Food Science Text Series

The ***Food Science Text Series*** provides faculty with the leading teaching tools. The Editorial Board has outlined the most appropriate and complete content for each food science course in a typical food science program and has identified textbooks of the highest quality, written by the leading food science educators.

B. Pam Ismail • S. Suzanne Nielsen
Editors

Nielsen's Food Analysis

Sixth Edition

Springer

Editors
B. Pam Ismail
Food Science and Nutrition
University of Minnesota
St. Paul, MN, USA

S. Suzanne Nielsen
Department of Food Science
Purdue University
West Lafayette, IN, USA

ISSN 1572-0330 ISSN 2214-7799 (electronic)
Food Science Text Series
ISBN 978-3-031-50642-0 ISBN 978-3-031-50643-7 (eBook)
https://doi.org/10.1007/978-3-031-50643-7

This Springer imprint is published by the registered company Springer Nature Switzerland AG
The registered company address is: Gewerbestrasse 11, 6330 Cham, Switzerland

For this sixth edition of the Food Analysis textbook, Springer has added "Nielsen's" at the beginning of the title. This is meant to recognize Dr. Suzanne Nielsen's efforts to edit this book over the past 30 years, starting with publication of the first edition in 1994. The sixth edition of the textbook will be her last. We are grateful for her years of dedication and tenacity which made Food Analysis the gold standard. Most importantly, we admire and applaud her dedication to students. Tentative plans are in place for editors of the future edition of the textbook. Thank you, Dr. Nielsen.

Preface and Acknowledgments

The intent of this sixth edition book is the same as that described in the Preface to the first five editions – a text primarily for undergraduate students majoring in food science, currently studying the analysis of foods. However, comments from users of the first five editions have indicated that the book is also a valuable text for persons in the food industry who either do food analysis or interact with analysts.

The focus of this edition was to do a general update, including addition and deletion of methods, as appropriate. The chapters included and the order of the chapters have not changed from the fifth edition. However, the lead author on a number of chapters has changed. As is common practice for textbooks, a number of long-time contributors to the book are being shifted to a secondary role, with more junior faculty members/scientists taking the lead on chapter authorship and revisions.

As stated for the first five editions, the chapters in this textbook are not intended as detailed references, but as general introductions to the topics and the techniques. Course instructors may wish to provide more details to students on a particular topic. Chapters focus on principles and applications of techniques. Procedures given are meant to help explain the principles and give some examples, but are not meant to be presented in the detail adequate to actually conduct a specific analysis. As in the first five editions, all chapters have summaries and study questions, and key words or phrases are in bold type, to help students focus their studies. The grouping of chapters by category is the same as in the fifth edition. Instructors are encouraged to cover the topics from this text in whatever order is most suitable for their course. Also, instructors are invited to contact Baraem P. Ismail for access to a website maintained with additional teaching materials related to this textbook and the accompanying laboratory manual.

Starting with the third edition, the competency requirements established by the Institute of Food Technologists were considered. Those requirements relevant to food analysis are as follows: (1) understanding the principles behind analytical techniques associated with food, (2) being able to select the appropriate analytical technique when presented with a practical problem, and (3) demonstrating practical proficiency in food analysis laboratory. This textbook should enable instructors to meet the requirements and develop learning objectives relevant to the first two of these requirements. The laboratory manual, now in its fourth edition, should be a useful resource to help students meet the third requirement.

The editors are grateful to all chapter authors for agreeing to be a part of this project. The authors have drawn on their experience of teaching students and/or working in the food industry and/or content expertise and experience to give chapters the appropriate content, relevance, and ease of use. Dr. Nielsen expresses special thanks to all the chapter authors who have worked with her since the first edition of this textbook: Sam K.C. Chang, Yong Hang, Dennis Miller, Michael Penner, Oscar Pike, Gary Reineccius, Denise Smith, and J. Scott Smith. The editors also wish to thank the authors of articles and books, and well as the publishers and industrial companies, for their permission to reproduce materials used here.

This is the last edition of the Food Analysis textbook with Dr. Nielsen serving as editor.

This sixth edition is co-edited by Dr. Nielsen and her former post-doctoral research associate, Dr. Baraem P. Ismail, who by now has taught Food Analysis at the University of Minnesota for over 15 years. Dr. Ismail has written multiple chapters for both the textbook and laboratory manual and provided input on the two previous editions of both books.

West Lafayette, IN, USA
St. Paul, MN, USA

S. Suzanne Nielsen
B. Pam Ismail

Contents

Contributors

William R. Aimutis North Carolina Food Innovation Lab, North Carolina State University, Kannapolis, NC, USA

Huseyin Ayvaz Department of Food Engineering, Canakkale Onsekiz Mart University, Canakkale, Turkey

James N. BeMiller Department of Food Science, Purdue University, West Lafayette, IN, USA

Subramanyam Bhadriraju Department of Grain Science and Industry, Kansas State University, Manhattan, KS, USA

Mirko Bunzel Department of Food Chemistry and Phytochemistry, Karlsruhe Institute of Technology, Karlsruhe, Germany

Sam K. C. Chang Department of Food Science, Nutrition, and Health Promotion, Mississippi State University, Starkville, MS, USA

Bingcan Chen Department of Plant Sciences, North Dakota State University, Fargo, ND, USA

Derek R. Dee Faculty of Land and Food Systems, University of British Columbia, Vancouver, BC, Canada

Hulya Dogan Department of Grain Science and Industry, Kansas State University, Manhattan, KS, USA

Jinping Dong Cargill Research and Development Center, Cargill, Inc., Plymouth, MN, USA

Wayne C. Ellefson Formerly at Covance Laboratories, Madison, WI, USA

M. Monica Giusti Department of Food Science and Technology, The Ohio State University, Columbus, OH, USA

M. Lourdes González-Miret Facultad de Farmacia, Unisersidad de Sevilla, Sevilla, Spain

Belén Gordillo Facultad de Farmacia, Unisersidad de Sevilla, Sevilla, Spain

Yong D. Hang Department of Food Science and Technology, Cornell University, Geneva, NY, USA

G. Keith Harris Department of Food, Bioprocessing and Nutritional Sciences, North Carolina State University, Raleigh, NC, USA

Kerry C. Huber Department of Animal and Food Science, Brigham Young University – Idaho, Rexburg, ID, USA

B. Pam Ismail Department of Food Science and Nutrition, University of Minnesota, St. Paul, MN, USA

Helen S. Joyner Premier Nutrition Company, Emeryville, CA, USA

Lisa J. Mauer Department of Food Science, Purdue University, West Lafayette, IN, USA

Dennis D. Miller Department of Food Science, Cornell University, Ithaca, NY, USA

Michael A. Mortenson Cargill Research and Development Center, Cargill, Inc., Plymouth, MN, USA

S. Suzanne Nielsen Department of Food Science, Purdue University, West Lafayette, IN, USA

Prakash R. Pandeya Department of Animal & Food Sciences, University of Kentucky, Lexington, KY, USA

Ronald B. Pegg Department of Food Science and Technology, The University of Georgia, Athens, GA, USA

Michael H. Penner Department of Food Science and Technology, Oregon State University, Corvallis, OR, USA

Tim D. Perez Research & Development, Conagra Brands, Omaha, NE, USA

Oscar A. Pike Department of Nutrition, Dietetics, and Food Science, Brigham Young University, Provo, UT, USA

Michael C. Qian Department of Food Science and Technology, Oregon State University, Corvallis, OR, USA

Qinchun Rao Department of Nutrition, Food and Exercise Sciences, Florida State University, Tallahassee, FL, USA

Gary A. Reineccius Department of Food Science and Nutrition, University of Minnesota, St. Paul, MN, USA

Bradley L. Reuhs Department of Food Science, Purdue University, West Lafayette, IN, USA

José I. Reyes De Corcuera Department of Food Science and Technology, The University of Georgia, Athens, GA, USA

Luis Rodriguez-Saona Department of Food Science and Technology, The Ohio State University, Columbus, OH, USA

Michael A. Rutzke School of Integrative Plant Science, Cornell University, Ithaca, NY, USA

Rachel R. Schendel Department of Animal & Food Sciences, University of Kentucky, Lexington, KY, USA

Shelly J. Schmidt Department of Food Science and Human Nutrition, University of Illinois at Urbana-Champaign, Urbana, IL, USA

Senay Simsek Department of Food Science, Purdue University, West Lafayette, IN, USA

Indarpal Singh Analytical & Applied Sciences, Conagra Brands, Omaha, ME, USA

Denise M. Smith School of Food Science, Washington State University, Pullman, WA, USA

J. Scott Smith Department of Animal Sciences and Industry, Kansas State University, Manhattan, KS, USA

Leonard C. Thomas DSC Solutions, Smyrna, DE, USA

Catrin Tyl Faculty of Chemistry, Biotechnology and Food Science, Norwegian University of Life Sciences, Ås, Norway

Robert E. Ward Department of Nutrition and Food Sciences, Utah State University, Logan, UT, USA

Vincent Yeung Department of Animal Science, California Polytechnic State University, San Luis Obispo, CA, USA

Umet Yucel Department of Animal Sciences and Industry, Kansas State University, Manhattan, KS, USA

Abbreviations

2-D	Two-dimensional
3-D	Three-dimensional
3-MCPD	3-Monochloropropane 1,2-diol
AACC	American Association of Cereal Chemists
AACCI	AACC International
AAS	Atomic absorption spectroscopy
AAPH	2,2′-Azobis (2-amidinopropane)
ABTS	2,2′-Azino-bis-(3-ethylbenzeneothazoline-6-sulfonic acid)
AC	Alternating current
ADI	Acceptable daily intake
ADP	Adenosine-5′-diphosphate
AE-HPLC	Anion exchange high-performance liquid chromatography
AES	Atomic emission spectroscopy
AFM	Atomic force microscopy
AI	Artificial intelligence
AMS	Accelerator mass spectrometer
AMS	Agricultural Marketing Service
ANSI	American Nation Standards Institute
AOAC	Association of Official Analytical Chemists
AOCS	American Oil Chemists' Society
AOM	Active oxygen method
APCI	Atmospheric-pressure chemical ionization
APHA	American Public Health Association
API	Atmospheric pressure ionization
APPI	Atmospheric pressure photoionization
AQC	6-Aminoquinolyl-*N*-hydroxysuccinimidyl carbamate
AQL	Acceptable quality limit
ASAP	Atmospheric pressure solid analysis probe
ASE	Accelerated solvent extraction
ASLS	Accelerated shelf-life studies
ASTA	American Spice Trade Association
ASTM	American Society for Testing Materials
ATCC	American Type Culture Collection
ATP	Adenosine-5′-triphosphate
ATR	Attenuated total reflectance
AUC	Area under the curve
AV	Acid value
a_w	Water activity
B_0	External magnetic field
BAW	Base and acid washed
BCA	Bicinchoninic acid
BCR	Community Bureau of Reference
Bé	Baumé modulus
BHA	Butylated hydroxyanisole
BHT	Butylated hydroxytoluene
BOD	Biochemical oxygen demand
BPA	Bisphenol A
BSA	Bovine serum albumin
BSDA	*Bacillus stearothermophilus* disk assay
BSE	Backscattered electrons
BSTFA	*N,O*-Bis(trimethylsilyl) trifluoroacetamide
CAD	Collision-activated dissociation
CAM	Chemical analytical methods
CAST	Calf antibiotic and sulfa test
CAT	Computerized axial tomography
CBP	Customs and Border Protection
CCD	Charge-coupled device
CD	Conjugated diene
CDC	Centers for Disease Control
CE	Capillary electrophoresis
CFR	Code of Federal Regulations
CFSAN	Center for Food Safety and Applied Nutrition
cGMP	Current Good Manufacturing Practices
CI	Confidence interval
CI	Chemical ionization
CID	Collision-induced dissociation
CID	Commercial item description
CID	Charge injection device
CIE	Commission Internationale d'Eclairage
CLA	Conjugated linoleic acid

CLND	Chemiluminescent nitrogen detector
CLSM	Confocal laser scanning microscopy
CMC	Critical micelle concentration
CMOS	Complementary metal oxide semiconductor
CNN	Convolutional neural networks
COA	Certificate of analysis
COD	Chemical oxygen demand
C-PER	Protein efficiency ratio calculation method
CPG	Compliance policy guidance
CP-MAS	Cross-polarization magic angle spinning
CQC	2,6-Dichloroquinonec hloroimide
CRC	Collision reaction cells
CSLM	Confocal scanning laser microscopy
CT	Computed technology
CT	Computed tomography
CV	Coefficient of variation
CVM	Center for Veterinary Medicine
Da	Dalton
DAL	Defect action level
DART	Direct analysis in real time
DC	Direct current
DDT	Dichlorodiphenyltrichloroethane
DE	Degree of esterification
dE*	Total color difference
DESI	Desorption electrospray ionization
DF	Dilution factor
DFE	Dietary folate equivalent
DHHS	Department of Health and Human Services
DIAAS	Digestible indispensable amino acid score
DIC	Differential interferential contrast
DL	Deep learning
DMA	Dynamic mechanical analysis
DMC	Dry matter coefficient
DMF	Dimethylformamide
DMD	D-Malate dehydrogenase
DMSO	Dimethyl sulfoxide
DNA	Deoxyribonucleic acid
DNFB	1-Fluoro-2,4-dinitrobenzene
dNTPs	Deoxynucleotide triphosphates
DON	Deoxynivalenol
DRI	Dietary references intake
DRIFTS	Diffuse reflectance Fourier transform spectroscopy
DRV	Daily reference value
DSC	Differential scanning calorimetry
DSHEA	Dietary Supplement Health and Education Act
DSPE	Dispersive solid-phase extraction
DTGS	Deuterated triglycine sulfate
DV	Daily value
DVB	Divinylbenzene
DVS	Dynamic vapor sorption
dwb	Dry weight basis
E_a	Activation energy
EAAI	Essential amino acid index
EAM	Elemental analysis method
EBT	Eriochrome black T
ECD	Electron capture detector
ECD	Electron capture dissociation
ECD	(Pulsed) electro-chemical detector
EDGE	Energized dispersive guided extraction
EDL	Electrodeless discharge lamp
EDS	Energy dispersive spectroscopy
EDTA	Ethylenediaminetetraacetic acid
EEC	European Economic Community
EFSA	European Food Safety Authority
EG	Ethylene glycol
EI	Electron impact (ionization)
EIE	Easily ionized elements
ELCD	Electrolytic conductivity detector
ELISA	Enzyme-linked immunosorbent assay
EM	Electron microscopy
EPA	Environmental Protection Agency
EPSPS	5-Enolpyruvyl-shikimate-3-phsophate synthase
Eq	Equivalents
ERH	Equilibrium relative humidity
ES	Electrospray
E-SEM	Environmental scanning electron microscopy
ESI	Electrospray interface
ESI	Electrospray ionization
ETD	Electron transfer dissociation
ETO	Ethylene oxide
EU	European Union
Fab	Fragment antigen-binding
FAIMS	Field-asymmetric ion mobility

FAME	Fatty acid methyl ester
FAO/WHO	Food and Agricultural Organization/World Health Organization
FAS	Ferrous ammonium sulfate
FBs	Fumonisins
Fc	Fragment crystallizable
FCC	Food Chemicals Codex
FD&C	Food, Drug and Cosmetic
FDA	Food and Drug Administration
FDAMA	Foods and Drug Administration Modernization Act
FDNB	1-Fluoro-2,4-dinitrobenzene
FFA	Free fatty acid
FID	Free induction decay
FID	Flame ionization detector
FIFRA	Federal Insecticide, Fungicide, and Rodenticide Act
FNB/NAS	Food and Nutrition Board of the National Academy of Sciences
FOS	Fructooligosaccharide
FPA	Focal plane array
FPD	Flame photometric detector
FPIA	Fluorescence polarization immunoassay
FPLC	Fast protein liquid chromatography
FPPS	Federal Purchase Program Specification
FRAP	Ferric reducing antioxidant power
FSIS	Food Safety and Inspection Service
FT	Fourier transform
FTC	Federal Trade Commission
FT-ICP	Fourier transform-inductively coupled plasma
FT-ICR	Fourier transform-ion cyclotron resonance
FTIR	Fourier transform infrared
FTMS	Fourier transform mass spectrometry
G6PDH	Glucose-6-phosphate dehydrogenase
GATT	General Agreement on Tariffs and Trade
GC	Gas chromatography
GC-AED	Gas chromatography-atomic emission detector
GC-FTIR	Gas chromatography-Fourier transform infrared
GC×GC	Comprehensive two-dimensional gas chromatography
GC-MS	Gas chromatography-mass spectrometry
GC-MS/MS	Gas chromatography-tandem mass spectrometry
GC-O	Gas chromatography-olfactory
GC-TOF	Gas chromatography-time of flight
GFC	Gel-filtration chromatography
GIPSA	Grain Inspection, Packers and Stockyard Administration
GLC	Gas-liquid chromatography
GLuDH	Glutamate dehydrogenase
GMA	Grocery Manufacturers of America
GMO	Genetically modified organism
GMP	Good manufacturing practices (also current good manufacturing practice in manufacturing, packing, or holding human food)
GOPOD	Glucose oxidase/peroxidase
Gox	Glucose oxidase
GPC	Gel-permeation chromatography
GRAS	Generally recognized as safe
HACCP	Hazard analysis critical control point
HAT	Hydrogen atom transfer
HCL	Hollow cathode lamp
HETP	Height equivalent to a theoretical plate
HFS	High fructose syrup
HIC	Hydrophobic interaction chromatography
HILIC	Hydrophilic interaction liquid chromatography
HIS	Hyperspectral imaging
HK	Hexokinase
H-MAS	High-resolution magic angle spinning
HMDS	Hexamethyldisilazane
HPLC	High-performance liquid chromatography
HPTLC	High-performance thin-layer chromatography
HQI	Hit quality index
HRGC	High-resolution gas chromatography
HRMS	High-resolution accurate mass spectrometry
HS	Headspace
HVP	Hydrolyzed vegetable protein
IC	Ion chromatography
IC_{50}	Median inhibition concentration
ICP	Inductively coupled plasma

ICP-MS	Inductively coupled plasma-mass spectrometer
ICP-OES	Inductively coupled plasma-optical emission spectroscopy
ICR	Ion cyclotron resonance
ID	Inner diameter
IDF	Insoluble dietary fiber
IDK	Insect damaged kernels
IEC	Inter-element correction
Ig	Immunoglobulin
IgE	Immunoglobulin E
IgG	Immunoglobulin G
IMS	Interstate Milk Shippers
InGaAs	Indium-gallium-arsenide
INT	Iodonitrotetrazolium
IR	Infrared
IRMM	Institute for Reference Materials and Measurements
IRMPD	Infrared multi-photon dissociation
ISA	Ionic strength adjustor
ISE	Ion-selective electrode
ISFET	Ion-sensitive field-effect transistor
ISO	International Organization for Standardization
ISO/IEC	International Organization for Standardization/International Electrotechnical Commission
IT	Ion trap
ITD	Ion trap detector
ITMS	Ion trap mass spectrometry
IU	International Units
IUPAC	International Union of Pure and Applied Chemistry
JECFA	Joint FAO/WHO Expert Committee on Food Additives
kcal	Kilocalorie
KDa	Kilodalton
KFR	Karl Fischer reagent
KFReq	Karl Fischer reagent water equivalence
KHP	Potassium acid phthalate
LALLS	Low-angle laser light scattering
LC	Liquid chromatography
LC-MS	Liquid chromatography-mass spectroscopy
LC-MS/MS	Liquid chromatograph-tandem mass spectrometry
LDH	Lactate dehydrogenase
LFS	Lateral flow strip
LIBS	Laser induced breakdown spectroscopy
LIMS	Laboratory information management system
LM	Light microscopy
LOD	Limit of detection
LOQ	Limit of quantitation
LRI	Linear retention index
LTM	Low thermal mass
LTP	Low-temperature plasma probe
LTPD	Lot tolerance percent defective
m/z	Mass-to-charge ratio
mAb	Monoclonal antibody
MALDI	Matrix-assisted laser desorption ionization
MALDI-TOF	Matrix-assisted laser desorption time-of-flight
MALLS	Multi-angle laser light scattering
MAS	Magic angle spinning
MASE	Microwave-assisted solvent extraction
MCL	Maximum contaminant level
MCT	Mercury-cadmium-telluride
MDGC	Multidimensional gas chromatography
MDL	Method detection limit
MDSC™	Modulated Differential Scanning Calorimeter™
mEq	Milliequivalents
MES-TRIS	2-(*N*-Morpholino)ethanesulfonic acid-tris(hydroxymethyl) aminomethane
MLR	Multiple linear regression
MRI	Magnetic resonance imaging
MRL	Maximum residue level
MRM	Multiple reaction monitoring
MRM	Multiresidue method
MS	Mass spectrometry (or spectrometer)
MS/MS	Tandem MS (multiple stages of MS)
Ms^n	Multiple stages of mass spectrometry
MTDSC	Modulated temperature differential scanning calorimetry
MW	Molecular weight
NAD	Nicotinamide adenine dinucleotide
NADH	Nicotinamide adenine dinucleotide hydride
NADP	Nicotinamide adenine dinucleotide phosphate
NADPH	Reduced NADP
NAS	National Academy of Sciences
NCM	*N*-Methyl carbamate

NCWM	National Conference on Weights and Measures
NDL	Nutrient Data Laboratory
NFDM	Nonfat dry milk
NIR	Near-infrared
NIRS	Near-infrared spectroscopy
NIST	National Institute of Standards and Technology
NLEA	Nutrition Labeling and Education Act
NMFS	National Marine Fisheries Service
NMR	Nuclear magnetic resonance
NOAA	National Oceanic and Atmospheric Administration
NOAEL	No observed adverse effect level
NPD	Nitrogen phosphorus detector or thermionic detector
NPN	Nonprotein nitrogen
NSF	National Science Foundation
NSSP	National Shellfish Sanitation Program
NVC	Nonvolatile compounds
NVOC	Nonvolatile organic compounds
OC	Organochlorine
OD	Outer diameter
ODS	Octadecylsilyl
OES	Optical emission spectroscopy
OMA	Official Methods of Analysis
OP	Organophosphate/organophosphorus
OPA	*o*-Phthalaldehyde
ORAC	Oxygen radical absorbance capacity
ORCA	Optimized Rowland circle alignment
OSI	Oil stability index
OT	Orbitrap
OTA	Ochratoxin A
PA	Polyacrylate
PAb	Polyclonal antibody
PAD	Pulsed-amperometric detector
PAGE	Polyacrylamide gel electrophoresis
PAM I	*Pesticide Analytical Manual, Volume I*
PAM II	*Pesticide Analytical Manual, Volume II*
P_c	Critical pressure
PCA	Photodiode array
PCBs	Polychlorinated biphenyls
PCR	Principal components regression
PCR	Polymerase chain reaction

PDA	Photodiode array
PDCAAS	Protein digestibility-corrected amino acid score
PDMS	Polydimethylsiloxane
PEEK	Polyether ether ketone
PER	Protein efficiency ratio
PFAS	Per-and polyfluoroalkyl substances
PFPD	Pulsed flame photometric detector
pI	Isoelectric point
PID	Photoionization detector
PLE	Pressurized liquid extraction
PLOT	Porous-layer open-tabular
PLS	Partial least squares
PME	Pectin methylesterase
PMO	Pasteurized Milk Ordinance
PMT	Photomultiplier tube
POD	Peroxidase
ppb	Parts per billion
PPD	Purchase product description
ppm	Parts per million
ppt	Parts per trillion
PSPD	Position-sensitive photodiode detector
PRV	Programmed temperature vaporization
PUFA	Polyunsaturated fatty acids
PV	Peroxide value
PVPD	Polyvinylidene difluoride
PVPP	Polyvinylpolypyrrolidone
Q	Quadrupole mass analyzer/filter
QA	Quality assurance
QC	Quality control
qMS	Quadruple mass spectrometry
QqQ	Triple quadrupole
Q-TOF	Quadrupole-time-of-flight
Q-trap	Quadruple-ion trap
QuEChERS	Quick, easy, cheap, effective, rugged and safe
RAC	Raw agricultural commodity
RACC	Reference amount customarily consumed
RAE	Retinol activity equivalents
RASFF	Rapid alert system for food and feed
RDA	Recommended daily allowance
RDI	Reference daily intake
RE	Retinol equivalent
R_f	Relative mobility
RF	Radio-frequency
RF	Response factor
R_m	Relative mobility

RI	Refractive index
RIA	Radioimmunoassay
R_m	Relative mobility
RMCD	Rapidly methylated ß-cyclodextrin
ROSA	Rapid one-step assay
RPAR	Rebuttable Presumption Against Registration
RS	Resistant starch
RVA	Rapid Visco Analyser
SAFE	Solvent-assisted flavor evaporation
SASO	Saudi Arabian Standards Organization
SBSE	Stir-bar sorptive extraction
SD	Standard deviation
SDE	Simultaneous distillation and extraction
SDF	Soluble dietary fiber
SDS	Sodium dodecyl sulfate
SDS-PAGE	Sodium dodecyl sulfate-poly-acrylamide gel electrophoresis
SEC	Size-exclusion chromatography
SEM	Scanning electron microscopy
SERS	Surface-enhanced Raman scattering
SET	Single electron transfer
SFC	Solid fat content
SFC	Supercritical fluid chromatography
SFC-MS	Supercritical fluid chromatography-mass spectrometry
SFE	Supercritical fluid extraction
SFE-GC	Supercritical fluid extraction-gas chromatography
SFI	Solid fat index
SI	International system of units
SIM	Selected ion monitoring
SIMCC	Soft independent modeling of class analogy
SKCS	Single kernel characteristics system
SMEDP	*Standard Methods for the Examination of Dairy Products*
SO	Sulfite oxidase
SOI	Standard of Identity
SOP	Standard operating procedures
SPDE	Solid-phase dynamic extraction
SPE	Solid-phase extraction
SPME	Solid-phase microextraction
SRF	Sample response factor
SRM	Standard reference materials
SRM	Selected reaction monitoring
SRM	Single-residue method
SSA	Significant scientific agreement
SSD	Solid-state detector
STOP	Swab test on premises
SVC	Semi-volatile compound
SVOC	Semi-volatile organic compounds
SXI	Soft x-ray imaging
TBA	Thiobarbituric acid
TBARS	TBA reactive substances
TCA	Trichloroacetic acid
TCD	Thermal conductivity detector
TCP	Tocopherols
TD	Time domain
TDA	Total daily intake
TDF	Total dietary fiber
TDI	Total daily intake
TDU	Thermal desorption unit
T-DNA	Transfer of DNA
TD-NMR	Time domain nuclear magnetic resonance
TEAC	Trolox equivalent antioxidant capacity
TEM	Transmission electron microscopies
TEMED	Tetramethylethylenediamine
Tg	Glass transition temperature
TGA	Thermogravimetric analysis
Ti	Tumor-inducing
TIC	Total ion current
TLC	Thin-layer chromatography
TMA	Thermomechanical analysis
TMCS	Trimethylchlorosilane
TMS	Trimethylsilyl
TOF	Time-of-flight
TOF-MS	Time-of-flight mass spectrometry
TPA	Texture profile analysis
TPC	Total polar compounds
TPTZ	2,4,6-Tripiyridyl-*s*-triazine
TQ	Triple quadrupole
TS	Total solids
TSQ	Triple stage quadrupole
TSS	Total soluble solids
TSUSA	Tariff Schedules of the United States of America
TTB	Alcohol and Tobacco Tax and Trade Bureau
TWI	Total weekly intake
UHPC	Ultra-high pressure chromatography
UHPLC	Ultra-high performance liquid chromatography

UPLC	Ultra-performance liquid chromatography
US	United States
USA	United States of America
USDA	United States Department of Agriculture
USDC	United States Department of Commerce
USP	United States Pharmacopeia
UV	Ultraviolet
UV-Vis	Ultraviolet-visible
Vis	Visible
VC	Volatile compounds
VOC	Volatile organic compounds
WDS	Wavelength dispersive x-ray
WHO	World Health Organization
wt	Weight
wwb	Wet weight basis
XMT	X-ray microtomography
XRD	X-ray diffraction
XRF	X-ray fluorescence spectroscopy
ZEA	Zearalenone

Part I

General Information

1 Introduction to Food Analysis

S. Suzanne Nielsen

Contents

1.1 INTRODUCTION

Investigations in food science and technology, whether by the food industry, government agencies, or universities, often require determination of food composition and characteristics. Trends and demands of consumers, national and international regulations, and realities of the food industry challenge food scientists as they work to monitor food composition and to ensure the quality and safety of the food supply. As summarized by McGorrin [1] in a review of food analysis history, "the growth and infrastructure of the model food distribution system heavily relies on food analysis (beyond simple characterization) as a tool for new product development, quality control, regulatory enforcement, and problem solving." All food products require analysis of various characteristics (i.e., chemical composition, microbial content, physical properties, sensory properties) as part of a **quality management** program, from raw ingredients, through processing, to the final product. Of course, food analysis is also used extensively for research on foods and food ingredients. The nature of the sample and the specific reason for the analysis commonly dictate the choice of analytical methods. **Speed, precision, accuracy, robustness,**

S. S. Nielsen (✉)
Department of Food Science, Purdue University,
West Lafayette, IN, USA
e-mail: nielsens@purdue.edu

B. P. Ismail, S. S. Nielsen (eds.), *Nielsen's Food Analysis*, Food Science Text Series,
https://doi.org/10.1007/978-3-031-50643-7_1

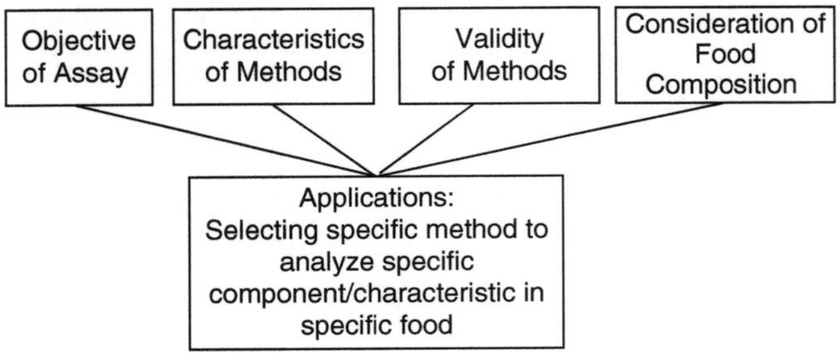

Fig. 1.1 Method selection in food analysis

Table 1.1 Reasons for analyzing foods

1. Food safety
2. Government regulations (a) Nutrition labeling (b) Standards: Mandatory and voluntary (c) Food inspection and grading (d) Authenticity
3. Quality control
4. Research and development

specificity, and **sensitivity** are often factors in this choice. *Validation* of the method for the specific **food matrix** being analyzed is necessary to ensure usefulness of the method.

Making an appropriate choice of analytical technique for a specific application requires good knowledge of the various techniques (Fig. 1.1). For example, the choice of method to determine the salt content of potato chips would differ depending on whether it is for nutrition labeling or quality control. The success of any analytical method relies on the proper selection and preparation of the food sample, carefully performing the analysis, and doing the appropriate calculations and interpretation of the data. Methods of analysis developed and endorsed by several nonprofit scientific organizations allow for standardized comparisons of results between different laboratories and for evaluation of less standard procedures. Such **official methods** are critical in the analysis of foods, to ensure that they meet the legal requirements established by government agencies. **Government regulations and international standards** most relevant to the analysis of foods are mentioned here but are covered in more detail in Chap. 2, and nutrition-labeling regulations in the United States, as related to food analysis, are covered in Chap. 3.

1.2 REASONS FOR ANALYZING FOODS AND TYPES OF SAMPLES ANALYZED

1.2.1 Overview

Consumer trends and demands, national and international regulations, and the food industry's need to manage product quality dictate the need for analysis of food ingredients and products (Table 1.1) and explain the types of samples analyzed.

1.2.2 Consumer Trends and Demands

Consumers have many choices regarding their food supply, so they can be very selective about the products they purchase. They demand a wide variety of products that are safe, nutritious, and are of high quality and offer a good value. Consumer demand has driven significant growth in products making claims, many of which bring an increased need for food analysis. For example, the consumer-driven, gluten-free claim has led to increased raw material/ingredient testing and finished product testing, required to comply in the United States with the definition of this claim established by the Food and Drug Administration (FDA). Many consumers are interested in the relationship between diet and health, including functional foods that may provide health benefits beyond basic nutrition. The use of social media by consumers has changed expectations and raised questions about food supply. Such trends and demands by consumers increase the need for food analysis and present some unique challenges regarding analytical techniques [2, 3].

1.2.3 Government Regulations and International Standards and Policies

For food companies to market safe, high-quality foods effectively in a national and global marketplace, they must pay increasing attention to government regulations and guidelines and to the policies and standards of international organizations. Food scientists must be aware of these regulations, guidelines, and policies related to food safety and quality and must know their implications for food analysis. Government regulations related to the composition of foods include nutrition labeling, product claims, standards, inspection and grading, and authenticity. This latter issue of authenticity is a challenge for the food industry, given the constant threat of economic adulteration of food products and ingredients. Detecting untargeted compounds in foods and determining their identity are challenges that require advanced analytical techniques that are powerful, sensitive, and fast [2]. The industry is also challenged and forced to "chase zeros" when laws are written stating zero as the level of certain compounds allowed. US government regulations relevant to food analysis are covered in Chaps. 2 and 3. Also covered in Chap. 2 are organizations active in developing international standards and safety practices relevant to food analysis.

1.2.4 Food Industry Management of Product Quality

1.2.4.1 Raw Ingredients to Final Product

To compete in the marketplace, food companies must produce foods that meet consumer demands, comply with government regulations, and meet the quality standards of the company. Whether it is new or existing food products or ingredients, the key concern for the food industry is safety of the food, but quality management goes well beyond safety. The management of product quality by the food industry is of increasing importance, beginning with the raw ingredients and extending to the final product eaten by the consumer. Analytical methods must be applied across the entire food supply chain to achieve the desired final product quality (Figs. 1.1 and 1.2). It is obvious that the food processor must understand the entire supply chain to successfully manage product quality.

In some cases, the cost of goods is linked directly to the composition as determined by analytical tests. For example, in the dairy field in the United States, the butterfat content of bulk tank raw milk determines how much money the milk producer is paid for the milk. For flour, the protein content can determine the price and food application. These examples point to the importance of accurate results from analytical testing.

Traditional quality control and quality assurance concepts are only a portion of a comprehensive quality management system. Food industry employees responsible for quality management work together in teams with other individuals in the company responsible for product development, production, engineering, maintenance, purchasing, marketing, and regulatory and consumer affairs.

One of the advantages that food scientists have over other disciplines in the food industry is their understanding of food analysis methods and principles. A solid understanding of food analysis methods and systems is critical to evaluate quality assurance systems across the value chain. Analytical information must be obtained, assessed, and integrated with other relevant information about the food system to address quality-related problems. Making appropriate decisions depends on having knowledge of the analytical methods and equipment utilized to obtain the data related to the quality characteristics. To design experiments in product and process development, and to assess results, one must know the operating principles and capabilities of the analytical methods. Upon completion of these experiments, one must critically evaluate the analytical data collected to determine whether product reformulation is needed or what parts of the process need to be modified for future tests. The situation is similar in the research laboratory, where knowledge of analytical techniques is necessary to design experiments and the evaluation of data obtained determines the next set of experiments to be conducted.

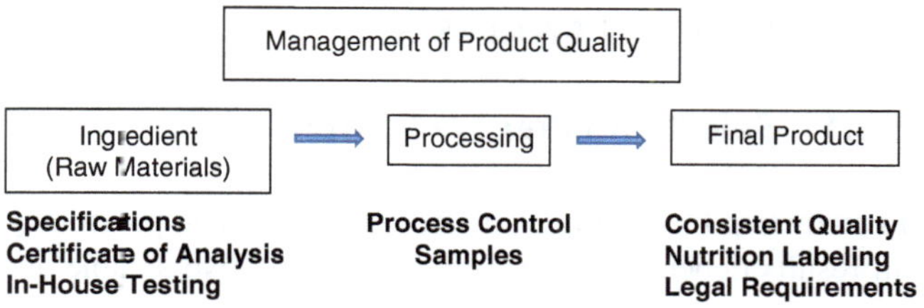

Fig. 1.2 Food industry management of quality

1.2.4.2 Types of Samples Analyzed

Chemical and physical analysis of foods is an important part of a quality assurance program in food processing, from ingredients and raw materials, through processing, to the finished products [4]. Chemical and physical analyses are also important in formulating and developing new products, in evaluating new processes for making food products, and in identifying the source of problems with unacceptable products (Table 1.2). Competitor's samples are increasingly relevant, e.g., store brands versus national

Table 1.2 Types of samples analyzed in a quality assurance program for food products

Sample type	*Critical questions*
Raw materials	Do they meet your specifications?
	Do they meet the required legal specifications?
	Are they safe and authentic?
	Will a processing parameter have to be modified because of any change in the composition of raw materials?
	Are the quality and composition the same as for previous deliveries?
	How does the material from a potential new supplier compare to that from the current supplier?
Process control samples	Did a specific processing step result in a product of acceptable composition or characteristics?
	Does a further processing step need to be modified to obtain a final product of acceptable quality?
Finished product	Does it meet the legal requirements?
	What is the nutritive value so that label information can be developed? Or is the nutritive value as specified on an existing label?
	Does it meet product claim requirements (e.g., "low fat," "gluten free")?
	Will it be acceptable to the consumer?
	Will it have the appropriate shelf life?
	If unacceptable and cannot be salvaged, how do you handle it (trash? rework? seconds?)
Competitor's sample	What are its composition and characteristics?
	How can we use this information to develop new products?
Complaint sample	How do the composition and characteristics of a complaint sample submitted by a customer differ from a sample with no problems?

Adapted and updated from Pearson [5] and Pomeranz and Meloan [6]

brands. For each type of sample listed in Table 1.2 to be analyzed, it may be necessary to determine either just one or many components. The nature of the sample and the way in which the information obtained will be used may dictate the specific method of analysis. For example, process control samples are usually analyzed by rapid methods, whereas nutritive value information for nutrition labeling generally requires the use of more time-consuming methods of analysis endorsed by scientific organizations. Critical questions, including those listed in Table 1.2, can be answered by analyzing various types of samples in a food processing system.

1.2.4.3 Increasing Dependence on Suppliers

The food industry takes many steps to reduce costs so that they can hold the price of food flat for as long as possible. With increasing competition in the food industry, one practice that has reduced costs is downsizing by pushing the responsibility for ingredient quality to the suppliers. Companies increasingly rely on others to supply high-quality and safe raw ingredients and packaging materials. Many companies have *select suppliers*, on whom they rely to perform the analytical tests to ensure compliance with detailed specifications for ingredients/raw materials. These specifications, and the associated tests, target various chemical, physical, and microbiological properties as appropriate based on the nature of the ingredient. Such specifications for raw materials/ingredients come in various forms within the food industry, with the three commonly used forms described below:

1. **Technical/Product Data Sheet**: supplier uses when salesperson is selling ingredients; gives maximum, minimum, and/or range of values, as appropriate, and methods of analysis.
2. **Specifications**: internal company document that defines company (processor) requirements (minimum, maximum, and/or target values) and links these to the specific methods of analysis; much of the data come from the technical/product data sheet; provides the requirements for the certificate of analysis.
3. **Certificate of Analysis** (COA): includes results of analytical tests related to predetermined specifications for specific shipment sent to a customer; gives actual values and methods of analysis.

As an example of specification information, Table 1.3 provides such information for the ingredient semolina, which is coarse-ground flour made from durum wheat (high protein), used especially to make pasta but also other cereal grain-based products. For the purposes of this table, specific numbers are not provided, since they would be in the actual documents for an ingredient.

Ingredient specifications and COAs are important for making specific food products. For example, with the wrong starch (i.e., with wrong specifications), a specific food product may not process correctly nor have the desired finished product quality attributes. Also, for example, if the COA indicates that the granulation size of a specific lot of rolled oats is "out of spec," the finished granola bars may not have the desired properties, which can result in increased consumer complaints. Setting the specifications for raw ingredients is commonly the responsibility of product developers, but it is often the production and quality control staff at the processing facility who must deal with challenges that arise when there are problems associated with ingredient specifications. Companies must have in place a means to maintain control of the COAs and react to them. With careful control over the quality of raw ingredients/materials, less testing is required during processing and on the final product.

1.2.4.4 Properties Analyzed

Analyzing foods for the reasons described in Table 1.1 to manage product quality commonly involves testing the following: chemical composition/characteristics, physical properties, sensory properties, and microbial quality. Table 1.4 shows the quality management tests typically done on dried pasta. Each test indicated would be done at some specified frequency and by a specified method. Note the relationship between the COA information for semolina in Table 1.3 and the quality tests conducted in-house on semolina by the processor, as reported in Table 1.4. While the COA generally requires considerable testing, done by official methods, routine in-house testing of ingredients is typically limited in scope and often uses rapid methods. The nature of testing required is largely determined by the nature of the food ingredient/product, but, not surprisingly, sensory properties (including taste, smell, appearance) are commonly tested across all foods and ingredients. This book focuses only on methods of analysis for testing chemical composition/characteristics and physical properties and not sensory properties or microbial quality.

1.3 STEPS IN ANALYSIS

1.3.1 Select and Prepare Sample

In analyzing food samples of the types described previously, all results depend on obtaining a representative sample and converting that sample to a form that can be analyzed.

Table 1.3 Example of information included in technical data sheet, specification, and certificate of analysis (COA) for semolina

Properties	*Technical/product data sheet*	*Specification*	*Certificate of analysis*
Chemical			
Moisture	Maximum value; AACCI	Maximum value; AOAC/AACCI	Actual value; AACCI
Protein	Minimum value; AACCI	Minimum value; AOAC/AACCI	Actual value; AACCI
Ash	Maximum value; AACCI	Maximum value; AOAC/AACCI	Actual value; AACCI
Falling number value	Target value, +/−; AACCI		
Enrichment			
Niacin		Maximum, minimum, target value; AOAC/AACCI	
Thiamine mononitrate		Maximum, minimum, target value; AOAC/AACCI	
Riboflavin		Maximum, minimum, target value; AOAC/AACCI	
Ferrous sulfate		Maximum, minimum, target value; AOAC/AACCI	
Folic acid		Maximum, minimum, target value; AOAC/AACCI	
Physical			
Bran specks	Maximum value; internal	Maximum value; AOAC/AACCI	Actual value; internal
Black specks	Maximum value; internal	Maximum value; internal	Actual value; internal
Color L, a, b		Maximum L value, minimum b value; AOAC/AACCI	Actual value (hunter)
Color (linear E)			Actual value (calculated from hunter lab)
Extraneous matter	Complies with FDA regulations; AACCI		
Insect fragments		Maximum value; AOAC/AACCI	
Rodent hair		Maximum value; AOAC/AACCI	
Granulation	Value, +/−; % over #40, 60, 80, 100 sieve; % through #100 sieve; Rotap	Minimum, maximum, target value; % over #40, 60, 80, 100 sieve; % thru #100 sieve; AOAC/AACCI	Actual value. % over #40, 60, 80, 100 sieve; % thru #100 sieve; Rotap
Microbiological			
Standard plate count; total plate count	Product is considered not ready to eat and requires further processing, so no microbiological guarantees are provided	Target value; FDA BAM	
Yeast		Target value; FDA BAM	
Mold		Target value; FDA BAM	
Vomitoxin	Complies with FDA's advisory max level	Maximum value; FDA BAM	
Shelf life	Number of days under recommended storage conditions		

Actual, minimum, maximum, and/or target value; source/type of method
Source of method: AOAC, AOAC International; AACCI, AACC International; FDA BAM, FDA's Bacteriological Analytical Manual; Internal, company method; Hunter, Hunter colorimeter; Rotap, machine to measure granulation

Neither of these is as easy as it sounds! **Sampling and sample preparation** are covered in detail in Chap. 5.

Sampling is the initial point for sample identification. Analytical laboratories must keep track of incoming samples and be able to store the analytical data from the analyses. This analytical information often is stored on a **laboratory information management system** (LIMS), which is a computer database program. Especially beneficial to commercial analytical laboratories, this system can capture all data for any specific sample and make it accessible to customers so that they can import, review, and analyze as they wish. Customers have real-time access to these data, which also enables them to review the testing status. There is no need for hand typing data, which saves times and decreases the chances for error, and the system settings can ensure standardized data entry for consistency and compliance with the analytical requirements.

1.3.2 Perform the Assay

Performing the assay is unique for each component or characteristic to be analyzed and may be unique to a specific type of food product. Certain chapters in this book address sampling

Table 1.4 Quality management test for dried pasta

Component/property being measured	*Name of the test*
Quality test done in-hours on semolina	
Moisture content	Rapid moisture analyzer
Color (L*, a*, b* determined to calculate the linear E value)	Colorimeter
Quality tests done in-process	
Moisture	Rapid moisture analyzer
Dimensions (after extrusion)	Micrometer and tape measurements by trained personnel
Metal detection	In-line metal detection (ferrous, nonferrous, stainless steel)
Package weight	In-line check with weight scale
Quality tests done on the final product	
Moisture	Rapid moisture analyzer
Color (L*, a*, b* determined to calculate the linear E value)	Colorimeter
Dimensions (diameter and shape)	Micrometer and tape measurements by trained personnel
Cooking quality	Sensory test by trained personnel (descriptive test; biting of samples)
Label	Visual inspection (with probability sampling methods)

and sample preparation (Chap. 5) and data handling (Chap. 4), while the remainder of this book addresses the step of actually performing the assay. Chapters 6–11 cover spectroscopy and spectrometry and Chaps. 12–14 cover chromatography. Since many analytical methods utilize spectroscopic and chromatographic techniques, these major general topics are covered before chapters that cover specific methods for chemical composition and characterization. For guidance in actually performing the assays, details regarding chemicals, reagents, apparatus, and step-by-step instructions are found in the books and articles referenced in each chapter. Numerous chapters in this book, and other books devoted to food analysis [7–9], make the point that for food analysis, we increasingly rely on expensive equipment, some of which require considerable expertise. Also, it should be noted that numerous analytical methods utilize automated instrumentation, including autosamplers and robotics to speed up the analyses.

1.3.3 Calculate and Interpret the Results

To make decisions and take actions based on the results obtained from performing the assay that determined the composition or characteristics of a food product, one must make the appropriate calculations to interpret the data correctly. **Data handling**, covered in Chap. 4, includes important statistical principles.

1.4 METHOD SELECTION

1.4.1 Objective of the Assay

Selection of a method depends largely on the objective of the measurement (Fig. 1.1). For example, methods used for rapid online processing measurements may be less accurate than official methods (see Sect. 1.5) used for nutrition labeling purposes. Methods referred to as reference, definitive, official, or primary are most applicable in a well-equipped and staffed analytical laboratory. Quality control testing of raw ingredients at the processing facility, in-process testing, and final product testing often rely on secondary/rapid quality control methods. This is likely in contrast to primary/official methods used on raw ingredient specifications and COAs and the testing for nutrition labeling. Both **primary** and **secondary methods** may be necessary and appropriate, with a secondary method calibrated against the primary method. The more rapid secondary or field methods may be more applicable on the manufacturing floor in a food processing facility. For example, the refractive index may be used as a rapid, secondary method for sugar analysis (see Chaps. 15 and 19), with results correlated with those of the primary method, high-performance liquid chromatography (HPLC) (see Chaps. 13 and 19). Moisture content data for a product being developed in the pilot plant may be obtained quickly with a rapid moisture analyzer that has been calibrated using a more time-consuming forced air oven method (see Chap. 15). Many companies commonly use unofficial, rapid methods for testing in processing plants but validate them against official methods. Calibration between these methods is critical. When evaluating potential/existing suppliers, it is critical to evaluate their calibration practices. A COA based on a rapid method that is not calibrated is of no value.

1.4.2 Characteristics of the Method

Numerous methods are often available to assay food samples for a specific characteristic or component. To select or modify methods used to determine the chemical composition and characteristics of foods, one must be familiar with the principles underlying the procedures and the critical steps. Certain properties of the methods and criteria described in Table 1.5 are useful to evaluate the appropriateness of a method in current use or a new method being considered. As highlighted in a review article on food analysis by Cifuentes [2], there is a continual need for the development of methods that can be characterized as being more robust, efficient, sensitive, and cost-effective. Many of the older "wet chemistry" methods have evolved into powerful and commonly used instrumental techniques. This has led to significant increases in accuracy, precision, detection limits, and sample throughput.

1.4.3 Validity of the Method

1.4.3.1 Overview

Numerous factors affect the usefulness and validity of the data obtained using a specific analytical method. One must consider certain characteristics of any method, such as specificity, precision, accuracy, and sensitivity (see Table 1.5 and Chap. 4). However, one also must consider how the variability of data from the method for a specific characteristic compares to the differences detectable and acceptable to a consumer and the variability of the specific characteristic inherent in processing of the food. One must consider the nature of the samples collected for the analysis, how representative the samples were of the whole, and the number of samples analyzed (Chap. 5). One must ask whether details of the analytical procedure were followed adequately, such that the results are accurate, repeatable, and comparable to data collected previously. For data to be valid, equipment to conduct the analysis must be standardized and appropriately used and the performance limitations of the equipment must be recognized.

Table 1.5 Criteria for choice of food analysis methods: characteristics of a method

Characteristic	*Critical questions*
Inherent properties	
Specificity/selectivity	Is the property being measured the same as that claimed to be measured, and is it the only property being measured? Are there interferences? What steps are being taken to ensure a high degree of specificity?
Precision	What is the precision of the method?* Is there within-batch, batch-to-batch, or day-to-day variation?* What step in the procedure contributes the greatest variability?*
Accuracy	How does the new method compare in accuracy to the old or a standard method? What is the percent recovery? Is it reproducible between labs?
Applicability of method to laboratory	
Reagents	Can you properly prepare them? What equipment are needed? Are they stable? For how long and under what conditions?
Equipment	Is the method highly sensitive to slight or moderate changes in the reagents? Do you have the appropriate equipment? Are personnel competent to operate equipment?
Cost Applicability to food/sample	What is the cost in terms of equipment, reagents, and personnel? Destructive or nondestructive? Online or offline? Official method/approval? Nature of food matrix?
Usefulness	
Time required	How fast is it? How fast does it need to be?
Reliability	How reliable is it from the standpoints of precision and stability?
Need	Does it meet a need or better meet a need? Is there simplicity of operation?
Personnel	
Safety	Is any change in method worth the trouble of the change?
Procedures	Who will do any required calculations?

*In-process samples may not accurately represent the finished product; must understand what variation can and should be present

1.4.3.2 Standard Reference Materials

A major consideration for determining method validity is the analysis of materials used as controls, often referred to as **standard reference materials** or **check samples** [10]. Analyzing check samples concurrently with test samples is an important part of quality control. Many large companies use internal check samples on each shift. These data are then plotted to help identify out-of-control situations (e.g., an under-trained technician; an HPLC column that is beginning to fail).

Standard reference materials (SRMs) can be obtained in the United States from the National Institute of Standards and Technology (NIST) and from the US Pharmacopeia (USP), in Europe from the European Commission's Joint Research Center (JRC) Institute for Reference Materials and Measurements (IRMM), and from specific organizations for other countries/regions. Besides government-related groups, numerous organizations offer check sample services that provide test samples to evaluate the reliability of a method [10].

Two examples of US-based organizations that offer check sample services are **American Association of Cereal Chemists** (AACC) and **American Oil Chemists' Society** (AOCS). (Note that **AACC** was initially known as the American Association of Cereal Chemists, then as AACC International, and is now known as the Cereals and Grains Association.) In this chapter, it will be referred to as AACC, since the methods are referred to as "AACC methods." AACC has a "laboratory proficiency rating program" in which a subscribing laboratory receives samples from AACC pertaining to specific analyses. The subscribing laboratory performs the specified analyses on the samples and returns the results to AACC. The AACC then provides a statistical evaluation of the analytical results and compares the subscribing laboratory's data with those of other laboratories to inform the subscribing laboratory of its degree of accuracy. The AACC offers check samples in specific categories that include various chemical analyses (e.g., moisture, ash, proteins, fats, total dietary fibers, specific vitamins). Samples also are available for testing physical properties and for microbiological and sanitation analyses. Very similar to AACC, the AOCS has a laboratory proficiency program and has available certified and quality reference materials related to various fat and oil products. Laboratories from many countries participate in these programs, to check the accuracy of their work, their reagents, and their laboratory apparatus against the statistical norm derived from the group data.

Standard reference materials are important tools to ensure reliable data. However, such materials need not necessarily be obtained from outside organizations. Control samples internal to the laboratory (i.e., **internal reference materials**) can be prepared by carefully selecting an appropriate type of sample, gathering a large quantity of the material, mixing and preparing to ensure homogeneity, packaging the sample in small quantities, storing the samples appropriately, and routinely analyzing the control sample when test samples are analyzed. Irrespective of the standard reference materials used, these should closely match the matrix of the samples to be analyzed by a specific method.

1.4.3.3 ISO Accreditation

To help ensure the validity of results generated by methods of analysis, commercial, private, and government laboratories are increasingly using **ISO** (International Organization for Standardization) accreditation (currently 17025:2017) to help assure their customers of the quality of their work (i.e., validity of data) [4, 11]. Accreditation can be for the entire laboratory and/or just individual assays (i.e., methods and procedures). Once accredited, these laboratories are audited for reaccreditation on a regular basis. ISO accreditation involves review of standard operating procedures, forms, records, work instructions, document control process, and test methods. While ISO accreditation does not deal with safety or the business, it does address questions about the following: (1) results validity, (2) equipment, (3) training, (4) analysts operating the equipment, (5) reagents and chemicals, and (6) customer communication.

Method accreditation requires comparison to other laboratories and the use of control samples/standards. For a food company with a corporate laboratory and multiple plant operations, the corporate laboratory conducts proficiency testing, using check samples (e.g., USP, NIST) and develops control samples that can be tested at both the corporate and plant laboratories for comparison of results.

A benefit of accreditation can include increased business for the laboratories (i.e., some customers choose ISO-accredited laboratories over nonaccredited laboratories, due to increased confidence in the data). Also, accreditation tends to create a mindset of continuous improvement within the laboratory. The greatest challenges for laboratories regarding accreditation are often updating procedures and balancing the need to get results to customers versus the need to check equipment and follow all required procedures. If a company fails an audit, the approach used is root cause analysis and corrective action investigation and implementation.

1.4.4 Consideration of Food Composition

Proximate analysis of foods refers to determining the major components of moisture (Chap. 15), ash (total minerals) (Chap. 16), lipids (Chap. 17), proteins (Chap. 18), and carbohydrates (Chap. 19). For proximate analysis, carbohydrate content is not analyzed directly but is rather calculated after analyzing the other four components of proximate composition as follows:

$$\%\ \text{carbohydrate} = 100\% - \begin{bmatrix} \%\ \text{moisture} + \%\ \text{ash} \\ +\%\ \text{lipids} + \%\ \text{proteins} \end{bmatrix} \quad (1.1)$$

The performance of many analytical methods is affected by the **food matrix**. The term "food matrix" has been used in reference to a number of things within the food industry. As related to food analysis, we talk in terms of a food's major chemical components, especially lipids, proteins, and carbohydrate contents. In food analysis, it is usually the food matrix that presents the greatest challenge to the analyst [12]. For example, high-fat or high-sugar foods can cause different types of interferences than can low-fat or low-sugar foods. Digestion procedures and extraction steps necessary for accurate analytical results can be very dependent on the food matrix. The complexity of various food systems often requires having not just one technique available for a specific food component but rather multiple techniques and procedures as well as the knowledge about which to apply to a specific food matrix.

A task force of **AOAC International** (formerly known as the Association of Official Analytical Chemists, AOAC) suggested a "triangle scheme" for dividing foods into matrix categories [13–20] (Fig. 1.3). The apexes of the triangle contained food groups that were 100% fat, 100% protein, or 100% carbohydrate. Foods were rated as "high," "low," or "medium" based on the levels of fats, carbohydrates, and proteins, which are the three nutrients expected to have the strongest effect on analytical method performance. This created nine possible combinations of high, medium, and low levels of fats, carbohydrates, and proteins. Complex foods were positioned spatially in the triangle according to their content of fats, carbohydrates, and proteins, on a normalized basis (i.e., fats, carbohydrates, and proteins normalized to a total 100%). General analytical methods ideally would be geared to handle each of the nine combinations, replacing more numerous matrix-dependent methods developed for specific foods. For exam-

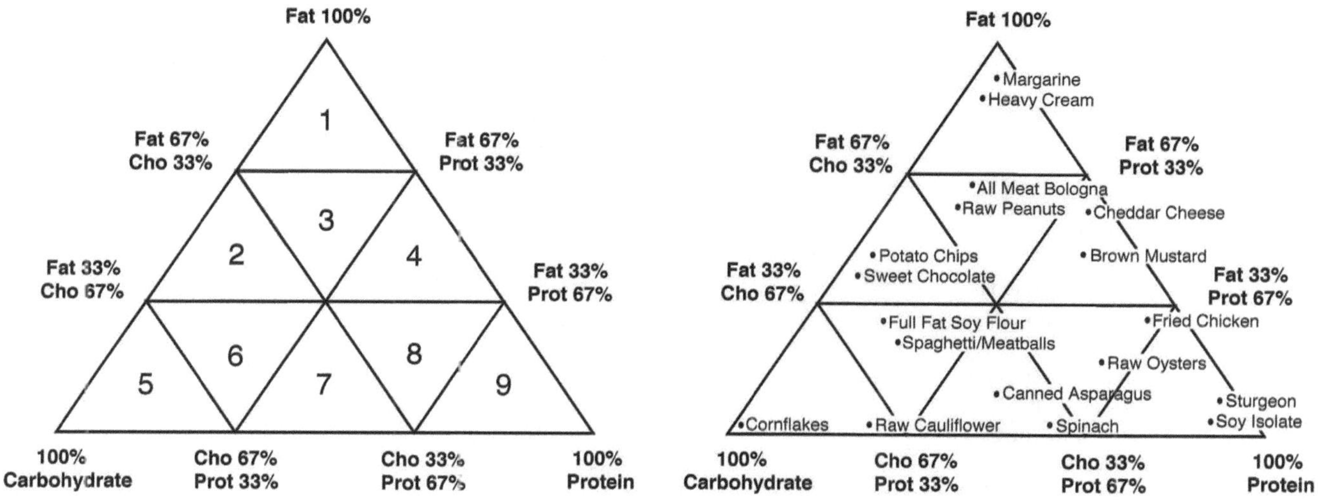

Fig. 1.3 Schematic layout of food matrixes based on protein, fat, and carbohydrate content, excluding moisture and ash. (Reprinted with permission from Ellis et al. [15], *Inside Laboratory Management*, September 1997, p. 33. Copyright 1997, by AOAC International)

ple, using matrix-dependent methods, one method might be applied to potato chips and chocolates, which are both low-protein, medium-fat, medium-carbohydrate foods, but another might be required for a high-protein, low-fat, high-carbohydrate food such as nonfat dry milk. In contrast, a robust general method could be used for all of the food types.

The AACC has approved a method studied using the triangle approach [16]. This approach also has been adopted by NIST for categorizing their SRMs [18–20] (see also Sect. 1.4.3.2). These SRMs can be used by laboratories to optimize methods for samples with specific matrices and to confirm that a method can be used reliably for a specific matrix. The food triangle categorization of both samples and reference materials enables analysts to choose the appropriate SRM for method development, optimization, or confidence in routine testing. NIST has, as a goal, the development of SRMs and other reference materials in all nine sections of the food matrix triangle [18].

1.5 OFFICIAL METHODS

The choice of method for a specific characteristic or component of a food sample is often made easier by the availability of *official methods*. Several nonprofit scientific organizations have compiled and published these methods of analysis for food products, which have been carefully developed and standardized. They allow for comparing results between different laboratories that follow the same procedure and for evaluating the results obtained using new or more rapid procedures.

1.5.1 AOAC International

AOAC International, which dates back to 1884, is well known for its program to provide official methods of analysis that fit their intended purpose (i.e., will perform with the necessary accuracy and precision under usual laboratory conditions). The official methods developed are intended for use by regulated industries, regulatory agencies, contract research organizations and testing laboratories, and academic institutions. AOAC International also has programs to develop standards, test the performance of commercial and analytical methods, and test laboratory proficiencies.

AOAC International uses advisory panels, stakeholder panels, working groups, and expert review panels in their collaborative system of voluntary consensus building to develop fit-for-purpose methods and services to ensure quality measurements. The *Official Methods of Analysis* of AOAC International [21] is the widely used compilation of methods adopted over the years. Details of the method validation program are given online and in the front matter of that book. Methods validated and adopted by AOAC International and the data supporting the method validation are published in the *Journal of AOAC International*. AOAC International's *Official Methods of Analysis* has been recently published as the 22nd edition in 2023, but an online version of the book is available as a "continuous edition," with new and revised methods posted as soon as they are approved. The *Official Methods of Analysis* of AOAC International includes methods appropriate for a wide variety of products and other materials (Table 1.6). These methods often are specified by the FDA with regard to legal requirements for food products regulated by the FDA. They are generally the methods followed by the FDA to check the nutritional

Table 1.6 Table of contents of the 2023 *Official Methods of Analysis of AOAC International* [21]

Chapter	*Title*
1	Agriculture liming materials
2	Fertilizers
3	Plants
4	Animal feed
5	Drugs in feeds
6	Disinfectants
7	Pesticide formulations
8	Hazardous substances
9	Metals and other elements at trace levels in foods
10	Pesticide and industrial chemical residues
11	Waters; and salt
12	Microchemical methods
13	Radioactivity
14	Veterinary analytical toxicology
15	Cosmetics
16	Extraneous materials: isolation
17	Microbiological methods
18	Drugs: Part I
19	Drugs: Part II
20	Drugs: Part III
21	Drugs: Part IV
22	Drugs: Part V
23	Drugs and feed additives in animal tissues
24	Forensic sciences
25	Baking powders and baking chemicals
26	Distilled liquors
27	Malt beverages and brewing materials
28	Wines
29	Nonalcoholic beverages and concentrates
30	Coffee and tea
31	Cacao bean and its products
32	Cereal foods
33	Dairy products
34	Eggs and egg products
35	Fish and other marine products
36	Flavors
37	Fruits and fruit products
38	Gelatin, dessert preparations, and mixes
39	Meat and meat products
40	Nuts and nut products
41	Oils and fats
42	Vegetable products, processed
43	Spices and other condiments
44	Sugars and sugar products
45	Vitamins and other nutrients
46	Color additives
47	Food additives: Direct
48	Food additives: Indirect
49	Natural toxins
50	Infant formulas, baby foods, and enteral products
51	Dietary supplements

labeling information and also to check foods for the presence or absence of undesirable residues or residue levels.

The **Food Safety and Inspection Service** (FSIS) of the **United States Department of Agriculture** (USDA) specifies, for nutrition analysis of USDA-regulated foods by USDA laboratories, the methods of the *Chemistry Laboratory Guidebook* [22]. However, if no USDA method is available and appropriate for the nutrients, the methods in the *Official Method of Analysis* of AOAC International [21] are to be used (see also Chap. 3, Sect. 3.3).

1.5.2 Other Endorsed Methods

The AACC publishes a set of approved laboratory methods, applicable mostly to cereal products (e.g., baking quality, gluten, physical dough tests, staleness/texture). The AACC process of adopting the *Approved Methods of Analysis* [23] is consistent with the process used by the AOAC International and AOCS. Approved methods of the AACC are continuously reviewed, critiqued, and updated and are now available online.

The AOCS publishes a set of official methods and recommended practices, applicable mostly to fat and oil analysis (e.g., vegetable oils, glycerin, lecithin) [24]. AOCS is a widely used methodology source on the subjects of edible fats and oils, oilseeds and oilseed proteins, soaps and synthetic detergents, industrial fats and oils, fatty acids, oleochemicals, glycerin, and lecithin.

Standard Methods for the Examination of Dairy Products [25], published by the American Public Health Association, includes methods for the chemical analysis of milk and dairy products (e.g., acidity, fat, lactose, moisture/solids, added water). *Standard Methods for the Examination of Water and Wastewater* [26] is published jointly by the American Public Health Association, the American Water Works Association, and the Water Environment Federation. *Food Chemicals Codex* (FCC) [27], published by USP, contains methods for the analysis of certain food additives.

Some trade associations publish standard methods for the analysis of their specific products. For example, the American Spice Trade Association (ASTA) publishes methods for use on spices in *The Official Analytical Methods of the American Spice Trade Association* [28].

The FDA refers analysts to manuals for general laboratory quality practices [29], elemental analysis methods (EAM) [30], pesticide analytical methods (PAM) [31], and macroanalytical procedures (e.g., evaluation by sight, smell, or taste) [32]. The FDA also refers analysts to manuals for chemical analytical methods (CAM), which are lists of validated procedures that the FDA regulatory laboratories use to determine food and feed safety [33].

1.6 SUMMARY

Food scientists and technologists routinely determine the chemical composition and physical characteristics of foods as part of their quality management, product development, or research activities. For example, the types of samples analyzed in a quality management program of a food company can include raw materials, process control samples, finished products, competitors' samples, and consumer complaint samples. Consumer, government, and food industry concerns for food quality and safety has increased the importance of analyses that determine composition and critical product characteristics.

To successfully base decisions on the results of any analysis, one must correctly conduct all three major steps in the analysis: (1) select and prepare samples, (2) perform the assay, and (3) calculate and interpret the results. The choice of analysis method is usually based on the objective of the analysis, the characteristics of the method itself (e.g., specificity, accuracy, precision, speed, cost of equipment, and training of personnel), and the food matrix involved. Validation of the method is important, as is the use of standard reference materials to ensure quality results. Rapid methods used for quality assessment in a production facility may be less accurate but much faster than official methods used for nutrition labeling. Endorsed methods for the chemical analyses of foods have been compiled and published by AOAC International, AACC, AOCS, and certain other nonprofit scientific organizations. These methods allow for comparison of results between different laboratories and for evaluation of new or more rapid procedures.

1.7 STUDY QUESTIONS

1. Identify six specific reasons you might need to determine certain chemical characteristics of.
2. A food product (or ingredient) as part of a quality management program.
3. In terms of number and nature of tests, in general, differentiate the testing of ingredients required of the supplier (e.g., certificate of analysis information) versus what is done by the food processor upon receipt of the ingredients from the supplier.
4. What is the difference between "specifications" and the "certificate of analysis" for an ingredient? Who in a food company is responsible for setting the specifications for an ingredient, and who must most commonly deal with problems when an ingredient does not meet those specifications?
5. Differentiate the nature and use of "primary methods" and "secondary methods" for food analysis.
6. Differentiate the nature of "standard reference materials," "check samples," and "internal reference materials." Identify how and why they are used.
7. In your work in an analytical laboratory, differentiate what is the meant when you are told that:
   (a) You need to do "proximate analysis" on a food sample.
   (b) The "food matrix" will affect an extraction procedure you need to do on a food sample.
8. You are considering the use of a new method to measure compound X in your food product. List six factors you will consider before adopting this new method in your quality assurance laboratory.
9. For each type of product listed below, identify a publication in which you can find standard methods of analysis appropriate for the product:
   (a) Ice cream
   (b) Enriched flour
   (c) Wastewater (from food processing plant)
   (d) Margarine
10. In your work at a food company, you mentioned to a coworker something about the *Official Methods of Analysis* published by AOAC International. The coworker asks you what AOAC International does and what the *Official Methods of Analysis* is. Answer your coworker's questions.

Acknowledgments The author thanks Garry Hellmich (retired from General Mills) and Chris Fosse and Indarpal Singh (Conagra Brands) for their input on the entire chapter in this edition. The author also thanks the following for their help with specific topics updated in the previous edition: Ryan Lane (TreeHouse Foods) for specifications for semolina; Samantha Park (TreeHouse Foods) and Yuezhen He (former student at Purdue University) for quality tests for pasta; Sandy Zinn (General Mills) for ISO accreditation; Ryan Ellefson (Covance) for LIMS; Karen Andrews (USDA) for food matrix and standard reference materials; and Julie Culp (General Mills) for food analysis linked to consumer demand and government regulations.

REFERENCES

1. McGorrin RJ (2009) One hundred years of progress in food analysis. J. Agric. Food CEhem. 57(18):8076–8088
2. Cifuentes A (2012) Food analysis: Present, future, and foodomics. ISRN Analytical Chemistry. Article ID 801607
3. Spence JT (2006) Challenges related to the composition of functional foods. J Food Compost Anal 19 Suppl 1: S4–S6
4. Alli I (2016) Food quality assurance: principles and practices, 2nd edn. CRC, Boca Raton, FL
5. Pearson D (1973) Introduction – some basic principles of quality control, Ch. 1. In: Laboratory techniques in food analysis. Wiley, New York, pp. 1–26
6. Pomeranz Y, Meloan CE (1994) Food analysis: theory and practice, 3rd edn. Chapman & Hall, New York
7. Cruz RMS, Khmelinskii I, Vieira M (2014) Methods in food analysis. CRC, Boca Raton, FL
8. Nollett LML, Toldra F (2012) Food analysis by HPLC, 3rd edn. CRC, Boca Raton, FL
9. Ötleş S (2011) Methods of analysis of food components and additives, 2nd edn. CRC, Boca Raton, FL
10. Latimer GW, Jr (1997) Check sample programs keep laboratories in sync. Inside Lab Manage 1(4): 18–20

11. ISO/IEC 17025:2017. General requirements for the competence of testing and calibration laboratories (last updated in 2018).
12. Wetzel DLB, Charalambous G (eds) (1998) Instrumental methods in food and beverage analysis. Elsevier Science, Amsterdam, The Netherlands
13. Wolf WR, Andrews KW (1995) A system for defining reference materials applicable to all food matrices. Fresenius J Anal Chem 352:73–76
14. Lovett RA (1997) U.S. food label law pushes fringes of analytical chemistry. Inside Lab Manage 1(4): 27–28
15. Ellis C, Hite D, van Egmond H (1997) Development of methods to test all food matrixes unrealistic, says OMB. Inside Lab Manage 1(8):33–35
16. DeVries JW, Silvera KR (2001) AACC collaborative study of a method for determining vitamins A and E in foods by HPLC (AACC Method 86–06). Cereal Foods World 46(5):211–215
17. Sharpless KE, Greenberg RR, Schantz MM, Welch MJ, Wise SA, Ihnat M (2004) Filling the AOAC triangle with food-matrix standard reference materials. Anal and Bioanal Chem 378:1161–1167
18. Wise SA, Sharpless KE, Sander LC, May WE (2004) Standard reference materials to support US regulations for nutrients and contaminants in food and dietary supplements. Accred Qual Assur 9:543–550
19. Sharpless KE, Lindstrom RM, Nelson BC, Phinney KW, Rimmer CA, Sander LC, Schantz MM, Spatz RO, Thomas JB, Turk GC, Wise SA, Wood LJ (2010) Preparation and characterization of standard reference material 1849 Infant/adult nutritional formula. 93(4):1262–1274
20. Thomas JB, Yen JH, Sharpless KE (2013) Characterization of NIST food-matrix standard reference materials for their vitamin C content. Anal Bioanal Chem 405:4539–4548
21. AOAC International (2023) Official methods of analysis, 22nd edn., 2023 (online). AOAC International, Rockville, MD
22. USDA Chemistry laboratory guidebook (2023) (online; updated continuously) Food Safety and Inspection Service, US Department of Agriculture, Washington, DC. https://www.fsis.usda.gov/news-events/publications/chemistry-laboratory-guidebook
23. Cereal & Grains Association (formerly AACC International) (2010) AACC Approved methods of analysis, 11th edn (online), Cereals & Grains Assoc., St. Paul, MN
24. AOCS (2020) Official methods and recommended practices, 7th edn, 2nd printing, American Oil Chemists' Society, Champaign, IL
25. Wehr HM, Frank JF (eds) (2004) Standard methods for the examination of dairy products, 17th edn. American Public Health Association, Washington, DC
26. American Public Health Association, American Water Works Association, Water Environment Federation (2023) Standard methods for the examination of water and wastewater, 24th edn. American Public Health Association, Washington, DC
27. US Pharmacopeia (USP) (2023) Food chemicals codex (FCC), 13th edn. United Book Press, Baltimore, MD (online subscription) https://www.foodchemicalscodex.org/
28. American Spice Trade Association (1997) Official analytical methods of the American spice trade association, 4th edn., Washington DC. https://www.astaspice.org/food-safety-technical-guidance/astas-analytical-methods/
29. FDA (2019) CFSAN laboratory quality manual (last updated 12/10/14) http://www.fda.gov/downloads/Food/FoodScienceResearch/UCM216952.pdf
30. FDA (2022) Elemental analysis manual for food and related products (last updated 01/19/2022) http://www.fda.gov/food/foodscienceresearch/laboratorymethods/ucm2006954.htm
31. FDA (1994) Pesticide analytical manual, vol I (PAMI) (updated Oct 1999) (content current as of 08/01/2018) (Methods which detect multiple residues) and vol II (PAMII) (updated Jan 2002) (Methods for individual pesticide residues), 3rd edn. National Technical Information Service, Springfield, VA. Available on Internet. (content current as of 08/01/2018) PAMI: https://www.fda.gov/food/laboratory-methods-food/pesticide-analytical-manual-volume-i-pam-3rd-edition. PAMII: https://www.fda.gov/food/laboratory-methods-food/pesticide-analytical-manual-volume-ii
32. FDA (2023) Macroanalytical Procedures Manual (last updated 03/27/2023) https://www.fda.gov/food/laboratory-methods-food/macroanalytical-procedures-manual-mpm Food Safety and Inspection Service, US Department of Agriculture, Washington, DC
33. US Food & Drug Administration. 2023. FDA Foods program compendium of analytical laboratory methods: Chemical analytical manual (CAM) Foods program compendium of analytical laboratory methods. (content current as of 05/31/2023) https://www.fda.gov/food/laboratory-methods-food/foods-program-compendium-analytical-laboratory-methods#CAM

US Government Regulations and International Standards Related to Food Analysis

2

S. Suzanne Nielsen

Contents

2.1 INTRODUCTION

Knowledge of government regulations relevant to the chemical analysis of foods is extremely important to persons working in the food industry. Federal laws and regulations reinforce the efforts of the food industry to provide wholesome foods, to inform consumers about the nutritional composition of foods, and to eliminate economic frauds. In some cases, they dictate what ingredients a food must contain, what must be tested, and the procedures used to analyze foods for safety factors and quality attributes. This chapter describes the US federal regulations related to the composition of foods. The reader is referred to references [1–3] for a comprehensive coverage of US food laws and regulations. Many of the regulations referred to in this chapter are published in the various titles of the *Code of Federal Regulations* (CFR) [4]. This chapter also includes information about food standards and safety practices established by international organizations. Nutrition labeling regulations are not covered in this chapter, but Chap. 3 is devoted to this topic, as it relates to food analysis.

S. S. Nielsen (✉)
Department of Food Science, Purdue University,
West Lafayette, IN, USA
e-mail: nielsens@purdue.edu

B. P. Ismail, S. S. Nielsen (eds.), *Nielsen's Food Analysis*, Food Science Text Series,
https://doi.org/10.1007/978-3-031-50643-7_2

2.2 US FEDERAL REGULATIONS AFFECTING FOOD COMPOSITION

2.2.1 US Food and Drug Administration (FDA)

The FDA is a government agency within the **Department of Health and Human Services** (DHHS) and is responsible for regulating, among other things, the safety of foods, cosmetics, drugs, medical devices, biologics, tobacco, and radiation-emitting products. It acts under laws passed by the US Congress to monitor the affected industries and ensure consumers of the safety of such products. A comprehensive collection of federal laws, guidelines, and regulations relevant to foods and drugs has been published by the Food and Drug Law Institute [1, 2].

2.2.1.1 Legislative History

Since the original Food and Drug Act of 1906, various acts, amendments, and regulations have been put in place to regulate foods. Table 2.1 summarizes these acts, amendments, and regulations with food analysis implications. The FDA implements and enforces the **Federal Food, Drug, and Cosmetic Act (FD&C) Act** of 1938 as amended. Select aspects of this and other regulations are further described in the subsections that follow.

2.2.1.2 Food Definitions and Standards

The food definitions and standards established by the FDA are published in 21 CFR 100–169 and include the standards of identity, quality, and fill. The **standards of identity (SOI)**, which have been set for a wide variety of food products, are most relevant to the chemical analysis of foods because they

Table 2.1 Food and Drug Administration acts, amendments, and regulations with food analysis implications

Act, amendment, or regulations (year)	*Intent*	*Food analysis relevance*
Federal Food, Drug, and Cosmetic Act (FD&C) (1938)	To assure consumers that foods are safe, wholesome, produced under sanitary conditions, and packaged and labeled truthfully and to prohibit adulteration and misbranding of foods	Authorized food definitions and standards of identity, many of which require testing food composition
Food Additive Amendment (1958)	To protect the health of consumers by requiring a food additive to be proven safe before addition to a food and used at only a safe level in foods. Delaney Clause (rider to amendment) prohibits the FDA from setting any tolerance as a food additive for substances known to be carcinogens	Requires testing food additives (type, amount, safety)
Color Additive Amendment (1960)	To define color additives, set rules for both certified and uncertified colors, provide approval of color additives, allow listing of color additives for specific uses, and set quantity limitations. Contains the Delaney clause (see above)	Requires testing food additives (type, amount, safety)
Current Good Manufacturing Practices (cGMP) (1969) (21 CFR 110 – General. Specific GMPs also exist for certain foods)	To prevent adulterated foods in the marketplace	Requires testing for adulterants (including extraneous matter)
Infant Formula Act (1980)	To establish minimum nutrient requirements, define adulteration, and establish nutrient and quality control procedures	Requires testing of formula composition
Nutrition Labeling and Education Act (NLEA) (1990)	To make nutrition labeling mandatory on most food products under FDA jurisdiction and establish definitions for health and nutrient claims	Requires testing for food composition (specific components and their levels)
Dietary Supplement Health and Education Act (DSHEA) (1994)	To define supplements as "dietary ingredients," set criteria to regulate claims and labeling and establish government agencies to handle regulation. Changed the definition and regulations for dietary supplements so that they are no longer considered food	Eliminated testing of dietary supplements as required when defined as food
Hazard Analysis Critical Control Point (HACCP) (1996): 21 CFR Part 123 – Seafood. 21 CFR Part 120 – Juice. 9 CFR Part 417 – Meat and poultry)	To improve food safety and quality	Requires testing for microbial, chemical, and physical hazards
Food Safety Modernization Act (FSMA) (2011)	To ensure that the food supply is safe by shifting the focus from responding to contamination to preventing it	Requires testing for hazards and preventive controls, including physical and chemical hazards, food allergens, sanitation controls, and supply chain controls
Food Labeling: Revision of the Nutrition and Supplement Facts Label (2016)	To provide updated information on the label to help consumers maintain health dietary practices	Changes the nutrients tested and reported on the label

specifically establish which ingredients a food must contain. Initially, the focus of the SOI program was to help address economic adulteration occurring in the marketplace. For example, some SOIs limit the amount of water permitted in certain products. The minimum levels for expensive ingredients are often set, and the maximum levels for inexpensive ingredients are sometimes set. The standards of identity for some foods include a list of optional ingredients. The kind and amount of certain vitamins and minerals that must be present in foods labeled "enriched" are specified. For example, the SOI for "enriched flour" specifies the content of the following: moisture, ash, protein, thiamine, riboflavin, niacin, iron, calcium, and folic acid (21 CFR 137.165). As a second example, the SOI for sour cream (21 CFR 131.160) specifies the product milkfat content and titratable acidity level, calculated as lactic acid. The standard of identity for a product often includes the recommended analytical method for determining the chemical composition, typically referring to methods published by AOAC International (Chap. 1, Sect. 1.5.1), as the *Official Methods of Analysis* (5).

The FDA began establishing SOIs within a year after the FD&C Act was established in 1938 and since then has established more than 250 SOIs. The FDA is currently updating the SOIs by focusing on the following: (1) "establish principles to determine whether to revise, eliminate, or establish food standards, (2) update SOIs to allow for continued innovation in the marketplace, and (3) update SOIs to support the production and innovation of healthier foods" [6]. For example, the FDA has amended the SOI for yoghurt (21 CFR 131.200) (compliance date of January 1, 2024) to require a pH of 4.5 or lower to ensure its safety. Also, all yoghurt must contain at least 8.25% solids not fats. Also, the fat contents requirement of "full-fat," "low-fat," and "nonfat" yoghurts are each defined.

Before leaving the topic of SOIs, the special attention the FDA pays to baby formula requirements and testing should be mentioned. The SOIs associated with the 1980 FDA Infant Formula Act have recently received careful review in part because of FDA's finding that a number of infant formula products had the same or similar statement of identity but were different in composition or intended use [7].

Compared to SOIs, the standards of quality and standards of fill generally are less related to the chemical analysis of foods, but they are important for economic and quality control considerations. The **standards of quality**, established by the FDA for some canned fruits and vegetables, set minimum standards and specifications for factors such as color, tenderness, weight of units in the container, and freedom from defects. The FDA has also set a standard of quality for bottled water, which includes set allowable levels for the following: coliform, turbidity, color, odor, radium-226 and radium-228 activity, beta-particle and photon radioactivity, and more than 70 chemical contaminants (21 CFR 165.110(b)). The FDA has established the **standards of fill** for some canned fruits and vegetables, tomato products, and seafood, stating how full a container must be to avoid consumer deception.

2.2.1.3 Inspection and Enforcement

The FDA has the broadest regulatory authority over most foods (generally, all foods other than meat, poultry, and egg products and water supplies). However, the FDA shares the responsibilities with other regulatory agencies for certain foods, as described in later sections of this chapter. Table 2.2 summarizes the food analysis-related scope of responsibility of the FDA, compared to and often in cooperation with other federal agencies, to ensure the quality and safety of foods in the USA. Table 2.3 summarizes which federal agency has the responsibility for ensuring the quality and safety of specific types of foods.

When violations of the FD&C Act are discovered by the FDA through periodic inspections of facilities and products and through analysis of samples, the FDA can issue warning letters, administrative detention, suspension of registration, seizures, injunctions, or recalls, depending on the circumstances. The FDA cannot file criminal charges but can rather recommend to the Justice Department that court action be taken that might result in fines or imprisonment of offenders. Details of these enforcement activities of the FDA are given in references [1–3].

2.2.2 US Department of Agriculture (USDA)

The USDA administers several federal statutes relevant to food standards, composition, and analysis. These include the standards of identity for meat products, grade standards, and inspection programs. Some programs for fresh and processed food commodities are mandatory, and others are voluntary.

2.2.2.1 Standards of Identity for Meat Products

The standards of identity have been established by the **Food Safety Inspection Service** (FSIS) of the USDA for many meat products (9 CFR 319). These commonly specify percentages of meat, fat, and water. Analyses are to be conducted using an AOAC method, if available.

2.2.2.2 Grade Standards

Grade standards developed for many foods by the USDA classify products in a range from substandard to excellent in quality. While most grade standards are not mandatory requirements (but they are mandatory for certain grains), they are widely used, voluntarily, by food processors and distributors as an aid in wholesale trading because the quality of a product affects its price. Such grade standards often

Table 2.2 Food analysis-related scope of responsibility of federal agencies

US Food and Drug Administration (FDA)	• Standards of Identity (most foods, except meats) • Standards of quality (canned fruits and vegetables) • Standards of fill (select foods) • Nutrition labeling (all foods but meats) (see Chap. 3 for details) • Standards for some alcoholic beverages and cooking wines • Enforces tolerances set on pesticides (note: Tolerances set by the EPA) • For seafood, works with NMFS to ensure safety; sets and enforces allowable levels of contaminants and pathogenic microorganisms in seafood • For shellfish, through the National Shellfish Ssanitation Program (NSSP) (part of the FDA): Sanitation for products shipped interstate; ensures that shellfish-growing areas are free of swage pollution and toxic industrial waste • For imported foods, works with CBP • For dairy products, works with USDA and states for safety and wholesomeness • Food additives (all foods) • Color additives (all foods) • Inspects food processing facilities for compliance with cGMP regulations and mandatory HACCP inspection programs
US Department of Agriculture (USDA)	• Standards of identity for meat products • Grade standards (voluntary; for many foods) • Inspection programs (meats, poultry, eggs, grains) • For select imported products, works with the CBP
US Department of Commerce, National Oceanic and Atmospheric Administration – National Marine Fisheries Service (NOAA-NMFS)	• Grading, standardization, and inspection program for seafood (works with the FDA and EPA)
US Department of Justice, Alcohol and Tobacco Tax and Trade Bureau (TTB)	• Standards and labeling of most alcoholic beverages
US Environmental Protection Agency (EPA)	• Sets tolerances for pesticide residues (note: Tolerances enforced by the FDA) • Sets standards for drinking water and lowed levels of contaminants • Sets standards and guidelines for effluents from food processing plants
US Customs and Border Protection (CBP)	• Ensures that imported products are taxed properly, are safe for human consumption, and are not economically deceptive (works with the FDA)
US Federal Trade Commission (FTC)	• Authority over advertising and sales promotion practices for foods

Table 2.3 Federal agencies with quality and safety responsibility of specific foods

Water	*Dairy products*	*Meats and meat products*	*Seafood*	*Shellfish*	*Alcoholic beverages*	*Fruits and vegetables*	*Harvested food or feed crops: pesticide residues*	*Infant formula*
EPA (drinking; effluent) FDA (bottled)	FDA USDA States	USDA	Dept. Commerce: NOAA-NMFS: FDA EPA	FDA: NSSP	Dept. Treasury: TTB (most products) FDA (select products)	FDA	EPA (sets tolerances) FDA (enforces tolerances)	FDA

are used as quality control tools. Consumers are familiar generally with grade standards for beef, butter, and eggs, but buyers for the retail market utilize grade standards for a wide variety of foods. Major users of standards include institutions such as schools, hospitals, restaurants, prisons, and the Department of Defense (see also Sect. 2.5).

The USDA has issued grade standards for more than 300 food products under the authority of the Agricultural Marketing Act of 1946 and related statutes. Standards for grades are not required to be stated on the label, but if they are stated, the product must comply with the specifications of the declared grade. Official USDA grading services are provided, for a fee, to pickers, processors, distributors, and others who seek official certification of the grades of their products.

While complete information regarding grade standards was published previously in the CFR, currently, only some standards are published in the CFR because they are USDA Agricultural Marketing Service (AMS) Administrative Orders. All grade standards are accessible on the Internet.

Grade standards, issued by the AMS of the USDA for agricultural products and by the Department of Commerce for fishery products, must not be confused with standards of quality set by the FDA or standards of identity set by the FDA or FSIS of the USDA, as discussed previously. Grade standards exist for many types of meats, poultry, dairy products, fruits, vegetables, and grains, along with eggs, domestic rabbits, certain preserves, dry beans, rice, and peas. Additional information about grade standards for dairy products is given in Sect. 2.3, but examples of grade standards for several other types of foods follow here.

The standards for grades of processed fruits and vegetables often include factors such as color, texture or consistency, defects, size and shape, tenderness, maturity, flavor, and a variety of chemical characteristics. Sampling procedures and methods of analysis are commonly given. As an example, the quality and analytical factors that determine the grade standards of frozen concentrated orange juice [8] are given in Table 2.4.

Grades for various grains (e.g., wheat, corn, soybeans, oats) are determined by factors such as test weight per bushel and percentages of heat-damaged kernels, broken kernels, and foreign material. Also, a grade limit is set commonly for moisture content. Grade standards for rice, beans, peas, and lentils are determined commonly by factors such as defects, the presence of foreign material, and insect infestation, and sometimes moisture content is specified.

2.2.2.3 Inspection Programs

The USDA administers some programs on inspection and certification that are mandatory, and some inspection programs are voluntary. Comprehensive inspection manuals specific to various types of foods have been developed to assist inspectors and industry personnel in interpreting and utilizing the regulations. Under the Federal Meat Inspection Act, the Poultry Products Inspection Act, and the Egg Products Inspection Act, the FSIS of the USDA inspects all meat, poultry, and egg products in interstate commerce (9 CFR 200-end). This includes a review of foreign inspection systems and packing plants that export meat and poultry to the USA. Imported products are reinspected at ports of entry. The Hazard Analysis Critical Control Point (HACCP) is a major component of FSIS rules for all slaughter and processing plants to improve the safety of meat and poultry. A program within the **Grain Inspection, Packers and Stockyards Administration** (GIPSA) of the USDA administers the mandatory requirements of the US Grain Standards Act (7 CFR 800). Regulations to enforce this act provide for a national inspection system for grains and mandatory official grade standards of numerous types of grains. Another program of the USDA standardizes, grades, and inspects fruits and vegetables under various voluntary programs. The inspection programs rely heavily on the HACCP concept.

Table 2.4 USDA standards for grade determinants of frozen concentrated orange juice [8]

Quality	*Analytical*
Appearance	Concentrate:
Reconstitution	Brix
Color	Brix/acid ratio
Defects	Reconstituted juice:
Flavor	Brix
	Soluble orange solids
	Recoverable oil

2.2.3 US Department of Commerce

2.2.3.1 Seafood Inspection Service

The **National Oceanic and Atmospheric Administration's** (NOAA) **National Marine Fisheries Service** (NMFS), a division of the **US Department of Commerce** (USDC), provides a seafood inspection service. The USDC Seafood Inspection Program ensures the safety and quality of seafoods consumed in the USA and certified for export through **voluntary grading, standardization, and inspection programs**, as described in 50 CFR 260. The inspection programs rely heavily on the HACCP concept. The US Standards for Grades of Fishery Products (50 CFR 261) are intended to help the fishing industry maintain and improve quality and thereby increase consumer confidence in seafoods. Standards are based on attributes such as color, size, texture, flavor, odor, workmanship defects, and consistency.

2.2.3.2 Interaction Between FDA and Environmental Protection Agency (EPA)

The FDA and the Environmental Protection Agency (EPA) work with the NMFS for the assurance of seafood safety. The FDA, under the FD&C Act, is responsible for ensuring that seafood shipped or received in interstate commerce is safe, wholesome, and not misbranded or deceptively packaged. The FDA has primary authority in setting and enforcing allowable levels of contaminants and pathogenic microorganisms in seafood. The EPA assists the FDA in identifying the range of chemical contaminants that pose a human health risk and are most likely to accumulate in seafood. A tolerance of 2.0 parts per million (ppm) for total polychlorinated biphenyls (PCBs) (21 CFR 109.30) is the only formal tolerance specified by the FDA for seafood. However, the EPA has established tolerances for certain pesticide residues, and the FDA has established a guidance level for methyl mercury [9].

2.2.4 US Alcohol and Tobacco Tax and Trade Bureau

2.2.4.1 Regulatory Responsibility for Alcoholic Beverages

Beer, wines, liquors, and other alcoholic beverages are termed "food" according to the FD&C Act of 1938. However, regulatory control over their quality, standards, manufacture, and other related aspects is specified by the **Federal Alcohol Administration Act**, which is enforced by the **Alcohol and Tobacco Tax and Trade Bureau** (TTB) of the **US Department of Treasury**. Issues regarding the composition and labeling of most alcoholic beverages are handled by the Bureau. However, the FDA has jurisdiction over certain other alcoholic beverages and cooking wines. The FDA also deals with questions of sanitation, filth, and the presence of deleterious substances in alcoholic beverages.

2.2.4.2 Standards and Composition of Beer, Wine, and Distilled Beverage Spirits

Information related to the **definitions, standards of identity**, and certain **labeling requirements** for beer, wine, and distilled beverage spirits is given in 27 CFR 1–30. The standards of identity for these types of beverages stipulate the need for analyses such as percentage alcohol by volume, total solids content, volatile acidity, and calculated acidity. For example, the fruit juice used for the production of wine is often specified by its °Brix and total solids content. The maximum volatile acidity (calculated as acetic acid and exclusive of sulfur dioxide) for grape wine must not be more than 0.14 g/100 mL (20 °C) for natural red wine and 0.12 g/100 mL for other grape wine (27 CFR 4.21). The percentage alcohol by volume is often used as a criterion for class or type designation of many alcoholic beverages. For example, "dessert wine" is grape wine with an alcohol content of greater than 14% but not over 24% by volume, while table wines must have an alcohol content not in excess of 14% by volume.

2.2.5 US Environmental Protection Agency (EPA)

The EPA regulatory activities most relevant to this book are control of pesticide residues in foods, drinking water safety, and the composition of effluents from food processing plants.

2.2.5.1 Pesticide Registration and Tolerance Levels

Pesticides are chemicals intended to protect our food supply by controlling harmful insects, diseases, rodents, weeds, bacteria, and other pests. However, most pesticide chemicals can have harmful effects on people, animals, and the environment if they are improperly used. The three federal laws relevant to protection of food from pesticide residues are (1) certain provisions of the Federal FD&C Act, (2) the Federal Insecticide, Fungicide, and Rodenticide Act (FIFRA), as amended, and (3) the **Food Quality Protection Act** of 1996. FIFRA, supplemented by the FD&C Act, authorizes a comprehensive program to regulate the manufacturing, distribution, and use of pesticides, along with a research effort to determine the effects of pesticides.

The Food Quality Protection Act amends both the FD&C Act and FIFRA to take pesticides out of the section of the FD&C Act that includes the Delaney Clause. This was done by changing the definition of a "food additive"to exclude pesticides. This redefinition leaves the Delaney Clause greatly reduced in scope and less relevant.

The EPA registers approved pesticides and sets tolerances for pesticide residues. The EPA is authorized to register approved pesticides and to establish an **allowable limit** or **a tolerance level** for any detectable pesticide residues that might remain in or on a harvested food or feed crop. While the EPA establishes the tolerance levels, the FDA enforces the regulations by collecting and analyzing food samples, mostly agricultural commodities. Livestock and poultry samples are collected and analyzed by the USDA. Pesticide residue levels that exceed the established tolerance levels are considered in violation of the FD&C Act.

Regulations regarding pesticide tolerances in foods are given in 40 CFR 180, which specifies the general categories of products and specific commodities with tolerances or exemptions and, in some cases, which part of the agricultural product is to be examined. Unless otherwise noted, the specific tolerances established for the pesticide chemicals apply to residues resulting from their application prior to harvest or slaughter. The tolerance set for a specific pesticide is sometimes revoked by the EPA based on safety data, but the testing of foods for that pesticide needs to continue for many years. For example, the insecticide chlorpyrifos was allowed for use on certain fruits, vegetables, and plants used for animal feed. Tolerances for chlorpyrifos were set for specific food commodities (e.g., apples, 0.01 ppm; corn oil, 0.25; meat from cattle, 0.05), but, in 2022, all food tolerances for chlorpyrifos were revoked.

The analytical methods to be used for determining whether pesticide residues are in compliance with the tolerances established are identified among the methods contained or referenced in the *Pesticide Analytical Manual* [10] maintained by and available from the FDA. The methods must be sensitive and reliable at and above the tolerance level. Pesticides are generally detected and quantitated by gas chromatographic or high-performance liquid chromatographic methods (see Chaps. 13 and 14, respectively). More details about pesticide analysis are included in Chap. 33, Sect. 33.3.

2.2.5.2 Drinking Water Standards and Contaminants

While the FDA regulates bottled water with a standard of identify (i.e., a detailed standard of quality) and specific current Good Manufacturing Practices (cGMP) regulations, the EPA regulates drinking (i.e., tap) water. The EPA administers the **Safe Drinking Water Act** of 1974, which is to provide for the safety of drinking water supplies in the USA and to enforce national drinking water standards. The EPA has identified potential contaminants of concern and established their maximum acceptable levels in drinking water. The EPA has primary responsibility of establishing the standards, while the states enforce them and otherwise supervise public water supply systems and sources of drinking water. **Primary and secondary drinking water regulations** (40 CFR 141 and 143, respectively) have been established. Concerns have been expressed regarding the special standardization of water used in the manufacturing of foods and beverages.

The **maximum contaminant levels** (MCLs) for primary drinking water are set for certain **inorganic and organic chemicals, turbidity, certain types of radioactivity**, and **microorganisms**. Sampling procedures and analytical methods for the analysis of chemical contaminants are specified, with common reference to *Standard Methods for the Examination of Water and Wastewater* [11] published by the American Public Health Association; *Methods of Chemical Analysis of Water and Wastes* [12], published by the EPA; and *Annual Book of ASTM Standards* [13], published by the American Society for Testing and Materials (ASTM). Methods commonly specified for the analysis of inorganic contaminants in water include atomic absorption (direct aspiration or furnace technique), inductively coupled plasma (see Chap. 9), ion chromatography (see Chap. 13), and ion-selective electrodes (see Chap. 21).

2.2.5.3 Effluent Composition from Food Processing Plants

In administering the **Federal Water Pollution and Control Act**, the EPA has developed effluent guidelines and standards that cover various types of food processing plants. Regulations prescribe effluent limitation guidelines for existing sources, standards of performance for new sources, and pretreatment standards for new and existing sources. Point sources of discharge of pollution are required to comply with these regulations, where applicable. Regulations are prescribed for specific foods under the appropriate point source category: dairy product processing (40 CFR 405), grain mills (40 CFR 406), canned and preserved fruit and vegetable processing (40 CFR 407), canned and preserved seafood processing (40 CFR 408), sugar processing (40 CFR 409), and meat and poultry products (40 CFR 432). The **effluent characteristics** commonly prescribed for food processing plants are **biochemical oxygen demand** (BOD) (see Chap. 28), **total soluble solids** (TSSs) (see Chap. 15), and **pH** (see Chap. 22). The test procedures for measurement of effluent characteristics are prescribed in 40 CFR 136. The effluent levels allowed are specific to the type of food processed in a plant.

2.2.6 US Customs and Border Protection (CBP)

Over 100 countries export food, beverages, and related edible products to the USA. The CBP assumes the central role in ensuring that imported products are taxed properly, safe for human consumption, and not economically deceptive. The CBP receives assistance from the FDA and USDA as it assumes these responsibilities. The major regulations promulgated by the CBP are given in Title 19 of the CFR.

All goods imported into the USA are subjected to duty or duty-free entry according to their classification under applicable items in the **Harmonized Tariff Schedule of the United States** (TSUSA). The US tariff system has official tariff schedules for more than 400 edible items exported into the USA [14]. The TSUSA specifies the food product in detail and gives the general rate of duty applicable to that product coming from most countries and any special higher or lower rates of duty for certain other countries.

The **rate of duty** for certain food products is determined by their chemical composition. For example, the rate of duty on some dairy products is determined in part by their fat content. The tariff for some syrups is determined by their fructose content, for some chocolate products by their sugar or butterfat content, for butter substitutes by their butterfat content, and for some wines by their alcohol content (percentage by volume).

2.2.7 US Federal Trade Commission (FTC)

The FTC is the most influential of the federal agencies that have authority over various aspects of advertising and sales promotion practices for foods in the USA. The major role of the FTC is to keep business and trade competition free and fair.

2.2.7.1 Enforcement Authority

The **Federal Trade Commission Act** of 1914 authorizes the FTC to protect both consumers and business people from anticompetitive behavior and unfair or deceptive business and trade practices. The FTC periodically issues industry guides and trade regulations and rules that tell businesses what they can and cannot do. These issuances are supplemented with advisory opinions given to corporations and individuals upon request. The FTC not only has guidance and preventive functions but is also authorized to issue complaints or shut down orders and sue for civil penalties for violation of trade regulation rules. The **Bureau of Consumer Protection** is one of the FTC bureaus that enforce and develop trade regulation rules.

2.2.7.2 Food Labels, Food Composition, and Deceptive Advertising

While the **Fair Packaging and Labeling Act** of 1966 is administered by the FTC, that agency does not have specific authority over the packaging and labeling of foods. The FTC and FDA have agreed upon responsibilities: the FTC has primary authority over advertising of foods and FDA has primary authority over labeling of foods.

Grading, standards of identity, and labeling of foods regulated by several federal agencies as described previously have eliminated many potential problems in the advertising of foods. Such federal regulations and voluntary programs have reduced the scope of advertising and other forms of product differentiation. Misleading, deceptive advertising is less likely to be an issue and is more easily controlled. For example, foods such as ice cream, mayonnaise, and peanut butter have standards of identity that set minimum ingredient standards. If these standards are not met, the food must be given a different generic designation (e.g., salad dressing instead of mayonnaise) or be labeled an "imitation." Grading, standards, and labeling of food aid consumers in making price–quality comparisons. Once again, analyses of chemical composition play an important role in developing and setting these grades, standards, and labels. In many cases in which the FTC intervenes, data from a chemical analysis become central evidence for all parties involved.

2.3 REGULATIONS AND RECOMMENDATIONS FOR MILK

The safety and quality of milk and dairy products in the USA are the responsibility of both the federal (FDA and USDA) and state agencies. The FDA has regulatory authority over the dairy industry's interstate commerce, while the USDA's involvement with the dairy industry is voluntary and service-oriented. Each state has its own regulatory office for the dairy industry within that state. The various regulations for milk involve several types of chemical analyses.

2.3.1 FDA Responsibilities

The FDA has responsibility under the FD&C Act, the Public Health Service Act, and the Import Milk Act to assure consumers that the US milk supply and imported dairy products are safe, wholesome, and not economically deceptive. As described in Sect. 2.1.2, the FDA promulgates standards of identity and labeling, quality, and fill-of-container requirements for milk and dairy products moving in interstate commerce.

For **Grade A milk and dairy products**, each state shares with the FDA the responsibility of ensuring safety, wholesomeness, and economic integrity. This is done through a **Memorandum of Understanding with the National Conference on Interstate Milk Shipments**, which is comprised of all 50 states. In cooperation with the states and the dairy industry, the FDA has also developed model regulations for state adoption regarding sanitation and the quality aspects of producing and handling Grade A milk. These regulations are contained in the *Grade A Pasteurized Milk Ordinance* (PMO) [15], which all states have adopted as minimum requirements. The PMO specifies that "all sampling procedures, including the use of approved in-line samples, and required laboratory examinations shall be in substantial compliance with the most current edition of *Standard Methods for the Examination of Dairy Products* (SMEDP) of the American Public Health Association [16], and the most current edition of *Official Methods of Analysis* of the *AOAC INTERNATIONAL (OMA)*" [5]. Examples of chemical analyses required by the PMO for Grade A pasteurized milk and milk products include the phosphatase level (test for adequacy of pasteurization) (Chap. 26) and any presence of drug residues (requires negative results) (Chap. 33).

2.3.2 USDA Responsibilities

Under the authority of the Agricultural Marketing Act of 1946, the **Dairy Quality Program** of the USDA offers **voluntary grading services** for manufactured or processed dairy products (7 CFR 58). If USDA inspection of a dairy manufacturing plant shows that good sanitation practices are being followed to meet the requirements in the *General Specifications for Dairy Plants Approved for USDA Inspection and Grading Service* [17], the plant qualifies for USDA services of grading, sampling, testing, and certification of its products. A product such as nonfat dry milk is graded based on flavor, physical appearance, and various laboratory analyses (Table 2.5) [18].

The USDA, under an arrangement with the FDA, assists states in establishing safety and quality regulations for manufacturing grade milk. Much as described previously for the FDA with Grade A milk, the USDA has developed model regulations [19] for state adoption regarding the quality and sanitation aspects of producing and handling manufacturing grade milk.

Table 2.5 US standards for grades of nonfat dry milk (spray process) [18]

Laboratory tests[a]	US extra grade	US standard grade
Bacterial estimate, standard plate count per gram	10,000	75,000
Milk fat content, percentage	1.25	1.50
Moisture content, percentage	4.0	5.0
Scorched particle content, mg	15.0	22.5
Solubility index, mL	1.2	2.0
US high heat	2.0	2.5
Titratable acidity (lactic acid), percentage	0.15	0.17

[a]All values are maximum allowed http://www.ams.usda.gov/sites/default/files/media/Nonfat_Dry_Milk_%28Spray_Process%29_Standard%5B1%5D.pdf

2.3.3 State Responsibilities

As described previously, individual states have enacted safety and quality regulations for Grade A and manufacturing grade milk that are essentially identical to those in the PMO and the USDA Recommended Requirements, respectively. The department of health or agriculture in each state is normally responsible for enforcing these regulations. The states also establish their own standards of identity and labeling requirements for milk and dairy products, which are generally similar to the federal requirements.

2.4 REGULATIONS AND RECOMMENDATIONS FOR SHELLFISH

Shellfish include fresh or frozen oysters, clams, and mussels. They may transmit intestinal diseases such as typhoid fever or act as carriers of natural or chemical toxins. This makes it very important that they be obtained from unpolluted waters and handled and processed in a sanitary manner.

The growing, handling, and processing of shellfish must comply not only with the general requirements of the FD&C Act but also with the requirements of state health agencies cooperating in the **National Shellfish Sanitation Program** (NSSP), a federal, state, industry voluntary cooperative program, administered by the FDA [20]. The FDA has no regulatory power over shellfish sanitation unless the product is shipped interstate. However, the Public Health Service Act authorizes the FDA to make recommendations and to cooperate with state and local authorities to ensure the safety and wholesomeness of shellfish. Through the NSSP, state health personnel continually inspect and survey bacteriological conditions in shellfish-growing areas. Any contaminated location is supervised or patrolled so that shellfish cannot be harvested from the area.

A major concern is the ability of shellfish to concentrate radioactive material, insecticides, and other chemicals from their environment. Thus, one aspect of the NSSP is to ensure that shellfish-growing areas are free from sewage pollution and toxic industrial waste. **Pesticide residues** in shellfish are usually quantitated by the gas chromatographic technique (Chap. 14), and **heavy metals** such as mercury are commonly quantitated by inductively coupled plasma mass spectrometry (ICP-MS) (Chap. 9, Sects. 9.5 and 9.6). Another safety problem with regard to shellfish is the control of **natural toxins**, which is a separate issue from sanitation. The naturally occurring toxins are produced by planktonic organisms, and testing is conducted using a variety of assays. Control of this toxicity is achieved by a careful survey followed by prohibition of harvesting from locations inhabited by toxic shellfish.

2.5 SPECIFICATIONS FOR FOODS PURCHASED BY GOVERNMENT AGENCIES

Large amounts of food products are purchased by federal agencies for use in domestic (e.g., school lunch) and foreign programs, prisons, veterans' hospitals, the armed forces, and other organizations. Specifications or descriptions developed for many food products are used by federal agencies for procurement of foods to ensure the safety and quality of the product specified. Such specifications or descriptions often include information that requires assurance of chemical composition and/or specified microbial quality. Often specified with the chemical composition requirements is the method of analysis required (e.g., AOAC or AOCS methods). Many such specification documents are a **commercial item description** (CID) or a **federal purchase program specification** (FPPS), often established by the USDA [21] but used across multiple federal agencies. Several examples follow: (1) canned tuna FPPS specifies the content of salt/sodium, methylmercury, and histamine [22]; (2) macaroni and cheese mix CID specifies fat and sodium contents [23], and (3) syrup CID specifies Brix, ash, and color [24].

2.6 International Standards and Policies

With the need to compete in the worldwide market, employees of food companies must be aware that allowed food ingredients, names of food ingredients, required and allowed label information, and standards for foods and food ingredients all differ between countries. For example, colorings and preservatives allowed in foods differ widely between countries, and nutrition labeling is not universally required. To develop foods for, and market foods in, a global economy, one must seek such information from international organizations and from organizations in specific regions and countries.

2.6.1 Codex Alimentarius

The **Codex Alimentarius Commission** (*Codex Alimentarius* is Latin for "code concerned with nourishment") was established in 1962 by two United Nations organizations, the Food and Agriculture Organization (FAO) and the World Health Organization (WHO), to develop international standards and safety practices for foods and agricultural products [25]. The standards, published in the *Codex Alimentarius*, are intended to protect consumers' health, ensure fair business practices in food trade, and facilitate international trade of foods.

The *Codex Alimentarius* is published in 13 volumes: 1 on general requirements (including labeling, food additives, contaminants, irradiated foods, import/export inspection, and food hygiene), 9 on the standards and codes of practice compiled on a commodity basis, 2 on residues of pesticides and veterinary drugs in foods, and 1 on the methods of analysis and sampling (Table 2.6). Codex has made efforts to validate and harmonize methods of food safety analysis among countries and regions, help maintain the smooth flow of international commerce, and ensure appropriate decisions on food exports and imports. The setting of international standards on food quality by Codex has been a high priority in world trade to minimize "nontariff" trade barriers. International trade of food and raw agricultural products has increased due to reduced economic trade restrictions and tariffs imposed. Codex has an online listing to search for Codex standards of various food products [25].

Table 2.6 Content of the *Codex Alimentarius* [25]

Volume	*Subject*
1A	General requirements
1B	General requirements (food hygiene)
2A	Pesticide residues in foods (general text)
2B	Pesticide residues in foods (maximum residue limits)
3	Residues in veterinary drugs in foods
4	Foods for special dietary uses
5A	Processed and quick-frozen fruits and vegetables
5B	Fresh fruits and vegetables
6	Fruit juices
7	Cereals, pulses (legumes) and derived products, and vegetable proteins
8	Fats and oils and related products
9	Fish and fishery products
10	Meat and meat products, soups and broths
11	Sugars, cocoa products and chocolate, and miscellaneous products
12	Milk and milk products
13	Methods of analysis and sampling

2.6.2 Other Standards

Other international, regional, and country-specific organizations publish standards relevant to food composition and analysis. For example, the **Saudi Arabian Standards Organization** (SASO) publishes standards documents (e.g., labeling, testing methods) important in the Middle East (except Israel), and the **European Commission** sets standards for foods and food additives for countries in the European Economic Community (EEC). In the USA, the Food Ingredients Expert Committee, which operates as part of the US Pharmacopeia, sets standards for the identification and purity of food additives and chemicals, published as the **Food Chemicals Codex** (FCC) [26]. For example, a company may specify in the purchase of a specific food ingredient that it be "FCC grade." Numerous countries other than the USA adopt FCC standard. At an international level, the **Joint FAO/WHO Expert Committee on Food Additives** (JECFA) sets standards for purity of food additives [27]. The Codex Alimentarius Commission is encouraged to utilize the standards established by JECFA. Standards established by FCC and JECFA are used by many countries as they develop their own standards.

2.7 SUMMARY

Various kinds of standards set for certain food products by federal agencies make it possible to get essentially the same food product whenever and wherever purchased in the USA. The standards of identity set by the FDA and USDA define what certain food products must consist of. The USDA and NMFS of the Department of Commerce have specified grade standards to define the attributes for certain foods. Grading programs are voluntary, while inspection programs may be either voluntary or mandatory, depending on the specific food product.

While the FDA has the broadest regulatory authority over most foods, responsibility is shared with other regulatory agencies for certain foods. The USDA has significant responsibilities for meat and poultry, the NOAA and the NMFS for seafood, and the TTB for alcoholic beverages. The FDA, the USDA, state agencies, and the dairy industry work together to ensure the safety, quality, and economic integrity of milk and milk products. The FDA, the EPA, and state agencies work together in the NSSP to ensure the safety and wholesomeness of shellfish. The EPA shares responsibility with the FDA for control of pesticide residues in foods and has the responsibility for drinking water safety and the composition of effluents from food processing plants. The CBP receives assistance from the FDA and USDA in its role to ensure the safety and economic integrity of imported foods. The FTC works with the FDA to prevent deceptive advertising of food products, as affected by food composition and labels. The chemical composition of foods is often an important factor in determining the quality, grade, and price of a food. Government agencies that purchase foods for special programs often rely on detailed specifications that include information on food composition.

International organizations have developed food standards and safety practices to protect consumers, ensure fair business practices, and facilitate international trade. The Codex Alimentarius Commission is the major international standard-setting group for food safety and quality. Certain regional and country-specific organizations also publish standards related to food composition and analysis.

2.8 STUDY QUESTIONS

1. Differentiate "standards of identity," "standards of quality," and "grade standards" with regard to what they are and which federal agency establishes and regulates them.
2. Government regulations regarding the composition of foods often state the official or standard method by which the food is to be analyzed. Give the full name of three organizations that publish commonly referenced sources of such methods.
3. For each type of product listed below, identify the government agency (or agencies) that has regulatory or other responsibilities for quality assurance. Specify the general nature of that responsibility and, if given, the specific types of analyses that would be associated with that responsibility.
   (a) Frozen fish sticks
   (b) Contaminants in drinking water
   (c) Dessert wine
   (d) Grade A milk
   (e) Frozen oysters
   (f) Imported chocolate products
   (g) Residual pesticide on wheat grain
   (h) Corned beef
4. Upon completing your college degree, you are employed by a major US food company that processes fruits and vegetables.
   (a) Where, specifically, would you look to find if a standard of identity exists for each of your processed products? What kind of information does such a standard include?
   (b) What US government agency sets the standards of identity for such products?
   (c) What are the minimum standards called that are set for some fruit and vegetable products?
   (d) What government agency sets the grade standards that you may want to use as a quality control tool and in marketing your products?
   (e) You are concerned about pesticide tolerances for the fruits and vegetables you process. What government agency sets those tolerances?
   (f) What government agency enforces pesticide tolerances?
   (g) For nutrition labeling purposes for your products, you want to check on official methods of analysis. Where, specifically, should you look? (see Chap. 3).
   (h) You want to check the detailed rules on nutrition labeling that would apply to your products. Where, specifically, would you look to find those rules?
   (i) You are considering marketing some of your products internationally. What resource could you check to determine whether there are international standards and safety practices specified for those products?
5. For each of the following, give the complete name of the US federal agency or organization that has the responsibility described relevant to food analysis:
   (a) Has primary authority over labeling of foods.
   (b) Has primary authority over advertising of foods.
   (c) Develops effluent guidelines and standards for food processing plants.
   (d) Administers voluntary grading services for dairy products.
   (e) Works with the FDA and EPA to ensure seafood safety.
   (f) Handles issues regarding the composition and labeling of most alcoholic beverages.
   (g) Works with the FDA and the USDA to ensure the safety of imported products.
   (h) Works with a voluntary cooperative program responsible for requirements to grow, handle, and process shellfish.

REFERENCES

1. Kanovsky SM, Pines, WL (2020) A practical guide to FDA's food and drug law and regulation, 7th edn. Food and Drug Law Institute, Washington, DC
2. Adams DG, Cooper RM, Hahn MJ, Kahan JS (2015) Food and drug law and regulation, 3rd edn. Food and Drug Law Institute, Washington, DC
3. Sanchez MC (2018) Food law and regulation for non-lawyers. A US perspective. 2nd edn, Springer, New York
4. Anonymous (2022) Code of federal regulations. Titles 7, 9, 21, 27, 40, 50. US Government Printing Office, Washington, DC. Available on Internet.
5. AOAC International (2023) Official methods of analysis, 22th edn, 2023 (Online). AOAC International, Rockville, MD
6. FDA (2023). Standards of identity for food. (04/07/2023) Food and Drug Administration. Washington DC. Available on Internet. https://www.fda.gov/food/food-labeling-nutrition/standards-identity-food#:~:text=SOIs%20were%20developed%20to%20help,consistent%20with%20what%20consumers%20expect

7. FDA (2023). Infant formula guidance documents & regulatory information. (02/28/2023) Food and Drug Administration, Washington, DC. Available on Internet. https://www.fda.gov/food/guidance-documents-regulatory-information-topic-food-and-dietary-supplements/infant-formula-guidance-documents-regulatory-information
8. USDA (1983) U.S. standards for grades of orange juice. 10 Jan 1983. Processed Products Branch, Fruit and Vegetable Division, Agricultural Marketing Service, US Department. of Agriculture, Washington, DC. Available on Internet. https://www.ams.usda.gov/sites/default/files/media/Canned_Orange_Juice_Standard[1].pdf
9. FDA (2022) Fish and fisheries products hazards and controls, June 2022 edn. Center for Food Safety and Applied Nutrition, Office of Food Safety, Food and Drug Administration, Washington, DC. Available on Internet. https://www.fda.gov/media/80637/download
10. FDA (1994) Pesticide analytical manual, vol I (PAMI) (updated Oct 1999) (content current as of 09/19/2018) (Methods which detect multiple residues) and vol II (PAMII) (updated Jan 2002) (Methods for individual pesticide residues), 3rd edn. National Technical Information Service, Springfield, VA. Available on Internet. (content current as of 08/01/2018) https://www.fda.gov/food/laboratory-methods-food/pesticide-analytical-manual-volume-i-pam-3rd-edition, https://www.fda.gov/food/laboratory-methods-food/pesticide-analytical-manual-volume-ii
11. American Public Health Association (APHA), American Water Works Association, Water Environment Federation (2023) Lipps WC, Braun-Howard EB, Baxter TE (eds) (2023) Standard methods for the examination of water and wastewater, 24th edn. APHA Press, Washington CD
12. EPA (1983) Methods of chemical analysis of water and wastes. EPA-600/4-79-020, March 1979. Reprinted in 1983. EPA Environmental Monitoring and Support Laboratory, Cincinnati, OH. Available on Internet.
13. American Society for Testing Materials (ASTM) International (2023) Annual book of ASTM standards, section 11, water and environmental technology vol 11.02, water (II). ASTM International, West Conshohocken, PA. Available on Internet.
14. US International Trade Commission (USITC) (2023) Official harmonized tariff schedule of the United States. USITC. Available on Internet.
15. US Department of Health and Human Services, Public Health Service, Food and Drug Administration (2019) Grade A pasteurized milk ordinance. Washington, DC. Available on Internet.
16. Wehr HM, Frank JF (eds) (2004) Standard methods for the examination of dairy products, 17th edn. American Public Health Association, Washington, DC. Available on Internet.
17. USDA (2012) General specifications for dairy plants approved for USDA inspection and grading service. Dairy Program, Agricultural Marketing Service, US Department of Agriculture, Washington, DC. Available on Internet.
18. USDA (2001) US standards for grades of nonfat dry milk (spray process) February 2, 2001. Agricultural Marketing Service, US Department of Agriculture, Washington DC. Available on Internet. http://www.ams.usda.gov/sites/default/files/media/Nonfat_Dry_Milk_%28Spray_Process%29_Standard%5B1%5D.pdf
19. USDA (2011) Milk for manufacturing purposes and its production and processing, recommended requirements. Dairy Program, Agricultural Marketing Service, US Department of Agriculture, Washington, DC. Available on Internet.
20. FDA (1995) National shellfish sanitation program. Food and Drug Administration, Washington, DC. Available on Internet.
21. USDA (2023) Product specifications & requirements. Specifications by products. US Department of Agriculture, Washington, DC. Available on Internet.
22. USDA (2016) Federal purchase program specification (FPPS) for canned/pouch tuna. Agricultural Marketing Service, Livestock and Poultry Program, Standards and Specification Division, US Department of Agriculture, Washington, DC. Available on Internet.
23. USDA (2019) Commercial item description (CID). Macaroni and cheese mix. A-A-20308B. May 23, 2019. General Services Administration, Specifications Section, Washington, DC. Available on Internet.
24. USDA (2017) Commercial item description. Syrup (not including maple syrup). A-A-20124E. General Services Administration, Specifications Section, Washington, DC. Available on Internet.
25. Joint FAO/WHO Codex Alimentarius. International Food Standards. Food standards programme. Codex Alimentarius Commission, Food and Agriculture Organization of the United Nations/World Health Organization, Rome, Italy. Online description of standards: https://www.fao.org/fao-who-codexalimentarius/en/ Online listing, by food product: https://www.fao.org/fao-who-codexalimentarius/codex-texts/list-standards/en/
26. US Pharmacopeia (USP) (2023) Food chemicals codex (FCC), 13th edn. United Book Press, Baltimore, MD (online subscription) https://www.foodchemicalscodex.org/
27. FAO and WHO (2023) Online edition: Combined compendium of food additive specifications. Joint FAO/WHO Committee on Food Additives (JECFA). Web version. FAO, Rome, Italy. https://www.fao.org/food/food-safety-quality/scientific-advice/jecfa/jecfa-additives/en/

Nutrition Labeling

3

S. Suzanne Nielsen

Contents

3.1 INTRODUCTION

Nutrition labeling regulations differ in countries around the world. The focus of this chapter is on nutrition labeling regulations in the USA, as specified by the **Food and Drug Administration** (FDA), with a brief summary of regulations for the **Food Safety and Inspection Service** (FSIS) of the **United States Department of Agriculture** (USDA). A major reason for analyzing the chemical components of foods in the USA is nutrition labeling regulations. Nutrition label information is not only legally required in many countries but is also of increasing importance to consumers as they focus more on health and wellness.

The FDA was authorized under the 1906 Federal Food and Drug Act and the 1938 **Federal Food, Drug, and Cosmetic** (FD&C) *Act* to require certain types of food labeling [1, 2]. This labeling information includes the amount of food in a package, its common or usual name, and its ingredients. In 1973, the FDA promulgated regulations that permitted, and in some cases required, food to be labeled with regard to their nutritional value. The **1990 Nutrition Labeling and Education Act** (NLEA) [1, 3] modified the 1938 FD&C Act to regulate nutrition labeling. Additionally, the 1997 **Food and Drug Administration Modernization Act** (FDAMA) [4] also amended the FD&C Act and included provisions that sped up the process for approving health and nutrient content claims. The FDA amended in 2016 its nutrition labeling regulations for conventional foods and dietary supplements [5]. The current FDA regulations [6], as they relate to food analysis, are the focus of this chapter.

The FDA and FSIS of the USDA have coordinated their regulations for nutrition labeling. FDA regulations, as related to food analysis, will be described in some detail in Sect. 3.2, focusing on the following:

1. What nutrients must be analyzed
2. How samples are to be collected
3. What methods of analysis are to be used
4. How data are to be reported
5. How data can be used to calculate caloric content
6. How data can be used for claims made on the food label

S. S. Nielsen (✉)
Department of Food Science, Purdue University,
West Lafayette, IN, USA
e-mail: nielsens@purdue.edu

B. P. Ismail, S. S. Nielsen (eds.), *Nielsen's Food Analysis*, Food Science Text Series,
https://doi.org/10.1007/978-3-031-50643-7_3

Following coverage of FDA regulations in Sect. 3.2, Sect. 3.3 will give a general discussion of the similarities and differences between FDA and USDA regulations. Complete details of the current nutrition labeling regulations of both FDA and USDA are available in the *Federal Register* (the chronological publication of proposed regulations and final regulations) and the *Code of Federal Regulations* (CFR) (a subject arrangement of regulations).

In developing a nutrition label for a food product, it is important to review the details of the regulations in the CFR and utilize other routinely updated resources available via the Internet. During the product development process, the effect of formulation changes on the nutrition label may be important. As an example, a small change in the amount of an ingredient may determine whether a product can be labeled low fat. As a result, the ability to immediately approximate how a formulation change will impact the nutrition label can be very useful. Food labels are often generated using a database calculation method to calculate the nutrient values based on the recipe provided, so no physical sample is analyzed. However, laboratory testing generally offers greater accuracy since the results reflect the actual product on the market. Analytical testing is especially important to obtain accurate nutrient values for certain types of foods (e.g., fried, fermented, brewed, brined, marinated). The use of nutrient databases and computer programs designed for preparing and analyzing nutrition labels can be valuable and can simplify the process of preparing a nutrition label (see Sect. 3.2.2.3 for more on FDA nutrient databases). The use of computer programs to prepare nutrition labels is beyond the scope of this chapter. However, an example computer program (TechWizard™, Owl Software) and a description of how this program can be used to prepare a nutrition label are found in the laboratory manual that accompanies this text [7].

3.2 U.S. FOOD AND DRUG ADMINISTRATION FOOD LABELING REGULATIONS

For each aspect of nutrition labeling regulations related to the food analysis described below, only FDA labeling requirements are covered. While the focus here is on **mandatory nutrition labeling** (21 CFR 101.9), it should be noted that the FDA has guidelines for **voluntary nutrition labeling** of raw fruits, vegetables, and fish (21 CFR 101.45) [8].

3.2.1 Mandatory Nutrition Labeling

3.2.1.1 Format

FDA regulations require nutrition labeling for most foods offered for sale and regulated by the FDA (21 CFR 101.9).

Fig. 3.1 Mandatory information shown on an example nutrition label. Food and Drug Administration (FDA) [21 CFR 101.9 (d)(12)]. Standard vertical label. Nutrition Labeling and Education Act of 1990, amended 2016. (Courtesy of the FDA, Washington, DC)

Certain nutrient information is required on the label, and other information is voluntary. The standard vertical format label showing mandatory nutrition information on food labels [21 CFR 101.9 (d)(12)] is displayed in Fig. 3.1. A similar standard vertical format exists for listing both mandatory and voluntary nutrients on a label. Note that all nutrients, including vitamins and minerals, must be reported by weight, in addition to relevant rules about expressing as a percentage of the daily value (DV). A simplified format for nutrition information may be used under certain specific conditions. Also, certain foods are exempt from mandatory nutrition labeling requirements.

3.2.1.2 Daily Values and Serving Size

"Daily value" (DV) is a generic term used to describe two separate terms, which are (1) "reference daily intake" (RDI) and (2) "daily reference value" (DRV). The term "RDI" is used for essential vitamins and minerals (Table 3.1), while the term "DRV" is used for select other food components (Table 3.2). The DRVs are based on a 2000 reference calorie intake. Nutrient content values and percentage DV calculations for the nutrition label are based on serving size [21 CFR 101.12 (b)]. The serving size is expressed in terms of "reference amounts customarily consumed" (RACC) per

Table 3.1 Reference daily intakes (RDIs) for vitamins and minerals essential for human nutrition

Nutrient	*RDI*
Vitamin A	900 µg
Vitamin C	90 mg
Calcium	1300 mg
Iron	18 mg
Vitamin D	20 µg
Vitamin E	15 mg
Vitamin K	120 µg
Thiamin	1.2 mg
Riboflavin	1.3 mg
Niacin	16 mg
Vitamin B_6	1.7 mg
Folate	400 µg
Vitamin B_{12}	2.4 µg
Biotin	30 µg
Pantothenic acid	5 mg
Phosphorus	1250 mg
Iodine	150 µg
Magnesium	420 mg
Zinc	11 mg
Selenium	55 µg
Copper	0.9 mg
Manganese	2.3 mg
Chromium	35 µg
Molybdenum	45 µg
Chloride	2300 mg
Potassium	4700 mg
Choline	550 mg

From Code of Federal Regulations (2023) [6]
Values are for adults and children aged 4 or more years. RDI values have also been established for infants, children under 4 years of age, and for pregnant and lactating women

Table 3.2 Daily reference values (DRVs) of food components[a]

Food component	*DRV*
Fat	78 g
Saturated fatty acids	20 g
Cholesterol	300 mg
Total carbohydrate	275 g
Fibers	28 g
Sodium	2300 mg
Protein	50 g
Added sugars	50 g

From Code of Federal Regulations (2023) [6]
[a]Based on the reference calorie intake of 2000 calories for adults and children ≥4 years

eating occasion. It is given by product category and expressed as a common household measure [e.g., cup(s), piece(s), fluid ounces, tablespoon(s)] [21 CFR 101.12 (b)]. The labeled serving size and reference amount are important since the use of nutrient content claims (Sect. 3.2.3) is dependent on the serving size and the reference amount.

3.2.1.3 Rounding Rules

Increments for the numerical expression of quantity per serving are specified for all nutrients (Table 3.3) [21 CFR 101.9 (c)]. Values obtained from nutrient analysis are all rounded for reporting on the nutrition label, following highly specific rules. For example, sodium content is to be reported to the nearest 5 mg amount up to and including 140 mg and to the nearest 10 mg above 140 mg. Sodium content can be reported as 0 if there are less than 5 mg per serving.

3.2.1.4 Caloric Content

Calories on the label can be calculated in numerous ways. A **calorie** is a measure of energy defined as the amount of heat required to raise the temperature of 1 g of water to 1 °C (1 calorie = 4.184 joules). The unit used to express the energy content of food and the body's energy requirements is Calorie or "kilocalorie" (kcal) (also called "Cal"), which equals 1000 calories. In this chapter, the term Calorie is used to express caloric content. FDA regulations specify multiple methods by which caloric content may be calculated [21 CFR 101.9 (c) (1)]:

1. Specific Atwater factors for calories per gram of protein, total carbohydrate, and total fat
2. The general factors of 4, 4, and 9 Cal/g of proteins, total carbohydrates, and total fats, respectively
3. The general factors of 4, 4, and 9 Cal/g of proteins, total carbohydrates (less the amount of nondigestible carbohydrates and sugar alcohols), and total fats, respectively. (Note: Regulations specify a general factor of 2 Cal/g for soluble nondigestible carbohydrates shall be used, and general factors for the caloric values of sugar alcohols is provided in 21 CRF 101.9 (c)(1)(i)(F), i.e., 0–3.0, depending on the specific sugar alcohol.)
4. Data for specific food factors for particular foods or ingredients approved by the FDA
5. Bomb calorimetry data subtracting 1.25 Cal/g protein to correct for incomplete digestibility
6. Using general factors for the caloric value of specific sugar alcohols (value given for each sugar alcohol)

3.2.1.5 Carbohydrate Food Components

Nutrition labeling regulations include special details regarding various carbohydrates [21 CFR 101.9 (c) (6) (i–iv)], summarized in Chap. 19, Sects. 19.1 and 19.6.1. The CFR details cover "total carbohydrates," dietary fibers (including "soluble fibers" and "insoluble fibers"), "total sugars," "added sugars," and "sugar alcohols." Some of these carbohydrate components are measured by specific methods (e.g., see fiber analysis methods described in Ch. 19, Sect. 19.6.2), and some are simply calculated. The recent FDA guidelines [9] related to "added sugars" have allowed some accommodation for single-ingredient foods (honey, maple syrup, pure sugars, and syrups, etc.) and for

Table 3.3 Rounding rules for declaring nutrients on the nutrition label

Nutrient/serving	*Increment rounding*[a,b]	*Insignificant amount*
Calories, calories from saturated fats	< 5 Cal: Express as 0 ≤ 50 Cal: Express to the nearest 5 Cal increment > 50 Cal: Express to the nearest 10 Cal increment	<5 Cal
Total fat, trans fats, polyunsaturated fats, monounsaturated, saturated fats	< 0.5 g: Express as 0 < 5 g: Express to the nearest 0.5 g increment ≥ 5 g: Express to the nearest 1 g increment	<0.5 g
Cholesterol	< 2 mg: Express as 0 2–5 mg: Express as "less than 5 mg" > 5 mg: Express to the nearest 5 mg increment	<2 mg
Sodium, potassium	< 5 mg: Express as 0 5–140 mg: Express to the nearest 5 mg increment > 140 mg: Express to the nearest 10 mg increment	<5 mg
Total carbohydrate, total sugars, added sugars, sugar alcohols, dietary fibers, soluble fibers, insoluble fibers, proteins	< 2% of RDI: May be expressed as: Zero An asterisk that refers to the statement "contains less than 2% of the daily value of this (these) nutrient (nutrients)" (or use symbol < in place of "less than") For vitamin D, calcium, iron, potassium: Statement "not a significant source of _____ (listing the vitamins or minerals omitted)" ≤ 10% of RDI: Express to the nearest 2% increment > 10% to ≤50% of RDI: Express to the nearest 5% increment > 50% of RDI: Express to the nearest 10% increment	<2% RDI
Vitamins and minerals	<0.1 mg: Express as 0 ≤0.8 mg: Express to the nearest 0.1 mg >0.8 mg: Express to the nearest 0.2 mg	
Fluoride		

Summarized from Code of Federal Regulations (2023) [6]

Notes below taken from: 21 CRF 101.9 (c); Food Labeling Guide. Appendix H. FDA rounding rules. 2015. Center for Food Safety and Applied Nutrition, Food and Drug Administration, Washington, DC

[a]To express nutrient values to the nearest 1-g increment, for amounts falling exactly halfway between two whole numbers or higher (e.g., 2.5–2.99 g), the values round up (e.g., 3 g). For amounts less than halfway between two whole numbers (e.g., 2.01–2.49 g), the values round down (e.g., 2 g)

[b]Percentage DV shall be calculated by dividing either the amount declared on the label for each nutrient or the actual amount of each nutrient (i.e., before rounding) by the DRV for the nutrient, except that the percentage for protein shall be calculated as specified in 21 CFR (c)(7)(ii) (described in Sect. 3.2.1.6 of this textbook chapter)

When rounding %DV for nutrients other than vitamins and minerals, when the %DV values fall exactly halfway between whole numbers or higher (e.g., 2.5–2.99), the values round up (e.g., 3%). For values less than halfway between two whole numbers (e.g., 2.01–2.49), the values round down (e.g., 2%). (Note: Sodium %DV is rounded off like for other nutrients that have a DRV, rather than like other minerals that have an RDI)

foods such as cranberries (juices, dried cranberries), to which sugar has been added, since they are naturally very tart and require some added sugar to make them palatable. In both special cases, the percentage DV (%DV) number is followed by the "ϯ" symbol, which can refer to a truthful explanation located either directly below (i.e., single-ingredient foods) or outside of (i.e., noted cranberry food products) the nutrition fact panel.

3.2.1.6 Protein Quality

Reporting the amount of protein as a percentage of its daily value on FDA-regulated foods is optional, except if a protein claim is made for the product or if the product is represented or purported to be used by infants or children under 4 years of age, in which case the statement is required [21 CFR 101.9 (c) (7)]. For infant foods, the corrected amount of protein per serving is calculated by multiplying the actual amount of protein (g) per serving by the relative protein quality value. This relative quality value is the **protein efficiency ratio** (PER) value of the subject food product divided by the PER value for casein. For foods represented or purported for adults and children 1 year or older, the corrected amount of protein per serving is equal to the actual amount of protein (g) per serving multiplied by the **protein digestibility-corrected amino acid score** (PDCAAS). Both the PER and PDCAAS methods to assess protein quality are described in Chap. 24, Sect. 24.3.2. The FDA allows use of the general factor 6.25 to calculate the protein content from the measured nitrogen content, except when official AOAC procedures require use of a different conversion factor (described in Chap. 18, Sect. 18.2.1.3).

3.2.2 Compliance

3.2.2.1 Sample Collection

Random sampling techniques are used by the FDA to collect samples to be analyzed for compliance with nutrition labeling regulations. A "lot" is the basis for sample collection by the FDA, defined as "a collection of primary containers or units of the same size, type, and style produced under conditions as nearly uniform as possible, and designated by a common container code or marking, or in the absence of any common container code or marking, a day's production." The sample used by the FDA for nutrient analysis consists of a "composite of 12 subsamples (consumer units), taken 1 from each of 12 different randomly chosen shipping cases, to be representative of a lot" [21 CFR 101.9 (g)].

3.2.2.2 Methods of Analysis

The FDA recommends that the nutrient values for labeling be based on product composition, as determined by laboratory analysis of each nutrient. Unless a particular method of analysis is specified for nutrients listed in 21 CFR 101.9 (c), the FDA states that appropriate methods of AOAC International published in the *Official Methods of Analysis* [10] are to be used [21 CFR 101.9 (g) (2)]. Other reliable and appropriate methods can be used if no AOAC method is available or appropriate. If scientific knowledge or reliable databases have established that a nutrient is not present in a specific product (e.g., dietary fibers in seafood, cholesterol in vegetables), then the FDA does not require analyses for the nutrients.

Table 3.4 provides commonly used AOAC official methods associated with nutrition label components, as reported by analytical laboratories. Also listed for each nutrition label component is the book chapter in this textbook that describes the type of method identified, along with associated methods.

3.2.2.3 FDA-Approved Databases

Compliance with the food labeling regulations can be obtained and aided by the use of FDA-approved databases [21 CFR 101.9 (g) (8)], which have been computed using FDA guidelines, and using foods that have been handled under good manufacturing practice conditions to prevent nutritional losses. The FDA refers to the "FDA Nutrition Labeling Manual – A Guide for Developing and Using Data Bases" [11]. This FDA document describes the nature of the

Table 3.4 Nutrition label components: commonly used AOAC International official methods and food analysis textbook coverage

Nutrient	*Name of method*	*AOAC official method number*	*AOAC official method locator number*	*Food analysis book chapter*
Calories	One of multiple methods allowed by the FDA: Calories = (g of carbohydrate x 4 Cal/g) + (g of protein x 4 Cal/g) + (g of fat x 9 Cal/g)			1
Total carbohydrate	Calculation for proximate composition (100% = total fat + protein + total carbohydrate + moisture + ash)	Proximate analysis		1, 19
Dietary fibers	Enzymatic gravimetric method for total, soluble, and insoluble dietary fibers	991.43 2011.25	32.1.17 32.1.43	19 19
Total sugars	High-performance liquid chromatography (HPLC)	977.20	44.4.13	19
Added sugars	No analytical method. Must be calculated based on formulation	–	–	
Proteins	Dumas	992.15	39.1.16	18
Total fat	Gas chromatography	996.06	41.1.28A	17
Saturated fat	Gas chromatography	996.06	41.1.28A	17, 23
Trans fat	Gas chromatography	996.06	41.1.28A	17, 23
Cholesterol	Gas chromatography	976.26	45.4.06	23
Moisture (to calculate total carbohydrate)	Forced draft oven at 105 ± 1 °C for 4 h (note: Other methods would be official/more appropriate for certain foods)	925.10	32.1.03	15
Ash (to calculate total carbohydrate)	Dry ashing using a muffle furnace at 550 °C	923.03	32.1.05	16
Sodium	Inductively coupled plasma–optical emission spectroscopy	2011.14	50.1.37	9
Vitamin D	Liquid chromatography–mass spectrometry	2002.05	45.1.22A	20
Calcium	Inductively coupled plasma–optical emission spectroscopy	2011.14	50.1.37	9
Iron	Inductively coupled plasma–optical emission spectroscopy	2011.14	50.1.37	9
Potassium	Inductively coupled plasma–optical emission spectroscopy	2011.14	50.1.37	9

The AOAC's official methods listed are common use methods, based on information (in 2023) from commercial analytical laboratories that perform nutrition label analyses. Other AOAC official methods are possible for many nutrients and, in some cases, may be more appropriate, depending on the nature of the food or ingredient databases and the responsibility of the groups that develop them (e.g., manufacturers, organizations, or trade associations that represent a group of manufacturers). The FDA encourages the industry to submit nutrition labeling databases to the agency for review, but such submission for the purpose of nutrition labeling is voluntary. The nutrition labeling databases are mostly in the "finished food" category, as a collection of data from nutrient analysis of specific products or commodities. The nutrient data submitted to the FDA are supported by information on sampling strategies, analytical methods used, and statistical treatment of data. The FDA requests that the organization building the database for each product includes a table with the method used to analyze each nutrient. A second type of nutrition labeling database in an ingredient or "recipe" database, for which software is used to calculate the label values from the nutrient content of ingredients based on the product's recipe but taking into account nutrient losses due to processing.

3.2.2.4 Levels for Compliance

The FDA recognizes the natural variability in the nutrient content of food, which is a result of manufacturing processes and environmental conditions. Therefore, the agency monitors accuracy of nutrient content information for compliance based on two classes of nutrients, described in Table 3.5. Class I nutrients are those that have been added to foods, such as vitamin D fortification in dairy products. A Class I nutrient must be formulated to meet or exceed the declared value on the label throughout the product's shelf life. Class II nutrients are naturally present in the food and must be at a level equal to at least 80% but no more than 120% of the declared label value throughout the product's shelf life.

Table 3.5 Basis for compliance of nutrition labeling regulation by the US Food and Drug Administration and the Food Safety and Inspection Service of the US Department of Agriculture

Class of nutrients	*Purposes of compliance*	*Nutrients regulated*	*Percentage (%) required*[a]
I	Added nutrients in fortified or fabricated foods	Vitamins, minerals, proteins, dietary fibers	≥100%
II	Naturally occurring (indigenous) nutrients	Vitamins, minerals, proteins, total carbohydrates, dietary fibers, soluble fibers, insoluble fibers, polyunsaturated or monounsaturated fats	≥80%
II	Naturally occurring (indigenous) nutrients	Calories, total sugars, added sugars (when the only source in food is added sugars), total fat, saturated fat, trans fat, cholesterol, sodium	≤120%

Summarized from Code of Federal Regulations (2023) [6]
[a]Amount of nutrients required in a food sample as a percentage of the label declaration or else the product is considered misbranded

Compliance regulations point to the importance of a robust analytical nutrition maintenance plan. This plan should include statistically significant sampling, with consideration given to multiple manufacturing locations, lines, and production lots. Samples should be carefully handled and prepared for accurate chemical analysis to ensure that the nutrition label information is correct. For example, a product fortified with iron would be considered misbranded if it contained less than 100% of the label declaration. A product that naturally contains dietary fibers would be considered misbranded if it contained less than 80% of the label declaration. A product would be considered misbranded if it had a caloric content greater than 20% in excess of the label declaration. Reasonable excesses over labeled amounts (of a vitamin, mineral, protein, total carbohydrate, polyunsaturated or monounsaturated fat, or potassium) or deficiencies below label amounts (of calories, sugars, total fats, saturated fats, cholesterol, or sodium) are acceptable within current good manufacturing practices (cGMP). Noncompliance with regard to a nutrition label can result in warning letters, recalls, seizures, and prosecution (21 CFR 1.21).

3.2.3 Nutrient Content Claims

The FDA has defined *nutrient content claims* that characterize the level of a nutrient, according to a specific definition (21 CFR 101.13, 101.54–101.67). Some of the terms allowed include "free," "low," "lean," "light," "reduced," "less," "fewer," "added," "extra," "plus," "fortified," "enriched," "good source," "contains," "provides," "more," "high," "rich in," "excellent source of," and "high potency." Of these terms, "less" (or "fewer"), "more," "reduced," "added" (or "extra," "plus," "fortified," and "enriched"), and "light" or "lite" are relative terms and require label information about the food product that is the basis of the comparison. The percentage difference between the original food (reference food) and the food product being labeled must be listed on the label for comparison. Many of the nutrient content claims related to the calorie content of foods require comparison to an appropriate reference food (21 CFR 101.60).

The use of nutrient content claims on food labels is typically based on the content of specific nutrients per reference amount or per serving. For example, the terms "free" or "low" for total fat, saturated fat, cholesterol, and sodium relate to specific amounts of those nutrients. Also, terms such as "high," "excellent source," and "enriched" require that the nutrient identified contain a specific percentage of the DV per reference amount. The term "healthy" or its derivatives may be used on the label or in labeling of foods under conditions defined by the FDA, relying on highly specific levels of various nutrients. Note that the FDA requirements on nutrient content claims do not apply to infant formulas and medical foods.

3.2.4 Health Claims

The FDA has defined and will allow claims for certain relationships between a nutrient or a food and the risk of a disease or health-related condition (21 CFR 101.14). The FDA utilizes several types of oversights to determine which health claims may be used in the labeling of food or dietary supplements, leading to three categories of health claims:

NLEA-authorized health claims characterize a relationship between a food, food component, dietary ingredient, or dietary supplement and risk of a disease (Table 3.6). The agency is prompted to authorize these health claims as a result of a petition, followed by an extensive review of the scientific literature, using the significant scientific agreement (SSA) standard.

The second category is Health Claims Based on Authoritative Statements, permitted by the FDA Modernization Act of 1997 (FDAMA). These claims are approved based on authoritative statements from scientific bodies within the US government or the National Academy of Sciences (NAS).

Finally, the FDA considers Qualified Health Claims for which there is evidence relating a nutrient or food to a disease state, but the evidence does not meet the SSA standard.

Table 3.6 Nutrition Labeling Education Act (NLEA)-authorized health claims

Claim	*CFR*[a] *reference*
Calcium, vitamin D, and osteoporosis	21 CFR 101.72
Dietary lipids and cancer	21 CFR 101.73
Sodium and hypertension	21 CFR 101.74
Dietary saturated fat and cholesterol and	21 CFR 101.75 risk of coronary heart disease
Fiber-containing grain products, fruits, and vegetables and cancer	21 CFR 101.76
Fruits, vegetables, and grain products that contain fibers, particularly soluble fibers, and risk of coronary heart disease	21 CFR 101.77
Fruits and vegetables and cancer	21 CFR 101.78
Folate and neural tube defects	21 CFR 101.79
Dietary noncariogenic carbohydrate sweeteners and dental caries	21 CFR 101.80
Soluble fibers from certain foods and risk of coronary heart disease	21 CFR 101.81
Soy protein and risk of coronary heart disease	21 CFR 101.82
Plant sterol/stanol esters and risk of coronary heart disease	21 CFR 101.83

[a]*CFR* Code of Federal Regulations

If the FDA determines the evidence to be credible, they will issue a letter outlining their intention to exercise enforcement discretion related to use of the claim.

Most health claims are based on specific nutrients and therefore the chemical analysis of those food components. Manufacturers should diligently follow the Health Claim regulations and consider the overall positioning of their product to avoid inadvertently making an impermissible drug claim.

3.2.5 Structure/Function Claims

These claims describe the relationship between a nutrient or ingredient and the normal structure or function of the body, e.g., "calcium builds strong bones." The FDA does not pre-approve structure/function claims, but manufacturers must be able to substantiate them with credible scientific literature and nutrition analytical results to prove that the claims are truthful and not misleading.

3.3 U.S. DEPARTMENT OF AGRICULTURE FOOD LABELING REGULATIONS

The FDA and the FSIS of the USDA have generally coordinated their regulations for nutrition labeling, although there are some notable differences. USDA regulations require nutrition labeling of most meat or meat products (9 CFR 317.300–317.400) [12] and poultry products (9 CFR 381.400–381.500) [13]. However, there are some unique regulations on what the FDA regulates and what USDA regulates. For example, what the USDA covers regarding "egg products" is considered anything outside the shell, but FDA has the responsibility for "in-shell" eggs. The FDA regulates exotic meats and meat products (including closed-faced sandwiches), but an open-faced sandwich is regulated by the USDA.

Unlike products regulated by the FDA, there are some circumstances in which manufacturers of USDA-regulated products must submit labels for FSIS's approval prior to use. The FSIS maintains label approval regulations (9 CFR 412) [14, 15] and publishes a compliance guide to address the types of labels that must be submitted to FSIS staff for "sketch" approval, and those that do not need to be submitted are generically approved ("Sketch" is a concept of the label, showing all features, size, and location of the label.)

Historically, the two agencies maintained close harmony regarding interpretation of the regulations and changes made in regulations; however, the USDA has yet to adopt the changes to the nutrition facts panel found in the Revision of the Nutrition and Supplement Facts Label final rule published by the FDA in 2016 [5]. Consequently, manufacturers who wish to apply the updated nutrition facts panel format to products regulated by the USDA must submit labels to FSIS for "sketch" approval prior to use.

Some other differences that exist in the regulations are due at least partially to the inherent differences in the food products regulated by the FDA and USDA (USDA regulates only meat, poultry, and egg products). For example, (1) nutrients allowed on the label by voluntary declaration, (2) serving size regulations, (3) compliance procedures, and (4) nutrient content claims. One specific difference is that FSIS does not require trans fats as a mandatory nutrient but permits it to be declared voluntarily. Two other specific areas different for the USDA from FDA regulations are summarized below.

Regarding compliance, for nutritional analysis, the FSIS specifies the methods of the *USDA Chemistry Laboratory Guidebook* [16]. If no USDA method is available and appropriate for the nutrient, then methods in the *Official Methods of Analysis* of AOAC International [10] are to be used. If no USDA, AOAC International, or specified method is available and appropriate, the FSIS specifies the use of other reliable and appropriate analytical procedures as determined by the Agency. The FSIS provides information on how it collects samples for compliance analysis [9 CFR 317.309 (h), 381.409 (h)].

Regarding nutrient content claim differences between FDA- and USDA-regulated foods, the following are some examples:

1. "Enriched" and "Fortified" are not defined in the FSIS regulations.
2. "Lean" and Extra lean" are defined and approved for all USDA-regulated products but only for the FDA-regulated products of seafood, game meat, and meal products.
3. The term "___% Lean" is approved only for USDA-regulated products.

This section has only briefly summarized USDA nutrition labeling regulations and compared them to FDA regulations. The reader is referred to the CFR regulations listed above for all details of USDA nutrition labeling regulations.

3.4 SUMMARY

A major reason for analyzing the chemical components of food in the United States (and many other countries) is nutrition labeling regulation. The FDA and FSIS of the USDA have coordinated their regulations on nutrition labeling. Regulations that implement the NLEA of 1990 require nutrition labeling for most foods regulated by the FDA, and FSIS requires a similar label on most meat and poultry products. The FDA amended regulations in 2016 regarding nutrition information on the label in an attempt to help consumers maintain healthy dietary practices, but the USDA has yet to adopt these changes. Nutrition labeling regulations define the format for nutrition information and give the rules and methods to report specific information. Specifications include sample collection procedures, the method of analysis to be used, and the nutrient levels required to ensure compliance with nutrition labeling regulations. Specific nutrient content claims and health claims are allowed on the nutrition label. The nutrition labeling regulations covered in this chapter are only those closely linked to food analysis. Readers are referred to appropriate sections of the CFR for all details of FDA and USDA regulations.

3.5 STUDY QUESTIONS

1. Utilize the data in the table below that you obtained on the nutrient content of your cereal product (actual amount per serving) to help develop a nutrition label that meets FDA requirements under the NLEA, amended in 2016. Use appropriate rounding rules to complete the blank columns. If you wanted to report the protein content as a percentage of the daily value, what would you need to do?

	Actual amount per serving[a]	Amount per serving reported on the label	Percentage (%) daily value reported on the label
Calories	192		–
Total fat	1.1 g		
Saturated fat	0 g		
Trans fat	0 g		
Cholesterol	0 mg		
Sodium	267 mg		
Total carbohydrate	44.3 g		
Dietary fibers	3.8 g		
Total sugars	20.2 g		–
Incl. Added sugars	6.6 g		
Protein	3.7 g		–
Vitamin D	2 μg		
Calcium	210 mg		
Iron	4.3 mg		
Potassium	217 mg		

[a]Serving size is 1 cup (55 g)

2. To what do the terms "reference daily intake," "daily reference values," and "daily value" refer, and how are these related?
3. Identify by the categories listed below which food components must be listed on the standard nutrition label of foods.

Components of proximate composition	Vitamins	Minerals	Other components/ quantities

4. Regarding the nutrition label on foods:
   (a) Why will you not find a value such as 209.56 mg sodium on a nutrition label?
   (b) Why do you typically find protein content on a nutrition label expressed in grams only and not as percentage daily value?
   (c) What components of "proximate composition" are not included on the nutrition label? Why must these components of "proximate composition" be determined to complete the required nutrition label?
   (d) What nutrients/food components required on the label are calculated values and not actually measured?
5. The FDA and FSIS of the USDA have very similar regulations for nutrition labeling.
   (a) Identify the differences in regulations between the FDA and FSIS regarding the first choice of methods for nutritional analysis.
   (b) Identify one difference between the agencies in nutrient content claims that is consistent with the statement that the differences in regulations between the agencies are primarily due to the inherent differences in the food products they regulate.

Acknowledgments The author thanks Noel Larson, Kyler Held, and Wendy Ramaker (General Mills Inc.) for providing some text related to both FDA and USDA nutrition labeling regulations. The author also thanks Kerry Huber (author of Chap. 19) for his help with the new section on carbohydrate food components.

REFERENCES

1. Kanovsky SM, Pines WL (2020) A practical guide to FDA's food and drug law and regulation, 7th edn. Food and Drug Law Institute, Washington, DC
2. Adams DG, Cooper RM, Hahn HJ, Kahan JS (2015) Food and drug law and regulation. 3rd edn. Food and Drug Law Institute, Washington, DC
3. US Congress (1990) US public law 101–535. Nutrition labeling and education act of 1990. 8 Nov 1990. US Congress, Washington, DC
4. US Congress (1997) US public law 105–115. Food and Drug Administration modernization act of 1997. 21 Nov 1997. US Congress, Washington, DC
5. Federal Register (2016) Food labeling: Revision of the nutrition and supplemental facts labels (5/27/2016). https://www.federalregister.gov/documents/2016/05/27/2016-11867/food-labeling-revision-of-the-nutrition-and-supplement-facts-labels
6. Code of Federal Regulations (2023) US Food & Drug Administration. Title 21 Code and Drugs. Food labeling, Sec. 101.1-101.108; Nutrition labeling of food, Sec. 101.9 [For most current version see electronic Code of Federal Regulations (eCFR)]
7. Roland, AM (2024) Nutrition labeling using a computer program. Ch. 5. In: Ismail BP, Nielsen SS (eds) Nielsen's food analysis laboratory manual, 4th edn. Springer, New York
8. Code of Federal Regulations (2023) US Food & Drug Administration. Title 21 CFR 101.45, Guidelines for the voluntary nutrition labeling of raw fruits, vegetables, and fish. (For most current version see electronic Code of Federal Regulations, eCFR) https://www.accessdata.fda.gov/scripts/cdrh/cfdocs/cfcfr/cfrsearch.cfm?fr=101.45
9. Food and Drug Administration (2019) The declaration of added sugars on honey, maple syrup, other single-ingredient sugars and syrups, and certain cranberry products: Guidance for industry. June 2019. Center for Food Safety and Applied Nutrition. https://www.fda.gov/media/127928/download
10. AOAC International (2023) Official methods of analysis, 22nd edn. (Online). AOAC International, Rockville, MD. Food Safety and Inspection Service, US Department of Agriculture, Washington DC
11. Food and Drug Administration (2023) FDA Nutrition labeling manual – A guide for developing and using data bases. March 1998 (content current as of 9/20/2018) Office of Nutrition and Food Labeling (HFS-800), Center for Food Safety and Applied Nutrition, College Park, MN 20740 https://www.fda.gov/regulatory-information/search-fda-guidance-documents/guidance-industry-guide-developing-and-using-data-bases-nutrition-labeling, https://www.fsis.usda.gov/sites/default/files/media_file/documents/CLG-1.00.pdf
12. Code of Federal Regulations (2023) US Department of Agriculture. Nutrition labeling of meat and meat food products. 9 CFR 317.300—317.400. Nutrition labeling of poultry products. 9 CFR 381.400. [For most current version, see electronic Code of Federal Regulations (eCFR)] https://www.ecfr.gov/current/title-9/chapter-III/subchapter-A/part-317/subpart-B/section-317.300
13. Code of Federal Regulations (2023) US Department of Agriculture. Nutrition labeling of poultry products. 9 CFR 381 subpart Y 381.400-381.500. [For most current version, see electronic Code of Federal Regulations (eCFR)] https://www.ecfr.gov/current/title-9/chapter-III/subchapter-A/part-381/subpart-Y/section-381.400
14. Code of Federal Regulations (2023) US Department of Agriculture. Label Approval: 9 CFR 412. [For most current version, see electronic Code of Federal Regulations (eCFR)] https://www.ecfr.gov/current/title-9/part-412
15. US Department of Agriculture (2020) FSIS Compliance Guideline for Label Approval. https://www.fsis.usda.gov/sites/default/files/media_file/2020-10/Label-Approval-Guide.pdf
16. USDA (2023) Chemistry laboratory guidebook. https://www.fsis.usda.gov/news-events/publications/chemistry-laboratory-guidebook

Evaluation of Analytical Data

4

J. Scott Smith

Contents

Keywords
Eval. Anal. Data – StatisticsData analysis

4.1 INTRODUCTION

The field of food analysis, or any type of analysis, involves a considerable amount of time learning principles, methods, and instrument operations and perfecting various techniques. Although these areas are extremely important, much of our effort would be for naught if there were not some way for us to evaluate the data obtained from the various analytical assays. Several mathematical treatments are available that provide an idea of how well a particular assay was performed or how well we can reproduce an experiment. Fortunately, the statistics are not too involved and apply to most analytical determinations.

Whether analytical data are collected in a research laboratory or in the food industry, important decisions are made based on the data. Appropriate data collection and analysis help avoid bad decisions being made based on the data. Having a good understanding of the data and how to interpret the data (e.g., what numbers are statistically the same) are critical to good decision-making. Talking with a statistician

J. S. Smith (✉)
Department of Animal Sciences and Industry, Kansas State University, Manhattan, KS, USA
e-mail: jsschem@ksu.edu

B. P. Ismail, S. S. Nielsen (eds.), *Nielsen's Food Analysis*, Food Science Text Series,
https://doi.org/10.1007/978-3-031-50643-7_4

before designing experiments or testing the products produced can help ensure appropriate data collection and analysis, for better decision-making.

The focus of this chapter is primarily on how to evaluate replicate analyses of the same sample for accuracy and precision. In addition, considerable attention is given to the determination of best line fits for standard curve data. Keep in mind as you read and work through this chapter that there is a vast array of computer software to perform most types of data evaluations and calculations/plots.

Proper sampling and sample size are not covered in this chapter. Readers should refer to Chap. 5 and Garfield et al. [1] for sampling in general and statistical approaches to determine the appropriate sample size and to Chap. 33, Sect. 33.4 for mycotoxin sampling.

4.2 MEASURES OF CENTRAL TENDENCY

To increase accuracy and precision, as well as to evaluate these parameters, the analysis of a sample is usually performed (repeated) several times. At least three assays are typically performed, though, often, number can be much higher. Because we are not sure which value is closest to the true value, we determine the mean (or average) using all the values obtained and report the results of the **mean**. The mean is designated by the symbol $\bar{x}$ and is calculated according to the equation below:

$$\bar{x} = \frac{x_1 + x_2 + x_3 + \ldots + x_n}{n} = \frac{\Sigma x_i}{n} \quad (4.1)$$

where:

$\bar{x}$ = mean
x_1, x_2, etc. = individually measured values (x_i)
n = number of measurements

For example, suppose we measured a sample of uncooked hamburger for percent moisture content four times and obtained the following results—64.53%, 64.45%, 65.10%, and 64.78%:

$$\bar{x} = \frac{64.53 + 64.45 + 65.10 + 64.78}{4} = 64.72\% \quad (4.2)$$

Thus, the result would be reported as 64.72% moisture. When we report the mean value, we are indicating that this is the best experimental estimate of the value. We are not saying anything about how accurate or true this value is. Some of the individual values may be closer to the true value, but there is no way to make that determination, so we report only the mean.

Another determination that can be used is the **median**, which is the midpoint or middle number within a group of numbers. Basically, half of the experimental values will be less than the median and half will be greater. The median is not used often because the mean is such a superior experimental estimator.

4.3 RELIABILITY OF ANALYSIS

Returning to our previous example, recall that we obtained a mean value for moisture. However, we did not have any indication of how repeatable the tests were or how close our results were to the true value. The next several sections will deal with these questions and some of the relatively simple ways to calculate the answers. More thorough coverage of these areas is found in references [2–4].

4.3.1 Accuracy and Precision

One of the most confusing aspects of data analysis for students is grasping the concepts of "accuracy" and "precision." These terms are commonly used interchangeably in society, which only adds to this confusion. If we consider the purpose of the analysis, then these terms become much clearer. If we look at our experiments, we know that the first data obtained are the individual results and a mean value ($\bar{x}$). The next questions should be: "How close were our individual measurements?" "How close were they to the true value?" Both questions involve accuracy and precision. Now, let us turn our attention to these terms.

Accuracy refers to how close a particular measure is to the true or correct value. In the moisture analysis of hamburger, recall that we obtained a mean of 64.72%. Let us say the true moisture value was actually 65.05%. By comparing these two numbers, you could probably make a guess that your results were fairly accurate because they were close to the correct value. (The calculations of accuracy will be discussed later.)

The problem in determining accuracy is that most of the time we are not sure what the true value is. For certain types of materials, we can purchase known samples from, for example, the National Institute of Standards and Technology and check our assays against these samples. Only then can we have an indication of the accuracy of the testing procedures. Another approach is to compare our results with those of other labs to determine how well they agree, assuming the other labs are accurate.

A term that is much easier to deal with and determine is **precision**. This parameter is a measure of how reproducible or how close replicate measurements become. If repetitive testing yields similar results, then we would say the precision of that test was good. From a true statistical view, the precision often is called **error**, when we are actually looking at

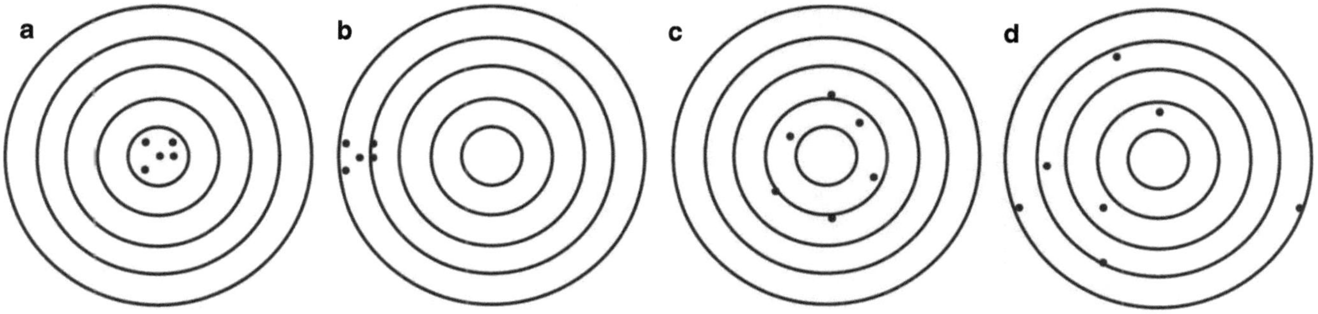

Fig. 4.1 Comparison of accuracy and precision: (**a**) good accuracy and good precision, (**b**) good precision and poor accuracy, (**c**) good accuracy and poor precision, and (**d**) poor accuracy and poor precision

experimental **variation**. So, the concepts of precision, error, and variation are closely related.

The difference between precision and accuracy can be illustrated best with Fig. 4.1. Imagine shooting a rifle at a target that represents experimental values. The bull's eye would be the true value, and where the bullets hit would represent the individual experimental values. As you can see in Fig. 4.1a, the values can be tightly spaced (good precision) and close to the bull's eye (good accuracy), or, in some cases, there can be situations with good precision but poor accuracy (Fig. 4.1b). The worst situation, as illustrated in Fig. 4.1d, is when both the accuracy and precision are poor. In this case, because of errors or variation in the determination, interpretation of the results becomes very difficult. Later, the practical aspects of the various types of error will be discussed.

When evaluating data, several tests are commonly used to give some appreciation of how much the experimental values would vary if we were to repeat the test (indicators of precision). An easy way to look at the variation or scattering is to report the range of the experimental values. The **range** is simply the difference between the largest and smallest observation. This measurement is not too useful and thus is seldom used in evaluating data.

Probably the best and most commonly used statistical evaluation of the precision of analytical data is the standard deviation. The **standard deviation** measures the spread of the experimental values and gives a good indication of how close the values are to each other. When evaluating the standard deviation, one has to remember that we are never able to analyze the entire food product. That would be difficult, if not impossible, and very time-consuming. Thus, the calculations we use are only estimates of the unknown true value.

If we have many samples, then the standard deviation is designated by the Greek letter sigma (σ). It is calculated according to Eq. 4.3, assuming all of the food product was evaluated (which would be an infinite amount of assays):

$$\sigma = \sqrt{\frac{\Sigma\left(x_{i} - \mu\right)^{2}}{n}} \tag{4.3}$$

where:

σ = standard deviation
x_i = individual sample values
μ = true mean
n = total population of samples

Because we do not know the value for the true mean, the equation becomes somewhat simplified so that we can use it with real data. In this case, we now call the σ term the standard deviation of the sample and designate it by SD or σ. It is determined according to the calculation in Eq. 4.4, where $\overline{x}$ replaces the true mean term μ and n represents the number of samples:

$$\mathrm{SD} = \sqrt{\frac{\Sigma\left(x_{i} - \overline{x}\right)^{2}}{n}} \tag{4.4}$$

If the number of replicate determinations is small (about 30 or less), which is common with most assays, then n is replaced by the n_1 term, and Eq. 4.5 is used. Unless you know otherwise, Eq. 4.5 is always used in calculating the standard deviation of a group of assays:

$$\mathrm{SD} = \sqrt{\frac{\Sigma\left(x_{i} - \overline{x}\right)^{2}}{n-1}} \tag{4.5}$$

Depending on which of the equations above is used, the standard deviation may be reported as SD_n or σ_n and SD_{n-1} or σ_{n-1}. (Different brands of software and scientific calculators sometimes use different labels for the keys, so one must be careful.) Table 4.1 shows an example of the determination of standard deviation. The sample results would be reported to average 64.72% moisture with a standard deviation of 0.293.

Once we have a mean and standard deviation, we must next determine how to interpret these numbers. One easy way to get a feel for the standard deviation is to calculate what is called the **coefficient of variation** (CV), also known as the **relative standard deviation**. This calculation is

Table 4.1 Determination of the standard deviation of percent moisture in uncooked hamburgers

Measurement	*Observed percentage moisture (%)*	*Deviation from the mean* $(x_i - \bar{x})$	$(x_i - \bar{x})^2$
1	64.53	−0.19	0.0361
2	64.45	−0.27	0.0729
3	65.10	+0.38	0.1444
4	64.78	+0.06	0.0036
	$\Sigma x_i = 258.86$		$\Sigma(x_i - \bar{x})^2 = 0.257$

$$\bar{x} = \frac{\Sigma x_i}{n} = \frac{258.86}{4} = 64.72$$

$$SD = \sqrt{\frac{\Sigma(x_i - \bar{x})^2}{n-1}} = \sqrt{\frac{0.257}{3}} = 0.2927$$

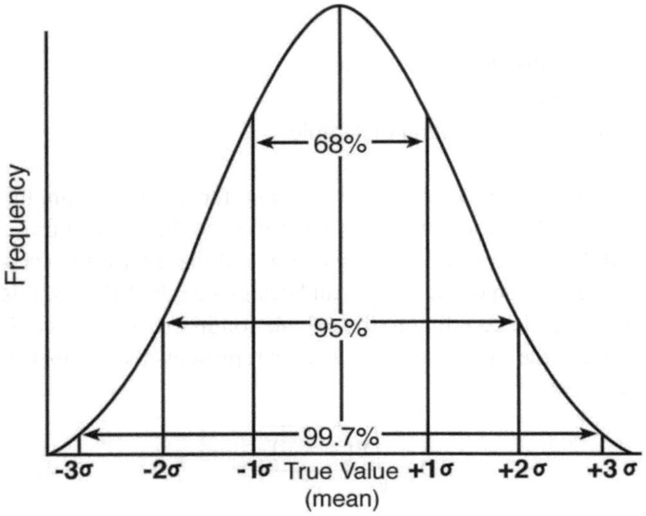

Fig. 4.2 A normal distribution curve for a population or a group of analyses

shown below for our example of the moisture determination of uncooked hamburgers:

$$\%\text{Coefficient of variation}(\%CV) = \frac{SD}{\bar{x}} \times 100 \quad (4.6)$$

$$\%CV = \frac{0.293}{64.72} \times 100 = 0.453\% \quad (4.7)$$

The CV tells us that our standard deviation is only 0.453% as large as the mean. For our example, that number is small, which indicates a high level of precision or reproducibility of the replicates. As a rule, a CV below 5% is considered acceptable, although it depends on the type of analysis.

Another way to evaluate the meaning of standard deviation is to examine its origin in statistical theory. Many populations (in our case, sample values or means) that exist in nature are said to have a normal distribution. If we were to measure an infinite number of samples, we would get a distribution similar to that represented by Fig. 4.2. In a population with a **normal distribution**, 68% of those values would be within ±1 standard deviation from the mean, 95% would be within ±2 standard deviations, and 99.7% would be within ±3 standard deviations. In other words, there is a probability of less than 1% that a sample in a population would fall outside ±3 standard deviations from the mean value.

Table 4.2 Values for *Z* for checking both upper and lower levels

Degree of certainty (confidence) (%)	*Z-value*
80	1.29
90	1.64
95	1.96
99	2.58
99.9	3.29

Another way of understanding the normal distribution curve is to realize that the probability of finding the true mean is within certain confidence intervals as defined by the standard deviation. For large numbers of samples, we can determine the **confidence limit** or **interval** around the mean using the statistical parameter called the **Z-value**. We do this calculation by first looking up the *Z*-value from statistical tables once we have decided the desired degree of certainty. Some *Z*-values are listed in Table 4.2.

The confidence limit (or interval) for our moisture data, assuming a 95% probability, is calculated according to Eq. 4.8. Since this calculation is not valid for small numbers, assume we ran 25 samples instead of 4:

$$\text{Confidence interval (CI)} = \bar{x} \pm \text{Z-value} \times \frac{\text{standard deviation (SD)}}{\sqrt{n}} \quad (4.8)$$

$$CI(\text{at } 95\%) = 64.72 \pm 1.96 \times \frac{0.2927}{\sqrt{25}} = 64.72 \pm 0.115\% \quad (4.9)$$

Because our example had only four values for the moisture levels, the confidence interval should be calculated using statistical *t*-tables. In this case, we have to look up the *t*-value from Table 4.3 based on the **degrees of freedom**, which is the sample size minus one ($n-1$), and the desired level of confidence.

The calculation for our moisture example with four samples (n) and three degrees of freedom ($n-1$) is given below:

$$CI = \bar{x} \pm t \text{ value} \times \frac{\text{standard deviation (SD)}}{\sqrt{n}} \quad (4.10)$$

Table 4.3 Values of t for various levels of probability[a]

	Levels of certainty		
Degrees of freedom (n−1)	*95%*	*99%*	*99.9%*
1	12.7	63.7	636
2	4.30	9.93	31.60
3	3.18	5.84	12.90
4	2.78	4.60	8.61
5	2.57	4.03	6.86
6	2.45	3.71	5.96
7	2.36	3.50	5.40
8	2.31	3.56	5.04
9	2.26	3.25	4.78
10	2.23	3.17	4.59

[a]More extensive t-tables can be found in statistics books

$$CI\left(at\,95\%\right) = 64.72 \pm 3.18 \times \frac{0.2927}{\sqrt{4}} = 64.72 \pm 0.465\% \tag{4.11}$$

To interpret this number, we can say that, with 95% confidence, the true mean for our moisture will fall within 64.72 ± 0.465% or between 65.185% and 64.255%.

The expression SD/$\sqrt{n}$ is often reported as the **standard error of the mean**. It is then left to the reader to calculate the confidence interval based on the desired level of certainty.

Other quick tests of precision used are the relative deviation from the mean and the relative average deviation from the mean. The **relative deviation from the mean** is useful when only two replicates have been performed. It is calculated according to Eq. 4.12, with values below 2% considered acceptable:

$$\text{Relative deviation from the mean} = \frac{x_i - \overline{x}}{\overline{x}} \times 100 \tag{4.12}$$

where:

x_i = individual sample value
$\overline{x}$ = mean

If there are several experimental values, then the **relative average deviation from the mean** becomes a useful indicator of precision. It is calculated similarly to the relative deviation from the mean, except that the average deviation is used instead of the individual deviation. It is calculated according to Eq. 4.13:

$$\text{Relative average deviation from the mean} = \frac{\frac{\Sigma\left|x_i - \overline{x}\right|}{n}}{\overline{x}} \times 1000 = \text{parts per thousand} \tag{4.13}$$

Using the moisture values discussed in Table 4.1, the $x_i - \overline{x}$ terms for each determination are −0.19, −0.27, +0.38, and + 0.06. Thus, the calculation becomes:

$$\text{Rel. avg. dev.} = \frac{\frac{0.19 + 0.27 + 0.38 + 0.06}{4}}{64.72} \times 1000 = \frac{0.225}{64.72} \times 1000 = 3.47\,\text{parts per thousand} \tag{4.14}$$

Up until now, our discussions of calculations have involved ways to evaluate precision. If the true value is not known, we can calculate only precision. A low degree of precision would make it difficult to predict a realistic value for the sample.

However, we may occasionally have a sample for which we know the true value and can compare our results with the known value. In this case, we can calculate the error for our test, compare it to the known value, and determine the accuracy. One term that can be calculated is the **absolute error**, which is simply the difference between the experimental value and the true value:

$$\text{Absolute error} = E_{abs} = x - T \tag{4.15}$$

where:

x = experimentally determined value
T = true value

The absolute error term can have either a positive or a negative value. If the experimentally determined value is from several replicates, then the mean ($\overline{x}$) would be substituted for the x term. This is not a good test for error because the value is not related to the magnitude of the true value. A more useful measurement of error is **relative error**:

$$\text{Relative error} = \text{E}_{rel} = \frac{\text{E}_{abs}}{\text{T}} = \frac{\text{x} - \text{T}}{\text{T}} \tag{4.16}$$

The results are reported as a negative or positive value, which represents a fraction of the true value.

If desired, the relative error can be expressed as percent relative error by multiplying by 100%. Then the relationship becomes the following, where x can be either an individual determination or the mean ($\overline{x}$) of several determinations:

$$\%E_{\text{rel}} = \frac{E_{\text{abs}}}{T} \times 100 = \frac{x - T}{T} \times 100 \tag{4.17}$$

Using the data for the percent moisture of uncooked hamburger, suppose the true value of the sample is 65.05%. The percent relative error is calculated using our mean value of 64.72% and Eq. 4.17:

$$\%E_{\text{rel}} = \frac{\bar{x} - T}{T} \times 100 = \frac{64.72 - 65.05}{65.05} \times 100 = -0.507\% \tag{4.18}$$

Note that we keep the negative value, which indicates the direction of our error, that is, our results were 0.507% lower than the true value.

Note that ISO/IEC 17025:2017 [9], "General requirements for the competence of testing and calibration laboratories," has requirements for estimating "Uncertainty of Measurements" included in their "General Requirements" section.

4.3.2 Sources of Errors [3]

As you may recall from the discussions of accuracy and precision, error (variation) can be quite important in analytical determinations. Although we strive to obtain correct results, it is unreasonable to expect an analytical technique to be entirely free of error. The best we can hope for is that the variation is small and, if possible, at least consistent. As long as we know about the error, the analytical method often will be satisfactory. There are several sources of error, which can be classified as: systematic error (determinate), random error (indeterminate), and gross error or blunders. Again, note that error and variation are used interchangeably in this section and essentially have the same meaning for these discussions.

Systematic or **determinate errors** produce results that consistently deviate from the expected value in one direction or the other. As illustrated in Fig. 4.1b, the results are spaced closely together, but they are consistently off the target. Identifying the source of this serious type of error can be difficult and time-consuming because it often involves inaccurate instruments or measuring devices. For example, a pipette that consistently delivers the wrong volume of reagent will produce a high degree of precision yet inaccurate results. Sometimes impure chemicals or the analytical method itself is the cause. Generally, we can overcome systematic errors by proper calibration of instruments, running blank determinations, or using a different analytical method.

Random or **indeterminate errors** are always present in any analytical measurement. This type of error is due to our natural limitations in measuring a particular system. These errors fluctuate in a random manner and are essentially unavoidable. For example, reading an analytical balance, judging the endpoint change in a titration, and using a pipette all contribute to random errors. Background instrument noise, which is always present to some extent, is a factor in random errors. Both positive and negative errors are equally possible. Although this type of error is difficult to avoid, fortunately, it is usually small.

Blunders are easy to eliminate, since they are so obvious. The experimental data are usually scattered, and the results are not close to an expected value. This type of error is a result of using the wrong reagent or instrument or of a sloppy technique. Some people call this type of error the "Monday morning syndrome" error. Fortunately, blunders are easily identified and corrected.

4.3.3 Specificity

Specificity of a particular analytical method means that it detects only the component of interest. Analytical methods can be very specific for a certain food component or, in many cases, can analyze a broad spectrum of components. Quite often, it is desirable for the method to be somewhat broad in its detection. For example, the determination of food lipids (fats) is actually the crude analysis of any compound that is soluble in an organic solvent. Some of these compounds are glycerides, phospholipids, carotenes, and free fatty acids. Since we are not concerned about each individual compound when considering the crude fat content of food, it is desirable that the method be broad in scope. On the other hand, determining the lactose content of ice cream would require a specific method. Because ice cream contains other types of simple sugars, without a specific method, we would overestimate the amount of lactose present.

There are no hard rules for what specificity is required. Each situation is different and depends on the desired results and type of assay used. However, it is something to keep in mind as the various analytical techniques are discussed.

4.3.4 Sensitivity and Limit of Detection [5]

Although often used interchangeably, the terms sensitivity and limit of detection should not be confused. They have different meanings, yet they are closely related. **Sensitivity** relates to the magnitude of change of a measuring device (instrument) with changes in compound concentration. It is an indicator of how little change can be made in the unknown material before we notice a difference on a needle gauge or a digital readout. We are all familiar with the process of tuning into a radio station on our stereo and know-how, at some point, once the station is tuned in, we can move the dial without disturbing the reception. This is sensitivity. In many situations, we can adjust the sensitivity of an assay to fit our needs, that is, whether we desire more or less sensitivity. We even may desire a lower sensitivity so that samples with widely varying concentration can be analyzed at the same time.

Limit of detection (LOD), in contrast to sensitivity, is the lowest possible amount that we can detect with some degree of confidence (or statistical significance). With every assay, there is a lower limit at which point we are not sure if something is present or not. Obviously, the best choice would be to concentrate the sample so we are not working close to the detection limit. However, this may not be possible, and we may need to know the LOD so we can work away from that limit.

There are several ways to measure the LOD, depending on the apparatus that is used. If we are using something like a spectrophotometer, gas chromatograph, or high-performance liquid chromatography (HPLC), the LOD often is reached when the signal-to-noise ratio is 3 or greater [5]. In other words, when the sample gives a value that is three times the magnitude of the noise detection, the instrument is at the lowest limit possible. Noise is the random signal fluctuation that occurs with any instrument.

A more general way to define the limit of detection is to approach the problem from a statistical viewpoint, in which the variation between samples is considered. A common mathematical definition of limit of detection is given below [3]:

$$X_{\mathrm{LD}} = X_{\mathrm{Blk}} + \left(3 \times \mathrm{SD}_{\mathrm{Blk}}\right) \tag{4.19}$$

where:

X_{LD} = minimum detectable concentration
X_{Blk} = signal of a blank
$\mathrm{SD}_{\mathrm{Blk}}$ = standard deviation of the blank readings

In this equation, the variation of the blank values (or noise, if we are talking about instruments) determines the detection limit. High variability in the blank values increases the limit of detection.

Another method that encompasses the entire assay method is the **method detection limit** (MDL). According to the US Environmental Protection Agency (EPA) [6], the MDL is defined as "the minimum concentration of a substance that can be measured and reported with 99 % confidence that the analyte concentration is greater than zero and is determined from analysis of a sample in a given matrix containing the analyte." What differentiates the MDL from the LOD is that it includes the entire assay and various sample types thus correcting for variability throughout. The MDL is calculated based on values of samples within the assay matrix and thus is considered a more rigorous performance test. The procedures on how to set up the MDL are explained in Appendix B of Part 136 (40 CFR, Vol 22) of the EPA regulations on environmental testing.

Though the LOD or MDL is often sufficient to characterize an assay, a further evaluation to check is the **limit of quantitation** (LOQ). In this determination, data are collected similar to the LOD, except the value is determined as $X_{\mathrm{Blk}} + (10 \times \mathrm{SD}_{\mathrm{Blk}})$ instead of $(X_{\mathrm{Blk}} + 3 \times \mathrm{SD}_{\mathrm{Blk}})$.

4.3.5 Quality Control Measures [1–3]

Quality control/assurance is desirable to evaluate analysis performance of a method or process. To explain how analytical data and control charts can be used in the food industry for **statistical process control**, this section will briefly describe quality control from the perspective of monitoring a specific process in making a food product (e.g., drying of a product, thereby affecting final moisture content). If the process is well defined and has known variability, the analytical data gathered can be evaluated over time. This provides set control points to determine if the process is performing as intended. Since all processes are susceptible to changes or drift, a decision can be made to adjust the process.

The best way to evaluate quality control is by control charting. This entails sequential plotting of the mean observations (e.g., moisture content) obtained from the analysis along with a target value. The standard deviation then is used to determine acceptable limits at the 95% or 99% confidence level and at what point the data are outside the range of acceptable values. Often, the acceptable limits are set as 2 standard deviations on either side of the mean, with the action limits set at 3 standard deviations. The charts and limits are used to determine if variation has occurred that is outside the normal variation for the process. If this occurs, there is need to determine the root cause of the variation and put in place corrective and preventive actions to further improve the process.

Two common types of **control charts** used are the Shewhart and CuSum charts described by Ellison et al. [2]. The CuSum chart is more involved and is better at highlighting small changes in the mean value. The Shewhart chart (Fig. 4.3) entails plots of the target value mean and both upper and lower limits for each measurement. An upper and lower warning limit and action limit are determined and added to the plot. The warning limit shows that the measurements may be moving out of the desirable limit (e.g., an

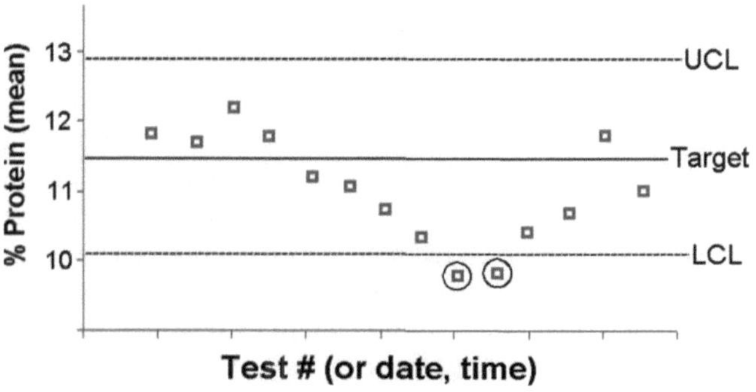

Fig. 4.3 An example of a Shewhart control chart for protein analysis. The upper control limit (UCL) and the lower control limit (LCL) are predetermined. Values that fall out of the limits (the circled values) indicate that the assay requires action and needs to be corrected or adjusted

upward trend). The action limit indicates that the measurements are past the acceptable limit, so the process needs to be evaluated for the causes of the drift. Examples of the calculations and charts are provided in references [2, 3].

4.4 CURVE FITTING: REGRESSION ANALYSIS [2–4]

Curve fitting is a generic term used to describe the relationship and evaluation between two variables. Most scientific fields use curve fitting procedures to evaluate the relationship two variables. Thus, curve fitting or curvilinear analysis of data is a vast area as evidenced by the volumes of material describing these procedures. In analytical determinations, we are usually concerned with only a small segment of curvilinear analysis, the standard curve, or regression line.

A **standard curve** or **calibration curve** is used to determine unknown concentrations based on a method that gives some type of measurable response that is proportional to a known amount of standard. It typically involves making a group of known standards in increasing concentration and then recording the particular measured analytical parameter (e.g., absorbance, area of a chromatography peak, etc.). What results when we graph the paired x and y values is a scatter plot of points that can be joined together to form a straight line relating concentration to observed response. Once we know how the observed values change with concentration, it is fairly easy to estimate the concentration of an unknown by interpolation from the standard curve.

As you read through the next three sections, keep in mind that not all correlations of observed values to standard concentrations are linear (but most are). There are many examples of nonlinear curves, such as antibody binding, toxicity evaluations, and exponential growth and decay. Fortunately, with the vast array of computer software available today, it is relatively easy to analyze any group of data.

4.4.1 Linear Regression [2–4]

So, how do we set up a standard curve once the data have been collected? First, a decision must be made regarding onto which axis to plot the paired sets of data. Traditionally, the concentration of the standards is represented on the x-axis, and the observed readings are on the y-axis. However, this protocol is used for reasons other than convention. The x-axis data are called **independent variables** and are assumed to be essentially free of error, while the y-axis data (**dependent variables**) may have an error associated with them. This assumption may not be true because an error could be incorporated as the standards are made. With modern-day instruments, the error can be very small. Although arguments can be made for making the y-axis data concentration, for all practical purposes, the end result is essentially the same. Unless there are some unusual data, the *concentration should be associated with the x-axis and the measured values with the y-axis*.

Figure 4.4 illustrates a typical standard curve used in the determination of caffeine in various foods. Caffeine is analyzed readily in foods by using HPLC coupled with an ultraviolet detector set at 272 nm. The area under the caffeine peak at 272 nm is directly proportional to the concentration. When an unknown sample (e.g., coffee) is run on the HPLC, a peak area is obtained, which can be related back to the sample using the standard curve.

The plot in Fig. 4.4 shows all the data points and a straight line that appears to pass through most of the points. The line almost passes through the origin, which makes sense because zero concentration should produce no signal at 272 nm. However, the line is not perfectly straight (and never is) and does not quite pass through the origin.

To determine the caffeine concentration in a sample that gave an area of say 4000, we could interpolate to the line and then draw a line down to the x-axis. Following a line to the x-axis (concentration), we can estimate the solution to be at about 42–43 ppm of caffeine.

We can mathematically determine the best fit of the line by using **linear regression**. Keep in mind the equation for a straight line, which is $y = ax + b$, where a is the slope and b is the y-intercept. To determine the slope and y-intercept, the regression equations shown below are used. We determine a and b, and, thus, for any value of y (measured), we can determine the concentration (x):

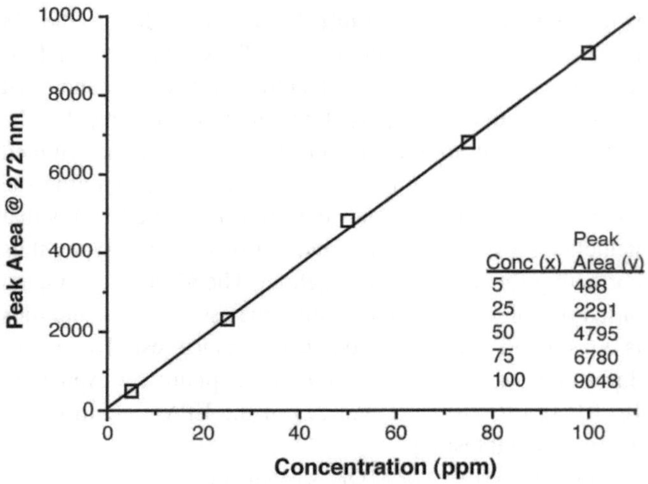

Fig. 4.4 A typical standard curve plot showing the data points and the generated best-fit line. The data used to plot the curve are presented on the graph

$$\text{slope } a = \frac{\Sigma(x_i - \bar{x})(y_i - \bar{y})}{\Sigma(x_i - \bar{x})^2} \quad (4.20)$$

$$y - \text{intercept } b = \bar{y} - a\bar{x} \quad (4.21)$$

where:

x_i and y_i = individual values
$\bar{x}$ and $\bar{y}$ = means of the individual values

Low-cost calculators and computer spreadsheet software can readily calculate regression equations, so no attempt is made to go through the mathematics in the formulas.

The formulas give what is known as the line of regression of y on x, which assumes that the error occurs in the y direction. The regression line represents the average relationship between all the data points and thus is a balanced line. These equations also assume that the straight line fit does not have to go through the origin, which at first does not make much sense. However, there are often background interferences so that even at zero concentration, a weak signal may be observed. In most situations, calculating the origin as going through zero will yield the same results.

Using the data from Fig. 4.4, calculate the concentration of caffeine in the unknown and compare with the graphing method. As you recall, the unknown had an area at 272 nm of 4000. Linear regression analysis of the standard curve data gave the y-intercept (b) as 90.727 and the slope (a) as 89.994 ($r^2 = 0.9989$):

$$y = ax + b \quad (4.22)$$

or

$$x = \frac{y - b}{a} \quad (4.23)$$

$$x(\text{conc}) = \frac{4000 - 90.727}{89.994} = 43.4393 \,\text{ppm caffeine} \quad (4.24)$$

The agreement is fairly close when comparing the calculated value to that estimated from the graph. Using high-quality graph paper with many lines could give us a line very close to the calculated one. However, as we will see in the next section, additional information can be obtained about the nature of the line when using computer software or calculators.

4.4.2 Correlation Coefficient

In observing any type of correlation, including linear ones, questions always surface concerning how to draw the line through the data points and how well the data fit to the straight line. The first thing that should be done with any group of data is to plot it to see if the points fit a straight line. By just looking at the plotted data, it is fairly easy to make a judgment on the linearity of the line. We also can pick out regions on the line where a linear relationship does not exist. The figures below illustrate differences in standard curves; Fig. 4.5a shows a good correlation of the data, and Fig. 4.5b shows a poor correlation. In both cases, we can draw a straight line through the data points. Both curves yield the same straight line, but the precision is poorer for the latter.

There are other possibilities when working with standard curves. Figure 4.6a shows a good correlation between x and y but in the negative direction, and Fig. 4.6b illustrates data that have no correlation at all.

The **correlation coefficient** defines how well the data fit to a straight line. For a standard curve, the ideal situation would be that all data points lie perfectly on a straight line. However, this is never the case because errors are introduced in making standards and measuring the physical values (observations).

The correlation coefficient and coefficient of determination are defined below. Essentially, all spreadsheet and plotting software will calculate the values automatically:

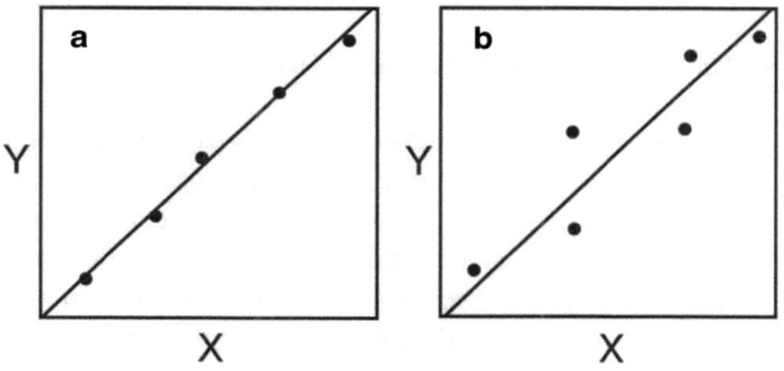

Fig. 4.5 Examples of standard curves showing the relationship between the x and y variables when there are (**a**) a high amount of correlation and (**b**) a lower amount of correlation. Both lines have the same equation

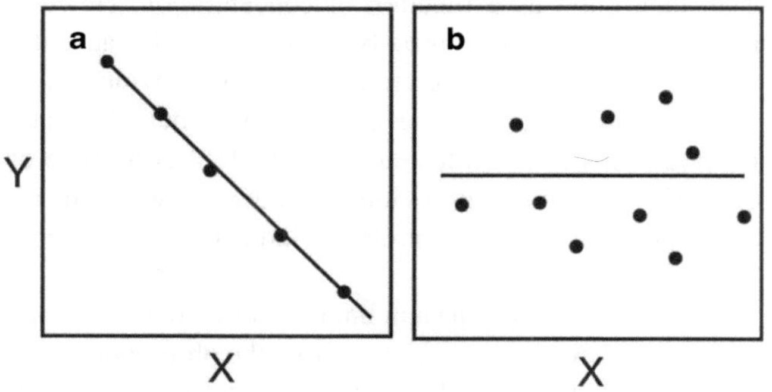

Fig. 4.6 Examples of standard curves showing the relationship between the x and y variables when there are (**a**) a high amount of negative correlation and (**b**) no correlation between the x and y values

correlation coefficient

$$= r = \frac{\Sigma(x_i - \bar{x})(y_i - \bar{y})}{\sqrt{\left[\Sigma(x_i - \bar{x})^2\right]\left[\Sigma(y_i - \bar{y})^2\right]}} \quad (4.25)$$

For our example of the caffeine standard curve from Fig. 4.3, $r = 0.99943$ (values are usually reported to at least four significant figures).

For standard curves, we want the value of r as close to +1.0000 or − 1.000 as possible because this value is a perfect correlation (a perfect straight line). Generally, in analytical work, the value of r should be 0.9970 or better (this does not apply to biological studies).

The **coefficient of determination** (r^2) is used often because it gives a better perception of the straight line even though it does not indicate the direction of the correlation. The r^2 for the example presented above is 0.99886, which represents the proportion of the variance of absorbance (y) that can be attributed to its linear regression on concentration (x). This means that about 0.114% of the straight line variation (1.0000–0.99886 = 0.00114 × 100% = 0.114%) does not vary with changes in x and y, and this is due to indeterminate variation. A small amount of variation is expected normally.

4.4.3 Errors in Regression Lines

While the correlation coefficient tells us something about the error or variation in linear curve fits, it does not always give the complete picture. Also, neither linear regression nor correlation coefficient will indicate that a particular set of data have a linear relationship. They only provide an estimate of the fit assuming the line is a linear one. As indicated before, plotting the data is critical when looking at how the data fit on the curve (actually, a line). One parameter that is used often is the ***y* residuals**, which are simply the differences between the observed values and the calculated or computed values (from the regression line). Advanced computer graphics software can actually plot the residuals for each data point as a function of concentration. However, plotting the residuals is usually not necessary because data that do not fit on the line are usually quite obvious. If the residuals are large for the entire curve, then the entire method needs to be evaluated carefully. However, the presence of one point that is obviously off the line while the rest of the points fit very well probably indicates an improperly made standard.

One way to reduce the amount of error is to include more replicates of the data such as repeating the observations with a new set of standards. The replicate x and y values can be entered into a calculator or spreadsheet as separate points for the regression and coefficient determinations. Another, probably more desirable, option is to expand the concentrations at which the readings are taken. Collecting observations at more data points (concentrations) will produce a better standard curve. However, increasing the data beyond seven or eight points is usually not beneficial.

Plotting **confidence intervals**, or **bands** or **limits**, on the standard curve along with the regression line is another way to gain insight into the reliability of the standard curve. Confidence bands define the statistical uncertainty of the regression line at a chosen probability (such as 95%) using the t-statistic and the calculated standard deviation of the fit. In some aspects, the confidence bands on the standard curve are similar to the confidence interval discussed in Sect. 4.3.1. However, in this case, we are looking at a line rather than a confidence interval around a mean. Figure 4.7 shows the caffeine data from the standard curve presented before, except some of the numbers have been modified to enhance the confidence bands. The confidence bands (dashed lines) consist of both an upper limit and a lower limit that define the variation of the y-axis value. The upper and lower bands are narrowest at the center of the curve and get wider as the curve moves to the higher or lower standard concentrations.

Looking at Fig. 4.7 again, note that the confidence bands show what amount of variation we expect in a peak area at a particular concentration. At 60 ppm concentration, by going up from the x-axis to the bands and interpolating to the y-axis, we see that with our data, the 95% confidence interval of the observed peak area will be 4000–6000. In this case, the variation is large and would not be acceptable as a standard curve and is presented here only for illustration purposes.

Error bars also can be used to show the variation of y at each data point. Several types of error or variation statistics can be used such as standard error, standard deviation, or percentage of data (i.e., 5%). Any of these methods give a visual indication of experimental variation.

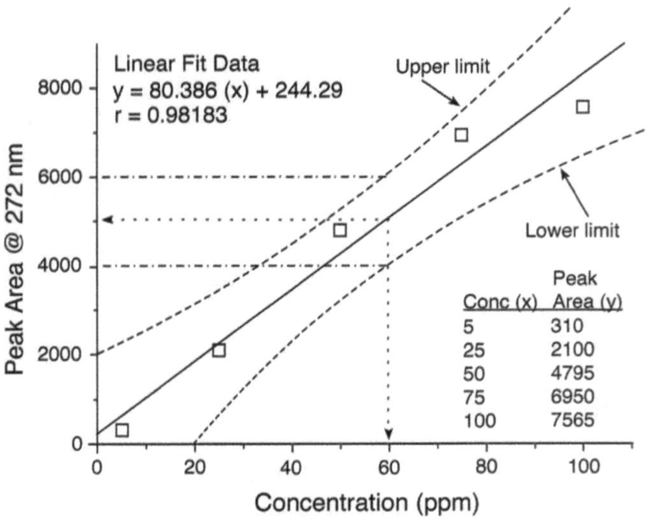

Fig. 4.7 A standard curve graph showing the confidence bands. The data used to plot the graph are presented on the graph as are the equation of the line and the correlation coefficient

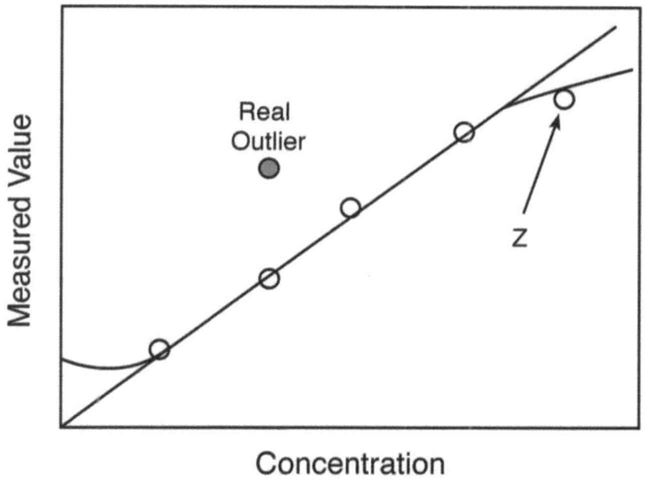

Fig. 4.8 A standard curve plot showing possible deviations in the curve in the upper and lower limits

Even with good standard curve data, problems can arise if the standard curve is not used properly. One common mistake is to extrapolate beyond the data points used to construct the curve. Figure 4.8 illustrates some of the possible problems that might occur when extrapolation is used. As shown in Fig. 4.8, the curve or line may not be linear outside the area where the data were collected. This can occur in the region close to the origin or especially at a higher concentration level.

Usually, a standard curve will go through the origin, but, in some situations, it may actually tail off as zero concentration is approached. At the other end of the curve, at higher concentrations, it is fairly common for a plateau to be reached where the measured parameter does not change much with an increase in concentration. Care must be applied at the upper limit of the curve to ensure that data for unknowns are not collected outside of the curve standards. Point Z in Fig. 4.8 should be evaluated carefully to determine if the point is an outlier or if the curve is actually tailing off. Collecting several sets of data at even higher concentrations should clarify this. Regardless, the unknowns should be measured only in the region of the curve that is linear.

4.5 REPORTING RESULTS

When dealing with experimental results, we are always confronted with reporting data in a way that indicates the sensitivity and precision of the assay. Ideally, we do not want to overstate or understate the sensitivity of the assay, and, thus, we strive to report a meaningful value, be it a mean, standard deviation, or some other number. The next three sections discuss how we can evaluate experimental values so as to be precise when reporting results.

4.5.1 Significant Figures

The term **significant figure** is used rather loosely to describe some judgment of the number of reportable digits in a result. Often, the judgment is not soundly based, and meaningful digits are lost or meaningless digits are retained. Exact rules are provided below to help determine the number of significant figures to report. However, it is important to keep some flexibility when working with significant figures.

Proper use of significant figures is meant to give an indication of the sensitivity and reliability of the analytical method. Thus, reported values should contain only significant figures. A value is made up of significant figures when it contains all digits known to be true and one last digit that is in doubt. For example, a value reported as 64.72 contains 4 significant figures, of which 3 digits are certain (64.7) and the last digit is uncertain. Thus, the 2 is somewhat uncertain and could be either 1 or 3. As a rule, numbers that are presented in a value represent the significant figures, regardless of the position of any decimal points. This is also true for values containing zeros, provided they are bounded on either side by a number. For example, 64.72, 6.472, 0.6472, and 6.407 all contain 4 significant figures. Note that the 0 to the left of the decimal point is used only to indicate that there are no numbers above 1. We could have reported the value as .6472, but using the 0 is better, since we know that a number was not inadvertently left off our value.

Special considerations are necessary for zeros that may or may not be significant:

1. Zeros after a decimal point are always significant figures. For example, 64.720 and 64.700 both contain 5 significant figures.
2. Zeros before a decimal point with no other preceding digits are not significant. As indicated before, 0.6472 contains 4 significant figures.
3. Zeros after a decimal point are not significant if there are no digits before the decimal point. For example, 0.0072 has no digits before the decimal point; thus, this value contains 2 significant figures. In contrast, the value 1.0072 contains 5 significant figures.
4. Final zeros in a number are not significant unless indicated otherwise. Thus, the value 7000 contains only 1 significant figure. However, adding a decimal point and another 0 yields the number 7000.0, which has 5 significant figures.

A good way to measure the significance of zeros, if the above rules become confusing, is to convert the number to the exponential form. If the zeros can be omitted, then they are not significant. For example, 7000 expressed in the exponential form is 7×10^3 and contains 1 significant figure. With 7000.0, the zeros are retained and the number becomes

7.0000×10^3. If we were to convert 0.007 into the exponential form, then the value is 7×10^{-3} and only 1 significant figure is indicated. As a rule, determining significant figures in arithmetic operations is dictated by the value having the least number of significant figures. The easiest way to avoid any confusion is to perform all the calculations and then round off the final answer to the appropriate digits. For example, $36.54 \times 238 \times 1.1 = 9566.172$, and, because 1.1 contains only 2 significant figures, the answer would be reported as 9600 (remember, the 2 zeros are not significant). This method works fine for most calculations, except when adding or subtracting numbers containing decimals. In those cases, the number of significant figures in the final value is determined by the numbers that follow the decimal point. Thus, when adding $7.45 + 8.725 = 16.175$, the sum is rounded to 16.18 because 7.45 has only 2 numbers after the decimal point. Likewise, 433.8–32.66 gives 401.14, which rounds off to 401.1.

A word of caution is warranted when using the simple rule stated above, for there is a tendency to underestimate the significant figures in the final answer. For example, take the situation in which we determined the caffeine in an unknown solution to be 43.5 ppm (see Eq. 4.24). We had to dilute the sample 50-fold using a volumetric flask in order to fit the unknown within the range of our method. To calculate the caffeine in the original sample, we multiply our result by 50 or 43.5 μg/mL × 50 = 2175 μg/mL in the unknown. Based on our rule above, we then would round the number to 1 significant figure (because 50 contains 1 significant figure) and report the value as 2000. However, doing this actually underestimates the sensitivity of our procedure because we ignore the accuracy of the volumetric flask used for the dilution. A class A volumetric flask has a tolerance of 0.05 mL; thus, a more reasonable way to express the dilution factor would be 50.0 instead of 50. We now have increased the significant figures in the answer by 2, and the value becomes 2180 μ/mL.

As can be seen, an awareness of significant figures and how they are adopted requires close inspection. The guidelines can be helpful, but they do not always work unless each individual value or number is closely inspected.

4.5.2 Outlier Data and Testing [2, 3]

Inevitably, during the course of working with experimental data, we will come across outlier values that do not match the others. Can you reject that value and thus not use it in calculating the final reported results?

The answer is "very rarely" and only after careful consideration. If you are routinely rejecting data to help make your assay look better, then you are misrepresenting the results and the precision of the assay. If the bad value resulted from an identifiable mistake in that particular test, then it is probably safe to drop the value. Again, caution is advised because you may be rejecting a value that is closer to the true value than some of the other values.

Consistently poor accuracy or precision indicates that an improper technique or an incorrect reagent was used or that the test was not very good. It is best to make changes in the procedure or change methods rather than try to figure out ways to eliminate the undesirable values.

There are several tests for rejecting outlier data. In addition, the use of more robust statistical estimators of the population mean can minimize the effects of extreme outlier values [2]. The simplest test for rejecting outlier data is the Dixon Q-test [7, 8], often called just the Q-test. The advantage is that this test can be easily calculated with a simple calculator and is useful for a small group of data. In this test, a ***Q*-value** is calculated as shown below and compared to values in a table. If the calculated value is larger than the table value, then the questionable measurement can be rejected at the 90% confidence level:

$$Q - \text{value} = x_2 - x_1 / \mathrm{W} \tag{4.26}$$

where:

x_1 = the questionable value
x_2 = the next closest value to x_1
W = the total spread of all values, obtained by subtracting the lowest value from the highest value

Table 4.4 provides the rejection Q-values for a 90% confidence level.

The example below shows how the test is used for the moisture level of uncooked hamburger for which four replicates were performed giving values of 64.53, 64.45, 64.78, and 55.31. The 55.31 value looks as if it is too low compared to the other results. Can that value be rejected? For our example, x_1 is the questionable value (55.31) and x_2 is the closest neighbor to x_1 (which is 64.45). The spread (W) is the high value minus the low measurement, which is 64.78–55.31:

$$\mathrm{Q} - \text{value} = \frac{64.45 - 55.31}{64.78 - 55.31} = \frac{9.14}{9.47} = 0.97 \tag{4.27}$$

Table 4.4 Q-values for the rejection of results

Number of observations	*Q of rejection (90% level)*
3	0.94
4	0.76
5	0.64
6	0.56
7	0.51
8	0.47
9	0.44
10	0.41

Adapted from Dean and Dixon [7]

From Table 4.4, we see that the calculated Q-value must be greater than 0.76 to reject the data. Thus, we make the decision to reject the 55.31% moisture value and do not use it when calculating the mean.

4.6 SUMMARY

This chapter focuses on the statistical methods to measure data variability, precision, etc. and basic mathematical treatment that can be used in evaluating a group of data. For example, it should be almost second nature to determine a mean, standard deviation, and CV when evaluating replicate analyses of an individual sample. In evaluating linear standard curves, best line fits should always be determined along with the indicators of the degree of linearity (correlation coefficient or coefficient of determination). Fortunately, most computer spreadsheet and graphics software will readily perform the calculations for you. Guidelines are available to enable one to report analytical results in a way that tells something about the sensitivity and confidence of a particular test. A section is included, which describes sensitivity and limit of detection as related to various analytical methods and regulatory agency policies. Additional information includes the proper use of significant figures, rules for rounding off numbers, and use of various tests to reject grossly aberrant individual values (outliers).

4.7 STUDY QUESTIONS

1. Method A to quantitate a particular food component was reported to be more specific and accurate than Method B, but Method A had lower precision. Explain what this means.
2. You are considering adopting a new analytical method in your lab to measure moisture content of cereal products. How would you determine the precision of the new method and compare it to the old method? Include any equations to be used for any needed calculations.
3. A sample known to contain 20 g/L glucose is analyzed by two methods. Ten determinations were made for each method and the following results were obtained:

Method A	Method B
Mean = 19.6	Mean = 20.2
Std. Dev. = 0.055	Std. Dev. = 0.134

   (a) Precision and accuracy:
      (i) Which method is more precise? Why do you say this?
      (ii) Which method is more accurate? Why do you say this?
   (b) In the equation to determine the standard deviation, n_1 was used rather than just n. Would the standard deviation have been smaller or larger for each of those values above if simply n had been used?
   (c) You have determined that the values obtained using Method B should not be accepted if outside the range of 2 standard deviations from the mean. What range of values will be acceptable?
   (d) Do the data above tell you anything about the specificity of the method? Describe what "specificity" of the method means as you explain your answer.
4. Differentiate "standard deviation" from "coefficient of variation," "standard error of the mean," and "confidence interval."
5. Differentiate the terms "absolute error" versus "relative error." Which is more useful? Why?
6. For each of the errors described below when performing an analytical procedure, classify the error as a random error, a systematic error, or a blunder and describe a way to overcome the error.
   (a) Automatic pipettor consistently delivered 0.96 mL rather than 1.00 mL.
   (b) Substrate was not added to one tube in an enzyme assay.
7. Differentiate the terms "sensitivity" and "limit of detection."
8. The correlation coefficient for standard curve A is reported as 0.9970. The coefficient of determination for standard curve B is reported as 0.9950. In which case do the data better fit a straight line?

4.8 PRACTICE PROBLEMS

Note that in Chap. 4 of the laboratory manual [10] that accompanies this textbook, there are additional sample problems regarding data handling.

1. How many significant figures are in the following numbers: 0.0025, 4.50, and 5.607?
2. What is the correct answer for the following calculation expressed in the proper amount of significant figures?

$$\frac{2.43 \times 0.01672}{1.83215} =$$

3. Given the following data on dry matter (88.62, 88.74, 89.20, 82.20), determine the mean, standard deviation, and CV. Is the precision for this set of data acceptable? Can you reject the value 82.20 since it seems to be different from the others? What is the 95% confidence level you would expect your values to fall within if the test were repeated? If the true value for dry matter is 89.40, what is the percent relative error?

4. Compare the two groups of standard curve data below for sodium determination by atomic emission spectroscopy. Draw the standard curves using graph paper or a computer software program. Which group of data provides a better standard curve? Note that the absorbance of the emitted radiation at 589 nm increases proportionally to sodium concentration. Calculate the amount of sodium in a sample with a value of 0.555 for emission at 589 nm. Use both standard curve groups and compare the results.

Sodium concentration (μg/mL)	Emission at 589 nm
Group A:—sodium standard curve	
1.0	0.050
3.0	0.140
5.0	0.242
10.0	0.521
20.0	0.998
Group B: Sodium standard curve	
1.0	0.060
3.0	0.113
5.0	0.221
10.0	0.592
20.0	0.917

Answers

1. 2, 3, and 4
2. 0.0222
3. mean = 87.19, SD_{n-1} = 3.34

$$\mathrm{CV} = \frac{3.34}{87.18} \times 100 = 3.83\%$$

Thus, the precision is acceptable because it is less than 5%.

$$\mathrm{Q\text{-}calc\ value} = \frac{88.62 - 82.20}{89.20 - 82.20} = \frac{6.42}{7.00} = 0.917$$

Q_{calc} = 0.92; therefore, the value 82.20 can be rejected because it is more than 0.76 from Table 4.4, using 4 as the number of observations.

$$\mathrm{CI}(\mathrm{at}\ 95\%) = 87.19 \pm 3.18 \times \frac{3.34}{\sqrt{4}}$$
$$= 87.19 \pm 5.31$$

Relative error = % E_{rel}, where the mean is 87.19 and the true value is 89.40:

$$\%E_{\mathrm{rel}} = \frac{\bar{x} - T}{T} \times 100 = \frac{87.19 - 89.40}{89.40} \times 100 = -2.47\%$$

4. Using linear regression, we get,

Group A: $y = 0.0504x - 0.0029$, $r^2 = 0.9990$.

Group B: $y = 0.0473x + 0.0115$, $r^2 = 0.9708$.

Group A r^2 is closer to 1.000 and is more linear and thus the better standard curve.

Sodium in the sample using Group A standard curve is:

$0.555 = 0.0504x - 0.0029$, $x = 11.1$ μg/mL

Sodium in the sample using Group B standard curve is:

$0.555 = 0.0473x + 0.0115$, $x = 11.5$ μg/mL.

Acknowledgments The author thanks Ryan Deeter for his contributions to the preparation of the content on quality control measures.

REFERENCES

1. Garfield F M, Klestra E, and Hirsch J (2000) Quality assurance principles for analytical laboratories, AOAC International, Gaithersburg, MD (This book covers mostly quality assurance issues, but Chap. 3 is a good chapter on sampling and on the basics of statistical applications to data.)
2. Ellison, SLR, Barwick, VJ, Farrant, TJD (2009) Practical Statistics for the Analytical Scientist: A Bench Guide, Chapter 5, 2nd ed. RSC Publishing, Cambridge, UK (This an excellent text for beginner and advanced analytical chemists alike. It contains a fair amount of detail, sufficient for most analytical statistics, yet works through the material starting at a basic introductory level. The authors have an entire chapter, Chap. 5, devoted to the various test used to evaluating and rejecting data outliers.)
3. Miller JN, Miller JC (2018) Statistics and chemometrics for analytical chemistry, 7th edn. Pearson Prentice Hall, Upper Saddle River, NJ (This is another excellent introductory text for beginner analytical chemists at the undergraduate and graduate level. It contains a fair amount of detail, sufficient for most analytical statistics, yet works through the material starting at a basic introductory level. Chapter 1 covers the types of experimental error. The assay evaluation parameters LOD, MDL, and LOQ are thoroughly discussed. The authors also cover outlier testing including the Q-test used for rejecting data.)
4. Skoog DA, West DM, Holler JF, Crouch SR (2021) Fundamentals of analytical chemistry, 10th edn. Cengage Brain, Independence, KY (Part I, Chaps. 5-8, have thorough coverage of the statistics needed by an analytical chemist in an easy-to-read style.)
5. Shrivastava A, Gupta VB (2011) Methods for the determination of limit of detection and limit of quantitation of the analytical methods. Chron Young Sci 2(1):21–25
6. Code of Federal Regulations (2011) Environmental Protection Agency. 40 CFR, Vol. 22. Chapter I part 136 – Guidelines Establishing Test Procedures for the Analysis of Pollutants. Appendix B to Part 136-Definition and Procedure for the Determination of the Method Detection Limit-Revision 1.11
7. Dean RB, Dixon WJ (1951) Simplified statistics for small numbers of observations. Anal Chem 23(4): 636–638
8. Rorabacher DB (1991) Statistical treatment for rejection of deviant values: critical values of Dixon's "Q" parameter and related subrange ratios at the 95 % confidence level. Anal Chem 63 (2):139–146
9. ISO/IEC 17025:2017, General requirements for the competence of testing and calibration laboratories, corrected version:2018-03, https://www.iso.org/standard/66912.html
10. Neilson, AP (2024) Statistics for food analysis. Ch. 4. In: Ismail BP, Nielsen SS (eds) Nielsen's food analysis laboratory manual, 4th edn. Springer, New York

Sampling and Sample Preparation

5

Indarpal Singh and Tim D. Perez

Contents

5.1 INTRODUCTION

Sampling has a direct effect on the evaluation of analytical data. One must ensure that sampling is conducted appropriately to take full advantage of the concepts described in Chap. 4 such as measuring central tendency and reliability of analysis. In fact, the theoretical foundation of statistics revolves around the concept of using a small subset from a larger group to make inferences about the group. The group of interest is typically referred to as the **population**. The population properties of interest are called **parameters**. **Statistics** are estimates of the population parameters calculated from the subset of the population called the **sample**.

As an example of sampling, suppose that a company is sent a 1000-pound shipment of ground beef. Suppose that the company wishes to know the mean protein value of the shipment. Here, the population is the 1000-pound shipment of ground beef, and the parameter of interest is the mean protein value. Some may not realize that it would require testing the entire 1000-pound shipment to know the true mean protein value of the shipment. However, this is almost never a practical solution because these tests are usually

I. Singh (✉)
Analytical & Applied Sciences, Conagra Brands,
Omaha, NE, USA
e-mail: indarpal.singh@conagra.com

T. D. Perez
Research & Development, Conagra Brands, Omaha, NE, USA
e-mail: tim.perez@conagra.com

B. P. Ismail, S. S. Nielsen (eds.), *Nielsen's Food Analysis*, Food Science Text Series,
https://doi.org/10.1007/978-3-031-50643-7_5

destructive. Moreover, the amount of testing would most likely be prohibitively large. It is in scenarios like this that sampling is employed to obtain an accurate estimate of the true mean protein for the shipment by testing a small fraction of the population. This chapter will outline some of the key principles for developing an appropriate sampling plan.

As discussed in Chap. 4, Sect. 4.3.1, **accuracy** is a measure of how close the data are to the true value. **Precision** is a measure of how close measurements are to one another. A great deal of attention is often given to the accuracy and precision of analytical methods. It is critical that attention is also given to the validity of sampling and sample preparation. Figure 5.1 is a graphical representation of the typical relative variability attributed to the four major sources of variability in food analysis. These are sampling, homogenization, sample preparation, and assay procedure. The graph illustrates that the greatest amount of error is typically introduced in the steps leading up to the actual analysis. Sampling can often be the greatest source of error in chemical analysis in general and can be especially problematic in food analysis [1–3]. This is an extremely intuitive concept. Imagine an analyst has an accurate and precise testing method. However, they are given an unrepresentative sample to test. They would then make accurate measurements of a sample that is not representative of the population. Clearly, the resulting statistics will not be representative of the population parameters. Consequently, the sample must be representative of the population to have any chance of accurately estimating the parameters.

A formal sampling plan is often required to ensure that sampling is conducted appropriately, because of the vital role it plays in food analysis. The **International Union of Pure and Applied Chemistry** (IUPAC) defines a sampling plan as "A predetermined procedure for the selection, withdrawal, preservation, transportation, and preparation of the portions to be removed from a lot as samples" [4]. The first step in developing a sampling strategy is understanding the key factors affecting the choice. Table 5.1 shows four major factors that one should consider with examples: the purpose of inspection, nature of population, nature of product, and nature of testing method. The purpose of inspection and nature of the testing method will determine what measurements the analyst should collect, the statistical analysis, and the acceptance criteria of the sampling plan. The nature of the population and product will affect the strategy for collecting the sample. Sampling plan development can start once these factors are determined.

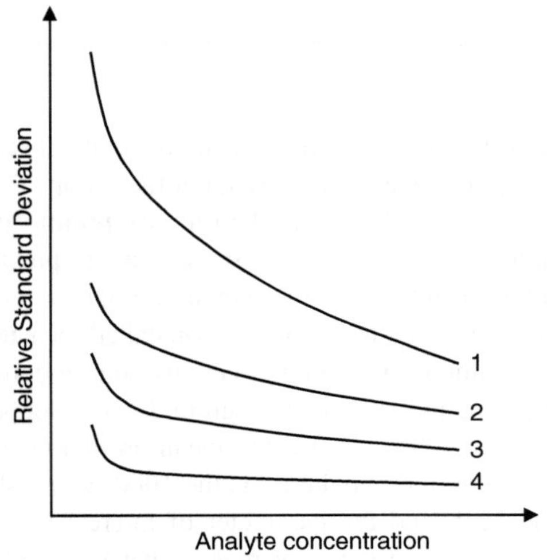

Fig. 5.1 Typical relative standard deviation of error components for a nonhomogeneous food matrix. (1) Sampling; (2) homogenization; (3) sample preparation; and (4) assay procedure. (Used with permission from Ref. Lichon [1])

Table 5.1 Factors that affect the choice of sampling plans

Factors to be considered	*Questions*
Purpose of the inspection	Is it to accept or reject the lot? Is it to measure the average quality of the lot? Is it to determine the variability of the product?
Nature of the population	Is the lot large but uniform? Does the lot consist of smaller, easily identifiable sub-lots? What is the distribution of the units within the population?
Nature of the product	Is it homogeneous or heterogeneous? What is the unit size? How consistently have past populations met the specifications? What is the cost of the material being sampled?
Nature of the testing method	Is the test critical or minor? Will someone become sick or die if the population fails to pass the test? Is the test destructive or nondestructive? How much does the test cost to complete?

Adapted from Ref. Puri et al. [2]

The principles described in this chapter are intended to provide the reader a basis for understanding, developing, and evaluating sampling plans and sample handling procedures for specific applications encountered. Analysts are encouraged to consult a statistician in the event they encounter situations that are not covered in this chapter. Keep in mind that sampling terminologies and procedures used may vary between companies and between specific applications.

5.2 DEFINITION AND PURPOSE OF INSPECTION

5.2.1 Types of Sampling

There must be a clear understanding of the purpose of the inspection prior to testing. The purpose of the inspection dictates the population of interest and the measurements taken. Sometimes the goal is estimation of population parameters. Other times, the goal is to determine whether the population parameter satisfies a predetermined quality threshold. These factors can affect the sample size and statistical analysis. Some of the most common purposes of inspection include:

- Nutrition labeling
- Detection of contaminants and foreign matter
- Acceptance of raw materials, ingredients, or products
- Process control samples
- Release of finished product
- Detection of adulterations
- Microbiological safety
- Authenticity of food ingredients

Sampling purposes tend to involve the examination of either variables or attributes [5].

Variable sampling is performed to numerically quantify the amount of a substance or characteristic. Results are measured on a continuous numeric scale in variable sampling. Examples include protein content or product weight. The analysis of these data is usually based on the normal distribution as described in Chap. 4, Sect. 4.3.1.

Attribute sampling is performed to determine the percentage of the population that possesses a certain characteristic. Often, attribute sampling results have a binary outcome such as the following:

1. Detection of contaminants and foreign matter: yes/no
2. Detection of adulterations: yes/no
3. Positive/negative
4. Conforming/nonconforming

The analysis of these data is based on discrete statistical distributions such as the binomial. In general, variable sampling requires smaller sample size than attribute sampling [6].

5.2.2 Sample Size Estimation: Variable Sampling

Precision analysis is often used by analysts for sample size determination in variable sampling. It uses the confidence interval approach and the assumption of approximate normality. As discussed in Chap. 4, Sect. 4.3.1, a confidence interval is an interval calculated from the sample data that should capture the population parameter with a given level of confidence. The example calculation shows that to test the concentration of sodium in a lot of ready-to-eat food product with a 10% margin of error at the 95% confidence level, the sample size needs to be 96 randomly selected trays. A confidence interval can be calculated using the following equation:

$$\bar{x} \pm z_{\alpha/2} \frac{s}{\sqrt{n}} \tag{5.1}$$

where:

$\bar{x}$ = sample mean
$z_{\alpha/2}$ = z-value corresponding to the level of confidence desired
s = known, or estimated, standard deviation of the population
n = sample size

In Eq. 5.1, $z_{\alpha/2}\frac{s}{\sqrt{n}}$ represents the margin of error for the sample at the desired level of confidence. One can utilize this equation to determine the sample size required to attain the desired margin of error. To do so, set the equation $E = z_{\alpha/2}\frac{s}{\sqrt{n}}$ and solve for n:

$$n = \left(\frac{z_{\alpha/2} s}{E}\right)^2 \tag{5.2}$$

In the formula, the desired margin of error, E, must be determined by the analyst. The margin of error can be a predetermined value or, in some cases, a percentage of the expected mean. However, that standard deviation must either be known for the population or estimated. The standard deviation can be estimated with data from a similar study or historical data.

As an example, assume that it is desired to test the concentration of sodium in a lot of a ready-to-eat food product with a 10% margin of error at the 95% confidence level. In a preliminary test, the average sodium content was 1000 mg per tray with a standard deviation of 500 mg. If the margin of error is desired to be 10% of the average, then assume that the preliminary sodium content estimate is accurate. Thus, the target margin of error would be 1000 mg x 10% = 100 mg. Use the following values in Eq. 5.2 to calculate the sample size:

- α = 0.05
- z = 1.96
- $\bar{x}$ = 1000
- s = 500
- Δ = 10%

$$n = \left(\frac{z_{\alpha/2} s}{E}\right)^2 = \left(\frac{z_{\alpha/2} s}{\Delta \times \bar{x}}\right)^2 = \left(\frac{1.96 \times 500}{0.1 \times 1000}\right)^2 = \left(\frac{980}{100}\right)^2 \approx 96\,\text{trays} \tag{5.3}$$

As a result, the sampled size to estimate the concentration of sodium in a lot of a ready-to-eat food product would be 96 randomly selected trays.

5.2.3 Sample Size Estimation: Attribute Sampling

5.2.3.1 Acceptance Sampling: Consumer Risk and Producer Risk

Acceptance sampling is a commonly used approach for attribute sampling and is an extremely broad topic. This discussion will cover only binary attributes such as whether a unit is defective or non-defective. Similarly, the acceptance sampling discussion will only include single-stage plans. The goal of acceptance sampling is to determine whether a population should be accepted or rejected. The decision of acceptance or rejection is based on quantified risks. Acceptance sampling plans control two major types of risks: consumer risk and producer risk.

Consumer risk (β) is the probability of incorrectly accepting a population that does not meet the predetermined quality threshold. As an example, assume that a food processor is receiving a lot or batch of materials from a supplier. The food processor requires a supply of materials with a defective unit rate below 10%. In scenarios like this, acceptance sampling can be used to help the food processor determine whether the lot should be accepted or rejected. Consumer risk is the probability of the food processor incorrectly accepting the lot from the supplier with a defective unit rate above 10%. A commonly used level of consumer risk is 5%. The threshold of percent defects at which a lot would be considered unacceptable is often called the **lot tolerance percent defective** (LTPD). In the example, the LTPD was 10%. The LTPD is typically set based on the consequences of accepting a lot. In scenarios in which the consequences are minor, one might use an LTPD of 5% or 10%. LTPD values of 1%, 0.1%, or lower are common when the consequences are severe. Remember, it would require testing all units in the population to claim that the population contains 0% defects. This is typically not a realistic option.

Producer risk (α) is the probability of incorrectly rejecting a population that meets the predetermined quality threshold. As an example, again assume that a food processor is receiving a lot or batch of materials from a supplier. The food processor still requires a supply of materials with a defective unit rate below 10%. The supplier is confident in the ability to produce lots with a defective unit rate below 10%. However, the supplier also knows it is likely that the supplied lot will have a defective unit rate around 1%. Producer risk is the probability of the food processor incorrectly rejecting the lot from the supplier with a defective unit rate below 1%. It is calculated at a value called the **acceptable quality limit** (AQL). An AQL is a percent of defects between zero and the LTPD. Typically, an AQL is no more than half of the LTPD. Producer risk is considered when consequences of a defective unit are minor or the producer's ability to manage the percentage of defects is limited. A commonly used level of consumer risk is 5%.

Both consumer risk and producer risk can be estimated using binomial distribution. Binomial distribution takes the following form and can be used to model the probability of observing d defective units in a sample of n units when the percentage of defects in the population is p:

$$P(D = d|n, p) = \binom{n}{d} p^d (1-p)^{n-d} \tag{5.4}$$

Figure 5.2 illustrates the basic framework of a single-stage acceptance sampling plan. The first step is to identify the target population. Step 2 is to collect a random sample of size n. Step 3 is to tally the observed defects d in the sample. Finally, if the observed defects in the sample d is greater than the predetermined threshold c, then the population is rejected. Otherwise, the population is accepted. The question remains, how does one derive the values of n and c? The short answer is that it will depend on the purpose of the sampling. The purpose will dictate the appropriate values for consumer risk, LTPD, producer risk, and AQL.

5.2.3.2 Zero Tolerance Plans

Consumer risk is often the only source of risk considered in the development of a sampling plan. When only consumer risk is considered, $c = 0$. In other words, a lot is only accepted if no defects are observed in the sample. As a result, plans that only consider consumer risk are often referred to as **zero tolerance plans**. In these cases, consumer risk is estimated by setting the percentage of defective units in the population p = LTPD. The probability of observing $d = c = 0$ defective units when p = LTPD and $n = 1$ is estimated using binomial distribution. If the resulting probability is greater than the desired consumer risk = β, then evaluate the probability when $n = 2$. Continue this process of adding to the sample size and

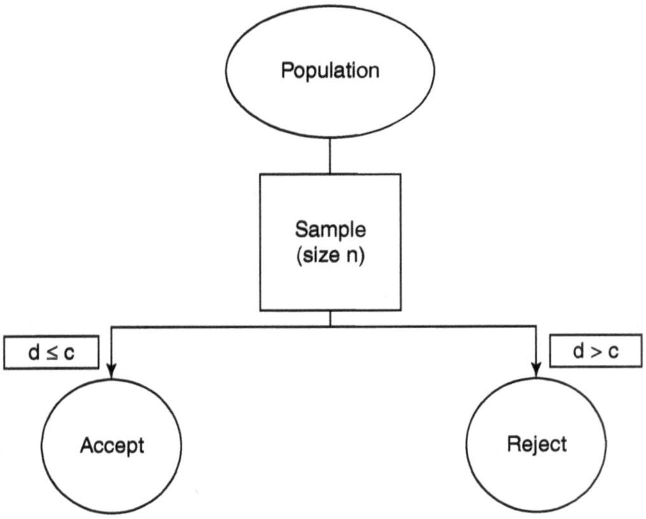

Fig. 5.2 Illustration of single-stage acceptance sampling plan's basic framework. n, sample size; c, rejection number; d, number of nonconformities

evaluating the probability until the probability of accepting at the LTPD is at or below the desired consumer risk = β.

As an example, assume that a food processor requires materials from a supplier to demonstrate that the rate of defects is below 10% with a consumer risk = 5%. Consumer risk in this case is the probability of accepting a lot with a rate of defect above 10%. The calculation below (Eq. 5.5) shows that if n = 29 and p = 10%, then the probability of observing $d = c = 0$ defects is only approximately 4.7%:

$$P(D = 0 \mid 29, 0.1) = \binom{29}{0} 0.1^0 (1-0.1)^{29} \approx 4.7\%. \quad (5.5)$$

As a result, there would only be a 4.7% chance of sampling 29 random units and observing zero defects if the actual rate of defects was 10%. Conversely, there would have been more than a 95% chance of observing at least one defective unit. Consequently, an appropriate sampling plan to demonstrate that the rate of defects is below 10% would be to test 29 randomly sampled units. If zero defects are observed, then the population would be accepted. Otherwise, the population would be rejected. Therefore, if a random sample of 29 units is collected and no defective units are observed, then it can be said that the true percent defects in the population is below 10% with 95% confidence.

The sample size calculation for zero tolerance plans can be approximated using the following formula:

$$N = \frac{\log \beta}{\log(1-p)} \quad (5.6)$$

where

β = the desired consumer risk
p = LTPD
N = rounded up to the nearest integer

Using the previous example of LTPD = 10%, consumer risk of α = 5%:

$$N = \frac{\log \beta}{\log(1-p)} = \frac{\log\ 0.05}{\log(1-0.1)} = \frac{\log\ 0.05}{\log(0.95)} \approx 28.4, \quad (5.7)$$

which rounds up to the nearest integer 29.

5.3 THE NATURE OF THE POPULATION AND PRODUCT

5.3.1 Types of Populations and Products

A clear definition and understanding of the nature of the population is key to the selection of an appropriate sampling plan. The ideal population and product would be uniform throughout and identical at all locations. Such a population would be **homogeneous**. However, homogeneous populations are extremely rare. Even a substance like water that may seem uniform can have suspended particles and sediments that render the population **heterogeneous**. As a result, an appropriate sampling plan should collect samples from various locations of the populations to be representative of the population. A larger sample size is often required for highly heterogeneous populations.

Another consideration is the distribution of the population. Populations tend to be either discrete or continuous. A **discrete** population is divided into distinct subunits. Discrete populations include things like cans of food, boxes of cereal, or bottles of juice. A **continuous** population has no such division into subunits [6]. Continuous populations include things like a semitruck of grain, a conveyer belt of potato chips, or a large tank of juice.

Populations also can vary widely in size. A population can be a day's production, a production lot, or even the contents of a container. The definition of a population is up to the analyst. However, results are only applicable to the population from which the sample is taken. If a sample is taken from the production on Monday, then conclusions are only applicable to that population. These conclusions cannot be extended to the production on Tuesday.

5.3.2 Sampling Methods

A representative sample is vital to attaining accurate estimates of population parameters. The collection of the sample can be conducted via manual or continuous sampling. **Manual sampling** is physically conducted by a human. **Continuous sampling** is collected automatically by a machine. Figure 5.3 shows an automatic sampling device

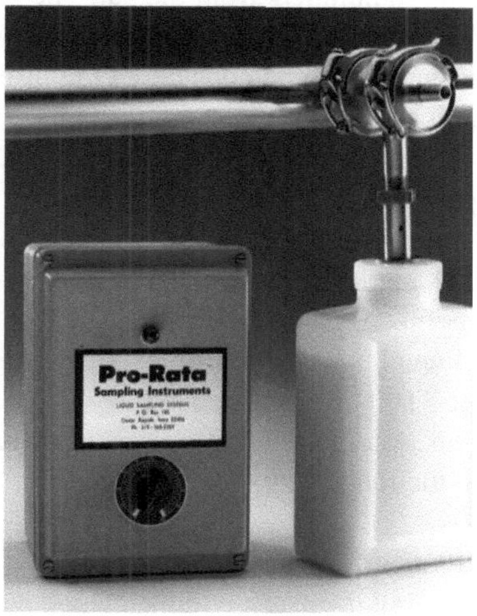

Fig. 5.3 An automatic liquid sampling device that uses air under high pressure to collect multiple 1.5-mL samples. The control box (*left*) regulates the sampling frequency. (Courtesy of Liquid Sampling Systems Inc., Cedar Rapids, IA)

that is used to collect liquid samples from a continuous production line. Continuous sampling should be less prone to human bias than manual sampling. If the sampling plan requires manual sampling, it is essential that the plan includes strategies to eliminate human bias. For example, when sampling from a container of liquid, the plan could require shaking the container well prior to sampling. However, mixing is impossible when sampling something like a rail car of grain. In this scenario, samples may be obtained with triers or probes that are inserted into the population at several locations. It should be noted that rounded particles may flow into the sampling compartments more easily than angular particles, resulting in some degree of sampling bias [7]. Nevertheless, probing methods are far less biased than simply sampling from the top of the car.

Sampling methods are the strategy utilized for attaining a sample. They fall into two general categories: non-probability sampling and probability sampling. Selection in **non-probability sampling** methods is based on convenience and ease. Typically, non-probability sampling should be avoided because the samples will tend to be biased. Selection in **probability sampling** methods is based on random chance. Probability sampling methods provide a statistically sound basis for obtaining representative samples with elimination of human bias [8]. The sampling methods in Sects. 5.3.2.1, 5.3.2.2, 5.3.2.3, 5.3.2.4 and 5.3.2.5 below are some of the most used probability sampling methods.

5.3.2.1 Simple Random Sampling

Simple random sampling is a method for which each unit in the population has an equal chance of being included in the sample. An example of simple random sampling is to assign each unit in the population an identification number. Then put all the identification numbers in a hat and randomly draw identification numbers from the hat. The final step is to include the selected identification numbers in the sample. The resulting sample can be considered unbiased. However, in most scenarios, this method is not feasible.

5.3.2.2 Systematic Sampling

Systematic sampling is a strategy for which every *n*th unit is sampled or samples are collected at evenly spaced time intervals. This method is ideal for scenarios in which samples are distributed evenly over time. An example of this is sampling every 100th unit produced as is comes along the production line. Another example is taking a sample every 15 min of production.

5.3.2.3 Stratified Sampling

Stratified sampling involves dividing the population into distinct subgroups called strata. and is used when the population consists of logical subpopulations for which units within a strata share common characteristics that are not shared among the strata. Random samples are then taken from each strata utilizing another sampling method, such as systematic sampling. An example of this is when production is conducted over the first and second shifts. Each shift can be considered as a strata. Random units would then be selected from each shift.

5.3.2.4 Cluster Sampling

Cluster sampling involves dividing population into subgroups called clusters. Cluster sampling is used when the population consists of logical subpopulations that share common characteristics. A random subset of clusters is then selected. In some cases, all of the units in the selected clusters are included in the sample. In other cases, random units are selected from the randomly select clusters for inclusion in the sample. As an example, consider a production that has been packaged and arranged on 100 pallets, and each pallet holds 360 units. Five units are randomly selected from each of ten randomly selected pallets.

5.3.2.5 Composite Sampling

Composite sampling involves collecting smaller samples from various points of the population called **primary samples**. The primary samples are combined into a larger sample called the **composite sample**. A sample is then taken from the composite sample called the **tested sample**. The key here is that the primary samples are taken randomly from the population and that the tested samples are taken randomly from the composite sample.

One scenario in which composite sampling is used is when working with a continuous population such as flour. Imagine a population for which flour is stored in multiple containers. In this case, primary samples can be taken from each container and combined into a composite sample. The composite sample then can be mixed thoroughly and sampled to produce the tested sample. Another example might be when the original sampling method is a systematic sample. Random samples are collected at prespecified time intervals. These primary samples are composited, and a random sample is taken from the composite. A final example of composite sampling is the sampling plan mandated by the Food and Drug Administration (FDA) and the Food Safety and Inspection Service (FSIS) of the United States Department of Agriculture (USDA) for nutrition labeling. They require a composite of 12 samples with at least 6 subsamples taken and analyzed for compliance with nutrition labeling regulations [9].

5.4 THE NATURE OF THE TESTING METHOD

Another major consideration is the nature of the testing method, which is the subject of many of the rest of the chapters in this book. Testing methods can vary widely in terms of cost, speed, accuracy, and precision. Another key characteristic to consider is whether the method is destructive.

Limitations on any of these characteristics will greatly impact the sampling plan.

5.5 PREPARATION OF SAMPLES

Food is a perishable material that provides the human body with nutrients for growth and development and, based on its composition, can fall into nine matrix categories on the AOAC food triangle as described in Chap. 1. Sample preparation is essential for analysis since the performance of many analytical methods is affected by the food matrix. Proper sample preparation includes considering storage conditions, sample size, particle size, and preservation techniques to stabilize a sample for food analysis.

5.5.1 Problems in Sample Storage

Unreliable data can be obtained by nonstatistical factors such as poor **sample storage**, resulting in sample degradation. Samples should be stored in a container that protects the contents from environmental factors (e.g., moisture, heat, light, air). To protect against changes in moisture content, samples should be stored in an airtight container. Light-sensitive samples should be stored in containers made of opaque glass or containers wrapped in aluminum foil. Oxygen-sensitive samples should be stored under nitrogen or an inert gas. Refrigeration or freezing may be necessary to protect chemically unstable samples. However, freezing should be avoided when storing unstable emulsions. Preservatives can be used to stabilize certain food substances during storage.

Mislabeling of samples causes mistaken sample identification. Samples should be clearly identified by marking the sample container in a manner such that markings are not removed or damaged during storage and transport. For example, plastic bags that are to be stored in ice should be marked with water-insoluble ink. Some laboratories use color-coded sample cups to help manage proper storage conditions.

The contents of **official** or **legal samples** must be sealed to protect against tampering and the seal mark easily identified. These samples must include the date of sampling with the name and signature of the sampling agent. The **chain of custody** of such samples must be documented and maintained from the time of sampling to completion of testing.

5.5.2 General Size Reduction Considerations

If the mass or particle size of the sample is too large for analysis, it must be reduced in bulk or particle size [6]. To obtain a smaller quantity for analysis, the sample can be spread on a clean surface and divided into quarters. The two opposite quarters are combined. If the mass is still too large for analysis, the process is repeated until an appropriate amount is obtained. This system of size reduction can be automated (Fig. 5.4) and adapted for homogeneous liquids by pouring into four containers. The samples need to be homogenized to ensure the negligible differences between each portion [8].

AOAC International [10] provides details on the preparation of specific food samples for analysis, which depends on the nature of the food and the analysis to be performed. For example, in the case of meat and meat products [10], it is specified in Method 983.18 that small samples should be avoided, as this results in significant moisture loss during preparation and subsequent handling. Ground meat samples should be stored in glass or similar containers, with air- and watertight lids. Fresh, dried, cured, and smoked meats are to be bone-free and passed three times through a food chopper with plate openings no more than 3-mm wide. The samples should be mixed thoroughly and analyzed immediately. If immediate analysis is not possible, samples should be chilled or dried for short-term and long-term storage.

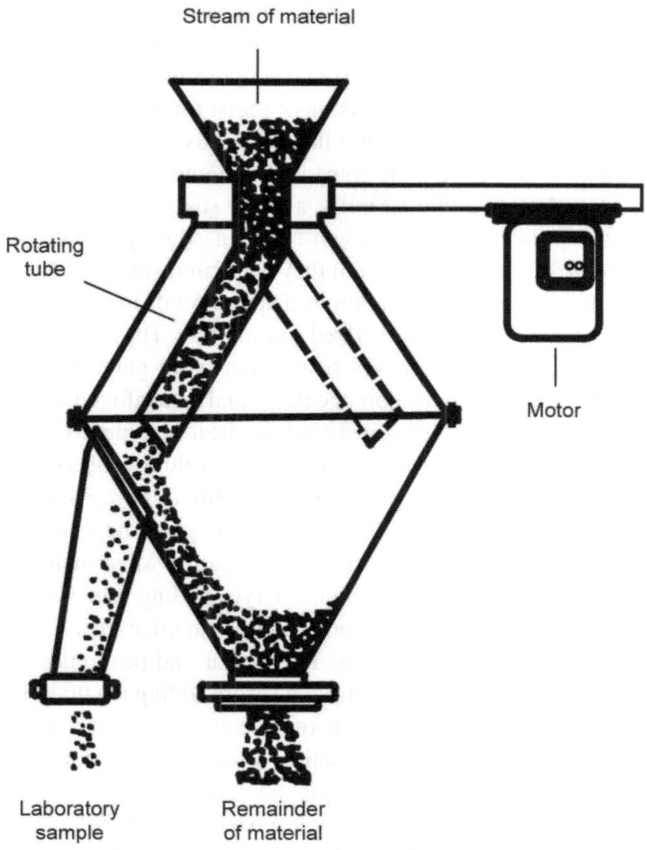

Fig. 5.4 A rotating tube divider for reducing a large sample (ca. 880 kg) of dry, free-flowing material to a laboratory-size sample (ca. 0.2 kg). (Courtesy of Glen Mills, Inc., Clifton, NJ)

5.5.3 Grinding

Grinding is important for sample preparation prior to analysis. Various mills are available for reducing particle size to achieve sample homogenization [9]. To homogenize moist samples, bowl cutters, meat mincers, tissue grinders, mortars and pestles, or blenders are used. Mortars and pestles and mills are best for dry samples. Some foods are more easily ground after drying in a desiccator or a vacuum oven. Grinding wet samples may cause significant loss of moisture and chemical changes. In contrast, grinding frozen samples reduces undesirable changes. The grinding process should not heat the sample; avoid overloading the grinder, which can produce heat through excess friction. For heat-sensitive samples, grinders can be cooled with liquid nitrogen and the ground samples stored at −80 °C. Contact of food with bare metal surfaces should be avoided if trace metal analysis is to be performed [11].

5.5.4 Applications for Grinding Equipment

To break up moist tissues, a number of slicing devices are available: bowl cutters can be used for fleshy tubers and leafy vegetables, while meat mincers may be better suited for fruits, roots, and meats [12]. Addition of sand as an abrasive can provide further subdivision of moist foods. Blenders are effective in mixing soft and flexible foods and suspensions. Rotating knives (25,000 rpm) will disintegrate a sample in suspension. In colloidal mills, a dilute suspension is flowed under pressure through a gap between slightly serrated or smooth-surfaced blades until they are disintegrated by shear. Sonic and supersonic vibrations disperse foods in suspension and in aqueous and pressurized gas solution. The Mickle disintegrator sonically shakes suspensions with glass particles while the sample is homogenized and centrifuged at the same time [12]. Alternatively, a low-shear continuous tissue homogenizer is fast and handles large volumes of sample. Another alternative is **cryogenic grinding** or **cryogrinding**. This method is ideal for biological samples and materials that are sensitive to oxygen or temperature. Most materials are suitable for this technique. Cryogrinding can be performed manually with a mortar and pestle after freezing the sample with liquid nitrogen. The mortar and pestle need to be prechilled with liquid nitrogen before adding the material. Also, there are several brands of specialized grinding equipment with an integrated cooling system that perform the cryogenic freezing and grinding automatically.

Mills differ according to their mode of action, being classified as a **burr**, a **hammer**, an **impeller**, a **cyclone**, an **impact**, a **centrifugal**, or a **roller mill** [12]. Methods for grinding dry materials range from a simple mortar and pestle to power-driven hammer mills. Hammer mills wear well and effectively grind cereals and dry foods, while small samples can be finely ground by ball mills. A ball mill grinds by rotating the sample in a container that is half filled with ceramic balls. A chilled ball mill can be used to grind frozen foods without pre-drying and reduces the likelihood of undesirable heat-initiated chemical reactions occurring during milling [12]. Dry materials can be ground using an ultracentrifugal mill by beating, impacting, and shearing. The food is fed from an inlet to a grinding chamber and is reduced in size by rotors. When the desired particle size is obtained, the particles are delivered by centrifugal force into a collection pan [12]. Large quantities can be ground continuously with a cyclone mill.

5.5.5 Determination of Particle Size

There are a variety of methods for measuring particle size, each suited for different materials. The simplest way to measure particle sizes of dry materials of less than 50 μm in diameter is by passing the sample through a series of vertically stacked sieves with increasing mesh number. The **mesh number** is the number of square screen openings per linear inch of mesh. As the mesh number increases, the apertures between the mesh are smaller and only finer and finer particles pass through subsequent sieves (see Table 5.2). The final particles of dried foods should be 20 mesh for moisture, total protein, or mineral determinations. Particles of 40 mesh size are used for extraction assays such as lipid and carbohydrate estimation. Sieve sizes have been specified for salt, sugar, wheat flour, cornmeal, semolina, and cocoa. The sieve method is inexpensive and fast, but it is not suitable for emulsions or very fine powders [13]. A standard in the industry is the W.S. Tyler® **Ro-Tap® Sieve Shaker** (Fig. 5.5).

A widespread technique to measure particle size distribution uses a principle known as **light scattering** or **laser diffraction**. When a coherent light source, such as a laser beam, is pointed to a particle, four interactions of the beam with the particle can take place: reflection, refraction, absorption, and diffraction. **Reflection** is the portion of light that is rejected by the particle. **Refraction** is the light that goes through transparent or translucent materials and exits the particle with an unchanged wavelength, at a different angle, and with a lower intensity. Some of the light can be **absorbed** by the particle and re-irradiated at a different frequency, which can be displayed as fluorescence or heat. The last interaction, **diffraction**, is the result of light interacting with the edges of the particle. This interaction makes the light spread out, or scatter, and produce wave fronts around the particle that act as secondary spherical waves. This phenomenon is similar to Young's double-slit experiment, which readers can find in a simple Google search.

Table 5.2 US standard mesh with equivalents in inches and millimeters

	Sieve opening	
US standard mesh	*Inches*	*Millimeters*
4	0.1870	4.760
6	0.1320	3.360
7	0.1110	2.830
8	0.0937	2.380
10	0.0787	2.000
12	0.0661	1.680
14	0.0555	1.410
16	0.0469	1.190
18	0.0394	1.000
20	0.0331	0.841
30	0.0232	0.595
40	0.0165	0.400
50	0.0117	0.297
60	0.0098	0.250
70	0.0083	0.210
80	0.0070	0.177
100	0.0059	0.149
120	0.0049	0.125
140	0.0041	0.105
170	0.0035	0.088
200	0.0029	0.074
230	0.0024	0.063
270	0.0021	0.053
325	0.0017	0.044
400	0.0015	0.037

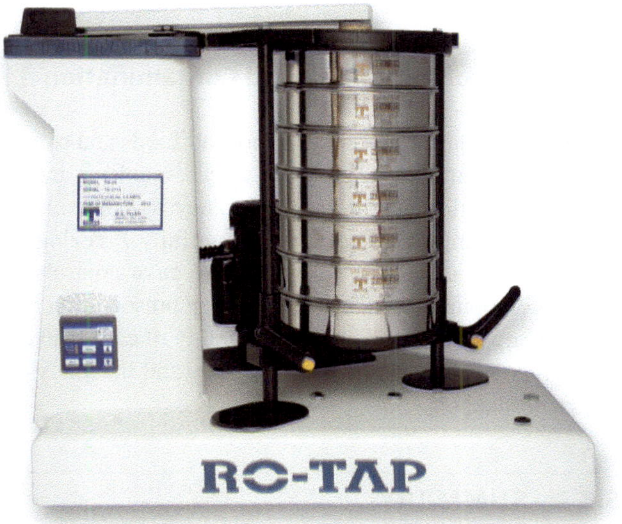

Fig. 5.5 W.S. Tyler® Ro-Tap® Sieve Shaker. (Courtesy of W.S. Tyler®, Mentor, OH)

The secondary waves generated at the edge of the particles interact with each other, in some areas constructively and in others destructively, producing interference patterns that correlate with the particle size. In general, larger particles scatter light at smaller angles and with a higher intensity. Smaller particles, on the other side, scatter light at wider angles and with less intensity. These scatter patterns can be correlated with particle size by using mathematical algorithms.

The size, shape, and optical properties affect the angle and intensity of the scattered light. When particles are capable of transmitting light, the refractive index plays a role in particle size determination. Therefore, light-scattering instruments take advantage of the refractive index to improve accuracy, especially for small particles.

A typical laser diffraction particle analyzer has a sample port, a medium for the sample transport, a flow cell, a laser beam, a detector, electronics, and software. The sample is introduced into the port, which contains some dispersion system to avoid particle agglomeration. The sample is then carried by a liquid (for wet samples) or compressed air (for dry samples) to a flow cell. As the sample goes through the flow cell, a laser beam irradiates the sample. This results in diffraction patterns that, after passing through a condensing lens, are projected on a detector where the diffraction angles and intensities are measured. This information is then transformed into a particle size distribution by the instrument electronics and a computer software algorithm (Fig. 5.6). The results are displayed as frequency distributions with particle size on the x-axis and frequency on the y-axis.

Determination of particle size by light scattering has many advantages, such as the following:

- A wide dynamic range—from nanometers to several millimeters
- Suitable for wet and dry materials—such as solid in liquid suspensions, dry powders, liquid–liquid emulsions, and pastes
- Fast measurement—in the order of seconds
- No calibration needed
- Easy to use and interpret results
- Proficient at detecting mixes of large and small particles
- Technique covered by the International Organization for Standardization (ISO) standard 13320:2009

The disadvantages include the following:

- Assumption that particles are spherical, which introduces error in particles that deviates greatly from sphericity
- Particle agglomeration
- Expression of results on a volume basis, which gives more weight to large particles
- Incomparability of the results with other size determination methods

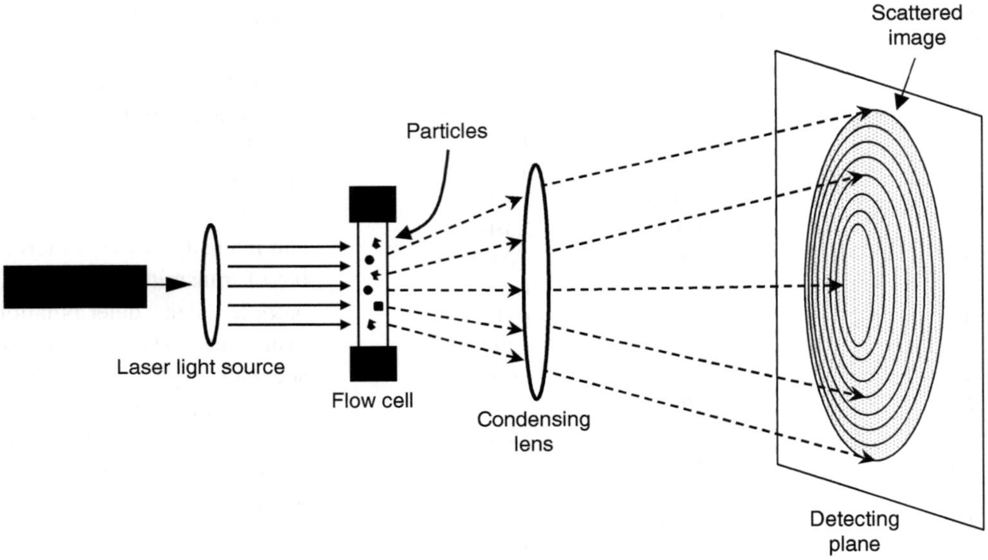

Fig. 5.6 Basics of an instrument to measure particle size distribution by light scattering (laser diffraction). (Adapted from Shimadzu Corporation)

For particles in the nanoscale range, a widely used technique for particle analysis is **dynamic light scattering**. These instruments determine the particle size and even the molecular weight of large molecules in solution. This is achieved by measurement of the frequency shifts of light scattered by particles due to Brownian motion.

Understanding of the principles of the instrument used to obtain size data is vital to appreciate the limitations of each method. For example, data obtained from the same sample using sieves and light scattering will differ [14]. Sieves separate particles using square holes, and, therefore, they distinguish size in the smallest dimension, independent of shape. Light-scattering techniques assume that the particle is spherical, and data are derived from the average of all dimensions. Particle size measurement is useful to maintain sample quality, but care must be taken when choosing the appropriate method and interpreting the data.

5.5.6 Enzymatic Inactivation

Food materials often contain enzymes that may degrade the food components being analyzed. Enzyme activity must be eliminated or controlled using methods that depend on the nature of the food. Heat denaturation to inactivate enzymes and freezer storage (−20 °C to −30 °C) for limiting enzyme activity are common methods. Some enzymes are more effectively controlled by changing the pH or by salting out [12]. Oxidative enzymes may be controlled by adding reducing agents.

5.5.7 Lipid Oxidation Protection

Lipids present particular problems in sample preparation. High-fat foods are difficult to grind and may need to be ground while frozen. Unsaturated lipids are sensitive to oxidative degradation and should be protected by storing under nitrogen or vacuum. Antioxidants help stabilize lipids and may be used if they do not interfere with the analysis. Light-initiated photooxidation of unsaturated lipids can be avoided by controlling lighting and using light-proof storage containers. Lipids are more stable when frozen in intact tissues rather than as extracts [12]. Ideally, unsaturated lipids should be extracted just prior to analysis. Low-temperature storage is generally recommended to protect most foods.

5.5.8 Microbial Growth and Contamination

Microorganisms are present in almost all foods and can alter the sample composition. Likewise, microorganisms are present on non-sterilized surfaces, so sample cross-contamination can occur if samples are not handled carefully. Freezing, drying, and chemical preservatives are effective controls, and often a combination of these is used. The preservation methods used are determined by the probability of contamination, the storage conditions, the storage time, and the analysis to be performed [12].

5.6 SUMMARY

Food quality is monitored at various processing stages, and 100% inspection is rarely possible or even desirable. Sampling techniques enable accurate estimation of the parameters of interest while testing only a small subset of the population of interest. However, appropriate sampling and sample reduction methods must be used to ensure that a representative sample is obtained for analysis. Great consideration should be given to the sampling plan because sampling and sample reduction can be the greatest sources of error. The selection of the sampling plan is determined by the purpose of the inspection, the nature of the population and product, and the nature of the testing method. Based on the purpose of the investigation, it should be determined if variable or attribute sampling is required. A sample size can be calculated to estimate population parameters with the desired precision if variable sampling is required. A sample size can be calculated to develop an acceptance sampling that will determine if a population is of acceptable quality while controlling consumer risk and producer risk. The nature of the population and product must be understood to select the most appropriate sampling method. Factors like the degree of heterogeneity in the population or if the population is discrete or continuous will help determine the most appropriate sampling method. Also, many of the logistical considerations like time and cost will be a function of the nature of the test. Remember that this chapter is an introduction to sampling. Analysts are encouraged to consult a statistician in development of appropriate sampling plans. The information in this chapter will help facilitate a productive conversation with a statistician.

Each sample must be clearly marked for identification and preserved during storage until completion of the analysis. Official and legal samples must be sealed and a chain of custody maintained. Only a portion of the sample is used for analysis and sample size reduction must ensure that the portion analyzed is representative of both the sample and population. Sample preparation and storage should account for factors that may cause sample changes. Samples can be preserved by limiting enzyme activity, preventing lipid oxidation, and inhibiting microbial growth/contamination.

5.7 STUDY QUESTIONS

1. As a supervisor in a quality assurance laboratory, you need to give a new employee instructions regarding choosing a sampling plan.
   (a) Which general factors would you discuss with the new employee?
   (b) Provide an example of a scenario in which one might conduct attribute sampling.
   (c) Provide an example of a scenario in which one might conduct variable sampling.
2. Your coworker wants you to develop and implement a sampling plan that demonstrates that there are no defective units in the population. What would you tell them? Why?
3. A production plant samples every 1000th meal produced for analytical testing. What sampling method is being used?
4. Samples: Collection, Preparation, and Precautions
   (a) Identify a piece of equipment that would be useful for *collecting* a representative sample for analysis. Describe the precautions to be taken to ensure that a representative sample is taken and a suitable food product that could be sampled with this device.
   (b) Identify a piece of equipment that would be useful for *preparing* a sample for analysis. What precautions should be taken to ensure that the sample composition is not changed during preparation?
5. For each of the problems identified below, which can be associated with collection and preparation of samples for analysis, state one solution to overcome the problem:
   (a) Heterogeneous population.
   (b) Change in composition during storage of the sample prior to analysis.
   (c) Metal contamination in grinding.
   (d) Microbial growth during storage of the product prior to analysis.
6. The instructions you are following for cereal protein analysis specify grinding a cereal sample to 10 mesh before you remove the protein by a series of solvent extractions.
   (a) What does 10 mesh mean?
   (b) Would you question the use of a 10-mesh screen for this analysis? Provide reasons for your answer.
7. You are to collect and prepare a sample of cereal produced by your company for the analyses required to create a standard nutritional label. Your product is considered "low-fat" and "high-fiber." (See reference [9] for information on nutrient content claims; see also Chap. 3, Sect. 3.2.2, on FDA compliance procedures for nutrition labeling.)
   (a) What kind of sampling plan will you use? Will you do attribute or variable sampling?
   (b) What are the risks associated with sampling in your specific case? What analysis might be used to calculate an appropriate sample size?
   (c) What specific problems would you anticipate in sample collection, storage, and preparation? How would you avoid or minimize each of these problems?

Acknowledgments The editors thank Rubén O. Morawicki (formerly at the University of Arkansas) for his involvement as the author of this chapter in the two previous editions of this textbook.

REFERENCES

1. Lichon, MJ (2000) Sample preparation for food analysis, general. In: Meyers RA (ed) Encyclopedia of analytical chemistry: applications, theory and instrumentation. John Wiley, New York
2. Lichon MJ (2004) Sample preparation, ch. 44. In: Nollet LML (ed) Handbook of food analysis, 2nd edn. Vol 3. Marcel Dekker, New York, pp. 1741–1755

3. Mälkki Y (1986) Collaborative testing of methods for food analysis. J Assoc Off Anal Chem 69(3):403–404
4. IUPAC (2019) Compendium of chemical terminology, 2nd edn. (the "Gold Book"). Compiled by McNaught AD Wilkinson A. Blackwell Scientific, Oxford. XML on-line corrected version: http://goldbook.iupac.org (2006-) created by Nic M, Jirat J, Kosata B; updates compiled by Jenkins A. ISBN 0–9678550–9–8. https://doi.org/10.1351/goldbook
5. Miller JC (1988) Basic statistical methods for analytical chemistry. I. Statistics of repeated measurements. A review. Analyst 113: 1351–1355
6. Horwitz W (1988) Sampling and preparation of samples for chemical examination. J Assoc Off Anal Chem 71: 241–245
7. Pehrsson P, Perry C, Daniel M (2013) ARS, USDA updates food sampling strategies to keep pace with demographic shifts. Procedia Food Sci 2:52–59
8. Puri SC, Ennis D, Mullen K (1979) Statistical quality control for food and agricultural scientists. G.K. Hall and Co., Boston, MA
9. FDA (2023) Code of federal regulations. 21 CFR 101.9 (g), 9 CFR 317.309 (h), 9 CFR 381.409 (h). US Government Printing Office, Washington, DC
10. AOAC International (2023) Official methods of analysis, 22nd edn. (On-line). AOAC International, Rockville, MD
11. Pomeranz Y, Meloan CE (1994) Food analysis: theory and practice, 3rd edn. Chapman & Hall, New York
12. Cubadda F, Baldini M, Carcea M, Pasqui LA, Raggi A, Stacchini P (2001) Influence of laboratory homogenization procedures on trace element content of food samples: an ICP-MS study on soft and durum wheat. Food Addit Contam 18: 778–787
13. Kenkel JV (2013) Analytical chemistry for technicians, 4th edn. CRC, Boca Raton, FL
14. Jordan JR (1999) Particle size analysis. Inside Lab Manage 3(7): 25–28

Part II

Spectroscopy and Mass Spectrometry

Basic Principles of Spectroscopy

6

Mirko Bunzel and Michael H. Penner

Contents

6.1 INTRODUCTION

Spectroscopy deals with the production, measurement, and interpretation of spectra arising from the **interaction of electromagnetic radiation with matter**. There are many different spectroscopic methods available for solving a wide range of analytical problems. The methods differ with respect to the species to be analyzed (such as molecular or atomic spectroscopy), the type of radiation–matter interaction to be monitored (such as absorption, emission, or diffraction), and the region of the electromagnetic spectrum used in the analysis. Spectroscopic methods are very informative and widely used for both quantitative and qualitative analyses. Spectroscopic methods based on the absorption or emission of radiation in the **ultraviolet** (UV), **visible** (Vis), **infrared** (IR), and radio-frequency **(nuclear magnetic resonance**, NMR) ranges are most commonly encountered in traditional food analysis laboratories. Each of these methods is distinct, in that it monitors different types of molecular or atomic transitions. The basis of these transitions is explained in the sections below. For detailed information on principles of spectroscopy refer to Refs. [1–14].

M. Bunzel (✉)
Department of Food Chemistry and Phytochemistry, Karlsruhe Institute of Technology (KIT), Karlsruhe, Germany
e-mail: mirko.bunzel@kit.edu

M. H. Penner
Department of Food Science and Technology, Oregon State University, Corvallis, OR, USA
e-mail: mike.penner@oregonstate.edu

6.2 LIGHT

6.2.1 Properties

Light may be thought of as particles of energy that move through space with wavelike properties. This image of light

B. P. Ismail, S. S. Nielsen (eds.), *Nielsen's Food Analysis*, Food Science Text Series,
https://doi.org/10.1007/978-3-031-50643-7_6

suggests that the energy associated with a ray of light is not distributed continuously through space along the wave's associated electric and magnetic fields but rather that it is concentrated in discrete packets. Light is therefore said to have a dual nature: **particulate** and **wavelike**. The phenomena associated with light propagation, such as interference, diffraction, and refraction, are most easily explained using the wave theory of electromagnetic radiation. However, the interaction of light with matter, which is the basis of absorption and emission spectroscopy, may be best understood in terms of the particulate nature of light. Light is not unique in possessing both wavelike and particulate properties. For example, the fundamental particles of matter, such as electrons, protons, and neutrons, are known to exhibit wavelike behavior.

The wave properties of electromagnetic radiation are described in terms of the wave's frequency, wavelength, and amplitude. A graphical representation of a plane-polarized electromagnetic wave is given in Fig. 6.1. The wave is plane-polarized, in that the oscillating electric and magnetic fields making up the wave are each limited to a single plane. The **frequency** (ν, the lower case Greek letter *nu*) of a wave is defined as the number of oscillations the wave will make at a given point per second. This is the reciprocal of the **period** (*p*) of a wave, which is the time in seconds required for successive maxima of the wave to pass a fixed point. The **wavelength** (λ) represents the distance between successive maxima on any given wave. The units used in reporting wavelengths will depend on the region of electromagnetic radiation used in the analysis. Spectroscopic data sometimes are reported with respect to **wave numbers** ($\bar{\nu}$), which are reciprocal wavelengths in units of cm^{-1}. Wave numbers are encountered most often in IR spectroscopy. The **velocity of propagation** (v_i) of an electromagnetic wave, in units of distance per second, in any given medium "i" can be calculated by taking the product of the frequency of the wave, in cycles per second, and its wavelength in that particular medium:

$$v_i = \nu\lambda_i \tag{6.1}$$

where:

v_i = velocity of propagation in medium *i*
ν = frequency (of the associated wave),
λ_i = wavelength in medium *i*.

The frequency of an electromagnetic wave is determined by the source of the radiation, and it remains constant as the wave traverses different media. However, the velocity of propagation of a wave will vary slightly depending on the medium through which the light is propagated. The wavelength of the radiation will change in proportion to changes in wave velocity as defined by Eq. 6.1. The **amplitude of the wave** (*A*) represents the magnitude of the electric vector at the wave maxima. The **irradiance** (radiant power) (*P*) and **intensity** (*I*) of a beam of radiation are proportional to the square of the amplitude of the associated waves making up that radiation. Figure 6.1 indicates that electromagnetic waves are composed of **oscillating magnetic** and **electric fields**, the two of which are mutually perpendicular, in phase with each other, and perpendicular to the direction of wave propagation. As drawn, the waves represent changes in the respective field strengths with time at a fixed location or changes in the respective field strengths over distance at a fixed time. The electrical and magnetic components of the waves are represented as a series of vectors whose lengths are proportional to the magnitude of the respective field. It is the oscillating electric field that is of most significance to spectroscopic phenomena such as absorption, transmission, and refraction. However, a purely electric field, without its associated magnetic field, is impossible.

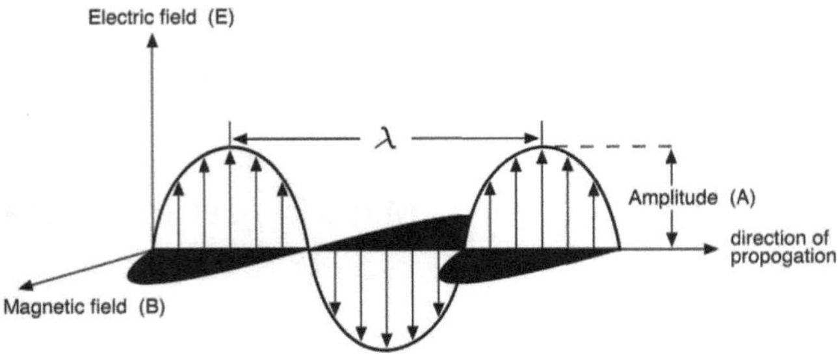

Fig. 6.1 Representation of plane-polarized electromagnetic radiation propagating along the *x*-axis. The electric and magnetic fields are in phase, perpendicular to each other and to the direction of propagation

6.2.2 Interference

"Interference" is the term used to describe the observation that when two or more coherent waves (with the same frequency and waveform) cross one another, they result in an instantaneous wave, at the point of intersection, whose amplitude is the algebraic sum of the amplitudes of the individual waves at the point of intersection. The law describing this wave behavior is known as the **principle of superposition**. Superposition of sinusoidal waves is illustrated in Fig. 6.2. Note that in all cases, the effective amplitude of the perceived wave at the point in question is the combined effect of each of the waves that crosses that point at any given instant. In spectroscopy, the amplitude of most general interest is that corresponding to the magnitude of the resulting electric field intensity. The **maximum constructive interference** of two waves occurs when the waves are completely in phase (i.e., the maxima of one wave align with the maxima of the other wave), while the **maximum destructive interference** occurs when waves are 180° out of phase (the maxima of one wave align with the minima of the other wave). This concept of interference is fundamental to the interpretation of diffraction data, which represents a specialized segment of qualitative spectroscopy. Interference phenomena also are widely used in the design of spectroscopic instruments that require the dispersion or selection of radiation, such as those instruments employing grating monochromators or interference filters, as described in Chap. 7.

Interference phenomena are best rationalized by considering the wavelike nature of light. However, phenomena such as the absorption and emission of radiation are more easily understood by considering the particulate nature of light. The particles of energy that move through space with wavelike properties are called **photons**. The energy of a photon can be defined in terms of the frequency of the wave with which it is associated (Eq. 6.2):

$$E = h\nu \tag{6.2}$$

where:

E = energy of a photon
h = Planck's constant
ν = frequency (of the associated wave)

This relationship indicates that the photons making up **monochromatic light**, which is electromagnetic radiation composed of waves having a single frequency and wavelength, are all of equivalent energy. Furthermore, just as the frequency of a wave is a constant determined by the radiation source, the energy of associated photons also will be unchanging. The brightness of a beam of monochromatic light, when expressed in terms of the particulate nature of light, will be the product of the photon flux and the energy per photon. **Photon flux** refers to the number of photons flowing across a unit area perpendicular to the beam per unit time. It follows that to change the brightness of a beam of monochromatic light, a change in photon flux is required. In spectroscopy, the term "brightness" is generally not used, but, rather, one refers to the **irradiance** (P) (radiant power or **intensity** (I) were formerly used terms, too) of a beam of light. Irradiance is the amount of radiant energy striking a given area per unit time. In terms of the International System of Units (SI), irradiance equals the number of joules of radiant energy impinging on a 1m² area of the detector per second (J $(m^2)^{-1}(s)^{-1}$ or W $(m^2)^{-1}$).

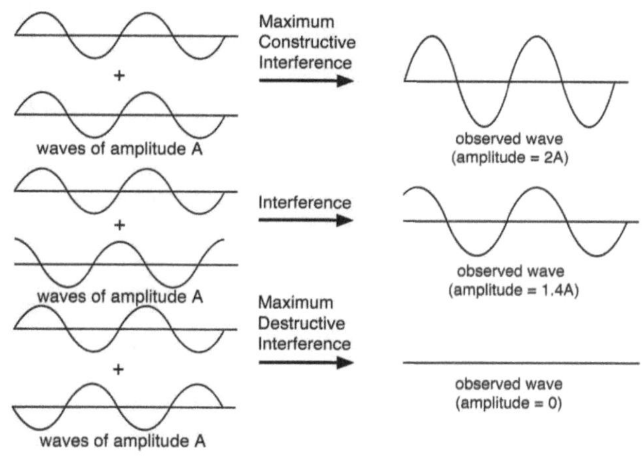

Fig. 6.2 Interference of coherent waves that are (**a**) in phase, (**b**) 90° out of phase, and (**c**) 180° out of phase

Table 6.1 Properties of light

Symbols/terms	*Relationship*	*Frequently used units*
λ = wavelength	$\lambda_i\nu = v_i$ ($v_i = c$ in a vacuum)	Nm (nanometers, 10^{-9} m) Å (angstrom units, 10^{-10} m) μm (microns, 10^{-6} m) mμ (millimicrons, 10^{-9} m)
ν = frequency		Hz (hertz, 1 Hz = 1 oscillation per second)
c = speed of light		2.9979×10^8m s^{-1} in vacuum
$\bar{\nu}$ = wave number	$\bar{\nu} = 1/\nu$	cm^{-1} kK (kilokayser, 1 kK = 1000 cm^{-1})
p = period	p = $1/\nu$	s
E = energy	E = $h\nu$ = hc/λ = $hc\bar{\nu}$	J (1 joule = 1 kg m^{2}s^{-2}) Cal (calorie, 1 cal = 4.184 J) Erg (1 erg = 10^{-7}J) eV (1 electron volt = 1.6022×10^{-19}J)
h = Planck's constant		6.6262×10^{-34} J s
P = irradiance (radiant power)	Amount of energy striking a given unit area per unit time	(joules) $(m^2)^{-1}(s)^{-1}$

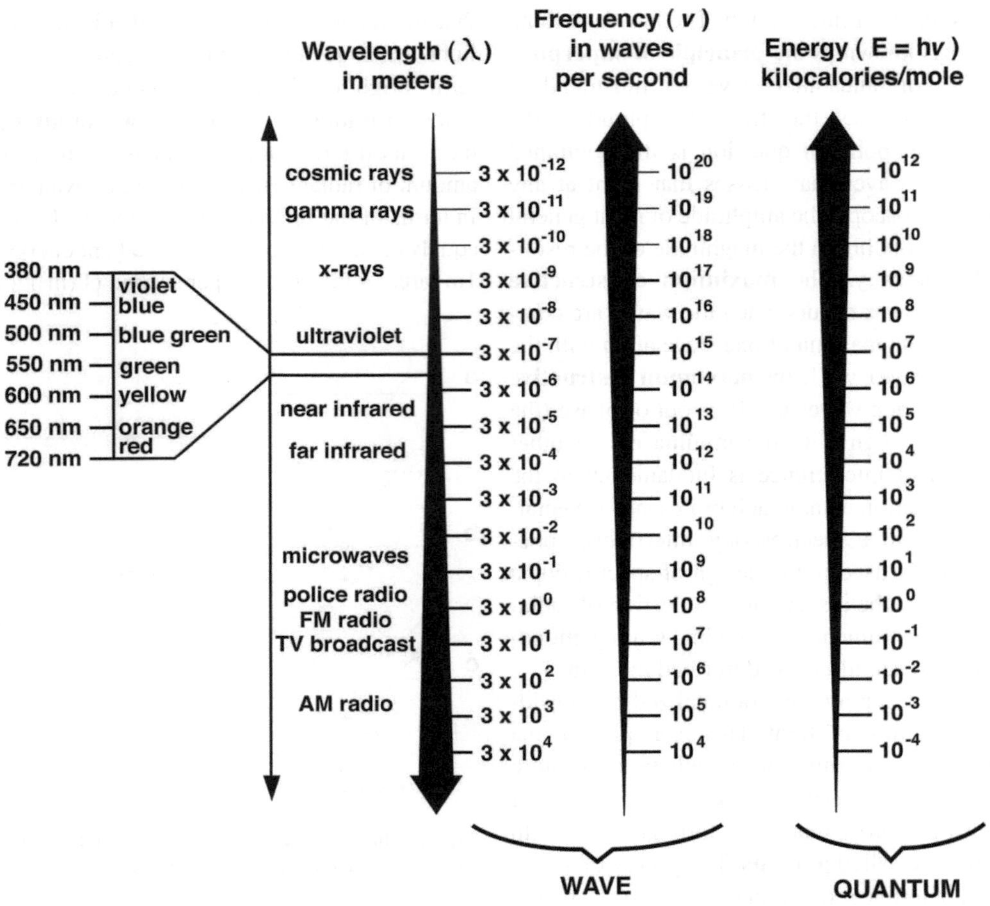

Fig. 6.3 The electromagnetic spectrum (from Milton Roy education manual for the SPECTRONIC® 20 & 20D spectrophotometers, p. 3 (1989). Courtesy of Milton Roy Company, Rochester, NY, a subsidiary of Sundstrand Corporation)

The basic interrelationships of light-related properties and a general scheme of the electromagnetic spectrum are presented in Table 6.1 and Fig. 6.3, respectively.

6.3 ENERGY STATES OF MATTER

6.3.1 Quantum Nature of Matter

The energy content of matter is quantized. Consequently, the potential or internal energy content of an atom or molecule does not vary in a continuous manner but rather in a series of discrete steps. Atoms and molecules, under normal conditions, exist predominantly in the **ground state**, which is the state of lowest energy. Ground-state atoms and molecules can gain energy, in which case they will be elevated to one of their higher energy states, referred to as **excited states**. The quantum nature of atoms and molecules puts limitations on the energy levels that are available to these species. Consequently, there will be specific "allowed" **internal energy levels** for each atomic or molecular species. Internal energy levels not corresponding to an allowed value for that particular species are unattainable. The set of available energy levels for any given atom or molecule will be distinct for that species. Similarly, the potential energy spacings between allowed internal energy levels will be characteristic of a species. Therefore, the set of potential energy spacings for a species may be used qualitatively as a distinct fingerprint. Qualitative absorption and emission spectroscopy make use of this phenomenon, in that these techniques attempt to determine an unknown compound's relative energy spacings by measuring transitions between allowed energy levels.

6.3.2 Electronic, Vibrational, and Rotational Energy Levels

The relative **potential energy** of an atom or molecule corresponds to the energy difference between the energy state in which the species exists and that of the ground state. Figure 6.4 is a partial molecular energy-level diagram depicting potential energy levels for an organic molecule. The lowest energy state in the figure, bottom line in bold, represents the ground state. There are three **electronic energy states**

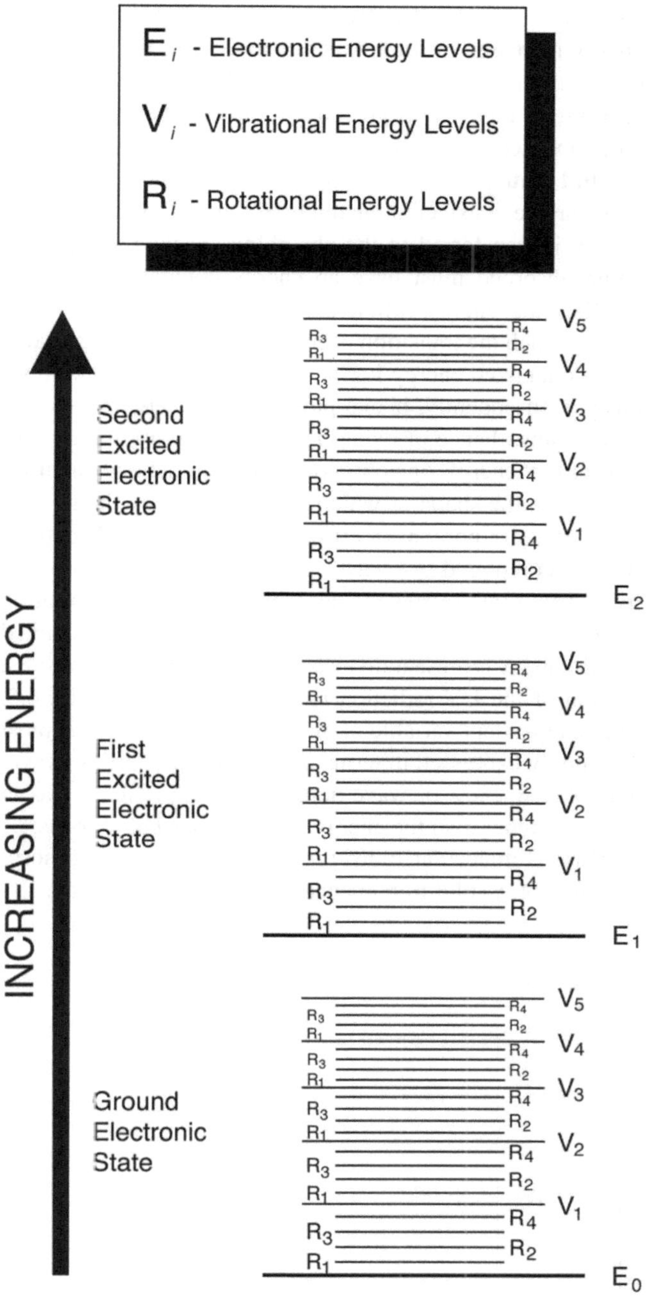

Fig. 6.4 Partial molecular energy-level diagram depicting three electronic states

depicted, each with its corresponding vibrational and rotational energy levels. Each of the electronic states corresponds to a given **electron orbital**. Electrons in different orbitals are of different potential energies. When an electron changes orbitals, such as when absorbing or emitting a photon of appropriate energy, it is termed an "electronic transition" since it is the electron that is changing energy levels. However, any change in the potential energy of an electron will, by necessity, result in a corresponding change in the potential energy of the atom or molecule that the electron is associated with.

Atoms are like **molecules**, in that only specific energy levels are allowed for atomic electrons. Consequently, an energy-level diagram of an atom would consist of a series of electronic energy levels. In contrast to molecules, the electronic energy levels of atoms have no corresponding vibrational and rotational levels and, hence, may appear less complicated. Atomic energy levels correspond to allowed electron shells (orbits) and corresponding subshells (i.e., 1 s, 2 s, 2p, etc.). The magnitude of the energy difference between the ground state and first excited states for valence electrons of atoms and bonding electrons of molecules is generally of the same range as the energy content of photons associated with UV and Vis radiation.

The wider lines within each electronic state of Fig. 6.4 depict the species' **vibrational energy levels**. The atoms that comprise a molecule are in constant motion, vibrating in many ways. However, in all cases, the energy associated with this vibrational motion corresponds to defined quantized energy levels. The energy differences between neighboring vibrational energy levels are much smaller than those between adjacent electronic energy levels. Therefore, it is common to consider that several vibrational energy levels are superimposed on each of the molecular electronic energy levels. Energy differences between allowed vibrational energy levels are of the same magnitude as the energy of photons associated with radiation in the IR region. Vibrational energy levels would not be superimposed on an atomic potential energy-level diagram since this vibrational motion does not exist in a single atom. In this respect, the potential energy diagram for an atom is less complex than that for a molecule, with the atomic energy-level diagram having fewer energy levels.

The potential energy of a molecule also is quantized in terms of the energy associated with the rotation of the molecule about its center of gravity. These **rotational energy levels** are yet more closely spaced than the corresponding vibrational levels, as depicted by the narrow lines within each electronic state shown in Fig. 6.4. Hence, it is customary to consider several rotational energy levels superimposed on each of the permitted vibrational energy levels. The energy spacings between rotational energy levels are of the same magnitude as the energy associated with photons of microwave radiation. Microwave spectroscopy is not commonly used in food analysis laboratories; however, the presence of these different energy levels will impact the spectrum observed in other forms of spectroscopy, as will be discussed later. Similar to the situation of vibrational energy levels, rotational energy levels are not of consequence to atomic spectroscopy.

In summation, the internal energy of an **atom** is described in terms of its **electronic energy levels**, while the internal energy of a **molecule** is dependent on its **electronic**, **vibrational**, and **rotational energies**. The algebraic form of these statements follows:

$$E_{atom} = E_{electronic} \quad (6.3)$$

$$E_{molecule} = E_{electronic} + E_{vibrational} + E_{rotational} \quad (6.4)$$

The spectroscopist makes use of the fact that each of these associated energies is quantized and that different species will have somewhat different energy spacings.

6.3.3 Nuclear Energy Levels in Applied Magnetic Fields

NMR spectroscopy makes use of yet another type of quantized energy level. The energy levels of importance to NMR spectroscopy differ with respect to those described above, in that they are relevant only in the presence of an applied external **magnetic field**. The basis for the observed energy levels may be rationalized by considering that the nuclei of some atoms behave as tiny bar magnets. Hence, when the atoms are placed in a magnetic field, their nuclear magnetic moment will have a preferred orientation, just as a bar magnet would behave. The NMR-sensitive nuclei of general relevance to the food analyst such as ^{1}H, ^{13}C, ^{15}N, and ^{31}P have two permissible orientations. The energy difference between these allowed orientations depends on the specific nuclei (it is, for example, different for ^{1}H and ^{13}C) and the effective magnetic field strength that the nuclei experience. The effective magnetic field strength will itself depend on the strength of the applied magnetic field and the chemical environment surrounding the nuclei in question. The applied magnetic field strength depends on the magnet of the NMR instrumentation (for example, a "500-MHz NMR spectrometer" employs a field of 11.75 T), and it is essentially equivalent for each of the nuclei in the applied field. Hence, differences in energy spacings of NMR-sensitive nuclei will depend solely on the identity of the nucleus and its environment. In general, the energy spacings between permissible nuclear orientations, under usable external magnetic field strengths, are of the same magnitude as the energy associated with radiation in the radio-frequency range.

6.4 ENERGY-LEVEL TRANSITIONS IN SPECTROSCOPY

6.4.1 Absorption of Radiation

The **absorption of radiation** by an atom or molecule is that process in which energy from a photon of electromagnetic radiation is transferred to the absorbing species. When an atom or molecule absorbs a photon of light, its internal energy increases by an amount equivalent to the amount of energy in that particular photon. Therefore, in the process of absorption, the species goes from a lower energy state to a more **excited state**. In most cases, the species is in the **ground state** prior to absorption. Since the absorption process may be considered quantitative (i.e., all of the photon's energy is transferred to the absorbing species), the photon being absorbed must have an energy content that exactly matches the energy difference between the energy levels across which the transition occurs. This must be the case due to the quantized energy levels of matter, as discussed previously. Consequently, if one plots photon energy versus the relative absorbance of radiation uniquely composed of photons of that energy, one observes a characteristic **absorption spectrum**, the shape of which is determined by the relative absorptivity of photons of different energies. The **absorptivity** of a compound is a wavelength-dependent proportionality constant that relates the absorbing species concentration to its experimentally measured absorbance under defined conditions. A representative absorption spectrum covering a portion of the UV radiation range is presented in Fig. 6.5. The independent variable of an absorption spectrum is most commonly expressed in terms of the wave properties (wavelength, frequency, or wave numbers) of the radiation, as in Fig. 6.5, rather than the energy of the associated photons.

Various molecular transitions resulting from the absorption of photons of different energies are shown schematically

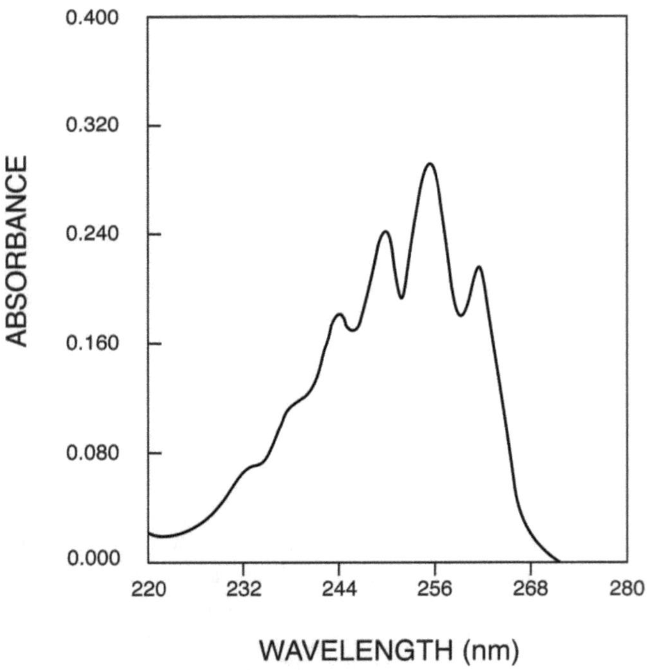

Fig. 6.5 Absorption spectrum of 0.005 M benzene in water solution

in Fig. 6.6. The transitions depicted represent those that may be induced by absorption of UV, Vis, IR, and microwave radiation. The figure also includes transitions in which the molecule is excited from the ground state to an exited electronic state with a simultaneous change in its vibrational or rotational energy levels. Although not shown in the figure, the absorption of a photon of appropriate energy also may cause simultaneous changes in electronic, vibrational, and rotational energy levels. The ability of molecules to have simultaneous transitions between the different energy levels tends to broaden the peaks in the UV–Vis absorption spectrum of molecules relative to those peaks observed in the absorption spectrum of atoms. This would be expected when one considers that vibrational and rotational energy levels are absent in an atomic energy-level diagram. The depicted transitions between vibrational energy levels, without associated electronic transitions, are induced by radiation in the IR region. Independent transitions between allowed rotational energy levels also are depicted, with these resulting from the absorption of photons of microwave radiation. A summary of transitions relevant to atomic and molecular absorption spectroscopy, including corresponding wavelength regions, is presented in Table 6.2.

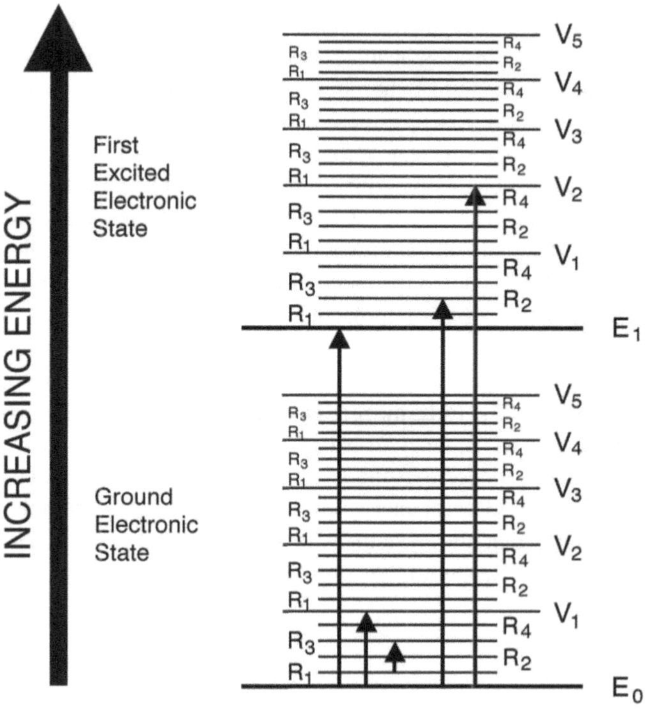

Fig. 6.6 Partial molecular energy-level diagram, including electronic, vibrational, and rotational transitions

6.4.2 Emission of Radiation

Emission is essentially the reverse of the absorption process, occurring when energy from an atom or molecule is released in the form of a photon of radiation. A molecule raised to an excited state will typically remain in the excited state for a very short period of time before relaxing back to the ground state. There are several **relaxation processes** through which an excited molecule may dissipate energy. The most common relaxation process is for the excited molecule to dissipate its energy through a series of small steps brought on by collisions with other molecules such as solvent molecules. The energy is thus converted to kinetic energy, with the net result being the dissipation of the energy as heat. Under normal conditions, the dissipated heat is not enough to measurably affect the system. In some cases, molecules excited by the absorption of UV or Vis light will lose a portion of their excess energy through the emission of a photon. This emission process is referred to as either **fluorescence** or **phosphorescence**, depending on the nature of the excited state. In molecular fluorescence spectroscopy, the photons emitted

Table 6.2 Wavelength regions, spectroscopic methods, and associated transitions

Wavelength region	*Wavelength limits*	*Type of spectroscopy*	*Usual wavelength range*	*Types of transitions in chemical systems with similar energies*
Gamma rays	0.01–1 Å	Emission	<0.1 Å	Nuclear proton/neutron arrangements
X-rays	0.1–10 nm	Absorption, emission, fluorescence, and diffraction	0.1–100 Å	Inner-shell electrons
Ultraviolet	10–380 nm	Absorption, emission, and fluorescence	200–380 nm	Outer-shell electrons in atoms, bonding electrons in molecules
Visible	380–780 nm	Absorption, emission, and fluorescence	380–780 nm	Same as ultraviolet
Infrared	0.078–1000 μm	Absorption	0.78–300 μm	Vibrational position of atoms in molecular bonds
Microwave	0.1–100 cm	Absorption Electron spin resonance	0.75–3.75 mm 3 cm	Rotational position in molecules Orientation of unpaired electrons in an applied magnetic field
Radio waves	1–1000 m	Nuclear magnetic resonance	0.6–10 m	Orientation of nuclei in an applied magnetic field

from the excited species generally will be of lower energy and longer wavelength than the corresponding photons that were absorbed in the excitation process. The reason is that, in most cases, only a fraction of the energy difference between the excited and ground states is lost in the emission process. The other fraction of the excess energy is dissipated as heat during vibrational relaxation. This process is depicted in Fig. 6.7, which illustrates that the excited species undergoes vibrational relaxation down to the lowest vibrational energy level within the excited electronic state and then undergoes a transition to the ground electronic state through the emission of a photon. The photon emitted will have an energy that equals the energy difference between the lowest vibrational level of the excited electronic state and the ground electronic state level it descends to. The fluorescing molecule may descend to any of the vibrational levels within the ground electronic state. If the fluorescence transition is to an excited vibrational level within the ground electronic state, then it will quickly return to the ground state (lowest energy level) via vibrational relaxation. In yet other cases, an excited species may be of sufficient energy to initiate some type of photochemistry that ultimately leads to a decrease in the system's potential energy. In all cases, the relaxation process is driven by the tendency for a species to exist at its lowest permissible internal energy level. The relaxation process that dominates a system will be the one that minimizes the lifetime of the excited state. Under normal conditions, the relaxation process is so rapid that the population of molecules in the ground state is essentially unchanged.

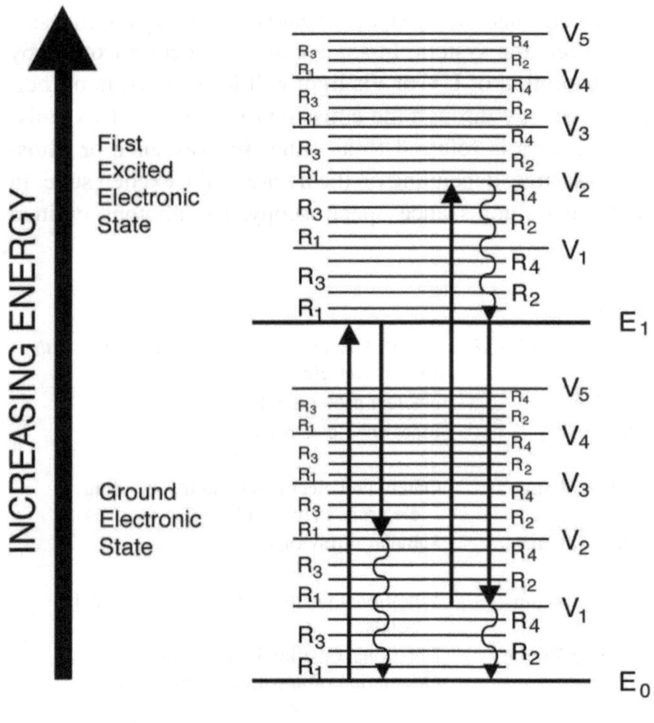

Fig. 6.7 Partial molecular energy-level diagram, including absorption, vibrational relaxation, and fluorescence relaxation

6.4.3 Population of Energy Levels

The preceding text discussed the existence and quantum nature of molecular, atomic, and nuclear energy levels. It is now relevant to ask how a population of particles is likely to be distributed among their permissible energy levels, i.e., the relative population of the different energy levels. This is important to consider because the intensity of the signal generated in a spectroscopic method is a function of the number of particles in the energy level corresponding to the origin of the signal. For example, when performing traditional molecular absorption spectroscopy, one measures the excitation of molecules in the ground state (Chap. 7); when performing atomic emission spectroscopy, one measures photons emitted by atoms in an excited state (Chap. 9). In the former case, the observed signal will be a function of the number of molecules in the ground state; in the latter case, the signal will be a function of the number of molecules in the appropriate excited state. For a given collection of particles, the relative population of the different energy levels is described by the Boltzmann distribution. The **Boltzmann distribution** is a probability distribution or a frequency distribution; when applied to a group of particles, it describes the average number of particles to be found in the different energy levels available to those particles (assuming that the system is in thermal equilibrium). The Boltzmann distribution law may be expressed as follows:

$$(p'/p) = e^{-(E'-E)/kT} \tag{6.5}$$

where:

p' = probability of finding a particle at energy level E'
p = probability of finding a particle at energy level E
E' = energy corresponding to a higher energy level
E = energy corresponding to a lower energy level
k = Boltzmann constant
T = absolute temperature

This equation indicates that the fraction of molecules in the higher energy state (E') decreases exponentially with increasing ΔE (i.e., $E' - E$). In UV–Vis spectroscopy, the vast majority of the molecules are in the ground state since the energy difference between the ground state and the first excited state is relatively large. In nuclear magnetic resonance spectroscopy, the populations in the two energy states are nearly identical due to the relatively small energy difference between the lower and higher energy states (with the lower energy state being slightly more populated). These relationships are consistent, with UV–Vis spectroscopy being significantly more sensitive than nuclear magnetic resonance spectroscopy. The equation also illustrates that increasing the temperature of a system will increase the fraction of molecules in the higher energy state. This is relevant to atomic emission spectroscopy in which molecules in the excited state emit the detected signal. At room temperature, there are insufficient neutral atoms in the excited state to perform atomic emission spectroscopy. However, as one continues to raise the temperature of the analyte mixture, the fraction of atoms in the excited state increases to the point that atomic emission spectroscopy becomes feasible.

6.5 INTRODUCTION TO METHODS OF MOLECULAR AND ATOMIC SPECTROSCOPY

Spectrochemical analysis, a branch of spectroscopy, encompasses a wide range of techniques used in analytical laboratories for the qualitative and quantitative analysis of the chemical composition of foods. Common methods of molecular spectroscopy include UV, Vis, and fluorescence spectroscopy (Chap. 7), infrared (IR) (mid-IR and near-IR, depending on the spectral range) (Chap. 8), absorption spectroscopy (Chap. 9), and NMR spectroscopy (Chap 10). Raman spectroscopy is another technique that is classified as vibrational spectroscopy (like mid-IR and near-IR spectroscopy); however, it is based on inelastic scattering (Chap. 8, Sect. 8.5). Moreover, in atomic spectroscopy, concentrations of atoms are measured by absorption or emission of characteristic wavelengths of light, often in the UV range. For an overview of these types of spectroscopies, see Table 6.3. In each of these methods of molecular and atomic spectroscopy, the analyst attempts to measure the amount of radiation either absorbed or emitted by the analyte, with the analyte being a molecule or an atom. All of these methods make use of the facts that the energy content of matter is quantized and that photons of radiation may be absorbed or emitted by matter if the energy associated with the photon equals the energy difference for allowed transitions of that given species. The above methods differ from each other with respect to the radiation wavelengths used in the analysis and/or the molecular vs. atomic nature of the analyte.

6.6 SUMMARY

Spectroscopic methods are of utmost importance in food analysis to analyze the structures of food constituents as well as to determine their concentrations. In a wider sense, spectroscopy deals with the production, measurement, and interpretation of spectra arising from the interaction of electromagnetic radiation (light) with matter. Spectroscopic methods have in common that the interaction of molecules or atoms in their ground state with electromagnetic radiation leads to an excited state of these molecules or atoms, followed by relaxation processes by which the molecules/atoms return to their ground state. Thus, to reach the excited state, a photon (particulate nature of light) must be absorbed, whose energy content is exactly equal to the difference between the energy levels of the ground state and the excited state. Depending on whether valence electrons, vibrations, or electron spins are excited, the energy of the photons must correspond to higher or lower energy differences; thus, the wavelengths of electromagnetic radiation (wavelike nature of light) required will be different depending on whether, for example, valence electrons are excited in UV spectroscopy (higher energy, shorter wavelengths) or electron spins are excited in NMR spectroscopy (lower energy, longer wavelengths). Monitoring of the energy/wavelengths of the absorbed electromagnetic radiation results in absorption spectra. The return of the excited molecules/atoms to their ground state (relaxation) results in a loss of energy, usually as heat; in specific cases, these relaxation processes may, however, be accompanied by the emission of electromagnetic radiation. This radiation is of lower energy than the radiation required for absorption (fluorescence spectroscopy). Thus, although the spectroscopic methods differ in detail, they share the same principles.

Table 6.3 Comparison of spectroscopic methods common in food analysis

Type of spectroscopy	*Principle*	*Application(s)*
Ultraviolet–visible (UV–vis)	Molecules absorb the electromagnetic radiation of specific wavelengths in the UV–vis range; the irradiance (radiant power, intensity) of a light beam is decreased, which is measured. The absorption of light causes the molecules to enter an excited electronic (and vibrational) state, followed by quickly relaxing back to their electronic ground state without emission of radiant energy (different from fluorescence)	Quantitation of molecules (by themselves or as the result of a chemical reaction) that will absorb the radiant energy of specific wavelengths in the UV or Vis region. For identification purposes, the full absorption spectra, e.g., 200–380 nm, are recorded. The principle is used for liquid chromatography detectors too
Fluorescence	Molecules absorb the electromagnetic radiation of specific wavelengths in the UV–vis range. The absorption of light causes molecules to enter an excited electronic (and vibrational) state. Following vibrational relaxation, the molecules relax from their excited electronic state to their corresponding ground state, thereby emitting electromagnetic radiation of specific wavelengths. The radiation emitted during relaxation is measured	Quantitation of molecules (by themselves or as the result of a chemical reaction) that will emit radiant energy in the UV or Vis region. The principle is used for liquid chromatography detectors too
Mid-IR (MIR)	Molecules in the ground state absorb mid-IR radiation and make a transition to the first excited state of vibration (fundamental vibration). The irradiance of a light beam is decreased (measuring principle), based on the absorption of the electromagnetic radiation of specific wavelengths triggering vibrations. The strength of absorption depends on the change of the dipole moment occurring during vibration. If there is no change in the dipole moment, the IR radiation is not absorbed	Characterization of unknown compounds (identification of functional groups). Verification of known compounds by comparing IR spectra with spectra of standards, e.g., polymers used for packaging. Quantitation of molecules. The principle is also used for gas chromatography detectors, although not as frequently
Near-IR (NIR)	Molecules absorb near-IR radiation (higher energy level than mid-IR) and make transitions to higher excited states of vibration (usually first and second overtone) and combination vibrations. The irradiance of a light beam is decreased (measuring principle). A net change in the dipole moment and large mechanical anharmonicity of the vibrating atoms are required	Quantitative analysis of samples. Several analytical parameters can be analyzed simultaneously (such as proteins, water, fats, etc.). Extensive calibration and application of chemometric methods are required. Authentication of food products based on chemometric models
Raman	Based on the inelastic scattering process of a monochromatic laser beam through its interaction with vibrating molecules, resulting in a change in the polarizability of the electron clouds of the molecules. Scattered photons, which are measured, are shifted to higher (anti-stokes shift) or lower (stokes shift) frequencies than the incident photons, giving information about the vibrational modes in the system. Complementary technique to IR spectroscopy	Qualitative and quantitative analysis of samples. Authentication of food products based on chemometric models
Nuclear magnetic resonance (NMR)	In the presence of an external magnetic field, spins of NMR-active nuclei (e.g., ^{1}H and ^{13}C) exist in slightly different energy states, and transitions between these states can be induced by radio-frequency pulses. Nonequilibrium magnetization states contain information about the nucleus, its chemical environment, and its dynamics. The detected time domain signal is the free induction decay, which is often transformed into the frequency domain by Fourier transformation. A multitude of NMR experiments allows for dedicated insight into materials properties	Identification of unknown food components (one-dimensional (1D) and two-dimensional (2D) NMR). Quantitation of molecules (qNMR). Purity determination of standard compounds. Authentication of food products based on chemometric models. Food quality control by NMR relaxation and diffusion imaging
Atomic absorption	By using a flame or an electrically heated graphite furnace, samples (molecules) are broken down into atoms. An element-specific radiation source emits electromagnetic radiation of atom-specific wavelengths (the commonly used wavelengths are in the UV range). Atoms absorb electromagnetic radiation, entering an excited electronic state. Thus, the irradiance of the light beam is decreased, which is measured	Quantitation of chemical elements
Atomic emission	Molecules in a plasma are converted into atoms. Collisions in the plasma promote atoms to excited electronic states. Excited atoms relax to their corresponding ground state, thereby emitting electromagnetic radiation of specific wavelengths. The radiation emitted during relaxation is measured	Quantitation of chemical elements

6.7 STUDY QUESTIONS

1. Which phenomena associated with light are most readily explained by considering the wave nature of light? Explain these phenomena based on your understanding of interference.
2. Which phenomena associated with light are most readily explained by considering the particulate nature of light? Explain these phenomena based on your understanding of the quantum nature of electromagnetic radiation.
3. What does it mean to say that the energy content of matter is quantized?
4. Molecular absorption of radiation in the UV–Vis range results in transitions between what types of energy levels?
5. Molecular absorption of radiation in the IR range results in transitions between what types of energy levels?
6. Why is an applied magnetic field necessary for NMR spectroscopy?
7. How do the allowed energy levels of molecules differ from those of atoms? Answer with respect to the energy-level diagram depicted in Fig. 6.4.
8. In fluorescence spectroscopy, why is the wavelength of the emitted radiation longer than the wavelength of the radiation used for excitation of the analyte?

REFERENCES

1. Atkins P, de Paulo J (2012) Elements of physical chemistry 6th ed, W.H. Freeman, New York
2. Ball DW (2001) The basics of spectroscopy. Society of Photo-optical Instrumentation Engineers, Bellingham, WA
3. Currell G (2000) Analytical instrumentation – performance characteristics and quality. Wiley, New York, pp. 67–91
4. Duckett 2 (2000) Foundations of spectroscopy. Oxford University Press, New York
5. Harris DC, Lucy CA (2019) Quantitative chemical analysis, 10th edn. WH Freeman, New York
6. Harris DC, Bertolucci MD (1989) Symmetry and spectroscopy. Dover, Mineola, NY
7. Harwood LM, Claridge TDW (1997) Introduction to organic spectroscopy. Oxford University Press, New York
8. Ingle JD Jr, Crouch SR (1988) Spectrochemical analysis. Prentice Hall, Englewood Cliffs, NJ
9. Meyers RA (ed) (2000) Encyclopedia of analytical chemistry: applications, theory, and instrumentation. 5, 2857–4332
10. Milton Roy educational manual for the SPECTRONIC® 20 & 20D spectrophotometers (1989) Milton Roy Co., Rochester, NY
11. Pavia DL, Lampman GM, Kriz GS Vyvyan JA (2015) Introduction to spectroscopy. 5th Edition, Cengage Learning, Independence, KY
12. Robinson JW, Frame EMS, Frame GM II (2014) Undergraduate instrumental analysis. 7th edn. Marcel Dekker, New York
13. Young HD, Freedman RA (2011) Sears and Zemansky's university physics, vol. 2, 13th edn. Addison-Wesley, Boston
14. Skoog DA, Holler FJ, Crouch SR (2017) Principles of instrumental analysis, 7th edn. Cengage Learning, Boston, MA

Ultraviolet, Visible, and Fluorescence Spectroscopy

7

Mirko Bunzel and Michael H. Penner

Contents

7.1 INTRODUCTION

Spectroscopy in the ultraviolet–visible (UV–Vis) range is one of the most commonly encountered laboratory techniques in food analysis. Diverse examples, such as the quantification of macrocomponents (total carbohydrate by the phenol sulfuric acid method), quantification of microcomponents (thiamine by the thiochrome fluorometric procedure), estimates of rancidity (lipid oxidation status by the thiobarbituric acid test), and surveillance testing (enzyme-linked immunoassays), are presented in this text. In each of these cases, the analytical signal for which the assay is based is either the emission or the absorption of radiation in the UV–Vis range. This signal may be inherent in the analyte, such as the absorbance of radiation in the visible range by pigments, or a result of a chemical reaction involving the analyte, such as the colorimetric copper-based Lowry method for the analysis of soluble proteins.

Electromagnetic radiation in the UV–Vis portion of the spectrum that is used in food analytical procedures ranges in wavelengths from approximately 200 to 780 nm. The UV range (200–380 nm) is colorless to the human eye, whereas different wavelengths in the visible range each have a characteristic color, ranging from violet at the short wavelength end of the spectrum to red-purple at the long wavelength end of the spectrum (Table 7.1). Spectroscopy utilizing radiation in the UV–Vis range may be divided into two general categories, namely, **absorbance** and **fluorescence** spectroscopies, based on the type of radiation–matter interaction that is being monitored. Each of these two types of spectroscopies may be

M. Bunzel (✉)
Department of Food Chemistry and Phytochemistry, Karlsruhe Institute of Technology (KIT), Karlsruhe, Germany
e-mail: mirko.bunzel@kit.edu; rachel.schendel@kit.edu

M. H. Penner
Department of Food Science and Technology, Oregon State University, Corvallis, OR, USA
e-mail: mike.penner@oregonstate.edu

B. P. Ismail, S. S. Nielsen (eds.), *Nielsen's Food Analysis*, Food Science Text Series,
https://doi.org/10.1007/978-3-031-50643-7_7

Table 7.1 The spectrum of visible radiation

Wavelength (nm)	*Color*	*Complementary hue*[a]
<380	Ultraviolet	
380–420	Violet	Yellow-green
420–440	Violet-blue	Yellow
440–470	Blue	Orange
470–500	Blue-green	Red
500–520	Green	Purple
520–550	Yellow-green	Violet
550–580	Yellow	Violet-blue
580–620	Orange	Blue
620–680	Red	Blue-green
680–780	Purple	Green
>780	Near-infrared	

[a]Complementary hue refers to the color observed for a solution that shows maximum absorbance at the designated wavelength assuming a continuous spectrum "white" light source

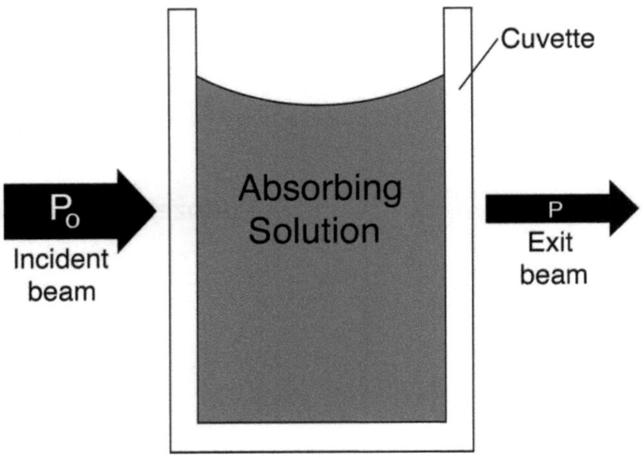

Fig. 7.1 Attenuation of a beam of radiation as it passes through a cuvette containing an absorbing solution

further subdivided into **qualitative** and **quantitative** techniques. In general, quantitative absorption spectroscopy is the most common of the subdivisions within UV–Vis spectroscopy. For detailed information on UV-Vis absorption and fluorescence spectroscopy refer to Refs. [1–22].

7.2 ULTRAVIOLET AND VISIBLE ABSORPTION SPECTROSCOPY

7.2.1 The Basis of Quantitative Absorption Spectroscopy

The objective of quantitative absorption spectroscopy is to determine the concentration of an analyte in a given sample solution. This determination is based on the measurement of the amount of light absorbed from a reference beam as it passes through the sample solution. In some cases, the analyte may naturally absorb radiation in the UV–Vis range, such that the chemical nature of the analyte is not modified during the analysis. In other cases, analytes that do not absorb radiation in the UV–Vis range are chemically modified during the analysis (derivatization) and are thus converted into a species that absorbs radiation of the appropriate wavelength. In either case, the presence of an analyte in the solution will affect the amount of radiation transmitted through the solution, and, hence, the relative transmittance or absorbance of the solution may be used as an index of analyte concentration.

In actual practice, the solution to be analyzed is contained in an absorption cell and placed in the path of radiation of a selected wavelength(s). The amount of radiation passing through the sample is then measured relative to a reference sample. The relative amount of light passing through the sample is then used to estimate the analyte concentration. The process of absorption may be depicted as in Fig. 7.1. The radiation incident on the absorption cell, P_0, will have significantly greater irradiance (radiant power) than the radiation exiting the opposite side of the cell, P. The decrease in irradiance (radiant power) as the beam passes through the solution is due to the capture (absorption) of photons by the absorbing species. The relationship between the power of the incident and exiting beams is typically expressed in terms of either the transmittance or the absorbance of the solution. The **transmittance** (T) of a solution is defined as the ratio of P to P_0, as given in Eq. 7.1. It may also be expressed as a percentage, as given in Eq. 7.2:

$$T = P / P_0 \tag{7.1}$$

$$\%T = \left(P / P_0\right) \times 100 \tag{7.2}$$

where:

T = transmittance.
P_0 = irradiance (radiant power) of the beam incident on the absorption cell
P = irradiance (radiant power) of the beam exiting the absorption cell.
$\%T$ = percentage transmittance.

The terms "T" and "$\%T$" are intuitively appealing, as they express the fraction of the incident light absorbed by the solution. However, T and $\%T$ are not directly proportional to the concentration of the absorbing analyte in the sample solution. The nonlinear relationship between transmittance and concentration is an inconvenience since analysts are generally interested in analyte concentrations. A second term used to describe the relationship between P and P_0 is "absorbance (A)." Absorbance is defined with respect to T, as shown in Eq. 7.3:

$$A = \log\left(P_0 / P\right) = -\log T = 2 - \log \%T \tag{7.3}$$

where:

A = absorbance.
T and $\%T$ = as in Eqs. 7.1 and 7.2, respectively.

Absorbance is a convenient expression, in that, under appropriate conditions, it is directly proportional to the concentration of the absorbing species in the solution. It should be noted that based on these definitions for A and T, the absorbance of a solution *is not* simply unity minus transmittance. In quantitative spectroscopy, the fraction of the incident beam that is not transmitted does not equal the solution's absorbance (A).

The relationship between the absorbance of a solution and the concentration of the absorbing species is known as **Beer's law** (Eq. 7.4):

$$A = abc \tag{7.4}$$

where:

A = absorbance
c = concentration of the absorbing species
b = pathlength through solution (in centimeters)
a = absorptivity

There are no units associated with absorbance, A, since it is the log of a ratio of beam powers. The concentration term, c, may be expressed in any appropriate unit (molar (M), millimolar (mM), milligrams per milliliter (mg/mL), percentage (%)). The pathlength, b, is in units of centimeters. The **absorptivity**, a, of a given species is a proportionality constant dependent on the molecular properties of the species. It is wavelength-dependent and may vary depending on the chemical environment (pH, ionic strength, solvent, etc.) to which the absorbing species is subjected. The units of the absorptivity term are $(\text{cm})^{-1}$ $(\text{concentration})^{-1}$. In the special case in which the concentration of the analyte is reported in units of molarity, the absorptivity term has units of $(\text{cm})^{-1}(M)^{-1}$. Under these conditions, it is designated by the symbol **ε**, which is referred to as the **molar absorption coefficient**. Beer's law expressed in terms of the molar absorption coefficient is given in Eq. 7.5. In this case, c specifically refers to the molar concentration of the analyte:

$$A = \varepsilon bc \tag{7.5}$$

where:

A and b = as in Eq. 7.4
ε = molar absorption coefficient
c = concentration in units of molarity

Quantitative spectroscopy is dependent on the analyst being able to accurately measure the fraction of an incident light beam that is absorbed by the analyte in a given solution. This apparently simple task is somewhat complicated in actual practice due to processes other than analyte absorption that also result in significant decreases in the power of the incident beam. A pictorial summary of reflection and scattering processes that will decrease the power of an incident beam is provided in Fig. 7.2. It is clear that these processes must be accounted for if a truly quantitative estimate of analyte absorption is necessary. In practice, a reference cell is used to correct for these processes. A **reference cell** is one that, in theory, exactly matches the sample absorption cell with the exception that it contains no analyte. Reference cells are often prepared by filling appropriate absorption cells with water (or other solvents used in a specific assay). A reference cell is placed in the path of the light beam, and the power of the radiation exiting the reference cell is measured and taken as P_o for the sample cell. This procedure assumes that all processes, except the selective absorption of radiation by the analyte, are equivalent for the sample and reference cells. The absorbance actually measured in the laboratory approximates Eq. 7.6:

$$A = \log\left(P_{\text{solvent}} / P_{\text{analyte solution}}\right) \cong \log\left(P_0 / P\right) \tag{7.6}$$

where:

P_{solvent} = irradiance (radiant power) of the beam exiting the cell containing the solvent (blank)
$P_{\text{analyte solution}}$ = irradiance (radiant power) of the beam exiting the cell containing the analyte solution
P_0 and P = as in Eq. 7.1
A as in Eq. 7.3.

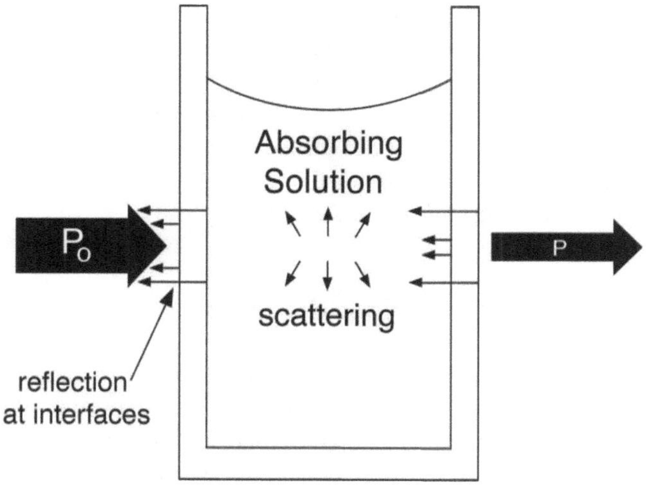

Fig. 7.2 Factors contributing to the attenuation of a beam of radiation as it passes through a cuvette containing an absorbing solution

7.2.2 Deviations from Beer's Law

It should never be assumed that Beer's law is strictly obeyed. Indeed, there are several reasons why the predicted linear relationship between absorbance and concentration may not be observed. In general, Beer's law is applicable only to dilute solutions, up to approximately 10 mM for most analytes. The actual concentration at which the law becomes limiting will depend on the chemistry of the analyte. As analyte concentrations increase, the intermolecular distances in a given sample solution will decrease, eventually reaching a point at which neighboring molecules mutually affect the charge distribution of the other. This perturbation may significantly affect the ability of the analyte to capture photons of a given wavelength, that is, it may alter the analyte's absorptivity (a). This causes the linear relationship between concentration and absorption to break down since the absorptivity term is the constant of proportionality in Beer's law (assuming a constant pathlength, b). Other chemical processes may also result in deviations from Beer's law, such as the reversible association–dissociation of analyte molecules or the ionization of a weak acid in an unbuffered solvent. In each of these cases, the predominant form of the analyte may change as the concentration is varied. If the different forms of the analyte (e.g., ionized versus neutral) have different absorptivities (a), then a linear relationship between concentration and absorbance will not be observed.

A further source of deviation from Beer's law may arise from limitations in the instrumentation used for absorbance measurements. Beer's law strictly applies to situations in which the radiation passing through the sample is monochromatic, since, under these conditions, a single absorptivity value describes the interaction of the analyte with all the radiation passing through the sample. If the radiation passing through a sample is polychromatic and there is variability in the absorptivity constants for the different constituent wavelengths, then Beer's law will not be obeyed. An extreme example of this behavior is when radiation of the ideal wavelength and stray radiation of a wavelength that is not at all absorbed by the analyte simultaneously pass through the sample to the detector.

7.2.3 Procedural Considerations

The goal of many quantitative measurements is to determine the concentration of an analyte with optimum precision and accuracy, in a minimal amount of time, and at minimal cost. To accomplish this, it is essential that the analyst considers potential errors associated with each step in a particular assay. Potential sources of errors for spectroscopic assays include inappropriate sample preparation techniques, inappropriate controls, instrumental noise, and errors associated with inappropriate conditions for absorbance measurements (such as extreme absorbance/transmittance readings).

Sample preparation schemes for absorbance measurements vary considerably. In the simplest case, the analyte-containing solution may be directly measured following homogenization and clarification. Except for special cases, homogenization is required prior to any analysis to ensure a representative sample. Clarification of samples is essential prior to taking absorbance readings in order to avoid the apparent absorption due to scattering of light by turbid solutions. The **reference solution** for samples in this simplest case will be the sample solvent, with the solvent being water or an aqueous buffer in many cases. In more complex situations, the analyte to be quantified may need to be chemically modified prior to taking absorbance measurements. In these cases, the analyte that does not absorb radiation in an appropriate spectral range is specifically modified, resulting in a species with absorption characteristics compatible with a given spectrophotometric measurement. Specific reactions such as these are used in many colorimetric assays that are based on the absorption of radiation in the Vis range. The reference solution for these assays is prepared by treating the sample solvent in a manner identical to that of the sample. The reference solution therefore will help correct for any absorbance due to the modifying reagents themselves and not the modified analyte.

A **sample-holding cell** or **cuvette** should be chosen after the general spectral region to be used in a spectrophotometric measurement has been determined. Sample-holding cells vary in composition and dimensions. A sample-holding cell should be composed of a material that does not absorb radiation in the spectral region being used. Cells meeting this requirement for measurements in the **UV range** may be composed of **quartz** or **fused silica**. For the **Vis range**, cells made of **silicate glass** are appropriate, and inexpensive **plastic** cells are also available for some applications. The dimensions of the cell will be important with respect to the amount of solution required for a measurement and with regard to the pathlength term used in Beer's law. A typical absorption cell is 1 cm^2 and is approximately 4.5 cm long. The pathlength for this traditional cell is 1 cm, and the minimum volume of solution needed for standard absorption measurements is approximately 1.5 mL. Absorption cells with pathlengths ranging from 1 to 100 mm are commercially available. Narrow cells, approximately 4 mm in width, with optical pathlengths of 1 cm, are also available. These narrow cells are convenient for absorbance measurements when limiting amounts of solution are available, e.g., less than 1 mL.

In many cases, an analyst must **choose an appropriate wavelength** at which to take absorbance measurements. If possible, it is best to choose the wavelength at which the analyte demonstrates maximum absorbance and where the absorbance does not change rapidly with changes in wavelength (Fig. 7.3). This position usually corresponds to the apex of the highest absorption peak. Taking measurements at this apex has two advantages: (1) maximum sensitivity, defined as the absorbance change per unit change in analyte concentration and (2) greater adherence to Beer's law since the spectral region making up the radiation beam is composed of wavelengths with relatively small differences in their molar absorptivities for the analyte being measured (Fig. 7.3). The latter point is important, in that the radiation beam used in the analysis will be composed of a small continuous band of wavelengths centered about the wavelength indicated on the instrument's wavelength selector.

The actual **absorbance measurement** is made by first calibrating the instrument for 0% and then 100% transmittance. The 0% transmittance adjustment is made while the photodetector is screened from the incident radiation by means of an occluding shutter, thus mimicking infinite absorption. This adjustment sets the base level current or "dark current" to the appropriate level, such that the readout indicates zero. The 100% transmittance adjustment is then made with the occluding shutter open and an appropriate reference cell/solution in the light path. The reference cell itself should be equivalent to the cell that contains the sample (i.e., a "matched" set of cells is used). In many cases, the same cell is used for both the sample and reference solutions. The reference cell is generally filled with the solvent that is used in the assay. The 100% T adjustment effectively sets $T = 1$ for the reference cell, which is equivalent to defining P_0 in Eq. 7.1 as equivalent to the irradiance (radiant power) of the beam exiting the reference cell. The 0% T and 100% T settings should be confirmed as necessary throughout the assay. The sample cell that contains the analyte is then measured without changing the adjustments. The adjustments made with the reference cell will effectively set the instrument to produce a sample readout in terms of Eq. 7.6. The readout for the sample solution will be between 0 and 100% T. Most modern spectrophotometers allow the analyst to take readout measurements in either absorbance units or as percentage transmittance. It is generally most convenient to take readings in absorbance units since, under optimum conditions, absorbance is directly proportional to concentration.

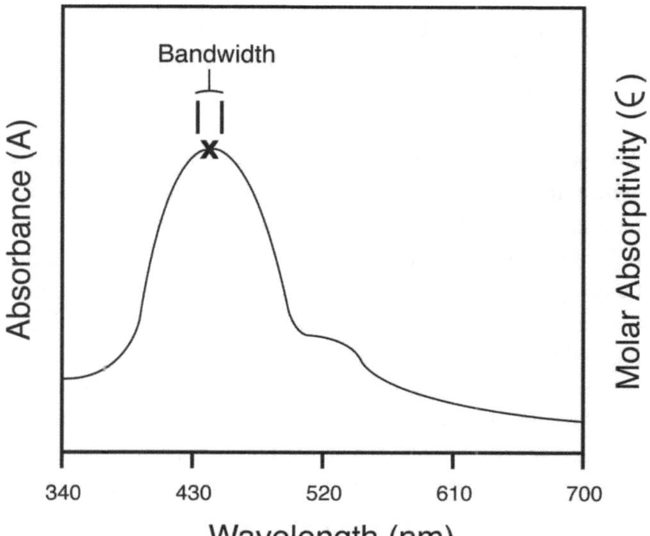

Fig. 7.3 A hypothetical absorption spectrum between 340 and 700 nm. The effective bandwidth of the radiation used in obtaining the spectrum is assumed to be approximately 20 nm. It should be noted that at the point indicated, there is essentially no change in molar absorptivity over this wavelength range

7.2.4 Calibration Curves

It is generally advisable to use calibration curves for quantitative measurements. Empirical assays that require the use of a calibration curve are common in food analyses. A calibration curve is used to establish the relationship between analyte concentration and absorbance. This relationship is experimentally established through the analysis of a series of samples of known analyte concentration. The standard solutions are best prepared with the same reagents and at the same time as the unknown. The concentration range covered by the standard solutions must include that expected for the unknown. Typical calibration curves are depicted in Fig. 7.4. **Linear calibration curves** are expected for those systems that obey Beer's law. **Nonlinear calibration curves** are used for some assays, but linear relationships are generally pre-

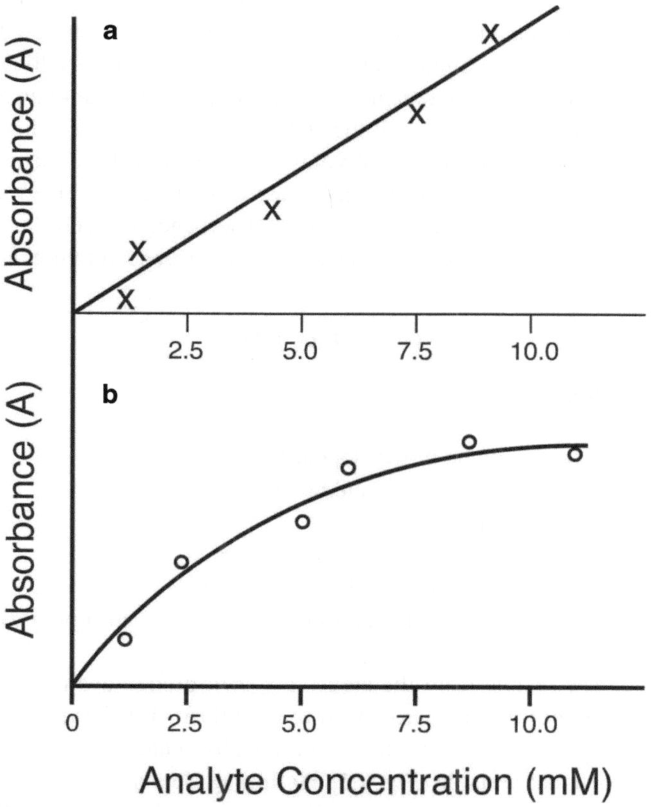

Fig. 7.4 Linear (**a**) and nonlinear (**b**) calibration curves typically encountered in quantitative absorption spectroscopy

ferred due to the ease of processing the data. Nonlinear calibration curves may be due to concentration-dependent changes in the chemistry of the system or the limitations inherent in the instruments used for the assay. The nonlinear calibration curve in Fig. 7.4b reflects the fact that the **calibration sensitivity**, defined as the change in absorbance per unit change in analyte concentration, is not constant. For the case depicted in Fig. 7.4b, the assay's concentration-dependent decrease in sensitivity obviously begins to limit its usefulness at analyte concentrations above 10 mM.

In many cases, truly representative calibration standards cannot be prepared due to the complexity of the unknown sample. This scenario must be assumed when insufficient information is available on the extent of interfering compounds in the unknown. **Interfering compounds** include those that absorb radiation in the same spectral region as the analyte, those that influence the absorbance of the analyte, and those that react with modifying reagents that are supposedly specific to the analyte. This means that calibration curves are potentially in error if the unknown and the standards differ with respect to pH, ionic strength, viscosity, types of impurities, and the like. In these cases, it is advisable to calibrate the assay system using a **standard addition protocol**. One such protocol goes as follows: to a series of flasks, add a constant volume of the unknown (V_u) for which you are trying to determine the analyte concentration (C_u). Next, to each individual flask, add a known volume (V_s) of a standard analyte solution of concentration C_s, such that each flask receives a unique volume of the standard. The resulting series of flasks will contain identical volumes of the unknown and different volumes of the standard solution. Next, dilute all flasks to the same total volume, V_t. Each of the flasks is then assayed, with each flask treated identically. If Beer's law is obeyed, then the measured absorbance of each flask will be proportional to the total analyte concentration as defined in Eq. 7.7:

$$A = k\left[\left(V_s C_s + V_u C_u\right)/\left(V_t\right)\right] \quad (7.7)$$

where:

V_s = volume of the standard
V_u = volume of the unknown
V_t = total volume
C_s = concentration of the standard
C_u = concentration of the unknown
k = proportionality constant (pathlength × absorptivity)

The results from the assays are then plotted against the volume of the standard added to each flask (V_s) as the independent variable and the resulting absorbance (A) as the dependent variable (Fig. 7.5). Assuming Beer's law, the line describing the relationship will be as in Eq. 7.8, in which all terms other than V_s and A are constants. Taking the ratio of the slope of the plotted line (Eq. 7.9) to the line's intercept

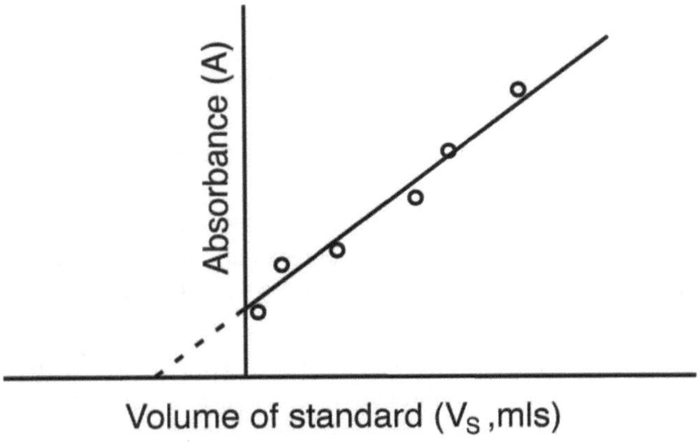

Fig. 7.5 A calibration curve for the determination of the analyte concentration in an unknown using a standard addition protocol. *A*, absorbance; V_s, volume of the standard analyte solution, as discussed in text

(Eq. 7.10) and rearranging yields Eq. 7.11, from which the concentration of the unknown, C_u, can be calculated since C_s and V_u are experimentally defined constants:

$$A = kC_s V_s / V_T + V_u C_u k / V_t \quad (7.8)$$

$$\text{Slope} = kC_s / V_t \quad (7.9)$$

$$\text{Intercept} = V_u C_u k / V_t \quad (7.10)$$

$$C_u = \left(\text{measured intercept / measured slope}\right)\left(C_s / V_u\right) \quad (7.11)$$

where:

V_s, V_u, V_t, C_s, C_u, and k = as in Eq. 7.7

7.2.5 The Effect of an Indiscriminate Instrumental Error on the Precision of Absorption Measurements

All spectrophotometric assays will have some level of **indiscriminate errors** associated with the absorbance/transmittance measurement itself. An indiscriminate error of this type is often referred to as **instrument noise**. It is important that the assay be designed such that this source of error is minimized, with the objective being to keep this source of error low relative to the variability associated with other aspects of the assay, such as sample preparation, subsampling, reagent handling, and so on. An indiscriminate instrumental error is observed with repeated measurements of a single homogeneous sample. The relative concentration uncertainty resulting from this error is not constant over the entire percentage transmittance range (0–100%). Measurements at intermediate transmittance values tend to have lower relative errors, and thus greater relative precision, than measurements made at either extremely high or extremely low transmittance. **Relative concentration uncertainty** or relative error may be defined as S_c/C, where S_c is the sample standard deviation and C is the measured concentration. Relative concentration uncertainties from 0.5% to 1.5% are to be expected for absorbance/transmittance measurements taken in the optimal range. The optimal range for absorbance measurements on simple, less expensive spectrophotometers is from approximately 0.2 to 0.8 absorbance units or 15–65% transmittance. On more sophisticated instruments, the range for optimum absorbance readings may be extended up to 1.5 or greater. To be safe, it is prudent to always take absorbance readings under conditions in which the absorbance of the analyte solution is less than 1.0. If there is an anticipated need to take measurements at absorbance readings greater than 1.0, then the relative precision of the spectrophotometer should be experimentally established by repetitive measurements of appropriate samples. Absorbance readings outside the optimal range of the instrument may be used, but the analyst must be prepared to account for the higher relative error associated with these extreme readings. When absorbance readings approach the limits of the instrumentation, then relatively large differences in analyte concentrations may not be detected.

7.2.6 Instrumentation

There are many variations of spectrophotometers available for UV–Vis spectrophotometry. Some instruments are designed for operation only in the visible range, whereas others encompass both the UV and Vis ranges. Instruments may differ with respect to design, quality of components, and versatility. A basic spectrophotometer is composed of five essential components: a **light source**, a **monochromator**, a **sample/reference holder**, a **radiation detector**, and a **readout device**. A power supply is required for instrument operation. A schematic depicting component interrelationships is shown in Fig. 7.6.

7.2.6.1 The Light Source

The **light sources** used in spectrophotometers must continuously emit a strong band of radiation encompassing the entire wavelength range for which the instrument is designed. The power of the emitted radiation must be sufficient for adequate detector response, and it should not vary sharply with changes in wavelength or drift significantly over the experimental time scale. The most common radiation source for Vis spectrophotometers is the **tungsten filament lamp**. These lamps emit adequate radiation covering the wavelength region from 350 to 2500 nm. Consequently, tungsten filament lamps are also employed in near-infrared spectroscopy. The most common radiation sources for measurements in the UV range are **deuterium electrical discharge lamps**. These sources provide a continuous radiation spectrum from approximately 160 nm through 375 nm. These lamps employ quartz windows and should be used in conjunction with quartz sample holders, since glass significantly absorbs radiation below 350 nm.

7.2.6.2 The Monochromator

The component that functions to isolate the specific, narrow, continuous group of wavelengths to be used in the spectroscopic assay is the **monochromator**. The monochromator is so named because light of a single wavelength is termed "monochromatic." Theoretically, **polychromatic radiation** from the source enters the monochromator and is dispersed according to wavelength, and **monochromatic radiation** of a selected wavelength exits the monochromator. In practice, light exiting the monochromator is not of a single wavelength, but, rather, it consists of a narrow continuous band of wavelengths. A representative monochromator is depicted in Fig. 7.7. As illustrated, a typical monochromator is composed of **entrance** and **exit slits**, **concave mirror(s)**, and a **dispersing element** (the grating in this particular example).

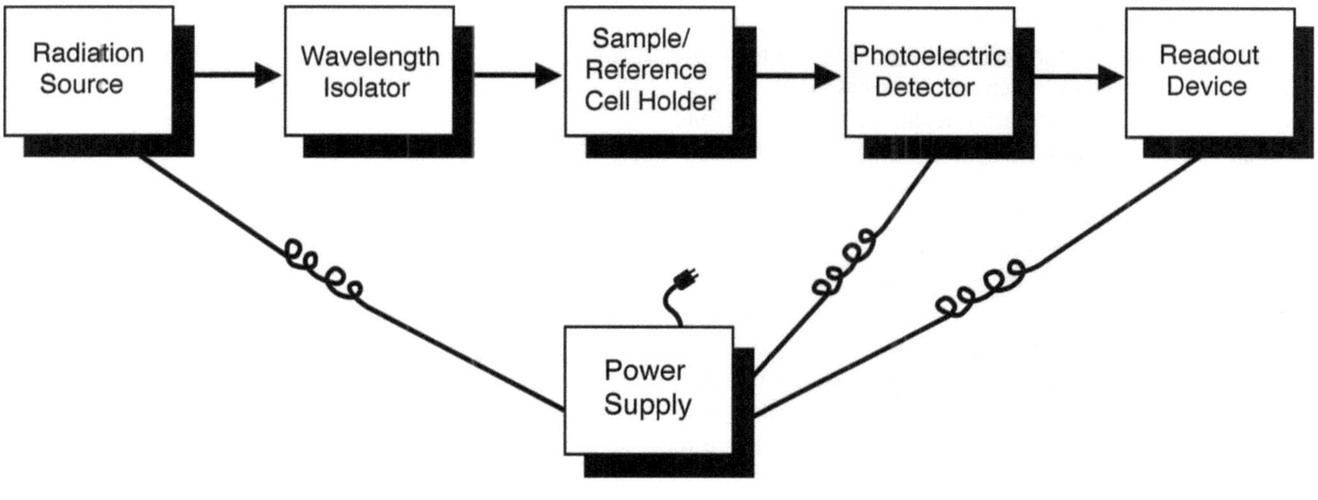

Fig. 7.6 Arrangement of components in a simple single-beam, UV–Vis absorption spectrophotometer

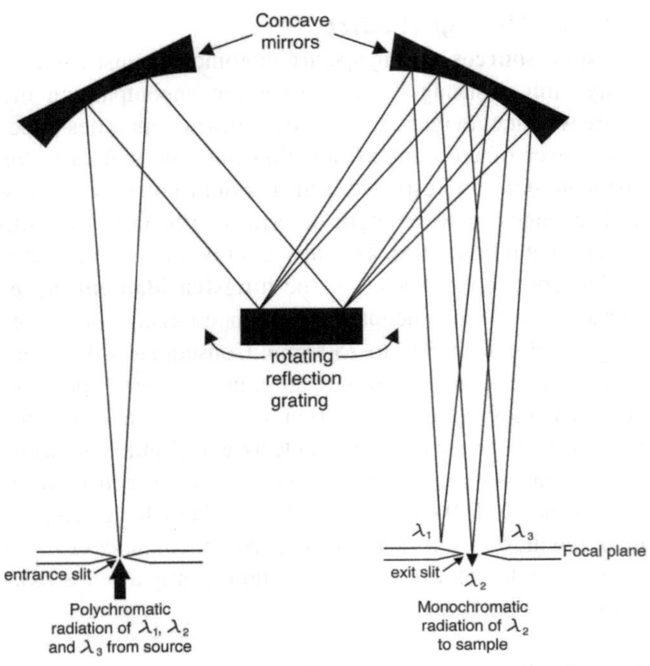

Fig. 7.7 A schematic of a monochromator employing a reflection grating as the dispersing element. The concave mirrors serve to culminate the radiation into a beam of parallel rays

Polychromatic light enters the monochromator through the entrance slit and is then culminated by a concave mirror. The culminated polychromatic radiation is then dispersed, with dispersion being the physical separation in the space of radiation of different wavelengths. The radiation of different wavelengths is then reflected from a concave mirror that focuses the different wavelengths of light sequentially along the focal plane. The radiation that aligns with the exit slit in the focal plane is emitted from the monochromator. The radiation emanating from the monochromator will consist of a narrow range of wavelengths presumably centered around the wavelength specified on the wavelength selection control of the instrument.

The size of the wavelength range passing out of the exit slit of the monochromator is termed the "bandwidth" of the emitted radiation. Many spectrophotometers allow the analyst to adjust the size of the monochromator exit slit (and entrance slit) and, consequently, the bandwidth of the emitted radiation. Decreasing the exit slit width will decrease the associated bandwidth and the irradiance (radiant power) of the emitted beam. Conversely, further opening of the exit slit will result in a beam of greater irradiance (radiant power) but one that has a larger bandwidth. In some cases in which resolution is critical, such as some qualitative work, a narrower slit width may be advised. However, in most quantitative work, a relatively open slit may be used since adsorption peaks in the UV–Vis range are generally broad relative to spectral bandwidths. Moreover, the signal-to-noise ratio associated with transmittance measurements is improved due to the higher irradiance (radiant power) of the measured beam.

The effective bandwidth of a monochromator is determined not only by the slit width but also by the quality of its dispersing element. The dispersing element functions to spread out the radiation according to the wavelength. **Reflection gratings**, as depicted in Fig. 7.8, are the most commonly used dispersing elements in modern spectrophotometers. Gratings are sometimes referred to as **diffraction gratings** because the separation of component wavelengths is dependent on the different wavelengths being diffracted at different angles relative to the grating normal. A reflection grating incorporates a reflective surface in which a series of closely spaced grooves has been etched, typically between 1200 and 1400 grooves per millimeter. The grooves themselves serve to break up the reflective surface such that each point of reflection behaves as an independent point source of radiation.

Referring to Fig. 7.8, lines 1 and 2 represent rays of parallel monochromatic radiation that are in phase and that strike the grating surface at an angle *i* to the normal. The maximum constructive interference of this radiation is depicted as occurring at an angle *r* to the normal. At all other angles, the two rays will partially or completely cancel each other. Radiation of a different wavelength would show maximum constructive interference at a different angle to the normal. The wavelength dependence of the diffraction angle can be rationalized by considering the relative distance that the photons of rays 1 and 2 travel and assuming that maximum constructive interference occurs when the waves associated with the photons are completely in phase. Referring to Fig. 7.8, prior to reflection, photon 2 travels a distance CD greater than photon 1. After reflection, photon 1 travels a distance AB greater than photon 2. Hence, the waves asso-

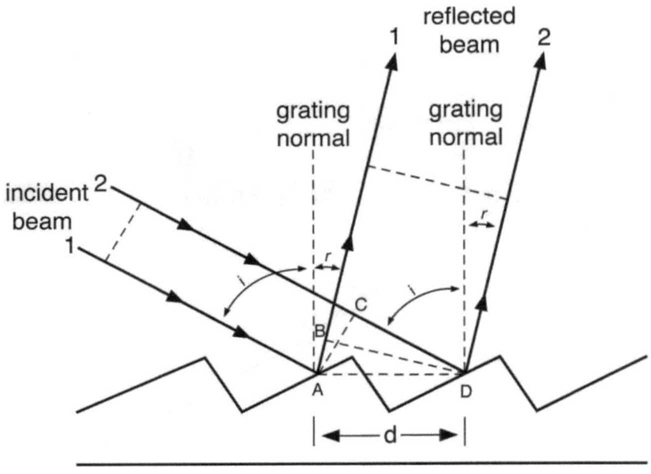

Fig. 7.8 A schematic illustrating the property of diffraction from a reflection grating. Each reflected point source of radiation is separated by a distance *d*

ciated with photons 1 and 2 will remain in phase after reflection only if the net difference in the distance traveled is an integral multiple of their wavelength. It should be noted that for a different angle *r*, the distance AB would change and, consequently, the net distance CD-AB would be an integral multiple of a different wavelength. The net result is that the component wavelengths are each diffracted at their own unique angles *r*.

7.2.6.3 The Detector

In a spectroscopic measurement, the light transmitted through the reference or sample cell is quantified by means of a **detector**. A detector is designed to produce an electrical signal when it is struck by photons. An ideal detector would emit a signal directly proportional to the irradiance (radiant power) of the beam striking it, it would have a high signal-to-noise ratio, and it would have a relatively constant response to light of different wavelengths, such that it was applicable to a wide range of the radiation spectrum. There are several types and designs of radiation detectors currently in use. The most commonly encountered detectors are the **phototube**, the **photomultiplier tube**, and the **photodiode detectors**. All of these detectors function by converting the energy associated with incoming photons into electrical current. The **phototube** consists of a semicylindrical cathode covered with a photoemissive surface and a wire anode, with the electrodes being housed under vacuum in a transparent tube (Fig. 7.9a). When photons strike the photoemissive surface of the cathode, there is an emission of electrons; the freed electrons are collected at the anode. The net result of this process is that a measurable current is created. The number of electrons emitted from the cathode and the subsequent current through the system are directly proportional to the number of photons, or irradiance (radiant power) of the beam, impinging on the photoemissive surface. The **photomultiplier tube** is of similar design. However, in the photomultiplier tube, there is an amplification of the number of electrons collected at the anode per photon striking the photoemissive surface of the cathode (Fig. 7.9b). The electrons originally emitted from the cathode surface are attracted to a dynode with a relative positive charge. At the dynode, the electrons strike the surface, causing the emission of several more electrons per original electron, resulting in an amplification of the signal. Signal amplification continues in this manner, as photomultiplier tubes generally contain a series of such dynodes, with electron amplification occurring at each dynode. The cascade continues until the electrons emitted from the final dynode are collected at the anode of the photomultiplier tube. The final gain may be as many as $10^6 - 10^9$ electrons collected per photon.

Photodiode detectors are now common in UV–Vis spectrophotometers. These are solid-state devices in which the light-induced electrical signal is a result of photons exciting electrons in the semiconductor materials from which they are fabricated, most commonly silicon. Spectrophotometers using photodiode detectors may contain a single diode detector or a linear array of diodes (diode array spectrophotometers). If a single photodiode detector is

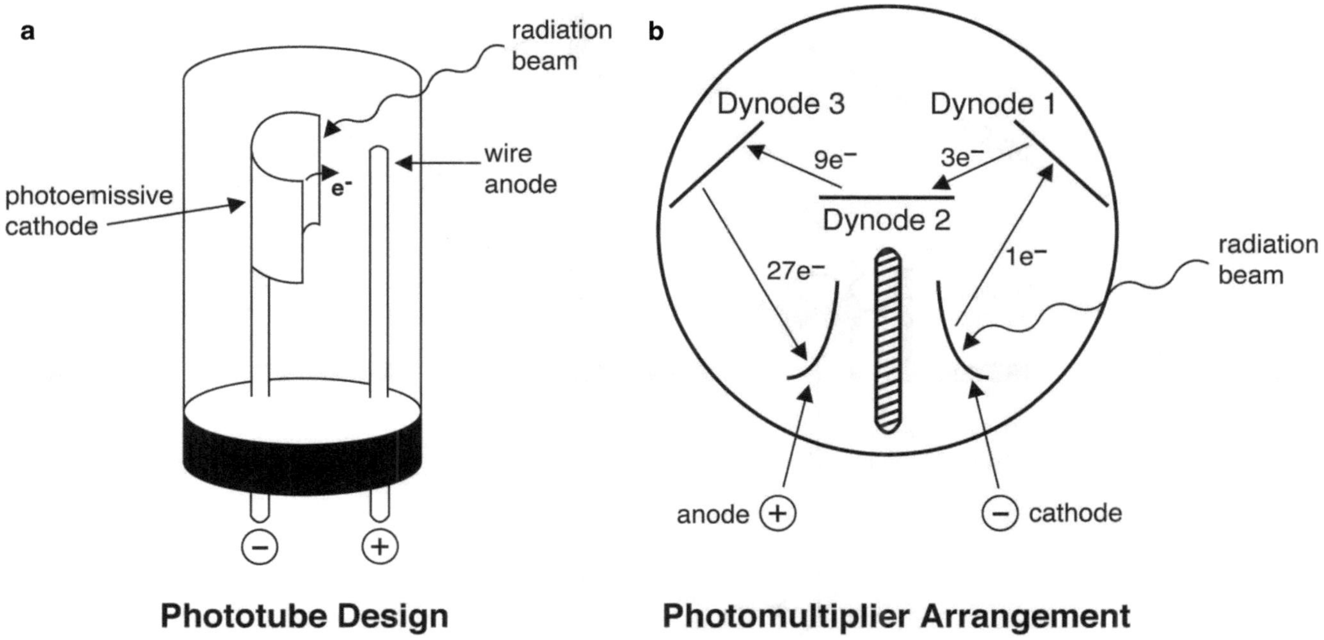

Fig. 7.9 A schematic diagram of a (**a**) typical phototube design and (**b**) the cathode–dynode–anode arrangement of a representative photomultiplier tube

used, then the arrangement of components is generally as depicted in Fig. 7.6. If an array of photodiode detectors is used, then the light originating from the source typically passes into the sample prior to it being dispersed. The light transmitted through the sample is subsequently dispersed onto the diode array, with each diode measuring a narrow band of the resulting spectrum. Dispersion of transmitted light is achieved using a polychromator. A polychromator can be described as a monochromator, but without the slit for wavelength selection. This design allows one to simultaneously measure multiple wavelengths, allowing nearly instantaneous collection of an entire absorption spectrum. Diode-based detectors are generally reported to be more sensitive than phototubes but less sensitive than photomultiplier tubes. Diode-based detectors using diode arrays are particularly important as detectors in liquid chromatography (Chap. 13) because they provide information about the identity of an analyte (via interpretation of the UV–Vis absorption spectrum) in addition to quantitative information.

7.2.6.4 The Readout Device

The signal from the detector is generally amplified and then displayed in a usable form to the analyst. The final form in which the signal is displayed will depend on the complexity of the system. Digital readouts are common and express the signal as numbers on the face of a meter. In these cases, there is a requirement for signal processing between the analog output of the detector and the final digital display. In virtually all cases, the signal processor is capable of presenting the final readout in terms of either absorbance or transmittance. Many instruments include microprocessors capable of more extensive data manipulations on the digitized signal. For example, the readouts of some spectrophotometers may be in concentration units, provided that the instrument has been correctly calibrated with appropriate reference standards.

7.2.7 Instrument Design

The optical systems of spectrophotometers fall into one of two general categories: they are either single-beam or double-beam instruments. In a **single-beam instrument**, the radiant beam follows only one path, which goes from the source through the sample to the detector (Fig. 7.6). When using a single-beam instrument, the analyst generally measures the transmittance of a sample after first establishing 100% T, or P_0, with a reference sample or blank. The blank and the sample are read sequentially since there is only a single light path going through a single cell-holding compartment. In a **double-beam instrument**, the beam is split

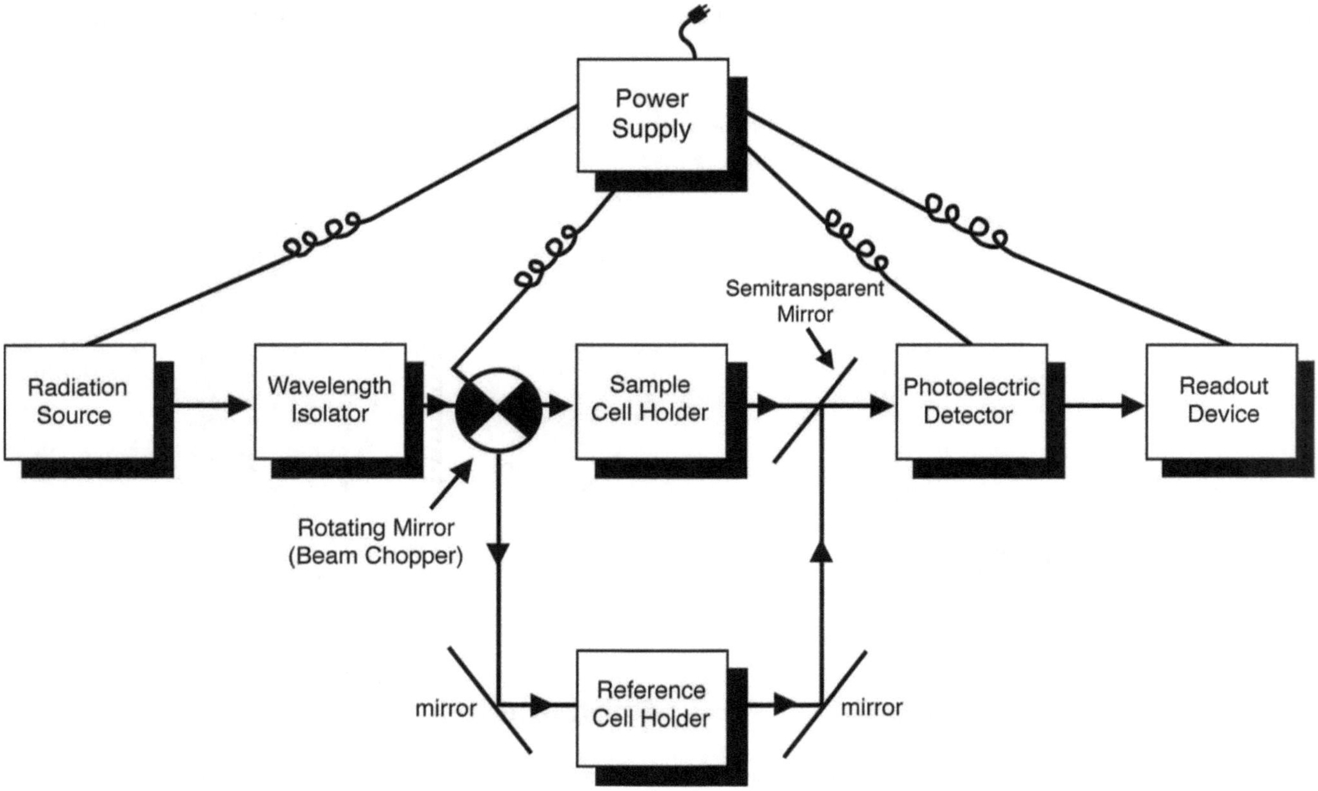

Fig. 7.10 Arrangement of components in a representative double-beam UV–Vis absorption spectrophotometer. The incident beam is alternatively passed through the sample and reference cells by means of a rotating beam chopper

such that one-half of the beam goes through one cell-holding compartment and the other half of the beam passes through a second. The schematic of Fig. 7.10 illustrates a double-beam optical system in which the beam is split in time between the sample and reference cell. In this design, the beam is alternately passed through the sample and reference cells by means of a rotating sector mirror with alternating reflective and transparent sectors. The double-beam design allows the analyst to simultaneously measure and compare the relative absorbance of a sample and a reference cell. The advantage of this design is that it will compensate for deviations or drifts in the radiant output of the source since the sample and reference cells are compared many times per second. The disadvantage of the double-beam design is that the irradiance (radiant power) of the incident beam is diminished because the beam is split. The lower energy throughput of the double-beam design is generally associated with inferior signal-to-noise ratios. Computerized single-beam spectrophotometers are available that claim to have the benefits of both the single- and double-beam designs. Their manufacturers report that previously troublesome source and detector drift and noise problems have been stabilized such that simultaneous reading of the reference and sample cell is not necessary. With these instruments, the reference and sample cells are read sequentially, and the data are stored, then processed, by the associated computer.

The Spectronic® 20 is a classic example of a simple single-beam visible spectrophotometer (Fig. 7.11). The white light emitted from the source passes into the monochromator via its entrance slit; the light is then dispersed into a spectrum by a diffraction grating, and a portion of the resulting spectrum then leaves the monochromator via the exit slit. The radiation emitted from the monochromator passes through a sample compartment and strikes the silicon photodiode detector, resulting in an electrical signal proportional to the intensity of impinging light. The lenses depicted in Fig. 7.11 function in series to focus the light image on the focal plane that contains the exit slit. To change the portion of the spectrum exiting the monochromator, one rotates the reflecting grating by means of a wavelength cam. A shutter automatically blocks light from exiting the monochromator when no sample/reference cell is in the instrument; 0% *T* adjustment is made under these conditions. The light control occluder is used to adjust the irradiance (radiant power) of the beam exiting the monochromator. The occluder consists of an opaque strip with a V-shaped opening that can be physically moved in or out of the beam path. The occluder is used to make the 100% *T* adjustment when an appropriate reference cell is in the instrument.

7.2.8 Characteristics of UV–vis Absorbing Species

The absorbance of UV–Vis radiation is associated with electronic excitations within atoms and molecules. Commonly encountered analytical methods based on UV–Vis spectroscopy do not use UV radiation below 200 nm. Hence, the excitations of interest in traditional UV–Vis spectroscopy are

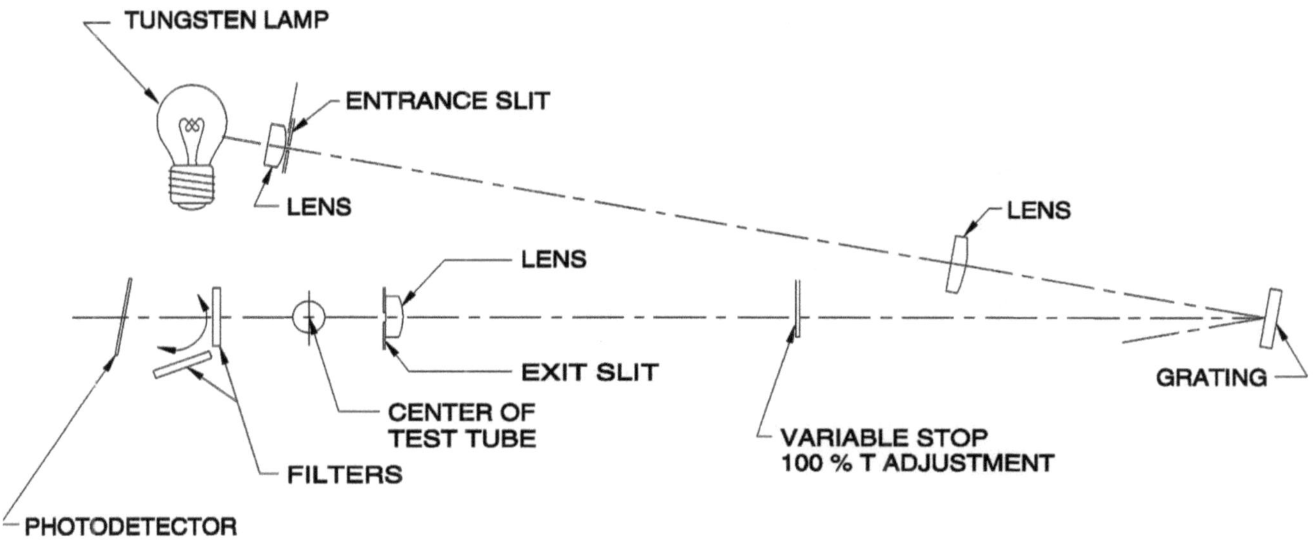

Fig. 7.11 An optical system for the Spectronic® 20 spectrophotometer. (Courtesy of Thermo Spectronic, Rochester, NY, Thermo Electron Spectroscopy)

the result of unsaturation and/or the presence of nonbonded electrons in the absorbing molecules. The UV–Vis absorption characteristics of several functional groups common to food constituents are tabulated in Table 7.2. The presented wavelengths of maximum absorbance and the associated molar absorption coefficients are only approximate since the environment to which the functional group is exposed, including neighboring constituents and solvents, will have an influence on the absorption properties of the functional group.

Table 7.2 Representative absorption maxima above 200 nm for select functional groups

Chromophore	*Example*	λ_{max}[a]	ε_{max}[b]	*References*
Nonconjugated systems				
R-CHO	Acetaldehyde	290	17	4
R_2-CO	Acetone	279	15	4
R-COOH	Acetic acid	208	32	4
R-$CONH_2$	Acetamide	220	63	24
R-SH	Mercaptoethane	210	1200	19
Conjugated systems				
$R_2C=CR_2$	Ethylene	<200	–	24
R-CH=CH-CH=CH-R	1,3-butadiene	217	21,000	24
R-CH=CH-CH=CH-CH=CH-R	1,3,5-Hexatriene	258	35,000	24
11 conjugated double bonds	β-Carotene	465	125,000	13
R_2C=CH-CH=O	Acrolein (2-propenal)	210	11,500	24
		315	14	24
HOOC-COOH	Oxalic acid	250	63	24
Aromatic compounds[c]				
C_6H_6	Benzene	256	200	24
C_6H_5OH	Phenol	270	1450	24
C_8H_7N	Indole	278	2500	NIST database[d]

[a]λ_{max}, wavelength (in nanometers) of a maximum absorbance greater than 200 nm

[b]ε_{max}, molar absorption coefficient, units of $(\text{centimeter})^{-1}$ $(\text{molarity})^{-1}$

[c]Spectra of the aromatic compounds generally contain an absorption band(s) of higher intensity at a lower wavelength (e.g., phenol has an absorption maxima of ~210 nm with a molar absorptivity of ~6200 $(cm)^{-1}$ $(M)^{-1}$; values are from Skoog et al. [1]). Only the absorption maxima corresponding to the longer wavelengths are included in this table

[d]NIST Standard Reference Database 69: *NIST Chemistry WebBook*. (The presented values were estimated from the UV–Vis spectrum for indole presented online: http://webbook.nist.gov/cgi/cbook.cgi?Name=indole&Units=SI&cUV=on. The website contains UV–Vis data for many compounds that are of potential interest to food scientists)

The type of information contained in Table 7.2 will likely be useful in determining the feasibility of UV–Vis spectroscopy for specific applications. For example, it is helpful to know the absorption characteristics of carboxyl groups if one is considering the feasibility of using UV–Vis absorption spectroscopy as a detection method to monitor non-derivatized organic acids eluting from liquid chromatography columns. With respect to this particular organic acid question, the table indicates that organic acids are likely to absorb radiation in the range accessible to most UV–Vis detectors (> 200 nm). However, the table also indicates that the sensitivity of such a detection method is likely to be limited due to the low molar absorption coefficient of carboxyl groups at such wavelengths. This explains why high-performance liquid chromatography methods for organic acid quantification sometimes make use of UV–Vis detectors tuned to ~210 nm [3] and why there are research efforts aimed at developing derivatization methods to enhance the sensitivity of UV–Vis-based methods for the quantification of organic acids [11].

The data of Table 7.2 also illustrate the effect of conjugation on electronic transitions. Increased conjugation leads to absorption maxima at longer wavelengths due to the associated decrease in the electronic energy spacing within a conjugated system (i.e., a lower energy difference between the ground and excited state). The aromatic compounds included in the table were chosen due to their relevance to protein quantification: benzene/phenylalanine, phenol/tyrosine, and indole/tryptophan (Chap. 18, Sect. 18.5.1). The table indicates that typical proteins will have an absorption maximum at approximately 278 nm (a high molar absorption coefficient of the indole side chain of tryptophan) as well as another peak at around 220 nm. This latter peak corresponds to the amide/peptide bonds along the backbone of the protein, with the rationale being deduced from the data for the simple amide included in the table (i.e., acetamide).

Figure 7.12 illustrates the effect of conjugation using the examples of the UV absorption spectra of guaiacol, vanillin, ferulic acid, and dihydroferulic acid. Extension of the conjugated system in guaiacol via vanillin to ferulic acid leads to a shift in absorption to higher wavelengths. In contrast, interruption of the conjugated system in dihydroferulic acid leads to absorption at shorter wavelengths (peak at 280.6 nm) compared to ferulic acid (peak at 318.4 nm).

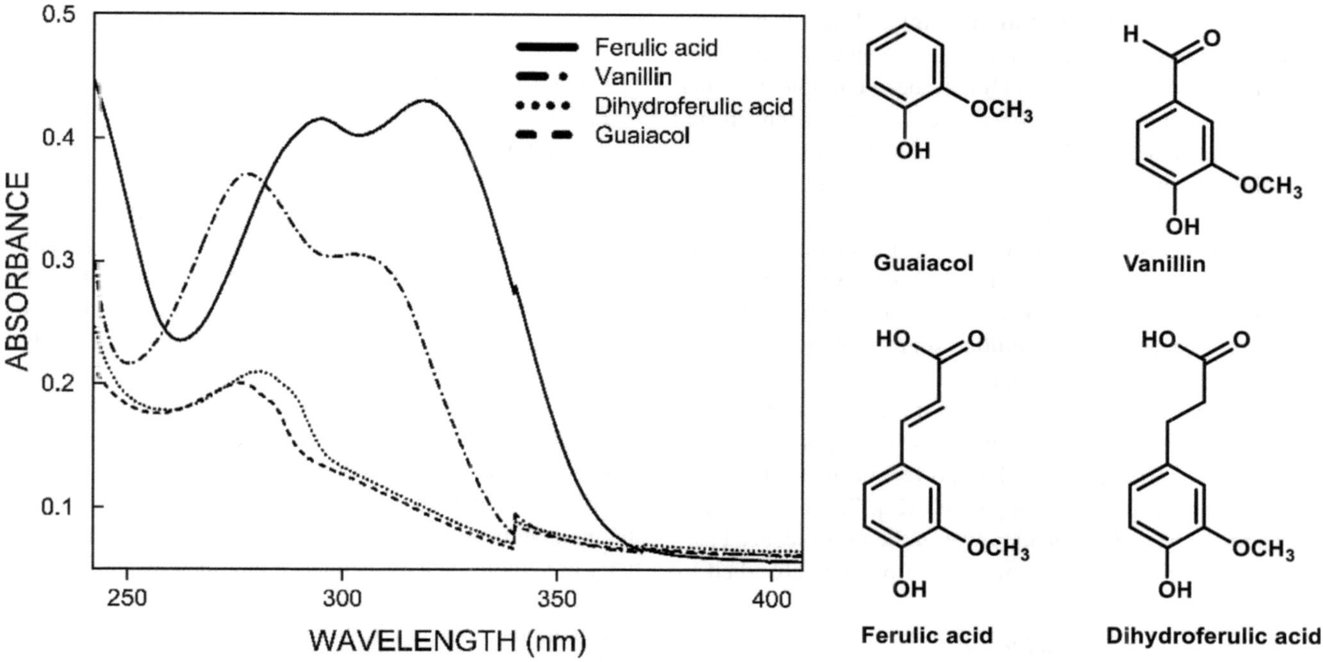

Fig. 7.12 UV absorption spectra (16 μM, in methanol) and chemical structures of guaiacol (maximum at 276.6 nm, shoulder at 283.5 nm), vanillin (maxima at 277.8 and 304.4 nm), ferulic acid (maxima at 294.8 and 318.4 nm), and dihydroferulic acid (maximum at 280.6 nm, shoulder at 288.3 nm) demonstrating the impact of conjugation on the shift of absorption to higher wavelengths

7.3 FLUORESCENCE SPECTROSCOPY

Fluorescence spectroscopy is generally one to three orders of magnitude more sensitive than the corresponding absorption spectroscopy. In **fluorescence spectroscopy**, the signal being measured is the electromagnetic radiation that is emitted from the analyte as it relaxes from an excited electronic energy level to its corresponding ground state. The analyte is originally activated to a higher energy level by the absorption of radiation in the UV or Vis range. The processes of activation and deactivation occur simultaneously during a fluorescence measurement. For each unique molecular system, there will be an optimum radiation wavelength for sample excitation and another, of longer wavelength, for monitoring fluorescence emission. The respective wavelengths for excitation and emission will depend on the chemistry of the system under study.

The instrumentation used in fluorescence spectroscopy is composed of essentially the same components as the corresponding instrumentation used in UV–Vis absorption spectroscopy. However, there are definite differences in the arrangement of the optical systems used for the two types of spectroscopies (compare Figs. 7.6 and 7.13). In fluorometers and spectrofluorometers, there is a need for two wavelength

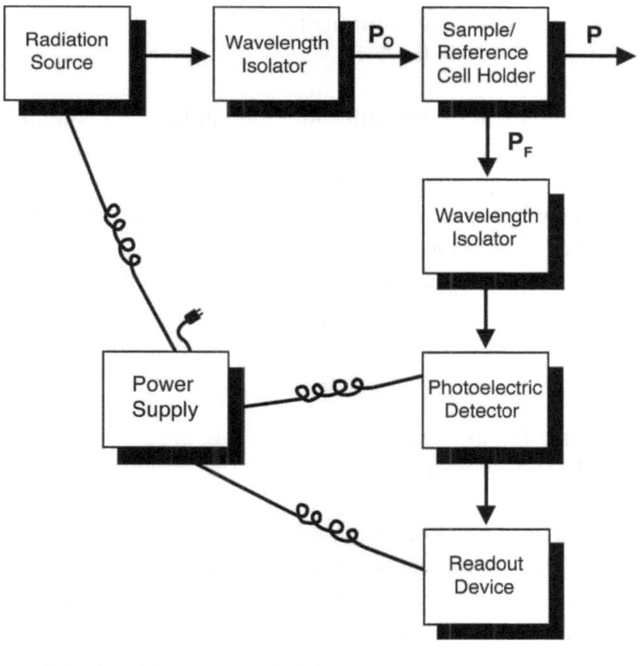

Fig. 7.13 A schematic diagram depicting the arrangement of the source, excitation and emission wavelength selectors, a sample cell, a photoelectric detector, and a readout device for a representative fluorometer or spectrofluorometer

selectors, one for the **excitation beam** and the other for the **emission beam**. In some simple fluorometers, both wavelength selectors are filters such that the excitation and emission wavelengths are fixed. In more sophisticated spectrofluorometers, the excitation and emission wavelengths are selected by means of grating monochromators. The photon detector of fluorescence instrumentation is generally arranged such that the emitted radiation that strikes the detector is traveling at an angle of 90° relative to the axis of the excitation beam. This detector placement minimizes signal interference due to transmitted source radiation and radiation scattered from the sample.

The **irradiance** (radiant power) of the fluorescence beam (P_F) emitted from a fluorescent sample is proportional to the change in the irradiance (radiant power) of the source beam as it passes through the sample cell (Eq. 7.12). Expressing this another way, the radiant power of the fluorescence beam will be proportional to the number of photons absorbed by the sample:

$$P_F = \varphi(P_0 - P) \quad (7.12)$$

where:

P_F = irradiance (radiant power) of the beam emitted from the fluorescent cell
φ = constant of proportionality
P_0 and P = as in Eq. 7.1

The constant of proportionality used in Eq. 7.12 is termed the "quantum efficiency" (φ), which is specific to any given system. The quantum efficiency equals the ratio of the total number of photons emitted to the total number of photons absorbed. Combining Eqs. 7.3 and 7.5 allows one to define P in terms of the analyte concentration and P_0, as given in Eq. 7.13:

$$P = P_0 10^{-\varepsilon bc} \quad (7.13)$$

where:

P_0 and P = as in Eq. 7.1
ε, b, and c = as in Eq. 7.5

Substitution of Eq. 7.13 into Eq. 7.12 yields an expression that relates the radiant power of the fluorescent beam to the analyte concentration and P_0, as shown in Eq. 7.14. At low analyte concentrations, $\varepsilon bc < 0.01$, Eq. 7.14 may be reduced to the expression of Eq. 7.15 (see Skoog et al. [1] for more on this). Further grouping of terms leads to the expression of Eq. 7.16, where k incorporates all terms other than P_0 and c:

$$P_F = \varphi P_0 \left(1 - 10^{\varepsilon bc}\right) \quad (7.14)$$

$$P_F = \varphi P_0 2.303\,\varepsilon bc \quad (7.15)$$

$$P_F = \mathrm{k} P_0 c \quad (7.16)$$

where:

k = constant of proportionality,
P_F = as in Eq. 7.12
c = as in Eq. 7.5.

Equation 7.16 is particularly useful because it emphasizes two important points that are valid for the conditions assumed when deriving the equation, particularly the assumption that analyte concentrations are kept relatively low. First, the fluorescent signal will be directly proportional to the analyte concentration, assuming that other parameters are kept constant. This is extremely useful because a linear relationship between signal and analyte concentration simplifies data processing and assay troubleshooting. Second, the sensitivity of a fluorescent assay is proportional to P_0, the power of the incident beam, with the implication being that the sensitivity of a fluorescent assay may be modified by adjusting the source output.

Equations 7.15 and 7.16 will eventually break down if analyte concentrations are increased to relatively high values. Therefore, the **linear concentration range** for each assay should be experimentally determined. A representative calibration curve for a fluorescence assay is presented in Fig. 7.14. The nonlinear portion of the curve at relatively high analyte concentrations results from decreases in the fluorescence yield per unit concentration. The fluorescence yield for any given sample is also dependent on its environment. Temperature, solvent, impurities, and pH may influence this parameter. Consequently, it is imperative that these environmental parameters be accounted for in the experi-

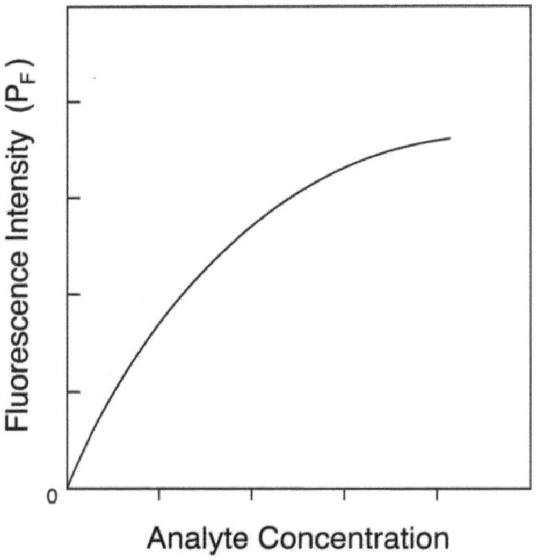

Fig. 7.14 The relationship between the solution concentration of a fluorescent analyte and that solution's fluorescence intensity. It should be noted that there is a linear relationship at relatively low analyte concentrations that eventually becomes nonlinear as the analyte concentration increases

mental design of fluorescence assays. This may be particularly important in the preparation of appropriate reference standards for quantitative work.

Although not very many analytes in food analysis meet the structural requirements for fluorescence (simplistically, a rigid conjugated structure), the possibility of derivatization should be considered. For example, (primary) amino acids can be modified to allow fluorescence measurement using *ortho*-phthaldialdehyde and 2-mercaptoethanol as derivatization reagents.

7.4 SUMMARY

UV and Vis absorption and fluorescence spectroscopy are widely used in food analysis. (See Chap. 6, Table 6.3, for a comparison of these types of spectroscopies, including their applications. The table also allows for comparison with other types of spectroscopies.) These techniques may be used for either qualitative and/or quantitative measurements. Qualitative measurements are based on the premise that each analyte has a unique set of energy spacings that will dictate its absorption/emission spectrum. Hence, qualitative assays are generally based on the analysis of the absorption or emission spectrum of the analyte. In contrast, quantitative assays most often are based on measuring the absorbance or fluorescence of the analyte solution at one wavelength. Quantitative absorption assays are based on the premise that the absorbance of the test solution will be a function of the solution's analyte concentration.

Under optimum conditions, there is a direct linear relationship between a solution's absorbance and its analyte concentration. The equation describing this linear relationship is known as Beer's law. The applicability of Beer's law to any given assay should always be experimentally verified by means of a calibration curve. The calibration curve should be established at the same time and under the same conditions that are used to measure the test solution. The analyte concentration of the test solution should then be estimated from the established calibration curve.

Molecular fluorescence methods are based on the measurement of radiation emitted from excited analyte molecules as they relax to lower energy levels. The analytes are raised to the excited state as a result of photon absorption. The processes of photon absorption and fluorescence emission occur simultaneously during the assay. Quantitative fluorescence assays are generally one to three orders of magnitude more sensitive than the corresponding absorption assays. Like absorption assays, under optimal conditions, there will be a direct linear relationship between the fluorescence intensity and the concentration of the analyte in the unknown solution. Most molecules do not fluoresce and, hence, cannot be assayed by fluorescence methods unless suitable derivatization approaches are available.

Instruments used for absorption and fluorescence methods have similar components, including a radiation source, wavelength selector(s), sample-holding cell(s), radiation detector(s), and a readout device.

7.5 STUDY QUESTIONS

1. Why is it common to use absorbance values rather than transmittance values when performing quantitative UV–Vis spectroscopy?
2. For a particular assay, the plot of absorbance vs. concentration is not linear; explain the possible reasons for this.
3. What criteria should be used to choose an appropriate wavelength at which to take absorbance measurements, and why is that choice so important?
4. In a particular assay, the absorbance reading on the spectrophotometer for one sample is 2.033 and for another sample 0.032. Would you trust these values? Why or why not?
5. Explain the difference between electromagnetic radiation in the UV and Vis ranges. How does quantitative spectroscopy using the UV range differ from that using the Vis range?
6. What is actually happening inside the spectrophotometer when the analyst "sets" the wavelength for a particular assay and why is it important?
7. Considering a typical spectrophotometer, what is the effect of decreasing the exit slit width of the monochromator on the light incident to the sample?
8. Describe the similarities and differences between a phototube and a photomultiplier tube. What is the advantage of one over the other?
9. Your laboratory has been using an old single-beam spectrophotometer that must now be replaced by a new spectrophotometer. You obtain the sales literature that describes single-beam and double-beam instruments. What are the basic differences between a single-beam and a double-beam spectrophotometer, and what are the advantages and disadvantages of each?
10. Explain the similarities and differences between UV–Vis spectroscopy and fluorescence spectroscopy with regard to instrumentation and principles involved. What is the advantage of using fluorescence spectroscopy?

7.6 PRACTICE PROBLEMS

1. A particular food coloring has a molar absorption coefficient of $3.8 \times 10^3 cm^{-1}M^{-1}$ at 510 nm.
   (a) What will be the absorbance of a 2×10^{-4}M solution in a 1-cm cuvette at 510 nm?
   (b) What will be the percentage transmittance of the solution in (a)?
2. You measure the percentage transmittance of a solution containing chromophore X at.
   (a) 400 nm in a 1-cm pathlength cuvette and find it to be 50%. What is the absorbance of this solution?
   (b) What is the molar absorption coefficient of chromophore X if the concentration of X in the solution measured in question 2 (a) is 0.5 mM?
   (c) What is the concentration range of chromophore X that can be assayed if, when using a sample cell of pathlength 1, you are required to keep the absorbance between 0.2 and 0.8?
3. What is the concentration of compound Y in an unknown solution if the solution has an absorbance of 0.846 in a glass cuvette with a pathlength of 0.2 cm? The absorptivity of compound Y is $54.2 cm^{-1}(mg/mL)^{-1}$ under the conditions used for the absorption measurement.
   (a) What is the molar absorption coefficient of compound Z at 295 nm and 348 nm, given the absorption spectrum shown in Fig. 7.15 (which was obtained using a UV–Vis spectrophotometer and a 1-mM solution of compound Z in a sample cell with a pathlength of 1 cm)?
   (b) Assume that you decide to take quantitative measurements of the amount of compound Z in different solutions. Based on the above spectrum, which wavelength will you use for your measurements? Give two reasons why this is the optimum wavelength.

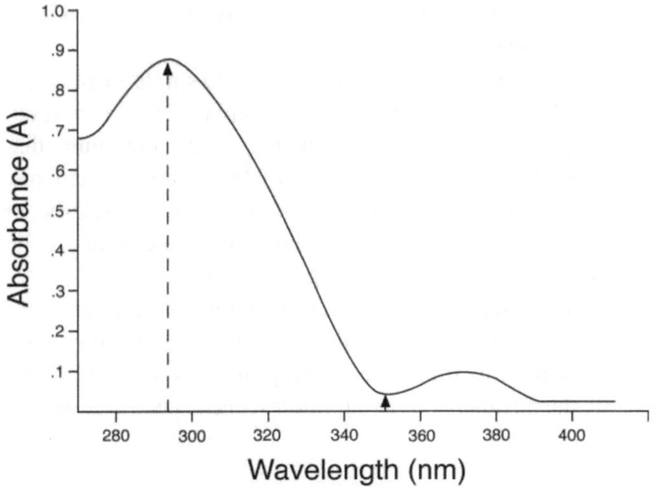

Fig. 7.15 The absorption spectrum of compound Z, to be used in conjunction with problems 4a and 4b

Answers

1. (a) = 0.76, (b) = 17.4

This problem requires knowledge of the relationship between absorbance and transmittance and the ability to work with Beer's law.

Given: Molar absorption coefficient = $3.8 \times 10^3 cm^{-1}M^{-1}$

(a) Use Beer's law: $A = \varepsilon bc$ (see Eq. 7.5) of the text).
- where,
  - $\varepsilon = 3.8 \times 10^3 cm^{-1}M^{-1}$
  - $b = 1$ cm,
  - $c = 2 \times 10^{-4}$M.

  Plugging into Beer's law gives the answer: Absorbance = 0.76

(b) Use definition of absorbance: $A = -\log T$ (see Eq. 7.3 of text).
- where
  - $T = P/P_0$
- Rearranging Eq. 7.3:
  - $-A = \log T$
  - $10^{-A} = T$
- $A = 0.76$ (from part (a) of question).
  - $10^{-.76} = .1737 = T$.
  - $\%T = 100 \times T$ (combining Eqs. 7.1 and 7.2 of text)
  - Answer: $\%T = 17.4$

2. (a) = 301, (b) = 602 $cm^{-1}M^{-1}$, (c) = 0.33×10^{-3}M to 1.33×10^{-3}M

This problem again requires knowledge of the relationship between absorbance and transmittance and the manipulation of Beer's law. Care must be taken when working with the appropriate concentration units.

(a) $T = 0.5$

  Use $A = -\log T = -\log .5 = .301$

  Answer: 0.301

(b) Given that the solution in part (a) is 0.5 mM (equivalent to 5×10^{-4}M).

  Rearranging Beer's law: $\varepsilon = A/(bc)$

  $\varepsilon = .301/[(1 \text{ cm}) \times (5 \times 10^{-4}\text{M})]$

  Answer: $\varepsilon = 602 cm^{-1}M^{-1}$

(c) To answer the problem, find the concentration that will yield an absorbance of 0.200 (lower limit) and the concentration that will yield an absorbance of 0.800 (upper limit). In both cases, use Beer's law to determine the appropriate concentrations:
- where
  - $c = A/\varepsilon b$
  - Lowest concentration = $0.2/[(602 cm^{-1}M^{-1})(1 \text{ cm})] = 3.3 \times 10^{-4}$M (i.e., 0.33 mM)
  - Highest concentration = $0.8/[(602 cm^{-1}M^{-1})(1 \text{ cm})] = 1.3 \times 10^{-3}$M (i.e., 1.33 mM)

3. 0.078 mg/mL

This problem illustrates (1) that concentration need not be expressed in units of molarity and (2) that the pathlength of the cuvette must be considered when applying Beer's law. In the present problem, the analyte concentration is given in milligrams per milliliter. Thus, the absorptivity must be in analogous units:

Apply: $c = A/\varepsilon b$

where

$A = 0.846$
$\varepsilon = 54.2\ \text{cm}^{-1}(\text{mg/mL})^{-1}$
$b = 0.2$ cm
Answer: 0.078 mg/mL

4. (a) = 860 at 295 nm, 60 at 348 nm; (b) = 295 nm; optimum sensitivity and more likely to adhere to Beer's law.

This problem presents the common situation in which one wants to use absorbance spectroscopy for quantitative measurements but is unsure what wavelength to choose for the measurements. Furthermore, the absorptivity of the analyte at the different wavelengths of interest is unknown. A relatively simple way to obtain the necessary information is to determine the absorption spectrum of the analyte at a known concentration.

(a) The arrows on the provided spectrum indicate the points on the spectrum corresponding to 295 nm and 348 nm. The problem notes that the absorption spectrum was obtained using a 1-mM solution (i.e., 1×10^{-3}M solution) of the analyte and that the pathlength of the cuvette was 1 cm. The answer to the problem is thus determined by taking the absorbance of the analyte at the two wavelengths in question and then plugging the appropriate data into Beer's law. It is somewhat difficult to get an exact absorbance reading from the presented spectrum, but we can estimate that the absorbance of the 1-mM solution is ~0.86 at 295 nm and ~ 0.06 at 348 nm.
   - Using $\varepsilon = A/bc$
   - Answer:
     - At 295 nm $\varepsilon = 0.86/[(1\ \text{cm})(0.001\ \text{M})] = 860\ \text{cm}^{-1}\text{M}^{-1}$.
     - At 348 nm $\varepsilon = 0.06/[(1\ \text{cm})(0.001\ \text{M})] = 60\ \text{cm}^{-1}\text{M}^{-1}$.

(b) In general, analysts strive to obtain maximum sensitivity for their assays, where sensitivity refers to the change in assay signal per unit change in analyte concentration (the assay signal in this case is absorbance). The absorbance values for the analyte at different wavelengths, taken from the absorption spectrum, and/or the relative absorptivity values for the analyte at different wavelengths, provide a good approximation of the relative sensitivity of the assay at different wavelengths (it is an approximation because we have not determined the variability/precision of the measurements at the different wavelengths). It can be seen from the given spectrum that absorbance "peaks" were at ~298 nm and ~ 370 nm. The sensitivity of the assay, relative to neighboring wavelengths, is expected to be maximum at these absorbance peaks. The peak at 295 nm is significantly higher than that at 370 nm, so the sensitivity of the assay is expected to be significantly higher at 295 nm. Thus, this would be the optimum wavelength to use for the assay. A second reason to choose 295 nm is because it appears to be in the middle of the "peak," and, thus, small changes in wavelength due to instrumental/operator limitations are not expected to appreciably change the absorptivity values. Therefore, the assay is more likely to adhere to Beer's law.

There are situations in which an analyst may choose to not use the wavelength corresponding to the overall maximum absorbance. For example, if there are known to be interfering compounds that absorb at 295 nm, then an analyst may choose to take absorbance measurements at 370 nm.

REFERENCES

1. Skoog DA, Holler FJ, Crouch SR (2017) Principles of instrumental analysis, 7th edn. Cengage Learning, Boston, MA
2. Brown CW (2009) Ultraviolet, visible, near-infrared spectrophotometers. In: Cazes J (ed) Ewing's analytical instrumentation handbook, 3rd edn. Marcel Dekker, New York
3. Currell G (2000) Analytical instrumentation – performance characteristics and quality. Wiley, New York, pp. 67–91
4. DeBolt S, Cook DR, Ford CM (2006) L-Tartaric acid synthesis from vitamin C in higher plants. Proc Natl Acad Sci 103: 5608–5613
5. Feinstein K (1995) Guide to spectroscopic identification of organic compounds. CRC, Boca Raton, FL
6. Hargis LG (1988) Analytical chemistry — principles and techniques. Prentice-Hall, Englewood Cliffs, NJ
7. Harris DC, Lucy CA (2019) Quantitative chemical analysis, 10th edn. W.H. Freeman, New York
8. Harris DC, Bertolucci MD (1989) Symmetry and spectroscopy. Dover, Mineola, NY
9. Ingle JD Jr, Crouch SR (1988) el, Englewood Cliffs, NJ
10. Lakowicz JR (2011) Principles of fluorescence spectroscopy, 3rd edn. Springer, New York
11. Milton Roy educational manual for the SPECTRONIC® 20 & 20D spectrophotometers (1989) Milton Roy Co., Rochester, NY
12. Miwa H (2000) High-performance liquid chromatography determination of mono-, poly- and hydroxycarboxylic acids in foods and beverages as their 2-nitrophenylhydrazides J Chromatogr A 881: 365–385

13. Owen T (2000) Fundamentals of UV–visible spectroscopy, Agilent Technologies. https://www.agilent.com/cs/library/primers/Public/59801397_020660.pdf
14. Pavia DL, Lampman GM, Kriz GS Jr (1979) Introduction to spectroscopy: a guide for students of organic chemistry. W. B. Saunders, New York
15. Pavia DL, Lampman GM, Kriz GS Vyvyan JA (2015) Introduction to spectroscopy. 5th Edition, Cengage Learning, Independence, KY
16. Perkampus H-H (1994) UV-Vis spectroscopy and its applications. Springer, Berlin, Germany
17. Robinson JW, Frame EMS, Frame GM II (2014) Undergraduate instrumental analysis, 7th edn. CRC Press, Inc., Boca Raton, FL
18. Shriner RL, Fuson RC, Curtin DY, Morrill TC (1980) The systematic identification of organic compounds – a laboratory manual. 6th edn. Wiley, New York
19. Smith BC (2003) Quantitative spectroscopy: theory and practice, Academic Press, Amsterdam
20. Thomas MJK, Ando DJ (1996) Ultraviolet and visible spectroscopy, 2nd edn. Wiley, New York
21. Valeur B, Berberan-Santos MN (2013) Molecular fluorescence: principles and applications, 2nd edn., Wiley-VCH, New York
22. Yadav LDS (2005) Organic spectroscopy. Kluwer Academic, Boston, MA

Infrared and Raman Spectroscopy

8

Luis Rodriguez-Saona and Huseyin Ayvaz

Contents

8.1 INTRODUCTION

Infrared (IR) spectroscopy refers to the measurement of the absorption of different frequencies of IR radiation by foods or other solids, liquids, or gases. IR spectroscopy began in 1800 with an experiment by Herschel [1]. When he used a prism to create a spectrum from white light and placed a thermometer at a point just beyond the red region of the spectrum, he noted an increase in temperature. This was the first observation of the effects of IR radiation. By the 1940s, IR spectroscopy had become an important tool used by

L. Rodriguez-Saona (✉)
Department of Food Science and Technology, The Ohio State University, Columbus, OH, USA
e-mail: rodriguez-saona.1@osu.edu

H. Ayvaz
Department of Food Engineering, Canakkale Onsekiz Mart University, Canakkale, Turkey
e-mail: huseyinayvaz@comu.edu.tr

B. P. Ismail, S. S. Nielsen (eds.), *Nielsen's Food Analysis*, Food Science Text Series,
https://doi.org/10.1007/978-3-031-50643-7_8

chemists to identify functional groups in organic compounds. In the 1970s, commercial near-IR (NIR) reflectance instruments that provided rapid quantitative determinations of moisture, protein, and fat in cereal grains and other foods were introduced. Today, IR spectroscopy is used widely in the food industry for both qualitative and quantitative analyses of ingredients and finished foods.

This chapter describes the three techniques of mid- and near-IR and Raman spectroscopy, including the principles by which molecules absorb IR radiation, the components, and configuration of commercial IR spectrometers, sampling methods of IR spectroscopy, and qualitative and quantitative applications of these techniques to food analysis. These three types of spectroscopies are briefly compared to each other and to other spectroscopic techniques in Table 6.3 of Chap. 6, which introduces all chapters related to spectroscopy. Infrared and Raman microspectroscopy will not be covered in this chapter but rather will be covered in Chap. 32, Sects. 32.3.2 and 32.3.3.

8.2 THE PRINCIPLES OF IR SPECTROSCOPY

8.2.1 The IR Region of the Electromagnetic Spectrum

Infrared radiation is electromagnetic energy with **wavelengths** (λ) longer than visible light but shorter than microwaves. Generally, wavelengths from 0.8 to 100 micrometers (μm) can be used for IR spectroscopy and are divided into the **near-IR** (0.8–2.5 μm; 12,500–4000 cm^{-1}), the **mid-IR** (2.5–15 μm; 4000–670 cm^{-1}), and the **far-IR** (15.4–100 μm; 650–100 cm^{-1}) regions. One μm is equal to 1×10^{-6} m. The near- and mid-IR regions of the spectrum are most useful for quantitative and qualitative analysis of foods.

IR radiation also can be measured in terms of its **frequency**, which is useful because frequency is directly related to the energy of the radiation by the following relationship:

$$E = h\nu \tag{8.1}$$

where

E = energy of the system
h = Planck's constant
ν = frequency in hertz

Frequencies are commonly expressed as **wave numbers** ($\bar{\nu}$, in reciprocal centimeters, cm^{-1}). Wave numbers are calculated as follows:

$$\bar{\nu} = 1/(\lambda \text{ in cm}) = 10^4/(\lambda \text{ in } \mu m) \tag{8.2}$$

8.2.2 Molecular Vibrations

A molecule can absorb IR radiation if it vibrates in such a way that its charge distribution, and therefore its electric dipole moment, changes during the vibration. Although there are many possible vibrations in a polyatomic molecule, the most important vibrations that produce a change in dipole moment are stretching (symmetric and asymmetric) and bending (scissoring, rocking, twisting, wagging) motions. Examples of these vibrations for the water molecule are shown in Fig. 8.1. Note that the stretching motions vibrate at higher frequencies than the scissoring motion. Also, asymmetric stretches are more likely to result in a change in dipole moment, with corresponding absorption of IR radiation, than are symmetric stretches.

8.2.3 Factors Affecting the Frequency of Vibration

The basic requirement for absorption of infrared radiation is that there must be a net change in dipole moment during the vibration of the molecule or functional group. A molecular vibration can be thought of as a **harmonic oscillator** (Fig. 8.2a), with the energy level for any molecular vibration given by the following equation:

$$E = \left(v + \tfrac{1}{2}\right)\left(\tfrac{h}{2\pi}\right)\sqrt{k \Big/ \frac{m_1 m_2}{m_1 + m_2}} \tag{8.3}$$

where

v = vibrational quantum number (positive integer values, including zero, only)
h = Planck's constant
k = force constant of the bond
m_1 and m_2 = masses of the individual atoms involved in the vibration

Note that the vibrational energy, and therefore the frequency of vibration, is directly proportional to the strength of the bond and inversely proportional to the mass of the molecular system. Thus, different chemical functional groups will vibrate at different frequencies. A vibrating molecular functional group can absorb radiant energy to move from the lowest (v = 0) vibrational state to the first excited (v = 1) state, and the frequency of radiation that will make this occur is identical to the initial frequency of vibration of the bond. This frequency is referred to as **fundamental absorption**. The harmonic oscillator provides a good fit to explain bond stretching vibrations for fundamental vibrations. However, molecules also can absorb radiation to move

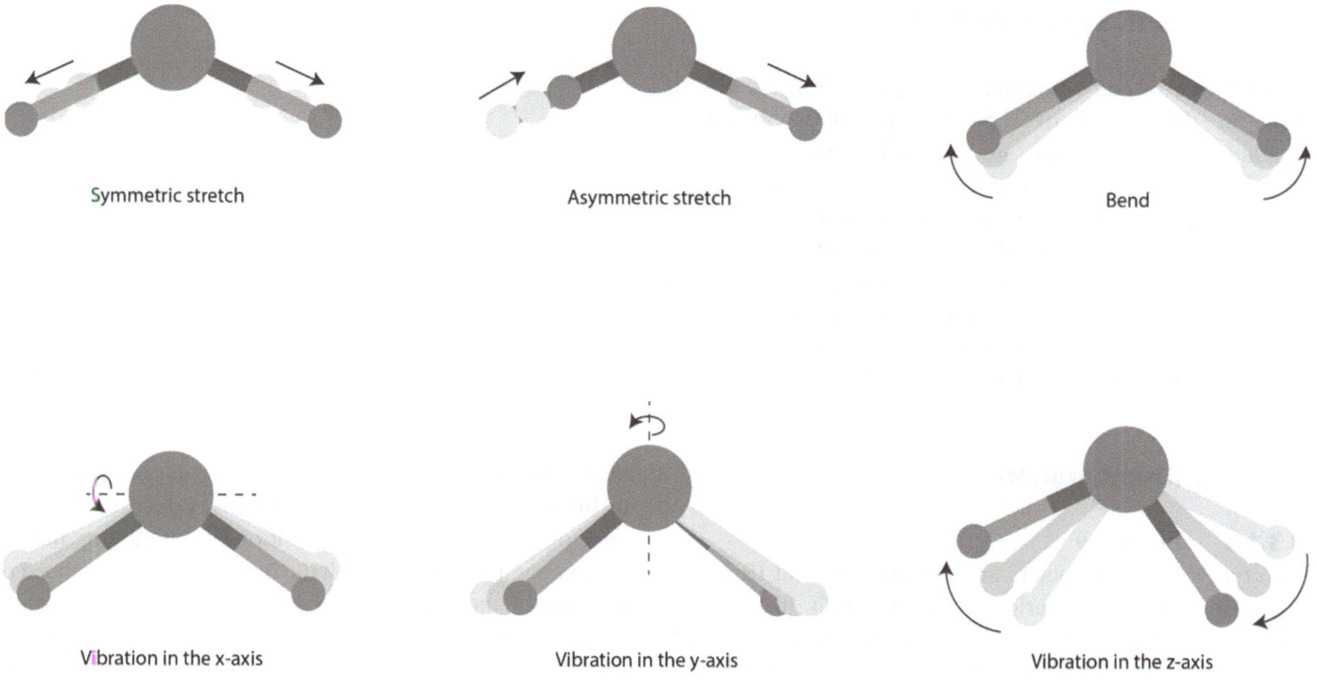

Fig. 8.1 Vibrational modes of a water molecule

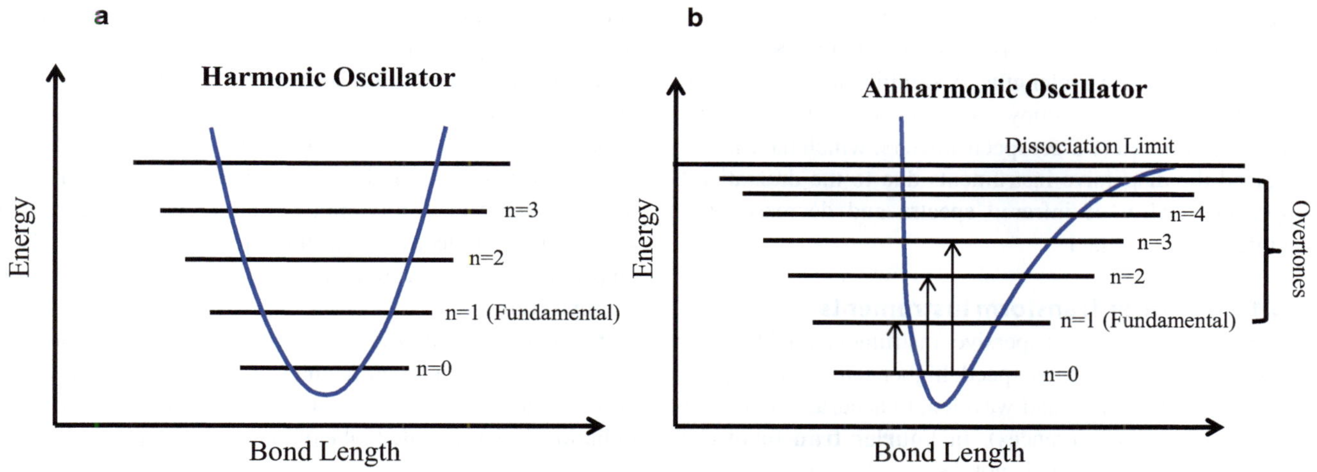

Fig. 8.2 A diagram of the differences in potential energy curves between a (**a**) harmonic and an (**b**) anharmonic oscillator model

to a higher ($v = 2$ or 3) excited state, such that the frequency of the radiation absorbed is two or three times that of the fundamental frequency. These absorptions are referred to as **overtones**, and the intensities of these absorptions are much lower than the fundamental since these transitions are less favored. The **anharmonic oscillator model** (Fig. 8.2b) accounts for the repulsion and attraction of the electron cloud and accommodates bond dissociation at higher energy levels. Overall, the fundamental vibrations are unaffected by the anharmonicity terms, but overtone transitions are influenced by anharmonicity, which must be taken into account when assessing the frequency of these higher-frequency vibrations. Combination bands can also occur if two or more different vibrations interact to give bands that are sums of their fundamental frequencies. The model of the harmonic oscillator and its modification to account for anharmonicity allows explanation of the origin of many of the characteristic frequencies that can be assigned to particular combinations of atoms within a molecule [2]. The overall result is that each functional group within the molecule absorbs IR radiation in distinct wavelength bands rather than as a continuum.

8.3 MID-IR SPECTROSCOPY

Mid-IR spectroscopy measures a sample's ability to absorb light in the region of 2.5–15 μm (4000–650 cm^{-1}). Fundamental absorptions are primarily observed in this spectral region. Mid-IR spectroscopy is very useful in studying organic compounds because the absorption bands are related to the vibrational modes of specific functional groups. The positioning of the band and its intensity are correlated with the energy of the bond, its environment, and its concentration in the matrix, making mid-IR spectroscopy ideal for both qualitative and quantitative applications.

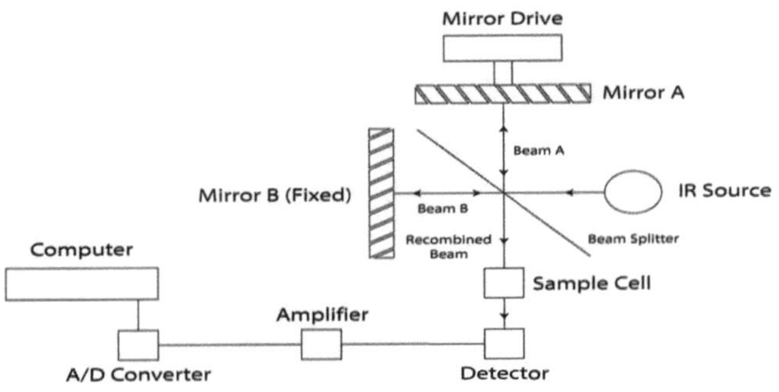

Fig. 8.3 A block diagram of an interferometer and associated electronics typically used in an FTIR instrument

8.3.1 Instrumentation

8.3.1.1 Overview

Two types of spectrometers are available for mid-IR analysis: dispersive and Fourier transform (FT) instruments. Dispersive systems have been available since the 1940s using prisms or gratings as dispersive elements. These systems contain components similar to ultraviolet–visible (UV–Vis) spectrometers, including a radiation source, a monochromator, a sample holder, and a detector connected to an amplifier system to record the spectra. In these systems, a filter, grating, or prism is used to separate the IR radiation into its individual wavelengths. A significant advance in the field of mid-IR spectroscopy was the development of Fourier transform infrared (FTIR) spectrometers, which have mostly replaced the dispersive instruments due to the dramatically improved quality of infrared spectra and decreased time required to obtain data.

8.3.1.2 Fourier Transform Instruments

Compared to mid-IR dispersive instruments, FTIR spectrometers allow for greater speed, higher sensitivity, superior wavelength resolution, and wavelength accuracy (details for advantages are in references). In **Fourier transform** (FT) instruments, the radiation is not dispersed; instead, all wavelengths arrive at the detector simultaneously. A mathematical treatment is used to convert the results into a typical IR spectrum. Instead of a monochromator, the instrument uses an **interferometer**. A Michelson interferometer is the most commonly used design, and its mechanism is simple (Fig. 8.3). The infrared radiation from the source is split into two beams by a beam splitter, and each half of the beam splitter goes to a mirror (either a fixed or moving mirror). The beams are reflected and recombined at the beam splitter, resulting in interference that is directed to the sample (or reference) and then the detector. The motion of the moving mirror results in the change of optical path length between the two split beams so that constructive, destructive, and intermediate interference states occur (with destructive interference being dominant). The resulting output is referred to as an **interferogram**, which is the intensity measured by the detector as a function of the position of the moving mirror. When a sample interacts with the recombined beam ahead of the detector, molecules absorb at their characteristic frequencies, and thus the radiation reaching the detector is modified (Fig. 8.4). Once the data are collected, a mathematical transformation called "Fourier transform" converts the interferogram from time domain (intensity versus time) to an IR spectrum in the frequency domain (intensity versus frequency). A computer allows the mathematical transformation to be completed rapidly.

The common **radiation sources** for a mid-IR spectrometer are inert solids heated electrically to 1000–1800 °C. Three popular types of sources are the Nernst glower (constructed of rare earth oxides), Globar (constructed of silicon carbide), and the Nichrome coil wrapped around a ceramic core that glows when an electrical current is passed through it. They all produce continuous radiations but with different radiation energy profiles.

Detectors include **thermocouples** for which output voltage varies with temperature changes caused by varying levels of radiation striking the detector. In a **Golay detector**, the radiation strikes a sealed tube of xenon gas, warming the gas and causing pressure changes within the tube. However, most current instruments use either **pyroelectric detectors**, such as deuterated triglycine sulfate (DTGS) crystals, or solid-state **semiconductor detectors**. Variation in the amount of radiation striking a DTGS detector causes the detector's temperature to change, which results in a change in the dielectric constant of the DTGS element. The resulting change in capacitance is measured. Semiconductor detectors, such as those made from a mercury–cadmium–telluride (MCT) alloy, have conductivities that vary according to the amount of radiation striking the detector surface. MCT detectors respond faster to smaller changes in radiation intensity than other detectors; however, they typically require cryogenic cooling. DTGS and MCT detectors are the most commonly used detectors in Fourier transform instruments.

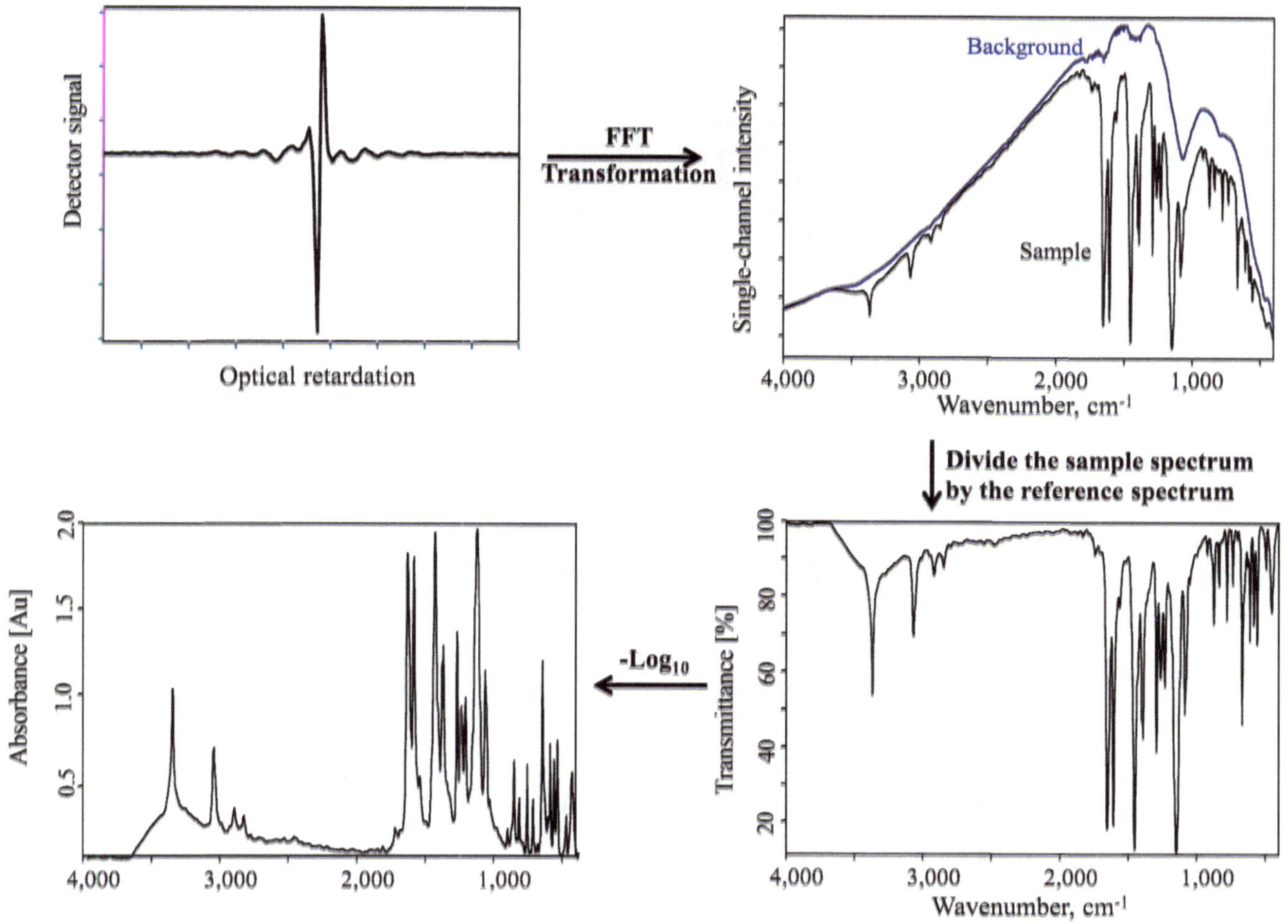

Fig. 8.4 An illustration of the process to convert an interferogram into an infrared spectrum using the Fourier transform algorithm

8.3.2 Sample Handling Techniques

Transmission mode is based on the IR beam passing through a sample placed in between two IR transparent windows. Liquids are often measured by **transmission IR spectroscopy**. Because absorptivity coefficients in the mid-IR are high, cells with path lengths of only 0.01–1.0 mm are commonly used. Quartz and glass absorb in the mid-IR region, so cell windows made of non-absorbing materials, such as halide or sulfide salts, are used. Halide salts are soluble in water, and care must be taken when selecting cells for use with aqueous samples. Cells are also available with windows made from more durable and less soluble materials, such as zinc selenide, but these are more expensive than those with halide salt windows. Liquid cells must be free of air bubbles, and extra care needs to be taken when cleaning between samples.

Transmission spectra of solids can be obtained by finely grinding a small amount of the sample with potassium bromide (KBr), pressing the mixture into a pellet under high pressure, and inserting the pellet into the IR beam. Limitations of this technique include the difficulty of handling and storing the hygroscopic KBr and the complexity and time required to make a good KBr pellet. An alternative technique is to disperse a finely divided solid in Nujol mineral oil to form a mull.

Transmission spectra can be obtained from gas samples using a sealed 2–10-cm glass cell with IR transparent windows. For trace analysis, multiple-pass cells are available that reflect the IR beam back and forth through the cell many times to obtain path lengths as long as several meters. FTIR instruments also can be interfaced with a gas chromatograph to obtain the spectra of compounds eluting from the chromatography column.

Attenuated total reflectance (ATR) is a widely applied sampling technique in infrared spectroscopy because it requires little or no sample preparation, eliminates cell path lengths, and provides consistent spectral collections. ATR allows for obtaining spectra from solid samples that are too thick for transmission measurements, e.g., pastes such as peanut butter and viscous liquids. ATR works based on the attenuation effect of infrared light (Fig. 8.5) when it is directed at an interface between an internal reflection element (a crystal) with high refractive index properties (i.e.,

a

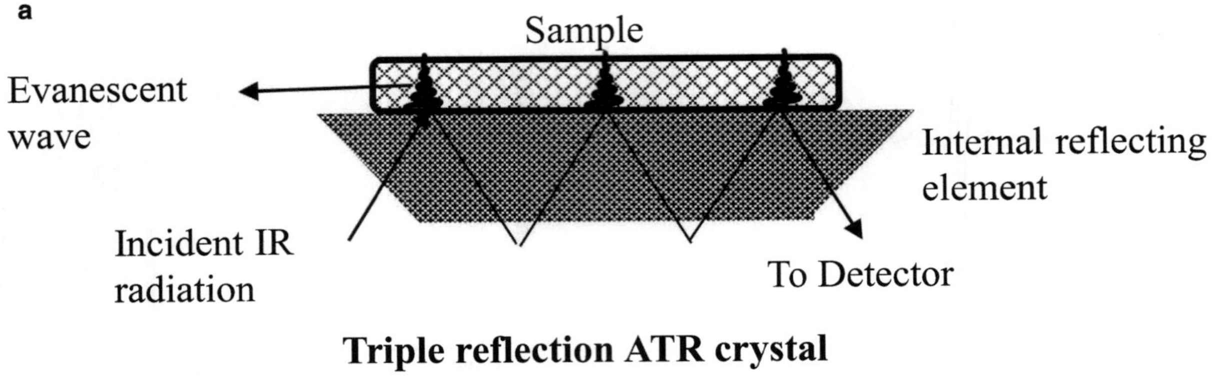

b

Guidelines for selection of ATR crystals

Material	Spectral range (cm^{-1})	Hardness (Kg/mm)	pH range	Refractive Index	Depth of penetration at 45°
Diamond	50,000 -2,500	9,000	1-14	2.4	1.66
Germanium	5,000 - 550	780	1-14	4.0	0.65
Silicon	8,333 - 33	1,150	1-12	3.4	0.81
KRS-5	17,900 - 250	40	5-8	2.4	0.85
ZnSe	20,000 - 500	130	5-9	2.4	1.66
AMTIR	11,000 - 725	170	1-9	2.5	1.46

Fig. 8.5 (**a**) An illustration of the reflection phenomena in a triple-reflection attenuated total reflection accessory and formation of the evanescence wave in the sample. (**b**) The characteristics of common ATR crystals

zinc selenide (ZnSe), thallium iodide-thallium bromide (KRS-5), germanium (Ge), silicon (Si), and diamond) and a low refractive index material (food sample) on its surface. Upon interaction with the reflecting surface, a radiation called an "evanescent wave" is formed, which exits the high refractive index material and penetrates into the sample. The sample material selectively absorbs, the intensity of the reflected radiation is decreased at wavelengths where the sample absorbs radiation, and the final attenuated radiation exiting the crystal is measured as being unique for the sample analyzed.

Radiation is not transmitted through the sample; therefore, there is no need for the sample to be thin enough to transmit the incident light. Since the penetration depth of the radiation is limited to a few micron (μm) thicknesses, the same spectra are obtained regardless of the amount of the sample placed on the surface, and there is no need to dilute the samples.

The physical state of the sample is an important factor because it must be in intimate contact with the ATR crystal to obtain a good ATR spectrum. Liquids and pastes usually exhibit better ATR spectra than solid samples. A pressure clamping system is used with solid samples to deform the sample, increasing the extent of contact between the ATR crystal and the sample.

Using multiple reflections of the beam, the intensity of the absorption spectrum can be enhanced, resolving minor bands in weakly absorbing samples. Multiple internal reflections (3 to up to 25) in a 4-mm diameter area result in effective path lengths of the sampling surface that equal the product of the number of reflections at the surface by the penetration depth (function of the crystal).

8.3.3 Applications of Mid-IR Spectroscopy

8.3.3.1 Absorption Bands of Organic Functional Groups

Infrared spectroscopy monitors the interaction of functional groups in chemical molecules with infrared light, resulting in predictable vibrations that provide a "fingerprint" characteristic of chemical substances in the sample. Spectra in the mid-IR region have well-resolved bands that can be assigned to functional groups of the components of foods. The positioning of the band facilitates structural characterization, and its intensity correlates with its concentration in the matrix, allowing for both qualitative and quantitative applications.

Spectra are commonly presented in wave numbers plotted on the *x*-axis and either percent transmittance or absorbance plotted on the *y*-axis. The mid-IR spectra of selected foods are shown in Fig. 8.6, displaying the major absorption bands associated with functional groups (Table 8.1) in fats and in protein- and carbohydrate-rich commodities.

The unique spectral profile can be used to identify specific functional groups present in an unknown substance. Comparing the mid-IR spectrum to a set of standard spectra and determining the closest match can accomplish the identification of chemical compounds. Spectral libraries are available from several sources, but probably the most extensive collection of standards is the Sadtler Standard Spectra (Sadtler Division of Bio-Rad, Inc., Philadelphia, PA) that contain more than 225,000 infrared spectra. Algorithms are used to compare the unknown spectrum and each spectrum in the reference database, and the hit quality index (HQI) is determined, indicating the similarity between spectra. Several HQI values can be generated for the unknown compounds, and the software will sort and display the best matches in a search report. Noise and spectral artifacts can impact the HQI and lead to mistakes in identification; thus, it is imperative to perform visual comparisons to confirm a good match. Spectral searches will most commonly be done on pure substances, rather than foods or commodities.

8.3.3.2 Applications of Mid-IR Spectroscopy

Mid-IR spectroscopic measurements obey **Beer's law**, although deviations may be greater than in UV–Vis spectroscopy due to the low intensities of IR sources, the low sensitivities of IR detectors, and the relative narrowness of mid-IR absorption bands. One of the first and most extensive uses of this technique is in **infrared milk analyzers**, which have the ability to analyze hundreds of samples per hour. The fat, protein, and lactose contents of milk can be determined simultaneously with one of these instruments. The ester carbonyl groups of lipid absorb at 5.73 μm (1742 cm^{-1}), the amide groups of protein absorb at 6.47 μm (1545 cm^{-1}), and the hydroxyl groups of lactose absorb at 9.61 μm (1045 cm^{-1}). These automated instruments homogenize the milk fat globules to minimize light scattering by the sample and then pump the milk into a flow-through cell through which the IR beam is passed. The instrument is calibrated using samples of known concentration to establish the slope and intercept of a Beer's law plot. Official methods have been adopted for IR milk analyzers, and specific procedures for operating these instruments are given [3, 4].

Mid-IR spectroscopy has many applications in food analysis, including quality control, authentication and adulteration detection, process monitoring, contaminant analysis, and nutritional analysis. Examples of food applications of mid-IR spectroscopy to foods are given in Table 8.2. In this table, quality control and macronutrient analysis (moisture, carbohydrates, proteins, and fats), common applications of IR spectroscopy in foods, are not given; instead, it focuses on specific applications for detection of minor components, contaminants, and ingredient authentication. Due to the complex nature of infrared spectra, multivariate statistical

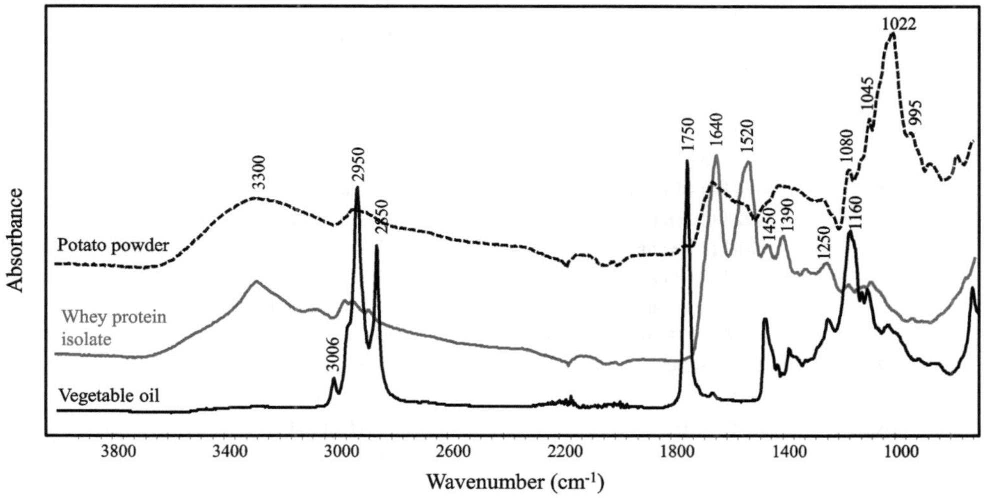

Fig. 8.6 Mid-IR spectra of corn oil, whey protein isolate, and potato flour measured by a triple-reflection ATR unit. Frequency in wave numbers is plotted on the *x*-axis, with intensity on the *y*-axis

Table 8.1 Characteristic mid-IR and near-IR absorption frequencies of major food components

Food components	*Infrared region*	*Frequency (cm⁻¹)*	*Assignment*
Fats	Mid-IR	3006	*Cis*-Olefinic groups
		3000–2800	C–H asymmetrical and symmetrical stretch of long-chain fatty acids
		1740	C=O stretch of fatty acid esters
		1711	C=O stretch of fatty acids
		1475–1435	Asymmetrical bending defor. $-CH_3$ in branched alkanes
		1465	=C–H *cis*-bending
		1350–1395	C – H symmetric bending in $-CH_3$
		1350–1150	C–H bending
		1240 and 1163	–C–O and $-CH_2-$ stretch bending
		1150–1000	–C=O stretch
		966	Isolated, out-of-plane *trans*-CuC bending
		914	*Cis* –C=C–H bending out of plane
	Near-IR	8700–8100 (1150–1235 nm)	C–H stretch, second overtone
		8563 (1168 nm)	*cis*-double bond
		7209 (1387 nm)	Combination C–H stretch and C–H bending
		5807 (1722 nm), 5681 (1760 nm)	C–H stretch, first overtone
		4705 (2125 nm)	C–H vibrations in isolated C=C bonds
		4336 (2306 nm), 4269 (2342 nm)	Combination bands of C–H and C–O stretch
Water	Mid-IR	3500	O–H stretch
		1650	O–H bending
	Near-IR	6900 (1450 nm)	First overtones stretch Free OH
		5150 (1940 nm)	OH combination bands
Proteins	Mid-IR	3300	Amide A
		1640	Amide I
		1540	Amide II
		1330–1230	Amide III
	Near-IR	5000–4550	Amides
		4855 (2016 nm)	Combination of amide A and amide II
		4580 (2180 nm)	Combination of amide A and amide III
Carbohydrates	Mid-IR	1745	Ester, pectin
		1630–1605	Carboxylate, pectin
		Shifts at 1617 and 1420	Metal coordination by the pectate chains
		1460–1340	C–C–H and C–O–H deformations
		1250–950	Endo–/or exocyclic C–C and C–O bonds
		1150	Characteristic of the pyranose ring
		1110	CO ring, C–4–O, C–6–O
		1080	Typical bending of C–1–H
		1060	C–1–OH (fructose residue)
		1030	Typical C–4–OH vibration
		1050–1020	Starch retrogradation, increased crystallinity
		995	Glycosidic linkage
		950–750	α- and β-anomeric region of saccharide
	Near-IR	6944 (1440 nm)	OH stretch, first overtone, crystalline structure
		5924 (1688 nm)	C–H stretch first overtone
		5882 (1700 nm)	C–H stretch methyl groups from cellulose or lignin
		4662 (2100 nm)	Combination C–O stretch/OH bend: Carbohydrates
		4280 (2336 nm)	Cellulose and starch
		4386 (2280 nm)–4292 (2330 nm)	Combination C–H stretch and CH_2 deformation
Aromatics	Mid-IR	3100–3000	Aromatic –CH stretch
		1600	–C=C– stretch
Alcohols	Mid-IR	3600–3200	–OH stretch
		1500–1300	–OH bend
		1220–1000	C–O stretch
	Near-IR	6850 (1460 nm)–6240 (1600 nm)	First overtone –OH
		6800 (1470 nm) and 7100 (1280 nm)	Typical first overtone bands of bonded OH–H– bonding
		4550 (1800 nm)–5550 (2200 nm)	Combination OH stretch and OH bending
Ethers	Mid-IR	1220–1000	C–O asymmetric stretch

Table 8.2 Some example food applications of mid-IR and near-IR spectroscopy

Application	*Food*	*Analysis*	*References*
Mid-IR spectroscopy	Milk	Determination of origin, storage time, adulterants (i.e., whey, hydrogen peroxide, urea, and melamine), acidity, pH, and lactose	[5–8]
	Cheddar cheese	Characterization of cheese ripening and flavor, classification of cheese samples based on their age, and detection of formaldehyde	[9–10]
	Yogurt	Determination of mineral elements (i.e., Ca, K, Mg, Na, P)	[11]
	Butter	Detection of adulterants (i.e., margarine and lard)	[12]
	Infant formula	Quantification of melamine and cyanuric acid	[13]
	Olive oil	Discrimination of olive oils by cultivar, geographic origin, and storage time and detection of adulteration (i.e., soybean oil)	[14–15]
	Wheat	DON (deoxynivalenol)	[16]
	Corn and corn oil	Fungal infection, DON	[17]
	Peanut and peanut oil	Fungal infection, aflatoxin B1, B2, G1, G2	[18]
	Pork	Total volatile basic nitrogen (TVB-N)	[19]
	Beef	Total viable count, specific microbial count, and detection of adulterants (i.e., pork, Turkey mince, horse meat)	[20–21]
	Chicken	Differentiation of fresh and frozen/thawed chicken meat	[22]
	Fish fillets	Detection of adulteration with cheaper species	[23]
	Sea bass	Differentiation of farmed and wild seabass	[24]
	Honey	Detection and quantification of adulterants and discrimination of honey samples from different botanical origins	[25–26]
	Coffee	Geographical origin and genotypic classification	[27]
	Roast ground coffee	Detection of various adulterants (i.e., coffee husks)	[28]
	Green Arabica coffee	Discrimination between defective and non-defective roasted beans	[29]
	Red wine	Authentication, geographical origin, and classification of wines	[30]
	Chocolate	Lard adulteration	[31]
	Paprika and chili powder	Detection of Sudan dye	[32]
	Saffron	Detection of adulterants (i.e., safflower, turmeric, tartrazine)	[33]
	Oregano	Detection of olive leaves	[34]
Near-IR spectroscopy	Milk	pH, acidity, lactose, protein, fat, SSC, total casein	[35–36]
	Wheat	Fungal infection, DON, ergosterol	[37]
	Wheat, corn, barley, rice, oats, peanut	Fungal infection, mycotoxins (i.e., deoxynivalenol, fumonisin, aflatoxin)	[38]
	Pork	Identification of repeatedly frozen meat	[39]
	Chicken	Bacterial species	[40]
	Fish fillets	Adulteration with cheaper species	[23]
	Eggs	Prediction of storage time at room temperature	[41]
	Grape	Brix, pH and anthocyanin, sugar	[42]
	Wine	Ethanol, malic, and tartaric acid and determination of mineral elements (i.e., Cu, Fe, K, Na, Mg, and Ca)	[43–44]
	Vinegar	Lactic acid and malic acid content	[45]
	Beer	Sugar, alcohol, and pH	[46]
	Alcoholic beverages	Quality analysis (i.e., alcohols, pH, total acidity, trace metals, age, geographical origin, varieties, and adulteration)	[36]
	Nonalcoholic beverages	Quality analysis (soluble solid content, pH, organic acids, sugars, sensory, and varieties)	[36]
	Black tea	Total polyphenol and caffeine	[47]
	Potato chips	Determination of acrylamide	[48]
	Mushrooms	Polyphenol oxidase	[49]
	Various fruits (i.e., bilberry, blackberry, and grapes)	Monomeric and total anthocyanins, anthocyanins, antioxidant capacity, total phenolics, total carotenoids, tannins, lycopene, total flavonoids, and various bioactive compounds (i.e., catechin, quercetin, chlorogenic acid, and resveratrol)	[50]
	Strawberry	Pesticide residues	[51]

analysis techniques (chemometrics) must often be used to extract information from the infrared spectra, allowing for the classification and quantitative analysis of multiple components in foods. Infrared calibration using chemometric techniques is discussed in more detail in Sect. 8.5.3.

8.4 NEAR-IR SPECTROSCOPY

Measurements in the **near-IR** (NIR) spectral region (0.7–2.5 μm, equal to 700–2500 nm) are more widely used for quantitative analysis of foods than are mid-IR measurements. Several commercial instruments are available for the compositional analysis of foods using near-IR spectroscopy. A major advantage of near-IR spectroscopy is its ability to directly measure the composition of solid food products using diffuse reflection techniques.

8.4.1 Principles

8.4.1.1 Principles of Diffuse Reflection Measurements

When radiation strikes a solid or granular material, part of the radiation is reflected from the sample surface. This mirrorlike reflection, called specular reflection, gives little useful information about the sample. Most of the specularly reflected radiation is directed back toward the energy source. Another portion of the radiation will penetrate through the sample's surface and be reflected off several sample particles before it exits the sample. This is referred to as **diffuse reflection**, and this diffusely reflected radiation emerges from the surface at random angles through 180°. Each time the radiation interacts with a sample particle, the chemical constituents in the sample can absorb a portion of the radiation. Therefore, the diffusely reflected radiation contains information about the chemical composition of the sample, as indicated by the amount of energy absorbed at specific wavelengths.

The amount of radiation penetrating and leaving the sample surface is affected by the size and shape of the sample particles. Compensation for this effect may be achieved by grinding solid or granular materials with a sample preparation mill to a fine, uniform particle size or applying mathematical corrections when the instrument is calibrated [52].

8.4.1.2 Absorption Bands in the Near-IR Region

The absorption bands observed in the near-IR region are primarily overtones and combinations. Therefore, the absorptions tend to be weak in intensity. However, this is actually an advantage since absorption bands that have sufficient intensity to be observed in the near-IR region arise primarily from functional groups that have a hydrogen atom attached to a carbon, nitrogen, or oxygen, which are common groups in the major constituents of food such as water, proteins, lipids, and carbohydrates. Table 8.1 lists the absorption bands associated with a number of important food constituents.

The absorption bands in the near-IR region tend to be broad and frequently overlap, yielding spectra that are quite complex. However, these broad bands are especially useful for quantitative analysis. Typical near-IR spectra of wheat, dried egg white, and cheese are shown in Fig. 8.7. Note that strong absorption bands associated with the -OH groups of water, which are the dominant features in the spectrum of cheese, containing 30–40% moisture, and they are still prominent even in the lower moisture wheat and egg white samples. Bands arising protein (2060 and 2180 nm) in the wheat sample are partially obscured by a strong starch absorption band and centered at 2100 nm. Relatively sharp absorption bands arising from –CH groups in lipids can be observed at 2310 nm, 2350 nm, and 1730 nm and are distinctly observable in the cheese spectrum.

8.4.2 Instrumentation

Two commercial near-IR spectrometers are shown in Fig. 8.8. The radiation source in most near-IR instruments is a tungsten halogen lamp with a quartz envelope, similar to a projector lamp. These lamps emit significant amounts of radiation in both the visible and near-IR spectral regions. Semiconductor detectors are most commonly used in near-IR instruments, with silicon detectors used in the 700–1100-nm range and lead sulfide used in the 1100–2500-nm region. In situations for which a rapid response to changing light intensity is needed, such as in online monitoring, indium–gallium–arsenide (InGaAs) detectors can be used. Many InGaAs detectors are limited to a maximum wavelength of 1700 nm, although commercial InGaAs detectors with a

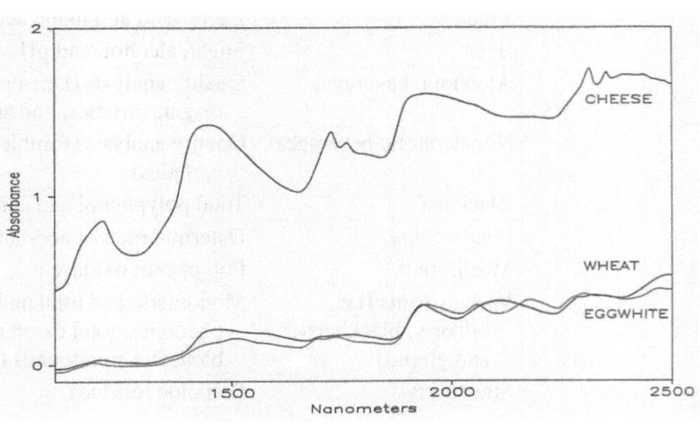

Fig. 8.7 Near-IR spectra of cheese, wheat, and dried egg white plotted as log (1/*R*) vs. wavelength in nanometers

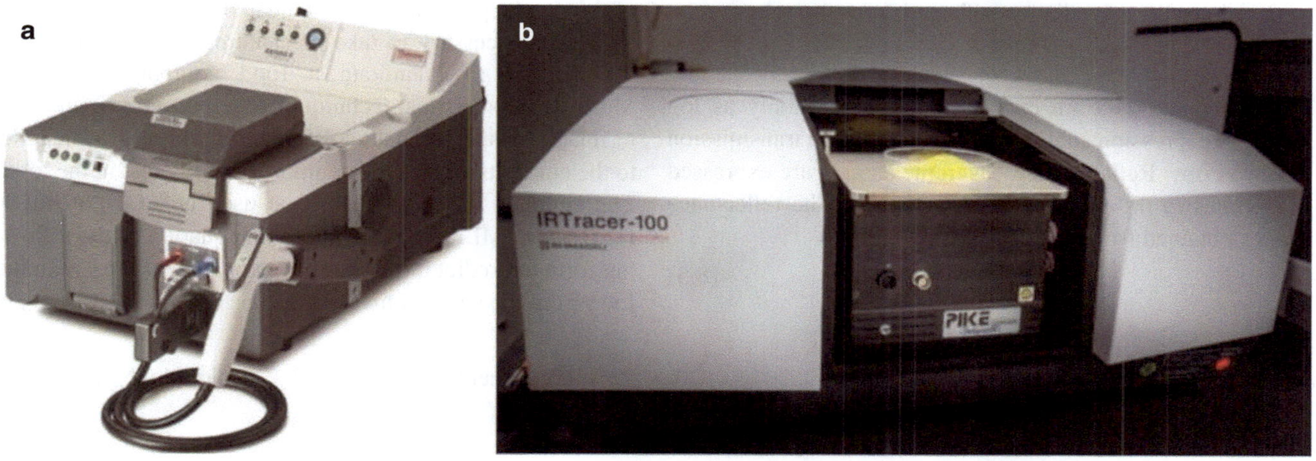

Fig. 8.8 Modern commercial near-IR instruments. (**a**) A Thermo Scientific Antaris II. (**b**) A Shimadzu IRTracer-100 equipped with a near-IR integrating sphere accessory from PIKE Technologies. (Photographs courtesy of Thermo Fisher Scientific and Shimadzu Scientific Instruments)

range extended to longer wavelengths are now available. Most commercial near-IR instruments use monochromators, rather than interferometers, although some commercial instruments now use FT technology. Monochromator-based instruments may be of the scanning type, in which a grating is used to disperse the radiation by wavelength, and the grating is rotated to impinge a single wavelength (or, more appropriately, a narrow band of wavelengths) onto a sample at any given time. Using this arrangement, it takes several seconds to collect a spectrum from a sample over the entire near-IR region. Some rapid scanning instruments impinge light over the entire near-IR region onto the sample. The reflected or transmitted light is then directed onto a fixed grating that disperses the light by wavelength and focuses it onto a multichannel array detector that measures all wavelengths simultaneously. These instruments can obtain a spectrum from a sample in less than 1 s.

Instruments dedicated to specific applications can use optical interference filters to select 6–20 discrete wavelengths that can be impinged on the sample. The filters are selected to obtain wavelengths that are known to be absorbed by the sample constituents. The instrument inserts a filter one at a time into the light beam to direct individual wavelengths of radiation onto the sample.

Either **reflection** or **transmission** measurements may be made in near-IR spectroscopy, depending on the sample type. In the reflection mode, which is used primarily for solid or granular samples, it is desirable to measure only the diffusely reflected radiation that contains information about the sample. In many instruments, this is accomplished by positioning the detectors at a 45° angle with respect to the incoming IR beam so that the specularly reflected radiation is not measured (Fig. 8.9a). Other instruments use an integrating sphere, which is a gold-coated metallic sphere with detectors mounted inside (Fig. 8.9b). The sphere collects the diffusely

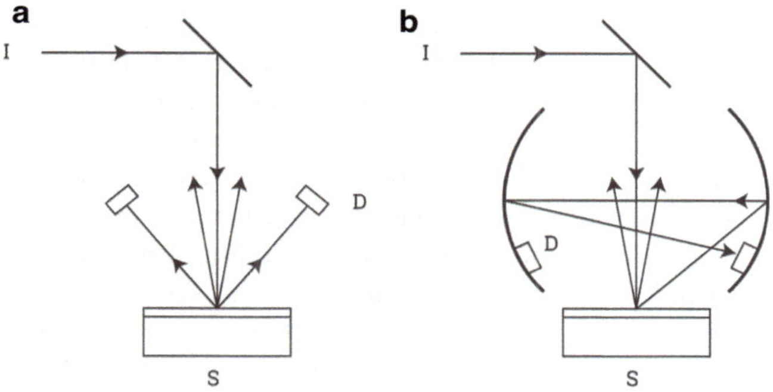

Fig. 8.9 Typical instrument geometries for measuring diffuse reflectance from solid food samples. Radiation from the monochromator (I) is directed by a mirror onto the sample (S). Diffusely reflected radiation is directly measured by the detectors (D) placed at a 45° angle to the incident beam (**a**) or is collected by an integrating sphere and focused onto the detectors (**b**). In both cases, the specularly reflected radiation is not measured

reflected radiation coming at various angles from the sample and focuses it onto the detectors. The specular component escapes from the sphere through the same port by which the incident beam enters and strikes the sample.

Samples are often prepared by tightly packing the food into a cell against a quartz window, providing a smooth, uniform surface from which reflection can occur. Quartz does not absorb in the near-IR region. At each wavelength, the intensity of light reflected from the sample is compared to that reflected from a non-absorbing reference, such as a ceramic or fluorocarbon material. Reflectance (R) is calculated by the following formula:

$$R = I / I_0 \tag{8.4}$$

where

I = intensity of radiation reflected from the sample at a given wavelength

I_0 = intensity of radiation reflected from the reference at the same wavelength

Reflectance data are expressed most commonly as log (1/*R*), an expression analogous to absorbance in transmission spectroscopy. Reflectance measurements also are expressed sometimes as differences, or derivatives, of the reflectance values obtained from adjacent wavelengths:

$$\left(\log R_{\lambda 2} - \log R_{\lambda 1}\right) \tag{8.5}$$

or

$$\left(2\log R_{\lambda 2} - \log R_{\lambda 1} - \log R_{\lambda 3}\right) \tag{8.6}$$

These derivative values are measures of the changes in the slope of the spectrum, where λ_1, λ_2, and λ_3 are adjacent wavelengths typically separated by 5–20 nm, with the higher numbers representing longer wavelengths.

Transmission measurements also can be made in the near-IR region, which is usually the method of choice for liquid samples. A liquid is placed in a quartz cuvette, and the absorbance is measured at the wavelengths of interest. Transmission measurements also can be taken from solid samples, but generally only in the 700–1100-nm range. In this wavelength region, the absorption bands are higher overtones that are very weak, allowing the radiation to penetrate through several millimeters of a solid sample. The use of transmission measurements can minimize the degree of sample preparation needed. Since the IR beam passes through the entire sample, the need for a smooth, homogeneous sample surface is reduced.

Near-IR energy can be transmitted through a fiber optic cable some distance from the monochromator or interferometer, allowing reflection or transmission measurements to be made remotely from the instrument. This is very useful for taking measurements in a processing plant environment. Commercial probes are available that can be inserted directly into bulk granular materials, or inserted into a pipe carrying a liquid.

As with mid-IR, near-IR imaging instruments are also now commercially available. These instruments use an array detector so that a digital image of a food sample can be obtained at various wavelengths or a spectrum can be obtained from a single pixel in a digital image. This technique is often referred to as hyperspectral imaging. It holds much potential for evaluating sample heterogeneity or for identifying small features or contaminants on an intact food sample.

8.4.3 Quantitative Methods Using Infrared Spectroscopy

Infrared instruments can be calibrated to measure various constituents in food and agricultural commodities. Because of the overlapping nature of the near-IR absorption bands, it is usually necessary to take measurements at two or more wavelengths to quantitate a food component reliably. Multivariate statistical techniques (**chemometrics**) are used to relate the spectral data collected at multiple wavelengths to the concentration of the component of interest in the food. The simplest statistical technique used is **multiple linear regression** (MLR), which applies an equation of the following form to predict the amount of a constituent present in the food from the spectral measurements:

$$\%\ \text{constituent} = z + a\log(1/R_{\lambda 1)} + b\log\left(1/R_{\lambda 2}\right) + c\log\left(1/R_{\lambda 3}\right) + \ldots \tag{8.7}$$

where each term represents the spectral measurement at a different wavelength multiplied by a corresponding coefficient. Each coefficient and the intercept (*z*) are determined by multivariate regression analysis.

Absorbance or derivatized reflectance data also can be used in lieu of the log(1/*R*) format. The use of derivatized reflectance data has been found to provide improved results in some instances, particularly with samples that may not have uniform particle sizes. Other mathematical techniques also are available that can be applied to the reflectance data to correct for the effects of nonuniform particle size [52].

To **calibrate** an infrared instrument for quantitative measurement, a set of samples is obtained that represents the product to be measured and contains the component of interest over the expected range of interest. The samples are then analyzed using the conventional method of analysis (e.g., for protein analysis, use Kjeldahl or Dumas methods; Chap. 18, Sects. 18.2.1 and 18.2.2), and the infrared spectrum of each sample is collected. A computer-assisted MLR analysis is then performed to determine the combinations of wavelengths that best predict the concentration of the component of interest and the coefficients associated with each wavelength, as shown in Eq. (8.7). In chemometric techniques such as MLR, the wavelengths are chosen based on the statistical correlation with the component being measured. However, the results should always be inspected to ensure that the selected wavelengths make sense from a spectroscopic perspective. Each calibration also should be tested using a second set of independent samples. Then, if the calibration yields satisfactory results, it can be used for routine analysis.

When using MLR, it may sometimes be difficult to include enough wavelengths to adequately define the relationship between the spectral and composition data. Adding too many wavelengths may "overfit" the calibration so that it does not apply well to samples that were not part of the original set. This can occur because the responses of individual wavelengths are highly intercorrelated. To overcome this problem and to obtain robust predictions, multivariate calibration methods such as **partial least squares** (PLS) regression and **principal component regression** (PCR) can be used. PLS and PCR are often referred to as "data compression" techniques, since they take the spectral variation from the entire wavelength range and express most of that variation with a smaller number of variables that are not correlated. These variables are used to develop a regression equation. PLS and PCR often provide improved results compared to MLR because they can use information from the entire spectrum with less risk of "overfitting" the results. For this reason, these two techniques are now the most widely used methods for calibrating mid-IR and near-IR instruments for quantitative analysis. Readers interested in a more detailed explanation of these techniques should consult the references [53].

8.4.4 Qualitative Analysis by Infrared Spectroscopy

Infrared spectroscopy also can be used to classify a sample into one of two or more groups, rather than to provide quantitative measurements. Classification techniques, such as **principal component analysis** (PCA), **soft independent modeling of class analogy** (SIMCA), or discriminant analysis, can be used to compare the infrared spectrum of an unknown sample to the spectra of samples from different groups. The unknown sample is then classified into the group to which its spectrum is most similar. While this technique has historically been used in the chemical and pharmaceutical industries for raw material identification, it is becoming more widely used for food applications, such as the identification of orange juice samples from different sources [54], authentication of the source of olive oils [55], and discrimination of beef [56]. Readers interested in a more detailed explanation of these classification techniques should consult the references [57].

8.4.5 Applications of Near-IR Spectroscopy to Food Analysis

The theory and applications of near-IR spectroscopy to food analysis have been discussed elsewhere [58]. The technique is widely used throughout the grain, cereal product, and oilseed processing industries. Near-IR techniques using measurements from ground or whole grain samples have been adopted as approved methods of analysis by American Association of Cereal Chemists (AACC) International [59] for measuring protein in barley, oats, rye, triticale, wheat, and wheat flour as well as moisture, protein, and oil in soybeans. These approved methods describe the instruments available for taking these measurements, including a list of current manufacturers with contact information in Method 39–30, as well as the proper techniques for preparing samples and calibrating the instruments. Near-IR instruments are now used by the official grain inspection agencies in the United States and Canada for measuring protein, moisture, and oil in cereals and oilseeds. Further specific examples of food applications of near-IR spectroscopy to foods are given in Table 8.2, excluding the well-known capability of near-IR spectroscopy in quality control and macronutrient analysis (moisture, carbohydrates, proteins, and fats).

If a substance absorbs in the near-IR region and is present at a level of a few tenths of a percent or greater, it has the potential to be measured by this technique. The primary advantage of near-IR spectroscopy is that once the instrument has been calibrated, several constituents in a sample can be measured rapidly (from 30 sec to 2 min) and simultaneously. To measure multiple constituents, a calibration equation for each constituent is stored in the instrument's memory. Measurements are taken at all wavelengths needed by the calibrations, and each equation then is solved to predict the constituents of interest. No sample weighing is required, and no hazardous reagents are used or chemical waste is generated. It also is adaptable for **online measurement systems** [60]. Disadvantages include the high initial cost of the instrumentation, which may require a large sample load to justify the expenditure, and the fact that specific calibrations may need to be developed for each product measured. Also, the results produced by the instrument can be no better than the data used to calibrate it, which makes careful analysis of the calibration samples of the highest importance.

Online and inline IR spectroscopy utilization in various industries provides valuable information and practical applications. Online implementation (i.e., real-time monitoring of moisture and temperature of dough in the baking industry as it moves through the baking process to prevent over- or under-baking, or moisture, starch, and protein contents of grains such as corn, rice, and wheat in the grain industry to optimize grain storage conditions better and preserve grain quality), involves incorporating the technique into a continuous monitoring system, enabling the analysis of samples in real-time during processing. By integrating infrared sensors into the processing equipment, manufacturers can continuously monitor these key parameters and make immediate adjustments to optimize product quality and consistency. Online IR spectrometers are especially beneficial in industries for which quick analysis and immediate feedback are critical requirements. Conversely, the inline application of infrared spectroscopy involves directly integrating the method into a production line. Inline instruments are specifically designed to measure samples without the necessity of manual sampling or preparation. For example, in the dairy industry, inline infrared spectroscopy can be used to monitor milk composition and detect any abnormalities or adulteration during processing. It provides rapid feedback on parameters such as milk fat, protein, lactose, and total solids, ensuring the quality and integrity of dairy products. Similarly, inline infrared spectroscopy can be employed in the beverage industry to monitor the sugar content and other quality parameters of juices, soft drinks, or alcoholic beverages. This capability enables rapid analysis of large sample volumes, providing immediate results that facilitate swift decision-making and process optimization. Inline infrared spectroscopy finds practical use in industries such as petroleum refining, polymer manufacturing, and pharmaceutical production, where the real-time monitoring of vital parameters and rapid quality control play a pivotal role. The utilization of infrared spectroscopy both online and inline provides substantial benefits, such as decreased analysis duration, enhanced process effectiveness, improved product excellence, and cost reduction. These methods enable industries to optimize their procedures, minimize wastage, ensure adherence to regulations, and deliver products of consistent high quality to customers.

8.5 RAMAN SPECTROSCOPY

8.5.1 Principles

Raman spectroscopy is a vibrational spectroscopic technique that is complementary to IR measurements. When a photon of light collides with a molecule, the collision can result in the photon being scattered. Molecules in the sample get excited and reach an unstable virtual energy state when they interact with the incident light, as shown in Fig. 8.10. However, this transition to a high-energy state in the molecule is a short-lived process, and most of the molecules relax back to their initial low energy level, resulting in the scattered photon having the same energy as the incident light. This is called elastic scattering process or **Rayleigh scattering**. However, a few molecules relax to a higher vibrational state with a change in the vibrational and rotational energy of the molecule, causing a shift in the wavelength of the scattered radiation. This is called the "inelastic scattering process" or **Raman scattering**. For Raman scattering to occur, a molecule must undergo a change in polarizability of the electron cloud of the molecule, but it does not need to undergo a change in dipole moment. Thus, Raman can observe symmetrical vibrations that cannot be detected by IR spectroscopy. Raman is complementary to IR spectroscopy, in that some vibrations are only Raman active, some are only IR active, and some are both.

During Raman scattering, scattered photons (approximately 1 in 10^7 photons) will shift to a longer wavelength (lower frequency), and this shift in frequency is called a **Stokes lines** or a Stokes shift. If the final energetic state is lower than the initial state, then scattered photons will shift to shorter wavelengths (higher frequency), and this is called an **anti-Stokes line** or an "anti-Stokes shift." Intensities of the Stokes lines are higher than those of the anti-Stokes lines, and, therefore, the Stokes lines are usually recorded as Raman spectra.

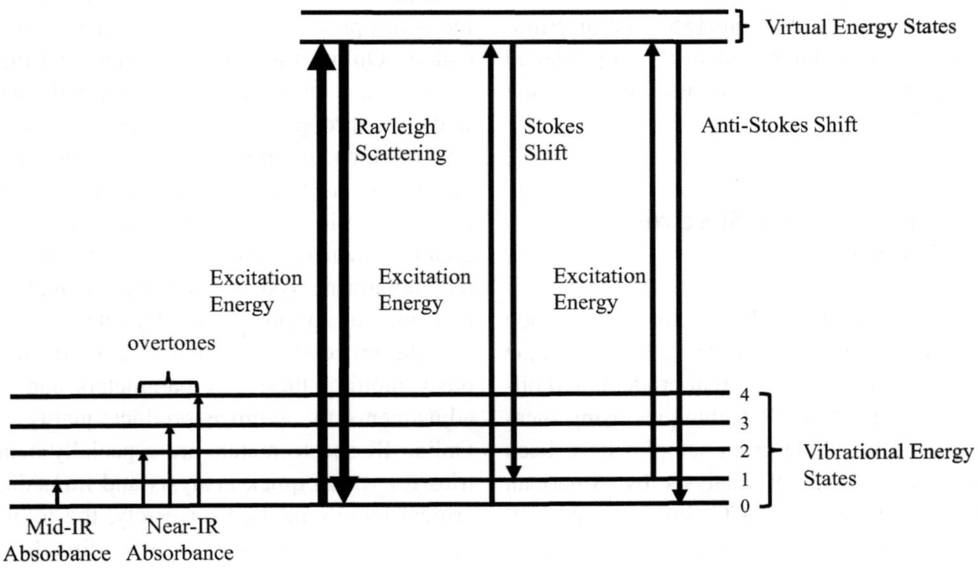

Fig. 8.10 An energy-level diagram of mid-IR, near-IR, Rayleigh, and Raman scattering. (Adapted from Boyaci et al. [89])

A typical Raman spectrum includes scattering intensity (photons per second) on the *y*-axis versus either increasing wavelength (nanometers) or Raman shift (centimeter^{-1}) on the *x*-axis. Each band in a Raman spectrum corresponds to chemical bonds and/or functional groups of the molecule, as illustrated in a representative Raman spectrum of selected foods in Fig. 8.11. Characteristic Raman intensities of different molecular bonds in various major sample/food matrices are summarized in Table 8.3. Similar to IR, a fingerprint of a molecule can be acquired by employing chemometric tools previously described in Sect. 8.4.3, and both qualitative and quantitative analyses can be generated since the intensity of the Raman bands is proportional to the concentration of the analyte.

Raman spectroscopy is positioned as an attractive technique because (1) it requires little or no sample preparation; (2) water and alcohols are weak Raman scatterers, allowing measurement of aqueous samples without any special accessories or sample preparation; and (3) it allows measurements through common transparent containers such as glass, quartz, and plastic, thus eliminating the need to open containers to analyze the contents. For food applications of Raman spectroscopy, the challenge of sample fluorescence may be a limiting factor. In general, the intensity of the Raman scatter is proportional to $1/\lambda^4$, so shorter excitation wavelengths deliver a much stronger Raman signal. The caveat is that when using short excitation wavelengths, **fluorescence** is more likely to occur under these conditions.

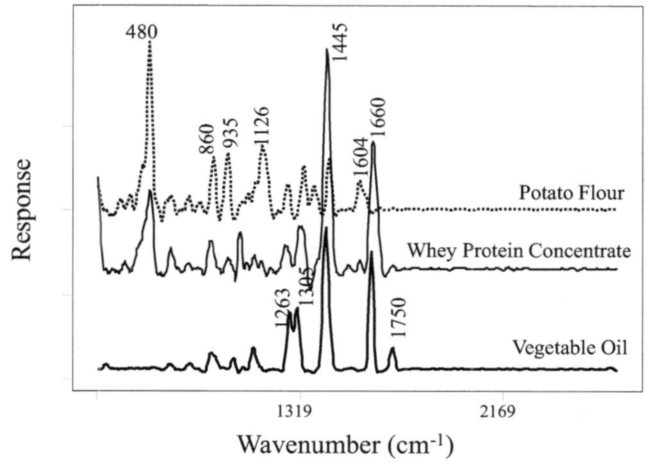

Fig. 8.11 Raman spectrum of vegetable oil, whey protein concentrate, and potato flour, measured by handheld Raman spectrometer equipped with a 1064 nm excitation laser. Raman shift in wavenumber is plotted on the x-axis, with intensity on the y-axis

8.5.2 Instrumentation

Raman spectrometers are based on **dispersive** and **Fourier Transform** technologies (Fig. 8.12). Each technique has unique advantages, and each is ideally suited to specific types of analyses.

A Raman dispersive system is composed of a laser source, a sample, a dispersing element (diffraction grating), a detector, and a computer. During a typical Raman spectrum collection process, a laser source gives a coherent beam of monochromatic light that is focused on the sample. The scattered light is passed through the notch filter that rejects the Rayleigh-scattered light, resulting in an important gain in sensitivity. Raman-scattered photons enter the monochromator, separated by wavelength, and are collected by a detector that records the intensity of the Raman signal at each wavelength. Lasers are typically used as the radiation source, as the strength of the Raman signal is proportional to the intensity of the incident light. Using the lasers as a radiation source to generate the Raman scattering has been a crucial development in Raman instrumentation. Typical wavelengths are 780 nm, 633 nm, 532 nm, and 473 nm. Raman detectors are frequently photodiode arrays (PDAs) or charge-coupled device (CCD) cameras. CCD detectors are extremely sensitive to light and contain thousands of picture elements (pixels) that acquire the whole spectrum at once in less than a second. CCD detectors allow the use of very low laser power, to prevent thermal or photochemical destruction of the sample.

FT–Raman spectrometers commonly use a near-IR laser, usually at 1030 or 1064 nm. Using lasers with excitation wavelengths in the near-IR region almost completely eliminates fluorescence; however, the Raman scattering intensity is weak. FT–Raman uses sensitive, single-element near-IR detectors such as **InGaAs** or liquid nitrogen-cooled germanium **(Ge)** detectors. An **interferometer** converts the Raman signal into an interferogram, permitting the detector to collect the entire spectrum simultaneously. Application of the Fourier transform algorithm to the **interferogram** converts the results into a conventional Raman spectrum. Besides the removal of the fluorescence interference, FT–Raman spectroscopy provides excellent wave number accuracy as a result of the internal interferometer calibration by the built-in helium–neon laser.

8.5.3 Surface-Enhanced Raman Scattering (SERS)

As mentioned earlier, one of the drawbacks of traditional Raman spectroscopy is the very low level of Raman sig-

Table 8.3 Characteristic Raman intensities of different molecular bonds in various major sample/food matrices

Food components	*Band position (cm^{-1})*	*Assignment*
Lipids	2885	C–H stretch (νCH_3)
	2855	C–H stretch (νCH_2)
	1750–1700	C=O stretching of esters v(C=O)
	1655	Double bond stretching ν(C=C)
	1660–1600	*cis*-double bond C=C stretching
	1442	Scissoring mode of methylene $\delta(CH_2)$
	1440–1390	CH_2 scissoring
	1302	In-phase methylene twist
	1270–1250	=C–H symmetric rock (*cis*)
	1213	Antisymmetric phosphoryl stretching corresponding to the vibrations of phospholipids and fatty acid chains
	1125	Aliphatic in-phase stretch ν(C–C)
	1080	Aliphatic stretch ν(C–C)
	1068	Aliphatic out-of-phase stretch ν(C–C)
	974	*Trans* RHC=CHR
Proteins	1645–1685 and 1200–1350	The secondary and tertiary structure of proteins, i.e., amide I and amide III bands
	1554	Tryptophan; CH_2 scissoring vibration
	1446	CH_2 bending modes of aliphatic amino acids
	1439	C − O stretching of aspartic and glutamic acids
	1334	Adenine
	1300–1270	α-Globular and α-fibrous helix formations
	1265	δ(C–H) bending at the *cis*-double bond in R–HC=CH–R
	1000	Phenylalanine
	937	α-Helical proteins, ν(C–CN) vibrations of Lys and glutamic acid
	879	ν(C–CN) coupled vibrations of Glu (849 cm^{-1}) + ala (852 cm^{-1}) + Val (850 cm^{-1}) + Lys (845 cm^{-1})
	850 and 826	Tyrosine doublet
	829	Out-of-plane γ(COO) modes of lactic acid (830 cm^{-1}), Phe (821 and 835 cm^{-1}), Leu (836 cm^{-1}), and Val (825 cm^{-1})
	766–740	Ring breathing vibration of tryptophan
	687 and 676	Melamine
	563–720	Amino acid side chains
Carbohydrates	1391 and 905	Amylopectin
	1260	CH_2 twisting mode of carbohydrates
	1253 and 850	Amylose
	1200	C–O–H stretching and –OH twisting
	1120–1064	CO stretching and CC and COH deformation
	950–870	COC deformation
	865	The vibration of CH
	824	The vibration of C(1)H, CH_2
	705	Stretching of CO and CCO, OCO bending
Water	3200–3080	OH stretching band
Alcohols	2980, 2890, and 2935	C–C stretching of ethanol
	1455, 1085, and 1045	CH_2 bending, $-CH_3$ rocking, and C–O stretching of ethanol
	880	C=C stretching vibration of ethanol

nals. One of the most popular techniques used to overcome this problem is surface-enhanced Raman scattering (SERS). Nanostructures on a metallic surface (typically gold and silver) can tremendously enhance the Raman signal of sample molecules on the order of 10^4–10^{11}, allowing detection in the ppb or single-molecule level [61]. Simple mechanisms of traditional Raman and SERS are compared in Fig. 8.13. Variations in the magnitude of the signal enhancement depend on the morphology of the particle (roughened surfaces can provide much more enhancements compared to flat (smooth) metal surfaces), the localization of the molecule in the system, and the selection of the excitation wavelength and light polarization [62].

The SERS phenomena result from an enhanced electromagnetic field produced at the surface of the metal and a

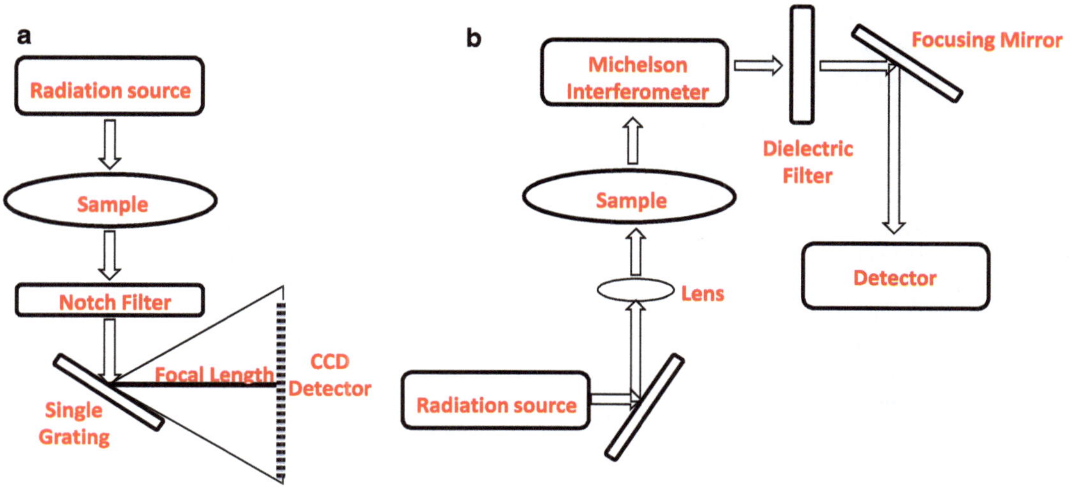

Fig. 8.12 Basic illustrations of the (**a**) dispersive Raman and (**b**) FT–Raman instrumentation. (Adapted from Das and Agrawal [90])

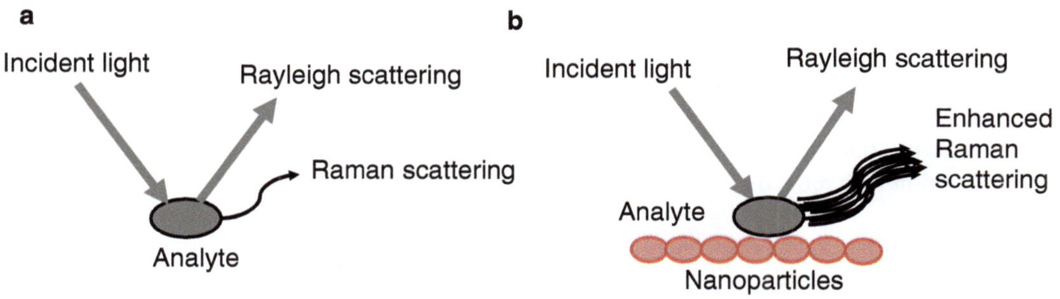

Fig. 8.13 Comparison of (**a**) traditional Raman and (**b**) the SERS mechanism. (Adapted from Zheng and He [91])

chemical enhancement due to a charge-transfer interaction between the metal and adsorbed molecules. When the wavelength of the incident light is close to a surface plasmon resonance (collective excitation of conductive electrons in small metallic structures), molecules adsorbed or in close proximity to the surface experience an exceptionally large electromagnetic field. In addition, the electronic transitions of many charge-transfer complexes between a metal surface and the molecule result in resonance enhancement.

The enormous signal enhancement achieved by SERS has positioned it as a very promising analytical tool for the biochemical, biomedical, and pharmaceutical fields. Food applications have been directed to food safety for the detection of foodborne pathogens as an alternative to current microorganism diagnostic tools, providing the possibility of developing portable pathogen sensors for on-site food inspection [61]. Other applications of SERS include the detection of food contaminants (pesticides and antibiotic residues) and adulteration (melamine, illegal food dyes, and mycotoxins) [61].

8.5.4 Applications of Raman Spectroscopy

Raman spectroscopy is a tool that is gaining attention for its applications in quality control, ingredient identification, and/or adulteration detection. Advances in Raman instrumentation have allowed this technique to be employed for qualitative and quantitative analyses of food and agricultural products, thus paving the way for exciting applications (Table 8.4). Features such as its ability to perform rapid analysis, be nondestructive, and be compatible with aqueous solutions make it a versatile tool for inline or online analysis. Furthermore, the characteristic fingerprint spectra of molecules allow one to identify food components in complex matrices, detect contaminants, provide geographical origin traceability, and screen for economically motivated adulteration. Raman spectroscopy has shown great potential in collecting biochemical information within intact cells, tissues, and even plants without extraction or using any other labeling agents, allowing to discriminate between transgenic and normal crops in crop breeding. Raman spectroscopy has been applied by the oil

Table 8.4 Application of Raman spectroscopy in foods

Application	*Sample/food matrix*	*Type of Raman spectroscopy*	*References*
Quality	Freshness of various fish	Micro-Raman	[63]
	Chemical changes during frozen storage of argentine red shrimp	FT–Raman	[64]
	Texture of frozen/thawed Chinese yellow bulls	Portable Raman	[65]
	Characterization of the lipid oxidation process in beef	Portable Raman	[66]
	Quality of ovine: Semimembranosus	Handheld Raman	[67]
	Microstructure of bread dough	Confocal Raman microscopy	[68]
Adulteration	Offal meat in beef burgers	Dispersive Raman	[69]
	Horsemeat in beef	Micro-Raman	[70]
	Atlantic salmon fish speciation	Portable Raman	[71]
	Melamine in milk powder	Handheld Raman	[72]
	Foreign fats and oils in cream and yogurt	Raman microscopy	[73]
	Whey addition to milk powder	FT–Raman	[74]
	Olive oil	Portable Raman	[75]
	Honey	Raman	[76]
Authentication	Wine: Cultivar, geographical origin, and vintage	FT–Raman	[77]
	Geographical origin of rice varieties	Raman	[78]
Aflatoxin contamination	Maize	Raman	[79]

industry to screen for economically motivated adulteration, especially the olive oil industry has evaluated the fingerprinting characteristics of Raman for detection of extra-virgin olive oil adulteration with refined olive oil or cheaper vegetable oils [80].

Raman spectroscopy also has allowed the nondestructive monitoring of quality in processed foods, such as meeting the wine industry requirements for process control from the start of grape ripening until bottling of the finished product [81], the kinetics of chemical transformation in the fermentation process from milk to yogurt [82], examining the oxidative degradation of edible oils [83], authenticating essential oils [84], detecting impurities such as erucic acid in canola oil [85], and monitoring the formation of histamine due to spoilage in fish [86]. Owing to its outstanding material identification capabilities, providing spatial information of vibrations from complex biological samples has rendered Raman spectroscopy a very accurate tool for analyzing highly complex matrices. Confocal dispersive Raman microscopy is an excellent technique for analyzing multi-thin-layer polymer materials, providing spatial resolution for detecting layers less than 1 μm in thickness and permitting rigorous examination of the multilayer films by cross-sectioning to identify manufacturing defects.

8.6 HANDHELD AND PORTABLE TECHNOLOGY

Vibrational spectroscopic techniques are exceptionally well suited to be used as portable or handheld instruments. Their simplicity, speed, selectivity, and ability to operate without sample preparation make them ideal to be used outside the laboratory for process monitoring in challenging environments. A single spectrometer can be used to verify the identity of bulk materials, check for contamination, process control, and confirm final product specifications. Field-based instruments have to tolerate harsh conditions, maintain reliability and accuracy, be easy to operate, be battery-powered, and be lightweight with an ergonomic design and intuitive user interface. Sample accessories must be robust, with limited or no sample preparation required, and capable of fast analysis (Fig. 8.14).

The advantages of portable mid-IR spectrometers (compared to near-IR) are their higher fundamental signal, allowing the detection of low analyte levels, and their unique fingerprinting capabilities. However, if a sample contains water, its strong mid-IR absorption may swamp useful information. Near-IR allows analysis without any sample preparation. The most commonly used sampling mode for solids is diffuse reflectance, whereas transflectance (combined transmission and reflectance) and transmission are suitable configurations for liquid analysis. Although a near-IR signal is 10–1000 times lower than mid-IR bands, the lower absorptivities permit the near-IR beam to penetrate deeper into the sample, resulting in a more representative analysis. Advantages of portable Raman spectroscopy instrumentation for field deployment include little or no sample preparation, noncontact and nondestructive capabilities, and a relatively weak Raman response of water to allow measuring aqueous solutions. Near-IR and Raman analyses permit measurements through glass and plastic films.

Portable and handheld vibrational spectrometers are attractive fingerprinting techniques for various applications, including food, pharmaceuticals, petrochemicals, and law enforcement. Ellis et al. [87] have summarized the applications of various commercially available spectroscopy-based

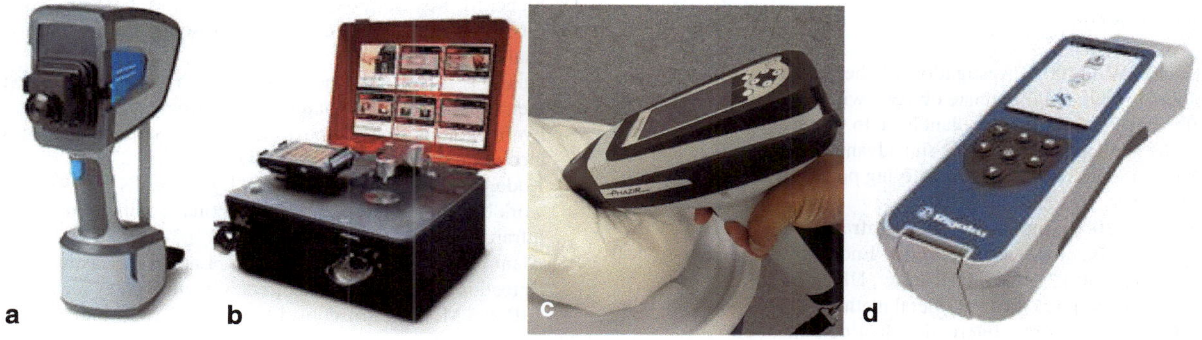

Fig. 8.14 Portable/handheld vibrational spectrometers commercially available. (**a**) An Agilent 4300 Handheld FTIR, (**b**) an Agilent 4500 Portable FTIR, (**c**) a Thermo microPhazir Handheld Near-IR, and (**d**) a Progeny™ 1064 nm handheld Raman analyzer by Rigaku Analytical Devices

approaches for rapid on-site food fraud analysis. In addition, Dos Santos et al. [88] have reviewed the applications of portable near-IR spectrometers in the agro-food industry.

8.7 SUMMARY

IR spectroscopy measures the absorption of radiation in the near-IR (λ = 0.8–2.5 µm) or mid-IR (λ = 2.5–15 µm) regions by molecules in food or other substances. IR radiation is absorbed as molecules change their vibrational energy levels. Mid-IR spectroscopy is especially useful for qualitative analysis, such as identifying specific functional groups present in a substance. Different functional groups absorb different frequencies of radiation, allowing the groups to be identified from the spectrum of a sample. Quantitative analysis can also be achieved by mid-IR spectroscopy, with milk analysis being a major application. Near-IR spectroscopy is used most extensively for quantitative applications, using either transmission or diffuse reflection measurements that can be taken directly from solid foods. Using multivariate statistical techniques, near-IR instruments can be calibrated to measure the amounts of various constituents of a food sample based on the amount of IR radiation absorbed at specific wavelengths. Mid-IR, near-IR, and Raman spectroscopy requires much less time to perform quantitative analysis than do many conventional wet chemical or chromatographic techniques.

8.8 STUDY QUESTIONS

1. Describe the factors that affect the frequency of vibration of a molecular functional group and, thus, the frequencies of the radiation that it absorbs. Also, explain how a molecule's fundamental absorption and overtone absorptions are related.
2. Describe the essential components of an FT mid-IR spectrometer and their function, and compare the operation of the FT instrument to a dispersive instrument. What advantages do FT instruments have over dispersive IR spectrophotometers?
3. Describe the similarities and differences between mid-infrared spectroscopy and Raman spectroscopy.
4. Of the three antioxidants, butylated hydroxytoluene (BHT), butylated hydroxyanisole (BHA), and propyl gallate, which would you expect to have a strong IR absorption band in the 1700–1750-cm^{-1} spectral region? Look up these compounds in a reference book if you are uncertain of their structure.
5. Describe the two ways in which radiation is reflected from a solid or granular material. Which type of reflected radiation is useful for making quantitative measurements on solid samples by near-IR spectroscopy? How are near-IR reflectance instruments designed to select for the desired component of reflected radiation?
6. Describe the steps involved in calibrating a near-IR reflectance instrument to measure the protein content of wheat flour. Why is it usually necessary to make measurements at more than one wavelength?

Acknowledgments We express our sincere gratitude to Dr. Ismail Hakki Boyaci and Dr. Siyu Yao for their invaluable contribution to this book chapter. Their expertise and insights have greatly enhanced the quality and depth of the content. We are truly thankful for their dedication and collaborative efforts throughout the writing process. The chapter authors and the book editors also acknowledge the contributions of Dr. Randy Wehling of the University of Nebraska, who was the sole author of this chapter in the first three editions and a contributor in the past two editions of this textbook.

REFERENCES

1. Herschel W (1800) Investigation of the powers of the prismatic colours to heat and illuminate objects; with remarks, that prove the different refrangibility of radiant heat. to which is added, an inquiry into the method of viewing the sun advantageously, with telescopes of large apertures and high magnifying powers. Philos Trans R Soc London Ser A 90:255-283
2. Coates J (2000) Interpretation of Infrared Spectra, A Practical Approach. In: Meyers RA (Ed.) Encyclopedia of Analytical Chemistry, Vol. 12, Wiley, Chichester, UK, pp. 10 815
3. AOAC International (2016) Official methods of analysis, 20th edn., Method 972.12. AOAC Internationals, Rockville, MD
4. Lahner BS (1996) Evaluation of Aegys MI 600 Fourier transform infrared milk analyzer for analysis of fat, protein, lactose, and solids nonfat: a compilation of eight independent studies. J AOAC Int 79: 1388
5. Santos PM, Pereira-Filho ER, Rodriguez-Saona LE (2013) Rapid detection and quantification of milk adulteration using infrared microspectroscopy and chemometrics analysis. Food Chem. 138:19
6. Jawaid S, Talpur FN, Sherazi STH, Nizamani SM, Khaskheli AA (2013) Rapid detection of melamine adulteration in dairy milk by SB-ATR-Fourier transform infrared spectroscopy. Food Chem 141: 3066
7. Behkami S, Zain SM, Gholami M, and Khir MFA (2019) Classification of cow milk using artificial neural network developed from the spectral data of single- and three-detector spectrophotometers. Food Chem 294:309
8. Wang YW, Ding W, Kou LP, Li L, Wang C, Jurick WM (2015) A non-destructive method to assess freshness of raw bovine milk using FT-NIR spectroscopy. J Food Sci Technol 52:5305
9. Subramanian A, Alvarez VB, Harper WJ, Rodriguez-Saona LE (2011) Monitoring amino acids, organic acids, and ripening changes in Cheddar cheese using Fourier-transform infrared spectroscopy. Int Dairy J 21: 434
10. Alkhalf MI, Mirghani ME (2017) Detection of formaldehyde in cheese using FTIR spectroscopy. Int Food Res J 24:496
11. Schmitt S, Garrigues S, de la Guardia M (2014) Determination of the mineral composition of foods by infrared spectroscopy: A review of a green alternative. Crit Rev Anal Chem 44: 186
12. Koca N, Kocaoglu-Vurma NA, Harper WJ, Rodriguez-Saona LE (2010) Application of temperature-controlled attenuated total reflectance-mid-infrared (ATR-MIR) spectroscopy for rapid estimation of butter adulteration. Food Chem 121:778
13. García-Miguel E, Meza-Márquez OG, Osorio-Revilla G, Téllez-Medina DI, Jiménez-Martínez C, Cornejo-Mazón M, Hernández-Martínez DM, Gallardo-Velazquez T (2018) Detection of Cyanuric Acid and Melamine in Infant Formula Powders by Mid-FTIR Spectroscopy and Multivariate Analysis. J Food Qual 2018: 1
14. Üçüncüoglu D, Küçük VA (2019) A multivariate data approach for FTIR-ATR monitoring of virgin olive oils: Lipid structure during autoxidation. OCL - Oilseeds fats, Crop. Lipids 26: 42.
15. Didham M, Truong VK, Chapman J, Cozzolino D (2020) Sensing the Addition of Vegetable Oils to Olive Oil: The Ability of UV–VIS and MIR Spectroscopy Coupled with Chemometric Analysis. Food Anal Methods 13: 601
16. Abramovic B, Jajic I, Abramovic B, Cosic J, Juric V (2007) Detection of deoxynivalenol in wheat by Fourier transform infrared spectroscopy. Acta Chim Slov 54: 859
17. Öner T, Thiam P, Kos G, Krska R, Schwenker F, Mizaikoff B (2019) Machine learning algorithms for the automated classification of contaminated maize at regulatory limits via infrared attenuated total reflection spectroscopy. World Mycotoxin J 12:113
18. Sieger M, Kos G, Sulyok M, Godejohann M, Krska R, Mizaikoff B (2017) Portable infrared laser spectroscopy for on-site mycotoxin analysis. Sci Rep 7: 1
19. Huang L, Zhao JW, Chen QS, Zhang YH (2014) Nondestructive measurement of total volatile basic nitrogen (TVBN) in pork meat by integrating near infrared spectroscopy, computer vision and electronic nose techniques. Food Chemy 145:228
20. Kodogiannis VS, Kontogianni E, Lygouras JN (2014) Neural network based identification of meat spoilage using Fourier-transform infrared spectra. J Food Eng 142:118
21. Alamprese C, Casale M, Sinelli N, Lanteri S, Casiraghi E (2013) Detection of minced beef adulteration with Turkey meat by UV-Vis, NIR and MIR spectroscopy. LWT- Food Sci Technol 53:225
22. Grunert T, Stephan R, Ehling-Schulz M, Johler S (2016) Fourier Transform Infrared Spectroscopy enables rapid differentiation of fresh and frozen/thawed chicken. Food Control 60:361
23. Alamprese C, Casiraghi E (2015) Application of FT-NIR and FT-IR spectroscopy to fish fillet authentication. LWT Food Sci. Technol 63: 720
24. Vidal NP, Goicoechea E, Manzanos MJ, Guillén MD (2014) Fourier transform infrared spectroscopy as a tool to study farmed and wild sea bass lipid composition. J Sci Food Agric 94: 1340
25. Se KW, Ghoshal SK, Wahab RA, Ibrahim RKR, Lani MN (2018) A simple approach for rapid detection and quantification of adulterants in stingless bees (Heterotrigona itama) honey. Food Res Int 105: 453
26. Gok S, Severcan M, Goormaghtigh E, Kandemir I, Severcan F (2015) Differentiation of Anatolian honey samples from different botanical origins by ATR-FTIR spectroscopy using multivariate analysis. Food Chem 170: 234
27. Link JV, Lemes ALG, Marquetti I, dos Santos Scholz MB, Bona E (2014) Geographical and genotypic classification of arabica coffee using Fourier transform infrared spectroscopy and radial-basis function networks. Chemom Intell Lab Syst 135: 150
28. Tavares KM, Pereira RGFA, Nunes CA, Pinheiro ACM, Rodarte MP (2012) Espectroscopia no infravermelho médio eanálise sensorial aplicada à detecção de adulteração de café torrado por adição de cascas de café. Quim Nova 35: 1164
29. Craig AP, Franca AS, Oliveira LS. Irudayaraj J. Ileleji K (2014) Application of elastic net and infrared spectroscopy in the discrimination between defective and non-defective roasted coffees. Talanta 128:393
30. Geana EI, Ciucure CT, Apetrei C, Artem V (2019) Application of spectroscopic UV-Vis and FT-IR screening techniques coupled with multivariate statistical analysis for red wine authentication: Varietal and vintage year discrimination. Molecules 24: 4166
31. Syahariza ZA, Che Man YB, Selamat J, Bakar J (2005) Detection of lard adulteration in cake formulation by Fourier transform infared (FTIR) spectroscopy. Food Chem 92:365
32. Lohumi S, Joshi R, Kandpal LM, Lee H, Kim MS, Cho H, Mo C, Seo Y-W, Rahman A, Cho B-K (2017) Quantitative analysis of Sudan dye adulteration in paprika powder using ftir spectroscopy. Food Addit Contam Part A 34:678
33. Petrakis EA, Polissiou MG (2017) Assessing saffron (Crocus Sativus L.) adulteration with plant-derived adulterants by diffuse reflectance infrared fourier transform spectroscopy coupled with chemometrics. Talanta 162:558
34. Wielogorska E, Chevallier O, Black C, Galvin-King P, Delêtre M, Kelleher CT, Haughey SA, Elliott CT (2018) Development of a comprehensive analytical platform for the detection and quantitation of food fraud using a biomarker approach. The oregano adulteration case study. Food Chem 239:32
35. Altieri G, Genovese F, Admane N, Di Renzo GC (2016) Online measure of donkey's milk properties by near infrared spectrometry. LWT Food Sci. Technol 69:348

36. Wang L, Sun DW, Pu H, Cheng JH (2017) Quality analysis, classification, and authentication of liquid foods by near-infrared spectroscopy: A review of recent research developments. Crit Rev Food Sci Nutr 57: 1524
37. Shen F, Zhao T, Jiang X, Liu X, Fang Y, Liu Q, Qiuhui Hu, Xingquan L, Yong F, Qing L, Qiuhui Hu, Xingquan L, Liu X (2019) On-line detection of toxigenic fungal infection in wheat by visible/near infrared spectroscopy. LWT Food Sci Technol 109: 2164
38. Freitag S, Sulyok M, Logan N, Elliott CT, Krska R (2022) The potential and applicability of infrared spectroscopic methods for the rapid screening and routine analysis of mycotoxins in food crops. Compr Rev Food Sci Food Safy 21:5199
39. Huang FR, Li YP, Wu J, Dong J, Wang Y (2016) Identification of repeatedly frozen meat based on near-infrared spectroscopy combined with self-organizing competitive neural networks. Int J Food Prop 19:1007
40. Xu Y, Kutsanedzie FYH, Sun H, Wang MX, Chen QS, Guo ZM, Wu JZ (2018) Rapid pseudomonas species identification from chicken by integrating colorimetric sensors with near-infrared spectroscopy. Food Anal Methods 11: 1199
41. Coronel-Reyes J, Ramirez-Morales I, Fernandez-Blanco E, Rivero D, Pazos A (2018) Determination of egg storage time at room temperature using a low-cost NIR spectrometer and machine learning techniques. Comput Electron Agric 145:1
42. Gomes V, Fernandes A, Martins-Lopes P, Pereira L, Faia AM, Melo-Pinto P (2017) Characterization of neural network generalization in the determination of pH and anthocyanin content of wine grape in new vintages and varieties. Food Chem 218:40
43. Martelo-Vidal MJ, Vazquez M (2015) Application of artificial neural networks coupled to UV-VIS-NIR spectroscopy for the rapid quantification of wine compounds in aqueous mixtures. Cyta – Journal of Food 13:32
44. Sauvage L, Frank D, Stearne J, Millikan MB (2002) Trace metal studies of selected white wines: an alternative approach. Anal Chim Acta 458, 223
45. Guan BB, Zhao JW, Jin HJ, Lin H (2014) The qualitative and quantitative analysis of aromatic vinegar produced during different seasons by near infrared spectroscopy. Anal Methods 6:9634
46. Viejo CG, Fuentes S, Torrico D, Howell K, Dunshea FR (2018) Assessment of beer quality based on foamability and chemical composition using computer vision algorithms, near infrared spectroscopy and machine learning algorithms. J Sci Food Agric 98:618
47. Pan WX, Zhao JW, Chen QS, Zhang DL (2015) Simultaneous and rapid measurement of main compositions in black tea infusion using a developed spectroscopy system combined with multivariate calibration. Food Anal Method 8:749
48. Segtnan VH, Kita A, Mielnik M, Jorgensen K, Knutsen SH (2006) Screening of acrylamide contents in potato crisps using process variable settings and near infrared spectroscopy. Mol Nutr Food Res 50:811
49. Baltacıoglu H, Bayındırlı A, Severcan M, Severcan F (2015) Effect of thermal treatment on secondary structure and conformational change of mushroom polyphenol oxidase (PPO) as food quality related enzyme: A FTIR study. Food Chem 187:263
50. Johnson JB, Walsh KB, Naiker M, Ameer K (2023) The Use of Infrared Spectroscopy for the Quantification of Bioactive Compounds in Food: A Review. Molecules 28: 3215
51. Yazici A, Tiryaki GY, Ayvaz H (2020) Determination of pesticide residual levels in strawberry (Fragaria) by near-infrared spectroscopy. J. Sci. Food Agric 100: 1980
52. Martens H, Naes T (2001) Multivariate calibration by data compression, Ch. 4. In: Williams PC, Norris KH (eds) Near infrared technology in the agricultural and food industries, 2nd edn. American Association of Cereal Chemists, St. Paul, MN, p 75
53. Bjorsvik HR, Martens H (2008) Data analysis: calibration of NIR instruments by PLS Regression, Ch. 8. In: Burns DA, Ciurczak EW (eds) Handbook of near-infrared analysis, 3rd edn. CRC Press, Boca Raton, FL
54. Goodner KL, Manthey JA (2005) Differentiating orange juices using Fourier transform infrared spectroscopy (FT-IR). P Fl St Hortic Soc 118:410
55. Rohman A, Man YBC, Yusof FM (2014) The use of FTIR spectroscopy and chemometrics for rapid authentication of extra virgin olive oil. J Am Oil Chem Soc 91:207
56. Meza-Márquez OG, Gallardo-Velázquez T, Osorio-Revilla G (2010) Application of mid-infrared spectroscopy with multivariate analysis and soft independent modeling of class analogies (SIMCA) for the detection of adulterants in minced beef. Meat Sci 86:511
57. Lavine BK (2000) Clustering and classification of analytical data. Robert A. Meyers (Ed). In "Encyclopedia of Analytical Chemistry". John Wiley & Sons Ltd, Chichester
58. Ozaki Y, McClure WF, Christy AA (2006) Near-infrared spectroscopy in food science and technology. Wiley, Hoboken, NJ.
59. AACC International (2010) Approved methods of analysis, 11th edn. (On-line). The American Association of Cereal Chemists, St. Paul, MN
60. Psotka J (2001) Challenges of making accurate on-line near-infrared measurements. Cereal Foods World 46:568
61. Craig AP, Franca AS, Irudayaraj J (2013) Surface-enhanced Raman spectroscopy applied to food safety. Annu Rew Food Sci Technol 4:369
62. Vlckova B, Pavel I, Sladkova M, Siskova K, Slouf M (2007) Single molecule SERS: perspectives of analytical applications. J Mol Struct, 834:42
63. Velioğlu, HM, Temiz HT, Boyaci IH (2015) Differentiation of fresh and frozen-thawed fish samples using Raman spectroscopy coupled with chemometric analysis. Food Chem 172: 283
64. Careche M, Herrero A, Carmona P (2002) Raman analysis of white spots appearing in the shell of Argentine red shrimp (Pleoticus muelleri) during frozen storage. J Food Sci 67:2892
65. Chen Q, Zhang Y, Guo Y, Cheng Y, Qian H, Yao W, Xie Y, Ozaki Y (2020) Non-destructive prediction of texture of frozen/thaw raw beef by Raman spectroscopy. J Food Eng 266:109693
66. Chen Q, Xie Y, Xi J, Guo Y, Qian HE, Cheng Y, Chen Y, Yao W (2018) Characterization of lipid oxidation process of beef during repeated freeze-thaw by electron spin resonance technology and Raman spectroscopy. Food Chem 243:58
67. Fowler SM, Schmidt H, van de Ven R, Wynn P, Hopkins DL (2015) Predicting meat quality traits of ovine m. semimembranosus, both fresh and following freezing and thawing, using a hand held Raman spectroscopic device. Meat Sci 108:138
68. Huen J, Weikusat C, Bayer-Giraldi M, Weikusat I, Ringer L, Lösche K (2014) Confocal Raman microscopy of frozen bread dough. J Cereal Sci 60:555
69. Zhao M, Downey G, O'Donnell CP (2015) Dispersive Raman spectroscopy and multivariate data analysis to detect offal adulteration of thawed beefburgers. J Agric Food Chem 63:1433
70. Boyacı IH, Temiz HT, Uysal RS, Velioğlu HM, Yadegari RJ, Rishkan MM (2014) A novel method for discrimination of beef and horsemeat using Raman spectroscopy. Food Chem 148: 37
71. Chen Z, Wu T, Xiang C, Xu X, Tian X (2019) Rapid identification of rainbow trout adulteration in Atlantic salmon by Raman spectroscopy combined with machine learning. Molecules 24:2851
72. Karunathilaka SR, Yakes BJ, He K, Brückner L, Mossoba MM (2018) First use of handheld Raman spectroscopic devices and onboard chemometric analysis for the detection of milk powder adulteration. Food Control 92:137
73. Karacaglar NNY, Bulat T, Boyaci IH, Topcu A (2019) Raman spectroscopy coupled with chemometric methods for the discrimination of foreign fats and oils in cream and yogurt. J Food Drug Anal 27:101

74. Almeida MR, Oliveira KDS, Stephani R, de Oliveira LFC (2011) Fourier-transform Raman analysis of milk powder: a potential method for rapid quality screening. J Raman Spectrosc 42:1548
75. Li Y, Fang T, Zhu S, Huang F, Chen Z, Wang Y (2018) Detection of olive oil adulteration with waste cooking oil via Raman spectroscopy combined with iPLS and SiPLS. Spectrochim Acta Part A 189:37
76. Li S, Shan Y, Zhu X, Zhang X, Ling G (2012) Detection of honey adulteration by high fructose corn syrup and maltose syrup using Raman spectroscopy. J Food Compos Anal 28:69
77. Dos Santos CAT, Páscoa RN, Sarraguça MC, Porto PA, Cerdeira AL, González-Sáiz JM, Pizarro C, Lopes JA (2017) Merging vibrational spectroscopic data for wine classification according to the geographic origin. Food Res Int 102:504
78. Feng X, Zhang Q, Cong P, Zhu Z (2013) Preliminary study on classification of rice and detection of paraffin in the adulterated samples by Raman spectroscopy combined with multivariate analysis. Talanta 115:548
79. Lee KM, Herrman TJ, Nansen C, Yun U (2013) Application of Raman spectroscopy for qualitative and quantitative detection of fumonisins in ground maize samples. J Reg Sci, 1:1
80. Barros IH, Paixão LS, Nascimento MH, Lacerda Jr V, Filgueiras PR, Romão W (2021) Use of portable Raman spectroscopy in the quality control of extra virgin olive oil and adulterated compound oils. Vib Spectrosc 116:103299
81. Martin C, Bruneel JL, Guyon F, Medina B, Jourdes M, Teissedre PL, Guillaume F (2015) Raman spectroscopy of white wines. Food Chem 181:235
82. Rodriguez R, Vargas S, Estevez M, Quintanilla F, Trejo-Lopez A, Hernández-Martínez AR (2013) Use of Raman spectroscopy to determine the kinetics of chemical transformation in yogurt production. Vib Spectrosc 68:133
83. Muik B, Lendl B, Molina-Díaz A, Ayora-Cañada MJ (2005) Direct monitoring of lipid oxidation in edible oils by Fourier transform Raman spectroscopy. Chem Phys Lipids 134:173
84. Daferera DJ, Tarantilis PA, Polissiou MG (2002) Characterization of essential oils from Lamiaceae species by Fourier transform Raman spectroscopy. J Agric Food Chem 50:5503
85. Velioglu SD, Temiz HT, Ercioglu E, Velioglu HM, Topcu A, Boyaci IH (2017) Use of Raman spectroscopy for determining erucic acid content in canola oil. Food Chem 221:87
86. Janči T, Valinger D, Kljusurić JG, Mikac L, Vidaček S, Ivanda,M (2017) Determination of histamine in fish by Surface Enhanced Raman Spectroscopy using silver colloid SERS substrates. Food Chem 224:48
87. Ellis DI, Muhamadali H, Haughey SA, Elliott CT, Goodacre R (2015) Point-and-shoot: rapid quantitative detection methods for on-site food fraud analysis – moving out of the laboratory and into the food supply chain. Anal Methods 7:9401.
88. Dos Santos CAT, Lopo M, Páscoa RN, Lopes JA (2013) A review on the applications of portable near-infrared spectrometers in the agro-food industry. Appl Spectrosc 67:1215
89. Boyaci IH, Temiz HT, Geniş HE, Soykut EA, Yazgan NN, Güven B, Uysal RS, Bozkurt AG, İlaslan K, Torun O, Şeker FCD (2015) Dispersive and FT-Raman spectroscopic methods in food analysis. RSC Adv. 5:56606
90. Das RS, Agrawal YK (2011) Raman spectroscopy: recent advancements, techniques and applications. Vib Spectrosc 57:163
91. Zheng J, He L (2014) Surface-Enhanced Raman Spectroscopy for the Chemical Analysis of Food. Compr Rev Food Scie Food Saf 13:317

Atomic Absorption Spectroscopy, Atomic Emission Spectroscopy, and Inductively Coupled Plasma Mass Spectrometry

9

Vincent Yeung, Dennis D. Miller, and Michael A. Rutzke

Contents

9.1 INTRODUCTION

As the names imply, atomic absorption spectroscopy (AAS) quantifies the **absorption** of electromagnetic radiation by well-separated neutral atoms, while atomic emission spectroscopy (AES) quantifies the **emission** of radiation from well-separated neutral atoms in the excited state. AAS and AES accurately measure individual elements even in the presence of other food components because atomic absorption and emission spectra are unique for each element. An example of emission spectra is shown in Fig. 9.1.

V. Yeung (✉)
Animal Science Department, California Polytechnic State University, San Luis Obispo, CA, USA
e-mail: ckyeung@calpoly.edu

D. D. Miller
Department of Food Science, Cornell University, Ithaca, NY, USA
e-mail: ddm2@cornell.edu

M. A. Rutzke
School of Integrative Plant Science, Cornell University, Ithaca, NY, USA
e-mail: mar9@cornell.edu

B. P. Ismail, S. S. Nielsen (eds.), *Nielsen's Food Analysis*, Food Science Text Series,
https://doi.org/10.1007/978-3-031-50643-7_9

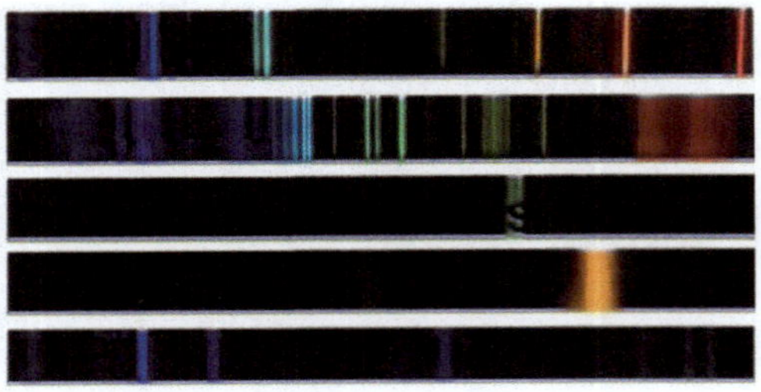

Fig. 9.1 Emission spectra of selected elements. (Photographed by Michael B. Rutzke)

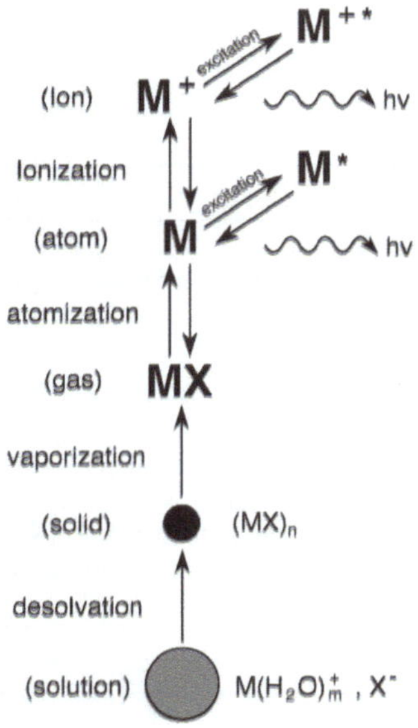

Fig. 9.2 A schematic representation of the atomization of an element in a flame or plasma. The *large circle* at the bottom represents a tiny droplet of a solution containing the element (M) as part of a compound. (From Boss and Fredeen [4], used with permission. Courtesy of the PerkinElmer Corporation, Shelton, CT)

In food matrices, most elements are present as compounds or complexes and therefore would have to go through the process of atomization (Fig. 9.2) before spectroscopic analysis. Table 9.1 shows the common sources of energy for atomization, including their temperature ranges. The development of inductively coupled plasma (ICP) in the 1960s [1, 2] as an energy source has allowed rapid measurement of multiple elements in a single sample. When combined with mass spectrometry (MS), ICP-MS is a powerful tool capable of measuring trace minerals with extremely low detection limits. Moreover, ICP-MS offers the added advantage of being able to separate and quantify multiple isotopes of the same element. Table 8.2 in Chap. 8 shows a comparison of different spectroscopy methods commonly available for food analysis, including AAS and AES.

This chapter deals with the basic principles of atomic spectroscopy and provides an overview of the instrumentation available. A discussion of ICP-MS is also included. Readers interested in a more thorough treatment of the topic are referred to references [3–7].

Table 9.1 Methods and temperature ranges for atomization of analytes

Source of energy for atomization	*Approximate atomization temperature range (K)*	*Analytical method*
Flame	2000–3400	AAS, AES
Electrothermal	1500–3300	AAS (graphite furnace)
Inductively coupled argon plasma	6000–7000	ICP-OES, ICP-MS

9.2 ATOMIC ABSORPTION SPECTROSCOPY

AAS is an analytical method developed by Alan Walsh in 1952 [8] based on the absorption of ultraviolet-visible (UV-Vis) radiation (200–800 nm) by free well-separated neutral atoms in the gaseous state. AAS can only measure one element at a time, and because it must operate in the air, AAS cannot measure nonmetals such as phosphorus as it requires vacuum UV radiation for excitation [5]. Despite being largely replaced by the more powerful ICP-based methods for commercial applications, AAS is still widely taught in university laboratories to help students understand better the principles of spectroscopy. Two types of atomization are commonly used in AAS: **flame atomization** and **electrothermal (graphite furnace) atomization**.

9.2.1 Principles of Flame Atomic Absorption Spectroscopy

A schematic diagram of a flame atomic absorption spectrometer is shown in Fig. 9.3. In flame AAS, the sample solution is **nebulized** (dispersed into tiny droplets), mixed with fuel and an oxidant, and burned in a flame produced by oxidation of the fuel. Atoms and ions are formed within the hottest portion of the flame as the sample solution goes through the process of desolvation, vaporization, atomization, and ionization (Fig. 9.2). Atoms and ions of the same element absorb radiation of different wavelengths and produce different spectra. Atomic absorption spectrometers are designed to measure *atomic* absorption, not *ionic* absorption. Therefore, it is desirable to choose a flame temperature hot enough to maximize atomization, while not being too hot to cause ionization.

Once the sample is atomized in the flame, the quantity of the analyte element is measured by determining the attenuation (decrease in intensity at the detector) of a beam of radiation passing through the flame, due to atomic absorption of incident radiation by the analyte element. For the measurement to be specific for the analyte element, the radiation source ideally should emit radiation of the exact discreet wavelengths that only the analyte element is capable of absorbing. This can be accomplished by using lamps with

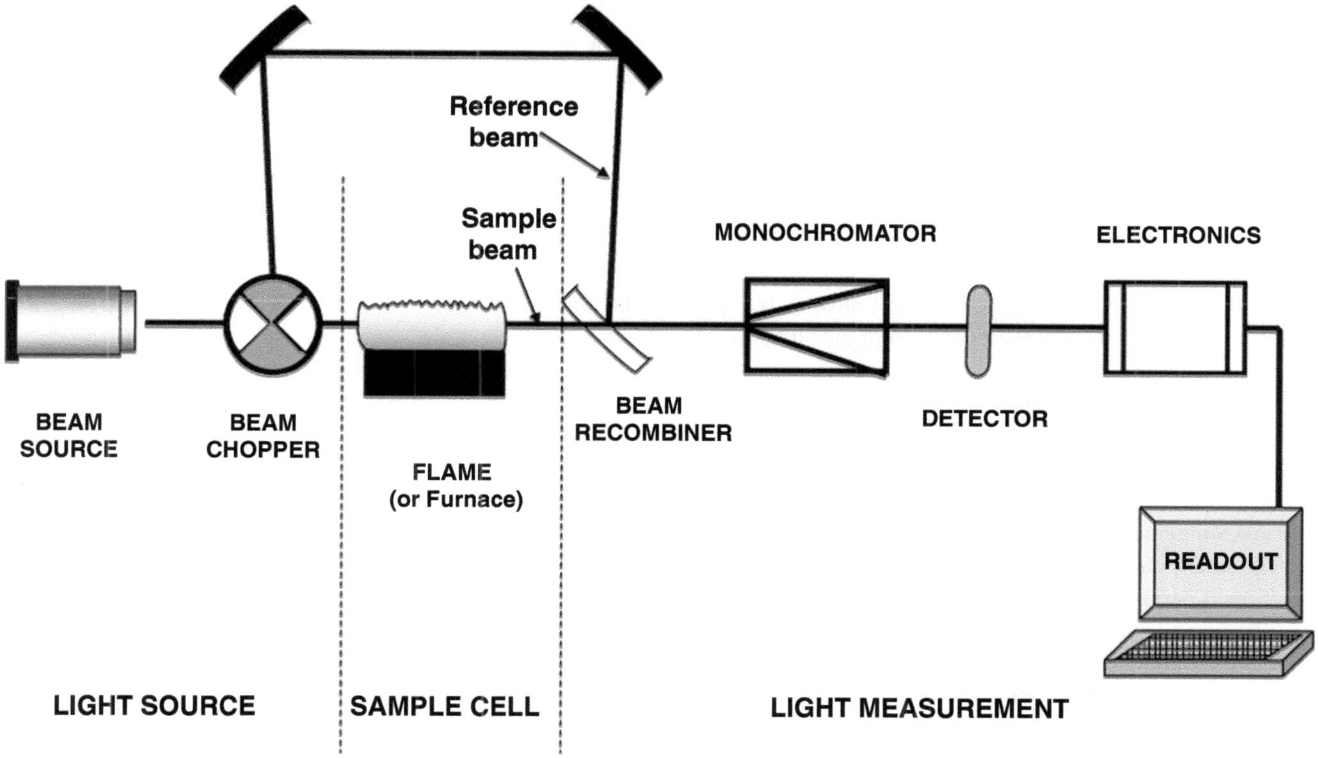

Fig. 9.3 Schematic representation of a double-beam atomic absorption spectrophotometer. (Adapted from Beaty and Kerber [3])

cathodes fabricated with the element to be determined. Thus, the beam of radiation emitted from the lamp is the emission spectrum of the element. The spectral line of interest is isolated by passing the beam through a monochromator so that only radiation of a very narrow bandwidth (0.002 nm) reaches the detector. Usually, one of the strongest spectral lines is chosen; for example, for sodium, the monochromator is set to pass radiation with a wavelength of 589.0 nm.

The amount of radiation absorbed (absorbance) is directly related to the concentration of the analyte element in the flame as governed by **Beer's law** (see Chap. 7, Sect. 7.2.1). Properly constructed calibration curves using pure standards are essential for accurate quantitative measurements (see Chap. 7, Sect. 7.2.4).

9.2.2 Principles of Electrothermal (Graphite Furnace) Atomic Absorption Spectroscopy

Electrothermal AAS is identical to flame AAS except for the atomization process. In electrothermal AAS, the sample is heated electrically in stages inside a graphite tube, commonly known as **graphite furnace**, to achieve atomization. The tube is aligned to the path of the radiation beam to be absorbed by the atomized sample, and absorbance is determined. Electrothermal AAS requires smaller sample sizes and offers lower detection limits. Disadvantages are the added expense of the graphite furnace, lower sample throughput, higher matrix interferences, and lower precision.

9.2.3 Instrumentation for Atomic Absorption Spectroscopy

Atomic absorption spectrometers, typically with a double-beam design (see Chap. 7, Sect. 7.2.7), consist of the following components (Fig. 9.3):

1. **Radiation source**, a hollow cathode lamp (HCL) or an electrode-less discharge lamp (EDL)
2. **Atomizer**, usually a nebulizer-burner system or a graphite furnace
3. **Monochromator**, usually a UV–Vis grating monochromator (200–800 nm)
4. **Detector**, a photomultiplier tube (PMT) or a solid-state detector (SSD)
5. **Readout device**, an analog or a digital readout

(Radiation source and atomizers will be further discussed in the following paragraphs. See Chap. 7, Sects. 7.2.6.2, 7.2.6.3, and 7.2.6.4 for a more detailed discussion of monochromators, detectors, and readout devices, respectively.)

9.2.3.1 Radiation Source

A **hollow cathode lamp** (HCL) consists of a hollow tube filled with argon or neon gas, an anode made of tungsten, and a cathode made of the metallic form of the element being measured (Fig. 9.4). When voltage is applied across the electrodes, the lamp emits radiation characteristic of the metal in the cathode. For example, if the cathode is made of iron, an iron spectrum will be emitted. As the radiation passes through the flame containing the atomized sample, only iron atoms (not atoms of other elements) will absorb this radiation because the emitted wavelengths from the HCL are specific for iron atoms. Of course, this means that it is necessary to use a different lamp for each element analyzed (there are a limited number of multielement lamps available that contain cathodes made of more than one element). HCLs for about 60 metallic elements are commercially available, suggesting that AAS may be used for the analysis of up to 60 elements.

The **beam chopper** is a disk with segments removed and positioned perpendicular to the light path between the HCL and the flame (Fig. 9.3). The disk rotates at a constant speed so that the light emitted from the HCL reaching the detector is either unimpeded or blocked at regular intervals, i.e., it is alternating. In contrast, emission from excited analyte atoms in the flame reaching the detector is continuous. The instrument electronics subtracts the continuous emission signal from the alternating emission signal, so only the signal from the attenuated HCL beam is recorded in the readout.

An **electrode-less discharge lamp** (EDL) contains no electrodes but a hollow glass vessel filled with an inert gas plus the element of interest. The discharge is produced by a high-frequency generator coil rather than an electric current [9]. EDLs are suitable for volatile elements such as arsenic, mercury, and cadmium.

9.2.3.2 Atomizers

Flame and **graphite furnace** atomizers are the two common types of atomizers used in AAS. When applicable, a **cold vapor technique** for mercury and a **hydride generation technique** for a few elements are used to enhance sensitivity.

The **flame atomizer** consists of a **nebulizer** and a **burner.** The nebulizer is designed to convert the sample solution into a fine mist or aerosol. This is accomplished by aspirating the sample through a capillary into a chamber through which oxidant and fuel are flowing. The chamber contains baffles that remove larger droplets, leaving a very fine mist. Only about 1% of the total sample is carried into the flame by the oxidant–fuel mixture. The larger droplets fall to the bottom of the mixing chamber and are collected as waste. The burner head contains a long, narrow slot that produces a flame that may be 5–10 cm in length. This gives a long path length that increases the sensitivity of the measurement.

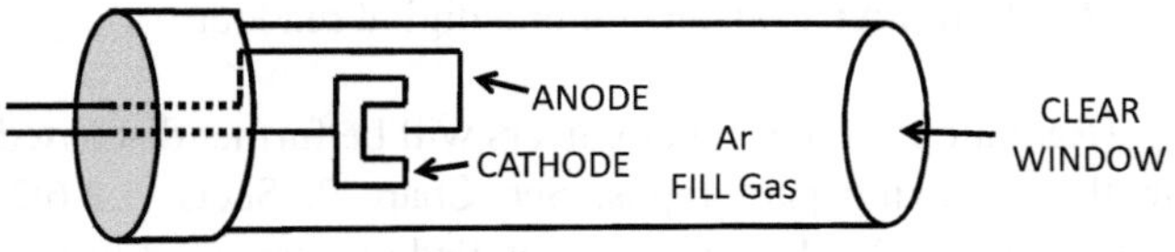

Fig. 9.4 Schematic representation of a hollow cathode lamp. (Adapted from Beaty and Kerber [3])

Flame characteristics may be manipulated by adjusting oxidant/fuel ratios and by choice of oxidant and fuel. **Air-acetylene** and **nitrous oxide-acetylene** are the most commonly used oxidant–fuel mixtures although other oxidants and fuels may be used for some elements. There are three types of flames:

1. **Stoichiometric**. This flame is produced from stoichiometric amounts (exact reaction ratios) of oxidant and fuel, so the fuel is completely burned and the oxidant is completely consumed. It is characterized by yellow fringes.
2. **Oxidizing**. This flame is produced from a fuel-lean (excess of oxygen) mixture. It is the hottest flame and has a clear blue appearance.
3. **Reducing.** This flame is produced from a fuel-rich mixture (excess of fuel compared to oxygen). It is a relatively cool flame and has a yellow color.

Analysts should follow manufacturer's guidelines or consult the literature for the proper type of flame for each element.

Flame atomizers have the advantage of being stable and easy to use. However, sensitivity is relatively low because much of the sample never reaches the flame and the residence time of the sample in the flame is short.

The **graphite furnace** is typically a cylindrical graphite tube connected to an electrical power supply. The sample is injected into the tube through an inlet using a microliter syringe with sample volumes ranging from 0.5 to 100 μL. During operation, the sample is confined to the small volume of the graphite tube and the system is flushed with an inert gas to prevent the tube from burning and to exclude air from the sample compartment. The tube is heated electrically in stages: First, the sample solvent is evaporated; then the sample is ashed; and finally, the temperature is rapidly increased to ~2000–3000 K to quickly vaporize and atomize the sample.

The **cold vapor technique** works only for mercury because mercury is the only mineral element that can exist as free atoms in the gaseous state at room temperature. Mercury compounds in a sample are first reduced to elemental mercury by the action of stannous chloride, a strong reducing agent. The elemental mercury is then carried in a stream of inert gas into an absorption cell without the need for further atomization. Cold vapor AAS is a popular technique for mercury analysis as it offers a much lower detection limit than ICP-OES. The **hydride generation technique** is limited to elements capable of forming volatile hydrides that include arsenic, lead, tin, bismuth, antimony, tellurium, germanium, and selenium. Samples containing these elements are reacted with sodium borohydride to generate volatile hydrides, which are carried into an absorption cell and decomposed by heat. Absorbance measurements with these two techniques are conducted in the same manner as with flame atomization, but sensitivity is greatly enhanced because there is very little sample loss [3].

9.2.4 Safety Precautions for Atomic Absorption Analysis

Standard operating procedures recommended by instrument manufacturers should be reviewed carefully, and safety precautions must be taken. The most commonly used fuel sources in flame AAS are mixtures of air/acetylene and nitrous oxide/acetylene. *ACETYLENE IS AN EXPLOSIVE GAS.* Proper ventilation must be in place before operation. The exhaust vent should be positioned directly above the burner to avoid the buildup of unburned fuel or any potentially hazardous toxic fumes. Flame atomic absorption spectrometers should never be left unattended while in operation.

9.2.5 Interferences in Atomic Absorption Spectroscopy

Two types of interferences are encountered in AAS: spectral and nonspectral. **Spectral interferences** are caused by the absorption of radiation by other elemental or molecular species at wavelengths that overlap with the spectral regions of the analyte present in the sample. **Nonspectral interferences** are caused by sample matrices and conditions that affect the atomization efficiency and/or the ionization of neutral atoms in the atomizer. General strategies for correcting interferences include using a deuterium lamp, Zeeman Splitting, and the Smith–Hieftje method. Please see reference [10] for more discussion.

9.2.5.1 Spectral Interferences

An element in the sample other than the element of interest may absorb at the wavelength of the spectral band being used. Such interference is rare because emission lines from the HCL are so narrow that only the element of interest is capable of absorbing the radiation in most cases. One example of when this problem does occur is with the interference of iron in zinc determinations. Zinc has a spectral line at 213.856 nm, which overlaps the iron line at 213.859 nm. The problem may be solved by choosing an alternative spectral line for measuring zinc or by narrowing the monochromator slit width.

The presence of alkaline earth oxide and hydroxide molecules may also lead to several specific spectral interferences. Spectra of calcium oxide and magnesium hydroxide will appear as **background absorption** for atomic absorption measurements of sodium and chromium, respectively [11]. These interferences are weak but must be taken into account when working with a complex sample matrix. If the flame is not hot enough to combust all of the organic molecules in the sample, serious background interferences can occur for elements absorbing at <250 nm [5].

9.2.5.2 Nonspectral Interferences

Transport interferences may occur when other components present in the sample matrix influence physical properties such as viscosity, surface tension, and vapor pressure of the sample solution, leading to differences in the rate of aspiration, nebulization, or transport between the sample solution and the standards during flame atomization. Such interferences often can be overcome by using the same solvent and by matching as closely as possible the physical properties of the sample solution and the standards. The standard addition protocol (see Chap. 7, Sect. 7.2.4) may also be used.

Matrix composition of the sample solution may affect the lateral migration of an analyte, resulting in **solute vaporization interferences**. For example, it has been observed in flame absorption and emission that alkaline earth metals are depressed by elevated levels of aluminum and phosphorus [12] and aluminum also suppresses the recovery of calcium [13]. **Chemical interferences** occur when an element forms thermally stable compounds, including the formation of metal oxides, that do not decompose during atomization. **Refractory metals** such as titanium and molybdenum may combine with oxygen to form stable oxides; metal oxide formation can be prevented by using higher-temperature nitrous oxide–acetylene flames (2900–3300 K) [5]. Also, phosphate in a sample matrix reacts with calcium to form calcium pyrophosphate which is not decomposed in the flame. This is particularly a problem for analyzing calcium in milk. A **releasing agent** such as lanthanum, which binds phosphate more strongly than calcium, may be added to the sample solution and the standards to free up calcium for atomization [13].

Ionization interferences are caused by ionization of analyte atoms in the flame and thereby reduce the concentration of neutral atoms for atomic absorption measurement. (Remember that atoms and ions of the same element absorb radiation of different wavelengths and produce different spectra.) Ionization increases with increasing flame temperature and normally is not a problem in cooler air–acetylene flames. It can be a problem in hotter nitrous oxide–acetylene flames, especially with elements that have ionization potentials of 7.5 eV or less (e.g., alkali metals). The ionization of atoms results in an equilibrium situation:

$$M \rightleftarrows M^{+} + e^{-} \quad (9.1)$$

Ionization interferences can be countered by spiking the sample solution and the standards with another **easily**

ionized element (EIE) such as potassium or cesium, known as an **ionization suppressant**, to create a pool of free electrons in the flame, which shifts the above equilibrium to the left and suppresses the ionization of the analyte atoms.

9.3 ATOMIC EMISSION SPECTROSCOPY

In contrast to AAS, the source of the measured radiation in AES is the excited atoms in the sample. Figure 9.5 shows a simplified diagram of an atomic emission spectrometer. Sufficient energy is first applied to the sample to excite atoms to higher energy levels. Emissions of wavelengths characteristic of individual elements are then measured when excited atoms fall back to the ground state or a lower energy state, which takes about 1×10^{-8} to 1×10^{-10} s [5]. The ratio of the number of excited atoms to ground-state atoms occurring in a flame or plasma is described by the Maxwell–Boltzmann equation (Chap. 6, Sect. 6.4.3), which predicts that a higher excitation temperature would lead to a substantial increase in the number of excited atoms.

Energy for excitation may be produced by **heat** (usually from a flame), **light** (from a laser), **electricity** (arcs or sparks), or **radio waves** (ICP). (*Note:* AES is also commonly called optical emission spectroscopy (OES), especially when combined with ICP. In this chapter, we will use ICP-OES rather than ICP-AES for our discussion, but the two terms are virtually interchangeable.)

The two most common forms of AES used in food analysis are **flame emission spectroscopy** and **inductively coupled plasma optical emission spectroscopy** (ICP-OES).

9.3.1 Principles of Flame Emission Spectroscopy

Flame emission spectrometers employ a nebulizer-burner system to atomize and excite the atoms of the elements being measured. The flame with the excited atoms serves as the radiation source, so an external source (i.e., the HCL) is not required. Otherwise, instrumentation for flame emission spectroscopy is essentially identical to that for AAS. Many modern atomic absorption spectrometers can also be operated as flame emission spectrometers.

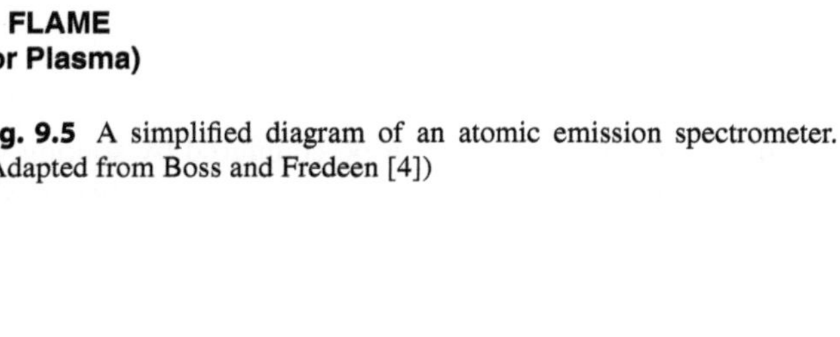

Fig. 9.5 A simplified diagram of an atomic emission spectrometer. (Adapted from Boss and Fredeen [4])

In some instruments, interference filters (instead of the more versatile monochromators typically found in absorption/emission spectrometers) are employed to isolate the spectral region of interest. **Flame photometers** are low-cost emission spectrometers equipped with interference filters and are specifically designed for the analysis of the alkali and alkaline earth metals in biological samples. Low flame temperatures are used so that only easily excited elements such as sodium, potassium, and calcium produce emissions. This results in a simpler spectrum and reduces interference from other elements present in the sample matrix.

9.3.2 Principles of Inductively Coupled Plasma Optical Emission Spectroscopy

The ICP concept, devised over a thirty-year period by Velmer Fassel and associates at Iowa State University, was explored as an atomization and excitation source for emission spectroscopy in the early 1960s [14]. (ICP was also explored independently by Stanley Greenfield's research group around the same time.) ICP-OES differs from flame emission spectroscopy in that it uses an argon plasma as the excitation source. A **plasma** is defined as a gaseous mixture containing significant concentrations of cations and electrons. Temperatures in argon plasmas could be as high as 10,000 K, with analyte excitation temperature typically ranging from 6000 to 7000 K.

The extremely high temperatures and the inert atmosphere of argon plasmas are ideal for the atomization, ionization, and excitation of the analyte atoms in the sample. The low oxygen content reduces the formation of oxides, which is sometimes a problem with flame methods. The nearly complete atomization of the sample also minimizes chemical interferences. The relatively uniform temperatures in plasmas (compared to nonuniform temperatures in flames) and long residence time give good linear responses over a wide concentration range (up to six orders of magnitude).

9.3.3 Instrumentation for Inductively Coupled Plasma Optical Emission Spectroscopy

ICP-OES instruments typically consist of the following components (Fig. 9.6):

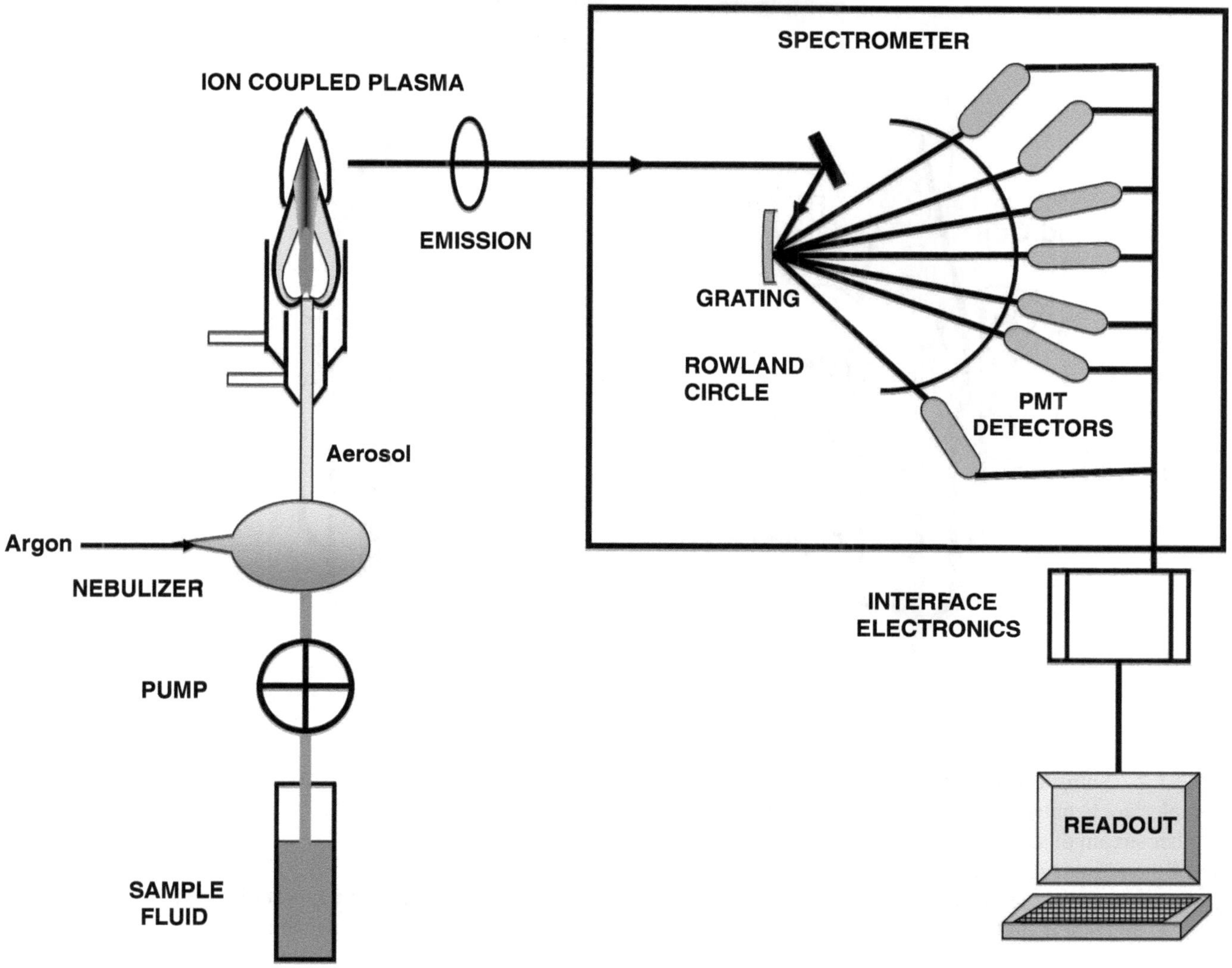

Fig. 9.6 Schematic of an inductively coupled plasma optical emission simultaneous spectrometer

1. **Argon plasma torch**
2. **Monochromator**, **polychromator**, or **echelle optical system**
3. **Detector(s)**, a single or multiple PMT(s) or solid-state array detector(s)
4. **Computer** for data collection and treatment

9.3.3.1 Argon Plasma Torch

9.3.3.1.1 Characteristics of an Argon Plasma Torch

The plasma torch (Fig. 9.7) consists of three concentric quartz tubes centered in a copper coil, called the **load coil**. During operation of the torch, a stream of argon gas flows through the outer tube and **radio frequency** (RF) power is applied to the load coil, creating a magnetic field oscillating at the frequency of the RF generator (usually 27 or 40 MHz). The plasma is started by ionizing argon atoms with an electric spark to form argon ions and electrons. The oscillating magnetic field couples with the argon ions and electrons, forcing them to flow in an annular (ring-shaped) path. Heating does not involve burning fuel to directly heat and atomize the sample, as is the case with flame AAS (argon is a noble gas and will not combust). Rather, heating is accomplished by transferring RF energy to free electrons and argon ions in a manner similar to the transfer of microwave energy to water in a microwave oven. These high-energy electrons in turn collide with argon atoms, generating even more electrons and argon ions and causing a rapid increase in temperature to approximately 10,000 K. The process continues until about 1% of the argon atoms are ionized. At this point, the plasma is very stable and self-sustaining for as long as the RF field is applied at constant power. The transfer of energy to a system through the use of electromotive forces generated by magnetic fields is known as **inductive coupling** [15], hence the name ICP.

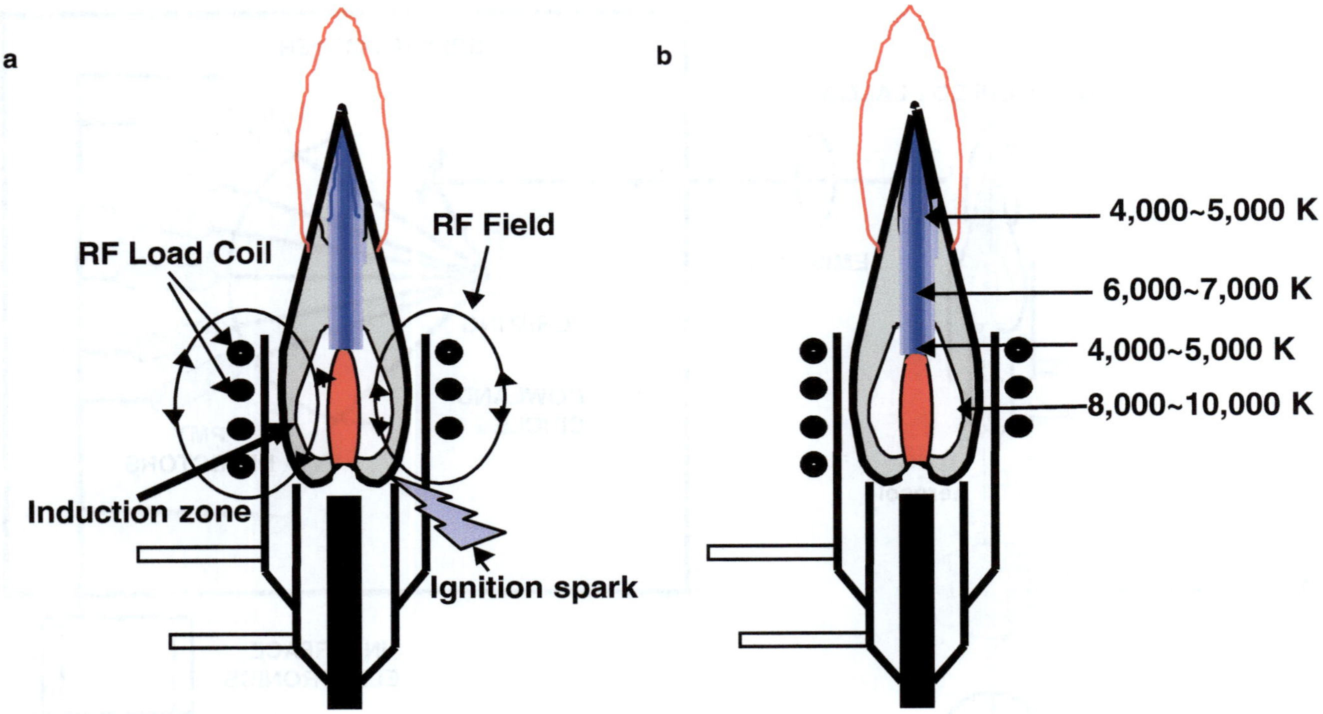

Fig. 9.7 The ICP plasma: (**a**) the process by which the plasma is formed and sustained and (**b**) the temperature distribution of the plasma

9.3.3.1.2 Sample Introduction and Analyte Excitation

Samples are nebulized and introduced as aerosols carried by another stream of argon gas in the inner tube inside the annulus of the plasma at the base of the RF load coil. The sample goes through the process of desolvation, vaporization, atomization, ionization, and excitation, as shown in Fig. 9.2. The exact mechanisms by which the analyte atoms and ions are excited in a plasma are not understood fully. Nevertheless, there is general agreement that excitation is dependent primarily on the number and temperature of the electrons in the plasma [16]. Presumably, when electrons are accelerated in a magnetic field, they acquire enough kinetic energy to excite analyte atoms and ions upon collision [16]. Exceptions to this mechanism include the excitation of neutral atoms of magnesium, copper, and a few other elements that are believed to be excited when an argon ion (Ar^+) extracts an electron from the analyte atom (M) to yield M^{+*} and Ar^0, where M is a generic abbreviation for a ground-state metal atom and M^{+*} is an excited metal ion. This mechanism is termed "charge transfer" [17, 18].

9.3.3.1.3 Radial and Axial Viewing

Emissions from ICP torches can be viewed either radially or axially. In the **radial view**, the optics are aligned perpendicular to the torch (Fig. 9.8a). In the **axial view**, the light is viewed by looking down the center of the torch (Fig. 9.8b). Axial viewing offers lower detection limits but is more prone to matrix interferences. Manufacturers of modern ICP-OES instruments have mostly combined the radial and axial configurations into single "dual-view" units, offering greater flexibility to the end user.

9.3.3.2 Detectors and Optical Systems

Older ICP-OES instruments were configured to focus emission lines from each of the analyte elements on separate PMTs arranged in a semicircle (the Rowland circle), as shown in Fig. 9.6. PMT-based instruments are relatively large and bulky, and while some are still in use today, they have been mostly replaced by modern instruments equipped with an **echelle optical system** and a solid-state array detector (Fig. 9.9), capable of measuring continuous emission spectra over a wide wavelength range.

The echelle optical system employs two dispersing components in series, a prism and a diffraction grating. The prism first disperses the radiation from the plasma torch without any wavelength overlap (in the x-direction). The radiation then strikes a low-density ruled grating (about 53 groves per mm). This further separates the radiation in a direction perpendicular to the direction of radiation dispersed by the prism (in the y-direction), producing a two-dimensional spectrum with a wavelength range of 166–840 nm. When the radiation passing through the echelle optical system is focused onto the solid-state array detector, electrons are liberated proportionally to the intensity of the incident radiation and trapped in the silicon-based, light-sensitive elements called pixels for signal processing. ICP-

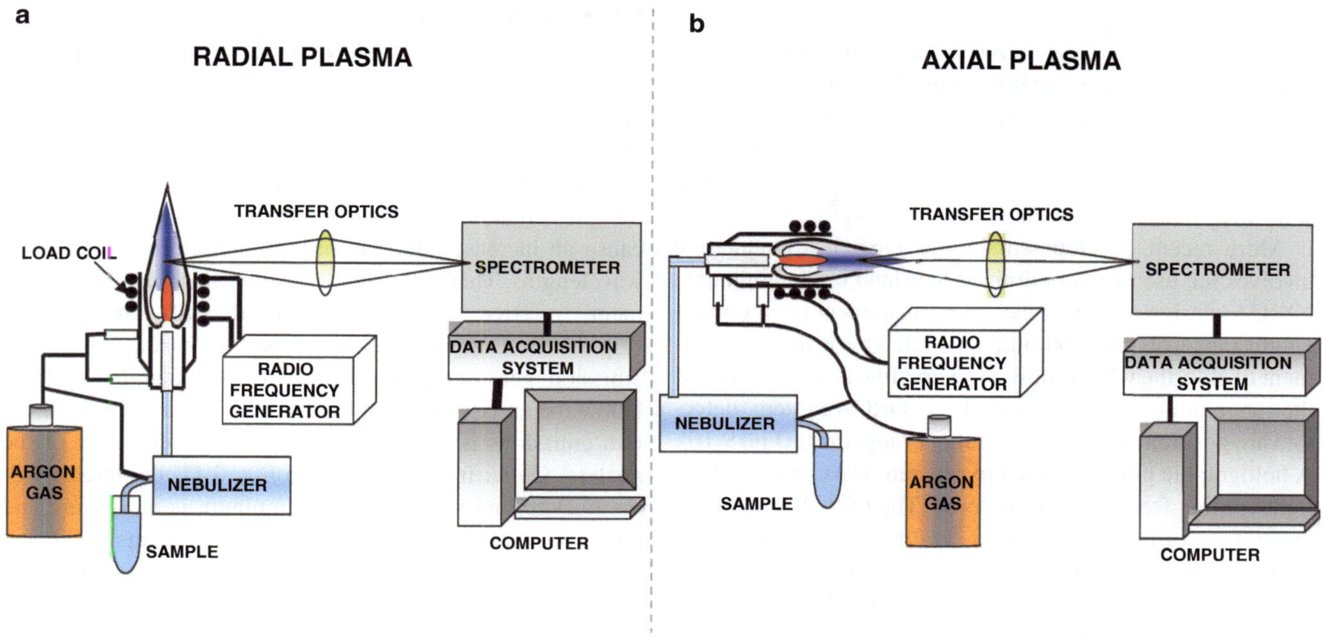

Fig. 9.8 Major components and typical layout of (**a**) a radially viewed ICP-OES instrument and (**b**) an axially viewed ICP-OES instrument

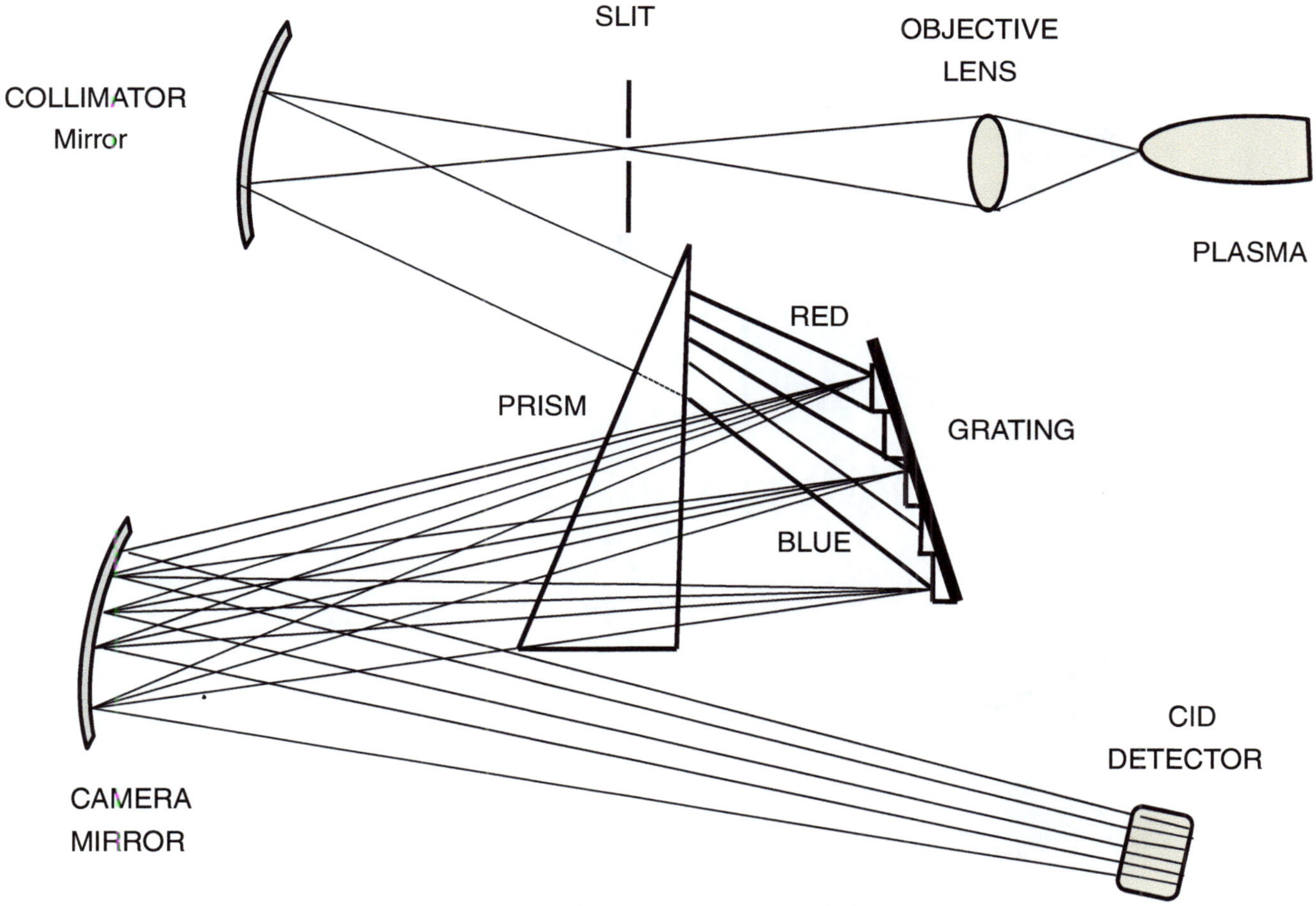

Fig. 9.9 A simplified example of an echelle spectrometer. The echelle spectrometer combines two light-dispersing elements in series. A prism first disperses the light in the x-direction without overlapping orders, followed by a grating which disperses the light in the y-direction, producing a two-dimensional spectrum on the detector

OES instruments typically use one of the three types of solid-state array detectors: **charge-coupled device** (CCD), **complementary metal oxide semiconductor** (CMOS), or **charge injection device** (CID). A description of these detectors is beyond the scope of this chapter. Interested readers are referred to the comparisons between different types of solid-state array detectors provided in references [19, 20].

More recent development in ICP-OES instrumentation involves the use of the optimized Rowland circle alignment (ORCA) design to further extend measurement over a wavelength range of 130–800 nm. This allows for the measurement of chlorine which emits a spectral line at 134 nm, below the range of an echelle system. In an ORCA system, detectors are positioned in a semicircular arrangement (Fig. 9.10). A holographic grating is used to separate wavelengths of the light coming from the plasma torch. The ORCA system has fewer light-dispersing elements (gratings and prisms) than an echelle system. This reduces light loss in the system and therefore increases sensitivity. It also allows for a more uniform resolution and greater stability.

9.3.4 Interferences in Inductively Coupled Plasma Optical Emission Spectroscopy

Generally, interferences in ICP-OES analyses are less of a problem than with AAS, but they do exist and must be taken into account. **Spectral interferences** are the most common. Samples containing high concentrations of certain ions may cause an increase (shift) in background emissions at some wavelengths. This will cause a positive error in the measurement, referred to as **background shift interference** (Fig. 9.11). Correction is relatively simple. Two additional emission measurements are made at a wavelength above and below the emission line of the analyte. The average of these two emissions is then subtracted from the emission of the analyte. (*Note*: In the example in Fig. 9.11, the intensity of the background shift from the aluminum is *increasing* and this is why it is necessary to make measurements above and below the emission line for lead. If the intensity of the background shift is constant, then only one measurement near the wavelength of the analyte's emission line is required for cor-

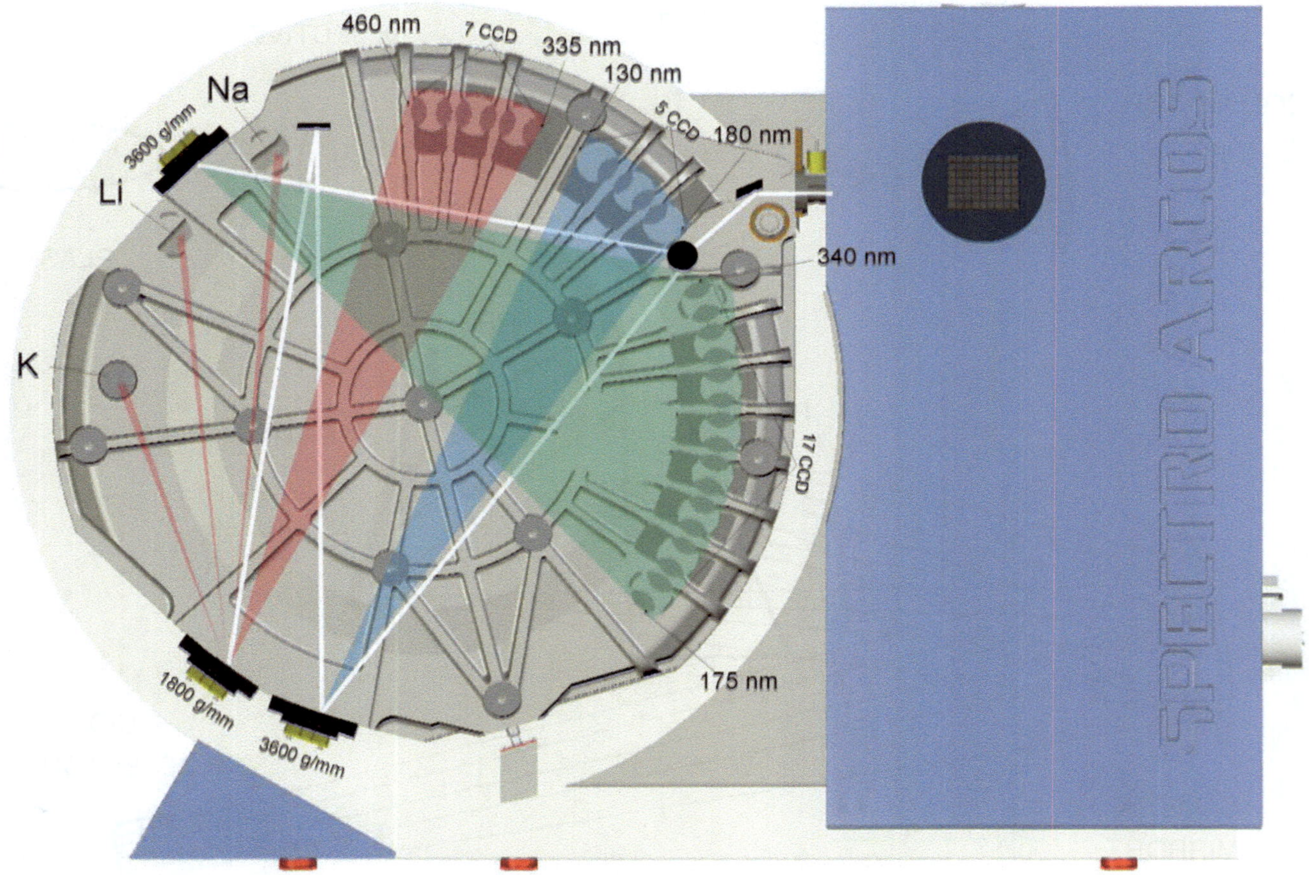

Fig. 9.10 The Optimized Rowland Circle Alignment (ORCA) design using three concave gratings. Each concave grating has the same curvature of the Rowland circle and will disperse the light according to its wavelength. The dispersed light will then be focused back onto one of the detectors placed along the diameter of the Rowland circle. The system uses CCD detectors in place of photomultiplier tubes (PMTs) and allows for the determination of nearly all the elements listed in the periodic chart. (Used with permission. Courtesy of SPECTRO Analytical Instruments GmbH Boschstr., Kleve, Germany)

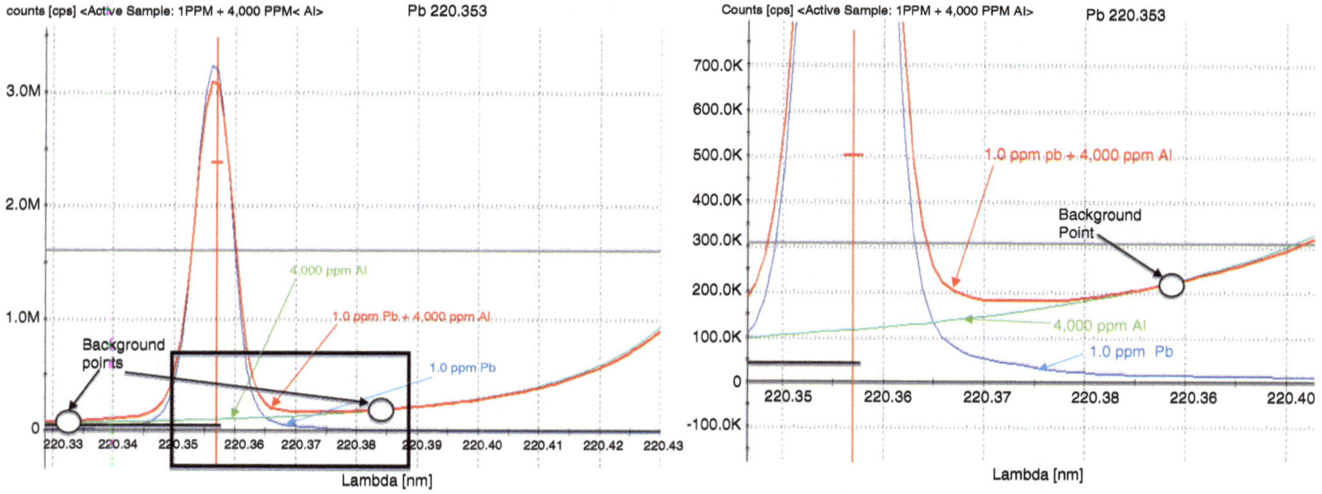

Fig. 9.11 Background shift interference of aluminum on lead. The two graphs show that aluminum will increase the background signal near the lead line at 220.35 nm. This is corrected by placing background subtraction points on both sides of the lead peak. The graph on the right is a blowup of the area that shows the background shift interference

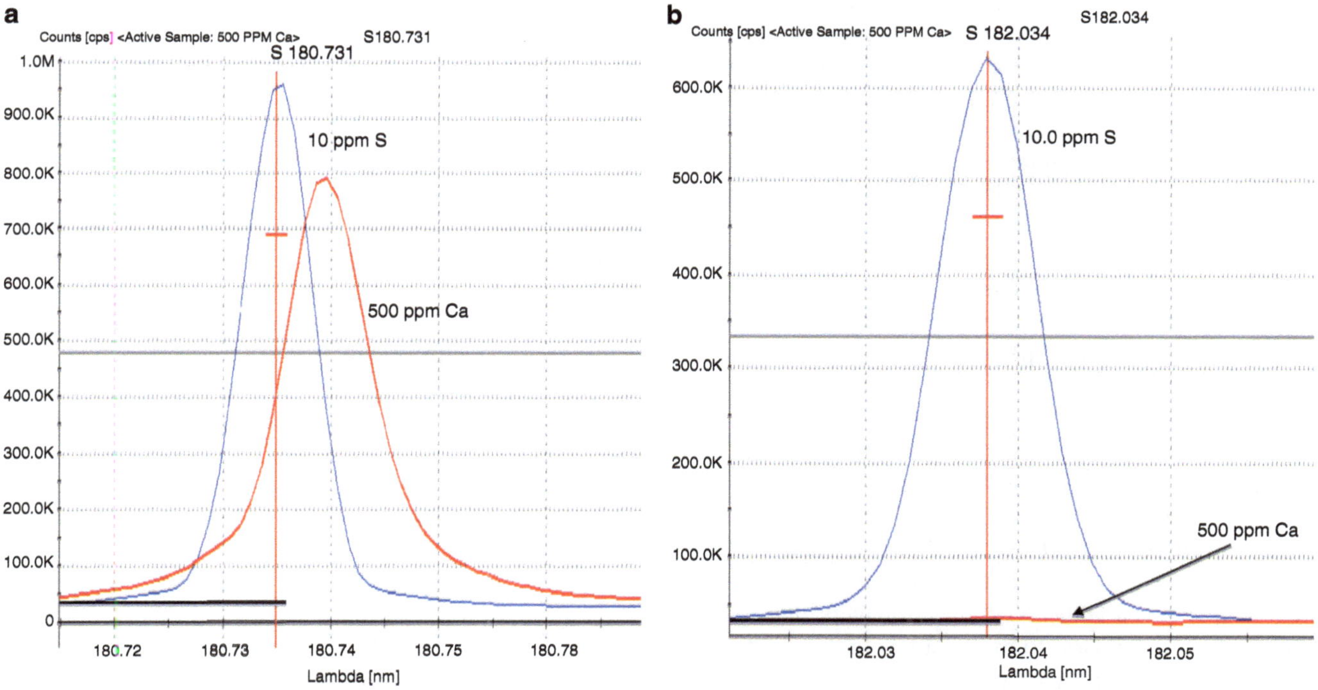

Fig. 9.12 Spectral overlap interference of calcium on sulfur. (**a**) The overlapping spectra of sulfur and calcium indicate that the resolution of the instrument is insufficient to resolve the emission from sulfur at 180.731 nm and calcium at 180.734 nm. (**b**) If the sample is being analyzed for sulfur, it is best to use the sulfur line at 182.034 nm where there is no spectral overlap

rection.) Alternatively, another emission line could be chosen in a region where there is no background shift.

A more problematic type of spectral interference, called **spectral overlap interference** (Fig. 9.12), occurs when the resolution of the instrument is insufficient to prevent overlap of the emission line (i.e., a very narrow bandwidth) of one element with that of another. For example, when determining sulfur in a sample containing high concentrations of calcium, some of the emission from one of the calcium lines overlaps with the sulfur emission line at 180.731 nm. This will cause an apparent increase in the measured concentration of sulfur. This problem may be overcome by either choosing a different sulfur emission line or calculating the **interelement correction** (IEC) factor. In the above example, this would require first measuring the emission of calcium at a different wavelength to determine its concentration in the sample. A standard solution of pure calcium is then prepared at that same concentration, and the apparent sulfur concentration in the pure calcium standard solution, presumably containing no sulfur, is determined. Finally, the sample is analyzed for sulfur, and the contribution of calcium to the sulfur signal is subtracted, thereby giving an accurate estimate of the true sulfur concentration.

9.4 APPLICATIONS OF ATOMIC ABSORPTION AND EMISSION SPECTROSCOPY

9.4.1 Uses

Atomic spectroscopy is widely used for the quantitative measurement of mineral elements in foods. In principle, any food may be analyzed with any of the atomic spectroscopy methods discussed. In most cases, it is necessary to ash the food to destroy organic matter and to dissolve the ash in a suitable solvent (usually water or dilute acid) prior to analysis. Readers are referred to Chap. 16 for a detailed discussion on the applications and procedures of dry/wet ashing as well as microwave ashing. Some liquid products such as vegetable oils may be analyzed without ashing by dissolving the oil in an organic solvent and aspirating the solution directly into a flame atomic absorption spectrometer. Alternatively, an aliquot of the oil sample may be introduced directly into a graphite furnace for atomization. The choice of method will depend on several factors, including instrument availability, cost, precision/sensitivity, and operator skill.

Two trending atomic spectroscopic techniques, namely **X-ray fluorescence spectroscopy** (XRF) and **laser induced breakdown spectroscopy** (LIBS), could be used with minimal sample preparation. XRF spectrometers utilize high-energy X-rays (primary) to excite atoms in the sample, and in response, the excited atoms emit fluorescent X-rays (secondary) that are unique to each element. For LIBS, pulsed laser beam is used as the energy source to produce a plasma on the surface of the sample, which contains excited atoms, ions, and electrons. Excited atoms emit unique spectra when they fall back to a lower energy state. Portable XRF and LIBS instruments support in situ applications and have been popular in the fields of metallurgy and geology for rapid analysis or identification. Food applications such as in-line quality control, screening, and inspection have made these two techniques potentially useful tools in the food industry [21, 22].

9.4.2 Practical Considerations

Quantitative atomic spectroscopy depends on a comparison of the sample measurement with a series of standards of known concentrations. Results should be routinely checked against standard reference materials (see Chap. 1, Sect. 1.4.3.2). Since concentrations of many mineral elements in foods are at the trace level, it is essential to use highly pure chemical reagents and water for preparation of samples and standard solutions. **Reagent blanks** should always be carried through the analysis. Vessels used for sample preparation and storage must be clean and free of the elements of interest. All glassware should be thoroughly washed with a detergent, carefully rinsed with distilled or deionized water, soaked in an acid solution (1 *N* HCl is sufficient for most applications), and rinsed again with distilled or deionized water.

9.5 INDUCTIVELY COUPLED PLASMA MASS SPECTROMETRY

As described above, atomic absorption and emission spectrometers are designed to quantify mineral elements in a sample by measuring either the absorption or emission of radiation by the element of interest at a wavelength unique to that element. Another approach is to directly measure the atoms (as ions in this case) of the element in the sample. In 1980, Houk et al. [23] reported the first account of the successful interfacing of an argon ICP with a mass spectrometer, paving the way for the development of **inductively coupled plasma mass spectrometry** (ICP-MS) instrument (Fig. 9.13) that allows extremely low detection limits at the single part per trillion (ppt) levels and quantification of individual isotopes. Note that isotopic analysis is not possible with atomic absorption or emission spectroscopy since absorption and emission lines are the same for all isotopes of a given element.

9.5.1 Principles of Inductively Coupled Plasma Mass Spectrometry

The principles of mass spectrometry and descriptions of different types of mass spectrometers are given in detail in Chap. 11. In ICP-MS mineral analyses, samples are prepared and aspirated into the ICP torch in the same manner as in ICP-OES. However, instead of having an optical system and a device for separating and detecting radiation of specific wavelengths, the ICP-MS uses a mass spectrometer to separate and detect *ions* of the elements directly based on their unique **mass-to-charge (*m/z*) ratios**. As shown in Fig. 9.13, the interface between the high-temperature ICP

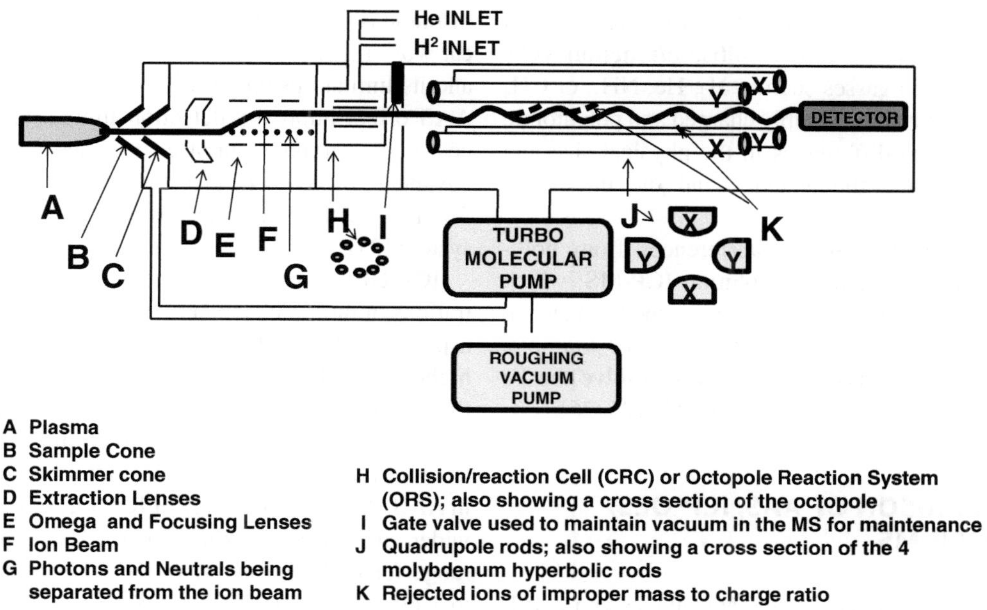

Fig. 9.13 A simplified diagram of an inductively coupled plasma mass spectrometer. The ions, electrons, photons, and neutral species generated in the plasma (**a**) are guided through an interface made up of a water-cooled sampling cone (**b**) and a skimmer cone (**c**) with a partial vacuum in between to remove argon gas and some neutral species. The remaining particles pass through the extraction lenses (**d**), which repel electrons and accelerate the positive ions further through to the off-axis ion omega lenses (**e**), bending the ion beam (**f**) as a result. The paths of the photons and neutral species (**g**) are unaffected and separated from the ion beam (see Chap. 11 for more details on the instrumentation of mass spectrometers)

operating at atmospheric pressure and the high vacuum mass spectrometer consists of (1) two funnel-shaped, water-cooled cones (the sampling cone and the skimmer cone) and (2) ion lenses (extraction lenses and omega lenses) to guide the analyte ions into the quadrupole while removing electrons, photons, and other neutral species from the ICP discharge.

9.5.2 Interferences in Inductively Coupled Plasma Mass Spectrometry

Interferences in ICP-MS arise when different ionic species from the sample have the same *m/z* ratio, leading to overlapping of signals. For example, iron has four naturally occurring stable isotopes: ^{54}Fe, ^{56}Fe, ^{57}Fe, and ^{58}Fe, with natural abundances of 5.8%, 91.75%, 2.1%, and 0.28%, respectively. Nickel has five stable isotopes: ^{58}Ni, ^{60}Ni, ^{61}Ni, ^{62}Ni, and ^{64}Ni, with natural abundances of 68.04%, 26.22%, 1.14%, 3.63%, and 0.93%, respectively. The signals for ^{58}Fe and ^{58}Ni will overlap, resulting in an **isobaric interference** because *m/z* = 58 for both isotopes. In this case, the analyst would select ^{56}Fe for the determination of iron and ^{60}Ni for the determination of nickel in the sample. (It is best to select the isotope with the highest natural abundance because the measurement precision is higher for more abundant isotopes. The concentration of the isotope in the sample is equal to the concentration of the element multiplied by the % natural abundance.) Most elements have at least one isotope with a unique mass number, thus allowing identification and quantification of elements.

The **doubly charged interference** occurs when ions of a particular element exist with double positive charges (instead of the normal single positive charge). Ions with charges greater than +1 typically have negligible impact except for ^{138}Ba, which may lose one or two electrons to produce singly and doubly charged ions with *m/z* = 138 and *m/z* = 69, respectively. The presence of doubly charged ^{138}Ba in a sample would interfere with ^{69}Ga. Another possible source of interference, referred to as **polyatomic interference**, comes from the formation of molecular species in the plasma between argon and elements from acids (e.g., H, N, O, Cl, etc.) used in dissolving the sample. Table 9.2 shows some examples of

Table 9.2 Examples of polyatomic interference in ICP-MS

Polyatomic species	*m/z*	*Interfered element/isotope*
$^{38}Ar^{1}H^{+}$	39	$^{39}K^{+}$
$^{35}Cl^{16}O^{+}$	51	$^{51}V^{+}$
$^{40}Ar^{12}C^{+}$	52	$^{52}Cr^{+}$
$^{38}Ar^{16}O^{1}H^{+}$	55	$^{55}Mn^{+}$
$^{40}Ar^{16}O^{+}$	56	$^{56}Fe^{+}$
$^{40}Ar^{35}Cl^{+}$	75	$^{75}As^{+}$
$^{40}Ar^{40}Ar^{+}$	80	$^{80}Se^{+}$

Adapted from Thomas [7]

polyatomic interferences. Modern ICP-MS instruments can be equipped with devices called **collision/reaction cells** (CRCs) through which gasses such as H_2, He, NH_3, or CH_4 may be introduced to remove doubly charged and polyatomic interferences based on differences in the physical size and kinetic energy between the analytes and the interfering species.

Another approach to reduce interferences from polyatomic species is to use a high-resolution ICP-MS, which utilizes a double-focusing sector-field mass spectrometer to separate ions generated by the plasma [24]. For example, high-resolution ICP-MS has been shown to resolve the Fe peak at mass 55.935 from ArO at mass 55.957 [25, 26].

9.6 COMPARISON OF AAS, ICP-OES, AND ICP-MS

Table 9.3 provides a summary of the advantages and disadvantages of AAS, ICP-OES, and ICP-MS. Flame AAS has enjoyed a long history of applications in mineral analysis. It is relatively inexpensive, easy to use, and ideal for analyzing a single element in a given sample. The major disadvantages of flame AAS include relatively low sensitivity, narrow linear working range, the use of potentially dangerous fuel gas, and its limitations on multielement analysis. Electrothermal AAS offers improved detection limits, but with the added cost of the graphite furnace, it somewhat negates the cost advantage of AAS. In fact, the cost of a high-end graphite furnace spectrometer overlaps with an entry-level ICP-OES system [27].

ICP-OES instruments are capable of determining concentrations of multiple elements in a single sample with a single aspiration. This offers a significant speed advantage and higher throughput over AAS when the objective is to quantify multiple elements (up to 70) in a given sample. The high temperature of the ICP torch also eliminates many nonspectral interferences (e.g., chemical interferences) encountered in AAS. Another advantage of ICP-OES over AAS is a much wider analytical working range. The analytical working range for ICP-OES is 6–7 orders of magnitude (i.e., 1 μg/L to 1 g/L without having to recalibrate the instrument), compared to about 3–4 orders of magnitude for AAS (i.e., 1 μg/L to 1 mg/L). All these advantages help explain the popularity of ICP-OES in commercial laboratories analyzing multiple elements in a large number of samples. ICP-OES is com-

Table 9.3 Advantages and disadvantages of AAS, ICP-OES, and ICP-MS

	Flame AAS	*Graphite furnace AAS*	*ICP-OES*	*ICP-MS*
Detection limit[a]	Good detection limits with many elements at the part per billion (ppb) level	Better than flame AAS and better than ICP-OES for some elements	Better than flame AAS	Overall best detection limits compared to other techniques
Elemental analytical capability	Single	Single	Multiple	Multiple
Approximate analytical working range	3–4 orders of magnitude	~2 orders of magnitude	6–7 orders of magnitude (could be higher with dual-view models)	9–10 orders of magnitude
Cost	Low	Low to medium	Medium	High
Use of explosive fuel gas	Yes (flame AAS instruments should not be unattended while in operation)	No	No	No
User-friendliness	Some skills required but relatively easy to use	Advanced skills required; extensive sample preparation	Easy to use once the computer interface is set up and operation is automated	Method development requires more expertise compared to other techniques
Interference	Both spectral and nonspectral interferences	Similar to flame AAS, but matrix (nonspectral) interferences likely higher	Spectral interferences including background shift and spectral overlap	Isobaric, doubly charged, and polyatomic interferences
Ideal application	A limited number of elements in a given sample; university instruction laboratories	A limited number of elements, and requiring better detection limits than flame AAS	Multiple elements in a large number of samples; routine mineral analysis in commercial laboratories with high throughput	Trace and ultra-trace mineral analysis; isotope information for food geographical identification
Isotopic analysis	N/A	N/A	N/A	Isotopic analysis possible because isotopes of the same element have different mass-to-charge ratios

[a]See Table 9.4 for more information on detection limit of each technique for different elements

monly used to obtain information for standard nutrition labeling.

ICP-MS retains the advantages offered by ICP-OES but, in conjunction with mass spectrometry, offers the lowest detection limits, enhanced multielemental capabilities, a wider analytical working range (9–10 orders of magnitude, i.e., 1 ng/L to 1 g/L), and isotopic information potentially for tracking the geographical origins of food products [28, 29]. Laboratories analyzing trace or ultra-trace concentrations of toxic heavy metals such as cadmium would be best served with an ICP-MS system. The major disadvantage for ICP-MS is perhaps the cost, which is about two to four times higher than its ICP-OES counterparts. It also costs more to maintain and operate than an ICP-OES.

Table 9.4 lists the detection limits with different techniques for mineral elements of interest in foods. It should be noted that these are approximate values and detection limits will vary depending on many factors such as the sample matrix and the stability of the instrument. A two- or threefold difference in detection limit is probably not meaningful, but an order of magnitude difference is noteworthy.

Table 9.4 Approximate detection limits[a] (μg/L) for selected elements analyzed with various instruments

Element	*Flame AAS*	*Graphite furnace AAS*	*ICP-OES*	*ICP-MS*
Al	45	0.1	0.5	0.00001
As	150	0.05	1.0	0.00005
Ca	1.5	0.01	0.03	0.00002
Cd	0.8	0.002	0.1	0.00009
Cr	3	0.004	0.17	0.00002
Cu	1.5	0.014	0.26	0.00001
Fe	5	0.06	0.08	0.00001
K	3	0.005	0.3	0.00002
Hg[b]	300	0.6	1.0	0.001
Mg	0.15	0.004	0.008	0.00001
Mn	1.5	0.005	0.026	0.00002
Mo	45	0.03	0.3	0.00001
Na	0.3	0.005	0.2	0.00001
Ni	6	0.07	0.4	0.00006
P	75,000	130	2.0	0.003
Pb	15	0.05	1.3	0.00001
Se	100	0.05	1.8	0.0003
Ti	75	0.35	0.04	0.00003
Tl	15	0.1	1.4	0.000004
Zn	1.5	0.02	0.07	0.0001

Adapted from Anonymous [27]

[a]Detection limit is defined as the lowest concentration of the element in a solution that can be distinguished from the blank with 98% confidence

[b]Detection limit for Hg with cold vapor AAS = 0.006 μg/L

9.7 SUMMARY

In comparison with traditional wet chemistry methods, AAS, AES, and ICP-MS methods are capable of measuring trace concentrations of elements in complex matrices rapidly and with excellent precision. For most applications, sample preparation involves the destruction of organic matter by dry or wet ashing, followed by dissolution of the ash in an aqueous solvent, usually a dilute acid. The sample solution is introduced as a fine mist into a flame atomizer or an ICP torch (or by injection into a graphite furnace) where it encounters very high temperatures (~2000–3000 K for flame or graphite furnace, and ~ 6000–7000 K for plasma). The sample goes through the process of desolvation, vaporization, atomization, and ionization. Analyte atoms, now in the gaseous state, are well separated and remain mostly neutral in a flame, but a significant fraction of them lose an electron and become charged in a plasma. The final step is to measure quantitatively the concentrations of the elements either by atomic spectroscopy or mass spectrometry.

Atomic spectroscopy depends on the absorption or emission of electromagnetic radiation (light) by the atoms in the gaseous state. In AAS, light of a discrete wavelength from the element-specific hollow cathode lamp will only be absorbed by the atoms of that element. Furthermore, the amount of light absorbed is directly related to the concentration of the atoms in the sample. By measuring the absorbance of light of a particular wavelength by an atomized sample, analysts can determine the concentration of an element even in the presence of other elements. In emission spectroscopy, the optical approach involves exciting the electrons in an element to a higher-energy state by a flame or plasma and then measuring the intensity of the light emitted when the electrons fall back to the ground state or a lower energy state.

In contrast to atomic spectroscopy, ICP-MS instruments are designed to measure ions of the element directly. This necessitates the ionization of atoms in the plasma. The ions of the element are then guided into a mass spectrometer which separates and detects ions according to their unique mass-to-charge (*m/z*) ratio. Quantification of elements with high sensitivity and specificity can be achieved because most elements have at least one isotope with a unique mass number.

Atomic spectroscopy is a powerful tool for the quantitative measurement of elements in foods. The development of these technologies over the past seven decades has had a major impact on food analysis. Today, accurate and precise measurements of a large number of mineral nutrients and nonnutrients in foods can be made rapidly using commercially available instrumentation. Analysts contemplating what instruments to acquire should make a decision based on an assessment of the required cost, user-friendliness, analytical working range, detection limit, multielemental capability, and the need for isotopic data.

9.8 STUDY QUESTIONS

1. Describe the process of "atomization" as it pertains to AAS and AES analyses.
2. AAS was the most widely used form of atomic spectroscopy in food analysis for many years. It has been largely replaced by the more powerful ICP-based spectroscopy in commercial laboratories.
   (a) Explain in simple terms the underlying principles of operation and instrumentation for ICP-OES (a simple drawing of the instrument might be helpful here).
   (b) Explain how AAS differs in principles of operation and instrumentation from what you described previously for ICP-OES.
   (c) Explain why AAS has been largely replaced by ICP-OES.
3. You are training a new technician in your analytical laboratory on mineral analysis by AAS and ICP-OES. Briefly describe the purpose of each of the following items.
   (a) HCL in AAS.
   (b) Plasma in ICP-OES.
   (c) Echelle optical system in ICP-OES.
   (d) Nebulizer in AAS and ICP-OES.
4. In the quantitation of Na by atomic absorption, KCl or LiCl was not added to the sample that also contained a relatively high concentration of K. Would you likely over- or underestimate the true Na content? Explain why either KCl or LiCl is necessary to obtain accurate results.
5. In preparing a sample (other than a clear liquid) for the analysis of specific minerals by AAS or ICP-OES:
   (a) What are the potential sources of error in sample preparation?
   (b) What ashing procedure would be the method of choice? Why?
6. You are performing iron analysis on a milk sample using AAS. Your results for the blank are high. What could be causing this problem and what is a possible remedy?
7. The detection limit for calcium is lower for ICP-OES than it is for flame AAS. How is the detection limit determined, and what does it mean?
8. When analyzing a sample for mineral elements using ICP-MS, the instrument is programmed to count the number of ions with a specific *m/z* ratio striking the detector. You decide to determine the concentrations of potassium and calcium in a sample of wheat flour. What *m/z* ratio would you use for potassium? For calcium? Why? (Hint, study the masses of all the naturally occurring and stable isotopes for the two elements and for argon (see table) and select isotopes with no interferences.) Why is it important to know the masses of argon isotopes as well as potassium and calcium?

Isotope	*Natural abundance (%)*
^{36}Ar	0.34
^{38}Ar	0.063
^{40}Ar	99.6
^{39}K	93.2
^{40}K	0.012
^{41}K	6.73
^{40}Ca	96.95
^{42}Ca	0.65
^{43}Ca	0.14
^{44}Ca	2.086
^{46}Ca	0.004

9.9 PRACTICE PROBLEMS

1. Your company manufactures and markets an enriched all-purpose flour product. You purchase a premix containing elemental iron powder, riboflavin, niacin, thiamin, and folate which you mix with your flour during milling. To comply with the standard of identity (see Chap. 2) for enriched flour, you specified to the supplier that the premix be formulated so that when added to flour at a specified rate, the concentration of added iron is 20 mg/lb. flour. However, you have reason to believe that the iron concentration in the premix is too low so you decide to analyze your enriched flour using your atomic absorption spectrometer. You follow the following protocol to determine the iron concentration.
   (a) Weigh out 10.00 g of flour, in triplicate (each replicate should be analyzed separately).
   (b) Transfer the flour to an 800-mL Kjeldahl flask.
   (c) Add 20 mL of deionized water, 5 mL of concentrated H_2SO_4, and 25 mL of concentrated HNO_3.
   (d) Heat on a Kjeldahl burner in a hood until white SO_3 fumes form.
   (e) Cool, add 25 mL of deionized water, and filter quantitatively into a 100-mL volumetric flask. Dilute to volume.
   (f) Prepare iron standards with concentrations of 2, 4, 6, 8, and 10 mg/L.
   (g) Install an iron hollow cathode lamp in your atomic absorption spectrometer, turn on the instrument, and adjust it according to instructions in the operating manual.

(h) Run your standards and each of your ashed samples and record the absorbances.

Calculate the iron concentration in each of your replicates, expressed as mg Fe/lb. flour.

Absorbance data for iron standards and flour samples

Sample	*Fe conc. (mg/L)*	*Absorbance*	*Corrected absorbance*
Reagent blank	–	0.01	–
Standard 1	2.0	0.22	0.21
Standard 2	4.0	0.40	0.39
Standard 3	6.0	0.63	0.62
Standard 4	8.0	0.79	0.78
Standard 5	10.0	1.03	1.02
Flour sample 1	?	0.28	0.27
Flour sample 2	?	0.29	0.28
Flour sample 3	?	0.26	0.25

2. Describe a procedure for determining calcium, potassium, and sodium in infant formula using ICP-OES. *Note:* Concentrations of Ca, K, and Na in infant formula are around 700 mg/L, 730 mg/L, and 300 mg/L, respectively.

Answers

1. The following steps may be used to determine the iron concentration in the flour samples.
   (a) Enter the data for the standards into Excel. Using the scatter plot function, plot the standard curve and generate a trend line using linear regression. Include the equation for the line and the R^2 value. Your results should look like the standard curve shown.

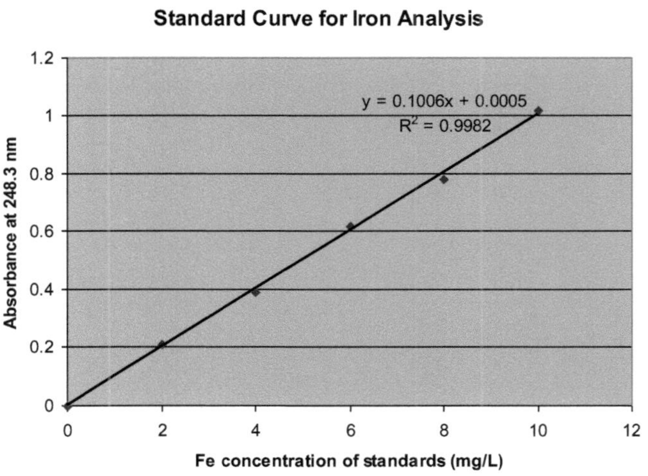

   (b) Using the equation, calculate the iron concentration in the solution in the volumetric flask for each of your samples. Your answers should be 2.68 mg/L, 2.79 mg/L, and 2.48 mg/L for Samples 1, 2, and 3, respectively. The mean is 2.65 mg/L; the standard deviation is 0.16.
   (c) Now determine the iron concentration in the flour. Recall that you transferred the solution from the Kjeldahl flask quantitatively into the 100-mL volumetric flask. Therefore, all of the iron in the flour sample should be in the volumetric flask. The mean concentration is 2.65 mg/L. The volume is 0.1 L. Therefore, the amount of iron in the 10 g of flour is 0.265 mg. To convert this to mg/lb., multiply by 454/10: 0.265 mg/10 g x 454 g/lb. = 12 mg Fe/lb. flour.
   (d) Your suspicions are confirmed; your supplier shorted you on iron in the premix. You need to correct this as soon as possible because your flour does not conform to the FDA's standard of identity for enriched flour and you may be subject to legal action by the FDA.
2. Consult AOAC Method 984.27 (see Chap. 1 for a description of AOAC International), and the following approach may be used:
   (a) Shake can vigorously.
   (b) Transfer 15.0 mL of formula to a 100-mL Kjeldahl flask. (Carry two reagent blanks through with sample.)
   (c) Add 30 mL of HNO_3-$HClO_4$ (2,1).
   (d) Leave samples overnight.
   (e) Heat until ashing is complete (follow AOAC procedure carefully—mixture is potentially explosive).
   (f) Transfer quantitatively to a 50-mL vol flask. Dilute to volume.
   (g) Calibrate instrument. Choose wavelengths of 317.9 nm, 766.5 nm, and 589.0 nm for Ca, K, and Na, respectively. Prepare calibration standards containing 200 μg/L, 200 μg/mL, and 100 μg/mL for Ca, K, and Na, respectively.
   (h) The ICP-OES computer will calculate concentrations in the samples as analyzed. To convert to concentrations in the formula, use the following equation:
   (i) Concentration in formula = Concentration measured by ICP × (50 mL/15 mL).

REFERENCES

1. Greenfield S, Jones IL, Berry CT (1964) High-pressure plasma as spectroscopic emission sources. Analyst 89:713–720
2. Wendt RH, Fassel VA (1965) Induction-coupled plasma spectrometric excitation source. Anal Chem 37:920–922
3. Beaty RD, Kerber JD (1993) Concepts, instrumentation and techniques in atomic absorption spectrophotometry, 2nd edn. The Perkin-Elmer Corporation, Norwalk, CT
4. Boss CB, Fredeen KJ (2004) Concepts, instrumentation and techniques in inductively coupled plasma atomic emission spectrometry, 3rd edn. The Perkin-Elmer Corporation, Norwalk, CT

5. Robinson JW, Frame ES, Frame II GM (2014) Undergraduate instrumental analysis, 7th edn. CRC, Taylor and Francis Group, Boca Raton, FL
6. Rutzke MA (2018) Atomic absorption, inductively coupled plasma optical emission spectroscopy, and infrared spectroscopy. In: White WM (ed) Encyclopedia of Geochemistry. Encyclopedia of Earth Sciences Series. Springer, Cham.
7. Thomas R (2013) Practical guide to ICP-MS: a tutorial for beginners, 3rd edn. CRC, Taylor and Francis Group, Boca Raton, FL
8. Dawson JB, Price WJ (1999) Alan Walsh - his influence on the development of atomic spectroscopy in the UK. Spectrochim Acta Part B At Spectrosc 14:2011-2015.
9. Ganeev A, Gavare Z, Khutorshikov VI, Hhutorshikov SV, Revalde G, Skudra A, Smirnova GM, Stankov NR (2003) High-frequency electrodeless discharge lamps for atomic absorption spectrometry. Spectrochimica Acta Part B 58:879–889
10. Anonymous (2023) Atomic Absorption Spectroscopy Learning Module. Maryville University. https://blogs.maryville.edu/aas/background-correction. Accessed February 2023
11. Koirtyohann SR, Pickett EE (1966) Spectral interferences in atomic absorption spectrometry. Anal Chem 38:585–587
12. West AC, Fassel VA, Kniseley RN (1973) Lateral diffusion interferences in flame atomic absorption and emission spectrometry. Anal Chem 45:1586–1594
13. Dockery CR; Blew MJ; Goode SR (2008) Visualizing the Solute Vaporization Interference in Flame Atomic Absorption Spectroscopy. J Chem Educ 85:854-858
14. Montaner A (1998) The Science of a Revolutionary Chemist and Spectroscopist: Velmer A. Fassel. Appl Spectrosc 52:406A-26A
15. Hou X, Jones BT (2000) Inductively coupled plasma-optical emission spectrometry. In: Meyers RA (ed) Encyclopedia of Analytical Chemistry. John Wiley and Sons Ltd, Chichester, England
16. Hieftje GM (1985) A steady-state approach to excitation mechanisms in the ICP. Spectrochem Acta 40:167-176
17. Hasegawa T, Umemoto M, Haraguchi H, Hsiech C, Montaser A (1992) Fundamental properties of inductively coupled plasma. In: Montaser A, Golightly DW (eds) Inductively Coupled Plasmas in Analytical Atomic Spectroscopy. VCH Publishers, New York
18. Lazar AC, Farnsworth PB (1999) Matrix effect studies in the inductively coupled plasma with monodisperse droplets. Part 1: The influence of matrix on the vertical analyte emission profile. Appl Spectrosc 53:457-464
19. Litwiller D (2005) CMOS vs. CCD: Maturing Technologies, Maturing Markets-The factors determining which type of imager delivers better cost performance are becoming more refined. Photonic Spectra 39:54-61
20. Sweedler JV, Jalkian RD, Pomeroy RS, Denton MB (1989) A comparison of CCD and CID detection for atomic emission spectroscopy. Spectrochimica Acta Part B 44:683-692
21. Stefas D, Gyftokostas N, Nanou E, Kourelias P, Couris S (2021) Laser-induced breakdown spectroscopy: an efficient tool for food Science and technology (from the analysis of Martian rocks to the analysis of olive oil, honey, milk, and other natural earth products). Molecules 26:4981.
22. Guild GE, Stangoulis JC (2022) Screening Ca concentration in staple food crops with energy dispersive x-ray fluorescence (EDXRF). Plant Soil 473:659-667.
23. Houk RS, Fassel VA, Flesch GD, Svec HJ, Gray AL, Taylor CE (1980) Inductively coupled argon plasma as an ion source for mass spectrometric determination of trace elements. Anal Chem 52:2283-9.
24. Jakubowski N, Moens L, Vanhaecke F (1998) Sector field mass spectrometers in ICP-MS. Spectrochimica Acta Part B 53:1739-1763
25. Thermo Scientific Application Note 30003 (2007) Determination of trace elements in clinical samples by high resolution ICP-MS. Thermo Fisher Scientific Inc., Waltham, MA
26. Thermo Scientific Application Note 30073 (2007) Determination of ultratrace elements in liquid crystal by high resolution ICP-MS. Thermo Fisher Scientific Inc., Waltham, MA
27. Anonymous (2022) Atomic spectroscopy: a guide to selecting the appropriate technique and system. PerkinElmer Inc. https://resources.perkinelmer.com/lab-solutions/resources/docs/BRO_WorldLeaderAAICPMSICPMS.pdf?_ga=2.186424106.283817021.1671227157-1322735730.1671227157. Accessed December, 2022
28. Luykx DM, Van Ruth SM (2008) An overview of analytical methods for determining the geographical origin of food products. Food Chem 107:897-911
29. Liu HL, Zeng YT, Zhao X, Tong HR (2020) Improved geographical origin discrimination for tea using ICP-MS and ICP-OES techniques in combination with chemometric approach. J Sci Food Agric 100:3507-3516

Nuclear Magnetic Resonance

10

Senay Simsek and Bradley L. Reuhs

Contents

10.1 INTRODUCTION

Nuclear magnetic resonance (NMR) **spectroscopy** is a powerful analytical technique with a wide variety of applications. It may be used for complex structural studies, for protocol or process development, or as a simple quality assay. NMR is non-destructive, and high-quality data may be obtained from milligram, even microgram, quantities of sample. Other spectroscopy techniques may be used to determine the nature of the functional groups present in a sample; only NMR spectroscopy can provide the data necessary to determine the complete structure of a molecule. Due to improved instrumentation and much lower costs, very complex and specialized NMR techniques can now be routinely performed by a student or technician. These experiments can be set up with the click of a button/icon, as all the basic parameters are included in default experiment files listed in the data/workstation software, and the results are obtained in a short time. These factors have led to NMR being increasingly used in quality control laboratories of food processing facilities (e.g., for fat or moisture analysis).

NMR instruments may be configured to analyze samples in solutions or in the solid state. In fact, these two types of analyses can be used in tandem to follow the fate of a given molecule within a specific system. For example, as a fruit ripens, many components will be released from the solid matrix around the plant cells into solution in the ripe fruit liquid. The development of this process can be followed by liquids versus solids analyses during the ripening time. As ripening progresses, some NMR signals will decrease in the solids NMR analyses and increase in the liquids NMR spectra. NMR instrumentation can also be utilized for routine analysis of moisture, protein, and fat contents of various

S. Simsek (✉) · B. L. Reuhs
Department of Food Science, Purdue University,
West Lafayette, IN, USA
e-mail: ssimsek@purdue.edu; breuhs@purdue.edu

B. P. Ismail, S. S. Nielsen (eds.), *Nielsen's Food Analysis*, Food Science Text Series,
https://doi.org/10.1007/978-3-031-50643-7_10

food samples. Additionally, there are many specific applications for NMR spectroscopy in food analysis.

Other food applications of NMR spectroscopy include structural analysis of food components, such as carbohydrates, to correlate the structure with the rheological properties. Routine analyses are used to determine the quality of a product or to test the purity of ingredients. Related techniques, such as **NMR relaxometry**, can be used to assess processing operations; for example, relaxometry is used to follow the solubilization of powdered ingredients in water, to optimize processing parameters. **Magnetic resonance imaging** (MRI) is a nondestructive technique that can be used to image product quality and changes during processing and storage. For example, MRI is used to image the freezing process, with the goal of increasing shelf life. **2-D NMR** methods can be utilized for structural analysis to identify thickeners used in food formulations. **Time-domain NMR** (TD-NMR) has many applications across a variety of foods for fat determination, moisture mobility, and through package analyses. **Cross-Polarization Magic-Angle Spinning NMR** (CP-MAS-NMR) has been utilized to characterize carbohydrate release in vegetables and starch molecular organization.

This chapter covers the basic principles and applications of NMR spectroscopy, as well as a brief description of relaxometry, MRI, and a recent development in instrumentation that uses NMR as part of a rapid moisture and fat analysis system. Specific applications to food analysis will be highlighted.

10.2 PRINCIPLES OF NMR SPECTROSCOPY

10.2.1 Magnetic Field

NMR differs from most other forms of spectroscopy in that it is the **atomic nuclei** that are the subject of study, and the measured energy is in the radio-frequency range. Many nuclei possess an angular momentum, which means that they have a characteristic spin quantum number (I) and may be analyzed using NMR. The most common nuclei analyzed by NMR are the proton (H), the ^{13}C isotope of carbon, ^{19}F, and ^{31}P, all of which have a spin $I = 1/2$. Nuclei with other spin quantum numbers will not be considered in this chapter, and the theoretical discussion will focus on the proton. These nuclei are charged, and a spinning charge generates a magnetic field. Simply put, the nuclei behave like tiny magnets that interact with an applied, external magnetic field.

Once the nuclei are placed within a strong **external magnetic field** (B_0), the spin of the nuclei will align with that field (Fig. 10.1). Because of quantum mechanical constraints (nuclei of spin I have $2I + 1$ possible orientations in the external magnetic field), there are only two orientations that the

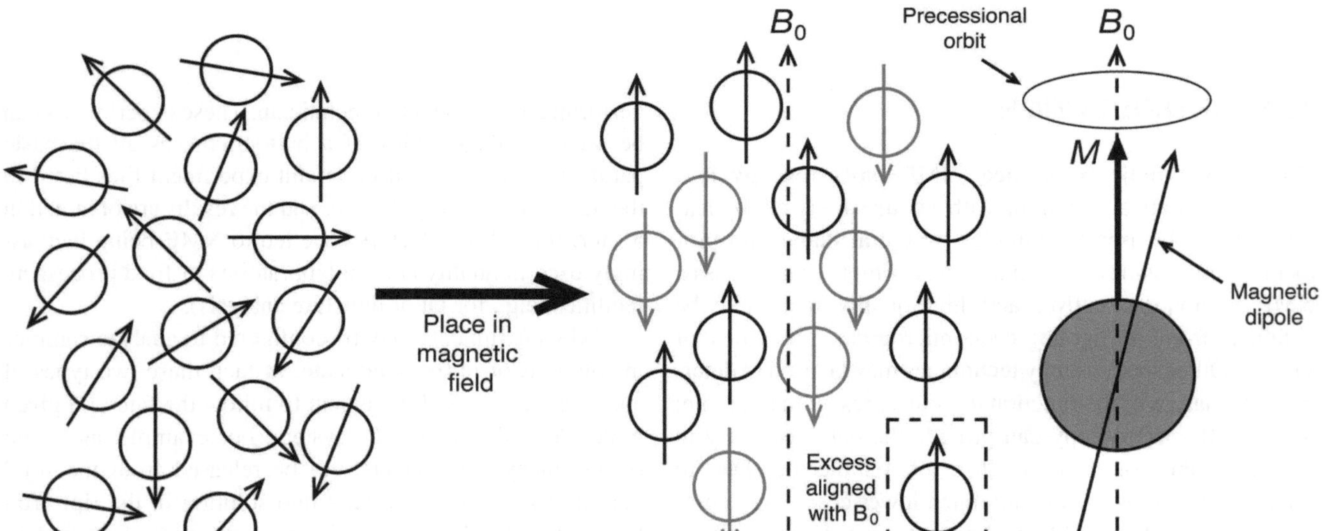

Fig. 10.1 Nuclear spin and magnetic vectors are randomly ordered outside of the NMR magnet. However, once placed in an applied magnetic field, the NMR magnet and the nuclei align either with the applied field, B_0, (parallel) or against it (antiparallel). There is a slight excess in the population aligned parallel to B_0. Although the magnetic dipole tracks a precessional orbit, the net magnetization (M) is aligned with B_0

spin 1/2 nuclei can adopt: either aligned with the applied magnetic field (parallel or spin +1/2) or aligned against the field (antiparallel or spin −1/2). The parallel orientation has a slightly lower energy associated with it and, therefore, has a slightly higher population. It is this excess of nuclei in the spin +1/2 state that produces the net magnetization that is manipulated and measured during an NMR experiment. The spin of the nuclei is not around the center axis but comparable to a gyration (Fig. 10.1). The motion of a spinning charged particle in an external magnetic field is similar to that of a spinning gyroscope in a gravitational field. This type of motion is known as **precession**, and there is a specific precessional orbit and frequency, the **Larmor frequency**, which is related to the magnetic properties of the nuclei. The magnitude and direction of the local magnetic field describe the magnetic moment or **magnetic dipole** of the system, and due to the precession and the lower energy state excess of nuclei, there is a net vector parallel to the applied field (Fig. 10.1).

All nuclei of the same element, H, for example, will have a nearly identical Larmor frequency in a magnetic field. The specific frequency is dependent on the strength of the external magnetic field, and the Larmor frequency of H in this field defines the NMR instrument. For example, a proton has a Larmor frequency of 500 MHz in an 11.7 Tesla magnet, so the instrument is termed a "500 MHz NMR spectrometer."

The position at which the magnetic moments spin on their precessional path at any given moment in time is referred to as the **phase**. The phase can be described as being in phase or **coherent** when the magnetic moments of the nuclei spins are all at the same point on the precessional path at any given moment. Conversely, when the spins are not at the same points at a given time, the phase is said to be out of phase or **incoherent**. In the resting state, the nuclei are incoherent and then become coherent when a pulse at the resonant frequency is applied.

10.2.2 Radio-Frequency Pulse and Relaxation

Early NMR instruments relied on electromagnets and a simple **radio-frequency** (RF) transmitter, and the analyses were performed by a sweep through the frequency range of the instrument. The collected spectra contain the frequency information; hence, it is termed frequency-domain NMR. Although this enabled the development of NMR spectroscopy, it was not sufficient to facilitate the modern NMR experiment. One of the major developments in NMR technology was the **RF pulse**, in which a large range of frequencies is excited by a short pulse of RF energy around a centered carrier frequency, which is at the Larmor, or resonance frequency of the nuclei under study. This pulse simultaneously excites all of the protons in the sample, and the NMR data for all the protons is collected during a short time after the pulse is applied. The excitation of a range of radio frequencies by a pulse is similar to the excitation of a range of audio frequencies when a bell is rung, and the size and construction of the bell determine the range that is emitted. In NMR, the carrier frequency, transmitter power, and duration of the RF pulse determine the frequency range of the pulse.

Once the sample is placed in the magnet, the protons align parallel or antiparallel to the applied, external magnetic field, $\boldsymbol{B}_0$, with an excess in parallel orientation. The net magnetization of the nuclei in the parallel orientation is aligned with the *z*-axis in an *xyz* graphical representation of the system (Fig. 10.2a). After a pulse of RF energy is applied to the system, the nuclei precess coherently and individual nuclei absorb energy and shift to a higher energy state. The pulse, which is applied by a transmitter coil perpendicular to the *z*-axis ($\boldsymbol{B}_1$), tilts the net magnetization vector away from the *z*-axis and toward the ***xy*-plane** (Fig. 10.2b).

Although the parameters that define a pulse include the transmitter power and the pulse duration, a specific pulse used in an NMR experiment is usually described by the degree to which the net magnetization is tilted. The most common pulse is the 90° pulse, which tilts the net magnetization exactly into the *xy*-plane, where the receiver coil is located, thereby maximizing the resulting signals. Many NMR experiments use a series of pulses, termed a **pulse sequence**, to manipulate the magnetization. Complex pulse sequences are essential for the two-dimensional (2-D) NMR experiments (Sect. 10.2.5) that are required for structural analyses of complex molecules.

Once the net magnetization has been tilted into the *xy*-plane by a 90° pulse, the magnetization begins to decay back to the *z*-axis. This process is termed **NMR relaxation**, and it involves both **spin-lattice** (*T*1) and **spin-spin** (*T*2) relaxation. ***T*1 relaxation** is associated with the interaction of the magnetic fields of the excited-state nuclei with the magnetic fields of other nuclei within the "lattice" of the total sample. ***T*2 relaxation** involves the interactions of neighboring nuclei that lead to a diminishment in the energy state of the excited-state nuclei and the loss of phase coherence. The mechanisms behind relaxation are complex, but the process can be utilized for some specific NMR experiments that, among other things, take advantage of the fact that samples in different forms, liquids vs. solids, for example, relax at different rates (this is very important for the instruments and experiments that are utilized for food processing and content analyses).

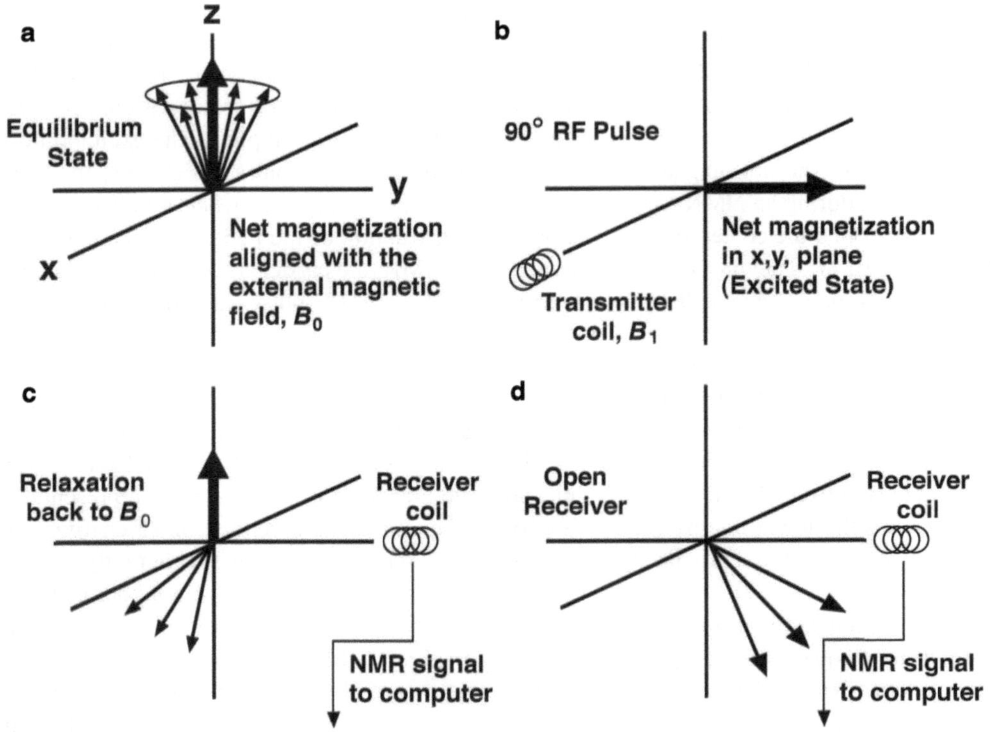

Fig. 10.2 (**a**) Prior to the RF energy pulse, the net magnetization (*M*) composed of all the component vectors is in the equilibrium state, aligned with B_0. (**b**) The 90° RF pulse, which covers the resonance frequencies of all relevant nuclei in the sample and originates perpendicular to the *z*-axis (B_1), causes the nuclei to move to a higher energy state, and the net magnetization rotates into the *xy*-plane. (**c**) Once in the *xy*-plane, the net magnetic vector separates into the component vectors for each unique population of nuclei. As these oscillate in the *xy*-plane, they emit RF signals that are detected by the NMR instrument after passing through the receiver coil, which is located perpendicular to both B_0 and the transmitter coil. (**d**) As the component vectors continue to oscillate in the *xy*-plane (and emit RF signals), the nuclei begin to relax back to the equilibrium state. The NMR instrument may be set up to repeat this process, with additional pulses, numerous times; the collected data are then added together to improve the signal/noise ratio and resolution

10.2.3 Chemical Shift and Shielding

The total H population in the sample determines the net magnetization of the system in the external magnetic field, $\mathbf{B}_0$. The exact frequency of a unique population of protons (i.e., all protons in a specific chemical configuration in the molecule), however, is also dependent on the immediate environment of the nucleus, principally the density of the electron cloud surrounding the nuclei, which determines the electronic environment of the nuclei. This is referred to as the "**shielding**" effect because the electrons create a secondary, induced magnetic field that opposes the applied field, shielding the nuclei from the applied field. The resulting frequency differences are so small, relative to the Larmor frequency, that they are commonly reported in parts per million (ppm). However, they are large enough to be clearly detected and resolved during an NMR experiment. The frequency differences that result from the differences in the electronic environments yield the **chemical shift** of the nuclei. Following the processing of the NMR data, these differences result in a series of resonance signals, each representing a unique proton population, along the *x*-axis in a one-dimensional (1-D) NMR spectrum. Those protons that have a relatively dense electron cloud are considered shielded, since the electron cloud works in opposition to the external magnetic field. The resonances of shielded protons will be found on the right, or upfield, side of the spectrum, at a lower chemical shift. As deshielding increases, the resonances are shifted further to the left, or downfield, at a progressively higher chemical shift (Fig. 10.3).

One of the most important determinants of the chemical shift for a specific population of protons (i.e., all the protons in the sample that are in an identical molecular environment) is the proximity to an electronegative group or atom, such as O (Fig. 10.3). For example, protons that are not located near (in a molecular context) any electronegative groups or atoms, such as those in the methyl group of a 6-deoxy sugar, are heavily shielded, and the resonances will be found on the far right side of a proton NMR spectrum (Fig. 10.3). In contrast, the proton on the C1 of a typical sugar (the anomeric proton) is near two O atoms, the O in the –OH group (or the O in the linkage to the next sugar in a polymer, such as starch) and the O that forms part of the hemiacetal ring structure of the sugar. Consequently, the resonances associated with the

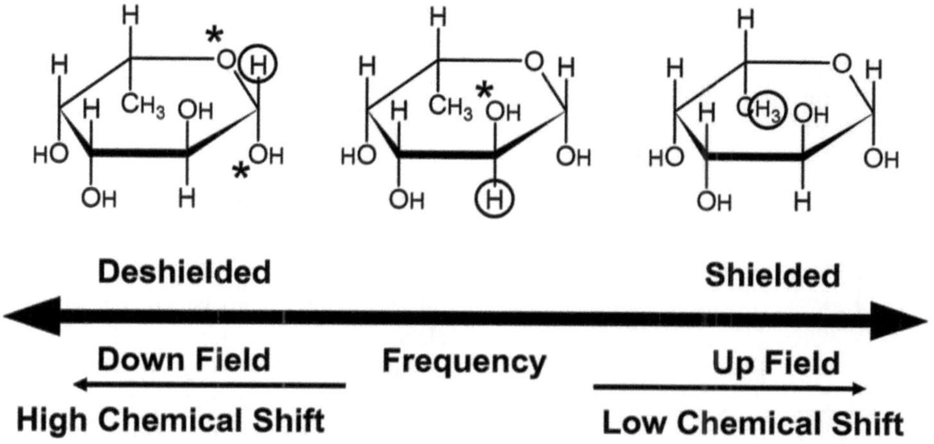

Fig. 10.3 The shielding effect is responsible for the small, but detectable, differences in the resonance frequencies of nuclei such as protons. Protons that are not close to an electronegative group in the molecule, such as the protons in the methyl group in fucose (*top right*), a common 6-deoxy sugar, will be shielded by the electrons surrounding it, and it will have a low chemical shift, upfield in the NMR spectrum. Protons near one oxygen atom (indicated by an asterisk), such as the ring protons in sugars (*top middle*), will be intermediately shielded and have a chemical shift toward the middle of the spectrum. A proton that is near two oxygen atoms, such as the anomeric proton in sugars (*top left*), will be relatively deshielded and have a high chemical shift downfield in the spectrum

highly deshielded anomeric protons are found on the left side of the proton spectrum. Protons near one O, such as the ring protons in a typical sugar, will be partially shielded, and the resonances associated with these protons will be in the central region of the spectrum.

10.2.4 1-D NMR Experiment

For solution ^{1}H-NMR spectroscopy, the sample is dissolved in deuterated solvents produced for NMR analysis, such as D_20 (where D = deuterium or ^{2}H; this is to avoid overloading the NMR signal with solvent protons), and pipetted into a NMR tube, which is then capped and placed into the magnet. The net magnetization of the nuclei in the parallel orientation in the external magnetic field, $\mathbf{B}_0$, is aligned with the z-axis; this is the equilibrium state. The 90° pulse, centered at the Larmor frequency of H, from the transmitter in the x-axis tilts the net magnetic vector into the xy-plane. There the component vectors (those representing each unique population of protons) will oscillate at their specific NMR resonance frequencies, separating the net vector into numerous component magnetic vectors, which then induce radio signals (the NMR signal) in the receiver coil located in the y-axis (Fig. 10.2c). At the same time, the magnitude of the vectors will decay in the xy-plane (relaxation) and return to the z-axis (Fig. 10.2d). This whole process is termed one **scan**.

The result of these actions is a combined signal for all protons in the sample, which rapidly decreases over time, yielding a **free induction decay** (FID) (Fig. 10.4a), which contains all the frequency and intensity information (as well as phase, which will not be considered) for each of the unique populations of nuclei in the sample. **Fourier transformation**, a mathematical operation that converts one function of a variable into another, is then applied to convert the time-domain FID to the frequency-domain **NMR spectrum** (an xy plot) (Fig. 10.4b). The frequency information for each unique proton population is presented on the x-axis of the NMR spectrum. The signal intensity is related to the y-axis; however, this is not labeled because it has no units; moreover, the linewidth differences of each resonance make the y-value imprecise for quantification in ^{1}H-NMR analyses. The best method to compare the signal intensity and, consequently, the relative abundance of specific protons in a 1-D NMR spectrum is to compare the integration values of the resonances.

In practice, the NMR spectrometer is set up so that the pulse is applied numerous times, usually in increments of 16 scans. For samples that are present in a high concentration, 16 or 32 scans per experiment are common and 256 or 512 scans are often used for more dilute samples. After each scan, the new data are added to the data already collected. The result of compiling the data from numerous scans is a significant improvement in the signal/noise ratio and resolution.

10.2.5 Coupling and 2-D NMR

Another essential concept to consider is "**coupling.**" Coupling is a result of the influence of electrons in covalent bonds on the local magnetic field of nearby nuclei. Through the intervening bonds, two nearby nuclei will affect the

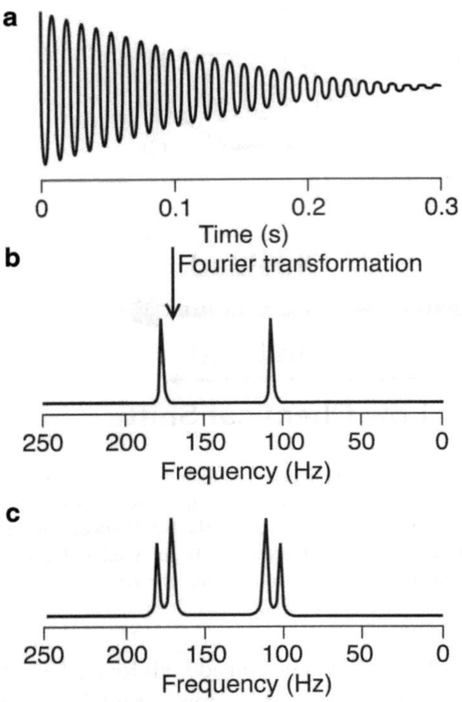

Fig. 10.4 After the 90° RF pulse moves the magnetization to a higher energy state in the *xy*-plane, the receiver is turned on and it collects the RF signals emitted by oscillating nuclei; the emitted RF signals rapidly decay over a short time as the magnetization relaxes back to the equilibrium state (see Fig. 10.2). (**a**) The result is NMR data that contains all of the frequency and intensity information for the nuclei under analysis and which diminishes as the signals decay; this data is termed a "free induction decay" (FID), and it represents the time-domain information obtained from the NMR experiment. (**b**) Once the FID has been processed by Fourier transformation, an NMR spectrum is obtained; this represents the frequency-domain NMR information. The resonance "peaks" found on the *x*-axis are due to unique populations of protons, two in this case. (**c**) If the two protons are coupled through the molecular bonds, they will each have the effect of splitting the resonance peaks into two distinct peaks. The degree of splitting, reported in Hz, is indicative of the strength of the coupling effect

chemical shift of one another, resulting in the splitting of the resonances from each unique population of nuclei into two distinct resonances (Fig. 10.4c). The coupling strength is affected by both the proximity of the nuclei to one another and the geometry of the intervening bonds. For example, protons on a carbon backbone that have a *trans* relationship ("across" from each other) have a much stronger coupling than those that have a *cis* relationship ("on the same side" as one another). Thus, the use of coupling data yields information about the geometry of a specific molecule.

A more important impact of coupling is that complex 2-D NMR experiments have been designed to take advantage of the coupling phenomenon, to produce the data necessary for the complete structure determination of a molecule. 2-D NMR experiments are essentially a series of 1-D experiments, in which the pulse sequence includes several pulses and a variable parameter, such as the delay time between two of the pulses. The computer collects all the spectra and plots them out as a 2-D plot, in which "**cross peaks**" show the coupling correlations of nearby nuclei.

10.3 NMR SPECTROMETER

The typical research NMR spectrometer consists of a powerful cryomagnet, into which the sample is placed, a set of electronics for transmitting and collecting radio signals, and a data/workstation (Fig. 10.5). Modern instruments use superconducting magnets that are cooled to a very low temperature by a jacket of liquid helium, which has a boiling point of 4.2 K. This jacket is, in turn, surrounded by an outer jacket of liquid nitrogen. The core of the superconducting magnet consists of coil windings of thin wires made of superconducting alloys, such as niobium-titanium or niobium-tin. The coil (i.e., the magnet) and the coolants are contained in an insulated Dewar, termed a cryostat, which includes a vacuum chamber around the liquid jackets.

Down the center of the magnet but external to the Dewar (i.e., at room temperature) is a tubular space, the **magnet bore**. A multifunction device, termed a **probe**, is placed inside the magnet bore from the bottom. Inside the probe, just inside of the main magnet coil, are small, secondary magnetic coils that receive power from the NMR instrument hardware. These small coils are manipulated by the operator to make fine adjustments to the magnetic field, to optimize the magnetic field; hence, they are called shims (in construction, a shim is a small piece of wood or metal that is placed between two layers of building material to obtain a better fit, such as when leveling a door frame). The probe also contains the coils for transmitting and receiving the RF energy. Thus, the probe is the central piece of hardware for the NMR experiment. Finally, at the top of the bore, and at the top of

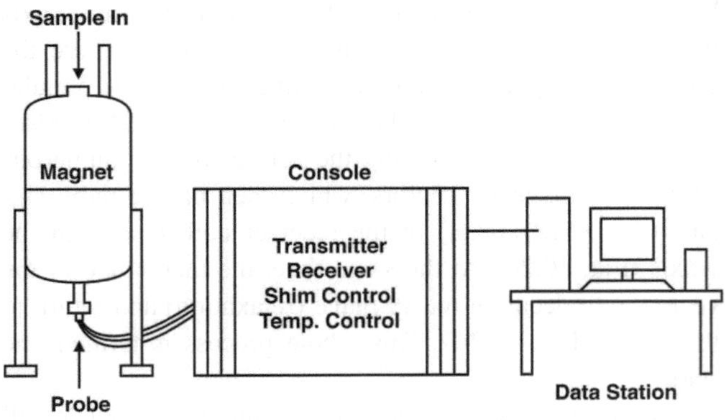

Fig. 10.5 A diagram of an NMR spectrometer. The instrument consists of a superconducting cryomagnet (the NMR magnet), an electronics console, and a data/workstation that also controls all the functions of the instrument

the whole system, is the sample insertion point. The sample tube, which is in a holder, is lowered down through the bore by a diminishing stream of forced air until it gently comes to rest at the top of the probe; there the sample is correctly aligned with the magnetic field and the probe hardware.

Connected to the magnet probe is an electronics console that includes the transmitter, the receiver, and other systems that control the NMR instrument, such as the sample temperature control unit. The transmitter includes systems to produce the pulse at the correct frequency for each nucleus that may be observed. For example, a 500-MHz NMR spectrometer requires a transmitter at 500 MHz for ^{1}H-NMR analyses and a second transmitter system at 125 MHz for ^{13}C-NMR. The console also houses the receiver electronics, which process the NMR signal from the probe receiver coils, the electronics that control the shim electromagnets, and the probe temperature control system. All of this is managed by a computer data/workstation and some NMR-specific peripherals.

In contrast to the large (and expensive) research instruments are the benchtop **time-domain (TD) NMR systems**, also termed **low-resolution NMR**. They do not provide frequency, or structural, information, but they can be applied to various types of content analyses. They are similar in size to a benchtop centrifuge and utilize a common magnet, with no need for coolants or a large Dewar flask. These instruments are relatively inexpensive and simple to use, and they have numerous food analysis applications, such as fat content.

10.4 APPLICATIONS

Table 10.1 lists recent applications of the NMR and related techniques to food analysis. The following discussion describes general applications and some specific applications to food research. See the references in Table 10.1 for detailed information on the various techniques.

Table 10.1 Recent food applications of magnetic resonance spectroscopy and related techniques

NMR method	*Food*	*Analysis*	*Ref.*
MRI	Par-baked bread	Bread inflation and gas cell size and distribution	[1]
^{1}H-MRI	Plum	Non-destructive assessment of bruising	[2]
^{1}H-MRI	Honey	Authenticity screening	[3]
HR ^{1}H NMR	Butter	Quantitative fatty acid composition	[4]
qHNMR	Commercial sweeteners	Quantification of rebaudioside A	[5]
^{1}H and spin-spin relaxation	Starches	Impact of proteins on starch gelatinization	[6]
2-D NMR	Processed food and beverages	Rapid identification of food thickeners	[7]
^{1}H NMR, TOCSY, HSQC and HMBC	Dry-cured hams	Metabolic profiling of dry-cured ham during processing	[8]
DOSY-NMR	Beverages	Aroma partitioning and release in beverage systems	[9]
^{1}H, ^{1}H-DPFGSE, F2-DPFGSE band-selective HSQC	Olive oil	Detection of aldehydes and phenolic compounds	[10]
^{13}C qNMR	Wine	In situ determination of fructose isomer concentration	[11]
HR-Solid-state ^{13}C MAS NMR	Wheat flour	Structural characterization of water-unextractable cell wall material	[12]
UF iSQC NMR	Viscous liquid foods	Sugar content, quality testing, and determination of adulteration	[13]
TD NMR	Caviar	Through-package water, fat, and protein determination	[14]
TD NMR	Wheat dough	Water mobility and gluten microstructure during mixing	[15]
TD NMR	Beef	Meat quality parameters related to aging	[16]
TD NMR	Vegetable emulsions	TD-NMR for water droplet distribution and oil droplet distribution	[17]
TD NMR (SMART6 ProFat™/ ORACLE™)	Chicken sausage	Moisture and fat contents	[18]
HR-MAS-NMR	Green tea leaves	Chemical profile and chemometric analysis for quality control	[19]
2D ^{1}H J-resolved NMR	Fish	Assessment of intact fish freshness	[20]
CP-MAS-NMR	Japanese radish	Release of rhamnogalacturonan during cooking	[21]
CP-MAS-NMR	Starch	Molecular organization	[22]

NMR nuclear magnetic resonance spectroscopy, *MRI* magnetic resonance imaging, *HR* high resolution, *qHNMR* quantitative proton NMR, *TOCSY* total correlation spectroscopy, *HSQC* heteronuclear single quantum coherence, *HMBC* heteronuclear multiple bond correlation, *DOSY-NMR* diffusion-ordered ^{1}H NMR, *DPFGSE* double pulsed field gradient spin echo, *qNMR* quantitative NMR, *MAS* magic-angle spinning, *UF iSQC NMR* ultrafast intermolecular single-quantum coherence, *TD NMR* time-domain NMR, *HR-MAS* high-resolution magic-angle spinning, *CP-MAS* cross polarization magic-angle-spinning NMR

10.4.1 NMR Techniques and General Applications

10.4.1.1 Liquids

Liquid NMR is used for relatively pure samples that readily go into solution in any of the many deuterated solvents produced for NMR analyses. Common samples include carbohydrates, proteins, lipids, phenolics, and many other classes of organic compounds. The experiment described in Sect. 10.2.4 is an example of a typical liquid 1-D ^{1}H-NMR spectroscopy analysis. The result is a 1-D NMR spectrum with the resonances plotted along the *x*-axis (the only plotted axis, hence, 1-D). This is the simplest application of NMR spectroscopy, but it can be very informative. For example, plant- and yeast-derived β-glucans are currently an important topic of research, as they have many health-related benefits and they are often discarded as waste from many food processing systems. NMR spectroscopy is the best analytical tool available to determine the purity and identity of the β-glucans as various food processors, particularly the cereal, baking, and brewing industries, work to extract these valuable byproducts from the waste stream in a cost-effective manner. Figure 10.6 shows a 1-D NMR spectrum of 1,3-1,4 mixed linkage β-glucans from cereal (oat) processing waste. From this spectrum, both the purity and relative ratio of 1,3 to 1,4 linkages could be determined. If additional structural information is needed, there are many powerful 2-D and 3-dimensional (3-D) NMR analyses available for the assignment of the chemical shifts of each ^{1}H and ^{13}C atom in an organic molecule.

NMR spectroscopy is also a valuable assay tool in batch ingredient analysis for quality assurance. In such assays, the structural assignments of the spectra would not be as important as the consistency of the spectra compared to a spectrum of a high-quality control product. This application can be used with many types of ingredients because NMR solvents are available for compounds with a range of solubility properties.

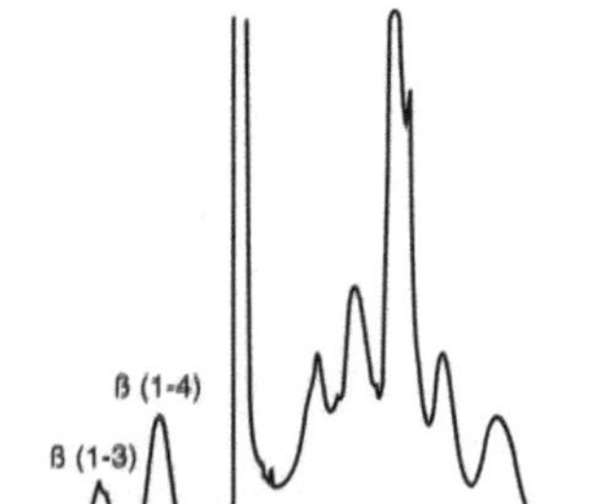

Fig. 10.6 A 1-D H-NMR spectrum of 1,3-1,4 mixed linkage β-glucans from oat processing waste. Both the purity of the sample and the ratio of 1,3 to 1,4 linkages could be determined from the spectrum

10.4.1.2 Solids

The principles that underlie solid-state NMR are similar to those discussed in Sect. 10.2; however, due to the fact that the sample is not freely tumbling about in solution, there is a "directional" aspect (anisotropic or orientation-dependent interactions) to the solid-state analysis. The anisotropic nature of solids results in very broad signals and yields spectra that lack the structural information obtained from samples in solution. One method to overcome this problem is **magic-angle spinning** (MAS), in which the line broadening due to the anisotropy is countered when the sample holder is spun at a specific "magic" angle relative to the external field, B_0, yielding much narrower lines. MAS is often combined with cross-polarization enhancement (CP-MAS), in which the magnetization from more easily detected nuclei is transferred to those that are less easily detected (such as from ^{1}H to ^{13}C).

Solid-state NMR analyses can be applied to many types of samples, such as powders and fresh vegetable tissue. Solid-state ^{13}C-CPMAS-NMR techniques can be used to monitor the chemical composition and the physicochemical properties in the solid portion of an intact food sample. This has been applied to composition studies of different mushroom species, and solid-state ^{13}C-CPMAS-NMR spectroscopy showed significant differences in the ratio of carbohydrate to protein resonances between different species. Also, high-resolution ^{1}H-MAS-NMR techniques enabled food researchers to discriminate between durum wheat flours from Southern Italy, which differ in composition depending on the region of origin. A similar application was used to correlate composition with origin in a study of Parmesan cheese.

10.4.1.3 Magnetic Resonance Imaging

Magnetic resonance imaging (MRI) is unique in that the sample can be placed into the magnet in the native form, and 2-D or 3-D images of the sample can be generated. MRI involves variations in field strength and the center frequency of the pulses over time and space, along with the application of field gradients in different geometric positions relative to the magnet bore (B_0). The end result is a spatial "encoding" of the sample protons with different phase and frequency values. After multidimensional Fourier transformations of multiple FIDs from different spatial "slices" of the sample, an image of the sample is produced that contains information about the state of the tissue or other material under study.

The sample can be a medical patient, a small test animal, a diseased plant stem, a ripened fruit, or even a complex food product in various steps of processing or its final form. For example, a packaged product could be analyzed over time in the package to track water movement or loss. The MRI analyses would not affect the product. There are many potential applications for MRI in the food industry; for example, it can be used to image the freezing process in frozen food production, with the goal of increasing shelf life. MRI may also be applied to analyze the composition and characteristics of pastes and sauces, to locate voids in products, or to examine the fat distribution in meats or the water/fat distribution in emulsions. It can also provide detailed information about the thickness of a filling or coating; structural changes, including water loss, in a product via heat transfer (cooking); or changes associated with hydration of a food product during processing. When combined with rheological analyses, sauce and paste flow in a processing system may be monitored.

MRI images of clementine fruit are shown in Fig. 10.7. One image shows freeze damage to the interior pericarp region of the fruit. The other image shows the presence of an unwanted seed. Such problems often are undetected by a simple visual inspection of the fruit. As the high costs of purchasing and maintaining a large-bore MRI instrument decrease over time, as they did for NMR spectroscopy, and as smaller bore instruments become more common, this important tool should become available to even small food companies and food science departments. These instruments may become a common sight in even modest research and development laboratories over the next decade.

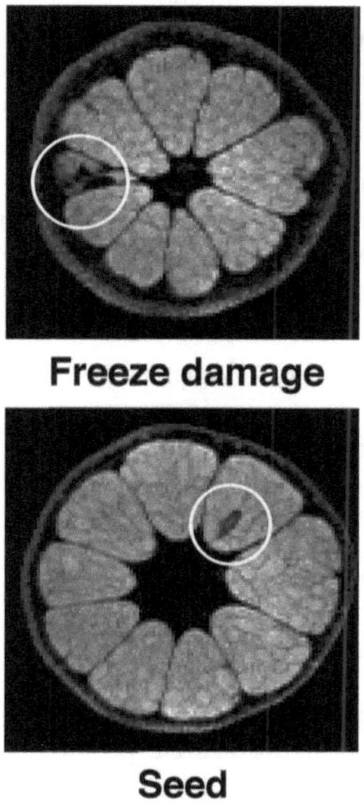

Fig. 10.7 MRI images (18 mm slice thickness) of clementine citrus fruit with defects. Freeze damage is shown in the image on the top, and an unwanted seed is shown in the image on the bottom. (Images courtesy of Michael McCarthy, Aspect AI Ltd., Netanya, Israel)

10.4.1.4 Relaxometry

Plasticizers are small molecules that act to make systems, composed of large polymers, more fluid. Water acts as a plasticizer in food processing; and the amount of water available to act as a plasticizer is a very important factor for food quality. An increase or a reduction in the amount of plasticizer can affect the glass transition process, which, in turn, affects the quality of the final product. Water exists in several states in food, and the interaction of water molecules with food components can be investigated by the measurement of NMR relaxation. This includes both the spin-lattice ($T1$) and spin-spin ($T2$) relaxation times (Sect. 10.2.2) of the water protons. The relaxation times are related to the magnetic interactions of water protons with the surrounding environment, and the effective relaxation time is related to the extent of the association between the water molecules and immobilized or slowly moving macromolecules. In general, as the macromolecular content increases, the relaxation times of the water protons also increase.

10.4.1.5 TD-NMR for Content Analyses

A recent development in NMR spectroscopy is instrumentation consistent with AOAC Method 2008.06 (moisture and fat in meats), using the FAST Trac NMR system (CEM Corporation), which combines rapid drying (microwave oven) with low-resolution (time domain) NMR in a benchtop instrument for moisture and fat analysis. CEM Corporation (Matthew, NC) makes two such instruments that combine a microwave plus NMR for fat content determination (ORACLE™, meets ISO/DIS 16756 for milk and milk products) or moisture/solids and fat contents (ORACLE™ paired with the SMART6™). These relatively inexpensive systems are operator friendly and easy to maintain. They provide rapid, reproducible information for quality control of food products, yielding information on the moisture and fat content of any product. The ORACLE™ is able to analyze products ranging from 0.05% to 100% fat in approximately 30 s. For example, the ORACLE™ Rapid NMR Fat Analyzer has been used to determine fat content in ice cream, chicken sausage, beef muscle meat, soybean oil oleogels, and upcycled mushroom protein.

10.4.2 Specific Food Application Examples

High-resolution NMR spectroscopy has been used for the analysis of complex systems such as food samples, biofluids, and biological tissues because it provides information on a wide range of compounds found in the food matrix in a single experiment. The short time frame needed to obtain NMR spectra (minutes), coupled with automation, enables the analysis of many samples with minimal operator input. There are two basic types of analysis in the application of NMR to the food industry: (1) identification of distinct resonances and, therefore, specific compounds and (2) the use of chemometric profile analysis, in which spectral profiles are compared without assigning particular resonances.

10.4.2.1 Oil/Fat

10.4.2.1.1 Fatty Acid Profile

Physical and chemical properties of fats, oils, and their derivatives are mainly influenced by their fatty acid profile. Even though gas chromatography (GC) is usually used for determining the fatty acid profile (Chap. 17, Sect. 17.2.7; Chap. 23, Sect. 23.6.2), the common unsaturated fatty acids, such as oleic, linoleic, and linolenic acids in an oil or fat sample can be quantified using ^{1}H-NMR, by integration of select signals in the spectra. Additionally, the fatty acid distribution on the glycerol anchors can be obtained from ^{13}C-NMR analysis. There are two groups of resonances in the carbonyl region of the spectrum: the first is due to fatty acids in positions 1 and 3 and the second is from fatty acids in position 2 of the glycerol moiety. Information about the fatty acid distribution on the glycerol anchors is important to determine the functionality of the ingredient in processing, such as the crystallization point or how it plasticizes the dough in a baked product. For example, the correct type of fat is essential for quality pie crusts or croissants.

10.4.2.1.2 Verification of Vegetable Oil Identity

Even though different oils or fats may be purposely mixed for specific reasons, the adulteration of high-value oils with oils of lesser value is an issue of economic and commercial importance. This is primarily a problem with olive oil because it is expensive and has superior nutritional value. Accordingly, many studies from major olive oil-producing Mediterranean countries, such as Greece, Italy, and Spain, deal with identifying lower-value oils, such as hazelnut oil, used for adulterating olive oil. The adulteration problem is complicated by the fact that lower-value oils usually have fatty acid profiles similar to olive oil. Among the methods used for analyzing potentially adulterated olive oil are ^{13}C-NMR and ^{1}H-NMR spectrometry. For example, NMR is utilized in conjunction with multivariate statistical analyses of specific resonances in NMR spectra of olive oil diluted with hazelnut or sunflower oil. These methods also can be used to identify the variety and geographical origin of the oil.

10.4.2.1.3 Monitoring of Oxidation

The oxidation of vegetable oils is a significant quality problem and can lead to further deterioration of the oil. Highly unsaturated fatty acids, with *bis*-allylic methylene groups, are particularly susceptible to oxidation. Primary and secondary oxidation products, such as hydroperoxides and aldehydes, are easily detected by ^{1}H-NMR analyses. ^{1}H-NMR is especially useful for such analyses because the samples do not require any additional treatments, such as derivatization, that could cause degradation.

10.4.2.1.4 Solid Fat Content (SFC)

While most analyses discussed in this chapter depend on high-resolution NMR instruments, a benchtop, low-resolution-pulsed NMR instrument can be used to determine the SFC of a sample (see also Chap. 23, Sect. 23.3.7.1). For example, the amount of solid triacylglycerols in the oil or fat at different temperatures can be determined. This method is based on the difference in relaxation times between solids and liquids, and after a delay, only the NMR signal of the liquid fat is measured. The solid content is then estimated. Crystallization mechanisms of fat blends also can be studied using SFC measurements.

10.4.2.2 Water

The glass transition is an important property of foods (see also Chap. 30), and the glass transition temperature (*Tg*), which is dependent on water content, impacts both the processing and the storage of food products. *Tg* can be determined with an **NMR state diagram**, which is a curve relating NMR relaxation time to glass transition temperature at different moisture contents. This information is important because processing and storage temperatures above *Tg* at any point during production and distribution of a product are associated with more rapid deterioration. Spin–spin relaxation time (T2) is commonly used as an indication of proton mobility, which is different above and below the *Tg* of a given product. Although the differential scanning calorimeter (DSC) (Chap. 33, Sect. 33.3.2) is most commonly used for simple *Tg* analyses, the ability to generate NMR state diagrams increases the value of NMR for many applications.

10.4.2.3 Ingredient Assays

Adulteration in fruit juice is not easy to detect by taste or color. For example, orange juice can be blended with relatively inexpensive grapefruit juice, but the presence of grapefruit juice in a commercially available orange juice product poses serious health risks for consumers with certain medical conditions. Grapefruit juice has a number of coumarin-like flavonoids and other powerful CYP450 inhibitors that negatively impact the metabolism of many prescribed drugs. Therefore, the detection and prevention of this kind of adulteration are especially important. NMR-based chemometric approaches using independent component analysis, a variant of principle component analysis, are now applied to this problem. Another common issue with juice preparation is the differentiation between freshly squeezed juices and those produced from pulp washes, which can be added to fresh-squeezed orange juice to reduce production costs. ^{1}H NMR, in combination with principal component analyses, can easily and accurately distinguish the fresh-squeezed and pulp-wash orange juice.

NMR is used in monitoring batch-to-batch quality and production site differences in beer. Large multinational breweries prepare their beers at many different geographic locations and require methods for quality control at a detailed molecular level. NMR can be used in conjunction with principal component analysis to distinguish beer from different production sites based on lactic acid, pyruvic acid, dextran, adenosine, inosine, uridine, tyrosine, and 2-phenylethanol content. Quantifying these compounds allows the producers to identify production sites where there is greater variability in these compounds (therefore, poorer quality control).

10.5 SUMMARY

Nuclear magnetic resonance technology provides powerful research instrumentation for a variety of applications, from structural elucidation of complex molecules to 3-D imaging of fresh tissue, to simple ingredient assays for quality assurance. NMR differs from most other forms of spectroscopy because the nucleus is the subject of analysis and the excitation step uses radio-frequency electromagnetic energy. The proton (H) and the ^{13}C isotope are the most commonly studied nuclei, and each has a characteristic charge and spin, which results in a small, local magnetic field. NMR analyses require an external magnetic field, which causes the local magnetic fields of the nuclei to align in a parallel or antiparallel orientation. There is a slight excess in the parallel orientation (in the z-axis aligned with B_0), and it is the net magnetic vector of this population that is detected during an NMR experiment. A pulse of RF energy moves this net magnetism into the xy-plane, where a reemitted radio signal (the NMR signal) is detected. This signal, which decays quickly, contains the intensity and frequency information for all the nuclei in the sample, and the resulting FID is converted by Fourier transformation into the NMR spectrum, which shows the various resonances spread along the x-axis based on differences in frequency.

The NMR instrument consists of a cryomagnet with the transmitter and receiver antennae in the central bore, an electronics console with the transmitter and receiver hardware, and a data/workstation that controls all the functions of the instrument. In addition to NMR spectrometers, with both solids and liquids applications, there are other related instruments, such as MRI, that are based on the same principles but yield different information. Among the common applications of NMR to food science are structural studies that examine the correlation between chemical structure and health benefits or functionality of food ingredients, studies of the effects of processing on food properties and quality, composition studies of food ingredients or even fresh vegetable tissue, imaging of food products, and determination of SFC or ingredient purity.

10.6 STUDY QUESTIONS

1. Explain the basic principles associated with NMR spectroscopy, including the function of the magnet and the concept of nuclear spin.
2. Describe the interaction of the net magnetization with the RF pulse (90°) and the subsequent NMR signals.
3. Explain the concept of shielding and chemical shift.
4. Describe the FID and the NMR spectrum, including the concepts of time domain, frequency domain, and data transformation.
5. List the components of the NMR spectrometer and their functions.
6. What kinds of samples are analyzed by (a) liquids NMR and (b) solid-state NMR?
7. What kind of final data does one obtain with an MRI? List two applications of MRI.
8. What is the primary use of relaxometry in food analysis?
9. List the general types of food applications of NMR and give an example of each.

REFERENCES

1. Grenier D, Lucas T, Le Ray D (2019) Enhanced aeration of part-baked bread using a novel combination of baking and partial vacuum. J Food Eng 248: 62–70
2. Fyfe S, Smyth HE, Schirra HJ, Rychlik M, Sultanbawa Y, Kurniawan ND (2023) Physical properties and magnetic resonance imaging of the Australian green plum (*Buchanania obovata*) through maturity and ripening and across locations. Postharvest Biol Technol 205:112494
3. Hategan AR, Guyon F, Magdas DA (2022). The improvement of honey recognition models built on ^{1}H NMR fingerprint through a new proposed approach for feature selection. J Food Compos Anal 114:104786
4. Fadzillah NA, Rohman A, Salleh RA, Amin I, Shuhaimi M, Farahwahida MY, Rashidi O, Aizat JM, Khatib A (2017) Authentication of butter from lard adulteration using high-resolution of nuclear magnetic resonance spectroscopy and high-performance liquid chromatography. Int J Food Prop 20(9):2147–2156
5. Paniagua-Vega D, Cavazos-Rocha N, Huerta-Heredia AA, Parra-Naranjo A, Rivas-Galindo VM, Waksman N, Saucedo AL (2019) A validated NMR method for the quantitative determination of rebaudioside A in commercial sweeteners. J Food Comp Anal 79:134–142
6. Luo H, Guo C, Lin L, Si Y, Gao X, Xu D, Jia R, Yang,W (2020) Combined Use of Rheology, LF-NMR, and MRI for Characterizing the Gel Properties of Hairtail Surimi with Potato Starch. Food Bioprocess Technol 13:637–647

7. Radziej S, Scherb-Forster J, Schlicht C, Eisenreich W (2021) Fast identification of food thickeners by nontargeted NMR-spectroscopy. J Agric Food Chem 69(12):3761–3775
8. Zhang J, Yi Y, Pan D, Zhou G, Wang Y, Dang Y, He J, Li G, Cao J (2019). ^{1}H NMR-based metabolomics profiling and taste of boneless dry-cured hams during processing. Food Res Int 122:114–122
9. Yeo H, Linforth R, MacNaughtan W, Williams H, Hewson L, Fisk ID (2023) Effect of sweeteners and carbonation on aroma partitioning and release in beverage systems. Food Res Int 164:112373
10. Klikarová J, Rotondo A, Cacciola F, Česlová L, Dugo P, Mondello L, Rigano F (2019) The phenolic fraction of Italian extra virgin olive oils: Elucidation through combined liquid chromatography and NMR approaches. Food Anal Methods 12:1759–1770
11. Colombo C, Aupic C, Lewis AR, Pinto BM (2015) In situ determination of fructose isomer concentrations in wine using 13C quantitative nuclear magnetic resonance spectroscopy. J Agric Food Chem 63(38):8551–8559
12. De Man WL, Chandran CV, Wouters AG., Radhakrishnan S, Martens JA, Breynaert E, Delcour JA (2022) Hydration of Wheat Flour Water-Unextractable Cell Wall Material Enables Structural Analysis of Its Arabinoxylan by High-Resolution Solid-State 13C MAS NMR Spectroscopy. J Agric Food Chem 70(34):10604–10610
13. Cai H-H, Chen H, Lin Y-L, Feng J-H, Cui X-H, Chen Z (2015) Feasibility of ultrafast intermolecular single-quantum coherence spectroscopy in analysis of viscous liquid foods. Food Anal Methods 8(7):1682–1690
14. Huang L, Cheng S, Song Y, Xia K, Xu X, Zhu BW, Tan, M (2017) Non-destructive analysis of caviar compositions using low-field nuclear magnetic resonance technique. J Food Meas Charact 11:621–628
15. Dufour M, Foucat L, Hugon F, Dugué A, Chiron H, Della Valle G, Kansou K, Saulnier L (2023) Water mobility and microstructure of gluten network during dough mixing using TD NMR. Food Chem 409:135329
16. Cônsolo NR, Silva J, Buarque VL, Samuelsson LM, Miller P, Maclean PH, Moraes TB, Barbosa LC, Higuera-Padilla A, Colnago LA, Netto AS (2021) Using TD-NMR relaxometry and 1D ^{1}H NMR spectroscopy to evaluate aging of Nellore beef. Meat Sci 181:108606
17. Linke A, Weiss J, Kohlus R (2020) Factors determining the surface oil concentration of encapsulated lipid particles: impact of the emulsion oil droplet size. Eur Food Res Technol 246:1933–1943
18. Rigdon M, Stelzleni AM, McKee RW, Pringle TD, Bowker B, Zhuang H, Thippareddi H (2021) Texture and quality of chicken sausage formulated with woody breast meat. Poultry Sci 100(3):100915
19. Maria de Fátima CS, Rech KS, Dutra LM, Menezes LR, Santos ADDC, Nagata N, Stefanello MÉA, Barison A (2023) ^{1}H HR-MAS NMR chemical profile and chemometric analysis as a tool for quality control of different cultivars of green tea (*Camellia sinensis*). Food Chem 408:135016
20. Tan C, Huang Y, Feng J, Li Z, Cai S (2018) Freshness assessment of intact fish via 2D ^{1}H J-resolved NMR spectroscopy combined with pattern recognition methods. Sens Actuators B 255:348–356
21. Tsuchida M, Yamaguchi H, Katayama N, Sato Y, Kawashima W, Kasai M (2021) Structural changes in cell wall of Japanese radish accompanied by release of rhamnogalacturonan during pressure cooker heating. Food Chem 349:129117
22. Okyere AY, Bertoft E, Annor GA (2019) Modification of cereal and tuber waxy starches with radio frequency cold plasma and its effects on waxy starch properties. Carbohydr Polym 223:115075

RESOURCE MATERIALS

Findeisen M, Berger S (2013) 50 and more essential NMR experiments: a detailed guide. Wiley–VCH, Weinheim, Germany

Capozzi F, Laghi L, Belton PS, (2015) Magnetic resonance in food science: defining food by magnetic resonance. The Royal Society of Chemistry, Cambridge, England

Günther H (2013) NMR Spectroscopy: Basic Principles, Concepts and Applications in Chemistry, 3rd edn. Wiley, New York

Lambert JB, Mazzola EP, Ridge CD (2019) Nuclear magnetic resonance spectroscopy: an introduction to principles, applications, and experimental methods. Wiley, New York

McMurry JE (2016) Organic chemistry, 9th edn. Brooks-Cole, Salt Lake City, UT

Skoog DA, Holler FJ, Crouch SR (2018) Principles of instrumental analysis, 7th edn. Brooks-Cole, Salt Lake City, UT

Van Duynhoven J, Van As H, Belton PS, Webb GA (eds) (2013) Magnetic resonance in food science: food for thought. The Royal Society of Chemistry, Cambridge, England

Mass Spectrometry

11

Umut Yucel and J. Scott Smith

Contents

ABBREVIATIONS

AMS	Accelerator mass spectrometer
APCI	Atmospheric pressure chemical ionization
API	Atmospheric pressure ionization
APPI	Atmospheric pressure photoionization
CAD	Collision-activated dissociation
CI	Chemical ionization
CID	Collision-induced dissociation
DART	Direct analysis in real time
ECD	Electron capture dissociation
EI	Electron impact ionization
ES	Electrospray
ESI	Electrospray ionization
ETD	Electron transfer dissociation
FT	Fourier transform
FT-ICR	Fourier transform-based ion cyclotrons
FTMS	Fourier transform mass spectrometry
GC	Gas chromatography
GC-MS	Gas chromatograph-mass spectrometry
HRMS	High-resolution mass spectrometry

U. Yucel (✉) · J. S. Smith
Department of Animal Sciences and Industry, Kansas State University, Manhattan, KS, USA
e-mail: yucel@ksu.edu; jsschem@k-state.edu

B. P. Ismail, S. S. Nielsen (eds.), *Nielsen's Food Analysis*, Food Science Text Series,
https://doi.org/10.1007/978-3-031-50643-7_11

ICP-MS	Inductively coupled plasma-mass spectrometry
IMS	Ion mobility mass spectrometry
IT	Ion trap
ITMS	Ion trap mass spectrometry
LC	Liquid chromatography
LC-MS	Liquid chromatography-mass spectrometry
m/z	Mass-to-charge ratio
MALDI	Matrix-assisted laser desorption ionization
MALDI-TOF	Matrix-assisted laser desorption ionization-time-of-flight
MRM	Multiple reaction monitoring
MS	Mass spectrometry
MS/MS	Tandem mass spectrometry
MS^n	Multiple stages of MS (tandem mass spectrometry)
OT	Orbitrap
Q	Quadrupoles mass filters
Q-TOF	Quadrupole time of flight
SFC	Super critical fluid chromatography
SRM	Selected-reaction monitoring
TIC	Total ion current
TOF	Time of flight
TQ	Triple quadrupole
TWIM	Travelling wave
UHPLC	Ultrahigh-performance liquid chromatography

11.1 INTRODUCTION

Over the past decade, **mass spectrometry** (MS) techniques have become indispensable for the identification, characterization, verification, and quantitation of small molecules (e.g., caffeine, 194 Da) to large complex biomolecules (e.g., immunoglobulin, 144,000 Da). The power of the MS technique is due to its ability to place a charge on a molecule, thereby converting it to an **ion** in a process called **ionization.** The generated ions are then separated according to their **mass-to-charge ratio** (*m/z*) by subjecting them to a combination of radio-frequency (RF) and electrostatic fields in a **mass analyzer.** The **ions** are finally detected by a highly sensitive **detector.** The resulting signals from the detector are digitized and processed by software to display the information as a mass spectrum, which reveals its molecular mass and its structural composition, leading to identification. These steps, in sequence, are given in Fig. 11.1 and are visualized in Fig. 11.2 as a schematic diagram of one type of mass spectrometer. The MS system shown has a quadrupole type of mass analyzer (described in Sect. 11.2.4.2), and the incoming sample comes from a gas chromatography (GC) instrument. A second stage of **ion fragmentation** may be included before detection to elicit structural information in a technique known as **tandem MS** (also known as **MS/MS**). In tandem MS, two or more mass analyzers are coupled together to increase selectivity, specificity, and sensitivity of the detection and structural elucidation of various compounds.

Two important developments led to the rapid rise in popularity of MS as an analytical technique. First was the development of **hyphenated** MS techniques, which coupled MS to the separation techniques of gas chromatography (GC) (Chap. 14) or liquid chromatography (LC) (LC is short for HPLC, high-performance liquid chromatography) (Chap. 13). This coupling of chromatography and MS dramatically lowered the detection limits for quantitative analysis while simultaneously increasing the confidence of measurement through high specificity. Second was the development of **hybrid** (i.e., using a combination of mass analyzers), benchtop, MS instruments that made high-resolution, accurate mass, LC-MS analysis routine. Hyphenated, hybrid MS techniques deliver robust, highly sensitive, precise measurements that withstand the rigor of statistical analysis for the purposes of quantitative analysis, while significantly reducing sample preparation time and effort. These advantages made MS a "must-have" technique when faced with complex

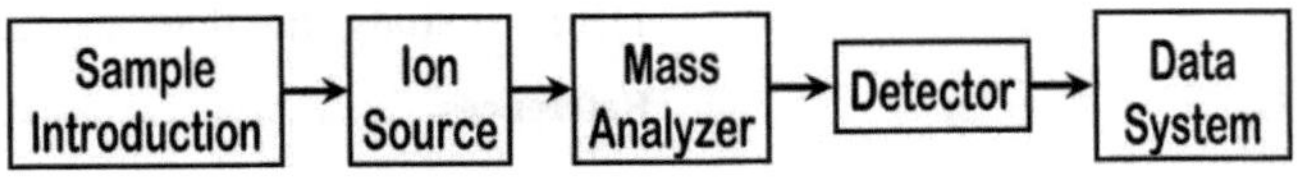

Fig. 11.1 A block diagram of the major components of a mass spectrometer

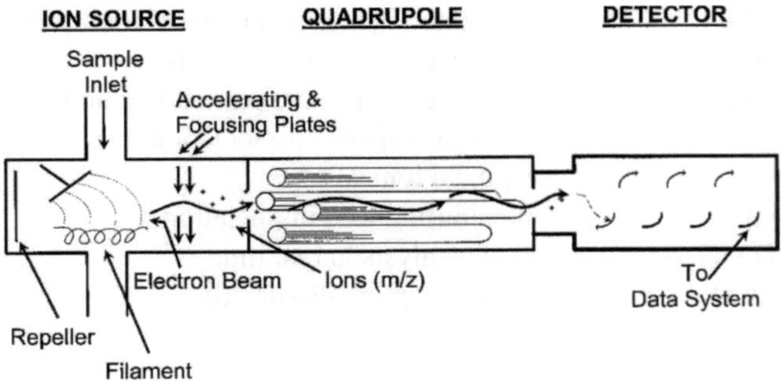

Fig. 11.2 Schematic of a typical mass spectrometer. The sample inlets (interfaces) at the *top* and *bottom* can be used for direct injection or interfacing to a GC

bioanalytical challenges such as pesticide screening in foods, trace analysis of environmental pollutants, composite characterization of natural products, or rapid identification of food-borne bacteria.

The GC-MS systems, which originated in the late 1960s, are simpler to operate and require less maintenance as compared to the more expensive and sophisticated LC-MS systems, which made ionization from liquids possible and started to gain adoption in the late 1980s. Both of these MS hyphenated techniques find common applications in many industries as prominent analytical methods. Due to its high speed and low-maintenance features, GC-MS is preferred for the analysis of thermally stable volatile molecules as described in Chap. 14, such as essential oils, flavors and off-flavors, fatty acids, alcohols, pesticides, and gases. Due to their versatility, LC-MS/MS systems are becoming common analytical workhorses for the analysis of a broader range of food constituents than can be analyzed by HPLC (Chap. 13). These include sugars and oligosaccharides (Chap. 19), amino acids and peptides (Chap 24), vitamins (Chap. 20), and chemical contaminants and compounds of concern including pesticide residues, mycotoxins, antibiotics, and sulfites (Chap. 33). Furthermore, discovered in 1988, the matrix-assisted laser desorption ionization (MALDI) unit is often coupled to the time-of-flight (TOF) mass analyzer. This "MALDI-TOF" system offers ionization from solid crystals and is popularly used for the analysis of peptides and proteins.

11.2 INSTRUMENTATION: THE MASS SPECTROMETER

11.2.1 Overview

Because there are so many acronyms associated with the MS instrumentation, a list of the acronyms used in this chapter is given just after Sect. 11.9. Many of those acronyms are first used in Table 11.1, which summarizes mass spectrometer components and types of instruments. This table also helps introduce the three basic functions an MS performs (also stated in Sect. 11.1): (1) There must be a way to **ionize** the molecules, which occurs in the ion source by a variety of techniques. (2) The charged molecular ion and its fragments must be **separated** according to their *m/z*, and this occurs in the **mass analyzer** section. (3) The separated, charged ions must be detected (electron multipliers, photomultipliers).

Sample introduction can be **static** or **dynamic,** the latter of which involves interfacing with GC or LC instruments. Since all mass spectrometers work in high vacuum, regardless of the state of the sample (gas, liquid, or solid), all ions are introduced into the MS as a gas. The MS **interface** converts the samples into an acceptable form for introduction into the vacuum chamber. Common MS interfaces will be discussed in more detail in the sections on GC-MS (Sect. 11.4) and LC-MS (Sect. 11.5).

As introduced in Sect. 11.1, Fig. 11.2 depicts the interior of a typical GC-MS instrument that uses a quadrupole mass analyzer. The region between ion generation and detection is maintained by different vacuum pumps. Each successive region from the ion source is kept at lower vacuum than the preceding region, with the mass analyzer/detector being in the region of strongest vacuum ($\approx 10^{-6}$–10^{-8} torr). A vacuum is necessary for two critical reasons: (1) to avoid ion–molecule reactions between the charged ions and other gaseous molecules before they reach the detector, and (2) for proper operation of ion lenses, mass analyzer electrodes, and ion detectors that require the use of high voltages. Vacuum performance determines the sensitivity and resolution of mass spectrometers.

11.2.2 Sample Introduction

11.2.2.1 Static Method

The initial step in operating the MS is to get the sample into the ion source chamber. Pure compounds or sample extracts of volatile or semivolatile compounds are injected directly into the source region without the requirement of a chromatographic separation. This **static method** of introducing the sample to the source is called **direct injection** and allows real-time MS analysis. Solutes with limited solubility can be analyzed using the **direct insertion probe method**, in which the sample is placed in a small cup at the end of a stainless steel rod or probe. The probe is inserted into the ion source through one of the sample inlets, and the source is heated until the solid vaporizes. The mass spectrum is then obtained on the vaporized solid material as with the direct injection method. Both direct injection and direct insertion probe methods work well with pure samples, but their use is very limited when analyzing complex mixtures of several compounds.

Direct analysis in real time (DART) is an example of a static sampling technique, in which ionization occurs at atmospheric pressure, as in atmospheric pressure chemical ionization (APCI). The ionization is achieved by using metastable helium (He) ions (19.8 eV energy) via the Penning process (much like EI), resulting in radical cations ($M^{+\cdot}$). A mixture of heated He and nitrogen is used to initiate the metastable ionization process, essentially creating a plasma-

Table 11.1 Summary of mass spectrometer components and types

	Types	*Applications and advantages*
Sample introduction		
Static method	Direct injection	Gas or volatile liquid
	Direct insertion probe	Solids
Dynamic	GC	Gas or volatile liquids
	LC	Nonvolatile solids or liquids
Ion source	Electron impact ionization (EI)	Primarily for GC-MS, for volatile compounds; electron energies at 70 eV
	Fast atom bombardment (FAB)	Soft ionization; argon or xenon atom beams used; lower S/N ratio; for more polar nonvolatile compounds
	Electrospray ionization (ESI)	Most popular method for LC-MS; normally for polar or slightly polar compounds
	Atmospheric pressure chemical ionization (APCI)	Primarily for LC-MS; normally for compounds of low polarity and some volatility; molecular weight determination of peptides
	Atmospheric pressure photoionization (APPI)	Same uses as APCI but has advantages in signal-to-noise ratio and detection limit
	Matrix-assisted laser desorption ionization (MALDI)	A "soft ionization," ideal for large biopolymers and other fragile molecules
	Chemical ionization (CI)	A "soft ionization," ideal for large biopolymers and other fragile molecules; provides less fragmentation
Mass analyzers	Quadrupoles mass analyzer/filter (Q)	Used in many types of instruments; compact; commonly used in benchtop instruments
	Ion trap (IT)	Capable of tandem MS (MS^n); higher resolution than quadrupoles
	Time of flight (TOF)	Useful to analyze biopolymers and large molecules
	Fourier transform-based mass analyzer (FT-ion cyclotrons, FT-ICR; FS-orbitrap)	Allows for easy-to-use benchtop LC-MS
	Magnetic sector	Specialized applications requiring ultrahigh resolution, e.g., dioxin analysis
	Isotope ratio MS	Useful in geochemistry and nutrition science; extreme specificity
	Accelerator mass spectrometer	Useful in geochemistry and nutrition science; extreme specificity
Hybrid MS: Common combinations of mass analyzers	Quadrupole TOF (e.g., Q-TOF, triple TOF)	Most LC-MS; provides for MS/MS, benchtop instruments
	Triple quadrupole (e.g., TQ; tandem MS)	Common for LC-MS. provides for MS/MS, benchtop instruments
	Ion trap (e.g., IT-FTMS, IT-orbitrap, Q-trap, IT-TOF)	Most LC-MS. provides for MS/MS. very high mass accuracy
Common MS instruments	Quadrupole MS (single quadrupole or TQ)	Quantitative and qualitative analysis
	ITMS	Qualitative analysis. Advantage of multistages of MS (MS^n)
	TOF/Q-TOF	Higher resolution than ITMS; high scan speed; allows structural analysis
	FTMS	High-resolution accurate mass needs; allows structural analysis

rich environment, wherein the metastable He reacts with ambient water, creating protonated water clusters, resulting ultimately in a charge transfer to the analyte of interest. The process has been well described by Hajslova et al. [1] for food QC and safety analysis. For example, in an earlier study, Schober and Peterson used an APCI-MS technique to monitor the concentrations of selected flavor molecules in a person's breath during consumption of hard candies [2].

11.2.2.2 Dynamic Method

For mixtures, sample introduction is a **dynamic method,** in which solutes need to be separated prior to MS analysis. This is done typically by GC or HPLC units connected to an MS by an **interface** (see Sects. 11.4 and 11.5). The interface removes excess GC carrier gas or HPLC solvent that would otherwise overwhelm the vacuum pumps of the MS.

11.2.3 Ionization

There are many methods used to produce ions for the compounds, depending on the type of chromatographic interface and nature of the compounds (Table 11.1). The major types of ion sources are briefly described in the subsections that follow.

11.2.3.1 Electron Impact Ionization (EI)

In GC-MS techniques, once the compound(s) coming from the GC enters the ion source, it is exposed to a beam of electrons emitted from a filament composed of rhenium or tungsten metal. When a direct current is applied to the filament (usually 70 electron volts, eV), it heats and emits electrons that move across the ion chamber toward a positive electrode. As the electrons pass through the source region, they come in close proximity to the sample molecule and extract an electron, forming an ionized molecule. Once ionized, the molecules contain such high internal energies they can further fragment into smaller molecular fragments. This entire process is called **electron impact** (EI) **ionization**, although the emitted electrons rarely hit a molecule.

11.2.3.2 Electrospray Ionization (ESI)

Electrospray ionization, the most popular LC-MS technique in use today, functions at atmospheric pressure and is a highly sensitive technique. Normally, polar compounds are amenable to ESI analysis, with the type of ion produced depending on the initial charge. That is, positively charged compounds yield positive ions, while negatively charged compounds such as those containing free carboxylic acid functional groups will produce negative ions.

The ESI source, as depicted in Fig. 11.3, consists of a nozzle that contains a fused-silica capillary sample tube (which serves to transfer the LC effluent) coaxially positioned within a metal capillary tube to which a variable electrical potential can be applied against a counter electrode,

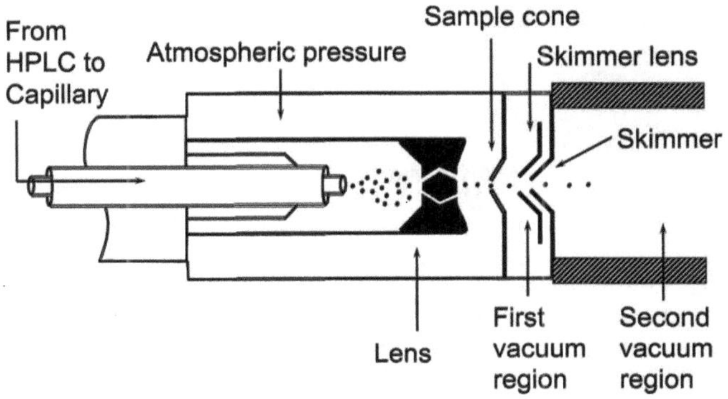

Fig. 11.3 Schematic of an electrospray LC-MS interface

which is usually the entrance to the MS. Compressed nitrogen gas at high velocity is coaxially introduced to aid in the nebulization of the LC effluent as it exits the tip of the metal capillary tube. The relative velocity difference between the streams of nitrogen gas (fast moving) and LC effluent (slow moving) at the ESI tip results in producing a fine spray of highly charged droplets. At nanoflow rates (<1 μL/min), the force of the electrical field is strong enough to break up the LC effluent into fine droplets without the use of nebulizing gas, in a process known as **nanospray**. For conventional HPLC flow rates (1–1000 μL/min), the sheer volume of liquid requires an initial droplet size reduction through the use of nebulizing nitrogen gas, creating the required microdroplets, which can now be influenced by the prevailing electrical field.

At this point, the repulsive forces due to the accumulation of "like" charges inside the rapidly reducing microdroplet volume create an imbalance with the forces of surface tension that are trying to conserve the spherical structure of the microdroplet. The positive charge is drawn out, but cannot escape the surface of the liquid, and forms what is known as a **Taylor cone**. Further reduction of the diameter of the droplets causes the Taylor cone to stretch to a critical point, at which the charge escapes the liquid surface and is emitted as a gas-phase ion in a process known as a **Coulombic explosion**.

One of the many advantages of the ESI process is its ability to generate multiple-charged ions and tolerate conventional HPLC flow rates. Proteins and other large polymers (e.g., between 2000 and 70,000 Da) can be easily analyzed on LC-MS systems having a mass limit of *m/z* 2000, due to this multiple charging phenomenon. Powerful software can process in excess of +50 charge states, to yield the molecular ion information for larger proteins. A limitation of the ESI process is the phenomenon of ion suppression/enhancement or matrix effects, which usually causes a variation in response for the analyte signal intensity in the presence of matrix components. Matrix factor corrections are used to account for ion suppression/enhancement effects, including the use of stable-labeled internal standards or matrix-assisted calibration curves for quantitative analysis.

11.2.3.3 Atmospheric Pressure Chemical Ionization (APCI)

The **APCI interface**, which operates at atmospheric pressure (like ESI), is normally used for compounds of low polarity and some volatility. In comparison to ESI, in which the liquid droplet of analyte is protonated, APCI is harsher and ionization occurs in the gas phase. Therefore, gas-phase chemistries of the analyte and solvent vapor play an important part in the APCI process (i.e., protonated solvent molecule donates a proton to the analyte).

Figure 11.4 shows the schematic diagram of an APCI interface. The LC effluent-carrying fused-silica capillary tube protrudes about halfway inside a silicon-carbide (ceramic) vaporizer tube. The vaporizer tube is maintained at approximately 400–500 °C and serves to vaporize the LC effluent. High voltage is applied to a corona needle positioned near the exit of the vaporizer tube. The high voltage creates a **corona discharge** that forms reagent ions from the mobile phase and nitrogen nebulizing gas. These ions react with the sample molecules (M) and convert them to ions. A common cascade of reactions occurring in the presence of water, nitrogen gas, and the high-voltage corona discharge is as follows:

$$e^- + N_2 \rightarrow N_2^{+\cdot} + 2e^- \quad (11.1)$$

$$N_2^{+\cdot} + H_2O \rightarrow N_2 + H_2O^{+\cdot} \quad (11.2)$$

$$H_2O^{+\cdot} + H_2O \rightarrow H_3O^+ + OH^\cdot \quad (11.3)$$

$$M + H_3O^+ \rightarrow (M+H)^+ + H_2O \quad (11.4)$$

The APCI interface is a robust interface and can handle high flow rates of up to 2 mL/ min. It is unaffected by minor changes in buffer strength or composition and is typically used to analyze molecules less than 2000 Da. It does not facilitate multiple charges and hence cannot be used to analyze large biomolecules/polymers. In terms of matrix effects, APCI usually shows "ion enhancement" rather than "ion suppression." This is due to the matrix components enriching the plasma generation process, thereby enhancing the efficiency of the ionization process. As a result, there is an increase in response for the analyte signal in the presence of matrix components, requiring matrix factor correction through the appropriate use of stable-labeled internal standards or matrix-assisted calibration curves for quantitative analysis.

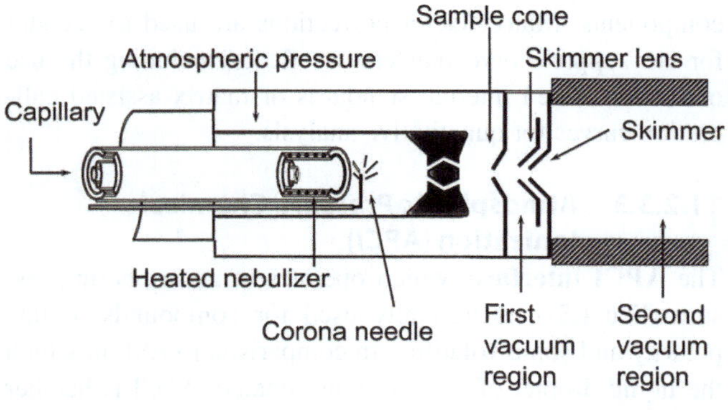

Fig. 11.4 Schematic of an atmospheric pressure chemical ionization LC-MS interface

11.2.3.4 Atmospheric Pressure Photoionization (APPI)

APPI is an ionization technique that improves on the interface possible with APCI. The APPI interface, which uses a krypton or xenon light source to generate a beam of photons instead of a corona discharge-generated plasma as in APCI. Compounds having ionization potentials lower than the wavelength of the light source will be ionized. Since most HPLC solvents do not ionize at the wavelengths generated by the commonly used photon sources, APPI improves the signal-to-noise ratio and hence detection limits.

11.2.3.5 Matrix-Assisted Laser Desorption Ionization (MALDI)

In MALDI, the sample is dissolved in an energy-absorbing matrix and ionized using a UV laser. The matrix plays an important role in ionization, acting both as the absorber of the laser energy, which causes it to vaporize, and as a proton donor and acceptor to initiate charge transfer to the analyte (Fig. 11.5). Since the sample is not directly ionized, MALDI is considered a "soft ionization" technique, and amenable to ionization of large biopolymers and other fragile molecules such as nucleic acids or carbohydrates [3]. The matrix used in MALDI is usually a weak organic acid with UV-absorbing properties [e.g., 2,5-dihydroxy benzoic acid (DHB), sinapinic acid (3,5-dimethoxy-4-hyroxycinnamic acid), gentisic acid (DHDA, 2,5-dihydroxybenzoic acid), or α-cyano-4-hydroxycinnamic acid (CHCA)]. Sensitivity is usually dependent on

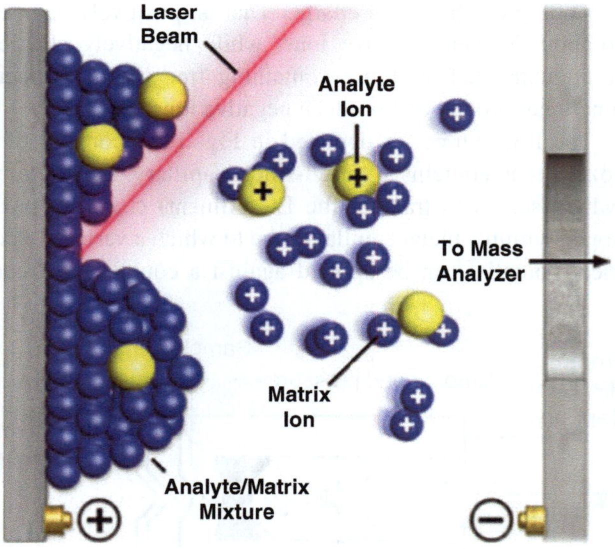

Fig. 11.5 Diagram of the MALDI desorption and ionization process used in some TOF instruments. (From Chaughtai and Heeren [3], used with permission of National High Magnetic Field Laboratory, www.magnet.fsu.edu)

critical pairing of the chemistries of matrix with the sample, especially for samples that are inherently nonvolatile or insoluble in most aqueous solvents.

The typical laser used for MALDI applications is neodymium-doped yttrium aluminum garnet (Nd-YAG) nitrogen laser operating at 337 nm or 355 nm (3.7–3.5 eV photon energies) in vacuum and pulsed at a repetition rate between 1 and 10 KHz. The laser beam size can be attenuated between 5 and 100 μm, which allows hundreds of laser shots to raster through a single sample spot. While the ionization mechanism is not fully understood, it is believed that a two-step process occurs; in step one, the matrix absorbs the UV energy from the laser and is consequently ionized $(M + H)^+$ and in step two a charge transfer to the sample $(S + H)^+$ is completed, allowing the charged sample to be focused into the mass analyzer. Infrared lasers are also used but are less popular, as is the case with atmospheric pressure-based MALDI ion sources.

11.2.3.6 Matrix Effects on Ionization

One key issue with all types of ionization is a phenomenon called **matrix effect**. This is when ionization of a molecule is either suppressed or enhanced by coeluting endogenous interferences contained in the matrix after sample cleanup. This effect has a direct impact on sensitivity; for the same level (e.g., 1 ng/mL), the ion intensity of the compound of interest will change in response to the coeluting matrix interferences such as salts, fatty acids, phospholipids, etc. For high-sensitivity quantitative MS analysis, a preliminary study to elucidate matrix effects is essential. For example, if pesticide residues in spinach and corn need to be analyzed, extracts of each material will have to be individually studied for matrix effects. In the process, a set of test pesticide standards at known levels are spiked in a clean matrix (spinach or corn not exposed to pesticides) and their peak intensities compared to the same level spiked in pure solvent. The difference in peak intensities between the pure standard versus matrix-spiked standard will determine the matrix effect during the final analysis of the samples to be tested. While all modes of ionization are susceptible to matrix effects, ESI seems to be most prone to ion suppression. Other techniques such as APCI can lead to ion enhancement, wherein the peak intensity is increased in the presence of matrix as compared to the pure standard.

11.2.3.7 Transition from Ion Source to Mass Analyzer

The eventual outcome of the ionization process, by any of the methods described in sections above, is the formation of both negatively and positively charged molecules of various sizes unique to each compound. When the repeller plate at the back of the ion source is positively charged, it repels the positive fragments toward the quadrupole mass analyzer. Thus, we look only at the **positive fragments**, although negative fragments are sometimes analyzed. As the positively charged fragments leave the ion source, they pass through holes in the accelerating and focusing plates. These plates serve to increase the energy of the charged molecules and to focus the beam of ions, so that a maximum amount reaches the mass analyzer.

11.2.4 Mass Analyzers

11.2.4.1 Overview

The heart of an MS is the **mass analyzer**. It performs the fundamental task of separating the charged molecules or their fragments based on their *m/z*, and it dictates the mass range, accuracy, resolution, and sensitivity. The sensitivity of a mass analyzer determines its limit of detection (LOD) and limit of quantification (LOQ) (terms introduced in Chap. 4, Sect. 4.3.4). Listed in Table 11.1 are the basic types of mass analyzers, the common combinations of basic mass analyzers (called **hybrid MS**), and the most common types of MS instruments, along with their typical applications. Described in the subsections below are only the four types of mass analyzers most commonly applied to food analysis.

11.2.4.2 Quadrupole Mass Analyzers (Q)

The word "quadrupole" of the quadrupole mass analyzer is derived from the Latin word for "fourfold" (quadruplus), and "pole," to describe the array of four rods that are used (Fig. 11.2). The four rods are used to generate two equal but out-of-phase electric potentials: one is alternating current (AC) frequency of applied voltage that falls in radio-frequency (RF) range, and one is direct current (DC). The potential difference can be varied to create an oscillating electrical field between two of the opposite rods, resulting in their having equal but opposite charges.

When, for example, a positive-charged ion enters the quadrupole field, it will be instantly attracted toward a rod maintained at a negative potential, and if the potential of that rod changes before the ion impacts, it will be deflected (i.e., change direction). Thus, every stable ion (i.e., ion with stable flight path) entering the quadrupolar region traces a sine wave-type pattern on its way to the detector. By adjusting the potentials on the rods, selected ions, a mass range, or only a single ion can be made stable and detected. The unstable ions impact one of the four rods, releasing them from the influence of the oscillating field, and they are pumped away by the vacuum pumps. A quadrupole mass analyzer is commonly referred to as a **mass filter**, because, in principle, the device filters ions that achieve stability from those that do not.

11.2.4.3 Ion Trap (IT) Mass Analyzers

Ion traps are essentially multidimensional quadrupole mass analyzers that store ions (trap) and then eject these trapped ions according to their *m/z* ratios. Once the ions are trapped, multiple stages of MS (MS^n) can be achieved, mass resolution can be increased, and sensitivity can be improved. The major difference between an ion trap and a quadrupole mass analyzer is as follows: in an ion trap, the unstable ions are ejected and detected while the stable ions are trapped (referred to as **MS in time**); in a quadrupole, the ions with a stable flight path reach the detector, and the unstable ions hit the rods and are pumped away (**MS in space**).

Figure 11.6 shows the cross-sectional view of a 3-D ion trap mass analyzer. It consists of a ring electrode sandwiched between a perforated entrance, end-cap electrode, and a perforated-exit, end-cap electrode. An AC (RF) voltage and variable amplitude is applied to the ring electrode, producing a 3-D quadrupole field within the mass analyzer cavity, which is responsible for trapping the ions, thus the efficiency of the MS [4].

Ions formed in the source are electronically injected into the ion trap, where they come under the influence of a time-varying RF field. The ions are trapped within the mass analyzer cavity, and the applied RF voltage drives ion motion in a figure eight toward the end-caps. Thus, for an ion to be trapped, it must have a stable trajectory in both the axial and radial directions. To detect the ions, the frequency applied to the ring electrode is changed and the ion trajectories are made unstable. Helium is continuously infused into the ion trap cavity and primarily serves as a dampening gas. The 3-D ion trap technology was improved with the introduction of **2-D ion traps**, which increased the ion trapping efficiency an order of magnitude by spreading the ion cloud in a quadrupole-like assembly [5].

11.2.4.4 Time-of-Flight Mass Analyzer (TOF)

Time-of-flight mass analyzers separate ions according to the time required to reach the detector while traveling over a known distance (Fig. 11.7). Ions are pulsed from the source with the same kinetic energy, which causes ions of different *m/z* ratios to acquire different velocities (lighter ions travel faster while heavier ions travel slower). The difference in velocities translates to difference in time reaching the detector, upon which the mass spectrum is generated. Theoretically, TOF instruments have no upper mass range, which makes them useful for the analysis of biopolymers and large molecules, and have fast cycles since they technically transmit all *m/z* ions (full scan mode). The resolution can be a challenge at high masses due to the time difference in the flight times of identical ions. The use of reflectrons (ion mirrors) can quickly increase mass resolution of TOF instruments by increasing ion drift path length by bouncing ions in a V or W pattern without drastically increasing the instrument footprint.

11.2.4.5 Fourier Transform-Based Mass Spectrometry (FTMS)

Fourier transform-based mass spectrometers deconvolute image currents produced by ion motion (harmonic oscillations or cyclotron motion) into mass spectra. FTMS technology is especially useful for extremely high-resolution mass spectroscopy (HRMS) applications, as explained in Sect. 11.7. A Fourier transform ion cyclotron resonance (FT-ICR) mass analyzer traps ions in a magnetic field (Penning traps), while a Fourier transform orbitrap mass analyzer traps ions in an electric field. Both analyzer types are unique as compared to the previously listed mass analyzers because the ions themselves are not detected by impinging upon a detector, but rather the frequency (cyclotron motion) is measured as a function of the applied electric (orbitrap) or magnetic field (ICR). Commercially launched in 2005, the orbitrap offers high resolution (400 K resolution at *m/z* 200) and is

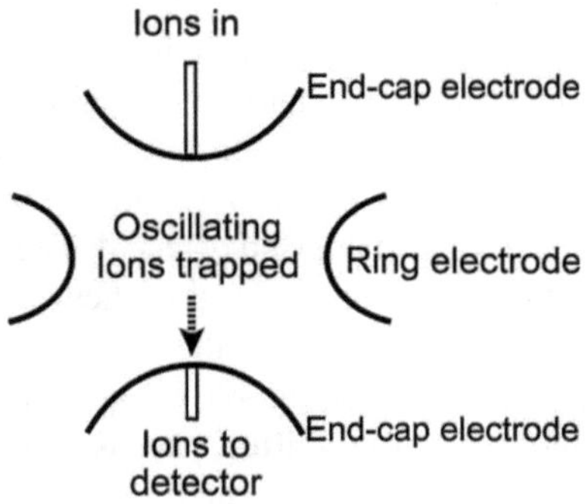

Fig. 11.6 Diagram of an ion trap mass analyzer

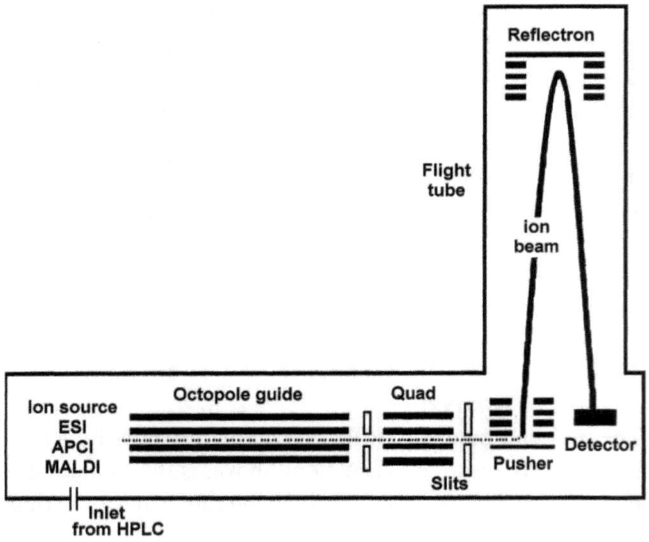

Fig. 11.7 Diagram of a single stage time-of-flight instrument

simpler to operate than an FT-ICP MS system (Fig. 11.8). While traditional FTMS (FT-ICP) delivers significantly higher resolution (7 M resolution at *m/z* 600), they require liquid helium-cooled superconducting magnets. This requirement makes them large in size, significantly more expensive than other mass analyzers, and difficult to operate without specialized operators—all of which limit their widespread adoption.

11.3 INTERPRETATION OF MASS SPECTRA

As previously indicated, a **mass spectrum** is a plot (or table) of the intensity of various mass fragments (*m/z*) produced when a molecule is subjected to one of the many types of ionization techniques. In classic GC-MS, the electron beam generated by a heated filament (used to ionize the molecules) is usually kept at a constant potential (70 eV), which produces sufficient ions without too much fragmentation. This fragmentation process would result in a loss of the higher-molecular-weight ions. Another advantage of using 70 eV for ionization is that the resulting mass spectra are usually very similar regardless of the make and model of the instrument.

In the full scan mode, all the mass fragments generated with ionization are detected to generate a total ion current (TIC) chromatogram, which allows for computer-assisted mass spectral matching to database libraries for unknown compound identification. This type of measurement is typically used in qualitative and nontargeted screening of the unknown molecules in composite materials. Typical MS setups come with a database of mass spectra and associated matching software, which provides possible matching of the unknown compounds as percentages (i.e., matching quality as correlation coefficient) to the ones listed in the database (not all chemicals are in the database). In fact, it is imperative to note that the accurate compound identification typically requires complementary analysis and considerations. A common approach involves verification by comparing the linear retention index (LRI) of analytical standards of the compounds suggested by the matching algorithm (i.e., LRI is further discussed in Chap. 14 on GC). Alternatively, the quantitative analyses of the targeted molecules with the known mass spectra are typically performed by using the selected ion monitoring (SIM) scan mode, in which the major ion fragments are selected to increase their relative abundances. Thus, the SIM mode lowers the LOQ by reducing the background noise and increasing the sensitivity of the instrument. The quantification process typically involves the use of an internal standard to account for artifacts from sample preparation and an external standard over a linear concentration range to produce a calibration curve (i.e., also known as a key curve), which correlates analyte concentrations with peak areas. It is worth noting that MS alone lacks the capability for complete resolution of molecular structures required for identification of unknown molecules among thousands of possibilities in biochemical systems. It is not uncommon that MS-hyphenated systems are also used in combination with detectors for LC and GC (Chaps. 13 and 14), to improve the qualitative power of the analysis.

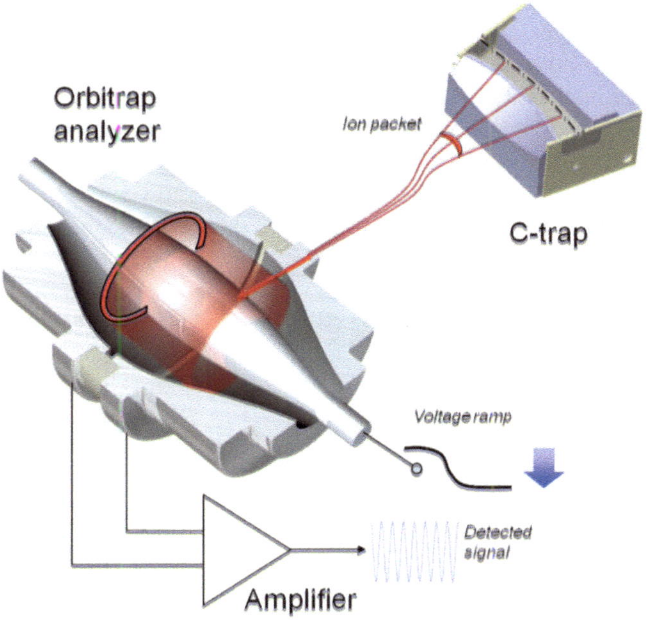

Fig. 11.8 Diagram of the orbitrap analyzer. Ions are captured in the C-trap after ionization during typical GC or LC-MS. They are then sent to the trap analyzer where the detected signal is then converted to mass https://en.wikipedia.org/wiki/Orbitrap#/media/File:OrbitrapMA&Injector.png

Typical mass spectra include only positive fragments that usually have a charge of +1. Thus, the **mass-to-charge ratio** is the molecular mass of the fragment divided by +1, which equals the mass of the fragment. The mass-to-charge ratio unit has no name and is currently abbreviated by the symbol *m/z* (older books use *m/e*). A mass spectrum for butane is illustrated in Fig. 11.9. The relative abundance is plotted on the *y*-axis, and the *m/z* is plotted on the *x*-axis. Each line on the bar graph represents a *m/z* fragment with the abundance unique to a specific compound. The spectrum always contains what is called the **base peak** or **base ion**. This is the fragment (*m/z*) that has the highest abundance or intensity, which is considered as 100%. The abundances of all the other *m/z* ions are adjusted relative to the base peak. The base peak is always presented as 100% relative abundance and is useful in compound identification. Butane has the base peak at an *m/z* of 43 (Fig. 11.9).

Another important fragment is the **precursor ion** (often called the **molecular ion** or **parent ion**), designated by the symbol $M^{+\bullet}$. This peak has the highest mass number and represents the positively charged intact molecule with an *m/z* equal to the molecular mass. The harsher ionizing techniques, such as the EI (Fig. 11.9), produce an ion (radical cation) by stripping an electron (*m/z* 58 for butane). The precursor ion is not always present in the mass spectra since it may fragment before passing through the mass analyzer. The lack of precursor ion in the mass spectra may present a problem for determining the molecular mass of an unknown molecule. The mass of a single electron is considered insignificant; thus, the molecular ion produced by EI-type ionization is equal to the molecular weight of that compound. All other molecular fragments originate from this charged ion; thus, it is called the **precursor** (molecular) **ion**. The remainder of the mass spectrum is a consequence of the stepwise cleavage of large fragments to yield smaller ones termed **product ions (daughter ions)**. The process is relatively straightforward for alkanes, such as butane, making possible identification of many of the fragments.

As indicated previously, the initial step in EI ionization is the abstraction of an electron from the molecule as electrons from the beam pass in close proximity. The equation below illustrates the first reaction that produces the positively charged product ion:

$$M + e\ (\text{from electron beam}) \Rightarrow \begin{array}{l} M^{+\bullet}(\text{molecular ion}) \\ + \\ 2\,e\ (\text{one electron from the electron beam and one from the molecular ion, } M) \end{array} \tag{11.5}$$

The M symbolizes the unionized molecule as it reacts with the electron beam and forms a radical cation. The cation will have a *m/z* equal to the molecular weight. The precursor (molecular) ion then sequentially fragments in a unimolecular fashion. (Note that the ion is often written as M^+, for which the free electron, symbolized by the dot, is assumed. Regardless, the molecule has lost one electron and still retains all the protons; thus, the net charge must be positive.)

The reactions of butane as it forms several of the predominant product ion (daughter) fragments are shown below:

$$CH_3-CH_2-CH_2-CH_3+e \Rightarrow CH_2-CH_2-CH_2-CH_3^{+\cdot}\ (m/z=58)+2e \tag{11.6}$$

$$CH_3-CH_2-CH_2-CH_3^{+\cdot} \Rightarrow CH_3-CH_2-CH_2-CH_2^{+}\ (m/z=57)+{}^{\cdot}H \tag{11.7}$$

$$CH_3-CH_2-CH_2-CH_3^{+\cdot} \Rightarrow CH_3-CH_2-CH_2^{+}\ (m/z=43)+\bullet CH_3 \tag{11.8}$$

$$CH_3-CH_2-CH_2-CH_3^{+\cdot} \Rightarrow CH_3-CH_2^{+\cdot}\ (m/z=29)+\cdot CH_3-CH_3 \tag{11.9}$$

$$CH_3-CH_2 \Rightarrow CH_3^{+\cdot}(m/z=15)+\bullet CH_2 \tag{11.10}$$

Many of the fragments for butane result from direct cleavage of the methylene groups. With alkanes, fragments that are produced by the sequential loss of CH_2 or CH_3 groups are traceable in the mass spectrum.

Close examination of the butane mass spectrum in Fig. 11.9 reveals a peak that is 1 *m/z* unit larger than the molecular ion at *m/z* = 58. This peak is designated by the symbol *M* + 1 and is due to the naturally occurring isotopes. The most abundant isotope of carbon has a mass of 12; however, a small amount of ^{13}C is also present (1.11%). Any ions that contained a ^{13}C or a deuterium isotope would be 1 *m/z* unit larger, although the relative abundance would be low.

Another example of MS fragmentation patterns is shown for methanol in Fig. 11.10. Similarly, the fragmentation pattern is straightforward. The precursor ion (CH_3-$OH^{+\bullet}$) is at a *m/z* of 32, which is the molecular weight. Other fragments include the base peak at a *m/z* of 31 due to CH_2-OH^+, the CHO^+ fragment at a *m/z* of 29, and the CH_3^+ fragment at a *m/z* of 15.

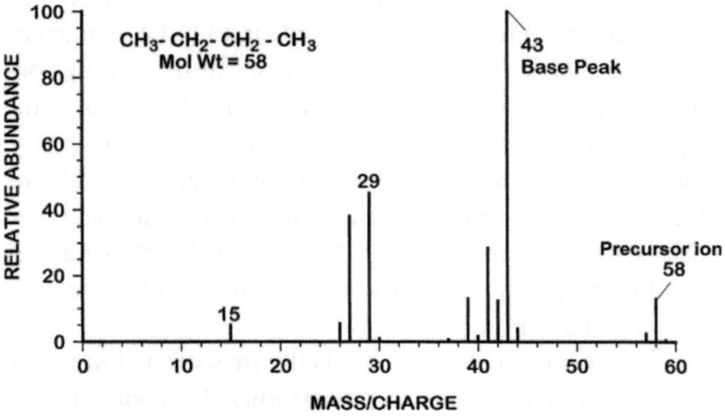

Fig. 11.9 Mass spectrum of butane obtained by electron impact ionization

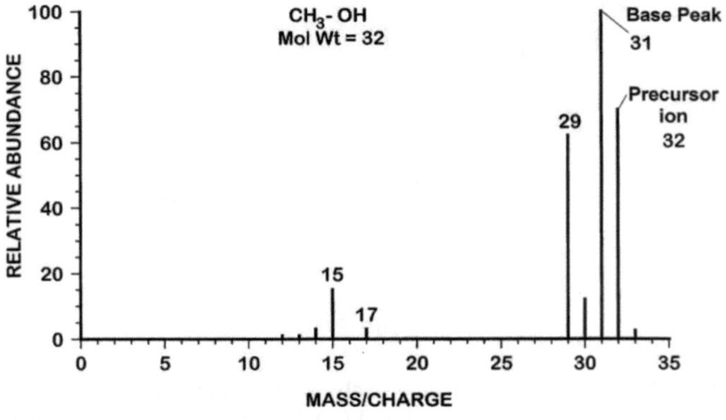

Fig. 11.10 Mass spectrum of methanol obtained by electron impact ionization

The **chemical ionization** (CI) method (see Table 11.1) is classified as a **soft ionization** since only a few fragments are produced. In this technique, a gas is ionized, such as methane (CH_4), which then directly ionizes the molecule. The most important use of CI is in the determination of the molecular ion since there is usually a fragment that is 1 *m/z* unit larger than that obtained with EI. Thus, a mass spectrum of butane taken by the CI method would have a quasimolecular (parent) ion at $m/z = 59$ (M + H). Many LC-MS interfaces use CI or electrospray ionization methods, so it is common to see the $(M + H)^+$ precursor ion. As can be seen in Eq. 11.6, 11.7, 11.8, 11.9, and 11.10, the reactions of the cleavage process can be quite involved. Many of the reactions are covered in detail in the book by McLafferty and Turecek listed in resource materials.

11.4 GAS CHROMATOGRAPHY-MASS SPECTROMETRY

Although samples can be introduced directly into the MS ion source, many applications require chromatographic separation before analysis. The rapid development of **gas chromatography-mass spectrometry** (GC-MS) has allowed for the coupling of the two methods for routine separation problems (see Chap. 14). A MS coupled to GC allows nontargeted screening of complex materials and identification of unknown molecules often with the use of complementary approaches (e.g., LRI and analytical standards) as explained in Sect. 11.3. In most cases, a **capillary GC column** is connected directly to the MS source via a **heated capillary transfer line**. The transfer line is kept hot enough so as to avoid condensation of the volatile component eluting from the GC column on its way into the low-pressure MS source. The sample flows through the GC column into the interface and then on to be processed by the MS. A computer is used to store and process the data from the MS.

Another critical function of GC-MS is to ascertain the purity of each peak as it elutes from the column. Peaks of the co-eluting molecules with similar retention times are merged. This information can be very useful for optimization of GC runs and temperature programming. An example is given for the separation of the methyl esters of several long-chain fatty acids (Fig. 11.11). Long-chain fatty acids must have the carboxylic acid group converted or blocked with a methyl group to make them volatile (i.e., fatty acid methyl esters, FAME, method; Chap. 17, Sect. 17.2.7). Figure 11.11 shows the TIC **chromatogram** of a sample containing methyl esters of palmitic (16:0), oleic (18:1), linoleic (18:2), linolenic (18:3), stearic (18:0), and arachidic (20:0) acids. The GC tracing showed only four peaks eluting off the column between 15.5 and 28 min, although the sample contained six different methyl esters. A logical explanation considers the poor resolution of one or two of the peaks of the methyl esters. The purity of the observed peaks is determined by taking mass spectra at very short increments of time (1 s or less). The mass spectra of a pure peak will be the same irrespective of the time point on the peak. The first peak at 15.5 min has the same mass spectrum throughout, indicating that only one compound is eluting. The mass spectrum of a pure peak is expected to show high correlation with the spectra database. A computer search of the MS library gives an identification of the peak to the methyl ester of palmitic acid. The mass spectra shown in Fig. 11.12 compare the material eluting from the column to the library

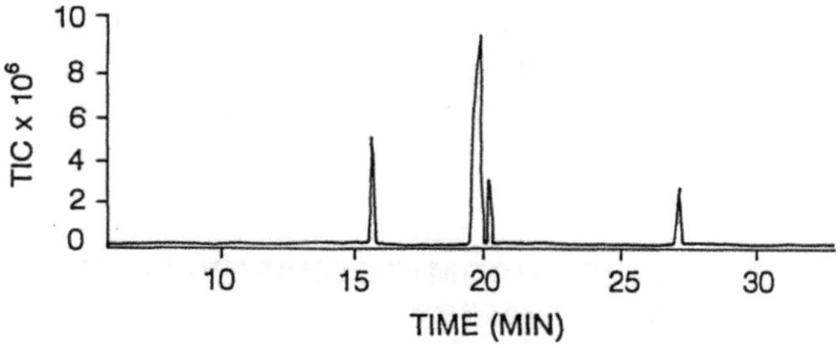

Fig. 11.11 Total ion current GC chromatogram of the separation of the methyl esters of six fatty acids. Detection is by electron impact ionization using a direct capillary interface

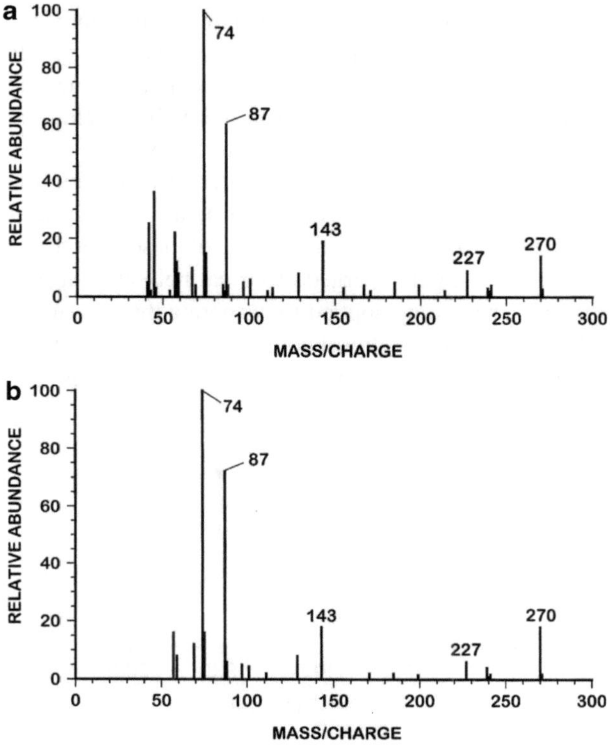

Fig. 11.12 Mass spectra of (**a**) the peak at 15.5 min in the TIC chromatogram shown in Fig. 11.11 and (**b**) the methyl ester of palmitic acid from a computerized MS library

mass spectrum. Most of the fragments match, although the GC-MS scan does have many small fragments not present on the library mass spectrum. This is a common background noise and usually does not present a problem. In a similar way, the peaks at 20 and 27 min observed to contain only one component. The computer match identifies the peak at 20 min as the methyl ester of stearic acid, and the peak at 27 min as the methyl ester of arachidic acid. In comparison, the peak located at 19.5 min is shown to have several different mass spectra, indicating impurity or coeluting compounds. Figure 11.13 shows the region enlarged around 19 min. The arrows indicate where different mass spectra were obtained. These molecules are identified based on library matching as linoleic acid methyl ester, oleic acid methyl ester, and linolenic acid methyl ester at 19.5, 19.7, and 19.8 min, respectively. This example illustrates the tremendous power of GC-MS used for targeted and nontargeted analyses.

11.5 LIQUID CHROMATOGRAPHY–MASS SPECTROMETRY

For a **high-performance liquid chromatography–mass spectrometry** (LC-MS) interface, the same overall requirements must be met as for GC-MS. There must be a way to remove the excess solvent, while converting a fraction of the liquid effluent into the gas phase, making it amenable for MS analysis. Furthermore, most compounds analyzed by HPLC are either nonvolatile or thermally labile, making the task of liquid-to-gas phase transition even more challenging, especially while maintaining compound integrity.

How does LC-MS work? A modern LC-MS ionization interface converts liquid (LC eluent) into gas-phase ions (sampled by the MS) by a process of desolvation in the presence of a highly charged electrical field at atmospheric pressure. The energy applied to evaporate the solvent (thermal and electrical) is almost completely used in the desolvation process, and it does not contribute to degradation

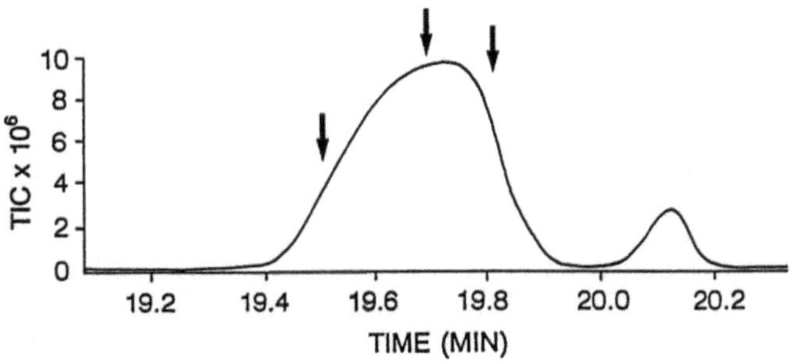

Fig. 11.13 Enlargement of the region 19.2–20.2 min from the TIC chromatogram shown in Fig. 11.11. *Arrows* indicate where mass spectra were obtained

(usually thermal) of any labile species present in the LC eluant. Of the many different types of LC-MS ionization interfaces developed over the years, it was the development of the atmosphere pressure-based ionization interfaces, ESI (Sect. 11.2.3.2) and APCI (Sect. 11.2.3.3) that made LC-MS a routine technique. More recently, an APPI (Sect. 11.2.3.4) has been developed as a complementary technique to APCI.

11.6 TANDEM MASS SPECTROMETRY

Tandem MS (MS/MS, MS^n) is used in both GC-MS and LC-MS but is especially helpful in LC-MS since it allows for characterization, verification, and quantification at ultra-high sensitivity. There are two basic types of tandem MS, one which is a result of **collision-induced dissociation** (CID) typically observed on beam-type instruments (triple quadrupoles, TQ), and the other is a result of **collision-activated dissociation** (CAD) or MS^n, typically observed in ion trap MS (MS in-time type instruments). Other fragmentation modes such as electron transfer dissociation (ETD), infrared multiphoton dissociation (IRMPD), and electron capture dissociation (ECD) are used for the analysis of compounds difficult to fragment.

Tandem MS using LC-**triple quadrupoles** (TQ) (Fig. 11.14) operated in the selected reaction monitoring (SRM) or multiple reaction monitoring (MRM) scan mode provides a factor of 100–1000× more sensitivity than ultraviolet (UV) or diode array detectors, making them indispensable for high-sensitivity quantitative LC-MS analysis. Using the mechanism of CID, the precursor molecule filtered by Q1 energetically collides with argon or nitrogen gas, buffered in the collision cell (Q2), and a specific product ion is transmitted through the third quadrupole (Q3). This mode offers the highest sensitivity critical for quantitative analysis that has been the mainstay of bioanalytical quantitative analysis for the past two decades. The high sensitivity of the SRM/MRM modes are due to two factors: (1) a dramatic improvement in signal-to-noise ratio by eliminating noise, and (2) TQ operation at nearly 100%, allowing seven scans or more to ensure capturing the top of the eluting chromatographic peak to accurately determine the area under the peak. The other advantages are simplicity of data analysis and the high specificity resulting from the MRM mode. The triple quadrupole remains the instrument of choice for high-sensitive, routine LC-MS quantitative analysis, even though it was first introduced in the early 1990s.

Multiple MS or "MS-to-the-nth" made ion traps popular for structural analysis and verification of molecules. Subtle changes can be identified by piecing together the various fragments, which result in the different stages of the MS/MS process. The difference between tandem MS on a TQ and

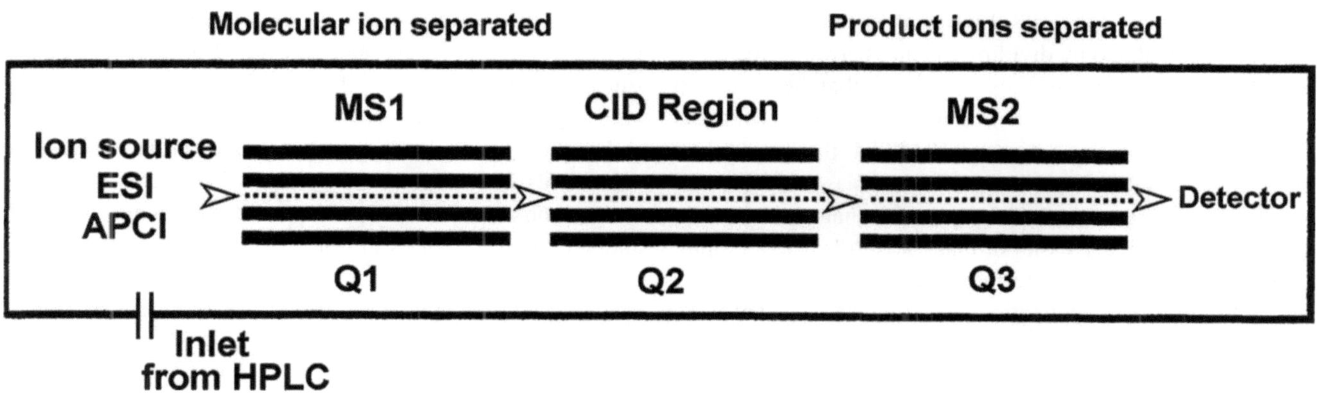

Fig. 11.14 Diagram of a triple quadrupole mass spectrometer capable of doing MS/MS. Q1 and Q2 are used to separate ions, and Q3 is the collision-induced dissociation area (CID). Once the compounds are ionized, they can be allowed to pass through without the CID activated, or the molecular ion can be further fragmented in the Q2 area via CID to yield product ions (fragments)

the MS^n mode on the IT is how the ion undergoes fragmentation. In the CID process on a TQ or Q-TOF, the precursor ion selected in Q1 is accelerated into the collision cell (Q2) filled with argon or nitrogen and smashed into its component pieces in one highly energetic step. In contrast, the CAD process occurring in an ion trap is energetically much gentler, involving a gradual ramping of voltages so that the internal energy of the specified trapped ion increases as it collides with the helium buffer gas, until the most fragile chemical bond breaks. As soon as this occurs, the *m/z* changes, and the product ions no longer experience the gradual ramping of the energy but remain trapped. This process allows for multiple steps of MS/MS analysis at high sensitivity. The most abundant fragment (typically) is then isolated and re-trapped, and the CAD process is repeated. In most cases, CAD can easily perform MS^4, but there have been cases where MS^{10} could be accomplished. Such a method of MS^n analysis makes it significantly easier to piece together the intact structure from its key fragments and determine or verify changes. Since ITMS is tandem MS in time, it has a significant advantage over TQ's operating Q3 in the "full scan MS/MS" mode. The main advantage of MS^n on ITMS is the high sensitivity of the MS^n scan mode, especially when connected to ultra-HPLC (UHPLC) (Chap. 13, Sect. 13.2.3.1).

11.7 HIGH-RESOLUTION MASS SPECTROMETRY (HRMS)

The widespread adoption of orbitrap and Q-TOF MS instruments has made high-resolution, accurate mass applications a growing trend. Resolution in a mass spectrometer is defined in terms of full width at half maxima (FWHM), which is the mass spectrum peak width at half height for a known *m/z* [6]. At unit mass resolution (nominal mass instruments such as quadrupoles and ion traps), FWHM is typically around 0.6 Da, so at *m/z* 300, the resolution would be 500 resolution (300÷0.6). An orbitrap or a Q-TOF, for example, can deliver FWHM of 0.01 Da, so it would deliver a resolution of 30,000 (300÷0.01) for *m/z* 300. An FTMS can deliver FWHM of 0.0001 Da, enabling it to deliver a resolution of 3,000,000 (300÷0.0001). High-resolution analysis has significant advantages in terms of S/N, fine isotopic structure analysis, and determining elemental composition from highly precise accurate mass assignment.

Accurate mass is important in mass spectrometry because it can give you elemental composition and enable identification of unknowns. With accurate mass and MS/MS fragmentation high-resolution data, uncertainty is significantly reduced in identification of "known" unknowns (i.e., compounds that are in a database, but not known to the analyst) and invaluable for screening type qualitative analysis. For mass accuracy below 5 ppm, elemental composition determination becomes straightforward through the application of sophisticated software algorithms.

The listing below for the compound caffeine ($C_8H_{10}N_4O_2$) illustrates the value of accurate mass:

- **Nominal mass:** 194 (used synonymously with molecular mass, is the sum of the integer mass of the most abundant isotope for each element, C = 12, H = 1, N = 14, O = 16)
- **Mono-isotopic mass:** 194.0804 (sum of the most abundant isotopic mass for each of the constituent elements, C = 12.0000, H = 1.007825, N = 14.003074, O = 15.994915)
- **Average mass:** 194.1906 (sum of the average atomic masses or sum of isotopes taking into account the relative abundances for the constitute elements. For example, O_{16} = 15.994915, but the average mass is 15.999405, which takes into account the O_{17} and O_{18} isotopes and their relative abundances)

Accurate mass is measured in terms of parts per million (ppm) and calculated by dividing the mass error by the theoretical mass. In the example above, assume that the measured mass on the high-resolution MS was 194.0811. Given the theoretical monoisotopic mass of 194.0804, this results in a mass error of 0.0007 (194.0811–194.0804), which would result in a mass accuracy of 3.6 ppm for that measurement ([194.0804/0.0007]*10^6). For mass accuracy below 5 ppm, elemental composition determination becomes straightforward through the application of sophisticated software algorithms.

The advancement of high-resolution MS coupled with availability of more databases has enhanced the determination of unknown compounds, a process termed nontargeted or "scouting" analysis. With accurate mass determinations, an unknown can be compared against LC-MS libraries for identification. The elemental composition can be determined and further verified by observing the fragment ions for definitive identification. There are several databases of compounds and their ionization patterns (i.e., ions produced). Most notably is the METLIN metabolite database by the Scripps Center for Metabolomics where the ion spectra are produced by ESI Q-TOF. In the case of caffeine (mass 194.080376), the spectrum contains a total of seven major ions, more than enough for positive identification. Another commonly used mass spectra database is ChemSpider. Unfortunately, the coverage of the databases is limited to the known compounds, and the mass spectrum depends on the type of ionization interface/method (i.e., recall that is not the case with EI-based GC-MS libraries). The limitations of these databases will only improve as more scientists provide additional data and with the use of complementary analysis (e.g., LRI, coupling with other detectors), as mentioned in Sect. 11.3. Both Milman [7] and Lehotay et al. [8] discuss MS libraries, screening of molecules, and nontarget identification.

11.8 APPLICATIONS

The use of MS in the field of food science is well established and growing rapidly as food exports from Asia increase yearly to the USA and Europe. MS-based analytical techniques are indispensable for the analysis of a diverse set of compounds in relation to food quality, safety, traceability, authenticity, and trademark protection. For example, hybrid MS systems are useful in identifying characteristic flavor profiles of traditional meat products, which, in return, assists with the development of plant-based meat alternatives with similar quality attributes. At the same time, MS-based techniques effectively determine the origin of meat protein sources or contamination during processing, thus preventing food adulteration. Regulatory agencies such as the Food and Drug Administration (FDA), Center for Food Safety and Applied Nutrition (CFSAN), European Food Safety Authority (EFSA), Health Canada, and Japan Food Safety Commission employ MS-based techniques to establish regulatory standards for banned substances, ensuring the safety of the food supply chain for human consumption. Chap. 33 provides explanation for the analysis of compounds of concern, such as chloramphenicol, nitrofurans, sulfonamides, tetracyclines, melamine, acrylamide, and malachite green in foods such as honey, fish, shrimp, and milk, for which LC-MS/MS is the most commonly used technique.

The MS applications for food analysis in combination with other techniques are too numerous to provide a comprehensive list. Interested readers can find a plethora of review articles on the subject (e.g., the work of Herrero et al., 2012 exemplifies various hybrid-MS techniques for the analysis of pesticides and antimicrobials in various food commodities [9]).

To give an appreciation of the usefulness of LC-MS, several examples are provided below. It is important to remember that there are a wide variety of methods now available to analyze just about any type of sample in a variety of matrices.

Due to the prevalence of consumption of caffeine-containing drinks throughout the world, the analysis of this small bioactive compound has been of interest for many years. Over 20 years ago, HPLC methods were published showing that caffeine and other alkaloids, theobromine, and theophylline could be analyzed by HPLC using an ultraviolet detector. While HPLC-UV analysis is quite acceptable, the use of LC-MS can verify and enhance identification in a variety of complex food systems.

Figure 11.15 illustrates a reversed-phase HPLC column separation and MS spectrum obtained using the ESI interface coupled with MS/MS [10]. An aqueous coffee extract was filtered and separated by HPLC using an acetic acid–acetonitrile mobile phase. For comparative purposes, a separate HPLC separation was achieved with the same HPLC column except detection was with a UV detector. The HPLC chromatogram in Fig. 11.15a shows the TIC, an indicator of total ions and thus compounds eluting, and matches the HPLC-UV chromatogram (not shown). Figure 11.15b is the selected ion trace of ions *m/z* = 180.7–181.7, which would correspond to the protonated molecular ions $(M + H)^+$ of theobromine and theophylline at 181.2 (both compounds are isomers and have identical masses of 180.2). The chromatogram in Fig. 11.15c is the selected ion trace of *m/z* = 194.2–196.2, which corresponds to the protonated molecular ion of caffeine (195.2). The MS/MS separations of caffeine and theobromine are presented in Figs. 11.15d and 11.15e, which show the respective protonated molecular ions and several ion fragments (*m/z*) for both caffeine and theobromine.

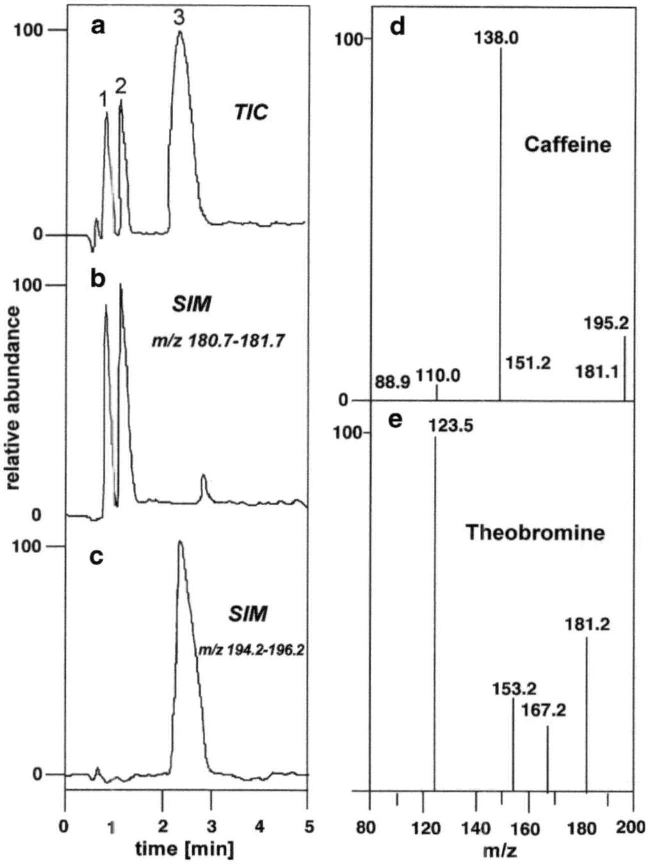

Fig. 11.15 Reversed-phase LC-MS (parts a-c) and MS/MS separation of an aqueous coffee extract: (**a**) TIC, total ion current of extract; (1) theobromine, (2) theophylline, (3) caffeine; (**b**) selected ion trace, *m/z* = 180.7–181.7; (**c**) selected ion trace, *m/z* = 194.2–196.2; (**d**) MS-MS ionization of caffeine; (**e**) MS/MS of theobromine. (From Huck et al. [10], used with permission of Elsevier, New York)

The **melamine** food contamination issue of 2007–2008 is another excellent example of the use of LC-MS and MS/MS in both a detective role and later as an official analytical method. Melamine is a six-membered cyclic nitrogenous ring compound with three amines attached to the carbons in the ring and contains 67% nitrogen (Fig. 11.16). Since a common test for protein measures nitrogen (Chap. 18, Sect. 18.2), the compound was used as an economic adulterant in wheat gluten and dried milk powder to artificially increase the apparent protein content. Due to the polar nature of the amine groups, melamine is not volatile and thus cannot be analyzed with GC unless a derivative is synthesized, thus LC-MS is a preferred method. Methods have been presented that entail the separation of melamine by HPLC with detection by ESI-MS/MS. ESI produces a strong protonated molecular ion at *m/z* = 127.1 with an MS/MS precursor ion (molecular) of 127.1 and a product ion of 85.1. One of the analytical methods suggested by the FDA entails monitoring

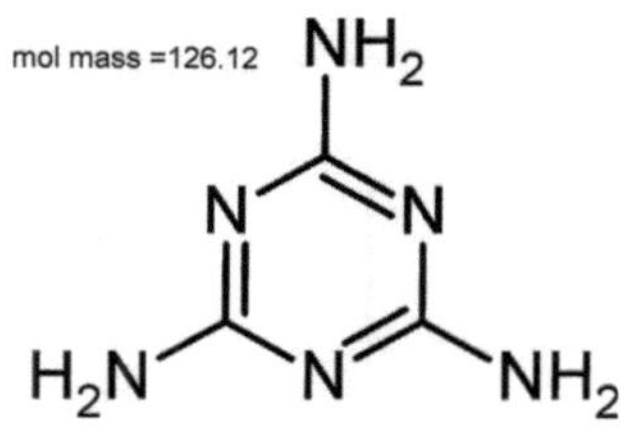

Fig. 11.16 Chemical structure of melamine

of all these ions, which produces very good specificity and sensitivity.

The area of bioactive food components has grown dramatically over the last 10 years in part due to the availability of LC-MS. Many bioactive compounds in fruits, vegetables, and spices are polar and are not volatile. Thus, identification and evaluations were very difficult without HPLC methods coupled with MS. A good example of the use of LC-MS is in the measurement of the polyphenolic flavonoids called the catechins. These antioxidant compounds present in catechu, green tea, cocoas, and chocolates appear to have several beneficial biological effects including enhanced heart and blood vessel health.

Figure 11.17 shows a separation of the major green tea catechins by HPLC with ESI-MS detection [11]. Depicted at the bottom of the figure is the chromatogram showing the separation of the catechins by HPLC. The ESI-MS detector was used to monitor all ions from *m/z* 120–2200 (TIC mode). The top left panel shows the MS obtained for the epicatechin peak eluting at about 12.1 min. As typical with ESI, there are few fragments, though the M + H molecular ion at 291.3 *m/z* is predominant. Further fragmentation (MS/MS) of the epicatechin yields two major protonated fragments, *m/z* 273.3 (loss of OH from the C ring) and 139.3 (oxidation and cleavage of the A ring). With these data, it is possible to elucidate isomers and also possible degradation pathways.

High-resolution analysis on orbitrap technology has found wide application in food analysis, with LC, GC, and supercritical fluid chromatography (SCF) as hyphenated front-end separation techniques before the mass analysis is performed. The examples below demonstrate the power of HRMS techniques for the analysis of pesticide contamination, which is covered in Chap. 33. Rajski et al. [12] performed a large multi-residue screening study (>250 pesticides) in fruits and vegetables using three different resolution settings on the orbitrap (R = 17.5 K, 35 K, and 70 K). The study revealed that using a ± 0.2 min retention time window, R = 70,000, and mass tolerance of 5 ppm gave less than 5% false-positive results at the 10 μg/kg level. Ishibashi et al. [13], used supercritical fluid (SFC) coupled to an orbitrap to simultaneously analyze 373 pesticides (10 μg/kg level which is the provisional MRL in Japan) in a QuEChERS spinach extract using an R = 70,000, and a mass tolerance of 5 ppm.

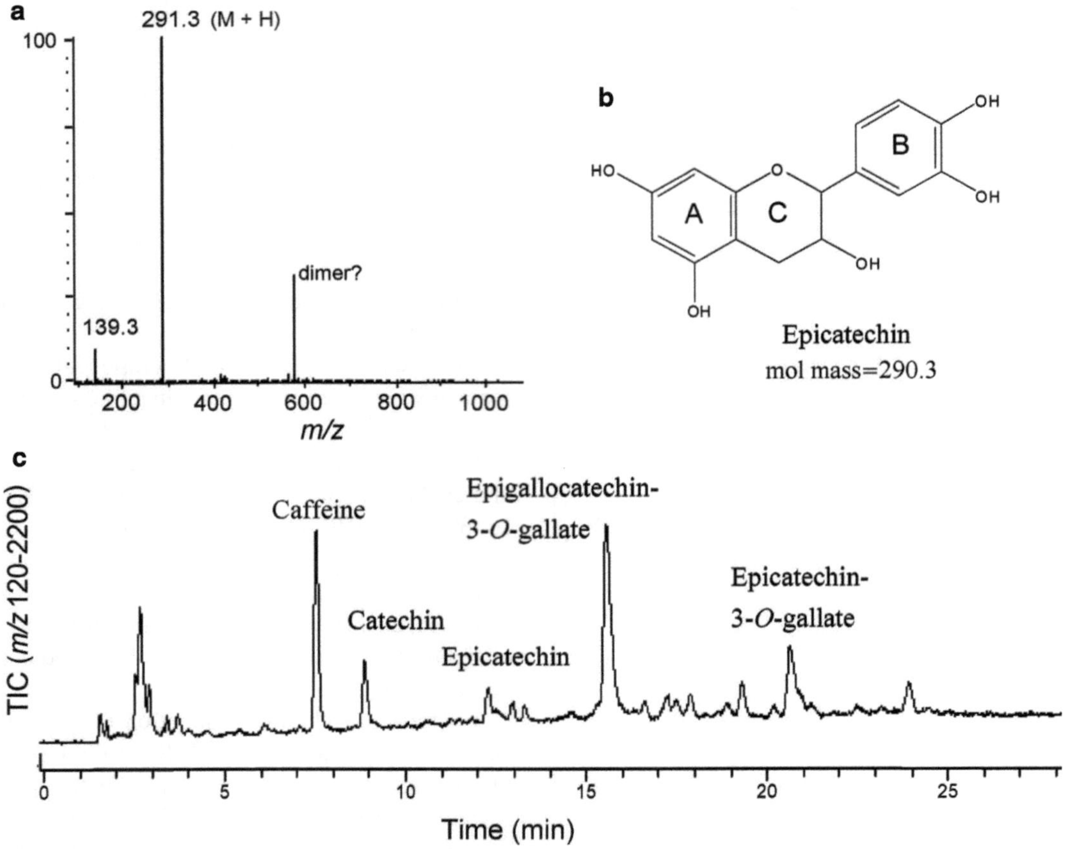

Fig. 11.17 LC-ESI-MS of green tea leaf extracts: (**a**) ESI mass spectra of epicatechin, (**b**) chemical structure of epicatechin, and (**c**) TIC scanned from *m/z* 120 to 2200 (From Shen et al. [11], used with permission of American Chemical Society, Washington, DC) https://en.wikipedia.org/wiki/Orbitrap#/media/File:OrbitrapMA&Injector.png

The high-throughput advantage of SFC separation allowed for 72 samples in approximately 45 min using SFC-OT technology, for compounds whose molecular weights ranged from *m/z* 99 to *m/z* 900.

A GC-capable version of the orbitrap [14] can enable high-resolution GC-MS at resolutions exceeding $R > 50{,}000$. In contrast to GC-TOF instruments currently available, which focused mainly MS acquisition speed at resolutions around 5000 (higher than quadrupole-based instruments), the GC coupling to the orbitrap truly allows for high-resolution, accurate mass analysis in both MS and MS/MS modes. The superior chromatographic separation power of GC, coupled to high resolution ($R = 60{,}000$ at *m/z* 200), allows for unambiguous analysis of 132 pesticides in even the most complex matrices at the 10 ng/g level, approaching sensitivity performance (IDL 10 ng/g) of routine GC-triple quadrupoles. These orbitrap-based techniques currently are used extensively for screening of pesticide residues and highly sensitive quantitative analysis of banned substances.

FTMS provides ultrahigh-resolution MS at the isotope fine-structure level, i.e., resolution of isobaric species of different elemental composition, which can yield elemental composition analysis. Essentially, isotope fine structure is the fingerprint of a small molecule because the *m/z* value, and their intensities in the fingerprint for a specific molecule, exactly reflects the atomic composition of the molecule. Because FTMS delivers resolutions in excess of $R > 1000.000$ (million), this allows for isotopologue analysis of the A + 1 and A + 2 ion signals. Using this capability, bioactive sulfur-containing compounds were identified in asparagus, purported to have angiotensin-converting enzyme (ACE) inhibitory function [15]. One such compound identified was aspartame (no UV-chromophores) detected at *m/z* 307.0893 (confirmed by high-resolution MS/MS) that required acquisition at $R = 1000{,}000$ to determine the elemental composition of $C_{10}\ H_{18}\ N_4O_3S_2$. These discoveries could only be achieved using FTMS, which allows for isotope fine-structure analysis.

Tandem and hybrid MS-hyphenated LC systems, which offer high resolution and often in combination with soft ionization techniques, such as MALDI, are the instruments of choice in omics-based workflows (i.e., foodomics, genomics, metabolomics, lipidomics, proteomics) [16]. Omics technologies largely rely on the generation of a large set of analytical data, which is imported into specific databases. Subsequently, knowledge graphs are constructed using statistical and machine learning algorithms. These data models play an indispensable role in biomedical, biochemical, and bioinformatics research. Thus, omics methodologies enable the simultaneous analysis of a large number of biochemical compounds and biomarkers in complex matrices within a couple of minutes. This analysis is related to metabolic and biochemical reaction pathways, allowing for compound profiling, authentication of food quality or safety, assessment of bioactivity and health benefits of food components, the development of new transgenic foods, evaluation of food contaminants and toxicity studies, and pharmaceutical applications.

This chapter concludes with several examples that demonstrate the versatility and power of MS tied to foodomics-based analyses. In the last few decades, proteomics work popularly involved MALDI-TOF to identify proteins and perform peptide fingerprinting, often combined with a suitable pretreatment, such as enzymatic treatment. In addition, sophisticated lipidomics and metabolomics approaches were used for authentication of meat products [17]. There is also interest in the analysis of intact proteins in relationship to their structure and micro-molecular interactions. In parallel, pioneering work involves the use of hybrid-MS and HRMS techniques for elucidating the effects that binding of small ligands (e.g., covalent binding of flavor molecules) have on the structure and functionality of intact proteins [18].

The MALDI-TOF was also used in food microbiology applications to rapidly identify bacteria and fungi through their protein signatures. In this technique, bacteria or fungi from the culture plates are directly spotted onto the MALDI target plate, sprayed with matrix, and then directly analyzed on the MALDI-TOF instrument. The resulting spectrum representative of the microorganisms proteomic fingerprint is matched against a known, verified spectrum in the library, and, if there is positive hit, the bacteria or fungi is rapidly identified. Wieme et al. [19] used MALDI-TOF to catalog 4200 mass spectra from 273 acetic acid bacteria and lactic acid bacteria covering 52 species responsible for beer spoilage, and then used the resulting library for routine quality control in the brewing industry. MALDI-TOF has been used for the rapid confirmatory identification (within 24 h) of more than 120 strains for *S. aureus* in milk, a pathogen that causes toxic shock syndrome toxin-1, a deadly form of food poisoning. This food-borne pathogen is usually a result of subclinical and clinical mastitis-affected dairy cattle [20]. Overall, analytical techniques based on MS play an integral role in advancing foodomics research with high potential for impact on food science and consumer safety. Valdes et al. provide a comprehensive list of the recently published review articles on foodomics and the use of MS [21].

11.9 SUMMARY

MS is a powerful analytical technique that can solve most complex problems faced by the food analytical chemist, both in a qualitative and quantitative manner. Its principles are fairly simple when examined closely. The basic requirements are to: (1) get the sample into an ionizing chamber where ions are produced; (2) separate the ions formed by magnets, quadrupoles, drift tubes, or electric fields; (3) detect the *m/z* of the precursor ion; (4) fragment the *m/z* selectively to derive more information if required; and (5) output the data to a computer for software evaluation.

Since the qualitative and quantitative aspects of MSs are so powerful, they are routinely coupled to a GC or HPLC and find growing use with static sample introduction techniques. The interface for GCs is versatile and easy to use; however, the extensive sample preparation required for GC-MS analysis makes its utility cumbersome. The adoption of LC-MS as an analytical technique has greatly increased because of far simpler sample preparation procedures, wider ionization ranges for different classes of compounds, faster analysis times, routine high sensitivity, access to accurate mass capability, and the advent of UHPLC.

11.10 STUDY QUESTIONS

1. What are the basic components of an MS?
2. Mass spectrometry is commonly used as a hyphenated technique with GC and HPLC.
   (a) What are the unique aspects of data obtained from GC-MS or HPLC-MS analysis, as compared to GC or HPLC with only traditional detectors?
   (b) What are the advantages of using tandem MS (MS/MS) versus just MS with GC or HPLC?
3. What is EI ionization? What is CI ionization?
4. What is the base peak on a mass spectrum? What is the precursor ion peak?
5. What is the difference between nominal mass and monoisotopic mass?
6. What are the major ions (fragments) expected in the EI mass spectrum of ethanol (CH_3-CH_2-OH)?
7. What are the major differences in how ionization occurs in the electrospray versus the APCI interface? What is ion suppression?
8. What does MALDI stand for and how does it differ from ESI?
9. What are the major differences between the quadrupole, ion trap, time of flight, and Fourier transform mass analyzer? What are the advantages of using each analyzer? What is especially unique about a Fourier transform-based mass analyzer?

10. What is the working principle behind the MALDI-TOF-based microbiology identification?
11. Which MS type is popular for quantitative analysis?
12. What is the difference between CAD and CID?

Acknowledgments The authors of this chapter thank Rohan A. Thakur (Bruker Daltonics, Billerica, MA, USA) for his contribution to the chapter in previous editions of the book.

REFERENCES

1. Hajslova J, Cajka T, Vaclavik L (2011) Challenging applications offered by direct analysis in real time (DART) in food-quality and safety analysis. Trends Anal. Chem. 30(2):204–218
2. Schober AL, Peterson DG (2004). Flavor release and perception in hard candy: Influence of flavor compound–flavor solvent interactions. J. Agric. Food Chem. 52, 2628–2631
3. Chughtai K, Heeren RMA (2010) Mass spectrometric imaging for biomedical tissue analysis. Chem. Rev. 110(5):3237–3277
4. Chang L, Shiying C, Siyuan T, Xinchi Y, You J, Xinhua D, Xiaoyun G, Xiang Fang, Di T (2021). Towards higher sensitivity of mass spectrometry: A perspective from the mass analyzers. Front. Chem. 9: 813359
5. Silveira JA, Ridgeway ME, Park MA (2014) High resolution trapped ion mobility spectrometry of peptides. Anal. Chem. 86(12):5624–56274
6. Balogh, MP (2004) Debating resolution and mass accuracy. LC-GC Europe 17(3):152–159
7. Milman BL, (2015) General principles of identification by mass spectrometry. Trends Anal. Chem. 69:24–33
8. Lehotay SJ, Sapozhnikova Y, Hans G.J., Mol, HGJ (2015) Current issues involving screening and identification of chemical contaminants in foods by mass spectrometry. Trends Anal. Chem. 69:62–75
9. Herrero M, Simo, C, Garcia-Canas V, Ibanez E, Cifuentes A (2012) Foodomics: MS-based strategies in modern food science and nutrition. Mass Spec. Rev. 31:49–69
10. Huck CW, Guggenbichler W, Bonn GK (2005) Analysis of caffeine, theobromine and theophylline in coffee by near infrared spectroscopy (NIRS) compared to high-performance liquid chromatography (HPLC) coupled to mass spectrometry. Analytica Chimica Acta 538(1-2):195–203
11. Shen D, Wu Q, Wang M, Yang Y, Lavoie EJ, Simon JE (2006) Determination of the predominant catechins in *Acacia catechu* by liquid chromatography/electrospray ionization-mass spectrometry. J Agric Food Chem 54(9):3219–3224
12. Rajski L, Gomez-Ramos MDM, Fernandez-Alba, AR (2014) Large pesticide multiresidue screening method by liquid chromatography-orbitrap mass spectrometry in full scan mode applied to fruit and vegetables. J. Chromatogr. A 1360:119–127
13. Ishibashi M, Izumi Y, Sakai M, Ando T, Fukusaki E, Bamba T. (2015) High-throughput simultaneous analysis of pesticides by supercritical fluid chromatography coupled with high-resolution mass spectrometry. J. Agric. Food Chem. 63(18):4457–4463
14. Peterson AC, Hauschild J-P, Quarmby ST, Krumwiede D, Lange O, Lemke RAS, Grosse-Coosmann F, Horning S, Donohue TJ, Westphall MS, Coon JJ, Griep-Raming J (2014) Development of a GC/quadrupole-orbitrap mass spectrometer, part I: Design and characterization. Anal. Chem. 86:10036–10043
15. Nakabayashi R, Yang Z, Nishizawa T, Mori T, Saito K (2015) Top-down targeted metabolomics reveals a sulfur-containing metabolite with inhibitory activity against angiotensin-converting enzyme in Asparagus officinalis. J. Nat. Prod. 78(5):1179–1183
16. Griffiths WJ, Wang Y (2009) Mass spectrometry: from proteomics to metabolomics and lipidomics. Chem. Soc. Rev. 38:1882–1896
17. Harlina PW, Maritha V, Musfiroh I, Huda S, Sukri N, Muchtaridi M (2022) Possibilities of liquid chromatography mass spectrometry (LC-MS)-based metabolomics and lipidomics in the authentication of meat products: A mini review. Food Sci. Anim. Resour. 42(5):744–761
18. Reineccius GA (2022) Flavor interactions with proteins. Curr. Op. Food Sci. 47:100884
19. Wieme AD, Spitaels F, Vandamme P, Landschoot AV (2014) Application of matrix-assisted laser desorption/ionization time-of-flight mass spectrometry as a monitoring tool for in-house brewer's yeast contamination: a proof of concept. J. Inst. Brewing. 120(4):438–443
20. El Behiry A, Zahran RN, Tarabees R, Marzouk E, Al-Dubaib M (2014) Phenotypical and genotypical assessment techniques for identification of some contagious mastitis pathogens. Int J. Med Health Biomed Bioeng Pharm Eng. 8(5):236–242
21. Valdes A, Alvarez-Rivera G, Socas-Rodriguez B, Herrero M, Ibanez E, Cifuentes A (2022) Foodomics: Analytical opportunities and challenges. Anal. Chem. 94:366–381

RESOURCE MATERIALS

Balogh, MP (2014) The mass spectrometer primer. Waters Series, Milford, MA. A very good introduction to modern mass spectrometry including newer developments in LC-MS technologies.

Cifuentes A (Ed) (2013) Foodomics: Advanced Mass Spectrometry in Modern Food Science and Nutrition. John Wiley & Sons, New York. Covers omics-based principles and applications in food and nutrition.

Gross JH (2017) Mass Spectrometry 3rd Ed. Springer. Comprehensive book with color illustrations including basics of MS and theories including ionization, fragmentation patterns, GC-MS, and LC-MS.

Ho C-T, Lin J-K, Shahidi F (Eds) (2009) Tea and tea products. CRC, Boca Raton, FL. A good review of current literature on the chemistry and health-promoting properties of tea. Includes several chapters that discuss analytical methods for analyzing bioactive compounds and flavonoids in teas.

Hoffmann E, Stroobant (2013) Mass Spectrometry: Principles and Applications. Wiley, New York. A good fundamental introduction to Mass Spectrometry with practical examples.

Niessen WMA (2006) Liquid chromatography–mass spectrometry. 3rd Ed. Taylor and Francis, New York. A thorough, though somewhat dated, review of LC-MS methods and interfaces for a variety of types of biological compounds.

Watson JT, Sparkman OD (2007) Introduction to Mass Spectrometry: Instrumntation, applications and strategies for data interpretation. 4th Ed. The fourth edition of a classic book on how molecules fragment in the ion source. Contains examples of different types of molecules, and provides systematic descriptions on several types of instruments and ionization techniques.

Part III

Chromatography

12 Basic Principles of Chromatography

B. Pam Ismail

Contents

12.1 INTRODUCTION

Chromatography has a great impact on most areas of constituents' analysis and, therefore, on the progress of analytical science in general. Chromatography differs from other methods of separation in that a wide variety of materials, equipment, and techniques can be used. (Readers are referred to references [1–31] for general and specific information on chromatography.) This chapter will focus on the principles of chromatography, mainly **liquid chromatography** (LC). Detailed principles and applications of **gas chromatography** (GC) will be discussed in Chap. 14. In view of its widespread use and applications, **high-performance liquid chromatography** (HPLC) will be discussed in a separate chapter (Chap. 13). The general principles of extraction are first described as a basis for understanding chromatography.

B. P. Ismail (✉)
Department of Food Science and Nutrition, University of Minnesota, St. Paul, MN, USA
e-mail: bismailm@umn.edu

B. P. Ismail, S. S. Nielsen (eds.), *Nielsen's Food Analysis*, Food Science Text Series,
https://doi.org/10.1007/978-3-031-50643-7_12

12.2 EXTRACTION

In its simplest form, extraction refers to the transfer of a solute from one liquid phase to another. Extraction in myriad forms is integral to food analysis—whether used for preliminary sample cleanup, concentration of the component of interest, or as the actual means of analysis. Extractions may be categorized as **batch, continuous**, or **countercurrent** processes. (Various extraction procedures are discussed in detail in other chapters: traditional solvent extraction in Chaps. 14, 17, and 33; accelerated solvent extraction and microwave-assisted solvent extraction in Chap. 33; and solid-phase extraction and solid-phase microextraction in Chaps. 14 and 33).

12.2.1 Batch Extraction

In **batch extraction**, the solute is extracted from one solvent by shaking it with a second immiscible solvent. The solute **partitions**, or distributes, itself between the two phases, and when equilibrium has been reached, the **partition coefficient**, *K*, is a constant:

$$K = \frac{\text{Concentration of solute in phase 1}}{\text{Concentration of solute in phase 2}} \qquad (12.1)$$

After shaking, the phases are allowed to separate and the layer containing the desired constituent is removed, for example, in a separatory funnel, or in a Mojonnier flask (Chap. 17, Sect. 17.2.6.1). In batch extraction, it is often difficult to obtain a clean separation of phases, owing to emulsion formation. Moreover, partition implies that a single batch extraction is usually incomplete. Multiple extractions are often performed, with fresh solvent each time, to ensure efficient extraction.

12.2.2 Continuous Extraction

Continuous extraction requires special apparatus but is more efficient than batch separation. One example is the use of a Soxhlet extractor (Chap 17, Sect. 17.2.5) for extracting fat from solids using organic solvents. Solvent is recycled, using a condenser system, so that the solid is repeatedly extracted with fresh solvent. Different types of equipment have been designed for the continuous extraction of substances from liquids and/or solids, and different extractors are used for solvents that are heavier or lighter than water.

12.2.3 Countercurrent Extraction

Countercurrent distribution refers to a serial extraction process. It separates two or more **solutes** (the solubilized constituents within the sample) with different partition coefficients from each other by a series of partitions between two immiscible liquid phases. **Liquid partition chromatography** (Sect. 12.4.2), also known as countercurrent chromatography, is a direct extension of countercurrent extraction. Years ago, the countercurrent extraction was done with a "Craig apparatus" consisting of a series of glass tubes designed such that the lighter liquid phase (**mobile phase**) was transferred from one tube to the next, while the heavy phase (**stationary phase**) remained in each tube [5]. The liquid–liquid (i.e., partition) extractions took place simultaneously in all tubes of the apparatus, which was usually driven electromechanically. Each tube in which a complete equilibration took place corresponded to one theoretical plate of the chromatographic column (refer to Sect. 12.5.1.2.1). The greater the difference in the **partition coefficients** of various substances, the better and quicker was the separation. Many more tubes were required to separate mixtures of substances with close partition coefficients, which made this type of countercurrent extraction very tedious and lengthy. Modern liquid partition chromatography (Sect. 12.4.2) is much more efficient and convenient.

12.3 CHROMATOGRAPHY

12.3.1 Historical Perspective

Modern chromatography originated in the late nineteenth and early twentieth centuries from independent works by David T. Day, an American geologist and mining engineer, and Mikhail Tsvet, a Russian botanist. Day developed procedures for fractionating crude petroleum by passing it through Fuller's earth, and Tsvet used a column packed with chalk to separate leaf pigments into colored bands. Because Tsvet recognized and correctly interpreted the chromatographic processes and named the phenomenon **chromatography**, he is generally credited with its discovery.

Chromatography began to evolve in the 1940s due to the development of column partition chromatography by Martin and Synge and the invention of paper chromatography. Meanwhile, GC had developed into the first sophisticated instrumental technique available commercially in the late 1960s because of its importance to the petroleum industry. HPLC, profiting from the theoretical and instrumental advances of GC, has extended the area of LC into an equally sophisticated and useful method. Supercritical fluid chromatography (SFC), first demonstrated in 1962, has been gaining popularity in food analysis [7]. Efficient chromatographic techniques, including automated systems, continue to be developed for utilization in the characterization and quality control of food ingredients and products [4, 7–13].

12.3.2 General Terminology

Chromatography is a general term applied to a wide variety of separation techniques based on the partitioning or distribution of a sample **(solute)** between a moving (i.e., mobile) phase and a fixed or stationary phase. Chromatography may be viewed as a series of equilibrations between the mobile and stationary phase. The relative interaction of a solute with these two phases is described by the **partition** (K) or **distribution** (D) **coefficient** (ratio of concentration of solute in mobile phase to concentration of solute in stationary phase). The mobile phase may be either a gas (for GC) or liquid (for LC) or a supercritical fluid (for SFC). The stationary phase may be a liquid or a solid. The field of chromatography can be subdivided according to the various techniques applied (Fig. 12.1) or according to the physicochemical principles involved in the separation. Table 12.1 summarizes some of the chromatographic procedures or methods that have been developed on the basis of different mobile–stationary phase combinations. Inasmuch as the nature of interactions between solute molecules and the mobile or stationary phases differ, these methods have the ability to separate different kinds of molecules. (The reader is urged to review Table 12.1 again after having read this chapter.)

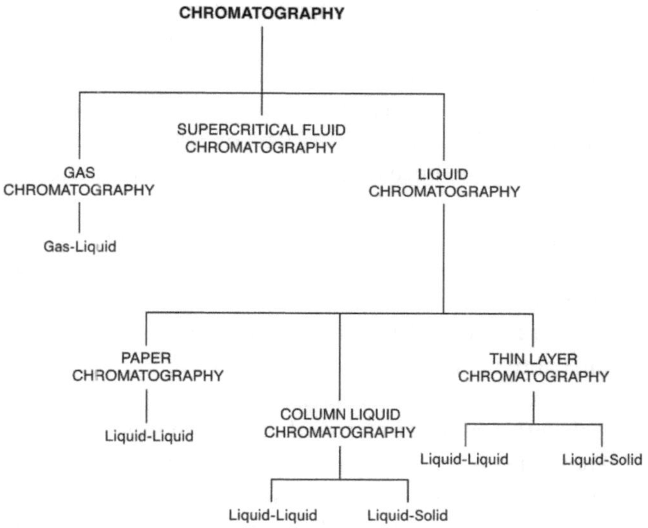

Fig. 12.1 A scheme for subdividing the field of chromatography, according to various applied techniques

12.3.3 Gas Chromatography

Gas chromatography is a column chromatography technique, in which the mobile phase is gas and the stationary phase can be solid, or mostly an immobilized liquid on an inert solid support in either a packed or capillary-type column. When the stationary phase is liquid, GC is commonly referred to as gas–liquid chromatography. Gas chromatography is used to separate thermally stable and volatile components of a mixture. Gas chromatography involves vaporizing a sample and injecting it onto the head of the column. Under a controlled temperature gradient, the sample is transported through the column by the flow of an inert, gaseous mobile phase. Volatiles are then separated based on several properties, including boiling point, molecular size, and polarity. Details of the chromatographic theory of separation as it applies specifically to GC, as well as detection and instrumentation of GC, are detailed in Chap. 14.

12.3.4 Liquid Chromatography

There are several **liquid chromatography** techniques applied in food analysis, involving **planar chromatography** (both paper and thin-layer chromatography) and column liquid chromatography, all of which involve a liquid mobile phase and either a solid or a liquid stationary phase. Separation of the solutes is based on their physicochemical interactions with the two phases, which is discussed in Sect. 12.4.

Table 12.1 Characteristics of different chromatographic methods

Method	*Mobile phase*	*Stationary phase*	*Retention varies with*
Gas-liquid chromatography	Gas	Liquid	Molecular size/polarity/boiling point
Gas-solid chromatography	Gas	Solid	Molecular size/polarity/boiling point
Supercritical fluid chromatography	Supercritical fluid	Solid	Molecular size/polarity
Reversed-phase chromatography	Polar liquid	Nonpolar liquid or solid	Molecular size/polarity
Normal-phase chromatography	Nonpolar liquid	More polar liquid or solid	Molecular size/polarity
Hydrophilic interaction Liquid chromatography	Less polar liquid	More polar liquid or solid	Molecular size/polarity
Ion-exchange chromatography	Polar liquid	Ionic solid	Molecular charge
Hydrophobic interaction chromatography	Polar liquid	Nonpolar liquid or solid	Molecular size/polarity
Affinity chromatography	Water	Binding sites	Specific structure
Size-exclusion chromatography	Liquid	Solid	Molecular size

Reprinted from Heftmann [1], p. A21, with kind permission from Elsevier Science-NL, Sara Burgerhartstraat 25, 1055 KV Amsterdam, The Netherlands

12.3.4.1 Planar Chromatography

12.3.4.1.1 Paper Chromatography

Paper chromatography was introduced in 1944 and since has been used in the food industry to analyze food additives (e.g., colorants) and identify contaminants. Today, it is mostly used as a teaching tool. Paper chromatography is a form of partition chromatography in which both the stationary phase (water) and the mobile phase (organic solvent) are liquid (see Sect. 12.4.2). Paper (usually cellulose) serves as a support for the liquid stationary phase. The support may also be impregnated with a nonpolar organic solvent and developed with water or other polar solvents. The dissolved sample is applied as a small spot about 1.5 cm from the edge of a strip or square of the paper, which is then allowed to dry. The dry strip is suspended in a closed container in which the atmosphere is saturated with the **developing solvent** (mobile phase) and the paper chromatogram is **developed**. The end closer to the sample is placed in contact with the solvent, which then travels up or down the paper by capillary action (depending on whether **ascending** or **descending** development is used), separating the sample components in the process. When the solvent front has traveled the length of the paper, the strip is removed from the developing chamber and the separated zones are detected by an appropriate method.

In the case of complex sample mixtures with a narrow range of partition coefficients, a two-dimensional technique may be used. The sample is spotted in one corner of a square sheet of paper, and one solvent is used to develop the paper in one direction. The paper is then dried, turned 90°, and developed again, using a second solvent of different polarity. Another means of improving resolution is the use of **ion-exchange** (Sect. 12.4.5) papers, i.e., paper with derivatized cellulose hydroxyl groups (with acidic or basic moieties).

In planar chromatography, components of a mixture are often characterized by their relative mobility (R_f) value, where:

$$R_f = \frac{\text{Distance moved by component}}{\text{Distance moved by solvent}} \tag{12.2}$$

Unfortunately, R_f values are not always constant for a given solute/sorbent/solvent but depend on many factors, such as the quality of the stationary phase, paper thickness, humidity, development distance, and temperature.

12.3.4.1.2 Thin-Layer Chromatography

Thin-layer chromatography (TLC), first described in 1938, is faster, more sensitive, and more reproducible than paper chromatography. The resolution in TLC is greater than in paper chromatography because the particles on the plate are smaller and more regular than paper fibers. There are several distinct advantages to TLC over column chromatography: low cost, high sample throughput, analysis of several samples and standards simultaneously, minimal sample preparation, and possibility to store the plate for later identification and quantification. Advances in TLC led to the development of **high-performance thin-layer chromatography** (HPTLC), which simply refers to TLC performed using plates coated with smaller, more uniform particles, allowing for enhanced resolution in a shorter time.

TLC, more so HPTLC, is applied in many fields, including environmental, clinical, forensic, pharmaceutical, food, flavors, and cosmetics. Within the food industry, TLC may be used for quality control. For example, corn and peanuts are tested for aflatoxins/mycotoxins prior to their processing into corn meal and peanut butter, respectively. Applications of TLC to the analysis of a variety of compounds, including lipids, carbohydrates, vitamins, amino acids, natural pigments, and sugar substitutes, are discussed in references [14, 17, 18].

1. **TLC General Procedures.** TLC utilizes a thin (ca. 250 μm thick) layer of **sorbent** or **stationary phase** bound to an **inert support**. The support is often a glass plate (traditionally, 20 cm × 20 cm), but plastic sheets and aluminum foil are commonly used since they are commercially available. Pre-coated plates, of different layer thicknesses, are available in a wide variety of sorbents, including chemically modified silica. Four frequently used TLC sorbents are silica gel, alumina, diatomaceous earth, and cellulose. Modified silica for TLC may contain polar or nonpolar groups, so both normal and reversed-phase (see Sect. 12.4.2.1) thin-layer separations may be carried out.

   If **adsorption** TLC is to be performed, the sorbent is first **activated** by drying for a specified time and temperature. As in paper chromatography, the sample (in carrier solvent) is applied as a spot about 1.5 cm from one end of the plate. After evaporation of the carrier solvent, the TLC plate is placed in a closed **developing chamber**, solvent migrates up the plate (**ascending development**) by capillary action, and sample components are separated. After the TLC plate has been removed from the chamber and solvent allowed to evaporate, the separated constituents are made visible or detected by other means. Specific **chemical reactions (derivatization)**, which may be carried out either before or after chromatography, often are used for this purpose. Two examples are reaction with sulfuric acid to produce a dark charred area (a **destructive chemical method**), and the use of iodine vapor to form a colored complex (a **nondestructive method** inasmuch as the colored complex is usually not permanent). Common **physical detection methods** include the measurement of absorbed or emitted electromagnetic radiation, such as measuring fluorescence when stained with 2,7-dichlorofluorescein, and measurement of β-radiation from radioactively labeled compounds.

Different reagents that can react selectively to generate colored products are also used [17]. **Biological methods** or biochemical inhibition tests can be used to detect toxicologically active substances. An example is measuring the inhibition of cholinesterase activity by organophosphate pesticides (Chap. 33, Sect. 33.3.3.2.1).

Quantitative evaluation of constituents separated by TLC may be performed [17]: (1) in situ (directly on the layer) by using a **densitometer** [18] or (2) scraping a zone off the plate, eluting compound from the sorbent, then analyzing the resultant solution (e.g., by liquid scintillation counting).

2. **Factors Affecting Thin-Layer Separations.** In both planar and column liquid chromatography, the nature of the compounds to be separated is considered when choosing the stationary phase. Separation of different classes of constituents (Table 12.2) can occur by various chromatographic mechanisms (Sect. 12.4).

   Solvents for TLC separations are selected for specific chemical characteristics and **solvent strength** (a measure of interaction between solvent and sorbent; see Sect. 12.4.1). In simple adsorption TLC, the higher the solvent strength, the greater the R_f value of the solute. An R_f value of 0.3–0.7 is typical. Mobile phases have been developed for the separation of various compound classes on the different sorbents (see Table 7.1 in reference [19]).

   In addition to the sorbent and solvent, several other factors must be considered when performing planar chromatography. These include the **type of developing chamber** used, **vapor phase conditions** (saturated vs. unsaturated), **development mode** (ascending, descending, horizontal, radial, etc.), and **development distance**. For additional reading refer to references [14–18].

12.3.4.2 Column Liquid Chromatography

Column liquid chromatography generally has enhanced resolution of solutes in a mixture and enables precise analysis and a straightforward quantification compared to planar chromatography. Separation of solutes occurs as a result of differential migration through a closed tube of stationary phase, and analytes can be monitored while the separation is in progress. In column liquid chromatography, the mobile phase is liquid and the stationary phase can be either solid or liquid supported by an inert solid. A system for **low-pressure** (i.e., performed at or near atmospheric pressure) column liquid chromatography is illustrated in Fig. 12.2. High-pressure column chromatography is described in Chap. 13.

Having selected a stationary and mobile phase suitable for the separation problem at hand, the analyst must first prepare the **stationary phase** (**resin, gel,** or **packing material**) for use according to the supplier's instructions (e.g., the stationary phase often must be **hydrated** or **preswelled** in the mobile phase). The prepared stationary phase is then **packed** into a column (usually glass or plastic), the length and diameter of which are determined by the amount of sample to be loaded, the separation mode to be used, and the degree of resolution required. Longer and narrower columns usually enhance resolution and separation (Sect. 12.5.1). The most common technique for column packing involves making a slurry of the stationary phase with the solvent and pouring it into the column to the desired bed height under a specified pressure and rate. Pouring uniform columns is an art that is mastered with practice. A poorly packed column leads to poor resolution. In **high-pressure** column chromatography (Chap. 13), commercially available columns are pre-packed under pressure with various types of stationary phases and have different dimensions.

The sample to be analyzed is dissolved in a minimum volume of the starting mobile phase, injected through a sample injection port, and carried by the mobile phase onto the column. Low-pressure chromatography utilizes only **gravity** flow or a high-precision **peristaltic pump** to maintain a constant flow of **mobile phase** (**eluent** or **eluting solvent**) through the column. If eluent is fed to the column by a peristaltic pump (see Fig. 12.2), then the flow rate is determined by the pump speed. Depending on the dimensions of the column, the flow rate is adjusted not to exceed the max pressure sustained by the pump.

The process of passing the mobile phase through the column is called **elution**, and the portion that emerges from the outlet end of the column is called the **eluate** (or effluent). Elution may be **isocratic** (constant mobile-phase composi-

Table 12.2 Thin-layer chromatography sorbents and mode of separation

Sorbent	*Chromatographic mechanism*	*Typical application*
Silica gel	Adsorption	Steroids, amino acids, alcohols, hydrocarbons, lipids, aflatoxins, bile acids, vitamins, and alkaloids
Silica gel RP	Reversed phase	Fatty acids, vitamins, steroids, hormones, and carotenoids
Cellulose, kieselguhr	Partition	Carbohydrates, sugars, alcohols, amino acids, carboxylic acids, and fatty acids
Aluminum oxide	Adsorption	Amines, alcohols, steroids, lipids, aflatoxins, bile acids, vitamins, and alkaloids
PEI cellulose[a]	Ion exchange	Nucleic acids, nucleotides, nucleosides, purines, and pyrimidines
Magnesium silicate	Adsorption	Steroids, pesticides, lipids, and alkaloids

Reprinted from Touchstone [19] by permission of Wiley, New York

[a]PEI cellulose refers to cellulose derivatized with polyethyleneimine (PEI)

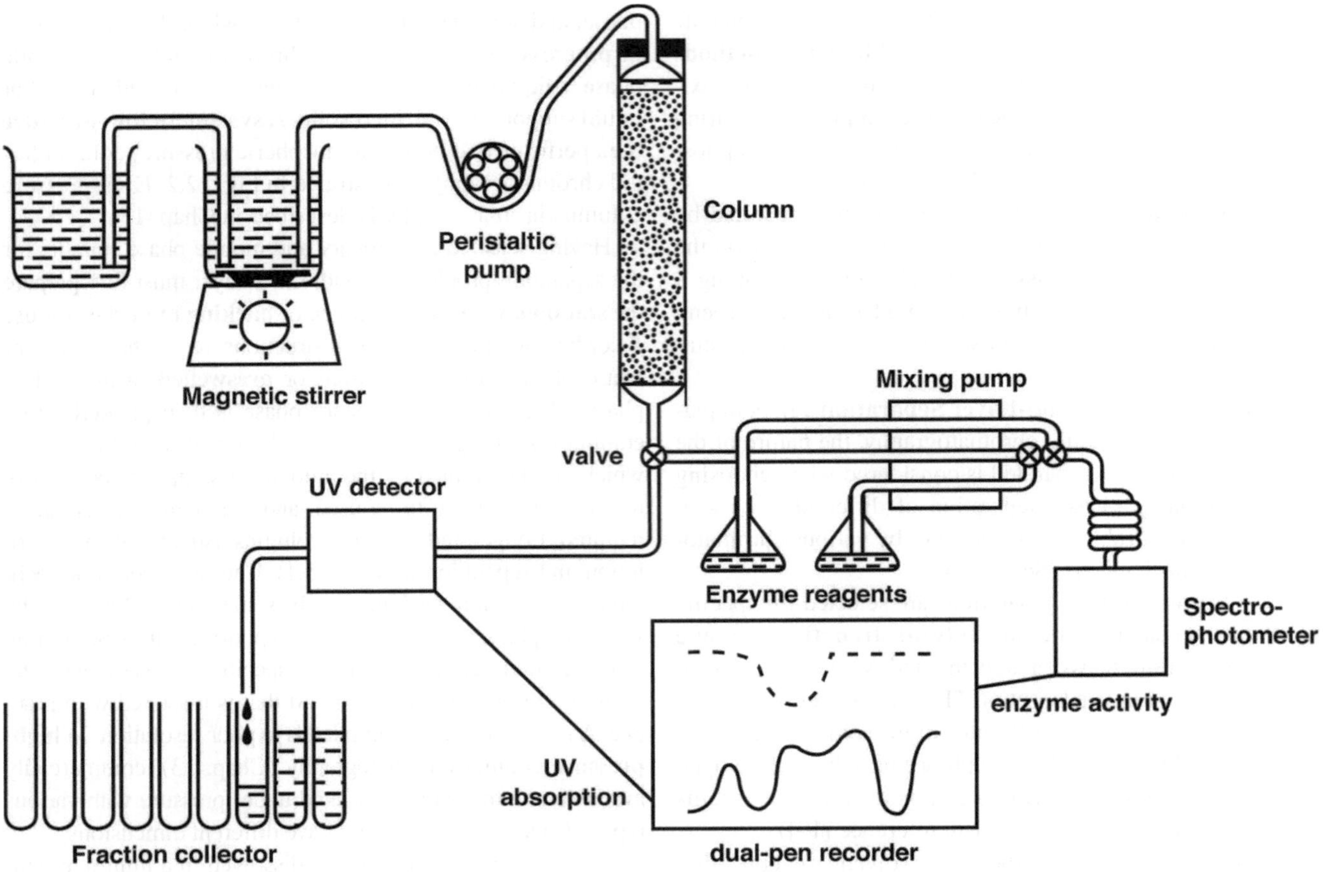

Fig. 12.2 A system for low-pressure column liquid chromatography. In this diagram, the column effluent is being split between two detectors in order to monitor both enzyme activity (*at right*) and UV absorption (*at left*). The two tracings can be recorded simultaneously by using a dual-pen recorder. (Adapted from Scopes [30], with permission)

tion) or a **gradient** (changing the mobile phase composition, i.e., increasing its strength). Increasing the strength of the mobile phase is done by changing its properties in a way that reduces the solutes interaction with the stationary phase and increases the solubilization of the solutes in the mobile phase. Gradient elution enhances resolution and decreases analysis time (see also Sect. 12.5.1). As elution proceeds, solutes are selectively retarded by the stationary phase based on the strength of interaction with the stationary phase and thus are eluted at different times.

The column eluate may be directed through a detector and then into tubes, changed at intervals by a fraction collector. The detector response, in the form of an electrical signal, may be recorded (the **chromatogram**) using a computerized software. Signals are then integrated for either qualitative (i.e., detection and identification) or quantitative analysis (Sects. 12.5.2 and 12.5.3). The fraction collector may be set to collect eluate at specified time intervals or after a certain volume has been collected. Solutes that have been chromatographically separated and collected can be further analyzed as needed.

12.3.5 Supercritical Fluid Chromatography

Supercritical fluid chromatography is performed above the **critical pressure** (P_c) and **critical temperature** (T_c) of the mobile phase. A supercritical fluid (or compressed gas) is neither a liquid nor a typical gas. The combination of P_c and T_c is known as the **critical point**. A supercritical fluid can be formed from a conventional gas by increasing the pressure or from a conventional liquid by raising the temperature. Carbon dioxide is frequently used as a mobile phase for SFC; however, it is not a good solvent for polar and high-molecular-weight compounds. A small amount of a polar, organic solvent such as methanol can be added to a supercritical fluid to enhance solute solubility, improve peak shape, and alter selectivity. Other supercritical fluids that have been used in food applications include nitrous oxide, trifluoromethane, sulfur hexafluoride, pentane, and ammonia.

Supercritical fluids confer chromatographic properties intermediate to LC and GC. The **high diffusivity** and **low viscosity** of supercritical fluids mean decreased analysis times and improved resolution compared to LC. An additional benefit of short analysis time is the reduced solvent consumption. SFC offers a wide range of **selectivity** (Sect. 12.5.2) adjustment, by changes in **pressure** and **temperature** as well as changes in **mobile phase composition** and the **stationary phase.** Compared to HPLC (Chap. 13), SFC can offer better resolution of compounds with a broad range of polarities. In addition, SFC makes possible the separation of **nonvolatile, thermally labile compounds**, which cannot be analyzed by GC without derivatization. In fact, SFC today is mainly used for the analysis of nonvolatiles.

Either **packed** or **capillary** columns can be used in SFC. Packed column materials are similar to those used for HPLC. Small particle, porous, high surface area, hydrated silica, either bare or bonded silica, is commonly used as the column packing material (Sect. 12.4.2.3 and Chap. 13). Capillary columns are generally coated with a polysiloxane (-Si-O-Si) film containing different functional groups with varying polarity such as methyl, phenyl, pyridine, or cyano.

The development of SFC benefited from advancement in the instrumentation for HPLC and GC. One major difference in instrumentation is the presence of a back pressure regulator to control the **outlet pressure** of the system. Without this device, the fluid would expand to a low-pressure, low-density gas. Detectors used in HPLC and GC (Chaps. 13 and 14) can be used with SFC, including coupling with mass spectrometry (MS) (Chap. 11).

SFC has been used for the analysis of a wide range of compounds. SFC is increasingly applied for the analysis of lipids, lipid soluble components such as fat-soluble vitamins and carotenoids, as well as polyphenols, carbohydrates, and food adulterants such as Sudan dyes [11]. SFC has also been used to detect pesticide residues, evaluate citrus essential oils, and characterize compounds extracted from microwave packaging [20]. SFC coupled with MS has been used for the analysis of the noncaloric fat substitute, Olestra®.

12.4 PHYSICOCHEMICAL PRINCIPLES OF CHROMATOGRAPHIC SEPARATION

Several physicochemical principles (illustrated in Fig. 12.3) are involved in chromatography mechanisms employed to separate or fractionate various compounds of interest, regardless of the specific techniques applied (discussed in Sect. 12.3). The mechanisms described below apply mainly to liquid chromatography; GC mechanisms will be detailed in Chap. 14. Although it is more convenient to describe each of these phenomena separately, it must be emphasized that more than one mechanism, or **separation mode**, may be involved in a particular chromatographic separation. Table 12.3 summarizes the different separation modes and associated stationary phases, mobile phases, and types of interactions.

12.4.1 Adsorption (Liquid–Solid) Chromatography

Adsorption chromatography is the oldest form of chromatography and originated with Tsvet experiments in 1903. In this chromatographic mode of separation, the stationary phase is a finely divided solid to maximize the surface area. The stationary phase (**adsorbent**) is chosen to permit differential interaction with the constituents of the sample to be resolved. The intermolecular forces primarily responsible for chromatographic adsorption include the following:

- Van der Waals forces
- Electrostatic forces
- Hydrogen bonds
- Hydrophobic interactions

Binding sites available for interaction with any given substance are heterogeneous. Binding sites with greater affinities, the most active sites, tend to be populated first, so that additional solutes are less firmly bound. The net result is that adsorption is a concentration-dependent process, and the **adsorption coefficient** is *not* a constant (in contrast to the **partition coefficient**). Sample loads exceeding the adsorptive capacity of the stationary phase will result in relatively poor separation (Sect. 12.5.1).

Classic adsorption chromatography utilizes mostly **silica** (slightly acidic), **alumina** (slightly basic), or charcoal (nonpolar). Both silica and alumina are polar adsorbents, possessing surface hydroxyl groups, and Lewis acid-type interactions determine their adsorption characteristics. The elution order of compounds from these adsorptive stationary phases can often be predicted on the basis of their relative **polarities** (Table 12.4). Compounds with the most polar functional groups are retained most strongly on polar adsorbents and, therefore, are eluted last. Nonpolar solutes are eluted first.

One model proposed to explain the mechanism of liquid–solid chromatography is that **solute** and **solvent** molecules are competing for active sites on the adsorbent. Thus, as relative adsorption of the mobile phase increases, adsorption of the solute must decrease. Solvents can be rated in order of their strength of adsorption on a particular adsorbent, such as silica. Such a **solvent strength** (or solvent polarity in this case) **scale** is called an **eluotropic series.** An eluotropic series for alumina is listed in Table 12.5. Silica has a similar rank ordering. Once an adsorbent has been chosen, solvents can be selected from the eluotropic series for that adsorbent. Mobile phase strength (polarity in this case) can be increased (often by admixture of more polar solvents) until elution of the compound(s) of interest has been achieved.

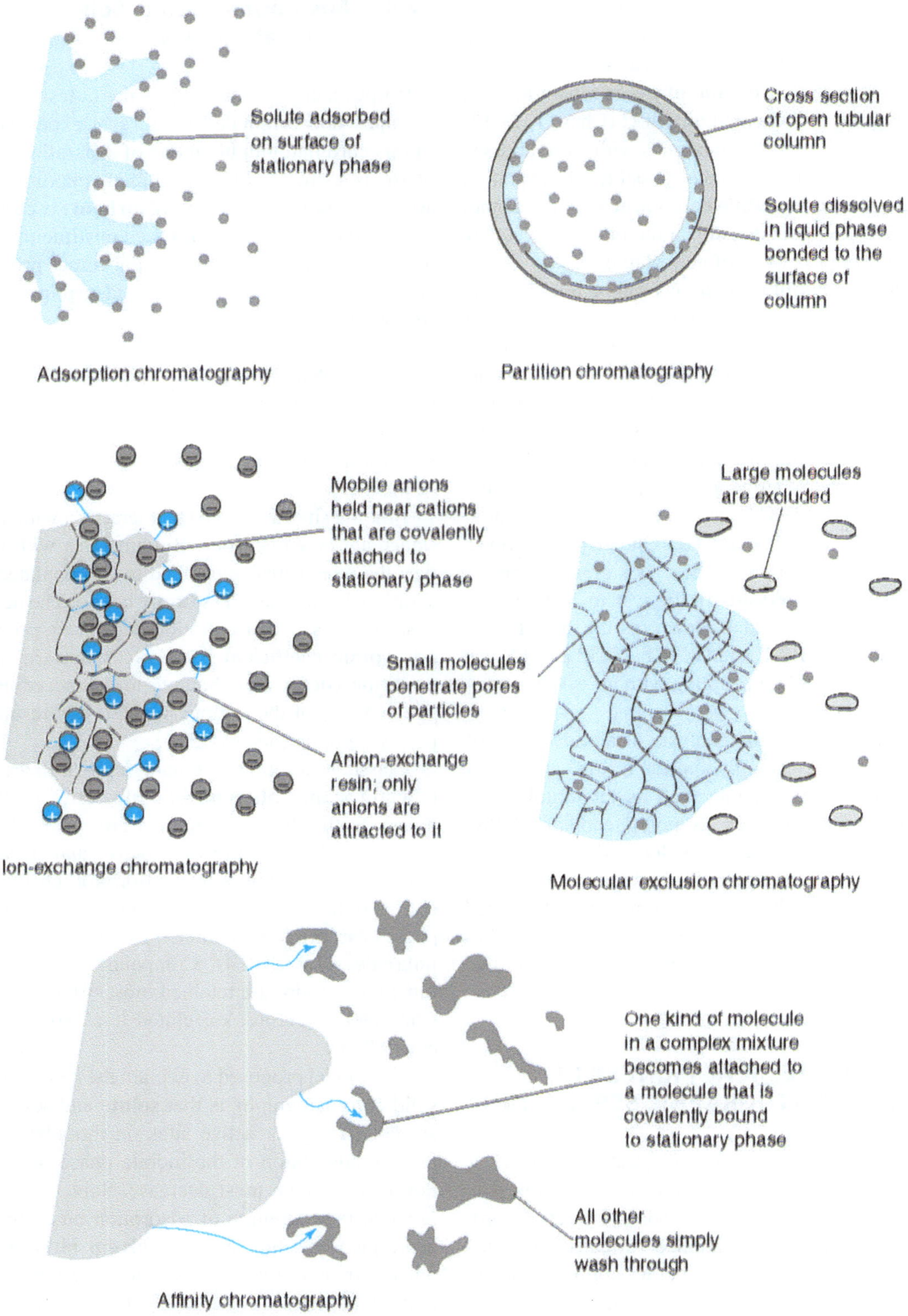

Fig. 12.3 Physicochemical principles of chromatography. (From Harris [2] *Quantitative Chemical Analysis, 10th ed.*, by D.C. Harris, copyright 2020 by W.H. Freeman & Co. Used with permission of publisher)

Table 12.3 Summary of different chromatographic separation modes

Separation mode	*Stationary phase*	*Mobile phase*	*Increasing mobile phase strength*	*Compounds eluting first/ eluting last*	*Type of interactions between solutes and stationary phase*
Normal phase (can be in the form of adsorption or partition chromatography)	Polar sorbent (e.g., silica, alumina, water)	Nonpolar solvent (e.g., aqueous methanol, acetonitrile)	Decreasing concentration of organic solvent (i.e., increasing polarity, making the mobile phase more like the stationary phase)	Least polar/most polar	H-bonding mostly
Reversed-phase (can be in the form of adsorption or partition chromatography)	Nonpolar sorbent (e.g., bonded silica, C8 or C18)	Polar solvent (e.g., water)	Increasing concentration of organic solvent (i.e., decreasing polarity, making the mobile phase more like the stationary phase)	Most polar/least polar	H-bonding, Van der Waals, hydrophobic interactions
Hydrophilic interaction	Polar-modified silica	Nonpolar solvent (e.g., 50% acetonitrile)	Decreasing concentration of organic solvent (i.e., increasing polarity, making the mobile phase more like the stationary phase)	Least polar/most polar	Van der Waals, electrostatic, H-bonding
Hydrophobic interaction	Nonpolar sorbent (e.g., butyl-Sepharose® and phenyl-Sepharose®)	Salt solution/buffer (e.g., 1 M ammonium sulfate; phosphate buffer)	Decreasing concentration of salt (result in reduced interaction of solute with the sorbent)	Least hydrophobic surface/most hydrophobic surface	Hydrophobic interactions
Cation exchange	Negatively charged functional groups (e.g., $RSO3^-$, $RCO2^-$)	Buffers of specific pH and ionic strength	Increasing the pH (in case of weak cation exchanger) or increasing salt concentration (e.g., increasing the pH will result in deprotonation, i.e., loss of positive charge of the solute so it no longer interacts with the functional group of the stationary phase, and increasing the salt concentration will provide counter ions that will displaces the solute on the functional groups of the stationary phase)	Solutes with the least positive charge density/ most positive charge density	Electrostatic
Anion exchange	Positively charged functional groups (e.g., RNR'3+, R-NHR'2+)	Buffers of specific pH and ionic strength	Decreasing the pH (in case of weak anion exchanger) or increasing salt concentration (e.g., decreasing the pH will result in protonation, i.e., loss of negative charge of the solute so it no longer interacts with the functional group of the stationary phase, and increasing the salt concentration will provide counter ions that will displace the solute on the functional groups of the stationary phase)	Solutes with the least negative charge density/ most negative charge density	Electrostatic
Affinity	Highly specific ligand bound to inert surface (e.g., antibodies, enzyme inhibitors, lectins)	Buffer	Changing the pH or ionic strength, or adding a ligand similar to the bound ligand of the stationary phase	Solutes with least affinity to bound ligands/most affinity to bound	H-bonding; van der Waals, hydrophobic interactions, electrostatic
Size exclusion	Porous inert material (e.g., Sephadex, a cross-linked dextran)	Mostly water or buffer	Not applicable	Largest in size/smallest in size	None

Table 12.4 Compounds class polarity scale[a]

Fluorocarbons
Saturated hydrocarbons
Olefins
Aromatics
Halogenated compounds
Ethers
Nitro compounds
Esters ≈ ketones ≈ aldehydes
Alcohols ≈ amines
Amides
Carboxylic acids

From Johnson and Stevenson [5], used with permission
[a]Listed in order of increasing polarity

Table 12.5 Eluotropic series for alumina[a]

Solvent
1-Pentane
Isooctane
Cyclohexane
Carbon tetrachloride
Xylene
Toluene
Benzene
Ethyl ether
Chloroform
Methylene chloride
Tetrahydrofuran
Acetone
Ethyl acetate
Aniline
Acetonitrile
2-Propanol
Ethanol
Methanol
Acetic acid

From Johnson and Stevenson [5], used with permission
[a]Listed in order of increasing polarity

Adsorption chromatography can be used to separate aromatic or aliphatic nonpolar compounds, based primarily on the type and number of functional groups present. The labile, fat-soluble chlorophyll and carotenoid pigments from plants have been studied extensively by adsorption column chromatography. Adsorption chromatography has also been used for the analysis of fat-soluble vitamins. Frequently, it is used as a batch procedure for the removal of impurities from samples prior to other analyses. For example, disposable solid-phase extraction cartridges (see Chap. 14) containing silica have been used for analyses of lipids in soybean oil, carotenoids in citrus fruit, and vitamin E in grain. Adsorption chromatography is also applied in specialized forms for the analysis of a wide range of compounds. Several of the chromatographic separation techniques described in the following sections are forms of specialized adsorption chromatography.

12.4.2 Partition (Liquid–Liquid) Chromatography

12.4.2.1 Introduction

In 1941, Martin and Synge undertook an investigation of the amino acid composition of wool, using a countercurrent extractor (Sect. 12.2.3) consisting of 40 tubes with chloroform and water flowing in opposite directions. The efficiency of the extraction process was improved enormously when a column of finely divided inert support material was used to hold one liquid phase (stationary phase) immobile, while the second liquid, an immiscible solvent (mobile phase), flowed over it, thus providing intimate contact between the two phases. Solutes partitioned between the two liquid phases according to their partition coefficients, hence the name **partition chromatography**.

In partition chromatography, depending on the characteristics of the compounds to be separated, the nature of the two liquid phases can be varied, usually in combination with solvents or pH adjustment of buffers. Often, the more polar of the two liquids is held stationary on the inert support, and the less polar solvent is used to elute the sample components (**normal-phase chromatography**). Reversal of this arrangement, using a **nonpolar stationary phase** and a **polar mobile phase**, has come to be known as **reversed-phase chromatography** (see Sect. 12.4.2.3).

Polar **hydrophilic** substances, such as amino acids, carbohydrates, and water-soluble plant pigments, are often separable by **normal-phase** partition chromatography. **Lipophilic** compounds, such as lipids, fat-soluble pigments, and **polyphenols**, may be resolved better with **reversed-phase** systems. Partition chromatography has been invaluable to carbohydrate chemistry. Column liquid chromatography on finely divided cellulose has been used extensively in preparative chromatography of sugars and their derivatives.

12.4.2.2 Coated Supports

In its simplest form, the stationary phase for partition chromatography consists of a liquid coating on a solid matrix. The solid support should be as inert as possible and have a large surface area in order to maximize the amount of liquid held. Some examples of solid supports that have been used are silica, starch, cellulose powder, and glass beads. All are capable of holding a thin film of water, which serves as the stationary phase. It is important to note that materials prepared for adsorption chromatography must be **activated** by drying them to remove surface water. Conversely, some of these materials, such as silica gel, may be used for partition chromatography if they are deactivated by impregnation with water or the desired stationary phase. One disadvantage of partition chromatographic systems is that the liquid stationary phase is often stripped off. This problem can be overcome by chemically bonding the stationary phase to the support material.

12.4.2.3 Bonded Supports

The liquid stationary phase may be covalently attached to a support by a chemical reaction. These **bonded phases** have become very popular for HPLC use, and a wide variety of both polar and nonpolar stationary phases is now available. It is important to note that mechanisms other than partition may be involved in the separation using bonded supports. Reversed-phase HPLC (see Chap. 13), with a nonpolar bonded stationary phase (e.g., silica bonded with C_8 or C_{18} groups) and a polar solvent (e.g., water-acetonitrile), is widely used. Bonded-phase HPLC columns have greatly facilitated the analysis of vitamins in foods and feeds, as discussed in Chap. 3 of reference [21]. Additionally, bonded-phase HPLC is widely used for the separation and identification of polyphenols such as phenolic acids (e.g., *p*-coumaric, caffeic, ferulic, and sinapic acids), carotenoids (e.g., lutein and zeaxanthin), and flavonoids (e.g., flavonols, flavones, isoflavones, anthocyanidins, catechins, and proanthocyanidins).

12.4.3 Hydrophilic Interaction Liquid Chromatography

Hydrophilic interaction liquid chromatography (HILIC) is a form of normal-phase chromatography, with a difference in the composition of the mobile phase. The stationary phase is a polar liquid, and the mobile phase in HILIC is composed of a high concentration of water miscible organic solvent (e.g., acetonitrile), 50–95%, in an aqueous buffer. In contrast, the mobile phase in normal-phase chromatography is a water-immiscible organic solvent. The stationary phase is mostly silica modified with a polar functional group such as amino, cyano, or cyclodextrin. Interactions between the stationary phase and solutes are polar interactions including van der Waals, electrostatic, or hydrogen bonding. HILIC is an alternative technique to normal-phase chromatography, with the advantage of interfacing well with MS, due to the water-miscible mobile phase [22]. It is commonly used for the determination of small-molecular-weight polar compounds such as amino acids, peptides, phospholipids, and organic acids [23].

12.4.4 Hydrophobic Interaction Chromatography

Hydrophobic interaction chromatography (HIC) is a type of adsorption chromatography, commonly used for the purification of compounds on a preparative and semi-analytical scale. In HIC, biomolecules adsorb to a weak hydrophobic surface at high salt concentration. Elution of adsorbed molecules is achieved by decreasing the salt concentration of the mobile phase over time. This technique takes advantage of hydrophobic moieties on the surface of a compound. Accordingly, HIC is commonly used for the purification of food proteins, enzymes, and antibodies while offering high resolution. The high salt concentration allows biomolecules with high surface charge to adsorb to the hydrophobic **ligands** by shielding the charges. Salt precipitation of proteins, in particular, onto the column does not cause denaturation, since the interaction with the hydrophobic ligands is weak.

The stationary phase in HIC consists of hydrophilic support bonded to hydrophobic ligands. Several polymeric materials can be used as support, including cellulose, agarose, chitosan, polymethacrylate, or silica. The support must be hydrophilic so as not to contribute additional hydrophobicity and hence strong interactions that may cause denaturation of the protein. Often the polymer used has a high degree of cross-linking to provide rigidity and high surface area. Most used ligands, chemically bonded to the support polymer, are linear chain alkanes. The size of the alkyl chains used depends on the surface hydrophobicity of the biomolecules to be separated, with the length of the chain being higher for more hydrophobic biomolecules. Sometimes, phenyl and other aromatic groups are also used. Often times though, ligands with intermediate hydrophobicity are used to avoid strong interactions. Butyl-Sepharose® and phenyl-Sepharose® are among the most commonly used stationary phases.

Different salts are used in HIC depending on their effects on protein precipitation. The most used salt is ammonium sulfate. Concentration of the salt is also a determining factor for the precipitation of the protein on the column. Often 1 M ammonium sulfate is used for initial screening. It is important to prepare the sample in the same salt solution/buffer as the mobile phase. However, this necessitates care in loading the sample to avoid precipitation of the protein prior to reaching the column. A wash step generally precedes loading the sample to allow for washing out impurities. Depending on the sample characteristics, elution can be performed gradually by decreasing salt concentration over time, which may allow for isolation of different proteins in the sample in less time. Changing of salt concentration over time requires optimization for best resolution and shortest analysis time. Switching directly to water after the salt wash may also be performed to elute all bound protein with minimal separation. This is dependent on the level of fractionation and purification required. If a compound resists elution even after reducing salt concentration to zero, other HIC ligands should be tried. Cleaning and regeneration of the column are required after several runs. Often 0.1–1 M NaOH is used to prevent fouling of the column, and sometimes detergents and alcohol washes are used.

The pH of the mobile phase can also influence retention on the column and elution. However, oftentimes buffers with pH 7 are used. Additives, such as water-miscible alcohols and detergents are sometimes used to help elute the protein faster. The hydrophobic parts of these additives will compete with the protein for binding to the ligand, causing desorption of the protein. Finally, temperature may play a role in HIC. As temperature increases, hydrophobic interaction increases allowing for better retention on the column; lowering the temperature aids in elution. Control of temperature during a run may enhance resolution and reduce analysis time (see also Sect. 12.5.1). For further details on HIC and applications, readers are directed to reference [24].

12.4.5 Ion-Exchange Chromatography

Separation by ion exchange can be categorized into three types: (1) ionic from nonionic, (2) cationic from anionic, and (3) mixtures of similarly charged species. In the first two cases, one solute binds to the ion-exchange medium, whereas the other solute does not.

Ion-exchange chromatography is viewed as a specialized type of adsorption chromatography in which interactions between the solute and stationary phase are primarily **electrostatic** in nature. The stationary phase (ion exchanger) contains fixed functional groups that are either negatively or positively charged (Fig. 12.4a). Exchangeable counterions

a

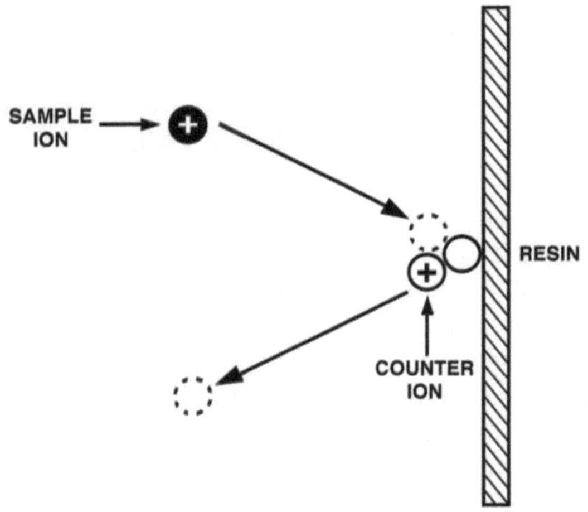

b

$M^+ + (Na^{+}\,{}^{-}O_3S\text{—Resin}) \rightleftharpoons (M^{+}\,{}^{-}O_3S\text{—Resin}) + Na^+$

M^+ = Cation

$X^- + (Cl^{-}\,{}^{+}R_4N\text{—Resin}) \rightleftharpoons (X^{-}\,{}^{+}R_4N\text{—Resin}) + Cl^-$

X^- = Anion

Fig. 12.4 The basis of ion-exchange chromatography. (**a**) Schematic diagram of the ion-exchange process; (**b**) ionic equilibria for cation- and anion-exchange processes. (From Johnson and Stevenson [5], used with permission)

preserve charge neutrality. A sample ion (or charged sites on large molecules) can exchange with the counterion for the interaction with the charged functional group. Ionic equilibrium is established as depicted in Fig. 12.4b. The functional group of the stationary phase determines whether cations or anions are exchanged. **Cation exchangers** possess covalently bound negatively charged functional groups (acidic residues), whereas **anion exchangers** possess bound positively charged groups (basic residues). The chemical nature of these acidic or basic residues determines how stationary-phase ionization is affected by the mobile-phase pH.

The strongly acidic sulfonic acid moieties (RSO3-) of "strong" cation exchangers are completely ionized at all pH values above 2. Strongly basic quaternary amine groups (RNR'3+) on "strong"-anion exchangers are ionized at all pH values below 10. Since maximum negative or positive charge is maintained over a broad pH range, the exchange or binding capacity of these stationary phases is essentially constant, regardless of mobile-phase pH. "Weak" cation exchangers contain weakly acidic carboxylic acid functional groups (RCO2-); consequently, their exchange capacity varies considerably between ca. pH 4 and 10. Weakly basic anion exchangers possess primary, secondary, or tertiary amine residues (R-NHR'2+) that are deprotonated in moderately basic solution, thereby losing their positive charge and the ability to bind anions. Thus, one way of eluting solutes bound to weak ion-exchange medium is to change the mobile phase pH. Changing the pH in the case of weak ion exchangers allows for enhanced resolution and better selectivity compared to strong ion exchangers when separating compounds with very similar charge densities. While strong ion exchangers are generally used for initial screening and optimization of the separation, oftentimes, weak ion exchangers are used to separate compounds that have similar adsorption coefficient, making use of a pH gradient. Changing the pH, however, may present a challenge when separating proteins. One must avoid the isoelectric point of the proteins during chromatographic separation; otherwise, they will fall out of solution. A second way to elute bound solutes from either strong or weak ion exchangers is to increase the ionic strength (e.g., use NaCl) of the mobile phase, to weaken the electrostatic interactions.

Chromatographic separations by ion exchange are based upon differences in affinity of the exchangers for the ions (or charged species) to be separated. The factors that govern **selectivity** of an exchanger for a particular ion include: (1) ionic valence, radius, and concentration, (2) nature of the exchanger (including its displaceable counterion), and (3) composition and pH of the mobile phase. To be useful as an ion exchanger, a material must be both ionic in nature and highly permeable. Synthetic ion exchangers are thus cross-linked polyelectrolytes, and they may be inorganic (e.g., aluminosilicates) or, more commonly, organic compounds. Two commonly used types of organic compound-based synthetic ion exchangers are polystyrene and polysaccharide.

Polystyrene, made by cross-linking styrene with divinyl benzene (DVB), may be modified to produce either anion- or cation-exchange resins (Fig. 12.5). Polymeric resins such as these are commercially available in a wide range of particle sizes and with different degrees of cross-linking (expressed as weight percent of DVB in the mixture). The extent of cross-linking controls the rigidity and porosity of the resin, which, in turn, determines its optimal use. Lightly cross-linked resins permit rapid equilibration of solute, but particles swell in water, thereby decreasing charge density and selectivity (relative affinity) of the resin for different ions. More highly cross-linked resins exhibit less swelling, higher exchange capacity, and selectivity, but longer equilibration times. The small pore size, high charge density, and inherent hydrophobicity of the older ion-exchange resins have limited their use to small molecules [molecular weight (MW) < 500].

Ion exchangers based on **polysaccharides**, such as cellulose, dextran, or agarose, have proven very useful for the separation and purification of large molecules, such as proteins and nucleic acids. These materials, called **gels,** are much softer than polystyrene resins, thus may be derivatized with strong or with weak acidic or basic groups via OH moieties on the polysaccharide backbone (Fig. 12.6). They have much larger pore sizes and lower charge densities than the older synthetic resins.

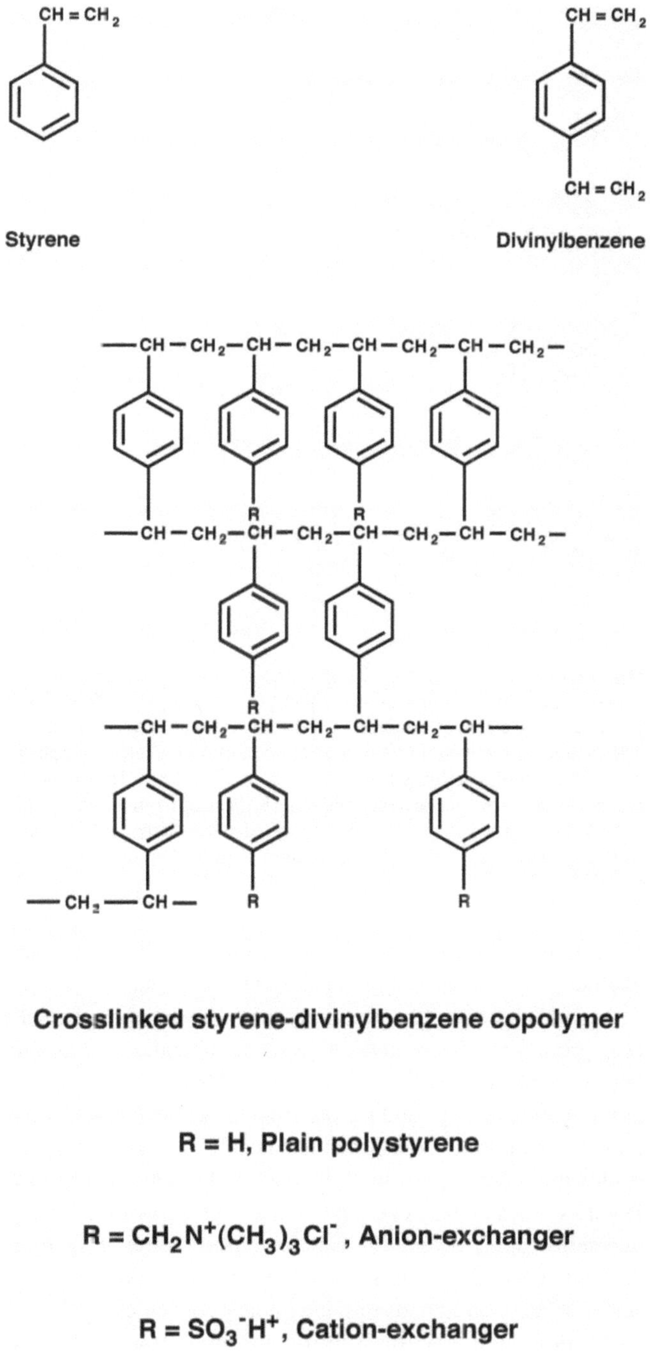

Fig. 12.5 Chemical structure of polystyrene-based ion-exchange resins

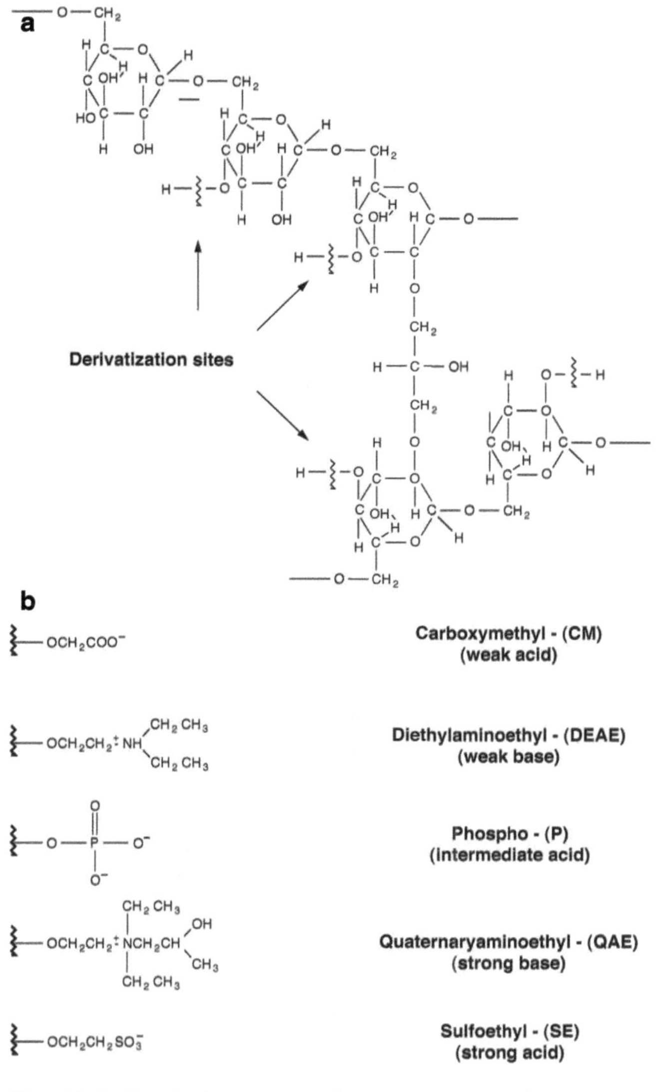

Fig. 12.6 Chemical structure of one polysaccharide-based ion-exchange resin. (**a**) Matrix of cross-linked dextran ("Sephadex," Pharmacia Biotech Inc., Piscataway NJ); (**b**) functional groups that may be used to impart ion-exchange properties to the matrix

Food-related applications of ion-exchange chromatography include the separation of amino acids, sugars, alkaloids, and proteins. Fractionation of amino acids in protein hydrolysates was initially carried out by ion-exchange chromatography; automation of this process led to the development of commercially produced amino acid analyzers (see Chap. 24). Many drugs, fatty acids, and organic acids, being ionizable compounds, may be analyzed by ion-exchange chromatography. For additional details on the principles and applications of ion-exchange chromatography, refer to reference [25].

12.4.6 Affinity Chromatography

Affinity chromatography is also another specialized form of adsorption chromatography, with the separation being based on a very specific, reversible interaction between a solute and a biological ligand immobilized on the chromatographic stationary phase (types of interactions are listed in Sect. 12.4.1). The immobilized ligands can be antibodies, enzyme inhibitors, lectins, or other molecules that selectively and reversibly bind to complementary analytes in the sample. Although both ligands and the analytes to be isolated are usually biological macromolecules, the term affinity chromatography also encompasses other systems, such as separation of small molecules containing *cis*-diol groups via phenylboronic acid moieties on the stationary phase.

The principles of affinity chromatography are illustrated in Fig. 12.7. A ligand, chosen based on its specificity/selectivity and strength of interaction with the analyte, is immobilized on a suitable support material. As the sample is passed through the column, analytes that are complementary to the bound ligand are adsorbed, while other sample components are eluted. Bound analytes are subsequently eluted via a change in the mobile phase composition as will be discussed below. After re-equilibration with the initial mobile phase, the stationary phase is ready to be used again. The ideal support for affinity chromatography should be a porous, stable, high-surface-area material that does not adsorb anything itself. Thus, polymers such as agarose, cellulose, dextran, and polyacrylamide are used, as well as controlled pore glass.

Affinity ligands are usually attached to the support or matrix by covalent bond formation, and optimum reaction conditions often must be found empirically. Immobilization generally consists of two steps: **activation** and **coupling**. During the activation step, a reagent reacts with functional groups on the support, such as hydroxyl moieties, to produce an activated matrix. After removal of excess reagent, the ligand is coupled to the activated matrix. (Preactivated supports are commercially available.) The coupling reaction most often involves free amino groups on the ligand, although other functional groups can be used. When small molecules such as phenylboronic acid are immobilized, a **spacer arm** (containing at least four to six methylene groups) is used to hold the ligand away from the support surface, enabling it to reach the binding site of the analyte.

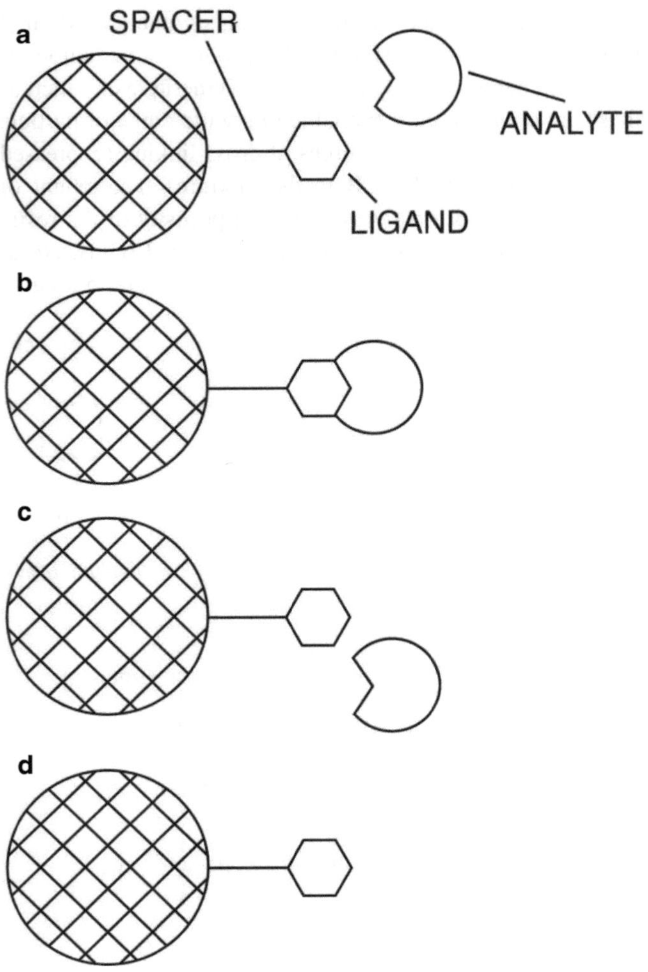

Fig. 12.7 Principles of bioselective affinity chromatography. (**a**) The support presents the immobilized ligand to the analyte to be isolated. (**b**) The analyte makes contact with the ligand and attaches itself. (**c**) The analyte is recovered by the introduction of an eluent, which dissociates the complex holding the analyte to the ligand. (**d**) The support is regenerated, ready for the next isolation. (Reprinted from Heftmann [1], p. A311, with kind permission from Elsevier Science NL, Sara Burgerhartstraat 25, 1055 KV Amsterdam, The Netherlands)

Ligands for affinity chromatography may be either **specific** (i.e., group specific) or **general**. Specific ligands, such as antibodies, bind only one particular solute. General ligands, such as nucleotide analogs and lectins, bind to certain classes of solutes. For example, the lectin concanavalin A binds to all molecules that contain terminal glucosyl and mannosyl residues. Bound solutes then can be separated as a group or individually, depending upon the elution technique used. Some of the more common general ligands are listed in Table 12.6. Although less selective, general ligands provide greater convenience.

Elution methods for affinity chromatography may be divided into **nonspecific** and **(bio)specific** methods.

Table 12.6 General affinity ligands and their specificities

Ligand	*Specificity*
Cibacron Blue F3G-A dye, derivatives of AMP, NADH, and NADPH	Certain dehydrogenases via binding at the nucleotide binding site
Concanavalin A, lentil lectin, wheat glycoproteins, germ lectin	Polysaccharides, glycolipids, and membrane proteins containing sugar residues of certain configurations
Soybean trypsin inhibitor, methyl esters of various amino acids, D-amino acids	Various proteases
Phenylboronic acid	Glycosylated hemoglobin, sugars, nucleic acids, and other *cis*-diol-containing substances
Protein A	Many immunoglobulin classes and subclasses via binding to the F_c region
DNA, RNA, nucleosides, nucleotides	Nucleases, polymerases, nucleic acids

Nonspecific elution involves disrupting ligand analyte binding by changing the mobile phase pH, ionic strength, dielectric constant, or temperature. If additional selectivity in elution is desired, for example, in the case of immobilized general ligands, a biospecific elution technique is used. Free ligand, either identical to or different from the matrix-bound ligand, is added to the mobile phase. This free ligand competes for binding sites on the analyte. For example, glycoproteins bound to a concanavalin A (lectin) column can be eluted by using buffer containing an excess of lectin. In general, the eluent ligand should display greater affinity for the analyte of interest than the immobilized ligand.

In addition to protein purification, affinity chromatography may be used to separate supramolecular structures such as cells, organelles, and viruses; concentrate dilute protein solutions; investigate binding mechanisms; and determine equilibrium constants. Affinity chromatography has been useful, especially in the separation and purification of enzymes and glycoproteins. In the case of the latter, carbohydrate-derivatized adsorbents are used to isolate specific lectins, such as concanavalin A, and lentil or wheat germ lectin. The lectin then may be coupled to agarose, such as concanavalin A–agarose or lentil lectin–agarose, to provide a stationary phase for the purification of specific glycoproteins, glycolipids, or polysaccharides. Other applications of affinity chromatography include purification and quantification of mycotoxins [26] and pesticide residues/metabolites [27] in food/crops (see Chap. 33). For additional details on affinity chromatography, refer to references [28, 29].

12.4.7 Size-Exclusion Chromatography

Size-exclusion chromatography (SEC), also known as molecular exclusion, gel permeation (GPC), and gel filtration chromatography (GFC), is probably the easiest mode of chromatography to perform and to understand. It is widely used in the biological sciences for the resolution of macromolecules, such as proteins and carbohydrates, and also is used for the fractionation and characterization of synthetic polymers. Unfortunately, nomenclature associated with this separation mode developed independently in the literature of the life sciences and in the field of polymer chemistry, resulting in inconsistencies.

In the ideal SEC system, molecules are separated solely on the basis of their size; no interaction occurs between solutes and the stationary phase. In the event that solute/support interactions do occur, the separation mode is termed nonideal SEC. The stationary phase in SEC consists of a column packing material that contains pores comparable in size to the molecules to be fractionated. Solutes too large to enter the pores travel with the mobile phase in the interstitial space (between particles) outside the pores. Thus, the largest molecules are eluted first from an SEC column. The volume of the mobile phase in the column, termed the column **void volume**, Vo, can be measured by chromatographing a very large species, such as Blue Dextran, a dye that has a molecular weight (MW) of 2000 kDa.

As solute dimensions decrease, approaching those of the packing pores, molecules begin to diffuse into the packing particles and, consequently, are slowed down. Solutes of low MW (e.g., glycyltyrosine) that have free access to all the available pore volume are eluted in the volume referred to as Vt, the **total permeation volume** of the column. The Vt is equal to the column void volume, Vo, plus the volume of liquid inside the sorbent pores (**internal pore volume**), Vi (Vt = Vo + Vi). These relationships are illustrated in Fig. 12.8.

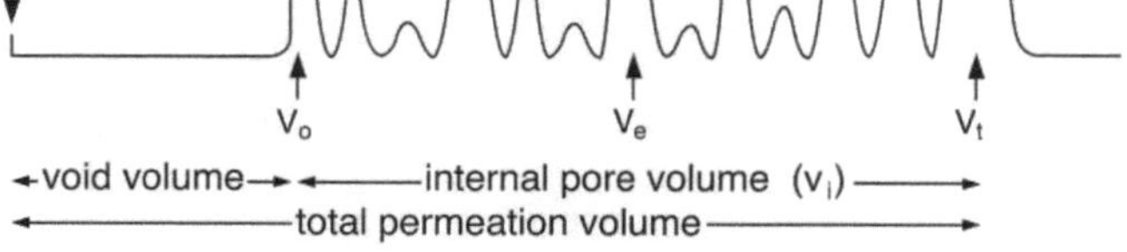

Fig. 12.8 Schematic elution profile illustrating some of the terms used in size-exclusion chromatography. (Adapted from Heftmann [1], p. A271, with kind permission from Elsevier Science—NL, Sara Burgerhartstraat 25, 1055 KV Amsterdam, The Netherlands)

Solutes are ideally eluted between the void volume and the total permeation volume of the column. Because this volume is limited, only a relatively small number of solutes (ca. 10), of a particular size range, can be completely resolved by SEC under ordinary conditions.

The behavior of a molecule in a size-exclusion column may be characterized in several different ways. Each solute exhibits an **elution volume**, Ve, as illustrated in Fig. 12.9. However, Ve depends on column dimensions, the way in which the column was packed, and the sorbent pore size. The **available partition coefficient**, Kav, is used to define solute behavior independent of these variables:

$$\mathrm{Kav} = (\mathrm{Ve} - \mathrm{Vo}) / (\mathrm{Vt} - \mathrm{Vo}) \quad (12.3)$$

where:

Kav = available partition coefficient
Ve = elution volume of solute
Vo = column void volume
Vt = total permeation volume of column

The value of Kav calculated from experimental data for a solute chromatographed on a given SEC column defines the proportion of pores that can be occupied by that molecule. For a large, totally excluded species, such as Blue Dextran or DNA, Ve = Vo and Kav = 0. For a small molecule with complete access to the internal pore volume, such as glycyltyrosine, Ve = Vt and Kav = 1.

For each size-exclusion packing material, a plot of Kav vs. the logarithm of the MW for a series of solutes, similar in molecular shape and density, will give an S-shaped curve (Fig. 12.9). In the case of proteins, Kav is actually better related to the Stokes radius, the average radius of the protein in solution. The central, linear portion of this curve describes the **fractionation range** of the matrix, wherein maximum separation among solutes of similar MW is achieved. This correlation between solute elution behavior and MW (or size) forms the basis for a widely used method for characterizing large molecules such as proteins and polysaccharides. A size-exclusion column is calibrated with a series of solutes of known MW (or Stokes radius) to obtain a curve similar to that shown in Fig. 12.9. The value of Kav for the unknown is then determined, and an estimate of MW (or size) of the unknown is made by interpolation of the calibration curve.

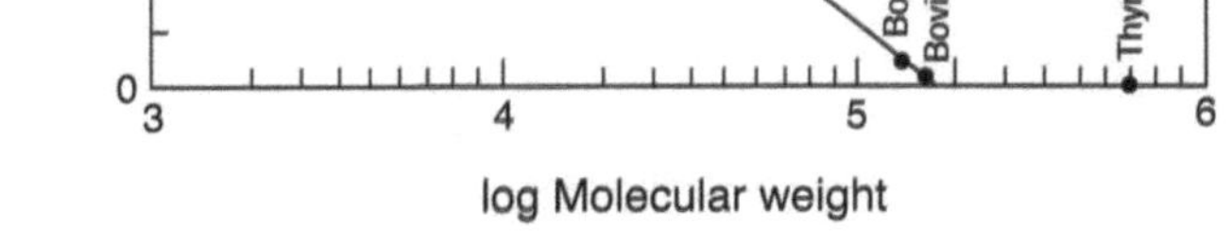

Fig. 12.9 Relationship between K_{av} and log (molecular weight) for globular proteins chromatographed on a column of Sephadex G-150 Superfine. (Reproduced by permission of Pharmacia Biotech, Inc., Piscataway, NJ)

Column packing materials for SEC can be divided into two groups: semirigid, hydrophobic media, and soft, hydrophilic gels. The former is usually derived from polystyrene and is used with organic mobile phases (GPC or nonaqueous SEC) for the separation of polymers, such as rubbers and plastics. Soft gels, polysaccharide-based packings, are typified by Sephadex, a cross-linked dextran (see Fig. 12.6a). These materials are available in a wide range of pore sizes and are useful for the separation of water-soluble substances in the MW range 1–2.5 × 10^7 Da. In selecting an SEC column packing, both the purpose of the experiment and size of the molecules to be separated must be considered. If the purpose of the experiment is group separation, for which molecules of widely different molecular sizes need to be separated, a matrix is chosen such that the larger molecules, e.g., proteins, are eluted in the void volume of the column, whereas small molecules are retained in the total volume. When SEC is used for separation of macromolecules of different sizes, the molecular sizes of all the components must fall within the fractionation range of the gel.

As discussed previously, SEC can be used, directly, to fractionate mixtures, purify, or, indirectly, to obtain MW/size information about a dissolved species. In addition to MW estimations, SEC is used to determine the MW distribution of natural and synthetic polymers, such as dextrans and gelatin preparations. Fractionation of and determination of the MW of peptides in a protein hydrolysate or of maltodextrins in hydrolyzed starch is another common application. Fractionation of biopolymer mixtures is probably the most widespread use of SEC, since the mild elution conditions employed rarely cause denaturation or degradation. Often times SEC is used as an early chromatographic separation step in a multidimensional chromatographic approach toward purification. It is also a fast, efficient alternative to dialysis for desalting solutions of macromolecules, such as proteins.

12.5 ANALYSIS OF CHROMATOGRAPHIC PEAKS

Once the chromatographic technique (Sect. 12.3) and chromatographic mechanism (Sect. 12.4) have been chosen, the analyst has to ensure adequate separation of constituents of interests from a mixture, in a reasonable amount of time.

After separation is achieved and chromatographic peaks are obtained, qualitative as well as quantitative analysis can be carried out. Basic principles of separation and resolution (separation of closely related analytes) will be discussed in the subsequent sections. Understanding these principles allows the analyst to optimize separation and perform qualitative and quantitative analysis.

12.5.1 Separation and Resolution

This section will discuss separation and resolution as it pertains mainly to LC; separation and resolution optimization as it pertains specifically to GC will be discussed in Chap. 14.

12.5.1.1 Developing a Separation

There may be numerous ways to accomplish a chromatographic separation for a particular compound. In many cases, the analyst will follow a standard laboratory procedure or published methods. In the case of a sample that has not been previously analyzed, the analyst begins by evaluating what is known about the sample and defines the goals of the separation. How many components need to be resolved? What degree of resolution is needed? Is qualitative or quantitative information needed? Molecular weight (or MW range)/size, polarity, and ionic characteristics of the sample will guide the choice of chromatographic separation mechanism (**separation mode**). Figure 12.10 shows that more than one correct choice may be possible. For example, small ionic compounds may be separated by ion-exchange, ion-pair reversed-phase (see Sect. 13.3.2.1), or reversed-phase LC. In this case, the analyst's choice may be based on convenience, experience, and personal preference.

Having chosen a separation mode for the sample at hand, one must select an appropriate stationary phase, elution conditions, and a detection method. Trial experimental conditions may be based on the results of a literature search, the analyst's previous experience with similar samples, or general recommendations from chromatography experts.

To achieve separation of sample components by all modes except SEC, one may utilize either isocratic or gradient elution. **Isocratic elution** is the simplest technique, in

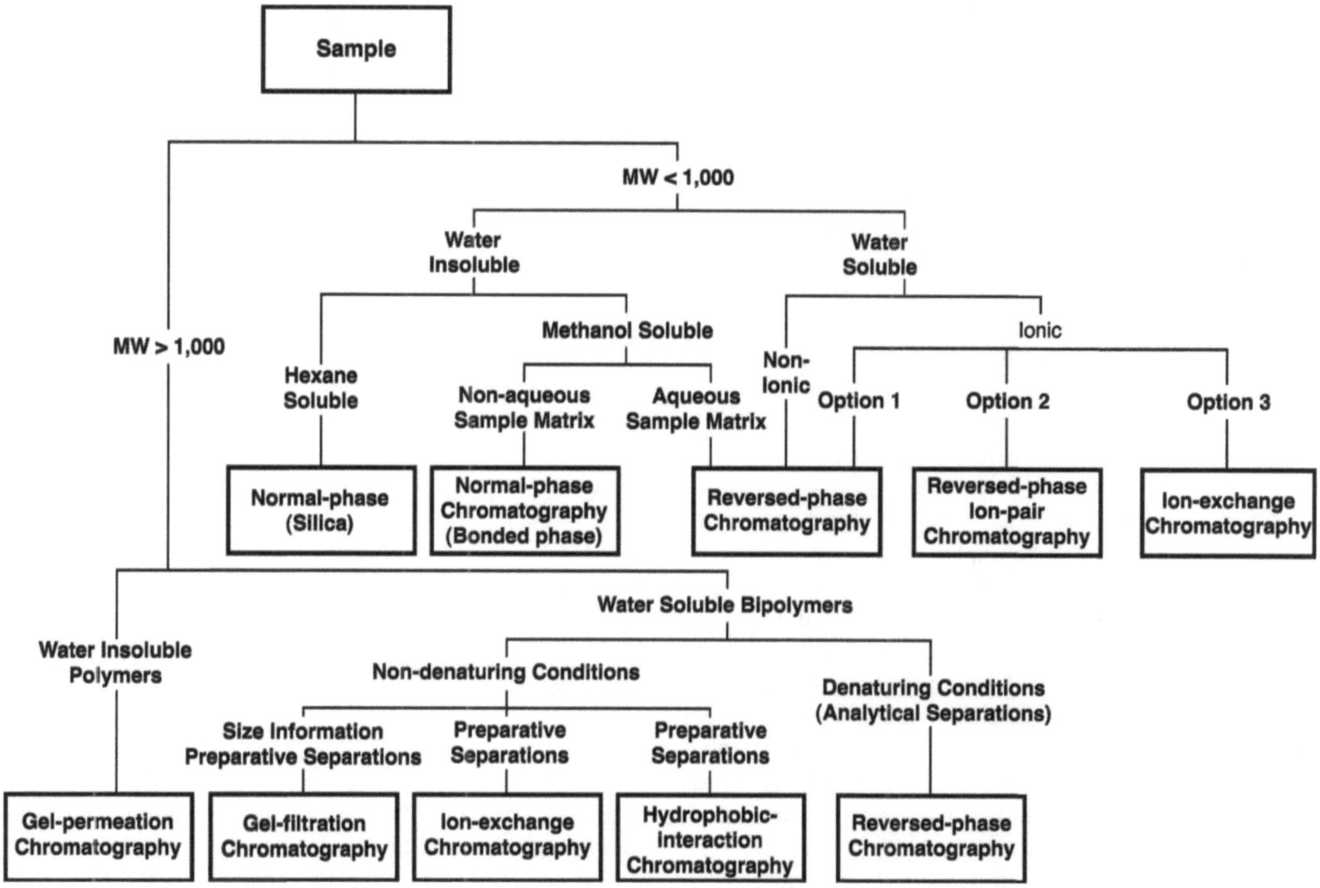

Fig. 12.10 A schematic diagram for choosing a chromatographic separation mode based on sample molecular weight and solubility. (From Lough and Wainer [31], used with permission)

which solvent composition, temperature, and flow rate are held constant throughout the chromatographic run. **Gradient elution** involves reproducibly varying the mobile phase composition or flow rate (flow programming) during the LC analysis. Gradient elution is used when sample components vary a lot in inherent characteristics such as polarity and/or charge density, so that an isocratic mobile phase does not elute all components within a reasonable time. The change may be continuous or stepwise. Gradients of increasing ionic strength, or changing pH, are extremely valuable in ion-exchange chromatography (see Sect. 12.4.5), whereas gradients of increasing or decreasing polarity are valuable in normal or reversed-phase chromatography, respectively (Sect. 12.4.2). Increasing the "strength" of the mobile phase (Sect. 12.4.1), either gradually (continuously) or in a stepwise fashion, shortens the analysis time.

Method development may begin with an isocratic mobile phase, possibly of intermediate solvent strength; however, using gradient elution for the initial separation may ensure that some level of separation is achieved within a reasonable time period and nothing is likely to remain on the column. Data from this initial run allows one to determine if isocratic or gradient elution is needed and to estimate optimal isocratic mobile phase composition or gradient range.

Once an initial separation has been achieved, the analyst can proceed to optimize resolution. This generally involves changing mobile phase variables, including the following: (1) nature and percentage of organic solvents (and type of organic solvents), (2) pH, (3) ionic strength, (4) nature and concentration of additives (such as ion-pairing agents), (5) flow rate, and/or (6) temperature. In the case of gradient elution, gradient steepness (slope, i.e., rate of change) is another variable to be optimized. However, the analyst must be aware of the principles of chromatographic resolution as will be discussed in the following section.

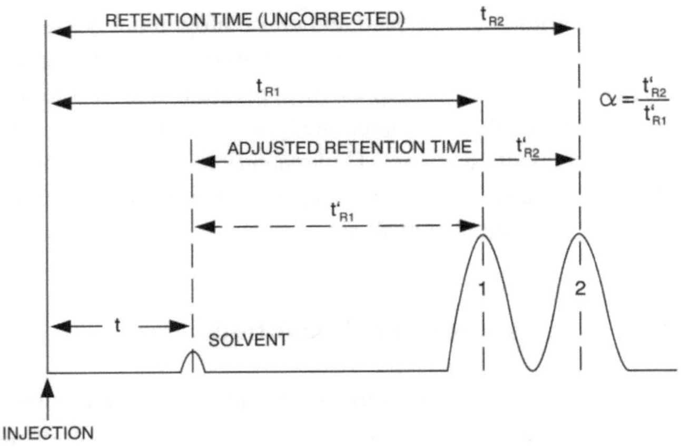

Fig. 12.11 Measurement of chromatographic retention. (Adapted from Johnson and Stevenson [5], with permission)

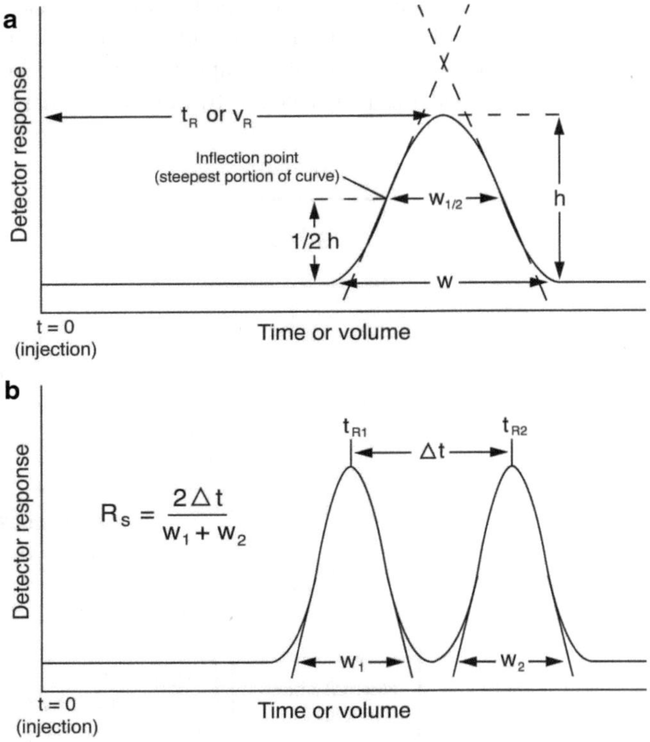

Fig. 12.12 Measurement of peak width and its contribution to resolution. (**a**) Idealized Gaussian chromatogram, illustrating the measurement of *w* and $w_{1/2}$; (**b**) the resolution of two bands is a function of both their relative retentions and peak widths. (Adapted from Johnson and Stevenson [5], with permission)

12.5.1.2 Chromatographic Resolution

12.5.1.2.1 Introduction

The main goal of chromatography is to segregate components of a sample into separate bands or peaks as they migrate through the chromatography column. A **chromatographic peak** is defined by several parameters including **retention time** (Fig. 12.11), **peak width,** and **peak height** (Fig. 12.12). The volume of the mobile phase required to elute a compound from a chromatography column is called the **retention volume**, V_R. The associated time is the **retention time**, t_R. Shifts in retention time and changes in peak width greatly influence **chromatographic resolution**.

Differences in column dimensions, stationary phase thickness, its particle and pore size, mobile phase flow rate, temperature, system dead volume, and detector geometry may lead to discrepancies in retention time. By subtracting the time required for the mobile phase or a non-retained solute (t_m or t_o) to travel through the column to the detector, one obtains an **adjusted retention time,** t'_R (or volume), as depicted in Fig. 12.11. The adjusted retention time (or volume) corrects for differences in system dead volume and signifies the time the analyte spends adsorbed on the stationary phase.

The **resolution** of two peaks from each other is related to the **separation factor**, α. Values for α (Fig. 12.11) depend on temperature, flow rate, stationary phase, and mobile phase used. Resolution is defined as follows:

$$R_s = \frac{2\Delta t}{w_2 + w_1} \tag{12.4}$$

where:

R_s = resolution
Δt = difference between retention times of peaks 1 and 2
w_2 = width of peak 2 at baseline
w_1 = width of peak 1 at baseline

Figure 12.12 illustrates the measurement of peak width [part (a)] and the values necessary for calculating resolution [part (b)]. (Retention and peak or band width must be expressed in the same units, i.e., time or volume.)

Chromatographic resolution is a function of column **efficiency, selectivity,** and the **capacity factor**. Mathematically, this relationship is expressed as follows:

$$R_s = \underbrace{1/4\sqrt{N}}_{a}\underbrace{\left(\frac{\alpha - 1}{\alpha}\right)}_{b}\underbrace{\left(\frac{k'}{k'+1}\right)}_{c} \tag{12.5}$$

where:

a = column **efficiency** term
b = column **selectivity** term
c = **capacity** term

These terms, and factors that contribute to them, will be discussed in the following sections.

12.5.1.2.2 Column Efficiency

If faced with the problem of improving resolution, a chromatographer should first examine the **efficiency** of the column. An efficient column keeps the solutes from spreading widely (by longitudinal diffusion) on the column and hence gives narrow peaks. Column efficiency can be calculated by:

$$N = \left(\frac{t_R}{\sigma}\right)^2 = 16\left(\frac{t_R}{w}\right)^2 = 5.5\left(\frac{t_R}{w_{1/2}}\right)^2 \tag{12.6}$$

where:

N = number of theoretical plates
t_R = retention time
σ = standard deviation for a Gaussian peak
w = peak width at baseline ($w = 4\sigma$)
$w_{1/2}$ = peak width at half height

The measurement of t_R, w, and $w_{1/2}$ is illustrated in Fig. 12.12. (Retention volume may be used instead of t_R; in this case, peak width is also measured in units of volume.) Although some peaks are not actually Gaussian in shape, normal practice is to treat them as if they were. In the case of peaks that are incompletely resolved or slightly asymmetric, peak width at half height is more accurate than peak width at baseline.

The value N calculated from the above equation is called the number of **theoretical plates**. The theoretical plate concept, borrowed from distillation theory, can best be understood by viewing chromatography as a series of equilibrations of solutes between mobile and stationary phases, analogous to countercurrent distribution. Thus, a column would consist of N segments (theoretical plates) with one equilibration occurring in each. As a first approximation, N is independent of retention time and is therefore a useful measure of column performance. One method of monitoring column performance over time is to chromatograph a standard compound periodically, under constant conditions, and to compare the values of N obtained. It is important to note that columns often behave as if they have a different number of plates for different solutes in a mixture. Different solutes have a different partition or distribution coefficient and thus have distinctive series of equilibrations between mobile and stationary phases. Peak broadening due to column deterioration will result in a decrease of N for a particular solute. Peak broadening is a result of an extended time for a solute to reach equilibrium between mobile and stationary phases (causing them to spread over a wider area on the column, i.e., diffuse within the stationary phase).

The number of theoretical plates is generally proportional to column length. The longer the column, the higher the number of theoretical plates, and hence the better is the resolution. Because columns are available in various lengths, it is useful to have a measure of column efficiency that is independent of column length. This may be expressed as:

$$\text{HETP} = \frac{L}{N} \tag{12.7}$$

where:

HETP = height equivalent to a theoretical plate
L = column length
N = number of theoretical plates

The so-called **HETP, height equivalent to a theoretical plate**, is sometimes more simply described as **plate height** (H). If a column consisted of discrete segments, HETP would be the height of each imaginary segment. Small HETP values (a large number of plates) indicate good efficiency of separation (minimal spread or diffusion of solute within the column, resulting in sharp peaks). Conversely, reduced number of plates results in poor separation due to the extended equilibrium time (spread of solutes on the column, resulting in wide peaks) in a deteriorating column.

In reality, columns are not divided into discrete segments and equilibration is not infinitely fast. The plate theory is used to simplify the equilibration concept. The movement of solutes through a chromatography column takes into account the finite rate at which a solute can equilibrate itself between stationary and mobile phases. Thus, peak shape depends on the rate of elution and is affected by solute diffusion. Any mechanism that causes a peak of solute to broaden will increase HETP and decrease column efficiency. The various factors that contribute to plate height are expressed by the **Van Deemter equation:**

$$\mathrm{HETP} = \mathrm{A} + \frac{\mathrm{B}}{u} + \mathrm{C}u \tag{12.8}$$

where:

HETP = height equivalent to a theoretical plate
A, B, C = constants
u = mobile phase rate

The constants A, B, and C are characteristic for a given column, mobile phase, and temperature. The A term represents the **eddy diffusion** or multiple flow paths. Eddy diffusion refers to the different microscopic flow streams that the mobile phase can take between particles in the column (analogous to eddy streams around rocks in a brook). Sample molecules can thus take different paths as well, depending on which flow streams they follow. As a result, solute molecules spread from an initially narrow band to a broader area within the column. The larger is the particle size of the packing material, the more paths a solute might take. Eddy diffusion may be minimized by good column packing techniques and the use of small-diameter particles of narrow particle size distribution to minimize the open space between the particles.

The B term of the Van Deemter equation, sometimes called the **longitudinal diffusion** term, exists because all solutes diffuse from an area of high concentration (the center of a chromatographic peak) to one of low concentration (the leading or trailing edge of a chromatographic peak). In LC, the contribution of this term to HETP is small except at low flow rate of the mobile phase. With slow flow rates, there will be more time for a solute to spend on the column; thus, its diffusion will be greater.

The C (mass transfer) term arises from the finite time required for the solute to equilibrate between the mobile and stationary phases. **Mass transfer** is practically the partitioning or distribution of the solute into the stationary phase, which does not occur instantaneously and depends on the solute's partition or distribution coefficients. If the stationary phase consists of porous particles (see Chap. 13, Sect. 13.2.3.2, Fig. 13.3), a sample molecule, entering a pore, ceases to be transported by the solvent flow and moves by diffusion only. Subsequently, this solute molecule may diffuse back to the mobile phase flow, or it may interact with the stationary phase. In either case, solute molecules inside the pores are slowed down relative to those outside the pores and peak broadening occurs. Contributions to HETP from the C term can be minimized by using porous particles of small pore diameter or pellicular packing materials (Chap. 13, Sect. 13.2.3.2.2). Also, using a narrower column with a smaller inner diameter or a capillary column reduces the C value, given that equilibrium time will be reduced since there is less packing material.

As expressed by the Van Deemter equation, **mobile phase flow rate**, u, contributes to plate height in opposing ways—increasing the flow rate increases the equilibration point (Cu) but decreases longitudinal diffusion of the solute particles (B/u). A Van Deemter plot (Fig. 12.13) may be used to determine the mobile phase flow rate at which plate height is minimized and column efficiency is maximized. Flow rates above the optimum may be used to decrease analysis time if adequate resolution is still obtained. However, at very high flow rates, there will be less time to approach equilibrium, which will lead to a shorter retention time and co-elution of closely related solutes.

In addition to flow rate, temperature can affect the longitudinal diffusion and the mass transfer. Increasing the temperature causes enhanced movement of the solute between the mobile phase and the stationary phase and within the column, thus leading to faster elution and narrower peaks.

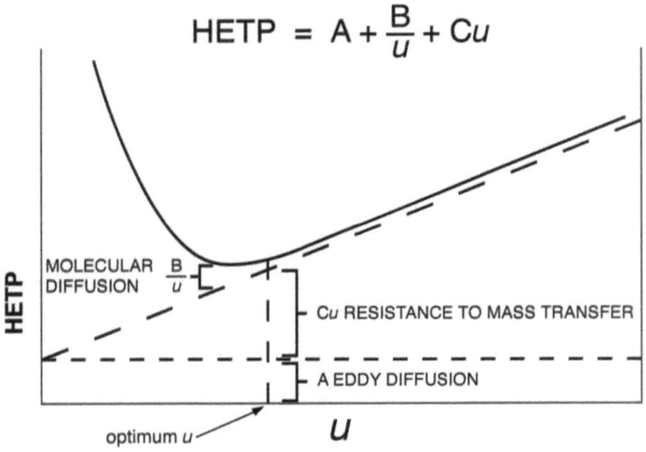

Fig. 12.13 Van Deemter plot of column efficiency (HETP) vs. mobile phase rate (u). Optimum u is noted. (Courtesy of Hewlett-Packard Co., Analytical Customer Training, Atlanta, GA)

12.5.1.2.3 Column Selectivity

Chromatographic resolution depends on column selectivity as well as efficiency. **Column selectivity** refers to the distance, or relative separation, between two peaks and is given by:

$$\alpha = \frac{t_{R2} - t_o}{t_{R1} - t_o} = \frac{t'_{R2}}{t'_{R1}} = \frac{K_2}{K_1} \qquad (12.9)$$

where:

- α = separation factor
- t_{R1} and t_{R2} = retention times of components 1 and 2, respectively
- t_o (or t_m) = retention time of unretained components (solvent front)
- t'_{R1} and t'_{R2} = adjusted retention times of components 1 and 2, respectively
- K_1 and K_2 = distribution coefficients of components 1 and 2, respectively

Retention times (or volumes) are measured as shown in Fig. 12.11. The time, t_o, can be measured by chromatographing a solute that is not retained under the separation conditions (i.e., travels with the solvent front). When this parameter is expressed in units of volume, V_o or V_m, it is known as the **dead or void volume** of the system. Selectivity is a function of the stationary and/or mobile phase. For example, selectivity in ion-exchange chromatography is not only influenced by the nature and number of ionic groups on the matrix but can also be affected by pH and ionic strength of the mobile phase. Changing the pH or the ionic strength of the mobile phase will impact the extent of interaction of the solutes with the column. Selectivity can be either static as in isocratic elution (the choice of the mobile phase from the beginning impact selectivity) or dynamic as in gradient elution (selectivity changing over time). Good selectivity is probably more important to a given separation than high efficiency (Fig. 12.14), since resolution is directly related to selectivity but is quadratically related to efficiency; thus, a fourfold increase in N is needed to double R_s (Eq. 12.5).

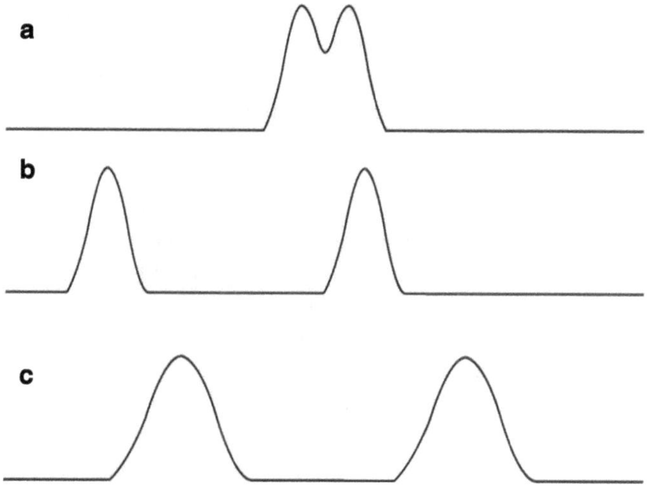

Fig. 12.14 Chromatographic resolution: efficiency vs. selectivity. (**a**) Poor resolution; (**b**) good resolution due to high column efficiency; and (**c**) good resolution due to column selectivity. (From Johnson and Stevenson [5], used with permission)

12.5.1.2.4 Column Capacity Factor

The **capacity** or **retention factor**, k', is a measure of the amount of time a chromatographed species (solute) spends in/on the stationary phase relative to the mobile phase. The relationship between capacity factors and chromatographic retention (which may be expressed in units of either volume or time) is shown below:

$$k' = \frac{KV_s}{V_m} = \frac{V_R - V_m}{V_m} = \frac{t_R - t_o}{t_o} \qquad (12.10)$$

where:

- k' = capacity factor
- K = distribution coefficient of the solute
- V_s = volume of stationary phase in column
- V_m = volume of mobile phase
- V_R = retention volume of solute
- t_R = retention time of solute
- t_o = retention time of unretained components (solvent front)

Small values of k' indicate little retention, and components will be eluted close to the solvent front, resulting in poor separations. Overuse or misuse of the column may lead to the loss of some functional groups, thus resulting in small k' values. Large values of k' result in improved separation but can also lead to broad peaks and long analysis times. On a practical basis, k' values within the range of 1–15 are generally used. (In the equation for R_s, k' is actually the average of k_1' and k_2' for the two components separated.)

12.5.2 Qualitative Analysis

Once separation and resolution have been optimized, identification of the detected compounds can be achieved. (Various detection methods are outlined in Chaps. 13 and 14) Comparing V_R or t_R to that of standards chromatographed under identical conditions often enables one to identify an unknown compound. When it is necessary to compare chromatograms obtained from two different systems or slightly different columns, it is better to compare **adjusted retention time**, t'_R (see Sect. 12.5.1.2). Oftentimes comparison across different systems and columns is not possible. Different compounds may have identical retention times using the same system. In other words, even if the retention time of an unknown and a standard are equivalent, the two compounds might not be identical. Therefore, other techniques are needed to confirm peak identity. For example:

1. Spike the unknown sample with a known compound and compare chromatograms of the original and spiked samples to see which peak has increased. Only the height of the peak of interest should increase, with no change in retention time, peak width, or shape.
2. A diode array detector can provide absorption spectra of designated peaks (see Sects. 13.2.4 and 13.2.4.1). Although identical spectra do not prove identity, a spectral difference confirms that sample and standard peaks are different compounds.
3. In the absence of spectral scanning capability, other detectors, such as absorption or fluorescence, may be used in a ratioing procedure. Chromatograms of sample and standard are monitored at each of two different wavelengths. The ratio of peak areas at these wavelengths should be the same if sample and standard are identical.
4. Peaks of interest can be collected and subjected to additional chromatographic separation using a different separation mode.
5. Collect the peak(s) of interest and establish their identity by another analytical method (e.g., mass spectrometry, which can give a mass spectrum that is characteristic of a particular compound; see Chap. 11).

12.5.3 Quantitative Analysis

Assuming that good chromatographic resolution and identification of sample components have been achieved, quantification involves mostly measuring peak area and comparing the data with those for standards of known concentration. Nowadays, all chromatography systems use data analysis software that recognizes the start, maximum, and end of each chromatographic peak, even when not fully resolved from other peaks. These values then are used to determine retention times, peak height, and peak areas. At the end of each run, a report is generated that lists these data and post-run calculations, such as relative peak areas, areas as percentages of the total area, and relative retention times. If the system has been standardized, data from external or internal standards can be used to calculate analyte concentrations (discussed below). In low-pressure, preparative chromatography, post-chromatography analysis of collected fractions is sometimes used to identify eluted samples. Examples of post-chromatography analysis include the BCA (bicinchoninic acid) protein assay (Chap. 18, Sect. 18.4.2.3) and the phenol-sulfuric acid assay for carbohydrate (Chap. 19, Sect. 19.3). After obtaining the absorbance reading on a spectrophotometer for such assays, the results are plotted as fraction number on the x-axis and absorbance on the y-axis to determine the presence of a particular component in each fraction.

Having determined peak areas, one must compare these data with appropriate standards of known concentration to determine sample concentrations. Comparisons may be by means of **external** standards, with or without the use of an **internal** standard (Table 12.7). Comparison of peak area of unknown samples with standards injected separately (i.e., **external standards**) is common practice. External standard solutions covering the desired concentration range (preferably diluted from one stock solution) are chromatographed, and the peak area is plotted versus concentration to obtain a standard curve. An identical volume of sample is then chromatographed, and area of the sample peak is used to determine sample concentration via the standard curve (Fig. 12.15a). This absolute calibration method requires precise analytical technique and requires that detector sensitivity be constant from day to day if the calibration curve is to remain valid.

The use of an **internal standard** can minimize errors due to losses incurred during sample preparation, apparatus noise (inherent apparatus error), and systematic errors in the operator's technique (including the volume of sample injected). In other words, the internal standard, upon addi-

Table 12.7 Differentiating internal vs. external standards

	Internal Standard	*External Standard*
Definition	A compound that is added to every sample and external standard solution in an equal amount or concentration	A compound that is not added to the sample, but rather run separately at different concentrations
Criteria	Structurally similar to the sample analyte but have different retention time and elute within the analysis time; not naturally present in the sample to be analyzed	Is the same compound present in the sample for peak identification and quantification
Purpose	Correct for variability due to sample loss during sample preparation to enhance accuracy; correct for systematic errors in the operator's technique (including the volume of sample injected)	Generate standard curve, with or without an internal standard, and identify and quantitate the compound based on retention time and detector response

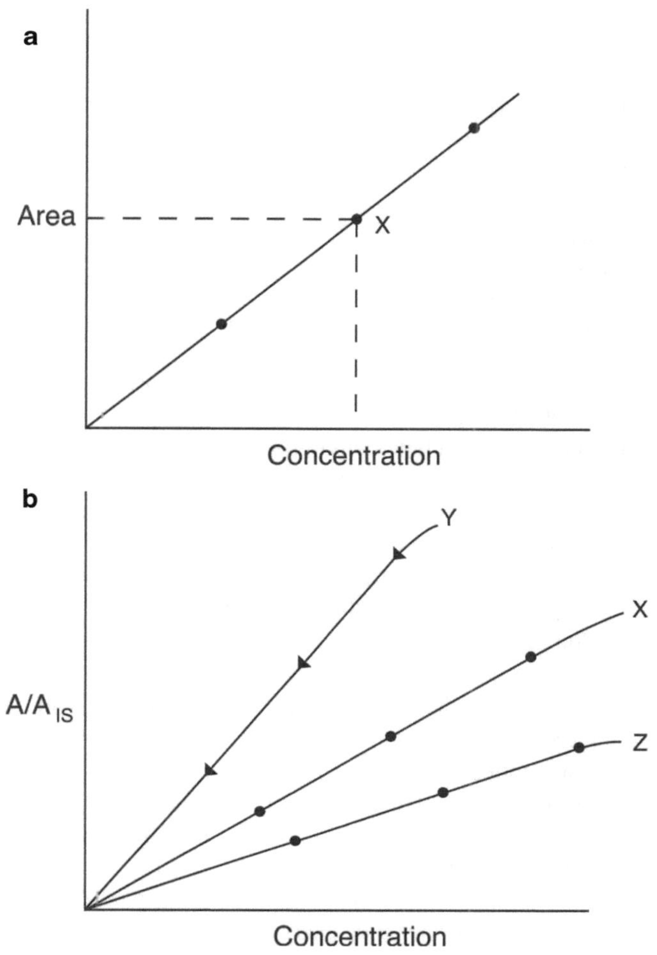

Fig. 12.15 Calibration curves for quantification of a sample component, *x*. (**a**): External standard technique; (**b**): internal standard technique. (Adapted from Johnson and Stevenson [5], with permission)

tion to the sample, will be subjected to the same sequence of events that may introduce changes to the actual content of the compounds of interest and the standard. In this technique, compared to the compounds of interest in the sample, the internal standard compound must: (1) be chemically/structurally related to the compounds of interest, (2) elute at a different time, and (3) not be naturally present in the sample. Basically, the amount of each component in the sample is determined by comparing the area of that component peak to the area of the internal standard peak. The same internal standard is also added to the external standard solutions. Each of these solutions is made to contain a known and constant amount of the internal standard, similar to what was added to the sample. These standard solutions are chromatographed, and peak area is measured. Ratios of peak area (compound of interest/internal standard) are calculated and plotted against concentration to obtain calibration curves such as those shown in Fig. 12.15b. A separate response curve must be plotted for each sample component to be quantified. Next, a known amount of internal standard is added to the unknown sample, and the sample is chromatographed. Peak area ratios (compound of interest/internal standard) are calculated and used to calculate the concentration of each relevant component using the appropriate calibration curve.

Standards should be included during each analytical session, since detector response may vary from day to day. Analyte recovery should be checked periodically. This involves addition of a known quantity of standard to a sample (usually before extraction) and determination of how much is recovered during subsequent analysis. During routine analyses, it is highly desirable to include a control or check sample, a material of known composition. This material is analyzed parallel to unknown samples. When the concentration of analyte measured in the control falls outside an acceptable range, data from other samples analyzed during the same period should be considered suspect. Carefully analyzed food samples and other substances are available from the National Institute of Standards and Technology (formerly the National Bureau of Standards) for use in this manner.

12.6 SUMMARY

Chromatography is a separation method based on the partitioning or distribution of a solute between a mobile phase and a stationary phase. The mobile phase may be liquid, gas, or a supercritical fluid. The stationary phase may be an immobilized liquid or a solid, in either a planar or column form. Based on the physicochemical characteristics of the analyte, and the availability of instrumentation, a chromatographic system is chosen to separate, identify, and quantify the analyte. Chromatographic separation modes include normal-phase, reversed-phase, hydrophilic interaction, hydrophobic interaction, ion exchange, affinity, and size-exclusion chromatography. Factors to be considered when developing a separation include mobile phase variables (polarity, ionic strength, pH, temperature, and/or flow rate) as well as column efficiency, selectivity, and capacity. Following detection, a chromatogram provides both qualitative and quantitative information via retention time, peak area data, and/or spectral analysis.

For an introduction to the techniques of HPLC and GC, the reader is referred again to Chaps. 13 and 14 in this text or to the excellent tenth edition of *Quantitative Chemical Analysis* by D.C. Harris [2]. The book by R.M. Smith [3] also contains information on basic concepts of chromatography and chapters devoted to TLC, LC, and HPLC, as well as an extensive discussion of GC. References [14–16] contain a wealth of information on TLC, and references [8, 9] cover SFC. *Chromatography* [1], the standard work edited by E. Heftmann (2004 and earlier editions), is an excellent source of information on both fundamentals (Part A) and applications (Part B) of chromatography. Part B includes chapters on the chromatographic analysis of amino acids, proteins, lipids, carbohydrates, and phenolic compounds. In addition, *Fundamental and Applications Reviews* published (in alternating years) by the journal *Analytical Chemistry* relate new developments in all branches of chromatography, as well as their application to specific areas, such as food. Recent books and general review papers are referenced.

12.7 STUDY QUESTIONS

1. Explain the principle of countercurrent extraction and how it developed into partition chromatography.
2. For each set of two (or three) terms used in chromatography, give a brief explanation as indicated to distinguish between the terms.

(a) Adsorption vs. partition chromatography

	Adsorption	*Partition*
Nature of stationary phase		
Nature of mobile phase		
How solute interacts with the phases		

(b) Normal-phase vs reversed-phase chromatography

	Normal-phase chromatography	*Reversed-phase chromatography*
Nature of stationary phase		
Nature of mobile phase		
What elutes last		

(c) Cation vs. anion exchangers

	Cation exchanger	*Anion exchanger*
Charge on column		
Nature of compounds bound		

(d) Internal standards vs. external standards

Nature of stds.	*How stds. are handled in relation to samples*	*What is plotted on std. curve*
Internal standard		
External standard		

(e) TLC vs. column liquid chromatography

	Thin-layer chromatography	*Column liquid chromatography*
Nature and location of stationary phase		
Nature and location of mobile phase		
How sample is applied		
Identification of solutes separated		

(f) HETP vs. N vs. L (from the equation HETP = *L/N*)

3. State the advantages of TLC as compared to paper chromatography.
4. State the advantages of column liquid chromatography as compared to planar chromatography.
5. Explain how SFC differs from LC and GC, including the advantages of SFC.
6. What is the advantage of bonded supports over coated supports for partition chromatography?
7. You are performing LC using a stationary phase that contains a polar nonionic functional group. What type of chromatography is this, and what could you do to increase the retention time of an analyte?
8. You applied a mixture of proteins, in a buffer at pH 8.0, to an anion-exchange column. On the basis of some assays you performed, you know that the protein of interest adsorbed to the column.
   (a) Does the anion-exchange stationary phase have a positive or negative charge?
   (b) What is the overall charge of the protein of interest that adsorbed to the stationary phase?
   (c) Is the isoelectric point of the protein of interest (adsorbed to the column) higher or lower than pH 8.0?
   (d) What are the two most common methods you could use to elute the protein of interest from the anion-exchange column? Explain how each method works (See also Chap. 24).
9. Would you use a polystyrene- or a polysaccharide-based stationary phase for work with proteins? Explain your answer.
10. Explain the principle of affinity chromatography, why a spacer arm is used, and how the solute can be eluted.
11. Explain how you would use SEC to estimate the molecular weight of a protein molecule. Include an explanation of what information must be collected and how it is used.

12. What is gradient elution from a column, and why is it often advantageous over isocratic elution?
13. A sample containing compounds A, B, and C is analyzed via LC using a column packed with a silica-based C18 bonded phase. A 1:5 solution of ethanol and H2O was used as the mobile phase. The following chromatogram was obtained:

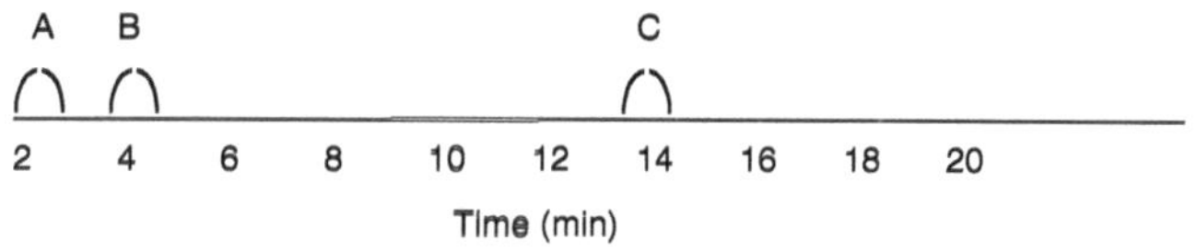

Assuming that the separation of compounds is based on their polarity:
   (a) Is this normal- or reversed-phase chromatography? Explain your answer.
   (b) Which compound is the most polar?
   (c) How would you change the mobile phase so compound C would elute sooner, without changing the relative positions of compounds A and B? Explain why this would work.
   (d) What could possibly happen if you maintained an isocratic elution mode at low solvent strength?
14. Using the Van Deemter equation, HETP, and N, as appropriate, explain why the following changes may increase the efficiency of separation in column chromatography:
   (a) Changing the flow rate of the mobile phase
   (b) Increasing the length of the column
   (c) Reducing the inner diameter of the column
   (d) Decreasing the particle size of the packing material
15. State the factors and conditions that lead to poor resolution of two peaks.
16. How can chromatographic data be used to quantify sample components?
17. When would you choose to use an internal standard? Describe how you would select an internal standard for use.
18. To describe how using internal standards works, answer the following questions.
   (a) What specifically will you do with the standards?
   (b) What do you actually measure and plot?
   (c) How do you use the plot?

REFERENCES

1. Heftmann E (ed) (2004) Chromatography, 6th edn. Fundamentals and applications of chromatography and related differential migration methods. Part A: fundamentals and techniques. Part B: applications. J Chromatog Library Ser vols 69A and 69B. Elsevier, Amsterdam
2. Harris DC (2020) Quantitative chemical analysis, 10th edn. W.H. Freeman, New York
3. Smith RM (1988) Gas and liquid chromatography in analytical chemistry, Wiley, Chichester, England
4. Fanali S, Haddad PR, Poole C, Schoenmakers P, Lloyd DK (2013) Liquid chromatography: fundamentals and instrumentation. Elsevier, Amsterdam.
5. Johnson EL, Stevenson R (1978) Basic liquid chromatography. Varian Associates, Palo Alto, CA
6. Craig LC (1943) Identification of small amounts of organic compounds by distribution studies. Application to Atabrine. J Biol Chem 150: 33–45
7. Bernal J, Martín M, Toribio L (2013) Supercritical fluid chromatography in food analysis. J Chromatogr A 1313:24–36
8. Lesellier E, West C (2015) The many faces of packed column supercritical fluid chromatography – A critical review. J Chromatogr A, 1382: 2–46
9. Caude M, Thiébaut D (1999) Practical supercritical fluid chromatography and extraction. Harwood Academic, Amsterdam
10. Radke W (2014) Polymer separations by liquid interaction chromatography: Principles – prospects – limitations. J Chromatogr A, 1335:62–79
11. Bernal J, Ares A, Pól J, Wiedmer S (2011) Hydrophilic interaction liquid chromatography in food analysis. J Chromatogr A 1218:7438–7452
12. Núñez O, Gallart-Ayala H, Martins C, Lucci P (2015) Fast Liquid Chromatography-Mass Spectrometry Methods in Food Analysis. Imperial College Press, London
13. Cordero C, Kiefl J, Schieberle P (2015) Comprehensive two-dimensional gas chromatography and food sensory properties: potential and challenges. Anal Bioanal Chem, 407: 169–191
14. Fried B, Sherma J (1999) Thin-layer chromatography, 4th edn. Marcel Dekker, New York
15. Hahn-Deinstrop E (2007) Applied thin-layer chromatography: best practice and avoidance of mistakes, 2nd edn. Wiley-VCH, Weinheim, Germany
16. Wall PE (2005) Thin-layer chromatography: a modern practical approach. The Royal Society of Chemistry, Cambridge, UK
17. Fuchs B, Süß R, Teuber K, Eibisch M, Schiller J (2011) Lipid analysis by thin-layer chromatography-A review of the current state. J Chromatogr A 1218:2754–2774
18. Morlock G, Meyer S, Zimmermann B, Roussel J (2014) High-performance thin-layer chromatography analysis of sterviol glycosides in Stevia formulations and sugar-free food products, and benchmarking with (ultra) high-performance liquid chromatography. J Chromatogr A 1350:102–111
19. Touchstone JC (1992) Practice of thin layer chromatography. Wiley, New York
20. Chester TL, Pinkston JD, Raynie DE (1996) Supercritical fluid chromatography and extraction (fundamental review). Anal Chem 68: 487R–514R
21. Snyder LR, Kirkland JJ (eds) (1979) Introduction to modern liquid chromatography, 2nd ed. Wiley, New York
22. Cavazzini A, Felinger A (2013) Hydrophilic interaction liquid chromatography, In Liquid chromatography: fundamentals and instrumentation, Eds. Fanali S, Haddad PR, Poole, Schoenmakers P, Lloyd DK. Elsevier, Amsterdam, p 105–119
23. Marrubini G, Appelblad P, Maietta M, Papetti A (2018). Hydrophilic interaction chromatography in food matrices analysis: An updated review. Food Chem 257: 53–66

24. Tomaz CT, Queiroz JA (2013) Hydrophobic interaction chromatography, In Liquid chromatography: fundamentals and instrumentation, Eds. Fanali S, Haddad PR, Poole, Schoenmakers P, Lloyd DK. Elsevier, Amsterdam, p 122–141
25. Weiss J (2004) Handbook of ion chromatography, 3rd edn. Wiley-VCH Verlag GmbH & Co. KGaA, Weinheim
26. Wacoo AP, Wendiro D, Vuzi, PC, Hawumba JF (2014) Methods for detection of aflatoxins in agricultural food crops. Applied Chem 2014, 1–15
27. Rollag JG, Beck-Westermeyer M, Hage DS (1996) Analysis of pesticide degradation products by tandem high-performance immunoaffinity chromatography and reversed-phase liquid chromatography. Anal Chem 68, 3631–3637
28. Walters RR (1985) Report on affinity chromatography. Anal Chem 57: 1099A–1113A
29. Zachariou M (2008) Affinity chromatography: methods and protocols. Humana, Totowa, NJ
30. Scopes RK (1994) Protein purification: principles and practice, 3rd edn. Springer-Verlag, New York
31. Lough WJ, Wainer IW (eds) (1995) High performance liquid chromatography: fundamental principles and practice. Blackie Academic & Professional, Glasgow, Scotland

High-Performance Liquid Chromatography

13

B. Pam Ismail and Bradley L. Reuhs

Contents

13.1 INTRODUCTION

High-performance liquid chromatography (HPLC) was developed during the 1960s as a direct offshoot of traditional low-pressure column liquid chromatography through improvements in the technology of columns and instrumental components (pumps, injection valves, column packing, and detectors). Originally, HPLC was the acronym for *high-pressure liquid chromatography,* reflecting the relatively high operating pressures generated by early columns. By the late 1970s, however, *high-performance liquid chromatography* had become the preferred term, emphasizing the enhanced resolution and throughput. In fact, newer columns and packing materials offer high performance at moderate pressure (but higher than atmospheric pressure). HPLC can be applied to the analysis of any compound with solubility in a liquid that can be used as the mobile phase. Although most often employed as an **analytical** technique, HPLC may also be used in **preparative** applications, commonly referred to as **semi-preparative** HPLC. There are many *advantages* of HPLC over traditional low-pressure column liquid chromatography:

1. A wide variety of stationary phases in prepacked columns
2. Improved resolution
3. Speed (many analyses can be accomplished in 30 min or less) leading to high throughput

B. P. Ismail (✉)
Department of Food Science and Nutrition, University of Minnesota, St. Paul, MN, USA
e-mail: bismailm@umn.edu

B. L. Reuhs
Department of Food Science, Purdue University, West Lafayette, IN, USA
e-mail: breuhs@purdue.edu

B. P. Ismail, S. S. Nielsen (eds.), *Nielsen's Food Analysis*, Food Science Text Series,
https://doi.org/10.1007/978-3-031-50643-7_13

4. Greater sensitivity (many different detectors can be employed)
5. Easy sample recovery (less eluent volume to remove)

Application of HPLC to the analysis of food began in the late 1960s, and its use increased with the development of column packing materials that would separate sugars. Using HPLC to analyze sugars was justified economically as a result of sugar price increases in the mid-1970s, which motivated soft drink manufacturers to substitute high-fructose corn syrup for sugar. Monitoring sweetener content by HPLC assured a good quality product. Other early food applications included the analysis of pesticide residues in fruits and vegetables, organic acids, lipids, amino acids, toxins (such as aflatoxins in peanuts), and vitamins [1]. HPLC continues to be applied to these, and many more, food-related analyses today [2–3]. Along with improved detectors and compatibility with mass spectrometry (MS) detection systems, these HPLC systems have had a significant impact on the use of HPLC in the food and drug industries.

13.2 COMPONENTS OF AN HPLC SYSTEM

A schematic diagram of a basic HPLC system is shown in Fig. 13.1. The main components of this system—**pump, injector, column, detector,** and **data system**—are discussed briefly in the sections below. Also important are the mobile phase (**eluent**) reservoirs and a fraction collector, the latter of which is used if further analysis of separated components is needed. Connecting tubing, tube fittings, and the materials out of which components are constructed also influence system performance and lifetime. References [1] and [4–9] include detailed discussions of HPLC equipment, with the book by Bidlingmeyer [1] especially appropriate for beginners. The unique organization of reference [7] is intended for those who may need to learn chromatography quickly in an industrial environment. Two useful references on HPLC troubleshooting are those by Sailaja et al. [10] and Snyder et al. [11]. In addition, much information on HPLC equipment, hardware, and troubleshooting hints may be found in publications such as *LCGC, American Laboratory, Chemical & Engineering News,* and similar periodicals. Manufacturers are also a source of practical information on HPLC instrumentation and columns/stationary phase material.

13.2.1 Pump

The **HPLC pump** delivers the mobile phase through the system, typically at a flow rate of 0.2–2 mL/min, in a controlled, accurate, and precise manner. Accuracy (±1% variation) and precision (<0.1% relative standard deviation) of the flow rate [12] are critical to ensure retention time and peak area reproducibility. In addition, the delivery of the mobile phase should be free from pressure pulsations since the baseline

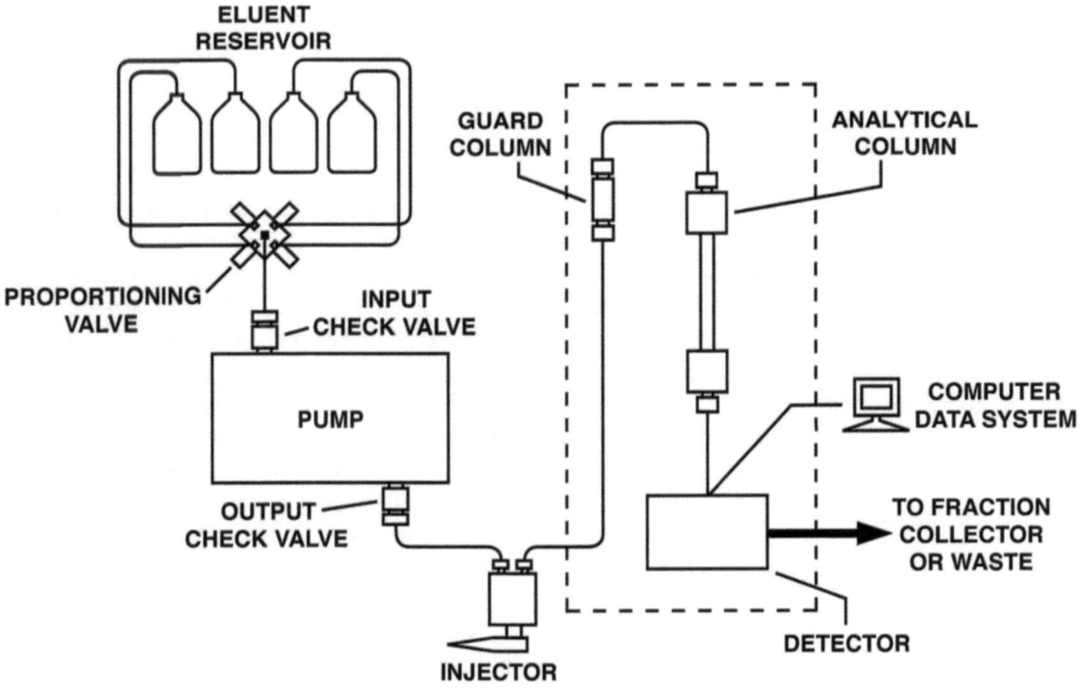

Fig. 13.1 Schematic representation of a system for high-performance liquid chromatography (not drawn to scale). Column(s) and detector may be thermostated, as indicated by the *dashed line*, for operation at elevated temperature

obtained using certain detectors, such as refractive index (Sect. 13.2.4.3), is sensitive to pump pulsations.

The majority of pumps currently used in HPLC (>90%) are reciprocating, piston-type pumps. The dual piston pump systems with ball check valves are the most efficient pumps available. One disadvantage of reciprocating pumps is that they produce a pulsating flow, requiring the addition of pulse dampers to suppress fluctuations.

HPLC pumps should allow rapid solvent changes in gradient elution. This is possible by designing pumps with a minimal dwell volume. **Gradient elution systems** are used to vary the mobile phase concentration during the run, by mixing mobile phase from two or more reservoirs. This is accomplished with low-pressure mixing, in which mobile phase components are mixed before entering the high-pressure pump, or high-pressure mixing, in which two or more independent, programmable pumps are used. For low-pressure gradient systems, a computer-controlled proportioning valve, followed by a mixing chamber at the inlet to the pump, is used, which results in extremely accurate and reproducible gradients. In such systems, 2–4 mobile phases can be used to build a gradient. In high-pressure gradient systems, the mixing chamber is placed at the exit of the pump. In such systems, typically two pumps are used, allowing for a gradient with a maximum of two mobile phases. Gradient HPLC is extremely important for the effective elution of all components of a sample and for optimal resolution (Chap. 12, Sect. 12.3.4.2). It is routinely applied to all modes of HPLC except size-exclusion chromatography.

The pump should be made of inert and chemically resistant materials. Many commercially available HPLC pumping systems and connecting lines are made of high-grade ANSI 316 stainless steel, which can withstand high pressure and is resistant to corrosion by oxidizing agents, acids, bases, and organic solvents (although mineral acids and halide ions will damage stainless steel).

All HPLC pumps contain moving parts such as check valves and pistons that are quite sensitive to dust and particulate matter in the eluent. Therefore, it is advisable to filter the mobile phase using 0.45 or 0.22-μm filters prior to use. Degassing HPLC eluents, by the application of a vacuum or by sparging with helium, is also recommended to prevent the problems caused by air bubbles in a pump or detector. Newer HPLC systems have a degasser unit installed ahead of the pump/s.

13.2.2 Injector

The role of the injector is to place the sample into the flowing mobile phase for introduction into the column. Virtually all HPLC systems use **valve injectors,** which separate sample introduction from the high-pressure eluent system. With the injection valve in the *load* position (Fig. 13.2a) [13], the sample is loaded into an **external, fixed-volume loop** using a syringe. Eluent, meanwhile, flows directly from the pump to the column at high pressure. When the valve is rotated to the *inject* position (Fig. 13.2b), the loop becomes part of the eluent flow stream and the sample is carried onto the column. Such injectors are generally trouble free and precise.

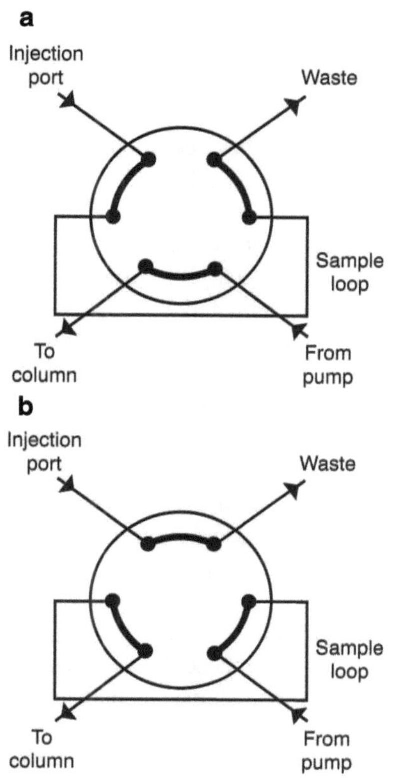

Fig. 13.2 Valve-type injector. The valve allows the sample loop to be (**a**) isolated from the pump eluent stream (LOAD position) or (**b**) positioned in it (INJECT position). (From Gertz [13], used with permission)

Changing the loop allows different volumes to be injected. Newer systems allow for injections of different volumes without the need to change the sample loop. Although injection volumes of 10–100 μL are typical, both larger (e.g., 1–10 mL) and smaller (e.g., ≤2 μL) sample volumes can be loaded. An important advantage of the loop valve design is that it is readily adapted to automated operation. Thus, automated sample injectors, or **autosamplers,** may be used to store and inject large numbers of samples. Samples are placed in uniform-size vials, sealed with a septum, and held in a (possibly refrigerated) tray. A cooler unit is sometimes part of the autosampler to avoid degradation of sensitive constituents during an automated sequence run. A computer-actuated syringe needle penetrates the septum to withdraw solution from the vial, and a mechanical system introduces it into the column. Autosamplers can reduce the tedium and labor costs associated with routine HPLC analyses and improve assay precision.

13.2.3 Column

The first two sections below describe the basic column hardware and packing material for HPLC, followed by a third section to describe how those components, along with operating pressure, differ for ultra-HPLC.

13.2.3.1 Column Hardware

An **HPLC column** is usually constructed of stainless steel with terminators and tubing that allow it to be connected between the injector and detector systems (Fig. 13.1). Columns are also made from glass, fused silica, titanium, and polyether ether ketone (PEEK) resin; the PEEK columns are essential for the high pH, high salt concentrations necessary for the powerful ion-exchange HPLC systems. Many types and sizes of columns are commercially available, ranging from 5 cm × 50 cm (or larger) preparative columns down to wall-coated capillary columns.

13.2.3.1.1 Precolumns

Auxiliary columns that precede the analytical HPLC column are termed **precolumns.** Short expendable columns, called **guard columns** (generally 0.5–3.0 cm in length)**,** often are used to protect the analytical column from strongly adsorbed sample components. A guard column (or cartridge) is installed between the injector and analytical column via short lengths of capillary tubing (or a cartridge holder). Guard columns, available commercially, are packed with the same stationary phase (or equivalent) to that of the analytical column. Typically, a guard column (or cartridge) is replaced before its binding capacity is exceeded and contaminants pollute the expensive analytical column. An increase in the system's pressure is often an indication of a blocked guard column. It is commonly recommended to change the guard column every 50–200 injections, depending on the type and dimensions of the guard column and on the injection volume.

13.2.3.1.2 Analytical Columns

The most commonly used **analytical** HPLC columns are 10, 15, or 25-cm long with an internal diameter of 4.6 or 5 mm [7]. Short (3-cm) columns, packed with ≤3-μm particles, are common in method development or process monitoring. The use of columns with smaller internal diameters (<0.5–2.0 mm), including wall-coated capillary columns, allows for a decreased use of mobile phase, an increased peak concentration, increased resolution, and the ability to couple HPLC with MS [14].

Various names have been used for the reduced-volume columns. Dorsey et al. [15] refer to columns with internal diameters of 0.5–2.0 mm as **microbore,** while packed or open tubular columns having internal diameters of <0.5 mm are termed **microcolumns** or **capillary columns** (a capillary column is a narrow-bore open tubular column, in which the inner surface is coated with a thin layer of stationary phase). In the case of the packed columns, the microbore or microcolumns contain very small particle-size packing material (≤2 μm).

13.2.3.2 HPLC Column Packing Materials

The development of a wide variety of column packing materials has contributed substantially to the success and widespread use of HPLC.

13.2.3.2.1 General Requirements

In most modes of chromatography, the column **packing material** serves as both the **support matrix** and the **stationary phase**. Requirements for HPLC column packing materials are good chemical stability, sufficient mechanical strength to withstand the high pressure generated during use, and a well-defined particle size with a narrow particle size distribution [11]. Two materials that meet the above criteria are porous silica and synthetic organic resins (see below Sects. 13.2.3.2.2 and 13.2.3.2.3, respectively).

13.2.3.2.2 Silica-Based Column Packings

Porous silica can be prepared in a wide range of particle and pore sizes, with a narrow particle size distribution. Both **particle size** and **pore diameter** are important: Small particles reduce the distance a solute must travel between stationary and mobile phases, which facilitates equilibration and results in good column efficiencies (Chap. 12, Sect. 12.5.1.2.2). However, small particles also yield greater flow resistance and higher pressure at equivalent flow rates. Spherical particles of 3-, 5-, or 10-μm diameter are utilized in analytical columns. One-half or more of the volume of porous silica consists of the **pores** [8]. Use of the smallest possible pore diameter will maximize **surface area** and **sample capacity**, which is the amount of sample that can be effectively separated on a given column. Packing materials with a pore diameter of 50–100 Å and a surface area of 200–400 m^2/g are used for low-molecular-weight (<500-Da) solutes. For increasingly larger molecules, such as proteins and polysaccharides, it is necessary to use materials with a wider pore (≥300 Å), so that the internal surface is accessible to the solute [8].

Bonded phases (Fig. 13.3a, [16]) are made by covalently bonding hydrocarbon moieties to -OH groups (silanols) on the surface of silica particles [8, 17]. Often, the silica is reacted with an organochlorosilane:

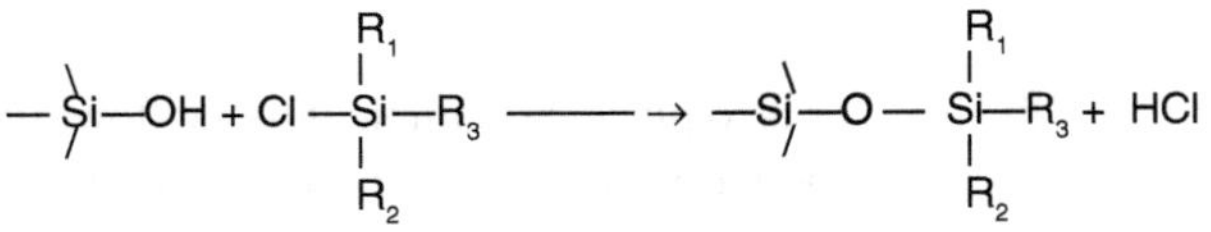

13.1

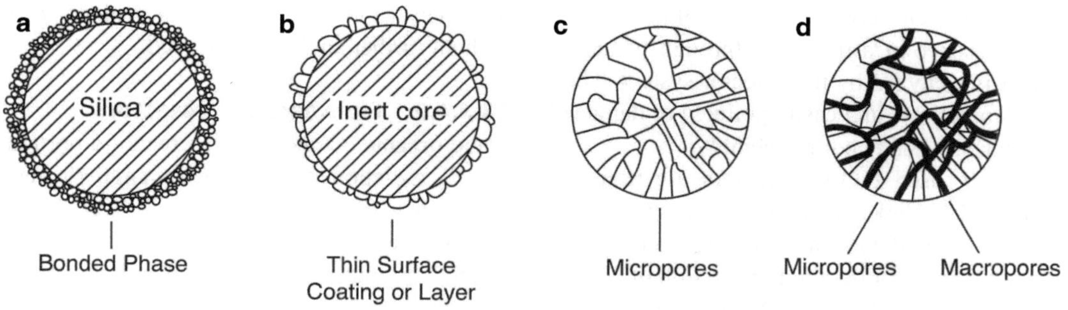

Fig. 13.3 Some types of packing materials utilized in HPLC. (**a**) Bonded-phase silica; (**b**) pellicular packing; (**c**) microporous polymeric resin; and (**d**) macroporous polymeric resin. (Adapted from Tomaz and Queiroz [16], p. 621, by courtesy of Marcel Dekker, Inc.)

Substituents R_1 and R_2 may be halides or methyl groups. The nature of R_3 determines whether the resulting bonded phase will exhibit normal-phase, reversed-phase, or ion-exchange chromatographic behavior. The main disadvantage of silica and silica-based bonded-phase column packings is that the silica skeleton slowly dissolves in aqueous solutions, and the rate of this process becomes prohibitive at pH <2 and >8.

A **pellicular packing material** (Fig. 13.3b) is made by depositing a thin layer or coating onto the surface of an inert, usually nonporous, microparticulate **core**. Functional groups, such as ion-exchange sites, are present at the surface only. Core material may be either inorganic, such as silica, or organic, such as poly(styrene-divinylbenzene) or latex. The rigid core ensures good physical strength, and the thin stationary phase provides for rapid mass transfer (Chap. 12, Sect. 12.5.1.2.2) and favorable column efficiency.

13.2.3.2.3 Porous Polymeric Column Packings

Synthetic organic resins offer the advantages of good chemical stability and the possibility to vary interactive properties through direct chemical modification. Two major categories of porous polymeric packing materials exist.

Microporous or gel-type resins (Fig. 13.3c) are composed of cross-linked copolymers in which the apparent porosity, evident only when the gel is in its swollen state, is determined by the degree of cross-linking. These gel-type packings undergo swelling and contraction with changes in the chromatographic mobile phase. Microporous polymers of less than ca. 8% cross-linking are not sufficiently rigid for HPLC use.

Macroporous resins are highly cross-linked (e.g., ≥50%) and consist of a network of microspheric gel beads joined together to form a larger bead (Fig. 13.3d). Large, permanent pores, ranging from 100 to 4000 Å or more in diameter, and large surface areas (≥100 m^2/g) are the result of interstitial spaces between the microbeads [17]. Rigid microparticulate poly(styrene-divinylbenzene) packing materials of the macroporous type are popular for HPLC use. They are stable from pH 1 to 14 and are available in a variety of particle and pore sizes. These resins can be used in the unmodified form for reversed-phase chromatography or functionalized for use in other HPLC modes.

13.2.3.3 Ultra-HPLC

Ultra-HPLC (UHPLC) and **ultra-performance liquid chromatography** (UPLC) are based on the same technique, and the terms can be used interchangeably, based on the manufacturer. UHPLC and UPLC instruments both use microbore or microcolumns with packing material of ≤2 µm and use much higher pressure than regular HPLC (HPLC max. of ~5800 psi; UHPLC/UPLC max. of 15,000–20,000 psi, depending on the manufacturer). The flow rate for UHPLC/UPLC is generally lower (0.2–0.7 mL/min) than that for regular HPLC (0.5–2 mL/min), but this is dependent on the column dimensions. Most UPLC/UHPLC systems are equipped with 50 × 2.1 mm columns with a total volume of 120 µL. With such a small column, sample injection volumes are only 1–5 µL. It is the combination of small packing material and increased pressure that results in reduced on-column dispersion (i.e., diffusion) of sample molecules [18]. With UPLC/UHPLC, efficiency of separation is increased, overall separation time is reduced, and less solvent is needed, compared to HPLC. The resolution and speed of UPLC/UHPLC instruments are well suited to be coupled with MS analysis. In general, UPLC/UHPLC requires unique detector technology with higher data acquisition rates [4], compared to the HPLC counterpart detectors (Sect. 13.2.4). UPLC/UHPLC is the basis for the AOAC official methods for numerous vitamins (i.e., A, B_6, B_{12}, C, D, and folate). Besides vitamins, UPLC/UHPLC is used for the separation of proteins (Chap. 24, Sect. 24.2.3.1), phenolic compounds (Chap. 25, Sect. 25.2.3.1), and antibiotic residues (Chap. 33, Sect. 33.5.2.2).

13.2.4 Detector

A **detector** translates sample concentration changes in the HPLC column effluent into electrical signals. Spectrochemical, electrochemical, or other properties of solutes may be measured by a variety of instruments, each of which has advantages and disadvantages. The choice of which to use depends on solute type and concentration, and on detector sensitivity, linear range, and compatibility with the solvent and elution mode to be used. Cost may also influence detector selection. One common feature of most HPLC detectors is the presence of a flow cell, through which the eluent flows as it is analyzed by the detector system. These flow cells are often delicate and easily polluted or damaged, so care must be taken when handling them.

The most widely used HPLC detectors are based on ultraviolet-visible (UV-Vis) and fluorescence spectrophotometry, refractive index determination, and electrochemical analysis (See Chap. 7 for detailed discussion of UV-Vis and fluorescence spectrophotometry). Many other methods, such as light scattering or MS, can also be applied to the detection of analytes in HPLC eluents. More than one type of HPLC detector may be used in series, to provide increased **specificity** and **sensitivity** for multiple types of analytes. In one food-related application, a multidetector HPLC system equipped with a diode array absorption detector coupled to fluorescence and electrochemical detectors was used to monitor a wide variety of Maillard reaction products.

13.2.4.1 UV-Vis Absorption Detectors

Many HPLC analyses are carried out using a **UV-Vis absorption** detector, which can measure the absorption of radiation by chromophore-containing compounds. The three main types of UV-Vis absorption detectors are **fixed wavelength, variable-wavelength**, and **diode array spectrophotometers** [8]. As its name implies, the simplest design operates at a single, **fixed wavelength**. A filter is used to isolate a single emission line (e.g., at 254 nm) from a source such as a mercury lamp. This type of detector is easy to operate and inexpensive but of limited utility.

The most popular general purpose HPLC detector today is the **variable-wavelength** detector, in which deuterium and tungsten lamps serve as sources of ultraviolet and visible radiation, respectively. Wavelength selection is provided by a **monochromator**, a device that acts to disperse light (Chap. 7, Sect. 7.2.6.2), and an exit slit that allows light from a limited range of wavelengths to pass through to the sample. Rotating the monochromator via a selector switch allows one to change the operating wavelength.

Diode array spectrophotometric detectors can provide much more information about sample composition than is possible with monochromatic detection. In this instrument, the light from a deuterium lamp is spread into a spectrum across an array of photodiodes (Chap. 7, Sect. 7.2.6.3) mounted on a silicon chip. These are read almost simultaneously by a microprocessor to provide the full absorption spectrum from 200 to 700 nm every 0.1 s, which may enable the components of a mixture to be identified. Although considerably more expensive than variable-wavelength detectors, they are useful in method development and in routine analyses in which additional evidence of peak identity, without further analysis, is needed.

13.2.4.2 Fluorescence Detectors

Some organic compounds can reemit a portion of absorbed UV-Vis radiation at a longer wavelength (lower energy). This is known as **fluorescence** (Chap. 7, Sect. 7.3), and measurement of the **emitted light** provides another useful detection method. Fluorescence detection is both selective and very sensitive, providing up to 1000-fold lower detection limits than for the same compound in absorbance spectrophotometry. Although relatively few compounds are inherently fluorescent, analytes often are converted into fluorescent derivatives (see Sect. 13.2.4.7). Ideal for trace analysis, fluorescence detection has been used for the determination of various vitamins in foods and supplements, monitoring aflatoxins in stored cereal products, and the detection of aromatic hydrocarbons in wastewater.

13.2.4.3 Refractive Index Detectors

Refractive index (RI) detectors measure change in the RI of the mobile phase due to dissolved analytes, which provides a nearly universal method of detection. However, because a bulk property of the eluent is being measured, RI detectors are less sensitive than other types. Another disadvantage is that they cannot be used with gradient elution, as any change in eluent composition will alter its RI, thereby changing the baseline signal. RI detectors are widely used for analytes that do not contain UV-absorbing chromophores, such as carbohydrates and lipids, when the analytes are present at relatively high concentration.

13.2.4.4 Electrochemical Detectors

Electroanalytical methods used for HPLC detection are based either on electrochemical oxidation–reduction of the analyte or on changes in conductivity of the eluent. **Amperometric detectors** measure the change in current as the analyte is oxidized or reduced by the application of voltage across electrodes in the flow cell. This method is highly selective (nonreactive compounds give no response) and very sensitive. A major application of electrochemical detection has been for the routine determination of catecholamines, which are phenolic compounds of clinical importance that are present in blood and tissues at very low levels. The development of a triple-pulsed amperometric detector (PAD), which overcame the problem of electrode poisoning (accumulation of oxidized product on the electrode surface), has allowed electrochemical detection to be applied to the analysis of carbohydrates (see Sect. 13.3.5.2). Pulsed electrochemical detection also has excellent sensitivity for the quantification of flavor-active alcohols, particularly terpenols.

Analytes that are ionized and carry a charge can be detected by measuring the change in eluent **conductivity** between two electrodes. **Conductivity detection** has been used mainly to detect inorganic anions and cations and organic acids upon elution from weak ion-exchange columns. Its principal application has been associated with ion exchange chromatography (Chap. 12, Sect. 12.4.5). An excellent overview of electrochemical detection is provided by Swadesh [7].

13.2.4.5 Other HPLC Detectors

Evaporative light scattering detector (ELSD) is the most widely used aerosol-based detector that is considered a universal detector for nonvolatile and nonchromophoric compounds [4]. In an ELSD, the mobile phase is sprayed into a heated air stream, evaporating volatile solvents and leaving nonvolatile analytes as aerosols. These droplets or particles can be detected because they will scatter a beam of light. HPLC with light scattering detection has been applied to the analysis of wheat flour lipids. Also, light scattering detectors are quite useful for the characterization of polymers (e.g., proteins and carbohydrates) by size-exclusion chromatography. Improvements in laser applications brought about the development of **low-angle laser light scattering** (LALLS) and **multi-angle laser light scattering** (MALLS) detectors. With these detectors, there is no need to evaporate the mobile phase, as the laser beam is directed at the flow cell and scattered laser light is then monitored by photodetectors set at specific angles to the cell. In MALLS, there may be many different photodetectors at discrete angles, each continuously collecting and analyzing the scattered light; from this data, the computer can determine the molecular weight of the eluting sample.

Radioactive detectors are widely used for pharmacokinetic and metabolism studies with radiolabeled drugs. Decay of a radioactive nucleus leads to excitation of a scintillator, which subsequently loses its excess energy by photon emission. Photons are counted by a photomultiplier tube, and the number of counts per second is proportional to radiolabeled analyte concentration [8].

A **chemiluminescent nitrogen detector** (CLND) allows nitrogen-containing compounds, such as amino acids, to be detected without using chemical derivatization (Sect. 13.2.4.7). This nitrogen-specific detection system has been used to quantify caffeine in coffee and soft drink beverages and to analyze capsaicin in hot peppers [19].

13.2.4.6 Coupled Analytical Techniques

To obtain more information about the analyte(s), eluent from an HPLC system can be passed on to a second analytical instrument, such as infrared (IR), nuclear magnetic resonance (NMR), or MS (see Chaps. 8, 10 and 11, respectively, or reference [5]). The coupling of spectrometers with liquid chromatography (LC) was initially slow to gain application, due to many practical problems. For example, in the case of **HPLC with mass spectrometric detection** (LC-MS), the liquid mobile phase affected the vacuum in the MS. This problem was addressed by the development of commercial interfaces that allow the solvent to be evaporated, so that only analyte is transferred to the spectrometer. Two commonly used interface techniques are discussed in detail by Harris [5]. The use of microbore or capillary HPLC columns with a low flow volume also facilitates direct coupling of the two instruments [14]. LC-MS systems continue to improve, and the applications are expanding to nearly every class of relatively low-molecular-weight compounds, including bioactives and contaminants.

13.2.4.7 Chemical Reactions

Detection sensitivity or specificity may sometimes be enhanced by converting the analyte to a **chemical derivative** with different or additional characteristics. An appropriate reagent can be added to the sample prior to injection (i.e., **precolumn derivatization**) or combined with column effluent before it enters the detector (i.e., **postcolumn derivatization**). Automated amino acid analyzers utilize postcolumn derivatization, usually with ninhydrin, for reliable and reproducible analysis of amino acids. Precolumn or postcolumn derivatization of amino acids with *o*-phthalaldehyde or similar reagents permits highly sensitive HPLC determination of amino acids using fluorescence (Chap. 24, Sect. 24.3.1.2) and absorption detection. In addition, fractions may be collected after passing through the detector and aliquots of each fraction analyzed by various means, including chemical/colorimetric assays, such as the bicinchoninic acid method for protein (Chap. 18, Sect. 18.4.2.3) or a total carbohydrate assay (Chap. 19, Sect. 19.3). The results can then be plotted and overlaid with the detector plot, yielding very important information about the compounds eluting in various peaks.

13.2.5 Data Station Systems

As a detector provides an electronic signal related to the composition of the HPLC column effluent, it is the function of the last element of the HPLC system to display the chromatogram and integrate the peak areas. **Data stations** and **software packages** are nearly ubiquitous with modern HPLC, and all come with very powerful tools for sample identification and quantitation. As an HPLC analysis progresses, the data from the HPLC detector(s) are digitized and saved to a computer hard drive. The data then can be manipulated (annotated) by assigning and integrating the peaks, for example, and then printed out, as plots and tables, for further assessment. Importantly, the software programs can be set up prior to the analysis to perform nearly all these functions, without further input from the operator. For example, retention times can be calculated relative to an internal standard in pesticide residue analysis and the results compared to a stored database of standards that the software automatically accesses when the analysis is complete. The software will then assign and integrate the peaks and construct a complete report that is displayed when the file is opened, even on a remote computer (e.g., in the office).

The current data stations are more comprehensive: the software packages include all the parameters needed to run the HPLC, including start and stop, autosampler injection of the sample, sequence run, and developing the gradient via control of the proportioning pump systems. The data station can carry out the entire operation, on hundreds of samples, in the absence of an operator, and, with networking, deliver the analysis file/report to any connected computer.

13.3 APPLICATIONS IN HPLC

The basic physicochemical principles underlying all liquid chromatographic separations—adsorption, partition, hydrophobic interaction, ion exchange, affinity, and size exclusion—are discussed in Chap. 12, and details will not be repeated here. The number of separation modes utilized in HPLC, however, is greater than that available in classic chromatography [20]. Examples of HPLC applications in food analysis are given in Table 13.1. This is attributable to the success of bonded phases, initially developed to facilitate partition chromatography (Chap. 12, Sect. 12.4.2). In fact, reversed-phase chromatography is the most widely used separation mode in modern HPLC.

13.3.1 Normal-Phase

13.3.1.1 Stationary and Mobile Phases

In **normal-phase** HPLC, the **stationary phase** is a **polar adsorbent,** such as bare silica or silica, to which polar nonionic functional groups—hydroxyl, nitro, cyano (nitrile), or amino—have been chemically linked. These bonded phases are moderately polar and the surface is more uniform, resulting in better elution profiles. The **mobile phase** for this mode consists of a **nonpolar solvent**, such as hexane, to which is added a more polar modifier, such as methylene chloride, to control solvent strength and selectivity. Solvent strength refers to solvent effects on the migration rate of the sample: Relatively **weak solvents** increase retention values (large k'), and **strong solvents** decrease retention values (small k').

13.3.1.2 Applications of Normal-Phase HPLC

In the past, normal-phase HPLC was used for the analysis of fat-soluble vitamins, although reversed-phase is currently applied more frequently for these analyses (see Table 13.1). Normal phase is currently used for the analyses of biologically active polyphenols from natural plant sources, such as grape and cocoa. It is also used for the analysis of relatively polar vitamins, such as vitamins A, D, E, and K (see Chap. 20), and also natural carotenoid pigments, which impart both color and health benefits to foods. Highly hydrophilic species, such as carbohydrates (see Chap. 19, Sect. 19.4.2.1), may also be resolved by normal-phase chromatography, using amino bonded-phase HPLC columns. Other applications include the analysis of antioxidants, such as butylated hydroxytoluene (BHT), butylated hydroxyanisole (BHA), and tert-butylhydroquinone (TBHQ, tertiary butylhydroquinone), and vitamin E compounds, such as tocopherols (TCP) [3]. Analysis of these compounds is increasingly necessary for the proper assessment of food products, as well as additives.

Table 13.1 Example applications of HPLC in the analysis of various food constituents

Analyte	*Separation mode*	*Method of detection*	*Chapter*	*Section*
Mono- and oligosaccharides	Ion exchange; normal or reversed phase	Electrochemical; refractive index; postcolumn analysis	19	19.4.2.1
Vitamins	Normal or reversed phase	Fluorescence; electrochemical; UV	20	20.4.1
Amino acids	Ion exchange; reversed phase	Post- or precolumn derivatization	24	24.3.1
Protein separation	Ion exchange, reversed phase, affinity, hydrophobic interaction	UV	24	24.2.3
Phenolics	Reversed phase	UV	25	25.2.3.1
Pesticides	Normal or reversed phase	UV; fluorescence; mass spectrometry	33	33.3.3.2.3
Mycotoxins	Reversed phase; immunoaffinity	UV; fluorescence	33	33.4.3.2.1
Antibiotics	Reversed phase	UV	33	33.5.2.2
Various food contaminants (e.g., acrylamide, melamine)	Reversed phase, ion exchange	UV; mass spectrometry	33	33.8
Sulfites	Ion exchange	UV; electrochemical	33	33.8.2

13.3.2 Hydrophilic Interaction

13.3.2.1 Stationary and Mobile Phases

The stationary phase in **hydrophilic interaction liquid chromatography** (HILIC) HPLC is similar to that of normal-phase HPLC in that the packing material is always polar (Chap. 12, Sect. 12.4.3), including unbonded silica, amino phases, diol bonded phases, and pentafluorophenyl phases [21]. Partition of solutes occurs between the water-enriched surface layer of the stationary phase and the mobile phase, which is mostly acetonitrile-rich. Methanol is not used in HILIC HPLC since it does not allow bilayer formation. Gradient elution is also common in HILIC HPLC, with gradually increasing the aqueous buffer concentration. The most used buffers are ammonium acetate and ammonium formate, which are soluble in acetonitrile and are compatible with MS detection. HILIC also complements reversed-phase HPLC but has an advantage with its ability to better retain hydrophilic compounds and its enhanced selectivity using aqueous/organic mobile phase [22].

13.3.2.2 Applications of Hydrophilic Interaction HPLC

The use of HILIC HPLC is expanding in various fields including pharmacology, biochemistry, and the food industry. While it is commonly used for drug and metabolites analysis, it is frequently used for carbohydrate analysis, peptide mapping, and detection and quantification of contaminants such as pesticides. In addition, HILIC HPLC is increasingly used for the analysis of lipids and lipid metabolites.

13.3.3 Reversed-Phase

13.3.3.1 Stationary and Mobile Phases

More than 70% of all HPLC separations are carried out in the reversed-phase mode, which utilizes a **nonpolar stationary phase** and a **polar mobile phase**. **Octadecylsilyl** (ODS) **bonded phases**, with an octadecyl (C_{18}) chain [-$(CH_2)_{17}CH_3$], are the most popular reversed-phase packing materials, although shorter chain hydrocarbons [e.g., octyl (C_8) or butyl (C_4)] or phenyl groups, and longer chain [e.g., tetracontanyl C_{40}] are also used. Many silica-based, reversed-phase columns are commercially available. Differences in their chromatographic behavior result from variation in the type of organic group bonded to the silica matrix or the chain length of organic moiety.

Reversed-phase HPLC utilizes **polar mobile phases**, usually water mixed with methanol, acetonitrile, or tetrahydrofuran. Solutes are retained due to **hydrophobic interactions** with the **nonpolar stationary phase** and are eluted in order of increasing hydrophobicity (decreasing polarity). Increasing the polar (aqueous) component of the mobile phase increases solute retention (larger k' values) (see Chap. 12, Sect. 12.5.1.2.4), whereas increasing the organic solvent content of the mobile phase decreases retention (smaller k' values). Various additives can serve additional functions. For example, although ionic compounds often can be resolved without them, **ion-pair reagents** may be used to facilitate chromatography of ionic species on reversed-phase columns. These reagents are ionic surfactants, such as octanesulfonic acid, which can neutralize charged solutes and make them more lipophilic. This type of chromatography is referred to as **ion-pair reversed-phase**.

A modification of reversed-phase systems has led to the development of hydrophobic interaction chromatography (HIC), which utilizes column matrices and elution conditions that minimize or eliminate protein denaturation (see Sect. 13.3.4).

13.3.3.2 Applications of Reversed-Phase HPLC

Reversed-phase has been the HPLC mode most used for analysis of plant proteins. Cereal proteins, among the most difficult of these proteins to isolate and characterize, are now routinely analyzed by this method. Both water- and fat-soluble vitamins (Chap. 20) can be analyzed by reversed-phase HPLC [2–5], and the availability of fluorescence detectors has enabled researchers to quantitate very small amounts of the different forms of vitamin B_6 (vitamins) in foods and biological samples. Figure 13.4 [16, 23] shows the separation of several of these vitamins in a rice bran extract achieved by reversed-phase ion-pair HPLC [24].

Reversed-phase ion-pair HPLC can be used to resolve carbohydrates on C_{18} bonded-phase columns [10], and the constituents of soft drinks (caffeine, aspartame, etc.) can be rapidly separated. Reversed-phase HPLC using a variety of detection methods, including RI, UV, and light scattering, has been applied to the analysis of lipids [3, 5, 11, 19]. Antioxidants, such as BHA and BHT, can be extracted from dry foods and analyzed with simultaneous UV and fluorescence detection [3]. Phenolic flavor compounds (such as vanillin) and pigments (such as chlorophylls, carotenoids, and anthocyanins) are also easily analyzed [3, 5, 11, 19]. A typical chromatogram of carotenoids present in a carrot extract is shown in Fig. 13.5 [25]. Reversed-phase ion-pair chromatography is also used for the separation of synthetic food colors (e.g., FD&C Red No. 40 and FD&C Blue No. 1).

Chlorogenic acid, an antioxidant with antidiabetic effects, is also analyzed by reversed-phase HPLC, as is the nutrient quality of sweet potatoes, which contain ß-carotene, in addition to many other beneficial components. The Scoville scale for hot peppers is based on capsicum content, which may also be measured by HPLC.

13.3.4 Hydrophobic Interaction

With the use of special matrices, hydrophobic interaction chromatography (HIC) offers the possibility to separate many compounds, including many nondenatured proteins,

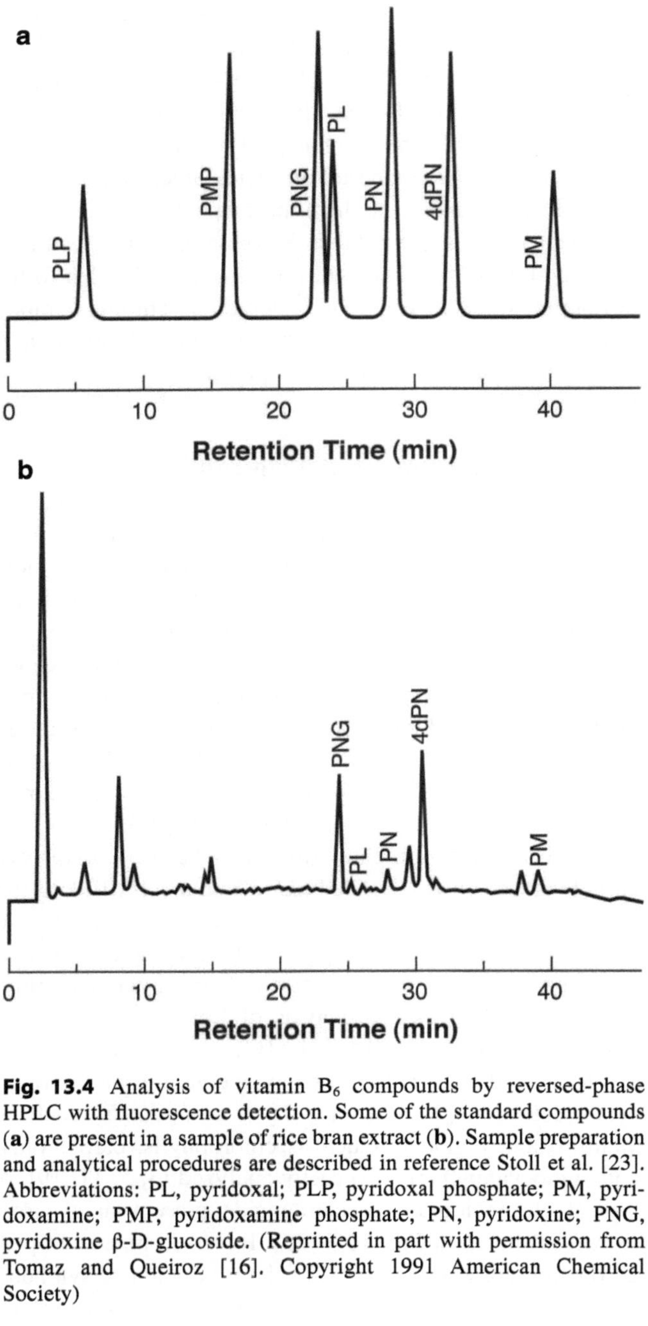

Fig. 13.4 Analysis of vitamin B_6 compounds by reversed-phase HPLC with fluorescence detection. Some of the standard compounds (**a**) are present in a sample of rice bran extract (**b**). Sample preparation and analytical procedures are described in reference Stoll et al. [23]. Abbreviations: PL, pyridoxal; PLP, pyridoxal phosphate; PM, pyridoxamine; PMP, pyridoxamine phosphate; PN, pyridoxine; PNG, pyridoxine β-D-glucoside. (Reprinted in part with permission from Tomaz and Queiroz [16]. Copyright 1991 American Chemical Society)

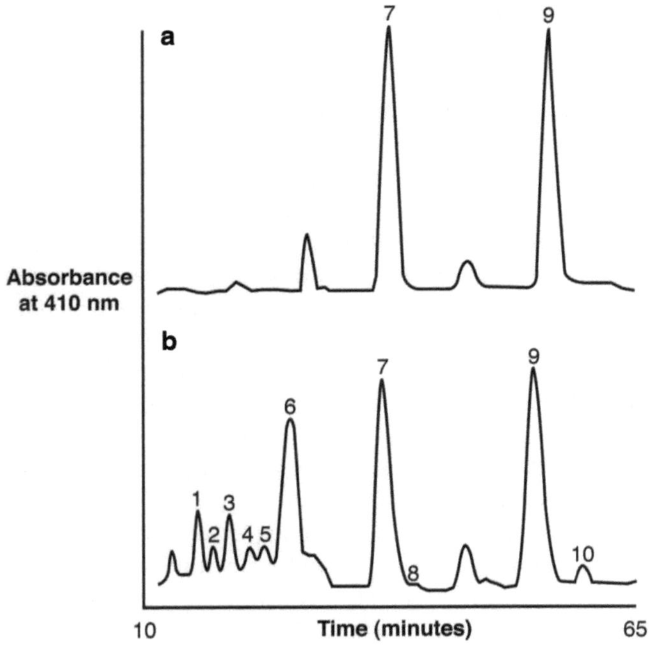

Fig. 13.5 Reversed-phase HPLC separation of α-carotene (AC) and β-carotene (BC) isomers in (**a**) fresh and (**b**) canned carrots using a 5-μm C_{30} stationary phase. Peak 1, 13-*cis* AC; 2, unidentified *cis* AC; 3, 13' –*cis* AC; 4, 15-*cis* BC; 5, unidentified *cis* AC; 6, 13-*cis* BC; 7, all-*trans* AC; 8, 9-*cis* AC; 9, all-*trans* BC; 10, 9-*cis* BC. (Reprinted with permission from Lessin et al. [25]. Copyright 1997 American Chemical Society)

by less harsh eluents and conditions. The technique is based on the presence of hydrophobic groups on or near the surface of the sample molecules that can interact with the hydrophobic column matrix under high saline (but nondenaturing) conditions (Chap. 12, Sect. 12.4.4). The sample is then eluted by lowering the salt concentration of the eluent. This allows for the collection of active proteins, such as enzymes, which can then be studied further in the lab [16]. HIC is often used in semi-preparative HPLC systems to allow not only for adequate resolution and analytical detection of biomolecules (mostly proteins), but for the collection of separated fractions in μg to mg quantities, thus facilitating further testing of the purified fractions.

13.3.5 Ion Exchange

13.3.5.1 Stationary and Mobile Phases

Packing materials for **ion-exchange HPLC** are usually **functionalized organic resins**, such as sulfonated or aminated poly(styrene-divinylbenzene), with either anion- or cation-exchange (Chap. 12, Sect. 12.4.5). **Macroporous resins** are most effective for HPLC columns due to their rigidity and permanent pore structure. **Pellicular packings** are also utilized, particularly in the CarboPac™ (Dionex) series, in which the nonporous, latex resin beads are coated with functionalized microbeads. The **mobile phase** in ion-exchange HPLC is usually an **aqueous buffer**, and solute retention is controlled by changing mobile phase ionic strength and/or pH. **Gradient elution** (gradually increasing ionic strength) is frequently employed.

13.3.5.2 Applications of Ion-Exchange HPLC

Ion-exchange HPLC has many applications, ranging from the detection of simple inorganic ions to analysis of carbohydrates and amino acids, to the preparative purification of proteins/peptides and oligosaccharides. Ion-exchange HPLC usually uses a conductivity detector. Because the mobile phase also contains ions, however, background conductivity can be relatively high. One step toward solving this problem is to use much lower capacity ion-exchange packing materials, so that more dilute eluents may be employed. Ion-exchange HPLC can be used to determine inorganic anions and cations, transition metals, organic acids, amines, phenols, surfactants, and sugars. Some specific examples applied to food matrices include the determination of organic and inorganic ions in milk, organic acids in coffee extract and wine, chlorine in infant formula, and trace metals, phosphates, and sulfites in foods. Figure 13.6 illustrates the simultaneous determination of organic acids and inorganic anions in coffee by ion chromatography.

Both cation- and anion-exchange HPLC have been used for carbohydrate analysis. The advantage of separating carbohydrates by anion exchange is that retention and selectivity may be altered by changes in eluent composition. Carbohydrate analysis has benefited greatly by the development of a system that involves anion-exchange HPLC at high pH (≥12) and detection by a PAD. Pellicular column packings (see Sect. 13.2.3.2.2), consisting of nonporous latex beads coated with a thin film of strong anion exchanger, provide the necessary fast exchange, high efficiency, and resistance to strong alkali. These systems may be used in a variety of applications, from routine quality control to basic research. One common application is the determination of oligosaccharide distributions in corn syrups and other starch hydrolysates (Fig. 13.7).

Amino acids have been resolved using ion-exchange HPLC for decades (see Chap. 24, Sect. 24.3.1.2). Ion-exchange HPLC is also one of the most effective modes for separation of proteins and of peptides in protein hydrolysates.

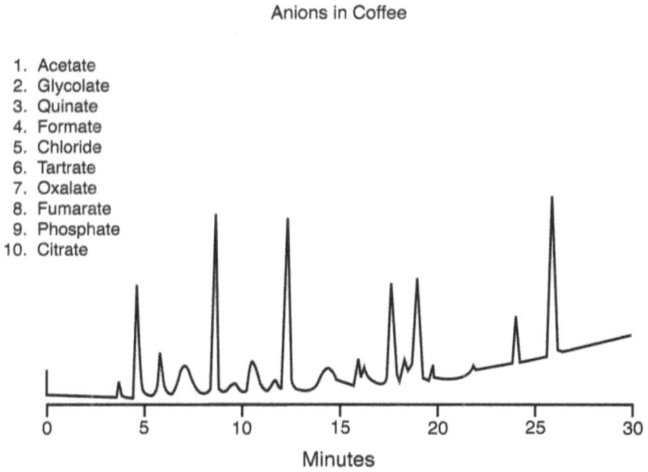

Fig. 13.6 Ion-chromatographic analysis of organic acids and inorganic anions in coffee. Ten anions (listed) were resolved on an IonPac AS5A column (Dionex) using a sodium hydroxide gradient and suppressed conductivity detection. (Courtesy of Dionex Corp., Sunnyvale, CA)

13.3.6 Size Exclusion

Size-exclusion HPLC (SE HPLC) fractionates solutes solely on the basis of size, with larger molecules eluting first (Chap. 12, Sect. 12.4.7). The resolution of SE HPLC columns has improved by using small particle packing materials.

13.3.6.1 Column Packings and Mobile Phases

Size-exclusion packing materials are selected so that matrix pore size matches the molecular weight range of the species to be resolved. Prepacked columns of microparticulate media are available in a wide variety of pore sizes. **Hydrophilic packings**, for use with water-soluble samples and aqueous mobile phases, may be surface-modified silica or methacrylate resins. **Poly(styrene-divinylbenzene) resins** are useful for nonaqueous size-exclusion chromatography of synthetic polymers.

The mobile phase in this mode is chosen for sample solubility, column compatibility, and minimal solute-stationary phase interaction. Otherwise, it has little effect on the separation. Aqueous buffers are used for biopolymers, such as proteins and polysaccharides, both to preserve biological activity and to prevent adsorptive interactions. Tetrahydrofuran or dimethylformamide is generally used for size-exclusion chromatography of other polymer samples, to ensure sample solubility.

13.3.6.2 Applications of SE HPLC

SE HPLC is used for the rapid determination of **average molecular weight** and **molecular weight range** of peptides, proteins, and polysaccharides including amylose, amylopectin, and other soluble gums such as xanthan, pullulan, guar, and water-soluble cellulose derivatives. **Molecular weight**

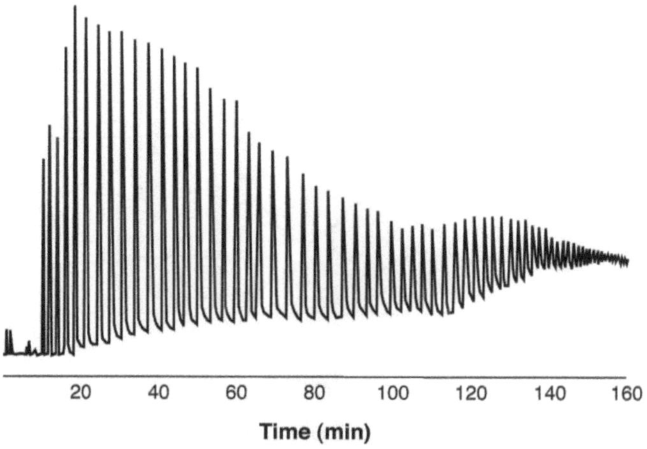

Fig. 13.7 Anion exchange analysis of isoamylase-treated waxy corn starch. The enzyme debranches the amylopectin, and the chromatogram represents the branch chain-length distribution, from four sugars in length and up. The analysis was performed with anion-exchange HPLC (Dionex™), with a pulsed amperometric detector

distribution can be determined directly from SEC HPLC, if LALLS or MALLS is used for detection [7]. The application of aqueous size-exclusion chromatography to two commercially important polysaccharides, xanthan and carboxymethyl cellulose, is discussed in detail in reference [7].

SE HPLC analysis is useful to better understand various food components and systems. SE HPLC analysis of tomato cell wall pectin from hot- and cold-break tomato preparations (Fig. 13.8) showed that the cell wall pectin was not differentially degraded by the different processing procedures. SE HPLC is a rapid, one-step method for assessing soybean cultivars on the basis of protein content (proteins in the extracts of nondefatted flours from five soybean cultivars were separated into six common peaks, and cultivars could be identified by the percent total area of the fifth peak). A SE HPLC method has also been applied to the determination of polymerized triacylglycerols in oils and fats [7].

13.3.7 Affinity

Affinity chromatography is based on the principle that the molecules to be purified can form a selective but reversible interaction with another molecular species that has been immobilized on a chromatographic support. Although almost any material can be immobilized on a suitably activated support, the major ligands are proteins, including lectins (Chap. 24, Sect. 24.2.3.2.2), nucleic acids, and dyes (Chap. 12, Sect. 12.4.6). Affinity chromatography is used to purify many glycoproteins. Affinity chromatography using immobilized folate-binding protein is an effective tool in purifying sample extracts for HPLC analysis of folates in foods (e.g., reference [26]).

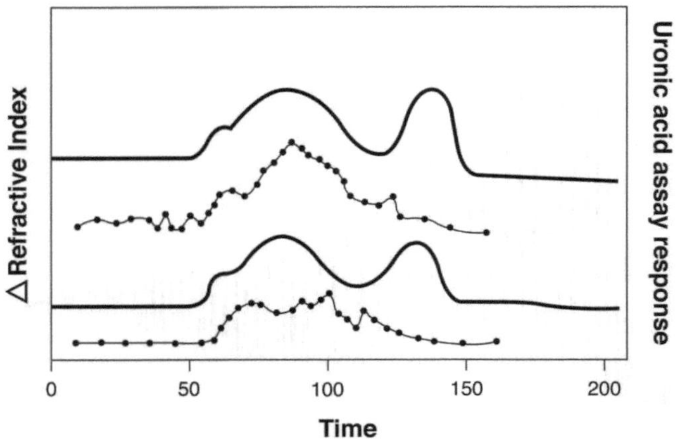

Fig. 13.8 Analysis of tomato cell wall pectin from hot- and cold-break tomato preparations by size-exclusion chromatography. The *solid lines* are from a refractive index detector response. The *lines with markers* result from a post-chromatography analysis. Aliquots of collected fractions were analyzed by a colorimetric chemical assay that is specific for pectic sugars

13.3.8 Multidimensional HPLC

Multidimensional HPLC, mostly 2D HPLC systems, is gaining popularity for automated and enhanced resolution, detection, and quantification of solutes with similar molecular characteristics. Multidimensional liquid chromatography (LC) has been performed for decades in the purification of components (e.g., peptides, carotenoids) for further structural and biological characterization. Fractions from the first dimension are collected and injected into a second- and third-dimension column and LC system for further purification. This process is tedious and requires intermediate sample preparation steps. 2D HPLC automated systems, in contrast, allow for fast separation and analysis of purified solutes. Such systems have embedded injectors and sample loops (Sect. 13.2.2), one ahead of the first column and one placed after the eluent exits the first detector. The second injector sends the fraction of interest onto the second column for additional resolution using a second-dimension pump system [27]. The isolated solutes proceed to the second detector within the system. 2D HPLC systems also allow detection using two different detectors; one of them can be an MS, allowing for robust qualitative and quantitative analysis. In addition, such systems can be available as UHPLC systems.

13.4 SUMMARY

HPLC is a chromatographic technique of great versatility and analytical power. A basic HPLC system consists of at least one pump, injector, column, detector, and data system. The pump delivers mobile phase through the system. An injector allows a sample to be placed into the flowing mobile phase for introduction onto the column. The HPLC column consists of stainless steel or polymer hardware filled with a separation packing material (stationary phase). Various guard columns may be used prior to the analytical column. Detectors used in HPLC include UV-Vis absorption, fluorescence, RI, electrochemical, and light scattering, as well as coupled analytical systems, such as a mass spectrometer. Detection sensitivity or specificity sometimes can be enhanced by chemical derivatization of the analyte. Computer-controlled data station systems offer data collection and processing capabilities and can run the instrument when an automated system is needed. A broad variety of column packing materials have contributed greatly to the widespread use of HPLC. These column packing materials may be categorized as silica-based (porous silica, bonded phases, pellicular packings) or polymeric (microporous, macroporous, or pellicular/nonporous). The success of silica-based bonded phases has expanded the applications of normal-phase and reversed-phase modes of separation in HPLC. Separations are also achieved with ion-exchange, size-exclusion, and affinity chromatography. The development of 2D systems enhanced the analytical detection and quantification by HPLC. HPLC is widely used for the analysis of small molecules and ions, such as sugars, vitamins, and amino acids, and is applied to the separation and characterization of macromolecules, such as proteins and polysaccharides.

13.5 STUDY QUESTIONS

1. Why might you choose to use HPLC rather than traditional low-pressure column chromatography?
2. What is a guard column and why is it used?
3. Give three general requirements for HPLC column packing materials. Describe and distinguish among porous silica, bonded phases, pellicular, and polymeric column packings, including the advantages and disadvantages of each type.
4. What is the primary function of an HPLC detector (regardless of type)? What factors would you consider in choosing an HPLC detector? Describe three different types of detectors and explain the principles of operation for each.
5. You are performing HPLC using a stationary phase that contains a polar nonionic functional group. What type of chromatography is this, and what could you do to increase the retention time of an analyte?
6. Describe one type of ion-exchange HPLC and one type of SE HPLC and give at least two examples for the use of each.

REFERENCES

1. Bidlingmeyer BA (1993) Practical HPLC methodology and applications. Wiley, New York
2. Núñez O, Gallart-Ayala H, Martins CPB, Lucci P (2012) New trends in fast liquid chromatography for food and environmental analysis. J Chromatogr A 1228: 298–323
3. Nollet LML, Toldra F (eds) (2012) Food analysis by HPLC, 3rd edn. Marcel Dekker, New York
4. Robards K, Ryan D (2022) High performance liquid chromatography: Instrumentation and techniques, Ch 5. In: Principles and Practice of Modern Chromatographic Methods, 2nd edn. Elsevier, Amsterdam
5. Harris DC (2020) Quantitative chemical analysis, 10th edn. W.H. Freeman and Co., New York
6. Hanai T (2004) HPLC: a practical guide. The Royal Society of Chemistry, Cambridge
7. Swadesh J (ed) (2000) HPLC: practical and industrial applications, 2nd edn. CRC, Boca Raton, FL
8. Lough WJ, Wainer IW (eds) (2008) High performance liquid chromatography: fundamental principles and practice. Springer, New York
9. LaCourse WR (2000) Column liquid chromatography: equipment and instrumentation (fundamental review). Anal Chem 72: 37R–51R
10. Sailaja A, Ghosh S, Reddy TPK, Deepthi PN, Banji D (2014) A Review on trouble shooting in HPLC and its solutions. International Journal of Pharmaceutical and Chemical Sciences 3: 625–635
11. Snyder LR, Kirkland JJ, Glajch JL (2012) Practical HPLC Method Development. John Wiley & Sons, Hoboken, New Jersey
12. Shackman JG (2013) General instrumentation. In: S. Fanali, et al. (Eds.) Liquid Chromatography: Fundamentals and Instrumentation, Elsevier, Amsterdam.
13. Gertz C (1990) HPLC tips and tricks. LDC Analytical, Riviera Beach, FL
14. Ishii D (ed) (1988) Introduction to microscale high-performance liquid chromatography. VCH Publishers, New York
15. Dorsey JG, Cooper WT, Siles BA, Foley JP, Barth HG (1996) Liquid chromatography: theory and methodology (fundamental review). Anal Chem 68:515R–568R
16. Tomaz CT, Queiroz JA (2013) Hydrophobic interaction chromatography, In Liquid chromatography: fundamentals and instrumentation, Eds. Fanali S, Haddad PR, Poole, Schoenmakers P, Lloyd DK. Elsevier, Amsterdam, p 122–141
17. Waters Corporation (2014) Beginners Guide to UPLC: Ultra-Performance Liquid Chromatography (Waters Series) 1st edn. Milford MA.
18. Swartz M (2005) UPLC™: An Introduction and Review. J. Liq. Chromatogr. Relat., 28:7–8, 1253–1263
19. Matissek R, Wittkowski R (eds) (1993) High performance liquid chromatography in food control and research. Behr's Verlag, Hamburg, Germany
20. Synder LR, Glajch JL, Kirkland JJ (1997) Practical HPLC method development, 2nd edn. Wiley, New York
21. Moldoveanu SC, David V (2017) Polar Analytical Columns, Ch 8. In: Selection of HPLC method in Chemical Analysis. Elsevier, Amsterdam
22. Liu Z, Xu M, Zhang W, Miao X, Wang PG, Li S, Yang S (2022) Recent development in hydrophilic interaction liquid chromatography stationary materials for glycopeptide analysis. Anal Methods 14: 4437–4448
23. Stoll DR, Wang X, Carr PW (2008) Comparison of the practical resolving power of one-and two-dimensional high-performance liquid chromatography analysis of metabolomics samples. Anal Chem 80(1):268–278
24. Gregory JF, Sartain DB (1991) Improved chromatographic determination of free and glycosylated forms of vitamin B_6 in foods. J Agric Food Chem 39:899–905
25. Lessin WJ, Catignani GL, Schwartz SJ (1997) Quantification of *cis-trans*isomers of provitamin A carotenoids in fresh and processed fruits and vegetables. J Agric Food Chem 45:3728–3732
26. Pfeiffer C, Rogers LM, Gregory JF (1997) Determination of folate in cereal-grain food products using tri-enzyme extraction and combined affinity and reverse-phase liquid chromatography. J Agric Food Chem 45: 407–413
27. Pirok BWJ, Stoll DR, Schoenmakers PJ (2018) Recent Developments in Two-Dimensional Liquid Chromatography: Fundamental Improvements for Practical Applications. Anal Chem 91: 240–263.

Gas Chromatography

14

Gary A. Reineccius and Michael C. Qian

Contents

14.1 INTRODUCTION

The first publication on gas chromatography (GC) was in 1952 [1] with the first commercial instruments manufactured in 1956. James and Martin [1] separated fatty acids by GC, collected the column effluent, and titrated the individual fatty acids for quantitation. GC has advanced greatly since that early work and is now considered to be a mature field that is approaching theoretical limitations.

The types of analysis that can be done by GC are very broad. GC has been used for the determination of fatty acids, triglycerides, cholesterol and other sterols, gases, solvent analysis, water, alcohols, and simple sugars, as well as oligosaccharides, amino acids and peptides, vitamins, pesticides, herbicides, food additives, antioxidants, nitrosamines, polychlorinated biphenyls (PCBs), drugs, flavor compounds, and many more. The fact that GC has been used for these various applications does not necessarily mean that it is the best method—often better choices exist. GC is ideally suited to the analysis of **thermally stable, volatile substances**. Substances that do not meet these requirements (e.g., sugars, oligosaccharides, amino acids, peptides, and vitamins) are more suited to analysis by a technique such as

G. A. Reineccius (✉)
Department of Food Science and Nutrition, University of Minnesota, St. Paul, MN, USA
e-mail: greinecc@umn.edu

M. C. Qian
Department of Food Science and Technology, Oregon State University, Corvallis, OR, USA
e-mail: michael.qian@oregonstate.edu

B. P. Ismail, S. S. Nielsen (eds.), *Nielsen's Food Analysis*, Food Science Text Series,
https://doi.org/10.1007/978-3-031-50643-7_14

high-performance liquid chromatography (HPLC) or supercritical fluid chromatography (SFC). Yet gas chromatographic methods appear in the literature for these substances after derivatization to render them volatile.

This chapter will discuss sample preparation for GC, GC hardware, columns, and chromatographic theory as it is uniquely applied to GC. Texts devoted to GC in general [2, 3] and food applications in particular [4, 5] should be consulted for more detail.

14.2 SAMPLE PREPARATION FOR GAS CHROMATOGRAPHY

14.2.1 Introduction

One cannot generally directly inject a food product into a GC without some sample preparation [6]. The high temperatures of the injection port will result in the degradation of nonvolatile constituents and create a number of false GC peaks corresponding to the volatile degradation products formed. In addition, very often the constituent of interest must be isolated from the food matrix simply to permit concentration such that it is at detectable limits for the GC or to isolate it from the bulk of the food. Thus, one must generally do some type of sample preparation, component isolation, and concentration prior to GC analysis.

Sample preparation often involves grinding, homogenization, or otherwise reducing particle size. There is substantial documentation in the literature showing that foods may undergo changes during sample storage and preparation. Many foods contain active enzyme systems that will alter the composition of the food product. This is very evident in the area of flavor work [7–9]. Inactivation of enzyme systems via high-temperature, short-time thermal processing, sample storage under frozen conditions, drying the sample, or homogenization with alcohol may be necessary.

Microbial growth or chemical reactions may occur in the food during sample preparation. Chemical reactions will often result in false peaks on the GC. Thus, the sample must be maintained under conditions such that degradation does not occur. Microorganisms are often inhibited by certain chemicals (e.g., sodium fluoride), thermal processing, drying, or frozen storage.

14.2.2 Isolation of Analytes from Foods

14.2.2.1 Introduction

The isolation procedure may be quite complicated depending on the constituent to be analyzed. For example, if one were to analyze the triglyceride-bound fatty acids in a food, one would first have to extract the lipids (free fatty acids; mono-, di-, and triglycerides; sterols; fat-soluble vitamins, etc.) from the food (e.g., by solvent extraction) and then isolate only the triglyceride fraction (e.g., by adsorption chromatography on silica). The isolated triglycerides would then have to be treated to first hydrolyze the fatty acids from the triglycerides and subsequently to form esters to improve gas chromatographic properties. The two latter steps might be accomplished in one reaction by transesterification (e.g., boron trifluoride in methanol) as described in Chap. 17, Sect. 17.2.7.2, and Chap. 23, Sect. 23.6.2.1. Thus, many steps involving several types of chromatography may be used in sample preparation for GC analysis.

The analysis of volatiles in foods (e.g., packaging or environmental contaminants, alcohols, and flavors or off-flavors) can be achieved by GC. These components may be isolated by headspace analysis (static or dynamic), simple solvent extraction, distillation, preparative chromatography (e.g., solid-phase extraction, column chromatography on silica gel), or some combination of these basic methods in preparation for GC analysis. Table 14.1 summarizes the various methods of isolation which are described in some detail in Sect. 14.2.2. The procedure(s) used for a particular food will depend on the food matrix as well as the compounds to be analyzed. The primary considerations are to isolate the compounds of interest from nonvolatile food constituents (e.g., carbohydrates, proteins, and vitamins) or those that would interfere with GC (e.g., lipids). Some of the chromatographic methods that might be applied to this task have been discussed in the basic chromatography chapter (Chap. 12) of this text. Methods for the isolation of volatile substances will be covered briefly as they pertain to the isolation of components for GC analysis.

It should be emphasized that the isolation procedure used is critical in determining the results obtained. An improper choice of method or poor technique at this step negates the best GC analysis of the isolated analytes. The influence of isolation technique on GC analysis of aroma compounds has been demonstrated [10]. These biases are discussed in the sections that follow and in more detail in books edited by Marsili [11] and Mussinan and Morello [12]. While these books relate to the analysis of aroma compounds in foods, the techniques for the isolation of these volatiles are the same as used in the analysis of other volatiles, perhaps less desirable components in foods.

14.2.2.2 Headspace Methods

One of the simplest methods of isolating volatile compounds from foods is by direct injection of the headspace vapors above a food product. There are two types of headspace sampling: static headspace sampling and dynamic headspace sampling.

Static (i.e., **direct**) **headspace sampling** has been used extensively when rapid analysis is necessary and a measur-

Table 14.1 Methods to isolate analytes (volatiles) from food for gas chromatographic analysis

General types	*Specific methods*	*Description*	*Advantage/applications*	*Disadvantages/limitations*
Headspace sampling	Direct headspace sampling	Headspace taken from sample in a closed vessel for direct injection into GC	Good for rapid analysis of very low-boiling point compounds	Low sensitivity
	Dynamic headspace sampling (purge and trap)	Sample is purged with inert gas to strip volatiles from the sample and retained by absorbent or cryogenic traps. The volatiles are then desorbed from the trap thermally or by organic solvent for GC analysis	Collects volatiles, no matter the polarity or boiling point. With adsorbent trap, no further extraction is required and the system is readily automated	With **cryogenic trap**, usually must further extract with organic solvent and dry to concentrate. With **adsorbent trap**, get differential adsorption affinity and limited capacity.
Distillation methods	Steam distillation (at normal pressure or in vacuum)	Use steam at atmospheric pressure or in vacuum to heat and codistill volatiles from the sample followed by solvent extraction of distillate	Convenient and efficient	Solvent extraction is required on dilute aqueous solution collected. Thermally generated artifacts
	Simultaneous distillation and extraction (SDE) AKA likens–Nickerson	Product steam and extracting solvent vapors are intermixed and condensed. The solvent extracts organic volatiles from condensed steam—conc. And GC	Time savings due to one-step isolation and concentration. Less solvent use vs what is needed to condense the sample in regular steam distillation	May get artifact formation due to elevated temperature (not necessarily a problem)
Solvent extraction	Simple batch extraction	Use typically organic solvent with, e.g., separatory funnel, to extract analytes of interest into the solvent	Efficient if multiple extractions and excessive shaking used	Cannot use if fat is present in the sample. Extraction may result in emulsion formation
	Continuous extraction	Use organic solvent or CO_2 with equipment to make continuous	More efficient than batch extraction	Need to remove nonvolatiles (e.g., fat) from the extracted sample
	Solvent-assisted flavor evaporation (SAFE)	Sample with extracting solvent are added slowly to the SAFE device. Volatiles and solvent are distilled and collected in the cold trap	High efficiency. Operate at low temperature. Minimum thermal artifact formation	Need special extraction device and high vacuum system. Tedious due to several steps in process
Solid-phase extraction (SPE)			Compared to traditional liquid-to-liquid extraction: less solvent, glassware, and time required, better precision and accuracy, minimum solvent evaporation for further analysis, readily automated	
	Simple solid-phase extraction (SPE)	A liquid sample is passed through a column or filter disk with chromatographic stationary phase. Solutes with affinity are retained on the phase. After rinsing the column with water or weak solvent, a strong eluent is used to elute analytes of interest		
	Solid-phase microextraction (SPME)	Stationary phase is bound to fine fused silica filament. The filament is immersed in the sample or headspace of the sample and then pulled into the metal sheath that is forced through the septum of GC. Volatiles are thermally desorbed from the filament	Simple operation. No solvent contamination or disposal. Many phases of fiber available, so analytes with a wide range of polarity and volatility can be analyzed	Poor reproducibility. Lot-to-lot variation. Fiber is highly selective, sample saturation, competitive absorption. Highly influenced by other volatiles in the sample
	Solid-phase dynamic extraction SPDE)	Similar to SPME, but the polymer is coated inside a special needle used to draw headspace of food. Volatiles are absorbed to the phase and then released when the needle is injected into GC	Similar to SPME, but less issues with analyte saturation and competition	Lack of commercial device

(continued)

Table 14.1 (continued)

General types	*Specific methods*	*Description*	*Advantage/applications*	*Disadvantages/limitations*
	Stir bar sorptive extraction (SBSE)	Magnetic stir bar is enclosed in a glass coated with absorbent film. The bar spins in the sample and absorbs analytes, and then the stir bar is transferred to thermal desorption unit and delivered to GC column	Higher sensitivity and accuracy than SPME. Highly automated. Can extract less volatile compounds	Limited phase available, instrumentation is expensive, high operation cost, not good for solid samples (use HS-SBSE)
Direct injection		Inject 2–3 μl sample directly into the GC column	Simple, if the sample and column allow for this	Thermal degradation of nonvolatiles. Damage to the column. Decreased separation if water in the sample. Contamination of the column and injection port by nonvolatiles

able, indicator component is present. In static headspace sampling, a food sample is placed in a vessel and the vessel is closed with an inert septum. At equilibrium, a sample of the headspace is taken using a gas-tight syringe (manually or auto-sampled) and then directly injected into the GC. Examples of method applications include measurement of hexanal as an indicator of oxidation [13, 14] and 2-methylpropanal, 2-methylbutanal, and 3-methylbutanal as indicators of nonenzymatic browning [15]. The determination of residual solvents in packaging materials may also be approached by this method. Unfortunately, this method does not provide the sensitivity needed for trace analysis. Instrumental constraints typically limit headspace injection volumes to 5 mL or less. Therefore, only volatiles present in the headspace at concentrations greater than 10^{-7} g/L headspace would be at detectable levels using a flame ionization detector (FID). Furthermore, the static headspace only allows the investigation of very low-boiling compounds.

Dynamic headspace sampling (i.e., **Purge and Trap**) has found wide usage in recent years (Fig. 14.1). In this dynamic headspace method, a sample is purged with an inert gas such as nitrogen or helium, which strips volatile constituents from the sample. This method may involve simply passing large volumes of headspace vapors through a cryogenic trap or, alternatively, an adsorption trap. A **cryogenic trap** (if properly designed and operated) will collect headspace vapors irrespective of compound polarity and boiling point. However, water is typically the most abundant volatile in a food product, and, therefore, this distillate must be extracted with an organic solvent, dried, and then concentrated for analysis. These additional steps add analysis time and provide opportunity for sample contamination and loss. A more commonly used technique is to pass the purge effluent through adsorbent traps. **Adsorbent traps** offer the advantages of providing a water-free volatile isolate (trap material typically has little affinity for water) and are readily automated. While Tenax is the most commonly used adsorbent, charcoal or synthetic porous polymers (Porapaks® and Chromosorbs®) may also be used as trapping materials.

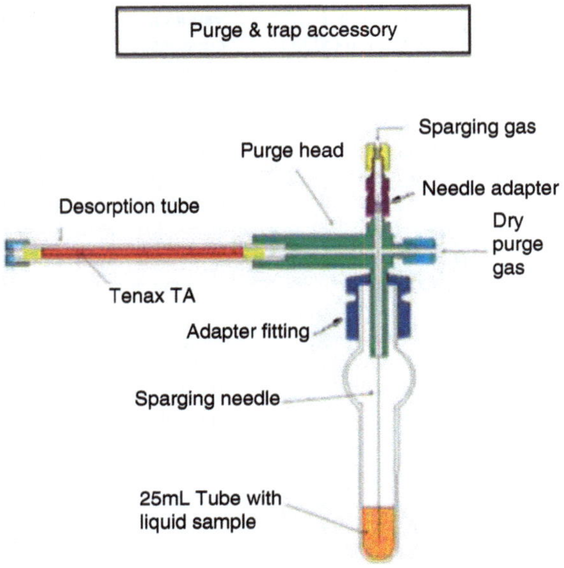

Fig. 14.1 Purge and trap accessory for headspace sampling

These polymers exhibit good thermal stability and have substantial adsorption capacity. The trapped volatiles are recovered from the trap with a suitable solvent or by thermal desorption. For automated purge-trap systems, the adsorbent traps are generally placed in a closed system and loaded, desorbed, and so on via the use of automated multiport valving systems. The automated closed system approach provides reproducible GC retention times and quantitative precision necessary for most studies. The primary disadvantages of adsorbent traps are their differential adsorption affinity for analytes and limited capacity.

This sampling technique is further challenged by some food samples, notably those high in protein, foaming on purging the sample. This foam may fill the purge chamber and enter the trap, resulting in contamination. The solution to foaming most commonly used is to purge with a tube (e.g., 1/16 inch metal) rather than a gas sparger. The tube does not produce small, stable bubbles and thus tends to not form stable foams. However, the large purge gas bubbles are inefficient at capturing sample volatiles, resulting in poorer sensitivity and less representative data. One could also consider the use of antifoaming agents, but these formulations typically contain volatile compounds that dominate the analytical profile, making the analysis complicated or impossible to interpret.

14.2.2.3 Distillation Methods

Distillation processes are quite effective at isolating volatile compounds from foods for GC analysis. Product moisture or outside steam is used to heat and codistill the volatiles from a food product. The most frequently used distillation method is steam distillation at atmospheric pressure. For most food analysis, a very dilute aqueous solution of volatiles results and a solvent extraction must be performed on the distillate to permit concentration for analysis. The distillation method most commonly used today is **simultaneous distillation and extraction** (SDE), modified from the original Likens–Nickerson distillation head (Fig. 14.2). In this apparatus, an aqueous sample is boiled in one side flask and a small amount of extracting solvent in another. The product steam and extracting solvent vapors are intermixed and condensed; the solvent extracts the organic volatiles from the condensed steam. The solvent and extracted distillate return to their respective flasks and are continuously distilled to extract the volatiles from the food. This one-step isolation concentration of flavor constituents allows a dramatic time saving over the separated operation and, because of their continuous recycling, a great reduction in organic solvents used. The drawback inherent in SDE is artifact formation. The elevated temperature applied during distillation may lead to lipid oxidation, Maillard browning, or Strecker reaction, which introduces artifacts. While the SDE system can be operated under

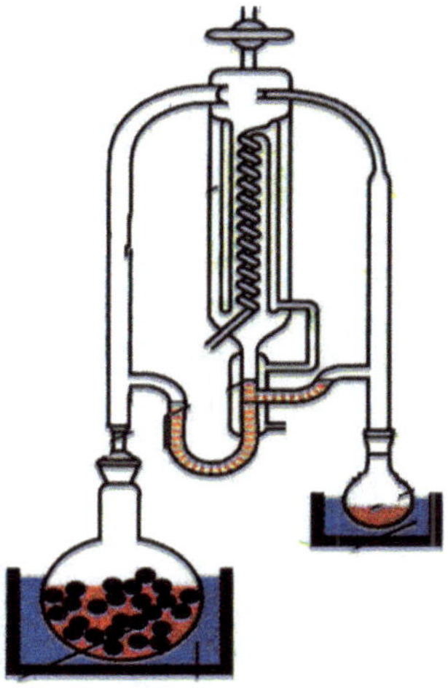

Fig. 14.2 Simultaneous distillation and extraction system for sample distillation

vacuum to reduce thermal exposure, it is not easy to regulate. Furthermore, some food aroma compounds, such as furaneol, may have low recovery by the SDE method [16]. While the distillation method is convenient and efficient, the volatile isolate can be contaminated by: (1) artifacts formed from solvents used in extraction, antifoam agents, steam supply (contaminated water), (2) thermally induced chemical changes, and (3) leakage of contaminated laboratory air into the system. This method finds greatest use today in detecting taints in foods since the tainting components are not likely to be formed as artifacts on sample heating.

14.2.2.4 Solvent Extraction

Solvent extraction is often the preferred method for the recovery of volatiles from foods. Recovery of volatiles will depend upon solvent choice (commonly dichloromethane or diethyl ether) and the relative solubility of the analytes in the sample vs extracting solvent. Solvent extraction typically involves the use of an organic solvent (unless sugars, amino acids, or some other water-soluble components are of interest).

Solvent extractions may be carried out in quite elaborate equipment, such as supercritical CO_2 extractors, or can be as simple as a batch process in a separatory funnel. Batch extractions can be quite efficient if multiple extractions and extensive shaking are used [17]. Continuous extractors (liquid–liquid) are more efficient but require more equipment and time.

Extraction with organic solvents limits the method to the isolation of volatiles from fat-free foods (e.g., wines; some breads, fruit, and berries; some vegetables; and alcoholic beverages). Even for the fat-free food systems, a small amount of fats, waxes, or other nonvolatile compounds may be extracted. These nonvolatile materials need to be removed; otherwise, they will interfere with subsequent concentration and GC analysis.

Nonvolatile compounds can be removed using **high vacuum distillation, molecular distillation,** or **solvent-assisted flavor extraction** (SAFE) distillation. SAFE is a compact and versatile distillation apparatus (Fig. 14.3) that can offer fast and reliable isolation of volatiles from complex matrices [18]. During SAFE distillation, the solvent and the solvent-extracted components are introduced slowly into the flask (top left) at low temperature under high vacuum and volatiles and solvents are evaporated instantly and condensed in a cold trap (right, typically liquid nitrogen), leaving nonvolatiles in the other flask. SAFE is highly efficient and can be used to isolate volatiles from solvent extracts or even directly from foods or beverages. Despite the drawback that compounds with high boiling points may not be completely recovered and it is tedious, *it tends to be the gold standard of volatile isolation procedures.*

Fig. 14.3 Solvent-assisted flavor evaporation (SAFE) distillation system

14.2.2.5 Solid-Phase Extraction

The extractions discussed above (Sect. 14.2.2.4) involve the use of two immiscible phases (water and an organic solvent). However, an alternative to such extractions is **solid-phase extraction** (SPE) [19, 20]. In one version of this technique, a liquid sample (most often aqueous based) is passed through a column (2–10 mL vol) filled with chromatographic packing or a filter disk (25–90 mm in diameter) that has the chromatographic packing embedded in it. The chromatographic packing (i.e., stationary phase coated on silica solid supports) may be any of a number of different materials (e.g., ion-exchange resins or a host of different reversed- or normal-phase HPLC column packings).

When a sample is passed through the cartridge or filter, analytes that have an affinity for the chromatographic phase will be retained on the phase, while those with little or no affinity will pass through. The trap phase is next rinsed with water, perhaps a weak solvent (e.g., pentane), and then a stronger solvent (e.g., dichloromethane). The strong eluent is chosen such that it will remove the analytes of interest.

SPE has been a popular method for sample extraction and cleanup. Many SPE columns are commercially available in different phases and formats. C18, amino, and silica gel are commonly used phases for food component analysis. C18 cartridges have high retention for nonpolar compounds, whereas highly cross-linked styrene-divinylbenzene (DVB) copolymers can extract a wide range of nonpolar and polar compounds. The new generation of polymers (Oasis® HLB, copolymer of divinylbenzene and N-vinylpyrrolidone) can even extract lipophilic, hydrophilic, acidic, and basic compounds with a single cartridge and perform fractionation based on the functional group [21].

Overall, solid-phase extraction has numerous advantages over traditional liquid–liquid extractions including: (1) less solvent is required, (2) speed, (3) less glassware is needed (less cost and potential for contamination), (4) better precision and accuracy, (5) minimum solvent evaporation before GC analysis, and (6) being readily automated. However, SPE method development requires tedious empirical procedures to achieve the best separation and recovery of an analyte.

Another version of this method is called **solid-phase microextraction** (SPME). This method was developed originally for environmental work [22, 23] but is now the most commonly used method for the isolation of flavor compounds from food systems. In this adaptation, the phase is bound onto a fine fused silica or metal filament (e.g., fiber) (approximately the size of a 10-μl syringe needle, Fig. 14.4). The fiber is immersed in a sample or in the headspace above a sample. After the desired extraction time, the fiber is pulled into a protective metal sheath, removed from the sample, and forced through the septum of a GC where the adsorbed volatiles are thermally desorbed from the fiber (Fig. 14.5) into the GC inlet.

Many different phases of fiber are commercially available, and compounds with a wide range of polarity or volatility can be analyzed. PDMS (polydimethylsiloxane) is a nonpolar phase coating and can be used to extract nonpolar compounds. Polar analytes can be extracted with polar phases (e.g., polyacrylate and Carbowax® coatings). Divinylbenzene (DVB) coating is good for many volatile compounds. The coating is available in various film thicknesses. Thicker film fibers (100 μm) are better for volatiles, whereas thinner film fibers (7 μm and 30 μm) are better for

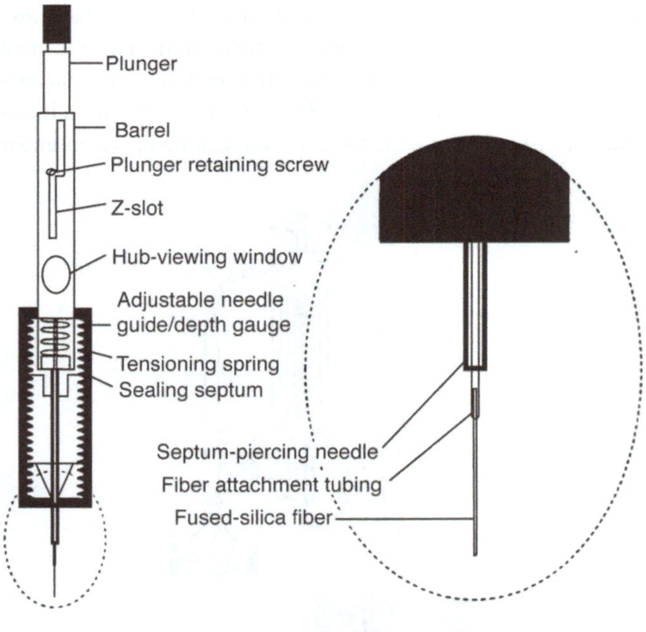

Fig. 14.4 Schematic of a solid-phase microextraction (SPME) device [22]. (Courtesy of Dr. Janusz Pawliszyn, Dept. of Chemistry, University of Waterloo, Waterloo, Ontario, Canada)

larger molecules. Multiphase fibers (such as Carboxen®/PDMS, Carboxen®/DVB/PDMS) are also available to extract both polar and nonpolar compounds. A Carboxen®/PDMS fiber is good for highly volatile compounds, particularly for volatile sulfur compounds [24]. A 2-cm 50/30 DVB/Carboxen®/PDMS fiber is frequently used for volatile and semivolatile flavor analysis in food systems.

SPME is widely used to extract volatile and semivolatile organic compounds from environmental, biological, and food samples [25–29]. The main advantages of this technique are the simplicity of operation and no solvent contamination or disposal. With an autosampler, high precision and sample throughput can be achieved. However, depending on the fiber used, the compounds extracted will be highly selective. Furthermore, SPME has limited absorption capacity, and the fiber can be easily saturated. Thus, other volatile compounds in the sample can compete for the active sites and cause competitive absorption. Other drawbacks are relatively poor reproducibility, fiber lot-to-lot variation, sensitivity to solvents, life time of the fiber, and difficulty in quantification. The simplicity of the method is often the basis for the wide use of this technique rather than its performance.

Solid-phase dynamic extraction (SPDE) is another technique for volatile extraction. SPDE is similar to SPME, except the phase is coated inside a special needle. A gas-tight syringe is used for SPDE to draw the headspace of food, and the volatiles are absorbed by the phase. The process can be repeated many times by moving the plunger up and down to achieve maximum absorption. The needle can then be inserted into a GC where the volatiles are thermally desorbed for analysis. Different phases are available to give the researcher some control over the types of compounds being absorbed. The volume of the trapping phase is about 4.5 μL compared with only 0.6 μL for SPME, so the SPDE has less issue with analyte saturation and competition.

A similar device works by drawing sample headspace repeatedly through a polymer trap (e.g., Tenax) rather than an SPDE needle (Fig. 14.5). Since this methodology uses a trap of much larger capacity than the SPDE, it will accumulate a greater quantity of volatiles, likely yielding better data.

In **stir bar sorptive extraction** (SBSE) I (Fig. 14.6), a magnetic stir bar is enclosed in a glass bar and then coated with a thick film of polymers such as polydimethylsiloxane (PDMS). The bar spins in the sample solution (or sample headspace) and absorbs analytes depending on mutual solubility. After "loading" the stir bar with volatiles, the stir bar is transferred to a compact thermal desorption unit (TDU) mounted on a GC. The analytes are thermally desorbed in the TDU and delivered to a GC column (Fig. 14.7).

The stir bar has 50–250 times more absorptivity than SPME. The PDMS volume is about 0.5 μL with SPME, compared to 24–126 μL with SBSE. Due to the increased volume of absorbent phase, SBSE has much higher sensitiv-

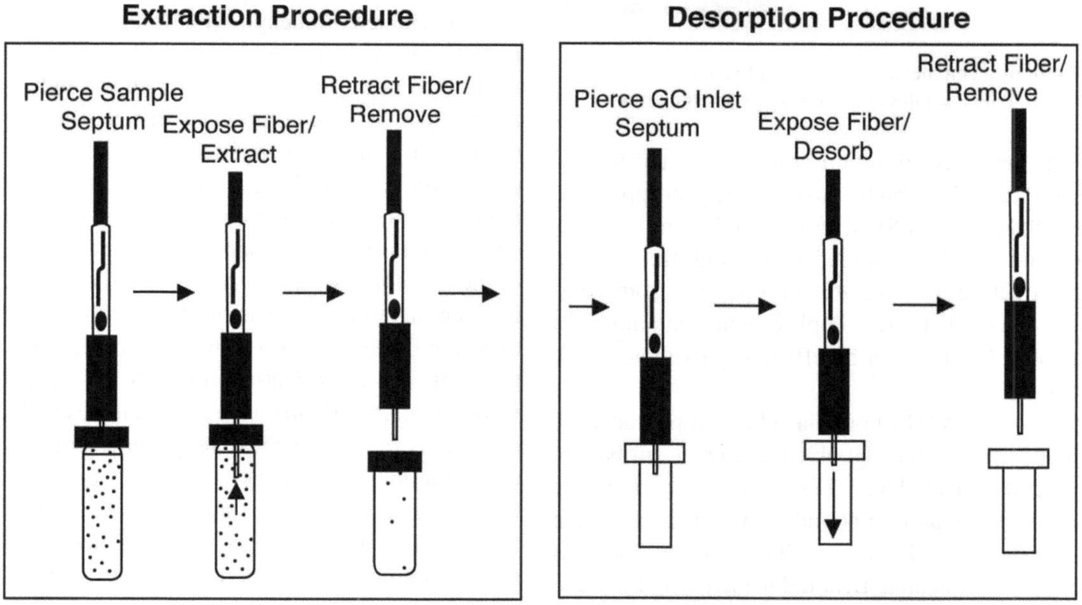

Fig. 14.5 Schematic showing the steps involved in the use of a solid-phase microextraction (SPME) device. (Reprinted with permission of Supelco, Bellefonte, PA)

Fig. 14.6 Diagram of stir bar sorptive extraction (SBSE) device. (Courtesy of Gerstel, Inc., Linthicum, MD)

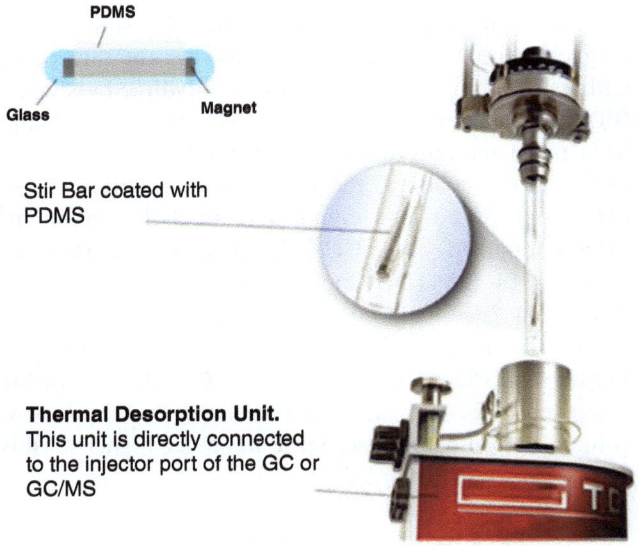

Fig. 14.7 Design of a commercial stir bar sorptive extraction device. (Courtesy of Gerstel, Inc., Linthicum, MD) (www.gerstel.com)

ity than SPME and has minimum competition and saturation effects [30, 31]. The high sensitivity (ppt to ppg) and flexibility of PDMS-based SBSE for nonpolar and medium polar compounds make it an effective and time-saving method for extracting trace volatile compounds from complex matrices [31–37]. Food samples even containing fat (<3%) or alcohol (<10%) can be efficiently extracted with this technique.

In addition to the PDMS nonpolar phase, a polyacrylate (PA) stir bar and an ethylene glycol (EG)-silicone stir bar are also commercially available. These phases, particularly EG-silicone, extract polar compounds more efficiently than the PDMS due to their polar nature. The use of stir bars has been applied to different analytical fields including food and wine [38].

14.2.2.6 Direct Injection

It is theoretically possible to analyze some foods by direct injection of the food into a GC. Assuming one can inject a 2–3-μl sample into a GC and the GC has a detection limit of 0.1 ng (0.1 ng/2 μl), one could detect analytes in the sample at concentrations greater than 50 ppb. Problems with direct injection arise due to: (1) thermal degradation of any nonvolatile food constituents in the injection port creating artifacts, (2) decreased separation efficiency due to water in the food sample, and (3) contamination of the GC column and injection port by nonvolatile materials. Despite these concerns, direct injection is sometimes used to determine oxidation in vegetable oils [39, 40]. A relatively large volume of oil (50–100 μl) can be directly injected into an injection port of a GC that has been packed with glass wool. Since vegetable oils are reasonably thermally stable and free of water, this method is particularly well suited to oil analysis.

There are numerous other approaches for the isolation of volatiles from foods. Some are simple variations of these methods, while others are unique. Several review articles are available that provide a more complete view of methodology [11, 12, 41].

14.2.3 Sample Derivatization

The compounds one wishes to determine by GC must be thermally stable under the GC conditions employed. Thus, for some compounds [e.g., pesticides, aroma compounds, polychlorinated biphenyls (PCBs), and volatile contaminants], the analyst can simply isolate the components of interest from a food as discussed above and directly inject them into a GC. For compounds that are thermally unstable, too low in volatility (e.g., sugars and amino acids), or yield poor chromatographic separation due to polarity (e.g., phenols or acids), a derivatization step must be included prior to GC analysis (see also Chaps. 19 and 24). A listing of some of the reagents used in preparing volatile derivatives for GC is given in Table 14.2. The most commonly used derivatizations are (1) silylation for alcohols, cholesterol, and carbohydrates, (2) esterification for fatty acids, and (3) oxime formation and derivatization for aldehydes and ketones. The conditions of use for these reagents are often specified by the supplier or can be found in the literature [42]. Derivatization can be performed in a liquid extract, or on a SPE extraction column, on an SPME fiber, or an SBSE, to allow automated extraction, derivatization, and GC analysis.

Table 14.2 Reagents used for making volatile derivatives of food components for GC analysis

Reagent	*Chemical group*	*Food constituent*
Silyl reagents	Hydroxy, amino carboxylic acids	Sugars, sterols, amino acids
Trimethylchlorosilane/hexamethyldisilazane		
BSA [*N,O*-bis(trimethylsilyl) acetamide		
BSTFA [*N,O*-bis (trimethylsilyl) trifluoroacetamide]		
t-BuDMCS (*t*-butyldimethylchlorosilyl/imidazole)		
TMSI (N-trimethylsilylimidazole)		
Esterifying reagents	Carboxylic acids	Fatty acids, amines, amino acids, triglycerides, wax esters, phospholipids, cholesteryl esters
Methanolic HCI		
Methanolic sodium methoxide		
N,N-Dimethylformamide dimethyl acetal		
Boron trifluoride methanol		
Miscellaneous		
Acetic anhydride/pyridine	Alcoholic and phenolic	Phenols, aromatic hydroxyl groups, alcohols
N-Trifluoroacetylimidizole/*N*-heptafluorobutyrlimidizole	Hydroxy and amines	Same as above
Alkylboronic acids	Polar groups on neighboring atoms	
O-Alkylhydroxylamine	Compounds containing both hydroxyl and carbonyl groups	Ketosteroids, prostaglandins

14.3 GAS CHROMATOGRAPHIC HARDWARE AND COLUMNS

The major parts of a GC are the **gas supply** system, **injection port, oven, column, detector, electronics,** and **recorder/data handling system** (Fig. 14.8). The hardware as well as operating parameters used in any GC analysis must be accurately and completely recorded. The information that must be included is presented in Table 14.3.

14.3.1 Gas Supply System

The gas chromatograph will require at least a supply of carrier gas and, most likely, gases for the detector (e.g., hydrogen and air for an FID). The gases used must be of high purity and all regulators, gas lines, and fittings of good quality. High-quality pressure regulators must be used to provide stable and continuous gas supply. The regulators should have stainless steel rather than polymer diaphragms since polymers will give off volatiles that may contribute peaks to the analytical run. All gas lines must be clean and contain no residual drawing oil. Nitrogen, helium, or hydrogen may be used as the **carrier gas** (i.e., mobile phase) to transport the analytes through the GC column. The carrier gas line should have traps (moisture trap, oxygen trap, and hydrocarbon trap) in line to remove any moisture and contaminants from the incoming gas. These traps must be periodically replaced to maintain effectiveness.

14.3.2 Injection Port

14.3.2.1 Hardware

The injection port serves the purpose of providing a place for sample introduction, its vaporization, and possibly some dilution and splitting. Liquid samples make up the bulk of materials analyzed by GC, and they are always done by syringe injection (manual or automated). The injection port contains a soft septum that provides a gas-tight seal but can be penetrated by a syringe needle for sample introduction.

Samples may be introduced into the injection port using a **manual syringe technique** or an **automated sampling system.** Manual sample injection is generally the largest single source of poor precision in GC analysis. Ten-microliter syringes are usually chosen since they are more durable than the microsyringes, and sample injection volumes typically range from 1 to 3 μl. These syringes will hold about 0.6 μl in the needle and barrel (this is in addition to that measured on the barrel). Thus, the amount of sample that is injected into the GC depends upon the proportion of this 0.6 μl that is included in the injection and the ability of the analyst to accurately read the desired sample volume on the syringe barrel. This can be quite variable for the same analyst and be grossly different between analysts. This variability between injections and the small sample volumes injected are the reasons internal (vs. external) standards are common for GC (see Chap. 12, Sect. 12.5.3).

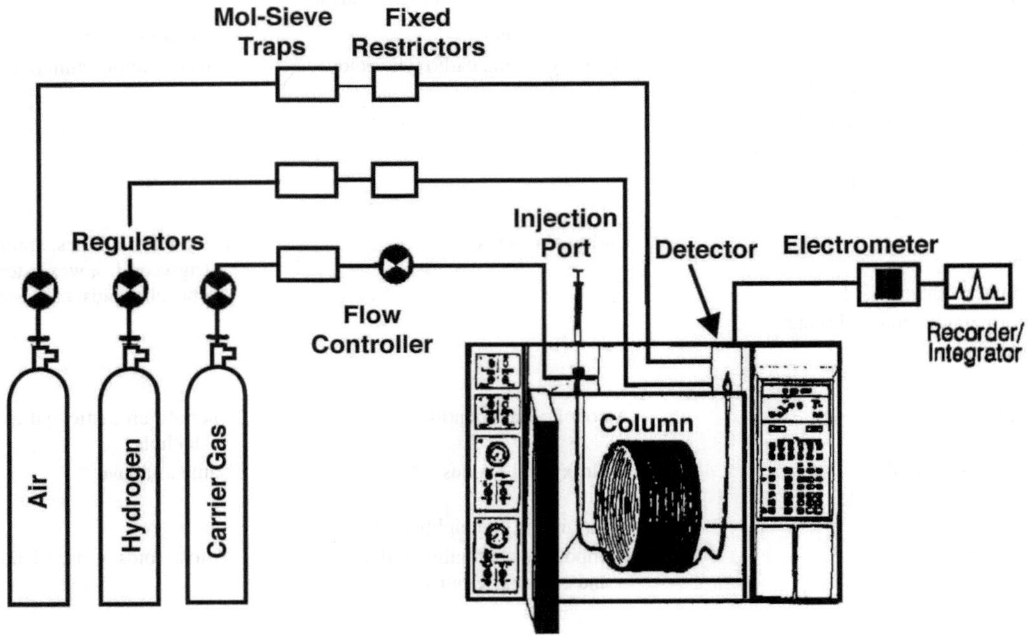

Fig. 14.8 Diagram of a gas chromatographic system. (Courtesy of Agilent Technologies, Inc., Santa Clara, CA)

Table 14.3 Gas chromatographic hardware and operating conditions to be recorded for all GC separations

Parameter	*Description*
Sample	Name and injection volume
Injection	Type of injection [e.g., split versus splitless and conditions (injection port flow rates)]
Capillary column	Phase, length, diameter, film thickness, and manufacturer
Packed columns	Solid support, size mesh, coating, loading (%)
Temperatures	Injector, detector, oven, and any programming information
Carrier gas	Flow rate (velocity) and type
Detector	Type

14.3.2.2 Sample Injection Techniques

The sample must be vaporized in the injection port to pass through the column for separation. This vaporization can occur quickly by flash evaporation (standard injection ports) or slowly in a gentler manner (temperature-programmed injection port or on-column injection). The choice depends upon the thermal stability of the analytes. Due to the various samples as well as instrumental requirements, there are several different designs of injection ports available.

14.3.2.2.1 Split Injection

Capillary columns have limited capacity, and the injection volume may have to be reduced to permit efficient chromatography. The injection port may serve the additional function of splitting the injection so that only a portion of the analyte goes on the column (i.e., **split injection**) (Fig. 14.9, split vent valve open). The injection port is operated at about 20 °C warmer than the maximum column oven temperature (commonly at 250 °C). The sample may be diluted with carrier gas to accomplish a split (1:50 to 1:100 preferred, split ratio = (column flow)/(column flow + venting flow), whereby only a small portion (1 part) of the analyte (more exactly, 1 part of gas flow) goes on the column, and the majority (49/50 or 99/100 parts) of the analytes are vented to the split vent. A high split ratio typically gives a sharp, narrow peak.

14.3.2.2.2 Splitless Injection

To increase the sensitivity, a splitless injection mode can be used. In **splitless injection**, the split vent valve is closed and all of the analyte goes on the column (Fig. 14.5). Similar to the split injection, the temperature of the injector is operated at 20 °C higher than the maximum column oven temperature. Splitless injection requires setting up the initial column temperature 10–20 °C lower than the boiling point of the sample solvent, so the solvent can recondense in the column for acceptable chromatography of early eluting compounds (called solvent effect).

14.3.2.2.3 Programmed Temperature Vaporization Injection

For **programmed temperature vaporization** (PTV) injection ports, the sample is introduced into an ambient temperature port and then it is temperature programmed to some desired temperature. Since the sample is not introduced to the hot injector, the technique is desired for temperature-

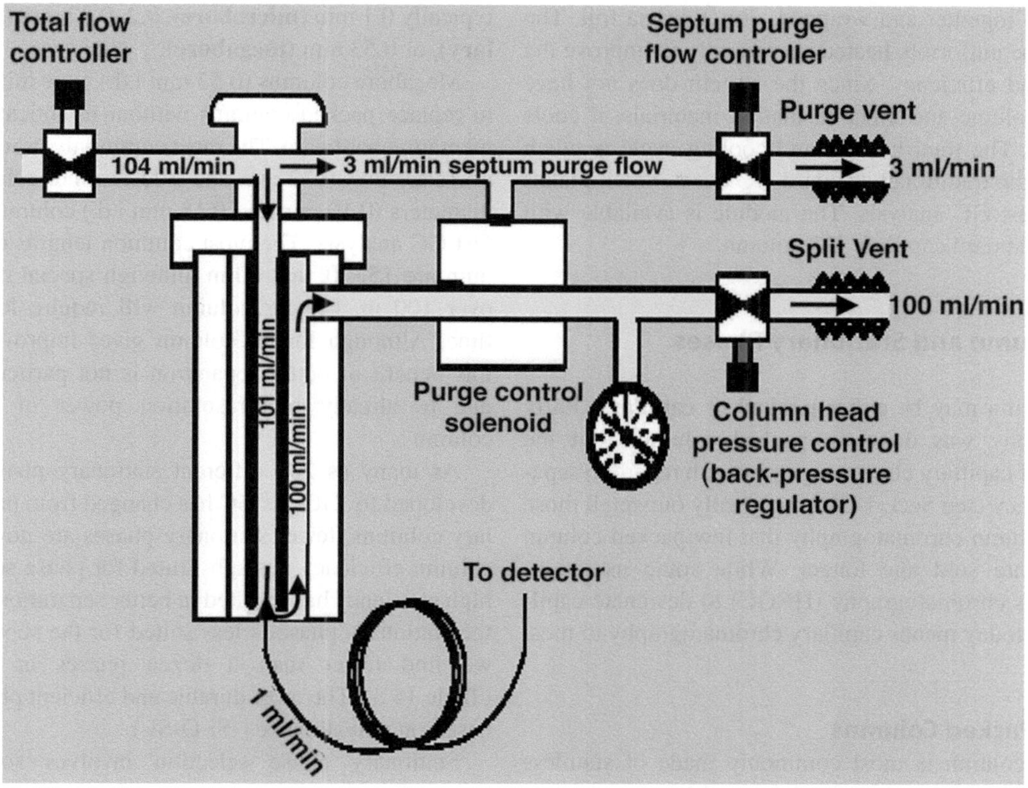

Fig. 14.9 Schematic of a GC injection port. (Courtesy of Agilent Technologies, Inc., Santa Clara, CA)

sensitive analytes. In addition, this technique is very useful to inject large amounts of sample when it is used together with split/splitless injection mode (solvent vent) to increase the sensitivity. For example, 10 µl of liquid sample can be injected at low temperature using a high split ratio to let the solvent vent out, and then the injection mode can be changed to "splitless" as the injector is heated up to evaporate and transfer analytes onto the column.

14.3.2.2.4 On-Column Injection

On-column injection is a technique whereby the sample is directly introduced into the column for which the temperature is at that of the GC oven. The sample is then slowly volatized as the oven heats up. The initial oven temperature needs to be below the boiling point of the solvent. This technique is good for thermally labile analytes.

14.3.2.2.5 Thermal Desorption Injection

The volatiles can be introduced onto the head of a GC column for chromatographic separation directly from food samples through **thermal desorption**. The sample is heated in a thermal desorption unit, and the volatiles are carried through a purge gas to a split/splitless injector. Cryofocusing with liquid nitrogen or CO_2 either in the injector or column is needed to attain sharp peaks. Alternatively, the volatiles can be retained using an absorbent such as a Tenax® trap during the purge stage and then thermally desorbed onto the column. The samples can be extracted with SPME or SBSE techniques described previously (Sect. 14.2.2.5) and then thermally desorbed onto the column for analysis. This technique has gained popularity to analyze volatile aroma compounds in foods including fruits [36, 43] and wine [44].

14.3.3 Oven

The oven controls the temperature of the column. In GC, one takes advantage of both an interaction of the analyte with the stationary phase and the boiling point for separation of compounds. Thus, the injection is often made at a lower oven temperature and is then temperature-programmed to some elevated temperature. While analyses may be done isothermally, temperature-programmed runs are most common. It should be obvious that higher temperatures will cause the sample to elute faster and, therefore, be at a cost of resolution.

Oven temperature program rates can range from as little as 0.1 °C/min to the maximum temperature heating rate that the GC can provide. A rate of 2–10 °C/min is most common.

The capillary column (Sect. 14.3.4.2) can also be directly heated with an insulated heating wire based on **low thermal mass** (LTM) **technology**. A temperature sensor is mounted on the column. The column, the heating wire, and the sensor are all coiled together and wrapped with alumina foil. The column can be uniformly heated very rapidly to improve the separation and efficiency. Since the system does not have much void volume and other insulation materials, it cools very quickly. The total heating and cooling cycle is much shorter than the traditional standard GC oven, which makes it ideal for fast GC analysis. The module is available with almost any standard capillary GC column.

14.3.4 Column and Stationary Phases

The GC column may be either **packed** or **capillary.** Early chromatography was done on packed columns, but the advantages of capillary chromatography with regard to separation efficiency (see Sect. 14.4.2) so greatly outweigh those of packed column chromatography that few packed column instruments are sold any longer. While some use high-resolution gas chromatography (HRGC) to designate capillary GC, GC today means capillary chromatography to most individuals.

14.3.4.1 Packed Columns

The packed column is most commonly made of stainless steel or glass and may range from 1.6 to 12.7 mm in outer diameter and be 0.5–5.0 m long (generally 2–3 m). It is packed with a granular material consisting of a "liquid" coated on an allegedly inert solid support. The **solid support** is most often diatomaceous earth (skeletons of algae) that has been purified, possibly chemically modified (e.g., silane treated), and then sieved to provide a definite mesh size (60/80, 80/100, or 100/120).

The liquid loading is usually applied to the solid support at 1–10% by weight of the solid support. While the liquid coating can be any one of the approximately 200 available, the most common are silicone-based phases (methyl-, phenyl-, or cyano-substituted) and Carbowax™ (ester based).

The liquid phase and the percent loading are determined by the analysis desired. The choice of liquid is typically such that it is of similar polarity to the analytes to be separated. Loading influences the time of analysis (retention time is proportional to loading), resolution (generally improved by increasing phase loading, within limits), and "bleed." The liquid coatings are somewhat volatile and will be lost from the column at high temperatures (this is dependent upon the phase itself). This results in an increasing baseline **(column bleeding)** during temperature programming.

14.3.4.2 Capillary Columns

The capillary column is a hollow fused silica glass (<100 ppm impurities) tube ranging in length from 5 to 100 m. The walls are so thin, ca. 25 μm, that they are flexible. The column outer walls are coated with a polyamide material to enhance strength and reduce breakage. Column inner diameters are typically 0.1 mm (**microbore**), 0.2–0.32 mm (**normal capillary**), or 0.53 mm (**megabore**).

Megabore columns (0.53 mm i.d.) were initially designed to replace packed columns without modification of instrumentation hardware. The most commonly used capillary columns are now 0.32 mm and 0.25 mm i.d. columns. Smaller diameters (0.10 mm and 0.18 mm i.d.) columns are used for fast GC analysis. The most common lengths of the GC column are 15, 30, and 60 m although special column can be over 100 m. Longer column will require longer analysis time. Although longer column gives improved resolution, this benefit of better separation is not particularly obvious due to already high-resolution power of capillary GC column.

As many as 200 different stationary phases have been developed for GC. As GC has changed from packed to capillary columns, fewer stationary phases are now in use since column efficiency has substituted for phase selectivity (i.e., high efficiency has resulted in better separations even though the stationary phase is less suited for the separation). Now, we find fewer than a dozen phases in common use (Table 14.3). The most durable and efficient phases are those based on polysiloxane (-Si-O-Si-).

Stationary phase selection involves some intuition, knowledge of chemistry, and help from the column manufacturer and the literature. There are general rules, such as choosing polar phases to separate polar compounds and the converse, or phenyl-based column phase to separate aromatic compounds. However, the high efficiency of capillary columns often results in separation even though the phase is not optimal. For example, a 5% phenyl-substituted methyl silicone phase applied to a capillary column will separate most polar and nonpolar compounds (Table 14.4).

Liquid coating is chemically bonded to the glass walls of capillary columns and internally cross-linked at phase thicknesses ranging from 0.1 to 5 μm. Film thickness directly affects separation. Thicker films retain compounds longer in the stationary phase; thus, the analytes will have longer interaction with the stationary phase to achieve separation. Generally, a thick-filmed column should be used to separate very volatile compounds. For example, an FFAP (polyethylene glycol treated with nitroterephthalic acid) column with 1-μm film thickness can effectively hold and separate hydrogen sulfide and other highly volatile sulfur compounds [24]. However, a thick film will also give higher baseline due to bleeding. A thin-film (0.25 μm) column is usually used to separate high-molecular-weight compounds; the analytes will stay in the stationary phase less time. A thin-film column also has less bleeding at high temperature, so it is used frequently for GC-MS.

Most compounds can be separated using nonpolar 5% phenyl, 95% dimethylpolysiloxane-based columns (e.g., DB-5, Agilent-5, RTX-5). This type of column has a very wide temperature range (−60 °C to 325 °C) and is very sta-

Table 14.4 Common stationary phases

Composition	*Polarity*	*Applications*[a]	*Phases with similar McReynolds constants*[b]	*Temperature limits*[c]
100% Dimethyl polysiloxane (gum)	Nonpolar	Phenols, hydrocarbons, amines, sulfur compounds, pesticides, PCBs	OV-1 SE-30	−60 °C–325 °C
100% Dimethyl polysiloxane (fluid)	Nonpolar	Amino acid derivatives, essential oils	OV-101, SP-2100	0 °C–280 °C
5% Phenyl 95% dimethyl polysiloxane	Nonpolar	Fatty acids, methyl esters, alkaloids, drugs, halogenated compounds	SE-52 OV-23 SE-54	−60 °C–325 °C
14% Cyanopropylphenyl methyl polysiloxane	Intermediate	Drugs, steroids, pesticides	OV-1701	−200 °C–280 °C
50% Phenyl, 50% methyl methyl polysiloxane	Intermediate	Drugs, steroids, pesticides, glycols	OV-17	60 °C–240 °C
50% Cyanopropylmethyl, 50% phenylmethyl polysiloxane	Intermediate	Fatty acids, methyl esters, alditol acetates	OV-225	60 °C–240 °C
50% Trifluoropropyl polysiloxane	Intermediate	Halogenated compounds, aromatics	OV-210	45 °C–240 °C
Polyethylene glycol-TPA modified	Polar	Acids, alcohols, aldehydes, acrylates, nitrites, ketones	OV-351 SP-1000	60 °C–240 °C
Polyethylene glycol	Polar	Free acids, alcohols, esters, essential oils, glycols, solvents	Carbowax 20 M	60 °C–220 °C

[a]Specific application notes from column suppliers provide information for choosing a specific column
[b]McReynolds constants are used to group stationary phases together on the basis of separation properties
[c]Stationary phases have both upper and lower temperature limits. Lower temperature limit is often due to a phase change (liquid to solid) and upper temperature limit to a volatilization of phase

ble. However, to separate very polar compounds such as alcohols and free fatty acids, a polar column is needed such as WAX (polyethylene glycol) or FFAP. A WAX type of column has superior separation power; however, it has a narrow usable temperature range (40 °C to 240 °C). It becomes solid (lost separation power) at low temperature and bleeds highly at high temperature. It is also sensitive to residue oxygen, so it deteriorates quickly if oxygen is not removed from the carrier gas. Cyanopropyl-based columns (SP-2560, CP-Sil 88) are good for *trans*-fatty acid esters. Other specialty phase columns have been developed to improve specific resolution. Ionic liquid-based GC columns [45, 46] have very good thermal stability and can be used to analyze very polar compounds. Ionic liquid column has been used to separate fatty acid methyl esters [47]. However, their applications in food systems are still limited. The ionic liquid column cannot easily replace a WAX-based column due to the strong interactions with acids (octanoic and decanoic acids), resulting in poor chromatography for these compounds. The β-cyclodextrin-based column is useful to separate chiral isomers of essential oil and other volatile compounds [48].

14.3.4.3 Gas–Solid (PLOT) Chromatography

Gas–solid chromatography is a very specialized area of chromatography accomplished without using a liquid phase—the analyte interaction is with a porous material. This material has been applied both to packed and capillary columns. For the capillary column, the porous material is chemically or physically (by deposition) coated on the inner wall of the capillary and the column is called a **porous-layer open-tabular** (PLOT) column. The most popular porous materials are alumina oxide, carbon, molecular sieve, and synthetic polymer such as Poropak® or Chromosorb® (trade names of polymers based on vinyl benzene). Separations usually involve water or other very volatile compounds such as headspace gas composition (N_2, O_2, CO_2, and CO) in packaged food and ethylene during fruit ripening and storage.

14.3.5 Detectors

There are numerous detectors available for GC, each offering certain advantages in either sensitivity (e.g., electron capture) or selectivity (e.g., atomic emission detector). The most common detectors are the **flame ionization detector** (FID), **thermal conductivity** (TCD), **electron capture** (ECD), **flame photometric** (FPD), **pulsed flame photometric detector** (PFPD), **photoionization** (PID) detector, and **mass spectrometry** (MS) (see Chap. 11 on MS). The operating principles and food applications of these detectors are discussed below. The characteristics of these detectors are summarized in Table 14.5.

Not described in detail below with traditional GC detectors, but mentioned here is **GC-olfactory** (GC-O). In GC systems with an olfactory detector outlet, the column effluent is split so that a portion of the effluent goes to a "sniffing port" and the remainder goes to a GC detector. The sniffing port typically consists of a glass cone for the operator to use their nose to identify "aroma-active" components eluting from the column (see Chap. 35, Food Forensic Investigation, Sect. 35.5.2.1).

Table 14.5 Characteristics of most common detectors for gas chromatography

Characteristic	*Thermal conductivity detector*	*Flame ionization detector*	*Electron capture detector*	*Flame photometric detector*	*Photoionization detector*
Specificity	Very little. Detects almost anything, including H_2O. Called the "universal detector"	Most organics	Halogenated compounds and those with nitro- or conjugated double bonds	Organic compounds with S or P (determined by which filter is used)	Depends on ionization energy of lamp relative to bond energy of analytes
Sensitivity limits	ca. 400 pg. Relatively poor. Varies with thermal properties of compound	10–100 pg for most organics. Very good	0.05–1 pg. Excellent	2 pg for S and 0.9 pg for P compounds. Excellent	1–10 pg depending on compound and lamp energy. Excellent
Linear range	10^4, poor. Response easily becomes nonlinear	10^6–10^7. Excellent	10^4. Poor	10^4 for P. 10^3 for S	10^7. Excellent

14.3.5.1 Thermal Conductivity Detector

14.3.5.1.1 Operating Principles

As the carrier gas passes over a hot filament (tungsten), it cools the filament at a certain rate depending on carrier gas velocity and composition. The temperature of the filament determines its resistance to electrical current. As a compound elutes with the carrier gas, the cooling effect on the filament is typically less, resulting in a temperature increase in the filament and an increase in resistance that is monitored by the GC electronics. Older-style TCDs used two detectors and two matching columns; one system served as a reference and the other as the analytical system. Newer designs use only one detector (and column), which employs a carrier gas switching value to pass alternately carrier gas or column effluent though the detector (Fig. 14.10). The signal is then a change in cooling of the detector as a function of which gas is passing through the detector from the analytical column or carrier gas supply (reference gas flow).

The choice of carrier gas is important since differences between its thermal properties and the analytes determine response. While hydrogen is the best choice, He is most commonly used since H_2 is flammable.

14.3.5.1.2 Applications

The most valuable properties of this detector are that it is *universal* in response and nondestructive to the sample. Thus, it is used in food applications for which there is no other detector that will adequately respond to the analytes (e.g., water, permanent gases, and CO) or when the analyst wishes to recover the separated compounds for further analysis [e.g., trap the column effluent for infrared, nuclear magnetic resonance (NMR), or sensory analysis]. It does not find broad use because it is relatively insensitive, and often the analyst desires specificity in detector response to remove interfering compounds from the chromatogram. The detector is most useful for gas composition analysis (CO_2, CO, O_2, and N_2) for food packaging and fruit ripening and shelf life.

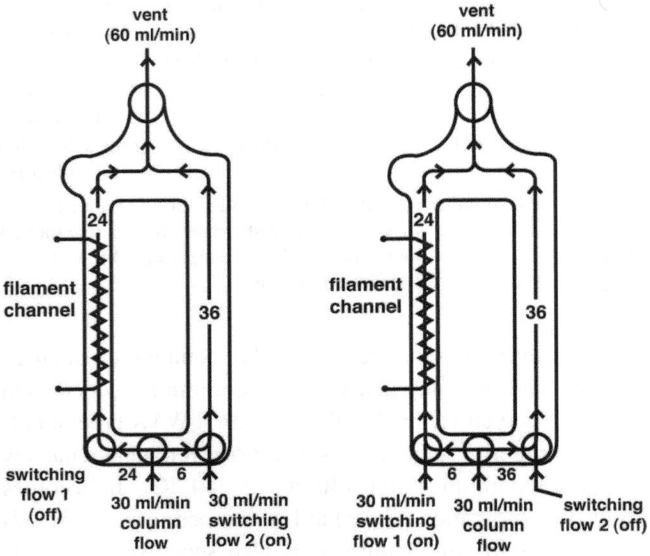

Fig. 14.10 Schematic of the thermal conductivity detector. (Courtesy of Agilent Technologies, Inc., Santa Clara, CA)

14.3.5.2 Flame Ionization Detector

14.3.5.2.1 Operating Principles

As compounds elute from the analytical column, they are burned in a hydrogen flame (Fig. 14.11). A potential (often 300 volts) is applied across the flame. The flame will carry a current across the potential which is proportional to the organic ions present in the flame from the burning of an organic compound. The current flowing across the flame is amplified and recorded. The FID responds to organics on a weight basis. It gives virtually no response to H_2O, NO_2, CO_2, and H_2S and limited response to many other compounds. Response is best with compounds containing C-C or C-H bonds. The FID can be modified to include a methanizer to convert CO and CO_2 to methane using Ni catalyst and then detected by FID. An alternative is to oxidize all compounds to CO_2 first and then reduce to methane before FID. This approach increases the sensitivity and detectability for many carbon-containing compounds. In addition, since all com-

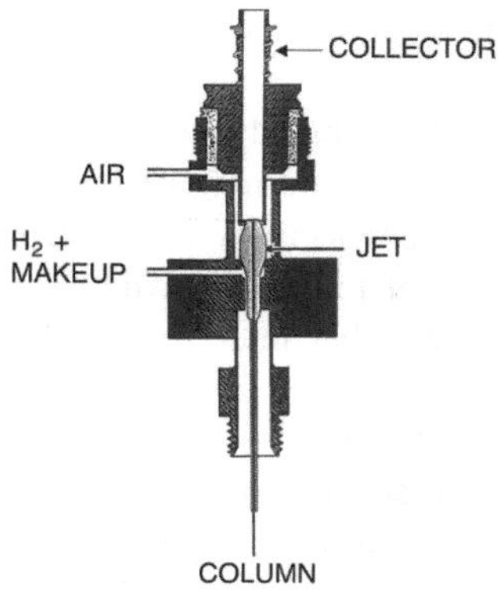

Fig. 14.11 Schematic of the flame ionization detector designed for use with capillary columns. (Courtesy of Agilent Technologies, Inc., Santa Clara, CA)

pounds are converted to methane and have the same response on FID, the detector allows for the analysis of compounds without using standards.

14.3.5.2.2 Applications

The food analyst is most often working with organic compounds, to which this detector responds well. Its' very good sensitivity, wide linear range in response (necessary in quantitation), and dependability make this detector the choice for most food work. Thus, this detector is used for virtually all food analyses for which a specific detector is not desired or sample destruction is acceptable (column eluant is burned in flame). This includes flavor studies, fatty acid analysis, carbohydrate analysis, sterols, contaminants in foods, and antioxidants.

14.3.5.3 Electron Capture Detector

14.3.5.3.1 Operating Principles

The ECD contains a radioactive foil coating that emits electrons as it undergoes decay (Fig. 14.12). The electrons are collected on an anode, and the standing current is monitored by instrument electronics. As an analyte elutes from the GC column, it passes between the radioactive foil and the anode. Compounds that capture electrons reduce the standing current and thereby give a measurable response. Halogenated compounds or those with high electrophilic compounds (diketones) give the greatest detector response. Unfortunately, this detector becomes saturated quite easily and thus has a very limited linear response range.

14.3.5.3.2 Applications

In food applications, the ECD has found its greatest use in determining PCBs and pesticide residues (see Chap. 33). The specificity and sensitivity of this detector make it ideal for this application. It is also used to analyze diacetyl and other

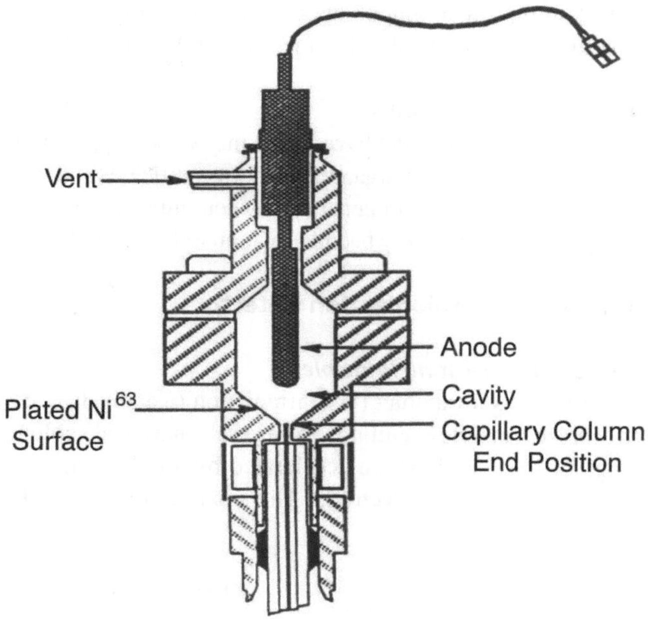

Fig. 14.12 Schematic of the electron capture detector. (Courtesy of Agilent Technologies, Inc., Santa Clara, CA)

vicinal diketone in beer because ECD is very sensitive for vicinal diketones.

14.3.5.4 Flame Photometric Detector and Pulsed Flame Photometric Detector

14.3.5.4.1 Operating Principles

The FPD works by burning all analytes eluting from the analytical column and then measuring specific wavelengths of light that are emitted from the flame using a filter and photometer (Fig. 14.13). The wavelengths of light that are suitable in terms of intensity and uniqueness are characteristic of sulfur (S) and phosphorus (P). Thus, this detector gives a greatly enhanced signal for these two elements (several thousand folds for S- or P-containing organic molecules versus non-S or P-containing organic molecules). Detector response to S-containing molecules is nonlinear, and thus quantification must be done with care.

The PFPD is very similar to the FPD. Unlike traditional flame photometric detection (FPD), which uses a continuous flame, the PFPD ignites, propagates, and self-terminates two to four times per second (Fig. 14.14). Specific elements have their own emission profile: hydrocarbons will complete emission early, while sulfur emissions begin at a relatively later time after combustion. Therefore, a timed "gate delay" can selectively allow for only emissions due to sulfur to be integrated, producing a clean chromatogram. This timed "gate delay" greatly improves the sensitivity. The PFPD can detect sulfur-containing compounds at a much lower detection limit than nearly all other methods of detection [49].

14.3.5.5.2 Applications

Both the FPD and PFPD have found major food applications in the determination of organophosphorus pesticides and volatile sulfur compounds in general. The determination of sulfur compounds has typically been in relation to flavor studies.

14.3.5.5 Photoionization Detector

14.3.5.5.1 Operating Principles

The PID uses ultraviolet (UV) irradiation (usually 10.2 eV) to ionize analytes eluting from the analytical column (Fig. 14.15). The ions are accelerated by a polarizing electrode to a collecting electrode. The small current formed is magnified by the electrometer of the GC to provide a measurable signal.

This detector offers the advantages of being quite sensitive and nondestructive and may be operated in a selective response mode. The selectivity comes from being able to control the energy of ionization, which will determine the classes of compounds that are ionized and thus detected.

14.3.5.5.2 Applications

The PID finds primary use in analyses for which excellent sensitivity is required from a nondestructive detector. This is most often a flavor application in which the analyst wishes to smell the GC effluent to determine the sensory character of the individual GC peaks. While this detector might find broader use, the widespread availability of the FID (which is suitable for most of the same applications) meets most of these needs.

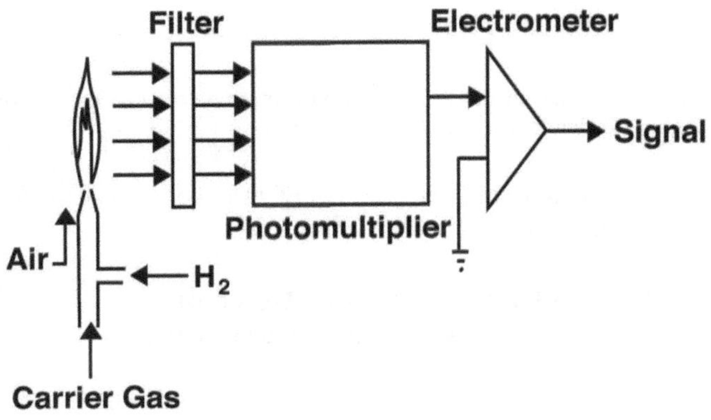

Fig. 14.13 Schematic of the flame photometric detector. (Courtesy of Agilent Technologies, Inc., Santa Clara, CA)

14.3.5.6 Electrolytic Conductivity Detector

14.3.5.6.1 Operating Principles

Compounds entering the **electrolytic conductivity detector** (ELCD) are mixed with a reagent gas (oxidizing or reducing depending on the analysis) in a nickel reaction tube producing ionic species. These products are mixed with a deionized solvent, interfering ions are scrubbed from the effluent, and the ionic analyte-transformation product is detected within the electrolyte conductivity cell. This detector can be used for the specific detection of sulfur-, nitrogen-, or halogen-containing molecules. For example, when operated in the nitrogen mode, the analyte is mixed with H_2 gas and hydrogenated over a nickel catalyst at

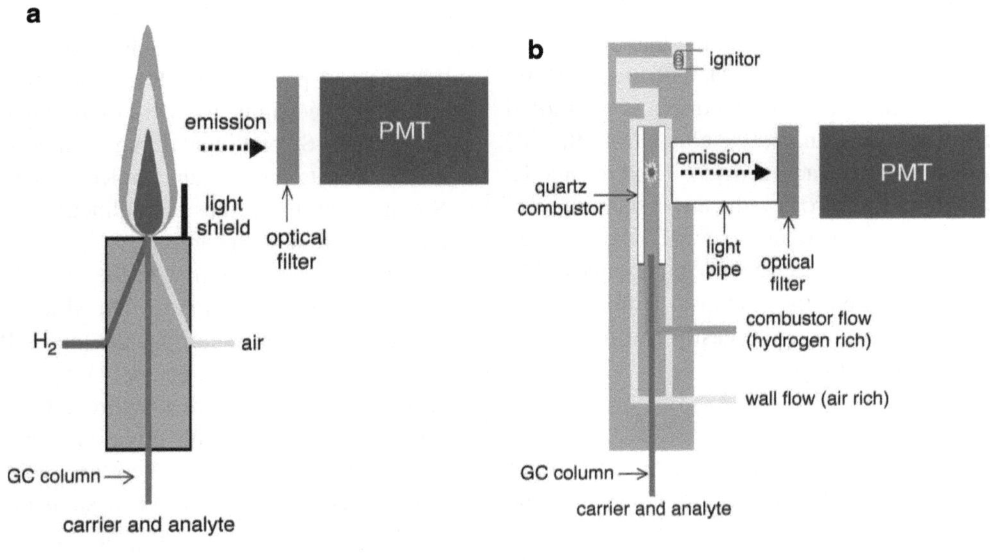

Fig. 14.14 Comparison of (**a**) flame photometric detector and (**b**) pulsed flame photometric detector. (Courtesy of Varian Inc., Palo Alto, CA)

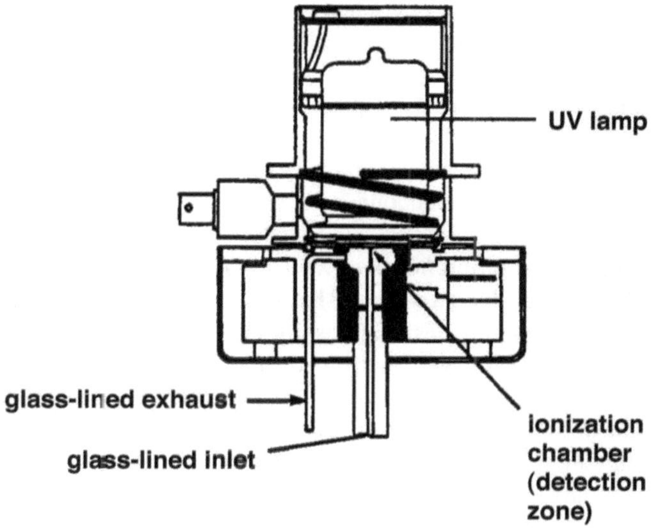

Fig. 14.15 Schematic of the photoionization detector. (Courtesy of Agilent Technologies, Inc., Santa Clara, CA)

850 °C. Acidic hydrogenation products are removed from the effluent by passage through an $Sr(OH)_2$ trap, and the NH_3 from the analyte passes to the conductivity cell where it is measured [50].

14.3.5.6.2 Applications

This detector can be used in many applications for which element specificity is desired. Examples would be pesticide, herbicide, nitrosamine, or flavor analysis. The ELCD is very selective and quite sensitive, having detection limits of 0.1–1 pg for chlorinated compounds, 2 pg for sulfur, and 4 pg for nitrogen.

14.3.5.7 Thermionic Detector

14.3.5.7.1 Operating Principles

The thermionic detector (also called the nitrogen phosphorus detector, NPD) is a modified FID in which a nonvolatile ceramic bead is used to suppress the ionization of hydrocarbons as they pass through a low-temperature fuel-poor hydrogen plasma. The ceramic bead is typically composed of rubidium that is heated to 300–350 °C. Most commonly, this detector is used for the selective detection of nitrogen- or phosphorus-containing compounds. It does not detect inorganic nitrogen or ammonia.

14.3.5.7.2 Applications

This detector is primarily used for the measurement of specific classes of flavor compounds, nitrosamines, amines, and pesticides.

14.3.5.8 Hyphenated Gas Chromatographic Techniques

Hyphenated gas chromatographic techniques are those that combine GC with another major technique. Examples are **GC-AED** (atomic emission detector), **GC-FTIR** (Fourier transform infrared detector), and **GC-MS** (mass spectrometry). While all of the techniques are established methods of analysis in themselves, they become powerful tools when combined with a technique such as GC. GC provides the separation, and the hyphenated technique provides the detector. GC-MS has long been known to be a most valuable tool for the identification of volatile compounds (see Chap. 11). The MS, however, may perform the task of serving as a specific detector for the GC by selectively focusing on ion fragments unique to the analytes of interest. The analyst can detect and quantify components without their gas chromatographic resolution in this manner. The same statements can be made about GC-FTIR (see Chap. 9). The FTIR can readily serve as a GC detector.

In GC-AED, the GC column effluent enters a microwave-generated helium plasma that excites the atoms present in the analytes. The atoms emit light at their characteristic wavelengths. This results in a very sensitive and specific elemental detector.

14.3.5.9 Multidimensional Gas Chromatography

Multidimensional gas chromatography (MDGC) greatly increases the separation ability of gas chromatography [51]. By simply coupling two GC columns, each of opposite polarity, an overall improvement in separation can be accomplished [52]. However, this tandem operation of GC columns does not actually represent multidimensionality but rather resembles the use of a mixed-stationary phase column [51]. True MDGC involves a process known as orthogonal separation for which a sample is first dispersed by one column, and the simplified subsamples are then applied onto another column for further separation. MDGC techniques can be generally divided into two classes: (1) conventional, or "heart-cut," MDGC and (2) **comprehensive two-dimensional gas chromatography** (GC × GC).

14.3.5.9.1 Conventional Two-Dimensional GC

Conventional two-dimensional GC is achieved by using coupled capillary columns for which a small portion, or heart-cut, of the effluent from the first ("pre-separation") column is transferred to the second ("analytical") column. The concept of conventional MDGC is almost identical to that of preparative GC operations, for which one column is used to obtain a partially separated fraction of a complex aroma mixture, which is then reinjected into another GC column, usually with an opposite stationary phase, for further separation. The only difference is that with MDGC, there are no requirements for manual collection of the effluent obtained from the pre-separation column since the two columns are directly connected.

Because the second column in the MDGC system is only injected with a small portion of the total sample at one time, a large quantity of the sample can be injected into the first column without the worry of chromatographic band smearing during analytical separations [53]. Therefore, trace compounds can be easily enriched for more successful detection and identification.

The MDGC technique is particularly useful to study enantiomers of flavor compounds. The interested compound can be "heart-cut" and transferred to an analytical column with an enantioselective stationary phase for good separation of targeted chiral compounds.

14.3.5.9.2 Comprehensive Two-Dimensional GC

Comprehensive two-dimensional GC is among the most powerful two-dimensional gas chromatographic techniques that have been developed to date (Fig. 14.16). Unlike conventional MDGC in which only particular segments are transferred from the pre-separation column onto the analytical column, comprehensive MDGC, or GC × GC, involves the transfer of the entire effluent from the first column onto a second column by way of a modulation interface so that complete two-dimensional data can be obtained for the entire run of the first column. The operation of the modulator involves the generation of narrow injection bands from the first column, which are continuously, but individually, sent to the secondary column for final separation. GC × GC requires that the second column can operate quickly enough to generate a complete set of data during the time that a single peak elutes from the first GC column, generally within 5 s [51,

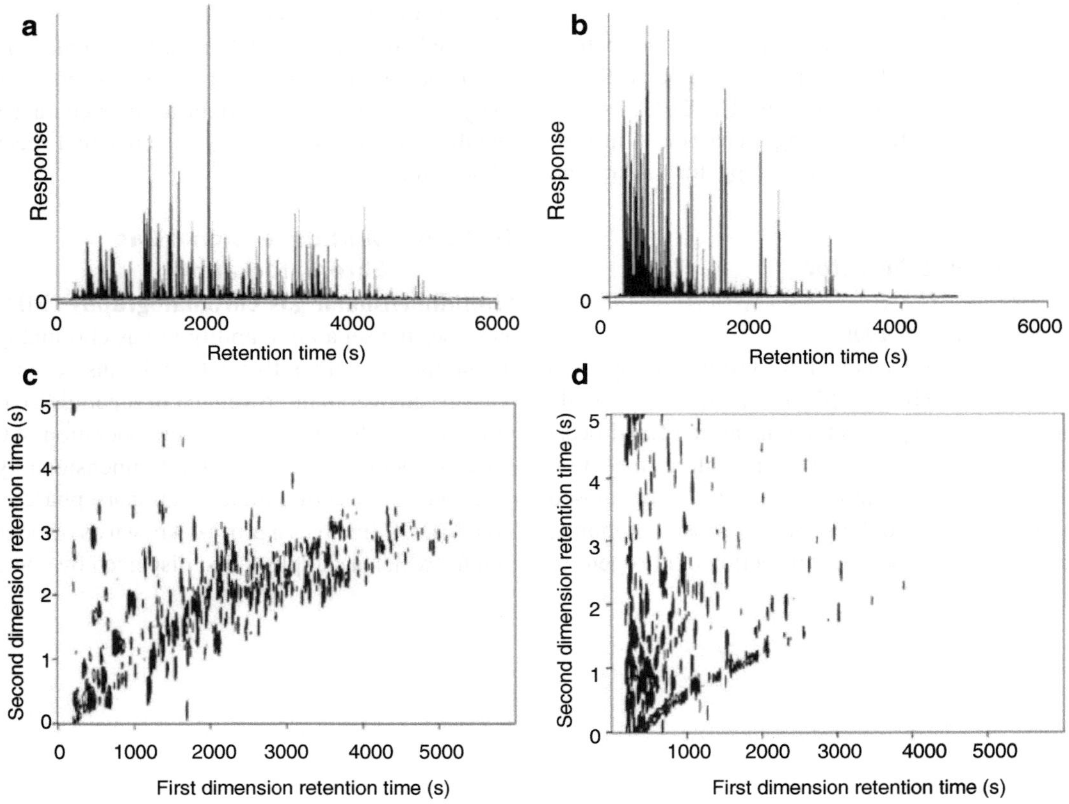

Fig. 14.16 Total ion chromatograms and their respective two-dimensional contour plots for an Arabica coffee extract separated by GC × GC using two different column sets: (**a** and **c**) polar × nonpolar and (**b** and **d**) nonpolar × polar. (Reprinted from Ref. Ryan et al. [54], used with permission)

55]. The data from both time axes are combined to create a set of coordinates for each peak so that the resultant chromatogram is actually a two-dimensional (2-D) plane rather than a straight line. Peak area information can be obtained by summing the integration over both dimensions.

In GC × GC, the two columns perform independently of each other; therefore, the overall peak capacity becomes the product of the capacities for each column. Because analytes elute from the second column so quickly, data acquisition must be adequately fast enough for proper detection. Time-of-flight mass spectrometry (TOF-MS) and rapid-scanning quadrupole mass spectrometry (qMS) (i.e., MS using time-of-flight or quadrupole mass analyzed) have both been used as effective detection methods for GC × GC to obtain mass spectral information [56, 57].

GC × GC has been used in the field of foods and beverages [58–60]. Although the instrumentation can be quite expensive, the use of comprehensive GC × GC for volatile aroma analysis has exponentially increased over the past few years as methodologies have become more established and systems have become commercially available. Overall, the application of MDGC, both conventional and comprehensive, has allowed for advanced separations of complex aromas to occur by using state-of-the-art instrumentation.

14.4 CHROMATOGRAPHIC THEORY

14.4.1 Introduction

GC may depend on several types (or principles) of chromatography for separation. The principles of chromatographic separations and chromatographic theory are discussed in Chap. 12, Sect. 12.4. For example, **size-exclusion chromatography** is used in the separation of permanent gases such as N_2, O_2, and H_2. A variation of size exclusion is used to separate chiral compounds on cyclodextrin-based columns; one enantiomorphic form will fit better into the cavity of the cyclodextrin than will the other form, resulting in separation. **Adsorption chromatography** is used to separate very volatile polar compounds (e.g., alcohols, water, and aldehydes) on porous polymer columns (e.g., Tenax^R phase). **Partition chromatography** is the workhorse for gas chromatographic separations. There are over 200 different liquid phases that have been developed for GC use over time. Fortunately, the vast majority of separations can be accomplished with only a few of these phases, and the other phases have fallen into disuse. GC depends not only upon adsorption, partition, and/or size exclusion for separation but also upon **analyte boiling point** for additional resolving powers. Thus, the separations accomplished are based on several properties of the analytes. This gives GC virtually unequaled resolution powers as compared to most other types of chromatography (e.g., HPLC, paper, or thin-layer chromatography).

A brief discussion of chromatographic theory will follow. The purpose of this additional discussion is to apply this theory to GC to optimize separation efficiency so that analyses can be done faster, less expensively, or with greater precision and accuracy. If one understands the factors influencing resolution in GC, one can optimize the process and gain in efficiency of operation.

14.4.2 Separation Efficiency

A good separation has narrow-based peaks and ideally, but not essential to quality of data, baseline separation of compounds. This is not always achieved. Peaks broaden as they pass through the column—the more they broaden, the poorer the separation and efficiency. As discussed in Chap. 12 (Sect. 12.5.1.2), a measure of this broadening is **height equivalent to a theoretical plate** (HETP). This term is derived from N, the number of plates in the column, and L, the length of the column. A good packed column might have $N = 5000$, while a good capillary column should have about 3000–4000 plates per meter for a total of 100,000 to 500,000 plates depending on the column length. HETP will range from about 0.1 to 1 mm for good columns.

14.4.2.1 Carrier Gas Flow Rates and Column Parameters

Several factors influence column efficiency (peak broadening). As presented in Chap. 12, these are related by the **Van Deemter equation** (Eq. 14.1): (HETP values should be small):

$$\text{HETP} = Au^{1/3} + B/u + Cu \tag{14.1}$$

where:

HETP = height equivalent to a theoretical plate
A = eddy diffusion
B = band broadening due to diffusion
u = velocity of the mobile phase
C = resistance to mass transfer

A is the eddy diffusion; this is a spreading of the analytes in the column due to the carrier gas having various pathways or nonuniform flow (Fig. 14.17). In capillary chromatography, the A term is relatively very small compared to packed column chromatography. However, as the diameter of the capillary column increases, the flow properties deteriorate and band spreading occurs. The most efficient capillary columns have small diameters (0.1 mm), and efficiency decreases rapidly as one goes to megabore columns (Fig. 14.18). Megabore columns are only slightly more efficient than packed columns. While column efficiency

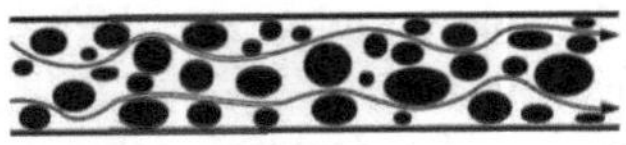

Fig. 14.17 Illustration of flow properties that lead to large eddy diffusion (Term A)

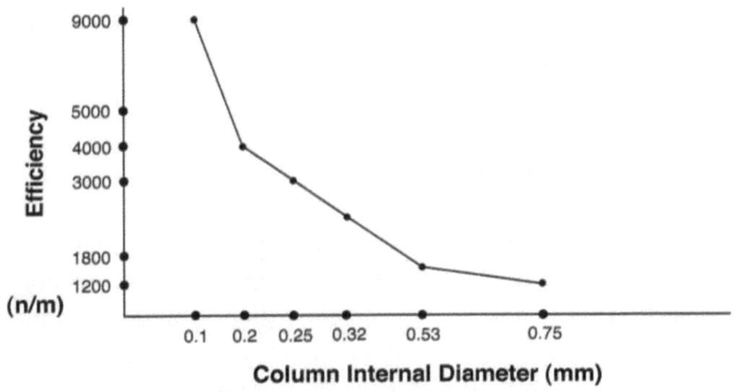

Fig. 14.18 Influence of column diameter on column efficiency (plates/meter). (Courtesy of Agilent Technologies, Inc., Santa Clara, CA)

increases as one goes to smaller columns, column capacity decreases rapidly. Microbore columns are easily overloaded (capacity may be 1–5 ng per analyte), resulting again in poor chromatography. Thus, column diameter is generally chosen as 0.2–0.32 mm to compromise efficiency with capacity.

B is band broadening due to diffusion; analytes will go from a high to a low concentration. The term *u* is the velocity of the mobile phase. Thus, very slow flow rates result in large amounts of diffusion band broadening, and faster flow rates minimize this term. The term *u* is influenced by the carrier gas choice. Larger-molecular-weight carrier gases (e.g., nitrogen) are more viscous than the lighter-molecular-weight gases (e.g., helium or hydrogen), and thus peak spreading is less for nitrogen than for helium or hydrogen carrier gases. This results in nitrogen having the lowest HETP of the carrier gases and theoretically being the best choice for a carrier gas. However, other considerations that will be discussed in Sect. 14.4.2.2 make nitrogen a very poor choice for a carrier gas.

C is the resistance to mass transfer. If the flow (*u*) is too fast, the equilibrium between the phases is not established and poor efficiency results. This can be visualized in the following way: if one molecule of solute is dissolved in the stationary phase and another is not, the undissolved molecule continues to move through the column, while the other is retained. This results in band spreading within the column. Another factor that influences this term is thickness of the stationary phase. Thick films give greater capacity (ability to handle larger amounts of a solute) but at a cost in terms of band spreading (efficiency of separation) since thick films provide more variation in diffusion properties in and out of the stationary phase. Thus, phase thickness is a compromise between maximizing separation efficiency and sample capacity (with too much sample, the column is overloaded and separation ability destroyed). Capillary columns with phase thicknesses of 0.25–1 μm are commonly used for most applications.

If the Van Deemter equation is plotted, giving the figure discussed in Chap. 12 (Fig. 12.13), we see an optimum flow rate due to the opposing effects of the *B* and the *C* terms. It should be noted that the GC may not be operated at a carrier flow velocity yielding maximum efficiency (lowest HETP). Analysis time is directly proportional to carrier gas flow velocity. If the analysis time can be significantly shortened by operating above the optimum flow velocity and adequate resolution is still obtained, velocities well in excess of optimum should be used.

14.4.2.2 Carrier Gas Type

The relationship between HETP and carrier gas flow velocity is strongly influenced by carrier gas choice (Fig. 14.19). Nitrogen is the most efficient (lowest HETP) carrier gas, as discussed in Sect. 14.4.2.1, but its minimum HETP occurs at a very low flow velocity. This low mobile phase velocity results in unnecessarily long analysis times. Considering the data plotted in Fig. 14.19, nitrogen has an HETP of about 0.25 at an optimum flow velocity of 10 cm/s. The HETP of He is only about 0.35 at 40 cm/s flow velocity. This is a small loss in resolution to reduce the analysis time fourfold (10 cm/s for nitrogen versus 40 cm/s for helium). One can potentially even push the flow velocity up to 60 or 70 cm/s and accomplish separation in even shorter times.

The plots in Fig. 14.19 suggest that hydrogen is an even better choice for a carrier gas than helium (i.e., has a flatter relationship between carrier gas flow velocity and HETP, thus providing the most rapid analysis). There have been some concerns voiced about hydrogen being flammable, but broad

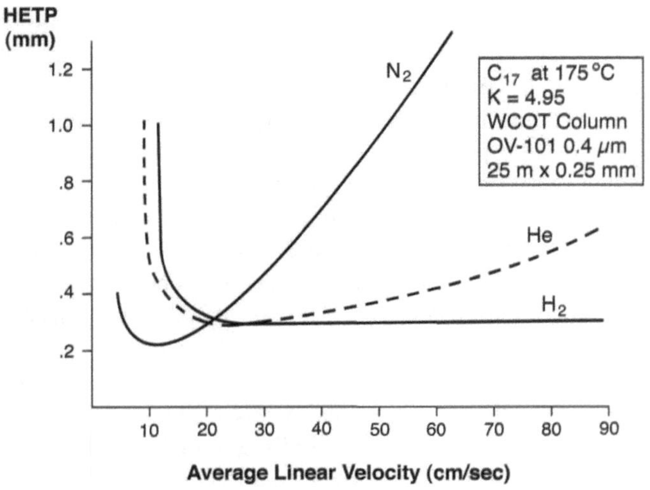

Fig. 14.19 Influence of carrier gas type and flow rate on column efficiency. (Courtesy of Agilent Technologies, Inc., Santa Clara, CA)

usage has not resulted in any fires/explosions being reported to date. H_2:AIR mixtures become flammable in very limited concentration ratios and are actually difficult to ignite.

14.4.2.3 Summary of Separation Efficiency

In summary, an important goal of analysis is to achieve the necessary separation in the minimum amount of time. The following factors should be considered:

1. **Column diameter.** In general, small diameter columns should be used since separation efficiency is strongly dependent on column diameter. While small diameter columns will limit column capacity, limited capacity can often be compensated for by increasing phase thickness. Increased phase thickness will also decrease column efficiency but to a lesser extent than increasing column diameter.
2. **Column operating temperature.** Lower column operating temperatures should be used—if elevated column temperatures are required for the compounds of interest to elute, use a shorter column if resolution is adequate.
3. **Column length.** One should keep columns as short as possible (analysis time is directly proportional to column length—resolution is proportional to the square root of length).
4. **Type of carrier gas.** Use hydrogen as the carrier gas if the detector permits. Some detectors have specific carrier gas requirements.
5. **Flow rate.** Operate the GC at the maximum carrier gas velocity that provides resolution.

The pyramid shown in Fig. 14.20 summarizes the compromises that must be made in choosing the analytical column and gas chromatographic operating conditions. One cannot optimize any given operating conditions and column choices to get one of these properties without compromising another property. For example, optimizing chromatographic resolution (small bore capillary diameter, thin phase coating, long column lengths, and slow or optimum carrier gas flow rate) will be at the cost of capacity (large bore columns and thick phase coating) and speed (thin-film coating, high carrier gas flow velocities, and short columns). Capacity will be at a cost of resolution and speed. The choice of column and operating parameters must consider the needs of the analyst and the compromises involved in these choices.

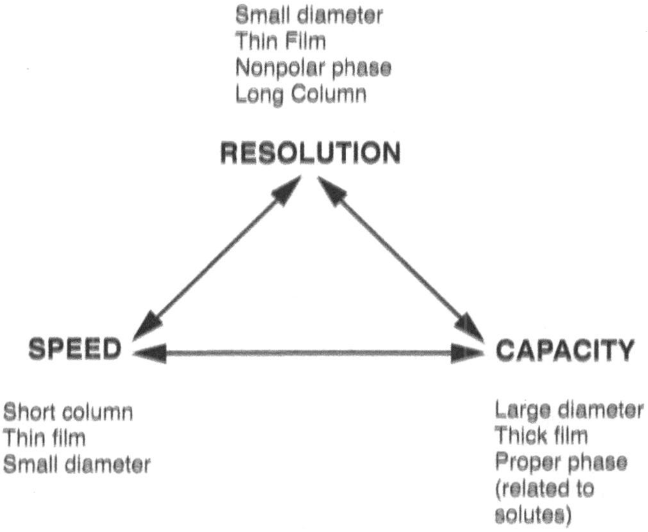

Fig. 14.20 Relationships among column capacity, efficiency, resolution, and analysis speed

14.5 APPLICATIONS OF GC

While Table 14.1 references some applications of GC in other chapters of this text. Two additional examples are presented below to illustrate separations and chromatographic conditions.

14.5.1 Residual Volatiles in Packaging Materials

Residual volatiles in packaging materials can be a problem both from health (if they are toxic) and quality standpoints (produce off-flavors in the food). As the industry has turned from glass to polymeric materials, there have been more problems in this respect. GC is most commonly used to determine the residual volatiles in these materials [61].

The GC chromatogram presented in Fig. 14.21 illustrates the analysis of styrene and ethylbenzene monomers, from polystyrene (PS) packaging, which migrate into foodstuffs (in the current example from yogurt). Peaks 6 and 9 represent ethylbenzene and styrene, respectively. The sample was

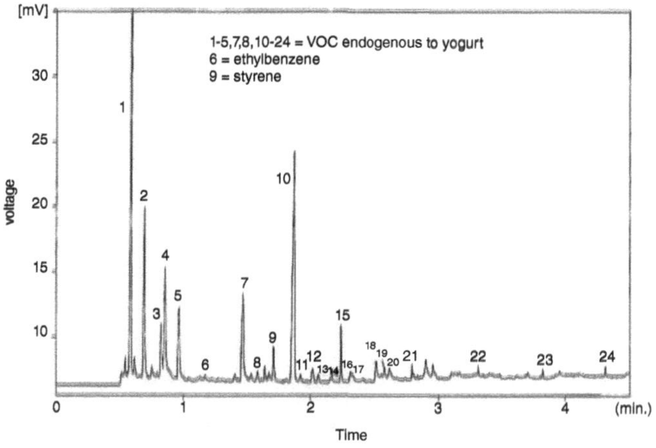

Fig. 14.21 SPME-Fast GC determination of packaging monomers in yogurt. (From Verzera et al. [62], used with permission)

extracted via HS-SPME with a divinylbenzene/carboxen/polydimethylsiloxane fiber and analyzed by fast GC to determine the targeted VOC by headspace isolation [62].

14.5.2 Identification of Unknown Compounds

Aroma is a complex sensation often composed of hundreds of molecules many having similar structures. Mass spectrometry is a key tool in compound identification, but MS data alone is generally not adequate to assure a correct identification. Thus, chromatographic data (compound elution time on a specific GC column) is generally needed as confirmation of compound identity. Ideally, one would make a tentative identification of an unknown compound based on MS data and then compare GC elution data (i.e., **linear retention index**, LRI) with that of the tentatively identified compound. Obtaining the elution time of every identified compound in a food would be tedious. Thus, an alternative approach to comparing retention times of compounds across literature studies has been devised using published "retention indices." The retention indices (often called Kovats Index after the creator of the index) of compounds are determined relative to the elution times of standard alkanes. To determine this number, a researcher would spike the aroma isolate with a homologous series of alkanes, e.g., C5–25, run the GC analysis again and then calculate where in the alkane series the unknown compound(s) elutes relative to the hydrocarbons (Fig. 14.22 shows the process and calculations).

For example, a compound eluting *exactly midway* between C7 and C8 hydrocarbons would be given an LRI (linear retention index) of 750. This is calculated by multiplying the number of carbons in the immediately lower hydrocarbon × 100 (700) and then calculating the fraction it is above the next larger hydrocarbon (0.5 × 100 = 50). In this example, the sum of the two values would be 750.

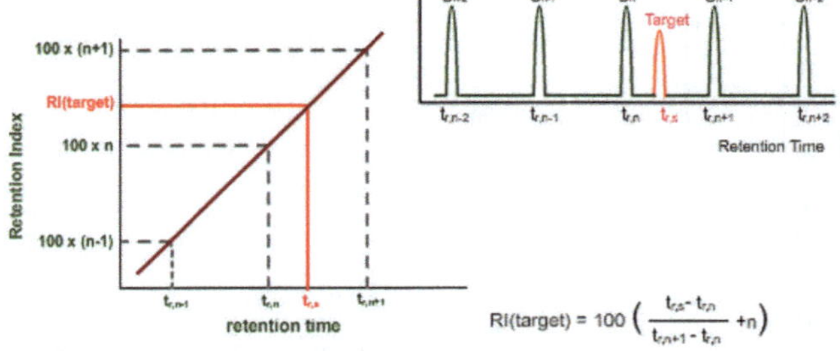

Fig. 14.22 Calculation of a linear retention index. https://www.ssi.shimadzu.com/service-support/technical-support/analysis-basics/gcms/fundamentals/retention/linear_retention_index.html

As noted earlier, one would use the LRI values to support or reject the identity suggested by MS data. The LRI of the compound suggested by the MS can be found in published databases [63]. The published values are gathered from publications from authors around the world so there is some variation in the values but generally, one considers the compound identified if the MS is judged similar and the LRI differs no more than by +/−10. Note that these indices are specific to the column phase (polar – Carbowax, semipolar and polar) but independent of GC oven temperature program.

14.6 SUMMARY

GC has found broad application in both the food industry and academia. It is exceptionally well suited to the analysis of volatile thermally stable compounds. This is due to the outstanding resolving properties of the method and the wide variety of detectors that can provide either sensitivity or selectivity in analysis.

Sample preparation generally involves the isolation of solutes from foods, which may be accomplished by headspace analysis, distillation, preparative chromatography (including solid-phase extraction), or extraction (liquid–liquid). Some analytes can then be directly analyzed, while others must be derivatized prior to analysis.

The gas chromatograph consists of a gas supply and regulators (pressure and flow control), injection port, column and column oven, detector, electronics, and a data recording and processing system. The analyst must be knowledgeable about each of these GC components: carrier and detector gases; injection port temperatures and operation in split, splitless, temperature-programmed, or on-column modes; column choices and optimization (gas flows and temperature profile during separation); and detectors (TCD, FID, NPD, ECD, FPD, and PID). The characteristics of these GC components and an understanding of basic chromatographic theory are essential to balancing the properties of resolution, capacity, speed, and sensitivity.

Unlike most of the other chromatographic techniques, traditional GC has reached the theoretical limits in terms of both resolution and sensitivity. Thus, this method will not change significantly in the future other than for minor innovations in hardware or associated computer software. New development and applications will be more related to multidimensional GC including GC × GC.

GC as a separation technique has been combined with AED, FTIR, and MS as detection techniques to make GC an even more powerful tool. Such hyphenated techniques are likely to continue to be developed and refined, especially in the area of GC × GC-TOF-MS. GC data are required to validate compound identification based on MS.

14.7 STUDY QUESTIONS

1. For each of the following methods to isolate analytes from food prior to GC analysis, describe the procedure, the applications, and the cautions in use of the method:
   (a) Headspace methods
   (b) Distillation methods
   (c) Solvent extraction
2. What is solid-phase extraction and why is it advantageous over traditional liquid–liquid extractions?
3. Why must sugars and fatty acids be derivatized before GC analysis, while pesticides and aroma compounds need not be derivatized?
4. Why is the injection port of a GC at a higher temperature than the oven temperature?
5. You are doing GC with a WAX column and notice that the baseline rises from the beginning to the end of each run. Explain a likely cause for this increase.
6. The most common detectors for GC are TCD, FID, ECD, FPD, and PID. Differentiate each of these with regard to the operating principles. Also, indicate below which detector(s) fits the description given:
   (a) Least sensitive
   (b) Most sensitive
   (c) Least specific
   (d) Greatest linear range
   (e) Nondestructive to sample
   (f) Commonly used for pesticides
   (g) Commonly used for volatile sulfur compounds
7. What types of chromatography does GC rely upon for separation of compounds?
8. In GC, explain why a balance has to be maintained between efficiency and capacity. Also, give an example situation in which you would sacrifice capacity for efficiency.
9. A fellow lab worker is familiar with HPLC for food analysis but not with GC. As you consider each component of a typical chromatographic system (and specifically the components and conditions for GC and HPLC systems), explain GC to the fellow worker by comparing and contrasting it to HPLC. Following that, state in general terms the differences among the types of samples appropriate for analysis by GC versus HPLC, and give several examples of food constituents appropriate for analysis by each (see also Chap. 13).

Acknowledgments The authors of this chapter thank Devin G. Peterson (Department of Food Science and Technology, The Ohio State University, Columbus, OH, USA) for his contributions to the chapter in previous editions of the book.

REFERENCES

1. James, AT, Martin, AJP (1952) Gas-liquid chromatography: the separation and microestimation of volatile fatty acids from formic acid to dodecanoic acid. Biochem J 50:679
2. Poole, CF (Ed.). (2021). Gas chromatography (2nd ed.). Elsevier. https://doi.org/10.1016/C2019-0-02050-X
3. Swinley, JP, De Coning, P., (2019). A practical guide to gas analysis by gas chromatography. Elsevier. eBook ISBN: 9780128188897
4. Wedad, Q L-Bukhaiti, A, Noman, A, Qasim, AS, AL-Farga, A (2017) Gas Chromatography: Principles, Advantages and Applications in Food Analysis. International J. Agriculture Innovations and Research 6(1) ISSN (Online) 2319-1473
5. Gordon, MH (1990) Principles and applications of gas chromatography in food analysis. E. Horwood, New York
6. O'Keeffe M (2000) Residue analysis in food: principles and applications. Harwood Academic, Amsterdam
7. Drawert F, Heimann W, Enberger R, Tressl R (1965) Enzymatische verandrung des naturlichen apfelaromass bei der aurfarbeitung. Naturwissenschaften 52:304
8. Fleming HP,Fore SP, Goldblatt LA (1968) The formation of carbonyl compounds in cucumbers. J Food Sci 33: 572
9. Kazeniak SJ, Hall RM (1970) Flavor chemistry of tomato volatiles. J Food Sci 35:519
10. Leahy MM, Reineccius GA (1984) Comparison of methods for the analysis of volatile compounds from aqueous model systems. In: Schreier P (ed) Analysis of volatiles: New methods and their application. DeGruyter, Berlin
11. Marsili R (1997) Techniques for analyzing food aroma. Marcel Dekker, New York
12. Mussinan CJ, Morello MJ (1998) Flavor analysis. American Chemical Society, Washington DC
13. Sapers GM, Panasiuk O, Talley FB (1973) Flavor quality and stability of potato flakes: Effects of raw material and processing. J Food Sci 38:586
14. Seo EW, Joel DL (1980) Pentane production as an index of rancidity in freeze-dried pork. J Food Sci 45:26
15. Buttery RG, Teranishi R (1963) Measurement of fat oxidation and browning aldehydes in food vapors by direct injection gas-liquid chromatography. J Agric Food Chem 11:504
16. Engel W, Bahr W, Schieberle P (1999) Solvent assisted flavor evaporation-a new and versatile technique for the careful and direct isolation of aroma compounds from complex food matrices. Z Lebensm Unters Forsch 209:237–241
17. Reineccius GA, Keeney PA, Weiseberger W (1972) Factors affecting the concentration of pyrazines in cocoa beans. J Agric Food Chem 20:202
18. Engel W, Bahr W, Schieberle P (1999) Solvent assisted flavour evaporation - a new and versatile technique for the careful and direct isolation of aroma compounds from complex food matrices. Eur Food Res Tech 209:237–241
19. Majors RE (1986) Sample preparation for HPLC and GC using solid-phase extraction. LC--GC 4:972
20. Markel C, Hagen DF, Bunnelle VA (1991) New technologies in solid-phase extraction. LC--GC 9:332
21. Li KM, Rivory LP, Clarke SJ (2006) Solid-phase extraction (SPE) techniques for sample preparation in clinical and pharmaceutical analysis: a brief overview. Curr Pharma Anal 2: 95–102
22. Pawliszyn J (1997) Solid phase microextraction: theory and practice. VCH Publishers, New York
23. Zhang Z, Yang ML, Pawliszyn J (1994) Solid phase-microextraction: a solvent-free alternative for sample preparation. Anal Chem 66:844A–857A
24. Fang Y, Qian MC (2005) Sensitive quantification of sulfur compounds in wine by headspace solid-phase microextraction technique. J Chromatogr A 1080:177–185

25. Verhoeven H, Beuerle T, Schwab W (1997) Solid-phase microextraction: artifact formation and its avoidance. J Chromatogr 46:63–66
26. Song J, Fan L, Beaudry RM (1998) Application of solid phase microextraction and gas chromatography/time-of-flight mass spectrometry for rapid analysis of flavor volatiles in tomato and strawberry fruits. J Agric Food Chem 46:3721–3726
27. Jetti RR, Yang EN, Kurnianta A, Finn C, Qian MC (2007) Quantification of selected aroma-active compounds in strawberries by headspace solid-phase microextraction gas chromatography and correlation with sensory descriptive analysis. J Food Sci 72:S487-S496
28. Kataoka H, Lord HL, Pawliszyn J (2000) Application of solid-phase microextraction in food analysis. J Chromatogr A 880:35–62
29. Vas G, Vekey K (2004) Solid-phase microextraction: A powerful sample preparation tool prior to mass spectrometric analysis. J Mass Spectrom 39:233–254
30. Pfanncoch E, Whitecavage J (2002) Stir bar sorptive extraction capacity and competition effects. Gerstel Global, Baltimore, MD, pp. 1–8
31. David F, Tienpont B, Sandra P (2003) Stir-bar sorptive extraction of trace organic compounds from aqueous matrices. LC-GC Europe 16:410
32. Ou C, Du X, Shellie K, Ross C, Qian MC (2010) Volatile compounds and sensory attributes of wine from cv. Merlot (Vitis vinifera l.) grown under differential levels of water deficit with or without a kaolin-based, foliar reflectant particle film. J Agric and Food Chem 58:12890–12898
33. Song J, Smart RE, Dambergs RG, Sparrow AM, Wells RB, Wang H, Qian MC (2014) Pinot noir wine composition from different vine vigour zones classified by remote imaging technology. Food Chem 153:52–59
34. David F, Sandra P (2007) Stir bar sorptive extraction for trace analysis. J Chromatogr A 1152:54–69
35. Kreck M, Scharrer A, Bilke S, Mosandl A (2001) Stir bar sorptive extraction (SBSE)-enantio-mdgc-ms - a rapid method for the enantioselective analysis of chiral flavour compounds in strawberries. Eur Food Res Technol 213:389–394
36. Malowicki SMM, Martin R,Qian MC (2008) Volatile composition in raspberry cultivars grown in the pacific northwest determined by stir bar sorptive extraction-gas chromatography-mass spectrometry. J Agric Food Chem 56:4128–4133
37. Malowicki SMM, Martin R, Qian MC (2008) Comparison of sugar, acids, and volatile composition in raspberry bushy dwarf virus-resistant transgenic raspberries and the wild type 'meeker' (Rubus idaeus l.). J Agric Food Chem 56:6648–6655
38. Zhou Q, Qian Y, Qian MC (2015) Analysis of volatile phenols in alcoholic beverage by ethylene glycol-polydimethylsiloxane based stir bar sorptive extraction and gas chromatography–mass spectrometry. J Chromatogr A 1390:22–27
39. Dupuy HP, Fore SP, Goldblatt LA (1971) Elution and analysis of volatiles in vegetable oils by gas chromatography. J Amer Oil Chem Soc 48:876
40. Legendre MG, Fisher GS,Fuller WH, Dupuy HP,Rayner ET (1979) Novel technique for the analysis of volatiles in aqueous and non-aqueous systems. J Amer Oil Chem Soc 56:552
41. Widmer HM (1990) Recent developments in instrumental analysis. In Flavor science and technology, Bessiere YThomas AF, Eds. John Wiley & Sons: Chichester, England, 1990; p 181.
42. Blau K, King GS (1993) Handbook of derivatives for chromatography, Vol 2. Wiley, New York.
43. Du X, Qian M (2008) Quantification of 2,5-dimethyl-4-hydroxy-3(2h)-furanone using solid-phase extraction and direct microvial insert thermal desorption gas chromatography-mass spectrometry. J Chromatogr. A 1208:197–201
44. Fang Y, Qian MC (2006) Quantification of selected aroma-active compounds in pinot noir wines from different grape maturities. J Agric Food Chem 54:8567–8573
45. Armstrong DW, He L, Liu Y-S (1999) Examination of ionic liquids and their interaction with molecules, when used as stationary phases in gas chromatography. Anal Chem 71:3873–3876
46. Berthod A, Ruiz-Angel M, Carda-Broch S (2008) Ionic liquids in separation techniques. J Chromatogr A 1184:6–18
47. Delmonte P, Kia A-RF, Kramer JK, Mossoba MM, Sidisky L, Rader JI (2011) Separation characteristics of fatty acid methyl esters using slb-il111, a new ionic liquid coated capillary gas chromatographic column. J Chromatogr A 1218:545–554
48. Bicchi C, Manzin V, D'Amato A, Rubiolo P (1995) Cyclodextrin derivatives in gc separation of enantiomers of essential oil, aroma and flavour compounds. Flavour and Fragrance J 10:127–137
49. Amirav A, Jing H (1995) Pulsed flame photometer detector for gas chromatography. Anal Chem 67:3305–3318
50. Buffington R, Wilson MK (1987) Detectors for gas chromatography. Hewlett-Packard Corp., Avondale, PA
51. Shellie R, Marriott P (2003) Opportunities for ultra-high resolution analysis of essential oils using comprehensive two-dimensional gas chromatography: a review. Flavour Fragrance J 18:179–191
52. Merritt C (1971) Application in flavor research. In: Zlatkis A, Pretorius V, (Eds) Preparative gas chromatography, Wiley-Interscience, New York, 1971; pp. 235–276
53. Kempfert KD (1989) Evaluation of apparent senitivity enhancement in gc/ftir using multidimensional gc techniques. J Chromatogr Sci 27:63–70
54. Ryan D, Shellie R, Tanchida P, Cassilli A, Mondello L, Marriott P (2004) Analysis of roasted coffee bean volatiles by using comprehensive two-dimensional gas chromatography-time-of-flight mass spectrometry. J Chromatography A 1054:57–65
55. Phillips JB, Xu J (1995) Comprehensive multi-dimensional gas chromatography (review). J Chromatogr A 703:327–334
56. Shellie R, Mondello L, Marriott P, Dugo G (2002) Characterisation of lavender essential oils by using gas chromatography-mass spectrometry with correlation of linear retention indices and comparison with comprehensive two-dimensional gas chromatography. J Chromatogr 970:225–234
57. Adahchour M, Brandt M, Baier H-U, Vreuls RJJ, Batenburg AM, Brinkman UAT (2005) Comprehensive two-dimensional gas chromatography coupled to a rapid-scanning quadrupole mass spectrometer: Principles and applications. J Chromatogr A 1067:245–254
58. Campo E, Cacho J, Ferreira V (2007) Solid phase extraction, multidimensional gas chromatography mass spectrometry determination of four novel aroma powerful ethyl esters. J Chromatogr A 1140:180–188
59. Campo E, Ferreira V, Lopez R, Escudero A, Cacho J (2006) Identification of three novel compounds in wine by means of a laboratory-constructed multidimensional gas chromatographic system. J Chromatogr A 1122:202–208
60. Adahchour M, Wiewel J, Verdel R, Vreuls RJJ, Brinkman UAT (2005) Improved determination of flavour compounds in butter by solid-phase (micro)extraction and comprehensive two-dimensional gas chromatography. J Chromatogr A 1086:99–106
61. Hodges K (1991) Sensory-directed analytical concentration techniques for aroma-flavor characterization and quantification. In: Risch SJ, Hotchkiss JH (Eds), Food packaging interactions ii, American Chemical Society, Washington, DC. p 174
62. Verzera A, Condurso C, Romeo V, Tripodi G, Ziino M. (2010). Solid-phase microextraction coupled to fast gas chromatography for the determination of migrants from polystyrene-packaging materials into yoghurt. Food Anal Methods 3(2):80–84
63. PubChem 2023 https://pubchem.ncbi.nlm.nih.gov/compound

Part IV

Compositional Analysis of Foods

Moisture and Total Solids Analysis

15

Lisa J. Mauer

Contents

15.1 INTRODUCTION

Moisture assays can be one of the most important analyses performed on a food product and yet one of the most difficult from which to obtain accurate and precise data. Water molecules are small and ubiquitous in the environment in which foods are produced, stored, and used. Moisture exchange between foods and the environment can lead to under- or overestimation from moisture assays, and water can be difficult to completely remove from foods. The first sections of this chapter describe various methods for moisture content analysis—their principles, procedures, applications, cautions, advantages, and disadvantages. Water activity measurement is also described later in this chapter, since it parallels the measurement of total moisture as an important stability and quality factor. Determining both the water con-

L. J. Mauer (✉)
Department of Food Science, Purdue University,
West Lafayette, IN, USA
e-mail: mauerl@purdue.edu

B. P. Ismail, S. S. Nielsen (eds.), *Nielsen's Food Analysis*, Food Science Text Series,
https://doi.org/10.1007/978-3-031-50643-7_15

tent and the water activity of a food provides a complete moisture analysis. With an understanding of the techniques described, one can apply appropriate moisture analyses to a wide variety of food products.

15.1.1 Importance of Moisture Assays

One of the most fundamental and important analytical procedures that can be performed on a food product is an assay for the amount of moisture, referred to as the **moisture or water content** of the food [1–4]. In this context, the words "water" and "moisture" are generally used interchangeably. The dry matter that remains after moisture removal is commonly referred to as **total solids**. This analytical value is of great economic importance to a food manufacturer, and there are legal limits as to how much water must or can be present in some foods. Some examples in which moisture content is important to the food processor are provided in Table 15.1.

Table 15.1 Importance of moisture content in the food industry

Moisture content is important for	*Food examples*
Preservation and stability	Dehydrated vegetables, potatoes, and fruits
	Dried milks and infant formulas
	Powdered eggs, coffees, and teas
	Spices and herbs
	Crispy fried or baked chips and crackers
	Cotton candy
Quality factors	Jams and jellies, to inhibit sugar crystallization
	Sugar syrups
	Prepared cereals: conventional 4–8% H_2O, puffed 7–8% H_2O
Convenience in packaging or shipping (usually reduced moisture)	Concentrated milks and fruit juices
	Liquid cane sugar (67% solids) and liquid corn sweetener (80% solids)
	Dehydrated products (which clump or become sticky and difficult to package if moisture content is too high)
Meeting compositional standards and standards of identity	Cheddar cheese must be ≤39% H_2O
	Enriched flour must be ≤15% H_2O
	Pineapple juice must have soluble solids of ≥10.5°Brix (conditions specified)
	Glucose syrup must have ≥70% total solids
	% H_2O in processed meats is commonly specified
Accurate computation of nutritional value	
Expressing results of other analytical determinations on a uniform basis	

In addition to quantifying the amount of water in foods, it is also important to document the energy status of the water in the food by determining the **water activity** [5]. It is the water activity, more so than the moisture content, that influences microbial growth, physical properties, and chemical and enzymatic reactions in foods. Additionally, it is differences in water activity, not moisture content, that drive moisture migration between different food components (such as between a crust and a filling) or between food and the environment. Water molecules move from regions of high water activity to regions of low water activity until equilibrium water activity is reached.

15.1.2 Water in Foods

The amount, physical state, and location of water in foods will affect the types of analyses best suited to a particular food product, the ease and rate of water removal, the time required for assay equilibration, and the sample handling.

15.1.2.1 Structure of the Water Molecule

Water molecules are composed of two hydrogen atoms covalently bound to an oxygen atom in a distorted tetrahedral arrangement (Fig. 15.1) [6]. There is a slight partially negative charge on the oxygen atom and a slight partially positive charge on the hydrogen atoms. Thus, water molecules are small and highly polar, have two hydrogen bond donor sites (the hydrogen atoms), and have two hydrogen bond acceptor sites (the two nonbonding electron pairs on the oxygen atom). Hydrogen bonds are relatively weak attractive interactions with short life spans picoseconds), readily breaking, and reforming as water molecules are constantly moving, particularly when temperature or humidity fluctuations occur [7, 8]. This poses challenges for accurately quantifying the amount or activity of water in a sample since the target can be moving from or escaping to the environment during sample handling.

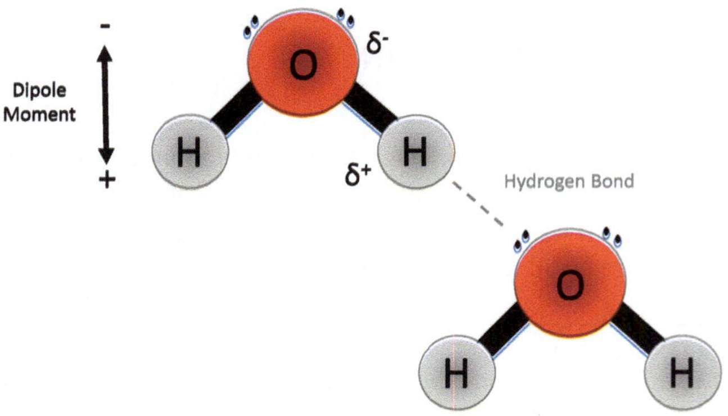

Fig. 15.1 Schematic of the water molecule and hydrogen bonding

15.1.2.2 Physical States and Properties of Water

The temperature, pressure, and extent of intermolecular hydrogen bonding between water molecules result in water existing in different physical states: as a solid (ice crystals), liquid (water), and gas (water vapor, humidity). Determining the total amount of water as well as the water activity in foods may involve analyzing water in all three of these states. Many direct moisture assays are based on the weight lost from evaporating all water from a sample, often by application of heat to overcome the enthalpy of vaporization. Other properties of water may be exploited for the indirect determination of moisture contents, including the dielectric constant, density, and freezing point. These and other properties of water are summarized in Table 15.2.

15.1.2.3 Water Interactions with Food Ingredients

Although water does not covalently bond with food ingredients, it can be difficult to remove from foods. Having some understanding of the water intermolecular interactions with food ingredients and the locations of water in foods can improve the application of moisture assays. The hydrogen bond donor and acceptor sites on water molecules interact with food ingredients via hydrogen bonds, dipole–dipole interactions, ionic attractions, and van der Waals forces [6]. Water molecules tend to cluster around ions and charged groups, as well as other hydrogen bond donor and acceptor groups such as hydroxyl groups (–OH), carbonyl oxygens (=O), and amine groups ($-NH_2$), adopting a more ordered structure in these hydration shell clusters than is found in bulk water [10]. To remove water from food, enough energy must be applied to overcome these intermolecular interactions. There are also different mechanisms by which water interacts with solids, including surface interactions (adsorption), condensed water (capillary condensation and deliquescence), and internalized water (absorption and crystal hydrate formation) [11]. Water can also be physically entrapped in food matrices, such as in gel structures or dense dehydrated or fried products, making it more difficult to remove than surface water. It is nearly impossible to remove all water from foods, particularly the internalized water for which heat and mass transfer rates and the properties of the food material can complicate complete moisture egress.

Table 15.2 Properties of water

Property	*Value*
Molecular weight	18.0153 g
Melting point (1 atm)	0.000 °C
Boiling point (1 atm)	100.000 °C
Enthalpy of vaporization (1 atm)	40.647 kJ/mol
Glass transition temperature (1 atm)	136 K
Critical temperature	373.99 °C
Density (20 °C)	0.99821 g/cm^3
Vapor pressure (20 °C)	2.3388 kPa
Heat capacity (20 °C)	4.1818 J/gK
Thermal conductivity (20 °C)	0.5984 W/mK
Thermal diffusivity (20 °C)	1.4×10^{-7} m^2/s
Dielectric constant (20 °C)	80.20
Water activity (20 °C)	1.000
Water activity (−20 °C)	0.82

Adapted from [5, 6, 9]

15.1.3 Sample Collection and Handling

General procedures for sampling, sample handling and storage, and sample preparation are given in Chap. 5. These procedures are perhaps the greatest potential source of error in any analysis. Precautions must be taken to minimize inadvertent **moisture losses** or **gains** that occur during these steps. Obviously, any exposure of a sample to the open atmosphere should be as short as possible. Any heating of a sample by friction during grinding should be minimized. Headspace in the sample storage container should be minimal because moisture transfer between the sample and the container environment will likely occur. It is critical to control temperature fluctuations since moisture will migrate in a sample to the colder part. To control this potential error, the entire sample should be removed from the container and reblended quickly and a new test portion should then be sampled [12, 13].

To illustrate the need for optimum efficiency and speed in weighing samples for analysis, Bradley and Vanderwarn [14] showed, using shredded Cheddar cheese (2–3 g in a 5.5 cm aluminum foil pan), that moisture loss within an analytical balance was a straight line function. The rate of loss was related to the relative humidity (RH). At 50% RH, it required only 5 s to lose 0.01% moisture. This time doubled at 70% RH for 0.01% moisture loss in 10 s. While one might expect a curvilinear loss, the moisture loss was actually linear over a 5-min study interval. Samples with lower a_w than the environmental RH, such as many powders and crispy fried products, may pick up moisture during handling leading to overestimation of moisture contents. These examples demonstrate the necessity of absolute control during collection of samples through weighing, before drying or other analysis.

15.2 MOISTURE/WATER CONTENT

15.2.1 Overview

The moisture content of foods varies greatly, as shown in Table 15.3 [3]. Water is a major constituent of many food products. The approximate, expected moisture content of a food can affect the choice of the method of measurement. It can also guide the analyst in determining the practical level

Table 15.3 Moisture content of selected foods

Food item	*Approximate percent moisture (wet weight basis)*
Cereals, bread, and pasta	
Wheat flour, whole grain	10.3
White bread, enriched (wheat flour)	13.4
Corn flakes cereal	3.5
Crackers saltines	4.0
Macaroni, dry, enriched	9.9
Dairy products	
Milk, reduced fat, fluid, 2%	89.3
Yogurt, plain, low fat	85.1
Cottage cheese, low fat or 2% milk fat	80.7
Cheddar cheese	36.8
Ice cream, vanilla	61.0
Fats and oils	
Margarine, regular, hard, corn, hydrogenated	15.7
Butter, with salt	15.9
Oil-soybean, salad, or cooking	0
Fruits and vegetables	
Watermelon, raw	91.5
Oranges, raw, California navels	86.3
Apples, raw, with skin	85.6
Grapes, American type, raw	81.3
Raisins	15.3
Cucumbers, with peel, raw	95.2
Potatoes, microwaved, cooked in skin, flesh, and skin	72.4
Snap beans, green, raw	90.3
Meat, poultry, and fish	
Beef, ground, raw, 95% lean	73.3
Chicken, broilers and fryers, light meat, meat and skin, raw	68.6
Finfish, flatfish (flounder and sole species), raw	79.1
Egg, whole, raw, fresh	75.8
Nuts	
Walnuts, black, dried	4.6
Peanuts, all types, dry roasted with salt	1.6
Peanut butter, smooth style, with salt	1.8
Sweeteners	
Sugar, granulated	0
Sugar, brown	1.3
Honey, strained or extracted	17.1

From the US Department of Agriculture, Agricultural Research Service (2023) USDA National Nutrient Database for Standard Reference. Release 28. Nutrient Data Laboratory Home Page, http://ndb.nal.usda.gov [3]

of accuracy required when measuring moisture content, relative to other food constituents. The method used for determining moisture content may measure more or less of the water present. This is the reason for official methods with stated procedures [15–17]. However, several official methods may exist for a particular product. For example, the AOAC International methods for cheese include Method 926.08, vacuum oven; 948.12, forced draft oven; 977.11, microwave oven; 969.19, distillation [15]. Usually, the first method listed by AOAC International is preferred over others in any section.

The different types of moisture content assays can be generally categorized into direct and indirect methods. **Direct methods** for moisture content are often done by removing water, although the method for moisture removal may vary. Drying, distillation, and extraction are commonly used for moisture removal followed by weighing, volumetry, or titration to determine moisture content. **Indirect methods** are based on properties of the food that are related to the presence of water, such as capacitance, specific gravity, density, refractive index, freezing point, and electromagnetic absorption.

15.2.2 Oven-Drying Methods

In **oven-drying methods**, which are direct methods, the sample is heated under specified conditions and the loss of weight is used to calculate the moisture content of the sample. The amount of moisture determined is highly dependent on the type of oven used, conditions within the oven, and the time and temperature of drying. Various oven methods are approved by AOAC International for determining the amount of moisture in many food products. The methods are simple, and many ovens allow for simultaneous analysis of large numbers of samples. The time required may be from ~1 h to over 24 h.

15.2.2.1 General Information

15.2.2.1.1 Removal of Moisture

Water evaporates more quickly at higher temperatures. Any oven method used to evaporate moisture has as its foundation the fact that the boiling point of water is 100 °C; however, this considers only pure water at sea level. According to Raoult's law, if 1 molecular weight (1 mol) of a solute is dissolved in 1.0 L of water, the boiling point would be raised by 0.512 °C. This boiling point elevation continues throughout the drying process as solute concentrations increase.

Moisture removal is sometimes best achieved in a two-stage process. Liquid products (e.g., juices and milk) are commonly predried over a **steam bath** before drying in an oven. Products such as bread and field-dried grain are often air-dried and then ground and oven-dried, with the moisture content calculated from moisture loss at both air and oven-drying steps. Particle size, particle size distribution, sample sizes, hygroscopicity, and surface area during drying influence the rate and efficiency of moisture removal.

15.2.2.1.2 Decomposition of Other Food Constituents

Moisture loss from a sample during analysis is a function of time and temperature. Decomposition enters the picture when time is extended too much or temperature is too high. Thus, most methods for food moisture content analysis involve a compromise between time and a particular temperature at which limited decomposition might be a factor. One major problem exists in that the physical process must separate all the moisture without decomposing any of the constituents that could release water. For example, carbohydrates decompose at elevated temperatures and release water to form dehydrated hydrocarbon compounds. The moisture generated in this decomposition leads to overestimation of the moisture content and is not the moisture that we want to measure. Certain other chemical reactions (e.g., sucrose hydrolysis) can result in utilization of moisture, which would reduce the moisture for measurement. The loss of **volatile constituents**, such as acetic, propionic, and butyric acids, and alcohols, esters, and aldehydes among flavor compounds, can also lead to errors. While weight changes in oven-drying methods are assumed to be due to moisture loss, weight gains can also occur due to oxidation of unsaturated fatty acids and certain other compounds.

Nelson and Hulett [18] determined that moisture was retained in biological products to at least 365 °C, which is coincidentally near the critical temperature for water. Their data indicate that among the decomposition products at elevated temperatures were CO, CO_2, CH_4, and H_2O. These were not given off at any one particular temperature but at all temperatures and at different rates at the respective temperature in question.

By plotting moisture liberated against temperature, curves were obtained that show the amount of moisture liberated at each temperature (Fig. 15.2). Distinct breaks were shown that indicated the temperature at which decomposition became measurable. None of these curves showed any break before 184 °C. Generally, proteins decompose at temperatures somewhat lower than that required for starches and celluloses. Extrapolation of the flat portion of each curve to 250 °C gave a true moisture content based on the assumption that there was no adsorbed water present at the temperature in question.

15.2.2.1.3 Temperature Control

Drying methods utilize specified drying temperatures and times, which must be carefully controlled. Moreover, there may be considerable variability of temperature, depending on the type of oven used for moisture analysis. One should determine the extent of variation within an oven before relying on data collected from its use.

Consider the temperature variation in three types of ovens: **convection (atmospheric), forced draft,** and **vacuum**. The greatest temperature variation exists in a convection oven. This is because hot air slowly circulates without the aid of a fan. Air movement is obstructed further by pans placed in the oven. When the oven door is closed, the rate of temperature recovery is generally slow. This is dependent also upon the load placed in the oven and upon the ambient temperature. A 10 °C temperature differential across a convection oven is not unusual. This must be considered in view of anticipated analytical accuracy and precision. A convection oven should not be used when precise and accurate measurements are needed.

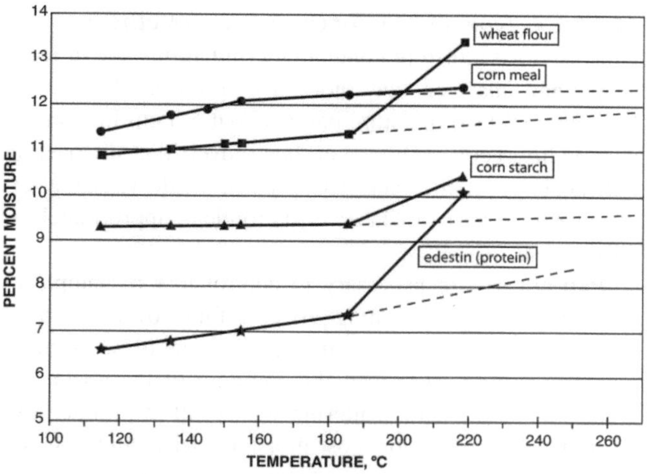

Fig. 15.2 Moisture content of several foods held at various temperatures in an oven. The hyphenated line extrapolates data to 275 °C, the true moisture content. (Reprinted with permission from [18]. Copyright 1920, American Chemical Society)

Forced draft ovens have the least temperature differential across the interior of all ovens, usually not greater than 1 °C. Air is circulated by a fan that forces air movement throughout the oven cavity. Forced draft ovens with air distribution manifolds have the added benefit of horizontal air movement across shelving. Thus, no matter whether the oven shelves are filled completely with moisture pans or only half filled, the result would be the same for a particular sample [14].

Two features of some **vacuum ovens** contribute to a wider temperature spread across the oven. One feature is a glass panel in the door. Although from an educational point of view it may be fascinating to observe some samples in the drying mode, the glass is a heat sink. The second feature is the way by which air is bled into the oven. If the air inlet and discharge are on opposite sides, conduct of air is virtually straight across the oven. Some newer models have air inlet and discharge manifolds mounted top and bottom. Air movement in this style of vacuum oven is upward from the front and then backward to the discharge in a broad sweep. The effect is to minimize cold spots as well as to exhaust moisture in the interior air.

15.2.2.1.4 Types of Pans for Oven-Drying Methods

Pans used for moisture content determinations are varied in shape and may or may not have a cover. The AOAC International [15] moisture pan is about 5.5 cm in diameter with an insert cover. Other pans have covers that slip over the outside edge of the pan. These pans, while reusable, are expensive, in terms of labor costs to clean appropriately to allow reuse.

Pan covers are necessary to control loss of sample by spattering during the heating process. If the cover is metal, it must be slipped to one side during drying to allow for moisture evaporation. However, this slipping of the cover also creates an area where spattering will result in product loss. Examine the interior of most moisture ovens and you will detect odor and deposits of burned-on residue, which, although undetected at the time of occurrence, produce erroneous results and large standard deviations [14].

Consider the use of **disposable pans** whenever possible; then purchase **glass fiber discs** for covers. At 5.5 cm in diameter, these covers fit perfectly inside disposable aluminum foil pans and prevent spattering while allowing the surface to breathe. Paper filter discs foul with fat and thus do not breathe effectively. Drying studies done on cheese using various pans and covers have shown that fat does spatter from pans with slipped covers. Fiberglass is the most satisfactory cover.

15.2.2.1.5 Handling and Preparation of Pans

The preparation and handling of pans before use require consideration. Use only **tongs** to handle any pan or wear gloves. Even the oils in fingerprints have weight. All pans must be oven-treated to prepare them for use. This is a factor of major importance unless disproved by the technologist doing moisture determinations with a particular type of pan. Disposable aluminum pans must be vacuum oven-dried for 3 h before use. At 3 h and 15 h in either a vacuum or forced draft oven at 100 °C, pans varied in their weight within the error of the balance (0.0001 g) and therefore longer drying times are not needed [14]. Store dried moisture pans in a functioning **desiccator.** The glass fiber covers should be dried for 1 h before use.

15.2.2.1.6 Control of Surface Crust Formation (Sand Pan Technique)

Some food materials tend to form a semi-permeable crust or lump together during drying, which will contribute to erratic and erroneous results. To control this problem, analysts use the **sand pan technique**. Clean, dry sand and a short glass stirring rod are preweighed into a moisture pan. Subsequently, after weighing in a sample, the sand and sample are admixed with the stirring rod left in the pan. The remainder of the procedure follows a standardized method if available; otherwise, the sample is dried to constant weight. The purpose of the sand is twofold: to prevent **surface crust** from forming and to disperse the sample so evaporation of moisture is less impeded. The amount of sand used is a function of sample size. Consider 20–30 g sand/3 g sample to obtain the desired distribution in the pan. Similar to the procedure, applications, and advantages of using sand, other heat-stable inert materials such as diatomaceous earth can be used in moisture content determinations, especially for sticky fruits.

The inert matrices such as sand and diatomaceous earth function to disperse the food constituents and minimize retention of moisture in the food products. However, the analyst must ascertain that the inert matrix used does not give erroneous results for the assay because of decomposition or entrapped moisture loss. Test the sand or other inert matrix for weight loss before using in any method. Add approximately 25 g of sand into a moisture pan and heat at 100 °C for 2 h and weigh to 0.1 mg. Add 5 mL water and mix with the matrix using a glass rod. Heat dish, matrix, cover, and glass rod for at least 4 h at 100 °C and then reweigh. The difference between weighing must be less than 0.5 mg for any suitable matrix [19].

15.2.2.1.7 Calculations

Moisture (wwb and dwb) and total solids contents of foods can be calculated as follows using oven-drying procedures:

$$\begin{aligned}\%\ \text{Moisture}(\text{wwb}) &= \frac{\text{wt of wet sample} - \text{wt of dry sample}}{\text{wt of wet sample}} \times 100 \\ &= \frac{\text{wt } H_2O \text{ in sample}}{\text{wt of wet sample}} \times 100\end{aligned} \tag{15.1}$$

$$\%\ \text{Moisture}(\text{dwb}) = \frac{\text{wt } H_2O \text{ in sample}}{\text{wt of dry sample}} \times 100 \tag{15.2}$$

$$\begin{aligned}\%\ \text{Total solids}(\text{wt}/\text{wt}) &= \frac{\text{wt of dry sample}}{\text{wt of wet sample}} \times 100 \\ &= 100 - \%\ \text{Moisture}(\text{wwb})\end{aligned} \tag{15.3}$$

15.2.2.2 Forced Draft Oven

When using a forced draft oven, the sample is rapidly weighed into a predried moisture pan, covered, and placed in the oven for an arbitrarily selected time if no standardized method exists. Drying time periods for this method are 0.75–24 h (Table 15.4), depending on the food sample and its pretreatment; some liquid samples are dried initially on a steam

Table 15.4 Forced draft oven temperature and times for selected foods

Product	*Dry on steam bath*	*Oven temperature (°C ± 2)*	*Time in oven (h)*
Buttermilk, liquid	X[a]	100	3
Cheese, natural type only		100	16.5 ± 0.5
Chocolate and cocoa		100	3
Cottage cheese		100	3
Cream, liquid and frozen	X	100	3
Egg albumin, liquid	X	130	0.75
Egg albumin, dried	X	100	0.75
Ice cream and frozen desserts	X	100	3.5
Milk Whole, low fat, and skim Condensed skim	X	100 100 100	3 3 3
Nuts: almonds, peanuts, walnuts		130	3
Fruit, dried		70	6
Coffee, roasted		70	16

From Wehr and Frank [16] p. 492, with permission, *Standard Methods for the Examination of Dairy Products,* 17th ed. H.M. Wehr and J.F. Frank (Eds.). Copyright 2004 by the American Public Health Association, Washington, DC

[a]*X* = samples must be partially dried on steam bath before being placed in oven

bath at 100 °C to minimize spattering. In these cases, drying times are shortened to 0.75–3 h. A forced draft oven is used with or without a steam table predrying treatment to determine the solids content of fluid milks (AOAC Method 990.19, 990.20).

An alternative to selecting a time period for drying is to weigh and reweigh the dried sample and pan until two successive weighings taken 30 min apart agree within a specified limit, for example, 0.1–0.2 mg for a 5 g sample. The user of this second method must be aware of sample transformation, such as browning which suggests moisture loss of the wrong form. Lipid oxidation and a resulting sample weight gain can occur at high temperatures in a forced draft oven. Samples high in carbohydrates should not be dried in a forced draft oven but rather in a vacuum oven at a temperature no higher than 70 °C.

15.2.2.3 Vacuum Oven

By drying under reduced pressure (25–100 mm Hg), the rate of evaporation is faster and one is able to obtain a more complete removal of water without decomposition within a 3–6-h drying time. Vacuum ovens need a dry air purge in addition to temperature and vacuum controls to operate within the method definition. In older methods, a vacuum flask is used, partially filled with concentrated sulfuric acid as the desiccant. One or two air bubbles per second are passed through the acid. Recent changes now stipulate an air trap that is filled with calcium sulfate containing an indicator to show water saturation (such as DrieRite™). Between the trap and the vacuum oven is an appropriately sized rotameter to measure air flow (100–120 mL//min) into the oven.

The following are important points in the use of a vacuum drying oven:

1. **Temperature** used depends on the product, such as 95–102 °C for some foods and lower temperatures (60–70 °C) for fruits and other high-sugar products. Even with reduced temperature, there can be some decomposition.
2. If the product to be assayed has a high concentration of **volatiles,** you should consider the use of a correction factor to compensate for the loss.
3. Analysts should remember that in a **vacuum,** heat is not conducted well. Thus, pans must be placed directly on the metal shelves to conduct heat.
4. **Evaporation** is an endothermic process; thus, a pronounced cooling is observed. Because of the cooling effect of evaporation, when several samples are placed in an oven of this type, you will note that the temperature will drop. Do not attempt to compensate for the cooling effect by increasing the temperature; otherwise, samples during the last stages of drying will be overheated.
5. The **drying time** is a function of the total moisture present, nature of the food, surface area per unit weight of sample, whether sand is used as a dispersant, and the relative concentration of sugars and other substances capable of retaining moisture or decomposing. The drying interval is determined experimentally to give reproducible results.

15.2.2.4 Microwave Analyzer

Determination of moisture contents in food products has traditionally been done using a standard oven, which, though accurate, can take many hours to dry a sample. **Microwave moisture analysis,** often called **microwave drying,** was the first precise and rapid technique that allowed some segments of the food industry to make in-process adjustment of the moisture content in food products before final packaging. For example, processed cheese could be analyzed, and the composition adjusted before the blend was dumped from the cooker. The ability to adjust the composition of a product in-process helps food manufacturers reduce production costs, meet regulatory requirements, and ensure product consistency. Such control could effectively pay for the microwave analyzer within a few months.

A particular microwave moisture/solids analyzer (CEM Corporation, Matthews, NC), or equivalent, is specified in the AOAC International procedures for total solids analysis of processed tomato products (AOAC Method 985.26) and moisture content analysis of meat and poultry products (AOAC Method 985.14).

The general procedure for use of a microwave moisture/solids analyzer has been to set the microprocessor controller to a percentage of full power to control the microwave output. Power settings are dependent upon the type of sample and the recommendations of the manufacturer of the microwave moisture analyzer. Next, the internal balance is tared with two sample pads on the balance. As rapidly as possible, a sample is placed between the two pads and then pads are centered on the pedestal and weighed against the tare weight. The time for the drying operation is set by the operator, and "start" is activated. The microprocessor controls the drying procedure, with percentage moisture indicated in the controller window. Some newer models of microwave moisture analyzers have a temperature control feature to precisely control the drying process, removing the need to guess appropriate time and power settings for specific applications. These new models also have a smaller cavity that allows the microwave energy to be focused directly on the sample.

There are some considerations when using a microwave analyzer for moisture determination: (1) the sample must be of a uniform, appropriate size to provide for complete drying under the conditions specified; (2) the sample must be centrally located and evenly distributed, so some portions are not burned and other areas under-processed; and (3) the amount of time used to place an appropriate sample weight between the pads must be minimized to prevent moisture loss or gain before weight determination. Sample pads should also be considered. There are several different types, including fiberglass and quartz fiber pads. For optimum results, the pads should not absorb microwave energy, as this can cause the sample to burn, nor should they fray easily, as this causes them to lose weight and can affect the analysis. In addition, they should absorb liquids well.

Another style of microwave oven that includes a vacuum system is used in some food plants. This vacuum microwave oven will accommodate one sample in triplicate or three different samples at one time. In 10 min, the results are reported to be similar to 5 h in a vacuum oven at 100 °C. The vacuum microwave oven is not nearly as widely used as conventional microwave analyzers but can be beneficial in some applications.

Microwave drying provides a fast (4–8 min), accurate method to analyze many foods for moisture content. The method is sufficiently accurate for routine assay. The distinct advantage of rapid analysis far outweighs its limitation of testing only single samples [20].

15.2.2.5 Infrared Drying

Infrared drying involves penetration of heat into the sample being dried, as compared to heat conductivity and convection with conventional ovens. Such heat penetration to evaporate moisture from the sample can significantly shorten the required drying time to 10–25 min. The infrared lamp used to supply heat to the sample results in a filament temperature of 2000–2500 K. Factors that must be controlled include distance of the infrared source from the dried material and thickness of the sample: infrared radiation has a limited penetration depth. The analyst must be careful that the sample does not burn or case-harden while drying. Infrared drying ovens may be equipped with forced ventilation to remove moist air and an analytical balance to read moisture content directly. No infrared drying moisture analysis techniques are approved by AOAC International currently. However, because of the speed of analysis, this technique is suited for qualitative in-process use.

15.2.2.6 Rapid Moisture Analyzer Technology

Many rapid moisture/solids analyzers based on thermogravimetric principles are used by the food industry. In addition to those based on infrared and microwave drying as described previously, compact instruments that depend on high heat created by different types of heaters are available. Two main categories of heating elements include halogen heaters (e.g., Halogen Moisture Analyzers, Mettler Toledo, Columbus, OH) and ceramic heaters (e.g., Computrac®, Arizona Instrument LLC, Chandler, AZ). These analyzers detect moisture levels from 50 ppm to 100% using sample weights of 150 mg–54 g. Smaller samples tend to dry more quickly, but it is important to use enough sample to be representative of the product. The test sample is placed on an aluminum pan or filter paper, and the heat control program (with a heating range of 25 °C – >200 °C) elevates the test sample to a constant temperature. As the moisture is driven from the sample, the instrument automatically weighs and calculates the percentage moisture or solids. The samples are not removed from the oven which minimizes weighing errors, and accurate results are obtained within minutes. These analyzers are utilized for both production and laboratory assays with results comparable to reference methods.

15.2.2.7 Thermogravimetric Analyzer

In a thermogravimetric analyzer (TGA), the mass of a sample is continuously measured as it is heated at a controlled rate in a controlled atmosphere. The sample (often 10–50 mg) is loaded into a pan that is then placed into the TGA instrument that contains a furnace and a precision balance. Foods can lose mass during heating through release of adsorbed compounds (such as water), chemical reactions, and decomposition. To determine moisture content, the sample chamber is purged with an inert gas (e.g., nitrogen) so the sample only reacts with temperature during heating and mass changes due to oxidation are minimized. The furnaces in TGA instruments cover a wider temperature range than many other moisture content determination techniques and may range from sub-ambient temperatures to >1500 °C. This enables not only moisture assessment (moisture content, temperature of dehydration, stoichiometry of a hydrate, and dehydration kinetics) but also at higher temperatures measurement of pyrolysis, decomposition, and weight % ash. More detailed information on TGA is provided in the chapter on thermal analysis (Chap. 30). To use TGA, a survey scan is often run on a sample from approximately 100 °C below to 100 °C above the transition of interest at a heating rate of 20 °C per minute. For water content determination of many foods, scans from ambient to 200 °C are generally sufficient and water content is determined by the mass loss up to ~100 °C. A derivative plot of the rate of mass loss will generate a peak showing the onset, midpoint, and endpoint temperatures for water loss. The precision balance and in situ mass monitoring result in precise and accurate moisture content determinations; however, decomposition or volatile loss at temperatures overlapping water evaporation can lead to errors in moisture content determination.

15.2.3 Distillation Procedures

15.2.3.1 Overview

Distillation is used as another direct measure of moisture content. Distillation techniques involve codistilling the moisture in a food sample with a high boiling point solvent that is immiscible in water, collecting the mixture that distills off, and then measuring the volume of water. Two distillation procedures are in use today: **direct** and **reflux distillations**, with a variety of solvents. For example, in direct distillation with immiscible solvents of higher boiling point than water, the sample is heated in mineral oil or liquid with a flash point well above the boiling point for water. Other immiscible liquids with boiling point only slightly above water can be used (e.g., toluene, xylene, and benzene). However, reflux distillation with the immiscible solvent toluene is the most widely used method.

Distillation techniques were originally developed as rapid methods for quality control work, but they are not adaptable to routine testing. The distillation method is an AOAC-approved technique for moisture content analysis of spices (AOAC Method 986.21), cheese (AOAC Method 969.19), and animal feeds (AOAC Method 925.04). It can also give good accuracy and precision for nuts, oils, soaps, and waxes.

Distillation methods cause less thermal decomposition of some foods than oven-drying at high temperatures. Adverse chemical reactions are not eliminated but can be minimized by using a solvent with a lower boiling point. This, however, will increase distillation times. Water is measured directly in the distillation procedure (rather than by weight loss), but reading the volume of water in a receiving tube may be less accurate than using a weight measurement.

15.2.3.2 Reflux Distillation with Immiscible Solvent

Reflux distillation uses either a solvent less dense than water (e.g., toluene, with a boiling point of 110.6 °C; or xylene, with a boiling range of 137–140 °C) or a solvent more dense than water (e.g., tetrachlorethylene, with a boiling point of 121 °C). The advantage of using this last solvent is that material to be dried floats; therefore it will not char or burn. In addition, there is no fire hazard with this solvent.

A **Bidwell–Sterling moisture trap** (Fig. 15.3) is commonly used as part of the apparatus for reflux distillation with a solvent less dense than water. The distillation procedure using such a trap requires about 1 h and involves using a brush to dislodge adhering water drops from the glassware, thereby minimizing error.

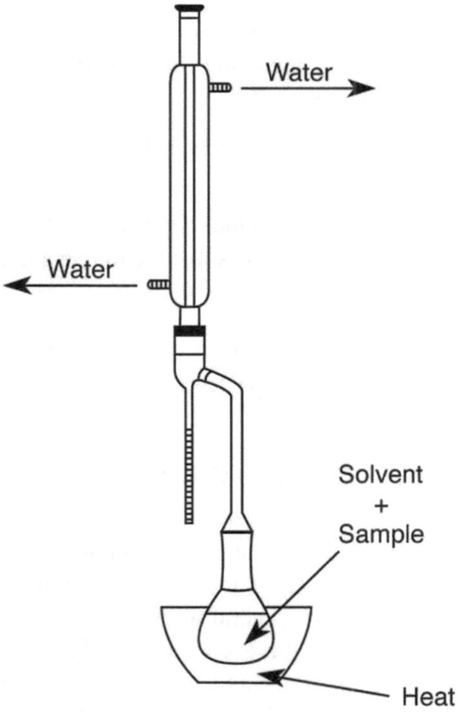

Fig. 15.3 Apparatus for reflux distillation of moisture from a food. Key to this setup is the Bidwell–Sterling moisture trap. This style can be used only where the solvent is less dense than water

Three potential sources of error with distillation should be eliminated if observed:

1. Formation of water–solvent emulsions that will not break. Usually, this can be controlled by allowing the apparatus to cool after distillation is completed and before reading the amount of moisture in the trap.
2. Clinging of water droplets to dirty apparatus. Clean glassware is essential, but even with this, a burette brush is needed to dislodge water droplets.
3. Decomposition of the sample with production of water. If this is a measurable problem, discontinue method use and find an alternative procedure.

15.2.4 Chemical Method: Karl Fischer Titration

The **Karl Fischer titration**, a direct measure of moisture content, is particularly adaptable to food products that show erratic results when heated or submitted to a vacuum. This is the method of choice for the determination of water content in many low-moisture foods such as dried fruits and vegetables (AOAC Method 967.19 E-G), candies, chocolate (AOAC Method 977.10), roasted coffee, oils and fats (AOAC Method 984.20), or any low-moisture food high in sugar or protein. The method is quite rapid and accurate and uses no heat. This method is based on the fundamental reaction described by Bunsen in 1853 [21] involving the reduction of iodine by SO_2 in the presence of water:

$$2H_2O + SO_2 + I_2 \rightarrow H_2SO_4 + 2HI \quad (15.4)$$

This was modified to include methanol and pyridine in a four-component system to dissolve the iodine and SO_2:

$$C_5H_5N{\cdot}I_2 + C_5H_5N{\cdot}SO_2 + C_5H_5N + H_2O \rightarrow 2C_5H_5N{\cdot}HI + C_5H_5N{\cdot}SO_3 \quad (15.5)$$

$$C_5H_5N{\cdot}SO_3 + CH_3OH \rightarrow C_5H_5N(H)SO_4{\cdot}CH_3 \quad (15.6)$$

These reactions show that for each mole of water, 1 mol of iodine, 1 mol of SO_2, 3 mol of pyridine, and 1 mol of methanol are used. For general work, a methanolic solution is used that contains these components in the ratio of 1 iodine:3 SO_2:10 pyridine and at a concentration so that 3.5 mg of water = 1 mL of reagent. A procedure for standardizing this reagent is given below.

In a **volumetric titration** procedure (Fig. 15.4 is manual titration unit; Fig. 15.5 is an example of automated titration unit), iodine and SO_2 in the appropriate form are added to the

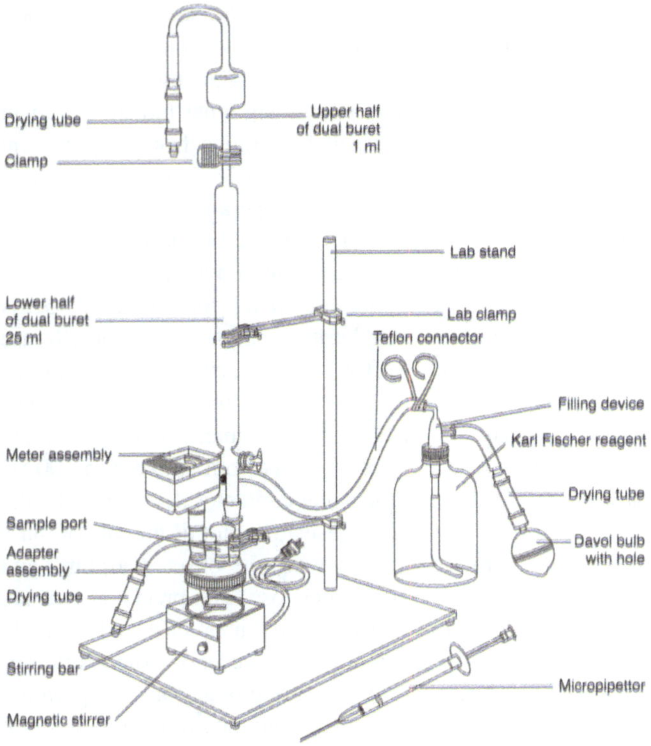

Fig. 15.4 Manual Karl Fischer titration unit. (Courtesy of Lab Industries, Inc., Berkeley, CA)

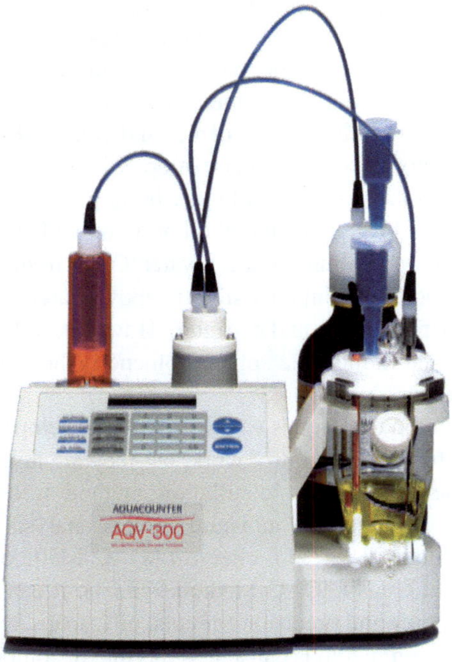

Fig. 15.5 Automated Karl Fischer volumetric titration unit. (Courtesy of Mettler-Toledo, Columbus, OH)

sample in a closed chamber protected from atmospheric moisture. The excess of I_2 that cannot react with the water can be determined **visually.** The endpoint color is dark red-brown. Some instrumental systems are improved by the inclusion of a potentiometer (i.e., **conductometric method**) to electronically determine the endpoint, which increases sensitivity and accuracy. The automated volumetric titration units (used for 100 ppm water to very high concentrations) use a pump for mechanical addition of titrant and use the conductometric method for endpoint determination (i.e., detection of excess iodine is by applying a current and measuring the potential).

The volumetric titration procedure described above is appropriate for samples with a moisture content greater than ~0.03%. A second type of titration, referred to as **coulometric titration,** is ideal for products with very low levels of moisture, from 0.03% down to parts per million (ppm) levels. In this method, iodine is electrolytically generated ($2I \rightarrow I_2 + 2e^-$) to titrate the water. The amount of iodine required to titrate the water is determined by the current needed to generate the iodine. Just like for volumetric titration, automated coulometric titration units are available commercially.

In a Karl Fischer volumetric titration, the **Karl Fischer reagent** (KFR) is added directly as the titrant if the moisture in the sample is accessible. However, if moisture in a solid sample is inaccessible to the reagent, the moisture is extracted from the food with an appropriate solvent (e.g., methanol). (Particle size affects the efficiency of extraction directly.) Then the methanol extract is titrated with KFR.

The noxious odor of pyridine makes it an undesirable reagent. Therefore, researchers have experimented with other amines capable of dissolving iodine and sulfur dioxide. Some aliphatic amines and several other heterocyclic compounds were found suitable. On the basis of these amines, **one-component reagents** (solvent and titrant components together) and **two-component reagents** (solvent and titrant components separate) have been prepared. The one-component reagent may be more convenient to use, but the two-component reagent has greater storage stability.

Before the amount of water found in a food sample can be determined, a **KFR water (moisture) equivalence** (KFReq) must be determined. The KFReq value represents the equivalent amount of water that reacts with 1 mL of KFR. Standardization must be checked before each use because the KFReq will change with time.

The KFReq can be established with **pure water**, a **water-in-methanol standard**, or **sodium tartrate dihydrate**. Pure water is a difficult standard to use because of inaccuracy in measuring the small amounts required. The water-in-methanol standard is premixed by the manufacturer and generally contains 1 mg of water/mL of solution. This standard can change over prolonged storage periods by absorbing atmospheric moisture. Sodium tartrate dihydrate ($Na_2C_4H_4O_6 \cdot 2H_2O$) is a primary standard for determining KFReq. This compound is very stable, contains 15.66% water under all conditions expected in the laboratory, and is the material of choice to use.

The KFReq is calculated as follows using sodium tartrate dihydrate:

$$\mathrm{KFReq}\left(\mathrm{mg\,H_2O/mL}\right) = \frac{36\,\mathrm{g\,H_2O/mol\,Na_2C_4H_4O_6 \cdot 2H_2O} \times S \times 1000}{230.08\,\mathrm{g/mol} \times A} \quad (15.7)$$

where:

KFReq = Karl Fischer reagent water (moisture) equivalence
S = weight of sodium tartrate dihydrate (g)
A = mL of KFR required for titration of sodium tartrate dihydrate

Once the KFReq is known, the moisture content of the sample is determined as follows:

$$\%\mathrm{H_2O} = \frac{\mathrm{KFReq} \times Ks}{S} \times 100 \quad (15.8)$$

where:

KFReq = Karl Fischer reagent water (moisture) equivalence
Ks = mL of KFR used to titrate sample
S = weight of sample (mg)

The major difficulties and sources of error in the Karl Fischer titration methods are as follows:

1. **Incomplete moisture extraction.** For this reason, fineness of grind (i.e., particle size) is important in the preparation of cereal grains and some foods.
2. **Atmospheric moisture**. External air must not be allowed to infiltrate the reaction chamber.
3. **Moisture adhering** to walls of unit. All glassware and utensils must be carefully dried.
4. **Interferences** from certain food constituents. **Ascorbic acid** is oxidized by KFR to dehydroascorbic acid to overestimate moisture content; **carbonyl compounds** react with methanol to form acetals and release water, to overestimate moisture content (this reaction may also result in fading endpoints); **unsaturated fatty acids** will react with iodine, so moisture content will be overestimated.

15.2.5 Physical Methods

Most physical methods are indirect measures of moisture content and do not separate the water from the sample for analysis. These techniques can be rapid and nondestructive, which has led to their widespread use in food production and quality control; however, they must be calibrated against data collected by a direct method to quantify the amount of water in samples.

15.2.5.1 Dielectric Method

The electrical properties of water are used in the **dielectric method** to determine the moisture content of certain foods, by measuring the change in **capacitance** or **resistance to an electric current** passed through a sample. These instruments require calibration against samples of known moisture content as determined by standard direct methods. Sample density or weight/volume relationships and sample temperature are important factors to control in making reliable and repeatable measurements by dielectric methods. These techniques can be very useful for process control measurement applications, where continuous measurement is required. These methods are limited to food systems that contain no more than 30–35% moisture.

The moisture content determination in dielectric-type meters is based on the fact that the dielectric constant of water (80.37 at 20 °C) is higher than that of most solvents. The **dielectric constant** is measured as an index of capacitance. As an example, the dielectric method is used widely for cereal grains. Its use is based on the fact that water has a dielectric constant of 80.37, whereas starches and proteins found in cereals have dielectric constants of 10. By determining this property on samples in standard metal condensers, dial readings may be obtained and the percentage of moisture determined from a previously constructed standard curve for a particular cereal grain.

15.2.5.2 Hydrometry

Hydrometry is the science of measuring **specific gravity** or **density,** which can be done using several different principles and instruments. While hydrometry is considered archaic in some analytical circles, it is still widely used and, with proper technique, is highly accurate. Specific gravity measurements with various types of **hydrometers** or with a **pycnometer** are commonly used for routine testing of moisture (or solids) content of numerous food products. These include beverages, salt brines, and sugar solutions. Specific gravity measurements are best applied to the analysis of solutions consisting of only one solute in a medium of water.

15.2.5.2.1 Hydrometer

One approach to measuring specific gravity is based on **Archimedes' principle**, which states that a solid suspended in a liquid will be buoyed by a force equal to the weight of the liquid displaced. The weight per unit volume of a liquid is determined by measuring the volume displaced by an object of standard weight. A hydrometer is a standard weight on the end of a spindle, and it displaces a weight of liquid equal to its own weight (Fig. 15.6). For example, in a liquid of low density, the hydrometer will sink to a greater depth, whereas in a liquid of high density, the hydrometer will not sink as far. Hydrometers are available in narrow and wide ranges of specific gravity. The spindle of the hydrometer is calibrated to read specific gravity directly at 15.5 °C or 20 °C. A **hydrometer** is not as accurate as a pycnometer, but the speed with which you can do an analysis is a decisive factor. The accuracy of specific gravity measurements can be improved by using a hydrometer calibrated in the desired range of specific gravities.

The rudimentary but surprisingly accurate hydrometer comes equipped with various modifications depending on the fluid to be measured:

1. The Quevenne and New York Board of Health **lactometer** is used to determine the density of milk. The Quevenne lactometer reads from 15 to 40 lactometer units and corresponds to 1.015 to 1.040 specific gravity. For every degree above 60 °F, 0.1 lactometer unit is added to the reading and 0.1 lactometer unit is subtracted for every degree below 60 °F.
2. The **Baumé hydrometer** was used originally to determine the density of salt solutions (originally 10% salt), but it has come into much wider use. From the value

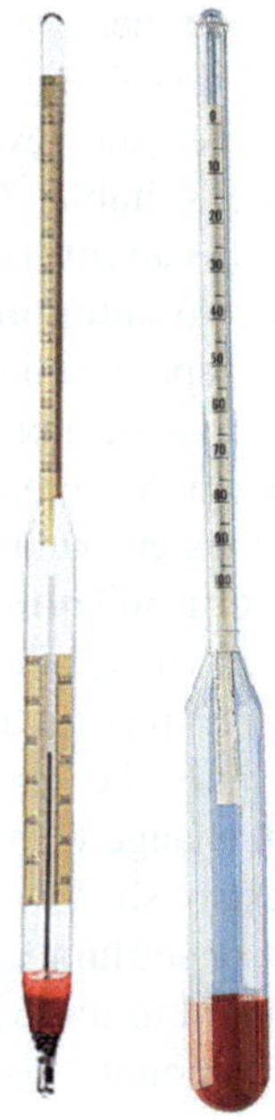

Fig. 15.6 Hydrometers. (Courtesy of Cole-Parmer Instrument Company, Vernon Hills, IL)

obtained in the Baumé scale, you can convert to specific gravity of liquids heavier than water. For example, it is used to determine the specific gravity of milk being condensed in a vacuum pan.

3. The **Brix hydrometer** is a type of **saccarometer** used for sugar solutions such as fruit juices and syrups, and one usually reads directly the percentage of sucrose at 20 °C. **Balling saccharometers** are graduated to indicate percentage of sugar by weight at 60 °F. The terms **Brix** and **Balling** are interpreted as the weight percentage of pure sucrose.
4. **Alcoholometers** are used to estimate the alcohol content of beverages. Such hydrometers are calibrated in 0.1° or 0.2° proof to determine the percentage of alcohol in distilled liquors (AOAC Method 957.03).
5. The **Twaddell hydrometer** is only for liquids heavier than water.

15.2.5.2.2 Pycnometer

Another approach to measuring specific gravity is a comparison of the weights of equal volumes of a liquid and water in standardized glassware, a **pycnometer** (Fig. 15.7). This will yield density of the liquid compared to water. In some texts and reference books, *20/20* is given after the specific gravity number. This indicates that the temperature of both fluids was 20 °C when the weights were measured. Using a clean, dry pycnometer at 20 °C, the analyst weighs it empty, fills it to the full point with distilled water at 20 °C, inserts the thermometer to seal the fill opening, and then touches off the last drops of water and puts on the cap for the overflow tube. The pycnometer is wiped dry in case of any spillage from filling and is reweighed. The density of the sample is calculated as follows:

$$\frac{\text{weight of sample filled pycnometer} \; \text{weight of empty pycnometer}}{\text{weight of water filled pycnometer} \; \text{-weight of empty pycnometer}} = \text{density of sample} \qquad (15.9)$$

This method is used for determining alcohol content in alcoholic beverages (e.g., distilled liquor, AOAC Method 930.17), solids in sugar syrups (AOAC Method 932.14B), and solids in milk (AOAC Method 925.22).

15.2.5.3 Refractometry

Moisture in liquid sugar products and condensed milks can be determined using a Baumé hydrometer (solids), a Brix hydrometer (sugar content), gravimetric means, or a **refractometer**. If it is performed correctly and no crystalline solids are evident, the refractometer procedure is rapid and surprisingly accurate (AOAC Method 932.14C, for solids in syrups). The refractometer has been valuable in determining the soluble solids in fruits and fruit products (AOAC Method 932.12; 976.20; 983.17).

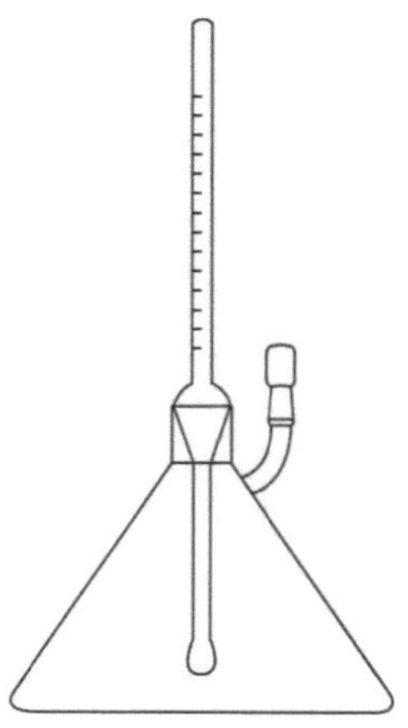

Fig. 15.7 Pycnometer

The **refractive index** (RI) of an oil, syrup, or other liquid is a dimensionless constant that can be used to describe the nature of the food. While some refractometers are designed only to provide results as refractive indices, others, particularly handheld, quick-to-use units, are equipped with scales calibrated to read the percentage of solids, percentage of sugars, and the like, depending on the products for which they are intended. Tables are provided with the instruments to convert values and adjust for temperature differences. Refractometers are used not just on the laboratory bench or as handheld units. Refractometers can be installed in a liquid processing line to monitor the °Brix of products such as carbonated soft drinks, dissolved solids in orange juice, and the percentage of solids in milk [22].

When a beam of light is passed from one medium to another and the density of the two differs, then the beam of light is bent or refracted. Bending of the light beam is a function of the media and the sines of the angles of incidence and refraction at any given temperature and pressure and is thus a constant (Fig. 15.8). The (RI) (η) is a ratio of the sines of the angles:

$$\eta = \frac{\text{sine incident ray angle}}{\text{sine refracted ray angle}} \qquad (15.10)$$

All chemical compounds have an index of refraction. Therefore, this measurement can be used for the qualitative identification of an unknown compound by comparing its RI with literature values. RI varies with **concentration** of the compound, **temperature**, and **wavelength of light**. Instruments are designed to give a reading by passing a light beam of a specific wavelength through a glass prism into a liquid, the sample. Benchtop or handheld units use **Amici prisms** to obtain the **D line of the sodium spectrum** or

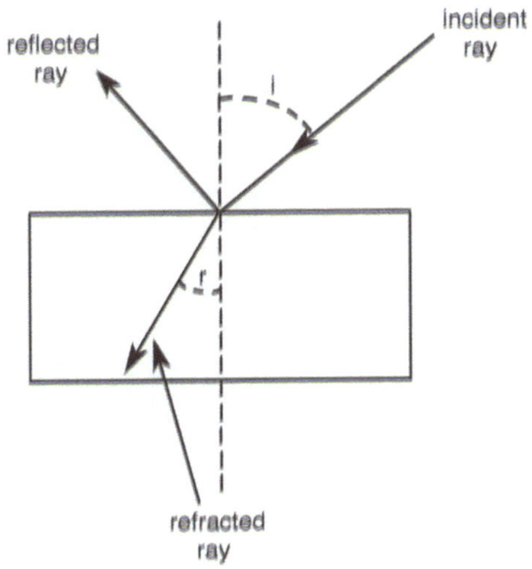

Fig. 15.8 Reflection and refraction concepts of refractometry

589 nm from white light. Whenever refractive indices of standard fluids are given, these are prefaced with η_D^{20} = a value from 1.3000 to 1.7000. The Greek letter η is the symbol for RI; the 20 refers to temperature in °C; and D is the wavelength of the light beam, the D line of the sodium spectrum.

Benchtop instruments are more accurate compared to handheld units mainly because of temperature control (Fig. 15.9). These former units have provisions for water circulation through the head where the prism and sample meet. Digital and **Abbe refractometers** are common for laboratory use. Care must be taken when cleaning the prism surface following use. The contact surface should be wiped clean with lens paper and rinsed with distilled water and then ethanol. The prism chamber should be closed and the instrument covered with a bag when not in use to protect the delicate prism surface from dust or other debris that might lead to scratches and inaccuracy.

The fact that the RI of a solution increases with concentration has been exploited in the analysis of total soluble solids of carbohydrate-based foods such as sugar syrups, fruit products, and tomato products. Because of this use, these refractometers are calibrated in °**Brix** (g of sucrose/100 g of sample), which is equivalent to percentage sucrose on a wt/wt basis. Refractive index measurements are used widely to approximate sugar concentration in foods, even though values are accurate only for pure sucrose solutions.

15.2.5.4 Infrared Analysis

Infrared spectroscopy (see Chap. 8) has attained a primary position in monitoring the composition of food products before, during, and following processing [23]. It has a wide range of food applications and has proven successful in the laboratory and online. Similar to the use of ultraviolet (UV) or visible (Vis) light in UV–Vis spectroscopy, in infrared

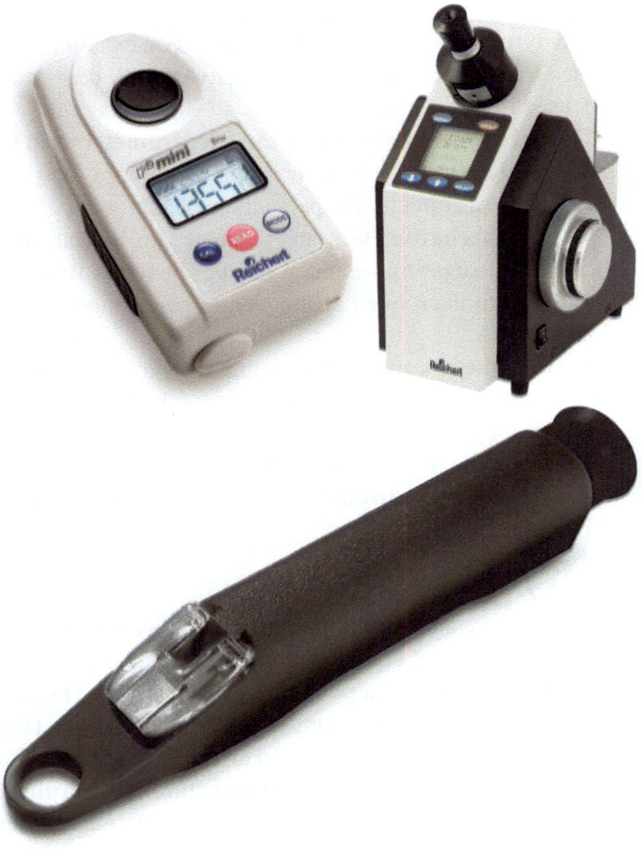

Fig. 15.9 Rhino Brix handheld refractometer, R^2 mini digital handheld refractometer, and Mark III Abbe refractometer. (Courtesy of Reichert Analytical Instruments, Depew, NY)

spectroscopy, a sample is exposed to IR radiation (near IR 700–2400 nm or mid-IR 2500–25000 nm), specific wavelengths are absorbed, and an IR spectrum is measured by calculating the intensity of the IR radiation before and after passing through the sample. The absorbance peaks are related to the type and amount of functional groups present. However, infrared spectrometers must be calibrated for each analyte to be measured, and the analyte must be uniformly distributed in the sample.

For water, near-infrared (NIR) bands (1400–1450; 1920–1950 nm) are characteristic of the -OH stretch of the water molecule and can be used to determine the moisture content of a food. NIR has been applied to moisture content analysis of a wide variety of food commodities and is an official method for moisture content determination in dried vegetables (AOAC Method 967.19).

The use of mid-infrared milk analyzers to determine fat, protein, lactose, and total solids in milk (AOAC Method 972.16) is covered in Chap. 8 of this text. The instrument must be calibrated using a minimum of eight milk samples that were previously analyzed for fat (F), protein (P), lactose (L), and total solids (TS) by standard methods. Then, a mean difference value, a, is calculated for all samples used in calibration:

$$a = \Sigma(TS - F - P - L)/n \quad (15.11)$$

where:

a = solids not measurable by the F, P, and L methods
n = number of samples
F = fat percentage
P = protein percentage
L = lactose percentage
TS = total solids percentage

Total solids then can be determined from any infrared milk analyzer results by using the formula

$$TS = a + F + P + L \quad (15.12)$$

The a value is thus a standard value mathematically derived. Newer instruments have the algorithm in their computer software to ascertain this value automatically. Although not directly measured, the moisture content can be calculated by subtracting the total solids content from 100.

15.2.5.5 Microwave Absorption

The absorption of microwaves, which are electromagnetic waves with wavelength of 0.001–1 m and frequency of 0.3–300 GHz, can be used to determine the moisture contents of a wide variety of food products. The approach is based on **dielectric constant** or **permittivity value** differences between water and dry materials. The permittivity of most dry materials is much lower than that of water, and small changes in the amount of water in a sample lead to measurable changes in permittivity. As microwave energy is passed through a sample (<40 mm thick) placed between a microwave transmitter and receiver, the absorption at 2.450 GHz (the most widely used frequency in microwaves) is linearly related to moisture content [24]. After calibration against laboratory standards, moisture contents can be determined in ~2 s using the microwave absorption method. Power absorption and attenuation may be affected by the dimensions, surface area, temperature, and dielectric properties of some food ingredients; however, the technique is rapid, nondestructive, and useful for some heterogeneous, powdered, multilayered, and frozen foods [4, 25].

15.2.5.6 Freezing Point

When water is added to a food product, many of the physical constants are altered. Some properties of solutions depend on the number of solute particles as ions or molecules present. These properties are vapor pressure, freezing point, boiling point, and osmotic pressure. Measurement of any of these properties can be used to determine the concentration of solutes in a solution. However, the most commonly practiced assay for milk is the change of the freezing point value. It has economic importance with regard to both raw and pasteurized milk. The **freezing point** of milk is its most constant physical property. While termed a physical constant, the freezing point varies within narrow limits and the vast majority of samples from individual cows fall between -0.503 °C and -0.541 °C, with the average very close to −0.521 °C. Herd or bulk milk will exhibit a narrower range unless the supply was watered intentionally or accidentally or if the milk is from an area where severe drought has existed.

The AOAC Method 961.07 for water added to milk uses a **cryoscope** to test for freezing points and assumes a freezing point for normal milk of −0.527 °C. The Food and Drug Administration will reject all milk with freezing points above −0.507 °C. Since the difference between the freezing points of milk and water is slight and since the freezing point can be used to calculate the amount of water added, it is essential that the method be as precise as possible. The thermistor used can sense temperature change to 0.001 °C. The general technique is to supercool the solution and then induce crystallization by a vibrating reed. The temperature will rise rapidly to the freezing point or eutectic temperature as the water freezes. In the case of pure water, the temperature remains constant until all the water is frozen. In the case of milk, the temperature is read when there is no further temperature rise. Time required for the automated instruments is 1–2 min per prechilled sample.

15.2.6 Comparison of Moisture Content Determination Methods

15.2.6.1 Principles

Characteristics of the various moisture content analysis methods described in Sects. 15.2.2, 15.2.3, 15.2.4, and 15.2.5 are summarized in Table 15.5. Direct moisture content determination methods often remove water from a sample and determine moisture contents by mass, volumetry, or titration. Oven-drying methods involve the removal of moisture from the sample and then a weight determination of the solids remaining to calculate the moisture content (and consequently also the total solids content). Non-water volatiles can be lost during drying, but their loss is generally a negligible percentage of the amount of water lost. Distillation procedures also involve a separation of the moisture from the solids, and the moisture is quantitated directly by volume. Karl Fischer titration is based on chemical reactions of the moisture present, reflected as the amount of titrant used.

Indirect moisture content determination methods analyze some property of the food that is related to the presence of water. The dielectric method is based on electrical properties of water. Hydrometric methods are based on the relationship

Table 15.5 Comparison of moisture analysis methods

Method	*Principle*	*What is actually measured?*	*How is/was removed/reacted/identified/etc.?*	*Cautions/things to control*	*Advantages*	*Disadvantages*	*Typical applications*
Forced draft oven	Sample is heated in oven to evaporate water. Weight loss equals moisture content	Weight change	Heat evaporates water when it boils at 100 °C	Control time, temperature, and particle size. Must predry some samples to avoid splattering	Easy to handle many samples at one time	Takes long time to get results. High temperature can cause loss of volatiles, lipid oxidation, Maillard browning, and sucrose hydrolysis, so not suitable for some types of foods	Official method for many types of samples. Not suitable for rapid quality control results. Not suitable for samples subject to loss of volatiles, lipid oxidation, Maillard browning, or sucrose hydrolysis
Vacuum oven	Sample is heated in oven under reduced pressure, so water evaporates at a lower temperature. Weight loss equals moisture content	Weight change	Heat sample under reduced pressure to evaporate water at −70 °C	Control time and temperature. Pull and release vacuum slowly	Easy to handle many samples at one time. Lower temperature for evaporating water reduces problems with high-sugar products	Takes long time to get results (though usually less time than with forced draft oven). More expensive than forced draft oven	An official method for many types of products. Not suitable for rapid quality control. Not suitable for powdered products, since they can blow around when vacuum is pulled and released
Microwave drying oven	Sample is heated with microwave energy to evaporate water. Weight loss equals moisture content	Weight change	Heat from microwave energy causes water evaporation	Control power and time to prevent sample decomposition. Spread sample evenly. Check calibration of analytical balance	Rapid	More expensive than other drying methods listed. Can only run one sample at a time	Suitable for rapid quality control, especially for liquid products, since us of pads avoids splattering
Infrared drying oven	Infrared lamp supplies heat that penetrates sample to evaporate water. Weight loss equals moisture content	Weight change	Heat from infrared lamp evaporates water	Control time and temperature. Spread sample evenly	Rapid	Expensive. Can only run one sample at a time	Suitable for rapid quality control, but not for high moisture products (would get splattering)
Rapid moisture analyzer	Sample is heated with heating elements to evaporate water. Weight loss equals moisture content	Weight change	Heat evaporates water when it boils at 100 °C	Control time and temperature. Spread sample evenly. Regular calibration of analytical balance	Rapid	Expensive. Can only run one sample at a time	Suitable for rapid quality control, but not for high moisture products (would get splattering)
Reflux distillation (with toluene)	When sample is heated in toluene (an immiscible liquid), the toluene and water are co-distilled. Collected moisture distills off, is condensed and collected, and volume of water is measured	Volume of after from sample collected after distillation and condensation	Co-distilled water from sample with toluene. Collect water and measure volume	Any emulsion formed must break to read volume of water. Need very clean glassware with no water. Use caution with solvents (fire hazards; toxic)	Causes less thermal decomposition of some foods than oven drying. Solvent protects sample from losing volatiles and minimized oxidation. Water is measured directly	Can only run one sample at a time. Solvent is likely flammable and toxic. Reading volume of water in receiving tube may be less accurate than by gravimetric method	AOAC method for spices

Karl Fischer	In titration of sample with Karl Fischer reagent, water in sample reacts with sulfur dioxide to cause reduction of iodine. Endpoint of titration is detected when excess iodine cannot react with water. Volume of titrant is used to calculate% moisture	Volume of Karl Fischer Reagent titrated	Water in sample reacts with iodine and sulfur dioxide to cause reduction of iodine	Control particle size of sample and humidity of room. Prevent any water in glassware. Must standardize KFR. Choose another methods if interferences from certain food constituents (e.g., ascorbic acid, carbonyl compounds, unsaturated fatty acids)	No heat, so no thermal decomposition. Rapid. Higher accuracy than many other methods for low-moisture foods	Can only run one sample at a time. Expensive, if using automated unit	Method of choice for many low-moisture foods (e.g., dried fruits and vegetables, candies, chocolate, roasted coffee, oils and fats, and many low-moisture foods that are high in sugar or protein). Good method to try if method with heating and/or vacuum gives erratic results
Hydrometer	Archimedes' principle. Compare relative density (specific gravity) of sample to that of water at same temperature	Volume displaced by hydrometer. Read specific gravity directly from hydrometer. Measuring solids content	Based on solids content of the solution, to determine specific gravity compared to pure water	Control temperature. Need clean hydrometer	Rapid. Easy. Inexpensive	Limited applications. Measures only solids content	Commonly used as rapid method to measure solids content of beverages, salt brines, and sugar solutions. Test applied to solutions with only one solute in a medium of water
Refractometer	Based on bending of light (i.e., refraction. Measure refractive index) as is hits surface of product. Refractive index can be used to determine concentration of compound of interest if nature of compound, temperature of sample, and wavelength of light are constant	Refractive index. Measuring solids content. Commonly calibrated in degree Brix (g of sucrose/100 g sample)	Based on solids content of solution, to determine refractive index	Control temperature. Need clean contact surface	Rapid. Easy. Inexpensive	Limited applications. Measures only solids content	Commonly used as rapid method to measure solids content of beverages and milk, and soluble solids of fruits and fruit products and of tomato products
Infrared analyzer	Measure absorption of infrared radiation at wavelength characteristic of the –OH stretch of the water molecule. Concentration of water is determined by energy that is reflected or transmitted, which is inversely proportional to energy absorbed	Amount of NIR light of specific wavelength reflected from sample	Molecular vibration of functional groups of water determines absorption, which is inversely related to reflected light that is measured	Must calibrate instrument for each type of product and each analyte being analyzed. Need to control sample particle size. Prevent scratches on glass container for sample. Remember that values obtained are only a prediction	Rapid. Easy. Can be used to estimate content of various food constituents	Expensive. Can only run one sample at a time. Values obtained are only estimates/predictions. Must calibrate the instrument for each analyte for each type of sample	Has wide range of food applications (in the laboratory, at-line, and on-line). NIR is heavily used in the grain/cereal industry to determine moisture, protein, and fat. HIR is heavily used in the dairy industry for total solids, fat, protein, and lactose in milk

Table 15.6 Common food industry uses of moisture content and water activity assays

Technique		Where used: Production, quality control	Product development	Basic research
Water content measurement				
Direct method	Forced draft oven drying	X	X	X
	Vacuum oven drying	X	X	X
	Microwave analyzer	X	X	
	Infrared drying	X	X	
	Rapid moisture analyzer technology	X	X	
	Thermogravimetric analyzer		X	X
	Lyophilization		X	X
	Chemical desiccation		X	X
	Karl Fischer titration	X	X	X
Indirect method	Dielectric capacitance	X		
	Hydrometer	X		
	Pycnometer	X		
	Refractometer	X	X	X
	NIR spectroscopy (absorbance or reflectance)	X	X	X
	Freezing point, cryoscope		X	X
	Microwave absorption	X	X	X
	Conductivity	X	X	
Water activity measurement				
	Dewpoint analyzer	X	X	X
	Electric (capacitance or electrolyte) hygrometer	X	X	X
	Freezing point depression			X
	Tunable diode laser sensor		X	X

between specific gravity and moisture content. The refractive index method is based on how water in a sample affects the refraction of light. Near-infrared analysis of water in foods is based on measuring the absorption at wavelength characteristic of the molecular vibration in water. Microwave absorption is based on the dielectric properties of water. Freezing point is a physical property of milk that is changed by a change in solute concentration.

15.2.6.2 Nature of Sample

While many foods will tolerate oven drying at high temperatures, some foods contain volatiles that are lost at such temperatures. Some foods have constituents that undergo chemical reactions at high temperatures to generate or utilize water or other compounds, and these reactions affect the calculated moisture content. Vacuum oven drying at reduced temperatures may overcome such problems for some foods. However, a distillation technique is necessary for some food to minimize volatilization and decomposition. For foods very low in moisture or high in fats and sugars, Karl Fischer titration is often the method of choice. The use of a pycnometer, hydrometer, and refractometer requires liquid samples, ideally with limited constituents.

15.2.6.3 Intended Purposes

Moisture content analysis data may be needed quickly for quality control purposes, in which high accuracy may not be necessary. Of the oven-drying methods, microwave drying, infrared drying, and the rapid moisture analyzer techniques are the fastest. Some forced draft oven procedures require less than 1 h drying, but most forced draft oven and vacuum oven procedures require a much longer time. The electrical, hydrometric, refractive index, and microwave absorption methods are very rapid but often require correlation to less empirical methods. Oven-drying procedures are official methods for a variety of food products. Reflux distillation is an AOAC method for chocolate, dried vegetables, dried milk, and oils and fats. Such official methods are used for regulatory and nutrition labeling purposes. A survey of food industry usage of the moisture content assays is summarized in Table 15.6, wherein the most commonly used techniques across a variety of companies for production, product development, and basic research applications are identified.

15.3 WATER ACTIVITY

15.3.1 Overview

Water content alone is not a reliable indicator of food stability, since foods with the same water content differ in their perishability [6]. It is the water activity (a_w) of foods that has been correlated with microbial growth, physical properties, and chemical and enzymatic reactions.

The water activity (a_w) of foods varies greatly, as shown in Table 15.7. For referring to water activity, the *a* is lower case and the *w* is a subscript because an activity coefficient of water is being used to describe its energy state. Generally, foods with higher moisture contents also have higher a_ws, although the relationship between moisture content and a_w is not linear. **Water activity** is a thermodynamic property of water in foods defined as the ratio of the fugacity (or escaping tendency) of water in the food to the fugacity of pure water at the same temperature and pressure [4, 6, 26]. Because fugacity cannot be directly measured, a_w is more commonly determined as the ratio of the vapor pressure of water in a food (p) to the vapor pressure of water (p_o) at the same temperature and barometric pressure, as shown in Eq. 15.13:

$$a_w = \frac{p}{p_o} \tag{15.13}$$

This concept is similar to how the relative humidity (RH) of the atmosphere is determined. The relationship between the a_w and equilibrium relative humidity (ERH) can be expressed as $a_w = \text{ERH}/100$. Water activity is a dimensionless number between 0 (absolutely no water) and 1 (pure water).

Table 15.7 Water activity (25°C) of foods and saturated solutions of food ingredients and salts[a]

Foods		Saturated solutions			
		Food ingredients		*a_w control salts*	
Type	*a_w*	*Type*	*a_w*	*Type*	*a_w*
Potato chips	0.07	*Single ingredients*		LiCl	0.11
Hard candy	0.12	Malic acid	0.58	CH_3CO_2K	0.23
Crisp crackers	0.13–0.20	Fructose	0.62	$MgCl_2$	0.33
Sugar-free hard candy	0.25	Sorbitol	0.67	K_2CO_3	0.43
Crisp cookie	0.25	Glucose	0.74	$Mg(NO_3)_2$	0.53
Chewy cookie	0.55	Citric acid	0.78	$CoCl_2$	0.65
Honey	0.56	Xylitol	0.79	NaCl	0.75
Beef jerky	0.61	Sucrose	0.85	KCl	0.84
Gummy candy	0.66	Lactose	0.97	K_2SO_4	0.97
High-fructose corn syrup	0.75	Maltose	0.97		
Condensed milk, strawberry preserves	0.84	*Ingredient blends*			
Soy sauce	0.87	NaCl + fructose	0.45		
Salted butter	0.90	Fructose + citric acid	0.50		
Bread	0.94	NaCl + sucrose	0.65		
Reduced sugar ketchup, unsalted butter	0.97	NaCl + glucose	0.71		
Juices, milk, fruits, vegetables	0.98–0.99				

[a]All measurements were done using an AquaLab 4TE instrument (Decagon Devices, Pullman, WA)

There are fewer official methods for determining the a_w of foods than there are for determining the moisture content [4, 15]. AOAC Method 978.18 describes techniques to determine the a_w of canned vegetables: in the regulation of acidified foods, an a_w of 0.85 is used as the cutoff for pathogen growth. In 21CFR Part 114 [27], low-acid foods are defined as foods with an equilibrium pH of >4.6 and a_w >0.85 and acidified foods are defined as low-acid foods to which acid has been added to create a finished equilibrium pH of ≤4.6 and that have an a_w >0.85. Foods with a_w <0.85 are not covered by 21 CFR Part 108, 113, or 114 [27]; however, it is important to recognize that lowering the a_w is not a kill step.

15.3.2 Importance of Water Activity

The a_w of foods affects important quality/acceptance (color, flavor, and texture), composition, and safety factors, thus the shelf-life of foods, and therefore a_w measurements are commonly used in food development and production. The a_w influences all aspects of microbial growth, and each microorganism has a threshold a_w below which it will not grow, although microorganisms may not be dead below this threshold a_w [28]. Understanding which spoilage or pathogenic microorganisms are a concern for a specific food and their a_w limits for growth can provide a foundation for formulation efforts to reduce the a_w of the food below that level. Because of the direct correlation between a_w and microbial growth, there are many food safety regulations that incorporate a_w guidelines (21CFR 110, 113, and 114 [27]; ANSI/NSF (American National Standards Institute/National Sanitation Foundation) Standard 75 [29]; ISO 21807:2004(E) [30]; AOAC Method 978.18 [15]), and a_w can be a critical control point in Hazard Analysis Critical Control Point (HACCP) plans.

Food stability maps have been created to display the relationships between water activity, microbial growth, chemical and biochemical reaction rates, and physical properties [31]. An example of a general map for an amorphous food is provided in Fig. 15.10, although the reaction rates and properties will vary across different a_w ranges for different foods. In addition to controlling microbial growth, maintaining the desired a_w of a product is a critical aspect in maintaining its color, flavor, texture, and composition throughout shelf life and there will be critical a_ws beyond which unwanted chemical and/or physical changes, or rates thereof, will occur in most products [32]. Chemical reactivity and reaction rates for many reactions that are associated with molecular mobility and the ability of water to act as a reactant and a solvent generally increase as a_w increases up to a maximum, followed by a decrease in reactivity with further a_w increases as

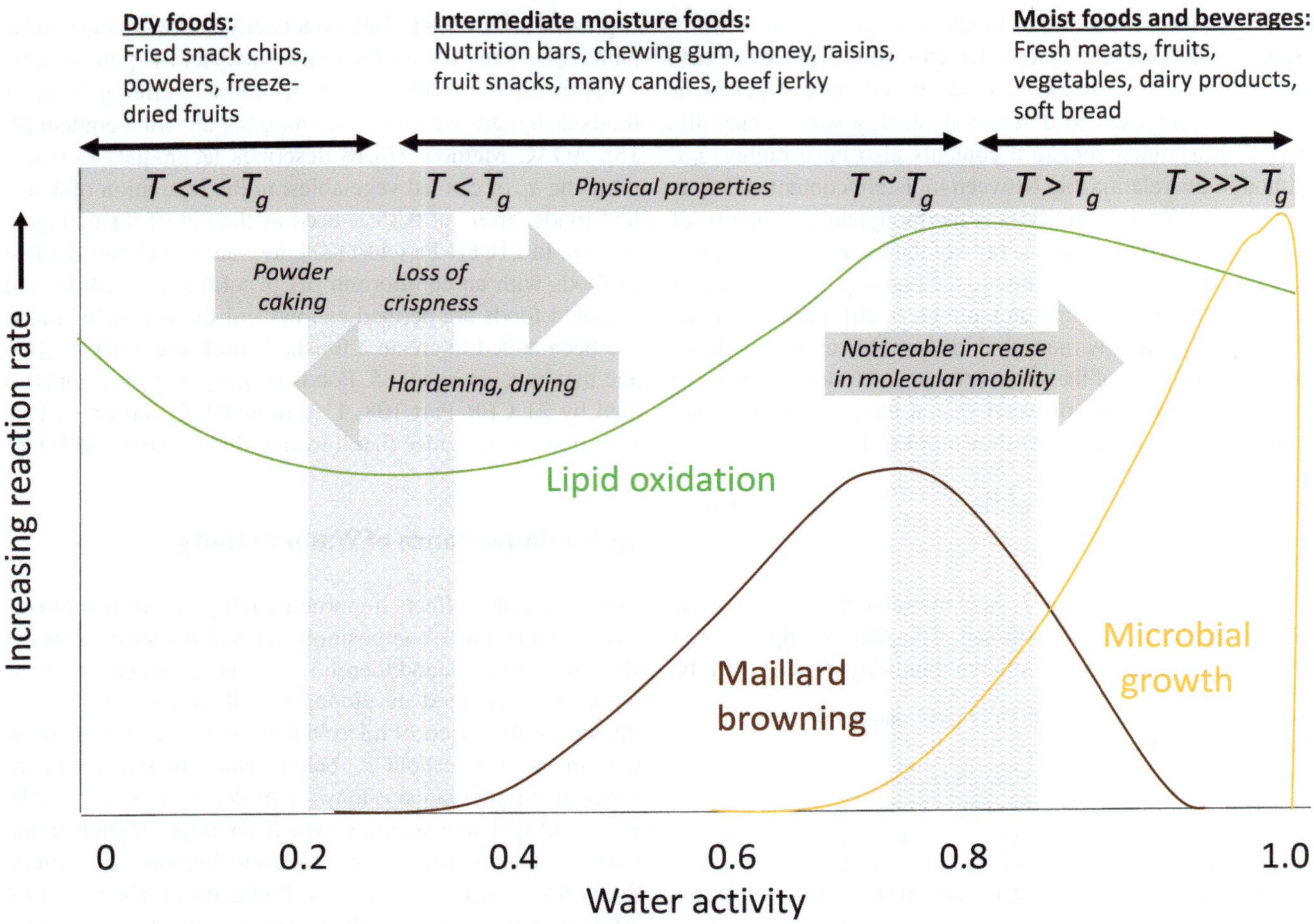

Fig. 15.10 Food stability map. (Adapted from Labuza et al. [31] and overlaid with data from Table 15.7)

dilution occurs (as shown for the Maillard reaction in Fig. 15.10) [33]. Such chemical changes can be associated with [33]: browning or other color change, which can be monitored as described in Chap. 31; vitamin degradation (including ascorbic acid, riboflavin, and thiamin), which has implications for compliance of nutrition labeling (refer to Chap. 3) and can be monitored as described in Chap. 20; and changes in flavor or taste (e.g., hydrolysis of aspartame). Lipid oxidation also influences flavor and shelf life but is affected differently by a_w (Fig. 15.10), occurring at high rates at very low a_ws (in the dry state) in which oxygen can diffuse freely and water does not inhibit the ability of metal ions to initiate free radical formation or quench free radicals. Analytical techniques for monitoring lipid oxidation are described in Chap. 23.

Because there is a thermodynamic driving force for water to move from areas of higher a_w to areas of lower a_w until equilibrium a_w is reached, if a crispy fried product (potato chip) is exposed to an environment with a higher RH than its a_w, moisture will migrate into the potato chip, increasing its a_w and leading to a softening of its texture once the critical a_w and moisture content are exceeded. Likewise, if a food product contains multiple components with different a_ws (cheese and cracker, pizza sauce on crust, pastry with filling, cereal flakes and fruit pieces, etc.), water migrates from the region of high a_w to the region of low a_w until a_w equilibration is attained, which can result in a hardening of one phase and a softening of the other. If these changes are unwanted, then reformulation of the phases to the same a_w (if possible) or physical separation of the phases in barrier packaging could extend shelf life. Increasing the a_w above a critical value can lead to clumping and caking of powders, recrystallization of sugars or salts in some products, a glassy-to-rubbery transition in amorphous solids, or even deliquescence of water-soluble deliquescent crystalline compounds. Changes in texture associated with water migration can be monitored by a variety of techniques, including the rheological methods described in Chap. 29.

The shelf life of a food is the period of time a food remains satisfactory in terms of quality, safety, composition, and general consumer acceptance. When the shelf life of a food can be defined by the extent of change of a measurable trait (e.g., vitamin content, status of lipid oxidation, crisp texture, microbial population, etc.), the relationships between reaction rates, a_w, and other factors such as temperature enable the prediction of shelf life by determining the rate constants of the reactions at several different storage conditions. Detailed descriptions of shelf life testing and prediction of shelf life for a wide range of food products can be found elsewhere [34, 35].

15.3.3 Water Activity Measurement

15.3.3.1 Principles

Since $a_w = p/p_o$, determining the a_w often involves the direct measurement of vapor pressures, although there are also indirect measurements for a_w. For the majority of a_w measurement techniques, a representative food sample is sealed in a container and allowed to reach equilibrium (i.e., the water will migrate out of the sample into the container headspace, or vice versa, until the a_w of the food and the RH of the headspace are equal) (Fig. 15.11). Enough sample must be used to assure that moisture migration during equilibration will not dry the sample but not so much sample that the sensor is compromised. A general recommendation is to fill half of the container with sample. More sample might be warranted for dense, dry products, while less sample could be used for wet products or liquids (AOAC Method 978.18 recommends canned vegetables fill >1/20 of the sample container volume).

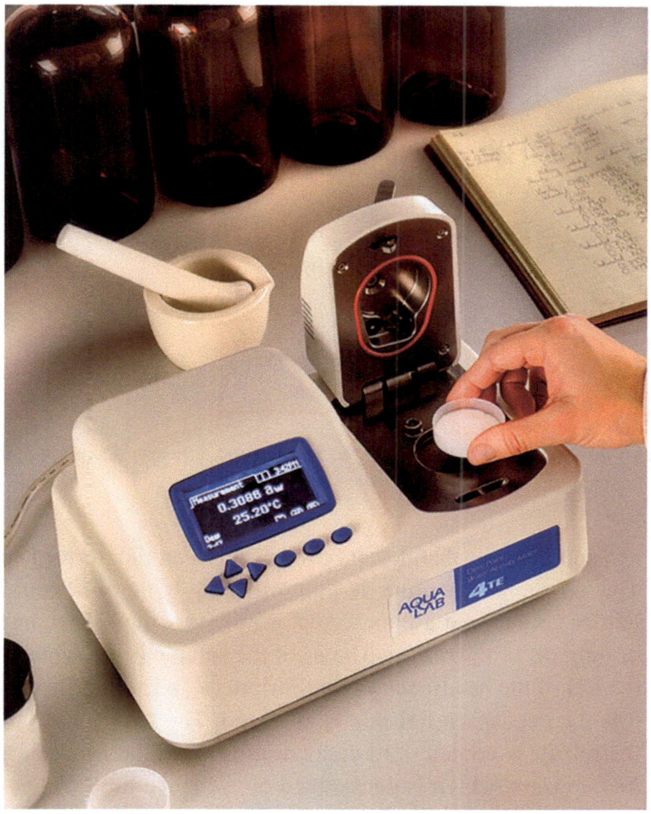

Fig. 15.11 Dewpoint water activity meter. (Courtesy of Decagon Devices, Pullman, WA)

The time required to attain headspace equilibrium can vary between samples and between assay types. Slow water-emitting samples (dense, dry, dehydrated, high fat, high viscosity) may require several hours for precise a_w measurement, while samples containing more water or those with water as the continuous phase may equilibrate much faster. For establishing the equilibrium criterion for a sample, the a_w could be monitored over time, such as at 15, 30, 60, and 120 min (AOAC Method 978.18), extending out in 60 min intervals until consecutive readings vary by <0.01 a_w (or other designated equilibrium criteria for a particular product). While some samples may require 24 h for equilibration, many foods (barring the slow water emitters) may reach equilibrium within 5 min for dewpoint analyses or 30–90 min for electric hygrometers [36].

Temperature is an important factor in equilibration and a_w—higher temperatures often lead to higher a_w readings for many foods if the moisture content is kept constant (water molecules move faster as temperature is increased). This is why temperature is reported along with a_w values. Additionally, inequalities in temperature between samples and containers or equipment can lead to unwanted water condensation and/or errors in determining the a_w. For example, a 1 °C temperature difference between sample and dewpoint leads to an a_w error of 0.06 in dewpoint instruments [4, 37]. The majority of reported a_w values were measured at 25 °C (as stipulated in AOAC Method 978.18) or 20 °C, although the temperatures of interest for a particular product may vary (particularly for refrigerated or frozen products).

It is important to use a calibrated sensor for a_w measurements. Calibration is often done using salt solution standards of known a_w (Table 15.7). It is important to comply with instrument manufacturer guidelines for cleaning, calibrating, and maintaining a_w instruments. Calibration may involve at least three measurement points (ISO 21807) or ≥5 salts (AOAC Method 978.18) that encompass the a_w range of interest for the sample and sensor to generate a standard curve and/or determine an offset adjustment. Once the instrument is calibrated and temperature and headspace equilibrium are reached, the headspace is analyzed to determine the a_w. There are a variety of sensors and methods to directly or indirectly determine the equilibrium water vapor pressure in the headspace, although only the most commonly used by the food industry (Tables 15.6 and 15.8) will be further described. Examples of a_w methods include dewpoint measurement, electric hygrometer sensors (including capacitance and electrolyte sensors), direct measurement of

Table 15.8 Comparison of water activity measurement methods

Method	*Instrument*	a_w *range*	*Accuracy*	*Repeatability*	*Resolution*	*Internal temperature control*	*Read time*	*Volatile interference*
Dewpoint	AquaLab4 (Decagon Devices)	0.030–1.000	0.003	0.001	0.0001	15–50 °C	2–5 min	Yes
Electric (resistive electrolytic based on electrolyte)	LabMaster-*aw* (Novasina)	0.03–1.000	0.003	0.001	0.001	0–50 °C	10–15 min	Yes[a]
Electric (capacitance)	HygroLab C1 (Rotronic)	0.0–1.000	0.008	0.001	0.001	no	4–6 min	Minimal[b]
Electric (capacitance)	Aw Therm (Rotronic)	0.005–1.000	0.005	0.001	0.001	10–60 °C	4–6 min	Minimal[b]
Tunable diode laser	TDL (Decagon Devices)	0.03–1.000	0.005	0.001	0.0001	15–50 °C	2–5 min	No

Adapted from Fontana and Campbell [4] and personal communications with Brady Carter (Decagon Devices), Markus Bernasconi (Novasina), and Harry Trainor and Rico Hasler (Rotronic)

[a]For the LabMaster-aw, filters can be used to block out low to medium levels of volatiles, or use the CM-3 sensor that is resilient to alcohol

[b]The sensors in the HygroLab C1 and Aw Therm instruments are affected by volatiles when subjected to high concentrations and long-term exposure. A single a_w measurement in 4–6 min is not enough exposure in most cases for volatiles to affect the analysis

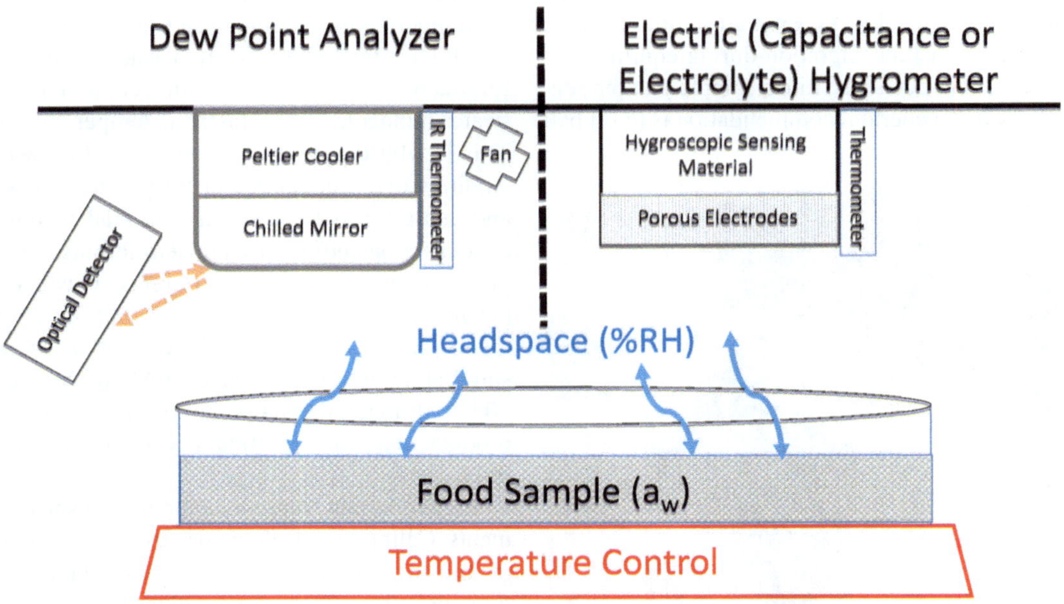

Fig. 15.12 Schematic of water activity measurement chambers for the two most common techniques used by the food industry: dewpoint and capacitance/electrolyte sensors

manometric pressure, change in length of a hair/thread, increase in sorbent mass in an isopiestic method, tunable diode laser sensor, thermocouple psychrometry, and freezing point determination [5, 38].

15.3.3.2 Chilled Mirror Dewpoint

In a chilled mirror dewpoint instrument, such as a Decagon Devices AquaLab (Fig. 15.12), a sample is equilibrated in a temperature-controlled sealed chamber that contains a fan to circulate air in the headspace, an infrared thermometer to measure sample temperature, a temperature-controlled mirror, and a sensor that detects condensation on the mirror [5]. The mirror temperature is controlled by a thermoelectric Peltier cooler, and a photodetector monitors the reflectance of light off of the mirror. The a_w of the sample is the same as the RH of the headspace at equilibrium (a_w = ERH/100), and it is the headspace RH that is analyzed. The temperature of the mirror is cooled until condensation from the headspace first appears, which is determined by a change in the reflectance. When condensation occurs, the temperatures of the sample and the mirror are recorded. The temperature of the sample is used to determine the water vapor pressure (p_o), and the dewpoint temperature is used to determine the vapor pressure of water in the headspace (p). The a_w of the sample is determined by applying Eq. 15.13, $a_w = p/p_o$. Volatiles that may condense on the mirror, such as alcohols, acetic acid, or propylene glycol, interfere with accurate a_w measurements using dewpoint techniques.

15.3.3.3 Electric Hygrometer

In an electric hygrometer, which for our purposes will include capacitance and electrolyte sensors such as Novasina and Rotronic instruments, a sample container is equilibrated in a chamber with a potentiometer and a sensor (such as an electrolytic moisture sensor) that measures the headspace RH and temperature [36]. Some models have internal temperature control features, while others may require an incubator or water-jacketed sample chamber for temperature control. The sensors monitor the electrical conductivity of their immobilized salt solutions or a hygroscopic polymer (the sensor composition is proprietary) that sorb or desorb water in response to the headspace RH. Changes in electric conductivity are calibrated to reflect changes in the headspace RH, and a_w is calculated as a_w = ERH/100. Volatile compounds may interfere with a_w measurements by electric hygrometers, and tunable diode laser techniques may be a better option for accurately determining the a_w of samples that contain alcohols, acetic acid, or propylene glycol.

15.4 MOISTURE SORPTION ISOTHERMS

15.4.1 Overview

A **moisture sorption isotherm** is a plot of the equilibrium relationship between a sample's moisture content and a_w at a constant temperature and pressure. The **hygrocapacity**, or water-holding capacity, of a sample is the amount of water it contains when equilibrated at a set a_w. Hygrocapacity depends on the affinity between water and the sample, temperature, RH, surface area, and water–solid interaction mechanism [39]. A key to establishing moisture sorption isotherms is to ensure that the samples have equilibrated at the set a_ws. Different samples exhibit different kinetics of moisture sorption and desorption. In the absence of equilibrium, a plot of moisture content versus a_w would simply be called a **sorption profile** (not an isotherm).

Generally as aw increases, moisture content also increases although the relationship is most certainly not linear. Different types of solids and different foods have different shapes of moisture sorption profiles, and these shapes have been organized into six major types (I–IV) [40]. Type II and III are most common for foods. The inflection points and slopes of the different isotherms provide useful information about hygroscopicity, potential for reformulating or drying to alter aw, how a sample might respond to different environments or formulations, and stability (microbial, physical, and chemical). Inflection points in sorption profiles occur during phase changes, such as a glassy-to-rubbery transition, hydrate formation, or deliquescence. Recrystallization events are often accompanied by expulsion of water from the matrix and a corresponding weight loss. Overlaying isotherms from different food types is a useful practice when considering co-formulation and multicomponent foods, keeping in mind that water does not move if the aws of the different components are equal. If differences in aw are initially present between the different food components, water will migrate until aw equilibrium is reached.

There are several approaches and instruments used to generate moisture sorption isotherms (or profiles), although there are no government regulations or AOAC official methods dictating techniques and parameters. A description of two common approaches for moisture sorption isotherm measurement is provided below.

15.4.2 Isopiestic Desiccator Method

In desiccator methods, samples are equilibrated in RH-controlled desiccators at set temperatures and, upon equilibrium, their moisture contents are determined by either a mass change from a known starting amount or any of the moisture content methods described above. The RH in the desiccators is most often controlled by saturated salt solutions (Table 15.7) with the exception of the 0% RH condition, which is created using calcium sulfate (DrieRite™) or phosphorous pentoxide (P_2O_5), the latter of which is more effective [41]. Enough saturated salt solution and headspace must be present to maintain a constant RH as moisture is exchanged with stored samples. General recommendations are to use ≥10% of the container volume filled with salt solution, >10:1 ratio of salt solution surface area to sample surface area, and a 20:1 ratio of headspace volume to sample volume. Vacuum may or may not be applied to the desiccators. Six to nine different RH chambers are recommended to generate a moisture sorption isotherm for a product [42]. Temperature is controlled by placing the desiccators in temperature-controlled incubators, ideally with temperature variation <1 °C. Samples should be stored and analyzed in triplicate over time, weighing to ±0.0001 g until some equilibrium criterion is reached. For example, equilibrium could be defined as <0.01% mass change across three consecutive days of weighings. The amount of time it takes for samples to reach equilibrium varies by sample type and RH condition, often from 10 to 21 days. Desiccator methods are time and labor intensive, there is potential for RH fluctuation as samples are removed from the desiccators for weighing, and challenges with chemical change or microbial growth in samples during extended equilibration may be encountered. There are a limited number RHs generated by saturated salt solutions; therefore, equations (such as the GAB and BET models) must be used to calculate sorption profiles from the measured data points [42]. However, the relative simplicity of the technique and ability to store multiple samples are attractive features of using the desiccator method to generate moisture sorption isotherms.

15.4.3 Automated Gravimetric Moisture Sorption Balance

In automated gravimetric moisture sorption balances, sometimes referred to as static dynamic vapor sorption (DVS) instruments, the mass of a sample is monitored as a function of time, while the sample is exposed to a series of tightly controlled RH step changes at constant temperature. An example of data collected using this technique is provided in Fig. 15.13. The initial moisture content of the sample must be known, or a sample can be dried in the instrument prior to analysis. The parameters used in generating automated moisture sorption isotherms (or profiles), including equilibrium criterion (% weight change in a set time period, e.g., 0.001% or 0.01%), step time (how long will a set RH be held before the next programmed RH step is initiated, in the absence of attaining the equilibrium criterion), and RH step size (1% RH, 5% RH, 10% RH, etc.), have a profound influence on the curves generated (Fig. 15.14).

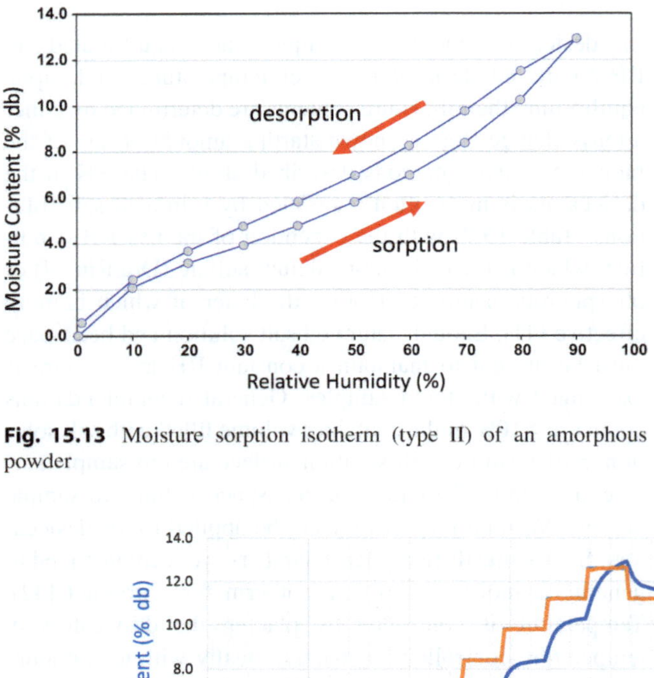

Fig. 15.13 Moisture sorption isotherm (type II) of an amorphous powder

When a DVS instrument moves to a programmed RH step, air with the programmed RH is moved into the sample chamber and the mass of the sample is monitored as it sorbs (or desorbs) water. Usually, the initial sorption is more rapid than the later stages of sorption, as seen in Fig. 15.14. If true equilibrium is reached between the sample and the surrounding environmental RH, then no more moisture uptake would occur and the equilibrium criterion would be 0.00% mass change over any time scale. However, due to the labile nature of some foods and the need to collect data for foods over a period of days not months, an equilibrium criterion of 0.01% weight change over some time scale (1–5 h) is more common. It is important to recognize that the shorter the step time, the more likely the sample is not in equilibrium with the surrounding RH resulting in sorption profiles and not isotherms.

Careful control and reporting of test parameters are important for reproducibility of moisture sorption isotherms. Advantages of DVS methods compared to desiccator methods include precise RH control and mass determinations, a smaller amount of sample (5 mg–1.5 g), smaller RH step sizes, and possibly a reduced time to generate a moisture sorption isotherm for a single sample. Disadvantages of DVS methods include high instrumentation costs, the upper RH limit is 95% RH (or lower RH at high temperatures), and a more limited sample size [many DVS instruments only analyze one sample at a time, although there is currently one that analyzes up to 23 samples simultaneously (SPS, ProUmid GmbH & Co., Germany)].

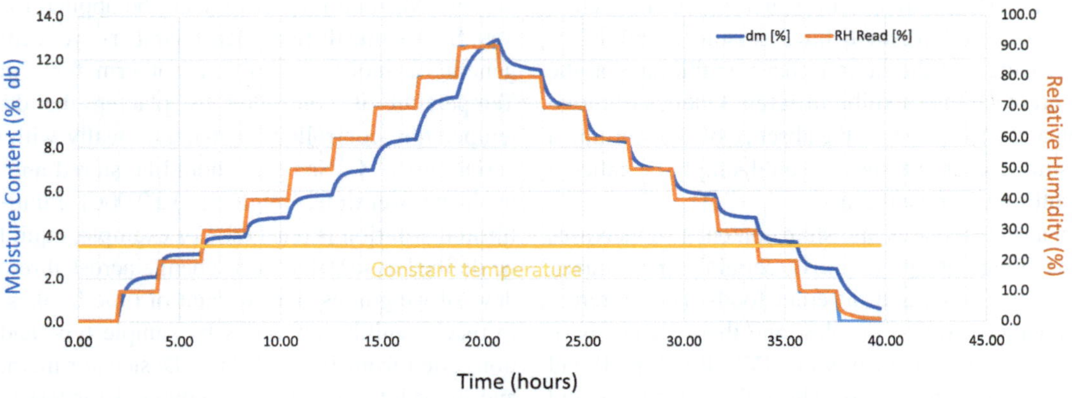

Fig. 15.14 Moisture sorption data for an amorphous powder collected on a DVS instrument. (SPS, ProUmid GmbH, Germany) and used to generate the moisture sorption isotherm shown in Fig. 15.13

A modified approach to the DVS instrumentation is found in the AquaLab Vapor Sorption Analyzer (Decagon Devices, Pullman, WA) operated in DDI mode. In the DDI mode, a small amount of saturated (100% RH) or dry (0% RH) air is introduced into the sample chamber, the chamber is sealed, the RH is continuously monitored by a capacitance RH sensor until a constant RH is reached (on average 5–6 min), and at this point both the mass and the water activity of the sample (based on the chilled mirror dewpoint technique) are recorded. While it is not possible to program RH steps or timing between measurements into the DDI technique, the DDI sorption profiles are very high resolution with numerous data points collected and may be collected in less time than the static DVS moisture sorption isotherms.

15.4.4 Phase Diagrams Containing a_w, Moisture Content, and T_g Relationships

In addition to routine moisture analyses for quality control, the food industry is increasingly taking a fundamental materials science approach to designing foods and processes and controlling food quality. An important practice here is to establish not only moisture sorption isotherm relationships for foods but also to establish the relationship between moisture content and glass transition temperature (T_g) as well as the relationship between water activity and glass transition temperature (T_g). More detailed information about T_g measurements is provided in Chap. 31.

15.5 SUMMARY

The moisture content and water activity of foods are important to food processors and consumers for a variety of reasons. While moisture determination may seem simplistic, it is often one of the most difficult assays in obtaining accurate and precise results. Direct moisture content analysis methods involve a separation of moisture in the sample from the solids and then quantitation by weight or volume. Indirect methods do not involve such a separation but instead are based on some physical or chemical property of the sample that varies with respect to its moisture content. A major difficulty with many methods is attempting to remove or otherwise quantitate all water present. This often is complicated by decomposition or interference by other food constituents. For each moisture analysis method, there are factors that must be controlled or precautions that must be taken to ensure accurate and precise results. Careful sample collection and handling procedures are extremely important and cannot be overemphasized. The choice of moisture analysis method is often determined by the expected moisture content, nature of other food constituents (e.g., highly volatile, heat sensitive), equipment available, speed necessary, accuracy and precision required, and intended purpose (e.g., regulatory or in-plant quality control).

15.6 STUDY QUESTIONS

1. Identify five factors that one would need to consider when choosing a moisture content analysis method for a specific food product.
2. Why is standardized methodology needed for moisture content determinations?
3. What are the potential advantages of using a vacuum oven rather than a forced draft oven for moisture content determination?
4. In each case specified below, would you likely overestimate or underestimate the moisture content of a food product being tested? Explain your answer.
   (a) Forced draft oven:
   - Particle size too large
   - High concentration of volatile flavor compounds present
   - Lipid oxidation
   - Sample very hygroscopic
   - Alteration of carbohydrates (e.g., Maillard browning)
   - Sucrose hydrolysis
   - Surface crust formation
   - Splattering
   - Desiccator containing DrieRite™ or P_2O_5 desiccant (~0% RH) with dried sample not sealed properly

   (b) Toluene distillation:
   - Emulsion between water in sample and solvent not broken
   - Water clinging to condenser

   (c) Karl Fischer:
   - Very humid day when weighing original samples
   - Glassware not dry
   - Sample ground coarsely
   - Food high in vitamin C
   - Food high in unsaturated fatty acids
5. The procedure for an analysis for moisture in a liquid food product requires the addition of 1–2 mL of deionized water to the weighed sample in the moisture pan. Why should you add water to an analysis in which moisture is being determined?
6. A new instrument based on near-infrared principles has been received in your laboratory to be used in moisture analysis. Briefly describe the way you would ascertain if the new instrument would meet your satisfaction and company standards.

7. A technician you supervise is to determine the moisture content of a food product by the Karl Fischer method. Your technician wants to know what is this "Karl Fischer reagent water equivalence" that is used in the equation to calculate percentage of moisture in the sample, why is it necessary, and how it is determined. Give the technician your answer.
8. Moisture Analysis Specific to Spices and Low-Moisture Fruit Product: For the food products listed below (i.e., spices, low moisture fruit products), a hot air oven would not be an appropriate method to determine the moisture content.
   (a) Would a forced draft oven drying cause an over- or underestimates of moisture content? Explain your answer on a chemical basis.

	Spices	Low-Moisture Fruit Product
Overestimate or Underestimate?		
Why?		

   (b) What method would you choose to use to more accurately determine for official purposes the moisture content of these products, and what is it about that method that makes it more suitable than use of a hot air oven?

	Spices	Low-Moisture Fruit Product
Method of choice?		
Why this method?		

9. Identify a moisture analysis method for each application.
   (a) One rapid, quality control moisture analysis method based on drying.
   (b) One rapid, quality control moisture analysis method **not** based on drying but suitable for a **solid** sample.
   (c) One rapid, quality control moisture analysis method **not** based on drying but suitable for simple, **liquid** sample.
10. You are fortunate to have available in your laboratory the equipment for doing moisture content analysis by essentially all methods—both official and rapid quality control methods. For each of the food products listed below (with the purpose specified as rapid quality control or official), indicate (a) the name of the method you would use, (b) the principle (not procedure) for the method, (c) a justification for use of that method (as compared to using a hot air-drying oven), and (d) two cautions in use of the method to ensure accurate results.
    (a) Ice cream mix (liquid)—quality control
    (b) Milk chocolate—official
    (c) Spices—official
    (d) Syrup for canned peaches—quality control
    (e) Oat flour—quality control
11. You are a manufacturer of processed cheese. The maximum allowed moisture content for your product is 40%. Your current product has a mean moisture content of 38%, with a standard deviation of 0.7. It would be possible to increase your mean moisture content to 39.5% if you could reduce your standard deviation to 0.25. This would result in a saving of $3.4 million per year. You can accomplish this by rapidly analyzing the moisture content of the cheese blend prior to the cooking step of manufacture. The cheese blend is prepared in a batch process, and you have 10 min to adjust the moisture content of each batch:
    (a) Describe the rapid moisture analysis method you would use. Include your rationale for selecting the method.
    (b) How would you ensure the accuracy and precision of this method (you need to be sure your standard deviation is below 0.25)?
12. You work in a milk drying plant. As part of the production process, you need to rapidly analyze the moisture content of condensed milk.
    (a) What rapid secondary method would you use, and what primary method would you use to calibrate the secondary method? Additionally, how would you ensure the accuracy and precision of your secondary method?
    (b) Your results with the secondary method are consistently high (about 1%), based on the secondary method you chose. What are some potential problems and how would your correct them?
13. During a 12 h period, 1000 blocks (40 lbs each) from ten different vats (100 blocks per vat) of Cheddar cheese were produced. It was later realized that the cooking temperature was too low during cheesemaking. You are concerned that this might increase the moisture content of the cheese above the legal requirement. Describe the sampling plan and method of analysis you would use to determine the moisture content of the cheese. You want the results within 48 h so you can determine what to do with the cheese.
14. Compare and contrast moisture content and water activity in terms of measurement approaches, effects on food texture and moisture migration with the environment, influence on microbial growth, and roles in vitamin stability, lipid oxidation, and the Maillard reaction.
15. Identify three factors that would lead to underestimation of the water activity of a food sample. Then identify three factors that would lead to overestimation of the water activity of a food sample.

16. You are put in charge of creating moisture sorption isotherms of several food products. Describe what method you will use and what approach you will take for assuring you have created moisture sorption isotherms and not moisture sorption profiles of:
    (a) Potato chips
    (b) An intermediate moisture nutrition bar containing oats, chocolate pieces, and a binder syrup
    (c) A gummy candy
    (d) Cheddar cheese

15.7 PRACTICE PROBLEMS

1. As an analyst, you are given a sample of condensed soup to analyze to determine if it is reduced to the correct concentration. By gravimetric means, you find that the concentration is 26.54% solids. The company standard reads 28.63%. If the starting volume was 1000 gal at 8.67% solids, and the weight is 8.5 lb per gallon, how much more water must be removed?
2. Your laboratory just received several sample containers of peas to analyze for moisture content. There is a visible condensate on the inside of the container. What is your procedure to obtain a result?
3. You have the following gravimetric results: weight of dried pan and glass disk is 1.0376 g, weight of pan and liquid sample is 4.6274 g, and weight of the pan and dried sample is 1.7321 g. What was the moisture content of the sample and what is the percent solids?

Answers

1. The weight of the soup initially is superfluous information. By condensing the soup to 26.54% solids from 8.67% solids, the volume is reduced to 326.7 gal [(8.67%/26.54%) × 1000 gal]. You need to reduce the volume further to obtain 28.63% solids [(8.67%/28.63%) × 1000 gal], or 302.8 gal. The difference in the gallons obtained is 23.9 gal (326.7 gal–302.8 gal) or the volume of water that must be removed from the partially condensed soup to comply with company standards.
2. This problem focuses on a real issue in the food processing industry—when do you analyze a sample and when don't you? It would appear that the peas have lost moisture that should be within the vegetable for correct results. You will need to grind the peas in a food mill or blender. If the peas are in a Mason jar or one that fits a blender head, no transfer is needed. Blend the peas to a creamy texture. If a container transfer was made, then put the blended peas back into the original container. Mix with the residual moisture to a uniform blend. Collect a sample for moisture analysis. You should note on the report form containing the results of the analysis that the pea samples had free moisture on container walls when they arrived.
3. Note Eqs. 15.1, 15.2, and 15.3 in Sect. 15.2.2.1.7. To use any of the equations, you must subtract the weight of the dried pan and glass disk. Then, you obtain 3.5898 g of original sample and 0.6945 g when dried. By subtracting these results, you have removed water (2.8953 g). Then (0.6945 g/3.5898 g) × 100 = 19.35% solids and (2.8953 g/3.5898 g) × 100 = 80.65% water.

Acknowledgment The author of this chapter thanks Robert L. Bradley, Jr (Professor Emeritus, Department of Food Science, University of Wisconsin, Madison, WI, USA) for his contribution to the chapter in the first four editions of the book.

REFERENCES

1. Pomeranz Y, Meloan C (1994) Food analysis: theory and practice, 3rd edn. Chapman & Hall, New York
2. Josyln MA (1970) Methods in food analysis, 2nd edn. Academic, New York
3. US Department of Agriculture, Agricultural Research Service (2016). USDA National Nutrient Database for Standard References, Release 28. https://data.nal.usda.gov/dataset/usda-national-nutrient-database-standard-reference-legacy-release
4. Fontana Jr AJ, Campbell CS (2004) Water activity (Chapter 3). In: Nollet L (ed.) Handbook of Food Analysis, 2nd edn. Marcel Dekker, Inc., New York p. 39–54
5. Fontana Jr AJ, Carter BP (2020) Measurement of water activity, moisture sorption isotherms, and moisture content of foods (Chapter 8). In: Barbosa-Canovas BV, Fontana AJ, Schmidt SJ, Labuza TP (eds). Water activity in foods: Fundamentals and applications. Second edition. John Wiley & Sons, Inc and Institute of Food Technologists, Chicago, IL, p. 207–226
6. Reid DS, Fennema OR (2007) Water and ice (Chapter 2). In: Damodaran S, Parkin K, Fennema OR (eds). Fennema's food chemistry, 4th edn. CRC, Boca Raton, FL
7. Arunan EDG, Klein RA, Sadlej J, Scheiner S, Alkorta I, Clary DC, Crabtree RH, Dannenbery JJ, Hobza P, Kjaergaard HG, Legon AC, Mennucci B, Nesbitt DJ (2011) Definition of the hydrogen bond (IUPAC recommendations 2011). Pure Applied Chemistry, 83(8):1637–1641
8. Martiniano H, Galamba, N (2013) Insights on hydrogen-bond lifetimes in liquid and supercooled water. J Physical Chem B, 117:16188–16195
9. CRC Handbook of Chemistry and Physics, 96th ed. (2015) CRC Press: Boca Raton, FL
10. Schmidt S (2004)Water and solids mobility in foods. Adv Food and Nutr Res., 48:1–101
11. Zografi G (1988) States of water associated with solids. Drug Development And Industrial Pharmacy, 14(14):1905–1926
12. Emmons DB, Bradley RL Jr, Sauvé J P, Campbell C, Lacroix C, Jimenez-Marquez SA (2001) Variations of moisture measurements in cheese. J AOAC Int., 84:593–604
13. Emmons DB, Bradley RL Jr, Campbell C, Sauve JP (2001) Movement of moisture in refrigeration cheese samples transferred to room temperature. J AOAC, 84:620–622

14. Bradley RL Jr, Vanderwarn MA (2001) Determination of moisture in cheese and cheese products. J AOAC Int., 84:570–592
15. AOAC International (2023) Official methods of analysis, 22nd edn. (online). AOAC International, Rockville, MD
16. Wehr HM, Frank JF (eds) (2004) Standard methods for the examination of dairy products, 17th edn. American Public Health Association, Washington, DC
17. AACC International (2010) Approved methods of analysis, 11th edn. AACC International, St. Paul, MN
18. Nelson OA, Hulett GA (1920) The moisture content of cereals. J Ind Eng Chem., 12:40–45
19. International Dairy Federation (1982) Provisional Standard 4A. Cheese and processed cheese: determination of the total solids content. International Dairy Federation, Brussels, Belgium
20. Bouraoui M, Richard P, Fichtali J (1993) A review of moisture content determination in foods using microwave oven drying. Food Res Inst Int., 26:49–57
21. Mitchell J Jr, Smith DM (1948) Aquametry. Wiley, New York
22. Giese J (1993) In-line sensors for food processing. Food Technol., 47(5):87–95
23. Wilson RH, Kemsley EK (1992) On-line process monitoring using infrared techniques. Food processing automation II proceedings of the American society of agricultural engineers, ASAE Publication 02-92. St. Joseph, MI: American Society of Agricultural Engineers
24. Pande A (1974) Handbook of moisture determination and control. New York: Marcel Dekker
25. Anonymous Kett Microwave Moisture Meters Santiago, CA: Kett; 2009 [cited 2023]. Available from: www.kett.com.
26. Reid DS (2007) Water activity: Fundamentals and relationships. In: Barbosa-Canovas BV, Fontana AJ, Schmidt SJ, Labuza TP (eds). Water activity in foods. Ames, IA: Blackwell Publishing; p. 15–28
27. Anonymous (2023) Code of Federal Regulations. Title 21 Parts 108, 113, 114. Washington DC: US Government Printing Office
28. Scott WJ (1957) Water relations of food spoilage microorganisms. Advances in Food Research, 7:83–127
29. International N. (2000) Non-potentially hazardous foods. Ann Arbor, MI: 2000 Nov. 10. Report No.: ANSI/NSF 75-2000
30. Anonymous (2017) International Standard ISO 18787: Microbiology of food and animal feeding stuffs - Determination of water activity. Switzerland: International Organization for Standardization
31. Labuza TP, Tannenbaum SR, Karel M (1970) Water content and stability of low moisture and intermediate moisture foods. Journal of Food Technology, 24:543–550
32. Barbosa-Canovas GV, Fontana Jr. AJ, Schmidt SJ, Labuza TP (2020) Water activity in foods: Fundamentals and applications. Second edition. John Wiley & Sons, Inc and Institute of Food Technologists, Chicago, IL
33. Bell LN (2020) Moisture effects on food's chemical stability (Chapter 9). In: Barbosa-Canovas BV, Fontana AJ, Schmidt SJ, Labuza TP (eds). Water activity in foods: Fundamentals and applications. Second edition. John Wiley & Sons, Inc and Institute of Food Technologists, Chicago, IL, p. 227–254
34. Taormina PJ, Hardin MD (2021) Food safety and quality-based shelf life of perishable foods. Springer Cham, Switzerland.
35. Steele R (2004) Understanding and measuring the shelf-life of foods. First edition. Woodhead Publishing.
36. Scott VN, Clabero RS, Troller JA (2001) Chapter 64. Measurement of water activity (a_w), acidity, and brix. In: Downes FP, Ito K, (eds). Compendium of Methods for the Microbiological Examination of Foods, 4th Edn: American Public Health Association, p. 649–657
37. Decagon Devices (2009) AquaLab water activity meter operator's manual, Version 4. Pullman, WA: Decagon Devices
38. Rodel W (2001)Water activity and its measurement in food (Chapter 16). In: Kress-Rogers E, Brimelow CJB (eds). Instrumentation and Sensors for the Food Industry. Cambridge: Woodhead, p. 453–843
39. Airaksinen S, Karjalainen M, Shevchenko A, Westermarck S, Leppanen E, Rantanen J, et al. (2005) Role of water in the physical stability of solid dosage formulations. J Pharm Sci., 2147–2165
40. Sing KSW, Everett DH, Haul RAW, Moscou L, Pierotti RA, Rouquerol J, et al. (1985) Reporting physisorption data for gas/solid systems with special reference to the determination of surface area and porosity. Pure and Applied Chemistry, 57(4):603–619
41. Trussel F, Diehl H (1963) Efficiency of chemical desiccants. Analytical Chemistry, 35:674–677
42. Bell LN, Labuza TP (2000) Moisture Sorption: Practical Aspects of Isotherm Measurement and Use. 2nd Edn. St. Paul, MN: American Association of Cereal Chemists

Ash Analysis

16

G. Keith Harris

Contents

16.1 INTRODUCTION

Ashing is an important step in proximate or specific mineral analysis. **Ash** refers to the inorganic (mineral) residue remaining after complete combustion or complete acid-facilitated oxidation of organic matter in food. A basic knowledge of the characteristics of ashing procedures and types of equipment is essential to ensure reliable results. Two major types of ashing are used: dry ashing and wet ashing. Dry ashing is used as part of proximate analysis to estimate total mineral content and as a preparatory step for specific mineral analyses. Dry ashing requires very high (500–600 °C) temperatures, which may be achieved by conventional or microwave heating. Wet ashing combines acid-facilitated oxidation and lower temperatures (100–120 °C) to eliminate organic material and is required for the analysis of minerals that might be volatilized and lost during dry ashing. Wet ashing methods often involve the use of pressure vessels to contain acids and samples, allowing for higher temperatures and faster processing. Microwave systems also can be used for both dry and wet ashing. In either case, microwave systems tend to speed up the ashing process, although sample throughput may be a limiting factor. Most dry samples (i.e., whole grain, cereals, and dried vegetables) need no preparation for ashing, while fresh vegetables and other high-moisture foods are generally dried prior to ashing to prevent rapid boiling and sample loss. High-fat products, such as meats, may need to be dried and fat-extracted before ashing to prevent smoke generation during heating steps. Flammable volatiles, like ethanol, should also be eliminated prior to ashing. Ashing is a gravimetric analysis, in which the final (ashed) weight is compared to the original weight of the sample. The ash content of foods can then be expressed on either a wet weight or a dry weight basis. For general and food-specific information on measuring ash content for humans and animals, see references [1–16].

G. K. Harris (✉)
Department of Food, Bioprocessing and Nutrition Sciences, North Carolina State University, Raleigh, NC, USA
e-mail: keith_harris@ncsu.edu

B. P. Ismail, S. S. Nielsen (eds.), *Nielsen's Food Analysis*, Food Science Text Series,
https://doi.org/10.1007/978-3-031-50643-7_16

16.1.1 Definitions

Dry ashing refers to the use of a muffle furnace capable of maintaining temperatures of 500–600 °C. Water and volatiles are vaporized, and organic substances are burned in the presence of oxygen to form CO_2 and oxides of N_2. In this way, organic materials are removed from the sample. The remaining minerals are converted to oxides, sulfates, phosphates, chlorides, and silicates. Some elements partially volatilize with this procedure, so other methods must be used if ashing is a preliminary step for the analysis of these minerals.

Wet ashing is a procedure used to chemically oxidize and remove organic substances using strong acids, oxidizing agents, and lower temperatures (100–120 °C) than for dry ashing. The acids and oxidizing agents used must also be capable of solubilizing the minerals to be analyzed. Hydrochloric, sulfuric, and nitricacids are commonly used. Perchloric acid was commonly used in the past for wet ashing (AOAC Method 975.03), but safety concerns over its formation of explosive peroxides have largely eliminated its use (so it will not be further discussed in this chapter). Since wet ashing is conducted at lower temperatures than dry ashing and may use sealed pressure vessels, minerals are not lost due to volatility. For this reason, wet ashing is preferable to dry ashing as a preparation for comprehensive mineral analysis of samples.

16.1.2 Importance of Ash in Food Analysis

Ash represents an estimate of the total mineral content of foods. Determining ash content is important for several reasons. First, it is used to determine the total mineral content of foods as part of proximate analysis. Second, ashing is used to prepare food samples for specific mineral analysis, whether for essential nutrients or for toxic heavy metals. Because certain foods are high in particular minerals, ash analysis can be important from nutritional, food quality, and toxicological standpoints. For example, dairy and beef are known to be rich sources of calcium and iron, respectively. The mineral content of plant foods tends to be more variable than that of animal products and is dependent on the soil in which it is grown. For example, rice grown in arsenic-containing soil effectively concentrates it. High levels of transition metals in lipid-rich foods may speed rancidity and limit shelf life. While ash is not reported directly on foods for human consumption, it is commonly part of the ingredient specifications for flours and wholegrains and may be reported on pet food labels.

Table 16.1 Ash content of selected foods

Food item	*Percent ash (wet weight basis)*
Cereals, bread, and pasta	
Rice, brown, long grain, raw	1.5
Corn meal, whole grain, yellow	1.1
Hominy, canned, white	0.9
White rice, long-grain, regular, raw,	0.6
enriched	1.6
Wheat flour, whole grain	0.9
Macaroni, dry, enriched	2.5
Rye bread	
Dairy products	
Milk, reduced fat, fluid, 2%	0.7
Evaporated milk, canned, with added	1.6
vitamin A	2.1
Butter, with salt	0.7
Cream, fluid, half and half	2.0
Margarine, hard, regular, soybean	1.1
Yogurt, plain, low fat	
Fruits and vegetables	
Apples, raw, with skin	0.2
Bananas, raw	0.8
Cherries, sweet, raw	0.5
Raisins	1.9
Potatoes, raw, skin	1.6
Tomatoes, red, ripe, raw	0.5
Meat, poultry, and fish	
Eggs, whole, raw, fresh	0.9
Fish fillet, battered or breaded and	2.5
fried	0.9
Pork, fresh, leg (ham), whole, raw	1.9
Hamburger, regular, single patty, plain	1.0
Chicken, broilers or fryers, breast meat	1.1
only, raw	6.8
Beef, chuck, arm pot roast, raw	
Beef jerky	

From the US Department of Agriculture, Agricultural Research Service [16], Food Data Central fdc.nal.usda.gov

16.1.3 Ash Contents in Foods

The ash content of food can range from 0 to 12%. The highest ash values are found in dry foods, such as beef jerky or dried vegetables, but rarely exceed 5% for fresh, moisture-containing foods. The average ash content for various food groups is given in Table 16.1.

16.2 METHODS

Principles, materials, instrumentation, general procedures, and applications are described below for various ash determination methods. Refer to the methods cited for detailed descriptions of the procedures.

16.2.1 Sample Preparation

Ash analysis does not require large sample sizes. A 2–10 g of sample is generally sufficient. Some newer instruments allow for sample sizes as low as 250 mg. For this reason, it is necessary to acquire a homogenous, representative sample for accurate results. Milling, grinding, and other methods used to reduce particle size and homogenize samples will not significantly alter ash values. In contrast, mineral contamination from the environment or from grinding equipment can introduce error, but the use of sample blanks can account for this. (This is particularly true of metals such as iron, since most grinders and mincers are of steel construction.) Crucibles and glassware also can be a source of contaminants unless they are carefully cleaned between samples. This can be addressed by soaking crucibles and glassware in an acid bath to solubilize mineral contaminants and rinsing repeatedly with distilled-deionized water. The crucibles themselves can also be "pre-ashed" in a muffle furnace to remove organic contaminants.

16.2.1.1 Plant Materials

Plant materials are generally oven-dried prior to grinding and ashing to remove moisture. Since dried samples often are used for multiple determinations (e.g., protein, fiber, and lipid), it may be necessary to keep oven temperatures at or below 100 °C to prevent the destruction or alteration of non-mineral analytes. Fresh stem and leaf tissue should be dried in two stages (i.e., first at a lower temperature of 55 °C and then at a higher temperature) to prevent formation of artifact lignin and other undesired products. Plant material with 15% or less moisture may be ashed without prior drying. Low-temperature drying can be accomplished via the use of a muffle furnace equipped with temperature gradient settings. Samples can thus be dried at low temperature and then ashed in the same crucible.

16.2.1.2 Macronutrient Composition Effects

Animal products, syrups, and spices require special consideration prior to ashing because of high fat, protein, moisture, or high sugar content that may result in loss of sample and underestimation of ash content due to spattering, swelling, or foaming. Conversely, high-sugar foods may form intractable tars that can lead to overestimation of ash. Depending on the application, crucible covers or the use of sealed crucibles may be used to contain samples. Meats, sugars, and syrups should be evaporated to dryness on a steam bath or with an infrared (IR) lamp prior to ashing. For products that tend to form a crust on heating, one or two drops of refined olive oil (which contains no ash) may be added to allow steam to escape as the crust forms.

Smoking and burning may occur when products high in fat and protein (e.g., cheese, spices, and peanuts) are dry-ashed. This process is allowed to finish slowly by keeping the muffle door slightly open until combustion stops, then the door is fully closed, and the normal ashing protocol is proceeded. High-fat samples may also be ashed after drying and fat extraction to minimize smoke production. In most cases, mineral loss is minimal during drying and fat extraction. Fat-extracted samples should not be heated until flammable extraction solvents (e.g., hexane and ether) have been evaporated completely to avoid the potential risk of solvent ignition or explosion. All of these procedures require the use of fume hoods to avoid release of smoke into the laboratory. Most of these macronutrient-related issues apply to dry ashing, which is described in more detail in the section that follows.

16.2.2 Dry Ashing

16.2.2.1 Principles and Instrumentation

Dry ashing is incineration at high temperature (525 °C or higher). Sample incineration is accomplished with a muffle furnace, an oven that can reach temperatures in excess of 1000 °C. Several models of muffle furnaces are available, ranging from large capacity units requiring either 208- or 240-V electric supplies to small benchtop units utilizing 110-V outlets. The muffle furnace must be placed in a fume hood and may have to be placed in a heat-proof room along with drying ovens. It is important that large muffle furnaces are equipped with a double-pole, single-throw switch. Heating coils are generally exposed, and care must be taken when taking samples in and out with metal tongs.

Crucibles are used to contain dry ashing samples. Lids are sometimes used to better contain samples. Crucible selection is critical in ashing because the type depends on the specific use. Primary considerations relate to chemical stability and resistance to high temperatures. **Quartz crucibles** are resistant to acids and halogens, but not alkali, at high temperatures. **Vycor® brand crucibles** are stable to 900 °C, but **Pyrex® Gooch crucibles** are limited to 500 °C. Ashing at a lower temperature of 500–525 °C may result in slightly higher ash values due to incomplete decomposition of carbonates. **Porcelain crucibles** resemble quartz crucibles in their properties but will crack with rapid temperature changes. Since they are relatively inexpensive, porcelain crucibles are often the crucible of choice. **Steel crucibles** are resistant to both acids and alkalis and are inexpensive, but they are composed of iron, chromium, and nickel, which can erroneously increase apparent ash content. **Platinum crucibles** are very inert and are probably the best overall, but they are currently far too expensive for routine use for large numbers of samples. **Quartz fiber crucibles** are disposable and unbreakable and can withstand temperatures of up to 1000 °C. They are porous, allowing air to circulate around the sample and speed combustion. This reduces ashing times significantly and makes them ideal for solids and viscous liquids. Quartz fiber also cools in seconds, virtually eliminating the risk of burns.

All crucibles should be marked for identification. Marks on crucibles with a felt-tip marking pen will disappear during ashing in a muffle furnace. Laboratory inks scribed with a steel pin are available commercially. Crucibles also may be etched with a diamond point and marked with a 0.5 M solution of $FeCl_3$, in 20% HCl. An iron nail dissolved in concentrated HCl forms a thick brown pigment that is a satisfactory marker. The crucibles should be fired and cleaned prior to use.

16.2.2.2 Procedures

AOAC International has several dry ashing procedures (e.g., AOAC Methods 900.02 A or B, 920.117, 923.03, 942.05) for certain individual human and animal foodstuffs. The general procedure includes the following steps:

1. Weigh a 5- to 10-g sample into a tared crucible. Account for the weight of crucible covers if they are used. Predry high-moisture samples.
2. Place crucibles in cool muffle furnace using tongs. Use tongs, gloves, and protective eyewear if the muffle furnace is warm.
3. Heat samples for 12–18 h (or overnight) at about 550 °C.
4. Turn off the muffle furnace. Wait until the temperature has dropped to 250 °C or lower before opening the furnace door. Open the door slowly to avoid the loss of the powdery ash that may result from rapid air movement.
5. Using tongs, quickly but carefully transfer crucibles to a desiccator with a porcelain plate and desiccant. Cover crucibles, close desiccator, and allow crucibles to cool to room temperature prior to weighing.

Note

Avoid placing hot samples into desiccators for two reasons. First, this could cause sudden displacement of the desiccator cover if sufficient pressure is generated. Second, as hot samples cool, a vacuum may form, making it difficult to remove the desiccator cover. Desiccator covers should always be removed gradually by sliding them to one side to prevent a sudden inrush of air. Covers with a ground-glass sleeve or fitted for a rubber stopper allow for a slower release of a vacuum.

The ash content is calculated as follows:

$$\%\text{Ash}\left(\text{dry basis}\right) = \frac{\left(\text{WAA} - \text{TWOC}\right)}{\left(\text{OSW x DMC}\right)} \times 100 \qquad (16.1)$$

where:

WAA = weight after ashing
TWOC = tare weight of crucible
OSW = original sample weight
DMC = dry matter coefficient (percent sample solids/100)

Consider one of three options for ash calculations. Use the DMC, or dry matter coefficient, to directly convert wet weight percent ash values to dry weight ash values. For example, if corn meal is known to be 87% dry matter, the dry matter coefficient would be 0.87. Dry weight ash content can be determined without the DMC if samples are dried at low temperatures in the same crucible prior to ashing. The denominator becomes (dry sample weight – tared crucible weight) and values are reported as dry weight. This method will be more accurate than the DMC method for dry weight determination but adds an extra, time-consuming step. If DMC values are unknown and samples have not been predried, the dry matter coefficient is deleted from the denominator and percent ash values are reported on an as-received, fresh, or wet weight basis, as appropriate.

16.2.2.3 Special Applications

Some AOAC procedures recommend steps in addition to those listed previously. If carbon is still present following the initial ashing (residual carbon will appear as a black, tar-like material), several drops of water or nitric acid should be added and samples should be re-ashed. If the carbon persists, which may occur with high-sugar samples, follow this procedure:

1. Suspend carbon-containing ash samples in water.
2. Filter samples through ashless filter paper because this residue tends to form a glaze.
3. Dry the filter and filtrate together.
4. Place paper and dried filtrate in muffle furnace and re-ash, subtracting the weight of the filter paper to determine ash percent.

Other suggestions that may be helpful to accelerate the ashing process:

1. High-fat samples should be solvent-extracted either using a crude fat determination procedure or, less ideally, by burning off lipids prior to closing the muffle furnace.
2. Mix high sugar, high-moisture samples, like jelly, with cotton wool to avoid spatter.
3. Salt-rich foods may require a separate ashing of water-insoluble components and salt-rich water extract. Use crucible covers to prevent spattering.
4. An alcoholic solution of magnesium acetate can be added to accelerate ashing of cereals. An appropriate blank determination is necessary.

The *advantages* of dry ashing are that it is a relatively safe and inexpensive method, that it requires no acids, added reagents or blank subtraction, that sample recovery is high, and that little attention is needed once ignition begins [2]. Depending on the oven size, a large number of crucibles can be handled at once, and the resultant ash can be used additionally in other analyses for most individual elements, acid-insoluble ash, and water-soluble and insoluble ash. The *disadvantages* of dry ashing are the length of time required (12–18 h or overnight) and the expense of equipment relative to traditional wet ashing. There will be some loss of select elements due to volatilization and to interactions between mineral components and crucibles. Volatile elements at risk of being lost include As, B, Cd, Cr, Cu, Fe, Pb, Hg, Ni, P, V, and Zn.

16.2.3 Wet Ashing

16.2.3.1 Principles, Materials, and Applications

Wet ashing is sometimes called **wet oxidation** or **wet digestion**. Its primary uses are to get a more accurate, comprehensive representation of percent ash than from dry ashing and to prepare for specific mineral analyses post-ashing.

The phasing out of perchloric acid for wet ashing (mentioned in Sect. 16.1.1) as a sole oxidizing solvent has led to the use of other acids. Unfortunately, using a single acid for wet ashing does not give complete and rapid oxidation of organic material. So, different concentrations and ratios of nitric and sulfuric acids, sometimes with hydrogen peroxide added, are often employed.

16.2.3.2 Procedures

This is a wet ash procedure using concentrated nitric and sulfuric acids (*must be performed in a fume hood*) (John Budin, Silliker Laboratories, Chicago, IL, personal communication):

1. Accurately weigh a dried, ground 1 g sample in a 125-mL Erlenmeyer flask (previously acid-washed and dried).
2. Prepare blanks (3 mL H_2SO_4 plus 5 mL HNO_3) to be run with all samples sets.
3. Add 3 mLH_2SO_4 followed by 5 mL HNO_3 to sample flasks.
4. Heat samples to boiling on a hot plate. Brown-yellow fumes will be observed.
5. Once brown-yellow fumes cease and white fumes from decomposing H_2SO_4 are observed, samples will darken. Remove flasks from the hot plate but do not allow them to cool to room temperature.
6. *Slowly* add 3–5 mLHNO_3.
7. Put flasks back on hot plates and allow HNO_3 to boil off. Proceed to the next step when all HNO_3 has been removed (samples will be clear or straw yellow in color). If the solutions remain dark in color, add another 3–5 mL HNO_3 to flasks and reboil. Repeat the process until the clear to straw yellow color is obtained.
8. Continue boiling to reduce the volume appropriately to allow for ease of final transfer. Allow samples to cool to room temperature, then quantitatively transfer samples to appropriately sized volumetric flasks.
9. Dilute samples to volume with ultrapure water and mix well. Dilute further, as appropriate, for the specific mineral analyses.

A combination dry and wet ash procedure is given in AOAC Method 985.35 "Minerals in Infant Formula, Enteral Products, and Pet Foods."

There are at least two advantages to using the wet ashing procedure over dry ashing: (1) minerals typically stay in solution throughout the procedure, making them easier to transfer for specific mineral analyses, and (2) wet ashing is faster than dry ashing. The disadvantages of wet ashing, by the open vessel hot plate method, over dry ashing are that: (1) this type of wet ashing requires virtually constant operator attention, (2) corrosive reagents are used, requiring additional safety precautions, caustic waste disposal, and personal protective equipment, and (3) only small numbers of samples can be handled at any one time.

16.2.4 Microwave Ashing

Both **dry** and **wet ashing** can be done using microwave heating, instead of the conventional dry and wet ashing procedures described previously. Companies like the CEM Corporation (Matthews, NC) and Anton-Paar (Graz, Austria) have developed a series of microwave-heated instruments for accelerated dry and wet ashing procedures. While ashing by conventional means can take many hours, the use of microwave heating can reduce sample preparation time to minutes, allowing laboratories to increase their sample throughput significantly (current models have a throughput of 5–24 samples per hour). This advantage has led to widespread use of microwave ashing, especially for wet ashing, within both analytical and quality control laboratories in food companies.

16.2.4.1 Microwave Wet Ashing

Microwave wet ashing (acid digestion) may be performed safely in either an open- or closed-vessel microwave system. The choice of open vs closed systems depends on the amount of sample and the temperatures required for digesting. Due to the ability of the closed vessels to maintain high pressures (up to 1500 psi), acids may be heated well past their boiling points. The samples generate the pressure based on decomposing in the acid and producing CO_2 and NOx gases. The vessel then holds this pressure, which allows for temperatures above the boiling point. This ensures fast, complete dissolution of hard-to-digest substances. It also allows the chemist to use nitric acid with samples that might normally require harsher acid mixtures, such as sulfuric or perchloric acids. Closed vessels specifically designed for high-temperatures/high-pressure reactions, nitric acid can reach a temperature of 240 °C (However, the methods specified by the equipment manufacturers, FDA, and AOAC use 200° C.). **Nitric acid** is often the acid of choice for microwave digestion, though hydrochloric, hydrofluoric, and sulfuric acids are also used, depending on the sample and the subsequent analysis being performed. **Closed-vessel microwave digestion systems** (Fig. 16.1) can process up to 40 samples at a time, with vessel liners available in Teflon®, TFM™ fluoropolymer, and quartz. Closed-vessel wet ashing systems allow for control of time, temperature, and pressure parameters. Some instruments enable the user to adjust these conditions in real time. This closed-vessel microwave ashing approach was used to improve AOAC Method 984.27, which had previously used conventional wet or dry ashing to prepare samples for ICP analyses [14].

Typically, in a closed-vessel microwave system, samples are placed in vessels with the appropriate amount of acid. Vessels are sealed and loaded into the carousel that is then placed in the microwave cavity. Floor-mounted sensors are used to measure the temperature of each vessel as it passes over the sensor. Adjustable settings may include time, temperature, and power in multiple stages. Digestions normally take less than 30 min. Because of the pressure generated by raising the temperature of a reaction, vessels must be allowed to cool before being opened. The ability to process multiple samples simultaneously provides the chemist with greater throughput than traditional methods. (Note that some closed-vessel microwave digestion systems may also be used for solvent extraction, protein hydrolysis, and chemical synthesis with the proper accessories.)

16.2.4.2 Microwave Dry Ashing

Compared to conventional dry ashing in a muffle furnace, which often takes many hours, **microwave muffle furnaces** (Fig. 16.2) can ash samples in minutes, decreasing analysis time by as much as 97%. Microwave muffle furnaces can reach temperatures of up to 1200 °C. These systems may be programmed with various heating and cooling methods. In addition, they are equipped with exhaust systems that circulate the air in the cavity to help decrease ashing times. Some also have scrubber systems to neutralize harmful fumes, thus eliminating the need for and space taken up by fume hoods. Any crucible that may be used in a conventional muffle furnace may be used in a microwave furnace, including those

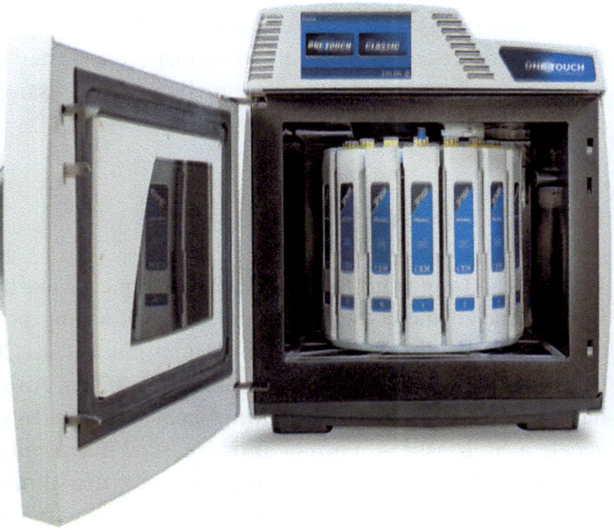

Fig. 16.1 Microwave closed-vessel digestion system. (MARS6; Courtesy of CEM Corporation, Matthews, NC)

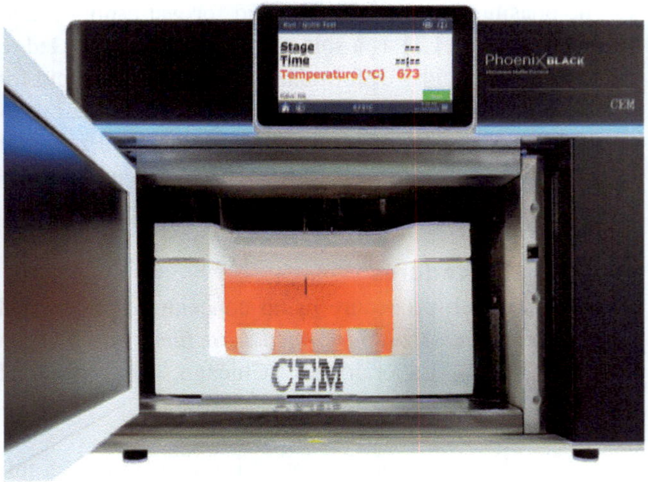

Fig. 16.2 Microwave muffle furnace. (Phoenix BLACK; Courtesy of CEM Corporation, Matthews, NC)

made of porcelain, platinum, quartz, and quartz fiber. Some systems can process up to 15 (25-mL) crucibles at a time.

Typically, in microwave dry ashing, a pre-ashed, dried crucible (one that has been stored in a desiccator) is weighed, sample is added, and the crucible is weighed again. The crucible and sample are placed in the microwave furnace, and time and temperature parameters are set. As with microwave-heated wet ashing, settings may include power, temperature, and pressure gradients. The system is started, and the program is run to completion. The crucible is carefully removed with tongs and reweighed. Tongs serve a dual purpose. They prevent contact with hot crucibles and prevent residues from the technician's hand or gloves from contaminating samples. The sample then may be further analyzed, if necessary. Some tests call for acid to be added to a dry-ashed sample, which may serve to enhance digestion or to solubilize minerals for further analysis.

A comparative study [15] showed that dry-ashing various plants for 40 min using a microwave system (CEM Corporation, Matthews NC) was similar to the 4 h time in a conventional muffle furnace. A time of 20 min was shown to be adequate for the plant material used except for Cu determinations, which required the full 40 min to obtain similar results. Other comparative examples include dried egg yolks, which can be ashed in 20 min in a microwave system but require 4 h in a conventional muffle furnace. It takes 16 h to ash lactose in a conventional muffle furnace, but only 35 min in a microwave furnace. Though microwave furnaces may not hold as many samples as a conventional furnace, their speed actually allows significantly more samples to be processed in the same amount of time. One additional advantage is that microwave furnaces typically do not require fume hood space.

16.2.5 Other Ash Measurements

The following are several special ash measurements and their applications:

1. **Soluble and Insoluble Ash** (e.g., AOAC Method 900.02). After dry-ashing a sample, ash is solubilized in boiling water and filtered. Water-soluble and water-insoluble fractions are then re-ashed. Soluble and insoluble ash are indices of the fruit content of preserves and jellies. Lower water-soluble ash indicates a higher percentage of fruit, since fruit contains a variety of minerals that could add to the insoluble fraction.
2. **Ash Insoluble in Acid**. After dry-ashing a sample, the ash is solubilized in 10% HCl, boiled, filtered, and the insoluble material is re-ashed. This may be used as a measure of the surface contamination of fruit, vegetables, wheat, and rice coatings. Contaminants are generally silicates and remain insoluble in acids, except HBr.
3. **Alkalinity of Ash** (e.g., AOAC Method 900.02, 940.26). Dry-ashed samples are treated with a known amount of 0.1 *N* HCl, then boiling water is added. Samples are titrated with 0.1 *N* NaOH using a methyl orange indicator. The volume of NaOH required indicates the alkalinity of the ash. The ash of fruits and vegetable is alkaline, while the ash of meats and some cereals is generally acidic.
4. **Sulfated Ash** (AOAC Method 900.02, 950.77). This method, applied mostly to sugars, syrups, and color additives, measures the amount of residual substance not volatilized when the sample is ignited in the presence of sulfuric acid. Samples are wetted with sulfuric acid and heated on a hot plate and then in a muffle furnace. This procedure is repeated until a constant weight is achieved, with the final weight expressed as % sulfated ash. (The procedure also can be performed in a microwave muffle furnace, e.g., CEM Corporation, Fig. 16.2).

16.3 COMPARISON OF METHODS

The four major ashing methods described in this chapter (dry and wet ashing using conventional vs microwave heating) are summarized and compared in Table 16.2. Dry ashing requires a muffle furnace, is more time-consuming, and uses more electricity but is less hazardous than wet ashing on a hot plate. Wet ashing using a hot plate requires a fume hood, use and disposal of corrosive reagents, and constant operator attention. While wet oxidation causes little volatilization of minerals, dry ashing will result in the loss of volatile elements. Both dry and wet ashing can be done using microwave systems that require relatively expensive instrumentation but greatly reduce the time needed for ashing and generally do not require use of a fume hood. If the intent is further elemental analyses, the specific elements being analyzed will dictate whether wet or dry ashing is used. Some minerals micronutrients and most volatile elements will require special equipment and procedures. Refer to Chaps. 9 and 21 for specific preparation procedures for mineral analyses.

Table 16.2 Summary of common ashing methods

Ashing method	*Principle*	*Sources of error*	*Advantages*	*Disadvantages*	*Applications*
Dry ashing	Sample heated to very high (500–600 °C) temperature. All organic matter incinerated. Remaining inorganic material quantitated gravimetrically	Microelement contamination (from grinder or water used to clean crucibles). Sample loss during pre-ash drying step. Volatilization of some elements. Incomplete combustion	Can analyze many samples at once. Requires little technician time. Safe. No blanks needed	Slow (takes 12–18 h). Some minerals volatilized. Minerals are difficult to resolubilize	Total ash content for proximate analysis. May be used as preparation for specific mineral analysis
Wet ashing	Organic matter oxidized using acids and oxidizing agents, leaving inorganic matter	Microelement contamination. Must run blanks to correct for organic matter in acid and oxidizing agent. Sample loss may occur due to spattering	Shorter time (~2 h) than dry ashing. Minerals stay in solution. Little or no volatilization	Requires constant attention. Lower sample throughput than dry ashing. Can be dangerous. Use of strong acids and oxidizers. Sample loss due to spattering	Ashing of samples prior to mineral analysis for official analyses
Microwave ashing (dry)	Microwave energy heats sample to very high temperatures, incinerates organic matter, leaves inorganic matter to be quantitated gravimetrically	Microelement contamination (from grinder or water used to clean crucibles). Sample loss during pre-ash drying step. Volatilization of elements. Incomplete combustion	More rapid (~30 min) than regular dry ashing	Costly. Lower sample throughput than regular dry ashing. Some minerals volatilized. May be hard to resolubilize	Determine total ash content for the determination of proximate composition or for quality control purposes
Microwave ashing (wet)	Microwave energy and acid (and sometimes oxidizing agent) are used to oxidize and incinerate organic matter, leaving inorganic matter	Microelement contamination. Must run blanks to correct for any organic matter in acid and oxidizing agent	Takes less time (~30 min) than regular wet ashing. Minerals stay in solution. Little or no volatilization	Expensive. Can handle fewer samples per run than standard wet or dry ashing procedures	Rapid ashing of samples prior to mineral analysis by rapid or official methods

16.4 SUMMARY

The two major types of ashing analysis, dry ashing and wet ashing, are used to determine total mineral content by conventional means or by the use of microwave systems. The procedure of choice depends on subsequent analyses to be performed and on cost, time, and sample number limitations. Conventional dry ashing is based on sample incineration at high temperatures in a muffle furnace. Except for certain volatile elements, which may be lost during dry ashing, the residue may be used for both proximate analysis and further specific mineral analyses. Conventional wet ashing is often used as a preparatory step for specific elemental analysis, by the spectroscopy methods described in Chap. 9, since it simultaneously dissolves minerals and removes organic material via oxidation. Wet ashing conserves volatile elements but requires more dedicated operator time than dry ashing if the process is not automated, is limited to a smaller number of samples, and requires the use of highly caustic solvents. Microwave ashing (dry or wet) is faster than conventional methods and requires little dedicated space, and if equipped with exhaust filters, may be used without a fume hood. Ashing may eventually be replaced by methods allowing for the direct determination of mineral identity and content from fresh samples. Due to its relative simplicity and cost-effectiveness, however, ashing remains an essential component of proximate analysis and is a key preparatory step for specific mineral analyses.

16.5 STUDY QUESTIONS

1. Identify four potential sources of error in the preparation of samples for ash analysis. Describe a way to overcome each error.
2. You are determining the total ash content of a product using the conventional dry ashing method. Your boss asks you to switch to a conventional wet ashing method because they have heard that it takes less time than dry ashing.
   (a) Do you agree or disagree with your boss concerning the time issue and why?

(b) If fast sample throughput is the primary concern and cost is less of an issue, what two methods might you suggest?

(c) Not considering the time issues, why might you want to continue using dry ashing, *and* why might you change to wet ashing?

3. Your lab technician was to determine the ash content of yogurt by conventional dry ashing. The technician weighed 5 g of yogurt into one preweighed platinum crucible, immediately put the crucible into the muffle furnace using a pair of all stainless steel tongs, and ashed the sample for 48 h at 800 °C. The crucible was removed from the muffle furnace and set on a rack in the open until it was cool enough to reweigh. What errors were made in the preceding method? What were the likely results of these errors? What instructions should you have given your technician before starting the procedure in order to avoid the errors you noted?
4. How would you recommend to your technician to overcome the following problems that could arise in conventional dry ashing of various foods?
   (a) You seem to be getting volatilization of phosphorus when you want to later determine the phosphorus content.
   (b) You are getting incomplete combustion of a product high in sugar after a typical dry ashing procedure (i.e., the ash is dark colored, not white or pale gray).
   (c) You have reason to believe the compound you want to measure after dry ashing may be reacting with the porcelain crucibles being used.
   (d) You want to determine the iron content of some foods but cannot seem to get the iron solubilized after the dry ashing procedure.
5. Identify the advantages and disadvantages of using wet and dry microwave ashing compared with conventional wet and dry ashing. Consider initial investments needed for equipment, the cost of supplies and reagents, as well as the relative energy costs, time invested to prepare, run, and clean up after analyses, and the use and disposal of hazardous chemicals.

16.6 PRACTICE PROBLEMS

1. A green coffee sample was found to contain 11.5% moisture. A 5.2146-g sample was placed into a crucible (28.5053 g tare weight). The ashed crucible weighed 28.5939 g. Calculate the percentage ash on (a) an as-received (aka. fresh or wet weight) basis and (b) on a dry matter basis. Make sure to adjust for significant figures in this and other practice problems.
2. A sweet potato sample (23.5000 g) was found to have 0.0850 acid-soluble ash and 0.0940 g acid-insoluble ash. What is the percentage acid-insoluble ash?
3. You wish to have at least 100 mg of ash from a sample of oats. Assuming 2.5% ash on average and no sample loss, what is the *minimum* weight of oats required for ashing?
4. You wish to have a coefficient of variation (CV) below 5% with your ash analyses. The following ash data are obtained: 2.15%, 2.12%, and 2.07%. Are these data acceptable, and what is the CV?
5. A sample of a plant-based protein burger was analyzed. Here are the results: sample weight, 2.034 g; weight after drying, 1.0781 g; weight after ether extraction of the dried sample to remove fat, 0.4679 g; and weight of ash, 0.0233 g. What is the percentage of ash on (a) a wet weight basis, (b) a fat-free basis (including water), and (c) a fat-free basis (without water)?

Answers

1. (a) 1.70%, (b) 1.92%

   Calculate ash from sample:

Crucible + ash	28.5939 g
Tared crucible	28.5053 g
Ash	0.0886 g

(a) Calculate for ash on a wet weight basis (a):

$$\frac{0.0886\ \text{g ash}}{5.2146\ \text{g sample}} \times 100\% = 1.70\%\ \text{or}\ 1.7\%$$

(b) Calculate for ash on a dry weight basis (b):

$$\frac{0.0886\ \text{g ash}}{5.2146\ \text{g sample} \times \left(\frac{100\% - 11.5\%}{100\%}\ \text{dry matter coeff.}\right)}$$

$$\times 100\% = 1.92\%$$

or

$$5.214\ \text{g sample} \times \frac{11.5\ \text{g water}}{100\ \text{g sample}} = 0.5997\ \text{g water}$$

$$5.214\ \text{g sample} - 0.5997\ \text{g water} = 4.6149\ \text{g sample dry wt}$$

$$\frac{0.0886\ \text{g ash}}{4.6149\ \text{g dry wt sample}} \times 100\% = 1.92\%$$

2. 0.4%

   Calculate % insoluble ash:

$$\frac{0.0940\ \text{g acid insoluble ash}}{23.5\ \text{g sample}} \times 100\% = 0.4\%$$

The content of acid-soluble ash is irrelevant to the calculation of insoluble ash percent.

3. 4 g

100 mg = 0.1 g ash

2.5% = 2.5 g ash/100g sample

$$\frac{2.5 \text{ g ash}}{100 \text{ g sample}} = \frac{0.1 \text{ g ash}}{x}$$

$$2.5x = 10$$

$$x = 4 \text{ g sample}$$

4. Yes, 1.9%

Calculate the mean:

$$\frac{2.15\% + 2.12\% + 2.07\%}{3} = 2.11\%$$

Calculation of mean and standard deviation was done using Excel:

1	2.15%
2	2.12%
3	2.07%
	Average = 2.11%
	Std deviation = 0.0404

$$\text{Coefficient of variation (CV)} = \frac{\text{SD}}{x} \times 100\%$$

$$\text{CV} = \frac{0.0404}{2.11} \times 100\% = 1.91\%$$

Yes, it is within the 5% level for CV.

5. (a) 1.1%, (b) 1.64%

Sample wet wt 2.034 g

Sample dry wt 1.0781 g

Wt after extraction 0.4679 g

Wt of ash 0.0233 g

(a) Calculate on a wet weight basis:

$$\frac{0.0233 \text{ g ash}}{2.034 \text{ g sample}} \times 100\% = 1.15\%$$

(b) Calculate on a fat-free, wet weight basis:

$$2.034 \text{ g wet sample} - 1.0781 \text{ g solids} = 0.9559 \text{ g water (this is 47\% moisture)}$$

$$1.0781 \text{ g solids dry wt} - 0.4679 \text{ g solids} = \text{after extraction} = 0.6102 \text{ g fat}$$

$$2.034 \text{ g wet sample} - 0.6102 \text{ g fat} = 1.4238 \text{ g wet sample wt without fat}$$

$$\frac{0.0233 \text{ g ash}}{(1.4238 \text{ g wet sample wt without fat})} \times 100\% = 1.64\% \text{ ash, fat-free basis}$$

(c) Calculate on fat-free, dry weight basis:

$$\frac{0.0233 \text{ g ash}}{(0.4679 \text{ g dry sample wt without fat})} \times 100\% = 4.98\% \text{ ash, fat-free, dry basis}$$

Acknowledgments The author of this chapter and the book editors thank Maurice R. Marshall (Department of FoodScience and Human Nutrition, University of Florida, Gainesville, FL, USA) for his contribution to the chapter in the previous edition of the book. Thanks is also expressed to Bob Lockerman and Eileen Stochl of the CEM Corporation for providing microwave ashing equipment images and checking relevant chapter text. The author of this chapter recognizes and thanks Ruth Watkins (Department of Food, Bioprocessing and Nutrition Sciences) for contributions to the current chapter. A special thanks is given to students, who encourage us to think more deeply about each topic we research and teach.

REFERENCES

1. Melini, V, Melini F (2021) Compositional and Nutritional Analysis. Galanakis, CM (ed). Innovative Food Analysis. Elsevier, San Diego, CA, USA. Provides a brief overview of wet and dryashing techniques and the relative advantage of microwave methods.
2. Akinyele, IO, Shokunbi, OS (2015) Comparative analysis of dry and wet digestion methods for the determination of trace and heavy metals in food. Food Chem 173: 682–684.
3. AOAC (Association of Official Analytical Chemists) International (2023) Official methods of analysis, 22nd edn. (On-line). AOAC International, Rockville, MD. This reference contains the official methods for many specific food ingredients. It may be difficult for the beginning student to follow.
4. ASTM (American Society for Testing and Materials) D4309-96 (2021) Standard sample digestion using closed vessel microwave technique for the determination of total metals in water.
5. Aurand LW, Woods AE, Wells MR (1987) Food composition and analysis. Van Nostrand Reinhold, New York. The chapters that deal with ash are divided by foodstuffs. General dry ashingprocedures are discussed under each major heading.
6. Bakkali K, Martos NR, Souhail B, Ballesteros E (2009) Characterization of trace metals in vegetables by graphite furnace atomic absorption spectrometry after closed vessel microwave digestion. Food Chem 116(2): 590–594.
7. EPA (Environmental Protection Agency) Method 3052 (1996) Microwave assisted digestion of siliceous and organically based matrices.
8. Kuboyama, K, Sasaki, N, Nakagome, Y, Kataoka, M (2005) Wet Digestion. Anal Chem 360: 184–191.
9. Mesko MF, De Moraes DP, Barin JS, Dressler VL, Knappet G (2006) Digestion of biological materials using the microwave-assisted sample combustion technique, Microchem J 82:183–188.
10. Neggers YH, Lane RH (1995) Minerals, ch. 8. In: Jeon IJ, Ikins WG (eds) Analyzing foodfor nutrition labeling and hazardous contaminants. Marcel Dekker, New York. This chapter compares wet and dry ashing, and summarizes the following in tables: losses of specific elements during dry ashing; acids used in wet oxidation related to applications; AOAC methods for specific elements related to food applications.

11. Palma M N N, Rocha G C, Valaderes Filho SC, Detman E (2015) Evaluation of Acid Digestion Procedures to Estimate Mineral Contents in Materials from Animal Trials. Asian-Australas J Anim Sci 11: 1624–1628.
12. Pomeranz Y, Meloan C (1994) Food analysis: theory and practice, 3rd edn. Chapman & Hall, New York. Chapter 35 on ash and minerals gives an excellent narrative on ashing methods and is easy reading for a student in food chemistry. A good reference list of specific mineral losses is given at the end of the chapter. No stepwise procedures are given, however.
13. Quirino DF, Palma MNN, Franco MO, Detmann E (2022) Variations in Methods for Quantification of Crude Ash in Animal Feeds. J AOAC Int. 22 (106):6–13.
14. Poitevin E, Nicolas M, Graveleau L, Richoz J, Andrey D, Monard F. (2009) Improvement of AOAC Official Method 984.27 for the determination of nine nutritional elements in food products by Inductively coupled plasma-atomic emission spectroscopy after microwave digestion: single-laboratory validation and ring trial. J AOAC Int. 2009 Sep-Oct; 92(5):1484–518.
15. Zhang H, Dotson P (1994) Use of microwave muffle furnace for dry ashing plant tissue samples. Commun Soil Sci Plant Anal 25 (9/10): 1321–1327.
16. US Department of Agriculture, Agricultural Research Service (2019), FoodData Central fdc.nal.usda.gov

Fat Analysis

17

Bingcan Chen and Wayne C. Ellefson

Contents

17.1 INTRODUCTION

17.1.1 Definitions

Lipids, proteins, and carbohydrates constitute the principal structural components of foods. Lipids are a group of substances that, in general, are soluble in ether, chloroform, or other organic solvents but are sparingly soluble in water. However, there exists no clear scientific definition of a lipid,

B. Chen (✉)
Department of Plant Sciences, North Dakota State University, Fargo, ND, USA
e-mail: bingcan.chen@ndsu.edu

W. C. Ellefson
Formerly at Covance Laboratories, Madison, WI, USA

B. P. Ismail, S. S. Nielsen (eds.), *Nielsen's Food Analysis*, Food Science Text Series,
https://doi.org/10.1007/978-3-031-50643-7_17

primarily due to the water solubility of certain molecules that fall within one of the variable categories of food lipids [1]. Some lipids, such as triacylglycerols, are very hydrophobic. Other lipids, such as di- and monoacylglycerols, have both hydrophobic and hydrophilic moieties in their molecules and are soluble in relatively polar solvents [2]. Short-chain fatty acids such as C1–C4 are completely miscible in water and insoluble in nonpolar solvents [1].

As implied above, the most widely accepted definition is based on solubility as previously stated. While most macromolecules are characterized by common structural features, the designation of "lipid" being defined by solubility characteristics is unique to lipids [2]. Lipids comprise a broad group of substances that have some common properties and compositional similarities [3]. Triacylglycerols are fats and oils that represent the most prevalent category of the group of compounds known as lipids. The terms lipids, fats, and oils are often used interchangeably.

The term "lipid" commonly refers to the broad, total collection of food molecules that meet the definition previously stated. Fats generally refer to those lipids that are solid at room temperature, and oils generally refer to those lipids that are liquid at room temperature. While there may not be an exact scientific definition, the US Food and Drug Administration (FDA) has established a regulatory definition for nutrition labeling purposes. The FDA has defined total fat as the sum of fatty acids from C4 to C24, calculated as triglycerides. This definition provides a clear path for resolution of any nutrition labeling disputes.

17.1.2 General Classification

The general classification of lipids that follows is useful to differentiate lipids in foods [1]:

1. **Simple lipids**: Ester of fatty acids with alcohol (i.e., fats, waxes)
2. **Compound lipids:** Compounds containing groups in addition to an ester of a fatty acid with an alcohol (i.e., phospholipids, cerebrosides, and sphingolipids)
3. **Derived lipids:** Substances derived from neutral lipids or compound lipids (e.g., fatty acids, long-chain alcohols, sterols, fat-soluble vitamins, and hydrocarbons)

17.1.3 Content of Lipids in Foods

Foods may contain any or all types of the lipid compounds previously mentioned. The lipid content in bovine milk (Table 17.1) illustrates the complexity and variability of lipids in a food system, having lipids that differ in polarity and concentrations.

Table 17.1 Lipids of bovine milk

Kinds of lipids	*Percent of total lipids*
Triacylglycerols	97–99
Diacylglycerols	0.28–0.59
Monoacylglycerols	0.016–0.038
Phospholipids	0.2–1.0
Sterols	0.25–0.40
Squalene	Trace
Free fatty acids	0.10–0.44
Waxes	Trace
Vitamin A	(7–8.5 μg/g)
Carotenoids	(8–10 μg/g)
Vitamin D	Trace
Vitamin E	(2–5 μg/g)
Vitamin K	Trace

Adapted from Ref. Patton and Jensen [4] with permission of S. Patton, and from Ref. Jenness and Patton [5] *Principles of Dairy Chemistry*, with permission. Copyright 1959, John Wiley & Sons, Inc.

Foods contain many types of lipids, but those that tend to be of greatest importance are the triacylglycerols and the phospholipids. **Liquid triacylglycerols** at room temperature are referred to as **oils**, such as soybean oil and olive oil, and are generally of plant origin. **Solid triacylglycerols** at room temperature are termed **fats**. Lard and tallow are examples of fats, which are generally from animals. The term *fat* is applicable to all triacylglycerols whether they are normally solid or liquid at ambient temperatures. Table 17.2 shows the wide range of lipid content in different foods.

17.1.4 Importance of Analysis

An accurate and precise quantitative and qualitative analysis of lipids in foods is important for accurate nutritional labeling, to determine whether the food meets the standard of identity, and to ensure that the product meets manufacturing specifications. Inaccuracies in analysis may prove costly for manufacturers and could result in a product of undesirable quality and functionality.

17.1.5 General Considerations

By definition, lipids are soluble in organic solvents and insoluble in water. Therefore, water insolubility is the essential analytical property used as the basis for the separation of lipids from proteins, water, and carbohydrates in foods. Glycolipids are soluble in alcohols and have a low solubility in hexane. In contrast, triacylglycerols are soluble in hexane

Table 17.2 Fat content of selected foods

Food item	*Percent fat (wet weight basis)*
Baked products	
Bread, white, commercially prepared	3.59
Bread, wholewheat, commercially prepared	3.55
Cookies, oatmeal, soft, with raisins	14.3
Cereal grains and pasta	
Flour, bread, white, enriched, unbleached	1.65
Flour, corn, yellow, fine meal, enriched	1.74
Flour, oat, whole grain	6.31
Flour, pastry, unenriched, unbleached	1.64
Dairy and egg products	
Milk, reduced fat, fluid, 2% milk fat, with added vitamin A and vitamin D	1.9
Milk, nonfat, fluid, with added vitamin A and vitamin D (fat free or skim)	0.08
Cheddar cheese	34
Yogurt, Greek, plain, whole milk	4.39
Fats and oils	
Oil, canola	94.5
Oil, peanut	93.4
Oil, olive, extra virgin	93.7
Soups, sauces, and gravies	
Sauce, pasta, spaghetti/marinara, ready-to-serve	1.48
Sauce, salsa, ready-to-serve	0.19
Fruits and fruit juices	
Apples, fuji, with skin, raw	0.16
Blueberries, raw	0.31
Kiwifruit, green, raw	0.44
Strawberries, raw	0.22
Legumes and legume products	
Beans, dry, black (0% moisture)	1.45
Flour, soy, full-fat	20.7
Hummus, commercial	17.1
Peanut butter, creamy	49.4
Pork products	
Ham, sliced, restaurant	3.68
Pork, cured, bacon, cooked	36.5
Nut and seed products	
Almond butter, creamy	53
Flaxseed, ground	37.3
Nuts, pecans, halves, raw	73.3
Sesame butter, creamy	62.4
Sausages and luncheon meats	
Frankfurter, beef, unheated	28
Sausage, Turkey, breakfast links, mild, raw	10.4

From the US Department of Agriculture, FoodData Central (2023) USDA Nutrient Data Laboratory. Home Page. https://fdc.nal.usda.gov/

and petroleum ether, which are nonpolar solvents. The wide range of relative hydrophobicity of different lipids makes the selection of a single universal solvent impossible for lipid extraction of foods. Some lipids in foods are components of complex lipoproteins and liposaccharides; therefore, successful extraction requires that bonds between lipids and proteins or carbohydrates are broken so that the lipids can be freed and solubilized in the extracting organic solvents [6].

17.2 SOLVENT EXTRACTION METHODS

17.2.1 Introduction

For routine quality control purposes, the total lipid content of a food is commonly determined by simple organic solvent extraction methods or by alkaline (Sect. 17.2.6.1) or acid (Sect. 17.2.6.2) hydrolysis followed by solvent extraction in a Mojonnier flask. For multicomponent food products, acid hydrolysis is often the method of choice. Solvent extraction methods may be used as a first step in the gas chromatographic determination of fatty acid content for nutrition labeling as required by FDA regulation in the United States for the determination of total fat (see Chap. 3). Solvent extraction methods described below are classified as continuous (Sect. 17.2.4), semicontinuous (Sect. 17.2.5), or discontinuous (Sect. 17.2.6).

The accuracy of solvent extraction methods (i.e., without prior acid or alkaline hydrolysis) greatly depends on the solubility of the lipids in the solvent used and the ability to separate the lipids from complexes with other macromolecules. The lipid content of a food determined by extraction with one solvent may be quite different from the content determined with another solvent of different polarity. In addition to solvent extraction methods, there are nonsolvent wet extraction methods and several instrumental methods that utilize the physical and chemical properties of lipids in foods for total lipid content determination.

Many of the methods cited in this chapter are official methods of AOAC International. Refer to these methods and other original references cited for detailed instructions on procedures. There are many methods available for the determination of lipid content. This chapter will focus on some of the primary methods in common use.

17.2.2 Sample Preparation

The validity of the fat analysis of a food depends on proper sampling and preservation of the sample before the analysis (see also Chap. 5). An ideal sample should be as close as possible in all of its intrinsic properties to the material from which it is taken. However, a sample is considered satisfactory if the properties under investigation correspond to those of the bulk material within the limits of the test [7].

The sample preparation for lipid analysis depends on the type of food and type and nature of lipids in the food [4]. The extraction method for lipids in liquid milk is generally different from that for lipids in solid soybeans. To analyze the lipids in foods effectively, knowledge of the structure, the chemistry, and the occurrence of the principal lipid classes and their constituents is necessary. Therefore, there is no single standard method for the extraction of all kinds of lipids in different foods. For the best results, sample preparation should be carried out under an inert atmosphere of nitrogen at low temperature to minimize chemical reactions such as lipid oxidation.

One or more preparatory steps are common in lipid analysis to aid in extraction: (1) removal of water, (2) reduction of particle size, and (3) separation of the lipid from bound proteins and/or carbohydrates through the use of techniques such as alkaline hydrolysis (Sect. 17.2.6.1) or acid hydrolysis (Sect. 17.2.6.2). The first two of these steps are described in sections immediately below.

17.2.2.1 Predrying Sample

Lipids cannot be effectively extracted with ethyl ether from moist food because the solvent cannot easily penetrate the moist food tissues due to the hydrophobicity of the solvents used or the hydroscopic nature of the solvents. The ether, which is hygroscopic, becomes saturated with water and inefficient for lipid extraction. Drying the sample at elevated temperatures is undesirable because some lipids become bound to proteins and carbohydrates, and bound lipids are not easily extracted with organic solvents. Vacuum oven drying at low temperature or lyophilization increases the surface area of the sample for better lipid extraction. Predrying makes the sample easier to grind for better extraction, breaks fat–water emulsions to make fat dissolve easily in the organic solvent, and helps to free fat from the tissues of foods [7].

17.2.2.2 Particle Size Reduction

The extraction efficiency of lipids from dried foods depends on particle size; therefore, adequate grinding is very important. The classical method of determining fat in oilseeds involves the extraction of the ground seeds with selected solvent after repeated grinding at low temperature to minimize lipid oxidation. For better extraction, the sample and solvent are mixed in a high-speed comminuting device such as a blender. It can be difficult to extract lipids from whole soybeans because of the limited porosity of the soybean hull and its sensitivity to dehydrating agents. The lipid extraction from soybeans is easily accomplished if the beans are broken mechanically by grinding. Extraction of fat from finished products can be a challenge, based on the ingredients (e.g., energy bars with nuts, caramel, protein, granola, soybean oil). Such products may best be ground after freezing with liquid nitrogen.

17.2.3 Solvent Selection

Ideal solvents for fat extraction should have a high solvent power for lipids and low or no solvent power for proteins, amino acids, and carbohydrates. They should evaporate readily and leave no residue, have a relatively low boiling point, and be nonflammable and nontoxic in both liquid and vapor states. The ideal solvent should penetrate sample particles readily, be in single component form to avoid fractionation, and be inexpensive and nonhygroscopic [8]. It is difficult to find an ideal solvent to meet all of these requirements. Ethyl ether and petroleum ether are the most commonly used solvents, but pentane and hexane are used to extract oil from soybeans.

Ethyl ether has a boiling point of 34.6 °C and is a better solvent for fat than petroleum ether. It is generally expensive compared to other solvents, has a greater danger of explosion and fire hazards, is hygroscopic, and forms peroxides [9]. **Petroleum ether** is the low boiling point fraction of petroleum and is composed mainly of pentane and hexane. It has a boiling point of 35–38 °C and is more hydrophobic than ethyl ether. It is selective for more hydrophobic lipids, cheaper, less hygroscopic, and less flammable than ethyl ether. The detailed properties of petroleum ether for fat extraction are described in AOAC Method 991.36 [10].

A combination of two or three solvents is frequently used. The solvents should be purified and peroxide-free, and the proper solvent–solute ratio must be used to obtain the best extraction of lipids from foods [9].

17.2.4 Continuous Solvent Extraction Method: Goldfisch Method

17.2.4.1 Principle and Characteristics

For continuous solvent extraction, solvent from a boiling flask continuously flows over the sample held in a ceramic thimble. Fat content is measured by weight loss of the sample or by weight of fat removed.

The continuous methods give faster and more efficient extraction than semicontinuous extraction methods. However, they may cause channeling which results in incomplete extraction. The Goldfisch test is an example of a continuous lipid extraction method [8, 9]. This method involves significant fire risk and has been discontinued in many laboratories, but it is briefly described here for comparison to semicontinuous and discontinuous extraction methods (Sects. 17.2.5 and 17.2.6).

17.2.4.2 General Procedure

The sample is weighed into a thimble. Fat is extracted from the sample with boiling ethyl ether in the Goldfisch apparatus (Fig. 17.1). Fat content is calculated as given in Eq. 17.1:

$$\begin{aligned}&\%\ \text{Fat on dry weight basis}\\&=\left(\text{g of fat in sample} / \text{g of dried sample}\right)\times 100\end{aligned} \qquad (17.1)$$

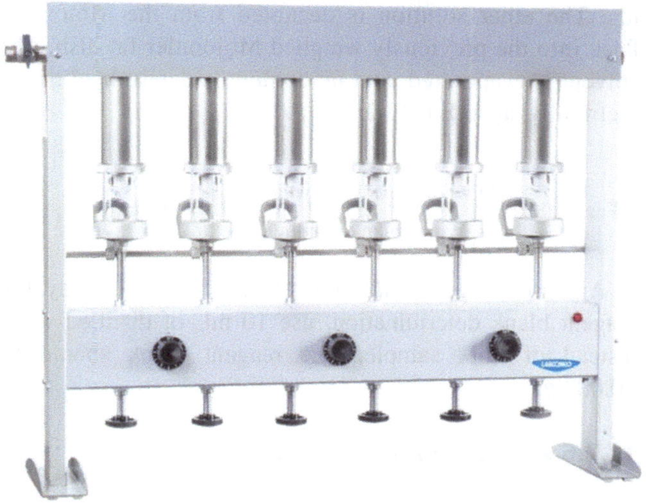

Fig. 17.1 Goldfisch fat extractor. (Courtesy of Labconco Corp., Kansas City, MO) (https://www.labconco.com/product/goldfisch-fat-extractor/982)

17.2.5 Semicontinuous Solvent Extraction Method: Soxhlet Method

The Soxhlet method (AOAC Method 948.22 for Nuts and nut products; AOAC Method 960.39 for Meat) [10] is an example of the semicontinuous extraction method and is described below.

17.2.5.1 Principle and Characteristics

For semicontinuous solvent extraction, **the solvent builds up in the extraction chamber for 5–10 min, completely surrounds the sample and then siphons back to the boiling flask. Fat content is measured by weight loss of the sample or by weight of fat removed.

This method provides a soaking effect of the sample and does not cause channeling. However, this method requires more time than the continuous method. Instrumentation for more rapid and automated versions of the Soxhlet method is available (e.g., Ankom XT15 Extractor, Ankom Technology, Macedon, NY; Soxtec™ 8000, FOSS in North America, Eden Prairie, MN).

17.2.5.2 General Procedure (See Fig. 17.2)

For the Soxhlet or Goldfisch methods, if the sample contains more than 10% H_2O, dry the sample to constant weight at 95–100 °C under pressure ≤ 100 mm Hg for about 5 h (AOAC Method 934.01). If analyzing a variety of food products, it may be more efficient to dry all samples, regardless of moisture content.

Samples are weighed into thimbles, placed in the Soxhlet apparatus (Fig. 17.2) and extracted with an appropriate solvent. As this procedure uses heat, it is more dangerous to use ethyl ether than other solvents. Many laboratories now use petroleum ether or hexane. Extraction time is 16 h in most cases. Certain products may lend themselves to shorter extraction time. The extract is then evaporated, and fat is determined gravimetrically. Fat content is calculated as in Eq. 17.2:

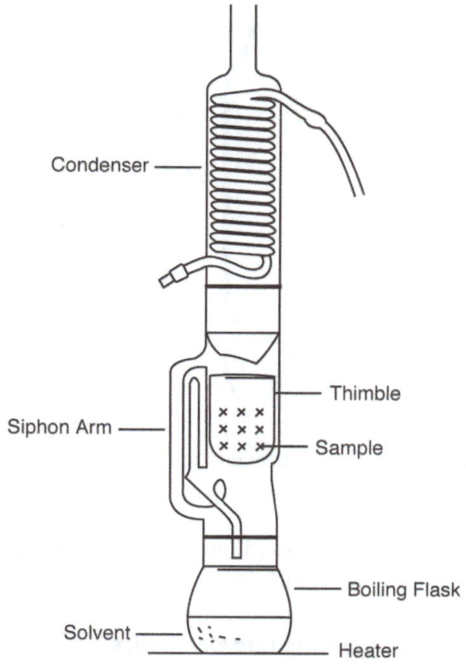

Fig. 17.2 Soxhlet extraction apparatus

$$\% \text{ Fat on dry weight basis} = (\text{g of fat in sample} / \text{g of dried sample}) \times 100 \quad (17.2)$$

17.2.6 Discontinuous Solvent Extraction Methods

Alkaline hydrolysis (also referred to as the Mojonnier method) and acid hydrolysis are both types of discontinuous solvent extraction methods. They involve treatment with acid or base, followed by solvent extraction.

17.2.6.1 Alkaline Hydrolysis Method (Mojonnier Method)

The terms Mojonnier fat, base hydrolysis, and alkaline hydrolysis are often used interchangeably to describe the method below. The term alkaline hydrolysis will be used in this chapter. After the use of ammonia to precipitate protein and free any bound fat, extraction is conducted with a mixture of ethyl ether and petroleum ether in a Mojonnier flask

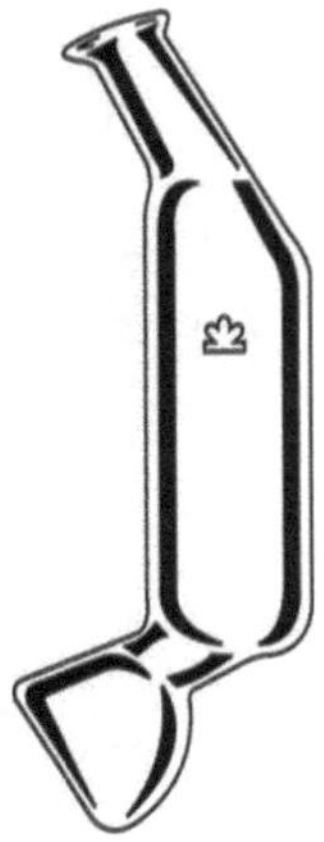

Fig. 17.3 Mojonnier fat extraction flask. (Courtesy of Kontes Glass Co., Vineland, NJ)

(Fig. 17.3), and the extracted fat is dried to a constant weight and expressed as percent fat by weight.

17.2.6.1.1 Principle and Characteristics

The alkaline hydrolysis test does not require removal of moisture from the sample, so it can be applied to both liquid and solid samples. The alkaline hydrolysis method was developed for and is applied primarily to dairy foods and remains the primary standard for the dairy industry. If petroleum ether is used to purify the extracted fat, this method is very similar in principle and practice to the **Roese–Gottlieb method** (AOAC Method 905.02) (which, like the "Mojonnier method," was developed for the dairy industry) [10]. The Mojonnier flasks (Fig. 17.3) are used not only for alkaline hydrolysis methods but may also be used to conduct hydrolysis (acid, alkaline, or combination) prior to fat extraction and gas chromatography (GC) analysis for the determination of fat content via fatty acid profile analysis (Sect. 17.2.6), according to the US nutrition labeling regulations. In place of using Mojonnier flasks for fat analysis, many laboratories now use separatory funnels or other glassware, for ease of use.

17.2.6.1.2 General Procedure

For AOAC Method 989.05 (fat in milk), the sample is prepared with appropriate warming and handling to ensure a homogenous sample, and then it is weighed into a Mojonnier flask [10]. Ammonium hydroxide is added to precipitate the milk protein, and then ethanol is added to prevent gel formation. Following these treatments, fat is extracted first with ethyl ether to extract most lipids present, followed by petroleum ether to assist with removal of water from the ethyl ether and to assist with complete extraction of nonpolar lipids. Extractions are commonly repeated twice more to ensure complete extraction of lipids. The ether solution is decanted from the Mojonnier flask into the previously weighed Mojonnier fat dish. The solvent is evaporated, and fat content is determined gravimetrically, as in Eq. 17.3:

$$\%\text{Fat} = 100 \times \left\{ \begin{matrix} \left[(\text{wt dish} + \text{fat}) - (\text{wt dish}) \right] \\ -(\text{avg wt blank residue}) \end{matrix} \right\} \Big/ \text{wt sample} \quad (17.3)$$

A pair of reagent blanks must be prepared every day. For reagent blank determination, use 10 mL of distilled water instead of milk sample. The reagent blank should be <0.002 g.

17.2.6.2 Acid Hydrolysis Procedure

17.2.6.2.1 Principle and Characteristics

A significant portion of the lipids in foods such as dairy, bread, flour, and animal products is bound to proteins and carbohydrates, and direct extraction with nonpolar solvents is inefficient. Such foods must be prepared for lipid extraction by acid hydrolysis. This includes a significant percentage of finished food products. Table 17.3 shows the inaccuracy that can occur if samples are not prepared by acid hydrolysis. Acid hydrolysis can break both covalently and ionically bound lipids into easily extractable lipid forms. Specifically, there are a variety of AOAC methods (AOAC Method 922.06 Fat in flour; 935.38 Fat in bread) for fat that involve an acid hydrolysis with HCl, followed by extraction with a combination of ethyl ether and petroleum ether [10]. Ethanol and solid hexametaphosphate may be added to facilitate separation of lipids from other components before food lipids are extracted with solvents [8, 9]. For example, the acid hydrolysis of two eggs requires 10 mL of HCl (AOAC Method 925.32) and heating in a water bath at 65 °C for 15–25 min or until the solution is clear [10].

17.2.6.2.2 General Procedure

For AOAC Method 922.06 (Fat in Flour), the sample is weighed into a Mojonnier flask. Hydrochloric acid digestion is carried out. Fat is extracted first with ethyl ether to extract most lipids present, followed by petroleum ether to assist

Table 17.3 Effects of acid hydrolysis on fat extraction from foods

	Percent fat	
	Acid hydrolysis	*No acid hydrolysis*
Dried egg	42.39	36.74
Yeast	6.35	3.74
Flour	1.73	1.20
Noodles	3.77–4.84	2.1–3.91
Semolina	1.86–1.93	1.1–1.37

Adapted from Ref. Joslyn [6], p. 154, with permission

with removal of water from the ethyl ether and to assist with complete extraction of nonpolar lipids. Extractions are commonly repeated twice more to ensure complete extraction of lipids. The ether solution is decanted from the Mojonnier flask into the previously weighed Mojonnier fat dish. The solvent is evaporated, and then fat is determined gravimetrically using Eq. 17.4:

$$\% \text{ Fat} = 100 \times \left\{ \begin{matrix} \left[(\text{wt dish} + \text{fat}) - (\text{wt dish}) \right] \\ - (\text{avg wt blank residue}) \end{matrix} \right\} \Big/ \text{wt sample} \quad (17.4)$$

17.2.6.3 Chloroform–Methanol Procedure

17.2.6.3.1 Principle and Characteristics

The combination of chloroform and methanol has been used commonly to extract lipids. The "Folch extraction" [11] is applied to small samples, while the "Bligh and Dyer extraction" [12] is applied to large samples of high moisture content. Both methods utilize this combination of solvents to recover lipids from foods. These methods have been reviewed, and procedures were modified by Christie [13] and others. The Bligh and Dyer procedure [12] is a modification of the Folch extraction [11], designed for more efficient solvent usage for low-fat samples. The Christie modification [13] of these former methods replaced water with 0.88% potassium chloride aqueous solution to create two phases.

In both the modified Folch extraction and Bligh and Dyer procedure, food samples are mixed/homogenized in a chloroform–methanol solution, and the homogenized mixture is filtered into a collection tube. A 0.88% potassium chloride aqueous solution is added to the chloroform–methanol mixture containing the extracted fats. This causes the solution to break into two phases: the aqueous phase (top) and the chloroform phase containing the lipid (bottom). The phases are further separated in a separatory funnel or by centrifugation. After the evaporation of the choloroform, the fat can be quantitated by weight.

The various methanol-chloroform extraction procedures are rapid, well suited to low-fat samples, and can be used to generate lipid samples for subsequent fatty acid compositional analysis. The procedure has been more applied to basic commodities, rather than to finished product samples. For consistent results, the procedures must be followed carefully, including the ratio of chloroform and methanol. A cautionary note is that chloroform and methanol are highly toxic, so the extraction procedure must be done in well-ventilated areas.

17.2.6.3.2 General Procedure

For AOAC Method 983.23 (Fat in Foods), samples are extracted first with methanol, and then chloroform is added on top, and extraction is continued. Up to two additional such extractions may be performed. Potassium chloride is added to aid in the separation of the layers. Solvent is evaporated, and fat is determined gravimetrically as in Eq. 17.5:

$$\% \text{ Fat} = 100 \times \left\{ \begin{matrix} \left[(\text{wt dish} + \text{fat}) - (\text{wt dish}) \right] \\ - (\text{avg wt blank residue}) \end{matrix} \right\} \Big/ \text{wt sample} \quad (17.5)$$

17.2.7 Total Fat by Gas Chromatography for Nutrition Labeling

17.2.7.1 Principle

After adding an internal standard and an antioxidant, the sample is treated by acid and/or alkaline hydrolysis, and then fat is extracted with ether. Fatty acids are converted to fatty acid methyl esters (FAMEs), then separated by gas chromatography (GC), and quantitated against the internal standard, with the sum equal to total fat (AOAC Method 996.06). (The principles of GC are described in Chap. 14.)

The saturated and monounsaturated fats are calculated as the sum of the respective fatty acids. Monounsaturated fat includes only *cis* form. Trans fat can be quantified utilizing this method in conjunction with identification criteria established by the American Oil Chemists' Society (AOCS Method Ce 1 h-05) [14] and Talebi et al. [15].

17.2.7.2 General Procedure

The GC method for total fat (AOAC 996.06) is summarized as follows:

1. Add the antioxidant pyrogallic acid to sample.
2. Add the internal standard (heneicosanoic acid methyl ester,C $_{21:0}$-ME) to sample (see Chap. 12, Sect. 12.5.3).
3. Subject sample to acid and/or alkaline hydrolysis.
4. Extract fat from sample with ethyl ether and/or petroleum ether.
5. Methylate fatty acids with boron trifluoride (in methanol) to form fatty acid methyl esters (FAMEs) to make the fatty acids volatile for GC analysis.
6. Separate FAMEs with GC capillary column and use detector to quantify peaks, compared to internal standard (Fig. 17.4).
7. Total fat is calculated as the sum of individual fatty acids expressed as triglyceride equivalents.

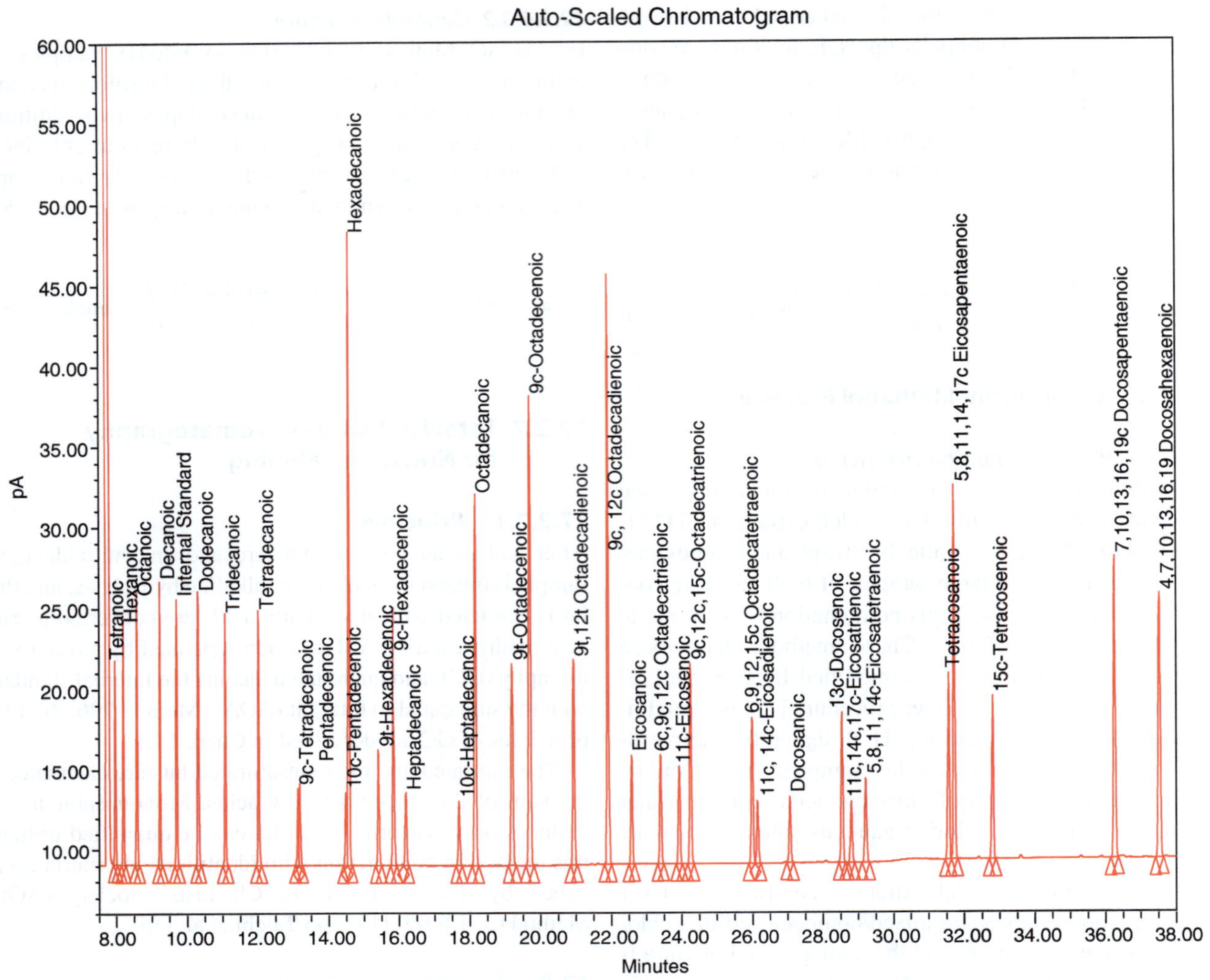

Fig. 17.4 Fatty acid chromatogram

17.3 NONSOLVENT WET EXTRACTION METHODS

17.3.1 Babcock Method for Milk Fat

17.3.1.1 Principle

In the Babcock method, H_2SO_4 is added to a known amount of milk in the Babcock bottle (Fig. 17.5). The sulfuric acid digests protein, generates heat, and releases the fat. Centrifugation and hot water addition isolate fat for quantification in the graduated portion of the test bottle. The fat is measured volumetrically, but the result is expressed as percent fat by weight.

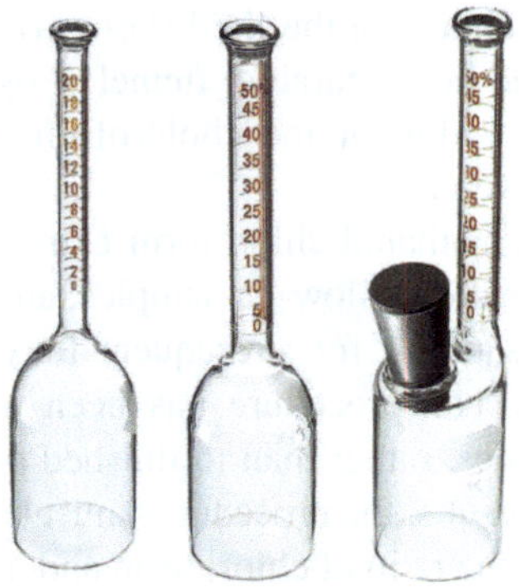

Fig. 17.5 Babcock milk test bottles for (**a**) milk, (**b**) cream, and (**c**) cheese (Paley bottle) testing. (Courtesy of Kimble Glass Co., Vineland, NJ)

17.3.1.2 Applications

The Babcock method, which is a common official method for the determination of fat in milk (AOAC Method 989.04), takes about 45 min, and duplicate tests should agree within 0.1%. The Babcock method does not determine the phospholipids in the milk products. It is not applicable to products containing chocolate or added sugar without modification

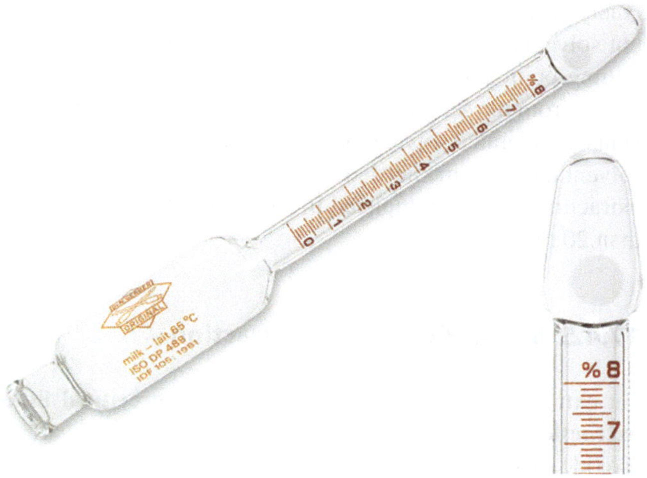

Fig. 17.6 Gerber fat butyrometer. (Courtesy of Gerber Instruments, Switzerland)

because of charring of chocolate and sugars by sulfuric acid. A modified Babcock method is used to determine essential oil in flavor extracts (AOAC Method 932.11) and fat in seafood (AOAC Method 964.12).

17.3.2 Gerber Method for Milk Fat

17.3.2.1 Principle

The principle of the Gerber method is similar to that of the Babcock method, but it uses sulfuric acid and amyl alcohol and a Gerber bottle (Fig. 17.6) (AOAC Method 2000.18). The sulfuric acid digests proteins and carbohydrates, releases fat, and maintains the fat in a liquid state by generating heat.

17.3.2.2 Applications

The Gerber method is comparable to the Babcock method but is simpler and faster and has wider application to a variety of dairy products [16]. The isoamyl alcohol generally prevents the charring of sugar found with the regular Babcock method. This test is more popular in Europe than in America.

17.4 INSTRUMENTAL METHODS

Instrumental methods offer numerous attractive features compared to the previously described extraction methods. In general, they are rapid, some are nondestructive, and require minimal sample preparation and chemical consumption. These methods can provide significant labor savings when one must perform analysis of many samples on a daily basis. However, the equipment can be expensive, and some methods require the establishment of calibration curves specific to various compositions. Despite these drawbacks, several of the following instrumental methods are very widely used in quality control as well as research and product development applications. The following section describes several of these instrumental methods.

17.4.1 Infrared Method

The infrared (IR) method is based on absorption of IR energy by fat at a wavelength of 5.73 μm. The more the energy absorption at 5.73 μm, the higher the fat content of the sample [17]. (See Chap. 8 for a discussion of IR spectroscopy.) Both near-infrared (NIR) or mid-infrared (mid-IR) spectrophotometers are used in this method. For example, AOAC Method 2007.04 (fat, moisture, and protein in meat and meat products; specifies use of a FOSS Food Scan™ NIR spectrophotometer) is based on correlating NIR data with the values obtained from analysis by conventional methods, to then predict the concentration of fat, moisture, and protein in a sample being tested. Mid-IR spectroscopy is used in infrared milk analyzers to determine milk fat content (AOAC Method 972.16). NIR spectroscopy has been used to measure the fat content of commodities such as meats, cereals, and oilseeds in the laboratory and can be adapted for online measurement. NIR has been applied to near-line batch monitoring for food processing plants. NIR is an analytical technique that provides a prediction of chemical measurement. The NIR models are developed and validated to provide equivalent results to the wet chemical measurement. This method requires a comprehensive database of constituent values to create the model along with ongoing verification.

17.4.2 X-Ray Absorption Method

X-ray absorption is a rapid analysis method that has received interest in application to lipid analysis in meats [18]. This method can be adapted to in-line analysis of meats. Determination of the amount of fat in meat and meat products is based on the fact that the X-ray absorption of lean meat is higher than that of fat. This method has been used for the rapid determination of fat in meat and meat products using the standard curve of the relationship between X-ray absorption and fat content determined by a standard solvent extraction method [9]. For example, the MeatMaster™II fat analysis instrument (Foss, Eden Prairie, MN), commonly used to rapidly determine the percent fat of meat products, is based on X-ray absorption.

17.4.3 Nuclear Magnetic Resonance

Nuclear magnetic resonance (NMR) can be used to measure lipids in food materials in a nondestructive way. It is one of the most popular methods for use in determining lipid melting curves to measure solid fat content (see Chap. 23), and more affordable instruments are becoming more popular for measuring total fat content. Total fat content can be measured using low-resolution pulsed NMR (AOAC Method 2008.06). (The principles and applications of NMR are described in Chap. 10.) NMR analysis is a very rapid and accurate method, and while the principles of NMR are relatively complex, the use of NMR can be quite simple, especially due to the high degree of automation and computer control. One example system is CEM's ORACLE unit (CEM Corp., Matthews, NC) to rapidly measure fat content, often combined with CEM's SMART 6 unit to rapidly measure moisture content.

17.4.4 Accelerated Solvent Extraction

Accelerated solvent extraction (ASE) was developed to replace Soxhlet and other extraction techniques for many samples, such as lipids in fours and industrial hemps [19, 20]. The automation and rapid extraction time of accelerated solvent extraction are its key advantages. Accelerated solvent extraction (Fig. 17.7) uses solvents at increased temperatures and pressures to accelerate the extraction process.

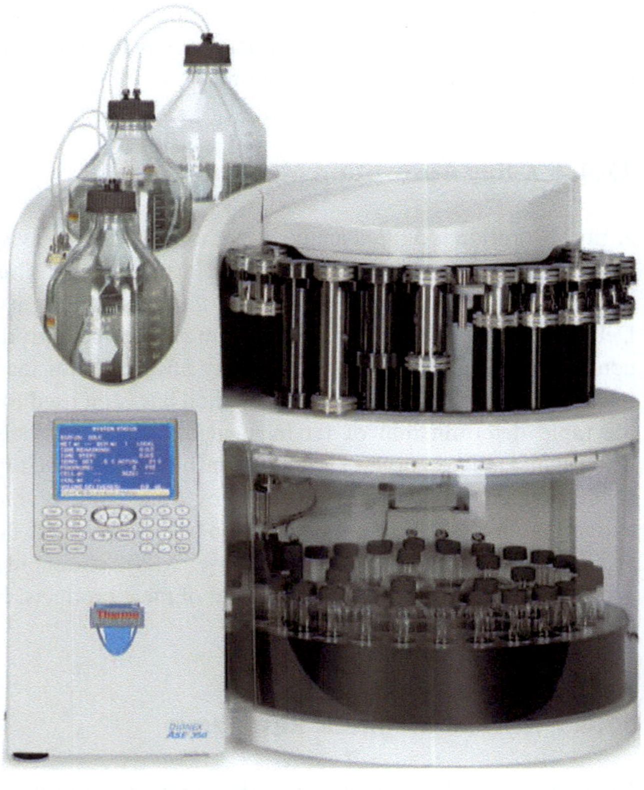

Fig. 17.7 Accelerated solvent extractor. (Courtesy of Thermo Fisher Scientific, Waltham, MA)

Analytes are more soluble and thus dissolve more quickly in hot solvents. Pressurized solvent is heated to well above its boiling point and introduced under high pressure to extract the fat from the sample. By using hot, pressurized liquid solvents, accelerated solvent extraction decreases the amount of solvent and the time needed to complete an extraction. Most extractions can be completed in less than 20 min, using less than 20 mL of solvent.

17.4.5 Supercritical Fluid Extraction

Fat extraction can be carried out using **supercritical fluid extraction** (SFE) instruments that use supercritical carbon dioxide (CO_2) as the solvent. Samples are weighed into extraction cells, the extraction chamber is heated and pressurized to a set value, and the supercritical fluid is then pumped through the extraction cell facilitating extraction of the target analytes from the sample matrix. This technique is finding greater use because of the cost and environmental problems associated with the use and disposal of organic solvents. When pressurized CO_2 is heated above a certain critical temperature, it becomes a supercritical fluid that has some of the properties of a gas and some of a liquid. The fact that it behaves like a gas allows it to easily penetrate into a sample and extract the lipids, while the fact that it behaves like a fluid helps it dissolve lipids (especially at higher pressures). The pressure and temperature of the solvent are later reduced, which causes the CO_2 to turn to a gas, leaving the lipid fraction remaining. The lipid content of a food is determined by weighing the percentage of lipid extracted from the original sample [9]. Automated SFE systems are available commercially.

17.5 COMPARISON OF METHODS

Various fat analysis methods are summarized and compared in Tables 17.4 and 17.5. Dried (or very low moisture content) samples are required for fat determination by the Soxhlet extraction method (traditional version or automated), which is applicable to many food commodities. If the samples are moist or liquid foods, the Mojonnier method is generally applicable to determination of the fat content. Acid hydrolysis or alkaline hydrolysis is widely used on many finished food products. The instrumental methods such as IR and NMR are very simple, reproducible, and fast and are gaining in popularity. The application of instrumental methods for fat determination generally requires a standard curve between the signal of the instrument analysis and the fat content obtained by a standard solvent extraction method. However, a rapid instrumental method can be used as a quality control method for fat determination of a specific food.

Table 17.4 Summary of fat analysis methods

Method	*Principle*	*Advantages*	*Disadvantages*	*Applications*
Soxhlet	Fat is extracted, semicontinuously, with organic solvent. Fat content is measured by weight loss of sample or weight of fat removed	Provides for soaking effect. Does not cause channeling. Sample remains cool	Time to result is long. Errs low on many products with bound fat. Manual test	Grains, raw meat
Soxhlet (automated)	These methods take the traditional Soxhlet methodology described above and automate the process, removing some of the manual labor necessary to obtain results	Requires less labor. Faster time to result	Higher equipment cost	Same as Soxhlet method
Mojonnier	Fat is extracted (discontinuous) with mixture of organic solvents. Extracted fat is dried to constant weight	Faster than Soxhlet. Hydrolysis and extraction done in same container. No need to dry sample	Errs low on many products with bound fat. Manual test	Same as above and on some nonprotein beverages
Mojonnier – base hydrolysis	Protein is precipitated with ammonia. Fat is extracted with mixture of organic solvents. Extracted fat is dried to constant weight	Best traditional method for dairy products. Hydrolysis and extraction done in same container. No need to dry sample	Manual test. Narrow range of scope	Intended for use on milk and milk-derived products
Mojonnier – acid hydrolysis	Organic matter (nonfat) is digested with acid. Fat is extracted with mixture of organic solvents. Extracted fat is dried to constant weight	Wide range of applicability for many food item. Hydrolysis and extraction done in same container. No need to dry sample	Manual test. More labor required than some methods	Applicable to most foods. Not applicable to raw soybeans
Chloroform–methanol	Fat is extracted with a combination of chloroform and methanol. Addition of potassium chloride causes the solution to break into two phases, with the fat in the chloroform phase evaporation of the chloroform allows for fat to be quantitated by weight	Wide range of applicability for many food items	Chloroform and methanol are both highly toxic. Measures all fat-soluble materials. Manual test, so significant labor expenditure. Need to deal with emulsions formed during extraction	Most foods
Gas chromatography (GC)	After adding an internal standard and a reagent to prevent oxidation, sample is treated by acid and/or alkaline hydrolysis, and then fat is extracted with ether. Fatty acids are converted to fatty acid methyl esters (FAMEs). FAMEs are separated by GC and quantitated. Sum of fatty acids equals fat content	Measures only true fat, as defined by FDA (not other fat-soluble compounds)	Time-consuming. Expensive. Many compounds to separate	Official method for nutrition labeling
Babcock	Sulfuric acid added digests protein, generates heat, and releases fat. After centrifugation of sample, fat content is measured volumetrically	Takes only ~45 min	Manual test. Get charring with high-sugar products, so difficult to read volume of fat	Commonly applied to milk and other dairy products
Gerber	Similar to Babcock method but uses sulfuric acid and amyl alcohol	Compared to Babcock, it is simpler and faster, and the isoamyl alcohol prevents charring of sugar	Manual test	Commonly applied to milk and other dairy products, especially high-sugar products
Infrared spectroscopy	NIR provides a prediction of fat measurement based on absorption by fat of IR energy at a specific wavelength. Instruments are developed and validated to provide equivalent results to classical methods	Rapid	Provides only an estimate of fat content. Instrument must be calibrated against official method	Mid-IR spectroscopy is commonly used in infrared milk analyzers. NIR commonly used to measure fat content of meats, cereals, and oilseeds

(continued)

Table 17.4 (continued)

Method	*Principle*	*Advantages*	*Disadvantages*	*Applications*
Specific gravity	Fat is extracted from sample with the solvent perchloroethylene. Specific gravity of sample solvent extract is related to fat content	Rapid	Values are estimates, based on correlation chart created using an official method	Meat
NMR	Certain nuclei will absorb and re-emit RF energy over a narrow band of frequencies when placed in a static magnetic field. The frequency at which the NMR effect occurs for a given nuclear isotope is dependent on the magnetic field strength of the magnet, and the phenomenon is caused by the interaction between the nuclear magnetic dipole of a nucleus and the magnetic field it experiences. Fat is determined by pulsed radio frequency (RF) energy while within a static 0.47 tesla magnetic field. The resulting NMR signal is recorded and analyzed for the total proton activity of fat present in the sample [8]	Rapid. Very low labor cost. Tremendous potential in high-throughput fat testing laboratories	Expensive instrument. Range of applicability has not yet been totally demonstrated	Potential application to many foods
Accelerated solvent extraction (ASE)	Pressurized solvent is heated to well above its boiling point and introduced under high pressure to extract the fat from the sample	Very rapid	Instrument cost. May require some work to match classical results	Use in place of Soxhlet method. Use for other fat-related tests
Supercritical fluid extraction (SFE)	The sample is heated and pressurized to a set value, and the supercritical fluid is then pumped through the extraction cell facilitating extraction of the fat from the sample matrix	Very rapid	Instrument cost. May require some work to match classical results	Use in place of Soxhlet method. Use for other fat-related tests.

Table 17.5 Comparison of classical methods for fat analysis

Methods	*Similarities*	*Differences*	
Soxhlet vs Goldfisch	Use for same applications (see list in Table 17.4) Use extraction with organic solvents; Measure fat by wt. loss or wt. of fat removed (gravimetric) Both take hours to do Sample must be dry or low in moisture Use same type of cellulose "thimble" to hold sample for extraction	*Soxhlet* Semicontinuous Soaking effect Sample stays cool No channeling May be slower (takes 4–16 h, depending on drip rate)	*Goldfisch* Continuous method No soaking effect Sample gets heated Channeling likely Extract for 4 h (so may be faster)
Babcock vs Mojonnier	Can use on liquid or solid sample Both developed originally for dairy	*Babcock* Use no solvents (uses sulfuric acid) Dairy products only Measure fat volume	*Mojonnier* Use multiple solvents Any food product Measure fat weight
Mojonnier vs Soxhlet	Use for same applications Gravimetric methods	*Mojonnier* Use multiple solvents Solid or liquid sample Discontinuous method Use same container for hydrolysis and extraction	*Soxhlet* Use single solvent Solid sample only (dry or low in moisture) Semicontinuous method Need separate containers for hydrolysis and extraction
Babcock vs Gerber	Same applications (dairy) Volumetric measurement	*Babcock* Use only sulfuric acid Use sulfuric acid to release fat Would get charring with high-sugar dairy product Slower; more complex	*Gerber* Use sulfuric acid plus isoamyl alcohol Can use on high sugar dairy product simpler and faster
Mojonnier vs GC method	GC method commonly uses Mojonnier for initial fat extraction	*Mojonnier* Common component of GC method Measure by wt. of fat Used for applications in summary table	*GC* Requires acid and/or alkaline hydrolysis Extraction is commonly done with Mojonnier-type flasks but has post-steps to measure fat as sum of FAs Used for nutrition labeling

Major uses of the Soxhletand Mojonnier methods include the following: (1) extract fat prior to GC analysis, (2) quality control of formulated products, (3) determine fat content during product development, (4) verify when fat content is <0.5 g/serving (so nutrient content claim can be made), and (5) defat samples prior to fiber analysis. Compared to GC analysis of fat content by AOAC Method 996.06, these three methods are faster and cheaper, but they may give a higher fat content (which must be recognized when using these methods for product development).

17.6 SUMMARY

Lipids have often been defined by their solubility characteristics rather than by some common structural feature. Lipids in foods can be classified as simple, compound, or derived lipids. The lipid content of foods varies widely, but quantitation is important because of regulatory requirements, nutritive value, and functional properties. To analyze food for the fat content accurately and precisely, it is essential to have a comprehensive knowledge of the general compositions of the lipids in the foods, the physical and chemical properties of the lipids as well as the foods, and the principles of fat determination. There is no single standard method for the determination of fats in different foods. The validity of any fat analysis depends on proper sampling and preservation of the sample prior to analysis. Predrying of the sample, particle size reduction, and acid/alkaline hydrolysis prior to analysis also may be necessary. The total lipid content of foods is commonly determined by organic solvent extraction methods, such as semicontinuous (e.g., Soxhlet) and discontinuous (e.g., Mojonnier), or by GC analysis for nutrition labeling. Nonsolvent wet extraction methods, such as the Babcockor Gerber, are commonly used for certain types of food products. Instrumental methods, such as NMR, infrared, ASE, SFE, and X-ray absorption, are also available for fat determination. These methods are rapid and may be useful for quality control but may require additional refinement in order to develop consistent correlation to a standard solvent extraction method.

17.7 STUDY QUESTIONS

1. Itemize the procedures that may be required to prepare a food sample for accurate fat determination by a solvent extraction method (e.g., Soxhlet method). Explain why each of these procedures may be necessary.
2. What are some important considerations when selecting solvents to be used in continuous, semicontinuous, and discontinuous solvent extraction methods?
3. To extract the fat from a food sample, you have the choice of using ethyl ether or petroleum ether as the solvent, and you can use either a Soxhlet or a Goldfisch apparatus. What combination of solvent and extraction would you choose? Give all the reasons for your choice.
4. Differentiate the principles of continuous, semicontinuous, and discontinuous fat extraction methods.
5. What is the purpose of the following chemicals used in the Mojonnier method?
   (a) Ammonium hydroxide
   (b) Ethanol
   (c) Ethyl ether
   (d) Petroleum ether
6. You performed fat analysis on a new super energy shake (high carbohydrate and protein) using standard Soxhlet extraction. The value obtained for fat content was much lower than that expected. What could have caused the measured fat content to be low and how would you modify the standard procedure to correct the problem?
7. What is a key application of the GC method and what does it specifically quantify?
8. Regarding total fat content analysis, the GC method is now required for nutrition labeling, based on FDA's definition of fat. Below is a list (out of order) of some of the basic steps of this analysis. For each step, briefly explain what would happen/how results would likely be affected if this step was not done.
   Description of step Why necessary?
   Methylate fatty acids with BF_3 to FAMEs
   Add internal standard (triundecanoin)
   Do acid and/or alkaline hydrolysis
   Add pyrogallic acid
9. Which of the following methods are volumetric and which are gravimetric determinations of lipid content: Babcock, Soxhlet, Mojonnier, Gerber?
10. Gas chromatography (GC) vs. Soxhlet vs. Mojonnier Methods for Fat Analysis
    (a) What is the rationale for the FDA requiring the GC method (which includes a solvent extraction step) vs. a simple solvent extraction method for total fat content on a nutrition label?
    (b) What characteristics of a discontinuous solvent extraction method (e.g., Mojonnier method), as compared to the Soxhlet method make it especially commonly used as a component of the overall GC method?

17.8 PRACTICE PROBLEMS

1. To determine the fat content of a semimoist food by the Soxhlet method, the food was first vacuum oven-dried. The moisture content of the product was 25.0%. The

fat in the dried food was determined by the Soxhlet method. The fat content of the dried food was 13.5%. Calculate the fat content of the original semimoist product.

2. The fat content of 10.00 g of commercial ice cream was determined by the Mojonnier method. The weights of extracted fat after the second extraction and the third extraction were 1.21 g and 1.24 g, respectively. How much fat, as a percentage of the total, was extracted during the third extraction?

Answers

1. If the sample weight of a semimoist food is 10 g and the moisture content is 25%, the dried weight of the original food is 7.5 g (10 g × 75% = 7.5 g). If the fat content of the dried food is 13.5%, the 7.5 g of dried sample has 1.0125 g fat (7.5 g dried food × 13.5% fat = 1.0125 g fat). The 10 g of semimoist food contains the same amount of fat, i.e., 1.0125 g. Therefore, the fat content of the semimoist food is 10.125% (1.0125 g fat/10 g semimoist food).
2. [(1.24 g – 1.21 g)/10 g] × 100 = 0.3%

REFERENCES

1. McClements DJ, Decker EA (2017) Lipids, chap 4. In: Fennema's Food Chemistry, 5th edn, CRC press, New York, pp 173–175
2. Belitz HD, Grosch W, Schieberle P (2009) Food chemistry, 5th edn, Springer, Berlin, pp 158–245
3. Caballero B, Allen L, Prentice A (2013). Encyclopedia of Human Nutrition, 3rd edn, Academic Press, Cambridge, MA
4. Patton S, Jensen RG (1976) Biomedical aspects of lactation. Pergamon, Oxford, p 78
5. Jenness R, Patton S (1959) Principles of dairy chemistry. Wiley, New York
6. Joslyn MA (1970) Methods in food analysis, 2nd edn. Academic, New York
7. Melini V, Melini F (2021) Compositional and nutritional analysis, Chapter 1. In: Galanakis CM (ed) Innovative Food Analysis, Academic Press, Cambridge, MA, pp 1–39
8. Akon CC (2017) Food Lipids: Chemistry, Nutrition, and Biotechnology, CRC Press, Boca Raton, FL
9. Poole CF (2019) Liquid-Phase Extraction, Elsevier, Cambridge, MA
10. Official Methods of Analysis, current on line edition (22nd edn). AOAC International, Gaithersburg, MD (2023)
11. Folch J, Lees M, Stanley GHS (1957) A simple method for the isolation and purification of total lipids from animal tissues. J Biol Chem 226: 497–509
12. Bligh EG, Dyer WJ (1959) A rapid method of total lipid extraction and purification. Can J Physiol 37: 911–917
13. Christie WW, Han X (2010) Lipid analysis. Isolation, separation, identification, and Lipidomic Analysis, 4th edn, Woodhead publishing, Oxdord
14. AOCS (2020) Official methods and recommended practices, 7th edn. American Oil Chemists' Society, Champaign, IL
15. Talebi M, Patil RA, Sidisky LM, Berthod A, Armstrong DW (2020) ranched-chain dicationic ionic liquids for fatty acid methyl ester assessment by gas chromatography. Anal. Bioanal. Chem. 410: 4633–4643
16. Wehr HM, Frank JF (eds) (2004) Standard methods for the examination of dairy products, 17th edn. American Public Health Association, Washington, DC
17. Gandhi K, Sharma N, Gautam PB, Sharma R, Mann B, Pandey V (2022). Infrared (IR) Spectroscopy. In: Advanced Analytical Techniques in Dairy Chemistry. Springer Protocols Handbooks. Springer, New York, NY, pp. 177–198
18. Devine C, Dikeman M (2014) Encyclopedia of Meat Sciences, Academic press, Cambridge, MA
19. Lan Y, Zha F, Peckrul A, Hanson B, Johnson B, Rao J, Chen B (2019) Genotype x Environmental Effects on Yielding Ability and Seed Chemical Composition of Industrial Hemp (Cannabis sativa L.) Varieties Grown in North Dakota, USA. J. AOCS 12: 1417–1425
20. Gu Z, Chen X, Rao J, Chen B (2022) Statistical Evaluation to Validate Matrix-Matched Calibration for Standardized Beany Odor Compounds Quantitation in Yellow Pea Flour Using HS-SPME-GC-MS. Food & Function 13: 3968–3981

Protein Analysis

18

Derek R. Dee and Sam K. C. Chang

Contents

D. R. Dee (✉)
Faculty of Land and Food Systems, University of British Columbia, Vancouver, BC, Canada
e-mail: ddee@mail.ubc.ca

S. K. C. Chang
Department of Food Science, Nutrition, and Health Promotion, Experimental Seafood Processing Laboratory, Coastal Research and Extension Center, Mississippi State University, Mississippi State, MS, USA
e-mail: sc1690@msstate.edu

B. P. Ismail, S. S. Nielsen (eds.), *Nielsen's Food Analysis*, Food Science Text Series,
https://doi.org/10.1007/978-3-031-50643-7_18

18.1 INTRODUCTION

18.1.1 Classification and General Considerations

Proteins are abundant in all cells, and almost all except storage proteins are important for biological functions and cell structure. Food proteins are very complex. Many have been purified and characterized. Proteins vary in molecular mass, ranging from approximately 5000 to more than a million Daltons. They are composed of elements including hydrogen, carbon, nitrogen, oxygen, and sulfur. Twenty α-amino acids are the building blocks of proteins; the amino acid residues in a protein are linked by peptide bonds. Nitrogen is the most distinguishing element present in proteins. However, nitrogen content in various food proteins ranges from 13.4% to 19.1% [1] due to the variation in the specific amino acid composition of proteins. Generally, proteins rich in basic amino acids contain more nitrogen.

Proteins can be classified by their composition, structure, biological function, or solubility properties. For example, simple proteins contain only amino acids upon hydrolysis, but conjugated proteins also contain non-amino acid components. Proteins have unique conformations that can be altered by denaturants such as heat, acid, alkali, urea, guanidine-HCl, organic solvents, and detergents. The solubility and functional properties of proteins can be altered by denaturants. The analysis of proteins is complicated by the fact that some food components possess similar physicochemical properties. Nonprotein nitrogen (NPN) could come from free amino acids, small peptides, nucleic acids, phospholipids, amino sugars, porphyrin, some vitamins and alkaloids, uric acid, urea, and ammonium ions. Therefore, the total organic nitrogen in foods would represent nitrogen primarily from proteins and to a lesser extent from all organic nitrogen-containing nonprotein substances. Depending on methodology, other major food components, including lipids and carbohydrates, may interfere physically with the analysis of food proteins.

Numerous methods have been developed to measure protein content. The basic principles of these methods include the determination of nitrogen, peptide bonds, aromatic amino acids, dye-binding capacity, ultraviolet (UV) absorptivity of proteins, and light scattering properties. In addition to factors such as sensitivity, accuracy, precision, speed, and cost of analysis, what is actually being measured must be considered in the selection of an appropriate method for a particular application.

18.1.2 Importance of Analysis

Protein analysis is important for:

1. **Nutrition labeling.**
2. **Pricing.** The cost of certain commodities is based on the protein content as measured by nitrogen content (e.g., cereal grains, milk for making certain dairy products, e.g., cheese).
3. **Functional property investigation**. Proteins in various types of food have unique food functional properties: for example, gliadin and glutenins in wheat flour for breadmaking, casein in milk for coagulation into cheese products, and egg albumen for foaming (see Chap. 24, Sect. 24.3.4).
4. **Biological activity determination**. Many proteins, including enzymes or enzyme inhibitors, are relevant to food science and nutrition: for instance, the proteolytic enzymes in the tenderization of meats, pectinases in the ripening of fruits, and trypsin inhibitors in legume seeds are proteins. To compare between samples, enzyme activity often is expressed in terms of specific activity, meaning units of enzyme activity per mg of protein.

Protein analysis is required when you want to know:

1. Total protein content
2. Content of a particular protein in a mixture
3. Protein content during isolation and purification of a protein
4. Nonprotein nitrogen
5. Amino acid composition (see Chap. 24, Sect. 24.3.1)
6. Nutritive value of a protein (see Chap. 24, Sect. 24.3.2)

18.1.3 Content in Foods

Protein content in food varies widely. Foods of animal origin and legumes are excellent sources of proteins. The protein contents of selected food items are listed in Table 18.1 [2].

18.1.4 Introduction to Methods

Principles, general procedures, and applications are described below for various protein determination methods. See Table 18.2 for a summary of the methods described, including more details about their applications and AOAC method numbers [3]. The advantages and disadvantages of methods are included in the summary table, rather than in the text. Please refer to the references cited within the text for detailed instructions on the procedures. Many of the methods covered in this chapter are described in somewhat more detail in books on food proteins [4–6] and proteins generally [7]. The Kjeldahl, Dumas (N combustion), infrared spectroscopy, and anionic dye-binding methods described are from the *Official Methods of Analysis* of AOAC International [3] and are used commonly in nutrition

Table 18.1 Protein content of selected foods

Food item	*Percent protein (wet weight basis)*
Cereals and pasta	
Rice, brown, long-grain raw	7.54
Rice, white, long-grain, regular, raw, enriched	7.13
Wheat flour, wholegrain	13.2
Corn flour, wholegrain, yellow	6.93
Pasta, dry, enriched	13
Cornstarch	0.26
Dairy products	
Milk, reduced fat, fluid, 2%	3.3
Milk, nonfat, dry, regular, with added vit. A and D	36.2
Cheese, cheddar, sharp, sliced	24.2
Yogurt, plain, low fat	5.25
Fruits and vegetables	
Apple, raw, with skin	0.26
Asparagus, raw	2.2
Strawberries, raw	0.67
Lettuce, iceberg, raw	0.9
Potatoes, flesh and skin, raw	2.05
Legumes	
Soybeans, mature seeds, raw	36.5
Beans, kidney, all types, mature seeds, raw	23.6
Tofu, raw, firm	17.3
Tofu, raw, regular	8.08
Meats, poultry, fish	
Beef, chuck, arm pot roast, lean, choice, raw	22
Beef, cured, dried beef	31.1
Chicken, broilers or fryers, breast meat only, raw	22.5
Ham, sliced, regular	16.6
Egg, raw, whole, fresh	12.6
Finfish, cod, Pacific, raw	15.3
Finfish, tuna, white, canned in oil, drained solids	26.5

From the US Department of Agriculture, FoodData Central (2023) USDA Nutrient Data Laboratory. Home Page [2]. https://fdc.nal.usda.gov/

labeling and/or quality control. The other methods described are used commonly in research laboratories working on proteins.

18.2 NITROGEN-BASED METHODS

18.2.1 Kjeldahl Method

18.2.1.1 Principle

In the Kjeldahl procedure, proteins and other organic food components in a sample are digested with sulfuric acid in the presence of catalysts. The **total organic nitrogen** is converted to ammonium sulfate. The digest is neutralized with alkali and distilled into a boric acid solution. The borate anions formed are titrated with standardized acid, which is converted to nitrogen in the sample. The result of the analysis represents the crude protein content of the food since nitrogen also comes from nonprotein components (note that the Kjeldahl method also measures nitrogen in any ammonia and ammonium sulfate).

18.2.1.2 Historical Background

In 1883, Johann Kjeldahl developed the basic process of today's Kjeldahl method to analyze organic nitrogen. An excellent book to review the Kjeldahl method for total organic nitrogen was written by Bradstreet [8]. Several important modifications have improved the original Kjeldahl process, but the original method and the current procedure (as described in detail below) both include the same basic steps: (1) digestion, (2) neutralization and distillation, and (3) titration.

18.2.1.3 General Procedures and Reactions

18.2.1.3.1 Sample Preparation

Solid foods are ground to pass a 20-mesh screen. Samples for analysis should be homogeneous. No other special preparations are required.

18.2.1.3.2 Digestion

Place sample (accurately weighed) in a Kjeldahl flask. Add acid and catalyst; digest until clear to get complete breakdown of all organic matter. Nonvolatile ammonium sulfate is formed from the reaction of nitrogen and sulfuric acid:

$$\text{Protein} \xrightarrow[\text{Sulfuric acid}]{\text{Heat, catalyst}} (NH_4)_2SO_4 \quad (18.1)$$

During digestion, protein nitrogen is liberated to form ammonium ions; sulfuric acid oxidizes organic matter and combines with ammonium formed; carbon and hydrogen elements are converted to carbon dioxide and water.

18.2.1.3.3 Neutralization and Distillation

The digest is diluted with water. Alkali-containing sodium thiosulfate is added to neutralize the sulfuric acid. The ammonia formed is distilled into a boric acid solution containing the indicators methylene blue and methyl red (AOAC Method 991.20):

$$(NH_4)_2SO_4 + 2NaOH \rightarrow 2NH_3 + Na_2SO_4 + 2H_2O \quad (18.2)$$

Table 18.2 Protein analysis method comparison

Method	*Chemical basis*	*Principle*	*Advantages*	*Disadvantages*	*Applications*
Kjeldahl	Nitrogen (total organic)	Determine N by method that involves digestion, neutralization, distillation, and titration. Use N content to calculate protein content	Inexpensive (if not automated system). Widely used and accepted for over a century	Measure total organic N, and not just protein N. Uses corrosive reagents. Lower precision than some other methods	Applicable to all foods. Little used now, due to availability of automated Dumas systems
Dumas	Nitrogen (total organic and inorganic)	N is released upon combustion of sample at very high temperature. N gas is quantitated by gas chromatography using a thermal conductivity detector. Use N content to calculate protein content	Requires no hazardous chemicals. Rapid (few minutes). Automated instruments allow for analyzing many samples without attention	Expensive equipment. Measure total organic and inorganic N, and not just protein N	Applicable to all foods. Widely used now, compared to Kjeldahl method, for both official and quality control purposes
Infrared spectroscopy	Peptide bond	Presence of peptide bond in protein molecules causes absorption of radiation at specific wavelength in mid- or near-infrared region	Rapid way to estimate protein content. Requires minimal training	Expensive equipment. Only provides an estimate of protein content. Instrument must be calibrated against results from official method	Applicable to wide range of food products (grains, cereal, meat, and dairy). Used as rapid, quality control method
Anionic dry-binding	Basic amino acid residues (of histidine, arginine, and lysine) and N-terminus of protein molecule	Residues identified react with anionic sulfonic acid dye to form an insoluble complex. Unbound soluble dye is measured by absorbance and related to protein concentration	Rapid (15 min. or less for non-automated method; much less for automated method). Relatively accurate. No corrosive reagents. Does not measure nonprotein N. More precise than Kjeldahl method. Can be used to estimate changes in available lysine content, since the dye does not bind altered, unavailable lysine	Not as sensitive as some other colorimetric methods. Requires a calibration curve for a given food commodity, since proteins differ in basic amino acid content, so they differ in dye-binding capacity. Not suitable for hydrolyzed proteins due to dye-binding to N-terminal amino acids. Some nonprotein components bind dye or protein, to cause error	Automated version used for quality control purposes, especially as a method to compare results against a nitrogen-based method (to check for economic adulteration)
Biuret	Peptide bond	Peptide bond is complexed with cupric ions under alkaline conditions to give color that is quantitated by spectroscopy	Less expensive, faster, and simpler than Kjeldahl method. Does not detect nonpeptide or nonprotein sources. Few interferences	Less sensitive than the Lowry and BCA methods	Not widely used now but is the partial basis of other methods based on peptide bond (e.g., BCA method)
Bicinchoninic acid (BCA)	Peptide bond (biuret reaction) and specific amino acids (cysteine, cystine, tryptophan, and tyrosine)	Peptide bond is complexed with cupric ions under alkaline conditions. Cuprous ions are chelated by BCA reagent to give color measured by spectroscopy	Good sensitivity, and micro-BCA method is even better (0.5–10 μg). Nonionic detergents and buffer salts do not interfere with the reaction, nor do medium concentrations of denaturing reagents	Color is not stable with time. Any compound capable of reducing Cu^{+2} to Cu^{+} will lead to color formation. Reducing sugars and high concentrations of ammonium sulfate interfere. Get color variation among proteins	Widely used method for protein isolation and purification. Has largely replaced other quantitative research colorimetric methods

Table 18.2 (continued)

Method	*Chemical basis*	*Principle*	*Advantages*	*Disadvantages*	*Applications*
Absorbance at 280 nm	Tyrosine and tryptophan	Aromatic amino acids, tryptophan and tyrosine, cause proteins to absorb at 280 nm. Absorbance can be used to estimate protein content	Rapid. Relatively sensitive (100 μg protein required). No interference from ammonium sulfate and other buffer salts. Nondestructive (so samples can be used after protein determination)	Nucleic acids can absorb at 280 nm. Aromatic amino acids in proteins vary between food sources, so results are qualitative. Requires relatively pure, clear, and colorless samples	Best used in purified protein systems (e.g., post-column detection of intact proteins)
Absorbance at 220 nm	Peptide bond	Peptide bonds cause proteins to absorb at 220 nm. Absorbance can be used to estimate protein content	Rapid. Nondestructive (so samples can be used after protein determination)	Many things other than peptide bonds absorb at 220 nm. Requires relatively pure, clear, and colorless samples	Best used with purified, hydrolyzed protein samples (i.e., post-column detection of hydrolyzed proteins)

$$NH_3 + H_3BO_3\left(\text{boric acid}\right) ® NH_4 + H_2BO_3^-\left(\text{borate ion}\right) \quad (18.3)$$

18.2.1.3.4 Titration

Borate anion (proportional to the amount of nitrogen) is titrated with standardized HCl:

$$H_2BO_3^- + H^+ ® H_3BO_3 \quad (18.4)$$

18.2.1.3.5 Calculations

$$\text{Moles of HCl} = \text{moles } NH_3 = \text{moles N in the sample} \quad (18.5)$$

A reagent blank should be run to subtract reagent nitrogen from the sample nitrogen:

$$\%N = N\text{HCl} \times \frac{\text{Corrected acid volume}}{g\text{ of sample}} \times \frac{14\,g\,\text{N}}{\text{mol}} \times 100 \quad (18.6)$$

where:

N HCl = normality of HCl, in moles/1000 mL
Corrected acid vol. = (mL std. acid for sample) – (mL std. acid for blank)
14 = atomic weight of nitrogen

A factor is used to convert percent N to percent crude protein. Most proteins contain 16% N, so the conversion factor is 6.25 (100/16 = 6.25):

$$\%\,N / 0.16 = \%\text{ protein}$$
$$\text{or} \quad (18.7)$$
$$\%\,N \times 6.25 = \%\text{ protein}$$

Table 18.3 Nitrogen-to-protein conversion factors for various foods

	Percent N	
	In protein	*Factor*
Egg or meat	16.0	6.25
Milk	15.7	6.38
Wheat	18.76	5.33
Corn	17.70	5.65
Oat	18.66	5.36
Soybean	18.12	5.52
Rice	19.34	5.17

Data from [1, 8]

Conversion factors for various foods are given in Table 18.3 [1, 9].

18.2.1.4 Applications

The Kjeldahl method is an AOAC official method for crude protein content and has been the basis for evaluation of many other protein methods. The Kjeldahl method is still used for some applications but now has limited use in many countries due to the availability and advantages of automated nitrogen combustion (Dumas) systems (Sect. 18.2.2) (see also Table 18.2 for advantages and disadvantages of the methods).

18.2.2 Dumas (Nitrogen Combustion) Method

18.2.2.1 Principle

The combustion method was introduced in 1831 by Jean-Baptiste Dumas. It has been modified and automated to improve accuracy since that time. Samples are combusted at high temperatures (700–1000 °C) with a flow of pure oxygen. All carbon in the sample is converted to carbon dioxide during the flash combustion. Nitrogen-containing components produced include N_2 and nitrogen oxides. The nitrogen oxides are reduced to nitrogen in a copper reduction column at a high temperature (600 °C). The total nitrogen (including both organic and inorganic compounds, i.e., nitrate and nitrite) released is carried by pure helium and quantitated by **gas chromatography** using a **thermal conductivity detector** (TCD) [10]. Ultrahigh purity acetanilide and ethylenediaminetetraacetate (EDTA) may be used as the standards for the calibration of the nitrogen analyzer. The nitrogen determined is converted to protein content in the sample using a protein conversion factor (Table 18.3).

18.2.2.2 Procedure

Samples (approximately 100–500 mg) are weighed into a tin capsule and introduced to a combustion reactor in automated equipment. The nitrogen released is measured by a built-in gas chromatograph. Figure 18.1 shows the flow diagram of the components of a Dumas nitrogen analyzer.

18.2.2.3 Applications

The combustion method is a faster and safer alternative to the Kjeldahl method [11] and is suitable for all types of foods. As an AOAC method, the Dumas method is widely used for official purposes, but its speed also allows for quality control applications.The industry uses different units/systems, depending on sample size and protein content. Freeze drying can be used to concentrate diluted liquid samples, e.g., waste stream samples.

18.3 INFRARED SPECTROSCOPY

18.3.1 Principle

Infrared spectroscopy measures the **absorption of radiation** (near- or mid-infrared regions) by molecules in food or other substances. Different functional groups in a food absorb different frequencies of radiation. For proteins and peptides, various **mid-infrared** bands (6.47 μm) and **near-infrared** (NIR) bands (e.g., 3300–3500 nm, 2080–2220 nm, and 1560–1670 nm) characteristic of the **peptide bond** can be used to estimate the protein content of a food. By irradiating a sample with a wavelength of infrared light specific for the constituent to be measured, it is possible to determine the concentration of that constituent by measuring the energy that is reflected or transmitted by the sample (which is inversely proportional to the energy absorbed) [12].

18.3.2 Procedure

See Chap. 8 for a detailed description of instrumentation, sample handling, and calibration and quantitation methodology.

18.3.3 Applications

Mid-infrared spectroscopy is used in infrared milk analyzers to determine milk protein content, while near-infrared spectroscopy is applicable to a wide range of food products (e.g., grains, cereal, meat, and dairy products) [3, 13, 14], especially as a rapid method to test nonstandard milk.

18.4 COLORIMETRIC METHODS

When protein reacts with specific reagents under certain conditions, colorful compounds are generated, and the absorbance is measured using spectrophotometry. The protein content is expressed on the basis of a standard protein, such as bovine serum albumin (BSA), and thus these methods do not report absolute protein content. Selection of the proper standard is key (ideally, the standard is as similar to the sample as possible), since protein-to-protein variation causes some proteins to be over- or underestimated [15]. However, because of high sensitivity, these methods have the advantage of requiring a small sample size.

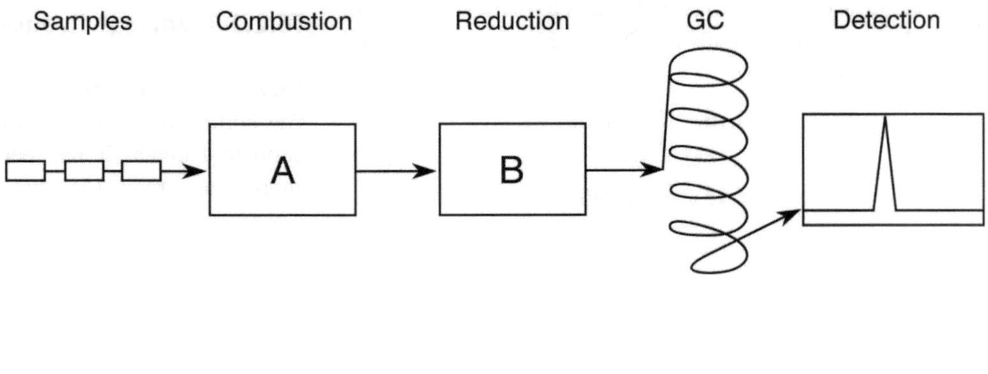

Fig. 18.1 General components of a Dumas nitrogen analyzer: the incinerator (**A**), copper reduction unit for converting nitrogen oxides to nitrogen (**B**), gas chromatography (GC) column, and detector

18.4.1 Dye-Binding Methods

18.4.1.1 Anionic Dye-Binding Method

18.4.1.1.1 Principle

The protein-containing sample is mixed with a known excess amount of **anionic dye** in a buffered solution. Proteins bind the dye to form an insoluble complex. The unbound soluble dye is measured after equilibration of the reaction and the removal of insoluble complex by centrifugation or filtration.

$$\begin{aligned}&\text{Protein} + \text{excess dye} \rightarrow \\ &\text{Protein} - \text{dye insoluble complex} + \text{unbound soluble dye}\end{aligned} \quad (18.8)$$

The anionic sulfonic acid dye, including acid orange 12, orange G, and Amido Black 10B, binds cationic groups of the **basic amino acid residues** (imidazole of histidine, guanidine of arginine, and ε-amino group of lysine) and the **free amino terminal group** of the protein [16]. The amount of the unbound dye is inversely related to the protein content of the sample [16].

18.4.1.1.2 Procedure

1. The sample is finely ground (60 mesh or smaller sizes) and added to an excess dye solution with known concentration.
2. The content is vigorously shaken to equilibrate the dye-binding reactions and filtered or centrifuged to remove insoluble substances.
3. Absorbance of the unbound dye solution in the filtrate or supernatant is measured and dye concentration estimated from a dye standard curve.
4. A straight calibration curve can be obtained by plotting the unbound dye concentration against the total nitrogen (as determined by the Kjeldahl method) of a given food covering a wide range of protein content.
5. Protein content of the unknown sample of the same food type can be estimated from the calibration curve or from a regression equation calculated by the least squares method.

18.4.1.1.3 Applications

Anionic dye-binding has been used to estimate proteins in milk [17, 18], wheat flour [19, 20], soy products [20, 21], and meats [22]. The anionic dye-binding method may be used to estimate the changes in available lysine content of cereal products during processing, since the dye does not bind altered, unavailable lysine. Since lysine is the limiting amino acid in cereal products, the available lysine content represents protein nutritive value of the cereal products [23]. An automated Sprint™Rapid Protein Analyzer has been developed by the CEM Company (Matthews, NC) based on the anionic dye-binding method. This rapid protein analyzer combines and automates all steps (sample weighing, homogenization, predetermined dye addition, filtration, absorption measurement, and protein calculation) in a compact benchtop unit. This dye-binding method has been approved in AOAC Method 2011.04 for raw and processed meats. This analyzer has shown great agreement with the Kjeldahl method in protein analysis of milk, cream, soybean, and soymilk [24, 25]. This method is safe (the proprietary dye iTag® is nonhazardous), easy to use, and efficient. Because different types of sample matrixes and proteins exhibit different responses on the Sprint Rapid Protein Analyzer, this automated method requires calibration for each type of food protein determined using other official methods.

18.4.1.2 Bradford Dye-Binding Method

18.4.1.2.1 Principle

When **Coomassie Brilliant Blue G-250** binds to protein, the **dye changes color** from reddish to bluish, and the absorption maximum of the dye is shifted from 465 to 595 nm. The change in the absorbance at 595 nm is proportional to the protein concentration of the sample [26]. Like other dye-binding methods, the Bradford relies on the **amphoteric nature of proteins**. When the protein-containing solution is acidified to a pH less than the isoelectric point of the protein(s) of interest, the dye added binds electrostatically. Binding efficiency is enhanced by hydrophobic interaction of the dye molecule with the polypeptide backbone adjoining positively charged residues in the protein [4]. In the case of the Bradford method, the dye bound to protein has a change in absorbance spectrum relative to the unbound dye.

18.4.1.2.2 Procedure

1. Coomassie Brilliant Blue G-250 is dissolved in 95% ethanol and acidified with 85% phosphoric acid.
2. Samples containing proteins (1–100 μg/mL) and standard BSA solutions are mixed with the Bradford reagent.
3. Absorbance at 595 nm is read against a reagent blank.
4. Protein concentration in the sample is estimated from the BSA standard curve.

18.4.1.2.3 Applications

The Bradford method has been used successfully to determine protein content in worts and beer products [27] and in potato tubers [28]. This procedure has been improved to measure microgram quantities of proteins [29]. Due to its rapidity, sensitivity, and fewer interferences than the Lowry method, the Bradford method has been used widely for the analysis of low concentrations of proteins and enzymes in their purification and characterizations. Improved Coomassie dye-based protein assays based on the Bradford method have been developed (e.g., CB™ and CB-X™, G-Biosciences, Geno Technology, Inc.) for increased compatibility with buffers and conditions commonly used in protein isolation. Other dye-based protein assays (e.g., RED 660™ Protein Assay, ThermoScientific Pierce) have been developed to improve the linearity, color stability, and interferences of the Bradford and improved Coomassie/Bradford assays.

18.4.2 Copper Ion-Based Methods

Since the biuret method was first developed to measure protein content based on the reaction with copper ions, several improved methods have been developed. Following the description of the basic biuret method below are the modified Lowry and the bicinchoninic acid (BCA) methods, which are both based in part on the biuret method.

18.4.2.1 Biuret Method

18.4.2.1.1 Principle

A violet-purplish color is produced when **cupric ions** are complexed with **peptide bonds** (substances containing at least two peptide bonds, i.e., oligopeptides, large peptides, and all proteins) under **alkaline conditions** (Fig. 18.2). The absorbance of the color produced is read at 540 nm. The color intensity (absorbance) is proportional to the protein content of the sample [30].

18.4.2.1.2 Procedure

1. A 5-mL biuret reagent is mixed with a 1-mL portion of protein solution (1 to 10 mg protein/ mL). The reagent includes copper sulfate, NaOH, and potassium sodium tartrate, which is used to stabilize the cupric ion in the alkaline solution.
2. After the reaction mix is allowed to stand at room temperature for 15 or 30 min, the absorbance is read at 540 nm against a reagent blank.
3. Filtration or centrifugation before reading absorbance is required if the reaction mixture is not clear.
4. A standard curve of concentration versus absorbance is constructed using **bovine serum albumin** (BSA).

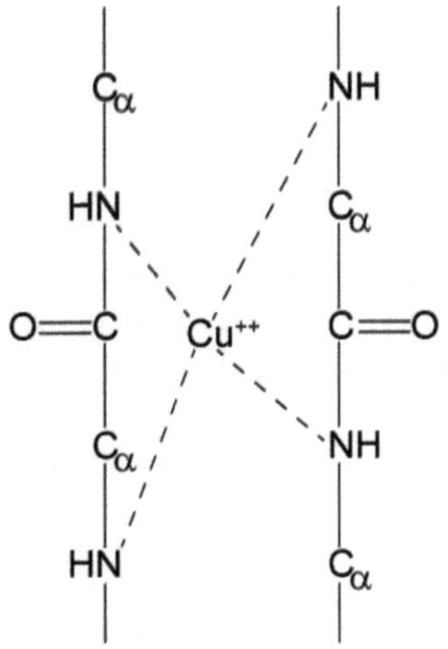

Fig. 18.2 Reaction of peptide bonds with cupric ions

18.4.2.1.3 Applications

The biuret method has been used to determine proteins in cereal [31, 32], meat [22], soybean proteins [21], and as a qualitative test for animal feed [33]. The biuret method also can be used to measure the protein content of isolated proteins, but it has been largely replaced for this use by more sensitive methods such as the modified Lowry and the BCA methods for the reasons described below (Sect. 18.4.2.2 and Sect. 18.4.2.3).

18.4.2.2 Lowry Method

18.4.2.2.1 Principle

The Lowry method [34, 35] combines the **biuret reaction** with the reduction of the **Folin–Ciocalteu phenol reagent** (phosphomolybdic-phosphotungstic acid) by **tyrosine** and **tryptophan** residues in the proteins (Fig. 18.3). The bluish color developed is read at 750 nm (high sensitivity for low protein concentration) or 500 nm (low sensitivity for high protein concentration). The original procedure has been modified by Miller [36] and Hartree [37] to improve the linearity of the color response to protein concentration and replace the use of two unstable reagents with one stable reagent.

18.4.2.2.2 Procedure

The following procedure is based on the modified procedure of Hartree [37]:

1. Proteins to be analyzed are diluted to an appropriate range (20–100 μg).
2. K Na Tartrate-Na_2CO_3 solution is added after cooling and then incubated at room temperature for 10 min.
3. $CuSO_4$-K Na Tartrate-NaOH solution is added after cooling and then incubated at room temperature for 10 min.

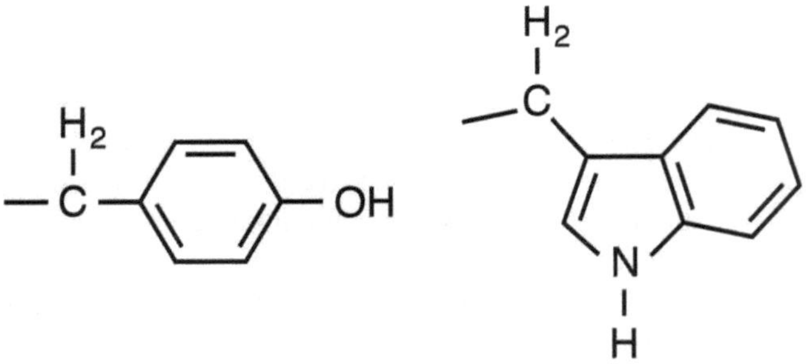

Fig. 18.3 Side chains of amino acids tyrosine (**a**) and tryptophan (**b**)

4. Freshly prepared Folin's reagent is added, and then the reaction mixture is mixed and incubated at 50 °C for 10 min.
5. Absorbance is read at 650 nm.
6. A standard curve of BSA is constructed for estimating protein concentration of the unknown.

18.4.2.2.3 Applications

Because of its simplicity and sensitivity, the Lowry method has been widely used in protein biochemistry. However, it has not been widely used to determine proteins in food systems without first extracting the proteins from the food mixture, where it may be used, for example, to examine extracted seed storage proteins [38].

18.4.2.3 Bicinchoninic Acid Method

18.4.2.3.1 Principle

Proteins and peptides (as short as dipeptides) reduce **cupric ions** to **cuprous ions** under **alkaline conditions** [39], which is similar in principle to that of the biuret reaction. The cuprous ion then reacts with the apple-greenish **bicinchoninic acid reagent** (BCA) to form a purplish complex (one cuprous ion is chelated by two BCA molecules) (Fig. 18.4). The color measured at 562 nm is near linearly proportional to protein concentration over a wide range of concentration from micrograms up to 2 mg/mL. Peptide bonds and four amino acids (cysteine/cystine, tryptophan, and tyrosine) contribute to the color formation with BCA.

18.4.2.3.2 Procedure

1. Mix (one step) the protein solution with the BCA reagent, which contains BCA sodium salt, sodium carbonate, NaOH, and copper sulfate, pH 11.25.
2. Incubate at 37 °C for 30 min or room temperature for 2 h or 60 °C for 30 min.

   The selection of the temperature depends upon the sensitivity desired. A higher temperature gives a greater color response.

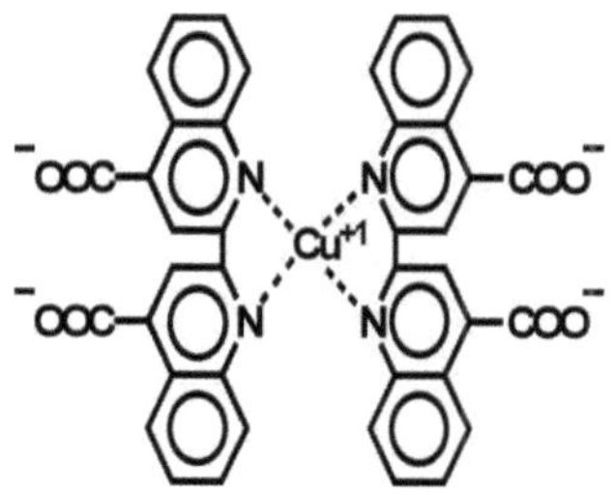

Fig. 18.4 Protein reaction with cupric ions under alkaline conditions to form cuprous ions, which react with bicinchoninic acid (BCA) to form purple color, measured at 562 nm. (Figure courtesy of Pierce BiotechnologyTechnical Library, Thermo Fisher Scientific, Inc., Rockford, IL)

3. Read the solution at 562 nm against a reagent blank.
4. Construct a standard curve using BSA.

18.4.2.3.3 Applications

The BCA method is widely used in protein isolation and purification due to its advantage (over the modified Lowry method and any Coomassie dye-based assay) of being compatible with samples containing up to 5% detergents. While most dye-binding methods are faster, the BCA method is less affected by protein compositional differences, so there is better protein-to-protein uniformity. The BCA method has been used to examine purified soy and egg white proteins [40].

18.5 ULTRAVIOLET ABSORPTION METHODS FOR PROTEINS AND PEPTIDES

18.5.1 Ultraviolet 280 nm Absorption for Protein

18.5.1.1 Principle

Proteins show strong absorption in the **ultraviolet** (UV) region at **UV 280 nm**, primarily due to **tryptophan** and **tyrosine** residues in the proteins. Because the content of tryptophan and tyrosine in proteins from each food source is fairly constant, the absorbance at 280 nm can be used to estimate the concentration of proteins, using **Beer's law.** Since each protein has a unique aromatic amino acid composition, the absorptivity or extinction coefficient (E_{280}) must be determined for individual proteins for protein content estimation. If the protein sequence is known, the E_{280} can be calculated using online tools such as ProtParam [41].

18.5.1.2 Procedure

1. Proteins are solubilized in buffer or alkali.
2. Absorbance of protein solution is read at 280 nm against a reagent blank.
3. Protein concentration is calculated according to the following equation:

$$A = abc \tag{18.9}$$

where:

A = absorbance
a = absorptivity
b = cell or cuvette path length
c = concentration

18.5.1.3 Applications

The UV 280 nm method has been used to determine the protein contents of milk [42] and meat products [43]. It has not been used widely in food systems. This technique is best used for pure proteins, especially those containing at least one Trp residue, or proteins that have been extracted in alkali or denaturing agents such as 8 *M* urea.

18.5.2 Peptide Measurement at 190–220 nm

Peptides without or with low level of tyrosine or tryptophan residues can be quantified at 190–220 nm at which peptide bonds have maximum absorption. The extinction coefficients in the far UV range can be calculated with the consideration of contribution of tyrosine and tryptophan to the absorption [44]. Absorbance at 205 nm was used to study peptide contents from various food hydrolysates [45]. Protein can also be measured in this UV range. For example, NanoDrop™ 2000/2000c from Thermo Scientific can be used for protein and peptide quantification at 205 nm with extinction coefficient being 31.

18.6 NONPROTEIN NITROGEN DETERMINATION

18.6.1 Principle

Protein is precipitated by trichloroacetic acid (TCA) and separated from nonprotein nitrogen (NPN)-containing compounds.

18.6.2 Procedure

1. Add appropriate amount of TCA into a sample solution to make the final TCA concentration around 10%. Alternatively, prepare a 10% TCA solution and add dry sample powder into the solution.
2. Mix the reaction mixture thoroughly and let precipitate settle for 5 min.
3. Filter the mixture through Whatman No. 1 filter paper. Alternatively, centrifugation at 30,000 g can be used to separate precipitated protein from the supernatant that contains NPN.
4. Determine nitrogen content of the filtrate by the Kjeldahl method or Dumas method.
5. Convert nonprotein nitrogen to protein equivalent with conversion factor.

18.6.3 Applications

The NPN determination is used to check economic adulteration with nitrogen-rich compounds such as urea, ammonia, and melamine. NPN for milk can be tested by AOAC methods (991.21 and 991.22). The results obtained are not absolute but are relative to the standardized conditions used in the procedure [46].

18.7 COMPARISON OF METHODS

1. **Sample Preparation.** The Kjeldahl, Dumas, and infrared spectroscopy methods require little preparation. Sample particle size of 20 mesh or smaller generally is satisfactory for these methods. Some of the newer NIR instruments can make measurements directly on wholegrains and other coarsely granulated products without grinding or other sample preparation. Other methods described in this chapter require fine particles for the extraction of proteins from the complex food systems.
2. **Principle.** The Dumas and Kjeldahl methods measure directly the nitrogen content of foods. However, the Kjeldahl method measures only organic nitrogen plus ammonia, while Dumas measures total nitrogen, including the inorganic fraction. (Therefore, Dumas gives a higher value for products that contain nitrates/nitrites.) Other methods of analysis measure the various properties of proteins. For instance, the biuret method measures peptide bonds, and the Lowry and BCA methods measure a combination of peptide bonds and specific amino acids. Infrared spectroscopy is an indirect method to estimate protein content, based on the energy absorbed when a sample is subjected to a wavelength of infrared radiation specific to the peptide bond.
3. **Sensitivity.** Kjeldahl, Dumas, and biuret methods are less sensitive than Bradford, Lowry, BCA, or UV methods.
4. **Speed.** After the instrument has been properly calibrated, infrared spectroscopy is likely the most rapid of the methods discussed. In most other methods involving spectrophotometric (colorimetric) measurements, one must separate proteins from the interfering insoluble materials before mixing with the color reagents or must remove the insoluble materials from the colored protein–reagent complex after mixing. However, the speed of determination in the colorimetric methods and in the Dumas method is faster than with the Kjeldahl method.
5. **Applications.** Although both the Kjeldahl and Dumas methods can be used to measure N content in all types of foods; in recent years, the Dumas method has largely replaced the Kjeldahl method for nutrition labeling (since

the Dumas method is faster, has a lower detection limit, and is safer). However, the Kjeldahl method is the preferred method for high-fat samples/products, since fat may cause an instrument fire during the incineration procedure in the Dumas method. The Kjeldahl method or the nitrogen combustion method is specified to correct for protein content in the most recent AOAC official method to measure the fiber content of foods (see Chap. 19, Sect. 19.6). Melamine, a toxic nitrogen adulterant, is included in the total nitrogen content if measured by the Kjeldahl or Dumas methods. NIR, CEM-Sprint dye-binding, and Dumas methods are good for quality control in food processing plants.

18.8 SPECIAL CONSIDERATIONS

1. To select a particular method for a specific application, sensitivity, accuracy, and reproducibility as well as physicochemical properties of food materials must be considered. The data should be interpreted carefully to reflect what actually is being measured.
2. Food processing methods, such as heating, may reduce the extractability of proteins for analysis and cause an underestimation of the protein content measured by methods involving an extraction step.
3. Except for the Dumas and Kjeldahl methods, and the UV method for purified proteins, all methods require the use of a standard or reference protein or a calibration with the Kjeldahl method. In the methods using a standard protein, proteins in the samples are assumed to have similar composition and behavior compared to the standard protein. The selection of an appropriate standard for a specific type of food is important.
4. **Nonprotein nitrogen** is present in practically all foods. To determine **protein nitrogen**, the samples usually are extracted under alkaline conditions and then precipitated with trichloroacetic acid or sulfosalicylic acid. The concentration of the acid used affects the precipitation yield. Therefore, nonprotein nitrogen content may vary with the type and concentration of the reagent used. Heating could be used to aid protein precipitation by acid, alcohol, or other organic solvents. In addition to acid precipitation methods used for nonprotein nitrogen determination, less empirical methods such as dialysis and ultrafiltration and column chromatography could be used to separate proteins from small nonprotein substances.
5. In the determination of the nutritive value of food proteins, including **protein digestibility** and **protein efficiency ratio** (PER), the Kjeldahl or Dumas method with a 6.25 conversion factor is used to determine crude protein content. The PER could be underestimated if a substantial amount of nonprotein nitrogen is present in foods. A food sample with a higher nonprotein nitrogen content (particularly if the nonprotein nitrogen does not have many amino acids or small peptides) may have a lower PER than a food sample containing similar protein structure/composition and yet with a lower amount of nonprotein nitrogen.

18.9 SUMMARY

Methods based on the unique characteristics of proteins and amino acids have been described to determine the protein content of foods. The Kjeldahl and Dumas methods measure nitrogen. Infrared spectroscopy is based on the absorption of a wavelength of infrared radiation specific to the peptide bond. Copper–peptide bond interactions contribute to the analysis by the biuret, Lowry, and BCA methods. Amino acid side chains are involved in the Lowry, BCA, dye-binding, and UV 280-nm methods. The BCA method also utilizes the reducing power of proteins in an alkaline solution. The various methods differ in their speed and sensitivity.

In addition to the commonly used methods discussed, there are other methods available for protein quantification. Because of the complex nature of various food systems, problems may be encountered to different degrees in protein analysis by available methods. Rapid methods may be suitable for quality control purposes, while a sensitive method is required for work with a minute amount of protein. Indirect colorimetric methods usually require the use of a carefully selected protein standard or a calibration with an official method.

18.10 STUDY QUESTIONS

1. Complete the table below to contrast the Dumas and Kjeldahl methods for protein analysis, making clear the advantages of the Dumas method over the Kjeldahl method.

	Dumas	*Kjeldahl*
What is measured?		
How is N measured?		
Speed		
Reagents		
Automation		
Cost		

2. What factors should one consider when choosing a method for protein determination?

3. The Kjeldahl method of protein analysis consists of three major steps. List these steps in the order they are done and describe in words what occurs in each step.
   Make it clear why milliliters of HCl can be used as an indirect measure of the protein content of a sample.
4. Why is the conversion factor from Kjeldahl nitrogen to protein different for various foods, and how is the factor of 6.25 obtained?
5. Differentiate the principles of protein determination by dye-binding with an anionic dye such as Amido Black vs the Bradford method, which uses the dye Coomassie Blue G-250.
6. With the anionic dye-binding method, would a sample with a higher protein content have a higher or a lower absorbance reading than a sample with a low protein content? Explain your answer.
7. For each of the situations described below, identify a protein assay method most appropriate for use and indicate the chemical basis of the method (i.e., what does it really measure?)
   (a) Nutrition labeling
   (b) Intact protein eluting from a chromatography column; qualitative or semiquantitative method
   (c) Intact protein eluting from a chromatography column; colorimetric, quantitative method
   (d) Rapid, quality control method for protein content of cereal grains
8. The FDA found melamine (see structure below) in pet food linked to deaths of pets in the United States. The FDA also found evidence of melamine in wheat gluten imported from China used as one of the ingredients in the production of pet food. Melamine is a nitrogen-rich chemical used to make plastic and is sometimes used as a fertilizer.

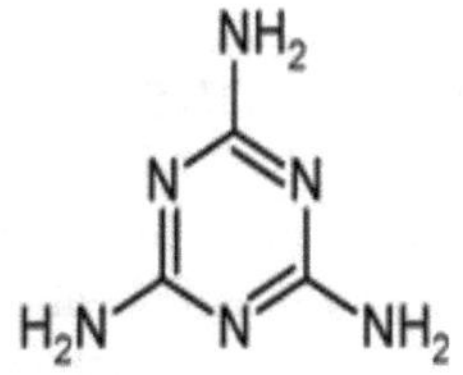

   (a) Knowing that each ingredient is tested and analyzed when imported, explain how melamine in wheat gluten could have escaped detection.
   (b) How can the adulteration of wheat gluten be detected (not necessarily detecting melamine specifically) using a combination of protein analysis methods? Explain your answer.
9. Identify methods that can quantitate protein content based on each of the following characteristics of proteins, and give a typical application for use of each method.

Characteristic	*Method*	*Application*
Peptide bond(s)		
Nitrogen content		
Presence of tyrosine and tryptophan		

10. When estimating the protein content of hydrolyzed proteins eluting from a chromatography column, should one use the absorbance method measured at 220 nm or at 280 nm? Explain your answer.
11. What assumptions are made in using nitrogen-based methods for protein analysis?

Explain your answer.

18.11 PRACTICE PROBLEMS

1. A dehydrated precooked pinto bean was analyzed for crude protein content in duplicate using the Kjeldahl method. The following data were recorded:
   Moisture content = 8.00%
   Wt of Sample 1 = 1.015 g
   Wt of Sample 2 = 1.025 g
   Normality of HCl used for titration = 0.1142 N
   HCl used for Sample 1 = 22.0 mL
   HCl used for Sample 2 = 22.5 mL
   HCl used for reagent blank = 0.2 mL

   Calculate crude protein content on both wet and dry weight basis of the pinto bean, assuming pinto bean protein contains 17.5% nitrogen.
2. A 20-mL protein fraction recovered from a column chromatography was analyzed for protein using the BCA method. The following data were the means of a duplicate analysis using BSA as a standard:

BSA mg/mL	*Mean absorbance at 562 nm*
0.2	0.25
0.4	0.53
0.6	0.74
0.8	0.95
1.0	1.15

   The average absorbance of a 1-mL sample was 0.44. Calculate protein concentration (mg/mL) and total protein quantity of this column fraction.

3. A turkey frankfurter was analyzed for crude protein content using the Kjeldahl method.

   The following data were recorded:

   Wt of sample = 0.5172 g
   N of HCl = 0.1027 *N*
   Vol of HCl used for sample = 8.8 mL

Vol of HCl used for reagent blank = 0.2 mL
Calculate % crude protein content on wet weight basis
For calculating the conversion factor, assume that amino acids in meat protein contain 16% nitrogen (or look up the conversion factor).

$$\%\text{Nitrogen} = \frac{(\text{mL HCl sample} - \text{mL HCl blank}) \times N\,\text{HCl} \times 1.4007}{\text{Sample weight (g)}}$$

% N × Conversion Factor = % Protein
100/% N = Conversion Factor.

4. Why is it difficult to accurately determine protein content of a composite food such as a sausage pizza when using the Kjeldahl and combustion methods for protein?
5. A standard procedure for the Dumas combustion method for protein states that a sample should contain between 10 and 50 mg N. Using the USDA Nutrient Database [2], calculate the weight of sample to use for analysis. Sample: turkey thigh meat, no skin, raw.
6. Estimate the weight of sample (Wheat Thins) you should use for Kjeldahl protein determination if you want to titrate the sample with at least 7 mL 0.1 *N* acid. Show calculation. Use the nutritional label as a guide (serving size: 29 g; protein: 2 g).
7. A stock solution of bovine serum albumin (BSA) was prepared previously by a coworker. Before using the BSA to prepare a standard curve for biuret determination, you need to confirm that the stock solution contains 20 mg protein/mL using UV spectroscopy. The $E^{1\%}$ for BSA is 6.3 at 280 nm. If the stock solution was correctly prepared, calculate the absorbance you would expect to find when the stock solution is diluted ten fold (0.3 mL stock with 2.7 mL water) and absorbance of this solution determined at 280 nm in a 3-mL cuvette with a 0.5-cm path length.
8. Colorimetric protein assay. Calculate the concentration of protein in an unknown solution X using the biuret test for protein. The stock solution contains 20-mg protein/mL. The standard curve was prepared using the stock solution as follows (used 1-cm path length cuvette):

Standard curve	*1*	*2*	*3*	*4*	*5*	*6*
Vol. H_20 (mL)	1	0.8	0.6	0.4	0.2	0.0
Vol. protein standard (mL)	0.0	0.2	0.4	0.6	0.8	1.0
Vol. biuret reagent (mL)	4.0	4.0	4.0	4.0	4.0	4.0
Absorbance at 540 nm	0.0	0.174	0.343	0.519	0.691	0.823

Solution X was prepared for analysis as follows:

Sample	*Tube 1*	*Tube 2*
Vol. H_20 (mL)	0.8	0.2
Vol. solution X (mL)	0.2	0.8
Vol. biuret reagent (mL)	4.0	4.0
Absorbance at 540 nm	0.451	0.857

(a) Explain why the absorbance of the biuret assay is measured at 540 nm.
(b) Calculate the concentration of protein (mg/tube) in each tube used to prepare the standard curve.
(c) Prepare a standard curve (in mg/tube). Label the graph properly.
(d) What sample dilution (tube 1 or tube 2) should you use to determine the concentration of the unknown sample? Why?
(e) Calculate the concentration of protein in solution X. List the concentration as mg protein/cuvette and then express results as mg protein/mL of solution X.

Answers

1. Protein content = 19.75% on a wet weight basis; 21.47% on a dry weight basis.

Calculations:

$$\%\text{N} = N\,\text{HCl} \times \frac{\text{Corrected acid volume}}{\text{g of sample}} \times \frac{14\,\text{g N}}{\text{mol}} \times 100$$

where:

N HCl = normality of HCl in mol/1000 mL
Corrected acid vol. = (mL std. acid for sample) − (mL std. acid for blank)
14 = atomic weight of nitrogen

Corrected acid volume for Sample 1

= 22.0 mL − 0.2 mL = 21.8 mL

Corrected acid volume for Sample 2

= 22.5 mL − 0.2 mL = 22.3 mL

$$\%\text{N for Sample 1} = \frac{0.1142\,\text{mole}}{1000\,\text{mL}} \times \frac{21.8\,\text{mL}}{1.015\,\text{g}} \times \frac{14\,\text{g N}}{\text{mol}} \times 100\%$$

$$= 3.433\%$$

$$\%\text{N for Sample 2} = \frac{0.1142\,\text{mole}}{1000\,\text{mL}} \times \frac{22.3\,\text{mL}}{1.025\,\text{g}} \times \frac{14\,\text{g N}}{\text{mol}} \times 100\%$$

$$= 3.478\%$$

Protein conversion factor

= 100%/17.5% N = 5.71

Crude protein content for Sample 1

= 3.433% × 5.71 = 19.6%

Crude protein content for Sample 2

= 3.478% × 5.71 = 19.9%

The average for the duplicate data

= (19.6% + 19.9%)/2 = 19.75%
= ~19.8% wet weight basis

To calculate protein content on a dry weight basis: Sample contains 8% moisture, and therefore, the sample contains 92% dry solids, or 0.92 g out of 1 g sample. Therefore, protein on a dry weight basis can be calculated as follows:

$$19.75\% / 0.92\,\text{g dry solids} = 21.47\%$$
$$= \sim 21.5\% \text{dry weight basis.}$$

2. Protein content = 0.68 mg/mL. Total protein quantity = 6.96 mg

Calculations:

Plot absorbance (*y*-axis, absorbance at 562 nm) versus BSA protein concentration (*x*-axis, mg/mL) using the data above. Determine the equation of the line ($y = 1.11\,x + 0.058$) and then use this equation and the given absorbance ($y = 0.44$) to calculate the concentration ($x = 0.344$ mg/mL). Since 1 mL of sample gives a concentration of 0.344 mg/mL and we have a total of 20 mL collected from column chromatography, we will have a total of (0.344 mg/mL × 20 mL) = 6.88 mg protein in this collected column fraction.

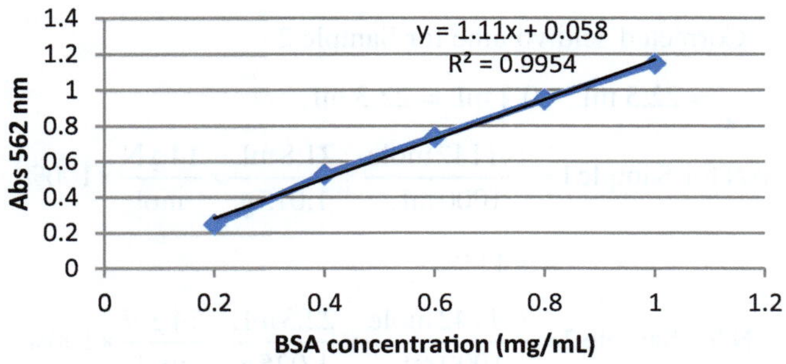

3. $$\frac{100}{16\%} = 6.25$$

$$\%N = \frac{(8.8 - 0.2\,\text{mL})(0.1027\,\text{N})(1.4007)}{0.5172}$$
$$= 2.39\%\text{N} \times 6.25 = 14.95\%\text{Protein}$$

4. The composite food contains many different proteins with different proportions of amino acids of varying nitrogen contents. The following is the conversion factors of some ingredients:
   Meat: 6.25
   Dairy: 6.38
   Wheat flour: 5.7

5. Database: 20.6 g protein/100 g
   Estimate:

$$\%N = \frac{20.6\%\text{protein}}{6.25} = 3.30\%\text{N}$$

$$\frac{3.30\,\text{g}}{100\,\text{g}} = \frac{0.010\,\text{g}}{\text{x}}$$

$$\text{x} = 0.303\,\text{g}$$
$$5\text{x} = 5 \times 0.303\text{g} = 1.515\,\text{g}$$
$$\text{Weight of sample} = 0.30 - 1.52\,\text{g}$$

6. $$\%\text{Protein} = \frac{2\,\text{g}}{29\,\text{g}} \times 100 = 6.9\%$$

   Conversion factor for wheat flour is 5.7, so

$$\%N = \frac{6.9\%}{5.7} = 1.2\%$$

$$1.2\% = \frac{(7\text{mLs})(0.1N)(1.4007)}{\text{sample weight}}$$

$$\text{Sample weight} = 0.82\,\text{g}$$

7. Concentration of BSA after ten fold dilution = 2 mg/mL = 0.002 g/g = 0.2%

$$A = a \times b \times c = 6.3 \times 0.5 \times 0.2 = 0.63$$

8. (a) The complex has maximum absorbance at this wavelength.
   (b) Tube 1: 0 mg/tube
       Tube 2: 4 mg/ tube
       Tube 3: 8 mg/ tube
       Tube 4: 12 mg/ tube
       Tube 5: 16 mg/ tube
       Tube 6: 20 mg/ tube
   (c) Standard curve prepared as in Problem 2.
   (d) Tube 1 is used because its absorbance is within the range of the standard curve.
   (e) Based on the standard curve and absorbance of X solution:
       Protein concentration calculated is about 10 mg/ tube. 10 mg/0.2 mL = 50 mg/mL.

Acknowledgments The authors thank Yan Zhang for his contribution to the chapter in the previous edition of the book. This work is supported in part by the Mississippi Agricultural and Forestry Experiment Station Project MIS 081710.

REFERENCES

1. Jones DB (1931) Factors for converting percentages of nitrogen in foods and feeds into percentages of proteins. US Dept. Agric. Circular No. 183, August. USDA, Washington, DC
2. US Department of Agriculture, Agricultural Research Service (2018). USDA National Nutrient Database for Standard Reference, Legacy. FoodData Central Home Page. https://fdc.nal.usda.gov/index.html
3. AOAC International (2023) Official methods of analysis, 22nd edn. (On-line). AOAC International, Rockville, MD
4. Yada RY, Jackman RL, Smith JL, Marangoni AG (1996) Analysis: quantitation and physical characterization, (Chapter 7). In: Nakai S, Modler HW (eds) Food proteins. Properties andcharacterization. VCH, New York, pp. 333–403
5. Kolakowski E (2001) Protein determination and analysis in food system, (Chapter 4). In: Sikorski ZE (ed) Chemical and functional properties of food protein. Technomic Publishing, Lancaster, PA pp. 57–112
6. Owusu-Apenten RK (2002) Food protein analysis. Quantitative effects onprocessing. Marcel Dekker, New York
7. Walter JM (Ed.) (2009) The Protein Protocols Handbook. Humana Press, Totowa NJ.
8. Bradstreet RB (1965) The Kjeldahl method for organic nitrogen. Academic, New York
9. Mossé J (1990) Nitrogen to protein conversion factor for ten cereals and six legumes or oilseeds. A reappraisal of its definition and determination. Variation according to species and to seed protein content. J Agric Food Chem 38: 18–24
10. Wilson PR (1990) A new instrument concept for nitrogen/protein analysis. A challenge to the Kjeldahl method. Aspects Appl Biol 25: 443–446
11. Wiles PG, Gray I, Kissling RC (1998) Routine analysis of proteins by Kjeldahl and Dumas methods: review and interlaboratory study using dairy products. JAOAC Int 81: 620–632
12. O'Sullivan A, O'Connor B, Kelly A, McGrath MJ (1999) The use of chemical and infrared methods for analysis of milk and dairy products. Int J DairyTechnol 52: 139–148
13. Luinge HJ, Hop E, Lutz ETG, van Hemert JA, de Jong EAM (1993) Determination of the fat, protein and lactose content of milk using Fourier transform infrared spectrometry. Anal Chim Acta 284: 419–433
14. Krishnan PG, Park WJ, Kephart KD, Reeves DL, Yarrow GL (1994) Measurement of protein and oil content of oat cultivars using near-infrared reflectance. Cereal Foods World 39(2): 105–108
15. Noble JE, Bailey MJ (2009) Quantitation of protein. Methods Enzymol 463: 73–95
16. Fraenkel-Conrat H, Cooper M (1944) The use of dye for the determination of acid and basic groups in proteins. J Biol Chem 154: 239–246
17. Udy DC (1956) A rapid method for estimating total protein in milk. Nature 178: 314–315
18. Janahar JJ, Marciniak A, Balasubramaniam VM, Jimenez-Flores R, Ting E (2021) Effect of pressure, shear, temperature, and their interactions on selected milk quality attributes. J Dairy Sci 1-4:1531–1547
19. Udy DC (1954) Dye-binding capacities of wheat flour protein fractions. Cereal Chem 31: 389–395
20. AACC International (2010) Approved Methods of American Association of Cereal Chemists International. 11th edn. (On-line), American Association of Cereal Chemists, St. Paul, MN
21. Pomeranz Y (1965) Evaluation of factors affecting the determination of nitrogen in soya products by the biuret and orange-G dye-binding methods. J Food Sci 30: 307–311
22. Moser C, Herman K (2011) Method for the rapid determination of protein in meats using the CEM™ Spring protein analyzer: Collaborative study. JAOAC Int 94: 1555–1561
23. Hurrel RF, Lerman P, Carpenter KJ (1979) Reactive lysine in foodstuffs as measured by a rapid dye-binding procedure. J Food Sci 44: 1221–1227
24. Amamcharla JK, Metzger LE (2010) Evaluation of a rapid protein analyzer for determination of protein in milk and cream. J Dairy Sci 93: 3846–3857
25. Ou YQ, Chang SKC (2011) Comparison of a rapid dye-binding method with the Kjeldahl and NIR methods for determining protein content in soybean and soymilk. Annual Meeting of the Institute of Food Technologists. June 11–14, 2011. New Orleans, LA
26. Bradford M (1976) A rapid and sensitive method for the quantitation of microgram quantities of protein utilizing the principle of protein-dye binding. Anal Biochem 72: 248–254
27. Lewis MJ, Krumland SC, Muhleman DJ (1980) Dye-binding method for measurement of protein in wort and beer. J Am Soc Brew Chem 38: 37–41
28. Snyder J, Desborou S (1978) Rapid estimation of potato tuber total protein content with Coomassie Brilliant Blue G-250. Theor Appl Genet 52: 135–139
29. Bearden Jr JC (1978) Quantitation of submicrogram quantities of protein by an improved protein-dye binding assay. Biochim Biophys Acta 533: 525–529
30. Robinson HW, Hodgen CG (1940) The biuret reaction in the determination of serum protein. 1. A study of the conditions necessary for the production of the stable color which bears a quantitative relationship to the protein concentration. J Biol Chem 135: 707–725
31. Jennings AC (1961) Determination of the nitrogen content of cereal grain by colorimetric methods. Cereal Chem 38: 467–479
32. Pinckney AJ (1961) The biuret test as applied to the estimation of wheat protein. Cereal Chem 38: 501–506
33. AOAC (1965) Official methods of analysis, 10th edn. Association of Official Analytical Chemists, Washington, DC
34. Lowry OH, Rosebrough NJ, Farr AL, Randall RJ (1951) Protein measurement with the Folin phenol reagent. J Biol Chem 193: 265–275
35. Peterson GL (1979) Review of the Folin phenol protein quantitation method of Lowry, Rosebrough, Farr, and Randall. Anal Biochem 100: 201–220
36. Miller GL (1959) Protein determination for large numbers of samples. Anal Chem 31: 964
37. Hartree EF (1972) Determination of protein: a modification of the Lowry method that gives a linear photometric response. Anal Biochem 48: 422–427
38. Aderinola TA, Alashi AM, Nwachukwu TN, Enujiugha VN, Aluko RE (2020) In vitro digestibility, structural and functional properties of *Moringa oleifera* seed proteins. Food Hydrocoll 101: 105574
39. Smith PK, Krohn RI, Hermanson GT, Mallia AK, Gartner FH, Provensano MD, Fujimoto EK, Goeke NM, Olson BJ, Klenk DC (1985) Measurement of protein using bicinchoninic acid. Anal Biochem 150: 76–85
40. Zheng J, Gao Q, Ge G, Wu J, Tang C-he, Zhao M, Sun W (2021) Heteroprotein complex coacervate based on β-conglycinin and

lysozyme: dynamic protein exchange, thermodynamic mechanism, and lysozyme activity. J Agric Food Chem 69: 7948–7959
41. Gasteiger E, Hoogland C, Gattiker A, Duvaud S, Wilkins MR, Appel RD, Bairoch A (2005) Protein identification and analysis tools on the ExPASy server. In Walker JM (eds) "The Proteomics Protocols Handbook." Humana Press. p. 571–607
42. Nakai S, Wilson HK, Herreid EO (1964) Spectrophotometric determination of protein in milk. J Dairy Sci 47: 356–358
43. Gabor E (1979) Determination of the protein content of certain meat products by ultraviolet absorption spectrophotometry. Acta Alimentaria 8(2): 157–167
44. Scopes RK (1974) Measurement of protein by spectrophotometry at 205 nm. Anal Biochem 59: 277–282
45. Nwachukwu ID, Aluko RE (2019) A systematic evaluation of various methods for quantifying food protein hydrolysate peptides. Food Chem 270: 25–31
46. Regenstein JM, Regenstein CE (1984) Protein functionality for food scientists. In "Food Protein Chemistry." Chapter 27. Academic Press Inc. p. 274–334

Carbohydrate Analysis

19

Kerry C. Huber and James N. BeMiller

Contents

19.1 INTRODUCTION

Carbohydrates are important in foods as a major supplier of energy, a key contributor to textural properties, and a predominant source of dietary fiber which contributes to overall health. There is interest in the analysis of food products and ingredients for the various types of carbohydrates (not only the different structural types but also types differing in physiological effects, e.g., digestible vs. nondigestible,

K. C. Huber (✉)
Department of Animal and Food Science, Brigham Young University – Idaho, Rexburg, ID, USA
e-mail: huberk@byui.edu

J. N. BeMiller
Department of Food Science, Purdue University, West Lafayette, IN, USA
e-mail: bemiller@purdue.edu

B. P. Ismail, S. S. Nielsen (eds.), *Nielsen's Food Analysis*, Food Science Text Series,
https://doi.org/10.1007/978-3-031-50643-7_19

metabolizable vs. non-metabolizable, caloric vs. reduced caloric vs. noncaloric, prebiotic vs. non-prebiotic). However, definitions of the types are not always agreed upon, and analytical methods do not always measure exactly what is included in the definition. These challenges result in controversies about what should be measured and how. This chapter covers analysis of carbohydrates primarily by structural type.

Digestible carbohydrates are converted into monosaccharides, which are absorbed and provide metabolic energy and satiety. Nondigestible polysaccharides (all those other than starch) comprise the major portion of dietary fiber (Sect. 19.6). Carbohydrates also provide other attributes, including bulk, body, viscosity, stability to emulsions and foams, water-holding capacity, stability to freezing and thawing, browning (including generation of flavors and aromas), and a range of desirable textures (e.g., firmness, resilience, crispness). They also may lower water activity to thereby inhibit microbial growth. Basic carbohydrate structures, chemistry, and terminology can be found in references [1, 2].

Occurrences of major carbohydrates in foods are presented by structural classes in Table 19.1. Ingested carbohydrates are almost exclusively of plant origin, with milk

Table 19.1 Occurrences of some major carbohydrates in foods

Carbohydrate	*Source*	*Constituent(s)*
Monosaccharides[a]		
D-glucose (dextrose)	Naturally occurring in honey, fruits, and fruit juices. Added as a component of glucose syrups and high-fructose syrups. Produced during processing by hydrolysis (inversion) of sucrose	
D-fructose	Naturally occurring in honey, fruits, and fruit juices. Added as a component of high-fructose syrups. Produced during processing by hydrolysis (inversion) of sucrose	
Sugar alcohol[a]		
Sorbitol (D-glucitol)	Added to food products, primarily as a humectant and low-calorie sweetener	
Disaccharides[a]		
Sucrose	Widely distributed in fruit and vegetable tissues and juices in varying amounts. Added to food and beverage products	D-fructose D-glucose
Lactose	In milk and products derived from milk	D-galactose D-glucose
Maltose	In malt. In varying amounts in various glucose syrups and maltodextrins	D-glucose
Higher oligosaccharides[a]		
Maltooligosaccharides	Maltodextrins. In varying amounts in various glucose syrups	D-glucose
Raffinose	Small amounts in beans	D-glucose D-fructose D-galactose
Stachyose	Small amounts in beans	D-glucose D-fructose D-galactose
Polysaccharides		
Starches[b]	Widespread in cereal grains and tubers. Added to foods as functional ingredients	D-glucose
Food gums/hydrocolloids[c] Algins Carboxymethylcelluloses Carrageenans Gellan Guar gum Gum arabic Hydroxypropylmethyl-celluloses Inulin Konjac glucomannan Locust bean gum Methylcelluloses Pectins Xanthan	Obtained from plants, algae, microorganisms, etc. Added to foods as functional ingredients	[d]
Cell-wall polysaccharides[c] Pectin (native) Cellulose Hemicelluloses Beta-glucan	Naturally occurring in plant tissues	

[a] For analysis, see Sect. 19.4.2
[b] For analysis, see Sect. 19.5.1
[c] For analysis, see Sect. 19.5.2
[d] For compositions, characteristics, and applications, see references [1, 2]

lactose being the major exception. Of the **monosaccharides** (sometimes called **simple sugars**), only D-glucose and D-fructose are found in other than minor amounts. Monosaccharides are the only carbohydrates that can be absorbed from the small intestine. Higher saccharides (**oligo-** and **polysaccharides**) must first be digested (i.e., hydrolyzed to monosaccharides) before absorption and utilization can occur. [Note: The Food and Agriculture Organization (FAO) and the World Health Organization (WHO) [3] recommend that carbohydrates be classified by molecular size into sugars [degree of polymerization (DP) 1-2], oligosaccharides (DP 3-9), and polysaccharides (DP >9), but carbohydrate chemists (according to international nomenclature rules) consider an oligosaccharide to be a carbohydrate composed of 2–10 (or 2–20) sugar (saccharide) units.] Polysaccharides usually contain from about 30–60,000 or more monosaccharide units. Humans can digest only sucrose, lactose, maltodextrins (maltooligosaccharides), and starch. All are digested with enzymes present in the mouth and small intestine.

At least 90% of the carbohydrate in nature is in the form of polysaccharides. As stated above, starch polymers are the only polysaccharides that humans can digest and use as a source of calories and carbon. All other polysaccharides are nondigestible. **Nondigestible polysaccharides** can be divided into **soluble** and **insoluble** classes. Along with lignin and other nondigestible, non-absorbed substances, they make up **dietary fiber** (Sect. 19.6.1). As dietary fiber, they regulate normal bowel function, reduce the postprandial hyperglycemic response, and may lower serum cholesterol. However, nondigestible polysaccharides most often are added to processed foods because of the functional properties they impart. Nondigestible oligosaccharides serve as prebiotics and are, therefore, increasingly used as ingredients in functional foods and nutraceuticals. The foods in which dietary fiber components can be used, and particularly the amounts that can be incorporated, are limited because addition above a certain level usually adversely alters the characteristics of the food product. Indeed, as already stated, they are most often used as ingredients because of their ability to impart important functional properties at a low level of usage, but occasionally are added to achieve a physiological effect.

Carbohydrate analysis is important from several perspectives. Qualitative and quantitative analyses are used to determine the composition of foods, beverages, and their ingredients. **Qualitative analysis** ensures that ingredient labels present accurate compositional information. **Quantitative analysis** ensures that added components are listed in proper order on ingredient labels and that Nutrition Facts information (stated amounts of specific components and their caloric contents) is appropriately determined. Table 19.2 summarizes some of the methods described in this chapter that are commonly used for nutrition labeling, quality assurance, or research of food ingredients and/or products. Of increasing importance are analyses to determine authenticity and origin of foods, beverages, and ingredients. Both qualitative and quantitative analyses can be used to authenticate (i.e., to detect adulteration of) food ingredients and products and for quality assurance.

The most commonly used methods of carbohydrate determination are presented here. However, methods often must be made specific to a particular food product due to the nature of the food matrix and the presence of other constituents. Approved methods are referenced where appropriate, but method approval has not kept pace with method development. Thus, in some cases, secondary methods are presented. Methods that have been in longtime use, while not giving as precise or as much information as newer methods, may nevertheless be useful for product standardization in some cases.

In general, evolution and succession of analytical methods for low-molecular-weight carbohydrates has followed the sequence: qualitative color tests, adaptation of the color test for reducing sugars based on reduction of Cu(II) to Cu(I) (Fehling test) to quantitation of reducing sugars, qualitative paper chromatography, quantitative paper chromatography, gas chromatography (GC) of derivatized sugars, qualitative and quantitative thin-layer chromatography, enzymic methods, and high-performance liquid chromatography (HPLC). Some older methods are still in use, and multiple official methods for the analysis of mono- and disaccharides in foods are currently approved by AOAC International [4]. Methods employing mid-infrared/Fourier transform infrared (FTIR) spectroscopy (Sect. 19.7.4 and Chap. 8, Sect. 8.3), near-infrared (NIR) spectroscopy (Sect. 19.7.5 and Chap. 8, Sect. 8.4), capillary electrophoresis (Sect. 19.4.2.4 and Chap. 24, Sect. 24.2.6.3), and mass spectrometry (MS) (Sect. 19.7.3 and Chap. 11) have been published but are not yet in general use for carbohydrate analysis. Reference [5] also may be consulted for food carbohydrate analysis.

According to the nutrition labeling regulations of the US Food and Drug Administration (FDA) [6] [21 CFR 101.9 (c) (6) (i–iv)], the following are details regarding carbohydrates (all declared in relation to a serving, as defined by the FDA):

1. **Total carbohydrate** content of a food must be calculated by the subtraction of the sums of the crude protein, total fat, moisture, and ash in a serving from the total weight of the food (i.e., total carbohydrate is determined by difference). (Note that this calculation is not a direct measurement of carbohydrate content. Though this method is required by US regulations for nutrition labeling, its accuracy relies on the correct determinations of other food components. As described in Chap. 3, Sect. 3.2.1.4, the total carbohydrate value (with or without taking into account the insoluble dietary content of the food) may be utilized in the calculation of the total caloric content for the label.)
2. **Dietary fiber** (FDA definition given in Sect. 19.6.1, Table 19.4) content in a serving must also be stated on the label. Declaration of **soluble fiber** and **insoluble fiber** contents is voluntary.

Table 19.2 Summary of carbohydrate analysis methods

To determine:	*Description of method*	*Method measures*	*Advantages/disadvantages*
Total carbohydrate for nutrition label	Grams of carbohydrate per serving is calculated as total grams of serving minus (g of moisture + g of protein + g of lipid + g ash)	Total carbohydrate by difference	Not an actual measurement of carbohydrate. Depends on accuracy of determination of other components, but this method is required by US regulations. Should not be used to calculate caloric content because carbohydrate components of dietary fiber, such as cellulose, provide essentially no calories
Total carbohydrate[a]	Spectrophotometric, phenol-sulfuric acid	Measures all carbohydrates except sugar alcohols	Solution must be clear, i.e., carbohydrates must be soluble, so the method may not measure all carbohydrates. Method requires a standard curve made with the same exact mixture of carbohydrates in the same ratio that occurs in the sample
Total reducing sugars[a]	Spectrophotometric. Somogyi–Nelson and related methods	Primarily used to measure glucose/dextrose, maltose, and other low-molecular-weight oligosaccharides in glucose syrups	If carbohydrates are not already in solution, requires extraction. Solution must be clear. Fructose gives some response
Glucose/dextrose[a]	(1) Enzymic assay using GOPOD reagent (spectrophotometric)[b] (2) HPLC	Both methods specifically determine the amount of glucose in a mixture of sugars	Extraction required. Enzymic method can be automated
Fructose[a]	(1) Enzymic assay using hexokinase (spectrophotometric) (2) HPLC	Specific determination of fructose in a mixture of sugars	Extraction required
Sucrose[a]	(1) Enzymic assay using GOPOD reagent (spectrophotometric)[b] (2) HPLC	Both methods specifically determine the amount of sucrose in a mixture of sugars	Extraction required. Enzymic method can be automated. For enzymic method, solution must be clear
Lactose[a]	(1) Enzymic assay (spectrophotometric)[b] (2) HPLC. Enzymic assay employs galactose oxidase[b]	Both methods specifically determine the amount of lactose in a mixture of sugars, except that the enzymic method will also measure free galactose (uncommon) or other galactose-containing substances	Extraction required. Enzymic method can be automated
Concentration of syrups[a]	(1) Measurement of specific gravity using a hydrometer (2) Refractive index using a refractometer	Concentration of solids in the solution	Solutions must be pure and of a simple substance
Starch[a]	(1) Hydrolysis of starch to glucose using a mixture of amylases and determination of glucose using the GOPOD reagent (2) Hydrolysis of starch with glucoamylase and determination of glucose with glucose oxidase[b]	Specific for starch, including modified starches	May not measure all forms of resistant starch[c]. Sample must be free of glucose or a correction made for it. Amylases must be purified to remove any interfering activities
Pectin[a]	Spectrophotometric. m-Hydroxydiphenyl-sulfuric acid method	Uronic acids	Extraction may be required. Standard curve is required. Other hydrocolloids containing uronic acids will interfere
Dietary fiber	Gravimetric (residue after removal of lipids, digestible starch, and protein, and subtraction of residual ash content)	"Total dietary fiber"	This method does not include low-molecular-weight soluble dietary fiber. It is not a measure of the physiological efficacy of the particular dietary fiber. Soluble and insoluble dietary fiber can be determined by specific methods for them

[a] For research and quality assurance. Not required for nutrition labeling
[b] The YSI Life Sciences instrument method uses an electrode that detects hydrogen peroxide
[c] A Megazyme kit provides an adaptation for inclusion of resistant starch in the measurement

3. **Total sugars** are defined for labeling purposes as the sum of all free monosaccharides and disaccharides (e.g., glucose, fructose, lactose, and sucrose).
4. **Added sugars** are defined by the FDA as follows: "Added sugars are either added during the processing of foods, or are packaged as such, and include sugars (free, mono- and disaccharides), sugars from syrups and honey, and sugars from concentrated fruit or vegetable juices that are in excess of what would be expected from the same volume of 100% fruit or vegetable juice of the same type, except …." The statement continues by stating the exceptions of what is not included as "added sugar." Additional accommodations are made for containers as single-ingredient foods (honey, maple syrup, pure sugars, etc.) and cranberry products (Chap. 3, Sect. 3.2.1.5).
5. **Sugar alcohols** "…are defined as the sum of saccharide derivative in which a hydroxyl group replaces a ketone or aldehyde group and whose use in the food is listed by FDA (e.g., mannitol or xylitol) or is generally recognized as safe (e.g., sorbitol)." Their declaration on the nutrition label is generally voluntary but becomes required if "a claim is made on the label or in labeling about sugar alcohol or total sugars, or added sugars when sugar alcohols are present in the food..." If only one sugar alcohol is present in a food (e.g., xylitol), the specific name of the sugar alcohol may be used in place of "sugar alcohol" in the ingredient statement.

Reference values for total carbohydrate, sugar, and total dietary fiber (TDF) values for food products can be accessed through the USDA Nutrient Database for Standard Reference [7].

19.2 SAMPLE PREPARATION

19.2.1 General Information

Sample preparation is specific to the carbohydrate being determined (because carbohydrates have such a wide range of solubilities) and the specific raw material, ingredient, or food product being analyzed. However, some generalities can be presented (Fig. 19.1).

For most foods, the first step is drying, which also can be used to determine moisture content. For other than beverages, drying is done by placing a weighed amount of material in a vacuum oven and drying to constant weight (Chap. 15, Sect 15.2.2.3). Then, the material is ground to a fine powder, and lipids are extracted (Chap. 17, Sect. 17.2). Without prior extraction of lipids and other lipid-soluble substances, extraction of water-soluble carbohydrates will likely be incomplete.

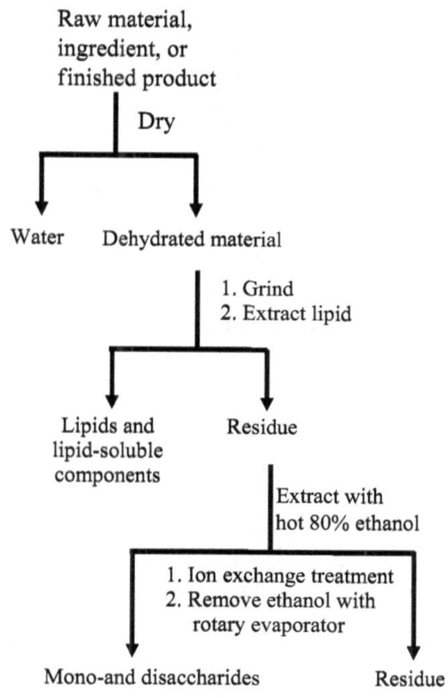

Fig. 19.1 Flow diagram for sample preparation and extraction of mono- and disaccharides

Specific sample preparation schemes often may be required. For example, the AOAC International [4] method for presweetened, ready-to-eat breakfast cereals calls for removal of fats by extraction with petroleum ether (hexane) and extraction of sugars with 50% ethanol (AOAC Method 982.14) [4], rather than the method described below.

19.2.2 Extraction and Cleanup for Determination of Mono- and Oligosaccharides

Food raw materials, products, and some ingredients are complex, heterogeneous, biological materials. Thus, it is quite likely that they may contain substances that interfere with measurement of the mono- and oligosaccharides present, especially if a spectrophotometric method is used. Interference may arise either from compounds that absorb light of the same wavelength used for the carbohydrate analysis or from insoluble, colloidal materials that scatter light (since light scattering will be measured as absorbance). Also, the aldehydic or keto group of the sugar can react with other components, especially amino groups of proteins, via the **nonenzymatic browning (Maillard) reaction**, which produces color and destroys the sugar. Even if chromatographic methods, such as HPLC (Sect. 19.4.2.1), are used for analysis, the mono- and oligosaccharides first must be isolated from the other components of the food before analysis

For the determination of any mono- (glucose, fructose), di- (sucrose, lactose, maltose), tri- (raffinose), tetra- (stachyose), or other oligosaccharides (e.g., **maltodextrins**) present, the dried, lipid-free sample (Sect. 19.2.1) is extracted with **hot 80% ethanol** in the presence of precipitated calcium carbonate to neutralize any acidity (AOAC Method 922.02, 925.05) (Fig. 19.1) [4]. Most of these low-molecular-weight carbohydrates are soluble in hot 80% ethanol, including some of the higher oligosaccharides from added maltodextrins or **fructooligosaccharides** (FOS), which may be likewise extracted. However, much of the composition of a food (other than water) is in the form of polymers, and almost all polysaccharides and proteins are insoluble in hot 80% ethanol. Thus, this extraction is rather specific. Extraction is done by a batch process: Refluxing 1 h, cooling, and filtering are standard practices. (A Soxhlet apparatus cannot be used because aqueous ethanol undergoes azeotropic distillation as 95% ethanol.) Hot 80% ethanol treatment should be done at least twice to ensure completeness of extraction. If a foodstuff or food product is particularly acidic (e.g., a low-pH fruit), neutralization may be necessary to prevent hydrolysis of sucrose, which is particularly acid-labile; thus, precipitated calcium carbonate is routinely added.

The 80% ethanol extract will contain components other than carbohydrates, in particular ash, pigments, organic acids, and perhaps free amino acids and low-molecular-weight peptides. Because the mono- and oligosaccharides are neutral and the contaminants are charged, the contaminants can be removed by **ion-exchange** techniques. Because reducing sugars can be adsorbed onto and be isomerized by strong anion-exchange resins in the hydroxide (OH^-) form, a weak anion-exchange resin in the carbonate (CO_3^{2-}) or hydrogen carbonate (HCO_3^-) form is often used. [**Reducing sugars** are those mono- and oligosaccharides that contain a free carbonyl (aldehydic or keto) group and, therefore, can act as reducing agents; see Sect. 19.4.1.] Because sucrose and sucrose-related oligosaccharides are very susceptible to acid-catalyzed hydrolysis, the anion-exchange resin should be used before the cation-exchange resin. However, because the anion-exchange resin is in a carbonate or hydrogen carbonate form, the cation-exchange resin (in the H^+ form) cannot be used in a column because of CO_2 generation. Mixed-bed columns are not recommended for the same reason. AOAC Method 931.02 [4] reads as follows for cleanup of ethanol extracts: Place a 50-mL aliquot of the ethanol extract in a 250-mL Erlenmeyer flask. Add 3 g of anion-exchange resin (OH^- form) and 2 g of cation-exchange resin (H^+ form). Let it stand for 2 h with occasional swirling.

Aqueous ethanol is not the preferred solvent for subsequent determination of the extracted carbohydrates and,

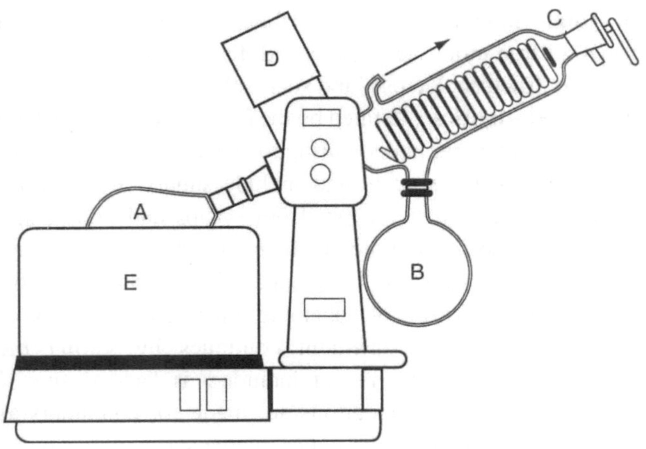

Fig. 19.2 Diagram of a rotary evaporator. The solution to be concentrated is placed in the round-bottom flask (***A***) in a water bath (***E***) at a controlled temperature. The system is evacuated by means of a water aspirator or vacuum pump; connecting tubing is attached at the arrow. Flask **A** turns (generally slowly). Because of the reduced pressure, evaporation is relatively rapid from a thin film on the inside walls of flask **A** produced by its rotation, the large surface area, and the elevated temperature. ***C*** is a condenser. ***D*** is the motor. Condensate collects in flask ***B***. The stopcock at the top of the condenser is for releasing the vacuum

therefore, is removed under reduced pressure using a **rotary evaporator** (Fig. 19.2) at a temperature of 45–50 °C. The carbohydrate residue is dissolved in an accurately measured amount of distilled water or buffer for analysis by HPLC (Sect. 19.4.2.1.2 or 19.4.2.1.4). Filtration should not be required but should be used if necessary. Some methods employ a final passage through a hydrophobic column such as a Sep-Pak C18 cartridge (Waters Associates, Milford, MA) as a final cleanup step to remove any residual lipids, proteins, and/or pigments. However, this should not be necessary if the lipids and lipid-soluble components were properly removed prior to extraction.

19.3 TOTAL CARBOHYDRATE: PHENOL-SULFURIC ACID METHOD

19.3.1 Principle and Characteristics

Carbohydrates are destroyed by strong acids and/or high temperatures. Under these conditions, a series of complex reactions takes place, beginning with a simple dehydration reaction as shown in Eq. 19.1:

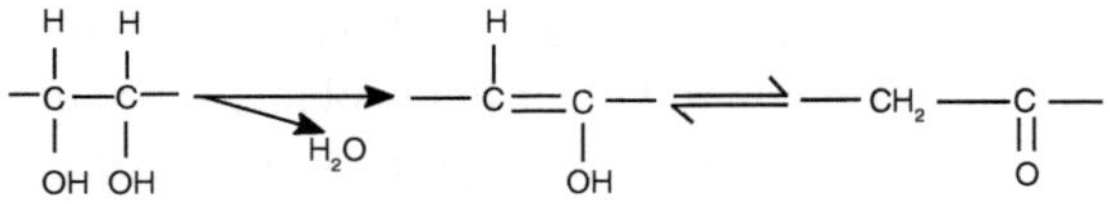

(19.1)

Continued heating in the presence of acid produces various furan derivatives that react with phenolic compounds, such as phenol, resorcinol, orcinol, α-naphthol, and naphthoresorcinol, and with various aromatic amines, such as aniline and *o*-toluidine, to produce colored compounds [1]. The most often used condensation is with phenol itself [8] with which a stable, yellow-gold color is produced. This widely used method is simple, rapid, sensitive, accurate, and specific for carbohydrates. The reagents are inexpensive, readily available, and stable. Virtually, all mono-, oligo-, and polysaccharides can be determined using the phenol-sulfuric acid method. (Oligo- and polysaccharides react because they undergo hydrolysis in the hot, strong acid, releasing monosaccharides.) However, neither sorbitol nor any other **sugar alcohol (alditol, polyol, and polyhydroxyalcohol)** gives a positive test. Under proper conditions, method results are reproducible and accurate to ±2%.

The reaction is not stoichiometric, and the extent of color formation is, in part, a function of the structure of the sugar. Therefore, a standard (calibration) curve (Chap. 4, Sect. 4.4 and Chap. 7, Sect. 7.2.4) must be used. Ideally, the standard curve will be prepared using mixtures of the same sugars present in the same ratio as they occur in the sample being analyzed. If this is not possible (e.g., if a pure preparation of the sugar being measured is not available or if more than one sugar is present either as free sugars in unknown proportions or as constituent units of oligo- or polysaccharides or mixtures of them), D-glucose is used to prepare the standard curve. In these cases, accuracy is a function of conformity of the standard curve made with D-glucose to the curve that would be produced from the exact mixture of carbohydrates being determined.

19.3.2 Outline of Procedure

1. A clear, aqueous solution of carbohydrate(s) is transferred using a pipette into a small tube.
2. An aqueous solution of phenol is added, and the contents are mixed.
3. Concentrated sulfuric acid is added rapidly to the tube so that the stream produces good mixing. The tube is agitated. (Adding the sulfuric acid to the water produces considerable heat.) A yellow-gold color results.
4. Absorbance is measured at 490 nm.
5. The average absorbance of blanks (sample alone and reagents alone) is subtracted, and the amount of sugar is determined by reference to a standard curve.

19.4 MONO- AND OLIGOSACCHARIDES

19.4.1 Total Reducing Sugars

19.4.1.1 Somogyi–Nelson Method

19.4.1.1.1 Principle

Oxidation is a loss of electrons; reduction is a gain of electrons. Reducing sugars are those sugars that have an aldehydic group (aldoses) that acts as a reducing agent by giving up electrons to an oxidizing agent, which is reduced by receiving the electrons. Oxidation of the aldehydic group produces a carboxylic acid group (Eq. 19.2):

$$R-\overset{O}{\overset{\|}{C}}-H + 2Cu(OH)_2 + NaOH \longrightarrow R-\overset{O}{\overset{\|}{C}}-O^-Na^+ + Cu_2O + 3H_2O \quad (19.2)$$

The most often used method to determine amounts of reducing sugars is the Somogyi–Nelson method [9], also at times referred to as the Nelson–Somogyi method. This and other reducing sugar methods (Sect. 19.4.1.2) can be used in combination with enzymic methods (Sect.19.4.2.3) for the determination of oligo- and polysaccharides. In enzymic methods, specific hydrolases are used to convert the oligo- or polysaccharide into its constituent monosaccharide or repeating oligosaccharide units, whose total amounts are measured using a reducing sugar method. The Somogyi–Nelson method is based on **reduction of Cu(II) ions to Cu(I) ions by reducing sugars**. The reaction is conducted in an alkaline solution containing tartrate or citrate ions, which function to keep the copper ions in solution. The Cu(I) ions then reduce an arsenomolybdate complex prepared by reacting ammonium molybdate [$(NH_4)_6Mo_7O_{24}$] and sodium arsenate (Na_2HAsO_7) in sulfuric acid. Reduction of the arsenomolybdate complex produces an intense, stable blue color that is measured spectrophotometrically. The extent of color formation is, in part, a function of the sugars present, so the method must be used with a standard curve (Chap. 4, Sect. 4.4 and Chap. 7, Sect. 7.2.4) of the sugars in the same ratio as they occur in the sample being analyzed or D-glucose (if a constituent sugar is unavailable or the constituents are unknown).

19.4.1.1.2 Outline of Procedure

1. A solution of copper(II) sulfate and an alkaline buffer solution are added by pipette to a solution of reducing sugars(s) (prepared as per the sample preparation procedure described in Sect. 19.2.2) and a water blank.
2. The resulting solution is heated in a boiling water bath.
3. A reagent prepared by mixing solutions of acidic ammonium molybdate and sodium arsenate is added.

4. After mixing, dilution, and remixing, absorbance is measured at 520 nm.
5. After subtracting the absorbance of the reagent blank, the A_{520} is converted into glucose equivalents using a standard plot of μg of glucose vs. absorbance.

19.4.1.2 Other Methods

An alternative method to the Somogyi–Nelson method that is also based on **reduction of Cu(II) ions in alkaline solution to Cu(I) ions** is the **Lane–Eynon method** (AOAC Method 945.66) [4]. To perform the Lane–Eynon method, the solution to be analyzed is added (using a burette) to a flask containing a boiling, alkaline solution of cupric sulfate of known concentration containing potassium sodium tartrate and methylene blue. Any reducing sugars in the solution being analyzed reduce Cu(II) ions to Cu(I) ions. When all the Cu(II) ions have been reduced, further addition of reducing sugars results in the indicator losing its blue color. The volume of the solution required to reach the endpoint is used to calculate the amount of reducing sugar present in the sample. Again, because this reaction is not stoichiometric and because each reducing sugar reacts differently, this method must be used with a standard curve (Chap. 4, Sect. 4.4 and Chap. 7, Sect. 7.2.4).

A keto group cannot be oxidized to a carboxylic acid group, and thus ketoses are not reducing sugars. However, under the alkaline conditions employed, ketoses are isomerized to aldoses [1] and, therefore, are measured as reducing sugars. Because the conversion is not 100%, the response is less with ketoses, so a standard curve (Chap. 4, Sect. 4.4 and Chap. 7, Sect. 7.2.4) made with D-fructose as one of the sugars in the mixture of sugars should be used if it is present.

The **dinitrosalicylic acid method** [10] will measure reducing sugars naturally occurring in foods or released by enzymes but is less often used. In this reaction, 3,5-dinitrosalicylate is reduced to the reddish monoamine derivative.

19.4.2 Specific Analysis of Mono- and Oligosaccharides

Determination of contents of specific mono- and oligosaccharides is often done chromatographically. The most commonly used method is high-performance liquid chromatography (HPLC) (Sect. 19.4.2.1). The method is simple and determines sugar alcohols in addition to reducing sugars. Gas chromatography (GC) (Sect. 19.4.2.2) is more time-consuming in that it requires derivatization of the sugars. In GC, sugars are determined as their reduced forms (sugar alcohols).

19.4.2.1 High-Performance Liquid Chromatography

19.4.2.1.1 Overview

HPLC (Chap. 13) is the method of choice for analysis of mono- and oligosaccharides in foods and can be used for the analysis of polysaccharides after hydrolysis (Sect. 19.5.2.2) to their constituent monosaccharides. HPLC gives both qualitative analysis (identification of the carbohydrate) and, with peak integration, quantitative analysis. HPLC analyses are rapid, can tolerate a wide range of sample concentrations, and provide a high degree of precision and accuracy. HPLC requires micron-filter filtration prior to injection. Complex mixtures of mono- and oligosaccharides can be separated and analyzed. The basic principles and important parameters of HPLC (the stationary phase, the mobile phase, and the detector) are presented and discussed in Chap. 13. Some details related to carbohydrate analysis are discussed here. The use of HPLC to determine food and other carbohydrates has been reviewed many times; some reviews can be found in references [11–20]. Specific details of methods of analysis for specific food ingredients or products should be obtained from the literature. Sample preparation for HPLC analysis is discussed in references [17, 19]. Various column packing materials and detectors have been used (Chap 13, Sects. 13.2.3 and 13.2.4, respectively). Only the most often employed column packing materials and detectors are presented here. The aforementioned references should be consulted for other columns and detectors.

19.4.2.1.2 Anion-Exchange HPLC

Separation of carbohydrates by HPLC is most often done using anion-exchange (AE-HPLC) columns (Chap. 13, Sect. 13.3.5.2). Carbohydrates have pKa values in the pH range 12–14 and are, therefore, very weak acids. In a solution of high pH, some carbohydrate hydroxyl groups are ionized, allowing sugars to be separated on columns of anion-exchange resins. Special resins have been developed for this purpose. The general elution sequence is sugar alcohols (alditols), monosaccharides, d*isa*ccharides, and higher oligosaccharides

19.4.2.1.3 Pulsed Electrochemical Detection

The **pulsed electrochemical detector** (ECD) [formerly called a **pulsed amperometric detector** (PAD)] (Chap. 13, Sect. 13.2.4.4), which relies on oxidation of carbohydrate hydroxyl and aldehydic groups, is universally used with anion-exchange chromatography [11–19, 21–25]. ECD requires a high pH. Both gradient and graded elutions can be used. The solvents employed (sodium hydroxide solutions, with or without sodium acetate) are simple and inexpensive. (Water may be used, but when it is, post-column addition of a sodium hydroxide solution is required.) The detector is suitable for both reducing and nonreducing monosaccharides. Lower detection limits are approximately 1.5 ng for monosaccharides and 5 ng for di-, tri-, and tetrasaccharides. ECD responses vary from sugar to sugar and change continuously, so standards must be run and response factors calculated at least daily.

AE-HPLC coupled to an ECD has been used to examine the complex oligosaccharide patterns of many food components and products. The method has the ability to achieve baseline separation within each class of carbohydrates (Fig. 19.3) and of providing separation of homologous series of oligosaccharides into their components [21, 26]. A newer detector with increasing use is the evaporative light scattering detector [27] (Chap. 13, Section 13.2.4.5).

19.4.2.1.4 Other HPLC Methods

There are other HPLC methods for carbohydrate analysis. Among them is the widely used normal-phase chromatography. In normal-phase chromatography (Chap. 13, Sect. 13.3.1), the stationary phase is polar, and elution is effected by employing a mobile phase of increasing polarity. Silica gel that has been derivatized with one or more of several reagents to incorporate amino groups is used. These so-called amine-bonded stationary phases are generally used with acetonitrile-water as the eluent. The elution order is monosaccharides and sugar alcohols, disaccharides, and higher oligosaccharides [28]. Amine-bonded silica gel columns have been used successfully to analyze the low-molecular-weight carbohydrate contents of foods [29].

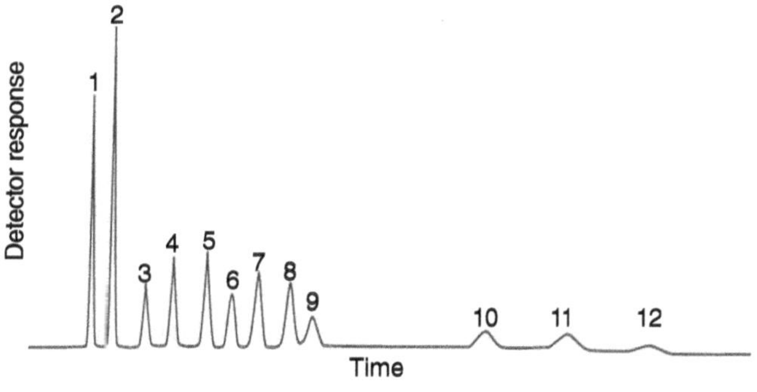

Fig. 19.3 High-performance liquid chromatogram of some common monosaccharides, disaccharides, alditols, and the trisaccharide raffinose at equal w/v concentrations separated by anion-exchange chromatography and detected by pulsed electrochemical detection. Peak *1*, glycerol; *2*, erythritol; *3*, L-rhamnose; *4*, D-glucitol (sorbitol); *5*, mannitol; *6*, L-arabinose; *7*, D-glucose; *8*, D-galactose; *9*, lactose; *10*, sucrose; *11*, raffinose; *12*, maltose

19.4.2.2 Gas Chromatography

19.4.2.2.1 Overview

GC (i.e., **gas-liquid chromatography,** GLC) (Chap. 14), like HPLC, provides both qualitative and quantitative analysis of carbohydrates [17, 30]. For GC, sugars must be converted into volatile derivatives. The most commonly used derivatives are the alditol peracetates [31, 32]. These derivatives are prepared as illustrated in Fig. 19.4 for D-galactose. A **flame ionization detector** (FID, Chap. 14, Sect. 14.3.5.2) is most often used to analyze peracetylated carbohydrate derivatives, but mass spectrometers are increasingly used as detectors. A mass spectrometric (MS) detector (Chap. 11, Sect. 11.4) lowers detection limits, and an MS/MS detector (Chap. 11, Sect. 11.6) lowers them even more [33]. A technique known as gas chromatography–combustion–isotope ratio mass spectrometry has been applied to determine the origin and adulteration of food and food ingredients [34].

The most serious problem with GC for carbohydrate analysis is that two preparation steps are involved: reduction of aldehydic groups to primary hydroxyl groups and conversion of the reduced sugars (alditol) into volatile peracetate esters; and of course, for the analysis to be successful, each of these steps must be 100% complete. The basic principles and important parameters of GC (the stationary phase, temperature programming, and detection) are presented and discussed in Chap. 14.

19.4.2.2.2 Neutral Sugars: Outline of Procedure [31]

1. **Reduction to Alditols.** Neutral sugars from the 80% ethanol extract (Sect. 19.2.2) or from hydrolysis of a polysaccharide (Sect. 19.5.2.2) are reduced at 40 °C with an excess of sodium or potassium borohydride dissolved in dilute ammonium hydroxide solution. After reaction, glacial acetic acid is added to destroy excess borohydride. The acidified solution is evaporated to dryness. A potential problem exists: If fructose is present, either as a naturally occurring sugar, from the hydrolysis of inulin, or as an additive [from a high-fructose syrup (HFS), invert

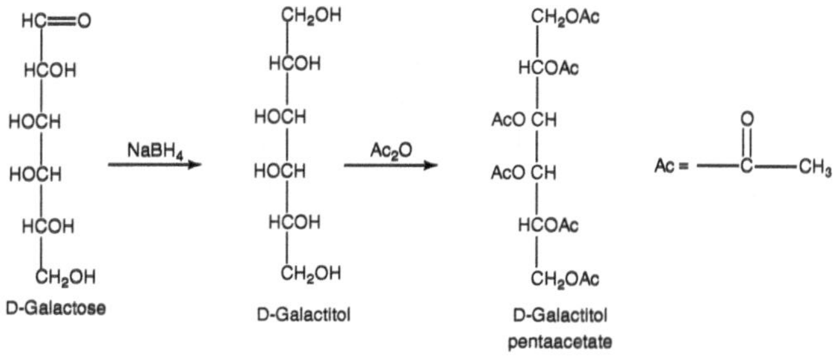

Fig. 19.4 Modification of D-galactose in preparation for gas chromatography

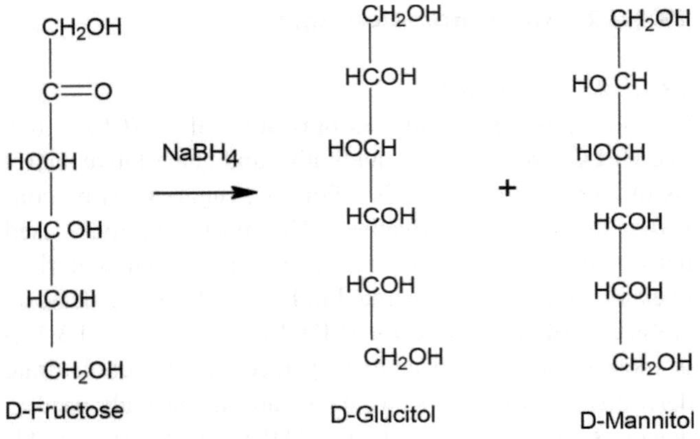

Fig. 19.5 Reduction of D-fructose to a mixture of alditols

sugar, or honey], it will be reduced to a mixture of D-glucitol (sorbitol) and D-mannitol (Fig. 19.5)

2. **Acetylation of Alditols.** Acetic anhydride and a catalyst are added to a dry mixture of alditols. After 10 min at room temperature, water and dichloromethane are added. After mixing, the dichloromethane layer is washed with water and evaporated to dryness. The residue of alditol peracetates is dissolved in a polar organic solvent (usually acetone) for chromatography.
3. **GC of Alditol Peracetates.** Alditol peracetates may be chromatographed isothermally and identified by their retention times relative to that of inositol hexaacetate, inositol being added as an internal standard prior to acetylation. It is essential to run standards of the alditol peracetates of the sugars being determined with inositol hexaacetate as an internal standard to determine elution times and relative responses.

19.4.2.3 Enzymic Methods

19.4.2.3.1 Overview

Enzymic methods (Chap. 26) generally have great specificity for the carbohydrate being determined, do not require high purity of the sample being analyzed, have very low detection limits, do not require expensive equipment, and are easily automated [35, 36]. However, the methods are spectrophotometric and thereby require clear solutions, so extraction and cleanup may be required (Sect. 19.4.2.3.2)

Enzymic methods for the determination of carbohydrates have been developed into commercially available kits (Table 19.3). They are often, but not always, specific for the substance being measured. The kits contain specific enzymes, required reagents, buffer salts, and detailed instructions that must be followed because enzyme concentration, substrate concentration, concentration of other required reagents, pH, and temperature all affect reaction rates and results. A good method description will point out any interferences from other substances and acknowledge experimental limitations

Table 19.3 Selected enzymic methods of carbohydrate analysis

Carbohydrate	*Reference*	*Kit form*[a]
Monosaccharides		
Pentoses		
L-Arabinose	[35, 36]	x
D-Xylose	[35, 36]	x
Hexoses		
D-Fructose	[35, 36]	x
D-Galactose	[35, 36]	x
D-galacturonic acid	[35]	x
D-glucose.		
Using glucose oxidase	[36], Sect. 19.4.2.3.3	x
Using glucose dehydrogenase	[35, 36]	
Using glucokinase (hexokinase)	[35, 36]	x
D-mannose	[35, 36]	x
Monosaccharide derivatives		
D-gluconate/ D-glucono-δ-lactone	[35, 36]	x
D-glucitol/sorbitol	[35, 36]	x
D-mannitol	[35, 36]	x
Xylitol	[35, 36]	x
Oligosaccharides		
Lactose	[35, 36]	x
Maltose	[35, 36]	x
Sucrose	[35, 36]	x
Raffinose, stachyose, verbascose	[35, 36]	x
Polysaccharides		
Amylose, amylopectin (contents and ratio)		x
Cellulose	[35, 36]	
Galactomannans (guar and locust bean gums)	[35]	x
β-Glucan (mixed-linkage)	[35]	x
Glycogen	[35, 36]	x
Hemicellulose	[35, 36]	
Inulin	[35, 36]	x
Pectin/poly(D-galacturonic acid)	[35, 36]	x
Starch	[35, 36], Sect. 19.5.1.1[a]	x

[a]Available in kit form from companies such as R-Biopharm, Megazyme, and/or Sigma-Aldrich

19.4.2.3.2 Sample Preparation

It is sometimes recommended that the **Carrez treatment** [37], which breaks emulsions, precipitates proteins, and absorbs some colors, be applied to food products prior to determination of carbohydrates by enzymic and other methods. The Carrez treatment involves addition of a solution of potassium ferrocyanide ($K_4[Fe(CN)_6]$, potassium hexacyanoferrate), followed by the addition of a solution of zinc sulfate ($ZnSO_4$), followed by the addition of a solution of sodium hydroxide. The suspension is filtered, and the clear filtrate is used directly in enzyme-catalyzed assays. Carrez solutions are commercially available

19.4.2.3.3 Enzymic Determination of D-Glucose (Dextrose)

Glucose oxidase oxidizes D-glucose quantitatively to D-glucono-1,5-lactone (glucono-delta-lactone), the other product being hydrogen peroxide (Fig. 19.6). To measure the amount of D-glucose present, **peroxidase** and a leuco dye (a colorless compound) are added. In the reaction catalyzed by peroxidase, the leuco dye is oxidized to a colored compound, which is measured spectrophotometrically. Different dyes are used in various commercial kits. The method using this combination of two enzymes and an oxidizable colorless compound is known as the **glucose oxidase/peroxidase (GOPOD) method** and is an example of a coupled-enzyme reaction (Chap. 26, Sect. 26.2.3.2)

Enzymic methods that measure a specific monosaccharide may be adapted to quantify polysaccharides comprised of the same constituent sugar. For example, starch can be solubilized and enzymically hydrolyzed to glucose, after which the GOPOD method may be used to quantify glucose to provide a measure of total starch (Sect. 19.5.1.1)

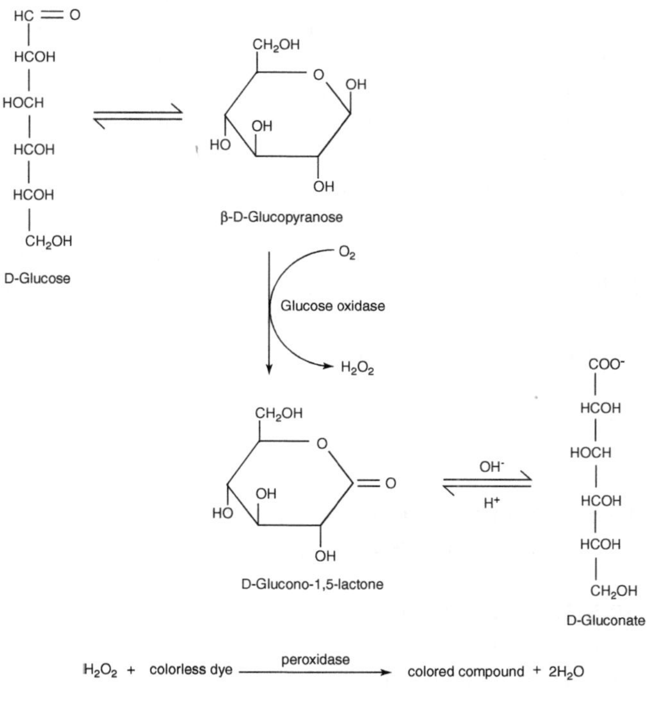

Fig. 19.6 Coupled-enzyme-catalyzed reactions for the determination of D-glucose

YSI Life Sciences makes a commercial instrument that utilizes the glucose oxidase enzyme immobilized between two membranes and an electrode that measures the released hydrogen peroxide. Results are obtained in less than 60 s. Using other immobilized enzymes, the instrument will determine amounts of D-galactose (using galactose oxidase), sucrose (using invertase and glucose oxidase), lactose (using galactose oxidase), and starch (using glucoamylase/amyloglucosidase and glucose oxidase).

Another **coupled-enzyme** enzymic method, also available in kit form, but less often used, involves reaction of D-glucose with ATP in the presence of hexokinase to form glucose 6-phosphate (G6P) + ADP. The reaction mixture also contains glucose 6-phosphate dehydrogenase (G6PDH) and $NADP^+$. G6PDH catalyzes the oxidation of G6P to D-gluconate 6-phosphate and reduction of $NADP^+$ to NADPH, so the amount of NADPH formed is equivalent to the amount of D-glucose present. The amount of NADPH formed is determined by measuring its absorbance at 340 nm (a wavelength which $NADP^+$ does not absorb).

19.4.2.4 Capillary Electrophoresis

Capillary zone electrophoresis (see also Chap. 24, Sect. 24.2.6.3) has also been used to separate and measure carbohydrates, but because carbohydrates lack chromophores, pre-column derivatization, and detection with a UV or fluorescence detector are required [11, 16, 38–43].

19.5 POLYSACCHARIDES

19.5.1 Starch

Starch is second only to water as the most abundant component of food. Starch is found in all parts of plants (leaves, stems, roots, tubers, and seeds). A variety of commercial starches are available worldwide as food additives. These include corn (maize), waxy maize, high-amylose corn (amylomaize), wheat, rice, potato, tapioca (cassava), yellow pea, sago, and arrowroot starches. In addition, starch is the main component of wheat, rye, barley, oat, rice, corn, mung bean, and pea flours and certain roots and tubers such as potatoes, sweet potatoes, and yams.

19.5.1.1 Total Starch

19.5.1.1.1 Principle and Procedure

The only reliable method for the determination of total starch is based on complete conversion of starch into D-glucose by purified starch hydrolyzing enzymes, followed by subsequent determination of the total released D-glucose using an

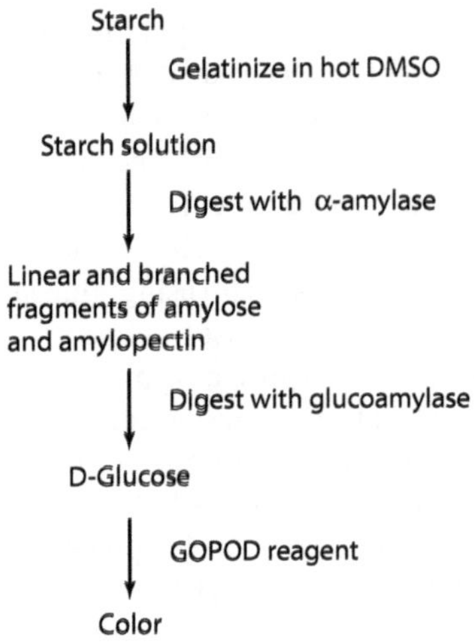

Fig. 19.7 Flow diagram for determination of total starch

enzyme-coupled reaction (Sect. 19.4.2.3.3). In the procedure outlined in Fig. 19.7, ***α*-amylase** catalyzes hydro*lys*is of unbranched segments of 1,4-linked *α*-D-glucopyranosyl units, forming primarily maltooligosaccharides composed of 3–6 units. **Debranching enzymes** (both **pullulanase** and **isoamylase**) catalyze hydro*lys*is of the 1,6 linkages constituting the branch points of starch po*lys*accharide chains and maltooligosaccharides (derived from the hydro*lys*is of starch po*lys*accharide molecules) to thereby produce short linear molecules. **Glucoamylase (amyloglucosidase)** acts at the nonreducing ends of starch oligo- and polysaccharide chains and releases D-glucose, one unit at a time; it will also catalyze hydro*lys*is of both 1,4 and 1,6 α-D-glucopyranosyl linkages. In the assay, glucose and a starch low in protein and lipid content (such as potato starch) are used as standards after determining their moisture contents

19.5.1.1.2 Potential Limitations

Starch-hydrolyzing enzymes (amylases) must be purified to eliminate any other enzymic activity (e.g., cellulases, invertase, sucrase, and β-glucanase) that would release D-glucose from non-starch substrates or destroy the hydrogen peroxide (i.e., catalase) generated for the enzymic determination of D-glucose. The former contamination would give false high values and the latter, false low values.

Even with purified enzymes, challenges still can be encountered. The method may not be quantitative for high amylose or other forms of starch that remain at least partially resistant to enzyme-catalyzed hydrolysis, sometimes even after cooking. Such starch material is characterized as **resistant starch** (RS), which, by definition, is composed of starch and starch degradation products that escape digestion in the small intestine [44]. Five types of starch are generally considered to be resistant to digestion in the small intestine or so slowly digested that they pass essentially intact through the small intestine, i.e., without conversion into D-glucose:

1. Starch that is physically inaccessible to amylases because it is trapped within a food matrix, even though it is gelatinized (**RS1**)
2. Starch that resists enzyme-catalyzed hydrolysis because it is uncooked, i.e., not gelatinized (**RS2**)
3. Retrograded starch [i.e., starch polymers that have recrystallized after gelatinization of the granule (**RS3**); cooled, cooked potatoes and other starchy foods, such as pasta, contain resistant starch]
4. Starch that has been chemically modified structurally in such a way as to make it less susceptible to digestion (**RS4**)
5. Starch involved in V-type complexes with lipids (e.g., fatty acids, fatty alcohols), making it resistant to amylase digestion (**RS5**)

RS is at best only partially converted into D-glucose by this method; rather, most of it constitutes dietary fiber (Sect. 19.6). Methods for the specific determination of RS have been reviewed [45].

One approach for the analysis of total starch (AOAC Method 996.11, AACCI Method 76-13.01) [4, 67] overcomes many of the aforementioned challenges associated with RS. Starch is dispersed in dimethyl sulfoxide (DMSO) and then is converted quantitatively to D-glucose by first treating the solution with a **thermostable α-amylase** to effect depolymerization and solubilization of the starch (Fig. 19.7). Addition of **glucoamylase (amyloglucosidase)** then effects complete conversion of the starch chain fragments (produced by the action of α-amylase) into D-glucose. Total D-glucose is measured using the GOPOD reagent (Sect. 19.4.2.3.3) to determine total starch. However, this method is not generally appropriate for the analysis of chemically modified starch derivatives (RS4), which may not be completely hydrolyzed to glucose by amylolytic treatment. Neither does it reveal the botanical source of the starch nor whether it is a native starch or a modified food starch. The botanical source of the starch may be determined by microscopic examination (Chap. 32, Sect. 32.2.2.4) of the material before it is cooked.

19.5.2 Non-starch Polysaccharides (Hydrocolloids/Food Gums)

19.5.2.1 Overview

Non-starch polysaccharides are almost always added as ingredients, although there are exceptions. These added polysaccharides, along with the protein gelatin, comprise a group of ingredients known as **hydrocolloids** or **food gums**. The non-starch polysaccharides used as additives in food products are obtained from land plants, seaweeds (marine algae), and microorganisms, as well as by chemical derivatization of cellulose. Their use is widespread and extensive.

Analytical methods are required for these polysaccharides to enable both suppliers and food processors to determine the purity of a hydrocolloid product, to ensure that label declarations of processors are correct, and to confirm that hydrocolloids have not been added to standardized products in which they are not allowed. For example, it may be desirable to determine the **β-glucan** content of an oat or barley flour – or to assess a breakfast cereal for a label claim of a specific dietary fiber or the **arabinoxylan** content of a wheat flour to set processing parameters for bakery products.

Determination of polysaccharides classified as hydrocolloids is problematic because of the inherent diversity of chemical structures, solubilities, and molecular weights among gum types. Further, the molecules comprising even a single polysaccharide preparation from a plant or microorganism, with very few exceptions, not only vary from molecule to molecule in chemical structure but also possess a range of molecular weights. Some polysaccharides are neutral; some are anionic. Some contain ether, ester, and/or cyclic acetal groups in addition to sugar units, either naturally or as a result of chemical modification. Some are soluble only in hot water; some are soluble only in room temperature or colder water; some are soluble in both hot and cold water; and some require aqueous solutions of acids, bases, or metal ion-chelating compounds to release them from plant tissues. This structural diversity complicates determination of both the types and amounts of polysaccharides in a food product [46]. As a result, there is no single approach that will either qualitatively or quantitatively determine all individual hydrocolloids present within a food. Other potential problems are that hydrocolloids are usually added to foods in very small amounts (0.01–1%), and blends of hydrocolloids are often used to extend functionalities.

Current methods depend on extraction of the hydrocolloid(s), followed by deproteinization of the extract and precipitation of the hydrocolloids by the addition of ethanol, acetone, or 2-propanol (isopropanol). However, low-molecular-weight (low-viscosity-grade) hydrocolloids may not be precipitated under these conditions. Because blends of hydrocolloids are often used in food products, fractionation may be required to separate a hydrocolloid blend into its individual components. Fractionation, like extraction and precipitation, invariably results in some loss of material. Most often, an isolated polysaccharide is identified by identifying and quantitating its constituent sugars after acid-catalyzed hydrolysis. However, sugars are hydrolyzed from polysaccharides at different rates and are destroyed by hot acids at different rates, so even the exact monosaccharide composition of a polysaccharide preparation may be difficult to determine and may not be achieved. A hydrolytic enzyme specific for the polysaccharide being determined (if available) is useful if the specific hydrocolloid present is known. Analytical strategies for and problems associated with the determination of hydrocolloids in foods have been reviewed [46, 47].

19.5.2.2 Hydrocolloid Content Determination

Most schemes for the analysis of food gums are targeted to a specific group of food products, as it is difficult, perhaps impossible, to develop a universal scheme. A general scheme for isolation and purification of non-starch, water-soluble polysaccharides is presented in Fig. 19.8 [48]. Letters within parentheses below correspond to the same letters in Fig. 19.8.

(a) The sample first must be dried. Freeze-drying is recommended. If the dried material contains lumps, it must be ground to a fine powder. It is usually difficult to extract polysaccharides quantitatively when fats, oils, waxes, and proteins are present. Therefore, any lipid-soluble substances are then removed. A known weight of dry sample is placed in a Soxhlet apparatus, and the lipid-soluble substances are removed with 19:1 v/v chloroform-methanol (*n*-hexane has also been used.). Solvent is removed from the sample by air-drying in a hood and then by placing the sample in a desiccator, which is then evacuated.
(b) Although not in the published scheme, soluble sugars, other low-molecular-weight compounds, and ash can be removed at this point using hot 80% ethanol as described in Sect. 19.2.2.
(c) Protein is removed by enzyme-catalyzed hydrolysis. The cited procedure [48] uses papain as the protease. However, bacterial alkaline proteases are recommended to prevent the action of contaminating carbohydrases – all of which have acidic pH optima. (Essentially all commercial enzyme preparations, especially those from bacteria or fungi, have carbohydrase activities in addition to proteolytic activity.) In this procedure, proteins are denatured for easier digestion by dispersing the sample in sodium acetate buffer containing sodium chloride and heating the mixture.
(d) Any solubilized polysaccharides are precipitated by the addition of sodium chloride to the cooled dispersion, followed by addition of four volumes of absolute ethanol. The mixture is centrifuged.
(e) The pellet is suspended in acetate buffer. To this suspension is added a freshly prepared solution of glucoamy-

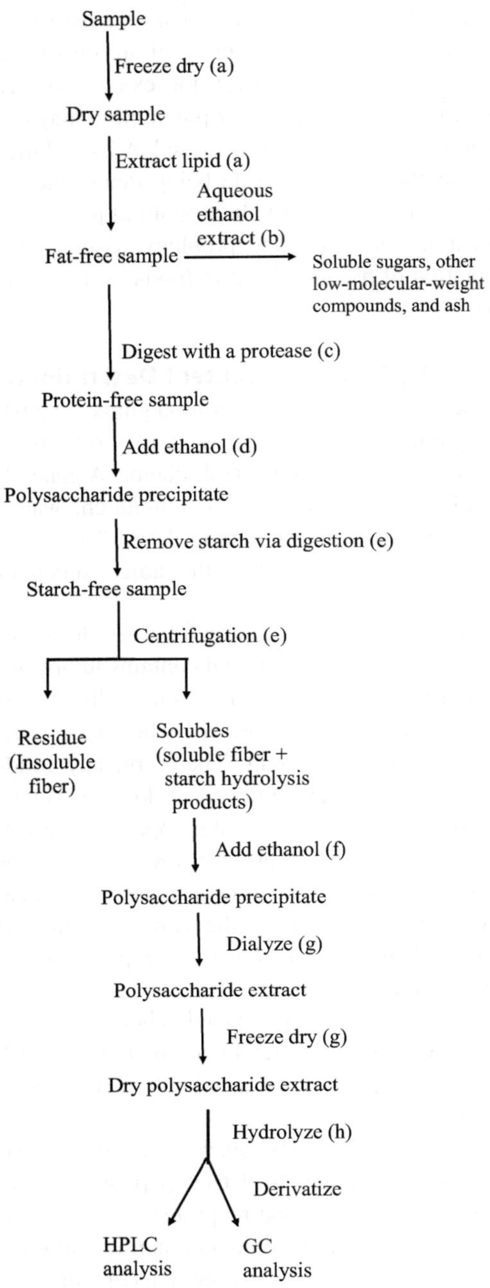

Fig. 19.8 Flow diagram for isolation and analysis of polysaccharides

lase in the same buffer. This suspension is then incubated. Just as in the analysis of starch, highly purified enzyme must be used to minimize hydrolytic breakdown of other polysaccharides. [This step may be omitted in future analyses of the same product if no glucose is found in the centrifugate (supernatant) from step f, indicating that no starch is present.] Centrifugation after the removal of starch polysaccharides isolates insoluble dietary fiber (IDF) (Sect. 19.6).

(f) Solubilized polysaccharides are reprecipitated by the addition of sodium chloride and four volumes of absolute ethanol to the cooled dispersion. The mixture is centrifuged. The precipitate (pellet) of water-soluble polysaccharides (often added hydrocolloids) is soluble dietary fiber (SDF) (Sect. 19.6).

(g) The pellet is suspended in deionized water, transferred to dialysis tubing, and dialyzed against frequent changes of sodium azide solution (sodium azide used to prevent microbial growth). Finally, dialysis against deionized water is done to remove the sodium azide. The retentate is recovered from the dialysis tubing and freeze-dried.

(h) Polysaccharide identification relies on hydrolysis to constituent monosaccharides and identification of these sugars. For hydrolysis, polysaccharide material is added to a Teflon-lined, screw-capped vial. Trifluoroacetic acid solution is added (usually 2 *M*), and the vial is tightly capped and heated (usually for 1–2 h at 121 °C) [49]. After cooling, the contents are evaporated to dryness in a hood under a stream of air or nitrogen. Sugars are determined by HPLC (Sect. 19.4.2.1) or GC (Sect. 19.4.2.2). If GC is used, inositol is added as an internal standard. Qualitative and quantitative analysis of the polysaccharides present can be determined by sugar analysis. For example, guaran, the polysaccharide component of guar gum, yields D-mannose and D-galactose in an approximate molar ratio of 1.00:0.56.

The described acid-catalyzed hydrolysis procedure does not release uronic acids quantitatively. Some examples of polysaccharides known to possess uronic acid units include pectin (D-galacturonic acid) and alginate (D-mannuronic acid and L-guluronic acid). The general presence of **uronic acids** (independent of their monosaccharide type) within a given polysaccharide material can be indicated or measured by the ***m*-hydroxydiphenyl (3-phenylphenol) assay** [50–52]. This and similar methods are based on the same principle as the phenol-sulfuric acid assay (Sect. 19.3) but involve condensation of uronic acid (rather than total sugar [as in the case of the phenol-sulfuric acid assay]) dehydration products with a phenolic compound to produce colored compounds that can be measured quantitatively by means of spectrophotometry. If present, specific uronic acids can be identified by a specific GC procedure for them.

19.5.2.3 Pectin

19.5.2.3.1 Nature of Pectin

Commercial pectin is primarily poly(α-D-galacturonic acid methyl ester). Even though **pectin** is a very important food hydrocolloid, no official method for its determination has been established. What few methods have been published basically involve its precipitation (by addition of ethanol) from jams, jellies, etc., in which it is the only added polysaccharide present.

19.5.2.3.2 Pectin Content Determination

Conditions for the extraction of pectin from various plant tissues, followed by its precipitation, have been studied for many years and continue to be investigated – not for analytical interests but because of pectin's commercial value. Several different extractants, extraction conditions, precipitants, and precipitation conditions (with variations of each) have been investigated and optimized. Product characteristics vary with the source material, isolation conditions, and in the case of fruit sources, the degree of ripeness. Because the principal, but not the sole, constituent of pectins is **D-galacturonic acid**, determination of the pectin content of a food or a food product is often based on it suronic acid content (as long as pectin is the only significant source of uronic acid in the food product, since this method is not specific for D-galacturonic acid, but will detect all or total uronic acid present) using the ***m*-hydroxydiphenyl(3-phenylphenol) method** [50–52] (Sect. 19.5.2.2). This method further requires isolation of the crude pectin and the use purified pectin of the same type to construct the standard curve. For reviews of methods for the determination of pectin, see references [53, 54]. A procedure involving methanolysis followed by reverse-phase HPLC has been published [55]. Interference by other hydrocolloids in the determination of pectin has been reviewed [56].

19.6 DIETARY FIBER

19.6.1 Definition

Because labeling of food products for dietary fiber content is mandated, an official analytical method(s) for its determination is required. The first step in adopting a method must be agreement on what constitutes dietary fiber. Then, there must be a method that measures what is described in the definition. However, while no single definition of **dietary fiber** has been agreed upon by all domestic and international organizations, there is progress toward harmonization [57]. One hurdle is that dietary fiber not only needs a chemical definition for the development of an assay procedure, but a measurement of it is important because of its positive physiological effects, which differ from one fiber source to another.

Four official definitions are given in Table 19.4. AACC International (now known as Cereal & Grains Association) was the first organization to develop an official definition for dietary fiber [58, 59]. According to the AACC International definition, and that of others, dietary fiber is essentially the sum of the nondigestible components of a food ingredient or product. No polysaccharide other than starch is digested in the human small intestine; so all polysaccharides other than nonresistant starch are included in this definition. Of the oli-

Table 19.4 Definitions of dietary fiber

AACC International (now known as Cereal and Grains Association)	"Dietary fiber is the edible parts of plants or analogous carbohydrates that are resistant to digestion and absorption in the human small intestine with complete or partial fermentation in the large intestine. Dietary fiber includes polysaccharides, oligosaccharides, lignin and associated plant substances. Dietary fiber promotes beneficial physiological effects, such as, laxation, and/or blood cholesterol attenuation, and/or blood glucose attenuation" [58, 59]
US Institute of Medicine (now known as National Academy of Medicine)	"Dietary fiber consists of nondigestible carbohydrates and lignin that are intrinsic and intact in plants. Functional fiber consists of isolated, nondigestible carbohydrates that have beneficial physiologic effects in humans. Total fiber is the sum of dietary fiber and functional fiber" [60]
Codex Alimentarius Commission	"Dietary fiber denotes carbohydrate polymers with 10 or more monomeric units, which are not hydrolysed by the endogenous enzymes in the small intestine of humans and belonging to the following categories: edible carbohydrate polymers naturally occurring in the food consumed; carbohydrate polymers obtained from food raw materials by physical, enzymatic, or chemical means; synthetic carbohydrate polymers" [61] (Footnote: Decision on whether to include carbohydrates from 3 to 9 monomeric units as fiber is left to national authorities.)
Food and Drug Administration	"Dietary fiber is defined as nondigestible soluble and insoluble carbohydrates (with 3 or more monomeric units), and lignin that are instrinsic and intact in plants; isolated or synthetic non-digestible carbohydrates (with 3 or more monomeric units) determined by FDA to have physiological effects that are beneficial to human health" [6]

gosaccharides, only sucrose, lactose, and those derived from starch (maltodextrins) are digested. **Analogous carbohydrates** are defined as those carbohydrate-based food ingredients that are nondigestible and nonabsorbable, but these are not natural plant components. Wax (suberin and cutin) is included within associated substances. The definition also includes some of the health benefits known to be associated with ingestion of dietary fiber.

Since publication of the AACC International definition of dietary fiber, modified versions have been adopted by both governmental and nongovernmental organizations around the world, such as the US Institute of Medicine (now known as National Academy of Medicine) [60], the Codex Alimentarius Commission [61], and the FDA (Table 19.4). The fiber definition of the Codex Alimentarius Commission (an international body) requires a demonstrated health benefit for all polysaccharides extracted or isolated from food/plant materials and for synthetically derived fibers – the decision whether or not to include nondigestible short-chain carbohydrates (3 to 9 monomeric units) as fiber is left up to individual governments. The FDA designates nondigestible carbohydrates (≥3 monomeric units) as fiber but similarly distinguishes between "intrinsic" and "intact" fibers native to plants (fruits, vegetables, cereal grains, etc.) versus "isolated" or "synthetic" nondigestible fibers (carbohydrate gums, polydextrose, etc.) that are added to foods. While all intrinsic and intact fibers qualify as dietary fiber, isolated and synthetic fibers must first be demonstrated to possess at least one beneficial physiological health effect based on documented scientific evidence (i.e., submission of a citizen petition to the FDA) to gain approval for a fiber designation. Once approved, a given isolated or synthetic fiber is deemed to meet the FDA definition and is included on a list of approved isolated dietary fibers for general use [62]. Definitions adopted by some other regulatory bodies, commissions, and organizations can be found in reference [63].

Formulating a definition acceptable to all is difficult because dietary fiber from different sources is often composed of diverse mixtures of nondigestible and nonabsorbable carbohydrates and other substances that impart different physiological effects. However, there is general agreement that dietary fiber consists of oligo- and polysaccharides, lignin, and other substances not acted upon by the digestive enzymes of the human stomach or small intestine and that most, but not all, dietary fiber is plant cell wall material (cellulose, hemicelluloses, and lignin) composed primarily of polysaccharide molecules. Because only amylose and amylopectin molecules in cooked or other nonresistant forms of starch are digestible, all other polysaccharides are theoretically components of dietary fiber. Some are components of **insoluble fiber**; some make up **soluble fiber**. Major components of soluble and insoluble dietary fiber are listed in Table 19.5. **Total dietary fiber** (TDF) is the sum of insoluble and soluble dietary fiber.

Table 19.5 Components of dietary fiber

Insoluble dietary fiber
Cellulose, including microcrystalline and powdered cellulose added as ingredients
Lignin
Insoluble hemicelluloses and soluble hemicelluloses entrapped in the lignocellulosic matrix
Resistant starch
Soluble dietary fiber
Soluble hemicelluloses not entrapped in the lignocellulosic matrix
Native pectin (most)
Hydrocolloids (most)
Nondigestible oligosaccharides, such as those derived from inulin (FOS)

19.6.2 Methods

19.6.2.1 Overview

Measurement of insoluble fiber is important not only in its own right but also for calculating the caloric content of a food. According to nutrition labeling regulations, one method allowed to calculate total calories involves subtracting the amount of insoluble dietary fiber from the value for total carbohydrate, before calculating the calories based on protein, fat, and carbohydrate content (approximately 4, 9, and 4 calories per gram, respectively) (Chap. 3, Sect. 3.2.1.4). This scheme ignores the fact that soluble fiber, like insoluble fiber, is also essentially noncaloric. [Fiber components can contribute some calories via absorption of their fermentation products (mostly short-chain fatty acids) within the colon.]

The food component that may be most problematic in fiber analysis is **starch**. In any method for the determination of dietary fiber, it is essential that all digestible starch be removed, since incomplete removal of digestible starch increases the residue weight and inflates the estimate of fiber. Some fiber methods include a heating step (e.g., 95–100 °C for 35 min) to **gelatinize starch granules** and make them susceptible to hydrolysis, which is effected using a thermostable α-amylase and glucoamylase (Sect. 19.5.1.1.1). While some forms of resistant starch granules and/or molecules (Sect. 19.5.1.1.2) remain essentially intact through the heating step and, therefore, are components of dietary fiber, other nondigestible resistant starch products (e.g., RS2) in ready-to-eat foods may be rendered partially or fully digestible during this heating step and, thus, may not be counted as dietary fiber by approved methods. Both AOAC Method 2017.16 (AACCI Method 32-60.01) [4, 67] and AOAC Method 2022.01 [64] have been adapted to better retain resistant starch components (Table 19.6).

Nondigestible oligosaccharides such as those derived from inulin (a fructan) [64], certain maltodextrins designed to be nondigestible, and partially hydrolyzed guar gum [65]

Table 19.6 Some official methods of analysis for dietary fiber in food ingredients and products

AOAC Official Method No. [4]	*AACCI Approved Method No.* [67]	*Description of method and measured substance*
994.13	32-25.01	TDF determined as neutral sugar and uronic acid monomer units plus Klason lignin by a gas chromatographic-spectrophotometric-gravimetric method
993.21		Nonenzymic-gravimetric method for TDF applicable to determination of >10% TDF in foods and food products with <2% starch
985.29	32-05.01	Enzymic-gravimetric method for TDF in cereal grains and cereal grain-based products
	32-06.01	A rapid gravimetric method for TDF
992.16		Enzymic-gravimetric method for insoluble dietary fiber in vegetables, fruits, and cereal grains.
993.19		Enzymic-gravimetric method for soluble dietary fiber
991.43	32-07.01	Enzymic-gravimetric method for total, soluble, and insoluble dietary fiber in grain and cereal products, processed foods, fruits, and vegetables
2002.02	32-40.01	Enzymic method for RS2 and RS3 in food products and plant materials
	32-21.01	Enzymic-gravimetric method for insoluble and soluble dietary fiber in oats and oat products
	32-32.01	Enzymic-spectrophotometric method for total fructan (inulin) in foods (not applicable to FOS)
999.03		Enzymic-spectrophotometric method for fructan (inulin) in foods (not applicable to FOS)
997.08	32-31.01	AE-HPLC method for fructan in foods and food products applicable to the determination of added inulin in processed foods
2000.11	32-28.02	AE-HPLC method for polydextrose in foods
	32-22.01	Enzymic method for β-glucan in oat fractions and unsweetened oat cereals
	32-23.01	Rapid enzymic procedure for β-glucan content of barley and oats
2001.03	32-41.01	Enzymic-gravimetric and HPLC method for dietary fiber containing added resistant maltodextrin
2001.02	32-33.01	HPLC method for *trans*-galactooligosaccharides (TGOS) applicable to added TGOS in selected food products
2009.01	32-45.01	Determinestotal dietary fiber, including high-molecular-weight and low-molecular-weight soluble dietary fibers (Codex Alimentarius definition), by an enzymic-gravimetric method and HPLC (underestimates RS4 [phosphate cross-linked starch] and FOS; see AOAC 2017.16 and AACCI 32-60.01 for an improved methodology)
2011.25	32-50.01	Determines insoluble, soluble, and total dietary fiber (Codex Alimentarius definition) by an enzymic-gravimetric method and HPLC (underestimates RS4 [phosphate cross-linked starch] and FOS; see AOAC 2022.01 for an improved methodology)
2017.16	32-60.01	Determines total dietary fiber, including high-molecular-weight and low-molecular-weight soluble dietary fibers (Codex Alimentarius definition), by an enzymic-gravimetric method and HPLC. Replacement method for AOAC 2009.01 and AACCI 32-45.01 with improved methodology for quantifying RS4 [phosphate cross-linked starch] and FOS
2022.01		Determines insoluble, soluble (including the oligosaccharide fiber fraction), and total dietary fiber (Codex Alimentarius definition) by an enzymic-gravimetric method and HPLC. Replacement method for AOAC 2011.25 with improved methodology for quantifying RS4 [phosphate cross-linked starch] and FOS

may be problematic in an analytical sense, since they are in the soluble portion that is not precipitated with 78% ethanol. They should be measured in AOAC Method 2017.06 (AACCI Method 32-60.01) and AOAC Method 2022.01 [4, 67]. Methods for the determination of fructans in certain products have been reviewed [66].

It is essential that either all digestible materials be removed from the sample so that only nondigestible components remain or that a correction be applied for any remaining digestible contaminants. **Lipids** are removed easily from the sample with organic solvents (Sect. 19.2) and generally do not pose analytical problems. **Protein** and **salts/minerals** that are not removed from the sample during the solubilization steps should be corrected for by Kjeldahl (Chap. 18, Sect. 18.2.1) or Dumas (Chap. 18, Sect. 18.2.) nitrogen analysis and by ashing (Chap. 16) of the fiber residue.

To comply with FDA nutrition labeling requirements for the determination of total dietary fiber, the declarable fiber is calculated as the total fiber quantified by official analytical methods minus the amount of added isolated or synthetic fiber that "does not meet the dietary fiber definition" (i.e., lacks supporting evidence for a beneficial physiological effect). A list of approved isolated and synthetic fibers is maintained by the FDA [62].

19.6.2.2 Sample Preparation

Measures of fiber are most consistent when samples are low in fat (less than 10% lipid), dried, and finely ground. If necessary, the sample is ground to pass through a 0.3–0.5-mm mesh screen. If the sample contains more than 10% lipid, the lipid is removed by extraction with 25 parts (v/w) petroleum ether or hexane in an ultrasonic water bath. The mixture is then centrifuged, and the organic solvent is decanted. This extraction is repeated. The sample is air-dried to first remove flammable organic solvent and then dried overnight in a vacuum oven at 70 °C if a measure of lipid and moisture content is required. Loss of weight due to fat and moisture removal is recorded, and the necessary correction is made in the calculation of the percentage dietary fiber value determined in the analysis.

If samples contain large amounts of soluble sugars (mono-, di-, and trisaccharides), the samples should be extracted three times with 80% aqueous ethanol in an ultrasonic water bath at room temperature for 15 min. The supernatant liquid is discarded, and the residue is dried at 40 °C.

A variety of methods have been developed and used at different times for different products. AOAC International *Official Methods of Analysis* [4] and AACC International *Approved Methods* [67] are listed in Table 19.6. Listed methods are generally specific for the type of fiber or the fiber component desired to be measured. Several methods are available in kit form.

19.6.2.3 Enzymic-Gravimetric Method

Dietary fiber is most often determined **gravimetrically** after digestible carbohydrates, lipids, and proteins are selectively solubilized by chemical reagents or removed by enzyme-catalyzed hydrolysis. After such treatments, non-solubilized and/or undigested materials are collected by filtration, and the fiber residue is recovered, dried, and weighed.

19.6.2.3.1 Total, Soluble, and Insoluble Dietary Fiber

AOAC Method 991.43 (AACCI Method 32-07.01) [4, 67] determines total, insoluble, and soluble dietary fiber in cereal products, fruits, vegetables, processed foods, and processed food ingredients. It contains the features of a general analytical method for dietary fiber:

1. **Principle.** Starch and protein are hydrolyzed and removed from a sample by sequential treatment with a thermostable α-amylase, a protease, and glucoamylase. The insoluble residue is recovered and washed [**insoluble dietary fiber** (IDF)]. Ethanol is added to the soluble portion to precipitate soluble polysaccharides [**soluble dietary fiber** (SDF)]. To obtain **total dietary fiber** (TDF), alcohol is added after digestion with glucoamylase, and the IDF and SDF fractions are collected together, dried, weighed, and ashed.
2. **Outline of Procedure.** A flow diagram outlining the general procedure for the method is given in Fig. 19.9. Letters in parentheses for each step below correspond to the same letters in Fig. 19.9.
   (a) To samples devoid of significant lipids is added a basic buffer (pH 8.2), followed by incubation with a thermostable α-amylase (95–100 °C) to gelatinize any starch so that the α-amylase can break it down.
   (b) After α-amylase digestion, the mixture is cooled to 60 °C, an alkaline protease is added, and the mixture is incubated to break down the protein.
   (c) The pH is then adjusted in the range of 4.1–4.8, and glucoamylase is added, with further incubation of the mixture (60 °C) to complete the digestion of any starch. The next few steps differ depending on whether total, insoluble, or soluble fiber is to be determined.
   (d) To determine TDF, four volumes of 95% ethanol are added (to give an ethanol concentration of 78%). The mixture is vacuum-filtered through a preweighed, fritted crucible containing prewashed Celite (a siliceous filter aid). The residue in the crucibles is dewatered by washing with 78% ethanol, 95% ethanol, and acetone in that order. Then, the crucibles are air-dried (to remove all acetone), oven-dried at 103 °C, and weighed. Since some protein and salts/minerals are combined with plant. Cell wall constituents, protein [Kjeldahl (Chap. 18, Sect. 18.2.1) or Dumas (Chap. 18, Sect. 18.2.2)] and ash [muffle furnace procedure (Chap. 16)] are determined on separate duplicate samples, and fiber values are corrected for residual protein and ash. If resistant starch in the fiber residue is to be analyzed separately, it can be deter-

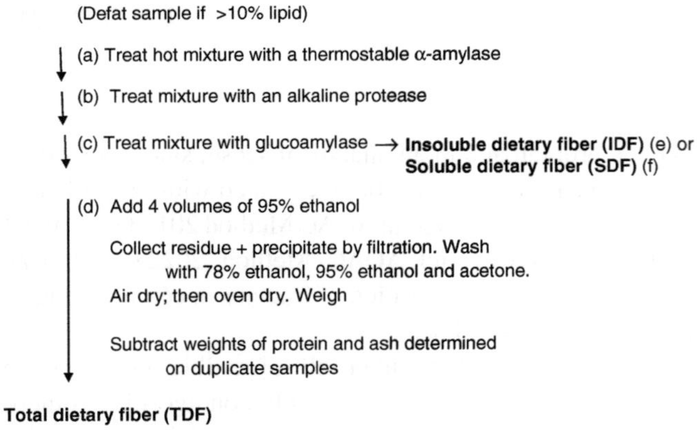

Fig. 19.9 Flow diagram of AOAC Method 991.43 (AACCI Method 32-07.01) [4, 67] – Please refer to Sect. 19.6.2.3.1 (Outline of the Procedure) for a more detailed description of experimental steps (**a**–**f**)

Table 19.7 Dietary fiber data sheet[a]

	SAMPLE				BLANK			
	Insoluble Fiber		*Soluble Fiber*		*Insoluble Fiber*		*Soluble Fiber*	
Sample wt (mg)	m_1	m_2						
Crucible + Celite wt (mg)								
Crucible + Celite + residue wt (mg)								
Residue wt (mg)	R_1	R_2	R_1	R_2	R_1	R_2	R_1	R_2
Protein (mg) **P**								
Crucible + Celite + ash wt (mg)								
Ash wt (mg) **A**								
Blank wt (mg) **B**[b]								
Fiber (%)[c]								

[a]Adapted with permission from *The Journal of AOAC International,* 1988, 71:1019. Copyright 1988 by AOAC International

[b]Blank (mg) $= \frac{R_1 + R_2}{2} - P - A$

[c]Fiber (%) $= \frac{\frac{R_1 + R_2}{2} - P - A - B}{\frac{m_1 + m_2}{2}} \times 100$

mined using AOAC Method 2002.02 (AACCI Method 32-40.01) [4, 67].

(e) To determine IDF, the mixture obtained from step (c) is vacuum-filtered through a tared, fritted crucible containing prewashed Celite. The residue retained by the filter is washed with water and then dewatered by washing in order with 78% ethanol, 95% ethanol, and acetone. The crucibles are air-dried (to remove all acetone), oven-dried at 103 °C, and weighed. Fiber weight is corrected for residual protein and ash as described for step (d) above. SDF is in the filtrate.

(f) To determine SDF, four volumes of 95% ethanol (to give an ethanol concentration of 78%) are added to the filtrate and water washes from step (e) to precipitate soluble fiber. The precipitate is collected by vacuum filtration through tared, fritted crucibles containing prewashed Celite. The residues are dewatered by washing with 78% ethanol, 95% ethanol, and acetone in that order. Then, the crucibles are air-dried (to remove all acetone), oven-dried at 103 °C, and weighed. Fiber weight is corrected for residual protein and ash as described for step (d) above. Duplicate reagent blanks must be run through the entire procedure for each type of fiber determination.

Table 19.8 Total, soluble, and insoluble dietary fiber in foods as determined by AOAC Method 991.43 [4]

Food	*Soluble*[a]	*Insoluble*[a]	*Total*[a]
Barley	5.02	7.05	12.25
High-fiber cereal	2.78	30.52	33.73
Oat bran	7.17	9.73	16.92
Soy bran	6.90	60.53	67.14
Apricots	0.53	0.59	1.12
Prunes	5.07	4.17	9.29
Raisins	0.73	2.37	3.13
Carrots	1.10	2.81	3.93
Green beans	1.02	2.01	2.89
Parsley	0.64	2.37	2.66

Adapted from *Official Methods of Analysis, 22nd* edn. Copyright 2023 by AOAC International

[a]Grams of fiber per 100 g of food on a fresh weight basis

Table 19.7 is a form used to calculate fiber percentages. Using the equations shown, percent dietary fiber is expressed on a dry weight basis if the sample weights are for dried samples. Representative values obtained using this method are given in Table 19.8.

Note: Neither this method for TDF nor that for SDF accounts for SDF that does not precipitate in 78% aqueous ethanol, including some or most inulin, polydextrose, digestion-resistant maltodextrins, and partially hydrolyzed guar gum and all fructo-, arabinoxylo-, xylo-, and galactooligosaccharides. AOAC Method 2017.16 (AACCI Method 32-60.01) and AOAC Method 2022.01 [4, 67] incorporate deionization and HPLC procedures to quantify these lower-molecular-weight, digestion-resistant materials in the filtrate so that all SDF is measured.

19.6.2.3.2 Dietary Fiber Components as Defined by Codex Alimentarius

The most recent method (AOAC Method 2022.01) [4] combines aspects of previously approved methods to measure individual components of dietary fiber (IDF, SDF, and TDF), with the added capability to quantify both polysaccharide (SDFP) and oligosaccharide (SDFS; DP ≈ 3-9) classes of soluble fiber as prescribed by Codex Alimentarius. This method is outlined in Fig. 19.10.

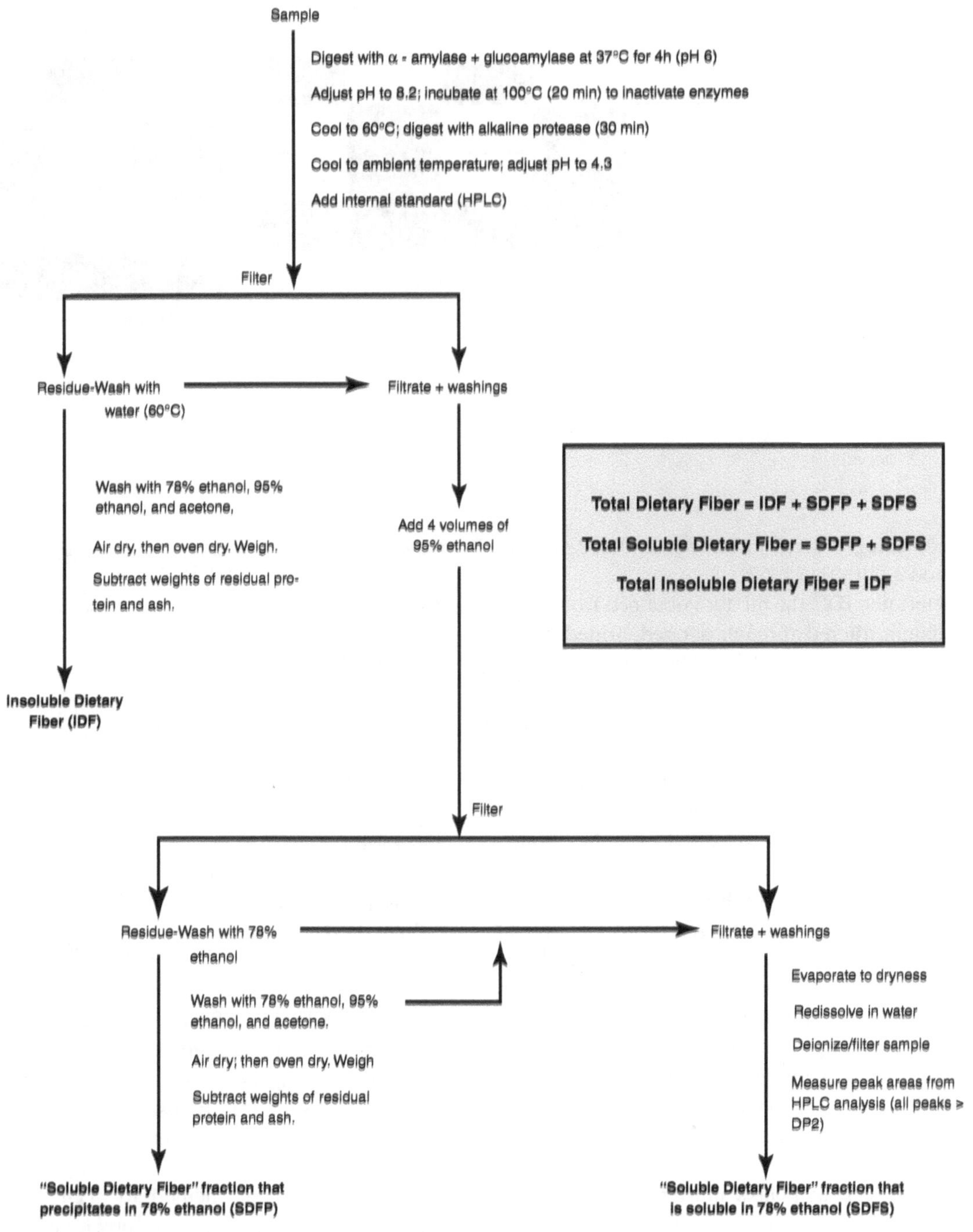

Fig. 19.10 Flow diagram of AOAC Method 2022.01 [4]

Note: **SDFP** is dietary fiber that is soluble in water but insoluble in 78% ethanol; it includes most hydrocolloids, and some of such compounds are polydextrose, nondigestible maltodextrins, inulin, and partially hydrolyzed guar gum. **SDFS** is dietary fiber that is soluble in both water and 78% ethanol; it includes various oligosaccharides, such as low-molecular-weight FOS and galactooligosaccharides.

19.7 PHYSICAL METHODS

19.7.1 Measurements of Sugar Concentrations in Solution

The concentration of a carbohydrate in solution can be determined by measuring the solution's specific gravity, refractive index (RI), or optical rotation. The **specific gravity** (Chap 15, Sect. 15.2.5.2) is the ratio of the density of a substance to the density of a reference substance (usually water) both at a specific temperature. The most common way to determine specific gravity is by the use of a hydrometer calibrated in °**Brix**, which corresponds to sucrose concentrations by weight, or in **Baumé modulus (°Bé)** (AOAC Method 932.14) [4]. The obtained values are then converted into concentrations using tables constructed for the substance being measured as a pure solution. Measurement of specific gravity as a means of determining sugar concentration is accurate only for pure sucrose or other solutions of a single pure substance, but it can be, and is, used for obtaining approximate values for liquid products for which appropriate specific gravity tables have been constructed.

When light passes from one medium to another, it changes direction, i.e., it is bent or refracted. The ratio of the sine of the angle of incidence to the sine of the angle of refraction is called the index of refraction, or **refractive index** (RI). RI varies with the nature and concentration of the dissolved compound, the temperature, and the wavelength of light used. By holding the nature of the compound, the temperature, and the wavelength constant, the concentration of the dissolved compound can be determined by measuring the RI (Chap. 15, Sect. 15.2.5.3). To determine RI, the solution must be clear. Like determination of specific gravity, the use of RI to determine concentrations is accurate only for pure sucrose or other solutions of a single pure substance. It is frequently used for obtaining approximate sugar concentrations in liquid products. Refractometers that read directly in sucrose concentration units are available.

Most compounds that contain a chiral carbon atom have optical activity, i.e., they will rotate the plane of polarization of polarized light. A **polarimeter** measures the extent to which a compound in solution rotates a plane of polarized light. Carbohydrates have chiral carbon atoms and, thus, optical activity. Carbohydrates can rotate the plane of polarized light through an angle that depends on the nature of the compound, the temperature, the wavelength of light, and the concentration of the compound. The concentration of the compound can be determined from a value known as the **specific optical rotation** if all other factors are held constant and if the solution contains no other optically active compounds. Determination of specific optical rotation can be used to measure sucrose concentration (AOAC Methods 896.02, 925.46, 930.37) [4]. Determination of sucrose concentration by **polarimetry** also requires a clear solution, and there are instruments available that read in units of the International Sugar Scale. Determination of specific optical rotation before and after hydrolysis of sucrose into its constituent sugars, D-glucose and D-fructose, a process called **inversion**, can be used to determine sucrose in the presence of other sugars (AOAC Methods 925.47, 925.48, 926.13, 926.14) [4].

19.7.2 Measurement of Starch Pasting Behavior and Thermal Properties

Starch gelatinization, pasting, and retrogradation properties are key factors that influence the functionality of both isolated starch ingredients (native or modified) and starch-containing food materials (e.g., cereal flours, potatoes, and legumes) that undergo heating in the presence of water. A detailed review of starch gelatinization, pasting, and retrogradation topics is found within these references [1, 2, 68].

19.7.2.1 Starch Pasting Properties

A starch pasting profile is generated by monitoring the viscosity of an aqueous slurry of starch (or starch-containing material) over time, as it is subjected to a controlled heating, hold, and cooling regimen in the presence of applied shear (i.e., stirring). In short, a pasting profile provides a functional fingerprint of a specific starch material that embodies viscosity changes in response to simulated gelatinization, pasting, and retrogradation events (Fig. 19.11). Defined pasting attribute values (i.e., Peak, Trough, and Final), which are either obtained or calculated from key points along the viscosity curve, provide key measures of starch behavior, such as extent of granule swelling upon gelatinization (Peak viscosity), hot paste stability/viscosity (Breakdown and Trough viscosities), and gelling tendency/strength (Setback and Final viscosities) [68]. These defined pasting attributes allow for quantitative comparison of different starch materials, based on measured differences in their viscosity profiles.

The primary instruments utilized for determining starch pasting profiles are the Rapid Visco Analyser (RVA, models 4500 or 4800), the Amylograph®-E (AG-E), and the Micro Visco-Amylo-Graph® (MVAG, a smaller, yet more nimble

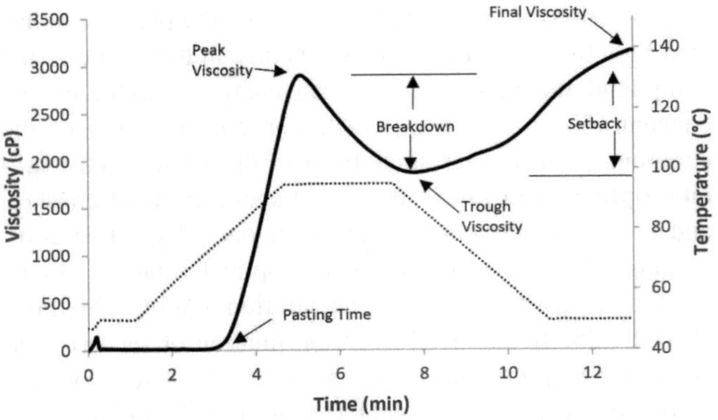

Fig. 19.11 Rapid Visco Analyser (RVA) example pasting profile (bold line = viscosity profile; dotted line = heating and cooling regimen) for native corn starch, depicting key pasting attributes. [Run parameters: initial hold at 50 °C (1 min); linear heating to 95 °C (3.7 min); intermediate hold at 95 °C (2.5 min); linear cooling to 50 °C (3.8 min); final hold at 50 °C (2 min)]

version of the AG-E) [69, 70]. Of the three instruments, the AG-E has the most limited operational capability (larger sample requirement, slower heating/cooling rates, fixed shear rate, longer run times), while both the RVA and VMAG offer greater adaptability (lesser sample requirement, variable and faster heating/cooling rates, adjustable shear/stir rates, shorter run times). Sample preparation involves creation of an aqueous starch slurry (7–10%, w/w), followed by heating to 95 °C, holding at 95 °C, cooling to 50 °C, and holding at 50 °C (refer to Fig. 19.11 for an example set of RVA run parameters). While 95 °C represents a practical temperature maximum for typical pasting applications (to avoid "boiling out" of the sample at 100 °C), the RVA 4800 is designed to achieve temperatures up to 140 °C, making it more relevant for high-temperature processing applications (e.g., retorting, extrusion).

Both the RVA and AG-E are the basis of multiple official methods for evaluating cooking/processing properties of milled rice flour (AACCI 61-02.01/AACCI 61-04.01 and AACCI 61.01.01, respectively), processing properties of milled/wholemeal oat flour (AACCI 76-22.01, RVA only), baking performance of wheat/rye flour (AACCI 76-22-0.01, RVA only), and sprout damage/amylase activity in grain flours (AACCI 22-08.02 and AACCI 22-10.01/AACCI 22-12.01, respectively) [67]. Beyond official methods, applications of starch pasting methods are diverse [71], finding use in quality assurance programs (raw ingredient/finished product testing), prediction of processing operations (e.g., extrusion) characterization of starch properties [72], investigation of starch–ingredient interactions [73, 74], and as a selection tool in plant breeding programs [75, 76].

For more sophisticated rheological analysis (beyond pasting behavior), starch pastes and gels may also be analyzed using a rotary (Chap. 29, Sect. 29.3.2) or dynamic oscillatory (Ch. 29, Sect. 29.3.3) rheometer [77, 78].

19.7.2.2 Starch Thermal Properties

Starch gelatinization temperature is most often measured by Differential Scanning Calorimetry (DSC, Chap. 30, Sect. 30.3.2). A typical procedure involves heating a small quantity of starch (10 mg, db) in a set volume of water (20 μL; water must be twofold [w/w] the amount of starch) within a sealed DSC pan at a rate of 10 °C/min over a temperature range of 20–180 °C [79]. Since the gelatinization event occurs over a range of temperature, data output includes the onset (T_o), peak (T_p), and conclusion (T_c) gelatinization temperatures, as well as a measure of gelatinization enthalpy (ΔH, energy associated with the loss of crystalline order) [70]. The DSC can also be used to detect/measure amylose V-type complexes [80] and monitor starch retrogradation [81].

19.7.3 Mass Spectrometry

There are many different variations of mass spectrometry (MS) (Chap. 11). Most of the techniques applied to carbohydrates are used for structural analysis. MS has been used for the determination of carbohydrates, but not in a routine manner. Particularly useful is the **matrix-assisted laser-desorption time-of-flight** (MALDI-TOF) technique for the analysis of a homologous series of oligosaccharides (Fig. 19.12). A comparison between anion-exchange HPLC (Sect. 19.4.2.1.2) (the most used carbohydrate analysis technique today), capillary electrophoresis (Sect. 19.4.2.4), and MALDI-TOF MS for the analysis of maltooligosaccharides led to the conclusion that the latter technique gave the best results [27].

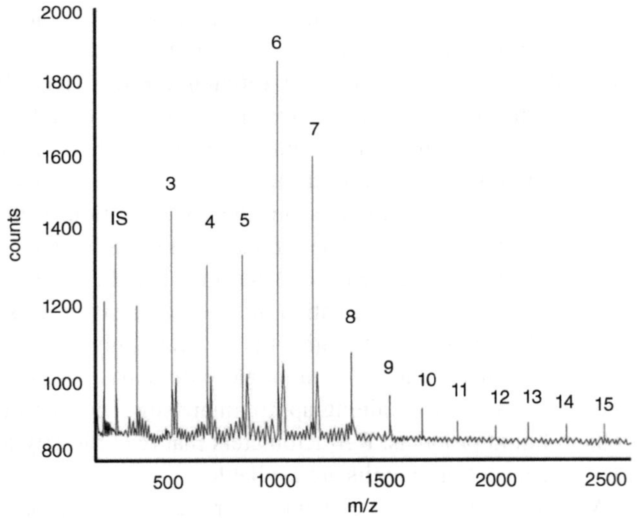

Fig. 19.12 MALDI-TOF mass spectrum of maltooligosaccharides produced by hydrolysis of starch. Numbers indicate DP. IS = internal standard. (From Kazmaier et al. [26], used with permission, Copyright Springer-Verlag, 1998)

19.7.4 Mid-Infrared/Fourier Transform Infrared (FTIR) Spectroscopy

FTIR (Chap. 8, Sect. 8.3.1.2) methods are simple and rapid. Detection limits are greater than those required for most other methods. Spectral libraries have been compiled for several hydrocolloids, including κ-, ι-, and λ-carrageenans [82–86], pectin [82, 84, 87], galactomannans [82, 84], and cellulose derivatives [88]. Dispersive mid-IR instruments are utilized as milk analyzers for rapid and straightforward determination of lactose content, along with fat and protein levels (Chap. 8, Sect. 8.3.3.2).

19.7.5 Near-Infrared (NIR) Spectroscopy

NIR (Chap. 8, Sect. 8.4) spectrometry has been used to determine dietary fiber [89] and sugar [90] contents and to identify cellulose derivatives [88].

19.8 SUMMARY

For the determination of low-molecular-weight carbohydrates, older colorimetric methods for total carbohydrate, various reducing sugar methods, and physical measurements have largely been replaced by chromatographic methods. Older chemical methods suffer from the fact that different sugars give different results, which makes them particularly problematic when a mixture of sugars is present. Older physical methods suffer from the fact that they work only with pure substances. However, some older methods continue to be used for simplicity, quality assurance, and product standardization. Chromatographic methods (HPLC and GC) separate mixtures into their component sugars, identify each component by retention time, and provide a measurement of the quantity of each component. HPLC is widely used for the identification and measurement of mono- and oligosaccharides. Enzymic methods are specific and sensitive, but seldom, except in the case of starch, is determination of only a single component desired.

Polysaccharides are important components of many food products. Yet, there is no universal procedure for their analysis. Generally, isolation must precede measurement. Isolation introduces errors because losses of constituents occur with extraction, recovery, and separation techniques. Identification is done by hydrolysis to constituent monosaccharides and their determination. An exception is starch, which can be digested to glucose using specific enzymes (amylases), followed by the measurement of the glucose released and, therefore, can be specifically measured.

Insoluble dietary fiber, soluble dietary fiber, and total dietary fiber are each composed primarily of non-starch polysaccharides. Methods for the determination of total dietary fiber and its components rely on removal of the digestible starch using amylases and often on removal of digestible protein with a protease, leaving a nondigestible residue. Improved methodologies for fiber analysis are closing the gap between prescribed, regulatory-based dietary fiber definitions and the previous inability to measure all designated fiber components.

19.9 STUDY QUESTIONS

1. Give three reasons why carbohydrate analysis is important.
2. "Proximate composition" refers to analysis for moisture, ash, fat, protein, and carbohydrate. Identify which of these components of "proximate composition" are actually required on a nutrition label. Also, explain why it is important to measure the non-required components quantitatively if one is developing a nutrition label.
3. Distinguish chemically between monosaccharides, oligosaccharides, and polysaccharides, and explain how solubility characteristics can be used in an extraction procedure to separate monosaccharides and oligosaccharides from polysaccharides.
4. Discuss why 80% ethanol (final concentration) is used to extract mono- and oligosaccharides, rather than using water. What is the principle involved?
5. You have a solid raw food product that contains cellulose, lignin, starch, sucrose, glucose, fructose, and pectin, along with many other noncarbohydrate food components. If the sample is ground then refluxed in 80% ethanol (final concentration), which of the noted components will be present in the filtered extract (i.e., not the residue)?
6. What are the principles behind total carbohydrate determination using the phenol-sulfuric acid method? Why is a standard curve employed? Why is a reagent blank used? What are limitations of the method?
7. What is the principle behind determination of total reducing sugars using the Somogyi–Nelson method?
8. In regard to carbohydrate analysis, briefly describe the use/application and the underlying principles for each described treatment or instrument:
   (a) 80% alcohol solution (final concentration) buffered with calcium carbonate
   (b) Carrez reagents
   (c) Glucose oxidase and peroxidase
   (d) Amylase and glucoamylase
   (e) A hydrometer
9. Describe the principle behind AE-HPLC separation of carbohydrates.

10. Describe the general procedure for the preparation of sugars for GC. What is required for this method to be successful?
11. Why has HPLC largely replaced GC for the analysis of carbohydrates?
12. What are the advantages of enzymic methods? What are the limitations (potential problems)?
13. Describe the principles behind the enzymic determination of starch. What are the advantages of this method? What are potential problems?
14. Briefly describe a method that could be used for each of the following:
    (a) To prevent hydrolysis of sucrose when sugars are extracted from fruits via a hot alcohol extraction
    (b) To remove proteins from solution for an enzymic determination of carbohydrates
    (c) To measure total carbohydrate
    (d) To measure total reducing sugars
    (e) To measure glucose enzymically
    (f) To measure simultaneously the concentrations of individual free sugars
15. Describe the principle behind each step in Fig. 19.8. What is the reason for each step (a–h)?
16. Describe the principles behind separation and analysis of cellulose, water-soluble gums, and starch.
17. Using pectin as an example, explain why the quantitative analysis of hydrocolloids in food products is so difficult.
18. What is a general definition of dietary fiber? Why is the definition of dietary fiber important? How does the definition of dietary fiber affect the development of an analytical procedure for it?
19. List the general component classes of dietary fiber that are usually determined for nutritional or research purposes.
20. List the constituents of dietary fiber.
21. Explain how measurement of dietary fiber relates to calculating the caloric content of a food product.
22. Explain the purpose(s) of each of the steps in the AOAC Method 991.43 for total dietary fiber listed below as applied to the determination of the dietary fiber content of a high-fiber snack food:
    (a) Heating sample and treating with α-amylase
    (b) Treating sample with protease
    (c) Treating sample with glucoamylase
    (d) Adding four volumes of 95% ethanol to sample after treatment with glucoamylase
    (e) After drying and weighing the filtered and washed residue, heating one duplicate final product to 525 °C in a muffle furnace and analyzing the other duplicate sample for protein
23. The nutrition labeling regulations require declaration of the dietary fiber content on food labels. The AOAC International Method 991.43 determines "Total Dietary Fiber" (TDF), with options for measuring both soluble (SDF) and insoluble (IDF) dietary fiber fractions.
    (a) Explain briefly why this method for determining TDF is described as an "enzymatic-gravimetric" method.
    (b) The method for TDF, as well as those for determining SDF and IDF fractions, each include a correction for any residual minerals and protein. What assays are used to determine residual mineral and protein in fiber fractions?
    (c) One way to obtain a value for TDF is to sum the individual values of the SDF and IDF fractions. (Soluble and insoluble fiber values may be voluntarily listed on nutrition labels.) Describe the methods for measuring the SDF and IDF values individually, as compared to the method for determining just TDF (without first separating SDF and IDF) (i.e., what procedures are done to make each possible, starting with the step following digestion of starch and protein.)
24. Describe the principles behind and the limitations of determining sugar (sucrose) concentrations by (a) specific gravity determination, (b) refractive index measurement, and (c) polarimetry.
25. As the new incoming QC Manager of a small, upstart manufacturer of ready-to-eat pudding products, you observe fluctuations in finished product texture upon receipt of each new lot of starch thickener. However, you learn that there is no established protocol for monitoring incoming ingredient quality. First, identify an appropriate test method that could be the basis of a certificate of analysis agreement that would ensure the quality of the starch ingredient. Second, describe specifically how the identified method could be utilized to monitor starch quality.

19.10 PRACTICE PROBLEMS

1. The following data were obtained when an extruded breakfast cereal was analyzed for total fiber by AOAC Method 991.43 (AACCI Method 32-07.01).

Sample wt, mg	1002.8
Residue wt, mg	151.9
Protein wt, mg	13.1
Ash wt, mg	21.1
Blank wt, mg	6.1

What is percent total fiber determined to the appropriate number of significant figures?

2. The following tabular data were obtained when a high-fiber cookie was analyzed for fiber content by AOAC Method 991.43 (AACCI Method 32-07.01).

	Sample			
	Insoluble		*Soluble*	
Sample wt, mg	1002.1	1005.3		
Crucible – Celite wt, mg	31,637.2	32,173.9	32,377.5	33,216.4
Crucible – Celite + residue wt, mg	31,723.5	32,271.2	32,421.6	33,255.3
Protein, mg	6.5		3.9	
Crucible – Celite + ash wt, mg		32,195.2		33,231.0
	Blank			
	Insoluble		*Soluble*	
Crucible + celite wt, mg	31,563.6	32,198.7	33,019.6	31,981.2
Crucible + Celite + residue wt, mg	31,578.2	32,213.2	33,033.4	33,995.6
Protein, mg	3.2		3.3	
Crucible + Celite + ash wt, mg		32,206.8		31,989.1

What is the (a) total, (b) insoluble, and (c) soluble fiber content of the cookie, each determined to the appropriate number of significant figures?

Answers

1. Number of significant figures = 2 (6.1 mg)

$$\frac{151.9-13.1-21.1-6.0}{1002.8}\times 100 = 11\%$$

2. (Calculations are done a little differently than those at the bottom of Table 19.7)

Insoluble dietary fiber
Number of significant figures = 2 (6.5 mg, 3.2 mg)
Blank residue = 31,578.2 mg – 31,563.6 mg = 14.6 mg
32,231.2 mg – 32,198.7 mg = 14.5 mg
Ave. = 14.6 mg
Blank ash = 32,206.8 – 32,198.7 = 8.1 mg
First sample residue = 31,723.5 mg – 31,637.2 mg = 86.3 mg
ash = 32,195.2 mg – 32,173.9 mg = 21.3 mg
86.3 mg (residue weight)
–14.6 mg (blank)
–3.3 mg (protein, 6.5 – 3.2 [blank])
–13.2 mg (ash, 21.3 – 8.1 [blank])
55.2 mg
(55.2 mg ÷ 1002.1 [sample wt.]) × 100 = 5.5%
Second sample residue = 32,271.2 mg – 32,173.9 mg = 97.3 mg
97.3 – 14.5 – 3.3 – 13.2 = 66.3 mg
(66.3 mg ÷ 1005.3 mg [sample wt.]) × 100 = 6.6%
Average of 5.5% and 6.6% = 6.1%

Soluble dietary fiber
Number of significant figures = 2 (3.9 mg, 3.3 mg)
Blank residue = 33,033.4 mg – 33,019.6 mg = 13.8 mg
33,995.6 mg – 31,981.2 mg = 14.4 mg
Ave. = 14.1 mg
Blank ash = 31,989.1 mg – 31,981.2 mg = 7.9 mg
First sample residue = 32,421.6 mg – 32,377.5 mg = 44.1 mg
ash = 33,231.0 mg – 33,216.4 mg = 14.6 mg
44.1 mg (residue weight)
–14.1 mg (blank)
–0.6 mg (protein 3.9 – 3.3 [blank])
–6.7 (ash, 14.6 – 7.9 [blank])
22.7 mg
(22.7 mg ÷ 1002.1 [sample wt]) × 100 = 2.3%
Second sample residue = 33,255.3 mg – 33,216.4 mg = 38.9 mg
38.9 mg (residue weight)
–14.1 mg (blank)
–0.6 mg (protein, 3.9 – 3.3 [blank])
–6.7 mg (ash, 14.6 – 7.9 [blank])
17.5 mg
(17.5 mg ÷ 1005.3 [sample wt.]) × 100 = 1.7%
Average of 2.3% and 1.7% = 2.0%
Total dietary fiber = 6.1% (insoluble fiber) + 2.0% (soluble fiber) = 8.1%

REFERENCES

1. BeMiller JN (2019) Carbohydrate Chemistry for Food Scientists, 3rd edn. Woodhead Publishing, Duxford, UK and AACC International, St. Paul, MN
2. Huber KC, JN BeMiller (2017) Carbohydrates (Chap 3). In: Damodaran S, Parkin KL, Fennema OR (eds), Food Chemistry, 5th edn. Marcel Dekker, New York
3. FAO/WHO expert consultation on carbohydrates in human nutrition. 14–18 April 1997, Rome
4. AOAC International (2023) Official methods of analysis, 22nd edn. (online) AOAC International, Rockville, MD
5. Peris-Tortajada M (2004) Carbohydrates and starch (Chap 13). In: Nollet LML (ed) Handbook of Food Analysis, 2nd edn. Marcel Dekker, New York
6. Anon. (2016) Code of Federal Regulations, Title 21, Part 101.9–Food Nutrition Labeling of Food. U.S. Government Printing Office, Washington, DC
7. USDA (2019) USDA Nutrient Database for Standard Reference. Legacy Release, 2019. http://data.nal.usda.gov/dataset/usda-national-nutrient-database-standard-reference-legacy-release
8. Dubois M, Gilles KA, Hamilton JK, Rebers PA, Smith F (1956) Colorimetric method for determination of sugars and related substances. Analytical Chemistry 28:350
9. Wood TM (1994) Enzymic conversion of cellulose into D-glucose. Methods in Carbohydrate Chemistry 10:219
10. Miller G, Blum R, Glennon WEG, Burton A (1960) Measurement of carboxymethyl cellulase activity. Analytical Biochemistry 1:127
11. El Rassi Z (ed) (1995) Carbohydrate analysis. Journal of Chromatography Library vol 58

12. Andersen R, Sørensen A (2000) Separation and determination of alditols and sugars by high-pH anion-exchange chromatography with pulsed amperometric detection. Journal of Chromatography A 897:195
13. Hanko VP, Rohrer JS (2000) Determination of carbohydrates, sugar alcohols, and glycols in cell cultures and fermentation broths using high-performance anion-exchange chromatography with pulsed amperometric detection. Analytical Biochemistry 283:192
14. Cataldi TRI, Campa C, DeBenedetto GE (2000) Carbohydrate analysis by high-performance anion-exchange chromatography with pulsed amperometric detection: The potential is still growing. Fresenius' Journal of Analytical Chemistry 368:7391
15. Zhang Y, Lee YC (2002) High-performance anion-exchange chromatography of carbohydrates on pellicular resin columns. Journal of Chromatography Library 66:207
16. Soga T (2002) Analysis of carbohydrates in food and beverages by HPLC and CE. Journal of Chromatography Library 66:483
17. Montero CM, Dodero MCR, Sánchez DAG, Barroso CG (2004) Analysis of low molecular weight carbohydrates in foods and beverages: A review. Chromatographia 59:15
18. Corradini C, Cavazza A, Bignardi C (2012) High-performance anion-exchange chromatography coupled with pulsed electrochemical detection as a powerful tool to evaluate carbohydrates of food interest: principles and applications. International Journal of Carbohydrate Chemistry 487564
19. Peris-Tortajada M (2012) HPLC determination of carbohydrates in foods (Chap 7) In: Nollet LM, Toldra F (eds) Food analysis by HPLC, 3rd edn. CRC Press, Boca Raton
20. Yan X (2014) High-performance liquid chromatography for carbohydrate analysis (Chap 3) In: Zuo Y (ed) High-performance liquid chromatography (HPLC). Nova Science Publishers, Hauppauge, NY
21. Ammeraal RN, Delgado GA, Tenbarge FL, Friedman RB (1991) High-performance anion-exchange chromatography with pulsed amperometric detection of linear and branched glucose oligosaccharides. Carbohydrate Research 215:179
22. Marioli JM (2001) Electrochemical detection of carbohydrates in HPLC. Current Topics in Electrochemistry 8:43
23. LaCourse WR (2002) Pulsed electrochemical detection of carbohydrates at noble metal electrodes following liquid chromatographic and electrophoretic separation. Journal of Chromatography Library 66:905
24. Baldwin RP (2002) Electrochemical detection of carbohydrates at constant potential after HPLC and CE separations. Journal of Chromatography Library 66:947
25. LaCourse WR (2009) Advances in pulsed electrochemical detection for carbohydrates. Advances in Chromatography 47:247
26. Kazmaier T, Roth S, Zapp J, Harding M, Kuhn R (1998) Quantitative analysis of malto-oligosaccharides by MALDI-TOF mass spectrometry, capillary electrophoresis and anion exchange chromatography. Fresenius' Journal of Analytical Chemistry 361:473
27. Dvořácková E, Šnóblová M, Hrdlička P (2014) Carbohydrate analysis: from sample preparation to HPLC on different stationary phases coupled with evaporative light-scattering detection. Journal of Separation Science 37:323
28. Churms SC (1995) High performance hydrophilic interaction chromatography of carbohydrates with polar sorbents (Chap 3). In: reference 11
29. Ball GFM (1990) The application of HPLC to the determination of low molecular weight sugar and polyhydric alcohols in foods: a review. Food Chem 35:117
30. Hernandez-Hernandez O, Moreno FJ, Sanz ML (2012) Analysis of dietary sugars in beverages by gas chromatography. Food and Nutritional Components in Focus 3:208
31. Ruiz-Matute AI, Hernandez-Hernandez O, Rodriguez-Sanchez S, Sanz ML, Martinez-Castro I (2011) Derivatization of carbohydrates for GC and GC-MS analyses. Journal of Chromatography B: Analytical Technologies in the Biomedical and Life Sciences 879:1226
32. Fox A, Morgan SL, Gilbart J (1989) Preparation of alditol acetates and their analysis by gas chromatography (GC) and mass spectrometry (MS) (Chap 5). In: Biermann CJ, McGinnis GD (ed) Analysis of carbohydrates by GLC and MS. CRC Press, Boca Raton
33. Fox A (2002) A current perspective on analysis of sugar monomers using GC-MS and GC-MS/MS. Journal of Chromatography Library 66:829
34. van Leeuwen KA, Prenzler PD, Ryan D, Camin F (2014) Gas chromatography-combustion-isotope ratio mass spectrometry for traceability and authenticity in foods and beverages. Comprehensive Reviews in Food Science and Food Safety 13:814
35. BeMiller JN (ed) (1994) Methods in carbohydrate chemistry, vol 10 Enzymic Methods. Wiley, New York
36. Bergmeyer HU (ed) (1984) Methods of Enzymatic Analysis, vol 6 Metabolites 1: Carbohydrates, 3rd edn. Verlag Chemie, Weinheim, Germany
37. Cabálková J, Žídková J, Přibyla L, and Chmelík J (2004) Determination of carbohydrates in juices by capillary electrophoresis, high-performance liquid chromatography, and matrix-assisted laser desorption/ionization-time of flight-mass spectrometry. Electrophoresis 25:487
38. Thibault P, Honda S (eds) (2003) Capillary electrophoresis of carbohydrates, Methods in Molecular Biology, vol 213
39. Cortacero-Ramírez S, Segura-Carretero A, Cruces-Blanco C, Hernáinz-Bermúdez de Castro M, Fernandez-Gutiérrez A (2004) Analysis of carbohydrates in beverages by capillary electrophoresis with precolumn derivatization and UV detection. Food Chemistry 87:471
40. Ramírez SC, Carretero S, Blanco CC, de Castro MHB, Gutiérrez AF (2005) Indirect determination of carbohydrates in wort samples and dietetic products by capillary electrophoresis. Journal of the Science of Food and Agriculture 85:517
41. Ma S, Lau W, Keck RG, Briggs JB, Jones AJS, Moorhouse K, Nashabeh W (2005) Capillary electrophoresis of carbohydrates derivatized with fluorophoric compounds. Methods in Molecular Biology 308:397
42. Momenbeik F, Johns C, Breadmore MC, Hilder EF, Macka M, Haddad PR (2006) Sensitive determination of carbohydrates labelled with p-nitroaniline by capillary electrophoresis with photometric detection using a 406 nm light-emitting diode. Electrophoresis 27:4039
43. Volpi N (ed) (2011) Capillary electrophoresis of carbohydrates. From monosaccharides to complex polysaccharides. Humana Press, New York.
44. BeMiller JN (2020) Resistant starch (Chap 7) In: Welti-Chanes J, Serna-Saldivar SO, Campanella O, Tejada-Ortigoza V (eds), Science and Technology of Fibers in Food Systems, Springer (Springer Nature Switzerland AG), Cham, Switzerland.
45. Perera A, Meda V, Tyler RT (2010) Resistant starch: a review of analytical protocols for determining resistant starch and of factors affecting the resistant starch content of foods. Food Research International 43:1959
46. BeMiller JN (2016) Gums/hydrocolloids: analytical aspects (Chap 6). In: Eliasson A-C (ed) Carbohydrates in Food, 3rd edn. CRC Press, Boca Raton
47. Baird JK (1993) Analysis of gums in foods (Chap 23). In: Whistler RL, BeMiller JN (eds) Industrial Gums, 3rd edn. Academic, San Diego
48. Harris P, Morrison A, Dacombe C (1995) A practical approach to polysaccharide analysis (Chap 18). In: Stephen AM (ed) Food polysaccharides and their applications. Marcel Dekker, New York
49. Biermann CJ (1989) Hydrolysis and other cleavage of glycosidic linkages (Chap 3). In: Biermann CJ, McGinnis GD (ed) Analysis of carbohydrates by GLC and MS. CRC Press, Boca Raton

50. Fillisetti-Cozzi TMCC, Carpita NC (1991) Measurement of uronic acid without interference from neutral sugars. Analytical Biochemistry 197:157
51. Ibarz A, Pagán A, Tribaldo F, Pagán J (2006) Improvement in the measurement of spectrophotometric data in the m-hydroxydiphenyl pectin determination methods. Food Control 17:890
52. Yapo BM (2012) On the colorimetric-sulfuric acid analysis of uronic acids in food materials: potential sources of discrepancies in data and how to circumvent them. Food Analytical Methods 5:195
53. Baker RA (1997) Reassessment of some fruit and vegetable pectin levels. Journal of Food Science 62:225
54. Walter RH (1991). Analytical and graphical methods for pectin (Chap 10). In: Walter RH (ed) Chemistry and technology of pectin. Academic Press, San Diego
55. Quemener C, Marot C, Mouillet L, Da Riz V, Diris J (2000) Quantitative analysis of hydrocolloids in food systems by methanolysis coupled to reverse HPLC. Part 2. Pectins, alginates, and xanthan. Food Hydrocolloids 14:19
56. Kyriakidis NB, Psoma E (2001) Hydrocolloid interferences in the determination of pectin by the carbazole method. Journal of AOAC International 84:1947
57. Jones JM (2013) Dietary fiber future directions: integrating new definitions and findings to inform nutrition research and communication. Advances in Nutrition 4:8–15
58. McCleary BV, Prosky L (eds) (2001) Advanced dietary fibre technology, Blackwell Science, London
59. Anon. (2001) The definition of dietary fiber. Cereal Foods World 46:112
60. Institute of Medicine, Food and Nutrition Board (2005) Dietary Reference Intakes: energy, carbohydrates, fiber, fat, fatty acids, cholesterol, protein and amino acids. National Academies Press, Washington, DC
61. Codex Alimentarius Commission, Food and Agriculture Organization, World Health Organization (2009) Report of the 30th session of the Codex Committee on Nutrition and Foods for Special Dietary Uses. http://www.codexalimentarius.net/download/report710/al132_26e.pdf
62. FDA (2021) Questions and answers on dietary fiber. https://www.fda.gov/food/food-labeling-nutrition/questions-and-answers-dietary-fiber
63. Jones JM (2014) CODEX-aligned dietary fiber definitions help to bridge the "Fiber gap". Nutrition Journal 13:34
64. McCleary BV, McLoughlin C (2022) Determination of insoluble, soluble, and total dietary fiber in foods using a rapid integrated procedure of enzymatic-gravimetric-liquid chromatography: First Action 2022.01. Journal of AOAC International 106:127–145
65. Juneja LR, Sakanaka S, Chu D-C (2001) Physiological and technological functions of partially hydrolysed guar gum (modified galactomannans). (Chap 30). In: McCleary BV, Prosky L (eds), Advanced dietary fibre technology. Blackwell Science, Oxford, UK
66. Austin S, Bhandari S, Cho F, Christiansen S, Cruijsen H, De GR, Deborde J-L, Ellingson D, Gill B, Haselberger P, et al. (2015) Fructans in infant formula and adult/pediatric nutritional formula. Journal of AOAC International 98:1038
67. AACC International (online) Approved Methods. AACC International, St. Paul, MN
68. BeMiller JN, Huber KC (2011) Starch. In: Elvers B (ed), Ullmann's Encyclopedia of Industrial Chemistry, 7th edn. Wiley-VCH, Weinheim
69. Mijland PJHC, Janssen AM, de Vries PAM (1999) Multivariate comparison of the Brabenderviscograph and the rapid visco analyzer using cross-linked starch. Starch-Staerke 51:33–39
70. Suh DS, Jane JL (2003) Comparison of starch pasting properties at various cooking conditions using the Micro Visco-Amylo-Graph and the Rapid Visco Analyser. Cereal Chemistry 80: 745–749
71. Balet S, Guelpa A, Fox G, Manley M (2019) Rapid Visco Analyser (RVA) as a tool for measuring starch-related physiochemical properties in cereals: a review. Food Analytical Methods 12:2344–2360
72. Ai Y, Jane J-L (2015) Gelatinization and rheological properties of starch. Starch/Staerke 67:213–224
73. Kim H-S, BeMiller JN (2012) Effects of hydrocolloids on the pasting and paste properties of commercial pea starch. Effects of hydrocolloids on the pasting and paste properties of commercial pea starch. Carbohydrate Polymers 88:1164–1171
74. Woodbury TJ, Grush E, Allan MC, Mauer LJ (2022) The effects of sugars and sugar alcohols on the pasting and granular swelling of wheat starch. Food Hydrocolloids 126: Article 107433
75. Crosbie GB, Chiu PC, Ross AS (2002) Shortened temperature program for application with a Rapid Visco Analyser in prediction of noodle quality in wheat. Cereal Chemistry 79:596–599
76. Higley JS, Love SL, Price WJ, Nelson JE, Huber KC (2003) The Rapid Visco Analyzer (RVA) as a tool for differentiating potato cultivars on the basis of flour pasting properties. American Journal of Potato Research 80:195–206
77. Biliaderis GB (2009) Structural transitions and related physical properties of starch (Chap 8). In BeMiller, JN, Whistler, R (eds), Starch: Chemistry and Technology, 3rd edn. Elsevier, New York
78. Kim H-S, Patel B, BeMiller JN (2013) Effects of amylose-amylopectin ratio on starch-hydrocolloid interactions. Carbohydrate Polymers 98:1438–1448
79. Geera BP, Nelson JE, Souza E, Huber KC (2006) Composition and properties of A- and B-type starch granules of wild-type, partial waxy, and waxy soft wheat. Cereal Chemistry 83:551–557
80. Seo T-R, Kim J-Y, Lim S-T (2015) Preparation and characterization of crystalline complexes between amylose and C18 fatty acids. LWT – Food Science and Technology 64:889–897
81. Allan MC, Mauer LJ (2022) Variable effects of twenty sugars and sugar alcohols on the retrogradation of wheat starch gels. Foods 11, 3008
82. Copikova J, Syntsya A, Cerna M, Kaasova J, Novotna M (2001) Application of FT-IR spectroscopy in detection of food hydrocolloids in confectionery jellies and food supplements. Czech Journal of Food Science 19:51
83. Chopin T, Whalen E (1993) A new and rapid method of carrageenan identification by FT IR diffuse reflectance spectroscopy directly on dried, ground algal material. Carbohydrate Research 246:51
84. Cerna M, Barros AS, Nunes A, Rocha SM, Delgadillo I, Copikova J, Coimbra MA (2003) Use of FT-IR spectroscopy as a tool for the analysis of polysaccharide food additives. Carbohydrate Polymers 51:383
85. Tojo E, Prado J (2003) Chemical composition of carrageenan blends determined by IR spectroscopy combined with a PLS multivariate calibration method. Carbohydrate Research 338:1309
86. Prado-Fernandez J, Rodriguez-Vazquez JA, Tojo E, Andrade JM (2003) Quantitation of κ-, ι-, and λ-carrageenans by mid-infrared spectroscopy and PLS regression. Analytica Chimica Acta 480:23
87. Monsoor MA, Kalapathy U, Proctor A (2001) Determination of polygalacturonic acid content in pectin extracts by diffuse reflectance Fourier transform infrared spectroscopy. Food Chemistry 74:233
88. Langkilde FW, Svantesson A (1995) Identification of celluloses with Fourier-transform (FT) mid-infrared, FT-Raman and near-infrared spectrometry. Journal of Pharmaceutical and Biomedical Analysis 13:409
89. Kays SE, Barton FE II (2002) Near-infrared analysis of soluble and insoluble dietary fiber fractions of cereal food products. Journal of Agricultural and Food Chemistry 50:3024
90. Mehrübeoglu M, Coté GL (1997) Determination of total reducing sugars in potato Samples using near-infrared spectroscopy. Cereal Foods World 42:409

Vitamin Analysis

20

Ronald B. Pegg

Contents

20.1 INTRODUCTION

20.1.1 Definition and Importance

Vitamins are defined as relatively low-molecular-weight organic compounds which humans, and for that matter, any living organism that depends on organic matter as a source of nutrients, require in small quantities for normal metabolism. With few exceptions, humans cannot synthesize most vitamins and therefore must obtain them from food and supplements. There are 13 vitamin activities identified, including the four fat-soluble vitamins [A, D, E, and K] and the nine water-soluble vitamins [B_1 (thiamine), B_2 (riboflavin), B_3 (niacin), B_5 (pantothenic acid), B_6 (pyridoxine), B_7 (biotin), B_9 (folate), B_{12} (cobalamin), and C (ascorbic acid)]. Insufficient levels of vitamins can result in lethal deficiency diseases [e.g., scurvy, beriberi, pellagra, and pernicious anemia, which are due to the lack of vitamins C, B_1, B_3, and B_{12}, respectively].

20.1.2 Importance of Analysis

Vitamin analysis of food and other biological samples has played a critical role in determining animal and human nutritional requirements. Furthermore, accurate food composition

R. B. Pegg (✉)
Department of Food Science and Technology, The University of Georgia, Athens, GA, USA
e-mail: rpegg@uga.edu

B. P. Ismail, S. S. Nielsen (eds.), *Nielsen's Food Analysis*, Food Science Text Series,
https://doi.org/10.1007/978-3-031-50643-7_20

information is required to determine dietary intakes to assess diet adequacy and improve human nutrition worldwide. From the consumer and industry points of view, reliable assay methods are required to ensure accuracy of food labeling. This chapter provides an overview of techniques for the analysis of the vitamin content of food.

20.1.3 Vitamin Units

When vitamins are expressed in units of mg or μg per tablet or food serving, it is very easy to grasp how much is present. Vitamins also can be expressed as **international units** (IU), **United States Pharmacopeia** (USP) **units**, and **% Daily Value** (DV). The IU is a unit of measurement for the amount of a substance, based on measured biological activity. Because each vitamin has its own level of biological efficacy, the IU measures the effectiveness rather than the mass. For details about IU and USP units of various vitamins, see the Vitamin Analysis chapter in the fourth edition of this textbook. For details about %DV for vitamins, see Chap. 3 in this textbook. When analysis of a foodstuff or dietary supplement is required for its content of vitamins, as might be the case for labeling and quality control purposes, being able to report the findings on different bases becomes important.

20.1.4 Extraction Methods

With the exception of some biological feeding studies, vitamin assays in most instances involve the extraction of a vitamin from its biological matrix prior to analysis. This generally includes one or several of the following treatments: **heat**, **acid**, **alkali**, **solvents**, and **enzymes**.

In general, extraction procedures are specific for each vitamin and designed to stabilize the vitamin. Some procedures are applicable to the combined extraction of more than one vitamin, for example, for thiamine and riboflavin as well as some of the fat-soluble vitamins [1, 2, 12]. Typical extraction procedures are as follows:

- **Ascorbic acid:** Cold extraction with meta-phosphoric acid/acetic acid.
- **Vitamin B_1 and B_2:** Boiling or autoclaving in acid plus enzyme treatment.
- **Niacin:** Autoclaving in acid (noncereal products) or alkali (cereal products).
- **Pantothenic acid:** Digestion with liver acetone powder from pigeon for the release ofthis vitamin from Coenzyme A or the acyl carrier protein.
- **Folate:** Enzyme extraction with α-amylase, protease, and γ-glutamyl hydrolase (conjugase).
- **Vitamins A, E, or D:** Organic solvent extraction, saponification, and re-extraction with organic solvents. For unstable vitamins such as these, antioxidants are routinely added to inhibit oxidation.

For fat-soluble vitamins, the initial extraction with a hydrophobic organic solvent removes all fat-soluble compounds from the food, including all of the triacylglycerols. The **saponification** step that follows (generally either overnight at room temperature or by refluxing at 70 °C, using an antioxidant that protects the sample from oxidation) renders liberated fatty acids from the triacylglycerols insoluble in an organic solvent (because they now exist as soap, typically as a potassium salt), but the fat-soluble vitamins remain soluble. These vitamins are then re-extracted with a hydrophobic organic solvent and concentrated as needed.

20.1.5 Overview of Methods

Vitamin assays can be classified as follows:

1. **Bioassays** involving humans and animals
2. **Microbiological assays** making use of protozoan organisms, bacteria, and yeast
3. **Chemical assays** that include spectrophotometric, fluorometric, chromatographic, enzymatic, immunological, and radiometric methods

In terms of ease of performance, but not necessarily with regard to accuracy and precision, the three systems follow the reverse order. It is for this reason that bioassays, on a routine basis at least, are very limited in their use to those instances in which no satisfactory alternative method is available.

The selection criteria for a particular assay depend on a number of factors, including accuracy and precision, but also economic factors and the sample load to be handled. Applicability of certain methods for a particular matrix also must be considered. It is important to bear in mind that many official methods presented by regulatory agencies are limited in their applicability to certain matrices, such as vitamin concentrates, milk, or cereals, and thus cannot be applied to other matrices without some procedural modifications, if at all.

Because of the sensitivity of certain vitamins to adverse conditions such as light, oxygen, pH, and heat, proper precautions need to be taken to prevent any deterioration throughout the analytical process, regardless of the type of assay employed. Such precautionary steps need to be followed with the test material in bioassays throughout the feeding period. They are required with microbiological and chemical methods during extraction as well as during the analytical procedure.

Just as with any type of analysis, proper sampling and subsampling as well as the preparation of a homogeneous sample are critical aspects of vitamin analysis. General guidelines regarding this matter are provided in Chap. 5 of this text.

The principles and procedures for select vitamin analysis methods are described in this chapter. Calculations for select vitamins are described with the Practice Problems (Sect. 20.8). Many of the methods cited are official methods of AOAC International [2], the British Standards Institution [3–10], or the US Pharmacopeial Convention [11]. Refer to these methods and other original references cited for detailed instructions on procedures. A summary of commonly used regulatory and other methods is provided in Table 20.1. Further details on standard reference materials (SRMs) and certified reference materials (CRMs) for the assays of the 13 vitamins are available [13]. The sections below on bioassay, microbiological, and chemical methods are not comprehensive but rather just give examples of each type of analysis.

Table 20.1 Commonly used regulatory methods for vitamin analysis

Vitamin	*Method designation*	*Application*	*Approach*
Fat-soluble vitamins			
Vitamin A (and precursors)			
Retinol	AOAC[a] 992.04	Vitamin A in milk and milk-based infant formula	HPLC[b] 340 nm
Retinol	AOAC 2001.13	Vitamin A in foods	HPLC 328 or 313 nm
All-*trans*-retinol	AOAC 2001.07	Vitamin A in infant formula and adult nutritionals	UHPLC[c] 326 nm
All-*trans*-retinol 13-*cis*-retinol	EN 1283-1 [3]	All foods	HPLC 325 nm or Fluorescence[d] E_x $\lambda = 325$ nm E_m $\lambda = 475$ nm
β-Carotene	AOAC 2005.07	β-Carotene in supplements and raw materials	HPLC 445 or 444 nm
β-Carotene	EN 1283-2 [3]	All foods	HPLC 450 nm
Vitamin D			
Cholecalciferol Ergocalciferol	AOAC 936.14	Vitamin D in foods	Bioassay
Cholecalciferol Ergocalciferol	AOAC 995.05	Vitamin D in infant formula and enteral products	HPLC 265 nm
Cholecalciferol Ergocalciferol	AOAC 2002.05	Vitamin D in selected foods	HPLC 265 nm
Cholecalciferol Ergocalciferol	AOAC 2011.11	Vitamin D in infant formula and adult/pediatric nutritional formula	UHPLC-MS/MS[e]
Cholecalciferol Ergocalciferol	AOAC 2012.11	Simultaneous determination of vitamins D_2 and D_3 in infant formula and adult/pediatric nutritional formula	ESI[f] LC-MS/MS
Cholecalciferol Ergocalciferol	EN 1282172 [5]	Vitamin D in foods	HPLC 265 nm
Vitamin E			
All-racemic α-Tocopherol	AOAC 2012.10	Simultaneous determination of vitamins E and A in infant formula and adult nutritionals	NP-HPLC[g] Fluorescence E_x $\lambda = 280$ nm E_m $\lambda = 310$ nm
α-Tocopherol	AOAC 2012.09	Vitamins A and E in infant formula and adult/pediatric nutritional formula	HPLC Fluorescence E_x $\lambda = 295$ nm E_m $\lambda = 330$ nm
R,R,R-Tocopherols	EN 12822 [6]	Vitamin E in foods	HPLC Fluorescence E_x $\lambda = 295$ nm E_m $\lambda = 330$ nm

(continued)

Table 20.1 (continued)

Vitamin	*Method designation*	*Application*	*Approach*
Vitamin K			
Phylloquinone	AOAC 999.15	Vitamin K in milk and infant formulas	HPLC postcolumn reduction Fluorescence E_x $\lambda = 243$ nm E_m $\lambda = 430$ nm
Phytonadione (K_1)	AOAC 2015.09 (50.2.08)	*Trans* and total (*cis* + *trans*) vitamin K1 in infant, pediatric, and adult nutritionals	NP-HPLC postcolumn reduction Fluorescence E_x $\lambda = 245$ nm E_m $\lambda = 440$ nm
Phylloquinone	EN 14148 [7]	Vitamin K in foods	HPLC postcolumn reduction Fluorescence E_x $\lambda = 243$ nm E_m $\lambda = 430$ nm
Water-soluble vitamins			
Ascorbic acid (vitamin C)			
Ascorbic acid	AOAC 967.21	Vitamin C in juices and vitamin preparations	2,6-Dichloroindophenol titration
Ascorbic acid	AOAC 967.22	Vitamin C in vitamin preparations	Fluorescence E_x $\lambda = 350$ nm E_m $\lambda = 430$ nm
Ascorbic acid	AOAC 2021.002	Vitamin C in all forms of infant formula, follow-up formula, baby foods, adult/pediatric formula, complementary food supplements, nutritional supplements for pregnant women and nursing mothers, and sports nutrition food	HPLC 254 nm or UHPLC 254 nm
Thiamine (vitamin B_1)			
Thiamine Thiamine•HCl	AOAC 942.23	Thiamine in foods	Thiochrome Fluorescence E_x $\lambda = 365$ nm E_m $\lambda = 435$ nm
Thiamine	AOAC 2015.14	Total vitamins B_1, B_2, B_3, and B_6 in infant formula and related nutritionals	Enzymatic digestion and UHPLC-MS/MS
Thiamine	EN 14122 [9]	Thiamine in foods	HPLC Thiochrome Fluorescence E_x $\lambda = 366$ nm E_m $\lambda = 420$ nm
Riboflavin (vitamin B_2)			
Riboflavin	AOAC 970.65	Riboflavin in foods and vitamin preparations	Fluorescence E_x $\lambda = 440$ nm E_m $\lambda = 565$ nm
Riboflavin	AOAC 2015.14	Total vitamins B_1, B_2, B_3, and B_6 in infant formula and related nutritionals	Enzymatic digestion and UHPLC-MS/MS
Riboflavin	EN 14152 [10]	Riboflavin in foods	HPLC Fluorescence E_x $\lambda = 468$ nm E_m $\lambda = 520$ nm
Niacin			
Nicotinic acid Nicotinamide	AOAC 944.13	Niacin and niacinamide in vitamin preparations	Microbiological
Nicotinic acid Nicotinamide	AOAC 985.34	Niacin and niacinamide in ready-to-feed milk-based infant formula	Microbiological
Nicotinic acid Nicotinamide	AOAC 2015.14	Total vitamins B_1, B_2, B_3, and B_6 in infant formula and related nutritionals	Enzymatic digestion and UHPLC-MS/MS
Vitamin B_6			

Table 20.1 (continued)

Vitamin	*Method designation*	*Application*	*Approach*
Pyridoxine Pyridoxal Pyridoxamine	AOAC 2004.07	Total vitamin B_6 in infant formula	HPLC Fluorescence E_x λ = 468 nm E_m λ = 520 nm
Pyridoxine Pyridoxal Pyridoxamine	AOAC 2015.14	Total vitamins B_1, B_2, B_3, and B_6 in infant formula and related nutritionals	Enzymatic digestion and UHPLC-MS/MS
Folic acid, folate			
Total folates	AOAC 2004.05	Total folates in cereals and cereal products – trienzyme procedure	Microbiological
Total folates	AOAC 2011.06	Total folates in infant formula and adult nutritionals	Trienzyme extraction and HPLC-MS/MS
Folic acid 5-methyl tetrahydrofolic acid	AOAC 2013.13	Folate in infant formula and adult/pediatric nutritional formula	Trienzyme extraction and UHPLC-MS/MS
Vitamin B_{12}			
Cyanocobalamin	AOAC 986.23	Cobalamin (vitamin B_{12}) in milk-based infant formula	Microbiological
Cyanocobalamin	AOAC 2011.10	Vitamin B_{12} in infant and pediatric formulas and adult nutritionals	HPLC 550 nm
Cyanocobalamin	AOAC 2014.02	Vitamin B_{12} in infant and pediatric formulas and adult nutritionals	UHPLC 361 nm
Biotin			
Biotin	USP29/NF24, Dietary Supplements Official Monograph [11]	Biotin in dietary supplements	HPLC 200 nm or microbiological
Pantothenic acid			
Calcium pantothenate	AOAC 992.07	Pantothenic acid in milk-based infant formula	Microbiological
Calcium pantothenate	AOAC 2012.16	Pantothenic acid (vitamin B5) in infant formula and adult/pediatric nutritional formula	UHPLC-MS/MS

[a] AOAC method [2]
[b] *HPLC* high-performance liquid chromatography (in some methods simply called liquid chromatography)
[c] *UHPLC* ultra-HPLC
[d] Fluorometric test, giving excitation (E_x) and emission (E_m) wavelengths
[e] *MS/MS* tandem mass spectrometry
[f] *ESI* electrospray ionization
[g] *NP* normal phase

20.2 BIOASSAY METHODS

Outside of vitamin bioavailability studies, bioassays at the present are used only for the analysis of **vitamins B_{12} and D**, and even for them, the bioassays have very limited use. For vitamin D, the bioassay reference standard method (AOAC Method 936.14) (specified for milk, vitamin preparations, and feed concentrates) is known as the **line test**, which is based on bone calcification. Rats are initially fed a diet that depletes rats of vitamin D and then groups of the rats are fed a diet with known (for standard curve) or unknown (sample) amounts of vitamin D. The rats are then sacrificed, and the sections of specific bones are stained to show the extent of bone calcification.

20.3 MICROBIOLOGICAL ASSAYS

20.3.1 Principle

The growth of microorganisms is proportional to their requirement for a specific vitamin, if all other nutritional needs of the microorganisms are met. Thus, in microbiological assays the growth of a certain microorganism in an extract of a vitamin-containing sample is compared against the growth of this microorganism in the presence of known quantities of that vitamin. Bacteria, yeast, or protozoans are used as test organisms. **Growth** can be measured in terms of **turbidity, acid production, gravimetry,** or by **respiration.** With bacteria and yeast, turbidimetry is the most commonly employed system. If turbidity measurements are involved, clear sample and standard extracts vs. turbid ones are essential. With regard to incubation time, turbidity measurement is also a less time-consuming method. The microorganisms are specified by ATCC™ numbers and are available from the *American Type Culture Collection* (ATCC™) (10801 University Blvd., Manassas, VA 20110).

20.3.2 Applications

Microbiological assays are limited to the analysis of water-soluble vitamins. The methods are very sensitive and specific for each vitamin. The methods are somewhat time-consuming, and strict adherence to the analytical protocol is critical for accurate results. All microbiological assays can use microtiter plates (96-well) in place of test tubes. Microplate usage results in significant savings in media and glassware, as well as labor.

20.3.3 Niacin

The microbiological analysis of niacin and nicotinamide, as an example of such an assay, is briefly described here (AOAC Method 944.13, 45.2.04) [2, 14]. *Lactobacillus plantarum* ATCC™ 8014 is the test organism. A stock culture needs to be prepared and maintained by inoculating the freeze-dried culture on Bacto *Lactobacilli* agar followed by incubation at 37 °C for 24 h prior to sample and standard inoculation. A second transfer may be advisable in the case of poor growth of the inoculum culture. The final inoculum is added to tubes of niacin assay medium that contain added known amounts of a USP niacin reference standard (for standard curve) and unknown amounts of niacin (food sample extract). The tubes are incubated at 37 °C for 16–24 h. The percent transmittance at a specific wavelength is measured to determine microbial growth as indicated by turbidity. Using *Lactobacillisp.* as the test organism, acidimetric measurements could be used instead of turbidity, but the required incubation time would be 72 h.

20.4 CHEMICAL METHODS

20.4.1 High-Performance Liquid Chromatography (HPLC) Methods

20.4.1.1 Overview

Because of their relative simplicity, accuracy, and precision, the chemical methods, in particular, the chromatographic methods using HPLC/UHPLC (ultra-HPLC) (see Chap. 13), are preferred for all vitamins overmicrobiological and bioassays. Numerous vitamins are now commonly measured by UHPLC (e.g., A, D, E, K, C, and various B vitamins), many as official methods and some unofficial. Liquid chromatography in combination with mass spectrometry (MS) (see Chap. 11) has added a new dimension to vitamin analysis, allowing for the determination of vitamers, which are different forms of a vitamin (sometimes just isomers) that possess activity of the vitamin in question. To estimate total vitamin activity, the respective vitamin activity of each of the vitamers must be understood and estimated, including assessments of bioaccessibility and bioavailablity. In general, LC-MS or electrospray ionization (ESI) LC-MS/MS (tandem mass spectrometry) methods are available for each fat- and water-soluble vitamin. Detection by MS leads to increased sensitivity as well as unequivocal identification and characterization of the vitamin. The LC-MS assays have become a mainstay of accurate, cost-effective vitamin analyses. For example, LC-MS is commonly employed for the verification of vitamin D content of products with difficult matrices (i.e., comparing results to those with standard LC analysis) (e.g., AOAC Method 2012.11, Simultaneous Determination of Vitamins D_2 and D_3 in Infant Formula and Adult/Pediatric Nutritional Formula) and LC-MS/MS for folate (AOAC Method 2013.13, Folate in Infant Formula and Adult/Pediatric Nutritional Formula by a UHPLC-MS/MS assay vs. the microbiological method).

Standard HPLC is commonly employed as an official method of analysis for vitamins A (e.g., AOAC Method 992.04, 50.1.02), E (e.g., AOAC Method 992.03, 50.1.04), and D (e.g., AOAC Method 2002.05, 45.1.22A), and as a quality control method for vitamin C. While HPLC/UHPLC involves a high capital outlay, it is applicable to most vitamins and lends itself in some instances to simultaneous analysis of several vitamins and/or vitamers. There is some concern about not getting information on the biological activities of vitamers using chemical methods. However, these vitamers can be recognized by microbiological methods. Implementation of multi-analyte procedures for the analysis of water-soluble vitamins can result in assay efficiency with savings in time and materials. To be useful, a simultaneous assay must not lead to loss of sensitivity, accuracy, and precision when compared to single-analyte methods. In general terms, multi-analyte methods for water-soluble vitamin assay of high-concentration products including pharmaceuticals, supplements, and vitamin premixes are quite easily developed. Though the applicability of HPLC has been demonstrated to a wide variety of biological matrices with no or only minor modifications in some cases, one must always bear in mind that all chromatographic techniques, including HPLC, are separation and not identification methods. Therefore, during adaptation of an existing

HPLC method to a new matrix, establishing evidence of peak identity and purity is an essential step of the method adaptation or development.

20.4.1.2 Vitamin A

Vitamin A or retinol refers to a group of closely related fat-soluble compounds present in our diet, namely all-*trans*-retinol, 13-*cis* retinol, retinyl esters, and retinaldehyde. Additionally, three carotenoids with provitamin A activity have been identified and include α-carotene, β-carotene, and β-cryptoxanthin, of which β-carotene is the predominant carotenoid in the human diet. Vitamin A is sensitive to ultraviolet (UV) light, air (and any pro-oxidants for that matter), high temperatures, and moisture. Therefore, steps must be taken to avoid any adverse changes in this vitamin due to such effects. Steps include the following: (1) using low actinic glassware, nitrogen, and/or vacuum, (2) avoiding excessively high temperatures, (3) working in subdued artificial light, and (4) adding pyrogallol as an antioxidant prior to saponification.

HPLC methods are considered the only acceptable methods to provide accurate food measurements of vitamin A activity. For example, in the HPLC method of vitamin A (i.e., retinol isomers) in milk and milk-based infant formula (AOAC Method 992.04, 50.1.02) [2], the test sample is saponified with ethanolic KOH, vitamin A (retinol) is extracted into organic solvent, and then concentrated. Vitamin A isomers – all-*trans*-retinol and 13-*cis*-retinol – levels are determined by HPLC on a silica column (i.e., normal phase). Vitamin A also can be analyzed using reversed-phase HPLC columns with photodiode array (PDA) or MS detection.

20.4.1.3 Vitamin D

Vitamin D-active compounds include vitamin D_3 (synthesized in the skin of vertebrates from 7-dehydrocholesterol by UVB light, also referred to as cholecalciferol), vitamin D_2 (synthesized from ergosterol in fungi, also referred to as ergocalciferol), 25-hydroxyvitamin D_2, and the corresponding metabolites of 1,25-dihydroxyvitamin D. Vitamin D is typically analyzed by HPLC with a UV/PDA detector (some version of AOAC Method 2002.05) and when needed for verification by LC-MS/MS. In the twenty-first century, however, LC-MS/MS methods are becoming more common place, especially in the vitamin analysis of meat, offal (i.e., variety meats), milk, infant formula, and adult nutritionals. Protection against oxidation is done as described for vitamin A above. For the HPLC-UV/PDA analysis, an internal standard (vitamin D_2) is added to the sample that is subjected to basic hydrolysis and then saponified in ethanolic KOH. This sample is extracted with heptane, and the heptane organic phase is evaporated to dryness. The reconstituted sample is subjected to a semi-preparative normal-phase HPLC column, from which the fractions are collected, concentrated, and diluted in acetonitrile-methanol. These samples are subjected to a reversed-phase HPLC column with UV detection to quantitate the D_3. A separate sample is tested in parallel to confirm the absence of endogenous D_2.

20.4.1.4 Vitamin E (Tocopherols and Tocotrienols)

Vitamin E is present in foods as eight different compounds: all are 6-hydroxychromans. The vitamin E family is comprised of α-, β-, γ-, and δ-tocopherol, characterized by a saturated side chain of three isoprenoid units and the corresponding unsaturated tocotrienols (α-, β-, γ-, and δ-). Like vitamins A and D, vitamin E must be protected from oxidation during sample preparation and is commonly analyzed by HPLC. Typically, a normal or reversed-phase HPLC column is connected to a fluorescence detector: $E_x\ \lambda = 290$ nm and $E_m\ \lambda = 330$ nm (E_x, excitation; E_m, emission) (for fluorescence spectroscopy, see Chap. 7, Sect. 7.3) [15]. An example chromatogram is depicted in Fig. 20.1. Vitamin E is quantitated by external standards from peak area by linear regression. If a fluorescence detector is not available on the HPLC system, saponification of the extracted lipids from the food sample in question is required to collect the unsaponifiable matter before HPLC-UV detection of the tocopherols can be performed.

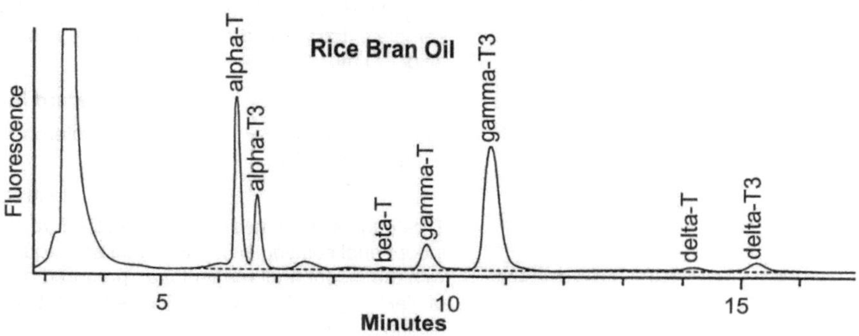

Fig. 20.1 Chromatogram of rice bran oil showing tocopherols and tocotrienols

20.4.2 Other Chemical Methods

20.4.2.1 Vitamin C

The vitamin (***L*-ascorbic acid** and ***L*-dehydroascorbic acid**) is very susceptible to oxidative deterioration, which is enhanced by high pH and the presence of ferric and cupric ions. For these reasons, the entire analytical procedure needs to be performed at low pH and, if necessary, in the presence of a chelating agent. Mild oxidation of ascorbic acid results in the formation of dehydroascorbic acid, which is also biologically active and is reconvertible to ascorbic acid by treatment with **reducing agents** such as β-mercaptoethanol and dithiothreitol. Two AOAC official methods for vitamin C are described below, but vitamin C also can be analyzed in infant formula and adult/pediatric nutritional formula by HPLC with UV detection (AOAC Method 2012.21) and UHPLC with UV detection (AOAC Method 2012.22).

20.4.2.1.1 2,6-Dichloroindophenol (DCIP) Titrimetric Method

This method is specified as an AOAC official method (AOAC Method 967.21, 45.1.14) [2] for vitamin preparation and juices (i.e., fruits) to monitor fortification, but it is sometimes used as a secondary method for other foods, because it is a more rapid assay than the microfluorometric method (described in Sec. 20.4.2.1.2) applicable to other foods. In the DCIP method, *L*-ascorbic acid is oxidized to *L*-dehydroascorbic acid by the oxidation–reduction indicator dye, DCIP. At the endpoint, excess unreduced dye appears rose pink in acid solution (see Figs. 20.2 and 20.3). With colored samples such as red beets or heavily browned products, the rose-pink endpoint is impossible to detect by the human eye. In such cases, it needs to be determined by observing the change of transmittance using a spectrophotometer with the wavelength set at $\lambda = 545$ nm.

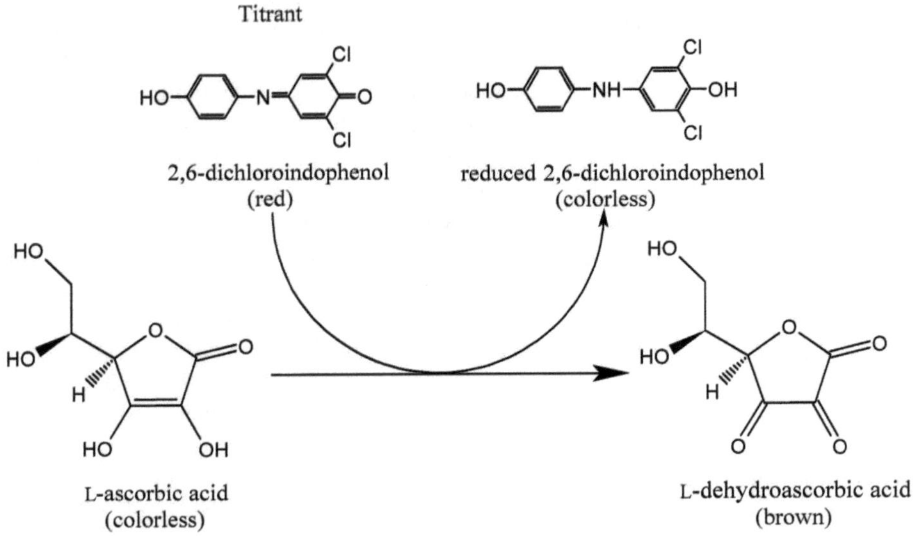

Fig. 20.2 Chemical reaction between L-ascorbic acid and the indicator dye, 2,6-dichloroindophenol

Fig. 20.3 Analysis of vitamin C by the 2,6-dichloroindophenol titration, AOAC Method 967.21, 45.1.14 (2). (Adapted from Pelletier [16])

VITAMIN C ASSAY PROCEDURE
2,6-DICHLOROINDOPHENOL (DCIP) TITRATION

Sample Preparation

Weigh and extract by homogenizing test sample in *meta*phosphoric acid-acetic acid solution (*i.e.*, 15 g of HPO_3 and 40 ml of HOAc in 500 ml of deionized H_2O). Filter (and/or centrifuge) sample extract, and dilute appropriately to a final concentration of 10-100 mg of ascorbic acid/100 ml.

Standard Preparation

Weigh 50 mg of USP L-ascorbic acid reference standard and dilute to 50 ml with HPO_3-HOAc extracting solution.

Titration

Titrate three replicates each of the standard (*i.e.*, to determine the concentration of the indophenol solution as mg ascorbic acid equivalents to 1.0 ml of reagent), test sample, and blank with the indophenol reagent (*i.e.*, prepared by dissolving 50 mg of DCIP sodium salt and 42 mg of $NaHCO_3$ to 200 ml with deionized H_2O) to a light but distinctive rose pink endpoint lasting ≥ 5 sec.

20.4.2.1.2 Microfluorometric Method

The vitamin C AOAC microfluorometric (AOAC Method 967.22, 45.1.15) assay is specified for vitamin preparations, but a semiautomated fluorometric AOAC method (AOAC Method 984.26, 45.1.16) is specified as applicable to all food products in the absence of erythorbate [2, 16]. Themicrofluorometric method measures both ascorbic acid and dehydroascorbic acid. All ascorbate forms are oxidized to dehydroascorbic acid (using a boric acid-sodium acetate solution), and then the dehydroascorbic acid is reacted with ***o*-phenylenediamine** to produce a **fluorescent quinoxaline compound.** The amount of fluorescence in the sample (compared to a standard and corrected with blanks) is used to quantitate the amount of vitamin C.

20.4.2.2 Thiamine (Vitamin B_1) Thiochrome Fluorometric Method

While thiamine can be quantitated by HPLC, it is still commonly analyzed by the longtime official thiochrome fluorometric procedure (AOAC Method 942.23) [2]. Following sample extraction with dilute acid, enzymatic hydrolysis of thiamine's phosphate esters, and chromatographic cleanup (i.e., purification), thiamine is oxidized to **thiochrome,** which is fluorescent. This method is based on the fluorescence measurement of thiochrome in the test solution compared to that from an oxidized thiamine standard solution.

20.4.2.3 Riboflavin (Vitamin B_2) Fluorometric Method

Like other B vitamins, riboflavin can be analyzed by HPLC, but its natural fluorescence allows for measurement based on this characteristic. Following sample extraction, cleanup, and compensation for the presence of interfering substances, riboflavin is determined fluorometrically, compared to a riboflavin standard (AOAC Method 970.65, 45.1.08) [2].

20.5 COMPARISON OF METHODS

Each type of method has its advantages and disadvantages. In selecting a certain method of analysis for a particular vitamin or vitamins, a number of factors need to be considered, some of which are listed below:

1. Method accuracy and precision
2. The need for bioavailability information
3. Time and instrumentation requirements
4. Personnel requirements
5. The type of biological matrix to be analyzed
6. The number of samples to be analyzed
7. Regulatory requirements – must official AOAC International methods be used?

At present, the applicability of microbiological assays is limited to water-soluble vitamins (most commonly B_3, B_5, and B_{12}). Though somewhat time-consuming, they generally can be used for the analysis of a relatively wide array of biological matrices without major modifications and can provide details about the importance of key vitamers of a vitamin being present or absent under the assay conditions. Furthermore, less sample preparation is often required compared to chemical assays; yet, with more and more official methods being developed for HPLC and UHPLC, the employment of these microbiological assays is expected to continue to decrease with time.

When selecting a system for analysis, at least initially, it is wise to consider the use of official methods that have been tested through interlaboratory studies and that are published by such organizations as AOAC International [2], the British Standards Institution [3–10], the US Pharmacopeial Convention [11], or the AACC International [17]. Again, one must realize that these methods are limited to certain biological matrices.

20.6 SUMMARY

Three types of methods for the analysis of vitamins – bioassays and microbiological and chemical assays – have been outlined in this chapter, with emphasis on the chemical methods. The methods are, in general, applicable to the analysis of more than one vitamin and several food matrices. However, the analytical procedures must be properly tailored to the analyte in question and the biological matrix to be analyzed; issues concerning sample preparation, extraction, and quantitative measurements are also involved. With the increase of vitamin premixes being used by the food industry for fortification, especially due to the rising consumption of functional foods, care is needed in vitamin analyses to ensure that over-fortification has not occurred. This is especially important when formula changes to different premixes have occurred. It is essential to validate any new application appropriately by assessing its accuracy and precision. Method validation is especially important with chromatographic methods such as UHPLC, because these methods basically accent separations rather than identification of compounds. For this reason, it is essential to ensure not only identity of these compounds but also, just as important, their purity.

20.7 STUDY QUESTIONS

1. What factors should be considered in selecting the assay for a particular vitamin?
2. To be quantitated by most methods, vitamins must be extracted from foods. What treatments are commonly used to extract the vitamins? For one fat-soluble vitamin and one water-soluble vitamin, give an appropriate extraction procedure.
3. What vitamin must be listed on the US standard nutrition label as of 2018 (see Chap. 3, Sect. 3.2.1.1), and what would be an official method for its analysis?
4. Identify the general steps in the procedure for extracting fat-soluble vitamins from food in preparation for analysis. Explain why each step is needed.
5. Due to the sensitive nature of vitamin A, what steps in sample preparation and storage are necessary to prevent any undesirable reactions?
6. Explain why it is possible to use microorganisms to quantitate a particular vitamin in a food product and describe such a procedure.
7. There are two commonly used AOAC methods to measure the vitamin C content of foods. Identify these two methods and then compare and contrast them with regard to the principles involved.
8. Would the vitamin C content as determined by the 2,6-dichloroindophenol method be underestimated or overestimated in the case of heat processed juice samples? Explain your answer.
9. What are the advantages and disadvantages of using HPLC for vitamin analysis?
10. Explain the similarity and difference between the chemical methods to measure thiamine and riboflavin? How is their analysis different from regular UV-visible spectroscopy? (See also Chap. 7, Sect. 7.3).

20.8 PRACTICE PROBLEMS

Please refer to the fourth edition of this Food Analysis textbook for practice problems.

Acknowledgments The author of this chapter thanks Ronald R. Eitenmiller (Professor Emeritus, Department of Food Science and Technology, the University of Georgia, Athens, GA, USA) for his contributions to this chapter in previous editions of the book.

REFERENCES

1. Eitenmiller RR, Ye L, Landen WO Jr (2008) Vitamin analysis for the health and food sciences, 2nd edn., p. 135. Taylor & Francis Group, CRC, Boca Raton, FL
2. AOAC International (2023) Official methods of analysis, 22nd edn., AOAC International, Oxford University Press, New York NY
3. British Standards Institution AW/275 (2014) Foodstuffs. Determination of vitamin A by high performance liquid chromatography. Part 1: Measurement of all-*E*-retinol and 13-*Z*-retinol, BS EN 12823–1:2014
4. British Standards Institution AW/275 (2000) Foodstuffs. Determination of vitamin A by high performance liquid chromatography. Part 2: Measurement of *β*-carotene, BS EN 12823-2:2000
5. British Standards Institution AW/275 (2009) Foodstuffs. Determination of vitamin D by high performance liquid chromatography. Measurement of cholecalciferol (D_3) and ergocalciferol (D_2), BS EN 12821:2009
6. British Standards Institution AW/275 (2014) Foodstuffs. Determination of vitamin E by high performance liquid chromatography. Measurement of α-, β-, γ- and δ-tocopherol, BS EN 12822:2014
7. British Standards Institution AW/275 (2003) Foodstuffs. Determination of vitamin K1 by HPLC, BS EN 14148:2003
8. British Standards Institution AW/275 (2014) Foodstuffs. Determination of vitamin B_1 by high performance liquid chromatography, BS EN 14122:2014
9. British Standards Institution AW/275 (2014) Foodstuffs. Determination of vitamin B_2 by high performance liquid chromatography, BS EN 14152:2014
10. British Standards Institution AW/275 (2014) Foodstuffs. Determination of vitamin B_6 by high performance chromatography, BS EN 14164:2014
11. United States Pharmacopeial Convention (2016) US Pharmacopoeia National Formulary, USP39–NF34, Nutritional Supplements, Official Monographs, United States Pharmacopoeial Convention, Rockville, MD
12. Blake CJ (2007) Analytical procedures for water-soluble vitamins in foods and dietary supplements: a review. Anal Bioanal Chem 389:63–76
13. Jakobsen J, Melse-Boonstra A, Rychlik M (2019) Challenges to quantify total vitamin activity: How to combine the contribution of diverse vitamers? Curr Dev Nutr 3(10):nzz086
14. Eitenmiller RR, Ye L, Landen WO Jr (2008) Niacin. In: Vitamin analysis for the health and food sciences, Second edition. CRC Press, Boca Raton, Florida pp. 361–400
15. Lee J, Ye L, Landen WO Jr, Eitenmiller RR (2000) Optimization of an extraction procedure for the quantification of Vitamin E in tomato and broccoli using response surface methodology. J Food Compos Anal 13:45–57
16. Pelletier O (1985) Vitamin C (L-ascorbic and dehydro-L-ascorbic acid). In: Augustin J, Klein BP, Becker DA, Venugopal PB (eds) Methods of vitamin assay, 4th edn. Wiley, New York, pp. 335–336
17. AACC International (2010) Approved methods of analysis, 11th edn. (On-line), AACC International, St. Paul, MN

Traditional Methods for Mineral Analysis

21

Robert E. Ward

Contents

21.1 INTRODUCTION

This chapter describes traditional methods for the analysis of minerals involving titrimetric and colorimetric procedures, ion-selective electrodes (ISEs), and instruments to measure salt content. Other traditional methods of mineral analysis include gravimetric titration (i.e., insoluble forms of minerals are precipitated, rinse, dried, and weighed) and redox reactions (i.e., mineral is part of an oxidation–reduction reaction, and product is quantitated). However, these latter two methods will not be covered because they currently are not used much in the food industry. The traditional methods that will be described have maintained widespread usage in the food industry despite the development of more modern instrumentation such as ion chromatography (Chap. 13), atomic absorption spectroscopy (AAS), and inductively coupled plasma-optical emission spectroscopy (ICP-OES) (Chap. 9). Traditional methods generally require chemicals and equipment that are routinely available in an analytical laboratory and are within the experience of most laboratory technicians. Additionally, traditional methods often form the basis for rapid analysis kits (e.g., AquaChek® for calcium and Quantab® for salt determination) and for automated benchtop analyzers for

R. E. Ward (✉)
Department of Nutrition and Food Sciences, Utah State University, Logan, UT, USA
e-mail: robert.ward@usu.edu

B. P. Ismail, S. S. Nielsen (eds.), *Nielsen's Food Analysis*, Food Science Text Series,
https://doi.org/10.1007/978-3-031-50643-7_21

salt content. In these cases, it is useful to understand the principles employed in the analysis. For additional examples of traditional methods, refer to references [1–5]. For analytical requirements for specific foods, see the *Official Methods of Analysis* of AOAC International [5].

21.1.1 Importance of Minerals in the Diet

Dietary **macrominerals** (calcium, phosphorus, sodium, potassium, magnesium, chlorine, and sulfur) are required at more than 100 mg per day by the adult [6]. Additional 10 minerals (iron, iodine, zinc, copper, chromium, manganese, molybdenum, fluoride, selenium, and silica) are required in milli- or microgram quantities per day and are referred to as **trace minerals. Ultra trace minerals**, including vanadium, tin, nickel, arsenic, and boron, are being investigated for possible biological function, but they currently do not have clearly defined biochemical roles. **Heavy metals** (lead, mercury, cadmium, and arsenic) have been documented to be toxic to the body and should, therefore, be avoided in the diet. Essential minerals such as fluoride and selenium also are known to be harmful if consumed in excessive quantities, even though they do have beneficial biochemical functions at proper dietary levels.

The Nutrition Labeling and Education Act of 1990 (NLEA) mandated labeling of **sodium**, **iron**, and **calcium** contents largely because of their important roles in controlling hypertension, preventing anemia, and impeding the development of osteoporosis, respectively (see Chap. 3, Fig. 3.1). The newest US food labeling regulations require the addition of **potassium** to the label due to its numerous health benefits (see Chap. 3). The content of these minerals in several foods are shown in Table 21.1. The content of other minerals may be included on the nutrition label at the producer's option, although this becomes mandatory if the mineral is the subject of a nutrient content claim on the label (Chap. 3, Sect. 3.2.3).

Table 21.1 Mineral content of selected foods

	mg/g (wet weight basis)			
Food Item	*Calcium*	*Iron*	*Sodium*	*Potassium*
Cereals, bread, and pasta				
Rice, brown, long grain, raw	9	1	5	250
Rice, white, long grain, regular, raw, enriched	28	4	5	115
Wheat flour, wholegrain	34	4	2	363
Wheat flour, white, all-purpose, unenriched	15	1	2	107
Pasta, dry, enriched	21	3	6	223
Dairy products				
Milk, whole, 3.25% milk fat with added vitamin D	113	<1	43	132
Butter, salted	24	<1	643	24
Cheese, cottage, low fat, 2% milk fat	111	<1	308	125
Fruits and vegetables				
Apples, raw, with skin	8	<1	1	107
Bananas, raw	5	<1	1	358
Raisins, seedless	50	2	11	749
Potatoes, raw, skin	30	3	10	413
Tomatoes, red, ripe, raw	10	<1	5	237
Meats, poultry, and fish				
Eggs, whole, raw, fresh	56	2	142	138
Fish fillet, battered or breaded, and fried	14	<1	561	251
Pork, fresh, leg (ham), whole, raw	6	1	55	369
Beef, chuck, arm pot roast, raw	16	2	62	290
Bologna, chicken, pork, beef	92	1	1120	313

From the US Department of Agriculture, Agricultural Research Service (2023) USDA National Nutrient Database for Standard Reference. Release 28
https://data.nal.usda.gov/dataset/usda-national-nutrient-database-standard-reference-legacy-release

21.1.2 Minerals in Food Processing

Minerals are of **nutritional** and **functional** importance and, for that reason, their levels need to be known and/or controlled. Some minerals are contained at high levels in natural foodstuffs. For example, milk is a good source of calcium, containing about 300 mg of calcium per 8-ounce cup. However, direct acid cottage cheese is very low in calcium because of the action of the acid, causing the calcium bound to the casein to be freed and consequently lost in the whey fraction. Similarly, a large portion of the phosphorus, zinc, manganese, chromium, and copper found in a grain kernel is lost when the bran layer is removed in the processing. The enrichment law for flour requires that iron be replaced in white flour to the level at which it occurred naturally in the wheat kernel before removal of the bran.

Fortification of some foods has allowed addition of minerals above levels ever expected naturally. Prepared breakfast cereals often are fortified with minerals such as calcium, iron, and zinc, formerly thought to be limited in the diet. Fortification of salt with iodine has almost eliminated goiter in the USA. In other cases, minerals may be added for functionality. Salt is added for flavor, to modify ionic strength that affects solubilization of protein and other food components, and as a preservative. This significantly increases the sodium content of products such as processed meats, pickles, and processed cheese. Phosphorus may be added as phosphates to increase water-holding capacity in meats and to change the texture of processed cheese. Calcium may be added to promote gelation of proteins and gums. Textural properties of

Table 21.2 Common water treatments and their rationales for use

	Treatments						
Quality parameter	*Filtration*	*Membrane filtration*	*Ion exchange*	*Chlorination/ozonization*	*UV radiation*	*pH adjustment*	*Activated carbon*
Solids	X	X					
Salts, including hardness		X	X				
pH correction							
Other chemical contaminants, e.g., organic residues	X	X	X				X
Bacteria		X		X	X		
Viruses		X		X	X		
Protozoa		X		X	X		
Algal bloom (toxin)							X

Adapted from European Hygenic Engineering and Design Group [11]

fruits and vegetables can be influenced by the "hardness" or "softness" of the water used during processing.

Water is an integral part of food processing, and **water quality** is a major factor to be considered in the food processing industry. Water is used for washing, rinsing, blanching, cooling, and as an ingredient in formulations. Microbiological safety of water used in food processing is very important. Also important, but generally not appreciated by the consuming public, is the mineral content of water used in food processing. Waters that contain excessive minerals can result in clouding of beverages.

Water quality standards in the USA are established by the Environmental Protection Agency as a result of the Safe Drinking Water Act (Chap. 2, Sect. 2.2.5.2), and local water suppliers must ensure regulated substances, such as heavy metals, are below maximum contaminant levels. Tests ensure that levels of total microorganisms, coliforms, inorganic contaminants, disinfection by-products (such as chlorine), and radioactive contaminants are below maximum contaminant levels. In addition, food processors should measure other water quality parameters such as pH, turbidity, hardness, heavy metals, iron, nitrates, and volatiles.

Water hardness is a measure of the dissolved calcium and magnesium salts in water and is commonly expressed in parts per million (ppm). Hardness of water is determined using the following scale: 0–60 ppm is soft, 60–120 ppm is moderately hard, 120–180 ppm is hard, and >180 ppm is very hard. Use of hard water can affect processing in several ways. For example, over time, calcium and magnesium salts may precipitate out of hard water and form scale in pipes and on other surfaces. In addition, hard water reduces the effectiveness of soaps and sanitizers. If the water source at a food production facility contains excess minerals, it can be softened using ion exchange resins. Table 21.2 lists other treatments that may be used to improve water quality for use in food production.

21.2 BASIC CONSIDERATIONS

21.2.1 Nature of Analyses

Mineral analysis is a valuable model for understanding the basic structure of analysis procedures to separate and measure. Separation of minerals from the food matrix is often specific, such as **complexometric titrations** (Sect. 21.3.1) or **precipitation titrations** (Sect. 21.3.2). In these cases of specific separation, nonspecific measurements such as volume of titrant are made and later converted to mass of mineral based on fundamental stoichiometric relationships. In other cases, separation of mineral involves nonspecific procedures such as **ashing** or **acid extraction**. These nonspecific separations require that a specific measurement be made as provided by **colorimetry** (Sect. 21.3.3), **ion-selective electrodes** (ISE) (Sect. 21.3.4), **atomic absorption spectroscopy** (AAS) (Chap. 9), or **inductively coupled plasma-optical emission spectroscopy** (ICP-OES) (Chap. 9).

Because determination mineral mass is the final objective of analysis, measures other than mass are considered to be surrogate, or stand-in, measures. **Surrogate measures** are converted into mineral mass via fundamental stoichiometric and physiochemical relationships or by empirical relationships. Empirical relationships are those associations that need be established by experimentation because they do not follow any well-established physiochemical relationship. An example of a surrogate measurement is the wavelength-specific absorbance of a chromogen–mineral complex (Sect. 21.3.3). It may be possible to convert absorbance into mass of mineral using the fundamental relationships defined by the molar absorptivity and stoichiometry of the chromogen–mineral complex. However, it is more commonly required that the absorbance vs. concentration relationship be empirically developed using a series of standards (i.e., a standard curve).

21.2.2 Sample Preparation

Some sample preparation is generally required for traditional methods of mineral analysis to ensure a well-mixed and representative sample and to make the sample ready for the procedure to follow. A major concern in mineral analysis is **contamination** during sample preparation. **Comminution** (e.g., grinding or chopping) and mixing using metallic instruments can add significant mineral to samples and, whenever possible, should be performed using nonmetallic instruments or instruments not composed of the sample mineral. For example, using an aluminum grinder is standard practice for comminution of meat samples undergoing iron analysis. **Glassware** used in sample preparation and analysis should be scrupulously cleaned using acid washes and triple rinsed in the purest water. The latter may necessitate installation of an **ultrapure water system** in the laboratory to further purify the general supply of distilled water.

Solvents, including water, can contain significant quantities of minerals. Therefore, all procedures involving mineral analysis require the use of the purest reagents available. In some cases, the cost of ultrapure reagents may be prohibitive. When this is the case, the alternative is to always work with a reagent blank. A **reagent blank** is a sample of reagents used in the sample analysis, quantitatively the same as used in the sample but without any of the material being analyzed. This reagent blank, representing the sum of the mineral contamination in the reagents, is then subtracted from the sample values to more accurately quantify the mineral.

A method such as near-infrared spectroscopy (Chap. 8, Sect. 8.4) allows for mineral estimation without destruction of the carbon matrix of carbohydrates, fats, protein, and vitamins that make up foods. However, traditional methods generally require that the minerals be freed from this organic matrix in some manner. Chapter 16 describes the various methods used to ash foods in preparation for the determination of specific mineral components of the food. In water samples, minerals may be determined without further preparation.

21.2.3 Interferences

Factors such as **pH**, **sample matrix**, **temperature**, and other **analytical conditions** and **reagents** can interfere with the ability of an analytical method to quantify a mineral. Often there are specific interfering substances that must be removed or suppressed for accurate analysis. Two of the more common approaches are to isolate the sample mineral or remove interfering minerals, using selective precipitations or ion exchange resins. Water may need to be boiled to remove carbonates that interfere with several traditional methods of mineral analysis.

If other interferences are suspected, it is a common practice to develop the standard curve using sample mineral dissolved in a background matrix containing interfering elements known to be in the food sample. For example, if a food sample is to be analyzed for calcium content, a **background matrix solution** of the known levels of sodium, potassium, magnesium, and phosphorus should be used to prepare the calcium standards for developing the standard curve. In this manner, the standard curve more closely represents the analysis response to the sample mineral when analyzing a food sample. Alternatively, the standard curve can be developed using a series of sample mineral spikes added to the food sample. A **spike** is a small volume of a concentrated standard that is added to the sample. The volume is small enough so as to not appreciably change the overall composition of the sample, except for the mineral of interest. Thus, measurements of both the standards and the sample are made in the presence of the same background. If the spikes are added before implementation of the analysis protocol, possible effects of incomplete extractions, sample mineral degradation, and other losses are integrated into the standard curve.

21.3 METHODS

21.3.1 EDTA Complexometric Titration

21.3.1.1 Background Information

The hexadentate ligand **ethylenediaminetetraacetate** (EDTA) forms stable 1:1 complexes with numerous mineral ions. This gives complexometric titration using EDTA broad application in mineral analysis. Stability of mineral–EDTA complexes generally increases with valence of the ion, although there is significant variation among ions of similar valence due to their coordination chemistry. Endpoints are detected using mineral chelators that have coordination constants lower than EDTA (i.e., indicator has less affinity for mineral ions than does EDTA) and that produce different colors in each of their complexed and free states. **Calmagite** and **Eriochrome Black T** (EBT) are such indicators that are pink when complexed with calcium or magnesium, but blue when no metal ions are complexed. The endpoint of a complexometric EDTA titration using either Calmagite or EBT as the indicator is detected when the color changes from pink to blue.

The pH affects a complexometric EDTA titration in several ways and must be controlled for best performance. The complexation equilibrium is strongly pH-dependent. With decreasing pH the chelating sites of EDTA become protonated, thereby decreasing its effective concentration. The pH must be 10 or more for calcium or magnesium to form stable complexes with EDTA. Also, the sharpness of the endpoint increases with increasing pH. However, magnesium and calcium precipitate as their hydroxides at pH 12, so titration pH should probably be no more than 11 to ensure their solubility. Considering all factors, EDTA complexometric titration of calcium and magnesium is specified at pH 10 ± 0.1 using an ammonia buffer [7].

21.3.1.2 Principle

Based on the background information above, a summary of the principle of the EDTA complexometric titration method follows (described for a sample containing just calcium): the EDTA in the titrant solution complexes with calcium, 1:1, at pH 10. The EDTA binds to calcium ions stronger than the indicator binds to the calcium. When all the calcium present in the sample has reacted with the EDTA titrant (i.e., there is no calcium left to react with the indicator), the indicator changes in color from pink to blue. The moles of calcium in the sample are equivalent to the moles of EDTA used to do the titration.

21.3.1.3 Procedure: Hardness of Water Using EDTA Titration

Water hardness is determined by EDTA complexometric titration of the total of calcium and magnesium, in the presence of Calmagite, and expressed as the equivalents of calcium carbonate (mg/L) (*Standard Methods for the Examination of Water and Wastewater*, Method 2340, Hardness) [7]. The calcium–Calmagite complex is not stable, and calcium alone cannot be titrated using the Calmagite indicator. However, Calmagite becomes an effective indicator for calcium titration if the buffer solution contains a small amount of neutral magnesium salt and enough EDTA to bind all magnesium. Upon mixing sample into the buffer solution, calcium in the sample replaces the magnesium bound to EDTA. The free magnesium binds to Calmagite, and the pink magnesium–Calmagite complex persists until all calcium in the sample has been titrated with EDTA. The first excess of EDTA removes magnesium from Calmagite and produces a blue endpoint.

21.3.1.4 Applications

The major application of EDTA complexometric titration is testing calcium plus magnesium as an indicator of water hardness [7]. However, EDTA complexometric titration is suitable for determining calcium in the ash of fruits and vegetables (AOAC Method 968.31) [5] and other foods that have calcium without appreciable magnesium or phosphorus.

Several of the methods for mineral analysis described in this chapter have been incorporated into portable test strips that can be used to test samples in real time. For example, the water hardness application of the EDTA complexometric titration is made easy using test strips impregnated with Calmagite and EDTA (e.g., **AquaChek**, Environmental Test Systems, Inc., a HACH Company, Elkhart, IN). The strips are dipped into the water to test for total hardness caused by calcium and magnesium. The calcium displaces the magnesium bound to EDTA, and the released magnesium binds to Calmagite, causing the test strip to change color. The color is compared to a reference standard to estimate the calcium concentration.

21.3.2 Precipitation Titration

21.3.2.1 Principles

When at least one product of a titration reaction is an insoluble precipitate, it is referred to as **precipitation titrimetry**. Major factors that have limited the usefulness and accuracy of these methods have been the long time necessary for complete precipitation, failure of the reaction to yield a single product of definite composition, and lack of an endpoint indicator for the reaction. Nonetheless, precipitation titration has resulted in at least two methods that are used widely in the food industry today: Mohr titration and Volhard titration.

The **Mohr method** for chloride determination is a direct or **forward titration** method, in which the chloride (in NaCl) is titrated with silver nitrate in the presence of potassium chromate (Fig. 21.1a). The arrow in the figure shows the titrant addition. Initially, the silver reacts with the chloride and the solution is cloudy white. When all the chloride is reacted, the excess silver reacts with chromate to form an orange-colored solid, silver chromate. The volume of silver nitrate titrant and the molarity of that solution are used to calculate the amount of chloride ions, which can be used to calculate the amount of NaCl:

$$Ag^{+} + Cl^{-} \rightarrow AgCl\left(\text{until all } Cl^{-} \text{ is complexed}\right) \quad (21.1)$$

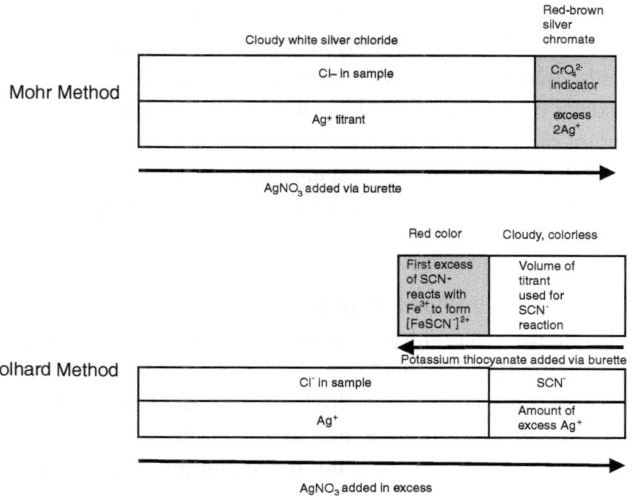

Fig. 21.1 Comparison of the Mohr (*forward*) and Volhard (*backward*) chloride titration methods

$$2Ag^+ + CrO_4^{2-} \rightarrow Ag_2CrO_4 \\ (\text{orange only after } Cl^- \text{ is all complexed}) \quad (21.2)$$

The **Volhard method** is an indirect or **back titration** method in which an excess of a standard solution of silver nitrate is added to a chloride-containing sample solution which is shown by the bottom arrow in Fig. 21.1b. The excess silver is then back-titrated (top arrow) using a standardized solution of potassium or ammonium thiocyanate, with ferric ion as an indicator. The volume of thiocyanate solution used in the titration is proportional to the amount of excess silver that did not react with the chloride in the sample. The moles of total silver are equal to the sum of the moles of chloride in the sample and moles of thiocyanate in the titrant:

$$Ag^+ + Cl^- \rightarrow AgCl(\text{until all } Cl^- \text{ is complexed}) \quad (21.3)$$

$$Ag^+ + SCN^- \rightarrow AgSCN \\ (\text{to quantitate silver not complexed with chloride}) \quad (21.4)$$

$$SCN^- + Fe^3 \rightarrow [FeSCN]^{2+} \\ (\text{red when there is any } SCN^- \text{ not complexed to } Ag^+) \quad (21.5)$$

21.3.2.2 Procedures

21.3.2.2.1 *Mohr Titration*

Salt in foods may be estimated by titrating the chloride ion with silver. For example, to measure the salt content of butter (AOAC Method 960.29), a butter sample is solubilized in boiling water, potassium chromate is added as the indicator, and then the solution is titrated with a standardized solution of silver nitrate. The orange endpoint in this reaction occurs only when all chloride ion is complexed, resulting in an excess of silver to form the colored silver chromate. The endpoint of this reaction is therefore at the first hint of an orange color. When preparing reagents for this assay, boiled water must be used to avoid interferences from carbonates in the water.

21.3.2.2.2 *Volhard Titration*

To measure the chloride content of cheese by the Volhard method (AOAC 935.43), excess silver nitrate is added to a cheese sample that is boiled in nitric acid solution. After cooling and filtering, the excess silver nitrate is titrated with potassium thiocyanate, using ferric aluminum as the indicator. Once chloride is determined by titration, the chloride weight is multiplied by 1.648 to obtain salt weight (based on molecular weight of Cl vs. NaCl). As in the Mohr titration method, water must be boiled to minimize errors due to interfering carbonates.

21.3.2.3 Applications

Precipitation titration methods are well suited for any foods that may be high in chlorides. Because of added salt in processed cheeses and meats, these products should certainly be considered for using this method to detect chloride; then salt content is estimated by calculation. A second example of a traditional mineral analysis method that has been adapted to a test strip format is the **Quantab® chloride titration** used in AOAC Method 971.19. This method is an adaptation of the principles involved in the Mohr titration method. This test strip adaptation allows for very rapid quantitation of salt in food products and is accurate to ±10% over a range of 0.3–10% NaCl in food products.

21.3.3 Colorimetric Methods

21.3.3.1 Principles

Chromogens are chemicals that, upon reaction with the compound of interest, form a colored product. Chromogens are available that selectively react with a wide variety of minerals. Each chromogen reacts with its corresponding mineral to produce a soluble colored product that can be quantified by absorption of light at a specified wavelength. The relationship between concentration and absorbance is given by **Beer's law** as detailed in Chap. 7. Generally, the concentration of a specific mineral in a sample is determined from a standard curve developed during the analysis, although in some cases it is possible to directly calculate concentration based on molar absorptivity of the chromogen–mineral complex.

Samples generally must be ashed or treated in some other manner to isolate and/or release the minerals from organic complexes that would otherwise inhibit their reactivity with the chromogen. The mineral of interest must be solubilized from dry ash and then subsequently handled in a manner that prevents its precipitation. The soluble mineral may need to be treated (e.g., reduced or oxidized) to ensure that all mineral is in a form that reacts with the chromogen [2]. Ideally, the chromogen reacts rapidly to produce a stable product, but if not, absorbance is read at a specific time. As with all mineral analysis of food, special efforts must be put in place to avoid contamination during sampling and analysis.

21.3.3.2 Procedures: Determination of Phosphorus in Milk

The total phosphorus content of foods can be quantified spectrophotometrically using modification of the Murphy–Riley method [8]. In this analysis, food-derived phosphorus reacts with ammonium molybdate at low pH to form phosphomolybdic acid. This product is subsequently reduced by ascorbic acid to a blue color, and antimony potassium tartrate is added to facilitate the reduction. The blue-colored phosphomolybdate complex formed has maximum absorptivity at 880 nm, a wavelength in the infrared range. However, this complex absorbs sufficient radiation at 700 nm, in the visible range, to be measured by most spectrophotometers. A requirement of the assay is that phosphorous must be soluble, and therefore solid food samples must first be ashed. To conduct the measurement, the ashed sample is solubilized in strong acid and then mixed with the Murphy–Riley reagent producing a colored product. A standard curve is prepared along with the samples with known amounts of phosphorus, and a regression equation is used to determine the concentration of phosphorus in the sample.

21.3.3.3 Applications

Colorimetry is used for the detection and quantification of a wide variety of minerals in food, and it is often a viable alternative to atomic absorption spectroscopy and other mineral detection methods. Colorimetric methods generally are very specific and usually can be performed in the presence of other minerals, thereby avoiding extensive separation to isolate the mineral of interest. The assays are particularly robust and often immune to matrix effects that can limit the usefulness of other methods for mineral analysis.

21.3.4 Ion-Selective Electrodes

21.3.4.1 Background Information

Many electrodes have been developed for the selective measurement of various cations and anions, such as bromide, calcium, chloride, fluoride, potassium, sodium, and sulfide [9, 10]. The pH electrode described in Chap. 22 is a specific example of an **ISE**. While the sensor of a pH electrode is specific to hydrogen ions, other sensors available are specific to individual mineral ions. For any ISE, an **ion-selective sensor** is placed such that it acts as a "bridging electrode" between two reference electrodes carefully designed to produce a constant and reproducible **potential**. In this manner, the potential within the sensor remains constant, while potentials develop at the sensor surfaces according to the **Nernst equation** (Chap. 22, Sect. 22.3.2.1), dependent on sample **ion activity** in the solutions contacting each surface. **Ion concentration** is generally substituted for ion activity, which is a reasonable approximation at low concentrations and controlled ionic strength environments. Indeed, this is observed within limitations set by electrode and instrumental capabilities (Fig. 21.2).

21.3.4.2 Principle

The principle for using ISE to measure a specific mineral is the same as for measuring pH (i.e., uses the Nernst equation), but the composition of glass in the sensing electrode is made to be specific for the element of interest. When sensing and reference electrodes (often as a combination electrode) are immersed in the sample solution, the electrical potential that develops at the surface of the sensing electrode is measured by comparing it to the reference electrode. The voltage that develops between the two electrodes is related to the ion activity for the element of interest, measured as mV. Ion activity is related to ion concentration via the activity coefficient, which is controlled by ionic strength of the sample. The concentration of the element of interest is determined using a standard curve of mV vs. log concentration.

21.3.4.3 General Methodology

For ISE analysis, one simply attaches an ISE for the sample ion to a **pH meter** set on the **mV scale** and follows instructions for determination. Detailed information regarding the performance of specific ISEs is available from vendor catalogs. Typical ISEs likely to be employed for analysis of foods operate in the range of 1–10^{-6} M (Fig. 21.2), although the electrode response may be distinctly nonlinear at the lower concentrations. Because the ISE responds to ionic activity, it is important that the **activity coefficient** be kept constant in samples and calibration standards. The activity

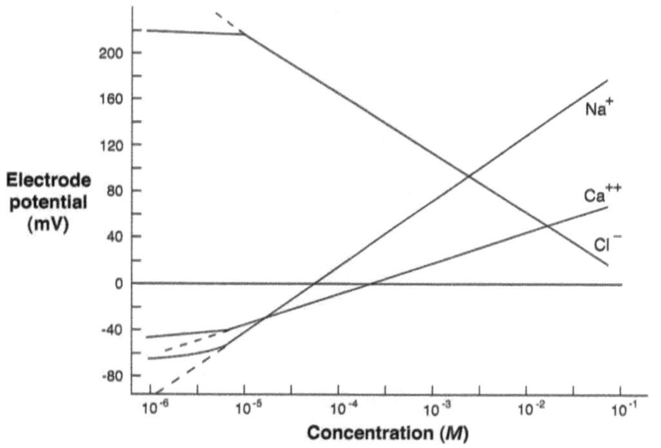

Fig. 21.2 Examples of ion-selective electrode calibration curves for ions important in foods. (Courtesy of Van London pHoenix Company, Houston, TX)

coefficient (γ) is used to relate ion activity (A) to ion concentration (C) ($A = \gamma C$). Activity coefficient is a function of ionic strength, so **ionic strength adjustment** (ISA) **buffers** are used to adjust the samples and standards to the same ionic strength. These ISA buffers are commercially available. The use of ISA buffers also adjusts the pH, which may be necessary if H^+ or OH^- activities affect the ion-specific sensor or if they interact with the analyte. In the case of metals having insoluble hydroxides, it is necessary to work at a pH that prevents their precipitation. Depending on the selectivity of the ISE, it may be necessary to remove interfering ions from the sample by selective precipitation or complexation.

In view of temperature effects on standard potentials and slopes of electrodes, it is important to keep the electrode and solutions at a constant temperature. This may involve working in a room that is thermostatically controlled to 25 °C (one of the internationally accepted temperatures for electrochemical measurements) and allowing sufficient time for all samples and standards to equilibrate to this temperature. Solutions should be gently stirred during the measurement to attain rapid equilibrium and to minimize concentration gradients at the sensor surface. Finally, it is important to allow sufficient time for the electrode to stabilize before taking a reading. ISEs may not completely stabilize within a practical time frame, so a decision needs to be made on when to take the reading. The reading may be taken when the rate of change has fallen below some predetermined value or at a fixed time after the electrode was placed in solution. A problem with the latter is that as samples are changed, many ISEs respond more rapidly to an increase in concentration of sample ion as compared to a decrease in concentration of sample ion.

21.3.4.4 Electrode Calibration and Determination of Concentration

In using an ISE, ion concentration can be determined using either a calibration curve or endpoint titration. It is common practice to develop a **calibration curve** when working with an ISE because it allows a large number of samples to be measured rapidly. The electrode potential (volts) is developed in a series of solutions of known concentration and plotted against the standard concentrations on a log scale. Examples of calibration curves for various ions are given in Fig. 21.2. Upon analysis of a test sample, the observed electrode potential is used to determine ion concentration by referring to the calibration curve. Note the nonlinear region of the curve at the lowest concentrations. Total ionic strength and the concentration of interfering ions are especially important factors limiting selective detection of low levels of ions.

ISEs can be used to detect the endpoints of potentiometric titrations by measuring the change in potential as the titration proceeds. One common use is for endpoint detection in the Mohr chloride assay previously described. In potentiometric silver chloride (Mohr) titrations, the ISE used for endpoint detection can either be specific for the titrant species (Ag^+) or the sample ion (Cl^-). If an ISE is used for Ag^+, a T-type titration curve results from the large increase in titrant activity detected at the equivalence point (see Fig. 21.3 for a cation titrant). Sample concentration is calculated from titrant volume to reach the equivalence point, taking into account the stoichiometric relationship between titrant species and sample ion.

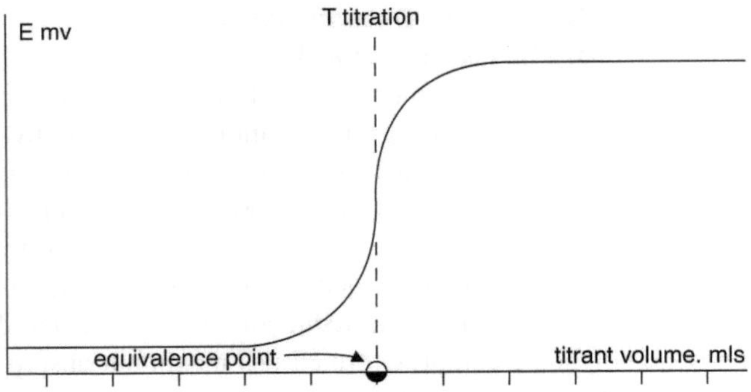

Fig. 21.3 A typical T-type titration. (From Kirk and Sawyer [2], used with permission)

21.3.4.5 Applications

Some examples of applications of ISEs are measuring salt and nitrate in processed meats, salt content of butter and cheese, calcium in milk, sodium in low-sodium ice cream, carbon dioxide in soft drinks, potassium and sodium levels in wine, and nitrate in canned vegetables. An ISE method applicable to foods containing <100 mg sodium/100 g is an official method of AOAC International (Method 976.25). This method employs a sodium combination ISE, pH meter, and magnetic stirrer. Several official methods exist for using ISEs to detect the endpoint in chloride and salt measurement, such as AOAC methods 971.27 (sodium chloride in canned vegetables), 976.18 (salt in seafood), and 986.26 (chloride in milk-based infant formula).

21.4 BENCHTOP RAPID ANALYZERS FOR SALT

There are several rapid methods of salt determination used in food analysis that are not necessarily based on the traditional methods described above in this chapter. Nonetheless, it is useful for food science students to be aware of the methods and to compare the principles to the other methods previously described.

Salt concentration can be measured by either hydrometry or refractometry (Chap. 15, Sects. 15.2.5.2 and 15.2.5.3). A

salometer is a hydrometer that measures brine strength using a scale that indicates the percent of salt saturation. A **salinity refractometer** can be either a manual or automatic device that gives a measurement analogous to that of sugar concentration, or brix, using a typical refractometer. An advantage of the automatic devices, over the less expensive manual models, is that they can correct for changes in the refractive index of salt solutions as a function of temperature.

Several instrument manufacturers sell devices that can be used to determine the salt content of a sample in under a minute using automatic titration. In these instruments, both the titrant dosing and the endpoint detection are controlled by the instrument which increases the precision over manual titrations where there is subjective error in both determining the titrant dose as well as the endpoint color change. These instruments work on the basic principle of precipitation titrations, as silver ions are still used to precipitate chloride ions, but there are a few differentiating features. For example, some instruments couple automated silver nitrate addition with endpoint determination using an ion-sensitive electrode. Several manufacturers offer models using this technique including Mettler-Toledo (Columbus, OH), Hanna Instruments (Woonsocket, RI), and Metrohm (Herisau, Switzerland). A second type of automatic chloride titrator does not rely on using a silver nitrate titrant but rather generates silver ions in the sample using the principle of coulometry. In instruments manufactured by Nelson-Jameson (Marshfield, WI), Sherwood Scientific (Cambridge, United Kingdom), and DKK-TOA (Toyko, Japan), silver ions are generated from a silver wire in situ via an electric current, and they complex with chloride ions in solution. The potential of the solution is monitored via an ISE, and the reaction is complete when the potential shifts dramatically as in Fig. 21.3. The electric current used to generate silver ions then may be converted to chloride and subsequently salt concentration. One advantage of this method is that no silver nitrate titrant is needed, and thus a standardization step is not necessary.

Another common method of salt determination in commercial laboratories is via conductance (e.g., DiCromat Salt Analyzer, Arrow Scientific, Gladesville, Australia). Salts in solution dissociate into negative and positive ions and thus increase conductance in proportion to their concentration. Conductance-based salt analyzers are rapid and also correct for differences in conductivity as a function of temperature.

21.5 COMPARISON OF METHODS

For labeling, processing, and even practical nutrition, the focus is on only a few minerals that generally can be analyzed by traditional methods. The traditional methods available for mineral analysis are varied, and only a very limited number of examples have been given in this chapter. Choice of methods for mineral analysis must be made considering method performance regarding accuracy, sensitivity, detection limit, specificity, and interferences. Information on method performance is available from the collaborative studies referenced with AOAC official methods. Other factors to be considered include cost per analysis completed, equipment availability, and analytical time compared to analytical volume.

Generally, for a small laboratory with skilled analytical personnel, the traditional methods can be carried out rapidly, with accuracy, and at minimal costs. If a large number of samples of a specific element are to be run, there is certainly a time factor in favor of using atomic absorption spectroscopy or emission spectroscopy, depending on the mineral being analyzed. Atomic absorption spectrophotometers with a graphite furnace or inductively coupled plasma-mass spectrometers are capable of sensitivity in the parts per billion range, which is much lower than the limits of traditional methods. Yet, instrumentation capable of such analysis is expensive and beyond the financial resources of many quality assurance laboratories, and often this degree of sensitivity is not required. This leaves the options of sending samples out to certified laboratories for analysis or utilizing one of the more traditional methods for analysis.

21.6 SUMMARY

The mineral content of water and foodstuffs is important because of their nutritional value, toxicological potential, and interactive effects with processing and texture of some foods. Traditional methods for mineral analysis include titrimetric and colorimetric procedures. The basic principles of these methods are described in this chapter, along with discussion of ISE methodology that has general application for mineral analysis, and some benchtop analyzers for salt content. Table 21.3 gives a summary of many methods covered in this chapter, with comparison to AAS and ICP-OES (Chap. 9).

Procedures are described in this chapter that illustrate use of these traditional methods to quantify minerals of concern in the food industry. These procedures generally require chemicals and equipment routinely available in an analytical laboratory and do not require expensive instrumentation. These methods may be suited to a small laboratory with skilled analytical personnel and a limited number of samples to be analyzed. The traditional procedures will often perform similarly to procedures requiring more instrumentation and may be more robust in actual practice.

Foods are typically ashed prior to traditional analyses because the methods generally require that the minerals be freed from the organic matrix of the foods. Sample prepara-

Table 21.3 Summary of mineral analysis methods

Methods	*Principle*	*Advantages*	*Disadvantages*	*Applications*	*AOAC method number*
EDTA titration method	Description for sample containing just calcium: EDTA of titrant solution complexes with calcium, 1:1, at pH 10. EDTA binds to calcium ions stronger than the indicator binds to calcium. When all the calcium present in the sample has reacted with the EDTA titrant, and no calcium is bound to the indicator, the indicator changes form pink to blue. Moles of calcium in the sample are equivalent to moles of EDTA used in titration.	Rapid, especially when done with test strip. Inexpensive	Can be interfering compounds. Endpoint is subjectively determined	Testing water hardness. Calcium in ash of fruits and vegetables	AOAC Method 968.31, calcium in canned vegetables
Mohr titration	A forward titration. Chloride (in NaCl) is titrated with silver nitrate in the presence of potassium chromate. Silver reacts with the chloride. When all chloride is reacted, the excess silver reacts with chromate to form an orange-colored solid, silver chromate. Volume and molarity of silver nitrate are used to calculate amount of chloride, which relates to amount of NaCl.	Does not require expensive equipment, highly trained personnel, or ashing of sample. Fewer reagents and less time-consuming than Volhard titration. Inexpensive (unless automated equipment)	Subjectivity of determining endpoint of titration (if manual titration)	Salt content of variety of foods. Test strip version and automated instruments available	AOAC Method 960.29, salt in dairy products and butter
Volhard titration	A backward titration. An excess of a standard solution of silver nitrate is added to a chloride-containing solution. Excess silver nitrate is back-titrated with a standardized solution of potassium or ammonium thiocyanate, with ferric ion as an indicator. Volume of thiocyanate solution used is proportional to the excess silver. Moles of total silver is equal to the sum of the moles of chloride in the samples and moles of thiocyanate in the titrant.	Does not require expensive equipment, highly trained personnel, or ashing of sample. Rapid. Inexpensive	Subjectivity of determining endpoint of titration. Requires more reagents and time than Mohr titration	Salt content of a variety of foods	AOAC Method 935.43 chloride (Total) in cheese; AOAC Method 915.01, chloride in plants
Colorimetric methods	Description for just mineral analysis: Chromogen in reagent reacts with mineral of interest to form a soluble colored compound that can be quantitated by absorption of light at a specific wavelength. Concentration of mineral of interest is determined from standard curve of absorbance vs. concentration, based on Beer's law.	Applicable to wide variety of minerals. Very specific. Other minerals usually do not interfere. Less expensive than AAS for ICP-OES, but similar accuracy and precision	Requires significant technician time	Low-cost method for analyzing single element	AOAC Method 944.02, iron in wheat flour
Ion-selective electrode	Principle is the same as for measuring pH (i.e., uses Nernst equation), but by varying he composition of the glass in the sensing electrode, the electrode can be sensitive to a specific mineral. Sensing and reference (often as combination electrode) are immersed in solution with element of interest. Electrical potential that develops at surface of sensing electrode is measured by comparing to reference electrode with fixed potential. Voltage between sensing and reference electrode is related to ion activity, measured in mV. Ion activity is related to ion concentration via the activity coefficient, which is controlled by ionic strength. Concentration of element is determined using standard curve of mV vs log concentration.	Can measure many anions and cations directly. Does not require expensive equipment (only pH meter), highly trained personnel, or ashing of sample. Analysis is independent of turbidity, color, or viscosity	Cannot measure at low concentration. Electrode response can be slow. Sensing and reference electrode must be specific to element being measured. High rate of premature failure for some electrodes	Quality control, especially for Na or K	AOAC Method 976.25, sodium in foods for special dietary use

Table 21.3 (continued)

Methods	*Principle*	*Advantages*	*Disadvantages*	*Applications*	*AOAC method number*
Atomic absorbance spectroscopy (AAS)	Heat energy from a flame converts molecules to atoms and then energy [at specific wavelength, from a hollow cathode lamp (HCL)] raises atoms of a specific element from ground to excited state. Measure absorption of energy, which is linearly related to concentration of element.	Can measure many elements (i.e., not just sodium). More sensitive than Mohr or Volhard titration for sodium	Expensive instrument. Requires highly trained personnel. Requires ashing of most types of samples. Requires a different HCL for each element. Uses potentially explosive fuel gas. More interferences than ICP-OES. Response is not linear over a broad concentration range	Single-element analysis in a given food or biological sample	AOAC Method 975.03, metals in plants and pet foods. AOAC Method 985.35, minerals in infant formula, enteral products, and pet foods
Inductively coupled plasma-optical emission spectrometry (ICP-OES)	Energy from plasma converts molecules to atoms and ions and then plasma energy raises atoms from ground to excited state. Measure emission of energy (at specific wavelength) as excited atoms return to a lower energy state. Amount of energy emitted is proportional to concentration of specific element.	Can measure many elements (i.e., not just sodium) simultaneously. More sensitive than Mohr or Volhard titration for sodium. Few interferences. Response linear over a large concentration range	Expensive instrument. Requires highly trained personnel. Requires ashing of most types of samples	Multiple element analysis in a large number of samples. Nutrition labeling	AOAC Method 984.27, calcium, copper, iron, magnesium manganese, phosphorus, potassium, sodium, and zinc in infant formula. AOAC 985.01, metals in plants and pet foods

tion and analysis must include steps necessary to prevent contamination or loss of volatile elements, and they must deal with a variety of potential interferences. Various approaches are described to account for these possible errors, including use of reagent blanks, addition of spikes, and development of standard curves using appropriate mineral matrix background.

Traditional methods for mineral analysis are often automated or adapted to test kits for rapid analysis. Tests for water hardness and the Quantab® for salt determination are examples currently being used. The basic principles involved in traditional methods will continue to be utilized to develop inexpensive rapid methods for screening mineral content of foods and beverages. Familiarity with the traditional principles will allow the food analyst to obtain the best possible performance with the kits and adapt to problems that may be encountered.

21.7 STUDY QUESTIONS

1. What is the major concern in sample preparation for specific mineral analysis? How can this concern be addressed?
2. If the ammonia buffer is pH 11.5 rather than pH 10 in the EDTA complexometric titration to determine the hardness of water, would you expect to overestimate or underestimate the hardness? Explain your answer.
3. In a back titration procedure, would overshooting the endpoint in the titration cause an over- or underestimation of the compound being quantified? Explain your answer.
4. How would the use of "lite salt" (i.e., half NaCl and half KCl, to reduce the sodium content) in a food affect the determination of "salt" content (i.e., NaCl content) of that sample if the salt determination is done with a silver nitrate titration method? Explain your answer, including the principle of the titration method.
5. Describe how and why to employ standards in background matrix, spikes, and reagent blanks.
6. Explain the principles of using an ISE to measure the concentration of a particular inorganic element in food. List the factors to control, consider, or eliminate for an accurate measure of concentration by the ISE method.
7. You have decided to purchase an ISE to monitor the sodium content of foods produced by your plant. List the advantages this would have over the Mohr/Volhard titration method. List the problems and disadvantages of ISE that you should anticipate.

8. What factors should be considered in selecting a specific method for mineral analysis for a food product?

21.8 PRACTICE PROBLEMS

1. If a given sample of food yields 0.750 g of silver chloride in a gravimetric analysis, what weight of chloride is present?
2. A 10-g food sample was dried, then ashed, and analyzed for salt (NaCl) content by the Mohr titration method ($AgNO_3 + Cl \rightarrow AgCl$). The weight of the dried sample was 2 g, and the ashed sample weight was 0.5 g. The entire ashed sample was titrated using a standardized $AgNO_3$ solution. It took 6.5 mL of the $AgNO_3$ solution to reach the endpoint, as indicated by the red color of Ag_2CO_4 when K_2CrO_4 was used as an indicator. The $AgNO_3$ solution was standardized using 300 mg of dried KCl. The corrected volume of $AgNO_3$ solution used in the titration was 40.9 mL. Calculate the salt (NaCl) content of the original food sample as percent NaCl (wt/wt).
3. A 25-g food sample was dried, then ashed, and finally analyzed for salt (NaCl) content by the Volhard titration method. The weight of the dried sample was 5 g, and the ashed sample weighed 1 g. Then 30 mL of 0.1 *N* $AgNO_3$ was added to the ashed sample, the resultant precipitate was filtered out, and a small amount of ferric ammonium sulfate was added to the filtrate. The filtrate was then titrated with 3 mL of 0.1 *N* KSCN to a red endpoint.
   (a) What was the moisture content of the sample, expressed as percent H_2O (wt/wt)?
   (b) What was the ash content of the sample, expressed as percent ash (wt/wt) on a dry weight basis?
   (c) What was the salt content of the original sample in terms of percent (wt/wt) NaCl (molecular weight Na = 23; molecular weight Cl = 35.5)?
4. Compound X in a food sample was quantified by a colorimetric assay. Use the following information and Beer's law to calculate the content of Compound X in the food sample, in terms of mg Compound X/100 g sample:
   (a) A 4-g sample was ashed.
   (b) Ashed sample was dissolved with 1 mL of acid and the volume brought to 250 mL.
   (c) A 0.75-mL aliquot was used in a reaction in which the total volume of the sample to be read in the spectrophotometer was 50 mL.
   (d) Absorbance at 595 nm for the sample was 0.543.
   (e) The absorptivity constant for the reaction (i.e., extinction coefficient) was known to be 1574 L g^{-1} cm^{-1}.
   (f) Inside diameter of the cuvette for spectrophotometer was 1 cm.
5. Colorimetric analysis
   (a) You are using a colorimetric method to determine the concentration of Compound A in your liquid food sample. This method allows a sample volume of 5 mL. This volume must be held constant but can be comprised of diluted standard solution and water. For this standard curve, you need standards that contain 0, 0.25, 0.50, 0.75, and 1.0 mg of Compound A. Your stock standard solution contains 5 g/L of Compound A.

   Devise a dilution scheme(s) for preparing the samples for this standard curve that could be followed by a lab technician. Be specific. In preparing the dilution scheme, use no volumes less than 0.5 mL.
   (b) You obtain the following absorbance values for your standard curve:

Sample(mg)	Absorbance (500 nm)
0.00	0.00
0.25	0.20
0.50	0.40
0.75	0.60
1.00	0.80

   Construct a standard curve and determine the equation of the line.
   (c) A 5-mL sample is diluted to 500 mL, and 3 mL of this solution is analyzed as per the standard samples. The absorbance is 0.50 units at 500 nm. Use the equation of the line calculated in part (b) and information about the dilutions to calculate what the concentration is of Compound A in your original sample in terms of g/L.
6. What is the original concentration of copper in a 100-mL sample that shows a potential change of 6 mV after the addition of 1 mL of 0.1 *M* $Cu(NO_3)_2$?

Answers

1. $$\frac{x\,\mathrm{g\,Cl}}{0.750\,\mathrm{g\,AgCl}} = \frac{35.45\,\mathrm{g/mol}}{143.3\,\mathrm{g/mol}}$$
$$x = 0.186\,\mathrm{g\,Cl}$$

2. $$N\ \mathrm{AgNO_3} = \frac{0.300\,\mathrm{g\,KCl}}{\mathrm{mL\,AgNO_3} \times 74.555\,\mathrm{g\,KCl/mol}}$$

$$0.0984\ N = \frac{0.300\,\mathrm{g}}{40.9\,\mathrm{mL} \times 74.555}$$

$$\text{Percent salt} = \left(\frac{0.0065\ \mathrm{L} \times 0.0984\,N\ \mathrm{AgNO_3} \times 58.5\,\mathrm{g/mol}}{10\,\mathrm{g}}\right) \times 100$$

Percent salt = 0.37%

3. (a) $\dfrac{25\,\text{g wet sample} - 5\,\text{g dry sample}}{25\,\text{g wet sample}} \times 100 = 80\%$

(b) $\dfrac{1\,\text{g ash}}{5\,\text{g dry sample}} \times 100 = 20\%$

(c) mol Ag added = mol Cl^- in sample + mol SCN^- added

$$\text{mol Ag} = (0.1\,\text{mol/L}) \times (0.03\,\text{L}) = 0.003\,\text{mol}$$
$$\text{mol}\,SCN^- = (0.1\,\text{mol/L}) \times (0.003\,\text{L}) - 0.0003\,\text{mol}$$
$$0.003\,\text{mol Ag} = \text{mol}\,Cl^- + 0.0003\,\text{mol}\,SCN^-$$
$$0.0027\,\text{mol} = \text{mol}\,Cl^-$$
$$(0.0027\,\text{mol}\,Cl^-) \times \frac{58.5\,\text{g NaCl}}{\text{mol}} = 0.1580\,\text{g NaCl}$$
$$\frac{0.1580\,\text{g NaCl}}{25\,\text{g wet sample}} = \frac{0.00632\,\text{g NaCl}}{\text{g wet sample}} \times 100$$
$$= 0.63\%\,\text{NaCl}\,(\text{w/w})$$

4. A = *abc*

$0.543 = (1574\ \text{L g}^{-1}\ \text{cm}^{-1})\ (1\ \text{cm})\ c$

$c = 3.4498 \times 10^{-4}$ g/L

$c = 3.4498 \times 10^{-4}$ mg/mL

$$\frac{3.4498 \times 10^{-4}\,\text{mg}}{\text{ml}} \times 50\,\text{ml} = 1.725 \times 10^{-2}\,\text{mg}$$
$$\frac{1.725 \times 10^{-2}\,\text{mg}}{0.75\,\text{ml}} \times \frac{250\,\text{ml}}{4\,\text{g}} = 1.437\,\text{mg/g}$$
$$= 143.7\,\text{mg}/100\text{g}$$

5. (a) Lowest dilution volume for 1 mL of stock:

$$\frac{0.25\,\text{mg}}{0.5\,\text{ml}} = \frac{1\,\text{ml}}{x\,\text{ml}} \times \frac{5\,\text{mg}}{\text{ml}}$$

x = 10 mL

Therefore, use a volumetric pipette to add 1 mL of stock to a 10-mL volumetric flask. Bring to volume with ddH_2O to give a diluted stock solution of 0.25 mg/0.5 mL. Use this to make up standards according to the following table.

mg A/5 mL	mL diluted stock solution	mL H_2O
0	0	5.0
0.25	0.5	4.5
0.50	1.0	4.0
0.75	1.5	3.5
1.0	2.0	3.0

(b)

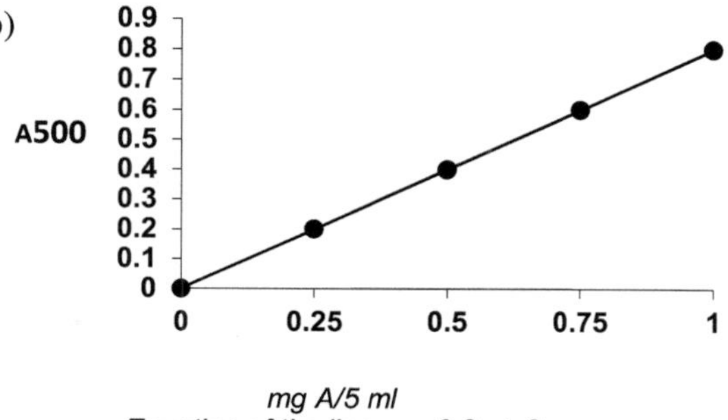

Equation of the line: y = 0.8x + 0

(c) $A_{500} = 0.50 = y$

$$0.50 = 0.8x + 0$$
$$x = 0.625$$
$$\frac{0.625\,\text{mg}}{5\,\text{ml}} \times \frac{5\,\text{ml}}{3\,\text{ml}} \times \frac{500\,\text{ml}}{5\,\text{ml}} = 20.8\ \text{mg/ml}$$
$$= 20.8\,\text{g/L}$$

6. $$C_O = \frac{0.001\,\text{L} \times \dfrac{0.001\,\text{moles}}{\text{L}} \times \dfrac{1}{0.100\,\text{L}}}{\left(10^{0.006/0.0285} - 1\right)} = 1.6\,\text{mM}$$

Acknowledgments The authors of this chapter thank Jerrad F. Legako (currently in the Department of Animal and Food Sciences, Texas Tech University, Lubbock, TX, USA) for his contribution in the chapter in the previous edition of the book.

REFERENCES

1. Schwendt G (1997) The essential guide to analytical chemistry, Wiley, New York
2. Kirk RS, Sawyer R (1991)Pearson's composition and analysis of foods, 9th edn. Longman Scientific and Technical, Essex, England
3. Skoog DA, West DM, Holler JF, Crouch SR (2022) Fundamentals of analytical chemistry,10th edn., Cengage, Boston, MA
4. Harris DC, Kucy, CA (2019) Quantitative chemical analysis, 10th edn. W.H. Freeman and Co., New York
5. AOAC International (2023) Official methods of analysis, 22ndedn., (On-line). AOAC International, Rockville, MD
6. Food and Nutrition Board, Institute of Medicine, National Academy of Sciences (2006)Dietary reference intakes: the essential guide for nutrient requirements. National Academies Press, Washington, DC
7. Eaton AD, Clesceri LS, Rice EW, Greenburg AE (eds) (2022) Standard methods for the examination of water and wastewater, 24thedn. Method 2340, hardness. American Public Health Association (APHA), Washington, DC.
8. Murphy J, Riley JP (1962) A modified single solution method for the determination of phosphate in natural waters. Anal Chim Acta 27:31–36
9. Mikhelson, KN (2013) Ion-selective electrodes, Springer Berlin, Heidelberg, Germany
10. Wang J (2023) Analytical electrochemistry, 4thedn. Wiley, New York
11. European Hygienic Engineering and Design Group (2007) Safe and hygienic water treatment in food factories. Trends in Food Sci Tech 18:S93–S100

Part V

Chemical Characterization and Associated Assays

pH and Titratable Acidity

22

Catrin Tyl

Contents

22.1 INTRODUCTION

There are two interrelated concepts in food analysis that relate to the presence of acids and bases: **pH** and **titratable acidity.** Each of these quantities is analytically determined in separate ways, and each provides its own particular insights on food quality. Both are regularly used when analyzing a wide range of foods, such as fruits, juices, fermented products, or coffee. This is tied to their relationship with food quality as well as safety. Microbial growth, enzymatic activity, and protein solubility are some of the main parameters affected by the pH of food. Sensory properties can be influenced by the pH and titratable acidity. However, titratable acidity is a better predictor of an acid's impact on flavor than pH and is often measured as an indicator of fruit ripening or to monitor the fermentation progress in dairy products.

This chapter will start with a review of terminology and definitions related to concentration measurements and acid–base reactions (Sect. 22.2) and then present information related to measuring pH (Sect. 22.3) and titratable acidity (Sect. 22.4). The text focuses on practical aspects.

For further general and food-specific information on measuring pH and titratable acidity, see Refs. [1–14]. For the pH and titratable acidity of select foods, see Ref. [8].

C. Tyl (✉)
Faculty of Chemistry, Biotechnology and Food Science, Norwegian University of Life Sciences, Ås, Norway
e-mail: catrin.tyl@nmbu.no

B. P. Ismail, S. S. Nielsen (eds.), *Nielsen's Food Analysis*, Food Science Text Series,
https://doi.org/10.1007/978-3-031-50643-7_22

22.2 CALCULATION AND CONVERSION FOR NEUTRALIZATION REACTIONS

22.2.1 Concentration Units

Before discussing and performing calculations related to acid–base reactions and pH, it can be helpful to review terminology and units that apply to concentrations and dilutions. A brief summary is given here; for a more detailed overview, see chapters in the food analysis laboratory manual that accompanies this textbook (Refs. [15–18]) (in Ref. [15], see Table 2.1; also see [16]).

Molarity (*M*) and normality (*N*) are the most common SI (International Scientific) terms used in food analysis, but concentrations also can be expressed as percentages, and analysts should be able to convert between units.

Molarity (*M*) is a concentration unit representing the number of moles of the solute per liter of solution. Molar concentrations are commonly signified by putting the value in square brackets. **Normality** (*N*) is a concentration unit representing the number of equivalents (Eq) of a solute per liter of solution. In acid and base solutions, normality represents the concentration or moles of H^+ or OH^- per liter that will be exchanged in a neutralization reaction when taken to completion. Normality also can be used for oxidation–reduction reagents, where it represents the concentration or moles of electrons per liter to be exchanged when the reaction is taken to completion. Conceptually, to convert from molarity to normality (and vice versa), consider the number of exchanged species per molecule, e.g., how many H^+ one acid molecule yields, or how many electrons one molecule can donate or accept. For most organic acids in food systems, the number of H^+ that can be donated at pH values relevant for foods equals the number of carboxyl groups (RCOOH). However, there are also organic acids in foods for which the H^+ is donated from a different group, such as in ascorbic acid. For chemical structures of acids, see Fig. 22.1. The following are some examples of molarity vs. normality (equivalents):

Acid–Base Reaction

1 *M* H_2SO_4 = 2 *N* sulfuric acid

Two equivalents of H^+ per mol of acid

1 *M* NaOH = 1 *N* sodium hydroxide

One equivalent of OH^- per mol of base

1 *M* CH_3COOH = 1 *N* acetic acid

One equivalent of H^+ per mol of acid

1 *M* HOOC-CH_2-CH(OH)-COOH = 2 *N* malic acid

Two equivalents of H^+ per mol of acid

Oxidation–Reduction Reaction

$HSO_3^- + I_2 + H_2O \rightleftarrows SO_4^{2-} + 2I^- + 3H^+$

1 *M* I_2 = 2 *N* iodine

Two equivalents of electrons gained per mol of I_2

1 *M* HSO_3^- = 2 *N* bisulfite

Two equivalents of electrons lost per mol of bisulfite

Many analytical determinations in food analysis use the concept of equivalents to measure the normality or equivalency of an unknown solution. Such tests use stoichiometry or the whole number ratios of atoms that react to produce a particular product. In **acid–base reactions,** hydrogen ions are exchanged and can be quantified through stoichiometric neutralization with a standard base. Examples for food analyses that use acid–base titration include nitrogen in the Kjeldahl protein determination (see Chap. 18, Sect. 18.2.1.3.4), benzoic acid in sodas, and percent titratable acidity determination.

Equivalent weight can be defined as the molecular weight divided by the number of equivalents in the reactions. For example, the molecular weight of H_2SO_4 is 98.08 g. Since there are two equivalents per mole of H_2SO_4, the equivalent weight of H_2SO_4 is 49.04 g. Table 22.1 provides a list of molecular and equivalent weights for acids important in food analysis. In working with normality and milliliters, the term **milliequivalents** (mEq) is usually preferred. Milliequivalent weight is the equivalent weight divided by 1000.

Percentage concentrations are the mass amount of solute or analyte per 100 mL or 100 g of material. Unlike molarity, normality, and equivalents, percent concentrations neither require nor provide insight into the stoichiometry of a reaction. Percentage can be expressed as volume of solution or for weight of solids and can be on a volume basis (v) or mass basis (w). Therefore, the particular basis is normally stated. For example, 10% (w/v), 10% (v/v), and 10% (w/w) express different concentrations in all cases for which the specific gravities of the two components are not the same. When the percentage becomes a number less than 1%, the terms **parts per million** (ppm), **parts per billion** (ppb), and even **parts per trillion** (ppt) usually are preferred. If percentage is defined as the mass of the solute or analyte per mass (or volume) of sample × 100, then ppm is simply the same ratio of mass of solute per mass of sample × 1000,000 (in Ref. [15], see Table 2.2).

22.2.2 Equation for Neutralization and Dilution

There are some general rules in evaluating equilibrium reactions that are helpful in most situations. At full neutralization, the equivalents of one reactant in the neutralization equal the equivalents of the other reactant. This can be expressed mathematically as:

$$(\text{vol of } X)(N \text{ of } X) = (\text{vol of } Y)(N \text{ of } Y) \qquad (22.1)$$

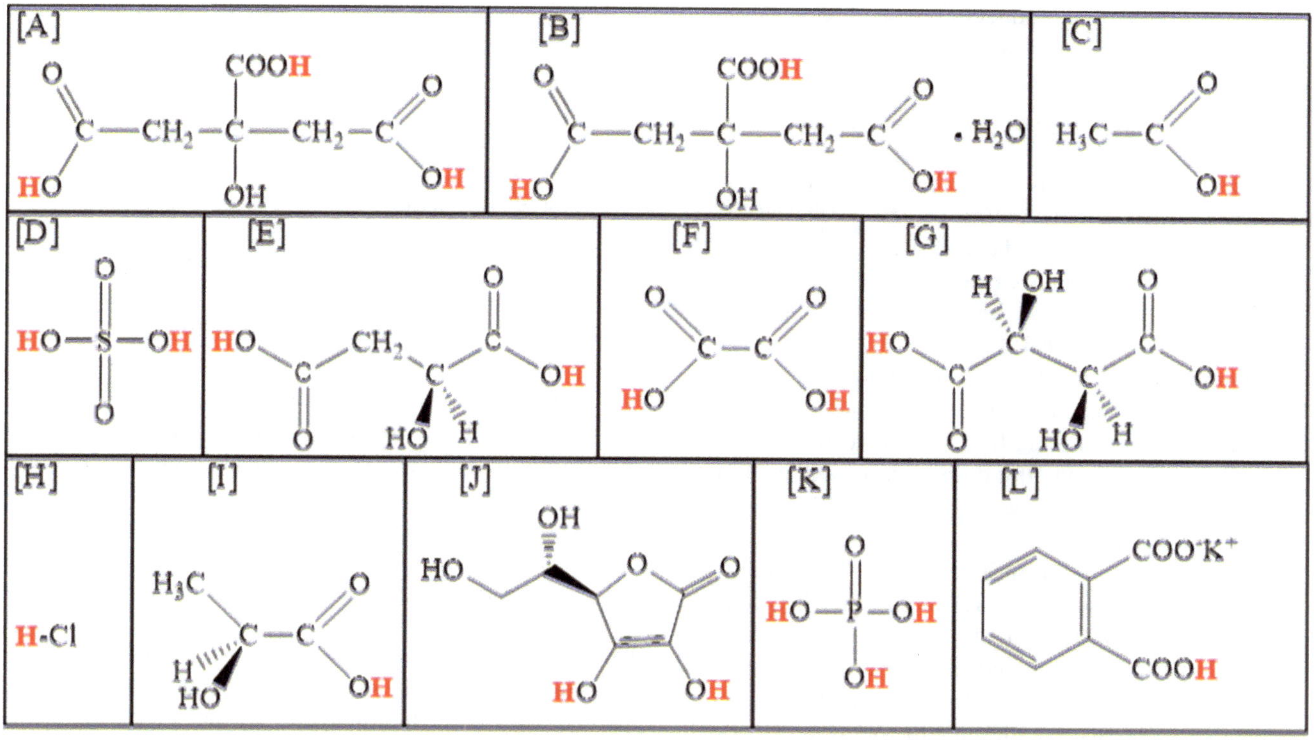

Fig. 22.1 Structures of acids listed in Table 22.1. H indicated in bold red font contribute to the equivalents per mole. Lactic, malic, and tartaric acid contain asymmetric carbon atoms and can thus occur in the form of stereoisomers (shown are isomers with stereocenters in S-configuration). However, this does not affect the properties listed in Table 22.1

Table 22.1 Molecular and equivalent weights of common food acids. Letters in square brackets next to an acid's name correspond to the position of the acid's structure in Fig. 22.1

Acid	*Molecular formula*	*Molecular weight*	*Equivalents per mole*	*Equivalent weight*
Citric (anhydrous) [A]	$C_6H_8O_7$	192.12	3	64.04
Citric (monohydrate) [B]	$C_6H_8O_7 \bullet H_2O$	210.14	3	70.05
Acetic [C]	$C_2H_4O_2$	60.06	1	60.05
Sulfuric [D]	H_2SO_4	98.08	2	49.04
Malic [E]	$C_4H_6O_5$	134.09	2	67.05
Oxalic [F]	$H_2C_2O_4$	90.04	2	45.02
Tartaric [G]	$C_4H_6O_6$	150.09	2	75.05
Hydrochloric [H]	HCl	36.47	1	36.47
Lactic [I]	$C_3H_6O_3$	90.08	1	90.08
Ascorbic [J]	$C_6H_8O_6$	176.12	2	88.06
Phosphoric [K]	H_3PO_4	98.00	3	32.67
Potassium acid phthalate [L]	$C_8H_5O_4K$	204.22	1	204.22

Equation 22.1 also can be used to solve dilution problems where X represents the stock solution and Y represents the working solution. When Eq. 22.1 is used for dilution problems, any value of concentration (grams, moles, ppm, etc.) can be substituted for N. Units should be recorded with each number, since cancelation of units provides a quick check on proper setup of the problem (See Ref. [15], Practice Problems 1–11).

22.3 PH

22.3.1 Acid-Base Equilibria

The Brønsted–Lowry theory of neutralization is based on the following definitions for acid and base:

Acid: A substance capable of donating hydrogen ions (H^+).
Base: A substance capable of accepting hydrogen ions.

Neutralization is the reaction of an acid with a base to form a salt, as shown below:

$$HCl + NaOH \rightleftarrows NaCl + H_2O \quad (22.2)$$

A hydrogen atom only contains one proton in the nucleus and one electron. Upon dissociation, the electron remains with the former binding partner (in foods, often an oxygen), which implies that a hydrogen ion consists of just one proton. Therefore, the terms hydrogen ion, proton, and H^+ are used interchangeably. However, dissociated H^+ become rapidly hydrated by surrounding water molecules, which leads to **hydronium ions** (H_3O^+). In food systems or processing operations, bases used are either salts that dissociate into **hydroxide ions** (OH^-) or compounds that abstract H^+ from water molecules, which also generate OH^-. Moreover, even in the absence of acids and bases, a small number of water molecules dissociate and thus produce H_3O^+ and OH^-. This process is referred to as self-ionization or autodissociation.

At any temperature, the product of the molar concentrations of H_3O^+ and OH^- in water is a constant referred to as the **ion product constant for water** (K_w):

$$[H_3O^+][OH^-] = K_w \quad (22.3)$$

K_w varies with the temperature. For example, at 25°C, K_w = 1.04 × 10^{-14} but at 100 °C, K_w = 58.2 × 10^{-14}. Experimentation has revealed that in pure water at 25 °C, the $[H_3O^+]$ and $[OH^-]$ is approximately 1.0 × 10^{-7} *M*. Because the concentrations of these ions are equal, pure water is referred to as being **neutral**.

Suppose that a drop of acid is added to pure water. The $[H_3O^+]$ would increase. However, as indicated in the name, K_w would remain constant (1.0 × 10^{-14}), revealing a decrease in the $[OH^-]$. Conversely, if a drop of base is added to pure water, the $[H_3O^+]$ would decrease, while the $[OH^-]$ would increase, maintaining the K_w at 1.0 × 10^{-14} at 25 °C.

How did the term pH derive from the above considerations? In approaching the answer to this question, consider the observed $[H_3O^+]$ and $[OH^-]$ in various foods, as shown in Table 22.2, which are bulky, and often the order of magnitude is more important than the exact value. This led a Swedish chemist, S.L.P. Sørensen, to develop the pH system in 1909.

pH is defined as the logarithm of the reciprocal of the hydrogen ion concentration, which is equal to the negative logarithm of the molar concentration of hydrogen ions:

$$\text{pH} = -\log[H^+] \quad (22.4)$$

Since there are no free H^+ in an aqueous system (because H_3O^+ forms), one may also write that the pH is equal to the negative decadic logarithm of $[H_3O^+]$, and it will be expressed as such throughout this chapter. In other words, one considers the exponent and inverts the sign, e.g., 1 × 10^{-6} is expressed simply as pH 6. Thus, a change in one pH unit is actually a tenfold change in the $[H_3O^+]$. While not used as extensively in food science, it is also possible to express the $[OH^-]$ as pOH. If the pH is 6, the pOH would be 8, since the values must add up to 14 based on the ion product constant of water (shown in Table 22.3).

Table 22.2 Concentrations of H_3O^+ and OH^- in various foods at 25 °C

Food	*[H_3O^+]*[a]	*[OH^-]*[a]	K_w
Cola	2.24 × 10^{-3}	4.66 × 10^{-12}	1 × 10^{-14}
Grape juice	5.62 × 10^{-4}	1.78 × 10^{-11}	1 × 10^{-14}
Seven-Up	3.55 × 10^{-4}	2.82 × 10^{-11}	1 × 10^{-14}
Schlitz beer	7.95 × 10^{-5}	1.26 × 10^{-10}	1 × 10^{-14}
Pure water	1.00 × 10^{-7}	1.00 × 10^{-7}	1 × 10^{-14}
Tap water	4.78 × 10^{-9}	2.09 × 10^{-6}	1 × 10^{-14}
Milk of magnesia	7.94 × 10^{-11}	1.26 × 10^{-4}	1 × 10^{-14}

From Pecsok et al. [12], used with permission. Copyright 1971 American Chemical Society
[a]Moles per liter. Note that the product of $[H^+][OH^-]$ is always 1 × 10^{-14}

The following calculation illustrates the conversion of $[H_3O^+]$ and pH, using beer with pH 4.30 as example. To find the $[H_3O^+]$:

Step 1. Substitute numbers into the pH equation:
pH = −log $[H_3O^+]$
4.30 = −log $[H_3O^+]$
−4.30 = log $[H_3O^+]$
Step 2. Calculate the antilog of −4.30: $[H^+]$ = 5.01 × 10^{-5} *M*.

This calculation is nowadays commonly done with a calculator or computer program (Practice Problem 13 at the end of this chapter shows how the calculation is done by hand).

Strictly speaking, the pH definition introduced above would only be valid in so-called ideal solutions (however, these do not exist). To account for nonideal behavior, the concept of activity was introduced. **Activity** is a measure of expressed chemical reactivity and is also referred to as effective concentration. It can be substituted for **concentrations** in equations. Because of the interactions of ions between themselves and with the solvent, activity is, in general, lower than the concentration. Activity and concentration tend to approach each other at infinite dilution and are related by the following equation:

$$A = \gamma C \quad (22.5)$$

where

A = activity
γ = activity coefficient
C = concentration

Table 22.3 Relationship of $[H^+]$ vs. pH and $[OH^-]$ vs. pOH at 25°C

[H^+][1]	*pH*	*[OH^-]*[a]	*pOH*
1 × 10^0	0	1 × 10^{-14}	14
10^{-1}	1	10^{-13}	13
10^{-2}	2	10^{-12}	12
10^{-3}	3	10^{-11}	11
10^{-4}	4	10^{-10}	10
10^{-5}	5	10^{-9}	9
10^{-6}	6	10^{-8}	8
10^{-7}	7	10^{-7}	7
10^{-8}	8	10^{-6}	6
10^{-9}	9	10^{-5}	5
10^{-10}	10	10^{-4}	4
10^{-11}	11	10^{-3}	3
10^{-12}	12	10^{-2}	2
10^{-13}	13	10^{-1}	1
10^{-14}	14	10^0	0

From Pecsok et al. [12], used with permission. Copyright 1971 American Chemical Society
[a]Moles per liter. Note that the product of $[H^+][OH^-]$ is always 1 × 10^{-14}

The **activity coefficient** is a function of ionic strength, which itself is a function of the concentration of, and the charge on, all ions in solution. Activity issues can become significant for hydronium ions below pH 1 and for hydroxyl ions at pH 13 and above. While such extreme pH values are not commonly encountered in food science, ionic strengths may be high, and thus different calculations would need to be used (refer to relevant literature, e.g., Ref. [1]).

22.3.2 pH Meter

The following is a brief overview of the principles behind pH measurements, which is likely a review from introductory chemistry courses, followed by some practical considerations. A detailed discussion is outside of the scope of this chapter. For more information on electrochemistry, the reader is referred to textbooks on general or analytical chemistry, such as Ref. [13].

22.3.2.1 General Principles

The pH meter is a **potentiometer,** i.e., a device that measures voltage at infinitesimal current flow. The basic principle of **potentiometry** involves the use of an electrolytic cell composed of two electrodes immersed into a test solution. A voltage develops, which is related to the ionic concentration of the solution. While this chapter focuses on the use of potentiometry to measure H^+, it can also be used to quantify the concentration of other ions (see Chap. 21, Sect. 21.3.4).

Four major parts of the pH system are needed: (1) reference electrode, (2) indicator electrode (pH-sensitive), (3) voltmeter or amplifier, and (4) the sample being analyzed (Fig. 22.2). Each of the two electrodes involved in the measurement is designed carefully to produce a constant, reproducible potential. Therefore, in the absence of other ions, the potential difference between the two electrodes is fixed. However, H_3O^+ ions in solution contribute to a new potential across an ion-selective glass membrane built into the indicating electrode. This alters the potential difference between the two electrodes in a way that is proportional to H_3O^+activity. The new potential resulting from the combination of all individual potentials is called the **electrode potential**. The **Nernst equation** relates the electrode response to the ion activity:

$$E = E_o + 2.303\frac{RT}{zF}\log A \quad (22.6)$$

where

E = measured electrode potential

E_o = standard electrode potential, a constant representing the sum of the individual potentials in the system at a standard temperature, ion concentration, and electrode composition

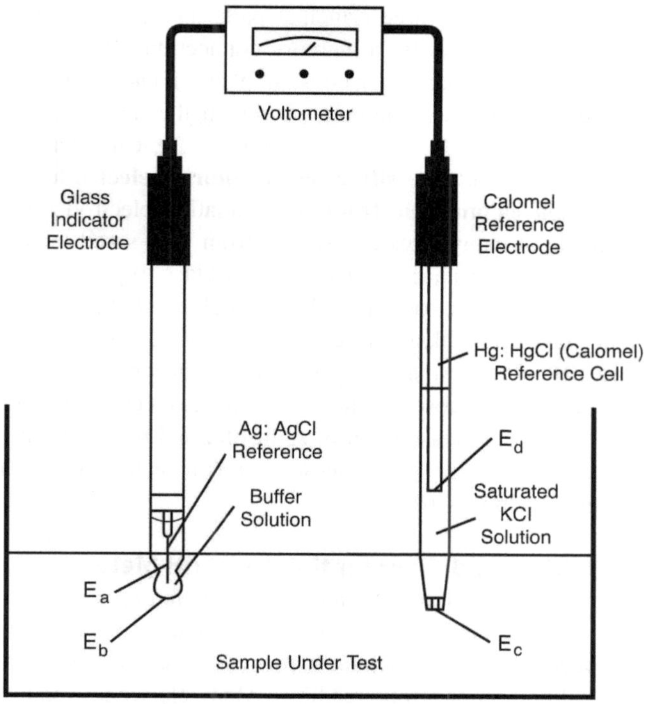

Fig. 22.2 The measuring circuit of the potentiometric system. E_a: contact potential between Ag/AgCl electrode and inner liquid. E_a is independent of pH of the test solution but is temperature-dependent. E_b, potential developed at the pH-sensitive glass membrane. E_b varies with the pH of the test solution and also with temperature. In addition to this potential, the glass electrode also develops an asymmetry potential, which depends on the composition and shape of the glass membrane. It also changes as the electrode ages. E_c, diffusion potential between saturated KCl solution and test sample. E_c is essentially independent of the solution under test. E_d, contact potential between calomel portion of electrode and KCl salt bridge. E_d is independent of the solution under test but is temperature-dependent. (From Bergveld et al. [3], used with permission)

R = universal gas constant, 8.313 $\left(\frac{\text{J}}{\text{mol K}}\right)$

F = Faraday constant, 96,490 coulombs (C) per mole $\left(\frac{\text{C}}{\text{mol}}\right)$

T = absolute temperature in Kelvin (K)

z = number of charges on the ion

A = activity of the ion being measured

Note that changes in temperature will alter the pH reading in accordance with Eq. 22.6. Therefore, modern pH meters have a sensitive attenuator (temperature compensator) built into them to account for this temperature effect. Today, most food analysis laboratories use **combination electrodes** that combine both the pH (i.e., indicator) and reference electrodes along with the temperature sensing probe in a common housing. The **indicator electrode** most commonly used in measuring pH today is referred to as the **glass electrode**

(Fig. 22.2). It has three principal parts: (1) a silver–silver chloride electrode with a mercury connection; (2) a buffer solution used to maintain a constant pH (E_a); and (3) a small pH-sensitive glass membrane for which the potential (E_o) varies with the pH of the test solution. Typical reference electrodes include the **silver–silver chloride electrode** and **saturated calomel electrode**. Combination electrodes are available in many sizes and shapes from very small microprobes to flat surface probes, from all glass to plastic, and from exposed electrode tip to jacketed electrode tips to prevent glass tip breakage. Microprobes may be used to measure pH of very small systems such as inside a cell or a solution on a microscope slide. Flat surface electrode probes can be used to measure pH of semisolid and high-viscosity substances such as meat, cheese, and agar plates and small volumes as low as 10 μl.

22.3.2.2 Guidelines for the Use of pH Meter

For proper operation and maintenance (including conditioning, cleaning, and storage), follow the specific instructions provided by the manufacturer to prevent issues such as incorrect calibration or erratic readings. Many electrodes require periodic refilling, and glass electrodes should not be allowed to dry out. Depending on the analyzed samples, specific cleaning procedures may be recommended. For example, residual protein can be removed by soaking in solutions containing proteases. User and reference manuals (which for newer instruments can usually also be found online) include technical data such as the pH and temperature measurement range. For maximum accuracy, the meter should be standardized using at least two buffers (**two-point calibration**). Select two buffers of pH values about 3 pH units apart, bracketing that of the anticipated sample pH. The three standardization buffers used most widely in laboratories are a pH 4.0 buffer, a pH 7.0 buffer, and a pH 9.0 buffer (at 25 °C). These are the typical pink, yellow, and blue solutions found adjacent to pH meters in many laboratories. Make sure to check the expiration date on the bottles, as they need to be periodically replaced.

22.4 TITRATABLE ACIDITY

22.4.1 Overview and Principle

The titratable acidity measures the total acid concentration in a food. Food acids are usually organic acids, with citric, malic, lactic, tartaric, and acetic acids being the most common. However, inorganic acids such as phosphoric and carbonic (arising from carbon dioxide in solution) acids often play an important and even predominant role in food acidulation. The organic acids present in foods influence the flavor (i.e., tartness), color (though their impact on anthocyanins and other pH-influenced pigments), microbial stability (via inherent pH sensitivity characteristics of organisms), and keeping quality (arising from varying chemical sensitivities of food components to pH). The titratable acidity of fruits is used, along with sugar content, as an indicator of maturity (Sect. 22.4.6). While organic acids may be naturally present in the food, they also may be formed through fermentation, or they may be added as part of a specific food formulation.

Titratable acidity is determined by neutralizing the acid present in a known quantity (weight or volume) of food sample using a standard base (i.e., 0.1 *N* NaOH). The endpoint for titration is usually either a target pH or the color change of a pH-sensitive dye (typically phenolpthalein). The volume of titrant used, along with the normality of the base and the volume (or weight) of sample, is used to calculate the titratable acidity, expressed in terms of the predominant organic acid (e.g., citric acid for orange juice). It is recommended that the reader look ahead to Fig. 22.6 to visualize this assay, done with either a potentiometric or colorimetric endpoint. Picturing this assay while reading the sections immediately below will help the reader see the importance of information on acid theory, understand the details of the titration, and appreciate the occasional problems that might arise in doing the titration.

22.4.2 General Considerations

22.4.2.1 Titration Curves of Strong and Weak Acids

Even though the concentration of acids in a system influences the pH, total acidity is not the same as pH. Since the pH signifies $-\log[H_3O^+]$, it is necessary that acid molecules dissociate to liberate their protons. As long as a proton is still bound to the acid molecule, it does not influence the $[H_3O^+]$. There are considerable differences among molecules when it comes to how easily a hydrogen ion dissociates, described via dissociation constants (see Table 22.4). In contrast to strong acids, for most organic acids commonly found in

Table 22.4 pK_a Values for some acids important in food analysis[a]

Acid	pK_a1	pK_a2	pK_a3
Oxalic	1.19	4.21	–
Phosphoric	2.12	7.21	12.30
Tartaric	3.02	4.54	–
Malic	3.40	5.05	–
Citric	3.06	4.74	5.40
Lactic	3.86	–	–
Ascorbic	4.10	11.79	–
Acetic	4.76	–	–
Potassium acid phthalate	5.40	–	–
Carbonic	6.10	10.25	–

[a]Note that some values differ from those listed in Table 2.2 of Ref. [15]. See that chapter for information on factors affecting dissociation constants

foods, only a small percentage of the acid dissociates in solution. The effect may be illustrated by comparing the pH and the titration curves of 0.1 *N* solutions of a strong monoprotic acid (e.g., hydrochloric) and a weak monoprotic acid (e.g., acetic) with a strong base (Figs. 22.3 and 22.4, respectively). Unless concentrations are high, strong acids are assumed to be fully disassociated. Therefore, in the titrations involving strong acids (e.g., at 0.1 *N*), the pH is equal to the hydronium ion concentration of the remaining acid or the hydroxyl ion concentration once all acid molecules have been neutralized and only the titrant base remains (Fig. 22.3).

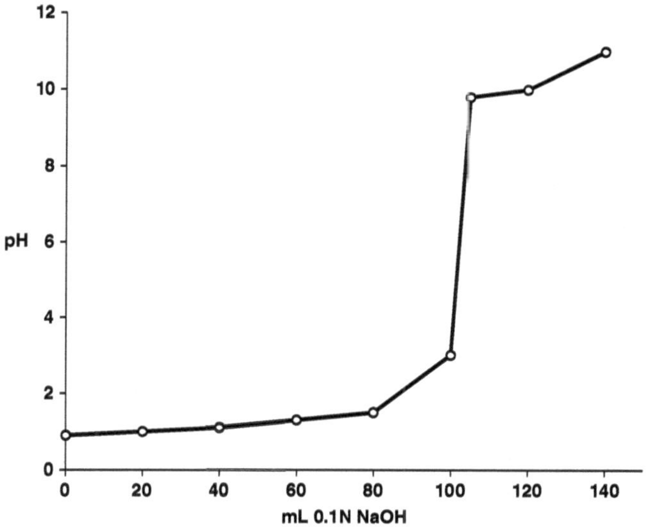

Fig. 22.3 Titration of a strong acid at a concentration of 0.1 N with sodium hydroxide (NaOH, a strong base) of the same concentration. As long as acid is present, the pH at any point in the titration is dictated by the hydrogen ion concentration of the remaining acid. Once all the acid has been consumed, the concentration of excess NaOH determines the pH

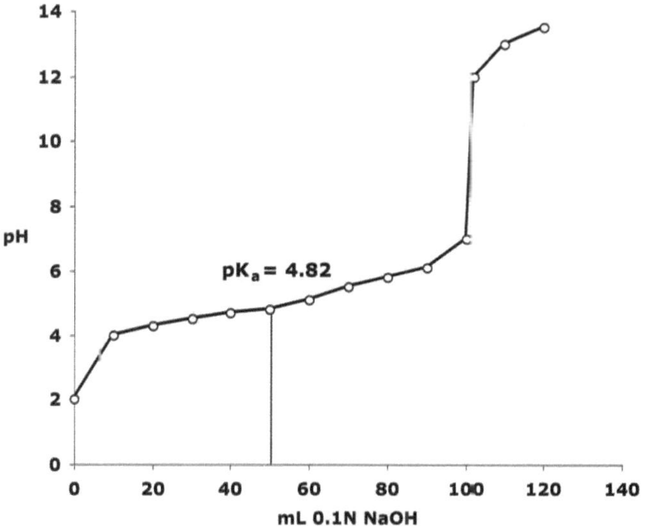

Fig. 22.4 Titration of a weak monoprotic acid with a strong base. A buffering region is established around the pK_a (4.82 for acetic acid). The pH at any point is described by the Henderson–Hasselbalch equation

In contrast to strong acid titration, when weak acids are titrated with a strong base, a mixture of this acid and its conjugated base is present in the system. This tends to cushion the solution from abrupt changes in pH. This property of a solution to resist change in pH is termed **buffering** and occurs whenever a weak acid and its conjugated base are present in the same medium. Because of buffering, a graph of pH vs. titrant concentration is more complex for weak acids than strong acids. However, this relationship can be estimated by the **Henderson–Hasselbalch equation**:

$$\text{pH} = \text{p}Ka + \log\frac{[\text{A}-]}{[\text{HA}]} \quad (22.7)$$

[HA] represents the concentration of undissociated acid, and [A$^-$] represents the concentration of its **conjugated base.** Table 22.4 lists the pK_a values for several acids common in foods. The **pK_a** is the negative decadic logarithm of an acid's dissociation constant. It also equals the pH at which equal quantities of undissociated acid and conjugated base are present because then the ratio of [A$^-$] to [HA] is 1, and the logarithm of 1 is 0. At this point, the equation buffering capacity is at a maximum. The graph showing the titration of 0.1 *N* acetic acid with 0.1 *N* NaOH illustrates this point (Fig. 22.4). Note the different shapes in Figs. 22.3 and 22.4. At the **equivalence point** in a titration, the number of acid equivalents exactly equals the number of base equivalents. For the strong acid, the equivalence point signifies that the concentrations of H^+ and OH^- are exactly the same, which means that the pH is neutral. If HCl was titrated with NaOH, aside from H^+ and OH^- only Na^+ and Cl^- would be present, neither of which affects the pH. In the case of titrating a weak acid, at the equivalence point all acid molecules have been converted to their conjugated base. For this reason, the denominator [HA] in the Henderson–Hasselbalch equation becomes insignificantly small as the equivalence point is approached, and the quotient [A$^-$]/[HA] increases exponentially. As a result, the solution pH rapidly increases and ultimately approaches the pH of the titrant. The exact equivalent point is the halfway mark on this slope of abrupt pH increase.

Di- and triprotic acids will have two and three buffering regions, respectively. A pH vs titrant graph of citric acid is given in Fig. 22.5. If the pK_a steps in polyprotic acids differ by three or more pK_a units, then the Henderson–Hasselbalch equation can predict the plateau corresponding to each step. However, the transition region between steps is complicated by the presence of protons and conjugate bases arising from other disassociation state(s). Consequently, the Henderson–Hasselbalch equation breaks down near the equivalence point between two pK_a steps. However, the pH at the equivalence point is easily calculated as (pK_a1 + pK_a2)/2.

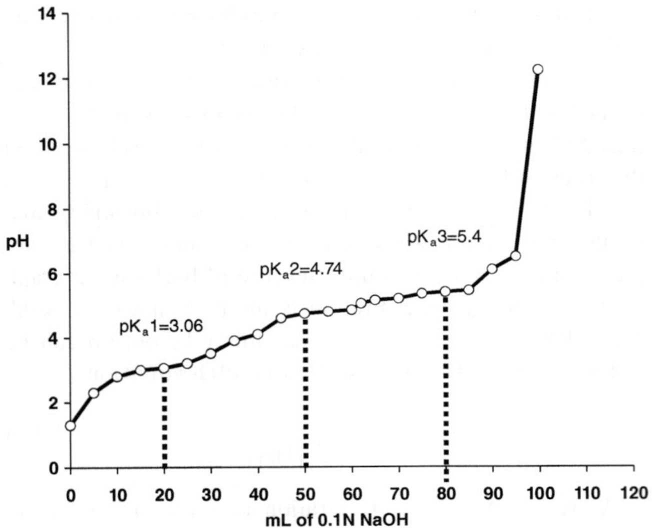

Fig. 22.5 Titration of a weak polyprotic acid with a strong base. Buffering regions are established around each pK_a. The Henderson–Hasselbalch equation can predict the pH for each pK_a value if pK_a steps are separated by more than three units. However, complex transition mixtures between pK_a steps make simple calculations of transition pH values impossible

Precise prediction of pH by the Henderson–Hasselbalch equation requires that all components form ideal solutions. An approximation to ideal solutions occurs at infinite dilution for all active components. However, real solutions may not behave ideally. For such solutions, the Henderson–Hasselbalch equation may only provide a good estimate of pH. For more details about buffers, and for practice problems using the Henderson–Hasselbalch equation to prepare buffers and calculate pH, see Ref. [15].

22.4.2.2 Potentiometric Titration

The use of a pH meter to identify the endpoint is called the **potentiometric method** for determining titratable acidity. The advantage of determining the equivalence point potentiometrically is that the precise equivalence point is identified. Since a rapid change in pH (and not some final pH value) signals the end of titration, accurate calibration of the pH meter is not even essential. However, to identify the equivalence point, a careful record of pH vs. titrant must be kept. This and the physical constraints of pH probes and slow response with some electrodes make the potentiometric approach somewhat cumbersome.

22.4.2.3 Indicators

For simplicity in routine work to determine TA, an indicator solution is often used to approximate the equivalence point. This approach tends to overshoot the equivalence point by a small amount. When indicators are used, the term **endpoint** or **colorimetric endpoint** is substituted for equivalence point. This emphasizes that the resulting values are approximate and dependent on the specific indicator. Phenolphthalein is the most common indicator for food use. It changes from clear to red in the pH region 8.0 to 9.6. Significant color change is usually present by pH 8.2. This pH is termed the **phenolphthalein endpoint.**

A review of pK_a values in Table 22.4 indicates that naturally occurring food acids do not buffer in the region of the phenolphthalein endpoint. However, phosphoric acid (used as an acidulant in some soft drinks) and carbonic acid (carbon dioxide in aqueous solution) do buffer at this pH. Consequently, taking the solution from the true equivalence point to the endpoint may require a large amount of titrant when quantifying these acids. Indistinct endpoints and erroneously large titration values may result. When these acids are titrated, potentiometric analysis is usually preferred over a phenolphthalein endpoint. Interference by CO_2 can be removed by boiling the sample and titrating the remaining acidity to a phenolphthalein endpoint.

Deeply colored samples also present a problem for endpoint indicators. When colored solutions obscure the endpoint, a potentiometric method is normally used. For routine work, pH vs. titrant data may not be plotted (as shown in Figs. 22.4 and 22.5), but samples are simply titrated to pH 8.2 (the phenolphthalein endpoint).

Dilute acid solutions (e.g., vegetable extracts) require dilute solutions of standard base for optimal accuracy in titration. However, a significant volume of dilute alkali may be required to take a titration from the equivalence point to pH 8.2. Bromothymol blue is sometimes used as an alternative indicator in low-acid situations. It changes from yellow to blue in the pH range 6.0–7.6. The endpoint is usually a distinct green. However, endpoint identification is somewhat more subjective than the phenolphthalein endpoint.

Indicator solutions rarely contain over a few tenths percent dye (w/v). All indicators are either weak acids or weak bases that tend to buffer in the region of their color change. If added too liberally, they can influence the titration by conferring their own buffering to the sample under analysis. Therefore, indicator solutions should be held to the minimum amount necessary to impart effective color. Typically, two to three drops of indicator are added to the solution to be titrated. The lower the indicator concentration, the sharper will be the endpoint.

22.4.3 Preparation of Reagents

22.4.3.1 Standard Alkali

Sodium hydroxide (NaOH) is the most commonly used base in titratable acidity determinations. In some ways, it appears to be a poor candidate for a **standard base**. Reagent grade NaOH is very hygroscopic and often contains significant quantities of insoluble sodium carbonate (Na_2CO_3). Consequently, the normality of working solutions is not precise, so each new batch of NaOH must be standardized against an acid of known normality. However, economy, availability, and long tradition of use for NaOH outweigh these shortcomings. Working solutions are normally made from a stock solution containing 50% sodium hydroxide in water (w/v) (see Refs. [15] and [18]). Sodium carbonate is essentially insoluble in concentrated alkali and gradually precipitates out of solution over the first 10 days of storage.

The NaOH titrant can react with dissolved and atmospheric CO_2 to produce new Na_2CO_3. This reduces alkalinity and sets up a carbonate buffer that can obscure the true endpoint of a titration. Water and CO_2 alone can also react to form buffering compounds and generate hydrogen ions, as shown in the following equations:

$$H_2O + CO_2 \rightleftarrows H_2CO_3 \text{ (carbonic acid)} \quad (22.8)$$

$$H_2CO_3 \rightleftarrows H^+ + HCO_3^- \text{ (bicarbonate)} \quad (22.9)$$

$$HCO_3^- \rightleftarrows H^+ + CO_3^{-2} \text{ (carbonate)} \quad (22.10)$$

Therefore, CO_2 should be removed from water prior to making the stock solution. This can be achieved by purging water with CO_2-free gas for 24 h or by boiling distilled water for 20 min and allowing it to cool before use. During cooling and long-term storage, air (with accompanying CO_2) will be drawn back into the container. Carbon dioxide can be stripped from reentering air with a soda lime (20% NaOH, 65% CaO, and 15% H_2O) or **ascarite trap** (NaOH-coated silica base). Air passed through these traps also can be used as purge gas to produce CO_2-free water.

Stock alkali solution of 50% in water is approximately 18 *N*. A working solution is made by diluting stock solution with **CO_2-free water**. There is no ideal container for strong alkali solutions. Glass and plastic are both used, but each has its drawbacks. If a glass container is used, it should be closed with a rubber or thick plastic closure. Glass closures should be avoided since, over time, strong alkali dissolves glass, resulting in permanent fusion of the contact surfaces. Reaction with glass also lowers the normality of the alkali. These liabilities also are relevant to long-term storage of alkali in burettes. NaOH has a low surface tension. This predisposes to leakage around the stopcock. Stopcock leakage during titration will produce erroneously high acid values. Slow evaporation of titrating solution from the stopcock valve during long periods of nonuse also creates a localized region of high pH with ensuing opportunities for fusion between the stopcock and burette body. After periods of nonuse, burettes should be emptied, cleaned, and refilled with fresh working solution.

Long-term storage of alkali in plastic containers also requires special vigilance because CO_2 permeates freely through most common plastics. Despite this shortcoming, plastic containers are usually preferred for long-term storage of stock alkali solutions. Whether glass or plastic is used for storage, working solutions should be restandardized weekly to correct for alkalinity losses arising from interactions with glass and CO_2.

22.4.3.2 Standard Acid

The impurities and hygroscopic nature of NaOH make it unsuitable as a primary standard. Therefore, NaOH titrating solutions must be standardized against a **standard acid. Potassium acid phthalate** (KHP) is commonly used for this purpose.

KHP's single ionizable hydrogen (pK_a = 5.4) provides very little buffering at pH 8.2. It can be manufactured in very pure form, is relatively nonhygroscopic, and can be dried at 120 °C without decomposition or volatilization. Its high molecular weight also favors accurate weighing.

KHP should be dried for 2 h at 120 °C and allowed to cool to room temperature in a desiccator immediately prior to use. An accurately measured quantity of KHP solution is titrated with a base of unknown normality. The base is always the titrant. CO_2 is relatively insoluble in acidic solutions. Consequently, stirring an acid sample to assist in mixing will not significantly alter the accuracy of the titration.

22.4.4 Sample Analysis

A number of official methods exist for determining titratable acidity in various foods [2]. However, determining titratable acidity on most samples is relatively routine, and various procedures share many common steps. An aliquot of sample (often 10 mL) is titrated with a standard alkali solution (often 0.1 *N* NaOH) to a phenolphthalein endpoint. Potentiometric endpoint determination is used when sample pigment makes use of a color indicator impractical.

Typical titration setups are illustrated in Fig. 22.6 for potentiometric and colorimetric endpoints. Erlenmeyer flasks are usually preferred for samples when endpoint indicators are used. A magnetic stirring bar may be used, but mixing the sample with hand swirling is usually adequate. When hand mixing is used, the sample flask is swirled with the right hand. The stopcock is positioned on the right side. Four fingers on the left hand are placed behind the stopcock valve, and the thumb is placed on the front of the valve. Titrant is dispensed at a slow, uniform rate until the endpoint is approached and then added dropwise until the endpoint does not fade after standing for some predetermined period of time, usually 5–10 s.

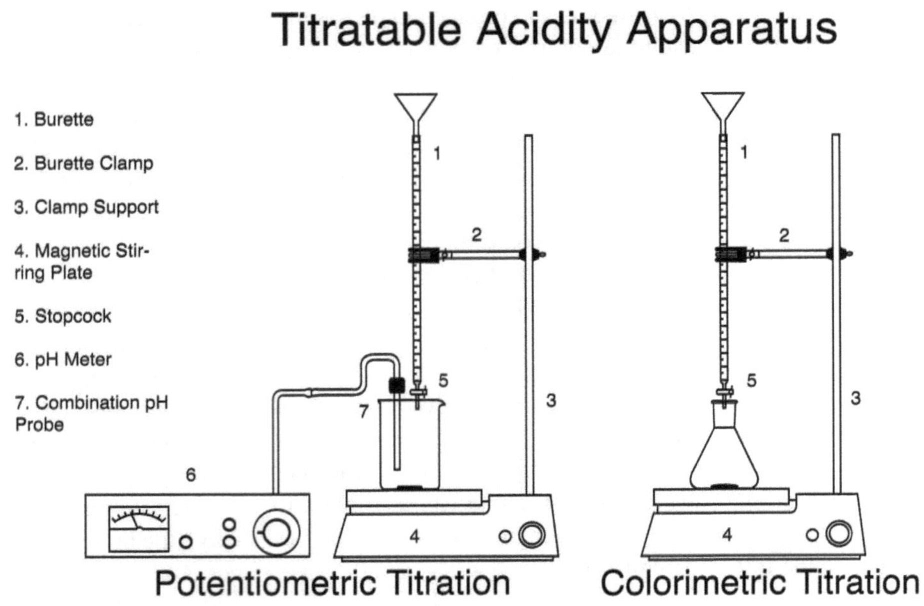

Fig. 22.6 Titratable acidity apparatus

The bulkiness of the pH electrode usually demands that beakers be used instead of Erlenmeyer flasks when samples are analyzed potentiometrically. Mixing is almost always achieved through magnetic stirring, and loss of sample through splashing is more likely with beakers than Erlenmeyer flasks. Otherwise, titration practices are identical to those described previously for colorimetric endpoint titrations.

Problems may arise when concentrates, gels, or particulate-containing samples are titrated. These matrices prevent rapid diffusion of acid from densely packed portions of sample material. This slow diffusion process results in a fading endpoint. Concentrates can simply be diluted with CO_2-free water. Then titration is performed, and the original acid content calculated from dilution data. Starch and similar weak gels often can be mixed with CO_2-free water, stirred vigorously, and titrated in a manner similar to concentrates. However, some pectin and food gum gels require mixing in a blender to adequately disrupt the gel matrix. Thick foams are occasionally formed in mixing. Antifoam or vacuum can be used to break the foams.

Immediately following processing, the pH values of particulate samples often vary from one particulate piece to another. Acid equilibration throughout the entire mass may require several months. As a result, particulate-containing foods should be liquified in a blender before titrating. The comminuting process may incorporate large quantities of air. Air entrapment makes the accuracy of volumetric measurements questionable. Aliquots often are weighed when air incorporation may be a problem.

22.4.5 Calculation of Titratable Acidity

In general chemistry, acid strength is frequently reported in normality (equivalents per liter) and can be calculated using Eq. 22.1. However, food acids are usually reported as percent of total sample weight. Thus, the equation for titratable acidity is as follows:

$$\%\ \text{acid}(\text{wt}/\text{wt}) = \frac{N \times V \times \text{Eq wt}}{W \times 1000} \times 100 \qquad (22.11)$$

where

N = normality of titrant, usually NaOH (mEq/mL)
V = volume of titrant (mL)
Eq. wt. = equivalent weight of predominant acid (mg/mEq)
W = mass of sample (g)
1000 = factor relating mg to grams (mg/g) (1/10 = 100/1000)

Note that the normality of the titrant is expressed in milliequivalents (mEq) per mL, which is a typical way of reporting normality for small volumes. This value is numericallythe same as equivalents/liter. Also note that it is easier to report sample mass in grams instead of milligrams, so multiplying sample mass by the factor of 1000 mg/g allows units to cancel.

For routine titration of fruit juices, milliliters can be substituted for sample weight in grams, as shown in Eqs. 22.12

and 22.13 and illustrated via an example using Eq. 22.14. Depending on the soluble solids content of the juice, the resulting acid values will be higher by 1–6%. However, this is common practice:

$$\% \text{ acid}(\text{wt}/\text{vol}) = \frac{N \times V_1 \times \text{Eq wt}}{V_2 \times 1000} \times 100 \quad (22.12)$$

or

$$\% \text{ acid}(\text{wt}/\text{vol}) = \frac{N \times V_1 \times \text{Eq wt}}{V_2 \times 10} \quad (22.13)$$

where

N = normality of titrant, usually NaOH (mEq/mL)
V_1 = volume of titrant (mL)
Eq. wt. = equivalent weight of predominant acid (mg/mEq)
V_2 = volume of sample (mL)
1000 = factor relating mg to grams (mg/g) (1/10 = 100/1000)

For example, if it takes 17.5 mL of 0.085 *N* NaOH to titrate a 15-mL sample of a juice, the total titratable acidity of that juice, expressed as percent citric acid (molecular weight = 192; equivalent weight = 64), would be 0.635%, wt/vol, citric acid:

$$\% \text{ acid}(\text{wt}/\text{vol}) = \frac{0.085 \times 17.5 \times 64}{15 \times 10} = 0.635\% \quad (22.14)$$

Notice that the equivalent weight of anhydrous citric acid always is used in calculating and reporting the results of titration.

22.4.6 Acid Content in Food

Most foods are as chemically complex as life itself. As such, they contain the full complement of Krebs cycle acids (and their derivatives), fatty acids, and amino acids. Theoretically, all of these contribute to titratable acidity. Routine titration cannot differentiate between individual acids. Therefore, titratable acidity is usually stated in terms of the **predominant acid**. For most foods, this is unambiguous. In some cases, two acids are present in large concentrations, and the predominant acid may change with maturity. In grapes, malic acid often predominates prior to maturity, while tartaric acid typically predominates in the ripe fruit. A similar phenomenon is observed with malic and citric acids in pears. Fortunately, the equivalent weights of common food acids are fairly similar (Table 22.1). Therefore, percent titratable acidity is usually not substantially affected by mixed predominance or incorrect selection of the predominant acid, unless equivalent weights are substantially different.

The range of acid concentrations in foods is very broad. Acids can exist at levels below detection limits or they can be the preeminent substance present in the food. The tartness of acids is reduced by sugars. Consequently, the **Brix/acid ratio** (often simply called ratio) is usually a better predictor of an acid's flavor impact than Brix [i.e., percent sugar (w/w) in a medium; see Chap. 15, Sect. 15.2.5.3) or titratable acidity alone. Acids tend to decrease with the maturity of fruit while sugar content increases. Therefore, the Brix/acid ratio is often used as an index of fruit maturity. For mature fruit, this ratio also can be affected by climate, variety, and horticultural practices. Table 22.5 gives typical acid composition and sugar levels for many commercially important fruits at maturity. Citric and malic acids are the most common acids in fruits and most vegetables; however, leafy vegetables also may contain considerable quantities of oxalic acid. Lactic acid is the most important acid in dairy foods for which titratable acidity is commonly used to monitor the progress of lactic acid fermentations in cheese and yogurt production [14].

Organic acids contribute to the refractometer reading of soluble solids. When foods are sold on the basis of pound solids, Brix readings are sometimes corrected for acid content. For citric acid, 0.20° Brix is added for each percent titratable acidity.

Table 22.5 Acid composition and °Brix of some commercially important fruits

Fruit	*Principal acid*	*Typical percent acid*	*Typical °Brix*
Apples	Malic	0.27–1.02	9.12–13.5
Bananas	Malic/citric (3: 1)	0.25	16.5–19.5
Cherries	Malic	0.47–1.86	13.4–18.0
Cranberries	Citric	0.9–1.36	
	Malic	0.70–0.98	12.9–14.2
Grapefruit	Citric	0.64–2.10	7–10
Grapes	Tartaric/malic (3: 2)	0.84–1.16	13.3–14.4
Lemons	Citric	4.2–8.33	7.1–11.9
Limes	Citric	4.9–8.3	8.3–14.1
Oranges	Citric	0.68–1.20	9–14
Peaches	Citric	1–2	11.8–12.3
Pears	Malic/citric	0.34–0.45	11–12.3
Pineapples	Citric	0.78–0.84	12.3–16.8
Raspberries	Citric	1.57–2.23	9–11.1
Strawberries	Citric	0.95–1.18	8–10.1
Tomatoes	Citric	0.2–0.6	4

22.4.7 Volatile Acidity

In acetic acid fermentations, it is sometimes desirable to know how much acidity comes from the acetic acid and how much is contributed naturally by other acids in the product. This can be achieved by first performing an initial titration to measure **titratable acidity** as an indicator of total acidity. The acetic acid is then boiled off, the solution is allowed to cool, and a second titration is performed to determine the **fixed acidity**. The difference between fixed and total acidity is the **volatile acidity**. A similar practice is used sometimes in the brewing industry to separate acidity due to dissolved CO_2 from fixed acids. Fixed acids are titrated after CO_2 is removed by low heat (40 °C) and gentle agitation.

22.4.8 Other Methods

Sometimes, it is of interest to quantify all acids in a food individually. High-performance liquid chromatography (HPLC) (see Chap. 13) and electrochemistry (such as voltammetry and polarography) both have been used to measure acids in food samples. One potential issue is interfering compounds, which often reduce the practicality of electrochemical approaches.

Unlike titration, chromatographic and electrochemical techniques do not differentiate between an acid and its conjugate base. As a result, acids determined by instrumental methods may be 50% higher than values determined by titration. It follows that Brix/acid ratios can be based only on acid values determined by titration.

New pH probe designs use an ion-sensitive field effect transistor (ISFET) to measure hydronium ion concentration. These may use standard calomel or Ag/AgCl reference electrodes. However, the development of a silicon-based reference electrode has met with mounting success. ISFET electrodes have faster response times than traditional electrodes. Also, compact chip technology makes miniaturized probes possible. ISFET probes do not require a fluid medium, making possible direct pH measurement of gels, meats, cheeses, and other dense foods. The higher cost of an ISFET probe and some tendency for reference electrode drift keep it from replacing traditional pH electrodes for routine analysis.

22.5 SUMMARY

Organic acids have a pronounced impact on food flavor and quality. Unlike strong acids that are fully dissociated, food acids are only partially ionized. Some properties of foods are affected only by this ionized fraction of acid molecules, while other properties are affected by the total acid content. It is impractical to quantify only free hydronium ions in solution by titration. Once the free ions are removed by chemical reaction, others arise from previously undissociated molecules. Indicator dyes, which change color depending on the hydronium ion environment, exist, but they only identify when a certain pH threshold has been achieved and do not stoichiometrically quantify free hydronium ions. The best that can be done is to identify the secondary effect of the hydronium ion environment on some property of the system such as the color of the indicator dyes or the electrochemical potential of the medium. The pH meter measures the change in electrochemical potential established by the hydronium ion across a semipermeable glass membrane on an indicator electrode. The shift in the indicator electrode potential is indexed against the potential of a reference electrode. The difference in millivolt reading between the two electrodes can be converted into pH using the Nernst equation. Indicator dyes are used commonly to identify the endpoint of acidity titrations, although pH meters can be used in critical work or when sample color makes indicators impractical.

The hydronium ion concentration can be back-calculated from pH using the definition of pH as the negative log of hydronium ion concentration. Buffer solutions of any pH can be created using the Henderson–Hasselbalch equation. However, the predictions of all these equations are somewhat approximate unless the activity of acids and conjugate bases is known and taken into account.

Titratable acidity provides a simple estimate of the total acid content of a food. However, most foods contain many acids and these cannot be differentiated through titration. Titratable acidity is not a good predictor of pH, since pH is a combined function of titratable acid and conjugate base. Instrumental methods such as HPLC and electrochemical approaches measure acids and their conjugate bases as a single compound and, therefore, tend to produce acid contents that are higher than those determined by titration. Titratable acidity is a better predictor of tartness than the concentration of free hydronium ions as reflected by pH. The perception of tartness is strongly influenced by the presence of sugars, and the Brix/acid ratio can be a useful indicator to take this into account.

22.6 STUDY QUESTIONS

1. Explain the theory behind using a pH meter to measure H_3O^+ concentration.
2. You return from a 2-week vacation and ask your lab technician about the pH of the apple juice sample you gave him or her before you left. Having forgotten to do it before, the technician calibrates a pH meter with one

standard buffer stored next to the meter and then reads the pH of the sample of unpasteurized apple juice immediately after removing it from the refrigerator (40 °F), where it has been stored for 2 weeks. Explain the reasons why this stated procedure could lead to inaccurate or misleading pH values.

3. How would you recommend determining the endpoint in the titration of tomato juice to determine the titratable acidity? Why?
4. The titratable acidity was determined by titration to a phenolphthalein endpoint for a boiled and unboiled clear carbonated beverage. Which sample would you expect to have a higher calculated titratable acidity? Why? Would you expect one of the samples to have a fading endpoint? Why?
5. Why and how is an ascarite trap used in the process of determining titratable acidity?
6. Regarding the use of a standard acid to standardize a NaOH solution to determine titratable acidity, what factors make (a) KHP a good choice and (b) HCl, HNO_3, or H_2SO_4 a poor choice?
7. For each of the food products listed below, what acid should be used to express titratable acidity?
   (a) Orange juice
   (b) Yogurt
   (c) Apple juice
   (d) Grape juice
8. What is a "Brix/acid ratio," and why is it often used as an indicator of flavor quality for certain foods, rather than simply Brix or acid alone?
9. Assume you are determining the titratable acidity of a fermented product and want to express the result in terms of the predominant acid. Why would the value be different if expressed as acetic acid vs. lactic acid?
10. An instructor was grading lab reports of students who had determined the titratable acidity of grape juice. One student had written that the percent titratable acidity was 7.6% citric acid. Give two reasons why the answer was marked wrong. What would have been a more reasonable answer?
11. Why is volatile acidity useful as a measure of quality for acetic acid fermentation products, and how is it determined?

22.7 PRACTICE PROBLEMS

1. You are performing a titration on duplicate samples and duplicate blanks that require 4 mL of 1 *N* NaOH per titration sample. The lab has 10% NaOH and saturated NaOH (assume 11.5 *M*). Choose one of these and describe how you would prepare the needed amount of NaOH solution.
2. You are performing duplicate titrations on five samples that require 15 mL of 6 *N* HCl each. How would you prepare the needed solution from reagent grade HCl?
3. What is the pH of a 0.057 *M* HCl solution?
4. Vinegar has a $[H_3O^+]$ of 1.77×10^{-3} *M*. What is the pH? What is the major acid found in vinegar?
5. Orange juice has a $[H_3O^+]$ of 2.09×10^{-4} *M*. What is the pH? What is the major acid found in orange juice?
6. A sample of vanilla yogurt has a pH of 3.59. What is the $[H_3O^+]$? What is the major acid found in yogurt?
7. An apple pectin gel has a pH of 3.30. What is the $[H_3O^+]$? What is the major acid found in apples?
8. How would you make 1 L of 0.1 *N* NaOH solution from an 18 *N* stock solution?
9. A stock base solution assumed to be 18 *N* was diluted to 0.1 *N*. KHP standardization indicated the normality of the working solution was 0.088 *N*. What was the actual normality of the solution?
10. A 20-mL sample of juice requires 25 mL of 0.1 *N* NaOH titrant. What would be the percent acid if the juice is: (1) apple juice, (2) orange juice, and (3) grape juice?
11. A lab analyzes a large number of orange juice samples. All juice samples will be 10 mL. It is decided that 5 mL of titrant should equal 1% citric acid. What base normality should be used?
12. A lab wishes to analyze apple juice. They would like each milliliter of titrant to equal 0.1% malic acid. Sample aliquots will all be 10 mL. What base normality should be used?
13. Using Table 22.2, calculate the pH of cola.

Answers

1. Four titrations of 4 mL each will be performed requiring a total of about 16 mL of 1 *N* NaOH. For simplicity, 20 mL of 1 *N* NaOH can be prepared. Use Eq. 22.1 to solve for the mL of stock solution, e.g., for a 10% NaOH:

$$10\,\text{g NaOH}/100\,\text{mL} = 100\,\text{g NaOH}/\text{L} = 2.5N\,\text{NaOH}$$
$$(20\,\text{mL})(1N\,\text{NaOH}) = (x\text{mL})(2.5N)$$
$$x\,\text{mL} = 8\,\text{mL of 10\% diluted to 20 mL with distilled water}$$

If saturated NaOH is used, approximately 8.7 mL of saturated NaOH diluted to 100 mL yields 1.0 *N*. Therefore, 1.87 mL or 2 mL of saturated NaOH diluted to 20 mL with distilled water will yield about 1 *N* NaOH.

2. A total of (5 samples) (2 duplicates) (15 mL) = 150 mL of 6 *N* HCl

$$(150\,\text{mL})(6N\,\text{HCl}) = (x\text{mL})(12N\,\text{HCl})$$
$$x\,\text{mL} = 75\,\text{mL concentrated HCl diluted with distilled water to 150 mL}$$

3. Since HCl is a strong acid, it will be almost completely dissociated. Therefore,

$$\left[\text{HCl}\right]=\left[\text{H}_3\text{O}^+\right]$$

$$\text{pH}=-\log\left[5.7\times10^{-2}\right]=1.24$$

4. 2.75; acetic acid.

   (Use Eq. 22.4 to solve Problems 4–7. For structures, see Fig. 22.1)

5. 3.68; citric acid.
6. 2.57×10^{-4} *M*; lactic acid.
7. 5.0×10^{-4} *M*; malic acid
8. Using Eq. 22.1 and solving for volume of concentrate, we get

$$\text{mL concentrated solution}=\frac{\text{final }N\times\text{final mL}}{\text{beginning }N}$$

$$=\frac{0.1\times1000}{18}$$

$$=5.55\,\text{mL}$$

   Consequently, 5.55 mL would be dispensed into a 1-L volumetric flask. The flask would then be filled to volume with distilled CO_2-free water. The normality of this solution will only be approximate since NaOH is not a primary standard. Standardization against a KHP solution or some other primary standard is essential. It is useful sometimes to back-calculate the true normality of the stock solution. Even under the best circumstances, the normality will decrease with time, but back-calculating will permit a closer approximation of the target normality the next time a working standard is prepared.

9. This answer is a simple ratio:

$$\frac{0.088}{0.100}\times18=15.85N$$

10. Table 22.5 indicates the principal acids in apple, orange, and grape juice are malic, citric, and tartaric acids, respectively. Table 22.1 indicates the equivalent weight of these acids are malic (67.05), citric (64.04), and tartaric (75.05). The *N*, v_1, and v_2 are given. Using Eq. 22.13, the percent acid for each of these juices would be:

$$\%\text{ malic acid}=\frac{0.1\times25\times67.05}{20\times10}=0.84\%$$

$$\%\text{ citric acid}=\frac{0.1\times25\times64.04}{20\times10}=0.80\%$$

$$\%\text{ tartaric acid}=\frac{0.1\times25\times75.05}{20\times10}=0.94\%$$

11. Quality control laboratories often analyze a large number of samples having a specific type of acid. Speed and accuracy are increased if acid concentration can be read directly from the burette. It is possible to adjust the normality of the base to achieve this purpose by rearranging Eq. 22.13.

$$\%\text{ acid}=\frac{N\times V_1\times\text{Eq wt}}{V_2\times10} \quad (22.15)$$

$$N=\frac{\%\text{ acid}\times V_2\times10}{V_1\times\text{Eq wt}}$$

   The desired % acid, v_1, and v_2, are 1%, 5, and 10 mL, respectively. The equivalent weight of citric acid is 64.04 (see Table 22.1). Thus,

$$N=\frac{1\times10\times10}{5\times64.04}=0.3123$$

   In actuality, the standard alkali solution used universally by the Florida citrus industry is 0.3123 *N*.

12. This can be solved analogously to Problem 11. This time, % acid is 0.1, v_1 is 1 mL, and v_2 is 10 mL. The equivalent weight is 67.05. Therefore,

$$N=\frac{0.1\times10\times10}{1\times67.05}=0.1491$$

13. Substitute the $[H_3O^+]$ into Eq. 22.4:

$$\text{pH}=-\log\left[\text{H}_3\text{O}^+\right]$$

$$\text{pH}=-\log\left[2.24\times10^{-3}\right]$$

   There are now different ways to obtain the right answer:

   (a) Type the concentration into a calculator, take the logarithm, and invert the sign.

   (b) Use a computer program such as Excel (the corresponding formula in Excel would be "-LOG10(2.24*10^-3)".

   (c) Calculate by hand: separate 2.24×10^{-3} into two parts; determine the logarithm of each part:

$$\log2.24=0.350$$

$$\log10^{-3}=-3$$

   Add the two logs together since adding logs is equivalent to multiplying the two numbers:

$$0.350+(-3)=-2.65$$

   Place the value into Eq. 22.1.

   All three of these methods should lead to a pH of 2.65.

Acknowledgments The author of this chapter thanks George D. Sadler (PROVE IT, LLC, Geneva, IL, USA) for his contribution to the chapter in previous editions of the book.

REFERENCES

1. Albert A., Serjeant EP (1984). The determination of ionization constants. A laboratory manual, 3rd edition. Chapman and Hall, New York
2. AOAC International (2023) Official methods of analysis, 22th edn. (Online). AOAC International, Rockville, MD
3. Bergveld, P(2003) Thirty years of ISFETOLOGY, What happened in the past 30 years and what may happen in the next 30 years. Sensors and Actuators B 88 1–20.
4. Efiok BJS, Eduok EE (2000) Basic calculations for chemical & biological analysis, 2nd edn. AOACInternational, Gaithersburg, MD
5. Thermo Scientific (2020) Thermo Scientific pH electrode handbook,https://assets.thermofisher.com/TFS-Assets/LSG/brochures/pH-Measurement-Handbook-S-PHREFBK-E.pdf
6. Gardner WH (1966) Food acidulants. Allied Chemical Co., New York
7. Harris DC,Lucy CA (2019) Quantitative chemical analysis,10th edn. W.H. Freeman & Co, New York
8. Joslyn MA (1970) pH and buffer capacity, ch. 12, and acidimetry, ch. 13. In: Methods in food analysis:Physical, chemical, andinstrumentalmethodsofanalysis, Academic, New York
9. Kenkel J (2014) Analytical chemistry for technicians. 4thedn. CRC Press, Boca Raton, FL
10. Beynon R, Easterby J (2003) Buffer solutions : the basics, Garland Science, New York
11. Nelson PE, Tressler DK (1980) Fruit and vegetable juice process technology, 3rd edn. AVI, Westport, CT
12. Pecsok RL, Chapman K, Ponder WH (1971) Modern chemical technology, vol 3, revised edn. American Chemical Society, Washington, DC
13. Skogg DA, West DM, Holler JF, Crouch SR (2021) Fundamentals of analytical chemistry 10th edn. Cengage Group, Boston, MA
14. Wehr HM, Frank JF (eds) (2004) Standardmethod for examination of dairy products, 17th edn. American Public Health Association, Washington, DC
15. Tyl C, Ismail BP (2024) Reagents and buffers, Ch. 2. In: Nielsen's food analysis laboratory manual, 4thedn., Ismail BP, Nielsen SS (eds), Springer, New York
16. Neilson, AP (2024) Dilutions and concentrations, Ch. 3. In: Nielsen's food analysis laboratory manual, 4thedn., Ismail BP, Nielsen SS (eds), Springer, New York
17. Ismail BP (2024) Preparation of solutions and buffers. Ch. 7 In: Nielsen's food analysis laboratory manual, 4thedn. Ismail BP, Nielsen SS (eds), Springer, New York
18. Ismail BP, Nielsen SS (2024) Standard solutions and titratable acidity. Ch. 22. In: Nielsen's food analysis laboratory manual. 4thedn. Ismail BP, Nielsen SS (eds), Springer, New York

Fat Characterization

23

Oscar A. Pike

Contents

O. A. Pike (✉)
Department of Nutrition, Dietetics, and Food Science, Brigham Young University, Provo, UT, USA
e-mail: oscar_pike@byu.edu

B. P. Ismail, S. S. Nielsen (eds.), *Nielsen's Food Analysis*, Food Science Text Series,
https://doi.org/10.1007/978-3-031-50643-7_23

23.1 INTRODUCTION

Numerous methods exist for the characterization of edible fats, oils, other lipids, and lipid-containing foods [1–8]. This chapter addresses those deemed appropriate for an undergraduate food analysis course, including methods required for nutritional labeling. Many traditional "wet chemistry" methods have been supplemented or superseded by instrumental methods such as gas chromatography (GC), high-performance liquid chromatography (HPLC), nuclear magnetic resonance (NMR), Fourier transform infrared (FTIR) spectroscopy, and differential scanning calorimetry (DSC). Nonetheless, an understanding of basic concepts derived from traditional methods is valuable in learning more sophisticated instrumental methods.

Many of the methods cited are official methods of the AOAC International [1], American Oil Chemists' Society (AOCS) [2], or International Union of Pure and Applied Chemists (IUPAC) [3]. The principles, general procedures, and applications are described for the methods. Refer to the specific methods cited in Table 23.1 for detailed information on procedures.

Table 23.1 Correlation of selected AOCS [2], AOAC [1], and IUPAC [3] methods

Method	*AOCS*	*AOAC*	*IUPAC*
Physical characterization			
Refractive index	Cc 7-25	921.08	2.102
Melting			
Capillary tube melting point	Cc 1-25	920.157	
Slip melting point	Cc 3-25		
	Cc 3b-92		
DSC melting properties	Cj 1-94		
Dropping point	Cc 18-80		
Wiley melting point	Cc 2-38[a]	920.156	
Smoke, flash, and fire points	Cc 9a-48		
	Cc 9b-55		
Cold test	Cc 11-53	929.08	
Cloud point	Cc 6-25		
Color			
Lovibond	Cc 13e-92		
	Cc 13j-97		
Spectrophotometric	Cc 13c-50		2.103
Solid fat index (SFI)	Cd 10-57[a]		2.141
Solid fat content (SFC)	Cd 16b-93		2.150
Consistency, penetrometer method	Cc 16-60		
Spreadability	Cj 4-00		
Chemical characterization			
Iodine value (IV)	Cd 1-25[a]	920.159	2.205
	Cd 1c-85		
	Cd 1b-87		
	Cd 1d-92	993.20	
	Cd 1e-01		
Saponification number (SV)	Cd 3-25	920.160	2.202
	Cd 3c-91		

Table 23.' (continued)

Method	*AOCS*	*AOAC*	*IUPAC*
	Cd 3a-94		
Free fatty acids (FFAs)	Ca 5a-40 Ca 5d-01	940.28	
Acid value	Cd 3d-63	969.17	2.201
Conjugated dienes (CD)	Ti 1a-64	957.13	2.206
	Ch 5-91		
Peroxide value (PV)	Cd 8-53[a]	965.33	2.501
	Cd 8b-90		
p-Anisidine value	Cd 18-90		2.504
Thiobarbituric acid (TBA) test	Cd 19-90		2.531
Hexanal (volatile organic compounds)	Cg 4-94		
Polar compounds in frying fats	Cd 20-91	982.27	2.507
Accelerated stability testing			
Oven storage test	Cg 5-97		
Oil stability index (OSI)	Cd 12b-92		
Active oxygen method (AOM)	Cd 12-57[a]		2.506
Accelerated oxidation test (Oxitest)	Cd 12c-16		
Lipid fractions			
Fatty acid composition (including saturated/unsaturated, *cis/trans*)	Ce 1-62[a] Ce 1a-13	963.22	2.302
	Ce 1b-89	991.39	
	Ce 1h-05	996.06	
	Ce 1i-07 Ce 1j-07		
Fatty acid methyl esters (FAMEs)	Ce 2-66	969.33	2.301
	Ce 2b-11 Ce 2c-11		
trans Isomer fatty acids using IR spectroscopy	Cd 14-95		2.207
	Cd 14d-99 Cd 14e-09 Cd 14f-14	2000.10	
Mono- and diacylglycerols	Cd 11-57[a]	966.18	
	Cd 11b-91		
	Cd 11d-96		
Triacylglycerols	Ce 5-86	986.19	
	Ce 5b-89		
	Ce 5c-93		
Cholesterol and other sterols	Ch 6-91	976.26	2.403

AOCS American Oil Chemists' Society, *AOAC* AOAC International, *IUPAC* International Union of Pure and Applied Chemists
[a]Although no longer current, these methods are included for reference because of their previous common use

The methods are divided into four sections. The first addresses methods that characterize the physical properties of fat/oil, such as melting point, smoke point, color, and consistency. The second section contains methods that describe the chemical properties of fat/oil, which include the amount of free fatty acids (FFAs), chain length (saponification value), and degree of unsaturation (iodine value). This section also includes methods characterizing fat/oil deterioration, including conjugated dienes (CDs), peroxides, aldehydes, and other organic compounds. Then, methods for monitoring the useful life of frying fats/oils are described. The following section contains methods evaluating lipid stability, as it relates to the shelf life of bulk fats/oils and foods containing fat/oil. The final set of methods is those that measure lipid fractions, e.g., fatty acids, mono-, di-, and triacylglycerols, and sterols.

Most characterization methods were developed to analyze the bulk fats/oils of commerce, but some methods can utilize intact foods. If enough lipid can be extracted from a food, methods developed for bulk fats/oils can be used to evaluate such food extracts.

Of the many methods available, some have limited use for edible fats/oils (in contrast to soaps and industrial oils). Other methods may require special apparatus not commonly available or have been antiquated by instrumental procedures. Methods to determine impurities, including moisture, unsaponifiable material (consisting of hydrocarbons, higher alcohols, oil-soluble vitamins, and sterols) in refined vegetable oil (see AOCS Methods Ca 6a-40 and Ca 6b-53), and insoluble impurities, also are not covered in this chapter. Defined methods exist for the sensory evaluation of fats/oils and foods containing fats/oils (see AOCS Methods Cg 1-83 and Cg 2-83) but are outside the scope of this text.

23.1.1 Definitions and Classifications

As explained in Chap. 17, the term **lipids** refers to a wide range of compounds soluble in organic solvents but only sparingly soluble in water. Chapter 17 also outlines the general classification scheme for lipids. Most lipids present in foods are of the following types: (1) fatty acids and their glycerides, including mono-, di-, and triacylglycerols, (2) phospholipids, (3) sterols (including cholesterol), (4) waxes, and (5) lipid-soluble pigments and vitamins. The composition of refined fats/oils is almost exclusively triacylglycerols. The commonly used terms monoglyceride, diglyceride, and triglyceride are synonymous with the proper nomenclature terms monoacylglycerol, diacylglycerol, and triacylglycerol, respectively.

In contrast to lipids, the terms fats and oils often refer to bulk products of commerce, crude or refined, that have already been extracted from animal products or oilseeds and other plants grown for their lipid content. The term **fat** signifies extracted lipids that are solid at room temperature, and **oil** refers to those that are liquid. This chapter will utilize "fat/oil" to refer to such bulk products. However, the three terms, lipid, fat, and oil, often are used interchangeably.

The FDA has defined **fat content** for nutritional labeling purposes as the total lipid fatty acids expressed as triglyceride, rather than the lipid content determined by extraction and gravimetric procedures used in the past (see also Chap.17, Sects. 17.1.1 and 17.2.7.1).

> *"Fat, total" or "Total fat": A statement of the number of grams of total fat in a serving defined as total lipid fatty acids and expressed as triglycerides where fatty acids are aliphatic carboxylic acids consisting of a chain of alkyl groups and characterized by a terminal carboxyl group. Amounts shall be expressed to the nearest 0.5 (1/2) gram increment below 5 grams and to the nearest gram increment above 5 grams. If the serving contains less than 0.5 gram, the content shall be expressed as zero* [21 CFR 101.9(c)(2)].

Fatty acids included in this definition may be derived from triacylglycerols, partial glycerides, phospholipids, glycolipids, sterol esters, or free fatty acids, but the concentration will be expressed as grams of triacylglycerols. This definition for nutritional labeling purpose requires that total fat be based on GC of **fatty acid methyl esters** (FAMEs). Although this requires more complex and expensive analytical equipment, the analysis provides a better estimation of the fat content of foods.

Saturated fat is defined for nutritional labeling purposes as the sum (in grams) of all fatty acids without carbon–carbon double bonds. Saturated fat is expressed to the closest 0.5 g below 5 g/serving and to the closest gram above 5 g. A food with less than 0.5 g of saturated fat per serving has the content expressed as zero. The optional category of **polyunsaturated fat** (PUFA) is defined as *cis, cis*-methylene-interrupted polyunsaturated fatty acids and has the same gram reporting requirements as saturated fats. Another optional category (unless certain label claims are made), **monounsaturated fat,** is defined as *cis*-monounsaturated fatty acids. The requirement that the fatty acids be *cis* prevents including fatty acids that contain *trans* isomers. ***Trans* fatty acids** must be included on the nutritional label in the United States. The definition for nutritional labeling requires that the *trans* acids are not conjugated. Most *trans* fatty acids found in foods are monounsaturated, and methods for monounsaturated fatty acid and *trans* fatty acid analysis must distinguish between *cis* and *trans* monoenes, as well as the conjugated *cis-trans* fatty acids such as **conjugated linoleic acid** (CLA). CLA isomers are thought to reduce the risk of cancer and other diseases.

> *"Trans fat" or "Trans": A statement of the number of grams of trans fat in a serving, defined as the sum of all unsaturated fatty acids that contain one or more isolated (i.e., nonconjugated) double bonds in a trans configuration, except that label declaration of trans fat content information is not required for products that contain less than 0.5 gram of total fat in a serving if no claims are made about fat, fatty acid or cholesterol content.* [21 CFR 101.9(c)(2)(ii)].

The presence of *trans* fat in foods has greatly declined due to this label requirement and because the FDA has determined that the major source of trans fat in the diet, **partially hydrogenated oils** (PHO), is no longer generally recognized as safe (GRAS). However, some foods will still contain small amounts of *trans* fat because it occurs naturally in various animal products and edible fats/oils.

23.1.2 Importance of Analyses

Such issues as the sourcing of fats/oils, the effect of dietary fat on health, and food labeling requirements necessitate that food scientists be able to not only measure the total lipid content of a food but also to characterize it. Edible fats/oils traded worldwide can be evaluated to ensure identity and quality, such as distinguishing extra virgin olive oil from other types of oils as well as other quality grades of olive oil. Health concerns require the quantitative measurement of such constituents as

Table 23.2 Typical values of selected parameters for fats/oils

Fat/oil source	*Refractive index (40 °C)*	*Melting point (°C)*	*Iodine value*	*Saponification value*
Butterfat (bovine)	1.453–1.457		25–42	210–254
Cocoa butter	1.456–1.458	30–35	32–42	190–200
Coconut	1.448–1.450	21–26	5–13	242–265
Corn	1.465–1.468	−18 to −10	107–135	156–196
Cottonseed	1.458–1.466	−2	96–121	189–199
Lard			45–168	192–203
Menhaden			150–200	192–199
Olive		−3 to 0	75–94	184–196
Palm (*Elaeisguineensis*)	1.453–1.460	27–42.5	45–56	190–209
Palm kernel (*Elaeisguineensis*)	1.448–1.452	24–30	14–24	230–257
Peanut	1.460–1.465	−5 to −2	73–107	184–196
Rapeseed (canola)	1.465–1.467		110–126	182–193
Safflowerseed	1.467–1.470	−5	136–151	186–203
Soybean	1.466–1.470		118–139	188–195
Sunflower seed	1.467–1.469		115–145	186–196
Tallow (beef)	1.450–1.458	45–48	33–50	190–202

Compiled from [9] with permission from AOCS Press, Copyright 2013

trans, n-3/ω3, saturated, mono- and polyunsaturated fatty acids, cholesterol, and phytosterols. Lipid stability impacts not only the shelf life of food products but also their safety, since some oxidation products (e.g., malonaldehyde, cholesterol oxides) have toxic properties. Another area of interest is the analysis of fats/oils used in deep-fat frying operations. Finally, the development of food ingredients composed of lipids that are not bioavailable (e.g., sucrose polyesters) or lipids not contributing the normal 9 Cal/g to the diet (e.g., short- and medium-chain triglycerides) accentuates the need to characterize the lipids present in food.

23.1.3 Lipid Content in Foods and Typical Values

Commodities containing significant amounts of lipids include butter; cheese; imitation dairy products such as margarine and spreads; shortening, frying fats/oils, cooking and salad oils; emulsified dressings such as mayonnaise; peanut butter; confections; and muscle foods such as meat, poultry, and fish [7, 8]. Information is available summarizing the total fat content of foods (see Chap. 17, Table 17.2) as well as their constituent fatty acids (e.g., [9]). Ongoing studies are characterizing the quantities of saturated and unsaturated fat, *trans* isomers, cholesterol, cholesterol oxides, phytosterols, and other specific parameters in foods.

Because of their usefulness as food ingredients, often it is important to know the physical, chemical, and quality characteristics of bulk fats/oils. Definitions and specifications for bulk fats/oils (e.g., soybean oil, corn oil, and coconut oil), including values for many of the tests described in this chapter, can be found in *Physical and Chemical Characteristics of Oils, Fats, and Waxes* [9], in the *Merck Index* [10], and in *Fats and Oils* [11]. Table 23.2 gives typical values for several of the tests for some of the common commercial bulk fats/oils. It must be remembered that bulk fats/oils can vary markedly in such parameters due to differences in source, composition, and susceptibility to deterioration. Foods containing even minor amounts of lipids (e.g., <1%) can have a shelf life limited by lipid oxidation and subsequent rancidity.

23.2 GENERAL CONSIDERATIONS

23.2.1 Crude Fat/Oil Processing

After extraction from their parent sources (e.g., palm, soybean, canola, and corn), crude fats/oils typically undergo purification steps: degumming (removes phospholipids), neutralization or physical refining (removes free fatty acids), bleaching (removes pigments), and deodorization. Modifications such as fractionation, winterization, interesterification, and hydrogenation also may be a part of processing, depending on the commodity being produced. Various methods discussed in this chapter can be used to monitor the processing of crude fats/oils, such as tests for color, solid fat content (SFC), and free fatty acids [12].

23.2.2 Chemical Changes During Processing, Storage, and Use

Chemical changes that lipids undergo during processing, storage, and use include lipolysis, oxidation, and polymerization. Some of these reactions result in the formation of molecules having unpalatable off-flavors and off-odors, as measured by sensory analysis, termed **rancid**. Rancid off-flavors and off-odors from various sources have multiple descriptors, including such terms as "painty," "fishy," "beany," "metallic," and "stale." These rancid flavors and odors influence the shelf life of fats/oils and foods, and the useful life of frying fats/oils. Knowing the pathways that cause rancidity in a fat/oil or food facilitates appropriate efforts to increase the product's shelf life.

23.2.2.1 Lipolysis

Lipolysis is the hydrolysis of fatty acids from the glyceride molecule (or other lipids). Hydrolysis may be catalyzed by enzymes (lipases), acids or bases, or may occur without catalysts. Because of their volatility, hydrolysis of short-chain fatty acids (shorter than C12) from triacylglycerols can result in off-odors, known as **hydrolytic rancidity**. Dairy products are susceptible to this problem because of the relatively high content of C4 (butyric) fatty acids in milk fat. Free C12 (lauric) fatty acids are often associated with a soapy taste but no aroma. **Free fatty acids** (FFAs) longer than C12 do not cause significant impairment in taste or odor.

The extent to which a fat/oil or food has undergone lipolysis is able to be measured using the Free Fatty Acids and Acid Value tests. Free fatty acids are more susceptible than triacylglycerols to lipid oxidation.

23.2.2.2 Lipid Oxidation

In microbially stable foods as well as bulk fats/oils, **lipid oxidation** is a primary cause of limited shelf life [13, 14]. Oxidation occurs as molecular oxygen (O_2) reacts with various lipids, including free fatty acids, triacylglycerols, phospholipids, and cholesterol. As shown in Fig. 23.1, lipid oxidation occurs through three main pathways: auto-oxidation, enzymatic oxidation, and photo-oxidation [14–16]. **Auto-oxidation** occurs in lipids via a self-sustaining free radical mechanism. Depending on the fatty acid composition, the prooxidants (e.g., metal ions, light, enzymes) to which it is exposed, the type and amount of antioxidants present, and other factors, lipid decomposition can proceed via various routes that produce many different compounds. The classical explanation is that the initial or primary products are hydroperoxides that undergo scission to form various secondary products, including aldehydes, ketones, organic acids, and hydrocarbons [17] (see Fig. 23.2). However, additional products, including epoxides, dimers, and polymers, can be produced simultaneously with hydroperoxides [18]. **Enzymatic oxidation** is caused by lipoxygenases, which add O_2 to lipid mole-

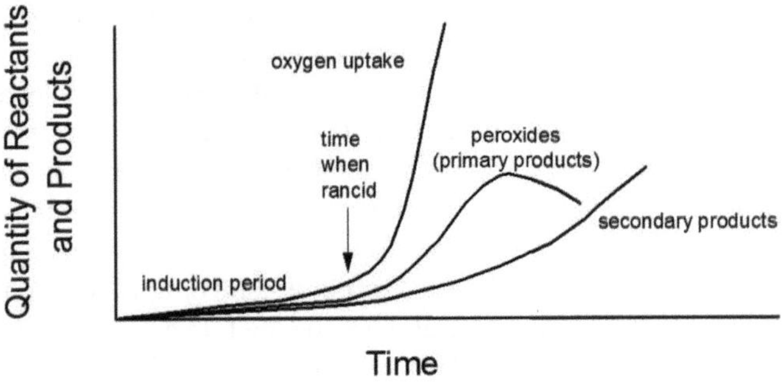

Fig. 23.2 Classical portrayal of changes in amounts of lipid oxidation reactants and products over time. (Adapted from [17], with permission. Copyright 1971, CRC Press, Boca Raton, FL)

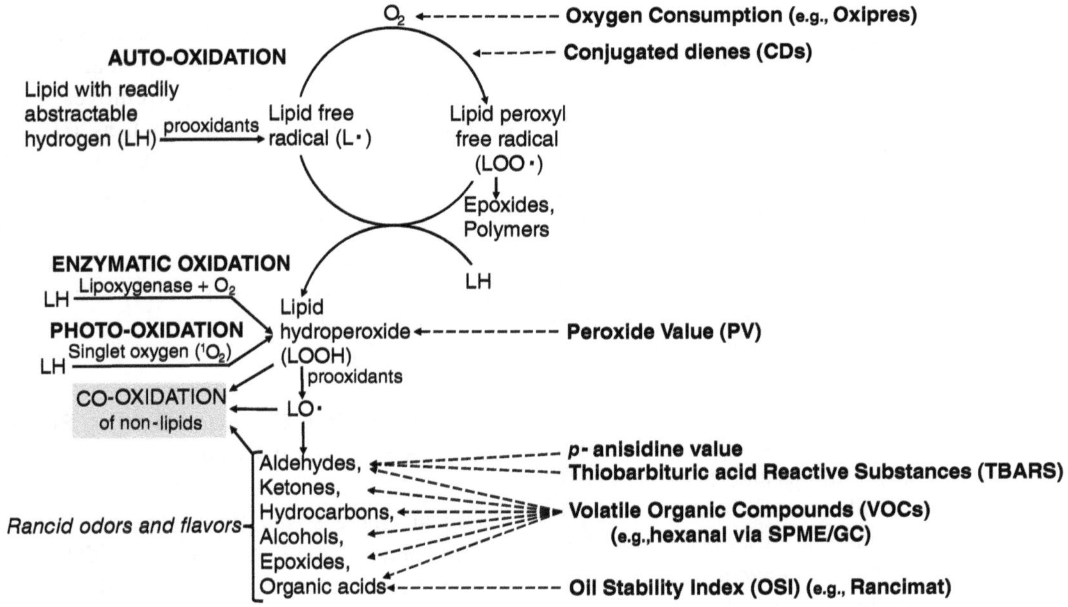

Fig. 23.1 Lipid oxidation major pathways and measurement methods. (Adapted and updated from [14–16])

cules to form hydroperoxides. **Photo-oxidation** occurs when light energy reacts with certain molecules (e.g., chlorophyll) in the presence of O_2 to form singlet oxygen ($^1O_2$), a high energy species that then reacts with lipid molecules to form hydroperoxides. All three oxidative pathways can result in the production of compounds having off-odors and off-flavors, known as **oxidative rancidity**.

The resistance of lipids to oxidation is known as **oxidative stability**. Because of their inherent properties (e.g., the amount of unsaturation and the presence of enzymes and natural antioxidants) as well as external factors (e.g., added antioxidants, processing, light, packaging, and other storage conditions), lipids and lipid-containing foods vary in their oxidative stability.

Many methods have been developed to measure the different compounds as they form or degrade during lipid oxidation. Figure 23.1 correlates common lipid oxidation measurement methods with the stage(s) of oxidation to which they correspond. In the early stages of deterioration, effective measures include oxygen consumption tests, conjugated dienes, epoxide, and hydroperoxide tests. Conjugated dienes, epoxides, and hydroperoxides degrade, however, and one test by itself (e.g., peroxide value) is not always an accurate measure of the degree of oxidation in fats/oils or foods, especially in stored products. Later stages of oxidation, during which rancid compounds are created, are monitored by methods measuring secondary products, including aldehydes, ketones, and organic acids. Since the system is dynamic and complex, multiple methods (e.g., methods measuring conjugated dienes, hydroperoxides, epoxides, and total carbonyls) should be used to obtain a more complete understanding of the oxidative status of a given fat/oil or food. This in turn provides valuable information regarding effective interventions to slow oxidation and extend shelf life. Schaich [18] discusses the principles, advantages, disadvantages, and applications of various methods for evaluating lipid oxidation.

In contrast to refined fats/oils, foods contain other non-lipid molecules that lipid free radicals can simultaneously oxidize, such as proteins, nucleic acids, polysaccharides, vitamins, and pigments. Measuring the **co-oxidation** of proteins and other compounds present in food provides a more complete understanding of the oxidative deterioration occurring in foods [18].

In some cases, it may be necessary to *visualize the location* of lipid molecules and lipid oxidation within a food or raw ingredient. **Fluorescence microscopy** with stains specific to lipids can be utilized (see Chap. 32, Sect. 32.2.2.3). Example applications include localizing oxidized lipids in a cereal product, visualizing interactions between lipids and emulsifiers, and localizing lipids in various foods.

23.2.2.3 Frying Fats/Oils and Lipid Polymerization

Fats/oils used to fry foods deteriorate and in discontinuous, batch frying operations must be replaced to maintain product quality. (Food fried in a continuous operation removes the fat/oil at a rate that typically exceeds quality deterioration.) Physical properties that change include viscosity, foaming, smoke point, and color. Chemical changes, which are the basis of effective methods for monitoring frying fat/oil quality, include the amount of **total polar compounds** and **polymerized triacylglycerols**. **Polymerization** of fats/oils occurs during frying operations where they are exposed to high temperatures for a prolonged time.

Rapid tests, based on the measurement of FFAs or oxidized lipids, are useful in quality assurance of frying operations. These rapid tests include colorimetric reaction sticks and portable instruments with digital readout [19, 20].

23.2.3 Shelf-Life Determination

The **shelf life** of fats/oils (or of foods containing fats/oils) can be defined as the time before rancidity or other characteristics (e.g., physical changes or safety concerns) reach an unacceptable level. Similarly, the **induction period** is the length of time before a rapid increase in some measure of lipid oxidation (see Fig. 23.2).

Shelf life based on rancidity is definitively determined using sensory analysis, which is expensive and time-consuming. Thus, various chemical characterization methods have been developed and their value in assessing rancidity depends on how well they correlate with sensory analysis.

Inasmuch as shelf-life determinations at ambient conditions of storage (usually room temperature) can require months or even years, accelerated tests have been developed to reduce the time required for the evaluation of the shelf life of fats/oils and foods. **Accelerated stability tests** artificially hasten lipid oxidation by exposing samples to heat, oxygen, or other catalysts. A major problem with accelerated tests is assuming reactions that occur at elevated temperatures or under other artificial conditions are the same as reactions occurring at the actual storage conditions of the product. Therefore, shelf-life determination must be done at ambient conditions of actual storage to determine a product expiration date. To the degree that accelerated stability tests correlate with ambient shelf-life studies, they may be useful in predicting changes in shelf life due to adjustments in product formulation, processing, and packaging. As described in Sect. 23.5, accelerated stability tests include those that measure oxygen consumption in a closed system (e.g., Oxipres, Oxitest) and that increase storage temperature (e.g., Oil Stability Index).

23.2.4 Sample Preparation

For bulk fats/oils, samples must be visually clear and free of sediment. When required (e.g., iodine value), samples are dried prior to testing (AOAC Method 981.11). Because exposure to heat, light, or air promotes lipid oxidation, avoiding these conditions during sample storage will retard rancidity. If samples are solid or semisolid at room temperature, they should be melted and thoroughly mixed before sampling. Sampling procedures are available for bulk fat/oils (AOCS Method C 1-47).

Various fat extraction solvents and methods are discussed in Chap. 17 and by Schaich [18]. For lipid characterization, extraction of fat/oil from foods can be accomplished by homogenizing with a solvent combination, such as hexane-isopropanol (3:2, vol/vol) or chloroform-methanol (2:1, vol/vol). The solvent then can be removed using a rotary evaporator or by evaporation under a stream of nitrogen gas. Lipid oxidation during extraction and testing can be minimized by adding antioxidants [e.g., 10–100 mg of t-butylated hydroxytoluene (BHT)/L] to solvents and by taking other precautions such as flushing containers with nitrogen and avoiding exposure to heat and light. Samples being prepared for the measurement of lipid oxidation are more effectively extracted with automated **accelerated solvent extraction** (ASE), using solvents at increased pressures (see Chap. 17, Sect. 17.4.4).

Sample preparation is hastened using **solid-phase extraction** (SPE), which consists of passing the lipid extract through a commercially available prepackaged absorbent (e.g., silica gel) that separates contaminants or various fractions based on polarity (see Chap. 14, Sect. 14.2.2.5, and Chap. 33, Sect. 33.2.2.3.2). Constituents present in lipid extractions that may present problems in lipid characterization include phosphatides, gossypol, carotenoids, chlorophyll, sterols, tocopherols, vitamin A, and metals.

23.3 PHYSICAL CHARACTERIZATION

23.3.1 Refractive Index

23.3.1.1 Principle

The **refractive index** (RI) of an oil is defined as the ratio of the speed of light in air (technically, a vacuum) to the speed of light in the oil. When a ray of light shines obliquely on an interface separating two materials, such as air and oil, the light ray is refracted in a manner defined by Snell's law, as shown in Eq. 23.1:

$$\theta_1 n_1 = \theta_2 n_2 \quad (23.1)$$

where:

θ_1 = angle of the incident light
n_1 = refractive index of material 1
θ_2 = angle of the refracted light
n_2 = refractive index of material 2

As can be seen in Fig. 23.3 and from Eq. 23.1, if the angles of incidence and refraction and the refractive index (n) of one of the two materials are known, the refractive index of the other material can be determined. In practice, the θ_1 and n_1 are constant, so n_2 is determined by measuring θ_2.

Because the frequency of light affects its refraction (violet light is refracted more than red light), white light can be dispersed or split after refraction through two materials of different refractive indexes. Refractometers often use monochromatic light to avoid errors from variable refraction of the different wavelengths of visible light.

23.3.1.2 Procedure

Samples are measured with a refractometer at 20 °C for oils and at specified higher temperatures for fats, depending on the temperature at which the fat is completely liquid.

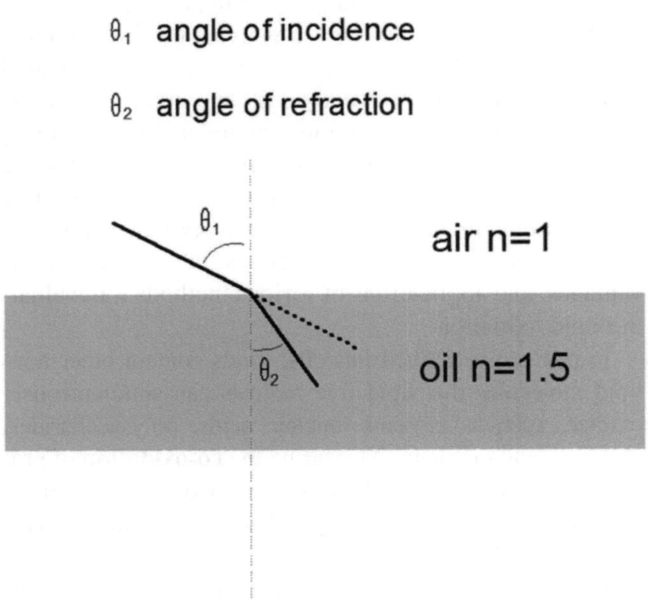

Fig. 23.3 Refraction of light in an air–oil interface

23.3.1.3 Applications

RI is related to the amount of saturation in a lipid; the RI decreases linearly as iodine value (IV) (a measure of total unsaturation) decreases. RI also is used as a measure of purity and as a means of identification, since each substance has a characteristic RI. However, RI is influenced by such factors as FFA content, oxidation, and heating of the fat or oil. The RI of various lipids is shown in Table 23.2. A relatively saturated lipid such as coconut oil has a different RI (n = 1.448–1.450) compared to a relatively unsaturated lipid such as menhaden oil (n = 1.472).

23.3.2 Melting Point

23.3.2.1 Principle

Melting point may be defined in various ways, each corresponding to a different residual amount of solid fat. The **capillary tube melting point**, also known as the **complete melting point** or **clear point**, is the temperature at which fat heated at a given rate becomes completely clear and liquid in a one-end closed capillary. The **slip melting point** is performed similarly to the capillary tube method and measures the temperature at which a column of fat moves in an open capillary when heated. The **dropping point** (also called the **dropping melting point** or **Mettler dropping point**) is the temperature at which the sample flows through a 0.11-in. hole in a sample cup placed in a specialized furnace. Fat melting properties can be evaluated using DSC (see Chap. 30, Sect. 30.3.2).

23.3.2.2 Applications

It appears the predominant method in the United States for measuring melting point is the dropping melting point. The procedure has been automated and therefore is not labor-intensive. The capillary tube method is less useful for fats/oils (in comparison to pure compounds), since they lack a sharp melting point due to their array of various components. The slip melting point often is used in Europe but has the disadvantage of a lengthy stabilization time.

23.3.3 Smoke, Flash, and Fire Points

23.3.3.1 Principle

The **smoke point** is the temperature at which the sample begins to smoke when tested under specified conditions. The **flash point** is the temperature at which a flash appears at any point on the surface of the sample; volatile gaseous products of combustion are produced rapidly enough to permit ignition. The **fire point** is the temperature at which evolution of volatiles (by decomposition of sample) proceeds with enough speed to support continuous combustion.

23.3.3.2 Procedure

A cup is filled with oil or melted fat and heated in a well-lighted container. The smoke point is the temperature at which a thin, continuous stream of bluish smoke is given off. The flash point and fire point are obtained with continued heating, during which a test flame is passed over the sample at 5 °C intervals. For fats/oils that flash at temperatures below 149 °C, a closed cup is used.

23.3.3.3 Applications

These tests reflect the volatile organic material in fats/oils, especially FFAs (Fig. 23.4) and residual extraction solvents. Frying fats/oils and refined fats/oils should have smoke points above 200 °C and 300 °C, respectively.

23.3.4 Cold Test

23.3.4.1 Principle

The **cold test** is a measure of the resistance of an oil to crystallization. The absence of crystals or turbidity indicates proper winterizing. Winterizing subjects an oil to cold temperatures and then separates the crystallized material from the bulk. This results in an oil that will not cloud at low temperature.

23.3.4.2 Procedure

Oil is stored in an ice bath (0 °C) for 5.5 h and observed for crystallization.

23.3.4.3 Applications

The cold test is a measure of success of the winterizing process. It ensures that oils remain clear even when stored at colder temperatures, useful in products such as salad dressings that will be stored in refrigeration after opening.

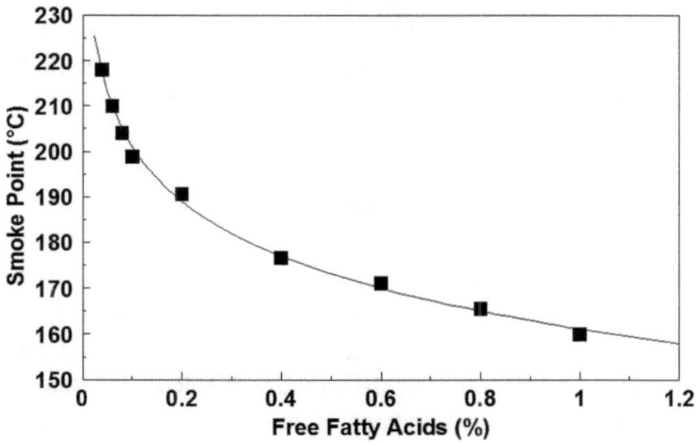

Fig. 23.4 Effect of free fatty acid content on smoke point of olive oil

23.3.5 Cloud Point

23.3.5.1 Principle

The **cloud point** is the temperature at which a cloud is formed in a liquid fat or oil due to the beginning of crystallization.

23.3.5.2 Procedure

The sample is heated to 130 °C and then cooled with agitation. The temperature of first crystallization is taken to be the point at which a thermometer in the fat/oil is no longer visible.

23.3.6 Color

Two methods for measuring the color of fats/oils are the **Lovibond** method and the **spectrophotometric** method.

23.3.6.1 Procedure

In the Lovibond method, oil is placed in a standard sized glass cell and visually compared with red, yellow, blue, and neutral color standards. Results are expressed in terms of the numbers associated with the color standards. Automated colorimeters are available.

For the spectrophotometric method, the sample is heated to 25–30 °C, placed in a cuvette, and absorbance read at the following wavelengths: 460, 550, 620, and 670 nm. The photometric color index is calculated as shown in Eq. 23.2 (AOCS Method Cc 13c-50):

$$\text{Photometric color index} = 1.29\left(A_{460}\right) + 69.7\left(A_{550}\right) + 41.2\left(A_{620}\right) - 56.4\left(A_{670}\right) \quad (23.2)$$

23.3.6.2 Applications

The color of fats/oils is commonly evaluated using the Lovibond method. Fats/oils from different sources vary in color. But if refined fat/oil is darker than expected, it is probably indicative of improper refinement or abuse [11]. Though specifically developed for testing the color of cottonseed, soybean, and peanut oils, the spectrophotometric method is probably applicable to other fats/oils as well.

23.3.7 Solid Fat Content

23.3.7.1 Principle

The amount of solids in a fat/oil, termed the **solid fat content** (SFC), can be determined using either continuous wave or pulsed NMR. Chapter 10 explains NMR as it relates to measuring the solid content of fats/oils and other foods. Chapter 30 discusses the use of DSC to measure SFC (see Chap 30, Sect. 30.2.3). Comparison between samples must be made using SFC values taken at the same temperature. The standard temperature used varies from country to country.

Originally, the amount of solids in a fat/oil was estimated using the **solid fat index** (SFI). SFI is measured using dilatometry, which determines the change in volume with change in temperature. Though the equipment is expensive, SFC is preferred over SFI because it measures the actual fat content (it is not an estimate), is less subject to error, and takes less time.

23.3.7.2 Applications

The amount of solid fat phase present in a plastic fat (e.g., margarine and shortening) depends on the type of fat, its history, and the temperature of measurement. The proportion of solids to liquids in the fat and how quickly the solids melt have an impact on functional properties, such as the mouthfeel of a food. An example of SFI use is shown in Fig. 23.5; the butter with high oleic and low saturated fatty acid composition has lower solid fat and is softer as it melts over the temperature range 10–35 °C and is more easily spreadable at refrigeration temperatures.

23.3.8 Consistency and Spreadability

The textural properties of plasticized fats (e.g., shortenings, margarine, butter) can be measured using such tests as consistency and spreadability. The **penetrometer method** of determining consistency measures the distance a cone-shaped weight will penetrate a fat in a given time period. The **spreadability** test delineates the parameters for using a Texture Technologies TA-XT2 Texture Analyzer® (or similar instrument) to determine the force needed to compress a sample. See Chap. 29 for general approaches to characteriz-

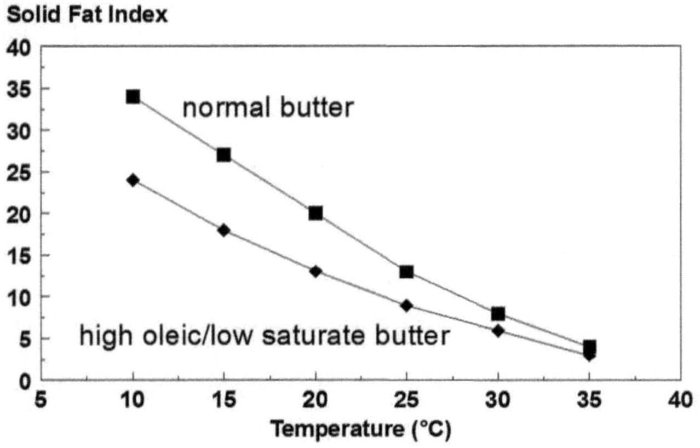

Fig. 23.5 SFI curves of butter with normal and high oleic/low saturated fatty acid compositions

ing the rheological properties of foods; many aspects can be applied to fats/oils.

The penetrometer method is useful for measuring the consistency of plastic fats and solid fat emulsions. Like SFC, consistency is dependent on the type of fat, its history, and the temperature during measurement. The spreadability method is applicable to fat/oil-containing solid suspensions, emulsion, and pastes that can maintain their shape at the temperature used for the analysis, including products such as peanut butter and mayonnaise.

23.4 CHEMICAL CHARACTERIZATION

23.4.1 Iodine Value

23.4.1.1 Principle

The iodine value (or iodine number) is a measure of degree of unsaturation, which is the number of carbon–carbon double bonds in relation to the amount of fat/oil. **Iodine value** (IV) is defined as the grams of iodine absorbed per 100 g of sample. The higher the amount of unsaturation, the more iodine is absorbed and the higher the iodine value.

Fat/oil dissolved in solvent is reacted with iodine or other halogen wherein addition to double bonds takes place (Eq. 23.3). Potassium iodide is added to reduce excess ICl to free iodine (Eq. 23.4). The liberated iodine interacts with starch to form a blue complex and is then titrated with standardized sodium thiosulfate (Eq. 23.5):

$$\underset{\text{(excess)}}{\text{ICl}} + \text{R}-\text{CH}=\text{CH}-\text{R} \rightarrow \text{R}-\text{CHI}-\text{CHCl}-\text{R} + \underset{\text{(remaining)}}{\text{ICl}} \quad (23.3)$$

$$\text{ICl} + \underset{\text{(clear)}}{2\text{KI}} \rightarrow \text{KCl} + \text{KI} + \underset{\text{(yellow)}}{\text{I}_2} \quad (23.4)$$

$$\text{I}_2 + \underset{\text{(blue)}}{\text{starch}} + 2\text{Na}_2\text{S}_2\text{O}_3 \rightarrow 2\text{NaI} + \underset{\text{(colorless)}}{\text{starch}} + \text{Na}_2\text{S}_4\text{O}_6 \quad (23.5)$$

A common practice is to determine **calculated iodine value** from the fatty acid composition (see Sect. 23.6.2) using AOCS Recommended Practice Cd 1c-85. The calculated iodine value is not meant to be a rapid method but instead gives two results (iodine value of triacylglycerols and FFAs) from one analysis (fatty acid composition).

23.4.1.2 Procedure

A quantity of melted fat or oil is dissolved in solvent. Excess iodine is added and allowed to stand for 30 min in the dark with occasional shaking. A solution of potassium iodide is added to reduce excess ICl to free iodine, which is yellow (Eq. 23.4). The liberated iodine is then titrated with a standardized sodium thiosulfate solution, with vigorous shaking, until the yellow color almost disappears. The starch indicator is added, and the titration is continued to a colorless endpoint (Eq. 23.5). Iodine value is calculated as shown in Eq. 23.6:

$$\text{Iodine value} = \frac{(B-S)\times N\times 126.9}{W\times 1000}\times 100 \quad (23.6)$$

where:

Iodine value = g iodine absorbed per 100 g of sample
B = volume of titrant (mL) for blank
S = volume of titrant (mL) for sample
N = normality of $Na_2S_2O_3$ (mol/L)
126.9 = MW of iodine (g/mol)
W = sample mass (g)
1000 = conversion of units (mL/L)

Calculated iodine value is obtained from fatty acid composition using Eq. 23.7 for triacylglycerols. A similar equation allows calculation of the iodine value of FFAs:

$$\begin{aligned}\text{Iodine value (triglycerides)} = &(\%\ \text{hexadecenoic acid}\times 0.950)\\ &+(\%\ \text{octadecenoic acid}\times 0.860)\\ &+(\%\ \text{octadecadienoic acid}\times 1.732)\\ &+(\%\ \text{octadecatrienoic acid}\times 2.616)\\ &+(\%\ \text{eicosenoic acid}\times 0.785)\\ &+(\%\ \text{docosenoic acid}\times 0.723)\end{aligned} \quad (23.7)$$

23.4.1.3 Applications

Iodine value is used to characterize fats/oils, to follow the hydrogenation process in refining, and as an indication of lipid oxidation, since there is a decline in unsaturation during oxidation.The calculated value tends to be low for materials with a low iodine value and for fats/oils with greater than 0.5% unsaponifiable material (e.g., fish oils). The Wijs iodine procedure uses ICl and the Hanus procedure uses IBr. The Wijs procedure may be preferable for highly unsaturated oils as it reacts faster with the doublebonds.

23.4.2 Saponification Value

23.4.2.1 Principle

Saponification is the process of breaking down or degrading a neutral fat/oil into glycerol and fatty acids by the treatment of the fat/oil with alkali (Eq. 23.8):

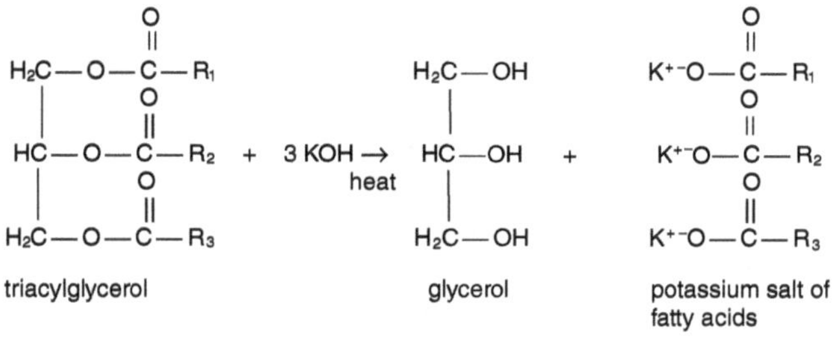

(23.8)

The **saponification value** (SV) (or saponification number) is defined as the amount of alkali necessary to saponify a given quantity of fat/oil. It is expressed as the mg KOH required to saponify 1 g of sample. The saponification value is an index of the mean molecular weight of the triacylglycerols in the sample. The mean molecular weight of the triacylglycerols may be divided by 3 to give an approximate mean molecular weight for the fatty acids present; the smaller the saponification value, the longer the average fatty acid chain length.

In common practice, the **calculated saponification value** is determined from the fatty acid composition (see Sect. 23.6.2) using AOCS Method Cd 3a-94.

23.4.2.2 Procedure

Excess alcoholic potassium hydroxide is added to the sample, and the solution is heated to saponify the fat (Eq. 23.8). The unreacted potassium hydroxide is back-titrated with standardized HCl using phenolphthalein as the indicator, and the saponification value is calculated (Eq. 23.9):

$$\text{Saponification value} = \frac{(B - S) \times N \times 56.1}{W} \tag{23.9}$$

where:

Saponification value = mg KOH per g of sample
B = volume of titrant (mL) for blank
S = volume of titrant (mL) for sample
N = normality of HCl (mmol/mL)
56.1 = MW of KOH (mg/mmol)
W = sample mass (g)

The **calculated saponification value** is obtained from fatty acid composition using Eq. 23.10. The **fractional molecular weight** of each fatty acid in the sample must be determined first by multiplying the fatty acid percentage (divided by 100) by its molecular weight. The **mean molecular weight** is the sum of the fractional weights of all the fatty acids in the sample:

$$\text{Calculated saponification value} = \frac{3 \times 56.1 \times 1000}{(\text{mean molecular weight} \times 3) + 92.09 - (3 \times 18)} \tag{23.10}$$

where:

Calculated saponification value = mg KOH per g of sample
3 = number of fatty acids per triacylglycerol
56.1 = MW of KOH (g/mol)
1000 = conversion of units (mg/g)
92.09 = MW of glycerol (g/mol)
18 = MW of water (g/mol)

23.4.2.3 Applications

Saponification value is useful for determining the average fatty acid chain length of a fat/oil. The calculated saponification value is not applicable to fats/oils containing high amounts of unsaponifiable material, FFAs (>0.1%), or mono- and diacylglycerols (>0.1%).

23.4.3 Free Fatty Acids and Acid Value

23.4.3.1 Principle

Measures of fat/oil acidity normally reflect the amount of fatty acids hydrolyzed from triacylglycerols (Eq. 23.11):

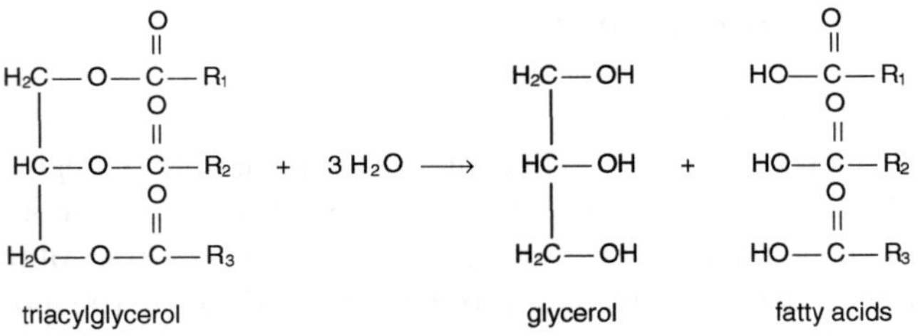

(23.11)

Free fatty acid (FFA) is the percentage by weight of a specified fatty acid (e.g., percent oleic acid). **Acid value** (AV) is defined as the mg KOH necessary to neutralize the free acids present in 1 g of fat/oil. In addition to FFAs, acid phosphates and amino acids also can contribute to acidity. In samples containing no acids other than fatty acids, FFA and acid value may be converted from one to the other using a conversion factor (Eq. 23.12). Acid value conversion factors for lauric and palmitic are 2.81 and 2.19, respectively:

$$\%\text{FFA}(\text{as oleic}) \times 1.99 = \text{acid value} \tag{23.12}$$

23.4.3.2 Procedure

To a liquid fat/oil sample, neutralized 95% ethanol and phenolphthalein indicator are added. The sample then is titrated with standardized KOH to a slight pink color endpoint. The percent FFA is calculated (Eq. 23.13):

$$\%\text{FFA}(\text{as oleic}) = \frac{V \times N \times 282}{W \times 1000} \times 100 \tag{23.13}$$

where:

% FFA = percent free fatty acid (g/100 g), expressed as oleic acid
V = volume of KOH titrant (mL)
N = normality of KOH titrant (mol/L)
282 = MW of oleic acid (g/mol)
W = sample mass (g)
1000 = conversion of units (mL/L)

23.4.3.3 Applications

In crude fat/oil, FFA or acid value estimates the amount of fat/oil that will be lost during refining steps designed to remove fatty acids. In refined fats/oils, a high acidity level means a poorly refined fat/oil or fat/oil breakdown after storage or use. However, if a fat/oil seems to have a high amount of FFAs, it may be attributable to acidic additives (e.g., citric acid added as a metal chelator) since any acid will participate in the reaction [11]. If the fatty acids liberated are volatile, FFA or acid value may be a measure of hydrolytic rancidity. The AV is often used as a quality indicator in frying fats/oils, where a limit of 2 mg KOH/g oil is sometimes used.

23.4.4 Conjugated Dienes

23.4.4.1 Principle

Double bonds in polyunsaturated fatty acids are changed from nonconjugated to conjugated bonds upon free radical formation and thus reflect the initiation of lipid oxidation in fats/oils. Fatty acid **conjugated dienes** (CDs) absorb UV light at about 232 nm and **conjugated trienes** at about 270 nm. (Conjugated products of lipid oxidation should not be confused with conjugated linoleic acid isomers, which are produced by incomplete hydrogenation in ruminants.)

23.4.4.2 Procedure

A homogeneous, clear lipid sample is weighed into a graduated flask and brought to volume with a suitable solvent (e.g., isooctane and cyclohexane), then absorbance is immediately measured using a spectrophotometer. For example, a 0.5% w/v solution (e.g., 50 mg oil sample in 10 mL solvent) is measured; it is diluted or concentrated as necessary to obtain a reading within the optimal range of the instrument, usually between 0.1 and 1.0 absorbance units (see Chap. 7, Sect. 7.2.5). Concentration of CDs can be reported as percent lipid weight or calculated on a molar basis from Beer's law using known molar extinction coefficients [18]. Interferences and background absorbances are minimized using appropriate steps such as carefully washed glassware, sample cleanup, and using the pure solvent as a blank.

A common practice is to calculate the amount of conjugated dienes from chromatographic analysis of fatty acids (see Chap. 14).

23.4.4.3 Applications

The conjugated diene method is useful for monitoring the early stages of lipid oxidation. The magnitude of the changes in absorption is not easily related to the extent of oxidation in advanced stages. Olive oil quality is monitored based on ultraviolet absorbance; low amounts of conjugated dienes and trienes are indicative of an unadulterated, high purity oil.

23.4.5 Peroxide Value

23.4.5.1 Principle

Lipid hydroperoxides are more stable than free radicals in the process of lipid oxidation (see Fig. 23.1) and as such they are commonly quantified to determine the extent of oxidation in fats/oils. **Peroxide value** (PV) is defined as the milliequivalents (mEq) of hydroperoxide per kg of fat/oil. It is a redox titrimetric determination wherein potassium iodide reduces lipid hydroperoxides to hydroxyl compounds (Eq. 23.14). The iodine produced reacts with starch to form a blue complex, then reacts with the sodium thiosulfate titrant until a colorless endpoint is reached (Eq. 23.15):

$$\mathrm{ROOH} + \mathrm{K^+I^-} \xrightarrow{\mathrm{H^+,heat}} \mathrm{ROH} + \underset{\text{(excess)}}{\mathrm{K^+OH^-}} + \mathrm{I_2} \tag{23.14}$$

$$\mathrm{I_2} + \underset{\text{(blue)}}{\mathrm{starch}} + 2\mathrm{Na_2S_2O_3} \rightarrow 2\mathrm{NaI} + \underset{\text{(colorless)}}{\mathrm{starch}} + \mathrm{Na_2S_4O_6} \tag{23.15}$$

23.4.5.2 Procedure

The fat/oil sample is dissolved in glacial acetic acid-isooctane (3:2). Excess potassium iodide is added along with starch as an indicator. The solution then is titrated with standardized sodium thiosulfate to a colorless endpoint. Because of the difficulty of visually detecting the endpoint, potentiometric determination of endpoint is advisable (see Chap. 22, Sect. 22.4.2.2). Precautions include sparging solvents with an inert gas to remove oxygen, performing the analysis in subdued lighting, using a magnetic stirrer, and performing the titration rapidly [18].

Peroxide value is calculated as (Eq. 23.16):

$$\text{Peroxide value} = \frac{(S-B)\times N}{W} \times 1000 \tag{23.16}$$

where:

Peroxide value = mEq peroxide per kg of sample
S = volume of titrant (mL) for sample
B = volume of titrant (mL) for blank
N = normality of $Na_2S_2O_3$ solution (mEq/mL)
1000 = conversion of units (g/kg)
W = sample mass (g)

23.4.5.3 Applications

Peroxide value measures a transient product of oxidation, i.e., hydroperoxides break down to form other products (see Fig 23.1). Though peroxide values measured on a given sample stored over time increase and then decrease (see Fig. 23.2), products are usually deemed rancid (via sensory analysis) during the period of increasing values. The AOCS methods require 5 g of fat/oil sample size, which may be difficult to obtain from foods low in fat. The method is empirical, and any modifications may change results. However, smaller sample quantities are routinely measured, with the amount required depending on the degree of oxidation [13,

18]. Despite limitations, peroxide value is one of the most common tests of lipid oxidation.

High-quality, freshly deodorized fats/oils will have a peroxide value of 0. Peroxide values >20 correspond to very poor quality fats/oils, which normally will have significant off-flavors. For soybean oil, peroxide values of 1–5, 5–10, and >10 correspond to low, medium, and high levels of oxidation, respectively (AOCS Method Cg 3-91).

23.4.6 *p*-Anisidine Value and Totox Value

23.4.6.1 Principle

The *p*-**anisidine value** estimates the amount of α- and β-unsaturated aldehydes (mainly 2-alkenals and 2,4-dienals), which are secondary oxidation products in fats/oils. The aldehydes react with *p*-anisidine to form a chromogen that is measured spectrophotometrically. The **totox value** tends to indicate the total oxidation of a sample using both the peroxide and *p*-anisidine values (Eq. 23.17):

$$\text{Totox value} = p\text{-anisidine value} + (2 \times \text{peroxide value}) \quad (23.17)$$

23.4.6.2 Procedure

The *p*-**anisidine value** by convention is defined as 100 times the absorbance at 350 nm of a solution containing precisely 1 g of test fat/oil diluted to 100 mL using a mixture of solvent (isooctane) and *p*-anisidine.

23.4.6.3 Applications

Since peroxide value measures hydroperoxides (which increase then decrease) and *p*-anisidine value measures aldehydes (decay products of hydroperoxides, which continually increase), the totox value usually rises continually during the course of lipid oxidation. Fresh soybean oil should have a *p*-anisidine value <2.0 and a totox value <4.0 (AOCS Method Cg 3-91). Though not common in the United States, *p*-anisidine and totox values are extensively used in Europe. Other tests for aldehydes may be preferable since *p*-anisidine is highly toxic [18].

23.4.7 Thiobarbituric Acid Reactive Substances

23.4.7.1 Principle

The **thiobarbituric acid reactive substances** (TBARS) test, also known as the **thiobarbituric acid** (TBA) test, measures secondary products of lipid oxidation, primarily **malonaldehyde**. It involves reaction of malonaldehyde (or malonaldehyde-type products including unsaturated carbonyls) with TBA to yield a colored compound that is measured spectrophotometrically. The food sample may be reacted directly with TBA but is often steam distilled, to eliminate interfering substances, and then the distillate is reacted with TBA. Other approaches include reacting TBA with lipids or aqueous extracts of the food. An alternative to spectrophotometric detection is to determine the content of malonaldehyde using HPLC. Many modifications of the test have been developed.

23.4.7.2 Procedure

In contrast to the direct method for fats/oils (see Table 23.1), the procedure outlined here requires distillation of the food prior to determining TBARS. A weighed sample is combined with distilled water and mixed. The pH is adjusted to 1.2, and the sample is transferred to a distillation flask. After addition of BHT (optional), antifoam reagent, and boiling beads, the sample is distilled rapidly, and the first 50 mL is collected. An aliquot of the distillate is combined with TBA reagent and heated in a boiling water bath for 35 min. Absorbance of the solution is determined at 530 nm, and, using a standard curve, absorbance readings are typically converted to mg of malonaldehyde (or TBARS) per kg of sample.

23.4.7.3 Applications

The TBA test correlates better with sensory evaluation of rancidity than does peroxide value but has many limitations. Despite common use, especially in muscle foods, tests for measuring malonaldehyde are only applicable to foods containing fatty acids with at least three double bonds and therefore are not useful in many foodsystems. Criticisms of the test include long reaction times, low sensitivity, and nonspecificity of reactive substances[18].

23.4.8 Volatile Organic Compounds

23.4.8.1 Principle

Volatile organic compounds (VOCs) present in fats/oils and foods are related to flavor, quality, and oxidative stability. These compounds include secondary products of lipid oxidation that are responsible for the off-flavors and odors of oxidized lipids (AOCS Method Cg 3-91). The compounds formed will vary depending on such parameters as the fatty acid composition of the sample and environmental storage conditions.

Prior to GC separation and detection of VOCs, various sampling methods are employed, which affect the number of volatiles detected and sensitivity of the test. **Static (direct) headspace analysis** entails the chromatographic analysis of a set volume of vapor obtained from the headspace above a sample held in a closed container. **Dynamic headspace sampling (purge and trap)** involves purging the sample with an inert gas that then passes through an absorbent material that retains the volatiles, which are then thermally desorbed onto the chromatographic column (see Chap. 14, Sect. 14.2.2.2). **Solid-phase microextraction** (SPME) improves the sensitivity of static headspace analysis of VOC because the volatiles are concentrated on a SPME fiber (see Chap. 14, Sect. 14.2.2.5).

23.4.8.2 Procedure

General guidelines regarding GC parameters are given in AOCS Method Cg 4-94, under the heading of static headspace; a review of research literature will indicate current practices used with various commodities. Typically, a small sample of the commodity is placed in a container having a septum cap. An internal standard is added, and the container is sealed and then heated for a given time to increase the concentration of headspace volatiles. An aliquot of the vapors in the container headspace is injected into a GC equipped with a flame ionization detector or mass selective detector. Automated headspace samplers both heat the sample and ensure a constant volume is being analyzed. The quantity of the volatiles of interest is calculated from peak areas (see Chaps. 12 and 14).

23.4.8.3 Applications

The formation of volatile aldehydes, such as hexanal, may correlate well with sensory determination of lipid oxidation since they are major contributors to off-odors in some food commodities. Commonly measured compounds include propanal, pentanal, hexanal, and 2,4-decadienal. However, collecting, identifying, and quantifying all volatiles, and combining this information with other oxidation tests, will provide a more complete understanding of the oxidation pathways (see Fig. 23.1) and extent of lipid oxidation in a fat/oil or food [18]. This, in turn, can suggest more effective means of inhibiting rancidity and lengthening shelf life.

An advantage of this approach is that since volatiles are collected from the headspace of the sample vial, lipid extraction is not required (i.e., intact foods can be analyzed). However, for SPME, various fibers are available, and the specific fiber used may not absorb all the volatiles produced.

Ongoing development of the **electronic nose** (E-nose) (Chap. 35, Sect. 35.5.2.2), which uses sensor arrays and pattern recognition systems to detect and identify odors and flavors, presents advantages and limitations in utilizing such instruments to identify edible fats/oils and evaluate the extent of deterioration in fats/oils and foods [21].

23.4.9 Total Polar Compounds

23.4.9.1 Principle

Deterioration of frying fats/oils can be monitored by measuring **total polar compounds** (TPCs) (see Sect. 23.2.2.3), which include monoacylglycerols, diacylglycerols, FFAs, and oxidation products formed during heating of foods. Nonpolar compounds are primarily unaltered triacylglycerols. The polar compounds in a sample can be separated from nonpolar compounds using chromatographic techniques.

23.4.9.2 Procedure

Polar compounds are measured by dissolving the fat/oil sample in light petroleum ether-diethyl ether (87:13), then applying the solution to a silica gel column. Polar compounds are adsorbed onto the column. Nonpolar compounds are eluted, the solvent evaporated, the residue weighed, and the total amount of polar compounds estimated by difference.

23.4.9.3 Applications

A suggested limit of approximately 25% polar compounds in frying fat/oil is a guide for when it should be discarded, but this limit varies by country. A limitation of this method is the time required. In contrast, portable digital devices rapidly estimate polar compounds based on the dielectric constant of the fat/oil[19].

23.4.10 Polymerized Triacylglycerols

Deterioration of frying fats/oils can be monitored by quantifying polymerized triacylglycerols (see Sect. 23.2.2.3). **Polymerized triacylglycerols** are measured by size exclusion chromatography [22] (see Chap. 13, Sect. 13.3.5). Sample oils are injected into the column and then developed with tetrahydrofuran. Polymerized triacylglycerols are eluted prior to unoxidized triacylglycerols and detected using a refractive index detector. Polymerized triacylglycerols are defined as (Eq. 23.18):

$$\text{Polymerized triacylglycerols}(\%) = C / \Sigma B \times 100 \quad (23.18)$$

where:

C = area of chromatographic peaks of polymerized triacyglycerols
ΣB = sum of area of all chromatographic peaks

In Europe, the limit of polymerized triacylglycerols in frying fats/oils is 10–16%.

23.5 ACCELERATED STABILITY TESTING

23.5.1 Overview

As discussed in Sect. 23.2.3, shelf-life determination should be done at ambient conditions of actual storage and can require months or even years. **Accelerated stability tests** hasten lipid oxidation by exposing samples to heat, oxygen, or other catalysts. However, reaction pathways occurring at elevated temperatures or other artificial conditions often are not the same as reactions occurring at actual storage conditions of the product. Nevertheless, accelerated stability tests that correlate well with ambient shelf-life studies may be helpful in predicting differences in shelf life due to product changes.

Induction period is the time before a rapid increase in lipid oxidation, as measured by various methods (see Fig. 23.2). The end of the induction period can be determined by such methods as calculating the maximum of the second derivative with respect to time or manually drawing tangents to the lines (Fig. 23.6). Measurement of the induction period allows a comparison of the stability of samples that contain differing ingredients or of samples held at varying storage conditions, and it provides an indication of antioxidant effectiveness in delaying lipid oxidation and thus prolonging the shelf life of fats/oils and foods. Guidelines for conducting such tests have been established (see AOCS Method Cg 7-05). (In contrast to measuring the effectiveness of antioxidants in prolonging food shelf life, antioxidant capacity refers to the ability of foods and ingredients to counter oxidative reactions in the human body and is discussed in Chap. 25.)

23.5.2 Accelerated Shelf-Life Tests

23.5.2.1 Principle

In **accelerated shelf-life tests** (ASLTs), the progression of lipid oxidation is evaluated by placing samples into carefully controlled storage conditions, including temperature, atmosphere, lighting (or darkness), and agitation (or without agitation). Samples are then removed at given time intervals for analysis by various methods, including the chemical tests discussed in Sect. 23.4 and other measures of oxidation. A more complete understanding is derived when multiple methods are used to determine the nature and extent of oxidation, including tests for initial products (e.g., CD, PV) and secondary products (e.g., VOCs).

23.5.2.2 Applications

This controlled approach to studying lipid oxidation is utilized for quality assurance as well as research purposes. It is useful in comparing product formulation (including antioxidants), processing, packaging, and other parameters on the stability and shelf life of fats/oils and foods. It is also useful in fundamental research studying the complexities of lipid oxidation.

ASLT has generally replaced the ill-defined Schaal oven test and the **oven storage test** (AOCS Method Cg 5-97). Like these previous tests, predicting shelf life using ASLT is problematic since even slightly elevated temperatures (40–60 °C) and other accelerating storage parameters can modify oxidation pathways occurring at lower temperatures and ambient conditions [18].

23.5.3 Oil Stability Index

23.5.3.1 Principle

The **oil stability index** (OSI) (AOCS Method Cd 12b-92) determines induction period by bubbling purified air through a fat/oil sample held at an elevated temperature (often 110 °C or 130 °C) and then passing the acidic volatiles (primarily formic acid) into a deionized water trap. The conductivity of the water is measured continuously, resulting in data similar to those shown in Fig. 23.6. Results should specify the air flow rate and temperature used, as well as induction period time. An instrument that automates this method is the Rancimat® (Metrohm USA, Inc.). The outdated and labor-intensive **active oxygen method** (AOM) is similar to the OSI except induction period is determined by discontinuous measurements of either peroxide value or sensory evaluation of rancid odor.

23.5.3.2 Applications

This method was originally designed to measure the effectiveness of antioxidants. Extrapolating OSI results to actual shelf life is difficult because the high temperatures used result in the formation of compounds not present in samples held at ambient or slightly elevated temperature conditions.

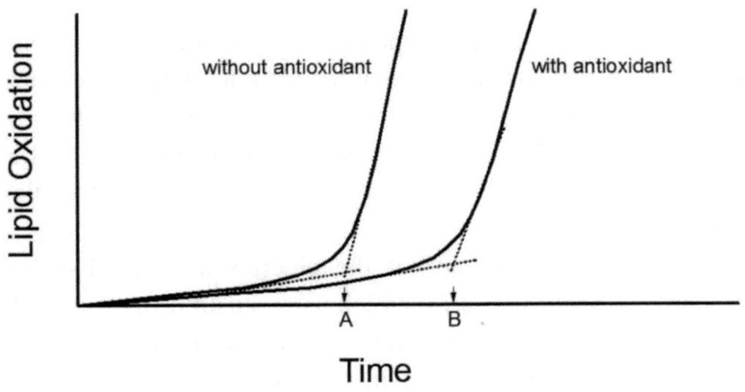

Fig. 23.6 A plot of lipid oxidation over time, showing the effect of an antioxidant on induction period. Time *A* is induction period of a less stable sample and Time *B* is induction period of more stable sample

Specification sheets for fats/oils sometimes report AOM values to accommodate those familiar with them. OSI values can be converted to AOM values.

Applicable to all fats/oils, the OSI has also been researched for applicability to certain low-moisture snack foods (e.g., potato chips and corn chips). Because of the continuous exposure to circulating air, samples that contain more than negligible amounts of water tend to dehydrate during the determination and are not likely to give reliable results because water activity can affect oxidation rates.

23.5.4 Oxygen ConsumptionTests

23.5.4.1 Principle

Inasmuch as lipid oxidation results in a decrease of oxygen in the surrounding environment (see Figs. 23.1 and 23.2), measuring the time required for the onset of rapid oxygen consumption (also referred to as absorbance, uptake or disappearance) in a closed system provides a means of determining oxidative stability. The devices have a sample cell pressurized (1–8 bars) with a controlled atmosphere (e.g., purified air or pure oxygen) and a controlled temperature (up to 200 °C) and then measure the change in pressure over time. Induction period is determined by measuring the time required for a sharp drop in pressure to occur, which corresponds with the rapid consumption of oxygen by the sample. Names of tests or devices correspond to instruments from various vendors, including the **Oxipres** (MikroLab Aarhus, Denmark), **RapidOxy** 100 (Anton Paar USA Inc., TX), and **Oxitest** (VELP Scientific, NY). The latter device is utilized in AOCS Method Cd 12c-16.

23.5.4.2 Applications

A major advantage, compared to the OSI, is that these tests may be used with intact foods instead of only extracted fats/oils. When used to measure stability or antioxidant effectiveness, conditions are typically high pressures (e.g., 5 bars) of pure oxygen and moderate temperatures (e.g., 60–80 °C). Tests of oxygen consumption may also be utilized to characterize the initial stage of oxidation, using lower pressures (e.g., 2 bars) of purified air. Comparison of oxygen consumption results, which monitors the initiation of lipid oxidation, with the OSI, which monitors secondary products of lipid oxidation (see Fig. 23.1), allows for distinguishing where a given treatment has its effect on lipid oxidation [18].

23.6 LIPID FRACTIONS

23.6.1 Overview

The lipid present in foods or bulk fats/oils can be characterized by measuring the amount of its various fractions, which include fatty acids, mono-, di-, and triacylglycerols, phospholipids, sterols (including cholesterol and phytosterols), and lipid-soluble pigments and vitamins. Another means of categorizing lipid fractions is inherent in nutrition labeling which involves the measurement of not only total fat but may also require quantification of saturated fat, monounsaturated fat, polyunsaturated fat, and *trans* isomer fatty acids. In addition, foods may contain lipids that do not contribute the same caloric content as normal lipids, for example, sucrose polyesters, medium-chain triglycerides, and triglycerides that contain short-chain fatty acids. Many of these fractions are determined readily by evaluating the component fatty acids. From the fatty acid composition, calculations are made to determine such parameters as total fat, saturated fat, calculated iodine value, and calculated saponification value.

GC (see Chap. 14) is ideal for the analysis of many lipid components. GC can be used for determinations such as total fatty acid composition, distribution and position of fatty acids in lipid, sterols, studies of fat/oil stability and oxidation, assessing heat or irradiation damage to lipids, and detection of adulterants and antioxidants. Methods exist that detail the analysis of various lipid fractions using GC [5]. GC combined with **mass spectrometry** (MS) (see Chap. 11) is a powerful tool used in the identification of compounds. **HPLC** (see Chap. 13) also is useful in lipid analyses, especially for components that are not readily volatilized, such as hydroperoxides and triacylglycerols [6]. **Thin-layer chromatography** (TLC) (see Chap. 12) has been used extensively in the past by lipid chemists. Partly due to low cost and ease, TLC is still useful, although many assays are more quantitative and have better resolution using GC or HPLC.

23.6.2 Fatty Acids

The **fatty acid composition,** or **fatty acid profile,** of a food product is determined by quantifying the kind and amount of fatty acids that are present, usually by extracting the lipids and analyzing them using GC (also described in Chap. 17, Sect. 17.2.7).

23.6.2.1 Principle

To increase the volatility before GC analysis, triacylglycerols are typically transesterified to form **fatty acid methyl esters (FAMEs)** (Eq. 23.19). Acyl lipids are readily transesterified using base such as sodium hydroxide and methanol. Sodium methoxide produced by this combination will create FAMEs from acyl lipids rapidly but will not react with free fatty acids. Acidic reagents such as methanolic HCl or boron trifluoride (BF_3) react rapidly with FFAs but more slowly with acyl lipids. Procedures such as AOCS Method Ce 1b-89 (a joint method with AOAC 991.39) use a two-step methylation, first reacting the lipid with 0.5*N* NaOH and then with excess BF_3/methanol. This allows a rapid methylation of FFAs, acyl lipids, and phospholipids. The sodium hydroxide step is not a saponification procedure (i.e., hydrolysis of acyl groups); it is a direct transmethylation:

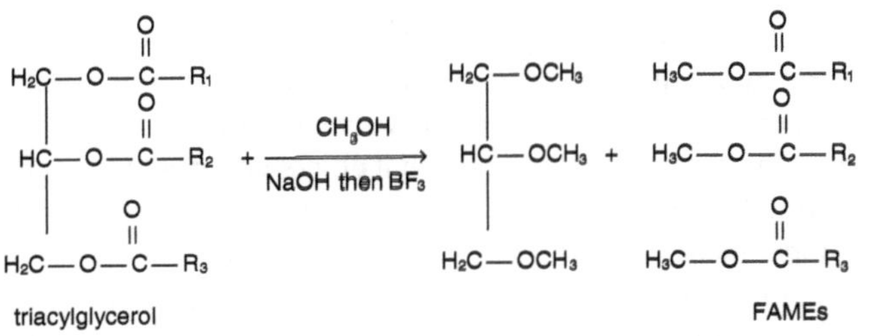

(23.19)

23.6.2.2 Procedure

The lipid is extracted from the food, for example, by homogenizing with a suitable solvent such as hexane-isopropanol (3:2, vol/vol) and then evaporating the solvent. The FAMEs are prepared by combining the extracted lipid with sodium hydroxide-methanol and internal standard in isooctane and then heating at 100 °C for 5 min. The sample is cooled, and then excess BF_3-methanol is added with further heating (100 °C for 30 min). After addition of saturated aqueous sodium chloride, additional isooctane, and mixing, the upper isooctane solution containing the FAMEs is removed and dried with anhydrous Na_2SO_4 and then diluted to a concentration of 5–10% for injection onto the GC.

Several methods (see Table 23.1) describe procedures and conditions for using GC to determine fatty acid composition. For examples, AOCS Method Ce 1b-89 is specific for marine oils.

23.6.2.3 Applications

Determination of the fatty acid composition of a product permits the **calculation** of the following categories of fats that pertain to health issues and food labeling: percent saturated fatty acids, percent unsaturated fatty acids, percent monounsaturated fatty acids, percent polyunsaturated fatty acids, CLAs, and percent *trans* isomer fatty acids. Calculation of fatty acids as a percentage is referred to as **normalization**, i.e., the areas of all of the FAMEs are summed and the percent area of each fatty acid is calculated relative to the total area. This is a reasonable procedure because with flame ionization detectors (FID), the weight of fatty acids in a mixture closely parallels the area on the chromatogram. However, this is not absolutely correct. Theoretical correction factors are needed to correct for the FID response, which is different depending on the level of unsaturation in FAMEs. A chromatogram showing separation of FAMEs of varying length and unsaturation is shown in Chap. 17 (Fig. 17.4).

The complexity of the fatty acids found in various foods will affect the details of the GC analysis that can be used. Analysis of FAMEs of a vegetable oil is quite simple and can easily be accomplished in less than 20 min using a column with a medium polarity liquid phase. The fatty acids present in most vegetable oils range from C14 to C24. Coconut and palm kernel oils also contain shorter-chain fatty acids such as C8–C12. Dairy fats contain butyric acid (C4) and other short-chain fatty acids, whereas peanut oil contains C26 at around 0.4–0.5% of the total FAMEs. Marine lipids contain a much wider range of fatty acids and require care in the separation and identification of FAMEs, many of which have no commercially available standards.

Trans fatty acids in foods originate from three main sources: biohydrogenation in ruminants, incomplete hydrogenation in the conversion of liquid oils to plastic fats, and high-temperature exposure during deodorization. The *trans* fatty acids formed from these three processes are quite different and require careful attention to achieve accurate analysis.

Separation of *trans* FAMEs is facilitated by selection of the most polar column phases available. Even with optimized temperature programming and column selection, resolution of *trans* isomers from partially hydrogenated vegetable oil mixtures is incomplete and facilitated by the use of a Fourier transform infrared (FTIR) detector (Chap. 8, Sect. 8.3.1.2) or mass spectrometry (Chap. 11).

23.6.3 *trans* Isomer Fatty Acids

Most natural fats/oils extracted from plant sources contain only isolated (i.e., methylene interrupted, nonconjugated) *cis* double bonds. Fats/oils extracted from animal sources may contain small amounts of *trans* double bonds. Inasmuch as the *trans* isomer is more thermodynamically stable, additional amounts of *trans* double bonds can be formed in fats/oils that undergo oxidation, or during processing treatments such as extraction, heating, and hydrogenation.

Measurement of *trans* isomer fatty acids is commonly done using GC techniques, such as AOCS Method Ce 1h-05 (see Sect. 23.6.2). However, this section describes the use of infrared (IR) spectroscopy for determining *trans* isomer fatty acids.

23.6.3.1 Principle

The concentration of *trans* fatty acids is measurable in lipids from an absorption peak at 966 cm^{-1} in the IR spectrum.

23.6.3.2 Procedure

AOCS Method Cd 14-95 requires liquid samples be converted to methyl esters and dissolved in an appropriate solvent that does not absorb in the IR region strongly due to planar (carbon disulfide) or tetrahedral (carbon tetrachloride) symmetry. The absorbance spectra between 1050 and 900 cm^{-1} are obtained using an infrared spectrometer (see

Chap. 8). Methyl elaidate is used as an external standard in calculating the content of *trans* double bonds. Alternately, AOCS Method Cd 14d-99 determines total *trans* fatty acids using attenuated total reflection-Fourier transform infrared (ATR-FTIR) spectroscopy (see Chap. 8, Sects. 8.3.1.2 and 8.3.2).

23.6.3.3 Applications

The methods described will only detect isolated (i.e., nonconjugated) *trans* isomers. This is especially important when oxidized samples are of interest since oxidation results in a conversion from nonconjugated to conjugated double bonds. Also, AOCS Method Cd 14-95 is restricted to samples containing at least 5% *trans* isomers, and AOCS Method Cd 14d-99 is limited to samples containing at least 0.8% *trans* isomer fatty acids. For samples containing less than 0.8% *trans* double bonds, a capillary GC method (e.g., AOCS Method Ce 1h-05) is recommended.

23.6.4 Mono-, Di-, and Triacylglycerols

Mono-, di-, and triacylglycerols may be determined using various techniques (see Table 23.1). Older methods use titrimetric approaches, whereas newer methods utilize chromatographic techniques. Short nonpolar columns and very high temperatures are needed for the analysis of intact triacylglycerols by GC. Section 23.6.6 describes the use of TLC to separate lipid classes, including mono-, di-, and triacylglycerols.

23.6.5 Cholesterol and Phytosterols

Many methods exist for the quantification of cholesterol and phytosterols in various matrices. Consulting research literature will give an indication of current practice and methods that may be less laborious or adapted for use with specific foods.

23.6.5.1 Principle

The lipid extracted from the food is saponified. The saponification process is a hydrolysis, with acyl lipids being converted to water-soluble FFA salts. Other components (called the unsaponifiable or nonsaponifiable matter) do not change in solubility after hydrolysis and thus remain soluble in organic solvents. Cholesterol (in the unsaponifiable fraction) is extracted and derivatized to form trimethylsilyl (TMS) ethers or acetate esters. This increases their volatility and reduces problems of peak tailing during chromatography. Quantitation is achieved using capillary GC peak area analysis.

23.6.5.2 Procedure

AOAC Method 976.26 outlined here is representative of the various procedures available for cholesterol determination. Lipids are extracted from the food and saponified, and the unsaponifiable fraction is extracted. This is accomplished by filtering an aliquot of the chloroform layer through anhydrous sodium sulfate and evaporating to dryness in a water bath using a stream of nitrogen gas. Concentrated potassium hydroxide and ethanol are added, and the solution is refluxed. Aliquots of benzene and 1 *N* potassium hydroxide are added and then shaken. The aqueous layer is removed, and the process is repeated with 0.5 *N* potassium hydroxide. After several washes with water, the benzene layer is dried with anhydrous sodium sulfate and an aliquot is evaporated to dryness on a rotary evaporator. The residue is taken up in dimethylformamide. An aliquot of this sample is derivatized by adding hexamethyl disilazane (HMDS) and trimethylchlorosilane (TMCS). Water (to react with and inactivate excess reagent) and an internal standard in heptane are added, and then the solution is centrifuged. A portion of the heptane layer is injected into a GC equipped with a nonpolar column. The HMDS and TMCS reagents are rapidly inactivated by water, and thus the reaction conditions must remain anhydrous.

23.6.5.3 Applications

GC quantitation of cholesterol is recommended, since many spectrophotometric methods are not specific for cholesterol. In the past, samples such as eggs and shrimp have had their cholesterol contents overestimated by relying on less specific colorimetric procedures. Other GC, HPLC, and enzymatic methods are available.

Cholesterol oxidation products as well as phytosterols can be quantified using the GC procedure outlined for cholesterol. A wide range of methods for analysis of sterols exist in the literature; most use TMS ether formation to increase volatility of the hydroxide-containing sterols and to improve chromatographic resolution (reduce peak tailing).

23.6.6 Thin-Layer Chromatography of Lipid Fractions

23.6.6.1 Procedure

Thin-layer chromatography (TLC) is performed using silica gel G as the adsorbent and hexane-diethyl ether-formic acid (80:20:2vol/vol/vol) as the eluting solvent system (Fig. 23.7). Plates are sprayed with 2′,7′-dichlorofluorescein in methanol and placed under ultraviolet light to view yellow bands against a dark background [5].

23.6.6.2 Applications

This procedure permits rapid analysis of the presence of the various lipid fractions in a food lipid extract. For small-scale preparative purposes, TLC plates can be scraped to remove various bands for further analysis using GC or other means. Many variations in TLC parameters are available that will separate various lipids.

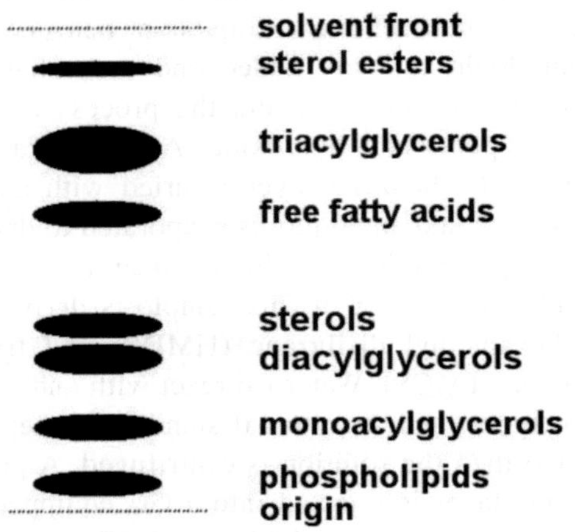

Fig. 23.7 Schematic of thin-layer chromatography (TLC) separation of lipid fractions on silica gel G. (Adapted from [5], with permission)

23.7 SUMMARY

The importance of fat characterization is evident in many aspects of the food industry, including ingredient technology, product development, quality assurance, product shelf life, and food regulations. Lipids are closely associated with health; the cholesterol or phytosterol compositions, and amounts of *trans*, saturated, and n-3/ω3 fatty acids are of great concern to consumers.

The methods described in this chapter help to characterize bulk fats/oils of commerce as well as the lipids in foods. A summary of some of the more common tests is given in Table 23.3. Methods described for bulk fats/oils can determine characteristics such as melting point, smoke point, color, degree of unsaturation, average fatty acid chain length, and amount of polar components. CD, PV, and VOC (e.g., hexanal) tests can be used to measure the extent of lipid oxidation, while the OSI can be used to predict the susceptibility of a fat/oil to oxidation and the effectiveness of antioxidants. Lipid fractions, including fatty acids, triacylglycerols, phospholipids, and cholesterol, are analyzed by chromatographic techniques such as GC and TLC.

The tests covered in this chapter represent only a few of the many methods that have been developed to characterize lipids. Consult the references cited for additional methods or more detailed explanations. Time, funding, availability of

Table 23.3 Summary of some of the common tests used to characterize fats/oils and foods containing fats/oils

Test	*Indicated*	*Actually measured*
Solid fat content (SFC)	Amount of solid vs. liquid in a fat/oil	Percent solid fat, by continuous wave or pulsed NMR
Iodine value (IV)	Degree of unsaturation	Iodine required for absorption by double bonds
Saponification value (SV)	Mean molecular weight of triacylglycerols	Alkali required to saponify fat/oil
Free fatty acids (FFAs)	Fatty acid hydrolysis from triglycerides (hydrolytic rancidity)	Potassium hydroxide necessary to neutralize free acids
Conjugated dienes (CD)	Initial stages of lipid oxidation (oxidative rancidity)	Change from nonconjugated to conjugated double bonds
Peroxide value (PV)	Initial stages of lipid oxidation (oxidative rancidity)	Hydroperoxides
Thiobarbituric acid reactive substances (TBARS)	Later stages of lipid oxidation (oxidative rancidity)	Malonaldehyde and similar secondary products of lipid oxidation
Volatile organic compounds (VOCs)	Later stages of lipid oxidation (oxidative rancidity)	Volatile organic compounds (e.g., hexanal, pentanal, pentane), secondary products of lipid oxidation
Accelerated Shelf-life Test (ASLT)	Oxidative stability and fundamental research in lipid oxidation	Progression of lipid oxidation during controlled accelerated conditions, measured using various lipid oxidation tests (e.g., CDs, PV, VOCs)
Oil stability index (OSI)	Oxidative stability	Induction period after exposure to air and high temperature, based on acidic volatiles, measured by conductivity
Fatty acid methyl esters (FAMES)	Fatty acid composition of fats/oils; calculated total fat (nutritional labeling)	Chromatographic separation and quantification of individual fatty acids after fat extraction and hydrolysis of triacylglycerols
trans Fatty acids	*trans* Fat (nutritional labeling)	Chromatographic separation and quantification of fatty acids containing trans bonds after fat extraction and hydrolysis of triacylglycerols
Cholesterol	Cholesterol (nutritional labeling)	Chromatographic determination of cholesterol after fat extraction and saponification

equipment and instruments, required accuracy, and purpose all will dictate the choice of method to characterize fats/oils and foods containing lipids.

23.8 STUDY QUESTIONS

1. You want to compare several fat/oil samples for the chemical characteristics listed below. For each characteristic, name one test (give full name, not abbreviation) that could be used to obtain the information desired:
   (a) Degree of unsaturation
   (b) Stability to lipid oxidation
   (c) Present status with regard to lipid oxidation (related to oxidative rancidity)
   (d) Average fatty acid molecular weight
   (e) Amount of solid fat at various temperatures
   (f) Degree of hydrolysis (related to hydrolytic rancidity)
2. Your analysis of anfat/oil sample gives the following results. What does each of these results tell you about the characteristics of the sample? Briefly describe the principle for each method used:
   (a) Large saponification value
   (b) Low iodine value
   (c) High TBARS
   (d) High FFA content
   (e) High OSI
3. Define solid fat content and explain the usefulness of this measurement.
4. Peroxide value, TBARS, and VOCs all can be used to help characterize a fat/oil sample.
   (a) What do the results of these tests tell you about a fat/oil sample?
   (b) Differentiate these three tests as to what chemical is being measured.
5. What methods would be useful in determining the effectiveness of various antioxidants added to anfat/oil?
6. You are responsible for writing the specifications for vegetable oil purchased from your supplier for use in deep-fat frying several foods processed by your company. Itemize the tests you should require in your list of specifications (specific values for the tests are not needed). For each test, briefly state what useful information is obtained.
7. The Nutrition Education and Labeling Act of 1990, as amended in 2016 (see Chap. 3), requires that the nutrition label on food products contains information related to lipid constituents. In addition to the amount of total fat (see Chap. 17), the label must state saturated fat, *trans* fat, and cholesterol contents. For a product such as potato chips, explain an appropriate method for the analysis of each of these lipid constituents.
8. You have developed a new butter containing added fish oil, which is high in PUFA. Before placing the product on the market, you need to determine its shelf life. What method or methods would you use and why?
9. You work in quality control for a company that makes peanut butter. You received information that between July and August several lots of peanut butter may have been improperly stored and you are concerned about potential lipid oxidation in the product. A total of 50 lots of peanut butter are involved.
   (a) What test(s) would you use to measure lipid oxidation? Include your rationale for selecting the method.
   (b) How would you decide what to do with the peanut butter?
10. You are developing a new food product for which the fat/oil is important for the ideal mouthfeel as perceived by consumers. For the various fats/oils you are considering for use, what property of the fats/oils would you appropriately measure and compare?
11. Before changing the type of fat used in the processing of a food product, the analytical lab of your company was asked to test samples of the current type of fat (sample A) and the newly proposed type of fat (sample B) for various chemical and physical properties under specified conditions. Results were as described below.
    (a) For each of these items, briefly state what the values tell you about the properties of A (vs. B).
        (1) A had a higher saponification number than B
        (2) A had a lower peroxide value than B
        (3) A had a lower hexanal content than B
        (4) A had a lower iodine value than B
        (5) A had a higher oil stability index than B
    (b) In total, are these values for sample A consistent with one another, or do they give a conflicting description of the oil? Explain your answer.
12. As shown on a certificate of analysis, the quality and authenticity of extra virgin olive oil (defined as cold-pressed and unrefined) can be determined using sensory analysis and various other tests, including free fatty acids, conjugated dienes and trienes (abbreviated as K_{232} and K_{270}, respectively), peroxide value, sterol content, and fatty acid profile. What do low values for K_{232} and K_{270} indicate regarding the oil's quality and authenticity?

23.9 PRACTICE PROBLEMS

1. A 5.00-g sample of oil was saponified with excess KOH. The unreacted KOH was then titrated with 0.500 *N* HCl (standardized). The difference between the blank and the sample was 25.8 mL of titrant. Calculate the saponification value.

2. A sample (5.0 g) of food grade oil was reacted with excess KI to determine peroxide value. The free iodine was titrated with a standardized solution of 0.10 *N* $Na_2S_2O_3$. The amount of titrant required was 0.60 mL (blank corrected). Calculate the peroxide value of the oil.
3. You analyze the saponification value of an unknown. You use 4.0 g oil and 0.5 *N* HCl to titrate. The difference between the blank and the sample was 43 mL of titrant. What is the average fatty acid molecular weight of the oil?
4. You analyze an oil's FAMEs by GC and find the following areas for your identified peaks:

16:0	2,853,369
18:0	1,182,738
18:1n-9	38,999,438
18:2n-6	14,344,172
18:3 n-3	2,148,207

Report the fatty acid composition as % and tentatively identify the oil.

Answers

1. Saponification value

$$= \frac{25.8\,\text{mL} \times 0.500\,\text{mmol/mL} \times 56.1\,\text{mg/mmol}}{5.00\,\text{g}} = 145$$

2. Peroxide value

$$= \frac{0.60\,\text{mL} \times 0.10\,\text{meq/mL} \times 1000}{5.0\,\text{g}} = 12\,\text{meq/kg}$$

3. Saponification value =

$$\frac{43\,\text{mL} \times 0.500\,\text{meq/mL} \times 56.1\,\text{mg/mmol}}{4.0\,\text{g}} = 302$$

$$302 = [3 \times 56.1 \times 1000] / \left\{ \begin{matrix} [(\text{Mean mol.wt} \times 3) + 92.09] \\ -(3 \times 18) \end{matrix} \right\}$$

Mean mol. wt.= 172

4. Sum of all areas = 58,416,924

Area % 16:0 = 100 × 2,853,369/58,416,924 = 4.9%

so,

16:0	4.9%
18:0	2.0%
18:1n-9	66.8%
18:2 n-6	22.7%
18:3n-3	3.7%

Based on the fatty acid composition, the oil is probably Canola oil.

Acknowledgment The author of this chapter thanks Sean O'Keefe (Department of Food Science and Technology, Virginia Tech, Blacksburg, VA, USA) for his contributions to this chapter in previous editions of the book.

REFERENCES

1. AOAC International (2023) Official methods of analysis of AOAC International, 22nd edn. AOAC International, Rockville, MD https://academic.oup.com/aoac-publications
2. AOCS (2020) Official methods and recommended practices of the AOCS, 7th edn. American Oil Chemists' Society, Champaign, IL
3. IUPAC (1987) Standard methods for analysis of oils, fats, and derivatives 7th edn. and supplements, International Union of Pure and Applied Chemistry, Commission on Oils, Fats and Derivatives, Paquot C and Hautfenne A (eds). Blackwell Scientific, Oxford
4. Christie WW, Han X (2012) Lipid analysis. Isolation, separation, identification and lipidomic analysis, 4th edn. Woodhead Publishing, Cambridge
5. Christie WW (1989) Gas chromatography and lipids. A practical guide. The Oily Press, Ayr, Scotland
6. Perkins EG (1991) Analyses of fats, oils, and lipoproteins. American Oil Chemists' Society, Champaign, IL
7. Shahidi F (ed) (2005) Bailey's industrial oil and fat products, 6th edn. John Wiley & Sons, New York
8. McClements DJ, Decker EA (2017) Lipids ch. 4. In: Damodaran S, Parkin KL (eds) Fennema's Food Chemistry, 5th edn. CRC Press, Boca Raton, FL
9. Firestone, D (2013) Physical and chemical characteristics of oils, fats, and waxes, 3rd ed., AOCS Press, Urbana, IL
10. Anon. (2022) TheMerck Index Online. Royal Society of Chemistry, London. https://www.rsc.org/merck-index
11. Stauffer CE (1996) Fats and oils. Eagan press handbook series. American Association of Cereal Chemists, St. Paul, MN
12. Lee Y-Y, Tang T-K, Phuah E-T, Lai O-M (Eds) (2022) Recent Advances in Edible Fats and Oils Technology, Springer Nature Singapore Pte Ltd., Singapore
13. Wrolstad RE, Acree TE, Decker EA, Penner MH, Reid DS, Schwarts SJ, Shoemaker CF, Smith DM, Sporns P (2005) Lipid oxidation/stability. In: Handbook of Food Analytical Chemistry: Water, Proteins, Lipids, and Carbohydrates, Vol. 1. John Wiley & Sons, Hoboken, JN, pp.513–564.
14. Hu M, Jacobsen C (eds) (2016) Oxidative stability and shelf-life of foods containing oils and fats. Academic Press, Cambridge, MA
15. Nawar WW (1996) Lipids ch. 5. In: Fennema OR (ed) Food Chemistry, 3rd edn. CRC Press, Boca Raton, FL
16. Frankel EN (1998) Lipid Oxidation. The Oily Press Ltd, Dundee, Scotland
17. Labuza TP (1971) Kinetics of lipid oxidation in foods. CRC Crit Rev Food Technol 2: 355–405
18. Schaich KM (2016) Analysis of lipid and protein oxidation in fats, oils, and foods, ch.1. In: Hu M, Jacobsen C (eds.) Oxidative stability and shelf-life of foods containing oils and fats, Academic Press, Cambridge, MA pp.1–131
19. Bansal G, Zhou W, Barlow PJ, Joshi PS, Lo HL, Chung YK. (2010) Review of rapid tests available for measuring the quality changes in frying oils and comparison with standard methods. Crit Rev Food Sci Nutr 50(6):503–14
20. Della Porta R, Aladedunye F (2022) Frying and stability of high-oleic oils. ch.8. In: Flider FJ (ed.) High Oleic Oils: Development, Properties, and Uses, Academic Press and AOCS Press, Cambridge, MA pp. 189–200
21. Majchrzak J, Wojnowski W, Dymerski T, Gebicki J, Namiesnik J (2018) Electronic noses in classification and quality control of edible oils: A review. Food Chem 246:192–201
22. Komoda M, Ishikawa O, Inoue M, Hatanaka H, Hattori T, Satoh C, Murakami C, Yamada, K, Tsutsumi T (2005) Quantitative determination of polymerized triacylglycerols in frying oils and fats by size exclusion chromatography: A collaborative study. J Oleo Sci54:529–536

Protein Separation and Characterization Procedures

24

B. Pam Ismail and Denise M. Smith

Contents

ABBREVIATIONS

$(NH_4)_2SO_4$	Ammonium sulfate
2D	Two dimensional
AQC	6-Aminoquinolyl-N-hydroxysuccinimidyl carbamate
C-PER	Protein efficiency ratio-calculation method
DIAAS	Digestible indispensable amino acid score
DNFB	1-Fluoro-2,3-dinitrobenzene (also called FDNB)
DSC	Differential scanning calorimetry
EAAI	Essential amino acid index
FAO	Food and Agriculture Organization
FDA	Food and Drug Administration
FPLC	Fast protein liquid chromatography
HIC	Hydrophobicity
HPLC	High-performance liquid chromatography
LGC	Least gelation concentration
MALS	Multiangle light scattering
MWCO	Molecular weight cutoff
NFDM	Nonfat dry milk
NLEA	Nutrition Labeling and Education Act
OBC	Oil-binding capacity
OPA	*o*-Phthalaldehyde
PAGE	Polyacrylamide gel electrophoresis
PDCAAS	Protein digestibility-corrected amino acid score
PDI	Protein dispersibility index
PER	Protein efficiency ratio
pI	Isoelectric point

B. P. Ismail (✉)
Department of Food Science and Nutrition, University of Minnesota, St. Paul, MN, USA
e-mail: bismailm@umn.edu

D. M. Smith
School of Food Science, Washington State University, Pullman, WA, USA
e-mail: denise.smith@wsu.edu

B. P. Ismail, S. S. Nielsen (eds.), *Nielsen's Food Analysis*, Food Science Text Series,
https://doi.org/10.1007/978-3-031-50643-7_24

R_m	Relative mobility
RVA	Rapid visco analyzer
SDS	Sodium dodecyl sulfate
SE HPLC	Size-exclusion HPLC
TEMED	Tetramethylethylenediamine
UPLC	Ultra-performance liquid chromatography
UV	Ultraviolet
WBC	Water binding capacity
WHC	Water holding capacity
V_e	Elution volume

24.1 INTRODUCTION

Many protein separation and isolation techniques are available to food scientists and the food protein industry. Several of the separation and isolation techniques described in this chapter are used commercially for the production of protein ingredients, whereas others are used to purify a protein from a food material for further study in the research laboratory. In general, separation techniques exploit the physical, biochemical, and structural differences in protein, including bulk density, solubility, charge, thermal stability, size, adsorption characteristics, and biological affinities for other molecules. These characteristics then are used to purify individual proteins from complex mixtures.

The nutritional, structural, and functional properties of food proteins can be characterized in a variety of ways. This chapter describes methods of amino acid analysis and several methods for protein nutritional quality analysis. In addition, key structural characteristics including molecular weight distribution, denaturation, and surface properties are covered to a certain extent. Finally, protein functionality, including binding with water or oil, solubility and dispersibility, emulsification, foaming, and protein network formation (e.g., gelation, dough formation, and texturization), is described.

This chapter has been revised to include techniques relevant to plant protein ingredients. In developed countries, the plant-based protein market is rapidly growing in response to changing consumer trends. Consumers are increasingly identifying as vegans, vegetarians, and flexitarians. These populations are largely motivated by the health benefits of plant-based foods, as well as their growing environmental consciousness, animal welfare concerns, and desires to diversify their diets [1] Accordingly, food and beverage manufacturers are increasingly utilizing plant protein ingredients in their products, and therefore are seeking efficient means to isolate and characterize plant proteins.

A list of abbreviations is included to summarize the terms used throughout this chapter.

24.2 METHODS OF PROTEIN SEPARATION AND ISOLATION

24.2.1 Initial Considerations

Usually, several separation techniques are used in sequence to purify a protein from a food material. In general, the purity of a protein preparation can be improved as more separation steps are added, although this will usually result in a lower recovery or yield. Food ingredients such as protein concentrates and isolates may be prepared using only one separation step, along with at least one concentration and/or isolation/purification step to produce ingredients with protein concentrations ranging from 30% to 90% protein. Three or more separation and purification steps are often used in sequence to prepare a pure protein for research. For example, a very common purification procedure for protein includes precipitation by ammonium sulfate, followed by hydrophobic interaction chromatography, ion-exchange chromatography, and size-exclusion chromatograph (in any order) (Chaps. 12 and 13).

Before starting a separation sequence, it is necessary to learn as much as possible about the physical, biochemical, and structural properties of a protein to determine any distinctive characteristics that will make separation easier. Characteristics include bulk density, molecular mass, isoelectric point (pI), surface properties, metal ion binding, denaturation temperature, and specific ligand recognition. The first separation step should be one that can easily be used with large quantities of material. This is often a technique that utilizes the differential solubility properties of a protein. Each succeeding step in a purification sequence will use a different mode of separation. Some of the most common methods of purification are described in this section and include dry fractionation, solubilization, precipitation, ion-exchange chromatography, hydrophobic interaction chromatography, affinity chromatography, size-exclusion chromatography, and membrane filtration. A summary of separation and isolation methods described in this chapter can be found in Table 24.1. More detailed information about the various purification techniques can be found in several published sources [2–4].

24.2.2 Dry Fractionation

Dry fractionation is achieved by air classification, which is a technique commonly used to concentrate proteins from **pulses** (non-oilseed legume, e.g., peas, chickpeas, beans, and lentils). Pulse proteins can be physically separated from starch and fiber based on the difference in particle sizes and bulk densities after milling the seeds into flour [5]. Milling

Table 24.1 Summary of protein separation methods

Method	*Basis of separation*	*Principle*
Dry fractionation	Physical separation	This separation is used mostly for pulses, where high protein fractions are separated from starch and fiber fraction by difference in bulk density
Alkaline pH (pH 8–10)	Solubilization	Proteins, namely globulin plant proteins, are mostly soluble under alkaline conditions, and thus can be separated from large molecular weight carbohydrates such as fiber and starch
Isoelectric precipitation	Precipitation	Protein has no net charge at pI, so it aggregates and precipitates from solution
Addition NaCl at relatively concentration (0.1–0.5 M NaCl)	Solubilization	This phenomenon is called salting in; proteins, namely globulin plant proteins are soluble at relatively low concentration salt solution, and thus can be separated from large molecular weight carbohydrates such as fiber and starch
Addition of ammonium sulfate	Precipitation	Proteins are precipitated from solution as ionic strength increases, using a neutral salt, such as ammonium sulfate
Solvent fractionation	Precipitation	Water-miscible organic solvents decrease the dielectric constant of an aqueous solution and decrease the solubility of most proteins, so proteins precipitate from solution
Protein denaturation	Precipitation	Proteins heated to high temperature or adjusted to extremes of pH precipitate from solution
Ion-exchange chromatography	Adsorption (based on charge)	Charged protein molecules in solution are reversibly adsorbed to a charged solid support matrix via electrostatic interactions. Bound proteins are eluted from a column by gradually changing the ionic strength or pH of the eluting solution
Affinity chromatography	Adsorption (based on specific biochemical characteristics)	Protein is adsorbed to chromatographic matrix that contains a ligand covalently bound to a solid support. The ligand used has reversible, specific, and unique binding affinity for the protein of interest. Protein that binds to ligand can be unbound by changing the pH, temperature, or concentration of salt or ligand in the eluting buffer
Hydrophobic interaction chromatography	Adsorption (based relative hydrophobicity)	Protein molecules in high concentration of salt solution are reversibly adsorbed to hydrophobic matrix by weak hydrophobic interactions. Bound proteins are eluted from a column by gradually reducing the salt concentration to 0%
Dialysis	Size	Semipermeable membrane permits passage of molecules based on molecular weight cutoff (MWCO) of the membrane's pores. Dialysis is an equilibrium-based process, and several washes are needed
Membrane processes (e.g., microfiltration, ultrafiltration, nanofiltration, and reverse osmosis)	Size	Pressure is applied to push the protein solution effectively through the semipermeable membrane with certain MWCO. Smaller molecules pass through, and larger ones are retained
Size-exclusion chromatography	Size	Proteins in solution flow through a column packed with beads that have different average pore sizes. Proteins larger than the pores are excluded and pass through the column quickly. Smaller proteins enter the pores of the beads, so elute more slowly from the column, at a rate dependent on their size
Sodium dodecyl sulfate-polyacrylamide gel electrophoresis (SDS-PAGE)	Size	Proteins bind SDS to become negatively charged. Under constant voltage, proteins move through an acrylamide gel matrix at a rate based on size alone since all molecules are highly negatively charged
Isoelectric focusing (IEF)	Charge	Proteins are separated by charge in an electric field on a gel matrix in which a pH gradient has been generated using ampholytes. Under constant voltage, proteins migrate to the location on the gradient at which pH equals the pI of the protein. Size is not a factor

helps separate the smaller protein bodies (<10 μm) from the larger starch granules and fiber components (20–70 μm) [6]. The flour is subjected to a stream of spiraling air in the separation chamber of an air classifier. Inside the chamber, there is a classifier wheel with a small slit that allows the fine fraction to float up and be collected, while the heavy/coarse fraction is kept in the lower portion of the chamber [6]. This process enables the recovery of both the concentrated protein (fine) fraction and the starch/fiber (heavy/coarse) fraction. This dry fractionation process can produce pulse protein concentrates of 50–60% protein with a protein recovery of up to 77% [6]. For higher protein purity, wet fractionation techniques are necessary, as will be described below.

24.2.3 Separation Based on Differential Solubility

24.2.3.1 Principle

Separation and purification by differential solubility exploit the differences in biochemical and structural properties of the protein (including ionizable groups and resulting charge at different pHs, molecular configuration, polarity, and denaturation temperature). pH-assisted solubilization and precipitation is one of the simplest methods for separating a protein of interest from other proteins or other components in a mixture. Proteins are polyelectrolytes; thus, solubility characteristics are determined by the type and charge of amino acids in the molecule. Proteins can be selectively solubilized or precipitated by changing **buffer pH, ionic strength, dielectric constant** (e.g., introducing organic solvents), or **temperature**. These separation techniques are advantageous when working with large quantities of material, are relatively quick, and are not usually influenced by other food components. Solubilization and precipitation techniques are used most commonly during early stages of a purification sequence.

24.2.3.2 Procedures

24.2.3.2.1 pH-Assisted Solubilization and Precipitation

Both optimal **solubilization pH** (the pH at which the protein is most soluble) and the **isoelectric point** (pI) pI are determined by the ionizable acidic and basic amino acids making up the protein. The pI is defined as the pH at which a protein has no net charge in solution. Proteins usually aggregate and precipitate at their pI because there is no electrostatic repulsion between molecules. Proteins have different pIs; thus, they can be separated from each other by adjusting the pH of the solution. When the pH of a solution is adjusted to the pI of a protein, the protein precipitates, while proteins with different pIs remain in solution. The precipitated protein can be resolubilized in another solution of different pH. Commercial plant protein isolates are mostly produced following a wet extraction process involving alkaline solubilization (pH 8–10) to separate the protein from carbohydrates such as fiber and/or starch. This is followed by precipitation at the average pI to purify the protein and remove soluble carbohydrates and small molecules [7]. The precipitated protein is washed, neutralized, pasteurized, and dried to produce an isolate with about 80–90% protein.

To produce soy protein concentrate, it is a common practice to disperse defatted flour in water and to reduce the pH to the pI of soy protein (pH 4.5). Both the protein and fiber will precipitate. The resulting precipitate is washed, neutralized, pasteurized, and dried to produce a concentrate with 60–80% protein. Dairy proteins are other examples of proteins that can be separated by differential pH solubility. For example, to separate casein from whey protein, the pH is adjusted to the pI of casein (pH 4.6). This pH adjustment is used in cheese curd formation. The separated liquid whey then can be used as the starting material to produce whey protein concentrates and isolates following membrane filtration (Sect. 24.2.5.2.2) or ion-exchange chromatography (Sect. 24.2.4.2.1).

24.2.3.2.2 Salting-In and Salting-Out

Proteins have unique solubility profiles in neutral salt solutions. Low concentrations of neutral salts usually increase the solubility of proteins; however, proteins are precipitated from solution as ionic strength is increased. This salt-based differential solubility is also dependent on the ionizable groups of the different amino acids within a protein. Accordingly, proteins in a complex mixture can be separated from non-protein components as well as from each other. Globulin proteins in plant material are soluble under dilute salt concentration (e.g., 0.1–0.5 M NaCl) and thus can be separated from insoluble starch and/or fiber. A protein mixture in solution can be separated from other solubles by using a high concentration of salt, typically **ammonium sulfate** [$(NH_4)_2SO_4$], Proteins in solution can be further separated from each other using salt solutions with varying concentration. For example, $(NH_4)_2SO_4$ is added at a concentration just below that necessary to precipitate the protein of interest. When the solution is centrifuged, less soluble proteins are precipitated while the protein of interest remains in solution. The next step is performed at an $(NH_4)_2SO_4$ concentration just above that necessary to precipitate the protein of interest. When the solution is centrifuged, the protein is precipitated, while more soluble proteins remain in the supernatant. One disadvantage of this method is that large quantities of salt contaminate the precipitated protein and often must be removed (often via dialysis, see Sect. 24.2.5.2.1) before the protein is resolubilized in buffer. Tables and formulas are available in many biochemistry books and online (type "ammonium sulfate calculator" into your web browser) to determine the proper amount of $(NH_4)_2SO_4$ to achieve a specific concentration.

24.2.3.2.3 Solvent Fractionation

Protein solubility at a fixed pH and ionic strength is a function of the dielectric constant of a solution. Thus, proteins can be separated based on solubility differences in **organic solvent-water** mixtures. The addition of water-miscible organic solvents, such as **ethanol** or **acetone**, decreases the dielectric constant of an aqueous solution and decreases the solubility of most proteins. Organic solvents decrease ionization of charged amino acids, resulting in reduced repulsive forces and enhanced protein aggregation and precipitation. The optimum quantity of organic solvent to precipitate a protein varies from 5% to 60%. Solvent fractionation is usually performed at low temperatures to prevent protein denaturation caused by temperature increases that occur when organic solvents are mixed with water. Soy protein concentrate can be prepared from defatted soybean flakes or flour using a 60–80% aqueous alcohol solution. Both protein and fiber components will precipitate, while soluble saccharides remain in the solvent. The precipitated concentrate is washed multiple times, pasteurized, and dried to result in a protein concentrate of 60–80% protein.

24.2.3.2.4 Thermal Denaturation

Many proteins are **denatured** and precipitated from solution when heated above a certain temperature or by adjusting a solution to highly acid or basic pHs. Proteins that are stable at high temperatures or at extremes of pH may be separated by this technique because many contaminating proteins will precipitate while the protein of interest remains in solution.

24.2.4 Separation by Adsorption Chromatography

24.2.4.1 Principle

After initial separation based on preferential solubility, purification of a protein usually includes one or more types of liquid chromatography. This section will cover the commonly used adsorptive chromatographic separations (Chap. 12, Sect. 12.4.1) that are based on the differential affinity of proteins in solution for the stationary phase. These chromatographic methods can be performed in columns at atmospheric pressure (by gravity flow) or under applied pressures as in **high-performance liquid chromatography** (HPLC) or **ultra-performance liquid chromatography** (UPLC) (Chap. 13). **Fast protein liquid chromatography** (FPLC) is a form of HPLC, with medium pressure, specific for protein purification with enhanced resolution compared to low pressure liquid chromatography [8], while allowing for high loading capacity. Ion-exchange chromatography, hydrophobic interaction chromatography, and affinity chromatography as applied for protein purification will be briefly described below.

24.2.4.2 Procedures

24.2.4.2.1 Ion-Exchange Chromatography

Ion-exchange chromatography (Chap. 12, Sect. 12.4.5) results in an average eightfold purification. The most commonly used exchangers for protein purification are anionic diethylaminoethyl-derivatized supports, followed by carboxymethyl- and phospho-cation exchangers. The protein of interest is first adsorbed to the selected ion exchanger under buffer conditions (ionic strength and pH) that maximize the affinity of the protein for the matrix. Components of different charges (or those not charged) pass through the exchanger unabsorbed. Proteins bound to the exchangers are selectively eluted from the column by gradually changing the ionic strength or pH of the eluting solution. As the composition of the eluting buffer changes, the affinity of the proteins for the ion-exchange matrix is decreased. Protein purification by ion-exchange chromatography must be done at a pH away from their pI to avoid protein precipitation.

Ion-exchange chromatography is commonly used to separate proteins in the laboratory as a first or second dimension of chromatographic purification [9] that precedes further characterization (e.g., molecular weight, amino acid sequence, enzyme activity or inhibition, and functional properties). Post complete protein digestion, amino acid compositional analysis can be performed using ion-exchange chromatography as described in Sect. 24.3.1. Ion-exchange chromatography is also used to produce protein ingredients such as whey protein isolate. Lactose, minerals, and residual fat are removed from whey in the process resulting in whey protein isolates (containing greater than 90% protein) [10, 11].

24.2.4.2.2 Affinity Chromatography

Affinity chromatography (Chap. 12, Sect. 12.4.6) is a very powerful technique and is a commonly used protein purification procedure. The average purification achieved by affinity chromatography is approximately 100-fold, although 1000-fold increases in purification have been reported, which is by far the highest among chromatographic separation protocols. Many covalently bound ligands and associated buffers are commercially available as kits. Preactivated solid supports used for covalent bonding of various ligands (Table 24.2) are also commercially available.

A specific protein of interest is isolated from a complex mixture based on its specific binding interaction with a ligand immobilized on a solid support. The protein is passed through a column containing the ligand bound to a solid support, under buffer conditions (pH, ionic strength, temperature, and protein concentration) that maximize binding of the

Table 24.2 Examples of common biological interactions used to separate proteins via affinity chromatography

Protein target	*Ligand*
Enzyme	Substrate, inhibitor, cofactor
Antibody	Antigen
Glycoprotein	Lectin, polysaccharide
Hormone	Hormone receptor
Proteins with histidine, cysteine, or tryptophan residues on the surface	Metal ions

protein to the ligand. Other components in the sample that do not bind the ligand are eluted. The bound protein is then desorbed or eluted from the column under conditions that decrease the affinity of the protein for the bound ligand, by changing the pH, temperature, or concentration of salt or ligand in the eluting buffer.

Affinity chromatography is commonly used for protein purification in the research laboratory and may be used for commercial preparation of proteins (e.g., antibodies, protein standards) by chemical suppliers. Recombinant proteins, for example, are often purified by immobilized metal ion affinity chromatography followed by size-exclusion chromatography (Sect. 24.2.5.2.3). Recombinant proteins can be engineered to contain multiple histidine residues (6–10 molecules of histidine). The proteins containing this polyhistidine tag will bind to an affinity column containing immobilized divalent metal ions, such as nickel. Once other constituents are washed from the column, the tagged protein is eluted using a gradient of imidazole counter ligands (the amino acid histidine contains an imidazole functional group). Affinity chromatography is not generally used for commercial production of food protein ingredients due to the high costs involved. However, this technique is used to purify some high-value bioactive peptides and proteins for nutritional applications.

24.2.4.2.3 Hydrophobic Interaction Chromatography

Hydrophobic interaction chromatography is another type of adsorptive chromatography often used in protein purification schemes (Chap. 12, Sect. 12.4.4). In this technique, proteins are separated due to their reversible interaction with a hydrophobic stationary phase under high salt concentration. The bound protein is usually eluted by decreasing the ionic strength of the mobile phase using a stepwise or gradient elution. This type of purification is often used for research purposes [12] and for pharmaceutical applications [13], but has not been scaled up for the production of protein ingredients.

24.2.5 Separation by Size

24.2.5.1 Principle

Protein molecular masses range from about 10 to over 1000 kDa; thus, size is a logical parameter to exploit for separations. Actual separation occurs based on the **Stokes radius** of the protein, not on the molecular mass. Stokes radius is the average radius of the protein in solution and is determined by protein conformation. For example, a globular protein may have an actual radius very similar to its Stokes radius, whereas a fibrous or rod-shaped protein of the same molecular mass may have a Stokes radius that is much larger than that of the globular protein. Thus, one limitation of these methods is that two proteins of the same molecular mass may occasionally separate differently.

24.2.5.2 Procedures

24.2.5.2.1 Dialysis

Dialysis is used to separate molecules in solution by the use of semipermeable membranes that permit passage of molecules based on the pore size of the membrane. Dialysis membranes commonly used for protein purification have a **molecular weight cutoff** (MWCO) of 1 kDa to 10 kDa. To perform dialysis, a protein solution is placed into dialysis tubing that has been tied or clamped at both ends of the tube. The bag is placed in a large volume of water or buffer (usually 500–1000 times greater than the sample volume inside the dialysis tubing), which is slowly stirred under refrigerated conditions. Solutes of molecular mass lower than the MWCO of the membrane diffuse from the bag, while buffer or water diffuses into the bag. Dialysis is an equilibrium-based process that requires at least 24 h with 3–4 changes of buffer or water. The protein solution inside the bag is often diluted during dialysis, due to osmotic strength differences between the solution and the buffer or water. Dialysis is often used to remove salt (e.g., from salt precipitated protein) and other impurities of low molecular mass between purification steps or to adjust the buffer composition of a final protein preparation.

24.2.5.2.2 Membrane Processes

Microfiltration, ultrafiltration, nanofiltration, and reverse osmosis all are processes that use a semipermeable membrane for the separation of solutes on the basis of size under an applied pressure. These methods are similar to dialysis but are much faster and are applicable to both small- and large-scale separations. Molecules larger than the membrane MWCO are retained and become part of the retentate, while smaller molecules pass through the membrane and become part of the filtrate (or permeate).

These membrane processes differ mainly in the porosity of the membranes and in the operating pressure used. The porosity of the membrane (membrane pore diameter) sequentially decreases, and the pressure used sequentially increases for microfiltration, ultrafiltration, nanofiltration, and reverse osmosis. The approximate pore size of each membrane process relative to the different components present in milk is shown in Fig. 24.1.

Ultrafiltration is commonly used in protein research laboratories, with various commercial units available. A stirred cell ultrafiltration unit is illustrated in Fig. 24.2. The protein solution in the stirred cell is filtered through the semipermeable membrane by gas pressure, leaving a concentrated solution of proteins larger than the membrane MWCO point inside the cell. Disposable centrifugal filtering units are available for small sample volumes with membrane MWCO ranging from 3 to 100 kDa. Solvents and molecules smaller than the membrane pore size are forced through the membrane by centrifugation resulting in concentration and purification of the protein in the retentate.

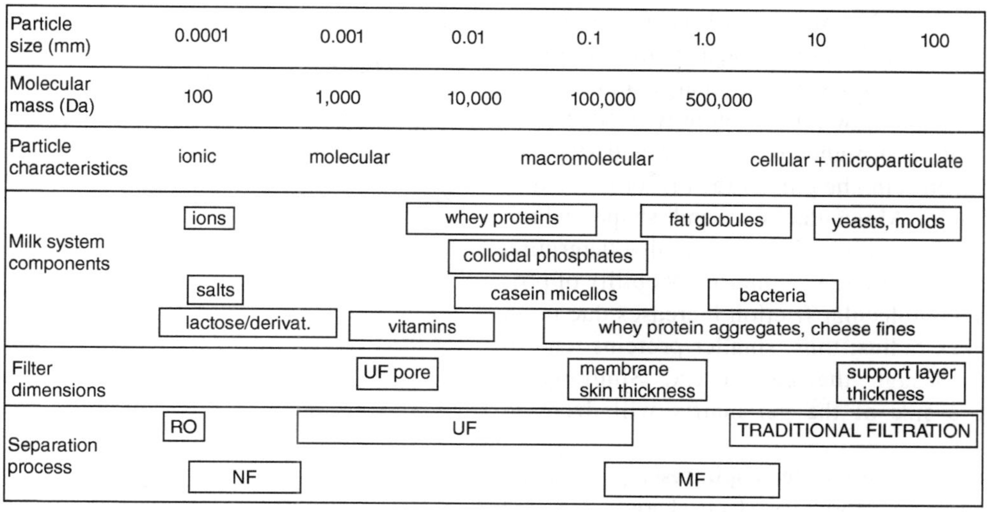

Fig. 24.1 Range of particle sizes used for various membrane filtration techniques compared to particle sizes of milk components and microorganisms. *RO* reverse osmosis, *NF* nanofiltration, *UF* ultrafiltration, *MF* microfiltration. (Adapted from Ref. Jelen [14], with permission of the International Dairy Federation)

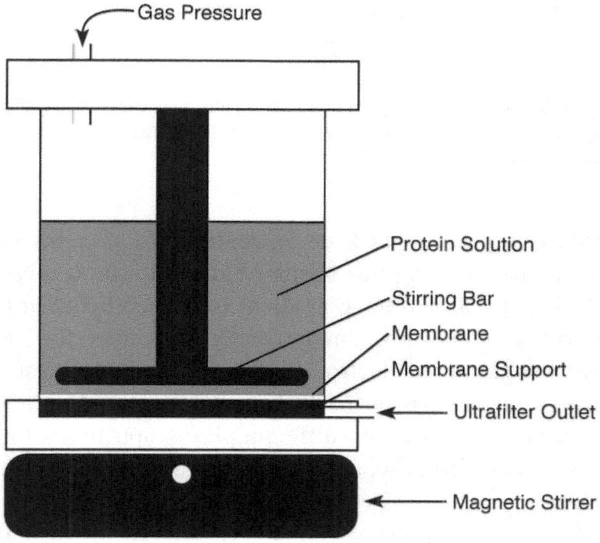

Fig. 24.2 Schematic diagram of a stirred cell ultrafiltration unit

Microfiltration can be used to remove particles and microorganisms and has been applied to wastewater treatment and to remove bacteria from milk and beer. Ultrafiltration and nanofiltration are used to remove salts, exchange buffer, or fractionate proteins on the basis of size. Ultrafiltration is used to concentrate milk for cheesemaking, whereas nanofiltration has been used to remove monovalent ions from salt whey. Reverse osmosis is often used to purify water and to remove aqueous salts, metal ions, simple sugars, and other small impurities with molecular mass below 2000.

Membrane filtration is commonly used in the food industry to produce protein ingredients (e.g., egg white concentrates and whey protein concentrates and isolates). Various membrane systems can be used in combination. For example, ultrafiltration and reverse osmosis in sequence are used to concentrate whey proteins, then remove salts and lactose, resulting in protein concentrates with 30–80% protein after drying, depending on the number of passes through the membrane. Microfiltration then can be performed on the retentate to remove large molecular weight components such as phospholipids and aggregated protein, resulting in whey protein isolate (permeate) with over 90% protein after drying.

24.2.5.2.3 Size-Exclusion Chromatography

Size-exclusion chromatography, also known as gel filtration or gel permeation chromatography, is a column technique that can be used to separate proteins on the basis of size (Chap. 12, Sect. 12.4.7). Size-exclusion chromatography is primarily used in the laboratory in a protein multidimensional separation sequence preceding or post ion-exchange chromatography. This technique is also used for desalting and removal of low molecular weight constituents to purify proteins.

24.2.6 Separation by Electrophoresis

24.2.6.1 Polyacrylamide Gel Electrophoresis

24.2.6.1.1 Principle

Electrophoresis is defined as the migration of charged molecules in a solution through an electrical field. The most common type of electrophoresis performed with proteins is **polyacrylamide gel electrophoresis** (PAGE). Proteins are positively or negatively charged, depending on solution pH and their isoelectric point (pI). A protein is negatively charged if solution pH is above its pI, whereas a protein is positively charged if solution pH is below its pI. The magnitude of the charge and applied voltage will determine how far a protein will migrate in an electrical field. The higher the voltage and stronger the charge on the protein, the greater the Migration within the electrical field. Molecular size and shape, which determine the Stokes radius of a protein, also determine migration distance within the gel matrix. Mobility of proteins decreases as molecular friction increases due to an increase in Stokes radius; thus, smaller proteins tend to migrate faster through the gel matrix. Similarly, a decrease in pore size of the gel matrix will decrease mobility.

In nondenaturing or **native electrophoresis**, proteins are separated in their native form based on charge, size, and shape of the molecule. Another form of electrophoresis commonly used for separating proteins is denaturing electrophoresis. PAGE with an anionic detergent, **sodium dodecyl sulfate** (SDS), is used to separate protein subunits by size (a process commonly referred to as SDS-PAGE). Proteins are solubilized and dissociated into subunits in a buffer containing SDS, which disrupts non-covalent interactions, with or without a reducing agent. **Reducing agents**, such as mercaptoethanol or dithiothreitol, are used to reduce disulfide bonds within a protein subunit or between subunits. Proteins bind SDS, become negatively charged, and are separated based primarily on size alone.

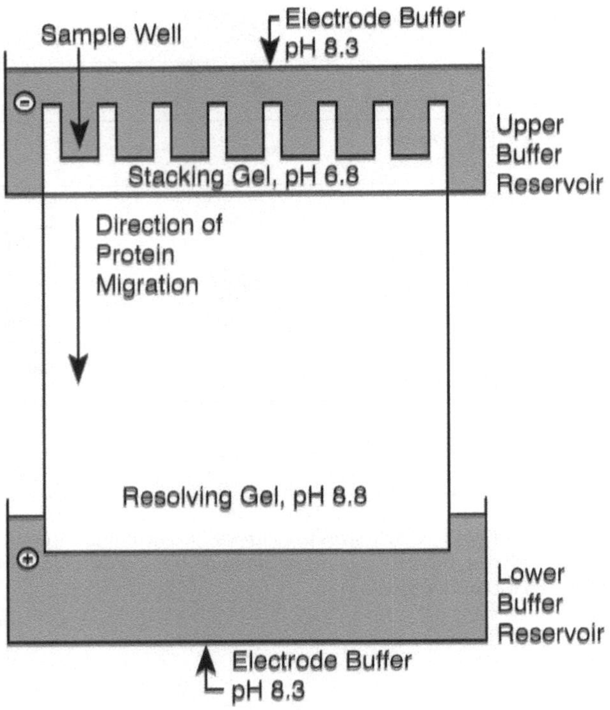

Fig. 24.3 Schematic diagram of a slab gel electrophoresis unit indicating the pHs of the stacking and resolving gels and the electrode buffer in an anionic discontinuous buffer system

24.2.6.1.2 Procedures

A power supply and electrophoresis apparatus containing the polyacrylamide gel matrix and two buffer reservoirs are necessary to perform a separation. A representative slab gel and electrophoresis unit are shown in Fig. 24.3. The power supply is used to generate the electric field by providing a source of constant current, voltage, or power. The electrode buffer controls the pH to maintain the proper charge on the protein and conducts the current through the polyacrylamide gel. Commonly used buffer systems include an anionic tris-(hydroxymethyl)aminomethane buffer with a resolving gel at pH 8.8 and a cationic acetate buffer at pH 4.3.

The polyacrylamide gel matrix is formed by polymerizing **acrylamide** and a small quantity (usually 5% or less) of the **cross-linking reagent, *N,N'*-methylenebisacrylamide**, in the presence of a **catalyst, tetramethylethylenediamine** (TEMED), and **source of free radicals, ammonium persulfate.** Precast, polyacrylamide gels within plastic cassettes are available commercially.

A discontinuous gel matrix is usually used to improve resolution of proteins within a complex mixture. The discontinuous matrix consists of a **stacking gel** with a large pore size (usually 3–4% acrylamide) and a **resolving gel** of a smaller pore size. The stacking gel, as its name implies, is used to stack or concentrate the proteins into very narrow bands prior to their entry into the resolving gel. At pH 6.8, a voltage gradient is formed between the chloride (high negative charge) and glycine ions (low negative charge) in the electrode buffer, which serves to stack the proteins into narrow bands between the ions. Migration into the resolving gel of a different pH disrupts this voltage gradient and allows separation of the proteins into discrete bands.

The pore size of the resolving gel is selected based on the molecular mass of the proteins of interest and is varied by altering the concentration of acrylamide in solution. Acrylamide concentrations of 15% may be used to separate proteins with molecular mass below 50 kDa. Proteins greater than 500 kDa are often separated on gels with acrylamide concentrations below 7%. A **gradient gel** in which the acrylamide concentration increases from the top to the bottom of the gel (most commonly 4–15% acrylamide) is often used to separate a mixture of proteins with a large molecular mass range.

To perform a separation, proteins in a buffer of the appropriate pH are loaded on top of the stacking gel. **Bromophenol blue tracking dye** is added to the protein solution. This dye is a small molecule that migrates ahead of the proteins and is used to monitor the progress of a separation. After an electrophoresis run, the separated protein bands in the gels can be visualized using a nonspecific **protein stain** such as **Coomassie Brilliant Blue stain, silver stain,** or a **fluorescent gel stain**. Specific enzyme stains or antibodies can be used to detect a particular protein or enzyme. Additional procedural details can be found in several sources [2–4].

24.2.6.1.3 Applications

Electrophoretic techniques can be used as one step in a purification process or to aid in the biochemical characterization of a protein. Electrophoresis can be used to help determine the purity of a protein extract. Commercially available preparative electrophoresis units are used to purify large quantities of protein. Small quantities of protein can be eluted and collected from electrophoresis gels using electroelution techniques. In other cases, proteins can be transferred from electrophoresis gels to a membrane and then stained with antibodies specific to a target protein in a process called electroblotting or western blotting (Chap. 27, Sect. 27.3.3). Separated protein subunits may also be excised from the gel for further evaluation by mass spectrometry (Chap. 11). Electrophoresis is often used to determine the protein composition of a food product. For example, differences in the protein composition of soy protein concentrates and whey protein concentrates produced by different separation techniques can be detected. The lanes on the left side of Fig. 24.4 illustrate the protein patterns observed when extracts of unheated and heated whey proteins are separated by native PAGE.

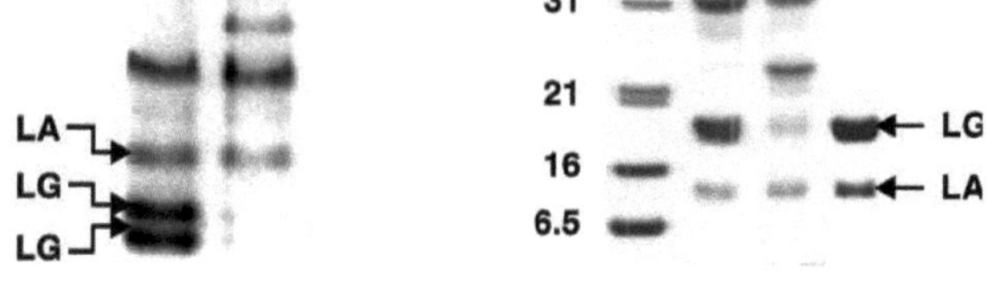

Fig. 24.4 Electrophoresis of whey proteins fractionated from raw and heated milk. *Left panel:* native-polyacrylamide gel electrophoresis (PAGE). Lane A = raw milk; lane B = heated milk. *Right panel*: sodium dodecyl sulfate-polyacrylamide gel electrophoresis (SDS-PAGE). Lane M = molecular weight marker; lane A = raw milk; lane B = heated milk; lane C β-lactoglobulin (LG) standard and α-lactalbumin (LA) standard. Ten micrograms of protein were loaded onto each lane. Note a decrease in two isoforms of LG after heating. (Reprinted from Ref. Chen et al. [15] with permission)

24.2.6.2 Isoelectric Focusing

24.2.6.2.1 Principle

Isoelectric focusing is a modification of electrophoresis, in which proteins are separated by charge in an electric field on a gel matrix in which a pH gradient has been generated using ampholytes. Proteins are focused or migrate to the location in the gradient at which pH equals the pI of the protein. Resolution is among the highest of any protein separation technique and can be used to separate proteins with pIs that vary less than 0.02 of a pH unit.

24.2.6.2.2 Procedure

A pH gradient is formed using **ampholytes**, which are small polymers (molecular masses of less than 1 kDa) containing both positively and negatively charged groups. An ampholyte mixture is composed of thousands of polymers that exhibit a range of pH values. Ampholyte mixtures are available that cover a narrow pH range (e.g., 1–2 pH units) or a broad range (e.g., 4–6 pH units) and should be selected for use based on properties of the proteins to be separated.

Ampholytes are added to the gel solution prior to polymerization. Once the gel is formed and a current applied, the ampholytes migrate to produce a linear pH gradient. Negatively charged ampholytes migrate toward the anode, while positively charged ampholytes migrate toward the cathode. The protein sample is loaded at one end of the gel and each protein migrates to its pI. There are commercial isoelectric focusing gel strips known as immobilized pH gradient strips.

24.2.6.2.3 Applications

Isoelectric focusing is an excellent method for determining the purity of a protein preparation. This method can be used to determine the isoelectric point of a protein by comparison to standard marker proteins of known pI. Genetic variants of many plant and animal proteins are visualized using isoelectric focusing. Isoelectric focusing is used to differentiate closely related animal and fish species based on protein patterns. The US Food and Drug Administration publishes the *Regulatory Fish Encyclopedia* [16], which contains isoelectric focusing patterns of about 1700 finfish and shellfish species. This guide helps state and federal officials to identify species substitution and thus detect product adulteration and economic fraud.

Isoelectric focusing is also used in combination with SDS-PAGE to increase the power of separation and purification. This technique is called **two-dimensional (2D) electrophoresis**. Proteins are first separated using isoelectric focusing strip. The gel strip, containing the proteins separated based on their pI, is then placed on top of an SDS-PAGE gel cassette, and proteins are further separated based on size. 2D electrophoresis is often used to obtain a high-purity protein from a mixture of a large number of proteins (up to 1000 proteins). The purified proteins are further characterized using analytical techniques such as mass spectrometry.

24.2.6.3 Capillary Electrophoresis

24.2.6.3.1 Principle

Capillary electrophoresis has been described as a hybrid technique incorporating aspects of traditional gel electrophoresis and liquid chromatography. Similar principles apply for the separation of proteins by both capillary and conventional electrophoretic techniques; proteins can be separated on the basis of charge or size in an electric field. The primary difference between capillary electrophoresis and conventional electrophoresis is that **capillary tubing** is used in place of acrylamide gels. As the separated proteins migrate along the capillary tubing, they are detected using detectors originally developed for liquid chromatography. Using standards, capillary electrophoresis can be used for quantification.

24.2.6.3.2 Procedure

A capillary electrophoresis system (Fig. 24.5) is comprised of a capillary column, power supply, detector, and two buffer reservoirs. The sample is introduced into the inlet side of the capillary tube by an injector placed prior to the capillary tube. Capillaries are composed of fused silica with internal diameters that commonly range from 10 to 100 μm. Column length varies from a few centimeters to 100 cm. High electric fields (5–30 kV/cm) can be used as the narrow columns dissipate heat very effectively, allowing for short run times of 10–30 min.

Protein bands are not visualized by staining as in conventional electrophoresis. Instead, protein peaks are detected as they migrate past a detector. Ultraviolet (UV)-visible detectors are most common, although fluorescence, conductivity, and light scattering detectors are available. The data obtained from a capillary electrophoresis run look like a typical HPLC chromatogram (see Chap. 13). Proteins can be labeled with a fluorescent derivative to increase sensitivity when a fluorescent detector is used.

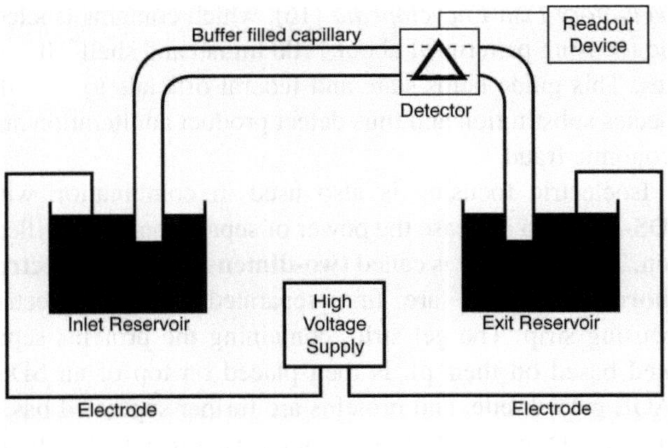

Fig. 24.5 Schematic diagram of a capillary electrophoresis system

There are three variations of capillary electrophoresis commonly used for protein separations. **Capillary zone electrophoresis** or **free-solution electrophoresis** is the most commonly used form of capillary electrophoresis. It is much like native PAGE, except proteins, are separated in free solution inside capillary tubes filled with buffer of the desired pH. Diffusion is prevented within the narrow diameter of the capillaries eliminating the need for a gel matrix. In capillary zone electrophoresis, electroosmotic flow also influences the separation of proteins within capillary tubes. The negatively charged fused silica capillary wall [containing silanol groups (SiO^-)] attracts positively charged ions (cations) from the buffer to form a double-ion layer at the interface between the capillary column wall and the buffer. When the electric field is applied, the cations forming the double layer are attracted toward the cathode and "pull" other molecules (independent of charge) in the same direction. Thus, in free-solution capillary electrophoresis, cations, anions, and uncharged molecules can be separated in a single run. Electroosmotic flow can be controlled by changing the pH or ionic strength of the buffer to alter the charge on the capillary wall and change the rate of protein migration.

SDS capillary gel electrophoresis techniques can be used to separate proteins by size and to determine molecular mass. In this technique, proteins are denatured and dissociated in the presence of SDS, with or without a reducing agent, then fractionation occurs in polyacrylamide gel-filled capillaries of specific pore sizes. Alternatively, linear polymers, such as methyl cellulose, dextrans, or polyethylene glycol, are added to the buffer within the capillary in a technique called dynamic sieving capillary electrophoresis. These entangled polymers act like the pores of the polyacrylamide gel to slow migration of the larger proteins and allow separation by size.

Proteins can also be separated on the basis of their isoelectric points, in a technique called **capillary isoelectric focusing**. Ampholytes (described in Sect. 24.2.6.2.2) are used to form a pH gradient within the capillary tube. A gel matrix is not needed. In this technique, electroosmotic flow is minimized by coating the capillary walls with buffer additives to prevent undesirable effects caused by surface charge.

24.2.6.3.3 Applications

Capillary electrophoresis is used primarily in analytical labs, although the use for routine quality control purposes is increasing. Capillary zone electrophoresis is used for a variety of applications, including the fractionation of milk, cereal, soybean, and muscle proteins [17]. Figure 24.6 illustrates the results obtained by capillary zone electrophoresis when used to fractionate the casein proteins of three mammalian milks. Microchip electrophoresis, a miniaturized ver-

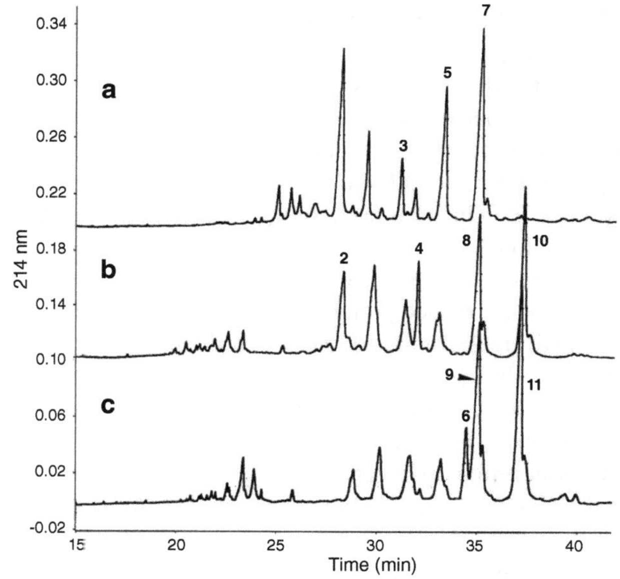

Fig. 24.6 Electropherograms illustrating differences in casein proteins isolated from (**a**) cow, (**b**) ewe, and (**c**) goat milk when separated by capillary zone electrophoresis. The peaks were identified as (*1*) bovine αS1-casein; (*2*) ovine αS1-casein; (*3*) bovine κ-casein; (*4*) ovine κ-casein; (*5*) bovine β-casein A1; (*6*) caprine κ-casein; (*7*) bovine β-casein A2; (*8*) ovine β2-casein; (*9*) caprine β2-casein; (*10*) ovine β1-casein; and (*11*) caprine β1-casein. Peaks eluting from the capillary column were detected by UV absorption. (Reprinted from Ref. Molina et al. [18] with permission)

sion of capillary electrophoresis using small volume microchannels, has been developed [19]. Electrophoretic separations on microchips can be completed in only a few minutes. This technique has been used to identify 15 different wheat varieties based on variations in the subunit composition of a major wheat protein, glutenin [20]. Results were comparable to those obtained using SDS-PAGE, but with an assay time of less than 1 min per analysis.

24.3 PROTEIN CHARACTERIZATION PROCEDURES

24.3.1 Amino Acid Analysis

24.3.1.1 Principle

Amino acid analysis is used to quantitatively determine the amino acid composition of a food, a protein ingredient, or relatively pure protein. Amino acid analysis is divided into three steps. First, the sample is hydrolyzed to release the amino acids. Amino acids are then separated using chromatographic techniques. Finally, the separated amino acids are detected and quantified. Several digestion procedures are reported [21–23], and others are AOAC methods [24]. Ion-exchange chromatography and reversed-phase liquid chromatography with pre- or post-column derivatization for detection are in widespread use and will be described in this section.

24.3.1.2 Procedures

In the most commonly used procedure, a sample is **hydrolyzed** in constant boiling 6 *N* HCl at 110 °C for 24 h to release amino acids prior to chromatography. Accurate quantification of some amino acids is difficult because they may be destroyed or converted into other reaction products during hydrolysis. Consequently, special hydrolysis procedures must be used to prevent errors. Faster digestion can be achieved using a microwave digestion unit (e.g., Mars 6, CEM, NC, USA) at 155 °C for 15 min.

Tryptophan is completely destroyed by acid hydrolysis. Methionine, cysteine, threonine, and serine are progressively destroyed during hydrolysis; thus, the duration of hydrolysis will influence results. Asparagine and glutamine are quantitatively converted to aspartic and glutamic acid, respectively, and cannot be measured. Isoleucine and valine are hydrolyzed more slowly in 6 *N* HCl than other amino acids, while tyrosine may be oxidized. AOAC Official Method 994.12 in the presence of performic acid is commonly used for the analysis of sulfur-containing amino acids (methionine and cysteine). Tryptophan can be analyzed after a basic digestion using concentrated NaOH (typically 4–5 *N*).

In the original methods developed in the 1950s, amino acids were separated by **cation-exchange chromatography** using a stepwise elution with three buffers of increasing **pH** and **ionic strength**. Variations of this method are still commonly used today and may include the use of gradient elution protocols. In a procedure called **post-column derivatization**, amino acids eluting from the column are derivatized and quantified by reaction with **ninhydrin** (reacts with primary amino group of amino acids) to produce a colored product that was measured spectrophotometrically. The method was automated in the late 1970s and adapted for use with high-performance liquid chromatographs in the 1980s as new ion-exchange resins were developed that could withstand high pressures. Amino acids eluting from the column also may be derivatized with ***o*-phthalaldehyde** (OPA) (reacts with the primary amino group of amino acids), then measured with a fluorescence detector. (Note: The ninhydrin and OPA methods can be used not only for amino acid analysis but also to monitor **hydrolysis of proteins** and to assay for **protease activity**, since these result in an increase of free primary amino groups.)

Other methods were developed in the 1980s using **precolumn derivatization** of the amino acids followed by reversed-phase HPLC. The hydrolyzed amino acids are derivatized prior to chromatography with **phenyl isothiocyanate** (reacts with primary amino group of amino acids), OPA, 6-aminoquinolyl-N-hydroxysuccinimidyl carbamate (AQC), or other compounds, separated by reversed-phase HPLC, and quantified by

UV or fluorescence spectroscopy. Methods using pre-column derivatizations may be more sensitive and can detect about 0.5–1.5 *uM* quantities of amino acids. Chromatographic runs usually take 30 min or less. The use of UPLC shortened further the time of analysis to no more than 10 min [23]. Quantification is done using external amino acid standards in the presence of an internal standard (e.g., L-Norvaline or alpha-aminobutyric acid) to correct for sample preparation errors.

24.3.1.3 Applications

Amino acid analysis is used to determine the amino acid composition of a protein or a mixture of proteins, determine quantities of essential amino acids to evaluate protein quality, identify proteins based on the amino acid profile, detect uncommon amino acids, and corroborate synthetic or recombinant protein structures. Amino acid analysis provides information for estimating the molecular mass of a purified protein. Amino acid analysis is also used to meet FDA nutritional labeling regulations for protein. In addition, the Food and Agriculture Organization (FAO) of the United Nations [25] recommends that amino acid analysis be used in place of the Kjeldahl method (N determination) for total protein, although they note that amino acid analysis requires the use of more sophisticated instrumentation and better trained technicians that are not always readily available in some countries. Proteins used in animal diets, infant formulas, sports nutrition products, and therapeutic human diets are often analyzed for protein quality to ensure adequate quantities of essential amino acids. A chromatogram illustrating the separation of amino acids from a casein hydrolysate on a reversed-phase column in under 10 min. is shown in Fig. 24.7. The amino acids were derivatized prior to separation (precolumn) using AQC and fractionated by UPLC using an internal standard such as alpha-aminobutyric acid [26].

24.3.2 Protein Nutritional Quality

24.3.2.1 Introduction

The nutritional quality of a protein is determined by the **amino acid composition** and the **digestibility** of that protein. **Antinutritional factors** can affect the nutritional quality of a protein. However, foods that contain heat-labile antinutritional factors (e.g., trypsin inhibitors) are usually cooked prior to consumption, thereby inactivating the inhibitor that might reduce protein digestibility. Some foods contain heat-stable antinutritional factors (e.g., tannins and saponins) that can decrease the nutritive value of a protein.

Many protein quality assessment methods utilize information about the essential amino acid content of a food. **Essential amino acids** are those that cannot be synthesized in the body and must be present in the diet. Although there are some special cases due to age and medical status of an individual, the amino acids generally categorized as essential (or indispensable) include histidine, isoleucine, leucine, lysine, methionine, phenylalanine, threonine, tryptophan, and valine. Requirements for these amino acids have been determined for various age groups (Table 24.3). The **first-limiting amino acid** of a food is defined as the essential amino acid present in the lowest amount compared to a reference protein or to human requirements.

A food scientist's concerns regarding protein nutritional quality include meeting the requirements of nutrition labeling, formulating products of high protein quality, and testing the effects of food processing on protein digestibility and essential amino acids such as lysine. The development of valid methods to measure protein quality of foods has been the focus of extensive research efforts over the past few decades [28, 29]. Protein nutritional quality assays may utilize animals in biological tests (*in vivo* assays), chemical or biochemical assays

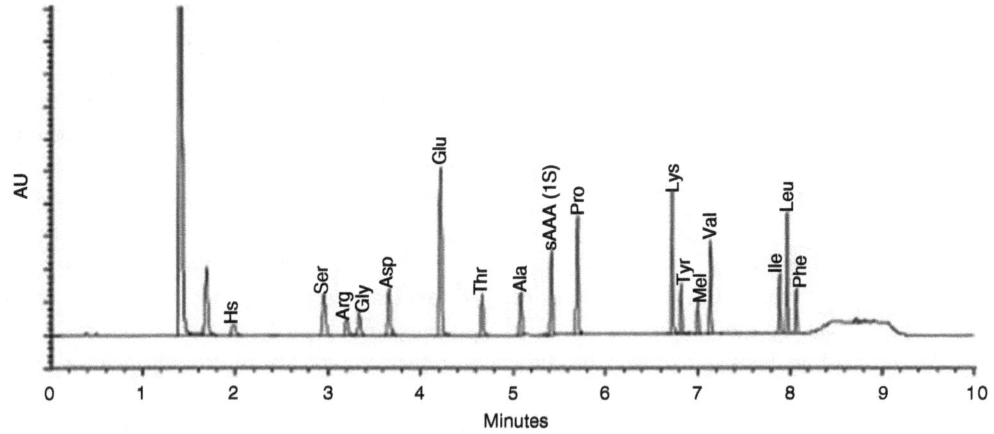

Fig. 24.7 High-performance liquid chromatographic analysis of amino acids from a casein hydrolysate using precolumn derivatization with 6-aminoquinolyl-N-hydroxysuccinimidyl carbamate (AQC) and separated on a reversed-phase column using an Acquity UPLC system (Waters Corp.) with a 2996 photodiode array detection system at 260 nm. A BEH C18 column (100 mm × 2.1 mm i.d., 1.7 μm) at a flow rate of 0.7 mL/min and column temperature of 55 °C was used. The injection volume was 1 μL. Gradient elution was performed using eluent (**a**) AccQ·Tag$_{ultra}$ concentrate (5%, v/v) and water (95%, v/v), and eluent. (**b**) AccQ·Tag$_{ultra}$. (Adapted from Ref. Boogers et al. [26] with permission)

(*in vitro* assays), and/or simply calculations. Because of the time and expense of *in vivo* methods, *in vitro* assays and calculations based on amino acid content often are used to estimate protein quality. This section of the chapter covers the tests and calculations required for nutrition labeling and mentions briefly several other protein quality methods for specialized applications. Please refer to Table 24.4 for a summary of methods used for measuring protein nutritional quality.

24.3.2.2 Protein Digestibility – Corrected Amino Acid Score

24.3.2.2.1 Principle

The **protein digestibility-corrected amino acid score** (PDCAAS method) is used to estimate protein nutritional quality by combining information from: (1) a calculation that compares the amount of the first-limiting amino acid in a protein to the amount of that amino acid in a reference protein and (2) an *in vivo* assay measuring the digestibility of the protein by rats. For nutrition labeling, the PDCAAS must be determined by methods described in 21 CFR 101.9 [30].

24.3.2.2.2 Procedure

1. Determine the amino acid composition of the food.
2. Calculate the **amino acid score** for the first-limiting amino acid, using the requirements of preschool age children as a reference pattern:

$$\text{Amino acid score} = \frac{\text{mg of amino acid in 1 g of test protein}}{\text{mg of amino acid in 1 g of reference protein}} \quad (24.1)$$

Table 24.3 Amino acid requirements of infants, preschool children, adolescents, and adults (males and females combined)

		His	*Ile*	*Leu*	*Lys*	*SAA*	*AAA*	*Thr*	*Trp*	*Val*
Age (years)		Amino acid requirements (mg/kg per day)								
0.5	(Infants)	22	36	73	64	31	59	34	9.5	49
1–2	(Preschool children)	15	27	54	45	22	40	23	6.4	36
11–14	(Adolescent)	12	22	44	35	17	30	18	4.8	29
>18	(Adult)	10	20	39	30	15	25	15	4.0	26
		Scoring pattern (mg/g protein requirement)								
0.5	(Infants)	20	32	66	57	28	52	31	8.5	43
1–2	(Preschool children)	18	31	63	52	26	46	27	7.4	42
11–14	(Adolescent)	16	30	60	48	23	41	25	6.5	40
>18	(Adult)	15	30	59	45	22	38	23	6.0	39

Adapted from Ref. Food and Agriculture Organization [27]
His histidine, *Ile* isoleucine, *Leu* leucine, *Lys* lysine, *SAA* sulfur amino acids, *AAA* aromatic amino acids, *Thr* threonine, *Trp* tryptophan, *Val* valine

Table 24.4 Summary of methods used to measure protein nutritional quality

Method	*What is measured*	*Application*
Protein digestibility- corrected amino acid score (PDCAAS)	Amino acid content of first- limiting amino acid compared to requirements of preschool-age children and true digestibility based on rat feeding experiment	Nutrition labeling for all but infant foods (to express g protein as % daily value)
Digestible indispensable amino acid score	This approach treats each essential amino acid as an individual nutrient, instead of only using the first-limiting amino acid as done in the PDCAAS method	Nutritional quality, but not approved yet for nutrition labeling
Protein efficiency ratio (PER)	Weight gain of rats per g protein consumed	Nutrition labeling for infant foods (to express g protein as % daily value)
pH shift *in vitro* digestibility	Change in pH due to enzymatic digestion of protein under standard conditions	Rapid test of protein digestibility
In vitro PDCAAS	*In vivo* digestibility using a commercial kit (Megazyme®), coupled with amino acid compositional analysis	Rapid test for comparative purposes among a number of samples
Amino acid score (AAS)	Amino acid content of first-limiting amino acid, compared to requirements of preschool age children	Part of PDCAAS assay
Essential amino acid index (EAAI)	Amino acid content of each of nine essential amino acids, compared to their content in a reference protein	Rapid calculation to determine optimum amounts of various proteins in food formulation
DNFB (1-fluoro-2,3-dinitrobenzene) method of available lysine	Amount of lysine that has not already reacted with other food constituents and thus become unavailable as an essential amino acid	Determine effect of heat treatment during food processing on available lysine

3. Feed male weanling rats standardized diets with 10% test protein or with no protein, following the procedure for true protein digestibility (AOAC Method 991.29) [24]. **True digestibility** is calculated based on nitrogen ingested and feed intake, corrected for metabolic losses in the feces. If available, published values of true digestibility for the test protein can be used.
4. Calculate PDCAAS:

$$\text{Amino acid score} \times \%\text{True digestibility} \tag{24.2}$$

5. For nutritional labeling: (50 g = Daily Value for protein)

$$\%\text{ Daily Value} = \frac{100 \times \left(\text{g protein / serving} \times \text{PDCAAS value}\right)}{50\,\text{g protein}} \tag{24.3}$$

24.3.2.2.3 *Applications*

The Nutrition Labeling and Education Act (NLEA) requires that the percent Daily Value for protein used on nutrition labels must be determined using the PDCAAS method, except for foods intended for consumption by infants (see also Chap. 3, Sect. 3.2.1.6. Because of the time and cost associated with the PDCAAS method, protein on the nutrition label is often expressed only as amount and not as a percent of the Daily Value. However, if a food label includes any claim regarding the protein, the nutrition label must include protein expressed as a percent of the Daily Value [30] [21CFR 101.9 © (7)].

The PDCAAS method generally is thought to better estimate protein quality for humans than the **protein efficiency ratio** (PER) method (Sect. 24.3.2.3.3). The actual digestion of protein by rats is thought to be fairly comparable to that by humans, so the protein digestibility portion of the PDCAAS method utilizes true digestibility as determined with rats. The PDCAAS method includes information on both amino acid composition and protein digestibility, since these are the factors that determine protein nutritional quality. However, perhaps as a limitation to the PDCAAS method, the amino acid score portion of the PDCAAS method includes only information about the first-limiting amino acid and no other essential amino acids. There is no differentiation in amino acid score between two proteins limiting to the same extent in one amino acid, but with the one protein only limiting in that amino acid and another protein limiting in many amino acids.

The continued use of the PDCAAS method remains under debate by the Food and Agriculture Organization (FAO) of the United Nations [27]. A newer protein quality measure, the **digestible indispensable amino acid Score** (DIAAS), has been recommended to replace the PDCAAS as the method of choice for dietary protein quality assessment for regulatory purposes. DIAAS is defined as

$$\text{DIAAS}(\%) = 100 \times \frac{\left(\begin{array}{l}\text{mg of digestible dietary}\\ \text{indispensable amino acid}\\ \text{in 1g of the dietary protein}\end{array}\right)}{\left(\begin{array}{l}\text{mg of the same dietary}\\ \text{indispensable amino acid in}\\ \text{1g of the reference protein}\end{array}\right)} \tag{24.4}$$

This approach treats each essential amino acid as an individual nutrient, instead of only using the first-limiting amino acid as done in the PDCAAS method. FAO is also recommending standardization of amino acid analysis methodologies. The DIAAS method is not yet approved by the FDA for nutritional labeling in the United States. The FDA, in a 2016 final rule, agreed that DIAAS is a better method, but did not approve it as a method to replace PDCAAS as the method for calculating protein % Daily Value on nutrition labels, citing that there was insufficient data to implement DIAAS.

24.3.2.3 Protein Efficiency Ratio

24.3.2.3.1 *Principle*

The PER method (AOAC Method 960.48) [24] estimates protein nutritional quality in an *in vivo* assay by measuring rat growth as weight gain per gram of protein fed.

24.3.2.3.2 *Procedure*

1. Determine the nitrogen content of the test protein-containing sample and calculate the protein content.
2. Formulate a standardized test protein diet and a casein control diet to each contain 10% protein.
3. Feed groups of male weanling rats the diet and water ad libitum for 28 days.
4. Record the weight of each animal at the beginning of the assay, at least every 7 days during the assay, and at the end of the 28 days.
5. Record the food intake of each animal during the 28-day feeding trial.
6. Calculate the PER using the average total weight gain and average total protein intake for each diet group at day 28:

$$\text{PER} = \text{Total weight gain of test group}(\text{g}) \,/\, \text{Total protein consumed}(\text{g}) \tag{24.5}$$

7. Normalize the PER value for the test protein (i.e., compare the quality of the test protein to that of casein) by assigning casein a PER of 2.5:

$$\text{Adjusted or corrected PER} = \text{PER of test protein} \,/\, \text{PER of casein control} \tag{24.6}$$

24.3.2.3.3 *Application*

The Food and Drug Administration (FDA) requires that the PER method, which measures rat growth [28], be used when determining protein % Daily Value on the nutrition label of foods intended for consumption by infants [30] (21 CFR 101.9 (c) (7)) (see also Chap. 3, Sect. 3.2.1.6. The PER method is limited to this application because the essential amino acid requirements for young rats are similar to those of human infants, but not those of other age groups. Rat growth is not comparable to that of adults, but it is more comparable to that of infants. The PER method is time consuming, and it does not give any value to a protein for its ability to simply maintain body weight (i.e., a protein that produces no weight gain in the assay has a PER of zero).

24.3.2.4 Other Protein Nutritional Quality Tests

24.3.2.4.1 *Essential Amino Acid Index*

Essential amino acid index (EAAI) estimates protein nutritional quality based on the content of all essential amino acids compared to a reference protein (or human requirements). The EAAI is a rapid method to evaluate and optimize the amino acid content of food formulations. Unlike the amino acid score component of the PDCAAS method that considers only the first-limiting amino acid, the EAAI method accounts for all essential amino acids. However, EAAI does not include any estimate of protein digestibility, which could be affected by the processing method. The essential amino acid content of the test protein (determined by amino acid analysis or from literature values) is compared to that of a reference protein (e.g., casein) or human requirements as follows: (Note: use values for methionine plus cystine and phenylalanine plus tyrosine because they can substitute for one another as essential amino acids.)

$$\text{Essential amino acid index} = 9\sqrt{\left[\left(\begin{matrix}\text{mg of lysine in}\\ \text{1 g of test protein}\end{matrix}\right) / \left(\begin{matrix}\text{mg of lysine}\\ \text{in 1 g reference protein}\end{matrix}\right)\right] \times [\text{etc. for other 8 essential amino acids}]} \quad (24.7)$$

24.3.2.4.2 *In Vitro Protein Digestibility*

The **pH shift method** is an *in vitro* protein digestibility assay used to estimate the digestibility of a protein by measuring the extent of protein hydrolysis under standardized conditions with commercial digestive enzymes. The digestion procedure is designed to simulate human digestion of protein by using the enzymes trypsin, chymotrypsin, peptidase, and a bacterial protease. The pH of the protein solution drops when proteases break peptide bonds, release carboxyl groups, and liberate hydrogen ions. The pH at the end of the digestion period is used to calculate protein digestibility.

The pH shift method is the enzyme digestibility part of the **protein efficiency ratio-calculation method** (C-PER) (AOAC Method 982.30) [24], which combines a calculation of the essential amino acid composition with a calculation based on the *in vitro* assay of digestibility. The C-PER assay is intended for routine quality control screening of foods and food ingredients, to estimate the PER as it would be determined with the rat bioassay method.

A newer *in vitro* digestibility assay is available commercially as a kit and is used for the determination of *in vitro* PDCAAS. The K-PDCAAS Megazyme® kit provides instructions for the enzymatic digestion (with pepsin, then trypsin + chymotrypsin) accompanied by detailed calculations guidelines [31], provided that amino acid compositional analysis is performed.

The *in vitro* digestibility assays can provide a rapid and inexpensive means to determine and compare the protein digestibility of food products. Such assays could be used to determine the effect of processing conditions on the protein digestibility of food products with the same formulation.

24.3.2.4.3 *Lysine Availability*

If the nutritional quality of a heat-processed product is lower than expected from its amino acid composition, one should assess the availability of lysine in the product. The **free ε-amino group** of the essential amino acid lysine can react with many food constituents during processing and storage to form biologically unavailable complexes with reduced nutritional quality. Lysine can readily complex with reducing sugars in the Maillard browning reaction, oxidized polyphenols, and oxidized lipids. Such reactions are accelerated upon heating and under alkaline conditions. **Lysinoalanine**, often found in alkali-treated proteins, and 2-furoyl-methyl-Lys (**furosine**), the Amadori compound formed upon the reaction of a lysine residue with a reducing sugar, decrease both the digestibility of the protein and the availability of lysine as an essential amino acid. The most commonly used method to measure **available lysine** is a spectrophotometric method that utilizes the reagent 1-fluoro-2,3-dinitrobenzene (DNFB, also referred to as FDNB). The DNFB method, detailed in AOAC Method 975.44 [24], can be used to determine if food processing operations have reduced the availability of lysine in a food.

24.3.3 Assessment of Protein Structural Properties

24.3.3.1 Introduction

Different sources of food proteins are made up of heterogeneous mixtures of proteins that vary in **structural proper-**

ties (molecular weight, configuration, and surface properties). Besides being important for nutritional considerations, a protein's primary structure (amino acid composition and sequence) also dictates its molecular weight, configuration, and surface properties, and thus its behavior (i.e., functionality) in a food system [32]. The protein's configuration (e.g., globular like whey protein, or open structure like casein) influences its thermal or nonthermal denaturation (i.e., structural stability). Changes in configuration impact surface properties. Protein structural properties are influenced by several external factors, including environmental factors (pH, temperature, and ionic strength), processing conditions (thermal treatment, solvent extraction), and protein modification (physical, chemical, enzymatic, biotechnological, and others).The following sections outline briefly the methods used to assess the most important structural properties that have a direct bearing on protein functionality in different food systems.

24.3.3.2 Molecular Weight Distribution

Molecular weight distribution allows for the determination of molecular size of peptides, protein subunits, monomers, dimers, quaternary structures, as well as high molecular weight polymers (mostly soluble aggregates) that form during processing. Estimation of molecular size can be done under nondenaturing and under denaturing conditions following techniques such as electrophoresis (Sect. 24.2.6), size-exclusion chromatography (Sect. 24.2.5.2.3), and mass spectrometry (Chap. 11).

SDS-PAGE is commonly used as a qualitative measure to visualize protein components within a sample and to estimate the molecular mass. Molecular mass is estimated by comparison to a molecular weight standard mix run on the same gel along with the sample (Fig. 24.4). Commercially prepared protein standards are available in several molecular mass ranges.

Size-exclusion HPLC (SE HPLC) is run under nondenaturing conditions, in which samples are solubilized in phosphate buffer at pH 7 to obtain the molecular weight of monomers, aggregates, and degradation products (peptides) and their relative concentration in a sample. SE HPLC also can be run under denaturing conditions, in which samples are solubilized in the presence of SDS with or without a reducing agent. Running the samples under different solubilization conditions allows for the determination of the type of interactions that contributed to the formation of polymers in the sample. Molecular mass can be calculated as described in (Chap. 12, Sect. 12.4.7).

The combination of SE HPLC with multiangle light scattering (MALS) detection (Chap. 13, Sect. 13.2.4.5) allows for a more accurate and detailed characterization of protein aggregation. In SE HPLC-MALS, light scattering and concentration are independently measured for each eluting fraction. Accordingly, the molar mass and molecular size can be accurately determined regardless of the retention time and the variation in the molecular structure (nonglobular structures).

Determining molecular weight distribution is useful when monitoring the formation of peptides, protein aggregates, and protein complexes with other constituents (e.g., carbohydrates) in different food systems as a result of various processing and modification conditions. For example, unheated and heated whey proteins are differentiated by SDS-PAGE (illustrated in Fig. 24.4, right side). Similar evaluation of changes in molecular weight distribution upon processing allows for the prediction of functionality (e.g.. solubility, emulsification, or gelation) in various food systems [33, 34]. Molecular weight distribution of proteins also allows for identifying differences among species (e.g., soybean compared to pea protein) and among varieties within a particular species (e.g., variation in protein fractions among different breeding lines of Faba bean, [35]).

24.3.3.3 Denaturation

Any change in a protein's quaternary, tertiary, and/or secondary structure results in protein denaturation [32]. **Protein denaturation** is a change in the native state of the protein structure. In simple terms, denaturation is often referred to as protein unfolding. Upon environmental stresses (e.g., heat, extreme pH, high salt concentration, organic solvents, and presence of denaturants such as SDS, urea, or a reducing agent such as β-mercaptoethanol), intramolecular interactions within the protein (ionic, hydrophobic, H-bonding, and/or disulfide linkages) may be disrupted, resulting in protein denaturation. The extent of denaturation can be determined following a number of instrumental and chemical measurements. **Differential scanning calorimetery** (DSC) (Chap. 33, Sect. 33.3.2) is among the most common methods used to determine the onset temperature of protein denaturation and the extent of denaturation through the calculation of the change in **enthalpy** (a thermodynamic quantity equivalent to the total heat content). When no change in enthalpy can be detected, then the protein is completely denatured.

Indirect chemical measurements of denaturation include measuring surface hydrophobicity (see Sect. 24.3.3.4), reduction in solubility (See Sect. 24.3.4.2) and increase in UV absorption or fluorescence. Upon protein unfolding, hydrophobic residues are exposed, including tyrosine and tryptophan, which can absorb light and can fluoresce.

Measuring protein denaturation is essential to determine the effects of processing on protein functionality. Depending on the extent of denaturation, the effect on protein functionality could be either positive or negative. Completely denatured protein may promote undesirable aggregation, leading to reduced solubility, which is detrimental for beverage applications. On the other hand, partially denatured protein may have enhanced surface activity for emulsification applications and may promote gelation for applications such as plant-based yogurt. Partial unfolding of the protein changes the hydrophobic to hydrophilic balance on the surface of the protein and hence may facilitate desirable protein/oil, protein/water, and/or protein/protein interactions.

24.3.3.4 Surface Properties

24.3.3.4.1 Surface Hydrophobicity

Typically, proteins fold in such a way that hides hydrophobic amino acid residues from the aqueous medium. However, all proteins have hydrophobic residues on the surface, resulting in a characteristic known as surface hydrophobicity. The **surface hydrophobicity** of a protein is defined by the portion of the nonpolar surface of the protein that makes contact with the surrounding bulk water. Proteins with high amounts of hydrophobic amino acids (>30%), such as most plant-derived food proteins, will have a relatively high surface hydrophobicity. Hydrophobic interactions between different monomers will result in quaternary structures involving three or more monomers. Such molecular configurations result in rigid structures of relatively high molecular weight, imparting unique functional properties, desired for some food applications (e.g., texturized vegetable proteins), and not desired in others.

The surface hydrophobicity of a protein is most commonly determined by measuring the amount of a hydrophobic fluorescent probe, such as 8-anilino-1-napthalenesulfonic acid ammonium salt (ANS) or 5,5′-dithiobis-(2-nitrobenzoic acid) (DTNB) that can bind to the protein via hydrophobic interactions. Fluorescence is measured using a spectrofluorometer (Chap. 7, Sect. 7.3) at excitation and emission wavelengths of 400/30 (wavelength/bandwidth) and 460/40 nm, respectively. A relative fluorescence index (RFI) is calculated relative to a blank at different protein concentrations (typically 0.01–0.05%), and the slope of the line is taken as the surface hydrophobicity index.

Measurement of surface hydrophobicity of a protein allows for comparison among different protein sources (e.g., plant vs animal protein) and for the prediction of changes in the protein structure upon processing. Changes in surface hydrophobicity contribute the protein's ability to interact with polar and nonpolar phases, or the potential for protein–protein interactions. Protein–protein interactions are desirable for protein network formation in gels, doughs, and extruded products. Determining surface hydrophobicity is also useful in explaining changes in protein such as solubility, foaming, and emulsification properties.

24.3.3.4.2 Surface Charge

Net surface charge is often estimated by measuring **zeta potential**, which is defined as the measure of magnitude of the electrostatic interactions between particles that are in solution. Zeta potential (ζ) provides the magnitude of the charge on the surface of a protein. The magnitude of the charge on a protein indicates the extent of potential repulsion between adjacent molecules in a dispersion. It is generally measured by placing the solution in a cell with electrodes at either end followed by applying a potential. The velocity of the particles is measured and expressed in unit field strength as their mobility. To measure the velocity of very high or low mobility samples, laser and light scattering technology may be used. An example of a commercial instrument for measuring Zeta potential is Malvern Zetasizer.

Zeta potential can be used to predict the stability of emulsions or suspensions, streamlining experimental design. Differences in Zeta potential among different protein sources can explain differences in protein solubility. Zeta potential is also impacted by pH and ionic strength. Proteins have a net positive charge below their pI and a net negative charge above their pI. Therefore, for comparison purposes among different proteins, zeta potential is measured over a range of pH. High ionic strength in a protein solution results in masking the magnitude of charge on the surface of the protein and may explain loss of solubility due to high salt content.

24.3.4 Assessment of Protein Functional Properties

24.3.4.1 Introduction

Protein functionality is defined as any physicochemical property of protein molecules that affects their behavior in food products during processing and storage, and consequently impacts overall quality attributes as perceived by the consumers. It is often desirable to characterize the functional properties of food proteins to optimize their use in different food applications. This section will briefly highlight several protein functional properties and methods of assessment. Although formulating high protein foods is on the rise, there remains a gap in standardized methods for determining protein functionality. Many methods exist in the literature, which makes it hard to compare results across different studies. The most commonly used methods will be discussed briefly in the following subsections. Functional properties that involve protein–water, protein–oil, protein–oil–water, and protein–water–protein interactions will be summarized. It should be pointed out that there is no single functional property test that is applicable to all food systems, so careful test selection is imperative.

24.3.4.2 Water Hydration/Binding Capacity and Viscosity

Water binding capacity (WBC), also sometimes called **water hydration capacity**, is defined as the grams of water bound per gram of protein when a dry powder is equilibrated with water vapor at 90–95% relative humidity. This is often measured by gradually adding water to a set amount of protein powder and mixing until the protein is completely wetted. The mixture is centrifuged at a low speed and the water that remains in the protein is weighed and calculated as the amount of water bound.

Several factors affect WBC, including the protein composition (charged, polar, nonpolar amino acid residues), protein denaturation, and environmental factors (pH, salt, and temperature) [32]. Many functional properties of proteins, such as dispersibility, solubility, emulsification, foaming, gelation, and water holding capacity, depend on water-protein interactions and hence on WBC. However, viscosity is the functionality that is most reliant on WBC.

Viscosity is defined as a solution's resistance to flow under an applied force or shear stress. This is generally measured by a viscometer (Chap. 29, Sect. 29.3.2). The viscosity of a protein solution is affected by the molecular configuration and denaturation state of the protein, protein concentration, and environmental conditions (pH, temperature, salt, and other constituents). To measure viscosity, a protein solution at a certain concentration is prepared and then placed under increasing amounts of shear at a certain rate and at varying temperatures. Viscosity measurement can be used to predict how a protein solution will perform under processing conditions when subjected to shear forces. Units of viscosity commonly reported are m2/s and stokes (St).

24.3.4.3 Dispersibility Index and Solubility

Protein dispersibility index (PDI) determines the amount of dispersible protein in water under the conditions of the test. It is a value representing the % protein that is dispersed in water relative to % total protein under standard conditions. Results are highly dependent upon the equipment used. The standardized AOCS method (Ba 10b-09) [36] calls for a vertical mixer with a variable speed (that is standardized for the test at 8500 rpm). A sample solution is prepared and mixed at the specified speed and centrifuged. The protein content of the supernatant is determined using Kjeldahl or Dumas procedures (Chap. 18, Sect. 18.2), and PDI is then calculated as percent water dispersed protein relative to the % total protein in the solution. PDI is mostly used for soy protein ingredients. For example, the PDI of soy flour correlates with the amount of heat to which the flour was subjected to during processing/production.

Given that PDI was developed exclusively for soy protein ingredients, **protein solubility** became the common measurement that replaced PDI. Proteins usually need to be soluble under the conditions of use for optimal functionality in food systems. Many other important functional attributes of proteins are influenced by protein solubility, such as emulsification, foaming, and gelation properties. Insoluble proteins have limited food applications.

Protein solubility is dependent on the molecular weight (large polymers tend to be insoluble) and the balance of hydrophobic to hydrophilic amino acids that make up the protein, especially those amino acids on the surface of the protein (i.e., surface hydrophobicity and charge). Protein solubility is influenced by pH, ionic strength, ion composition, and interactions with other food components, such as lipids or carbohydrates. Common food processing operations, such as heat and physical treatment (e.g., shearing), may influence protein solubility. Proteolysis by endogenous or exogenous proteases also may alter protein solubility.

Similar to PDI, for protein solubility measurement, a protein is dispersed in water or buffer at a specified pH, and the dispersion is centrifuged using defined conditions. Protein insoluble under the test conditions precipitates while soluble protein remains in the supernatant. Protein concentration, pH, salt content, and temperature can be varied to predict the protein solubility and stability in different food or beverage systems (e.g., the protein stays in solution during storage of high protein beverages). Total protein and protein in the supernatant are measured usually by the Dumas method or a colorimetric procedure, such as the Bradford assay or the bicinchoninic acid (BCA) assay (Chap. 18, Sects. 18.2.2 and 18.4). However, colorimetric methods have limitations due to matrix interferences and choice of standards used for calibration. Percentage protein solubility is measured as a percentage, similar to that of PDI, by dividing the protein in the supernatant (soluble protein) by the total protein and multiplying by 100.

24.3.4.4 Oil Binding Capacity

Oil binding capacity (OBC) is defined as the amount of oil that is bound to the protein, which is expressed in grams of oil held per gram of protein. The OBC of proteins is related to the protein's surface hydrophobicity and its overall solubility. In general, proteins with relatively high surface hydrophobicity have high OBC. In protein powders, OBC is influenced by the particle size. Powders with low density and small particle size adsorb and entrap more oil. OBC is generally measured by slowly adding oil to a set amount of protein and mixing until the protein is completely wetted. The mixture is centrifuged at a low speed and the oil that remains in the protein is weighed and calculated as the amount of oil that can be bound. OBC has a direct bearing on emulsification properties but is not the only factor to affect these properties. OBC is of great importance in developing food products as it affects texture and mouthfeel.

24.3.4.5 Emulsification

Emulsions are defined as a dispersion of two immiscible liquids that generate a high surface tension. Energy, in the form of homogenization, blending, or shaking, is used to disperse one immiscible liquid into the other, forming droplets [37]. Oil and water are the two most common immiscible liquids found in food emulsions, although many other food components are usually present. An emulsion is an unstable system and proteins are often used to form and stabilize emulsions. Proteins as emulsifiers lower the interfacial energy between phases to facilitate emulsion formation and to improve the stability of emulsions.

The **emulsification properties** of food proteins are affected by the protein surface properties, molecular configuration, denaturation state, protein concentration in the system, and environmental conditions (pH, temperature, salt, and other constituents). Emulsification properties are important for emulsion-type products (e.g., margarine, butter, cream, infant formulas, mayonnaise, processed cheese, salad dressing, meat products, baked goods, and frozen desserts).

Food emulsions usually are highly complex systems with multiple ingredients, so industrial scientists often choose to investigate the properties of emulsions in simplified model systems containing only a few of the most important ingredients. For a protein-based emulsion, this model system may contain only water or buffer, oil, and protein. It must be remembered that pH, salt concentration, temperature, type and amount of oil, protein concentration, energy input, and temperature during emulsion formation have a large effect on the properties of the final emulsion. These parameters must be selected prior to establishing a procedure.

The quality of an emulsion is dictated by many factors including droplet size, droplet size distribution, density differences between the two phases, viscosity of the continuous phase (the higher the viscosity, the more stable is the emulsion), electrostatic and steric interactions between molecules at the interface, and thickness of the adsorbed protein layer. The emulsification properties thus can be evaluated by determining the size distribution (often determined by turbidimetric techniques) of the droplets formed (the smaller the size the more stable the emulsion is against coalescence, flocculation, and creaming), Zeta potential on the surface of the droplets (higher surface charge reduces flocculation), viscosity, emulsion microstructure (by microscopy, Chap. 32, Sect. 32.2), emulsification activity index, emulsification stability, and emulsification capacity. **Emulsification activity index** is a measure of the ability of a protein to form an emulsion. The greater the activity of the protein, the smaller the oil droplets formed. **Emulsification stability index** is the measure of how quickly an emulsion will separate into two phases, expressed as a unit of time. **Emulsification capacity** is the measure of the maximum amount (volume or weight) of oil that can be stabilized by a unit weight of protein, expressed in grams or mL of oil per gram of protein or per gram powder. Details of all of the mentioned methods used for the evaluation of emulsification properties can be found in references [37–40].

24.3.4.6 Foaming

Foams are coarse dispersions of gas bubbles in a liquid or semisolid continuous phase. Like emulsions, foams require energy input during formation and are inherently unstable. Whipping, shaking, and sparging (gas injection) are three common methods of foam formation. Proteins in the continuous phase lower the surface tension between the two phases during foam formation and impart stability to films formed around the gas bubbles. Foams are found in cakes, breads, marshmallows, whipped cream, meringues, ice cream, soufflés, mousses, and beer.

Foam volume and foam stability are two important parameters used to evaluate foams. **Foam volume** is dependent on the ability of a protein to lower surface tension between the aqueous phase and gas bubbles during foam formation. **Foaming capacity** is the common method to evaluate foam volume. Foaming capacity is a measurement of the ability of a protein solution to produce a foam after agitation, generally expressed as the change in volume per gram of protein. It can be expresses in terms of overrun (foam volume/volume of initial liquid × 100), or foaming power (foam expansion calculated as volume of incorporated gas/volume of liquid × 100). The foaming power increases with protein concentration until a maximum volume is reached. Foaming capacity is generally measured by preparing a foam using a set volume of a protein solution of a predetermined protein concentration (dependent of the type of protein), using a blender/mixer at set parameters. Volume measurements are then recorded. Other methods for generating a foam include the use of a foamscan, which sparges air through the sample to generate a foam.

Foam stability depends on the properties of the protein film formed around the gas droplets. Free liquid is released as a foam breaks down. A more stable foam usually takes a longer time to collapse. Foaming stability refers to the ability of a protein foam to remain stable over time and against gravitational or mechanical stresses. It is generally expressed as the final volume of the foam after a certain period of time, divided by the initial volume and multiplied by 100. It is also often expressed as the time required for 50% reduction in foam volume (as a half-life). In one of the simplest methods, the foam (resulting from a sample whipped for a fixed time under standardized conditions) is placed in a funnel over a graduated cylinder. The time for half of the original weight or volume of the foam to drain away from the foam is recorded. The greater the half-life, the more stable the foam.

Foam volume and stability are influenced by energy input, pH, temperature, heat treatment, and the type and concentration of ions, sugars, lipids, and proteins in the foam. Hence, all of these variables, except the one being tested, must be standardized when designing a procedure to measure foam properties. Other methods can be used to better understand the molecular properties of the foam including (1) determination of interfacial properties such as surface pressure and film thickness, (2) characterization of the viscoelastic properties of the film, and (3) characterization of the bubble size and distribution.

24.3.4.7 Protein Network Formation

24.3.4.7.1 Gelation and Water Holding Capacity

A gel is a three-dimensional network formed between a solid and a liquid, entrapping water and small molecules. Protein gels are made by treating a protein solution with heat, enzymes (e.g., chymosin action on casein micelle to form cheese curd), divalent cations (e.g., adding Ca^{2+} or Mg^{2+} to form tofu from soy proteins), or by the lowering the pH to the pI under appropriate conditions. However, most food **protein gels** are made by heating protein solutions. Proteins are transformed from the "soluble" state to a cross-linked "gel-like" state in protein gelation. The continuous, cross-linked network structure formed may involve covalent (i.e., disulfide bond) and/or noncovalent interactions (i.e., hydrophobic interactions, hydrogen bonds, and electrostatic interactions). Some protein gel networks made by heating a protein solution are thermally reversible (e.g., gelatin gel that is formed by heating then cooling, but reverts to a liquid when reheated), while some protein gels are thermally irreversible (e.g., egg white). The strength of a protein gel is affected by a variety of factors, such as the structural properties and concentration of the protein, temperature, rates of heating and cooling, pH, ionic strength, and the presence of other food constituents [32]. Therefore, these variables must be predetermined and controlled to measure and compare gelation properties of proteins.

Least gelation concentration (LGC), i.e., the least amount of protein necessary to form a gel, is an important parameter that needs to be determined for each type of protein. It is measured by preparing a series of protein solutions at various concentrations, typically 2–20% protein, and applying heat at a moderate rate. The heating temperature is determined based on the denaturation temperature of the protein being tested. Often the heating temperature range is 7595 °C. The containers are inverted and the protein concentration at which the matrix does not slide down is determined as the LGC.

To measure the **gel strength**, which as another important gelation property, the gel is formed, a concentration above the LGC is chosen, and the gels are tested to evaluate firmness or other texture parameters. Common test methods include using an Instron, Texture Analyzer with a texture profile analysis (TPA) method, or following a cutting or shearing protocol. Often the extent of resistance is measured, as an indication of gel strength, when a force is applied, reported in Newton's. Techniques used to measure rheological properties are described in more detail in Chap. 29. In addition, many empirical tests have been developed to measure gel properties of specific proteins, such as the **Bloom test** designed to measure the strength of gelatin gels and the **folding test** to measure the elasticity of surimi gels.

24.3.4.7.2 Dough Formation

Wheat protein is unique in its ability to form a **viscoelastic dough** suitable for making bread and other bakery products. Gluten-forming proteins, the major storage proteins of wheat, are a heterogeneous mixture of gliadins and glutenins. The unique amino acid composition of these proteins makes possible the formation of a viscoelastic protein network within a dough, which can entrap carbon dioxide gas during yeast fermentation. Bread-making quality of wheat varieties, with and without dough enhancers (e.g., ascorbic acid), is commonly tested by measuring dough strength, viscosity, and extensibility. The effect of dough ingredients (e.g., whey or soy proteins, phospholipids, and surfactants) can be tested in the same way.

Dough strength is measured under standardized conditions with a variety of commercially available instruments. The **Mixograph®** is used to test the mixing properties of flour to ensure the dough will have the proper consistency, which is essential for automated manufacture of baked products. A **Farinograph®** is a torque meter to determine the water absorption of flour and the mixing properties of dough. The Extensograph®, often used in conjunction with the Farinograph, measures the extensibility of dough and thus can predict baking behavior. The **Rapid Visco Analyser®** (RVA) is a cooking viscometer to rapidly test starch-pasting properties. The RVA incorporates heating, cooling, and variable shear, to test the viscosity of starches, cereals, and other foods. The **Alveograph®** is used to measure gluten strength and extensibility by measuring the force required to blow and break a bubble of dough. Other tests of wheat quality include the **Glutomatic®** which measures the wet gluten content of flour.

Water holding capacity (WHC) is another important property related to gelation. WHC of a protein refers to the ability of a protein network to hold water and retain it against gravitational force within the protein gel. The term is often mistakenly used interchangeably with WBC. WHC is generally measured by forming a gel under predetermined conditions and then applying a gravitational force, followed by measuring the amount of water that is released. The water that remains in the protein matrix is reported as the WHC and is usually expressed as % water retained in the gel. WHC is of great importance in developing food products since it affects texture and mouthfeel, especially in meat products and baked goods. High water holding capacity will also increase profit margins since many food products contain 50% or more water.

24.3.4.7.3 Texturization

Texturization is a protein property of particular interest for the use of plant protein in meat alternative applications [41]. Specifically, texturization of plant proteins using extrusion technology produces fibers that mimic meat fibers through heat, shear, pressure, and cooling [42]. In this process, globular proteins unfold, associate, aggregate, and cross-link. Disulfide linkages, hydrophobic interactions, and hydrogen bonding are instrumental in forming these protein fibers. High moisture conditions often are used to create meat analogous with unique physical properties including chewiness, elasticity, softness, and juiciness [32]. Soy protein and gluten protein ingredients often are utilized for meat analogs applications due to their high degree of surface hydrophobicity, high molecular weight, and relatively high content of cysteine residues. These characteristics enable the formation of protein fibrous structures. Textural properties are evaluated using various texture analysis techniques (Chap. 29) and microscopy (Chap. 32).

24.3.4.8 Applications of Testing for Protein Functionality

Food product developers often need to understand the effect of ingredient substitutions on the functional properties and subsequent product quality. Variations in raw materials can lead to differences in protein functionality. Protein functional tests can be used to compare ingredients from different manufacturers or to verify the quality of protein ingredients in each batch purchased from a manufacturer. Also, if a product developer is trying to substitute soy protein for another plant protein ingredient in a formulation, it is crucial to know how the functional attributes of each ingredient differ under the unique processing and formulation conditions. Specific to plant proteins, it is relevant to evaluate how selective breeding and growing conditions may impact structural properties that consequently influence functionality.

24.4 SUMMARY

There are a variety of techniques used to separate, isolate, and characterize proteins. Separation and purification techniques rely on differences in the physical, biochemical, and structural differences in protein, including bulk density, solubility, charge, thermal stability, size, adsorption characteristics, and biological affinities for other molecules. Proteins can be separated by differential solubility under different pH, salt, solvent, and thermal conditions. Additionally, adsorptive chromatography can be used to separate proteins based on charge (ion-exchange chromatography), hydrophobicity (HIC), and unique affinity to specific ligands (affinity chromatography). Proteins can also be separated based on size, utilizing techniques such as dialysis, membrane filtration, and size-exclusion chromatography. Electrophoresis can be used to separate proteins from complex mixtures on the basis of size (SDS-PAGE) and charge (isoelectric focusing). Capillary electrophoresis is an adaptation of conventional electrophoresis in which proteins are separated in capillary tubes coupled with detectors commonly used in HPLC, allowing not only separation but also quantification. Protein characterization of isolated protein or a mixture of proteins starts with the analysis of amino acid composition, which involves complete protein hydrolysis and then HPLC analysis. The nutritional quality of a protein is determined by the amino acid composition, essential amino acid content, and protein digestibility. The PDCAAS is the method approved for reporting % protein daily value on the nutrition label. Protein structural properties, including molecular weight distribution, denaturation, and surface properties are impacted by several innate characteristics as well as by processing. Many tests are available to evaluate each of these structural properties, including instrument- and chemical-based tests. Protein functionality is influenced by the structural properties of the protein. Therefore, food processers need to evaluate the structural and functional properties of protein ingredients to determine their optimal food use. There are multiple tests available for evaluating protein functionality; however, these tests are not standardized, and no single test is applicable to all food systems.

24.5 STUDY QUESTIONS

1. You have a protein system with the following characteristics:

Protein	Solubility in (NH4)2SO4 (%)	Solubility in ethanol (%)	pI	Denaturation temperature (°C)
1	10–20	5–10	4.6	80
2	70–80	10–20	6.4	40
3	60–75	10–20	4.6	40
4	50–70	5–10	6.4	70

Describe how you would separate protein 4 from the others.

2. Compare and contrast the principles and procedures of SDS-PAGE vs. isoelectric focusing to separate proteins. Include in your explanation how and why it is possible to separate proteins by each method and what you can learn about the protein by running it on each type of system.

3. Explain how capillary electrophoresis differs from SDS-PAGE.
4. Briefly describe what each of the following tells you about the characteristics of the proteins of interest described in the statement (Note: The protein is not the same one in each statement):
   (a) When subjected to dialysis using tubing with a molecular mass cutoff of 3 kDa, a protein of interest is found in the retentate (i.e., not in the filtrate).
   (b) When subjected to ultrafiltration using a membrane with a molecular mass cutoff of 10 kDa, a protein of interest is found in the filtrate (i.e., not in the retentate).
   (c) When the protein was subjected to ion-exchange chromatography using an anion-exchange column and a buffer of pH 8.0, a protein of interest bound to the column.
   (d) When a protein of interest was subjected to isoelectric focusing, the protein migrated to a position of approximately pH 7.2 in the pH gradient of the gel.
   (e) When a protein of interest was subjected to SDS-PAGE in both the presence and absence of mercaptoethanol, the protein appeared as three bands at molecular mass 42, 45, and 48 kDa.
   (f) When a solution with various proteins was heated to 60 °C, the protein of interest was found in the precipitate obtained upon centrifugation of the solution.
5. You are submitting a soy protein sample to a testing laboratory with an amino acid analyzer (ion-exchange chromatography) so that you can obtain the amino acid composition. (Note: You want to quantify all the amino acids.)
   (a) Describe the procedure for how the sample will be treated initially.
   (b) Describe the procedure for how the amino acids will be quantified.
6. In amino acid analysis, a protein sample hydrolyzed to individual amino acids is applied to a cation-exchange column. The amino acids are eluted by gradually increasing the pH of the mobile phase.
   (a) Describe the principles of ion-exchange chromatography.
   (b) Differentiate anion vs cation exchangers.
   (c) Explain why changing the pH allows different amino acids to elute from the column at different times.
7. Briefly describe the differences between the following assay procedures:
   (a) Amino acid score vs essential amino acid index
   (b) *In vivo* vs. *in vitro* PDCAAS
8. You are helping to develop a new process for making a high-protein snack food from cereal grains and soy. You want to determine the protein quality of the snack food under various processing (toasting and drying) conditions. Considering the number of samples to be tested, you cannot afford an expensive *in vivo* assay, and you cannot wait more than a few days to get the results.
   (a) What method/s would you use to compare the protein quality of the snack food made under different processing conditions? Include an explanation of the principles involved.
   (b) You suspect that certain time–temperature combinations lead to overprocessed products. Your testing from (7a) shows that these samples have a lower nutritional quality. What amino acid(s) in the snack food would you suspect to be the most adversely affected by thermal abuse?
   (c) What test(s) could you use to confirm that amino acid(s) have become nutritionally unavailable by the overprocessing? How are these tests conducted?
9. Explain three structural properties of food proteins, state one factor that may influence each of these properties, and one analytical method for each structural property.
10. Define "protein functionality" and list three important functional properties of foods.
11. Describe a functional test used to measure:
    (a) Protein solubility
    (b) Emulsion stability
    (c) Foam volume
    (d) Gelation
    (e) Dough rheology
12. You need to reconstitute nonfat dry milk (NFDM) powder for use in a yogurt formula. You have experienced a large amount of variation in the hydration times required for different lots of NFDM. As the quality control manager, you need to develop a test to measure the solubility and hydration characteristics of each lot of NFDM before it is used (in hopes of avoiding future rehydration problems).
    (a) How would you measure solubility and what precautions would you take with the method you develop?
    (b) Assume that you did the following for the test you developed: (1) mix 20 g of NFDM with 200 g water at 60 °C, (2) blend for 2 min, (3) transfer 50 mL to a centrifuge tube, (4) centrifuge at 10,000× g for 5 min, and (5) measure the protein content of the supernatant after centrifugation. If the supernatant contains 2.95% protein and the starting NFDM contained 36.4% protein, what is the percent protein solubility?
13. You work for a plant protein product manufacturer, and you were asked to substitute soy protein for pea

protein for a meat analog application. With this application, the protein needs to form fibrous structures upon high moisture extrusion. What protein properties would you need to analyze to ensure good performance of pea protein compared to soy protein. Explain why each of these properties would need to be evaluated.

24.6 PRACTICE PROBLEMS

1. Using the data provided in the table below:
   (a) Calculate the EAAI for defatted soy flour.
   (b) Determine the amino acid score for the soy flour.
   (c) Calculate the PDCAAS, using the true digestibility value of 87% for defatted soy flour.

Amino acid	Soy (mg/g protein)	Reference pattern (mg/g protein)
Histidine	26	18
Isoleucine	46	31
Leucine	78	63
Lysine	64	52
Methionine/ cystine	26	26
Phenylalanine/ tyrosine	88	46
Threonine	39	27
Tryptophan	14	7.4
Valine	46	42

2. You work for a manufacturer of protein supplements sold to bodybuilders. You need to screen several proteins that may be used in a new protein supplement. You have three samples (A, B, and C) to evaluate. The amino acid profiles of these three samples and the reference profile (i.e., amino acid requirements of preschool age children) are shown below.

Amino acid	Reference profile	Sample A	Sample B	Sample C
Histidine	18	26	35	24
Isoleucine	31	50	55	35
Leucine	63	65	46	32
Lysine	52	80	92	80
Methionine/ cysteine	26	70	48	50
Phenylalanine/ tyrosine	46	70	90	85
Threonine	27	51	40	39
Tryptophan	7.4	16	22	25
Valine	42	60	64	42

   (a) Calculate the PDCAAS for each supplement. (Assume that the true digestibilities of Samples A, B, and C are 87%, 93%, and 64%, respectively.)
   (b) Which sample would you use if Sample A costs \$1.25/lb., Sample B costs \$3.25/lb, and Sample C costs \$1.15/lb?

Answers

1.
   (a) Essential amino acid index =

$$\sqrt[9]{(26/18)(46/31)(78/63)(64/52)(26/26)(88/46)(39/27)(14/7.4)(46/42)} =$$

$$\sqrt[9]{(1.44)(1.48)(1.24)(1.23)(1.00)(1.91)(1.44)(1.89)(1.10)} =$$

$$\sqrt[9]{18.5866} = 1.38$$

   (b) Amino acid score = 26/26 = 1.00; lowest ratio represents the limiting amino acid, methionine/cystine.
   (c) PDCAAS = amino acid score × true digestibility = 1.00 (0.87) = 0.87

2.
   (a) The first-limiting amino acid for all three samples is leucine. (Identify by determining the ratio of each amino acid compared to reference profile.)
   PDCAAS = amino acid score for first-limiting amino acid x true digestibility:

$$\text{Sample A} = (65/63) \times .87 = .90$$

$$\text{Sample B} = (46/63) \times .93 = .68$$

$$\text{Sample C} = (32/63) \times .64 = .33$$

   (b) The cost-to-protein quality ratio for each sample is as follows:

$$\text{Sample A} = (\$1.25/.90) = \$1.39$$

$$\text{Sample B} = (\$3.25/.68) = \$4.78$$

$$\text{Sample C} = (\$1.15/.33) = \$3.48$$

Sample A provides the highest amount of usable protein per dollar, suggesting it would be best to use.

REFERENCES

1. Formanski K (2020) Plant-based Proteins: Incl Impact of COVID-19 – US – May 2020 – Market Research Report. *Mintel.* Retrieved from https://reports-mintel-com.ezp3.lib.umn.edu/display/986972/?fromSearch=%3FcontentType%3Dreport%26freetext%3Dplant%2520protein
2. Coligan JE, Dunn BM, Speicher DW, Wingfield PT (eds) (2015) Current Protocols in Protein Science. Wiley, New York
3. Burgess RR, Deutscher MP (eds) (2009) Guide to Protein Purification. Vol. 436. Methods in Enzymology, 2nd edn, Academic Press, San Diego, CA
4. Janson JC (ed) (2011) Protein Purification. Principles, High Resolution Methods, and Applications. 3rd edn. Methods of Biochemical Analysis. Vol 54 Wiley, Hoboken NJ

5. Tulbek MC, Lam RSH, Wang YC, Asavajaru P, Lam A (2017) Pea: A Sustainable Vegetable Protein Crop. In Sustainable Protein Sources. Elsevier Inc. (pp. 145–164)
6. Pelgrom PJM (2015). Dry fractionation for sustainable production of plant protein concentrates. Doctoral dissertation, Wageningen University
7. Taherian A, Mondor M, Lamarche F (2012). Enhancing Nutritional Values and Functional Properties of Yellow Pea Protein via Membrane Processing. In: Peas: Cultivation, Varieties and Nutritional Uses. Comstock A, Lothrop B (eds). Pp. 1–48. Nova Science Publishers, Inc.
8. Prajapati H, Desai C, Narkhede S (2021) A Brief Review on Fast Protein Liquid Chromatography- FPLC. J Pharm Sci Bioscientific Res, 10: 235–237
9. Shee C, Sharma AK (2007) Purification and characterization of a trypsin inhibitor from seeds of Murraya koenigii. J. Enzyme Inhib. Med. Chem, 22: 115–120
10. Doultani S, Turhan KN, Etzel MR (2004) Fractionation of proteins from whey using cation exchange chromatography. Process Biochem 39: 1737–1743
11. Chen GQ, Qu Y, Gras SL, Kentish, SE (2023) Separation Technologies for Whey Protein Fractionation. Food Eng Rev, 15:438–465
12. Schneider A, Bu, F, Ismail, B (2023). Enhancement of Pea Protein Solubility and Thermal Stability for Acidic Beverage Applications via Endogenous Maillard-Induced Glycation and Chromatography Purification. Curr Res Food Sci, 6: 100452
13. Lu Y, Williamson B, Gillespie R (2009) Recent Advancement in Application of Hydrophobic Interaction Chromatography for Aggregate Removal in Industrial Purification Process. Curr Pharm Biotechnol, 10: 427–433
14. Jelen P (1991) Pressure-driven membrane processes: principles and definitions. In new applications of membrane processes. Document No. 9201. pp 6–41, International Dairy Federation, Brussels, Belgium
15. Chen WL, Hwang MT, Liau CY, Ho JC, Hong KC, Mao SJ. (2005) β-Lactoglobulin is a thermal marker in processed milk as studied by electrophoresis and circular dichroic spectra. J Dairy Sci 88:1618–1630
16. FDA (2015) Regulatory Fish Encyclopedia (updated 2015 May 12 2015; cited 2015 July 1. Silver Spring, MD. Available from: http://www.fda.gov/Food/FoodScienceResearch/RFE/.
17. Dolnik V (2008) Capillary electrophoresis of proteins 2005–2007. Electrophoresis 29: 143–156
18. Molina E, Marin-Alvarez PJ, Ramos M. (1999) Analysis of cows', ewes' and goats' milk mixtures by capillary electrophoresis: quantification by multivariate regression analysis. Int Dairy J 9:99–105
19. Breadmore MC (2012) Capillary and microchip electrophoresis: Challenging the common conceptions. J Chromatog A 1221:42–55
20. Marchett-Deschmann M, Lehner A, Peterseil V, Sovegjarto F, Hochegger R, Allmaier G (2011) Fast wheat variety classification by capillary gel electrophoresis-on-a-chip after single-step one-grain high molecular weight glutenin extraction. Anal Bioanal Chem 400:2403–2414
21. Otter DE (2012) Standardized methods for amino acid analysis of food. British J Nutr 108:S230-S237
22. La Cour R, Jorgensen H, Schjoerring J (2019). Improvement of Tryptophan Analysis by Liquid Chromatography-Single Quadrupole Mass Spectrometry Through the Evaluation of Multiple Parameters. Front Chem, 7:797
23. Temtrirath K (2022). A UHPLC-PDA method for the quantitative analysis of total amino acids in infant formula with microwave-assisted acid hydrolysis. Chem Pap, 76:3065–3076.
24. AOAC International (2023) Official methods of analysis (on line), 22nd edn. Official methods of analysis, 22nd edn. AOAC International, Gaithersburg, MD
25. Food and Agriculture Organization (2003) Food energy – Methods of analysis and conversion factors: report of a technical workshop. FAO Food and Nutrition Paper 77. Rome, Italy
26. Boogers I, Plugge W, Stokkermans YQ, Duchateau ALL (2008) Ultra-performance liquid chromatographic analysis of amino acids in protein hydrolysates using an automated pre-column derivatization method. J Chromatog A 1189(1–2):406–409 https://doi.org/10.1016/j.chroma.2007.11.052
27. Food and Agriculture Organization (2013) Dietary protein quality evaluation in human nutrition. FAO Expert Consultation, Food and Nutrition Paper 92. Rome, Italy
28. Boye J, Wijesinha-Bettoni R, Burlingame B (2012) Protein quality evaluation twenty years after the introduction of the protein digestibility corrected amino acid score method. Br J Nutr 108:S183-S211
29. Bandyopadhyay S, Kashyap S, Calvez J, Devi S, Azzout-Marniche D, Tomé D, Kurpad AV, Gaudichon C (2022) Evaluation of Protein Quality in Humans and Insights on Stable Isotope Approaches to Measure Digestibility – A Review. Adv Nutr, 13:1131–1143
30. Code of Federal Regulations (2023) US Food & Drug Administration. Title 21 Code of Federal Regulations. Nutrition labeling of food, Sec. 101.9 21 CFR 101.9 I (7) "Protein". [For most current version see electronic Code of Federal Regulations (eCFR)] 21 CFR 101.9: https://www.accessdata.fda.gov/scripts/cdrh/cfdocs/cfcfr/cfrsearch.cfm?fr=101.9 For easy to read form of 21 CFR101 – Food Labeling (which includes 21 CFR 101.9): https://www.ecfr.gov/current/title-21/chapter-I/subchapter-B/part-101
31. Plank DW (2017) In vitro method for estimating in vivo protein digestibility. US Patent 9,738,920 B2, issued Aug. 22, 2017
32. Damodaran S (2017) Amino Acids, Peptides, and Proteins. In: Damodaran S, Parkin KL (eds) Fennema's Food Chemistry, 5th edn. CRC Press, Boca Raton, FL, pp 235–356
33. Bu F, Nayak G, Bruggeman P, Annor G, Ismail BP (2022) Impact of plasma reactive species on the structure and functionality of pea protein isolate. Food Chemistry, 371, 131135
34. Hinnenkamp C, Ismail, BP (2021) A proteomics approach to characterizing limited hydrolysis of whey protein concentrate. Food Chemistry, 350, 129235
35. Warsame AO, Michael N, O'Sullivan DM, and Tosi P (2020) Identification and Quantification of Major Faba Bean Seed Proteins. J Agric Food Chem, 12, 68: 8535–8544
36. AOCS (2013) Official methods and recommended practices of the AOCS, 6th edn., 3rd printing. American Oil Chemists Society, Champaign, IL
37. McClements DJ (2015) Food emulsions: Principles, Practices, and Techniques. 3rd edn. CRC Press, Boca Raton, FL
38. Wrolstad RE, Acree TE, Decker EA, Penner MH, Reid DS, Schwartz SJ, Shoemaker CF, Smith D, Sporns P, (eds) (2005) Handbook of food analytical chemistry vol 1. Wiley, Hoboken, NJ
39. Bu F, Feyzi S, Nayak G, Mao Q, Kondeti VS, Bruggeman P, Chen C, Ismail BP (2023) Investigation of Novel Cold Atmospheric Plasma Sources and their Impact on the Structural and Functional Characteristics of Pea Protein. Innovative Food Science and Emerging Technologies, 83, 103248
40. Hinnenkamp C, Ismail BP (2021) Enhancing emulsion stability: The synergistic effect of combining Procream and partially hydrolyzed whey protein. Intl Dairy J, 119, 105059
41. Zhang J, Liu L, Liu H, et al (2019) Changes in conformation and quality of vegetable protein during texturization process by extrusion. Crit Rev Food Sci Nutr 59:3267–3280
42. Zahari I, Ferawati F, Helstad A, et al (2020) Development of high-moisture meat analogues with hemp and soy protein using extrusion cooking. Foods 9:1–13

Determination of (Total) Phenolics and Antioxidant Capacity in Food and Ingredients

25

Rachel R. Schendel and Prakash R. Pandeya

Contents

25.1 INTRODUCTION

The term **phenolics** refers to several thousand aromatic plant metabolites, which possess at least one hydroxyl group attached to the phenyl ring. In the plant, phenolic compounds are important structural components of the plant cell wall [1], and they help the plant quickly strengthen tissues damaged by wounding and infestation. Nonstructural phenolic compounds are also necessary for plants' defense strategies against biotic and abiotic stressors. In food products, phenolic compounds like tocopherols and gallates have long been valued for their ability to increase the oxidative stability of lipids, and they are therefore commonly incorporated as antioxidant ingredient additives. Some phenolics are used as food colorants (e.g., anthocyanins), and others contribute to the flavor of food products (e.g., vanillin). More recently, interest in phenolics was generated from studies suggesting potential health-promoting effects of phenolics against specific diseases such as coronary heart disease and certain forms of cancer [2]. However, many health-promoting effects of specific components were suggested based only on in vitro assays. A broadly protective health effect of phenolics is unlikely, and some food-based phenolics exert toxic effects at high doses [3].

Phenolics are often classified as either **simple phenolics**, which contain only one phenolic ring, for example, hydroxybenzoic acids and (monomeric) hydroxycinnamic acids, or **polyphenols**, which have two or more phenolic units, for example, flavonoids, stilbenes, hydrolyzable and non-hydrolyzable tannins, or lignans (Fig. 25.1). In their original plant-based form, both simple phenolics and polyphenols are often chemically linked to a non-phenolic component such as monosaccharides or organic acids.

R. R. Schendel (✉) · P. R. Pandeya
Department of Animal & Food Sciences, University of Kentucky, Lexington, KY, USA
e-mail: rachel.schendel@uky.edu; pandeya.prakash@uky.edu

B. P. Ismail, S. S. Nielsen (eds.), *Nielsen's Food Analysis*, Food Science Text Series,
https://doi.org/10.1007/978-3-031-50643-7_25

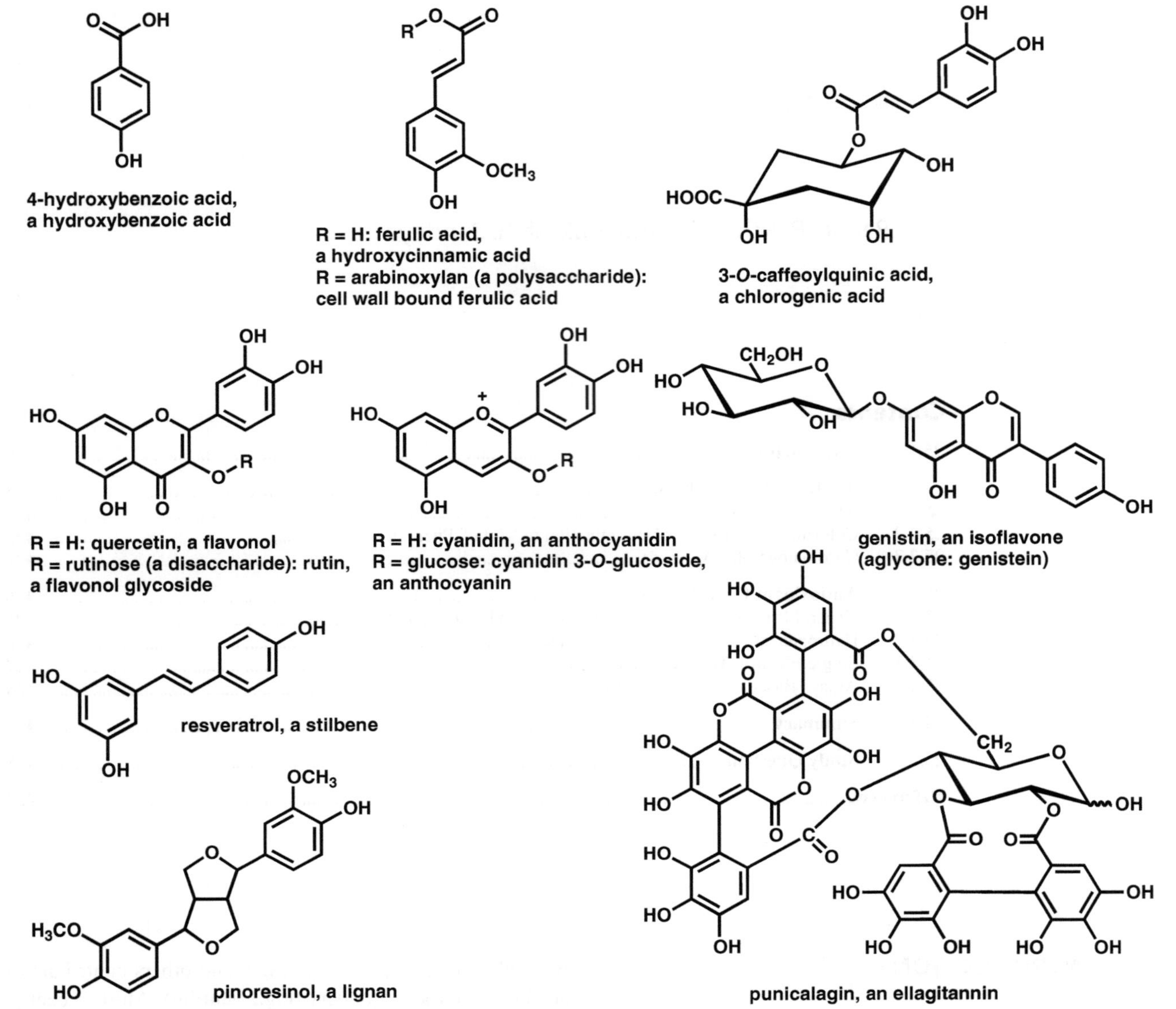

Fig. 25.1 Examples of different phenolic compounds in food products

Additionally, the previous examples are all comparably low-molecular-weight phenolic compounds, and high-molecular-weight phenolic compound polymers such as lignin or the phenolic domain of suberin are also found in plant-based foods. Finally, food processing leads to the formation of new phenolic compounds or significant modifications to the original plant-based phenolics. In coffee processing, for example, new phenolic compounds (melanoidins) are formed by the Maillard reaction, and naturally occurring phenolics (chlorogenic acids) are incorporated into these compounds [4].

Four types of assays are described in this chapter: (total) phenolics, hydrogen atom transfer-based antioxidant capacity assays, single electron transfer-based antioxidant capacity assays, and accelerated lipid oxidation assays. Official methods of analysis are emphasized when available, and limitations of the described methods are discussed. A summary of the described methods is provided in Table 25.1.

Table 25.1 Summary of common methods used to determine (total) phenolics and antioxidant capacity in foods

To determine	General assay types	Specific methods	Method principle	Method advantages	Method disadvantages
Total phenolics	Colorimetric methods	Folin-Ciocalteu assay	Phenolic compounds are transformed to phenolates under alkaline conditions. These phenolates reduce the initially yellow-colored reagent, producing a blue-colored product, which is measured spectrophotometrically.	This method is simple, fast, and inexpensive.	Poor sensitivity and selectivity. Interference by various non-phenolic reducing compounds such as ascorbic acid, reducing sugars, aromatic amines, thiols, etc., requiring sample cleanup to remove these interferences.
		Fast Blue BB assay	The Fast Blue BB reagent contains diazonium ions that couple specifically with phenolic hydroxyl groups under alkaline conditions, forming colored azo complexes, which are measured spectrophotometrically.	More selective and sensitive toward phenolics than the Folin-Ciocalteu assay. Reagent is inexpensive and stable.	The FBBB reagent is reactive with free tyrosine. The diazonium reagent is carcinogenic.
	Chromatographic methods	High performance liquid chromatography (HPLC)	Phenolic analytes are distributed through a liquid mobile phase and stationary phase.	Individual phenolic compounds can be identified and quantified.	Compounds with similar structures and polarities may fail to separate and co-elute. Accurate identification and quantification requires the acquisition of specific standard compounds.
		Gas chromatography (GC)	Volatilized analyte is carried by an inert gas stream into a stationary phase (GC column). Separation of the different components is achieved based on their relative vapor pressure and affinities for the stationary phase.	Individual phenolic compounds can be identified and quantified.	Nonvolatile phenolic compounds require prior derivatization to improve volatility. Standard compounds are needed for accurate identification. GC is not suitable for large conjugated phenolic compounds.

(continued)

Table 25.1 (continued)

To determine	General assay types	Specific methods	Method principle	Method advantages	Method disadvantages
Antioxidant capacity	Hydrogen atom transfer (HAT) assays	Oxygen radical absorbance capacity (ORAC)	Azo compounds are used, which upon heating at 37 °C and exposure to an oxygenated environment, form peroxyl radicals that attack the fluorescent probe (fluorescein). HAT-acting antioxidants quench the peroxyl radicals and delay the loss of fluorescein which is monitored spectrophotometrically over time.	Assay utilizes a controlled source of peroxyl radicals, which mimics real-world oxidation scenarios well.	Assay is complicated to perform. The reaction temperature and dissolved oxygen concentrations must be exactly controlled [33]. Both sample and reagent concentration must be optimized for different materials [33].
	Single electron transfer (SET) assays	Trolox equivalent antioxidant capacity (TEAC) assay, also called ABTS assay	This assay utilizes the stable and colored nitrogen-based $ABTS^{•+}$ radical. The analyte antioxidant transfers a single electron to the $ABTS^{•+}$ radical to form colorless ABTS. The disappearance of color after reaction with antioxidants is monitored spectrophotometrically at 734 nm.	Assay is inexpensive and simple to perform.	Only limited relevance to real-world oxidation reactions. Reaction rate is primarily determined by steric considerations, so small antioxidant compounds that can access the large ABTS•+ radical molecule react rapidly, but antioxidants with bulky structures react more slowly [33, 35]. Due to the slow reaction of this system with HAT-acting antioxidants, it will underestimate the antioxidant activity of food materials rich in HAT antioxidants.
		2,2-Diphenyl-1-picryhydrazyl radical (DPPH) assay	The DPPH radical has a blue-violet color that absorbs at 517 nm. Quenching of the radical by antioxidant compounds decolorizes the solutions and decreases the absorbance at 517 nm.	Assay is simple to perform.	This method is not suitable for quantitative comparison of mixtures. Mixtures of antioxidants display reduced antioxidant activity [33]. The reaction rate varies with the concentration of individual antioxidants in the samples [33, 37].
		Ferric reducing antioxidant power (FRAP) assay	The assay monitors the reduction of a ferric salt (Fe(III)-TPTZ) by antioxidants, generating a colored product that absorbs at 593 nm.	Method can be used to screen antioxidants for SET functionality.	This assay is not suitable for HAT-acting antioxidants.
	Antioxidant assays based on oxidation of lipids	Accelerated lipid oxidation assay	This assay is used to test the protective effect of antioxidants in food products. The food product is subjected to conditions that accelerate oxidation with or without the presence of the tested antioxidant, and the extent of lipid oxidation is compared.	The observation of oxidative reactions in some food products requires months to years if tested at normal storage conditions. Accelerated oxidation conditions increase sample throughput and reduce study length.	The oxidative reactions during accelerated storage conditions may not exactly coincide with the reactions at normal storage conditions.

25.2 ANALYSIS OF (TOTAL) PHENOLICS

Based on the structural diversity of phenolic compounds, it is impossible to determine all phenolic compounds with a single assay as the term "total phenolic assays" may suggest. Many tests that measure sum parameters are unselective, e. g., prone to false-positive results from matrix compounds, a fact that is also true for total phenolic assays. Thus, a critical evaluation of the results is necessary before claiming or comparing specific total phenolic contents of food products and ingredients. A key step in the application of total phenolic assays is choosing the extraction procedure. Extraction conditions define which phenolic compounds will be included in the test results and may reduce co-extraction of matrix compounds that will interfere with the assays.

25.2.1 Sample Preparation

Clear beverages such as white wine or clear apple juice may be used directly or after dilution. Fresh fruits, vegetables, cereals, and processed food products may be ground and extracted, or they can be dried first by lyophilization. Direct extraction of undried material still requires information about the food product's water content to reliably adjust the organic solvent/water ratio during the extraction step, a requirement that is often neglected in practice. Lyophilization is the gentlest way to dry food products, although small losses of phenolic compounds may still occur during this process. Sample grinding prior to extraction is necessary since many phenolic compounds are located in vacuoles and are best extracted after cell wall and membrane rupture. Simple solvent extraction is sufficient for low-molecular-weight non-structural phenolics, but phenolic compounds that are bound to plant polymers need to be liberated by hydrolytic procedures [5]. Additionally, after crushing fresh fruits or vegetables, the mash must be processed quickly to minimize phenolic losses from action of the enzyme polyphenol oxidase found in plant tissues.

25.2.1.1 Extraction

Phenolic compounds and their water-soluble conjugates such as sugar derivatives (phenolic glycosides) are extracted with water; polar organic solvents like methanol, ethanol, and acetone; or their mixtures [6]. Aqueous 80% ethanol or methanol solutions work well for many food products, solubilizing most phenolic compounds and, at the same time, precipitating out many polymers such as polysaccharides and proteins. Acidifying the extraction solutions may stabilize certain phenolic compounds such as anthocyanins [7]. Extraction temperatures from room temperature up to 90 °C are used, depending on the analyzed food product and the stability of its phenolic compounds. Ultrasonication during extraction, or ultrasound-assisted extraction, may be used to optimize phenolic extraction rates [7, 8]. In some cases, less polar organic solvents such as ethyl acetate are used, which extract less polar phenolic compounds but exclude common polar phenolics (especially phenolic glycosides and other polar conjugates).

25.2.1.2 Hydrolysis

Depending on the purpose of the analysis, measurement of insoluble phenolics in addition to soluble phenolics may be desired. Common types of linkages involving phenolic compounds are glycosidic linkages and ester linkages. Ester-linked phenolic acids are abundant in cereal grain products but are also found in other plant-based foods. Their liberation requires alkaline hydrolysis (e.g., 2 *M* NaOH, room temperature, 16 h) of the solvent-extracted residue (Sect. 25.2.1.1). Phenolic acids may oxidatively degrade under these alkaline conditions. To minimize degradation, NaOH solutions should be sparged with nitrogen and the headspace of the capped hydrolysis tubes should be flushed with nitrogen. Following alkaline hydrolysis, the hydrolysate is often acidified to protonate the liberated phenolic acids and make them extractable into less polar organic solvents such as ethyl acetate or diethyl ether.

25.2.2 Colorimetric Assays for Determination of "Total" Phenolics

25.2.2.1 Folin-Ciocalteu Assay

The foundation of many colorimetric assays, which determines total phenolic contents, is the ability of phenolic compounds to be oxidized. Permanganate and ferric ions were historically applied as oxidation reagents, but presently the **Folin-Denis reagent** [9, 10] and the **Folin-Ciocalteu reagent** [11] are more common. These reagents contain polymeric ions formed from phosphomolybdic and phosphotungstic heteropoly acids and were originally developed to measure the amino acids tyrosine and tryptophan. To apply these reagents to extracted phenolics, alkaline conditions are utilized, which transforms the phenolics into phenolates. The phenolates are oxidized, which reduces the initially yellow phosphotungstic-phosphomolybdic reagent and generates a blue color ("molybdenum-tungsten blue"), which can be measured spectrophotometrically. Color formation by a sample is compared to that of a standard compound, preferentially gallic acid, (+)- catechin, or more traditionally, tannic acid, and the results are reported as gallic acid, (+)-catechin or tannic acid equivalents.

The assays using both the Folin-Ciocalteu reagent and Folin-Denis reagent have shortcomings. Phenolic compounds are only oxidized under alkaline conditions, but both the oxidizing reagent and the formed molybdenum-tungsten blue have been described as unstable in alkaline solutions. So, several procedures to minimize this problem have been suggested, which involve varying the sequences of reagent addition and time periods in between reagent addition and spectrophotometric analysis. When using the Folin-Denis reagent, a precipitate may form. This issue can be addressed by adding lithium sulfate to the reagent. The molar absorptivity of the reagent (Chap. 7, Sect. 7.2.1) depends both on which reagent is used and which phenolic compounds are tested. For example, *ortho*-diphenols have greater absorptivities than monophenolic compounds or *meta*-diphenols, meaning that different phenolic compounds will not be equally quantified. Additionally, the Folin-Ciocalteu reagent has been described as being more sensitive (e.g., able to detect lower concentrations of phenolics) than the Folin-Denis reagent [12].

Poor selectivity is another shortcoming of both Folin reagent-based total phenolic assays. Other reducing compounds besides phenolics can reduce the Folin reagents and are thus incorrectly determined as phenolic compounds in this assay. Ascorbic acid and some other vitamins, thiols (for example, cysteine and glutathione), nucleotide bases, and redox-active metal ions are all active in this test [13]. Interference from reducing sugars has also been described. Due to their overall reactivity toward reducing compounds, the Folin-based phenolic assays were suggested as candidates for measuring the antioxidant or reducing activity of a sample rather than estimating total phenolics.

AOAC Official Method 2017.13 provides detailed method steps for the Folin-Ciocalteu procedure [14]. Briefly, an extract of the target food material is mixed with the Folin-Ciocalteu reagent, the pH is adjusted with sodium carbonate solution [15], and the absorbance at 765 nm is measured. The total phenolics content is determined using a standard curve and reported as equivalents of standard compounds such as gallic acid or (+)-quercetin.

25.2.2.2 Fast Blue BB Assay

The Fast Blue BB Assay is another colorimetric method that can detect and quantify phenolics in food and beverages [16, 17]. This method utilizes the Fast Blue BB diazonium salt reagent (FBBB), which contains diazonium ions ($-^{+}N\equiv N$). Under alkaline conditions, these ions react with an active hydroxyl group from phenolics' aromatic ring forming a stable azo complex that is measured spectrophotometrically at 420 nm. An alkaline reaction environment is required for a linear response [16].

The Fast Blue BB method has some clear advantages over the Folin-Ciocalteu procedure for the analysis of total phenolics. The major advantage of the Fast Blue BB method is its selectivity and sensitivity towards phenolics, since unlike the Folin-Ciocalteu method, the FBBB reagent is not susceptible to interference by non-phenolic reducing compounds [18–20]. FBBB has also been reported to be more sensitive than Folin-Ciocalteu for many food matrices. Food matrices containing substantial amounts of free tyrosine will require special sample preparation since tyrosine reacts with the FBBB reagent (as does the Folin-Ciocalteu reagent). Additionally, like the Folin-Ciocalteu method, the FBBB assay suffers from differing molar absorptivities between different phenolic compounds and measurement solvents [17–20].

The Fast Blue BB assay entails mixing a sample solution with the Fast Blue BB diazonium reagent and adding a basic solution to create an alkaline reaction environment. The absorbance is measured at 420 nm, and the total phenolics content is quantified using a standard curve and presented as the equivalents of the selected standard used, such as mg gallic acid equivalents per liter.

25.2.3 Chromatographic Methods

Chromatographic methods are widely used to quantify individual phenolic compounds or to monitor phenolic profiles of food products. However, they are ill-suited for quantifying the total phenolic contents of food products unless the food contains only a few phenolic compounds, which are all known and available as standard compounds; a rather unlikely scenario. The choice of the chromatographic method depends on the phenolic analyte. Where all extractable phenolics are theoretically appropriate for liquid chromatographic methods, few phenolic compounds are volatile enough to be analyzed by gas chromatography (GC) without prior derivatization.

25.2.3.1 High-Performance Liquid Chromatography

High-performance liquid chromatography (HPLC)/ultra-performance liquid chromatography (UPLC) (Chap. 13, Sect. 13.2.3.3) is a superb method for most phenolic compounds. Following extraction of the phenolic compounds as described in Sects. 25.2.1.1 and/or 25.2.1.2, the samples are either directly (after membrane filtration or sample centrifugation) injected into the HPLC system, or the extracts are further purified with liquid/liquid extraction or solid-phase extraction (SPE) (Chap. 14, Sect. 14.2.2.5). Reversed-phase SPE cartridges are conditioned and loaded with the sample. Unwanted matrix components are eluted with water or mixtures of water with low amounts of organic modifiers such as methanol and discarded, then the analytes are eluted with solvents or solvent-water mixtures containing larger portions of organic modifiers. The capacity of the selected SPE cartridges, the solvent choice for the elution of unwanted matrix components, and eluent volumes and flow rates are critical optimization parameters to achieve high recoveries. If SPE is used to separate phenolic acids from neutral phenolic compounds, the solvent pH may also be adjusted to optimize SPE conditions [21].

HPLC separation of purified phenolic compounds is usually done on reversed-phase columns (Chap. 13, Sect. 13.3.3), with C18- and phenyl-hexyl-stationary phases being preferentially used. Phenyl-hexyl-columns may, depending on the gradient system, provide extra selectivity for phenolic compounds due to π-π interactions (i.e., noncovalent attractive interactions that involve π-electron systems, often formed between two aromatic rings, also known as π-π stacking) between the stationary phase and the phenolic analytes. Mobile phases are typically water and varying amounts of polar organic modifiers, like methanol and acetonitrile. The separation of some phenolic compounds is improved by adding an acidic modifier, such as trifluoroacetic acid or formic acid, to the eluent. For example, eluent acidification is necessary to separate phenolic acids; without eluent acidification, peaks are broad due to partial deprotonation of the carboxyl group, whereas acid addition to the eluent protonates the carboxyl groups and results in narrow, symmetric peaks. If mass spectrometric (MS) detection is used (Chap. 11), trifluoroacetic acid should be avoided because it suppresses ionization using electrospray ionization (ESI); formic acid may be used instead [22].

All phenolic compounds are UV active, making UV detection an excellent detection mode [23]. Depending on the conjugation of the π-electron system, the maximum of UV absorption can vary considerably. For example, ferulic acid with its extended π-electron system (propenylic side chain) shows a UV maximum (depending on the solvent and pH) around 325 nm, whereas 3-(3-hydroxy-4-methoxyphenyl) propionic acid has a UV maximum around 275 nm (Fig. 25.2). With photodiode array detectors, capable of measuring the full UV-visible spectrum simultaneously (Chap. 7, Sect. 7.2.6.3), each compound can be measured at its individual UV absorption maximum simultaneously. When only single-wavelength UV detectors are available, 280 or 254 nm are often used to monitor all phenolic compounds. Because phenolic compounds are easily oxidized, electrochemical detection is another option which, however, has not seen broad application in the analysis of phenolics. Depending on the analyzed phenolic compounds, fluorescence detection may be used, with its obvious advantages of increased selectivity and sensitivity (Chap. 7, Sect. 7.3). With state-of-the-art MS detectors becoming more affordable, MS and MS/MS detection (Chap. 11) is increasingly used in both research and routine analytical testing laboratories.

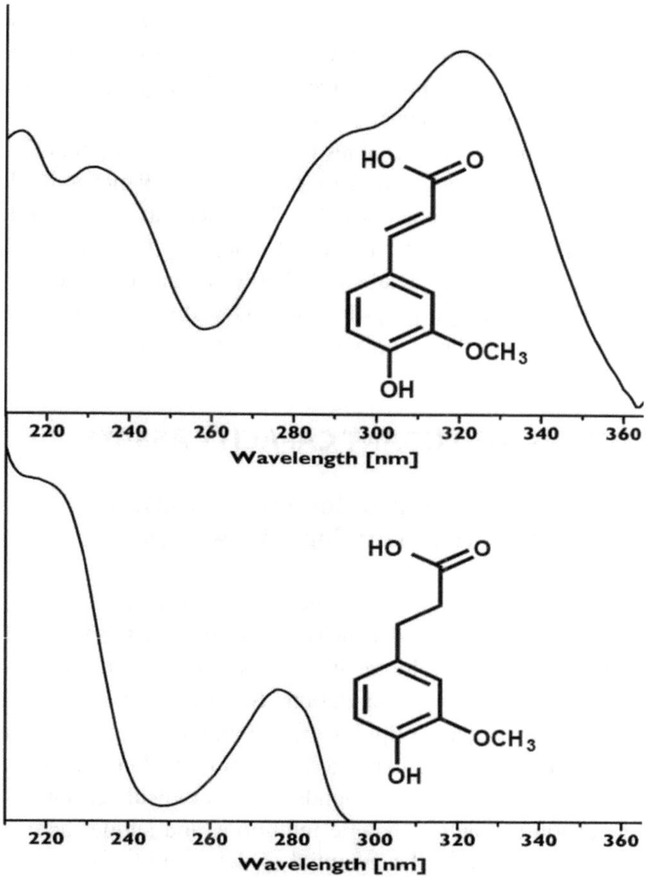

Fig. 25.2 UV spectra and structures of ferulic acid (*top*) and 3-(3-hydroxy-4-methoxyphenyl)propionic acid (*bottom*)

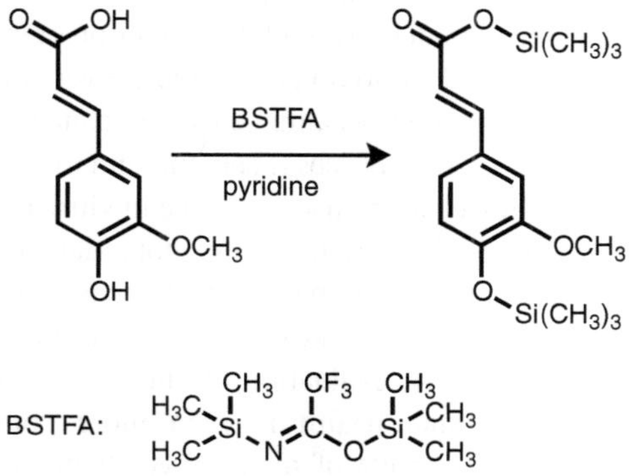

Fig. 25.3 Silylation of ferulic acid using *N,O*-bis(trimethylsilyl)trifluoroacetamide (BSTFA) as derivatization reagent and pyridine as catalyst. Carboxyl groups are converted into trimethylsilyl esters; hydroxyl groups are converted into trimethylsilyl ethers

25.2.3.2 Gas Chromatography

Gas chromatographic analysis of most phenolics requires prior derivatization to increase their volatility (Chap. 14, Sect. 14.2.3). **Silylation** using **derivatization reagents** such as *N,O*-bis(trimethylsilyl)trifluoroacetamide (BSTFA) is a commonly used derivatization method for phenolic compounds. BSTFA reacts with both hydroxyl and carboxyl groups of phenolic compounds, which results in trimethylsilyl (TMS) ethers and esters, respectively (Fig. 25.3). Silylation often can be performed directly in the silylation reagent, with or without addition of a catalyst [24], but the silylation of sterically hindered hydroxyl groups may be improved by using aprotic solvents and/or catalysts. Although derivatization sufficiently improves volatility of simple phenolic compounds and enables GC analysis, larger phenolics, especially if conjugated with sugars, are still not suitable analytes for GC procedures. The TMS derivatives of simple phenolics are well separated on nonpolar stationary GC phases (100% methyl-substituted polysiloxane or 5% diphenyl-95% dimethyl polysiloxane). Detection can be performed using a flame ionization detector (FID) or via MS.

25.3 ANTIOXIDANT CAPACITY ASSAYS

25.3.1 General Principles and Limitations of Antioxidant Capacity Assays

The antioxidant capacity of individual compounds or plant extracts was traditionally measured to identify suitable antioxidants for the development of food products with enhanced stability against lipid oxidation. Several mechanisms are involved in oxidative lipid degradation, including photooxidation, lipoxygenase-catalyzed oxidation, and autoxidation. Therefore, different compounds such as radical scavengers, metal ion chelators, enzyme inhibitors, and singlet oxygen quenchers all have the potential to delay lipid oxidation in food products. However, many antioxidant capacity studies published recently were not performed with the end purpose of identifying compounds or extracts that could limit food product oxidation. Rather, they targeted biological processes occurring in the human body, with the intention of evaluating the potential role of antioxidants in the prevention of diseases such as atherosclerosis, diabetes, and cancer (i.e., diseases in which oxidative processes are thought to be critically involved) [2]. However, the use of simple in vitro antioxidant tests (as described below) to assess potential antioxidant behavior in the human body has been criticized because it is not possible to mimic complex in vivo environments with these test systems. The evaluation of antioxidants from the perspective of human health is even more complicated because the health effects of many phytochemicals do not necessarily arise from their direct antioxidant capacity but, among other factors, from their induction of classical Phase II enzymes (antioxidant and detoxifying enzymes) [25, 26]. Because of these concerns, the USDA's Nutrient Data Laboratory (NDL) removed the "USDA Oxygen Radical Absorbance Capacity (ORAC) Database for Selected Foods" from the NDL website. They concluded that "there is no evidence that the beneficial effects of polyphenol-rich foods can be attributed to the antioxidant properties of these foods. The data for antioxidant capacity of foods generated by in vitro (test-tube) methods cannot be extrapolated to in vivo (human) effects, and the clinical trials to test benefits of dietary antioxidants have produced mixed results" [27].

If any of the antioxidant capacity tests described below are used, one must clearly define the goal of performing this test and ask whether this assay is suitable to obtain appropriate answers. It is particularly important to reflect whether the chosen test system simulates the target food system (or other biological systems) in sufficient detail [28]. A critical factor is whether the food system is an emulsion or bulk oil. If the food product of interest is an emulsion but the test system uses only water (to test for "hydrophilic antioxidants") or an organic solvent (to test for "lipophilic antioxidants"), the results from these assays are hard to interpret because the distribution of the antioxidants in an emulsion and at the interface is not reflected by the test system. If the test system is able to simulate an emulsion, it should be asked whether the oxidized substrate employed by the assay represents the compounds to be oxidized in the food product. For example, many test systems use free fatty acids, which form micelles and behave very differently from emulsified triglycerides.

Numerous variations of several antioxidant capacity assays have been applied [29, 30]. Efforts to standardize these test systems are underway, but even where standardization was achieved, different protocols are still used. Assays that measure radical scavenging capacities were broadly classified into **hydrogen atom transfer** (HAT) reaction-based assays and **single electron transfer** (SET) reaction-based assays [31]. **HAT assays** determine the ability of an antioxidant or a mixture of antioxidants to scavenge free radicals by hydrogen donation. **SET assays** measure the ability of antioxidants to transfer an electron, thereby reducing radicals and also other compounds such as redox-active metal ions. By accepting an electron, the radical is converted into an anion, which can accept a proton to form a stable compound.

25.3.2 Hydrogen Atom Transfer (HAT) Assays

25.3.2.1 Oxygen Radical Absorbance Capacity (ORAC) Assay

The ORAC assay utilizes azo compounds (i.e., compounds with the functional group R–N=N–R) to generate free radicals and monitors the progression of oxidation via disappearance of a fluorescent probe. When gently heated (37 °C), the azo compound [2,2′-azobis(2-amidinopropane) dihydrochloride, AAPH] disintegrates, releasing nitrogen gas and carbon-centered radicals. In an oxygenated environment, these carbon-centered radicals rapidly form peroxyl radicals, which attack the fluorescent probe (usually fluorescein). However, if HAT-acting antioxidants that quench the peroxyl radicals are present, the loss of fluorescein is delayed until the antioxidants are consumed. The loss of the fluorescent probe is monitored over time with a fluorescence spectrophotometer, and fluorescence intensity vs. time is plotted. The antioxidative capacity of tested food extracts is measured as the area under the curve (AUC) for the sample, minus the AUC for a blank. These unitless areas are converted into **Trolox** (a water-soluble vitamin E analog) **equivalents** based on the area generated by Trolox in the same assay. ORAC assay conditions may be modified to accommodate either lipophilic or hydrophilic antioxidants. AOAC Method 2012.23 for total antioxidant activity provides detailed method procedures [32], and Schaich et al. helpfully discuss technical challenges related to the ORAC assay [33].

25.3.3 Single Electron Transfer (SET) Assays

25.3.3.1 Trolox Equivalent Antioxidant Capacity (TEAC) Assay

The TEAC assay, also known as the 2,2′-azino-bis(3-ethylbenzothiazoline-6-sulfonic acid) ($ABTS^{•+}$) assay, utilizes the stable, nitrogen-based $ABTS^{•+}$ radical. The $ABTS^{•+}$ radical is strongly colored, but ABTS is colorless, which allows for easy monitoring of disappearance of the radical via UV-V is absorbance. Current TEAC/ $ABTS^{•+}$ methods first generate high $ABTS^{•+}$ concentrations, followed by addition of the food extract containing antioxidants, and then monitor $ABTS^{•+}$ decline with a UV-Vis spectrophotometer [34]. Samples containing SET-acting antioxidants will decrease the UV-Vis absorbance, since the colored $ABTS^{•+}$ radical is being transformed into the colorless ABTS.

The TEAC/ $ABTS^{•+}$ method is primarily governed by steric considerations, e.g., access to the large nitrogen-based $ABTS^{•+}$ radical molecule. Antioxidants that function primarily via SET or do not have bulky ring systems react rapidly, whereas unwieldy antioxidant structures or antioxidants functioning via HAT react slowly [33, 35]. Therefore, a strong response in the TEAC/ $ABTS^{•+}$ method does not necessarily correlate well with radical-quenching capabilities in real-world food or biological systems.

25.3.3.2 2,2-Diphenyl-1-Picryhydrazyl Radical (DPPH) Assay

The 2,2-diphenyl-1-picryhydrazyl radical (DPPH) assay was proposed over 60 years ago for “the antioxidant assay of biological materials” [36]. It is now the basis of AOAC Method 2012.04, for antioxidant activity in foods and beverages [32]. DPPH is a stable organic nitrogen radical and produces deeply blue-violet solutions that absorb at 517 nm. Upon quenching of the radical by antioxidant compounds, solutions are decolorized and the absorbance decreases. Quenching of the DPPH radical happens rapidly by the SET mechanism, but HAT-functioning antioxidant compounds are also active. However, hydrogen-bonding solvents, such as the methanol or ethanol often used for this assay, impede the HAT mechanism and thus favor SET-acting antioxidant compounds [37, 38].

Although the DPPH assay is simple to perform and can identify the presence of SET-acting antioxidant compounds, it is hampered by major drawbacks. Like the ABTS assay, the nitrogen-based, bulky DPPH radical is also a poor model for the radicals encountered in food and biological systems. Determining the total amount of DPPH consumed by the sample after several hours and its “total antioxidant capacity” ignores the short lifetimes (seconds or less) of free radicals in food systems [39]. Focusing on reaction rates in the DPPH assay to rank antioxidants has also proven futile because reaction rates are almost exclusively governed by the steric accessibility of the antioxidant molecule to the bulky DPPH’s radical site [37]. Additionally, both the “total antioxidant capacity” and reaction rate of individual antioxidants are affected by antioxidant concentration, with stoichiometry (moles of DPPH consumed/mole of antioxidant) decreasing with increased antioxidant concentration [37]. Mixtures of antioxidants display more activity depression than can be explained by steric interferences [33], meaning that the assay is completely irrelevant for ranking samples containing multiple components in varying concentrations, such as plant extracts. Finally, the assay does not capture antioxidants which function primarily via the HAT mechanism. Based on these limitations, the DPPH assay should not be used for quantitative ranking of the antioxidant capacity of individual compounds or mixtures.

25.3.3.3 Ferric Reducing Antioxidant Power (FRAP) Assay

In contrast to the TEAC/ $ABTS^{•+}$ and DPPH assays, which operate under mixed SET and HAT mechanisms, the FRAP assay measures exclusively SET-active compounds. The assay monitors reduction of a ferric salt (ferric 2,4,6,tripiridyl-*s*-triazine, TPTZ), producing a colored product that absorbs at 593 nm. The redox potential of Fe(III)-TPTZ (0.7 V) is similar to that of $ABTS^{•+}$ (0.68 V), so compounds that show activity in the TEAC/ $ABTS^{•+}$ assay will also react in the FRAP assay. However, because the FRAP assay must be performed at an acidic pH value of 3.6 to maintain iron solubility, which increases the redox potential, FRAP values are usually lower than TEAC/ $ABTS^{•+}$ assay values [29].

Because the assay’s mechanism is purely SET based, it can be used to screen antioxidants for SET functionality. Additionally, the assay is rapid and robust. However, the active mechanism of the FRAP assay [the ability to reduce Fe(III)] has little correlation to the radical-quenching mechanisms (HAT) displayed by many antioxidants.

25.3.4 Assays Based on Oxidation of Lipids

The effects of individual compounds, mixtures of antioxidants, or plant extracts on oxidative stability in more complex food systems can also be tested in **accelerated lipid oxidation tests**. These test systems are described in detail in Chap. 23, Sect. 23.5. Using, for example, the Rancimat® does not automatically require a lipid extraction, but direct measurement of many food products is possible. Using a whole food product is closest to what happens during real-world storage; however, foods with high water or protein contents such as salad dressings or sausages require a prior lipid extraction. If the lipid phase needs to be extracted from an emulsion-based food product such as salad dressings, the same limitations as described earlier occur (i.e., partition of the antioxidants between the phases and at the interface cannot be modeled). Limitations of using accelerated oxidation conditions are described in Chap. 23, Sect. 23.5.

25.4 SUMMARY

Because phenolic compounds have antioxidant characteristics, they have the potential to stabilize food products against lipid deterioration. Because many human diseases are linked to oxidative stress, phenolic compounds from food products and other plant sources are discussed in terms of disease prevention too. Thus, there is a demand to measure phenolic compounds in food products and to evaluate the antioxidant capacity of individual food constituents, extracts, and ingredients. Methods to determine the total phenolic contents of food products such as the Folin-Ciocalteu and Fast Blue BB assays are unspecific and do not, depending on the overall composition of the food product, necessarily reflect the phenolic contents. HPLC/UPLC or GC approaches enable analysis of individual phenolics. A plethora of antioxidant capacity assays has been described in the past that differ in the applied principle and procedural details such as probes, radical initiators, solvents, etc. These tests were designed as relatively high-throughput methods, so they can be used to screen many compounds/extracts, but suffer from oversimplification and do not reflect the complexity of oxidative degradation of lipids and other organic compounds in biological systems. Therefore, it is not advised to use these assays to screen for compounds with potential health benefits. If these assays are used to screen compounds to improve the shelf life of food, the assays should be critically evaluated for whether they are suitable models for the target food system. The efficacy of antioxidants identified in these test systems to delay lipid oxidation and extend shelf life needs to be confirmed in real food products under real storage conditions.

25.5 STUDY QUESTIONS

1. Why do compounds such as ascorbic acid and glutathione show a positive response in the Folin-Ciocalteu assay?
2. Name different ways to extract food products with or without prior hydrolysis for the analysis of phenolic compounds and give examples of phenolic compounds that are either included or excluded in the analysis depending on the extraction procedure.
3. Why does RP-HPLC analysis of phenolic acids usually require a pH adjustment of the mobile phase?
4. Why does the GC analysis of phenolic compounds often require a prior derivatization? What is a suitable derivatization procedure for many phenolic compounds?
5. Why did the USDA's Nutrient Data Laboratory remove the USDA ORAC database for Selected Foods from the NDL website?
6. Why do many antioxidant capacity assays fail to model food products that are emulsions?
7. What is the difference between HAT and SET antioxidant capacity assays?
8. What are some limitations of the DPPH assay due to the structure of the stable radical DPPH used in this assay?

Acknowledgments The chapter authors gratefully acknowledge Dr. Mirko Bunzel, who was the lead author of this chapter for the fifth edition of this textbook. The authors also wish to express their thanks to Baraem Ismail, Andrew Neilson, and Jairam Vanamala for reviewing the chapter and providing helpful comments when the chapter was first created for the fifth edition.

REFERENCES

1. Ralph J, Bunzel M, Marita JM, Hatfield RD, Lu F, Kim H, Schatz PF, Grabber JH, and Steinhart H (2004) Peroxidase-dependent cross-linking reactions of *p*-hydroxycinnamates in plant cell walls. Phytochem Rev 3: 79–96.
2. Halliwell B (2007) Dietary polyphenols: Good, bad, or indifferent for your health? Cardiovasc Res 73(2): 341–347.
3. Cladis DP, Weaver CM, and Ferruzzi MG (2022) (Poly)phenol toxicity *in vivo* following oral administration: A targeted narrative review of (poly)phenols from green tea, grape, and anthocyanin-rich extracts. Phytother Res 36(1): 323–335.
4. Gniechwitz D, Reichardt N, Ralph J, Blaut M, Steinhart H, and Bunzel M (2008) Isolation and characterisation of a coffee melanoidin fraction. J Sci Food Agric 88(12): 2153–2160.
5. Barberousse H, Roiseux O, Robert C, Paquot M, Deroanne C, and Blecker C (2008) Analytical methodologies for quantification of ferulic acid and its oligomers. J Sci Food Agric 88(9): 1494–1511.
6. Naczk M and Shahidi F (2004) Extraction and analysis of phenolics in food. J Chromatogr A 1054(1): 95–111.
7. Espada-Bellido E, Ferreiro-González M, Carrera C, Palma M, Barroso CG, and Barbero GF (2017) Optimization of the ultrasound-assisted extraction of anthocyanins and total phenolic compounds in mulberry (*Morus nigra*) pulp. Food Chem 219: 23–32.
8. He B, Zhang LL, Yue XY, Liang J, Jiang J, Gao XL, and Yue PX (2016) Optimization of Ultrasound-Assisted Extraction of pheno-

lic compounds and anthocyanins from blueberry (*Vaccinium ashei*) wine pomace. Food Chem 204: 70–76.
9. Folin O and Denis W (1912) On phosphotungstic-phosphomolybdic compounds as color reagents. J Biol Chem 12(2): 239–243.
10. Folin O and Denis W (1912) Tyrosine in proteins as determined by a new colorimetric method. J Biol Chem 12(2): 245–251.
11. Folin O and Ciocalteu V (1927) On tyrosine and tryptophane determinations in proteins. J Biol Chem 73(2): 627–650.
12. Singleton VL and Rossi JA (1965) Colorimetry of total phenolics with phosphomolybdic-phosphotungstic acid reagents. Am J Enol Vitic 16(3): 144–158.
13. Everette JD, Bryant QM, Green AM, Abbey YA, Wangila GW, and Walker RB (2010) Thorough study of reactivity of various compound classes toward the Folin–Ciocalteu reagent. J Agric Food Chem 58(14): 8139–8144.
14. Kupina S, Fields C, Roman MC, and Brunelle SL (2018) Determination of total phenolic content using the Folin-C Assay: Single-laboratory validation, First Action 2017.13. J AOAC Int 101(5): 1466–1472.
15. Singleton VL, Orthofer R, and Lamuela-Raventós RM, *Analysis of total phenols and other oxidation substrates and antioxidants by means of Folin-Ciocalteu reagent*, in *Methods Enzymol*. 1999, Elsevier. p. 152–178.
16. Medina MB (2011) Simple and rapid method for the analysis of phenolic compounds in beverages and grains. J Agric Food Chem 59(5): 1565–1571.
17. Medina MB (2011) Determination of the total phenolics in juices and superfruits by a novel chemical method. J Funct Foods 3(2): 79–87.
18. Lester GE, Lewers KS, Medina MB, and Saftner RA (2012) Comparative analysis of strawberry total phenolics via Fast Blue BB vs. Folin–Ciocalteu: Assay interference by ascorbic acid. J Food Compost Anal 27(1): 102–107.
19. Hinojosa-Nogueira D, Muros J, Rufián-Henares JA, and Pastoriza S (2017) New method to estimate total polyphenol excretion: Comparison of Fast Blue BB versus Folin–Ciocalteu performance in urine. J Agric Food Chem 65(20): 4216–4222.
20. Pico J, Pismag RY, Laudouze M, and Martinez MM (2020) Systematic evaluation of the Folin–Ciocalteu and Fast Blue BB reactions during the analysis of total phenolics in legumes, nuts and plant seeds. Food Funct 11(11): 9868–9880.
21. Robbins RJ (2003) Phenolic acids in foods: An overview of analytical methodology. J Agric Food Chem 51(10): 2866–2887.
22. Jilek ML and Bunzel M (2013) Dehydrotriferulic and dehydrodiferulic acid profiles of cereal and pseudocereal flours. Cereal Chem 90(5): 507–514.
23. Dobberstein D and Bunzel M (2010) Separation and detection of cell wall-bound ferulic acid dehydrodimers and dehydrotrimers in cereals and other plant materials by reversed phase high-performance liquid chromatography with ultraviolet detection. J Agric Food Chem 58(16): 8927–8935.
24. Bunzel M, Ralph J, Marita JM, Hatfield RD, and Steinhart H (2001) Diferulates as structural components in soluble and insoluble cereal dietary fibre. J Sci Food Agric 81(7): 653–660.
25. Stevenson DE and Hurst RD (2007) Polyphenolic phytochemicals–just antioxidants or much more? Cell Mol Life Sci 64(22): 2900–2916.
26. Dinkova-Kostova A (2002) Protection against cancer by plant phenylpropenoids: Induction of mammalian anticarcinogenic enzymes. Mini Rev Med Chem 2(6): 595–610.
27. USDA, *USDA Database for the Oxygen Radical Capacity (ORAC) of Selected Foods, Release 2*, in *USDA National Nutrient Database for Standard Reference*, USDA, Editor. 2010.
28. Frankel EN and Meyer AS (2000) The problems of using one-dimensional methods to evaluate multifunctional food and biological antioxidants. J Sci Food Agric 80(13): 1925–1941.
29. Prior RL, Wu X, and Schaich K (2005) Standardized methods for the determination of antioxidant capacity and phenolics in foods and dietary supplements. J Agric Food Chem 53(10): 4290–4302.
30. Moon J-K and Shibamoto T (2009) Antioxidant assays for plant and food components. J Agric Food Chem 57(5): 1655–1666.
31. Huang D, Ou B, and Prior RL (2005) The chemistry behind antioxidant capacity assays. J Agric Food Chem 53(6): 1841–1856.
32. *Official Methods of Analysis of AOAC INTERNATIONAL 22nd Edition*. 2023, Oxford University Press.
33. Schaich KM, Tian X, and Xie J (2015) Hurdles and pitfalls in measuring antioxidant efficacy: A critical evaluation of ABTS, DPPH, and ORAC assays. J Funct Foods 14: 111–125.
34. Re R, Pellegrini N, Proteggente A, Pannala A, Yang M, and Rice-Evans C (1999) Antioxidant activity applying an improved ABTS radical cation decolorization assay. Free Radic Biol Med 26(9): 1231–1237.
35. Tian X and Schaich KM (2013) Effects of molecular structure on kinetics and dynamics of the Trolox equivalent antioxidant capacity assay with $ABTS^{+\bullet}$. J Agric Food Chem 61(23): 5511–5519.
36. Blois MS (1958) Antioxidant determinations by the use of a stable free radical. Nature 181: 1199–1200.
37. Xie J and Schaich KM (2014) Re-evaluation of the 2,2-diphenyl-1-picrylhydrazyl free radical (DPPH) assay for antioxidant activity. J Agric Food Chem 62(19): 4251–4260.
38. Foti MC, Daquino C, and Geraci C (2004) Electron-transfer reaction of cinnamic acids and their methyl esters with the DPPH• radical in alcoholic solutions. J Org Chem 69(7): 2309–2314.
39. Pryor WA (1986) Oxy-radicals and related species: Their formation, lifetimes, and reactions. Annu Rev Physiol 48(1): 657–667.

Application of Enzymes in Food Analysis

26

José I. Reyes De-Corcuera

Contents

26.1 INTRODUCTION

Enzymes are protein catalysts capable of great specificity and reactivity under physiological conditions. Enzymatic analysis is the measurement of compounds with the aid of added enzymes or the measurement of endogenous enzyme activity to give an indication of the state of a biological system including foods. The fact that enzyme catalysis can take place under relatively mild conditions allows for measurement of relatively unstable compounds not amenable to some other techniques. In addition, the specificity of enzyme reactions can allow for measurement of components of complex mixtures without the time and expense of chromatographic separations.

There are several uses of enzyme analyses in food science and technology. In several instances, enzyme activity is a useful measure for adequate processing of a food product. The thermal stability of enzymes has been used extensively as a measure of heat treatment; for example, peroxidase activity is used as a measure of adequacy of blanching of vegetable products. Catalase activity is used as an indicator of raw or under cooked meat. Similarly, lactoperoxidase has been used to verify the adequacy of milk pasteurization. β-galactosidase is used as a kit in combination with other enzymes to determine any residual concentration of lactose in lactose-free dairy products. Enzyme activity assays are also used by the food technologist to assess potency of enzyme preparations used as processing aids such as pectinases and cellulases blends used for the clarification of fruit juices or amylases used for starch hydrolysis in breadmaking or grain fermentations. Proteinases like papain and bromelain are used as meat tenderizers. In other instances, residual enzyme activities negatively affect the flavor, color, or texture, and hence the shelf life of food products. For example, if not properly blanched, lipoxygenase is responsible for the development of off-flavors in stored frozen vegetables. Polyphenol oxidase catalyzes the discoloration of wheat flour and noodles as well as of several fruit juices and pastes (see Chap. 31). Polygalacturonase hydrolyzes pectin and is

J. I. Reyes De-Corcuera (✉)
Department of Agricultural and Biological Engineering, The University of Florida, Gainesville, FL, USA
e-mail: jireyes@ufl.edu

B. P. Ismail, S. S. Nielsen (eds.), *Nielsen's Food Analysis*, Food Science Text Series,
https://doi.org/10.1007/978-3-031-50643-7_26

responsible for the softening of fruit and vegetables. Residual lipase activity in vegetable oils hydrolyzes triacylglycerols producing undesirable free fatty acids (see Chap. 23). Also, residual lipase activity in milk is responsible for the development of off-flavors in milk chocolate.

The food scientist can also use commercially available enzyme preparations to measure constituents of foods that are enzyme substrates. For example, glucose content can be determined in a complex food matrix containing other monosaccharides by using readily available enzymes. A corollary use of commercially available enzymes is to measure enzyme activity as a function of enzyme-inhibitor content in a food. Organophosphate insecticides are potent inhibitors of the enzyme acetylcholinesterase, and hence the activity of this enzyme in the presence of a food extract is a measure of organophosphate insecticide concentration in the food. Also of interest is the measurement of enzyme activity associated with food quality. For example, catalase activity is markedly increased in milk from mastitic udders. Catalase activity also parallels the bacterial count in milk. Another use of enzyme assays to determine food quality is estimation of protein nutritive value by monitoring the activity of added proteases on food protein samples (see Chap. 24). Enzymes can be used to measure the appearance of degradation products such as trimethylamine in fish during storage. Enzymes are also used as preparative tools in food analysis. Examples include the use of amylases and proteases in fiber analysis (Chap. 19) and the enzymatic hydrolysis of thiamine phosphate esters in vitamin analysis.

Food scientists must realize that enzyme activity is greatly affected by its environment and that enzymes can degrade and lose their activity at high temperatures or under other conditions that deviate not too far from optimal. Therefore, processing enzymes must be stored and handled carefully, typically refrigerated or frozen. To some, stabilizing agents are added. To successfully carry out enzyme analyses in foods, an understanding of certain basic principles of enzymology is necessary. After a brief overview of these principles, examples of the use of enzymatic analyses in food systems are examined.

26.2 PRINCIPLES

26.2.1 Enzyme Kinetics

26.2.1.1 Overview

Enzymes are proteins with catalytic activity. A catalyst increases the rate (velocity) of a thermodynamically possible reaction. The enzyme does not modify the equilibrium constant of the reaction, and the enzyme catalyst is not consumed in the reaction. Because enzymes affect rates (velocities) of reactions, some knowledge of **enzyme kinetics** (study of rates) is needed for the food scientist to effectively use enzymes in analysis. To measure the rate of an enzyme-catalyzed reaction, typically one mixes the enzyme with the substrate under specified conditions (pH, temperature, ionic strength, etc.) and follows the reaction by measuring the amount of product that appears or by measuring the disappearance of substrate. Some of the fundamentals of chemical kinetics are summarized on Fig. 26.1A. However, because students of food analysis have typically already studied enzyme kinetics in biochemistry and food chemistry courses, many details of enzyme kinetics such as the derivation of reaction rate equations are not covered here, but can be found in the fourth edition of this textbook or elsewhere [1].

The formation of the enzyme-substrate complex (Fig. 26.1, Eq. 26.4), called the **pre-steady-state period,** is very rapid (on the millisecond scale) and is not normally seen in the laboratory. The slope of the linear portion of the curve following the pre-steady-state period gives us the **initial velocity** (v_0). After the pre-steady-state period, a **steady-state period** exists in which the concentration of the enzyme-substrate complex is constant. A time course needs to be established experimentally by using a series of points or a continuous assay to establish the appropriate time frame for the measurement of the initial velocity.

The rate of an enzyme-catalyzed reaction depends on the concentration of the enzyme and on the substrate concentration. With a fixed enzyme concentration, increasing substrate concentration results in an increased velocity (see Fig. 26.1B). As substrate concentration increases further, the increase in velocity slows until, with a very large concentration of substrate, no further increase in velocity is noted. The velocity of the reaction at this very large substrate concentration is the **maximum velocity** (V_{max}) of the reaction under the conditions of that particular assay. The substrate concentration at which one half V_{max} is observed is equal to the **Michaelis constant** (K_M). This constant is an indication of the relative binding affinity of the enzyme for a particular substrate. The lower the K_M, the greater the affinity of the enzyme for the substrate. Both K_M and V_{max} are affected by environmental conditions such as pH, temperature, ionic strength, and the nature of the solvent, that is, organic or aqueous (although most enzyme reactions occur in aqueous environments, some reaction occur in organic media such as in lipids) (Table 26.1).

26.2.1.2 Michaelis-Menten Equation

Michaelis-Menten equation (Fig. 26.1, Eq. 26.7) is that of a hyperbola that relates the reaction velocity to substrate concentration for an enzyme-mediated reaction (Fig. 26.1B).

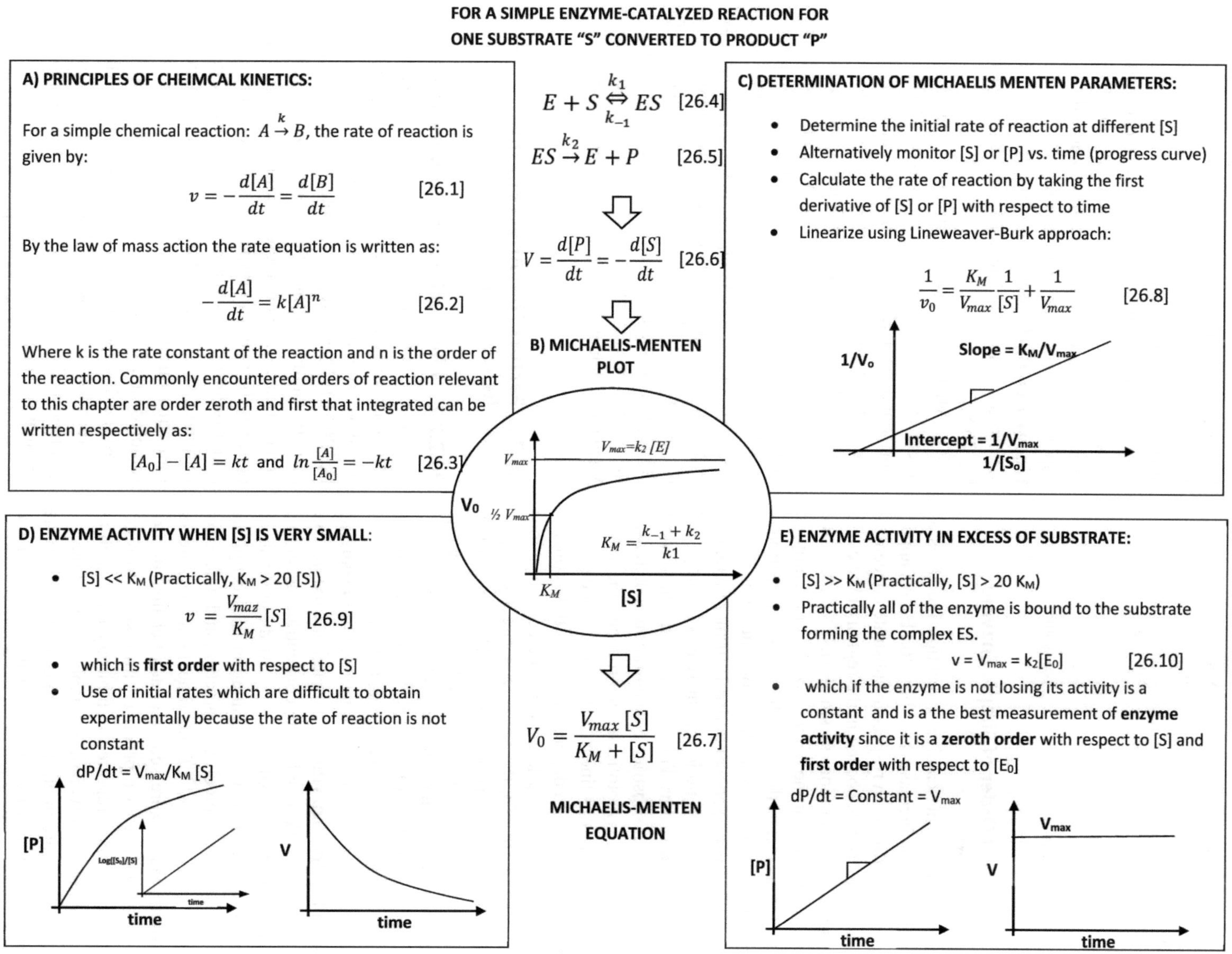

Fig. 26.1 Principles of chemical and enzyme kinetics

Table 26.1 Factors that affect enzyme reaction rate

Factors	*Michaelis constant* *(K_M)*	*Maximum velocity* *(V_{max})*
Source of enzyme	Yes	Yes
Substrate	Yes	Yes
Enzyme concentration	No	Yes
Substrate concentration	No	No
Temperature	Yes	Yes
pH	Yes	Yes
Inhibitors	Yes	Yes (some)

26.2.1.3 Apparent Order of Simple Enzyme Reactions

The velocity of an enzyme-catalyzed reaction increases as substrate concentration increases (see Fig. 26.1B). An apparent **first-order reaction** with respect to substrate concentration (Fig. 26.1, Eq. 26.9) is obeyed in the region of the curve where substrate concentration is small ($[S] < <K_M$, see Fig. 26.1D), assuming that the active enzyme concentration $[E]$ is constant. This means that the velocity of the reaction is directly proportional to the substrate concentration in this region. When the substrate concentration is further increased, the velocity of the reaction no longer increases linearly, and the reaction is **mixed order**. This is seen in the curvilinear portion of Fig. 26.1B. If substrate concentration is increased further, the velocity asymptotically approaches the maximum velocity (V_{max}). In this linear, nearly zero slope portion of the plot, the velocity is independent of substrate concentration. However, note that at large substrate concentrations ($[S]> > K_M$, see Fig. 26.1E), the velocity is directly proportional to enzyme concentration (Fig. 26.1E, Eq. 26.10). Thus, in this portion of the curve, where $[S]> > K_M$, the rate of the reaction is **zero order** with respect to substrate concentration (is independent of substrate concentration) but first order with respect to enzyme concentration.

For practical purposes, if we are interested in **measuring the amount of active enzyme** in a reaction mixture, we should, if possible, work at substrate concentrations so that the observed velocity approximates V_{max}. At these substrate concentrations, active enzyme concentration is directly rate limiting to the observed velocity. In practice, and unless not possible, V_{max} is used most of the time as a measure of **enzyme activity** and is expressed as a rate of change in concentration of a substrate or a product with respect to time. Enzyme activity is often reported in arbitrary units such as change in absorbance per weight of sample per unit time.

Conversely, if we are interested in **measuring substrate concentration** by measuring initial velocity, we must be at substrate concentrations less than K_M in order to have a rate directly proportional to substrate concentration. Knowledge of K_M and V_{max} is necessary, which in practice would require at least the experimental determination of V_{max} immediately prior to using the enzyme to account for any decrease of enzyme activity over time.

26.2.1.4 Determination of K_M and V_{max}

To properly design an experiment in which velocity is zero order with respect to substrate and first order with respect to enzyme concentration, or conversely an experiment in which we would like to measure rates that are directly proportional to substrate concentration, we must know the K_M. The most popular method for determining K_M is the use of a **Lineweaver-Burk plot** (Fig. 26.1C) drawn by using the reciprocal of the Michaelis-Menten equation (Fig. 26.1C, Eq. 26.8).

The Michaelis-Menten equation is that of a straight line where $1/v_0$ is the ordinate (vertical axis), $1/[S]$ is the abscissa (horizontal axis), K_M/V_{max} is the slope, and $1/V_{max}$ is the ordinate at the origin. Linear regression is used to determine K_M and V_{max}. Selection of experimental concentrations of substrate concentrations that result in evenly distributed values of $1/V_{max}$ results in more accurate determinations of K_M and V_{max} but is seldom found in the literature. Enzyme mechanisms are often more complex than what is presented here. For example, oxidases often require two substrates that bind to the enzyme simultaneously or in sequence. Detailed description of such mechanisms and rate equations is beyond the scope of this chapter. However, in practice, simple Michaelis-Menten kinetics is frequently assumed.

26.2.2 Factors That Affect Enzyme Reaction Rate

26.2.2.1 Effect of Enzyme Concentration

In excess of substrate, the velocity of an enzyme-catalyzed reaction is directly proportional to (increases linearly with) the enzyme concentration in the reaction mixture (Fig. 26.1E, Eq. 26.10). Therefore, if possible, determination of enzyme activity should be done at concentrations of substrate much greater than K_M. Under these conditions, a zero-order dependence of the rate with respect to substrate concentration and a first-order relationship between rate and enzyme concentration exist. It is critical that the substrate concentration is saturating during the entire period the reaction mixture is sampled and the amount measured of product formed or substrate disappearing is linear over the period during which the reaction is sampled. The activity of the enzyme is obtained as the slope of the linear part of the line of a plot of product or substrate concentration vs. time as depicted by the solid lines in Fig. 26.2. However, when the concentration of enzyme is very large, during the course of the reaction, the substrate concentration decreases and is no longer in excess, resulting in deviations from linearity as shown in the dashed lined in Fig. 26.2. If a large number of samples are to be assayed, a single aliquot is often taken and assayed at a single reaction time. This can be risky and will give good results only if the time at which the sample is taken falls on the linear portion of a plot of substrate or product concentration versus time of reaction (see Fig. 26.2). Normally, one designs an experiment in which enzyme concentration is estimated such that no more than 5–10% of the substrate has been converted to product within the time used for measuring the initial rate.

Sometimes it is not possible to carry out enzyme assays at $[S] >> K_M$. The substrate may be very expensive or relatively insoluble or K_M may be large (i.e., $K_M > 100$ m*M*). In these cases, the reaction rate is determined by measuring initial rates of the reactions, in which the change in substrate or product concentration is determined at times as close as possible to time zero. This is shown in Fig. 26.2 by the solid lines drawn tangent to the slopes of the initial parts of the curves. The slope of the tangent line gives an approximation of the initial rate. Enzyme concentration can also be estimated at substrate concentrations much less than K_M. When substrate concentration is much less than K_M, the substrate term in the denominator of the Michaelis-Menten equation can be ignored and $v = (V_{max}/K_M)[S]$ which is the equation for a first-order reaction with respect to substrate concentration

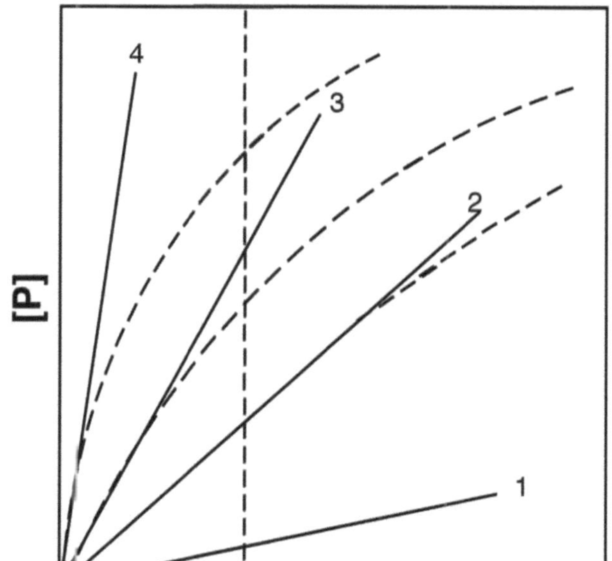

Fig. 26.2 Effect of enzyme concentration on time course of an enzyme-catalyzed reaction. The *dashed lines* are experimentally determined data with enzyme concentration increasing from 1 to 4. The *solid lines* are tangents drawn from the initial slopes of the experimental data. If a single time point, *a*, is used for data collection, a large difference between actual data collected and that predicted from initial rates is seen

(see Fig. 26.1D). Under these conditions, a plot of product concentration vs. time gives a nonlinear plot (Fig. 26.1D, left plot). A plot of log $([S_o]/[S])$ vs. time gives a straight-line relationship (Fig. 26.1D, left plot inset). The slope of the line of the log plot is directly related to the enzyme concentration. When the slopes of a series of these log plots are further plotted as a function of enzyme concentration, a straight-line relationship should result. If possible, the reaction should be followed continuously or aliquots removed at frequent time intervals and the reaction allowed to proceed to greater than 10% of the total reaction.

26.2.2.2 Effect of Substrate Concentration

The effect of substrate concentration on the velocity of an enzyme-catalyzed reaction in which enzyme concentration is constant is shown in Fig. 26.1b. As noted before, the rate of the reaction is first order with respect to substrate concentration when $[S] << K_M$ (Fig. 26.1D). At $[S] >> K_M$, the reaction is zero order with respect to substrate concentration and first order with respect to $[E]$ (Fig. 26.1E). At substrate concentrations between the first-order and zero-order regions, the enzyme-catalyzed reaction is **mixed order** with respect to substrate concentration. When initial rates are obtained, a linear relationship between v_0 and E_o should be seen. However, in practice in most common laboratories, it is very difficult to accurately determine initial rates because it is practically impossible to instantaneously mix substrates and enzyme and instrumentally determine product or substrate concentrations. Therefore, experimental determinations of initial rates typically underestimate the actual value.

26.2.2.3 Environmental Effects

The subsections below describe how various environmental factors such as temperature, pH, and the presence of inhibitors affect K_M and V_{max}. Although beyond the scope of this chapter, as the application of high-pressure processing of foods increases, food analysts must also be aware that like temperature, pressure also affects K_M and V_{max} [2]. Furthermore, high pressures up to about 300 MPa have stabilized several food enzymes against their thermal inactivation [3].

26.2.2.3.1 Effect of Temperature on Enzyme Activity

Increasing temperature typically increases the rate of the enzyme-catalyzed reactions up to a point beyond which the enzyme unfolds. In some cases, cooling also produces enzyme denaturation. Temperature affects the solubility of products and substrates, thus it may affect their effective concentration and the rate of reaction. Temperature also affects the equilibrium of species in solution, thus it may affect the pH of the system. For example, for buffering species Tris (tris [hydroxymethyl] aminomethane), the pK_a changes 0.031 per 1 °C change.

Temperature affects both the stability and the activity of the enzyme, as shown in Fig. 26.3. At relatively low temperatures, the enzyme is stable. However, at higher temperatures, denaturation dominates, and a markedly reduced enzyme activity represented by the negative slope portion of line 2 is observed:

$$E_{\text{active}} \xrightarrow{k_{\text{inact}}} E_{\text{inactive}} \quad (26.11)$$

Line 1 of Fig. 26.3 shows the effect of temperature on the velocity of the enzyme-catalyzed reaction. The velocity is expected to increase exponentially as the temperature is increased. As shown by line 1, the velocity approximately doubles for every 10 °C rise in temperature. The net effect of increasing temperature on the rate of conversion of substrate to product (line 1) and on the rate of the denaturation of enzyme (line 3) is line 2 of Fig. 26.3. The temperature optimum of the enzyme is at the maximum point of line 2. The temperature optimum is not unique for any given enzyme. Each enzyme has multiple optima depending on the type of substrate, pH, salt concentration, substrate concentration, and time of reaction. For this reason, investigators should fully describe a system in which the effects of temperature on observed enzyme activity are reported. Each optimum temperature occurs under conditions in which the enzyme is denaturing. Therefore, even under excess of substrate, plots of substrate or product concentration vs. time are often not linear since the enzyme concentration is constantly changing. For this reason, for analytical purposes, enzyme activity is assayed at temperatures colder than the optimal for activity.

The data of Fig. 26.3, line 1, can be plotted according to the **Arrhenius equation**:

$$k = Ae^{-E_a/RT} \quad (26.12)$$

which can be written as:

$$\log k = \log A - \frac{E_a}{2.3RT} \quad (26.13)$$

where:

k = a specific rate constant at some temperature, T (K)
E_a = activation energy, the minimum amount of energy a reactant molecule must have to be converted to product
R = gas constant
A = a frequency factor (preexponential factor)

The rate constant of inactivation k_{inact} also follows Arrhenius equation. The positive slope on the left side (high temperature) of the **Arrhenius plot** (Fig. 26.4) gives a measure of the **activation energy** (E_a) for the denaturation of the enzyme. Note that a small change in temperature has a very large effect on the rate of denaturation. The slope on the right side of Fig. 26.4 gives a measure of the E_a for the transformation of substrate to product catalyzed by the enzyme. If the experiment is carried out under conditions in which V_{max} is measured ([S]> > K_M), then the activation energy observed will be for the catalytic step of the reaction (k_2).

26.2.2.3.2 Effect of pH on Enzyme Activity

The observed rate of an enzyme-catalyzed reaction is greatly affected by the pH of the medium. Enzymes have pH optima

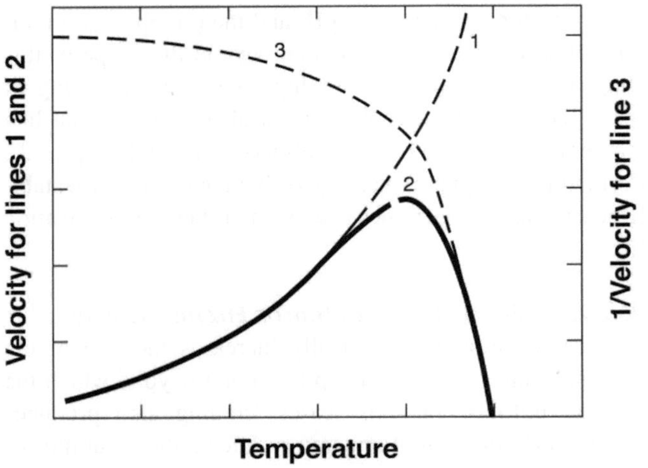

Fig. 26.3 Effect of temperature on velocity of an enzyme-catalyzed reaction. Temperature effect on substrate to product conversion is shown by line 1. Line 3 shows effect of temperature on rate of enzyme denaturation (right-hand y-axis is for line 3). The net effect of temperature on the observed velocity is given by line 2, and the temperature optimum is at the maximum of line 2

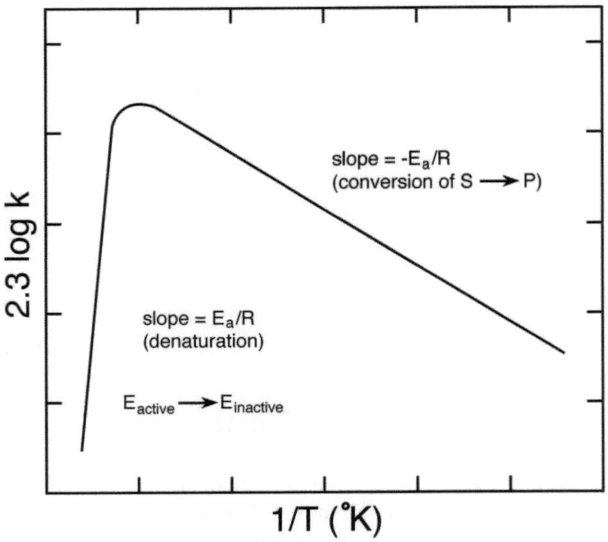

Fig. 26.4 Effect of temperature on rate constant of an enzyme-catalyzed reaction. The data are plotted 2.3 log k versus 1/T(°K) according to the Arrhenius equation, $k = Ae^{E_aRT}$

and commonly have bell-shaped curves for activity versus pH (Fig. 26.5). This pH effect is a manifestation of the effects of pH on enzyme stability and on rate of substrate to product conversion and may also be due to changes in ionization of substrate.

The rate of substrate to product conversion is affected by pH because pH may affect binding of substrate to enzyme and the ionization of catalytic groups such as carboxyl or amino groups that are part of the enzyme's active site. The stability of the tertiary or quaternary structure of enzymes is also pH dependent and affects the velocity of the enzyme reaction, especially at extreme acidic or alkaline pHs. The pH for maximum stability of an enzyme does not necessarily coincide with the pH for maximum activity of that same enzyme. For example, the proteolytic enzymes trypsin and chymotrypsin are stable at pH 3, while they have maximum activity at pH 7–8.

To establish the pH optimum for an enzyme reaction, the reaction mixture is buffered at different pHs and the activity of the enzyme is determined. To determine pH enzyme stability relationships, aliquots of the enzyme are buffered at different pH values and held for a specified period of time (e.g., 1 h). The pH of the aliquots is then adjusted to the pH optimum and each aliquot is assayed. The effect of pH on enzyme stability is thus obtained. These studies are helpful in establishing conditions for handling the enzyme and also may be useful in establishing methods for controlling enzyme activity in a food system. Note that pH stability and the pH optimum for the enzyme activity are not true constants. That is to say, these may vary with particular source of enzyme, the specific substrate used, the temperature of the experiment, or even the buffering species used in the experiment. In the use of enzymes for analysis, it is not necessary that the reaction be carried out at the pH optimum for activity, or even at a pH at which the enzyme is most stable (although it should be stable for the duration of the experiment), but it is critical to maintain a fixed pH during the reaction (i.e., use buffer) and to use the same pH in all studies to be compared.

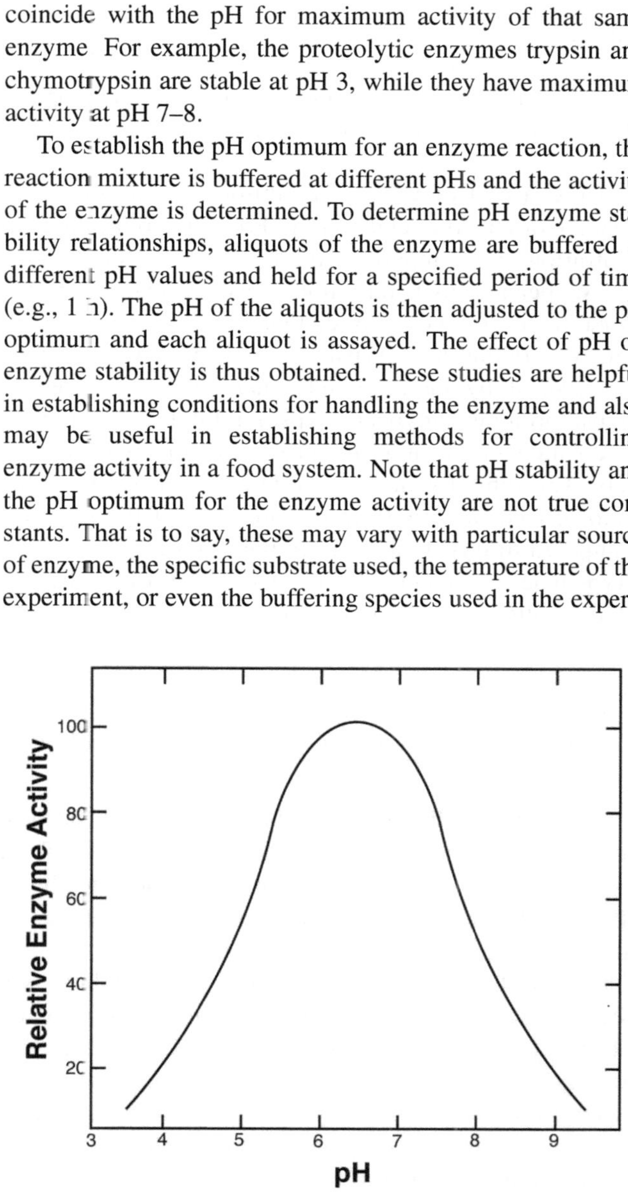

Fig. 26.5 Typical velocity-pH curve for an enzyme-catalyzed reaction. The maximum on the curve is the optimum for the system and can vary with temperature, specific substrate, and enzyme source

26.2.2.4 Activators and Inhibitors

26.2.2.4.1 Activators

Some enzymes contain, in addition to a protein portion, small molecules that are activators of the enzyme. Some enzymes show an absolute requirement for a particular inorganic ion for activity, while others show increased activity when small molecules are included in the reaction medium. These small molecules can play a role in maintaining the conformation of the protein, or they may form an essential component of the active site, or they may form part of the substrate of the enzyme.

In some cases, the activator forms a nearly irreversible association with the enzyme. These nonprotein portions of the enzyme are called **prosthetic groups**. The amount of enzyme activator complex formed is equal to the amount of activator present in the mixture. In these cases, activator concentration can be estimated up to concentrations equal to total enzyme concentration by simply measuring enzyme activity.

In most cases, dissociation constants for an enzyme activator complex are within the range of enzyme concentration. Dissociable nonprotein parts of enzymes are categorized as **coenzymes**. When this type of activator is added to an enzyme, a curvilinear relationship similar to a Michaelis-Menten plot results, making difficult the determination of an unknown amount of activator. A reciprocal plot analogous to a Lineweaver-Burk plot can be constructed using standards and unknown activator concentrations estimated from such a plot.

One food-related enzyme reaction involving an activator is measuring pyridoxal phosphate, a form of vitamin B_6. This reaction measures the reactivation of coenzyme-free yeast aminotransferase by coupling the transamination reaction with malate dehydrogenase. This is possible when malate dehydrogenase, NADH, alpha-ketoglutarate, aspartate, and the aminotransferase are in excess, and the pyridoxal 5-phosphate added is rate limiting:

$$\text{alpha} - \text{ketoglutarate} + \text{aspartate} \xrightleftharpoons{\text{aminotransferase}} \text{glutamate} + \text{oxalacetate} \quad (26.14)$$

$$\text{oxalacetate} + \text{NADH} + \text{H}^+ \xrightleftharpoons{\text{malate dehydrogenase}} \text{malate} + \text{NAD}^+ \quad (26.15)$$

Another example of an essential activator is the pyridine coenzyme NAD^+. NAD^+ is essential for the oxidation of ethanol to acetaldehyde by alcohol dehydrogenase:

$$\text{ethanol} + NAD^+ \xrightleftharpoons{\text{alcohol dehydrogenase}} \text{acetaldehyde} + NADH + H^+ \qquad (26.16)$$

In the reaction, NAD^+ is reduced to NADH and can be considered a second substrate. Another example of an activator of an enzyme is the chloride ion with α-amylase. In this case, α-amylase has some activity in the absence of chloride. With saturating levels of chloride, the α-amylase activity increases about fourfold. Other anions, including F^-, Br^-, and I^-, also activate α-amylase. These anions must not be in the reaction mixture if α-amylase stimulation is to be used as a method of determining chloride concentration.

26.2.2.4.2 Inhibitors

An enzyme inhibitor is a compound that when present in an enzyme-catalyzed reaction medium decreases the enzyme activity. Enzyme inhibitors can be categorized as **irreversible** or **reversible** inhibitors. Enzyme inhibitors include inorganic ions, such as Pb^{2+} or Hg^{2+}, which can react with sulfhydryl groups on enzymes to inactivate the enzyme, compounds that resemble substrate, and naturally occurring proteins that specifically bind to enzymes (such as protease inhibitors found in legumes).

(1) **Irreversible Inhibitors.** When the dissociation constant of the inhibitor enzyme complex is very small, the decrease in enzyme activity observed will be directly proportional to the inhibitor added. The speed at which the irreversible combination of enzyme and inhibitor reacts may be slow, and the effect of time on the reduction of enzyme activity by the addition of inhibitor must be determined to ensure complete enzyme-inhibitor reaction. For example, the amylase inhibitor found in many legumes must be preincubated under specified conditions with amylase prior to measurement of residual activity to accurately estimate inhibitor content [4]. Irreversible inhibitors decrease V_{max} as the amount of total active enzyme is reduced.

(2) **Reversible Inhibitors.** Most inhibitors exhibit a dissociation constant such that both enzyme and inhibitor are found free in the reaction mixture. Several types of reversible inhibitors are known: **competitive, noncompetitive**, and **uncompetitive.**

Competitive inhibitors usually resemble the substrate structurally and compete with substrate for binding to the active site of the enzyme, and only one molecule of substrate or inhibitor can be bound to the enzyme at one time. An inhibitor can be characterized as competitive by adding a fixed amount of inhibitor to reactions at various substrate concentrations and by plotting the resulting data by the Lineweaver-Burk method and noting the effect of inhibitor relative to that of control reactions in which no inhibitor is added. If the inhibitor is competitive, the slope and x-intercept of the plot with inhibitor are altered, while the y-intercept ($1/V_{max}$) is unaltered. It can be shown that the ratio of the **uninhibited initial velocity** (v_o) to the **inhibited initial velocity** (v_i) gives:

$$\frac{v_0}{v_i} = \frac{[I]K_M}{K_i\left(K_M + [S]\right)} \qquad (26.17)$$

where:

K_i = dissociation constant of the enzyme-inhibitor complex
[I] = concentration of competitive inhibitor.

Thus, a plot of v_o/v_i vs. inhibitor concentration will give a straight-line relationship. From this plot, the concentration of a competitive inhibitor can be found [1, 4, 5]. Soybean trypsin inhibitor, a well-known antinutritional factor, is a polypeptide that acts as reversible competitive inhibitor. Trypsin is a protease enzyme synthesized by humans and other animals needed to digest proteins. Soybean trypsin inhibitor is typically inactivated with heat during processing of soybean products. The inhibitor activity is determined based on the ratio of trypsin activity in the presence of the inhibitor compared to trypsin activity in the absence of inhibitor. Trypsin activity can be determined by following the rate of the following reaction by measuring the change in absorbance at 253 nm:

$$\text{N-benzoyl-L-arginine ethyl ester} + H_2O \xrightarrow{\text{Trypsin}} \text{N-benzoyl-L-arginine} + \text{ethanol} \qquad (26.18)$$

A **noncompetitive inhibitor** binds to enzyme independent of substrate and is bound outside the active site of the enzyme. A noncompetitive inhibitor can be identified by its effect on the rate of enzyme-catalyzed reactions at various substrate concentrations and the data plotted by the Lineweaver-Burk method. A noncompetitive inhibitor will affect the slope and the y-intercept as compared to the uninhibited system, while the x-intercept, $1/K_M$, is unaltered. Analogous to competitive inhibitors, a standard curve of v_o/v_i vs. inhibitor concentration may be prepared and used to determine the concentration of a noncompetitive inhibitor [1, 4, 5].

Uncompetitive inhibitors bind only to the enzyme-substrate complex. Uncompetitive inhibition is noted by adding a fixed amount of inhibitor to reactions at several substrate concentrations and plotting the data by the

Lineweaver-Burk method. An uncompetitive inhibitor will affect both the *x*- and *y*-intercepts of the Lineweaver-Burk plot as compared to the uninhibited system, while maintaining an equal slope to the uninhibited system (i.e., a parallel line will result). A plot of v_o/v_i vs. inhibitor concentration can be prepared to use as a standard curve for the determination of the concentration of an uncompetitive inhibitor [1, 4, 5].

26.2.3 Methods of Measurement

26.2.3.1 Overview

For practical enzyme analysis, it is necessary to be familiar with the methods of measurement of the reaction. Any physical or chemical property of the system that relates to substrate or product concentration can be used to follow an enzyme reaction. A wide variety of methods are available to follow enzyme reactions, including **absorbance spectrometry, fluorimetry, manometric methods, titration, isotope measurement, chromatography, mass spectrometry,** and **viscosity**. A good example of the use of spectrophotometry as a method for following enzyme reactions is use of the spectra of the pyridine coenzyme NAD(H) and NADP(H), in which there is a marked change in absorbance at 340 nm upon oxidation-reduction (Fig. 26.6). Many methods depend

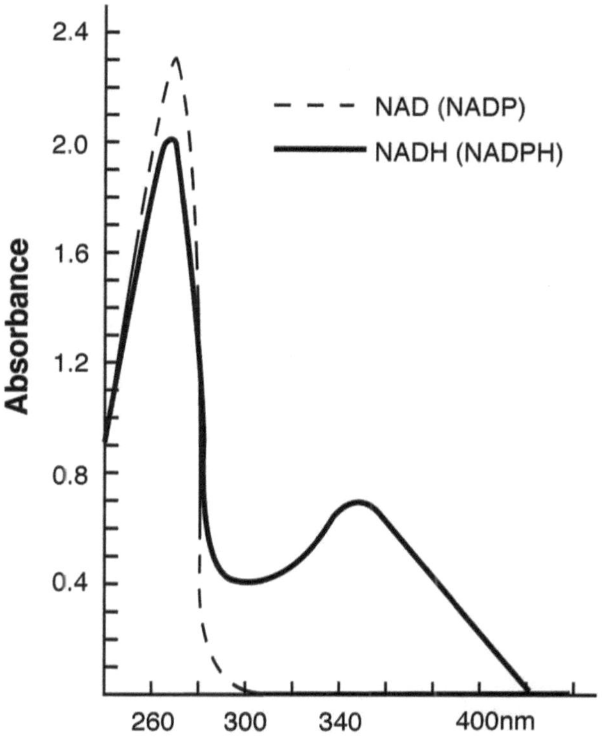

Fig. 26.6 Absorption curves of NAD(P) and NAD(P)H; λ = wavelength. Many enzymatic analysis methods are based on the measurement of an increase or decrease in absorbance at 340 nm due to NAD(H) or NADP(H)

on the increase or decrease in absorbance at 340 nm when these coenzymes are products or substrates in a coupled reaction.

An example of using several methods to measure the activity of an enzyme is in the assay of α-amylase activity [5]. *A*-Amylase cleaves starch at α-1,4 linkages in starch and is an endoenzyme. An endoenzyme cleaves a polymer substrate at internal linkages. This reaction can be followed by a number of methods, including reduction in viscosity (see Chap 29), increase in reducing groups upon hydrolysis, reduction in color of the starch-iodine complex, and polarimetry. However, it is difficult to differentiate the activity of α-amylase from β-amylase using a single assay. β-Amylase cleaves maltose from the nonreducing end of starch. While a marked decrease in viscosity of starch or reduction in iodine color would be expected to occur due to α-amylase activity, β-amylase can also cause changes in viscosity and iodine color if in high concentration. To establish whether α-amylase or β-amylase is being measured, the analyst must determine the change in number of reducing groups as a basis of comparison. Because α-amylase is an endoenzyme, hydrolysis of a few bonds near the center of the polymeric substrate will cause a marked decrease in viscosity, while hydrolysis of an equal number of bonds by the exoenzyme, β-amylase, will have little effect on viscosity.

In developing an enzyme assay, it is wise to first write out a complete, balanced equation for the particular enzyme-catalyzed reaction. Inspection of the products and substrates for chemical and physical properties that are readily measurable with available equipment will often result in an obvious choice of method for following the reaction in the laboratory.

If one has options in methodology, one should select the method that is able to monitor the reaction continuously, is most sensitive, and is specific for the enzyme-catalyzed reaction. For multiple routine analyses, the use of controlled temperature 96-well microplate readers with automatic injection and shaking with absorbance and fluorescence detectors has become widespread. These instruments are particularly useful to characterize enzyme kinetics using multiple enzyme and/or multiple substrate dilutions. These instruments are also commonly used in molecular biology assays such as enzyme-linked immunosorbent assays (Chap. 27, Sect. 27.3.2) and polymerase chain reaction (Chap. 33, Sect. 33.6.2) for the detection of bacteria, allergens, and genetically modified organisms.

26.2.3.2 Coupled Reactions

Enzymes can be used in assays via coupled reactions. **Coupled reactions** involve using two or more enzyme reactions so that a substrate or product concentration can be readily followed. In using a coupled reaction, there is an **indicator reaction** and a **measuring reaction**. For example:

$$\underset{\text{measuring reaction}}{S1 \xrightarrow{E1} P1} \quad (26.19)$$

$$\underset{\text{indicating reaction}}{P1 \xrightarrow{E2} P2} \quad (26.20)$$

The role of the indicating enzyme (E2) is to produce P2, which is readily measurable and, hence, is an indication of the amount of P1 produced by E1. Alternatively, the same sequence can be used in measuring S1, the substrate for E1. When a coupled reaction is used to measure the activity of an enzyme (e.g., E1 above), it is critical that the indicating reaction be fast, hence not rate limiting: the measuring reaction must always be the slowest, hence rate determining. Consequently, E2 activity should be much greater than E1 activity for an effective assay. Coupled enzyme reactions can have problems with respect to pH of the system if the pH optima of the coupled enzymes are quite different. It may be necessary to allow the first reaction (e.g., the measuring reaction catalyzed by E1 above, Eq. 26.19) to proceed for a time and then arrest the reaction by heating to denature E1. The pH is adjusted, the indicating enzyme (E2, Eq. 26.20) added, and the reaction completed. If an endpoint method is used with a coupled system, the requirements for pH compatibility are not as stringent as for a rate assay because an extended time period can be used to allow the reaction sequence to go to completion.

26.3 APPLICATIONS

As described previously, certain information is needed prior to using enzyme assays analytically. In general, knowledge of K_M, time course of the reaction, the enzyme's specificity for substrate, the pH optimum and pH stability of the enzyme, and effects of temperature on the reaction and stability of the enzyme are desirable. Many times this information is available from the literature and databases such as BRENDA (https://www.brenda-enzymes.org/). However, a few preliminary experiments may be necessary, especially in the case of experiments in which velocities are measured. A time course to establish linearity of product formation or substrate consumption in the reaction is a necessity. An experiment to show linearity of velocity of the enzyme reaction to enzyme concentration is recommended (see Fig. 26.2).

26.3.1 Substrate Assays

The following is not an extensive compendium of methods for the measurement of food components by enzymatic analysis. Instead, it is meant to be representative of the types of analyses possible. Table 26.2 summarizes substrate assays covered in this section. The reader can consult handbooks published by the manufacturers of enzyme kits that are

Table 26.2 Summary of substrate analysis examples

Substrate of interest	*Enzyme(s)*	*Compound measured*	*Measured property*	*Uses/comments*
Glutamate	Glutamate dehydrogenase (GluDH) and lactate dehydrogenase (LDH)	NADH	Endpoint decrease in A_{340}	Relevant to products containing monosodium glutamate (MSG). Coupled reaction + regeneration system: All glutamate is converted into α-ketoglutarate, all NADH is consumed with pyruvate + LDH. Enzymes are heat inactivated, then NADH + GluDH are added
Glutamate	Glutamate dehydrogenase (GluDH) and diaphorase	Formazan	Endpoint increase in A_{492}	Coupled reactions. NADH produced by GluDH reacts with iodonitrotetrazolium chloride
Sulfite	Sulfite oxidase (SO) and NADH-peroxidase (POD)	NADH	Endpoint decrease in A_{340}	Used to determine sulfites in wines. Coupled reaction
Glucose	Glucose oxidase (GOx) and POD	Oxidized o-diansidine	Endpoint increase in A_{420}	Used to monitor glucose concentration in yeast fermentations. Coupled reaction
	Hexokinase (HK) and glucose-6-phosphate dehydrogenase (G6PDH)	NADPH	Endpoint increase in A_{340}	Used to determine malic acid in apple juice as alternative to titratable acidity. Coupled reaction
D-Malic acid	D-malate dehydrogenase	NADH	Endpoint decrease in A340	

increasingly used in industry (e.g., Megazyme; https://secure.megazyme.com/Dynamic.aspx?control=CSCatalogue&categoryName=AssayKits), the review article by Whitaker [6], the handbook from Boehringer-Manheim [7], and the series by Bergmeyer [8] for a more comprehensive guide to enzyme methods applicable to foods.

26.3.1.1 Sample Preparation

Because of the specificity of enzymes, sample preparation prior to enzyme analysis is often minimal and may involve only extraction and removal of solids by filtration or centrifugation. Regardless, due to the wide variety of foods that might be encountered by the analyst using enzyme assays, a check should be made of the extraction and enzyme reaction steps by standard addition of known amounts of analyte to the food and extract, and then calculate recovery of that standard. If the standard additions are fully recovered, this is a positive indication that the extraction is complete, that sample does not contain interfering substances that require removal prior to the enzymatic analysis, and that the reagents are good. In some cases, interfering substances are present but can be readily removed by precipitation or adsorption. For example, polyvinylpolypyrrolidone (PVPP) powder can be used to decolorize juices or red wines. With the advent of small syringe minicolumns (e.g., C18, silica, and ion exchange cartridges), it is also relatively easy and fast to attain group separations to remove interfering substances from a sample extract.

26.3.1.2 Total Change/Endpoint Methods

While substrate concentrations can be determined in rate assays when the reaction is first order with respect to substrate concentration ($[S] < <K_M$), substrate concentration can also be determined by the total change or endpoint method. In this method, the enzyme-catalyzed reaction is allowed to go to completion so that concentration of product, which is measured, is directly related to substrate. An example of such a system is the measurement of glucose using glucose oxidase and peroxidase, described in Sect. 26.3.1.3.2.

In some cases, an equilibrium is established in an end point method in which there is a significant amount of substrate remaining in equilibrium with product. In these cases, the equilibrium can be altered. For example, in cases in which a proton-yielding reaction is used, alkaline conditions (increase in pH) can be used. Trapping agents can also be used, in which product is effectively removed from the reaction, and by mass action the reaction goes to completion. Examples include the trapping of ketones and aldehydes by hydrazine. In this way, the product is continually removed and the reaction is pulled to completion. The equilibrium also can be displaced by increasing cofactor or coenzyme concentration.

Another means of driving a reaction to completion is a regenerating system [8]. For example, in the measurement of glutamate, with the aid of glutamate dehydrogenase, the following can be done:

$$\text{glutamate} + NAD^+ + H_2O \xrightarrow{\text{glutamatedehydrogenase}} 2\ \text{oxoglutarate} + NADH + NH_4^+ \quad (26.21)$$

$$\text{pyruvate} + NADH + H^+ \xrightarrow{\text{lactate dehydrogenase}} NAD^+ + \text{lactate} \quad (26.22)$$

In this system, NADH is recycled to NAD^+ via lactate dehydrogenase until all the glutamate to be measured is consumed. The reaction is stopped by heating to denature the enzymes present, a second aliquot of glutamate dehydrogenase and NADH is added, and the α-ketoglutarate (equivalent to the original glutamate) measured via decrease in absorbance at 340 nm. An example in which the same equilibrium is displaced in the measurement of glutamate is as follows:

$$\text{glutamate} + NAD^+ + H_20 \xleftrightarrow{\text{glutamate dehydrogenase}} \alpha - \text{ketoglutarate} + NADH + NH_4^+ \quad (26.23)$$

$$NADH + INT \xrightarrow{\text{diaphorase}} NAD + \cdots + \text{formazan} \quad (26.24)$$

Iodonitrotetrazolium chloride (INT) is a trapping reagent for the NADH product of the glutamate dehydrogenase-catalyzed reaction. The formazan formed is measurable colorimetrically at 492 nm.

26.3.1.3 Specific Applications

26.3.1.3.1 Measurement of Sulfite

Sulfite is a food additive that can be measured by several techniques, including titration, distillation followed by titration, liquid chromatography, and an enzymatic method (Chap. 33, Sect. 33.8.2). In the enzymatic method, sulfite is specifically oxidized to sulfate by the commercially available enzyme sulfite oxidase (SO):

$$SO_3^{2-} + O_2 + H_2O \xrightarrow{SO} SO_4^{2-} + H_2O_2 \quad (26.25)$$

The H_2O_2 product can be measured by several methods including use of the enzyme NADH-peroxidase:

$$H_2O_2 + NADH + H^+ \xrightarrow{\text{NADH-peroxidase}} 2H_2O + NAD^+ \quad (26.26)$$

The amount of sulfite in the system is equal to the NADH oxidized, which is determined by decrease in absorbance at 340 nm. Ascorbic acid can interfere with the assay but can be removed by using ascorbic acid oxidase [9].

26.3.1.3.2 Colorimetric Determination of Glucose

The combination of the enzymes glucose oxidase and peroxidase can be used to specifically measure glucose in a food system [10] (see also Chap. 19, Sect. 19.4.2.3.3). Glucose is preferentially oxidized by glucose oxidase to produce gluconolactone and hydrogen peroxide. The hydrogen peroxide plus o-dianisidine in the presence of peroxidase produces a yellow color that absorbs at 420 nm (Eqs. 26.17 and 26.18). This assay is normally carried out as an endpoint assay, and there is stoichiometry between the color formed and the amount of glucose in the extract, which is established with a standard curve. Because glucose oxidase is quite specific for glucose, it is a useful tool in determining the amount of glucose in the presence of other reducing sugars:

$$\beta\text{-D-glucose} + O_2 \xrightarrow{\text{glucose oxidase}} \alpha\text{-gluconolactone} + H_2O_2 \quad (26.27)$$

$$H_2O_2 + \text{o-diansidine} \xrightarrow{\text{peroxidase}} H_2O + \text{oxidized dye (colored)} \quad (26.28)$$

26.3.1.3.3 Starch/Dextrin Content

Starch and dextrins can be determined by enzymatic hydrolysis using amyloglucosidase, an enzyme that cleaves α-1,4 and α-1,6 bonds of starch, glycogen, and dextrins, liberating glucose (see Chap. 19, Sect. 19.5.1.1). The glucose formed can be subsequently determined enzymatically. Glucose can be determined by the previously described colorimetric method. An alternative method of measuring glucose is by coupling hexokinase (HK) and glucose-6-phosphate dehydrogenase (G6PDH) reactions:

$$\text{glucose} + \text{ATP} \xrightarrow{\text{HK}} \text{glucose-6-phosphate} + \text{ADP} \quad (26.29)$$

$$\text{glucose-6-phosphate} + \text{NADP}^+ \xrightarrow{\text{G6PDH}} \text{6-phosphogluconate} + \text{NADPH} + H^+ \quad (26.30)$$

The amount of NADPH formed is measured by absorbance at 340 nm and is a stoichiometric measure of the glucose originating in the dextrin or starch hydrolyzed by amyloglucosidase. Note that HK catalyzes the phosphorylation of fructose as well as glucose. The determination of glucose is specific because of the specificity of the second reaction, catalyzed by G6PDH, in which glucose-6-phosphate is the substrate. Note as well that hydrolysis methods do not allow determining the degree of polymerization of the starting molecule.

This assay sequence can be used to detect the dextrins of corn syrup used to sweeten a fruit juice product. A second assay would be needed, however, without treatment with amyloglucosidase to account for the glucose in the product. The glucose determined in that assay would be subtracted from the result of the assay in which amyloglucosidase is used.

The same HK-G6PDH sequence used to measure glucose can also be used to measure other carbohydrates in foods. For example, lactose and sucrose can be determined via specific hydrolysis of these disaccharides by β-galactosidase and invertase, respectively, followed by the use of the earlier described HK-G6PDH sequence.

26.3.1.3.4 Determination of D-Malic Acid in Apple Juice

Two stereoisomeric forms of malic acid exist. L-Malic acid occurs naturally, while the D form is normally not found in nature. Synthetically produced malic acid is a mixture of these two isomers. Consequently, synthetic malic acid can be detected by a determination of D-malic acid. One means of detecting the malic acid is through the use of the enzyme decarboxylating D-malate dehydrogenase (DMD) [11]. DMD catalyzes the conversion of D-malic acid as follows:

$$\text{D malic acid} + \text{NAD}^+ \xrightarrow{\text{DMD}} \text{pyruvate} + CO_2 + \text{NADH} + H^+ \quad (26.31)$$

The reaction can be followed by the measurement of NADH photometrically. Because CO_2 is a product of this reaction and escapes, the equilibrium of the reaction lies to the right and the process is irreversible. This assay is of value because the addition of synthetic D–/L-malic acid can be used to illegally increase the acid content of apple juice and apple juice products.

26.3.2 Enzyme Activity Assays

Like the section above on substrate assays, this section on enzyme assays is not comprehensive, but rather is intended to give select assays important in the food industry. Table 26.3 summarizes the enzyme activity assays covered.

26.3.2.1 Peroxidase Activity

Peroxidase is found in most plant materials and is reasonably stable to heat. A heat treatment that will destroy all peroxidase activity in a plant material is usually considered to be more than adequate to destroy other enzymes and most microbes present. In vegetable processing, therefore, the adequacy of the blanching process can be monitored by following the disappearance of peroxidase activity [12]. Peroxidase catalyzes the oxidation of guaiacol (colorless) in the presence of hydrogen peroxide to form tetraguaiacol (yellow brown) and water (Eq. 26.32). Tetraguaiacol has an absorbance maximum around 450 nm. Increase in absorbance at 450 nm can be used to determine the activity of peroxidase in the reaction mixture:

$$H_2O_2 + \text{guaiacol} \xrightarrow{\text{peroxidase}} \underset{\text{(colored)}}{\text{tetraguaiacol}} + H_2O \quad (26.32)$$

Table 26.3 Summary of enzyme activity analysis examples

Enzyme of interest	*Substrate*	*Compound measured*	*Measured property*	*Uses/comments*
Peroxidase	Guaiacol	Tetraguaiacol	Dynamic increase in A_{450}	Used to assess the efficacy of blanching
Lipoxygenase	Linoleic acid	Dissolved O_2	Dynamic decrease in $[O_2]$, amperometric measurement	Used to assess the efficacy of blanching. Quantitative and does not require clear solutions
Alkaline phosphatase	Disodium phenyl phosphate	Indophenol	Endpoint measurement of A_{650}	Used to determine the adequacy of dairy products pasteurization. Coupled reaction that requires physical separation of indophenol
α-Amylase	β-Limit dextrin (produced in excess of β-amylase)	Starch-iodine color	Time to match color in a color standard disk after	Used to determine degree of sprouting in wheat flours as well as the activity of enzyme formulations added to flours
	Starch		Time for a plunger to fall after heating starch + water (falling number) or measurement of viscosity after ramping the temperature of a starch + water mix	
Rennet	Azocasein (casein to which a dye has been covalently bound)	Fragments of azocasein	Endpoint measurement of A_{345}	To measure the proteolytic activity of rennet by separation of fragments of azocasein that do not precipitate upon addition of trichloroacetic acid
Pectinmethylesterase	High methoxyl pectin	H^+	Time to return to initial pH (~7.7) after addition of 0.1 mL of 0.05 *N* NaOH solution	To determine residual PME activity in citrus juices after pasteurization. Also relevant in other fruit products

26.3.2.2 Lipoxygenase

Lipoxygenase may be a more appropriate enzyme to measure the adequacy of blanching of vegetables than peroxidase [13]. Lipoxygenase refers to a group of enzymes that catalyzes the oxidation by molecular oxygen of fatty acids containing a *cis*, *cis*, 1,4-pentadiene system producing conjugated hydroperoxide derivatives:

$$\left(-\mathrm{CH}=\mathrm{CH}\ \ \mathrm{CH_2}\ \ \mathrm{CH}=\mathrm{CH}-\right)+\mathrm{O_2} \xrightarrow{\text{lipoxygenase}} \left(-\underset{\text{(conjugated)}}{\mathrm{COOH}}-\mathrm{CH}=\mathrm{CH}-\mathrm{CH}=\mathrm{CH}-\right) \quad (26.33)$$

A variety of methods can be used to measure lipoxygenase activity in plant extracts. The reaction can be followed by measuring loss of fatty acid substrate, oxygen uptake, occurrence of the conjugated diene at 234 nm, or the oxidation of a cosubstrate such as carotene [14]. All these methods have been used, and each has its advantages. The oxygen electrode method is widely used and replaces the more cumbersome manometric method. The electrode method is rapid and sensitive and gives continuous recording. It is normally the method of choice for crude extracts, but secondary reactions involving oxidation must be corrected for or eliminated. Zhang et al. [15] have reported the adaptation of the O_2 electrode method to the assay of lipoxygenase in green bean homogenates without extraction. The Clark electrode can be used for amperometric measurement of the oxygen. More recently, oxygen sensors based on fiber optics and fluorescence quenching have been made commercially available and have the advantage of not needing the maintenance of Clark electrodes. The formation of conjugated diene fatty acids with a chromophore at 234 nm can also be followed continuously. However, optically clear mixtures are necessary. Bleaching of carotenoids has also been used as a measure of lipoxygenase activity. However, the stoichiometry of this method is uncertain, and all lipoxygenases do not have equal carotenoid bleaching activity. Williams et al. [13] have developed a semiquantitative spot test assay for lipoxygenase in which I^- is oxidized to I_2 in the presence of the linoleic acid hydroperoxide product and the I_2 detected as an iodine starch complex.

26.3.2.3 Phosphatase Assay

Alkaline phosphatase is a relatively heat-stable enzyme found in raw milk. The thermal stability of alkaline phosphatase in milk is greater than the nonspore-forming microbial pathogens present in milk. The phosphatase assay has been applied to dairy products to determine whether pasteurization has been done properly and to detect the addition of raw milk to pasteurized milk. A common phosphatase test is based on the phosphatase-catalyzed hydrolysis of disodium phenyl phosphate liberating phenol [16]. The phenol product is measured colorimetrically after reaction with CQC (2,6-dichloroquinonechloroimide) to form a blue indophenol. The indophenol is extracted into *n*-butanol and measured at 650 nm. This is an example of a physical separation of product to allow the ready measurement of an enzyme reaction. More recently, a rapid fluorometric assay was developed and commercialized for measurement of alkaline phosphatase in which the rate of fluorophore production can be monitored directly without butanol extraction used to measure indophenol when phenylphosphate is used as substrate [17]. The fluorometric assay was shown to give greater repeatability compared to the standard assay in which phenylphosphate is used as substrate and was capable of detecting 0.05% raw milk in a pasteurized milk sample. Similar chemistry has been applied to the measurement of acid phosphatase activity in meats as a means of ensuring adequate cooking via correlation of enzyme activity to end-point temperature [18].

26.3.2.4 α-Amylase Activity

Amylase activity in malt is a critical quality parameter. The amylase activity in malt is often referred to as diastatic power and refers to the production of reducing substances by the action of α- and β-amylases on starch. The measurement of diastatic power involves digestion of soluble starch with a malt infusion (extract) and following increase in reducing substances by measuring reduction of Fehling's solution or ferricyanide. Specifically measuring α-amylase activity (often referred to as **dextrinizing activity**) in malt is more complicated and is based on using a limit dextrin as substrate. **Limit dextrin** is prepared by the action of β-amylase (free of α-amylase activity) on soluble starch. The β-amylase clips maltose units off the nonreducing end of the starch molecule until an α-1,6- branch point is encountered. The resulting product is a β-limit dextrin that serves as the substrate for the endo cleaving α-amylase. A malt infusion is added to the previously prepared limit dextrin substrate and aliquots removed periodically to a solution of dilute iodine. The α-amylase activity is measured by changed color of the starch-iodine complex in the presence of excess β-amylase used to prepare the limit dextrin. The color is compared to a colored disc on a comparator. This is continued until the color is matched to a color on a comparator. The time to reach that color is **dextrinizing time** and is a measure of α-amylase activity, a shorter time representing a more active preparation.

Because α-amylase is an endoenzyme, when it acts on a starch paste, the viscosity of the paste is dramatically reduced, greatly influencing flour quality. Consequently, α-amylase activity is of great importance in whole wheat. Wheat normally has small amounts of α-amylase activity, but when wetted in the field, preharvest sprouting (pregermination) can occur in wheat, with a dramatic increase in α-amylase activity. Preharvest sprouting cannot be easily detected visually, so measurement of α-amylase activity can be used as a sensitive estimate of preharvest sprouting. The **falling number method** is a procedure in which ground wheat is heated with water to form a paste, and the time it takes for a plunger to fall through the paste is recorded [19, 20]. Accordingly, the time in seconds (the falling number) is inversely related to the α-amylase activity and the degree of preharvest sprouting. A rapid visco analyzer (RVA) can be used as a refined version of this method. Samples are placed in the RVA where the temperature is ramped and the decrease in viscosity over time is monitored. The viscosity after 20 min is a good indication of α-amylase activity. This method of measuring enzyme activity is a good example of using change in physical property of a substrate as a means of estimation of enzyme activity.

26.3.2.5 Rennet Activity

Rennet, an extract of bovine stomach, is used as a coagulating agent in cheese manufacture. Most rennet activity tests are based on noting the ability of a preparation to coagulate milk. For example, 12% nonfat dry milk is dispersed in a 10 m*M* calcium chloride solution and warmed to 35 °C. An aliquot of the rennet preparation is added and the time of milk clotting observed visually. The activity of the preparation is calculated in relationship to a standard rennet. As opposed to coagulation ability, rennet preparations can also be evaluated for proteolytic activity by measuring the release of a dye from azocasein (casein to which a dye has been covalently attached). In this assay, the rennet preparation is incubated with 1% azocasein. After the reaction period, the reaction is stopped by addition of trichloroacetic acid. The trichloroacetic acid precipitates the protein that is not hydrolyzed. The small fragments of colored azocasein produced by the hydrolysis of the rennet are left in solution and absorbance read at 345 nm [21, 22]. This assay is based on the increase in solubility of a substrate upon cleavage by an enzyme.

26.3.2.6 Pectinmethylesterease Activity

Pectinmethylesterase (PME) is present in citrus and other fruits. It is particularly important to the citrus juice industry because as it cleaves the methoxyl groups of pectin, producing methanol and a proton. It causes the precipitation of the juice cloud which is undesirable. In fact, the extent of thermal treatment of citrus juices is not dictated by microbial kill but by PME inactivation. PME activity is assayed by titration with NaOH at a pH near 7.7 by determining the time to return to initial pH (~7.7) after addition of 0.1 mL of 0.05 N NaOH to a high methoxyl pectin solution to which the enzyme is added. Due to the release of protons, the pH of the pectin solution decreases and, as a result, the activity of the enzyme changes as well. For that reason, the deviation from pH ~7.7 should be small. The activity units are in milliequivalents of NaOH per unit time, per mass of sample [23].

26.3.3 Biosensors/Immobilized Enzymes

The use of immobilized enzymes as analytical tools is currently receiving increased attention. An immobilized enzyme in concert with a sensing device is an example of a biosensor. A biosensor is a device comprised of a biological sensing element (e.g., enzyme and antibody) coupled to a suitable transducer (e.g., optical and electrochemical). The most widely used enzyme electrode is the glucose electrode in which glucose oxidase is combined with an oxygen electrode or with electrochemical detection of hydrogen peroxide to determine glucose concentration [24–27]. When the electrode is put into a glucose solution, the glucose diffuses into the membrane where it is converted to gluconolactone by glucose oxidase with the uptake of oxygen and production of hydrogen peroxide. The oxygen uptake and the production of hydrogen peroxide are measures of the glucose concentration. Similar systems have been commercialized in which lactate, ethanol, sucrose, lactose, and glutamate can be measured. In the case of some of these sensors, multiple enzymes are immobilized. For example, for sucrose analysis, invertase, mutarotase, and glucose oxidase are immobilized on the same membrane. A large number of other enzyme electrodes have been reported. For example, a glycerol sensor, in which glycerol dehydrogenase was immobilized, has been developed for the determination of glycerol in wine [28]. NADH produced by the enzyme was monitored with a platinum electrode. Currently, the main limitation of enzyme biosensors is their short operational life due to the lack of stability of most of the oxidases used in their construction [29]. Table 26.3 summarizes substrate assays covered in this section.

26.3.4 Structural Analysis of Enzymes

Tremendous progress has been made over the last two decades in the structural characterization of enzymes. The number of enzymes whose 3-D structures have been characterized by crystallography or nuclear magnetic resonance and reported in the protein data bank (https://www.rcsb.org/) has increased exponentially. Furthermore, the use of small-angle X-ray scattering and small-angle neutron scattering of enzymes in solution through Guinier analysis is increasing our fundamental understanding of protein stability [30]. The recent success of AlphaFold to predict protein folding from amino acid sequences [31] is likely to accelerate the production of enzyme variants useful to food analysis. While currently it is not possible to quantitatively calculate the stability of an enzyme based on its structure, the food technologist must stay abreast of these advanced that are likely to become of practical relevance within the next two decades.

26.4 SUMMARY

Enzymes, due to their specificity and sensitivity, are valuable analytical devices for quantitating compounds that are enzyme substrates, activators, or inhibitors. In enzyme-catalyzed reactions, the enzyme and substrate are mixed under specific conditions (pH, temperature, ionic strength, substrate concentration, and enzyme concentrations). Changes in these conditions can affect the reaction rate of the enzyme and thereby the outcome of the assay. The enzymatic reaction is followed by measuring either the amount of product generated or the disappearance of the substrate. Applications for enzyme analyses will increase as a greater number of enzymes are purified and become commercially available. In some cases, gene amplification techniques will make enzymes available that are not naturally found in great enough amounts to be used analytically. The measurement of enzyme activity is useful in assessing food quality and as an indication of the adequacy of heat processes such as pasteurization and blanching. In the future, as in-line process control (to maximize efficiencies and drive quality developments) in the food industry becomes more important, immobilized enzyme sensors, along with microprocessors, will likely play a prominent role.

26.5 STUDY QUESTIONS

1. The Michaelis-Menten equation mathematically defines the hyperbolic nature of a plot relating reaction velocity to substrate concentration for an enzyme-mediated reaction. The reciprocal of this equation gives the Lineweaver-Burk formula and a straight-line relationship as shown below.

$$\frac{1}{\nu_o} = \frac{K_M}{V_{max}} \frac{1}{[S]} + \frac{1}{V_{max}}$$

(a) Define what ν_o, K_M, V_{max}, and $[S]$ refer to in the Lineweaver-Burk formula.

ν_o,
K_M.
V_{max}.
$[S]$

(b) Based on the components of the Lineweaver-Burk formula, label the y-axis, x-axis, slope, and y-intercept on the plot.

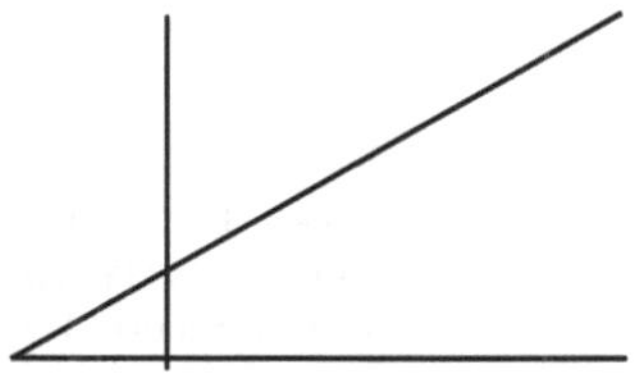

(c) What factors that control or influence the rate of enzyme reactions affect K_M and

V_{max}?

K_M.

V_{max}.

2. Explain, on a chemical basis, why extremes of pH and temperature can reduce the rate of enzyme-catalyzed reactions.
3. Differentiate among competitive, noncompetitive, and uncompetitive enzyme inhibitors.
4. You believe that the food product you are working with contains a specific enzyme inhibitor. Explain how you would quantitate the amount of enzyme inhibitor (I) present in an extract of the food. The inhibitor (I) in question can be purchased commercially in a purified form from a chemical company. The inhibitor is known to inhibit the specific enzyme E, which reacts with the substrate S to generate product P, which can be quantitated spectrophotometrically.
5. What methods can be used to quantitate enzyme activity in enzyme-catalyzed reactions?
6. What is a coupled reaction, and what are the concerns in using coupled reactions to measure enzyme activity? Give a specific example of a coupled reaction used to measure enzyme activity.
7. Explain how D-malic acid can be quantitated by an enzymatic method to test for adulteration of apple juice.
8. Why is the enzyme peroxidase often quantitated in processing vegetables?
9. Explain the purpose of testing for phosphatase activity in the dairy industry and explain why it can be used in that way.
10. The falling number value often is one of the quality control checks in processing cereal-based products. What is the falling number test, and what information does it provide? What other tests could be used to assay this quality factor?
11. Explain how glucose can be quantitated using a specific immobilized enzyme.

Acknowledgments The author of this chapter thanks Joseph R. Powers (Professor Emeritus, School of Food Science, Washington State University, Pullman, WA, USA) for his contributions to the chapter in early editions of the book.

REFERENCES

1. Cavalieri RP, Reyes-De-Corcuera JI (2005) Kinetics of chemical reactions in foods, In Barbosa-Canovas GV (ed) Food Engineering. Encyclopedia of Life Support Systems. UNESCO Publishing, Paris, pp. 215–239
2. Basak S, Chakraborty S (2023) Effect of high-pressure processing on the kinetic parameters of enzymes, In Leite Júnior BR, Artigiani Tribst A. (eds) Effect of High-Pressure Technologies on Enzymes. Academic Press, London, pp. 77–107
3. Eisemenger MJ, Reyes-De-Corcuera JI (2009) Enzyme Microb Technol 45:331
4. Powers JR, Whitaker JR (1977) Effect of several experimental parameters on combination of red kidney bean (*Phaseolus vulgaris*) α-amylase inhibitor with porcine pancreatic α-amylase. J Food Biochem 1:239
5. Bernfeld P (1955) Amylases, α and β. Methods Enzymol 1:149
6. Whitaker JR (1985) Analytical uses of enzymes. In: Gruenwedel D, Whitaker JR (eds) Food Analysis. Principles and Techniques, vol 3, Biological Techniques, Marcel Dekker, New York, pp 297–377
7. Boehringer-Mannheim (1987) Methods of Biochemical Analysis and Food Analysis. Boehringer Mannheim Gmb H Mannheim, W. Germany
8. Bergmeyer HU (1983) Methods of Enzymatic Analysis. Academic Press, New York
9. Beutler H (1984) A new enzymatic method for determination of sulphite in food. Food Chem 15:157
10. Raabo E, Terkildsen TC (1960) On the enzyme determination of blood glucose. Scand J Clin Lab Inves 12:402
11. Beutler H, Wurst B (1990) A new method for the enzymatic determination of D-malic acid in foodstuffs. Part I: Principles of the Enzymatic Reaction. Deutsche Lebensmittel-Rundschau 86:341
12. USDA (1975) Enzyme inactivation tests (frozen vegetables). Technical inspection procedures for the use of USDA inspectors. Agricultural Marketing Service, U.S. Department of Agriculture, Washington, DC
13. Williams DC, Lim MH, Chen AO, Pangborn RM, Whitaker JR (1986) Blanching of vegetables for freezing — Which indicator enzyme to use. Food Technol 40(6): 130
14. Surrey K (1964) Spectrophotometric method for determination of lipoxidase activity. Plant Physiology 39:65
15. Zhang Q, Cavalieri RP, Powers JR, Wu J (1991) Measurement of lipoxygenase activity in homogenized green bean tissue. J Food Sci 56:719
16. Murthy GK, Kleyn DH, Richardson T, Rocco RM (1992) Phosphatase methods. In: Richardson GH (ed) Standard methods for the examination of dairy products, 16th edn. American Public Health Association, Washington, DC, p 413
17. Rocco R (1990) Fluorometric determination of alkaline phosphatase in fluid dairy products: Collaborative study. J Assoc Off Anal Chem 73:842
18. Davis CE (1998) Fluorometric determination of acid phosphatase in cooked, boneless, nonbreaded broiler breast and thigh meat. J AOAC Int 81:887
19. Delwiche SR, Vinyard BT, Bettge AD (2015) Repeatability precision of the falling number procedure under standard and modified methodologies. Cereal Chem. 92 (2):177
20. Cereals & Grains Association (formerly AACC International) (2010) AACC Approved methods of analysis, 11th edn. (On-line), Cereals and Grains Assoc., St. Paul, MN
21. Christen, GL, and Marshall, R.T. 1984. Selected properties of lipase and protease of *Pseudomonas fluorescens* 27 produced in 4 media. Journal of Dairy Science, 67:1980
22. Kim, SM, Zayas, JF. 1991. Comparative quality characteristics of chymosin extracts obtained by ultrasound treatment. J Food Sci 56:406
23. Kimball (1996) Citrus Processing, a Complete Guide. 2nd ed. Aspen Publishers, Inc. Gaithersburg, MD, p 259–263
24. Shimizu Y, Morita K (1990) Microhole assay electrode as a glucose sensor. Anal Chem 62:1498
25. Reyes-De-Corcuera JI, Cavalieri RP (2003) Biosensors. In: Encyclopedia of agricultural, food, and biological engineering, p 119–123

26. Guilbault GG, Lubrano GJ (1972) Enzyme electrode for glucose. Anal Chim Acta 60:254
27. Borisov SM, Wolbeis OS (2009) Optical biosensors. Chem Rev 108:423
28. Matsumoto K (1990) Simultaneous determination of multicomponent in food by amperometric FIA with immobilized enzyme reactions in a parallel configuration. In: Schmid RD (ed) Flow injection analysis (FIA) based on enzymes or antibodies, GBF monographs, vol 14, VCH Publishers, New York, pp. 193–204
29. Reyes De Corcuera JI Olstad HE, Garcia-Torres R (2018) Stability and stabilization of enzyme biosensors: The key to successful application and commercialization. Ann Rev Food Sci Technol 9:293.
30. Chaudhuri, BN (2015) Emerging applications of small angle solution scattering in structural biology. Prot Sci. V24, 267–276
31. Jumper, J., Evans, R., Pritzel, A. et al. (2021) Highly accurate protein structure prediction with AlphaFold. Nature 596, 583–589. https://doi.org/10.1038/s41586-021-03819-2

Immunoassays

27

Qinchun Rao

Contents

27.1 INTRODUCTION

Immunochemistry is the study of biochemical and molecular aspects of immunology, especially the nature of antibodies, antigens, and their interactions [1]. One of the most useful analytical developments associated with this science is immunoassay. Originally, immunoassays were developed in medical settings to facilitate the study of immunology. Immunoassays are now finding widespread applications outside the clinical field because they are appropriate for a wide range of analytes ranging from proteins to small organic molecules. In the food analysis area, immunoassays are widely used for chemical residue analysis, identification of bacteria and viruses, and detection of proteins in food and agricultural products (Table 27.1). Protein detection is important for determining allergens and meat species content, identifying seafood species, and detecting genetically modified plant tissues. While immunoassays of all formats are too numerous to cover completely in this chapter, there are several assay formats that have become standard for food analysis because of their specificity, sensitivity, and simplicity (Fig. 27.1).

Q. Rao (✉)
Department of Nutrition and Integrative Physiology, Florida State University, Tallahassee, FL, USA
e-mail: qrao@fsu.edu

Table 27.1 Immunoassay application in food analysis

Applications	Format	Example
Allergens	Sandwich ELISA	[2]
Misbranded food components	Sandwich ELISA	[3]
Pesticides	Lateral flow assay	[4]
Antibiotics	Sandwich ELISA	[5]
Pathogens	Sandwich ELISA	[6]
Toxins	Dot blot	[7]

B. P. Ismail, S. S. Nielsen (eds.), *Nielsen's Food Analysis*, Food Science Text Series,
https://doi.org/10.1007/978-3-031-50643-7_27

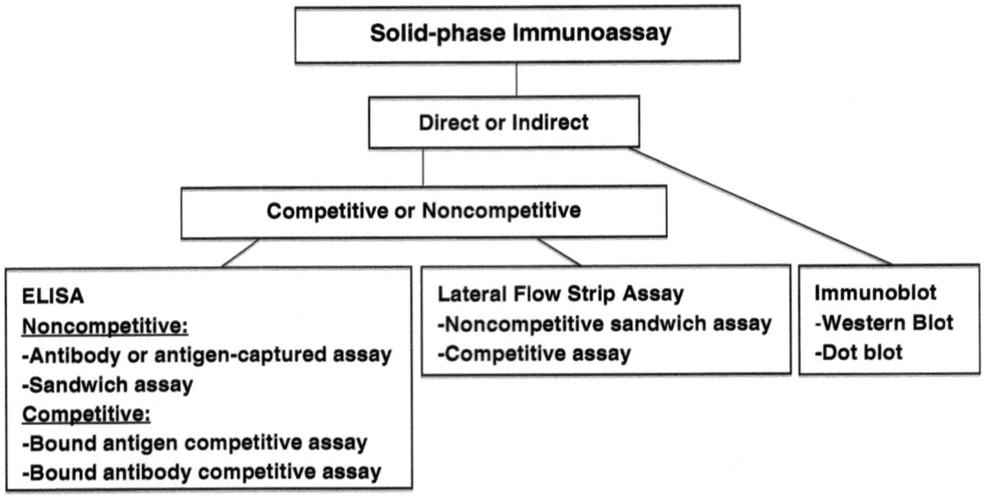

Fig. 27.1 Commonly used immunoassays for food analysis

27.1.1 Definitions

Immunoassay is a ligand-binding assay that uses a specific antigen or antibody, capable of binding to the analyte, to identify and quantify substances [1]. To fully understand immunoassays, some of these terms need to be defined. The two essential elements of any immunoassay are antigens and antibodies. In an immunoassay, antigens and antibodies are used either as the target or capture molecules. In other words, a particular antigen can be used to capture its specific antibody, or a specific antibody can be used to trap the target antigen in a sample. An **antigen** is a substance that stimulates the immune system to produce a set of specific antibodies and that combines with the antibody through a specific binding site, i.e., **epitope**, which is any part of a molecule that acts as an antigenic determinant [1]. An **antibody** is a protein (**immunoglobulin**) produced by the activated **B cells** in the immune system of an organism in response to exposure to a foreign molecule (antigen) [1]. The production of antibodies relies on the in vivo humoral response of animals. Generally, a molecule must be greater than 5000 Dalton, abbreviated as Da (unit of molecular weight), to be perceived as an antigen by a mammalian immune system. During antibody development using animal immunization technique, mainly because almost all proteins are large molecules, they have the ability to induce antibody formation in the body of host animals. However, many of the molecules analyzed in food are not as large as proteins but rather are small molecules such as toxins, antibiotics, and pesticides. When animals are injected with small molecules, they do not develop antibodies against them. To induce specific antibodies to recognize and bind to the small target molecule, one of the solutions is to link the small molecule covalently, or some appropriate derivative of the small molecule, to a **carrier molecule**. The most common molecules used as carriers are proteins that are fairly soluble for simplicity in chemical linking and are foreign to the host animal to properly stimulate an immune response. Typical carrier molecules include albumin proteins from different species, such as bovine serum albumin and hemocyanins obtained from crustaceans. The small molecule that must be linked to a large carrier protein before it can be used as an immunogen to induce antibodies is called a **hapten**. The carrier protein-linked hapten is called a **conjugate antigen**. Haptens react specifically with the appropriate antibodies but are not immunogenic. It should be noted that when a conjugate antigen is used for immunization of an animal, its immune system is stimulated to produce antibodies that bind to not only the hapten but also the carrier protein.

Antibody molecules are roughly Y-shaped molecules consisting of paired **heavy** and **light polypeptide chains** linked by inter- and intra-disulfide bonds. Based on their heavy chain structure, antibody molecules can be classified into five different immunoglobulins (Ig): IgA, IgE, IgG, IgM, and IgD. IgA is mainly found in mucous secretions and plays an important role in mucosal immunity. Animal blood contains trace amounts of IgD and IgE. The unique function of IgD is still unclear. IgE is responsible for allergic response in immediate hypersensitivity reactions. IgM is a huge molecule and can be regarded as a precursor of IgG. Among these five classes of antibodies, IgG has the highest concentration in blood and is the most important class used in food immunoassay. Since the antibody and antigen are central to any immunoassay, it is helpful to better understand the basic structure of the antibody and how it binds to the antigen. Figure 27.2 is an idealized diagram of an IgG antibody molecule. The IgG antibody molecule has four polypeptide chains, forming a flexible Y-shaped structure. Two of the heavy polypeptide chains are identical and roughly twice as large as the other two identical light polypeptide chains. Overall, an IgG antibody is a very large protein of approximately 150,000 Da.

Antigen is bound by two identical binding sites made up of the end portions (N terminals) of a heavy and light chain at the top of the Y-shaped IgG structure. These two fragments capable of binding with antigens are called **Fab** (**fragment antigen binding**). The third fragment with no antigen-binding capability is called **Fc** (**fragment crystallizable**). Different antibodies produced by different B cells can have many variations in amino acid sequences near the binding sites for both the heavy and light chains. This leads to a tremendous diversity of binding sites for different antibodies. For example, a mouse has 10^7–10^8 different antibodies (and at least this number of different B cells), each with a unique binding site. The rest of the antibody (away from the binding site) is quite consistent, and small variations in this region on the heavy chains result in different antibody classes.

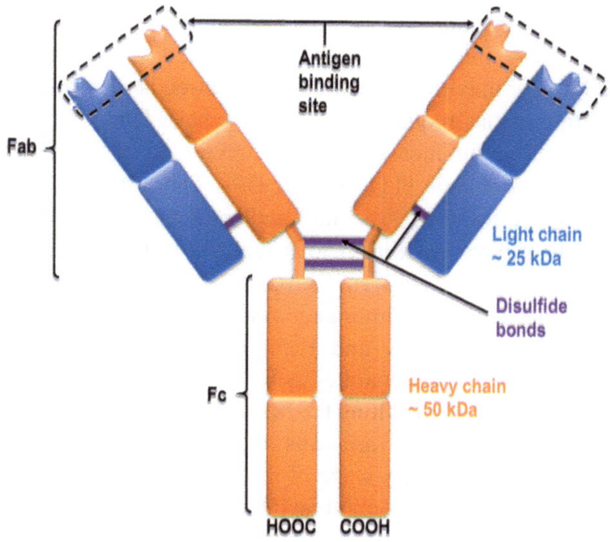

Fig. 27.2 Antibody (IgG) structure

27.1.2 Binding Between Antigen and Antibody

Antibodies can develop remarkably strong binding affinities for their antigens. These affinities are among the strongest noncovalent interactions known between molecules, including electrostatic interactions, hydrogen bonds, van der Waals forces, and hydrophobic and pi-cation interactions. The binding strength (affinity) between antibody and antigen is one of the most important factors that determine the sensitivity of an immunoassay. The antibody binds to the outside of the antigen molecule in a specific region. This specific region bound by a single antibody-binding site is known as an **epitope**. Two types of epitopes on an antigen can be formed. A **linear epitope** is formed by a continuous sequence of amino acid residues, and a **conformational epitope** is formed by noncontinuous amino acid sequences that are folded into close proximity from neighboring or overlapping peptide chains on the surface of the antigen (Fig. 27.3). If this three-dimensional conformation of the antigen is altered by some kind of environmental conditions, such as heating or pH changes, the conformational epitope will be destroyed, which means that the antigen cannot bind to the antibody.

27.1.3 Types of Antibodies

A major variable in an immunoassay is the type of antibody used. When antiserum from any immunized animal is used,

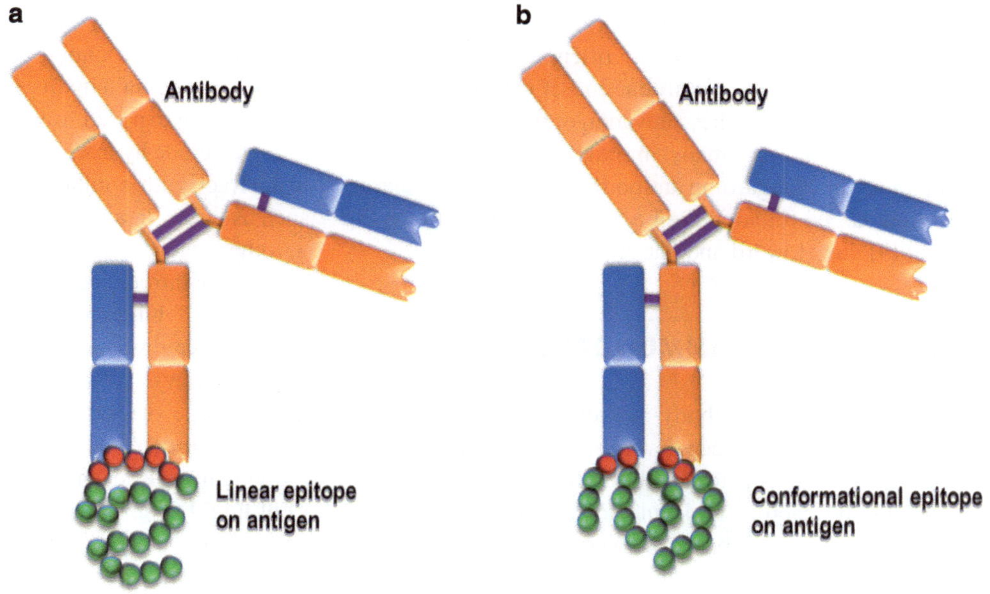

Fig. 27.3 Linear epitope (**a**) and conformation-dependent epitope (**b**)

there are many different antibodies that bind different epitopes on the same antigen. This collection of different antibodies is known as **polyclonal antibodies** (pAb). Although pAb is effective for many biological purposes, it has inherent disadvantages that relate to the heterogeneity of pAb. First, each batch of pAb is different from all other batches, even if these batches are raised in a genetically identical animal by using the identical preparation of immunogen with the same immunization protocol. Second, pAb is produced in limited volumes based on the longevity of the immunized animal, and thus it is impossible to use the identical pAb in the long run. Finally, each batch of pAb needs to be purified by affinity chromatography and standardized before use. This process is costly and time-consuming.

Scientists knew that individual B cells produced antibodies with only one binding site but were unable to culture B cells outside of the animal. However, in 1975, Köhler and Milstein [8] successfully fused cancer, or myeloma, cells with B cells. The new fused cells, or **hybridomas**, retained the properties of both parent cells. That is, they could be cultured, like cancer cells, and produce antibodies like the B cells. Hybridomas, thus, can be cloned and cultured individually to produce different antibodies with different epitopes. Antibodies produced with this procedure became known as **monoclonal antibodies** (mAb). mAbs produced from a single hybridoma are identical in every way and bind antigens with only one type of binding site; that is, a single epitope is bound; therefore, they can be used as standard reagents in immunoassays. Moreover, the hybridomas were "immortalized" by the procedure and, with proper care, could produce as many identical antibodies as required. It did not take the scientific community long to appreciate the tremendous advantages of these mAbs, so Köhler and Milstein were awarded the Nobel Prize in Physiology or Medicine for their work on this in 1984. While mAbs are initially much more expensive to produce, there is the possibility for limitless identical antibody production, often from non-animal sources such as large-scale production of the hybridomas in cell growth chambers. These advantages outweigh the initial development costs for many immunoassay manufacturers. Detail procedures regarding antibody production and characterization can be found in the books written by Howard and Bethell [9] and Howard and Kaser [10].

27.2 THEORY OF IMMUNOASSAYS

Based on the specific antibody-antigen affinity, various types of immunoassays have been developed to use either antibody as the capture molecule to search for the target antigen or the antigen as the capture molecule to trap the antibody in a complex sample. The basis of every immunoassay is the detection and measurement of the primary antibody-antigen reaction. In its simplest form, antibody capture of antigen can involve simple precipitation and be detected visually. Since all antibodies have at least two identical binding sites, they can cross-link epitopes from two identical antigens. If other antigen epitopes are further cross-linked by different antibodies, a large, insoluble network can form, which is seen as a precipitate. Immunoprecipitation techniques, including immunodiffusion and agglutination, formed the basis of the early development of immunoassay techniques and have been used widely for protein and cell identification using antisera. However, these methods only work for antigens with multiple epitopes.

To measure the quantity of soluble antibody-bound antigen molecules in a solution, all immunoassays require two things. First, there must be some method to separate or differentiate free antigens from bound antigens. Second, these antibody-bound antigens must be quantifiable at low concentrations for maximum sensitivity. Detection at very low concentrations has required very active labels. One of the first successful immunoassay procedures was developed by Yalow and Berson [11] in the 1950s. This procedure used radioactive iodine, I^{131}, a "hot" radioisotope with a half-life of only 8 days, as a label to reveal the primary antibody-antigen complex. This radioactive label allowed for the second requirement of immunoassays: quantification at low concentrations. Yalow and Berson used paper chromato-electrophoresis to separate their antibody-bound antigen from free antigen, fulfilling the first requirement of an immunoassay. With all the variations in separation and detection techniques in the early stage of immunoassay development; however, the radioactive iodine labeling remained, and these assays became known as **radioimmunoassays** (RIA). Rosalyn Yalow was awarded the Nobel Prize in Physiology or Medicine for the development of RIA in 1977.

One of the techniques for the separation of unbound from bound molecules in immunoassays involves immobilizing protein on a hydrophobic solid surface. Proteins have large regions that contain hydrophobic amino acid groups that prefer not to be exposed to water. These nonpolar hydrophobic groups include hydrocarbons and aromatic groups that prefer to interact with similar groups rather than a polar solvent such as water. In aqueous conditions, these regions will bind to other hydrophobic surfaces excluding water. Surfaces commonly introduced in immunoassays to take advantage of this type of binding include charcoal, nitrocellulose, and plastic. Plastic surface in many forms is commonly used for immunoassays. Among the most popular are **microtiter plates** made of plastics such as polystyrene or polyvinyl. These microtiter plates typically are formatted to contain 96 individual wells, each with a maximal capacity of about 300 μl of liquid (Fig. 27.4a). To differentiate the wells, the vertical rows are labeled A to H, and the columns are numbered 1 to 12. It is important to realize that proteins randomly bind to

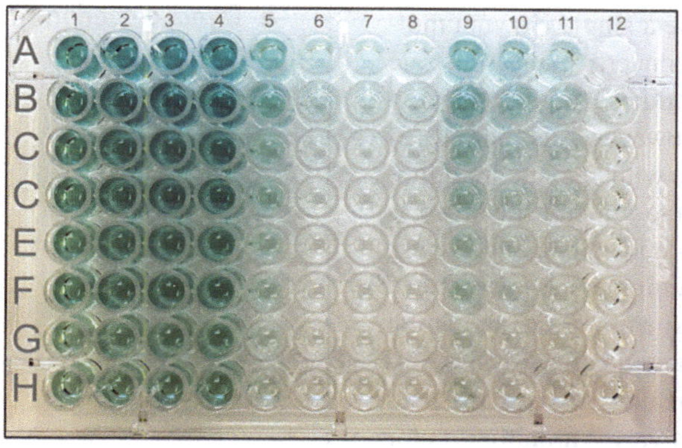

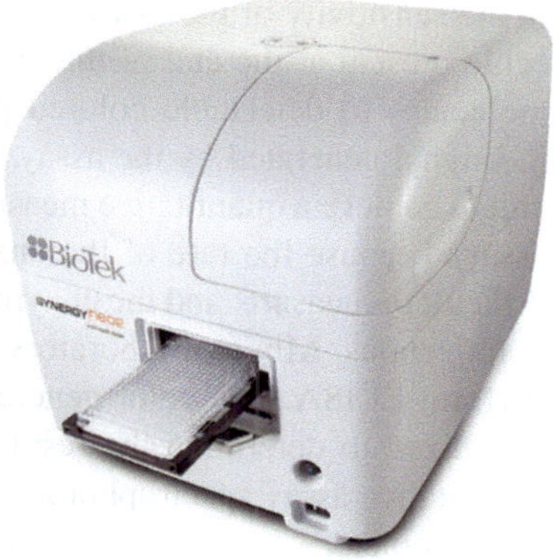

Fig. 27.4 Image of a 96-well plastic microtiter plate (**a**) and plate reader (**b**) used for ELISA

the bottom and sides of the wells in these plates through hydrophobic interactions. The hydrophobic interactions between proteins and these surfaces increase at lower temperatures (due to less molecular motion) and with increased ionic strength of the solvent (increasing solvent polarity). Other forms of solid surface commonly used in immunoassays include plastic vials, magnetic beads, and nitrocellulose or polyvinylidene difluoride (PVDF) membranes or strips.

27.3 SOLID-PHASE IMMUNOASSAYS

27.3.1 Overview

Every immunoassay technique developed is based on the selection of an **amplification method** that will improve the sensitivity of assays. Labeling of the detecting molecule, either antigen or antibody, is necessary to locate or quantify the target molecule in a solid-phase immunoassay that uses a solid surface for the separation of unbound from bound molecules. For example, having an easily detected label attached to the antibody allows the target antigen in a complex food sample matrix to be detected. A number of labeling compounds and materials have also been used, such as radioisotopes, fluorochromes, enzymes, biotin, and gold nanoparticles. These different label options become very useful for immunoassay development or in the use of antibodies for detection in different systems, such as the examination of tissue under a microscope or proteins separated using electrophoretic techniques followed by an immunoassay.

While RIAs worked well in the clinical field, they were confined to specially equipped laboratories because of the dangers associated with the use of radioactive material. Immunoassays did not develop for more general use, including field use, until enzyme labels were developed. Any immunoassay that uses an enzyme label to reveal the primary antibody-antigen binding is called an **enzyme immunoassay**. Pioneers in this development were Engvall and Perlmann [12], who developed a type of enzyme immunoassay in 1971 that they called an **enzyme-linked immunosorbent assay** or ELISA. ELISA involves the binding of a soluble antigen or antibody to a solid support (immunosorbent), typically in the form of a 96-well plastic microtiter plate. The bound and unbound molecules can be easily separated by a washing step of the plate. Microtiter plates of hydrophobic plastics are commonly used to immobilize proteins and separate unbound molecules. Similar immunoassays using solid supports other than plastic microtiter plates also have been developed and are commonly used. Examples are the **dot blot** assay, **Western blot** assay, and more recently developed lateral flow immunochromatographic assay, which use nitrocellulose or polyvinylidene difluoride (PVDF) membranes. These developments have expanded the use of immunoassays to a wider range of applications.

27.3.2 ELISA

27.3.2.1 Introduction

Because enzyme immunoassay, especially ELISA, has become the most popular immunoassay for food analysis applications, the general principle and protocols of various formats of ELISA are illustrated in subsequent sections using ELISA as examples.

The enzyme label used in an ELISA converts a colorless substrate to a colored soluble product in the solution, thus generating a detectable signal for the assay. The amount of target protein antigen present in the sample extract is determined from the intensity of color developed in the immunoassay. The ideal enzyme for an enzyme immunoassay is one that is stable, easily linked to antibodies or antigens, and rapidly catalyzes a noticeable change with a simple substrate. With the many enzymes available, two enzymes, **horseradish peroxidase** and **alkaline phosphatase,** are by far the most commonly used in immunoassays. Other enzymes used include β-galactosidase, glucose oxidase, and glucose-6-phosphate dehydrogenase. The use of an enzyme to catalyze a chemical reaction that generates a color signal contributes to the sensitivity of the assay because a single enzyme molecule present at the end of the test converts many substrate molecules to detectable colored products, thus amplifying the signal generated by the assay. However, it is more difficult to achieve a quantitative measure with enzyme immunoassays because the rate of the enzyme reaction involved is difficult to measure, and the enzyme-labeled reagents are not homogenous. Minimal laboratory equipment is required to perform ELISA. The color generated from an assay can be visualized to determine the result in a qualitative or semi-quantitative assay spectrophotometrically. The type of spectrophotometer used to quantitatively monitor color development caused by the enzyme action is called a **microplate reader** (Fig. 27.4b). Automated **microplate washer** also is available, although most assays can be washed manually.

All ELISA protocols include the following five general steps:

(1) *Coating* of antibody or antigen onto the wells of a microtiter plate (solid phase).
(2) *Blocking* the remaining uncoated surface on the solid phase with a blocking buffer containing a nonspecific protein such as bovine serum albumin (this is to minimize the nonspecific reactions and also protect the adsorbed antigen or antibody from surface denaturation).
(3) *Incubating* with different immunoassay reagents at a specified temperature and time.
(4) *Washing* the coated surface to separate free, unbound molecules from bound molecules.
(5) *Detecting* the color developed from the assay visually or spectrophotometrically.

Specific procedures vary with the different variations of ELISA. It is important to include both positive and negative controls in an assay along with any analyzed food sample because materials in the sample extract can vary widely, and these other components can have an effect on the competition for the antibody-binding site. This is to ensure that the immunoassay works well (positive control shows positive signal) and that there is no contamination or nonspecific reactions in the assay system (negative control shows negative signal).

27.3.2.2 Direct Versus Indirect Detection

All immunoassay signals can be detected directly or indirectly. In the **direct detection method**, the detecting molecules are purified and linked to the label to directly measure the amount of the antibody-antigen complex (Fig. 27.5a). Therefore, more purified immunoreagents need to be used for the label conjugation procedure. In contrast, the **indirect detection method** uses a commercially available intermediate reagent to link the capture molecule. Most often, a labeled **antispecies antibody** is used for an antibody-captured assay to indirectly measure the amount of antibody-antigen complex formed (Fig. 27.5b). Although an additional step is involved, indirect assays require fewer immunoreagents. In many cases, it could be more sensitive because more labeling molecules can be linked to the detection antigen or antibody for enhanced signal production. While the direct detection method is essential when an accurate quantification is required for the assay, the indirect detection method is used in most solid-phase immunoassays.

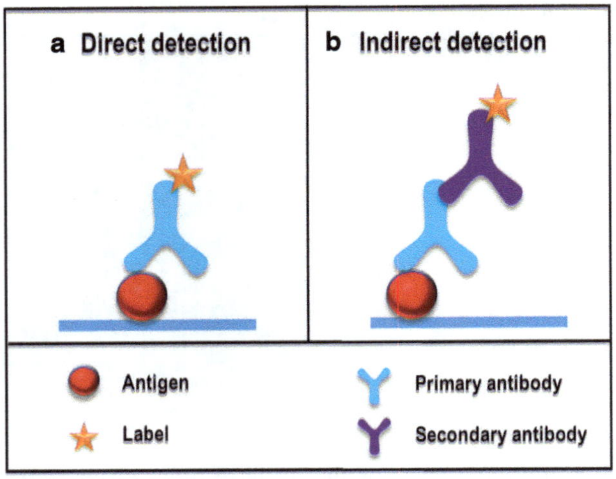

Fig. 27.5 Direct detection (**a**) and indirect (**b**) detection methods

A simple form of indirect antibody-captured ELISA is demonstrated in Fig. 27.6. The method is often used in the early stage of immunoassay development to detect primary antibodies in antisera or screen hybridoma supernatants for searching desired antibodies. **Primary antibody** refers to the antibody that binds the antigen. Theoretically, the antibody-captured ELISA can be made either as a direct or indirect ELISA. However, because the target molecule is the primary antibody that appears in the biological fluid in a very low quantity, it usually is impossible to obtain enough quantity of the primary antibody to prepare the enzyme-conjugated labeling reagent for subsequent detection. Therefore, antibody-captured ELISA is almost always configured in an indirect assay format. The soluble antigen is adsorbed (coating) onto the surface of the microplate wells and incubated. After blocking, diluted samples of antisera or hybridoma supernatants are then added to the wells and incubated to allow the immobilized antigen to bind primary antibodies in the sample. After washing away any unbound molecules, those bound antibodies can be detected by adding an enzyme-linked secondary antispecies antibody, which can easily bind to the primary antibody. After another incubation and washing steps, a solution containing substrate is added to generate

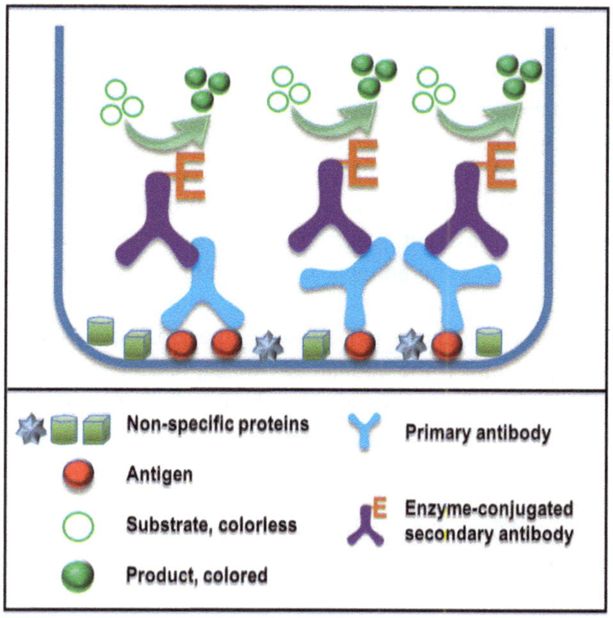

Fig. 27.6 Indirect noncompetitive ELISA

a color. The color is positively related to the amount of target antibody present in the sample.

The **secondary antibody** used in the indirect ELISA does not have the specificity to bind the antigen but only recognizes the primary antibody; thus, it links the enzyme label to the bound immunomolecules without interfering with the primary antigen-antibody binding. Since antibodies are proteins, they can act as antigens in other animal species. For example, rabbit antibodies injected into a goat can stimulate the goat's immune system to produce goat antibodies that bind to the rabbit antibodies. In this way, goat anti-rabbit antibodies can be produced to bind any antibody produced in a rabbit. There are many advantages to these secondary antispecies antibodies. For example, when antispecies antibodies are used in the above antibody-captured ELISA format (Fig. 27.6), there is no need to label the primary antibody with an enzyme. The primary antibody does not need to be chemically modified; thus, the loss of its activity is avoided. After excess material is washed away, goat anti-rabbit antibody labeled with an enzyme can be added to detect the presence of any primary antibody which is produced in rabbit antiserum and binds to the antigen coated on the microtiter plate. Although this procedure adds an additional step, there are many advantages: (1) antispecies antibodies of all types are commercially available from many manufacturers, (2) antispecies antibodies come with a variety of labels such as different enzymes, radioisotopes, or fluorescent compounds for different immunoassay systems, and (3) since the antibody is a very large protein, it has many sites for attachment of a labeled antispecies antibody. This multiplies the labels per antibody, increasing the ability to detect the antibody and

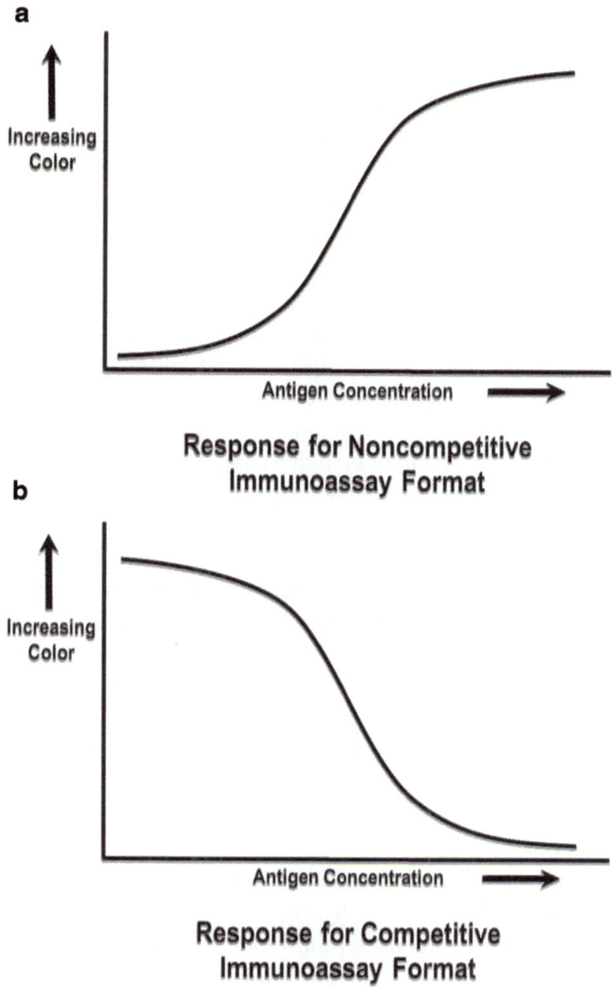

Fig. 27.7 Relationship between color development and antigen concentration for noncompetitive (**a**) and competitive (**b**) immunoassay formats

resulting in stronger signals with increased sensitivity in an immunoassay, so less primary antibody reagent is needed. Therefore, indirect methods are used in most detection applications.

27.3.2.3 Noncompetitive Versus Competitive Immunoassay Variations

Noncompetitive immunoassay is commonly employed to analyze large molecules, such as proteins in a food sample, while **competitive immunoassay** is competitive in nature and is mainly used for small molecule analysis. Both competitive and noncompetitive immunoassays can be detected directly or indirectly. For simplicity, these variations are illustrated in the following sections using ELISA as examples. In general, the color intensity for the noncompetitive ELISA positively correlates to the amount of antigen present in the sample (Fig. 27.7a). With any competitive ELISA format, there is an inversely proportional relationship between the color intensity developed and the amount of antigen present in the sample (Fig. 27.7b).

27.3.2.3.1 Noncompetitive ELISA

Noncompetitive ELISA variations involve revealing the amount of primary antibody-antigen complex immobilized on the solid phase by the amount of enzyme linked to the detection antigen or antibody molecules to produce a colored product in the assay solution. Therefore, at the end of the assay, the color intensity is positively related to the amount of the target molecules. The absence of the target molecules produces no color, and a high concentration of the target molecules produces intense color. This type of ELISA is often used to detect proteins in a food sample because protein molecules are large enough to link one or more antibodies or to an additional enzyme label on the surface of the protein.

One of the most popular formats for a noncompetitive enzyme immunoassay is the **sandwich immunoassay.** A sandwich ELISA model using direct and indirect detection methods is demonstrated in Fig. 27.8a and b–c, respectively. The "meat" in the antibody sandwich is the target antigen. In food analysis, this can involve identifying a protein adulterant such as undeclared pork meat in a beef product, or a protein allergen such as peanut protein, or wheat protein in a product that would be a problem for people suffering from celiac disease.

Generally, a primary antibody that binds to the antigen is first immobilized onto a hydrophobic solid phase such as plastic. Excess antibody is removed by washing with a washing solution or simply water, followed by a blocking step, and then the test is ready to analyze a food extract. The immobilized antibody is called a **capture antibody**. The food extract being tested contains many compounds with or without the target antigen. However, the antibody was prepared by immunization of an animal with a specific, purified protein antigen, and only this protein antigen in the food extract solution will bind to the capture antibody. Now the antigen and the capture antibody are immobilized, and the remaining unbound molecules can be washed away. After the washing step, another primary antibody labeled with an enzyme (Fig. 27.8a) is introduced. This antibody, called the **detection antibody**, also recognizes the antigen, thus forming an antibody-antigen-antibody complex. Again excess detection antibody is washed away, and then colorless enzyme substrate is added to develop a color if the bound enzyme is present. The enzyme will only be present if the detection antibody has been immobilized by binding to the antigen. The greater the color development, the greater the amount of antigen present. That is, there is a direct proportionality between the color intensity seen in the final step and the amount of antigen present in the extracted food sample. To increase the sensitivity of a sandwich immunoassay, one can use more antibodies to capture the antigen or link more enzyme molecules to the detection antibody through the use of intermediate reagents, such as biotin and labeled streptavidin. In the latter case, the detection antibody in the indirect sandwich assay (Fig. 27.8b) is purified and linked to biotin. Biotinylation of a primary antibody is relatively simple and rarely affects antibody activity. Then an additional step of adding enzyme-labeled avidin or streptavidin to the reaction system is needed. Avidin/streptavidin easily and tightly binds to the biotin groups, so more enzymes are linked to the immobilized molecules (Fig. 27.8b). In certain situations, a biotinylated secondary antibody against the detection antibody may be used (Fig. 27.8c). In general, this indirect sandwich immunoassay format can be made remarkably specific and very sensitive since two antibodies must detect the same antigen, and more labels are involved in

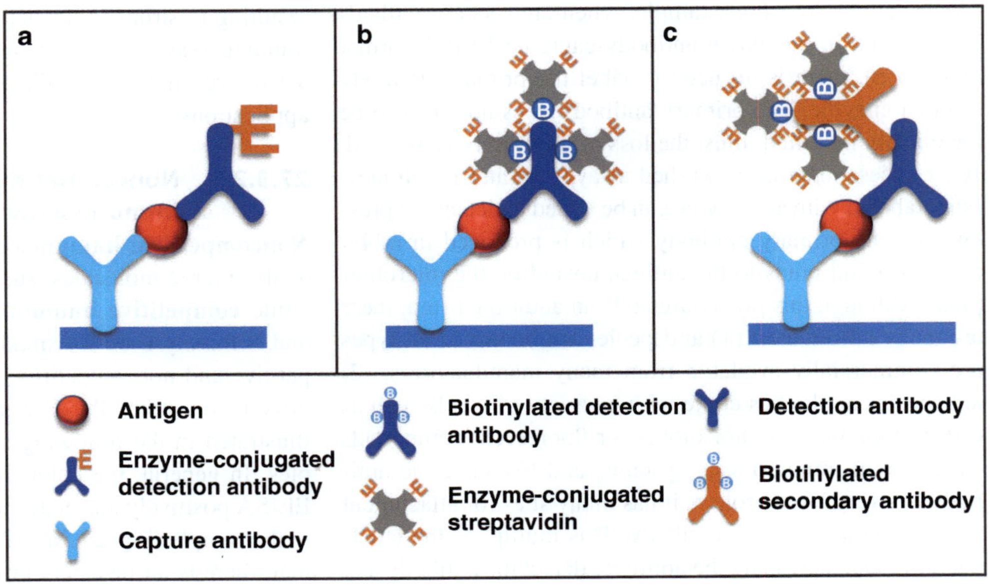

Fig. 27.8 Direct (**a**) and indirect (**b–c**) noncompetitive sandwich ELISA

amplifying the reaction signal. If the detection antibody in a sandwich ELISA cannot be labeled, such as IgE in human serum, a biotinylated secondary antibody directed against the host species of the detection antibody can be used to bind to the detection antibody (Fig. 27.8c). These types of biotinylated secondary antibodies are commercially available, which is advantageous as noted in Sect. 27.3.2.2. It should be noted, however, that the species used to generate the secondary antibody should always be different from the capture antibody host species.

When pAbs are used in the sandwich immunoassay, the pAb solution is divided into two parts. One part is bound to plastic to become the capture antibody. The second portion of the pAb solution is purified and conjugated to an enzyme or biotinylated and becomes the detection antibody. mAbs can also be used, but care must be exercised because a mAb cannot be used for both the capture and detection antibodies since only one unique epitope is recognized by any mAb. In other words, the antigen must be large enough to bind two antibodies simultaneously. It, therefore, must use at least two distinct epitopes recognized by different mAbs that recognize two distinct antigen epitopes.

27.3.2.3.2 Competitive Immunoassays

A problem in developing an immunoassay for detecting a small molecule is that a sandwich immunoassay format will not work since two different epitopes on the antigen are required for both antibodies to bind. A small molecule represents only one epitope or even only part of the epitope. Therefore, the competitive immunoassay format (Fig. 27.9) was developed to solve this problem. The first requirement in a competitive ELISA involves immobilizing the small molecule, often a hapten, or immobilizing the antibody. Subsequent procedures involve the competition between the free small antigen (from a sample) and the hapten (as an added reagent) for binding a limited amount of the specific antibody. To bind the hapten to a solid surface, such as nitrocellulose or plastic, it can again first be linked to a protein that binds to these hydrophobic surfaces. However, the protein used for binding the hapten to the surface differs from the carrier protein binding the hapten used for animal immunization since the animal also has developed antibodies against the carrier protein used for injection. Only the hapten-specific antibodies are desired for the competitive immunoassay. Since all types of competitive immunoassays involve a reduction in absorbance with respect to a control (containing no small molecule or analyte), data often are presented as a ratio of sample absorbance to the absorbance of the control. The concentration of the inhibitor (target antigen) required to reduce the assay absorbance by 50% (defined as **IC_{50}**) is a useful value to be determined for a competitive immunoassay because this is the region of greatest change in response compared to concentration changes and therefore the lowest coefficient of variation.

To increase the sensitivity of a competitive immunoassay, the amount of limiting antibody should be reduced. Note that this is the reverse of what one would do to increase the sensitivity of a noncompetitive immunoassay such as a sandwich immunoassay. Theoretically, the most sensitive competitive immunoassay would be between one antibody-binding site and one hapten, with either of the two labeled with an enzyme. It is for this reason that the ability to detect the presence of the enzyme is so important for a competitive immunoassay. The more sensitive the system is to detecting the enzyme, the more sensitive the competitive immunoassay. Two competitive ELISA procedures are described here.

(1) **Bound Hapten Format**. In the bound hapten competitive immunoassay format (Fig. 27.9a), the protein-bound hapten is first immobilized to a solid surface by hydrophobic interactions. Excess material is washed away.

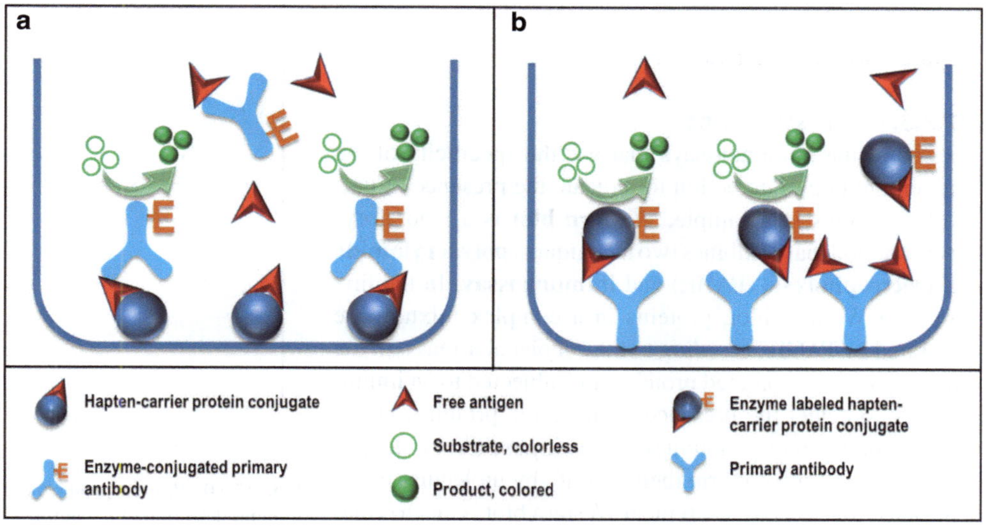

Fig. 27.9 Direct competitive ELISA in bound hapten (**a**) and bound antibody (**b**) formats

Next, a competition is created between the protein-bound hapten and the free small antigen molecule in a food extract, both competing for binding to the limited binding sites on the antibody labeled with an enzyme. It is important to realize that the free small molecule in the food extract is not completely identical to the immobilized hapten since the latter is covalently linked to a protein. However, if properly designed, the free molecule in the food extract is so chemically similar to the bound hapten that the competition for the limited number of antibody-binding sites is nearly equal. The primary antibody bound to immobilized hapten remains after a subsequent washing step. The more small molecules in the food extract, the more antibody is bound to these free small molecules, and this unbound antibody (and its attached enzyme) will be washed away in the subsequent washing procedure. Finally, the amount of bound antibody is identified by adding the enzyme substrate and observing the amount of color developed. Therefore, there is an inverse relationship between the concentration of small molecules or analytes in the food and the color intensity developed in the final step.

(2) **Bound Antibody Format.** The other variation for a competitive immunoassay is to bind a limited amount of antibody to the solid phase and create a competition between enzyme-labeled conjugate antigen and free small antigen molecules in the food extract (Fig. 27.9b). It is generally believed that this second format is somewhat superior to the first format for sensitivity, although it can require the use of more antibody reagents. Again after a washing step, the final procedure is a color development to determine the amount of antibody-bound conjugate antigen-enzyme. This competitive format also results in an inverse relationship between the color intensity developed and the amount of free small molecules in the food extract.

27.3.3 Immunoblots

27.3.3.1 Western Blot

As one of the immunoassays that use the specificity of the antigen-antibody interaction to indicate the presence of particular proteins in a sample, **Western blot** is a laboratory-based method that combines two techniques: **polyacrylamide gel electrophoresis** (PAGE) and **immunoassay**. In the first part of a Western blot, proteins in a complex mixture are separated by PAGE according to their molecular mass. In the second part, the separated proteins are subjected to an immunoassay to detect the presence of antigenic proteins. Using this combination of techniques makes it possible to identify target proteins and confirm their identity by molecular mass. The detection reagent in a typical Western blot is an enzyme-labeled antibody conjugate directly or indirectly labeled to the detection antibody. If the original protein mixture was labeled with a radioactive material, autoradiography is used to visualize the radioactive signal. The primary antibody used largely determines the specificity and sensitivity of the method. Specific proteins in picogram quantities can be detected in a highly sensitive Western blot.

To prepare protein samples for the initial separation by PAGE, they are typically boiled in a buffer solution containing a reducing agent (usually mercaptoethanol) and detergent (e.g., sodium dodecyl sulfate), to unfold the protein peptide chains. The treated sample is applied to a polyacrylamide gel and separated by electrophoresis based on **molecular mass**. To prepare for the immunoassay portion of the method, the separated protein bands are transferred from the polyacrylamide gel to a **nitrocellulose** or **PVDF membrane**. After blocking the nonspecific binding site, for a direct detection format, this membrane is then incubated with a solution of primary antibody-enzyme conjugate. After washing away the excess conjugate from the membrane, the enzyme substrate is added. A colored band forms at the site on the membrane where the protein that reacted with the antibody was immobilized (Fig. 27.10). The enzyme substrate used in Western blot is different from the ones used in

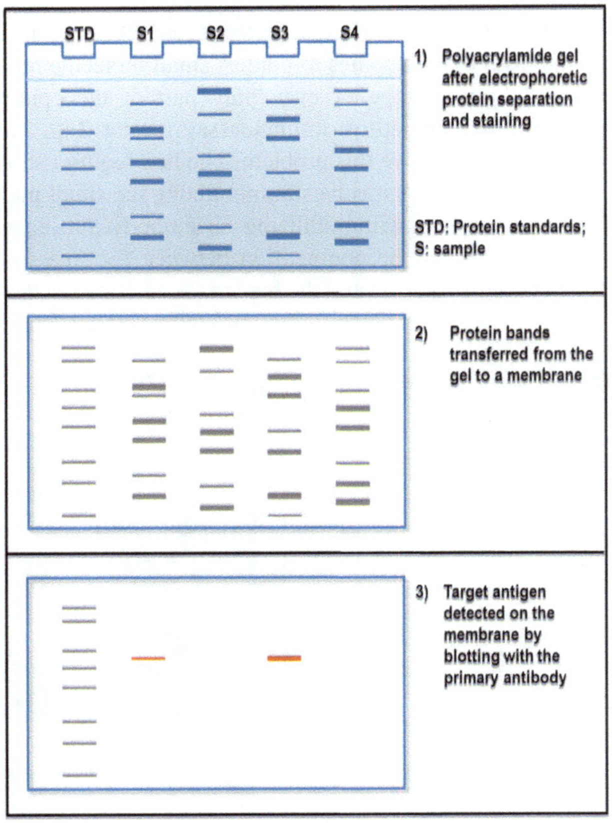

Fig. 27.10 Protein profile before (1) and after (2) transferring and immunodetection of the antigen on the membrane (3)

ELISA because the intent is to form an insoluble colored product that stays on the membrane. The color intensity and width of the protein band together indicate the concentration of the target protein in the sample extract. The molecular mass of the protein bound by the antibody can be estimated by its position relative to standard proteins of known molecular weight.

The Western blot procedure is technically complex, requiring highly trained personnel in a laboratory setting. However, the Western blot method is especially well suited to the analysis of food samples that have been subjected to processing conditions. The Western blot method requires that the antibody used binds to a linear epitope on the protein or is reactive to **denatured protein** (which would occur under the denaturing conditions in PAGE sample preparation). Because the Western blot method is highly sensitive and specific, and the antibodies used recognize denatured proteins, it is uniquely well suited to detect the presence of target food proteins at low concentrations in processed food.

27.3.3.2 Dot Blot

Dot blot represents a simplified version of Western blot. It is usually performed by depositing a drop of sample extract containing the molecule to be detected directly onto a piece of nitrocellulose or PVDF membrane as a dot. Unlike Western blot, protein samples in a dot blot assay do not need to be separated electrophoretically. The target protein molecule is directly detected by specific antibodies. The following general guideline describes a commonly used antigen bound dot blot assay for detecting a particular protein in food samples (Fig. 27.11):

(1) A drop of the food sample extract is applied to the membrane.
(2) The remaining binding sites on the membrane are "blocked" using a protein unrelated to the target antigen to minimize the nonspecific binding.
(3) Following a rinsing step, the bound molecules in the sample spot are incubated with the primary antibody (for indirect detection) or with the primary antibody-enzyme conjugate (for direct detection). An additional step using an antispecies antibody-enzyme conjugate is required for the indirect assay.
(4) After washing and adding the substrate solution, the reaction signal can be revealed either visually by color or measured by chemiluminescence imaging apparatus, depending on the enzyme label and the substrates used. The spot signal (i.e., color) intensity is positively related to the amount of target protein present in the sample.

This blot dot technique does not offer molecular weight information of the antigen. However, it is useful as a qualitative method for rapidly screening a large number of samples to probe the presence or absence of a protein target in food samples. This technique is also commonly used to evaluate the quality of the antibodies and test the suitability of experimental design parameters.

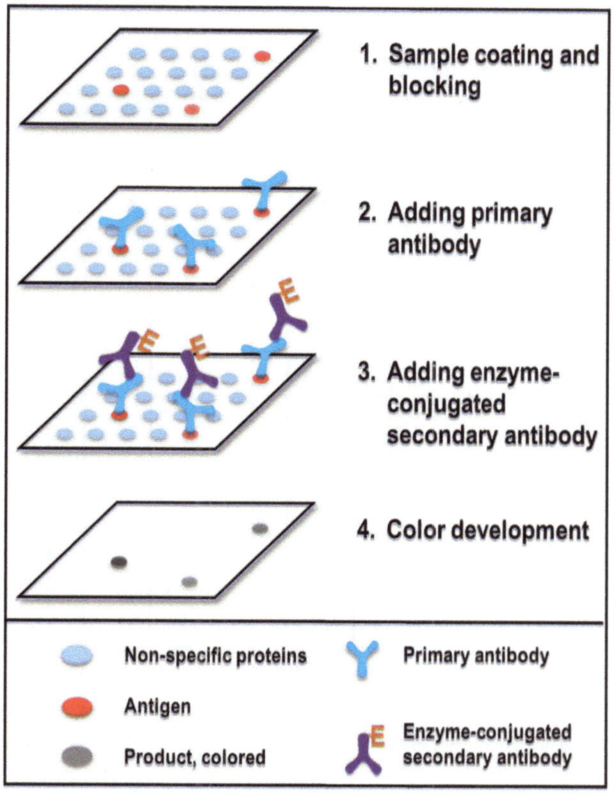

Fig. 27.11 Dot blot technique

27.3.4 Lateral Flow Strip Test

27.3.4.1 Overview

The **lateral flow strip** (LFS) test is a simple immunoassay format commonly used for proteins to determine if their concentration is above or below a specified threshold (cutoff) (Fig. 27.12). The home pregnancy test is the best-known LFS method. Results with LFS usually can be visualized in 10–20 min., and no washing steps are required to separate bound and unbound molecules. The characteristics of LFS methods – simplicity, low cost, ease of use, and reliability – make them ideal for use outside a laboratory setting, i.e., **field testing**, where supplies and equipment are limited.

Like other immunoassays, LFS methods are configured in a competitive assay format for the detection of small molecules such as toxins or chemical residues or in a noncompetitive sandwich immunoassay format for detecting large molecules. While the color in an enzyme immunoassay comes from enzyme action on its substrate to create a colored reaction product, the LFS methods instead use very

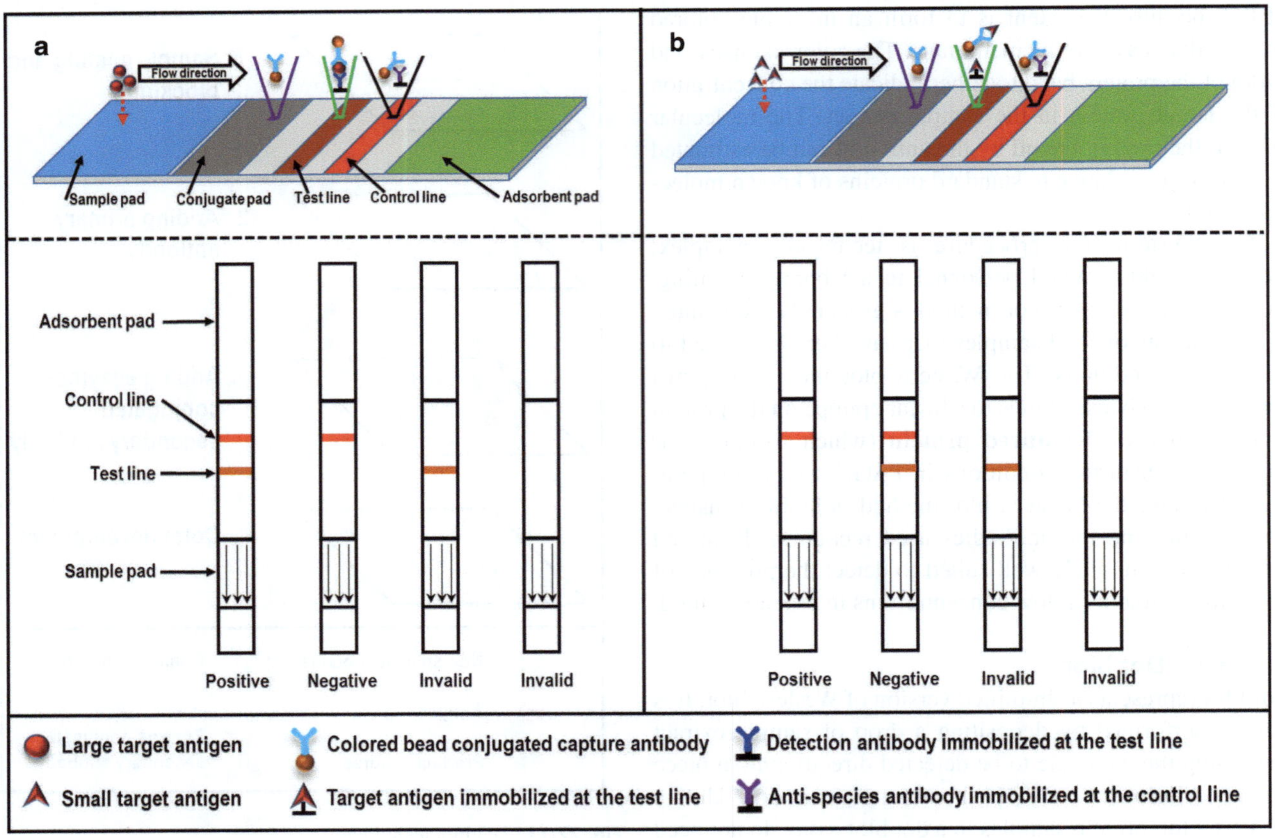

Fig. 27.12 Schematic view of a sandwich-type (noncompetitive) (**a**) and a competitive-type (**b**) lateral flow test strip

small, spherical, colored particles (colloidal gold or colored latex) attached to antibodies to generate a positive colored signal. The capture antibody is immobilized in a zone on a porous membrane (usually nitrocellulose). By capillary action of the membrane, the test sample travels past the zone of the immobilized antibody. The target protein in the test sample binds to the capture antibody. This type of sample movement and separation explains why LFS is called an **immunochromatographic assay**.

27.3.4.2 Procedure

Figure 27.12a depicts a typical LFS sandwich immunoassay and shows the various regions of the test strip. A primary antibody capable of binding to the target protein is coated onto the surface of very small, colored particles (usually 20–40 nm diameter, colloidal gold or colored latex). These antibody-coated colored particles are dried in a porous pad. When these particles come in contact with liquid samples, they get reconstituted and are able to flow with the sample, moving across the **sample pad** of the strip by capillary action. Any large particulates in the sample liquid can be filtered out by a fiber filter placed at the front of the strip in the sample pad area. A second primary antibody, also capable of binding to the target protein, is immobilized at the **test line** (zone) on the surface of the fibers of the porous nitrocellulose membrane. A **control line** above the test line, containing an antispecies antibody capable of binding antibody-colored particle conjugates, serves to indicate that the test ran appropriately. The sample gets drawn through the test strip by an **absorbent pad** placed at the back of the strip.

Both liquid and solid samples can be analyzed by the LFS assay. A solid sample must be dissolved or dispersed in a liquid solution to extract the target protein (antigen) from the sample. To do the assay, the strip is put in contact with the sample solution at the end of the sample pad. The sample solution is drawn into the test strip by capillary action. The sample first passes through the filter, and the target protein is bound by the colored particle-antibody conjugate. Then the antigen-antibody-colored particle conjugate complex gets drawn into the membrane and is captured by the immobilized antibody in the test zone. A colored line becomes evident as more colored complexes are captured. The color intensity generally correlates with the amount of antigen present in the sample. A dedicated device can be used to measure the color intensity of the test line if a quantitative result is required. Samples containing no target protein show no color at the test line. The control zone binds the excess antibody-colored particle conjugates that pass through the test zone and then forms a colored line for any complete test. If no colored line is formed at the control zone, the assay is invalid and needs to be repeated on a new strip.

In the case of competitive LFS immunoassays, the target antigen that is immobilized on the test line and the free antigen from the sample solution will compete for the antibody-colored particle conjugates. Therefore, no colored line is formed for samples containing an antigen concentration above the cutoff value of the LFS test (Fig. 27.12b).

27.3.4.3 Applications

Current generation of LFS methods are made into versatile formats and are useful in applications for which their primary attributes (i.e., speed, simplicity, and low cost) are important aspects of the testing (e.g., field testing). LFS methods have been developed for qualitative and, to some extent, quantitative monitoring of food allergens, foodborne pathogens, food toxins, hormones, and certain food protein ingredients to ensure food safety and quality. They also have been developed to detect the presence of genetically modified organisms in processed foods for consumers' interest and to detect prohibited ruminant proteins in ruminant feedstuffs for the surveillance of mad cow disease. These easy-to-use rapid tests allow food processors and regulators to comply with regulations governing the labeling of food and feed products with minimal user training and facilities. Various devices of LFS for food allergen detection have been described in a recent handbook [13]. The strengths, weaknesses, opportunities, and threats of LFS methods have been thoroughly discussed in a review article [14].

27.4 IMMUNOAFFINITY PURIFICATION

Besides the use of antibodies in immunoassays as described above, antibodies are often used in food analyses as complements to other analytical methods. This is due to the remarkable specificity of antibodies and their strong binding to antigens. The most common example of this is **immunoaffinity** purification, which is an antigen-capture technique. Basically, the antibody is immobilized on some support, most often using a covalent linking method so that there is no concern with "bleeding" of the antibody in later steps (Fig. 27.13). The antibody can be bound to a solid phase such as agarose (e.g., Sepharose®) gel. These antibody-bound solid phases can be used later for antigen purification via a chromatography method or by using these phases on the surface of magnetic beads that are separated using a magnet. A simple purification sequence would involve exposing the antibody-bound solid to a food extract to first bind antigen, then washing the solid phase free of all unbound material, and finally releasing the pure antigen. Even though antibodies have such remarkably strong binding constants, they can be treated to release antigens by simple procedures such as changes in pH or solvent. Since the antibody is a protein, pH changes or solvent changes result in denaturation that changes the conformation of the binding site, causing the release of the antigen. If these changes are carefully selected, denaturation can be reversed by reestablishing moderate conditions so that the valuable antibody-bound solid phase can be reused repeatedly. For sensitive antigens, like enzymes, these elution conditions also can be a concern.

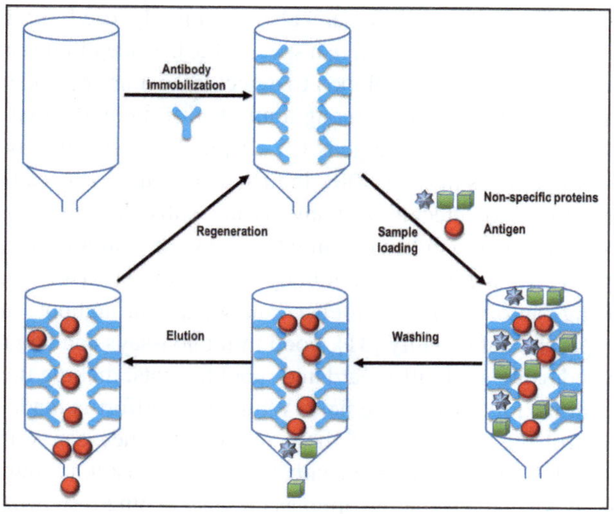

Fig. 27.13 An example of antibody-bound immunoaffinity chromatography

These immunoaffinity purification procedures have been used for small molecules like toxins (e.g., aflatoxins) and even materials as large as cells. Different microorganisms contain unique cell surface antigens that can be selectively bound to aid in purification and differentiation.

27.5 APPLICATIONS

Immunoassays are a well-developed area in food analysis, and there is a plethora of literature available. For all sorts of laboratory techniques, Harlow and Lane [9] wrote one of the best books on the use of antibodies. The theory and practice of immunoassays are well handled in several books [15–17]. There are even entire journals, such as *Food and Agricultural Immunology*, devoted to describing methods for preparing food immunoassays.

Because of the speed, simplicity, sensitivity, and selectivity of immunoassays, they are used widely as screening tests for pesticide [18, 19] and drug residues [20, 21] in food (see Chap. 33, Table 33.1). Besides chemical analysis, immunoassay techniques are used in microbiology to rapidly detect foodborne pathogens [22, 23] and bacterial and fungal toxins

[24, 25]. Immunoassays are also commonly used for the identification of meat and fish species [26]. Since immunoassays can easily be developed to detect trace amounts of specific proteins, they are among a number of methods used to detect hidden food allergens [27–29] and genetically modified organisms in foods [30]. In fact, an immunoassay can be developed to detect almost any organic substances in a food system. The flexibility is limited only by the availability and quality of the specific antibody used. Immunoassays are being automated for higher analytical throughput and improved data quality [31]. Food immunoassays can be prepared using straightforward and rapid formats, making them ideal for kits used in the field. While every effort is made to control the specificity of these rapid tests, they can suffer from false positives and negatives. For this reason, immunoassay kits are used most often as rapid screening tests, while food samples that test positive for the target analyte by immunoassay are often confirmed using another, more sophisticated analytical method. Immunoassays are also applicable in situations for which analysis by conventional methods is either not possible or is prohibitively expensive.

The research in the immunoassay area continues to develop the following: (1) specific antibodies, especially mAbs, for providing the ultimate specificity for the antibody-antigen interaction, (2) different solid-phase configurations to provide simpler means to carry out the assay, (3) new detection systems to further improve the assay sensitivity, and (4) alternative methodologies, such as immunosensors and immunoarray chips for multiple analyte detection and pattern recognition [2, 23].

27.6 SUMMARY

Almost any organic molecule in food can be determined using immunoassays as long as the specific antibodies are available. Both polyclonal and monoclonal antibodies or a combination of them can be used in an immunoassay. The remarkable selectivity and sensitivity of these assays result from the strong binding affinity between antibodies and their antigens. While the precise protocols of immunoassays can vary a great deal, all immunoassays use either a noncompetitive or a competitive format. The competitive format is the only one that can be used to quantify small (about 1000 Da or less) molecules. ELISA has become the most popular immunoassay that uses an enzyme as the label to reveal the primary antibody-antigen binding through a color reaction catalyzed by the enzyme. In a noncompetitive ELISA with enzyme-derived color development, the more antibody-bound molecules (analyte) in the food sample, the more color develops, while in a competitive ELISA, the reverse is true. The most common labels used for food immunoassays are enzymes, and the two most common enzymes used are alkaline phosphatase and horseradish peroxidase. The reaction signals of both competitive and noncompetitive immunoassays can be detected using direct or indirect labeling methods, although indirect methods are used in most detection applications due to their various advantages over direct methods.

Two other commonly used immunoassays are immunoblot and LFS test. Two types of immunoblots, Western blot and dot blot, are commonly used in food analysis. Western blot is a laboratory-based immunoassay that combines PAGE and immunoassay to reveal the presence and the molecular mass of the antigenic protein on a membrane. Western blot is the most commonly used method for identifying and characterizing an antigen. A positive result also indicates that the antigen confers a linear epitope with the antibody. Dot blot is a simplified version of Western blot without the protein separation procedure, hence does not offer the molecular weight information of the antigen. The LFS, on the other hand, is by far the simplest form of immunoassay. The assay is easy to use and is ideal for outside-the-laboratory applications for which access to equipment and supplies is limited. While the general immunoassay procedure of the immunoblot is similar to ELISA, the LFS assay is a one-step assay based on the one-direction movement of the antigen in a sample solution toward immobilized capturing antibodies at different zones. The separation of bound and unbound molecules by washing steps is not required. Because ELISA involves a signal amplification activity of the enzyme, it is generally regarded as inherently more sensitive than LFS methods. However, protein concentrations of less than a part per billion can be detected with a highly sensitive LFS.

Besides being the required constituent in immunoassays, antibodies also can be used to purify specific compounds in food for other analysis methods. These immunoaffinity purification methods allow for the rapid purification of analytes from complex food matrices.

27.7 STUDY QUESTIONS

1. What is the relationship between an antigen and an antibody?
2. What is an epitope? What are the two types of epitopes?
3. What is the difference between monoclonal and polyclonal antibodies?
4. All immunoassays have two conditions that they must satisfy; what are they?
5. What is a hapten, and what is a conjugate antigen?
6. What are the five general steps for an ELISA procedure?
7. What is the rationale for the blocking step in an immunoassay protocol?

8. Immunoassays. Differentiate between the following terms relevant to enzyme-linked immunosorbent assays:
   (a) Regarding competitive vs. noncompetitive assays:
      (i) Components and how it works
      (ii) What does high absorbance value (vs. low) mean, regarding level of antigen?
      (iii) Application
   (b) Regarding direct ELISA vs. indirect ELISA assays:
      (i) Components and how it works
      (ii) Advantage and disadvantage
9. Two common immunoassays are the sandwich assay and the competitive assay. Which molecules are best detected by each? Why?
10. What are two types of competitive ELISA procedures? Describe the main differences between these two.
11. Give four common applications of immunoassays in food analysis.
12. What is a Western blot? What is the major difference in reaction signals between Western blot and ELISA?
13. Compare and contrast ELISA and LFS, by identifying the similarities and differences in their characteristics, principles, and applications.
14. Describe, in general terms, how you would use immunoaffinity purification to isolate a protein for which you have developed antibodies.
15. An ELISA assay was devised and used by a research group for the routine screening of aflatoxin B1 in peanut butter. The standard aflatoxin Bl (to create the standard curve), antibody against aflatoxin Bl, and enzyme-labeled antibody against aflatoxin B1 were purchased from biochemical companies. Aflatoxin Bl in peanut butter was extracted with some organic solvents to create a sample extract. According to the procedure, the following were added to the microtiter plate in the order given, with washing steps in between:
    1. Antibody against B1
    2. Aflatoxin Bl standards (for the standard curve) or sample extract
    3. Enzyme-labeled antibody against aflatoxin B1
    4. Enzyme substrate (after incubation, the absorbance at 405 nm was determined)
    (a) Was this a competitive or noncompetitive ELISA assay? How do you know this?
    (b) Was this a direct or indirect ELISA assay? How do you know this? What would have been the major advantage of the other type of assay?
    (c) For this assay, what will be plotted on the x-axis and y-axis of a standard curve? Will the slope of the standard curve be positive or negative?
    (d) Would a high absorbance reading at 405 nm for the sample mean a high or low concentration of aflatoxin B1 in peanut butter?
16. All commercial potatoes contain the toxic glycoalkaloids α-solanine and α-chaconine. Both of these glycoalkaloids have the same large alkaloid portion, known as solanidine. Therefore, pAbs can be developed in rabbits against solanidine by chemically linking it to a foreign protein (foreign to the rabbit) and injecting the protein-conjugated hapten (solanidine linked using a succinic acid derivative) into rabbits. The antibodies that develop in the rabbit against the hapten bind to the alkaloid portion of both toxic glycoalkaloids. The rabbit antiserum containing primary polyclonal antibodies can be made highly specific to solanidine without cross-reactivity with other similar molecules by removing all cross-reactive antibody components using an immunoaffinity column procedure.

To develop an appropriate competitive ELISA, solanidine is again linked to a protein, but this time a different protein, and this conjugate is used to coat plastic microtiter plates. After excess conjugate is washed away, the plates are ready for the competitive ELISA procedure.

The glycoalkaloids in potatoes are extracted with methanol, and this extract is further diluted with water for use in the ELISA procedure. A standard curve is prepared by diluting standard solutions of α-chaconine at low, medium, and high concentrations with similar aqueous methanol solutions. In addition, a negative control is prepared using methanol and water at similar concentrations to the diluted potato extracts and standards but without any glycoalkaloid present. Now the various extracts, standards, and negative controls are placed in individual wells with equivalent amounts of diluted rabbit serum containing the specific polyclonal antibodies. After incubation for 30 min at room temperature, all of the wells on the plate are again washed. Next, a solution of commercially available goat anti-rabbit antibody conjugated to peroxidase is added to each well. After another 30 min incubation, the wells are again thoroughly washed.

Finally, phenylenediamine substrate solution is added to each well along with peroxide and again the plate is incubated for 30 min. After 30-min, the plate is rapidly read (in under 1 min) using an ELISA plate reader. The wells all contain different amounts of yellow color.

(a) Tomatidine is a glycoalkaloid found in tomatoes and contains the alkaloid portion tomatine. Would the polyclonal antibodies detect tomatidine?
(b) Why is the protein that the hapten is attached to different for the ELISA procedure than for the injection?
(c) Is the ELISA protocol direct or indirect?
(d) Which wells would you expect to contain the most color, standards, potato extracts, or negative controls?
(e) Would you be concerned if a potato extract gave almost no color at the end of the ELISA procedure?

Acknowledgments The author of this chapter thanks Y.-H. Peggy Hsieh (Professor Emeritus, Department of Nutrition and Integrative Physiology, Florida State University, Tallahassee, FL, USA) for her contribution to the chapter in previous editions of the book.

REFERENCES

1. IUPAC (1997) Compendium of chemical terminology. Blackwell Science, Oxford, UK
2. Shankaran DR, Gobi KV, Miura N (2007) Recent advancements in surface plasmon resonance immunosensors for detection of small molecules of biomedical, food and environmental interest. Sens Actuators B Chem 121: 158–177
3. Jiang X, Wu M, Dong W, Rao Q, Huo H, Han Q (2020) Monoclonal antibody-based sandwich enzyme-linked immunosorbent assay for porcine hemoglobin quantification. Food Chem 324: 126880
4. Hossain SZ, Luckham RE, McFadden MJ, Brennan JD (2009) Reagentless bidirectional lateral flow bioactive paper sensors for detection of pesticides in beverage and food samples. Anal Chem 81: 9055–9064
5. Huet A-C, Charlier C, Tittlemier SA, Singh G, Benrejeb S, Delahaut P (2006) Simultaneous determination of (fluoro)quinolone antibiotics in kidney, marine products, eggs, and muscle by enzyme-linked immunosorbent assay (ELISA). J Agric Food Chem 54: 2822–2827
6. Zhao X, Lin C-W, Wang J, Oh DH (2014) Advances in rapid detection methods for foodborne pathogens. J Microbiol Biotechnol 24: 297–312
7. Singh M, Agrawal RK, Singh BR, Mendiratta SK, Agarwal RK, Singh MK, Kumar D (2017) Development and evaluation of simple dot–blot assays for rapid detection of *Staphylococcal* enterotoxin-A in food. Indian J Microbiol 57: 507–511
8. Kohler G, Milstein C (1975) Continuous cultures of fused cells secreting antibody of predefined specificity. Nature 256: 495–497
9. Harlow E, Lane D (1999) Using antibodies: a laboratory manual. Cold Spring Harbor Laboratory Press, Cold Spring Harbor, New York
10. Howard GC, Kaser MR (2014) Making and using antibodies: a practical handbook. CRC Press, Boca Raton, FL
11. Berson SA, Yalow RS, Bauman A, Rothschild MA, Newerly K (1956) Insulin-I^{131} metabolism in human subjects: demonstration of insulin binding globulin in the circulation of insulin treated subjects. J Clin Invest 35: 170–190
12. Engvall E, Perlmann P (1972) Enzyme-linked immunosorbent assay, ELISA. 3. Quantitation of specific antibodies by enzyme-labeled anti-immunoglobulin in antigen-coated tubes. J Immunol 109: 129–135
13. Baumert JL, Tran DH (2015) Lateral flow devices for detecting allergens in food. In: Flanagan S, editor. Handbook of food allergen detection and control, Woodhead Publishing, Cambridge, UK. pp. 219–228
14. Posthuma-Trumpie GA, Korf J, van Amerongen A (2009) Lateral flow (immuno)assay: its strengths, weaknesses, opportunities and threats. A literature survey. Anal Bioanal Chem 393: 569–82
15. Crowther JR (2000) The ELISA guidebook. 2nd ed. Humana Press, New York
16. Deshpande SS (1996) Enzyme immunoassays: from concept to product development. Chapman and Hall, New York
17. Wild D (2013) The immunoassay handbook: Theory and applications of ligand binding, ELISA and related techniques. 4th Waltham, WA
18. Yao J, Wang Z, Guo L, Xu X, Liu L, Xu L, Song S, Xu C, Kuang H (2020) Advances in immunoassays for organophosphorus and pyrethroid pesticides. Trends Analyt Chem 131: 116022
19. Morozova VS, Levashova AI, Eremin SA (2005) Determination of pesticides by enzyme immunoassay. J Analyt Chem 60: 202–217
20. Ahmed S, Ning J, Peng D, Chen T, Ahmad I, Ali A, Lei Z, Abu bakr Shabbir M, Cheng G, Yuan Z (2020) Current advances in immunoassays for the detection of antibiotics residues: a review. Food Agric Immunol 31: 268–290
21. Reig M, Toldrá F (2008) Veterinary drug residues in meat: concerns and rapid methods for detection. Meat Sci 78: 60–67
22. Di Nardo F, Chiarello M, Cavalera S, Baggiani C, Anfossi L (2021) Ten years of lateral flow immunoassay technique applications: trends, challenges and future perspectives. Sensors 21: 5185
23. Banada PP, Bhunia AK (2008) Antibodies and immunoassays for detection of bacterial pathogens. In: Zourob M, Elwary S, and Turner A, editor. Principles of bacterial detection: biosensors, recognition receptors and microsystems, Springer, New York. pp. 567–602
24. Tripathi P, Upadhyay N, Nara S (2018) Recent advancements in lateral flow immunoassays: a journey for toxin detection in food. Crit Rev Food Sci Nutr 58: 1715–1734
25. Li W, Powers S, Dai S (2014) Using commercial immunoassay kits for mycotoxins: 'Joys and sorrows'? World Mycotoxin J 7: 417–430
26. Hsieh Y-HP (2005) Meat species identification. In: Hui YH and Sherkat F, editor. Handbook of Food Science, Technology, and Engineering, CRC, Boca Raton, FL. pp. 30-1–30-19
27. Immer U, Lacorn M (2015) Enzyme-linked immunosorbent assays (ELISAs) for detecting allergens in food. In: Flanagan S, editor. Handbook of food allergen detection and control, Woodhead Publishing, Cambridge, UK. pp. 199–217
28. Owusu-Apenten RK (2002) Determination of trace protein allergens in foods. Ch. 11. In: Owusu-Apenten RK, editor. Food protein analysis: quantitative effects on processing, CRC press, New York. pp. 297–339
29. Poms R, Klein C, Anklam E (2004) Methods for allergen analysis in food: a review. Food Addit Contam 21: 1–31
30. Holst-Jensen A (2009) Testing for genetically modified organisms (GMOs): past, present and future perspectives. Biotechnol Adv 27: 1071–1082
31. Bock JL (2000) The new era of automated immunoassay. Am J Clin Pathol 113: 628–646

Determination of Oxygen Demand

28

Yong D. Hang

Contents

28.1 INTRODUCTION

Food processing facilities work to minimize the level of organic pollutants in wastewater and monitor oxygen demand of the wastewater. Oxygen demand is a commonly used parameter to evaluate the potential effect of organic pollutants on either a wastewater treatment process or a receiving water body. Because microorganisms utilize these organic materials, the concentration of dissolved oxygen is greatly depleted from the water. The oxygen depletion in the environment can have a detrimental effect on fish and plant life in the receiving water body. Since wastewater treatment facilities must reduce the level of organic pollutants in the wastewater before releasing the water into rivers or lakes, food companies are charged based on the oxygen demand of the wastewater.

The two main methods used to measure the oxygen demand of water and wastewater are **biochemical oxygen demand** (BOD) and **chemical oxygen demand** (COD). This chapter briefly describes the principles, procedures, applications, and limitations of each method. Methods described in this chapter are adapted from *Standard Methods for the Examination of Water and Wastewater* [published by the American Public Health Association (APHA), the American Water Works Association (AWWA), and Water Environment Federation (WEF)] [1]. The book includes step-by-step procedures with equipment for BOD, COD, and other tests for water and wastewater.

28.2 METHODS

28.2.1 Biochemical Oxygen Demand (BOD)

28.2.1.1 Principle

The BOD determination is a measure of the amount of oxygen required by microorganisms to oxidize the biodegradable organic constituents present in water and wastewater. The method is based on the direct relationship between the concentration of organic matter and the amount of oxygen used to oxidize the pollutants to water, carbon dioxide, and

Y. D. Hang (✉)
Department of Food Science and Technology, Cornell University, Geneva, NY, USA
e-mail: ydh1@cornell.edu

B. P. Ismail, S. S. Nielsen (eds.), *Nielsen's Food Analysis*, Food Science Text Series,
https://doi.org/10.1007/978-3-031-50643-7_28

inorganic nitrogenous compounds. The **oxygen demand** of water and wastewater is proportional to the amount of **organic matter** present. The BOD method measures the biodegradable carbon (carbonaceous demand) and, under certain circumstances, the biodegradable nitrogen (nitrogenous demand).

28.2.1.2 Procedure

Place a known amount of a water or wastewater sample that has been seeded with an effluent from a biological waste treatment plant in an airtight BOD bottle and measure the initial dissolved oxygen immediately. Incubate the sample at 20 °C, and after 5 days, measure the dissolved oxygen content again (APHA Method 4500-0). The dissolved oxygen content can be determined by the membrane electrode method (APHA Method 4500-O G) or the azide modification (APHA Method 4500-0 C), permanganate modification (APHA Method 4500-0 D), alum flocculation modification (4500-0 E), or copper sulfate-sulfamic acid flocculation modification (APHA Method 4500-0 F) of the iodometric method (APHA Method 4500-0 B) to minimize interference by nitrite or ferrous or ferric iron. The iodometric method is a titrimetric procedure that is based on the oxidizing property of dissolved oxygen, while the membrane electrode method based on the diffusion rate of dissolved oxygen across a membrane. A dissolved oxygen meter with an oxygen-sensitive membrane electrode made by Fisher, Orion, YSI, or other companies is used to measure the diffusion current, which is linearly proportional to the concentration of dissolved oxygen under steady-state conditions. It is important to change frequently and calibrate the membrane electrode to eliminate the effect of interfering gases such as hydrogen sulfide. The azide-modified iodometric procedure, for example, is used to remove interference of nitrite, which is the most commonly interfering material in water and wastewater. The alum flocculation modification method is commonly used to minimize the interference caused by the presence of suspended solids. The BOD value, which is expressed as mg/L, can be calculated from the difference in the initial dissolved oxygen and the content of dissolved oxygen after the incubation period according to the following equation (APHA Method 5210 B):

$$\mathrm{BOD}(\mathrm{mg}/\mathrm{L}) = 100/P \times (\mathrm{DOB} - \mathrm{DOD}) \quad (28.1)$$

where:

DOB = initial oxygen in diluted sample, mg/L
DOD = oxygen in diluted sample after 5-day incubation, mg/L
P = mL sample × 100/capacity of bottle

28.2.1.3 Applications and Limitations

The BOD test is used most widely to measure the organic loading of waste treatment processes, to determine the efficiency of treatment systems, and to assess the effect of wastewater on the quality of receiving waters. The 5-day BOD test has some drawbacks because:

1. The procedure requires an incubation time of at least 5 days.
2. The BOD method does not measure all the organic materials that are biodegradable.
3. The test is not accurate without a proper seeding material.
4. Toxic substances such as chlorine present in water and wastewater may inhibit microbial growth.

28.2.2 Chemical Oxygen Demand (COD)

28.2.2.1 Principle

The COD determination is a rapid way to estimate the quantity of oxygen used to oxidize the organic matter present in water and wastewater. Most organic compounds are destroyed by refluxing in a strong acid solution with a known quantity of a strong oxidizing agent such as **potassium dichromate.** The excess amount of potassium dichromate left after digestion of the organic matter is measured. The amount of organic matter that is chemically oxidizable is directly proportional to the potassium dichromate consumed.

28.2.2.2 Procedure

A known quantity of sample of water or wastewater is refluxed at elevated temperatures for up to 2 h with a known quantity of potassium dichromate and sulfuric acid using an open reflux method (APHA Method 5220 B), a closed reflux titrimetric method (APHA Method 5220 C), or a closed reflux colorimetric method (APHA Method 5220 D). The amount of potassium dichromate left after digestion of the organic matter is titrated with a standard ferrous ammonium sulfate (FAS) solution using ortho-phenanthroline ferrous complex as an indicator. The amount of oxidizable organic matter, determined as oxygen equivalent, is proportional to the potassium dichromate used in the oxidative reaction. The COD value can be calculated from the following equation (APHA Method 5220 B):

$$\mathrm{COD}(\mathrm{mg}/\mathrm{L}) = (A - B) \times M \times 8000/D \quad (28.2)$$

where:

A = mL FAS used for blank
B = mL FAS used for sample
M = molarity of FAS
D = ml sample used
8000 = milliequivalent weight of oxygen × 1000 mL/L

28.2.2.3 Applications and Limitations

Potassium dichromate is widely used for the COD method because of its advantages over other oxidizing compounds in oxidizability, applicability to a wide variety of waste samples, and ease of manipulation. The dichromate reflux method can be used to measure the samples with COD values of greater than 50 mg/L.

The COD test measures carbon and hydrogen in organic constituents but not nitrogenous compounds. Furthermore, the method does not differentiate between biologically stable and unstable compounds present in water and wastewater. The COD test is a very important procedure for routinely monitoring industrial wastewater discharges and for the control of waste treatment processes. The test is faster and more reproducible than the BOD method. The obvious disadvantages of the COD method are as follows:

1. Aromatic hydrocarbons, pyridine, and straight-chain aliphatic compounds are not readily oxidized.
2. The method is very susceptible to interference by **chloride**, and thus the COD of certain food processing waste effluents such as pickle and sauerkraut brines cannot be readily determined without modification. This difficulty may be overcome by adding **mercuric sulfate** to the sample prior to refluxing. Chloride concentrations greater than 500–1000 mg/L may not be corrected by the addition of mercuric sulfate. A chloride correction factor can be developed for a particular waste by the use of proper blanks.

28.3 COMPARISON OF BOD AND COD METHODS

The BOD and COD analyses of water and wastewater can result in different values because the two methods measure different materials. As shown in Table 28.1, the COD value of a waste sample is usually higher than its BOD because:

1. Many organic compounds that can be chemically oxidized cannot be biochemically oxidized. For example, cellulose cannot be determined by the BOD method but can be measured by the COD test.
2. Certain inorganic compounds such as ferrous iron, nitrites, sulfides, and thiosulfates are readily oxidized by potassium dichromate. This inorganic COD introduces an error when computing the organic matter of water and wastewater.
3. The BOD test can give low values because of a poor seeding material. The COD test does not require an inoculum.
4. Some aromatics and nitrogenous (ammonium) compounds are not oxidized by the COD method. Other organic constituents such as cellulose or lignin, which are readily oxidized by potassium dichromate, are not biologically degraded by the BOD method.
5. Toxic materials present in water and wastewater that do not interfere with the COD test can affect the BOD results.

Table 28.1 Oxygen demand of tomato processing wastes

Item	*1973*	*1974*	*1975*
BOD, mg/L	2400	1300	1200
COD, mg/L	5500	3000	2800

From Splittsloesser and Downing [2]

Table 28.2 COD and BOD values of selected fruit and vegetable processing wastes

Product	*COD (mg/L)*	*BOD (mg/L)*	*Mean ratio (BOD/COD)*
Apples	395–37,000	240–19,000	0.55
Beets	445–13,240	530–6400	0.57
Carrots	1750–2,910	817–1927	0.52
Cherries	1200–3795	600–1900	0.53
Corn	3400–10,100	1587–5341	0.50
Green beans	78–2200	43–1400	0.55
Peas	723–2284	337–1350	0.61
Sauerkraut	470–65,000	300–41,000	0.66
Tomatoes	652–2305	454–1575	0.72
Wax beans	193–597	55–323	0.58
Wine	495–12,200	363–7645	0.60

From Splittsloesser and Downing [2]

The COD has value for specific wastes since it is possible to obtain a direct correlation between COD and BOD values. Table 28.2 shows the COD and BOD values of waste effluents from fruit and vegetable processing factories. The BOD/COD ratios of these processing waste effluents varied considerably and ranged from 0.50 to 0.72 [2]. The **BOD/COD ratio** can be a useful tool for rapid determination of the biodegradability of organic matter present in the wastes. A low BOD/COD ratio indicates the presence of a large amount of nonbiodegradable organic matter. Samples of wastewater with high BOD/COD ratios have a small amount of organic matter that is nonbiodegradable.

28.4 SAMPLING AND HANDLING REQUIREMENTS

Samples of water and wastewater collected for oxygen demand determinations must be analyzed as soon as possible or stored under properly controlled conditions until analyses can be made. Samples for the BOD test can be kept at low temperatures (4 °C or below) for up to 48 h. Chemical preservatives should not be added to water and wastewater because they can interfere with BOD analysis. Untreated wastewater samples for the COD test must be collected in glass containers and analyzed promptly. The COD samples can be stored at 4 °C or below for up to 28 days if these are acidified with a concentrated mineral acid (sulfuric acid) to a pH value of 2.0 or below.

28.5 SUMMARY

Oxygen demand is most widely used to determine the effect of organic pollutants present in water and wastewater on receiving streams and rivers. The two important methods used to measure oxygen demand are BOD and COD. Table 28.3 summarizes the principle, advantages, and disadvantages of BOD and COD. The BOD test measures the amount of oxygen required by microorganisms to oxidize the biodegradable organic matter present in water and wastewater. The COD method determines the quantity of oxygen consumed during the oxidation of organic matter in water and wastewater by potassium dichromate. Of the two methods used to measure oxygen demand, the BOD test has the widest application in measuring waste loading to treatment systems, in determining the efficiency of treatment processes, and in evaluating the quality of receiving streams and rivers because it most closely approximates the natural conditions of the environment. The COD test can be used to monitor routinely the biodegradability of organic matter in water and wastewater if a relationship between COD and BOD has been established.

Table 28.3 Comparison between biochemical oxygen demand and chemical oxygen demand

	Principal	*Advantages*	*Disadvantages*
Biochemical oxygen demand (BOD)	Measure amount of oxygen required by microorganisms to oxidize the biodegradable organic matter present in water and wastewater (i.e., correlation between amount of organic matter and amount of oxygen used to oxidize pollutants)	Measures true compounds of interest (i.e., organic matter). Less expensive. No interference from certain compounds that affect COD results (see text for details)	Slower. Less precise. Gives low values if poor seed material. Requires inoculum
Chemical oxygen demand (COD)	Organic compounds are destroyed by refluxing in strong acid solution with a known excess of an oxidizing agent (potassium dichromate) (i.e., correlation between amount of organic matter chemically oxidized and amount of oxidizing agent consumed)	Faster. More precise	Not a direct measurement of organic matter. More expensive. Can overestimate organic matter if sample is high in certain compounds, and can underestimate if high in other compounds (see text for details)

28.6 STUDY QUESTIONS

1. In your new job as supervisor of a lab that has previously been using the BOD method to determine oxygen demand of wastewater, you have decided to change to the COD method.
   (a) Differentiate the basic principle and procedure of the BOD and COD methods for your lab technicians.
   (b) In what case would they be instructed to use mercuric sulfate in the COD assay?
   (c) How do you justify making the change from the BOD method to the COD method?
2. In each case described below, indicate if you would expect the COD value to be higher or lower than the results from a BOD test. Explain your answer.
   (a) Poor seed material in BOD test
   (b) Sample containing toxic materials
   (c) Sample high in aromatics and nitrogenous compounds
   (d) Sample high in nitrites and ferrous iron
   (e) Sample high in cellulose and lignin

28.7 PRACTICE PROBLEMS

1. Determine the BOD value of a sample given the following data (see Eq. 28.1):
   DOB = 9.0 mg/L
   DOD = 6.6 mg/L
   P = 15 mL
   Capacity of bottle = 300 mL
2. Determine the COD value of a sample given the following data (see Eq. 28.2):
   mL FAS for blank = 37.8 mL
   mL FAS for sample = 34.4 mL
   Molarity of FAS = 0.025 M
   Sample = 5 mL

Answers

1. BOD = 48 mg/L

   Calculation:

   $$\begin{aligned} \text{BOD (mg/L)} &= 100/P \times (9.0\ \text{mg/L} - 6.6\ \text{mg/L}) \\ &= 100/P \times 2.4 \\ &= 240/P \\ &= 240/(15\ \text{mL} \times 100/300\ \text{mL}) \\ &= 240/5 \\ &= 48 \end{aligned}$$

2. COD = 136 mg/L

   Calculation:

   $$\begin{aligned} \text{COD (mg/L)} &= (37.8\ \text{mL} - 34.4\ \text{mL}) \times 0.025 \times 8000/D \\ &= 3.4 \times 0.025 \times 8000/D \\ &= 680/D \\ &= 680/5 \\ &= 136 \end{aligned}$$

References

1. Lipps WC, Braun-Howard EB, Baylor TE (eds) (2023) Standard methods for the examination of water and wastewater, 24th edn. American Public Health Association, American Water Works Association, Water Environment Federation, Washington, DC
2. Splittstoesser DF, Downing DL (1969) Analysis of effluents from fruit and vegetable processing factories. NY State Agr Exp Sta Res Circ 17. Geneva, NY

Part VI

Analysis of Physical Properties of Foods

Rheological Principles for Food Analysis

29

Helen S. Joyner

Contents

29.1 INTRODUCTION

29.1.1 Rheology and the Food Industry

Food scientists are routinely confronted with the need to measure physical properties related to sensory texture and behavior during processing. These properties are determined by **rheological methods**; **rheology** is a science devoted to the deformation and flow of all materials. Rheological properties should be considered a subset of the textural properties of foods because sensory evaluation of texture encompasses factors beyond rheological properties. Specifically, rheological methods accurately measure force, deformation, and flow, and food scientists and engineers must determine how best to apply this information. For example, the flow of salad dressing from a bottle, the snapping of a candy bar, or the pumping of cream through a homogenizer are each related to the rheological properties of these materials. In this chapter, we describe fundamental concepts of food rheology and discuss typical rheological tests for common foods, as well as how to select rheological tests for a given food product. A

H. S. Joyner (✉)
Premier Nutrition Company, Emeryville, CA, USA
e-mail: academicjoyner@gmail.com

B. P. Ismail, S. S. Nielsen (eds.), *Nielsen's Food Analysis*, Food Science Text Series,
https://doi.org/10.1007/978-3-031-50643-7_29

glossary is included to define and summarize the rheological terms used throughout the chapter.

29.1.2 Rheological Methods in the Food Industry

Rheological properties are determined by applying and measuring forces and deformations as a function of time. In the food industry, both fundamental and empirical measurement methods may be used. **Fundamental methods** account for the magnitude and direction of forces and deformations while placing restrictions on acceptable sample shapes and compositions. Fundamental tests have the advantage of being based on known concepts and equations of physics. Therefore, fundamental tests performed on different testing equipment with different sample geometries yield comparable results. However, food samples often have compositions and/or shapes that are too complex to be used in fundamental testing, and fundamental test results may not be able to be related to processing or sensory results. Therefore, **empirical methods** are often used for evaluating the rheological behaviors of food products. These empirical methods are typically descriptive in nature and ideal for rapid analysis. However, the results of empirical tests are dependent on the equipment and sample geometry, so it may be difficult to accurately compare data among different samples. Empirical tests are of value especially when they correlate with a property of interest, whereas fundamental tests determine true physical properties.

29.1.3 Basic Assumptions for Fundamental Rheological Methods

The precision and accuracy of fundamental measurements are dependent on multiple assumptions. Two important assumptions for fundamental methodologies are that the material is **homogeneous** and **isotropic**. **Homogeneity** implies a well-mixed and compositionally similar material, an assumption generally valid for fluid foods provided they are not a suspension of large particles, such as vegetable soup. For example, milk, mayonnaise, and apple juice are each considered homogeneous and isotropic. Homogeneity is more problematic in solid foods. For example, frankfurters without skins can be considered homogeneous. However, when particle size is significant, such as fat particles in some processed meats like salami, one must determine if homogeneity is a valid assumption. **Isotropic** materials display a consistent response to a load regardless of the applied direction. In foods such as a steak, muscle fibers make the material anisotropic so the response varies with the direction of the force or deformation. Another critical assumption for many tests is that the food composition and structure does not permanently change during testing. This assumption holds for traditional rheometry, but novel testing methods can violate this assumption, requiring careful interpretation of the data and numerous replicates to properly understand the measured behaviors.

29.2 FUNDAMENTALS OF RHEOLOGY

Rheology is concerned with how all materials respond to applied forces and deformations, and food rheology is the material science devoted to foods. Basic concepts of **stress** (force per area) and **strain** (relative deformation) are key to all rheological evaluations. Special constants of proportionality, called **moduli**, link stress with strain. Materials that are considered ideal solids (e.g., gelatin gels) obey **Hooke's Law**, for which stress is related directly with strain by a modulus. Materials that are considered ideal fluids (e.g., water, honey) obey **Newtonian** principles, and the proportionality constant is commonly referred to as **viscosity**, defined as an internal resistance to flow. These principles for solid and fluid behavior form the foundation for the entire chapter and are described in this section, with additional detail available in Steffe and Daubert [1].

29.2.1 Concepts of Stress, Strain, and (Shear) Strain Rate

Stress (σ) is always a measurement of **force**. Defined as the force (F, **Newtons**) divided by the area (A, **meters**^{2}) over which the force is applied, stress is generally expressed with units of **Pascals** (Pa). To illustrate the notion of stress, imagine placing a water balloon on a table as opposed to placing it on the tip of a pin (Fig. 29.1). The tip of the pin has a considerably smaller surface area, causing the stress, or weight (force due to gravity) per unit area of contact, to be much larger when compared with the stress of a balloon resting on a tabletop. Although the force magnitude (i.e., the weight of

Fig. 29.1 Concepts of stress

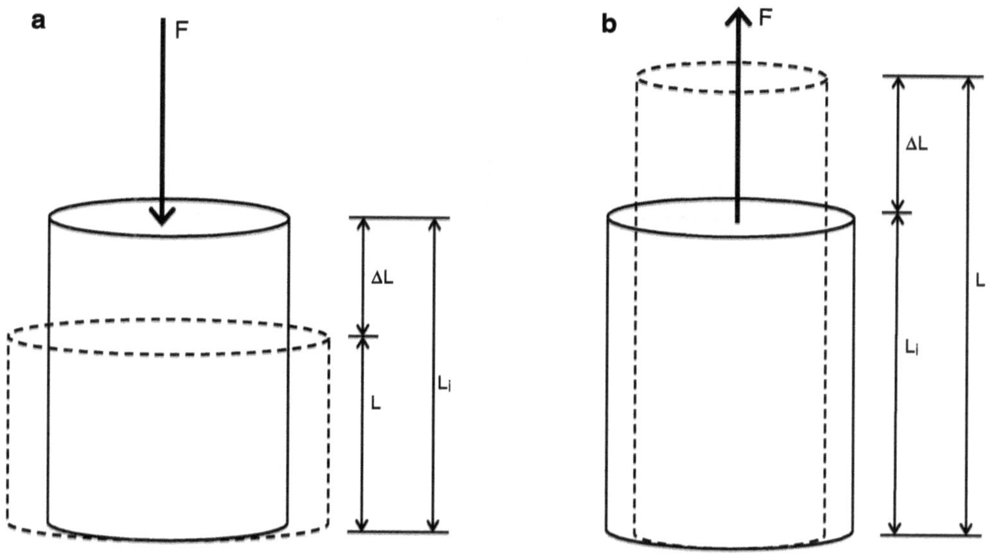

Fig. 29.2 Normal strain in a cylinder in (**a**) compression and (**b**) tension

the water balloon) is a constant value for each case, the final outcomes will be very different.

The direction of the force, with respect to the surface area impacted, determines the type of stress. For example, if the force is perpendicular to a surface, a **normal stress** results and can be achieved under tension or compression. If the force acts in parallel to the sample surface, a **shear stress** is experienced. Examples of normal stresses include pressing the edges of a piecrust together and biting through a solid food. Examples of shear forces include spreading butter over a slice of toast, brushing barbecue sauce on chicken, or stirring milk into a cup of coffee.

When a stress is applied to a food, the food deforms or flows. Strain is a dimensionless quantity representing the relative deformation of a material, and the direction of the applied stress with respect to the material surface will determine the type of strain. If the stress is normal (perpendicular) to a sample surface, the material will experience normal strain (ε) (Fig. 29.2). Foods show normal strains when they are compressed (compressive stress) or stretched (tensile stress). On the other hand, when a sample encounters a shear stress, such as the pumping of tomato paste through a pipe, a **shear strain** (γ) is observed. Figure 29.3 shows how a sample deforms when a shear stress is applied.

Normal strain ($\boldsymbol{\varepsilon}$) can be calculated using the original and change in length of the material:

$$\varepsilon = \ln\left(1+\frac{\Delta L}{L}\right) \tag{29.1}$$

Strain calculations give negative valves for compression and positive values for extension (tensile strains). Rather than expressing a negative strain, many rheologists typically record the absolute value of the strain and denote the compressive test mode.

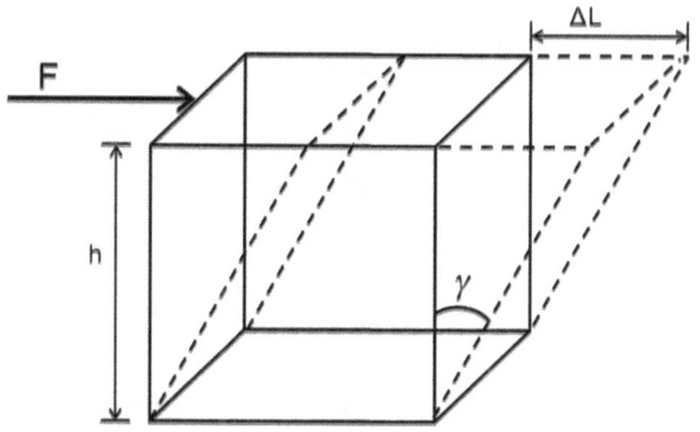

Fig. 29.3 Shear strain in a cube

When the material is a liquid, this approach for strain quantification is a bit more challenging. As coffee is stirred, water is pumped, or milk is pasteurized, these fluids all are exposed to shear and display irrecoverable deformation. Therefore, a **(shear) strain rate** ($\dot{\gamma}$), often called shear rate, is typically used to quantify strain during fluid flow. Shear rate is the amount of deformation (strain) per unit time with units of s^{-1}.

To picture the shear rate concept, consider a fluid filling the gap between two moveable, parallel plates separated by a known distance, h, as illustrated in Fig. 29.4. Now, set one plate in motion with respect to the other at a constant horizontal velocity, U. The shear rate for this system can be approximated by dividing the plate velocity by the fluid gap height, producing a value with units of s^{-1}.

$$\dot{\gamma} = \frac{U}{h} \tag{29.2}$$

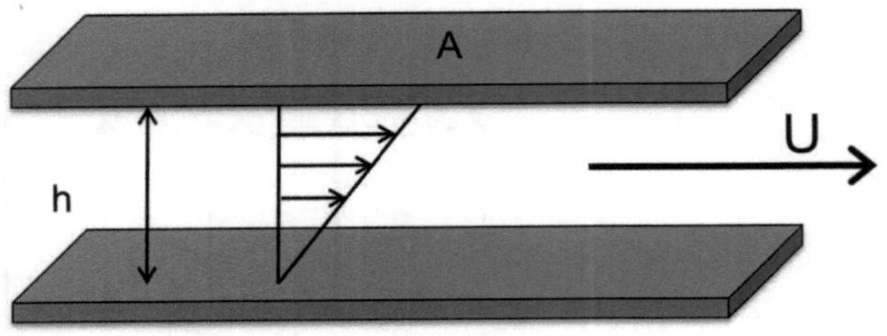

Fig. 29.4 Shear flow between parallel plates

Table 29.1 Elastic and shear moduli for common materials

Material	*E, elastic moduli (Pa)*	*G, shear moduli (Pa)*
Apple	1.0×10^7	0.38×10^7
Potato	1.0×10^7	0.33×10^7
Spaghetti, dry	0.27×10^{10}	0.11×10^{10}
Glass	7.0×10^{10}	2.0×10^{10}
Steel	25.0×10^{10}	8.0×10^{10}

This shear rate may be more easily understood through the deck of cards analogy. Imagine a stack of playing cards, with each card representing an infinitely thin layer of fluid. When the top card is pushed forward, the entire deck deforms to some degree proportional with the magnitude of the force. This type of movement is called **simple shear**, defined as a laminar deformation along a plane parallel to the applied force.

29.2.2 Elastic and Shear Moduli

Hooke's Law states that when a solid material is exposed to a stress, it experiences an amount of deformation or strain proportional to the magnitude of the stress. The constants of proportionality, used to equate stress with strain, are called **moduli**. If a normal stress is applied to a sample, the proportionality constant is known as **elastic modulus** (E), often called Young's modulus:

$$\sigma = \frac{F}{A} = E\varepsilon \quad (29.3)$$

Likewise, if the applied stress is shear stress, the constant is the **shear modulus** (G):

$$\sigma = G\gamma \quad (29.4)$$

These moduli are inherent properties of the material and have been used as indicators of quality. Moduli of selected foods and materials are provided in Table 29.1.

29.2.3 Viscosity Profiles

Very simple fluids, such as water, milk, and honey, have constant viscosity that is independent of shear rate and time. In other words, **Newton's postulate** is obeyed. This postulate states that if the shear stress is doubled, the velocity gradient (shear rate) within the fluid is also doubled. Similar to stress and strain for solids, the shear stress on fluids can be expressed as a function of shear rate to provide an indication of the internal resistance of a fluid to flow. For **Newtonian fluids**, shear stress and shear rate are related by the **coefficient of viscosity** or **Newtonian viscosity** (μ):

$$\sigma = \mu\dot{\gamma} \quad (29.5)$$

However, most liquids do not have constant viscosity due to compositional or structure effects. Instead, their viscosity changes as a function of shear rate, and the material is considered **non-Newtonian. Apparent viscosity** (η) is used for these fluids, defined as the shear stress divided by the shear rate:

$$\eta = f(\dot{\gamma}) = \frac{\sigma}{\dot{\gamma}} \quad (29.6)$$

When reporting the viscosity of a non-Newtonian fluid measured at a specific shear rate, it is important to report the shear rate at which that viscosity was measured with the viscosity value. However, it is more beneficial to show the entire viscosity profile of a non-Newtonian fluid to provide a better understanding of the overall flow behavior of the food. The flow behavior of a food can be described with fluid **rheograms** or appropriate rheological models. A rheogram is a graphical representation of flow behavior, showing the relationship between stress and strain or shear rate. The shape of the rheogram indicates the type of flow behavior. For example, if a plot of shear stress versus shear rate results in a straight line passing through the origin, the material is a Newtonian fluid, with the slope of the line equaling the Newtonian viscosity (μ), (Fig. 29.5). A number of common foods exhibit this ideal response, including water, milk, fruit juices, vegetable oils, and honey. Table 29.2 provides the Newtonian viscosity for common materials at 20 °C; this information is helpful for understanding into what range the viscosity of various food products might fall.

As established earlier, most fluid foods do not show Newtonian flow behavior. Their flow behavior changes with shear rate (i.e., mixing speed) or with time at a constant shear rate. Deviation from ideal Newtonian behavior results in a nonlinear relationship between shear stress and shear rate. Figure 29.5 shows example rheograms for several non-Newtonian flow behaviors.

If the viscosity decreases as shear rate increases, the material is referred to as **shear thinning** or **pseudoplastic**. Examples of pseudoplastic food items are beverages thickened with gums, mayonnaise, applesauce, and ice cream mix. If the viscosity of the material increases with increased shear rate, the sample is called **shear thickening** or **dilatent**. Cornstarch slurries are well-known for dilatent behavior.

Pseudoplasticity and dilatency are **time-independent properties**. Materials that thin and thicken with time are

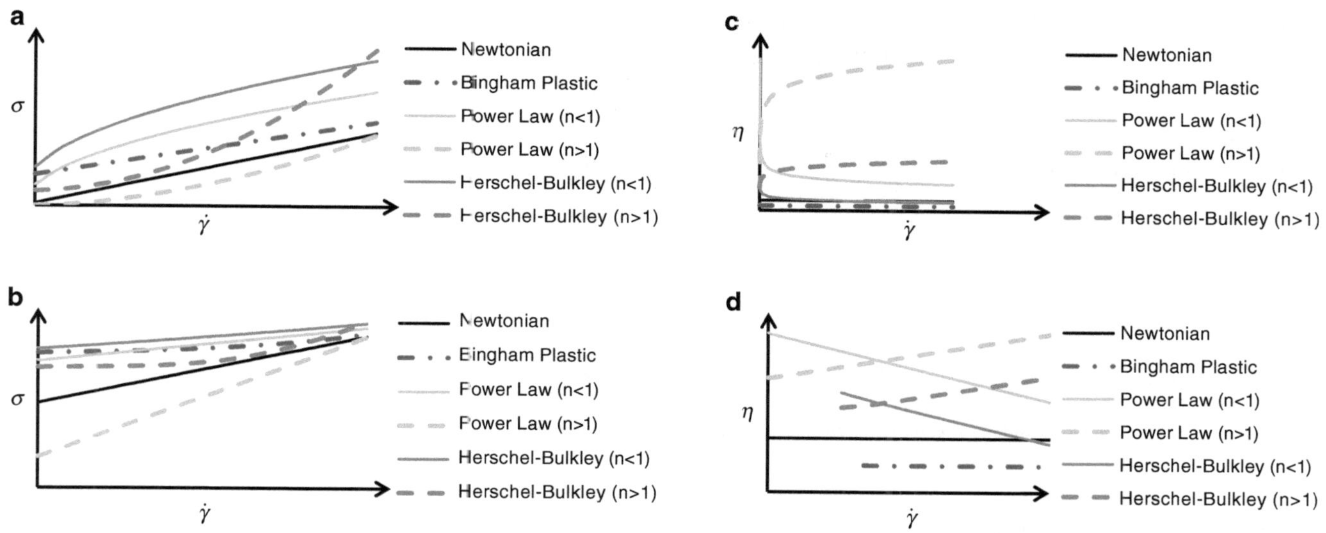

Fig. 29.5 Rheogram for fluids with different flow behaviors. (**a**) stress versus shear rate in normal scale, (**b**) stress versus shear rate in log scale, (**c**) viscosity versus shear rate in normal scale, and (**d**) viscosity versus shear rate in log scale

Table 29.2 Newtonian viscosities for common materials at 20 °C

Material	*Viscosity, μ (Pa s)*
Honey	11.0
Rapeseed oil	0.163
Olive oil	8.4×10^{-2}
Cottonseed oil	7.0×10^{-2}
Raw milk	2.0×10^{-3}
Water	1.0×10^{-3}
Air	1.81×10^{-5}

Adapted from Refs. Steffe [2] and Muller [3]

known as **thixotropic** and **rheopectic** liquids, respectively. These fluids are easily detected by monitoring the viscosity over time at a constant shear rate. If peanut butter is mixed at a constant speed, the material thins (thixotropy) with time due to the destruction of weak bonds linking the molecules. This effect can be observed in a jar of peanut butter: the area that has been disturbed multiple times with a knife or spoon is softer and more easily spread than an undisturbed area.

Many fluids do not flow at low stresses. In fact, a certain catsup brand once staked numerous marketing claims on the "anticipation" of flow from the bottle. Often, additional force needs to be applied to the catsup container to expedite the pouring. The minimum force, or stress, required to initiate flow is known as a **yield stress** (σ_0). Because Newtonian fluids require the stress–shear rate relationship to be a continuous straight line passing through the origin, any material with a yield stress is automatically non-Newtonian. A few common foods possessing yield stresses are catsup, yogurt, mayonnaise, and chocolate milk. Many foods are explicitly designed to include a certain yield stress. For example, if melted cheese did not have a yield stress, the cheese would flow off a cheeseburger or pizza. If salad dressings flowed at the lowest of applied stresses, the force of gravity would cause the dressing to run off the salad leaves. There are many fascinating rheological features of foods that consumers may never consider!

Once data for shear stress and shear rate are collected, **rheological models** can be used to gain a greater understanding of the flow response. Rheological models are mathematical expressions relating shear stress to shear rate, providing a "flow fingerprint" for a particular food. In addition, the models enable prediction of rheological behavior across a wide range of processing conditions. For practical purposes, the **Herschel–Bulkley model** can account for the steady-state rheological performance of many fluid foods:

$$\sigma = \sigma_o + K\dot{\gamma}^n \quad (29.7)$$

K and n represent material constants called the **consistency coefficient** and **flow behavior index**, respectively. The flow behavior index provides an indication of Newtonian or non-Newtonian flow, provided the material has no yield stress. Table 29.3 illustrates how the Herschel–Bulkley model is used to identify specific flow characteristics. It also shows how the Herschel–Bulkley model can be modified for foods that do not have yield stresses or have a yield stress but Newtonian flow behavior. Take, for example, the data for peanut oil. Table 29.4 reports no yield stress and a flow behavior index of 1.00 – the scenario for a Newtonian fluid. Accordingly, the Herschel–Bulkley model does in fact collapse to the special Newtonian case. Other equations are also available for modeling flow behaviors and are discussed in more detail in [2, 4].

Temperature is an important consideration when measuring flow behaviors. Viscosity generally decreases with

Table 29.3 Manipulation of Herschel–Bulkley model to describe flow behavior

Fluid type	*Shear dependent?*	σ_o	n	*Adjusted model*
Newtonian	No	0	1.0	$\sigma = \mu\dot{\gamma}$
Non-Newtonian				
Pseudoplastic, no yield stress	Yes	0	<1.0	$\sigma = K\dot{\gamma}^n$
Pseudoplastic, yield stress	Yes	>0	<1.0	$\sigma = \sigma_o + K\dot{\gamma}^n$
Dilatent, no yield stress	Yes	0	>1.0	$\sigma = K\dot{\gamma}^n$
Dilatent, yield stress	Yes	> 0	>1.0	$\sigma = \sigma_o + K\dot{\gamma}^n$
Yield stress only	No	>0	1.0	$\sigma = \sigma_o + \mu\dot{\gamma}$

σ_o yield stress, n flow behavior index, μ Newtonian viscosity, K consistency coefficient

Table 29.4 Herschel–Bulkley model data for common foods

Product	*Temp. (°C)*	*Shear Rate (s^{-1})*	*K (Pa s^n)*	*n (–)*	σ_o *(Pa s)*
Orange juice (13 °Brix)	30	100–600	0.06	0.86	–
Orange juice (22 °Brix)	30	100–600	0.12	0.79	–
Orange juice (33 °Brix)	30	100–600	0.14	0.78	–
Apple juice (35 °Brix)	25	3–2000	0.001	1.00	–
Ketchup	25	50–2000	6.1	0.40	–
Apple sauce	32	–	200	0.42	240
Mustard	25	–	3.4	0.56	20
Melted chocolate	46	–	0.57	1.16	1.16
Peanut oil	21.1	0.32–64	0.065	1.00	–

Adapted from Refs. Steffe [2] and Rao [5]

increased temperature. Accordingly, a small change in temperature, say 1–2 °C, can have a major impact on food flow behaviors. This is especially true when measuring food products near the melting point of the fats they contain, or near the gelation temperature of proteins. Generally, the viscosities of foods are measured at a temperature of interest, such as serving temperature, body temperature, or processing temperature. The selected temperature depends on how the data will be used. For example, 37 °C, or human body temperature, is advised when the viscosity data are to be compared with sensory attribute scores. However, it is a common practice to measure viscosity at 20 or 25 °C (room temperature), since many products are stored and/or consumed at this temperature.

29.3 STANDARD RHEOMETRY

Rheometers are devices used to determine viscosity profiles and other rheological properties of materials. Relationships between shear stress and shear rate are derived from values of system configurations, pressures, flow rates, and other applied conditions.

29.3.1 Compression, Extension, and Torsion Analyses

The rheological properties of solid foods are measured by compressing, extending, or twisting the material and can be accomplished by two general approaches called small- or large-strain testing. In **small-strain tests**, the goal is to apply the minimal amount of strain or stress required to measure the rheological behavior while at the same time preventing (or at least minimizing) damage to the sample. The goal in **large strain** and **fracture tests** is the opposite. Samples are deformed until the food matrix is significantly strained, damaged, or possibly fractured. Small-strain tests are used to understand the properties of an intact food network, whereas large-strain tests give an indication of sensory texture or product durability. In general, compressive and extensional tests are generally performed using large strains. Torsional (shear) tests may be performed using small or large strains, depending on the information desired.

29.3.1.1 Large-Strain Testing

Compression and tension (i.e., extension) tests are used to determine large strain and fracture properties. **Compression**

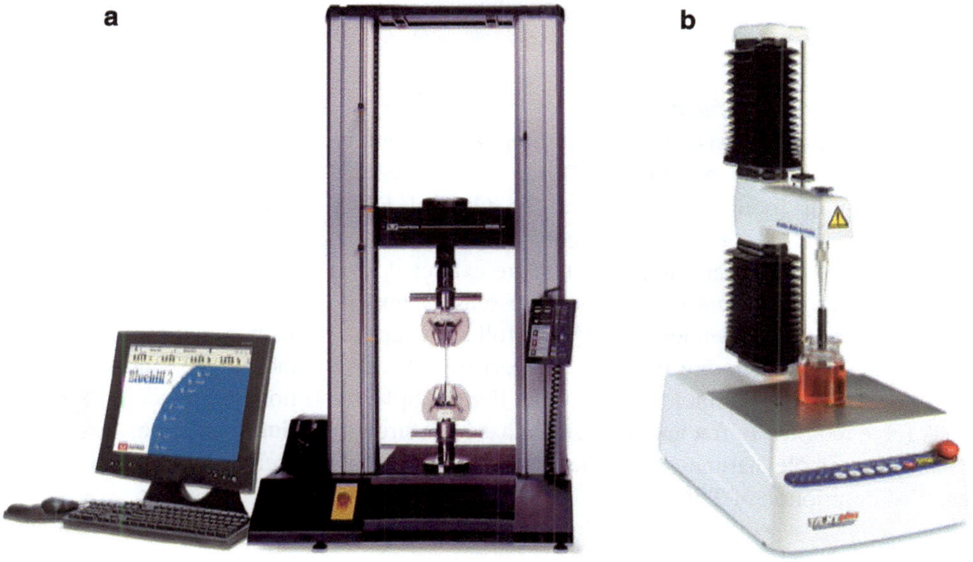

Fig. 29.6 Photos of (**a**) universal testing machine. (Courtesy of Instron®, Norwood, MA) and (**b**) Texture Analyzer. (Courtesy of Texture Technologies Corp., Hamilton, MA)

tests are generally selected for solid or viscoelastic solid foods when a tight attachment between sample, and the testing fixture is not required, thereby simplifying sample preparation. **Tension** and **torsion tests** are well suited for foods that require a high level of strain to fracture them. The main disadvantage to tension and torsion tests is that the sample must be attached to the test fixture [5]. Chen and Opara [6] and Hamann et al. [7] provide a detailed comparison and analysis of different methods of large deformation rheological testing.

Compression testing is the most common large-strain test performed in the food industry. It is typically performed using a **universal testing machine**, a **Texture Analyzer**, or a **penetrometer** (Fig. 29.6). Both the universal testing machines and the Texture Analyzer have a motor that applies a specific force or deformation on a sample, and the resulting deformation or force, respectively, is measured. Penetrometers use a calibrated weight in a rig that drops from a set height into the sample. The penetration depth of the geometry, usually a cone, into the sample is recorded. This depth is not directly convertible into a stress or strain but can be used to compare samples tested with the same weight and rig.

29.3.1.1.1 *Determining Stress, Strain, and Elastic Modulus* (E) in Compression

There are several assumptions to consider when doing compression testing. Along with the previously mentioned considerations of a homogeneous and isotropic material, the assumption that the food is an **incompressible material** greatly simplifies matters. An incompressible material changes in shape but not volume when compressed. Foods such as frankfurters, cheese, cooked egg white, and other high-moisture, gel-like foods are considered incompressible.

During compression, the initial cross-sectional area (A_i) increases as the length decreases. To account for this change, a correction term incorporating a ratio of the cylinder lengths (L/L_i) is applied to the stress calculation Eq. 29.8. Strain during compression can be calculated using Eq. 29.1.

$$\sigma = \frac{F}{A_i}\left(\frac{L}{L_i}\right) \tag{29.8}$$

In compression testing, one should use a cylindrical shaped sample with a length (L) to diameter ratio of >1.0. A cork borer is often used to create cylinders for testing. Gelled samples can be prepared in cylindrical molds. The sample should be compressed between two flat plates with diameters exceeding the lateral expansion of the compressed sample, and the entire sample cross section should be in contact with the plates during testing. The sample should also maintain a cylindrical shape with straight sides when compressed. If this is not the case, the contact surface between the plate and the sample may need lubrication. Water or oil can be used; the lubricant used should not be absorbed by or cause any deleterious effects to the sample.

If the material to be tested is a pure elastic solid, the compression rate does not matter. However, if the material is viscoelastic (as is the case with most foods), then the values for stress, strain, and elastic modulus can change with the speed of compression. A complete characterization of a viscoelastic material requires determining these values at a variety of compression rates. Another factor to consider is the level of compression. The sample can be compressed to fracture, or some level below fracture. The level of compression should be selected based on the desired information of the test. For example, compression to fracture is recommended for comparison to sensory data, while compression without fracture may be more desirable for understanding processing behaviors.

29.3.1.1.2 Texture Profile Analysis

Texture profile analysis (TPA) is an empirical technique using a two-cycle compression test. This test was developed by a group of food scientists from the General Foods Corporation and is compiled as force during compression and time (Fig. 29.7). Various parameters are extracted from the force–time curve and used to estimate sensory texture. Provided that the guidelines of TPA are followed and the appropriate parameters are used for comparison, TPA data can show good correlation to sensory data. For example, the peak force required to fracture a specimen has been strongly related to sample hardness. Bourne [8] provides a more detailed description of TPA.

29.3.1.2 Fracture Testing

Fracture tests are large-strain tests carried out to the point of sample failure. Generally, the failure point appears as a sudden decrease in stress as the sample fractures. Fracture tests may be performed in compressive, tensile, or shear (torsion) modes. Compressive or shear modes are usually used for food materials, since difficulties in gripping the sample without damaging it make tensile fracture testing difficult.

Several considerations are needed when performing fracture testing. The geometry of the sample must be controlled, especially in fundamental fracture testing. Certain fracture methodologies specify the shape of the sample, e.g., a cylinder, capstan, or beam shape. The homogeneity of the sample must also be considered. Anisotropic samples such as meats may have different fracture behaviors depending on their orientation. Testing parameters such as strain rate can affect fracture behavior as well. Chen and Opara [6] and Hamann et al. [7] provide a more detailed discussion of fracture testing.

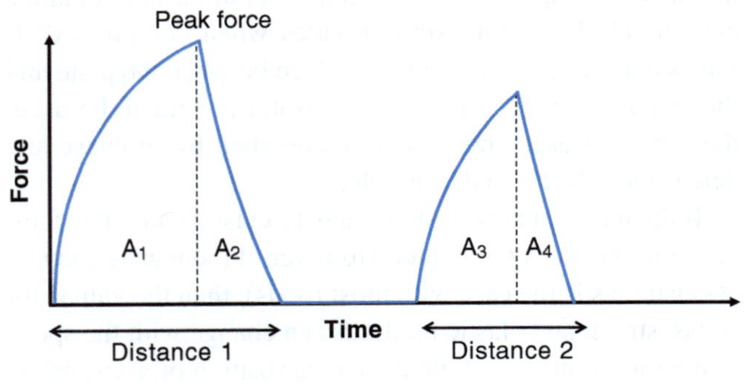

Hardness: Peak force
Cohesiveness: $(A_3+A_4)/(A_1+A_2)$
Springiness: Distance 2/Distance 1
Gumminess: Hardness x Cohesiveness
Chewiness: Hardness x Cohesiveness x Springiness
Resilience: A_2/A_1

Fig. 29.7 Double compression curve showing TPA parameters

29.3.2 Rotational Viscometry

For fluids, the primary mode for rheological measurement in the food industry is rotational viscometry, providing rapid and fundamental information. **Rotational viscometry** involves a known test fixture (geometrical shape) in contact with a sample, and through some mechanical, rotational means, the fluid is sheared by the fixture. Primary assumptions are made for the development of **constitutive equations** (relationships between shear stress and rate) and include the following:

- **Laminar Flow**. Laminar flow is synonymous with streamline flow. If we were to track the velocity and position of a fluid particle through a horizontal pipe, the path would only be in the horizontal direction without a shift toward the pipe wall.
- **Steady State**. There are no net changes to the system over time.
- **No-slip Boundary Condition**. When the test fixture is immersed in the fluid sample, the walls of the fixture and the sample container serve as boundaries for the fluid. At whatever speed either boundary is moving, an infinitely thin layer of fluid immediately adjacent to the boundary is moving at precisely the same velocity, so the boundary serves as a sort of anchor for the edge of the fluid.

When conducting rotational tests, it is important to use equipment and set test parameters that meet these assumptions. If the assumptions are violated, it is difficult to obtain accurate and precise rheological measurements. Since rheological software generally cannot determine whether the assumptions are met, the person running the test must check to ensure all assumptions are met through the duration of the test.

Rotational rheometers may operate in two modes: **steady shear** or **oscillatory**. **Steady shear** is a condition in which the sheared fluid velocity, contained between the boundaries, remains constant at any single position. Furthermore, the velocity gradient across the fluid is a constant.

29.3.2.1 Fundamental Geometries for Steady Shear Measurements

Three test fixtures most often used in steady shear rotational viscometry are the **concentric cylinder, cone and plate**, and **parallel plates**. The concentric cylinder attachment consists of a cylindrical fixture shape, commonly called a **bob**, suspended from a measuring device that is immersed in a sample fluid contained in a slightly larger cylinder, referred to as the **cup** (see Figs. 29.8 and 29.9).

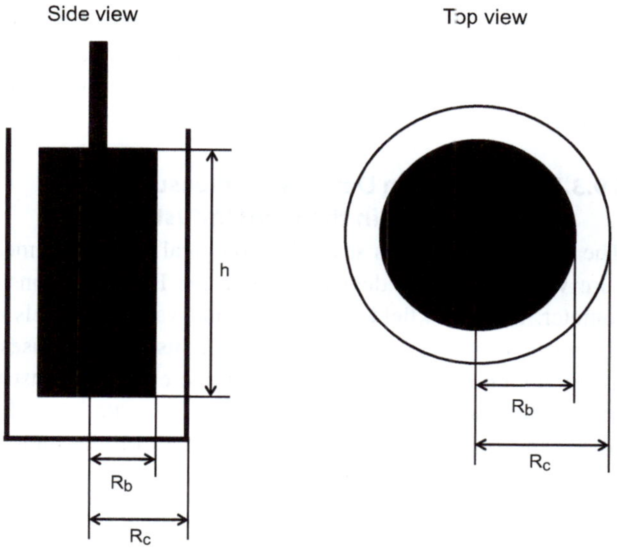

Fig. 29.8 Concentric cylinder geometry

Fig. 29.9 Photo of cup and bob test fixtures (concentric cylinder geometry) for rheological measurements. (Courtesy of TA Instruments, New Castle, DE)

To derive rheological data from experiments, shear stress, and shear rate are used. Calculations for these values require the torque (force) on the bob from the sample, the dimensions of the cup and bob (see Fig. 29.8), and the **rotational speed**, or **angular velocity**, of the bob. Typically, these calculations are performed by the software associated with the instrument. This software usually provides rheograms of shear stress (σ) versus shear rate $\left(\dot{\gamma}\right)$ or apparent viscosity versus shear rate. Example data for tomato catsup collected using a standard cup and bob system are shown in Table 29.5.

Table 29.5 Rheological data of tomato catsup collected using a concentric cylinder test fixture

rpm	*Torque (N m)*	*Shear rate* (s^{-1})	*Shear stress (Pa)*	*Apparent viscosity (Pa s)*
1.0	0.00346	2.09	22.94	10.98
2.0	0.00398	4.19	26.39	6.30
4.0	0.00484	8.38	32.10	3.83
8.0	0.00606	16.76	40.18	2.40
16.0	0.00709	33.51	47.02	1.40
32.0	0.00848	67.02	56.23	0.84
64.0	0.01060	134.04	70.29	0.52
128.0	0.01460	268.08	96.82	0.36
256.0	0.01970	536.16	130.63	0.24

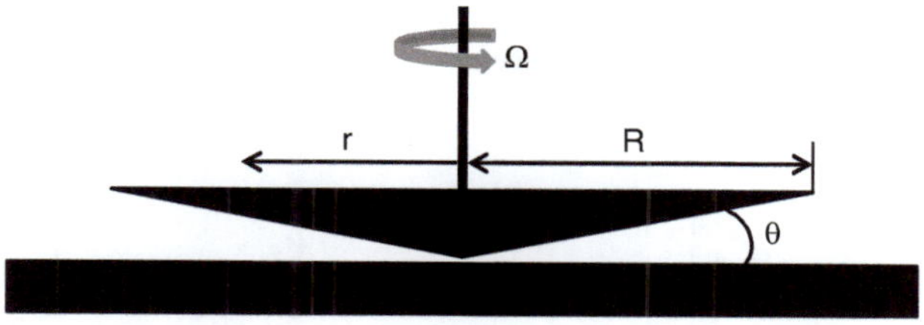

Fig. 29.10 Cone and plate geometry

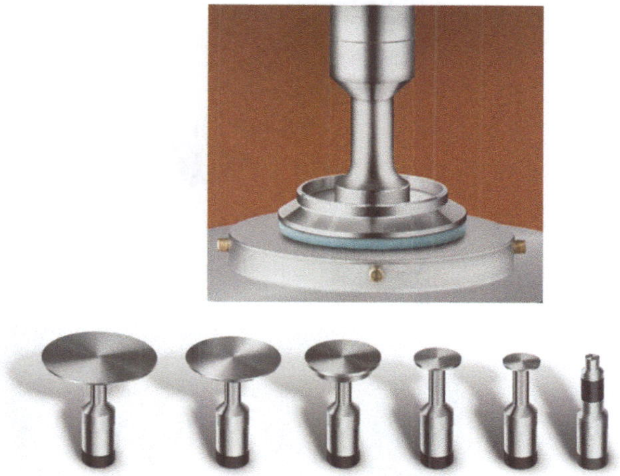

Fig. 29.11 Photos of cone and plate and parallel plate test fixtures for rheological measurements. (Courtesy of TA Instruments, New Castle, DE)

Another popular system for rotational measurement is the cone and plate configuration (Figs. 29.10 and 29.11). Its special design results in a constant shear stress and shear rate for any location of sample in the fluid gap. Test quality is best when the cone angle is small, and large errors may be encountered when the gap is improperly set or not well-maintained.

A primary advantage of the cone and plate test fixture is that shear stress and rate are independent of position, so they are constant throughout the sample. However, if there are large particles in the sample, they may become trapped under the point of the cone, invalidating the viscosity measurement. In this case, a parallel plate configuration may be used (Fig. 29.12), since the standard gap in this configuration is 1.0 mm. Unlike the cone and plate, shear stress and strain are not constant between the parallel plates. For consistency, the outer edge of the plate is set as the measurement location. Therefore, care must be taken to ensure the outer edge of the sample does not change during testing (e.g., no drying occurs).

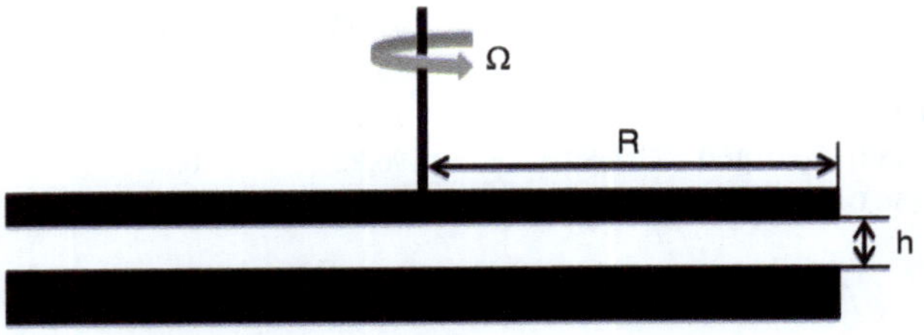

Fig. 29.12 Parallel plate geometry

29.3.2.2 Common Devices for Measuring Viscosity in the Food Industry

The food industry uses several rheological devices to measure viscosity. These devices include the Bostwick consistometer, the Brookfield viscometer, and various models of rheometers (Fig. 29.13). The **Bostwick consistometer** uses a given amount of sample poured into one end of the instrument behind a gate. A timer is started as the gate is quickly lifted, and the time to reach a certain mark on the ramp is measured. The distance the fluid travels is called the consis-

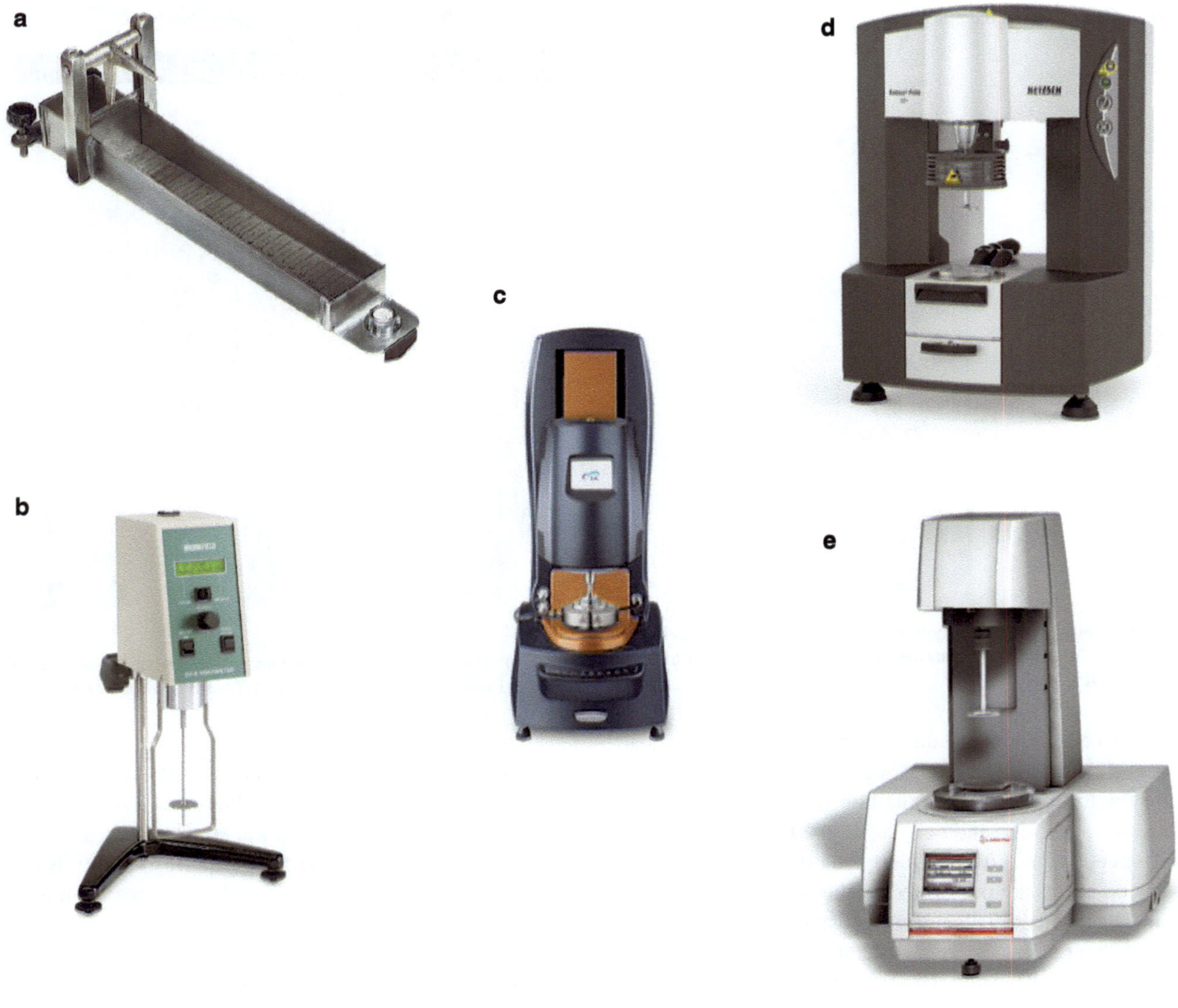

Fig. 29.13 Photo of (**a**) Bostwick consistometer. (Courtesy of Cole-Parmer, Vernon Hills, IL), (**b**) Brookfield viscometer. (Courtesy of Brookfield AMETEK, Middleboro, MA), (**c**) TA DHR3 rheometer, (**d**) KINEXUS rheometer, (**e**) MCR302 rheometer

tency. Note that consistency and viscosity are not interchangeable.

The **Brookfield viscometer** is one of the most common rheological devices found in the food industry. This apparatus uses a spring as a torque sensor. The operator selects a rotational speed (rpm) of the bob, attached to the spring. As the bob moves through the sample fluid, the viscosity impedes free rotation, causing the spring to wind. The degree of spring windup is a direct reflection of the torque magnitude (M), used to determine a shear stress at the bob surface. Newer Brookfield models convert torque to viscosity automatically, while older models provide conversion factors for calculating viscosity from torque.

Rheometers are devices that consist of a measuring head with a drive motor and a measuring platform. There are two main types of rheometer: strain-controlled and stress-controlled. Most rheometers used in the food industry are stress-controlled, meaning that the motor applies a torque on the sample and a sensor measures the angle of deflection of the measuring tool. Strain-controlled rheometers apply a deformation and measuring the resulting torque from the sample. Stettin and Läuger [9] provide an in-depth explanation of the principles of rheometer operation.

The Brookfield viscometer and the Bostwick consistometer are suitable for quick quality control measurements, as they provide rapid measurements and are easy to use and clean. However, these are empirical instruments and are not as precise as a rheometer. Furthermore, they are generally used to measure viscosity at a single shear rate, which can lead to incorrect assumptions about flow behavior. For example, what might happen if the Newtonian and a non-Newtonian fluid in Fig. 29.5 were tested only at the shear rate at which their viscosities are equivalent? It is recommended that these rheological devices be used with fluids that are Newtonian or close to Newtonian to remove shear rate effects from the measurements.

29.3.3 Oscillatory Rheometry

The goal of **oscillatory rheometry** is to characterize the **viscoelastic** properties of a material, generally under small stresses and strains. The term viscoelastic implies the material exhibits both fluid-like (viscous) and solid-like (elastic) behaviors simultaneously. Think of a piece of cheese: it springs back (elastic behavior) if you compress it gently, but not quite to its original height (viscous behavior shown as permanent deformation). Various viscoelastic behaviors are determined by applying: (1) a stress or strain in **oscillation** and measuring the respective strain or stress and phase angle between stress and strain, (2) a **constant strain** and measuring the decrease (relaxation) in stress, or (3) a **constant stress** and measuring the rate of deformation (creep). These tests are generally used to evaluate the viscoelastic behaviors of solid and semisolid foods, such as cheese and pudding, although they can also be used to evaluate fluid foods, such as salad dressing. They can be performed with the same geometries as fundamental rotational tests (see Sect. 29.3.2.1) and should be performed with a rheometer. A more detailed description of these techniques may be found in Steffe [2] and Rao [5].

29.3.4 Novel Rheological Methods

Rotational and oscillatory testing have been in use for over a century to evaluate the rheological properties of food products. However, the results of these tests do not usually give a complete picture of food processing and sensory behaviors. Over the past several decades, multiple novel rheological analyses have been developed to determine additional fundamental information about food products to generate a more complete picture of their behaviors and obtain improved correlation to sensory attributes and processing behaviors.

29.3.4.1 Tribology

Tribology is a subfield of rheology involving the study of friction, lubrication, and wear behaviors of materials. Originally used to study materials such as engine lubricants, tribology has gained interest from the food industry as a way to investigate friction-related aspects of food texture. Sensory attributes such as mouthcoat, chalkiness, or astringency relate poorly to traditional rheological measurements. However, these attributes all involve a sensation of friction. Tribology may be helpful in determining the mechanisms behind these friction-related textural attributes.

Tribological tests are performed by moving one surface against another, fixed surface. A thin layer of the test material is placed between the two surfaces and acts as a lubricant. Different geometries for tribological testing may be seen in Fig. 29.14. The moving surface is slid along the stationary surface at different speeds. Stribeck curves, or plots of friction coefficient versus sliding speed, are generated from the results (Fig. 29.15). Stribeck curves consist of three different behavioral regimes: the **boundary regime** (a), in which the sliding surfaces are in contact and friction is relatively constant; the **mixed regime** (b), in which the sliding surfaces are mostly separated by the lubricating material and friction decreases to a minimum value with increased sliding speed; and the **hydrodynamic regime** (c), in which the sliding surfaces are fully separated by the lubricating material. Friction behavior in the mouth is generally in the boundary to mixed regime, with oral sliding speeds estimated at about 10-30 mm/s.

As a system property rather than a physical property, friction is affected by many aspects of the system. For foods,

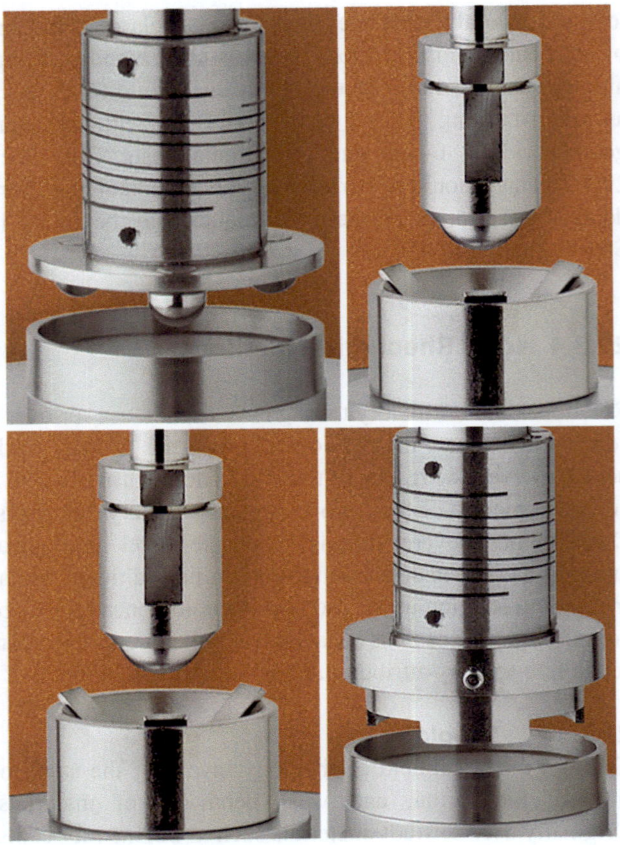

Fig. 29.14 Tribological testing geometries. (Courtesy of TA Instruments, New Castle, DE)

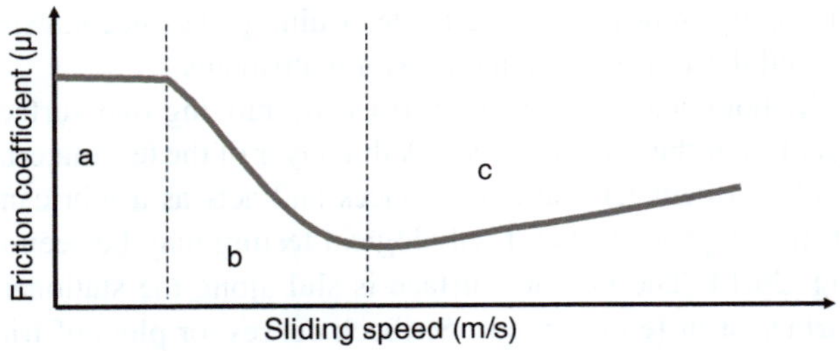

Fig. 29.15 Stribeck curve

friction can be impacted by the food composition and physicochemical properties, food viscosity, food particle size, oral surface composition and conditions, and saliva composition and amount. For example, whole wheat bread has a rougher mouthfeel (higher friction) than bread made with refined flour because the large, irregularly shaped bran particles increase the friction of the whole wheat bread as it moves against the oral surfaces during consumption. A more detailed review of tribology and its uses in the food industry may be found in [10, 11]

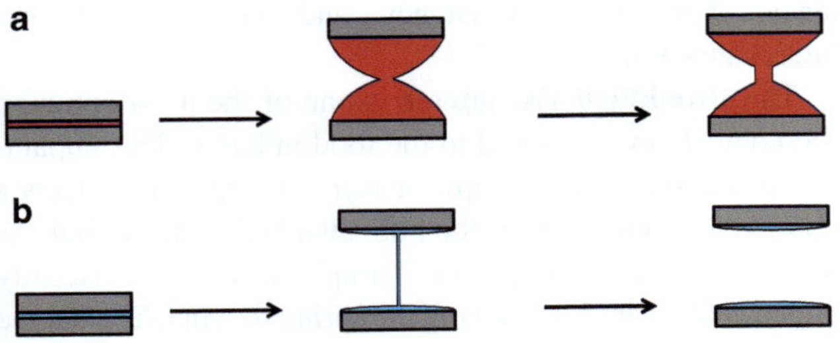

Fig. 29.16 Extensional viscosity measurement on a fluid with (**a**) high extensional viscosity and (**b**) low extensional viscosity

29.3.4.2 Extensional Viscosity

Extensional viscosity is the resistance to flow in the extensional direction. To measure extensional viscosity, the fluid is placed between two plates, and one plate is moved rapidly to a set height. A high-speed camera records the movement of the fluid during and after stretching. Fluids with a high extensional viscosity form strings between the two plates, while fluids with a low extensional viscosity rapidly separate (Fig. 29.16). Extensional viscosity has been found to be important in determining how easy it is to swallow a liquid and may be related to sensory parameters such as mouthcoat. More information on extensional viscosity and food behaviors may be found in [12–14].

29.3.4.3 Large Amplitude Oscillatory Shear

While traditional rheometry uses small strains for oscillatory test and requires no permanent deformation of the sample, **large amplitude oscillatory shear** is performed using strains large enough to cause significant changes in the material's microstructure. The stress response of the food to the imparted strain is plotted and analyzed to determine how the material's microstructure responds and recovers to the oscillating strain. Because the test conditions are more similar to industrial or oral processing conditions (i.e., the material undergoes some structural breakdown) compared to traditional oscillatory rheometry, the data from large amplitude oscillatory shear may relate more closely to material behaviors during processing or chewing. A more detailed description of this testing method and how the results can be related to food behaviors during industrial and oral processing is provided by [15].

29.4 SELECTING RHEOLOGICAL TESTS

Given the broad array of tests and test setups available for measuring food rheological properties, it can be difficult to select the appropriate test(s) for determining the information needed for a particular situation. However, the test selection process can be simplified by considering what information is needed, the sample itself, and the testing resources available. Table 29 6 provides a summary of appropriate tests for determining specific properties of a given material.

29.4.1 Determining Required Information

Determining what information is required of the food product is the first step in selecting a rheological test. This does not necessarily mean selecting a particular rheological parameter to measure. Rather, it means fully understanding what rheological information is needed to answer a question or solve a problem. For example, it is tempting to focus on yield stress measurements of a sauce that does not flow properly from a tank under gravity. However, the flow issue may be due to shear-dependent flow behavior: a shear-thinning sauce could flow well when pumped (high shear) but flow much more slowly under gravity (low shear). Therefore, the required information for this scenario would be the overall viscosity profile of the sauce. Required information may also include behavior under specific environmental conditions, such as a certain temperature or temperature change.

29.4.2 Sample Considerations

There are a number of factors to keep in mind when preparing a food product for a rheological test or performing that test on the food product. First, the classification of the food product as a solid, semisolid, or liquid is important for selecting appropriate tests. For example, materials undergoing fracture testing must be able to hold their shape and not sag or flow under their own weight. Table 29.6 provides a summary of appropriate tests for different material classifications.

In addition, it is important to maintain the composition of the food throughout the test. Some food products are highly susceptible to gaining or losing moisture during testing. These foods should be tested in a controlled environment or have their exposed edges protected to maintain constant composition. They should also be prepared and tested as quickly as possible without losing accuracy to minimize changes during testing.

Knowing how samples react to temperature is also an important consideration for testing. Samples containing lipids may be solid at refrigeration temperatures but fluid at room or body temperatures. Fluid samples containing protein may gel at certain temperatures. If measuring melt or gelation behavior is not the purpose of the rheological test, care should be taken to maintain the samples at a temperature that allows them to remain in their desired state. Even if the sample does not change state over a certain temperature range, temperature may still have a major impact on food rheological behaviors.

Sample age is another consideration. Some food products, such as cheese and protein bars, show large rheological changes over time. Depending on the product, significant rheological changes may happen over minutes, hours, days, weeks, or months. It is important to understand the timeframe of these rheological changes to properly compare datasets among food products. If the extent of rheological changes in the food over time is unknown, testing should be conducted to determine this information before comparing data from food products of varying ages.

Some food products, such as hot dogs, gelatin gels, and juices, are relatively insensitive to shear or compression and require quite a large force or deformation to damage their internal structure. Other foods, such as foams and potato chips, have much more delicate structures that require careful handling so as not to damage the internal structure before testing. Even transferring some foods from a storage container to the testing geometry can cause irreversible structural damage. Testing can be done to check for such structural sensitivities. In some cases, the sample can be tested in its storage container, such as using a vane on a cup of set yogurt to measure its yield stress.

29.4.3 Testing Considerations

Selection of a rheological test is based on the material to be tested and the desired rheological information from that material. The selected test should provide the required information as discussed in Sect. 29.4.1. If the required information is too complex to be obtained in a single test, multiple tests may need to be performed on the material. Different tests may provide similar information, but one test may be more appropriate for the material to be tested because of its composition or the amount of material available. For example, the difference in viscosity of a cheese sauce at refrigeration temperature (4 °C) and hot-hold temperature (60 °C) may be determined by shear rate sweeps at these two temperatures or by a temperature sweep at a constant shear rate. The former test would provide the entire viscosity profile but no information on how the rheological behaviors might change between the two temperatures. For example, the melting of fats at a certain temperature might have a dramatic effect on viscosity. The latter test would provide more information on how temperature impacts the cheese sauce, but only a single shear rate would be used so any changes in shear-dependent behavior would be unknown.

Once a test has been selected, the parameters for that test need to be chosen to best provide the information being sought. Test parameters include rate of compression, shear

Table 29.6 Overview of common rheological tests for solid, semisolid, and fluid foods

Material classification	*Rheological behavior*	*Rheological instrument*	*Testing geometry*	*Testing parameters*[a]
Solid: holds its shape when cut or formed, shows a clear fracture path, generally requires chewing to be able to swallow (e.g., crackers, string cheese, meats, jelly, tomatoes, cup-set yogurt, and French fries)	Compression	Instron or Texture Analyzer or Cone Penetrometer	Plate (fundamental testing) or specialized geometry (empirical testing)	Strain of 10–50% for most foods Ambient temperature Crosshead velocity of 0.5–4 mm/s
	Fracture	Instron or Texture Analyzer	Plate (fundamental testing) or specialized geometry (empirical testing)	Strain of >40% for most foods Ambient temperature Crosshead velocity of 0.5–4 mm/s
	Viscoelastic profile	Rheometer	Smooth parallel plates for samples that do not exude moisture or fat, sandblasted or crosshatched parallel plates for samples that exude moisture or fat	Strain of 0.01–100% Frequency of 0.01–100 Hz Ambient temperature (20–25 °C) or temperature reflective of storage, processing, or eating conditions (e.g., 4 °C or 37 °C)
	Creep			Low strain (<1%) Ambient temperature (20–25 °C) or temperature reflective of storage, processing, or eating conditions (e.g., 4 °C or 37 °C)
Semisolid: spoonable or spreadable, forms a mound on a spoon, sags under its own weight, generally requires palating to swallow (e.g., whipped cream, pudding, Greek yogurt, peanut butter, and mashed potatoes)	Viscosity profile	Rheometer or viscometer	Parallel plates or cone and plate	Shear rate of 0.01–100 s^{-1} Ambient temperature (20–25 °C) or temperature reflective of storage, processing, or eating conditions (e.g., 4 °C or 37 °C)
	Yield stress	Rheometer		Stress of 0.01–10,000 Pa Ambient temperature (20–25 °C) or temperature reflective of storage, processing, or eating conditions (e.g., 4 °C or 37 °C)
	Viscoelastic profile	Rheometer		Strain of 0.01–100% Frequency form 0.01 to 100 Hz Ambient temperature (20–25 °C) or temperature reflective of storage, processing, or eating conditions (e.g., 4 °C or 37 °C)
	Creep	Rheometer		Low strain (<1%) Ambient temperature (20–25 °C) or temperature reflective of storage, processing, or eating conditions (e.g., 4 °C or 37 °C)
Fluid: pourable, takes the shape of its container, generally requires little oral manipulation to swallow (e.g., milk, juice, honey, ketchup, and drinkable yogurt)	Viscosity profile	Rheometer or viscometer	Cone and plate, cup and bob, double-gap cup and bob, or appropriate viscometer spindle selected based on estimated viscosity range of sample. A Bostwick consistometer can be used on fluids that are Newtonian or close to Newtonian, but is not recommended on samples with significant shear-dependent behavior	Shear rate of 0.01–100 s^{-1} Ambient temperature (20–25 °C) or temperature reflective of storage, processing, or eating conditions (e.g., 4 °C or 37 °C)
	Yield stress	Rheometer		Stress of 0.01–1000 Pa Ambient temperature (20–25 °C) or temperature reflective of storage, processing, or eating conditions (e.g., 4 °C or 37 °C)
	Viscoelastic profile	Rheometer		Strain of 0.01%–100% Frequency of 0.01–100 Hz Ambient temperature (20–25 °C) or temperature reflective of storage, processing, or eating conditions (e.g., 4 °C or 37 °C)

[a]Testing parameters are general guidelines. Consult a rheologist and/or published literature for more precise testing parameters for a specific food product

rate range, strain and/or frequency range, and temperature. Pretest setup should also be considered. After they are placed onto the testing surface, samples may need time to reach the testing temperature or relax from the stresses placed on them during loading. Table 29.6 provides some suggested testing parameters for various products.

When considering which rheological tests and testing parameters are appropriate to obtain the desired information, it is important to understand the limitations of certain testing equipment. If the material needs to be tested under certain environmental conditions, the equipment used either needs to have an option to control these conditions or be moved into a location that is maintained at the desired conditions. For example, a rheometer would have the ability to measure a milk-based beverage at 4 °C, but a Brookfield viscometer would not unless the sample was placed in a temperature-controlled bath. Some instruments or geometries require the sample to be cut into certain shapes. Many geometries require the sample to be homogeneous and anisotropic. Products such as yogurt with fruit pieces or unstabilized salad dressings do not meet these requirements, making testing challenging.

The capability of the instrument to accurately and precisely apply or detect forces or deformations should also be considered. The geometry used for testing can be used to maximize the ability of the instrument to measure the rheological behaviors of the material. For example, a double-gap concentric cylinder system can be used to measure the viscosity of thin fluids such as milk because the higher surface area of the testing surfaces increases the torque signal to the rheometer. A Kramer shear cell can be used to measure compression, shear, and extrusion behaviors of small, irregularly shaped pieces of food, such as meat crumbles, cereal flakes, or berries, that are difficult to test with more standard geometries.

29.4.4 Other Considerations

Understanding what information is desired and what limitations the sample and available testing equipment exist are critical to selecting an appropriate rheological test. However, there are a few additional factors to consider when selecting these tests. First, the amount of available sample may dictate what and how many tests can be run. The amount of available sample is especially important when considering the number of replicates that need to be performed to obtain a good understanding of the average behavior of a sample. A sugar syrup may require only two replicates to determine its viscosity profile, while whole-muscle meats may require over twenty replicates in compressive testing due to the product's natural variation.

Additionally, the amount of time available to determine the required information is important. Some tests take only a few minutes to perform, while others need over an hour to run. Sample preparation, cleaning between samples and after testing is completed, and the number of replicates required all add to the testing time. Available personnel time and the tendency of samples to undergo rheological changes over time are additional considerations for selecting rheological tests.

Finally, a certain amount of expertise is needed to set up and run the tests, then analyze the resulting data. Accurately and precisely running a standard test requires relatively little training, while designing a new testing protocol and interpreting data requires significant experience in rheology and a knowledge of the food product's structure and composition. This knowledge is especially important when working with products that are not homogeneous and isotropic, as well as when working near the limits of the testing equipment.

29.5 SUMMARY

Rheological testing is simple in that it only requires the measurement of force, deformation, and time. To convert these measurements into fundamental physics-based rheological properties requires an understanding of the material and testing method. Materials should be homogeneous and isotropic for most fluid foods and many solid foods. Fundamental rheological properties are determined based on knowledge of the stress or strain applied to the sample and the geometry of the testing fixture. Once rheological properties are determined, they can be described by physical or mathematical models to gain a more complete understanding of the rheological properties. The advantage of determining fundamental, rather than empirical, rheological properties is the use of common units, independent of the specific instrument, to determine the rheological property. This approach not only allows for comparison among values determined on different instruments but it also permits comparisons like the flow behavior of honey vs. the flow of paint. Through rheological methods, food scientists have the ability to relate theoretical and experimental information from a range of disciplines, including polymer chemistry and materials sciences, to gain a greater understanding of the quality and behavior of food materials.

29.6 GLOSSARY

Bostwick consistometer	A rheological device used to measure viscosity in the food industry
Boundary regime	Tribological behavior characterized by high, constant friction due to surface-surface contact

Brookfield viscometer	A rheological device used to measure viscosity in the food industry
Compression	A force acting in a perpendicular (normal) direction toward the body
Concentric cylinder	A test fixture for rotational viscometry frequently called a cup and bob
Cone and plate	A test fixture for rotational viscometry
Constitutive equation	An equation relating stress with strain and sometimes other variables including time, temperature, and concentration
Dilatent	Shear-dependent thickening
Empirical test	Simple tests measuring poorly defined parameters but typically found to correlate with textural or other characteristics
Extensional viscosity	Viscosity measured by extending a material, typically but stretching between two surfaces
Fundamental test	A measurement of well-defined, physically based rheological properties
Homogeneous	Well mixed and compositionally similar regardless of location
Hydrodynamic regime	Tribological behavior characterized by complete surface-surface separation by the lubricating material
Incompressible	No change in material density
Isotropic	The material response is not a function of location or direction
Kinematic viscosity	The viscosity divided by the density of the material
Laminar flow	Streamline flow
Large amplitude oscillatory shear:	An oscillatory rheometric test that uses a strain large enough to cause permanent deformation to a material's structure
Mixed regime	Tribological behavior characterized by a reduction friction to a minimum as surface-surface decreases
Modulus	A ratio of stress to strain
Newtonian fluid	A fluid with a linear relationship between shear stress and shear rate without a yield stress
Non-Newtonian fluid	Any fluid deviating from Newtonian behavior
No slip	The fluid velocity adjacent to a boundary has the same velocity as the boundary
Oscillatory rheometry	Dynamic test using a controlled sinusoidally varying input function of stress or strain
Parallel plate	A test fixture for rotational viscometry
Penetrometer	An instrument that measures the depth of penetration of a testing geometry into a sample
Pseudoplastic	Shear thinning
Rheology	A science studying how all materials respond to applied stresses or strains
Rheogram	A graph showing rheological relationships
Rheometer	An instrument measuring a wide array of (usually fundamental) rheological properties
Rheopectic	Time-dependent thickening of a material
Shear (strain) rate	Change in (shear) strain with respect to time
Simple shear	The relative motion of a surface with respect to another parallel surface creating a shear field within the fluid contained between the surfaces
Steady shear	A flow field in which the velocity is constant at each location with time
Steady state	Independent of time
Strain	Relative deformation
Stress	Force per unit area
Tension	A force acting in a perpendicular direction away from the body
Test fixture	A rheological attachment, sometimes called a geometry, which shears the sample material
Texture Analyzer	An instrument that measures empirical and fundamental large-strain compression and tensile behaviors
Thixotropic	Time-dependent thinning of a material
Torque	A force-generating rotation about an axis, which is the product of the force and the perpendicular distance to the rotation axis
Torsion	A twisting force applied to a specimen

Tribology	A branch of rheology involving friction, lubrication, and wear
Universal Testing Machine	An instrument that measures empirical and fundamental large-strain compression and tensile behaviors
Viscoelastic	Exhibiting fluid-like (viscous) and solid-like (elastic behavior simultaneously
Viscometer	An instrument that measures viscosity of fluids
Viscosity	An internal resistance to flow
Yield stress	A minimum stress required for flow to occur

29.7 NOMENCLATURE

Symbol	Name	Units
A	Area	m^2
A_i	Initial sample area	m^2
E	Modulus of elasticity	Pa
F	Force	N
G	Shear modulus	Pa
h	Height	m
K	Consistency coefficient	Pa s^n
L	Length	m
L_i	Initial length	m
ΔL	Change in length	m
n	Flow behavior index	unitless
t	Time	s
U	Velocity	m s^{-1}
ϵ	Normal strain	unitless
γ	Shear strain	unitless
$\dot{\gamma}$	Shear (strain) rate	s^{-1}
η	Apparent viscosity	Pa s
μ	Newtonian viscosity	Pa s
σ	Stress	Pa
σ_o	Yield stress	Pa

29.8 STUDY QUESTIONS

1. How is stress different from force?
2. What is the difference between shear stress and normal stress?
3. What is the definition of apparent viscosity? How does apparent viscosity differ from Newtonian viscosity?
4. Pure maple syrup is a Newtonian fluid and imitation maple syrup is a pseudoplastic fluid. What are the differences in flow behavior of these fluids and how do these differences alter the processing and final texture of these foods?
5. The stress response of applesauce at 26 °C may be described by the following mathematical expression:

   $$\sigma = 5.6\dot{\gamma}^{0.45}$$

   The stress response of honey at 26 °C obeys a Newtonian model:

   $$\sigma = 8.9\dot{\gamma}$$

   In both equations, stress is in Pascals and shear rate is in s^{-1}.
   (a) Which rheological model is described by the equation for applesauce? What are the consistency coefficient and flow behavior index? Include the units on these quantities.
   (b) Calculate the apparent viscosities of applesauce and honey at a shear rate of 0.25, 0.43, 5.10, and 60.0 s^{-1}.
   (c) Compare how the viscosities of the two food products change with shear rate. Which food has a higher viscosity?
   (d) Explain the importance of multipoint testing when measuring viscosity.
6. You are designing a new chip dip. Describe at least three rheological behaviors that you would like your dip to display.
7. What sort of tribological differences might you expect between a full-fat product and a low-fat product?
8. Give an example of a food that would have a high extensional viscosity and a food that would have a low extensional viscosity. What are their textures like, and how might their extensional viscosity relate to their textures?
9. Rheological testing can be done using a variety of test parameters.
   (a) What are some things to consider when selecting parameters for a given rheological test?
   (b) You have been asked to determine which of three chocolate pudding formulas is most resistant to changes during processing. What types of rheological test could you run on the puddings to answer this question? Explain your answer.
   (c) Your supervisor wants to know which of the three puddings formulations in part b is most stable by the end of next week. You have three 6-oz containers of each pudding formulation currently in the lab. How does this information change your answer to part b?
10. What type(s) of instruments and geometries would be commonly used to measure the compression behavior

(a-c) or apparent viscosity (d-e) of the following for routine quality control purposes:

(a) Diced peaches
(b) Spreadable cream cheese
(c) Shredded wheat
(d) Stirred yogurt
(e) Pure maple syrup

Acknowledgments The author of this chapter thanks Christopher R. Daubert (currently in College of Agriculture, Food and Natural Resources, University of Missouri, MO, USA) and E. Allen Foegeding (professor emeritus in the Department of Food, Bioprocessing, and Nutrition Sciences, North Carolina State University) for their contribution to the chapter in previous editions of the book.

REFERENCES

1. Steffe JF, Daubert CR (2006) Bioprocessing pipelines: rheology and analysis. Freeman, East Lansing, MI
2. Steffe JF (1996) Rheological methods in food process engineering, 2nd edn. Freeman, East Lansing, MI
3. Muller HG (1973) An introduction to food rheology. Crane, Russak, Inc., New York
4. Macosko CW (1994) Rheology: principles, measurements, and applications. VCH, New York
5. Rao MA (1999) Rheology of fluid and semisolid foods: principles and applications. Aspen, Gaithersburg, MD
6. Chen, L., & Opara, U. L. (2013). Texture measurement approaches in fresh and processed foods—A review. *Food research international*, *51*(2), 823–835.
7. Hamann D, Zhang J, Daubert CR, Foegeding EA, Diehl KC (2006) Analysis of compression, tension and torsion for testing food gel fracture properties. J Texture Stud 37:620–639
8. Bourne MC (1982) Food texture and viscosity: concept and measurement. Academic, New York
9. Läuger J, Stettin H. (2010). Differences between stress and strain control in the non-linear behavior of complex fluids. Rheologica acta, 49(9), 909–930.
10. Sarkar, A., & Krop, E. M. (2019). Marrying oral tribology to sensory perception: A systematic review. Current Opinion in Food Science, 27, 64–73.
11. Rudge, R. E., Scholten, E., & Dijksman, J. A. (2019). Advances and challenges in soft tribology with applications to foods. Current Opinion in Food Science, 27, 90–97.
12. Padmanabhan M. (1995). Measurement of extensional viscosity of viscoelastic liquid foods. Journal of food engineering, 25(3), 311–327.
13. Hadde EK, Chen J. (2021). Texture and texture assessment of thickened fluids and texture-modified food for dysphagia management. Journal of Texture Studies, 52(1), 4–15.
14. Lv Z, Chen J, Holmes M. (2017). Human capability in the perception of extensional and shear viscosity. Journal of texture studies, 48(5), 463–469.
15. Joyner HS. (2021). Nonlinear (large-amplitude oscillatory shear) rheological properties and their impact on food processing and quality. Annual Review of Food Science and Technology, 12, 591–609.

Thermal Analysis

30

Shelly J. Schmidt and Leonard C. Thomas

Contents

30.1 INTRODUCTION

Thermal analysis is a term used to describe a broad range of analytical techniques that measure physical and chemical properties as a function of temperature, time, and atmosphere (inert or oxidizing gas, pressure, and relative humidity). Depending on the technique, test temperatures can range from −180 to 1000 °C or more, allowing investigation into a range of applications, including low-temperature stability and processing (e.g., freezing and freeze-drying) to high-temperature processing and cooking (e.g., extrusion, spray drying, and frying).

S. J. Schmidt (✉)
Department of Food Science and Human Nutrition, University of Illinois at Urbana-Champaign, Urbana, IL, USA
e-mail: sjs@illinois.edu

L. C. Thomas
Formerly at DSC Solutions, Smyrna, DE, USA

Thermal analysis results provide insight into the structure and thermal stability of starting materials, as well as finished products. The physical structure (amorphous, crystalline, and semi-crystalline) of a material creates a set of physical properties, which in turn define end-use properties, such as texture and storage stability. Areas of application include quality assurance, product development, and research into new materials, formulations, and processing conditions [1–4].

Specific instruments are typically used for characterization of a particular property. Instrumentation includes a transducer, used to measure the property of interest, and a temperature-measuring device, such as a thermocouple, thermopile, or platinum resistance thermometer, used to record the sample's temperature. Experiments are performed while heating, cooling, or at a constant temperature (isothermal), and measured signals are stored for analysis. Thermal analysis techniques of major interest to the food researcher and the properties they measure are listed in Table 30.1, with the first three widely used techniques discussed in detail in Sect. 30.3.

B. P. Ismail, S. S. Nielsen (eds.), *Nielsen's Food Analysis*, Food Science Text Series,
https://doi.org/10.1007/978-3-031-50643-7_30

Table 30.1 Thermal analysis techniques of major interest to the food researcher and the properties they measure

Techniques	*Abbreviation*	*Property Measured*
Thermogravimetric analysis	TGA	Weight change
Differential scanning calorimetry	DSC	Heat flow
Modulated temperature DSC	MTDSC	Heat flow and heat capacity
Thermomechanical analysis	TMA	Dimensional change
Dynamic mechanical analysis	DMA	Stiffness and energy dissipation
Rheology	Rheometer	Viscosity/flow behavior
Moisture sorption analysis	MSA	Moisture sorption

Thermogravimetric analysis (TGA) should be the first thermal analysis measurement done when characterizing a new material. TGA detects and quantifies volatile components, such as bulk and associated water, and identifies the temperature at which molecular decomposition (chemical change) begins. TGA data are also useful in selecting optimum **differential scanning calorimetry** (DSC) and **modulated temperature DSC** (MTDSC) experimental conditions and interpreting DSC and MTDSC data. DSC and MTDSC thermal analysis techniques provide fundamental and advanced information about material structure and changes in structure (transitions) as a function of temperature and time. By understanding structure and related physical properties, formulations can be developed that provide desired end-use properties, such as crispness, fast dissolution rate, and extended shelf life.

30.2 MATERIALS SCIENCE

Since the pioneering work of Slade and Levine (e.g., [5, 6]), food scientists have been actively applying the principles of materials science to the study of food materials (e.g., [7–10]). One of the main driving forces underlying the application of materials science principles to foods is that the end-use properties (functionality) of a material at a specific temperature are dependent on the structure of the components at that temperature. Therefore, it is necessary to measure structure as a function of temperature. The primary use of thermal analysis is to determine structure by measuring the physical properties (e.g., enthalpy, heat capacity, flow, expansion, and rigidity) associated with that structure.

30.2.1 Amorphous Structure

Many food products are amorphous or have a high amorphous content (semi-crystalline). Examples include extruded snacks and breakfast cereals, low-moisture cookies and crackers, sugar-based hard candies, and spray-dried powdered drink mixes. **Amorphous structure** has no regular or systematic long-range molecular order, which means that it has the highest energy and entropy content, the highest molecular mobility, and the fastest rate of dissolution, compared to crystalline structure. Recent research, however, does show that there are some differences between the short-range (or local) molecular ordering and associated physicochemical properties of amorphous materials made using different amorphization methods, such as melt-quenched sucrose compared to freeze-dried sucrose [11]. A feature and potential problem with all amorphous structure is that physical properties can change by orders of magnitude at a specific temperature, termed the **glass transition temperature** (*Tg*).

At temperatures below *Tg*, an amorphous material acts like a glass. It is rigid, has low molecular mobility, and has very high viscosity (low ability to flow). Above *Tg*, these materials act like a rubbery material, viscous liquid, or gel with greater free volume and much higher molecular mobility. Some amorphous materials (e.g., fats, oils, and water) have the ability to crystallize when cooled to lower temperatures. However, these materials may associate (e.g., hydrogen bond) with another material in the formulation and not crystallize, even at temperatures well below their freezing point. In the case of water, it is well known that not all the water in a food material freezes, even at very low storage temperatures.

Since properties change so significantly at *Tg*, it is important to be able to measure and control *Tg* by selecting appropriate ingredients in the correct weight ratio to other ingredients in a formulation or recipe. In foods, *Tg* can change by 50 °C or more via changes in moisture content. Crisp snack foods, such as crackers and potato chips, are classic examples of foods whose *Tg* and resulting physical properties are affected by changes in water content. For example, the *Tg* of freshly processed potato chips is much higher than room temperature; thus, the chips are crisp and have a pleasant texture. If the chips are exposed to ambient temperature and high humidity conditions for several hours, they become soft and pliable, which are textural characteristics typical of a low-moisture food held above *Tg*. Once the package of chips is opened, the low-moisture chips begin to absorb moisture from the air, which lowers *Tg* and creates a different set of physical properties at the consumption temperature.

Tg can be measured using most thermal analysis techniques due to significant changes in physical properties, such as heat capacity (for DSC), coefficient of thermal expansion (for **thermomechanical analysis**, TMA), and stiffness (for **dynamic mechanical analysis**, DMA), that accompany *Tg* [7]. However, DSC is usually the technique of choice because of easy sample preparation, short test times, straightforward data interpretation, and the ability to use sealed pans (hermetic), which prevent loss of moisture as the sample is heated. Because of the significant increase in molecular mobility and heat capacity that occurs as the sample is heated to a temperature above *Tg*, there is a corresponding increase in the heat flow rate measured by DSC, as illustrated in Fig. 30.1 for a sample of candy corn. Data analysis software measures the temperature and magnitude of the change in heat flow that occurs, which is proportional to the amount of amorphous material in the sample. For complete characterization of the *Tg*, best practice is to report all four *Tg* parameters – onset, midpoint, endpoint, and change in heat capacity (ΔCp, J/g °C), as well as the experimental conditions employed [7].

30.2.2 Crystalline Structure

Crystalline structure is different from amorphous structure in many ways. Molecules have long-range order, lower energy (heat content) and entropy, higher density, and a different set of physical properties. Molecular mobility is low, which means that heat capacity is low. Melting of the crystalline material creates an amorphous liquid. Upon cooling of the amorphous liquid, some materials (e.g., fats) will crystallize, while others (e.g., sucrose and fructose) transition into an amorphous glass below their *Tg*.

Because of the increased density and molecular order within the crystal, there is a reduction in the ability of crystalline material to form hydrogen bonds and, thus, a reduced tendency to "absorb" moisture into their bulk structure from the atmosphere. However, crystalline materials "adsorb" water onto their surface quite quickly, mirroring the % relative humidity of the surrounding environment [12]. The melting of crystalline material occurs at a higher temperature than the glass transition associated with amorphous material, which makes crystalline material more stable and much more rigid (often gritty like table sugar and salt) than amorphous material over a wide temperature range. Because the crystalline structure is more stable, physical properties change less with time.

Since crystalline material has lower heat content than amorphous material, crystalline material must absorb heat (endothermic process) to become amorphous. The absorption of heat during a DSC experiment is seen as an endothermic peak, termed the melting peak. Data analysis software can measure the onset, peak, and end temperatures of the endothermic melting peak and calculate the heat (J/g) required to melt the sample by determining the area under the peak. The area of the melting peak (J/g) increases as the percent crystallinity of the material increases.

Normally, crystalline structure is converted to amorphous structure by heating the sample to a temperature (an energy level) that is high enough to overcome the energy associated with the crystalline lattice (termed thermodynamic melting).

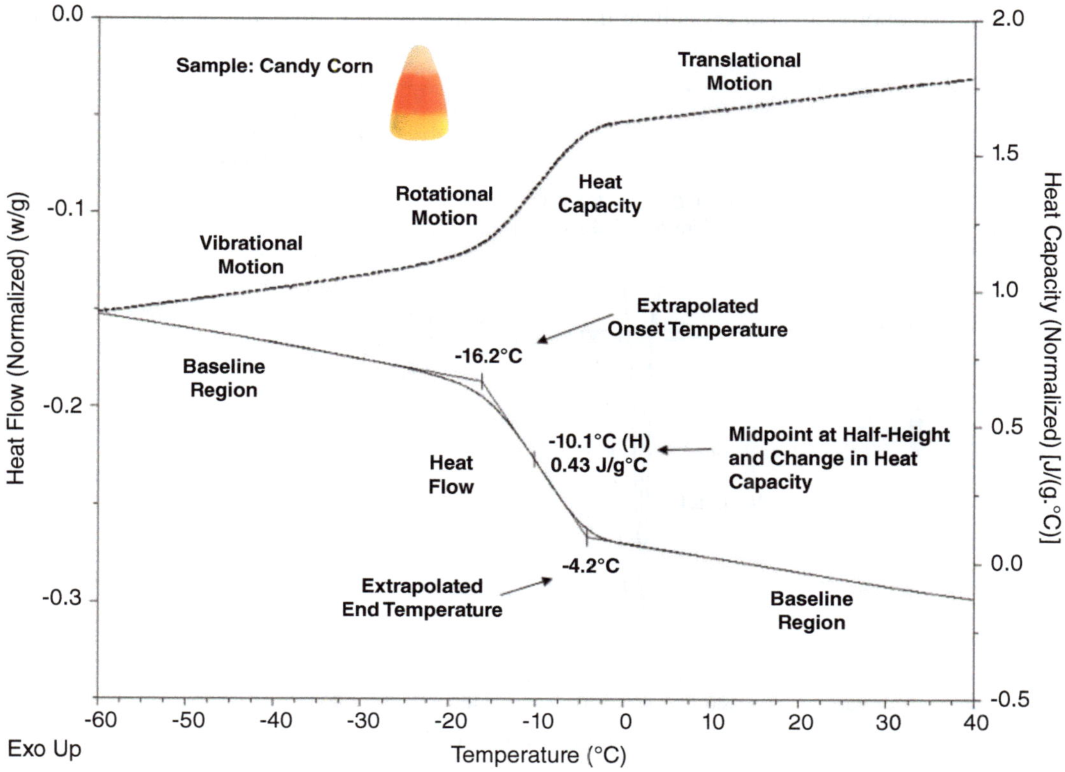

Fig. 30.1 The glass transition of amorphous structure can be measured using differential scanning calorimetry (DSC) due to the significant increase in heat capacity that occurs as the material is heated to a temperature above its glass transition temperature (*Tg*). This increase in heat capacity results in an increase in molecular mobility. Typical, *Tg* analysis includes the extrapolated onset, midpoint (temperature at one-half of the heat capacity change), extrapolated end temperature, and step change in heat capacity

However, crystalline structure can also be lost due to processes such as dissolving of crystals in a solvent (dissolution), dehydration of a hydrated crystalline form, chemical interaction of functional groups between two materials in a mixture, and breaking of chemical bonds (decomposition) at temperatures below the true melting point of the material [13]. Since these are time-dependent (kinetic) processes, the endothermic peak observed in DSC data shifts to a higher temperature as heating rate increases. Temperature onset values for mannitol and sucrose at 5 and 25 °C/min heating rates are compared in Fig. 30.2. Mannitol (a sugar alcohol) illustrates thermodynamic melting, while sucrose (a disaccharide simple sugar) shows a heating rate dependent or kinetic response. Figure 30.2 contains multiple Y-axes, as do other figures in this chapter. Since heat flow rate (W/g) is proportional to heating rate, it is necessary to scale the heat flow curves to different sensitivities to visually compare curves run at different heating rates, resulting in the need for more than one Y-axis.

30.2.3 Semi-crystalline Structure

Many foods contain both amorphous and crystalline structures. Therefore, DSC and MTDSC data are often complex, such as observed in Fig. 30.3 for soft caramels. Amorphous structure is detected by the presence of a glass transition, while crystalline structure is detected as an endothermic melting peak. When the amorphous and crystalline transitions overlap, careful analysis is required to accurately determine the associated glass transition and melting parameters.

In some materials, such as a lipid that melts over a temperature range, one ingredient can exist in both the solid and the liquid phases at the same temperature. A term that is sometimes used to describe a mixture of phases in a lipid is **solid-fat content** (SFC) (see Chap. 23, Sect. 23.3.7). At a particular temperature (usually room temperature, 72 °F or 22 °C), a certain fraction or percentage of the lipid material is solid (crystalline), and the remainder is liquid (melted). One lipid that illustrates this property at room temperature is cocoa butter, a common ingredient in chocolate. Cocoa butter can crystallize into six known forms (termed **polymorphs**), with the lowest stability form (form I) melting at approximately 17 °C and the highest stability form (form IV) melting at approximately 35 °C. Form V is the most desirable crystalline form of cocoa butter in chocolate, yielding a dark brown, glossy appearance, a satisfying snap when broken, and a melting temperature near mouth temperature.

The DSC thermogram of a soft caramel (Fig. 30.3) shows a glass transition over the temperature range of −20 to −7 °C, which overlaps with lipid melting transitions associated with a combination of butter, cream, and palm oil between −7 and 60 °C. The extrapolated end of the glass transition is −7.24 °C, and the extrapolated onset of melting is −6.86 °C. By measuring the percentage of melting below and above 22 °C, the ratio of liquid to solid phases can be

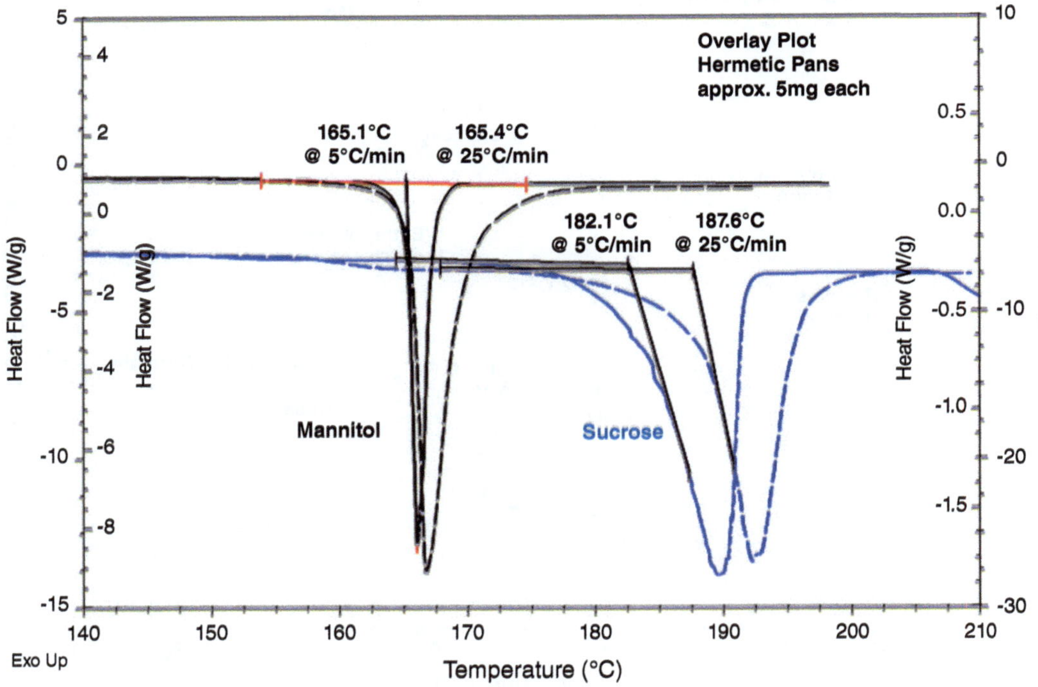

Fig. 30.2 Conversion of crystalline to amorphous structure is always seen as an endothermic peak because the enthalpy of amorphous structure is higher than crystalline structure. True melting is a thermodynamic event, which means that the melting temperature is independent of heating rate (mannitol). Loss of crystalline structure caused by kinetic processes is time dependent and the onset temperature of the endothermic peak increases with heating rate (sucrose)

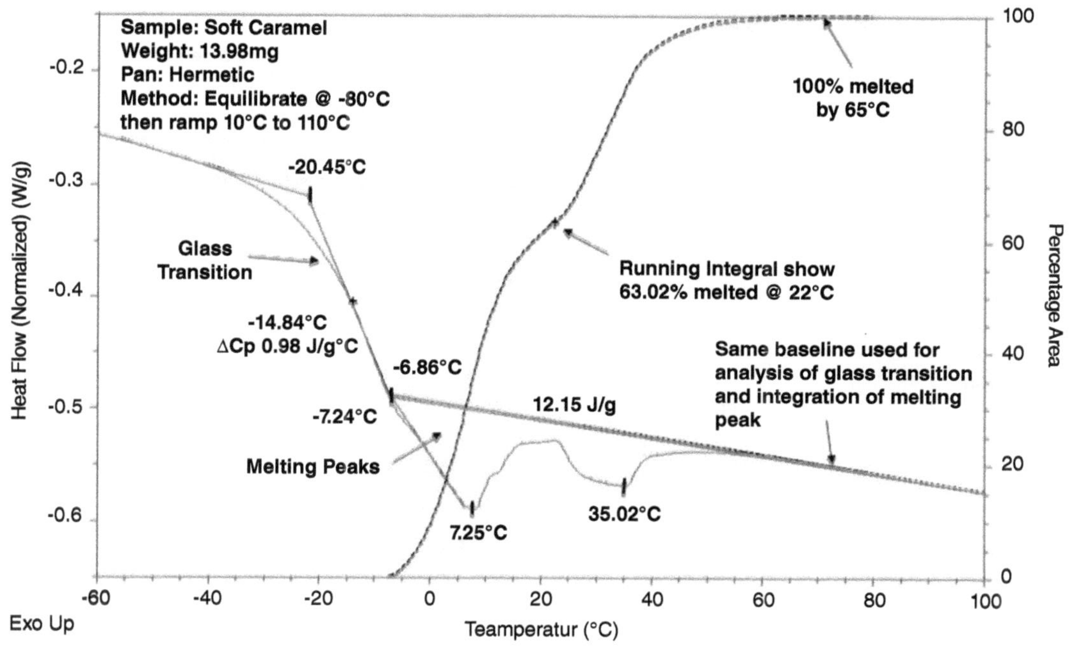

Fig. 30.3 DSC thermogram and associated data for a soft caramel candy illustrates a material with a semi-crystalline structure. The amorphous structure would be soft (viscous flow) because the glass transition occurs below room temperature. Lipid melting peaks show several crystalline components (butter, cream, and palm oil) that melt over a broad temperature range. Running integral analysis shows 63% of the crystalline structure has melted at room temperature (22 °C)

determined for room temperature. A feature of the DSC data analysis software, termed the "running integral," can plot percent melted versus temperature. Results show that 63% of the lipid is liquid (melted) at room temperature in this particular sample. The presence of the liquid lipid provides a creamy texture to the soft caramel candy.

30.2.4 Thermodynamic and Kinetic Properties

As discussed in Sect. 30.3.2.1, a natural limitation of DSC, and other thermal analysis techniques used to measure structure, is that heating of the sample is typically required. As temperature increases, molecular mobility increases, which permits the structure to change in ways that are not always obvious in the data. Thermodynamic properties (e.g., heat capacity, enthalpy, and density) have absolute values as a function of temperature, while kinetic properties are always a function of time and temperature. Some common examples of kinetic processes observed in foods are freezing of water during cold storage, crystallization of fats and amorphous sweeteners, adsorption and desorption of water, staling of bread, and decomposition and oxidation during processing, such as deep-frying.

The easiest way to determine if an observed event in the data is due to thermodynamics or kinetics is to change the heating rate. Since heating rate has the units of °C/min, the reciprocal is min/°C. The higher the heating rate, the less time the sample experiences at each temperature. Therefore, high heating rates decrease the probability of structural change, while low heating rates increase the probability of structural change. If a kinetic event is identified in the heating/cooling data, the time dependence (kinetics) of the process at specific temperatures can be evaluated with isothermal (constant temperature) experiments.

As stated in Sect. 30.2.1, amorphous structure is not stable at temperatures above *Tg*, where it is a viscous liquid with high molecular mobility. Above *Tg*, amorphous materials often have the ability to crystalize. Figure 30.4 is an example of how DSC can be used to evaluate formation of structure in freeze-dried amorphous sucrose. The first experiment (Fig. 30.4a) is typically done at 10 °C/min over a broad temperature range to detect the presence of a kinetic process. Analysis of Fig. 30.4a shows a glass transition in the temperature range of 20 to 28 °C, followed by crystallization starting near 63 °C and forming 52.8 J/g of crystalline structure. This amorphous-to-crystalline transition is referred to as cold crystallization, since it occurs while heating from the low-temperature side of the transitions [14]. Since crystallization is a kinetic process, it can occur at lower temperatures than 63 °C with additional time. If this product is expected to experience high temperatures during shipment and storage, subsequent experiments can then be done at a variety of isothermal temperatures to verify the time dependence of the observed event and the associated temperature stability of the product.

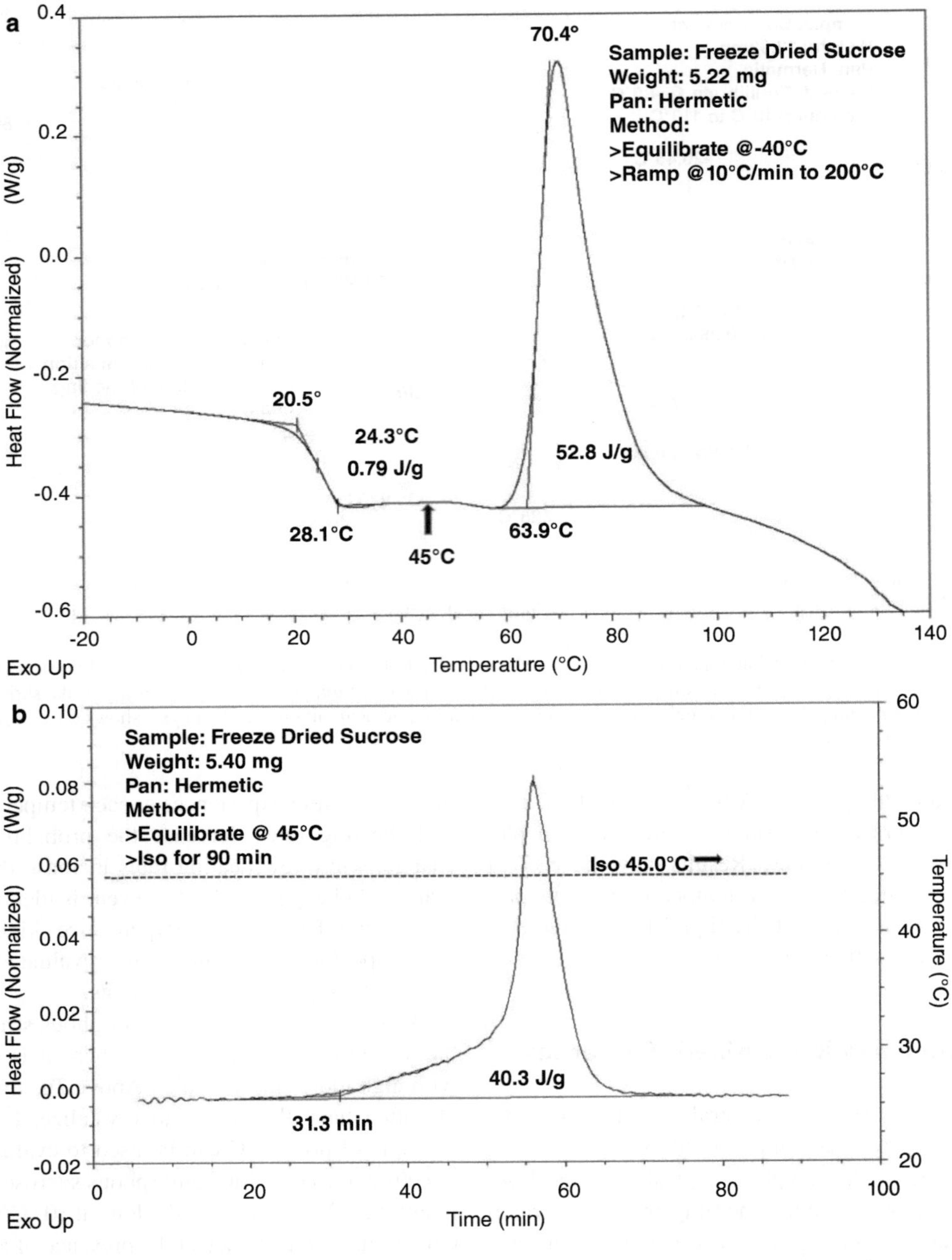

Fig. 30.4 (**a**) Heating of a sample of freeze-dried sucrose shows that the material is amorphous with a glass transition near room temperature (22 °C) and cold crystallization starting near 63 °C. The data could be incorrectly interpreted as the material is stable at temperatures below 63 °C. (**b**) The isothermal experiment at 45 °C for 90 mins verifies that the freeze-dried sucrose sample is not stable at 45 °C, and that crystallization can occur at lower temperatures given sufficient time. In this case, 45 °C is about 20 °C higher than the midpoint temperature of the glass transition (24.3 °C)

Figure 30.4b shows the results of an isothermal experiment of the freeze-dried amorphous sucrose at 45 °C for 90 minutes. The 10 °C/min heating rate data (Fig. 30.4a) indicated that crystallization begins near 63 °C. This could be incorrectly interpreted to mean that the material is stable at lower temperatures. However, the isothermal experiment (Fig. 30.4b) shows that cold crystallization begins in just 31 minutes at 45 °C. The amount of structure formed is lower than observed in the heating rate experiment (40.3 and 52.8 J/g respectively). This is as expected because molecular mobility increases with temperature, and the temperature of crystallization is higher in the heating rate experiment compared to the isotherm experiment.

30.3 PRINCIPLES AND METHODS

This section describes the working principles of the most frequently used thermal techniques and makes recommendations on optimum experimental conditions for characterizing common materials.

30.3.1 Thermogravimetric Analysis

30.3.1.1 Overview

Thermogravimetric analysis (TGA) should be the first thermal analysis technique used to characterize a new material. TGA provides information about the composition (number of components) of the material and the temperature where weight losses occur due to evaporation or decomposition (chemical breakdown). A significant benefit of that information is guidance in the selection of DSC experimental conditions. Weight losses would increase pressure in a DSC hermetic pan and can result in pan leakage and possible contamination of the DSC cell. TGA instruments use a specially designed and very sensitive analytical balance to measure weight changes, since samples are typically heated from room temperature to 1000 °C or more. A thermocouple is located close to the sample to continuously record the temperature as weight changes occur. The heated sample chamber is typically purged with an inert gas, such as nitrogen or helium; however, air or oxygen can be used when measuring oxidative stability. Most weight changes are weight losses due to volatilization or decomposition, but weight gain may be observed during early stages of oxidation. A specialized version of TGA is designed with humidity control so that the rate of moisture sorption (both absorption and desorption) can be measured as a function of time, temperature, and relative humidity.

Figure 30.5 shows a schematic of possible designs for both conventional TGA and a humidity-controlled sorption analysis TGA. Both contain sample and reference pans attached to the balance. The reference pan is empty, since its purpose is to offset the weight of the sample pan. With conventional TGA, the reference pan is typically not heated. The reference pan for sorption analysis is exposed to the same temperature and relative humidity as the sample. This greatly improves the stability and baseline performance of the measurement. Most modern instruments have autosamplers so that many samples can be run sequentially without need of operator presence.

Although TGA provides a quantitative measurement of weight change, it is often difficult to quantify the weight of a specific component because weight losses typically overlap in temperature (Fig. 30.6). Improved temperature resolution of these weight losses can usually be obtained by slowing the heating rate, reducing the sample weight, and increasing sample surface area. However, the lower heating rate increases test time (reduced productivity), and the smaller

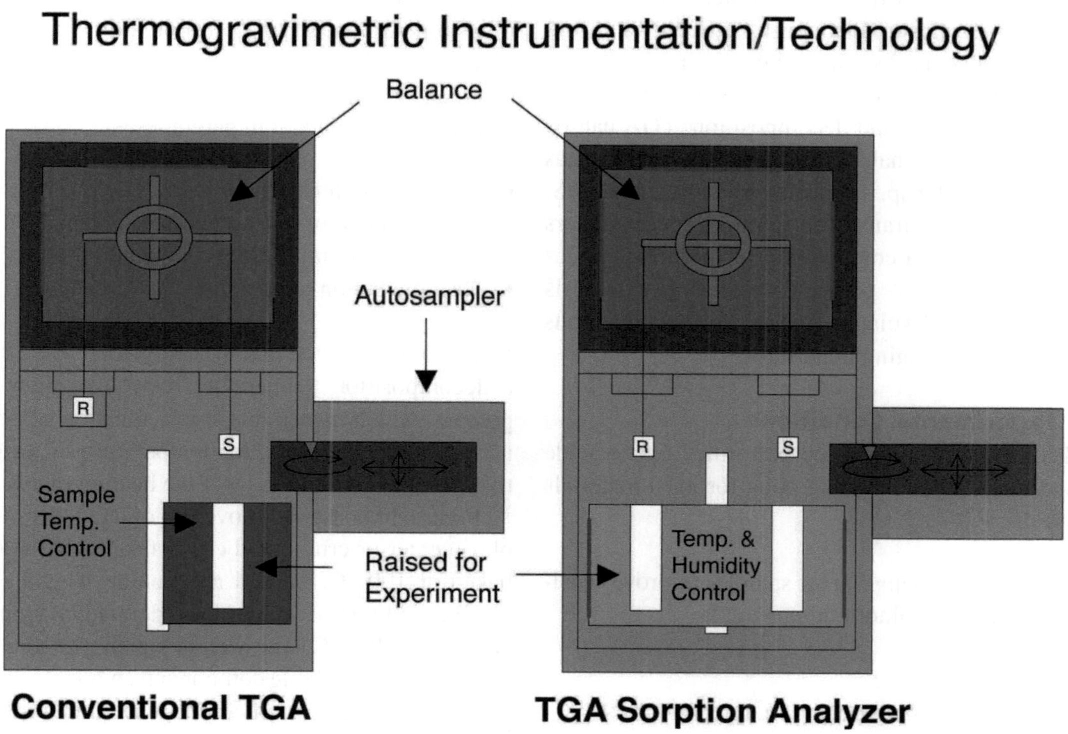

Fig. 30.5 Thermogravimetric instrumentation measures weight change using a sensitive analytical balance. The sample container is typically suspended into a temperature- and atmosphere-controlled chamber. (Adapted from figure by TA Instruments, New Castle, DE)

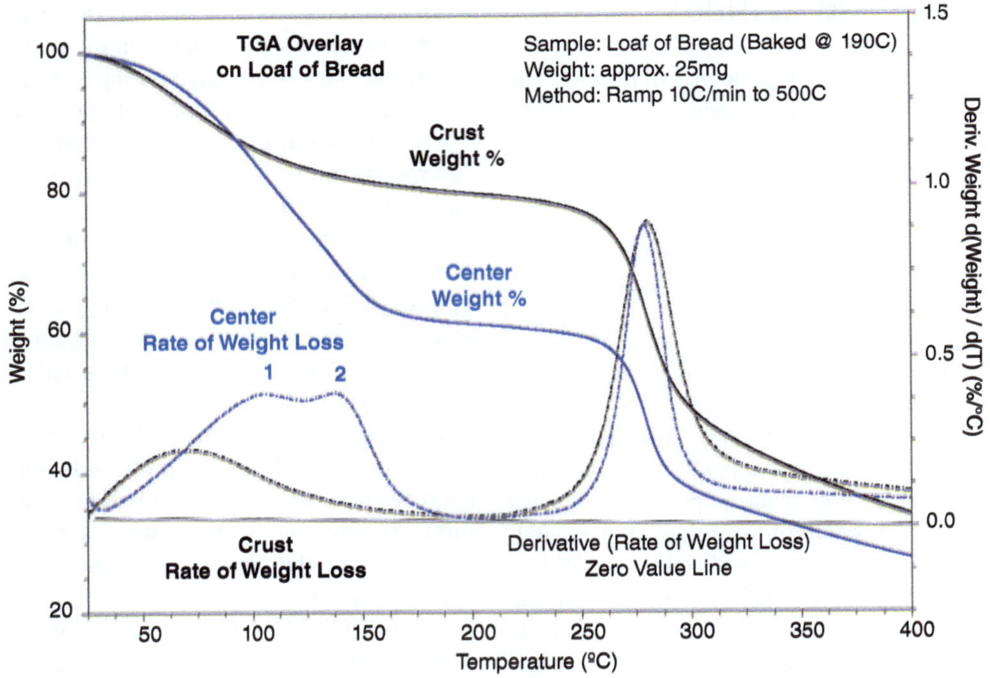

Fig. 30.6 A comparison of TGA data on the center (crumb) and crust of a loaf of bread baked at 190 °C shows substantial differences at temperatures below 200 °C. This is because the center of the loaf remains much cooler than the crust during a large portion the baking process. The bread's internal temperature profile is usually referred to as an "S" curve, as the temperature slowly increases initially, then rapidly rises as water is removed during baking

sample size reduces accuracy of small weight changes. Another limitation of TGA is that it cannot identify the chemistry of gases evolved from the sample. Knowledge of the gas composition helps to distinguish between water loss and loss of low molecular weight additives, such as flavors and fragrances, and helps to determine the chemical mechanisms involved in cooking and decomposition. TGA can be coupled to a number of analytical techniques, such as gas chromatographs (GC, Chap. 14), mass spectrometers (MS, see Chap. 11), Fourier transform infrared spectrometers (FTIR, see Chap. 8), or a combination of these techniques (GC/MS or MS/FTIR), to perform evolved gas analysis (EVG) for volatile, semi-volatile, and pyrolysis compounds from food and food packaging materials.

30.3.1.2 Experimental Conditions

Although TGA experiments can be performed over a wide range of conditions, a good starting point for most materials is as follows:

- Sample weight: 10–20 mg (larger samples improve sensitivity for detection of minor components)
- Pan type: platinum
- Purge gas: nitrogen
- Start temperature: room temperature, typically 20 °C
- Heating rate: 10 °C/min (lower rates improve resolution of overlapping weight losses)
- Final temperature: 300 °C

30.3.1.3 Common Measurements

TGA experiments are primarily heating experiments; however, isothermal (constant temperature) conditions can be used to determine drying rates or follow weight changes at processing/cooking temperatures. The most common measurements include the following (chapters containing associated material are noted in parentheses):

- Moisture content (Chap. 15)
- Water content of crystal hydrates (Chap. 15)
- Composition (multiple components)
- Decomposition temperature

It should be noted that in reality, there is no such thing as a decomposition temperature. Decomposition is a kinetic process, which means that it is a function of both time and temperature. Therefore, the temperature of weight loss due to decomposition increases if the heating rate is increased.

Figure 30.6 shows an overlay of TGA data for two samples (center or crumb and crust) taken from a loaf of bread baked at 190 °C. Plotted are sample weight (% of initial weight) and rate of weight loss (derivative) from room temperature to 400 °C. The weight signal provides quantitative information on volatile components, while the derivative signal gives a fingerprint of components (composition) remaining in the sample after baking. The larger the derivative signal (relative to zero value line drawn on the figure), the larger the component and faster the rate of weight loss.

Results show substantial differences below 200 °C, where weight loss is primarily due to evaporation of volatile components (water, flavors etc.). This is as expected because the center of the loaf (crumb) remains cooler than the crust during baking. Above 200 °C, weight loss is typically due to decomposition, and the two samples show similar results since they started with the same initial composition. The first weight loss starts immediately and is typical of unassociated or free water within the sample. The peak in the derivative signal continues above 100 °C (boiling point of water) because time is required for diffusion of the water through the sample. The large amount of volatiles in the samples would likely cause leakage in a typical aluminum hermetic pan used in DSC runs if heated above 100 °C. Even if pan leakage could be avoided by using very small samples (1–2 mg), no meaningful DSC data could be obtained above 200 °C where decomposition (chemical change) is occurring.

30.3.2 Differential Scanning Calorimetry

30.3.2.1 Overview

Differential scanning calorimetry (DSC) is the most frequently used thermal analysis technique, and for good reason. It is the universal detector for identifying and measuring changes in structure since *every* change in structure (transition) either absorbs (endothermic) or releases (exothermic) heat. It is important to remember that end-use properties of materials (foods, plastics, drugs, etc.) are a function of composition (number and amount of components; obtained using TGA) and structure (amorphous, crystalline, semi-crystalline; obtained using DSC).

The ability of DSC to measure very small rates of heat flow (microwatts, μJ/sec) is greatly enhanced because it uses a differential signal. An empty reference pan is subjected to the same thermal environment as the sample pan, and the measured signal from the DSC is the difference in heat flow rate between the sample and reference. This effectively eliminates signal noise and drift caused by heat exchanges with the environment or atmosphere around the pans.

There are two general approaches to making the differential measurement. One approach uses a common furnace for both the sample and reference (heat flux design), while the other uses individual furnaces (power compensation design). Each design has theoretical advantages and limitations, but performance is mostly based on the manufacturer's ability to build a completely symmetric system so that instrumental effects on the measurement are minimized. This provides the highest sensitivity (ability to detect weak transitions), resolution (ability to separate transitions close in temperature), accuracy, and precision. Heat flux is the most common approach to making DSC measurements and is discussed below.

Figure 30.7 is a cross-sectional view of a heat flux DSC. A common furnace is used for the sample and reference pans and is typically purged with high purity, ultra-dry nitrogen. The furnace and cooling accessory provide a wide tempera-

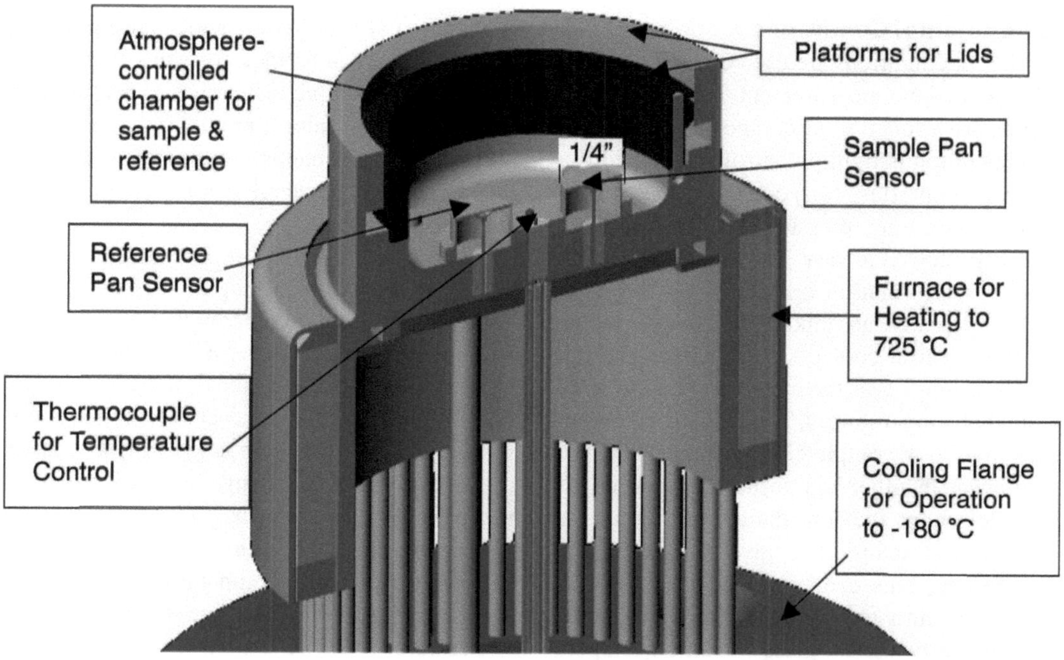

Fig. 30.7 Cross-sectional view of a heat flux DSC. The sample and reference pans are located in a common chamber that is temperature controlled over the range of −180 to 725 °C. The chamber is typically purged with high purity, ultra-dry nitrogen gas. (Courtesy of TA Instruments, New Castle, DE)

ture range from −180 to 725 °C. The sample and reference sensors are temperature sensors, such as thermocouples, thermopiles, or platinum resistance thermometers. They provide the basis of the differential heat flow measurement and directly measure temperature.

Even though DSC is the most useful of the thermal techniques for measuring structure and changes in structure, it has a number of natural limitations, including the following:

- DSC measures the *sum* of all heat flows within the calorimeter. It is sometimes difficult to interpret data because of overlapping events (multiple transitions occurring at the same time and temperature). These events can be as broad as 100 °C.
- Regardless of the type of material, all changes in structure are seen as endothermic or exothermic events. Therefore, interpretation of events requires the user to have some knowledge of the possible physical and chemical properties of the materials they are investigating.
- Most measurements involve heating the sample to higher temperatures. As temperature increases, mobility increases, permitting the structure to change in ways that are not always obvious. The measured structure may not be the original structure at the start of the experiment.
- DSC uses a single heating rate. However, higher heating rates provide better sensitivity, while lower heating rates provide better resolution. Therefore, it is not possible to optimize both sensitivity and resolution in a single DSC experiment.

30.3.2.2 Experimental Conditions

Significantly different conditions are used depending on the material and the goal of the measurement. As stated above, end-use properties are a function of composition and structure. Even if composition is tightly controlled, large differences in physical properties can occur if the process conditions (temperature, time, pressure, etc.) are intentionally or inadvertently allowed to vary. The first heat in a DSC experiment provides information on structure for a given formulation with typically an unknown thermal history (How fast was the material cooled? Was the material heat treated, if so for how long? Was the material stored, if so for how long and at what temperature?). The DSC thermograms for two samples may look completely different on the first heat due to differences in either composition or thermal history. The way to determine the contribution of thermal history to the measured structure is to give the sample(s) a known thermal history. This can be done in the DSC by cooling the sample from a temperature above its melting point to a temperature below its glass transition temperature. After cooling, a second heat is done and a comparison is made between the first and second heats. This "heat-cool-heat" experiment is illustrated for a gelatin gummy and discussed in Sect. 30.4, Understanding the Effect of Structure on Physical Properties.

Larger sample weights and higher heating rates improve sensitivity (ability to detect an event), while smaller sample weights and slower heating rates improve resolution. The conditions listed below provide a good starting point for conducting a DSC experiment, but they require optimization for the best results. The first decision is selecting the type of DSC pan that will be used. If the sample is dry (less than 0.5% volatile components, including water, at 100 °C in the TGA), then standard aluminum crimped pans will provide the best results. Crimped pans have lower mass and typically provide better heat transfer between the sample and sensor. However, they are not sealed, which permits evaporation of volatile components as the sample is heated. Hermetic pans are sealed and recommended for samples with volatile components. However, they are heavier than crimped pans and generally provide poorer contact between the sample and sensor, which can lower sensitivity and resolution. Regardless of the pan type, when preparing the sample pan, keep in mind that heat transfer between the sample and DSC sensor takes place through the bottom of the pan. Therefore, it is important to provide good contact between the sample and the bottom of the DSC pan to avoid artifacts in the data due to sample movement during the experiment. This can be done by lightly compacting the sample in the pan or, in some situations, using an inner crimped lid to force the solid sample to the bottom of the pan prior to sealing with the hermetic lid. One additional consideration is sample homogeneity. Many food products are heterogeneous, having large variations in composition and structure in their cross sections. Since thermal analysis techniques use relatively small sample sizes, users must look for ways to obtain representative samples. Two simple techniques to address this problem are to use the largest sample possible (that still fits in pan) and do all initial measurements in triplicate to detect compositional inconsistencies. General recommended starting conditions are as follows:

- Sample weight: 8–12 mg
- Pan type: Aluminum hermetic
- Start temperature: At least 25 °C below the first transition of interest. This gives the baseline time to stabilize before the temperature of the event is reached and permits better quantification of the change in heat content (**enthalpy**) resulting from the change in structure.
- Purge gas: Dry nitrogen
- Heating rate: 10 °C/min
- Final temperature: Temperature of 5% weight loss due to decomposition in TGA data should be the maximum temperature for a DSC experiment. In general, it is detrimental to decompose samples in the DSC cell because decomposition products can condense and affect the quality of future data.

30.3.2.3 Common Measurements

DSC can be used to measure the properties and structure of most ingredients used in the food industry. Typically, heating experiments are performed; however, measurements also are made while cooling or under constant temperature (isothermal). Examples of common measurements include the following (chapters containing associated material are noted in parentheses):

- Glass transition temperature of amorphous structure
- Melting temperature of crystalline structure (Chap. 23)
- Percent crystallinity of semi-crystalline material
- Crystallization of amorphous material
- Denaturation/renaturation of proteins (Chap. 24)
- Gelatinization of starch (Chap. 19)
- Analysis of frozen solutions used for freeze-drying
- Oxidative stability of fats and oils (Chap. 23)

Changes in structure (transitions) are either **endothermic**, i.e., the sample absorbs additional energy (heat), or **exothermic**, i.e., heat is released. Figure 30.8 shows endothermic transitions associated with denaturation of albumin from chicken egg at concentrations of 1 and 10% (w/w) solution with water. Unfolding of the protein results in an increase in enthalpy, free volume, and molecular mobility, and this requires heat to occur. Denaturation and gelatinization are low-energy processes and therefore require large sample sizes to provide sufficient sensitivity. In the case of Fig. 30.8, sample weights of approximately 80 mg were used in high-volume, hermetic, stainless steel pans.

30.3.3 Modulated Temperature DSC

30.3.3.1 Overview

Modulated Temperature DSC (MTDSC) is a special type of DSC measurement that applies two simultaneous heating rates to the sample [15]. A linear rate is used to obtain the same total heat flow information as provided by conventional DSC, while a nonlinear rate (such as sinusoidal or step) causes a response from just the **heat capacity** component of the total heat flow signal. Analysis software subtracts the heat capacity component from the total heat flow signal to create a third signal, which is the time-dependent, kinetic component. Each of the signals (total, heat capacity, and kinetic) can be seen in Eq. 30.1, which is often used to describe the heat flow signal from DSC:

$$\frac{dH}{dt} = Cp\frac{dT}{dt} + f(T,t) \qquad (30.1)$$

where

dH/dt = measured heat flow rate (mW = mJ/s)

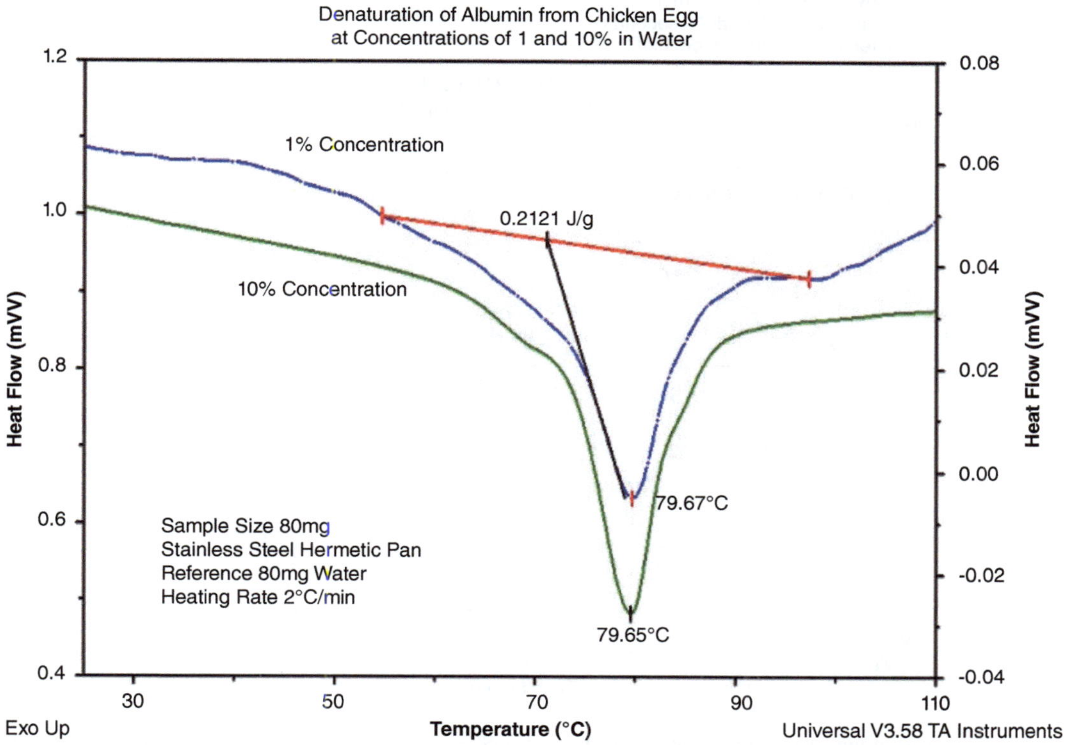

Fig. 30.8 DSC data showing protein denaturation for two concentrations (1 and 10% w/w in water) of albumin from chicken egg. Large samples (>50 mg) in high-volume pans are recommended to improve sensitivity of the weak transitions associated with protein denaturation and starch gelatinization

Cp = heat capacity (J/°C); product of specific heat (J/g°C) × sample weight (g)
dT/dt = heating rate (°C/min)
$f(T,t)$ = heat flow rate due to time-dependent, kinetic processes (mW)

Figure 30.9 shows temperature versus time and heating rate versus time for a MTDSC experiment using a sinusoidal (nonlinear) temperature change superimposed on a linear temperature change. Traditional DSC uses only a linear temperature rate (°C/min).

30.3.3.2 Experimental Conditions

Essentially every manufacturer of DSC provides an option for performing a modulated temperature experiment. However, there are significant differences (sinusoidal, step, constant frequency, variable frequency, etc.) in the way that the nonlinear temperature is applied. Regardless of the brand of instrument used, it is important to:

- Keep the size of the temperature modulation (°C) relatively small (typically less than 2 °C) so that there are at least 4–5 modulation cycles in the temperature region of a transition
- Keep the time period (seconds) of the modulation cycle long enough (typically greater than 60 seconds), so that there is sufficient time for the sample to follow the temperature modulation
- Use average heating rates less than 5 °C/min in order to obtain sufficient modulation cycles in the temperature region of a transition
- Keep sample size relatively small (typically less than 10 mg) in order to minimize thermal lag within the sample

30.3.3.3 Common Measurements

MTDSC is used on the same materials as DSC, but it has the significant advantage of being able to separate the heat flow signal into the heat capacity and kinetic components. However, lab productivity is significantly reduced due to the low average heating rates of 1–5 °C/min required in MTDSC versus 10–20 °C/min for DSC. For this reason, initial analysis of new materials should begin with DSC. However, if analysis of complex structural change is required, MTDSC is a good next step. One complex analysis example is shown in Fig. 30.10 for a sample of freeze-dried amorphous sucrose. As can be observed in Fig. 30.10, the **total heat flow** signal would be difficult to analyze because there are two overlapping events between 40 and 50 °C. The MTDSC data shows that the glass transition is due to a change in heat capacity and is clearly seen in the **reversing heat flow** signal, whereas the endothermic peak

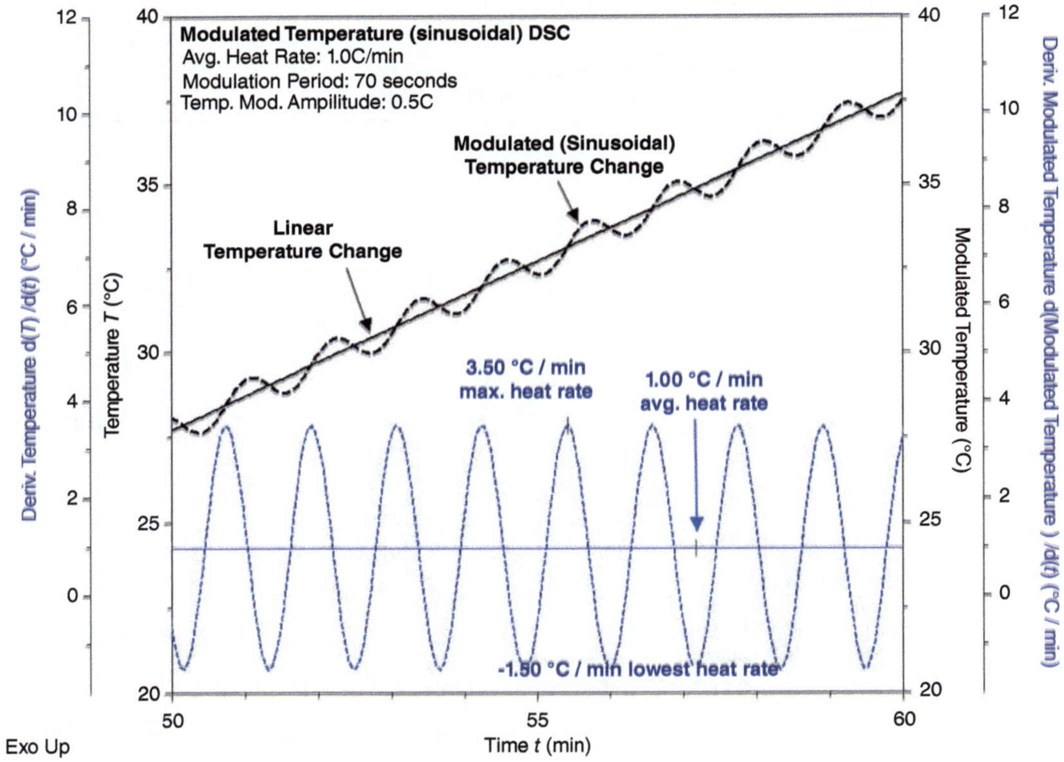

Fig. 30.9 Temperature change during a modulated temperature DSC (MTDSC) experiment has both linear (equivalent to conventional DSC) and nonlinear components. This complex temperature profile provides the ability to separate the heat capacity and kinetic components of the total heat flow signal

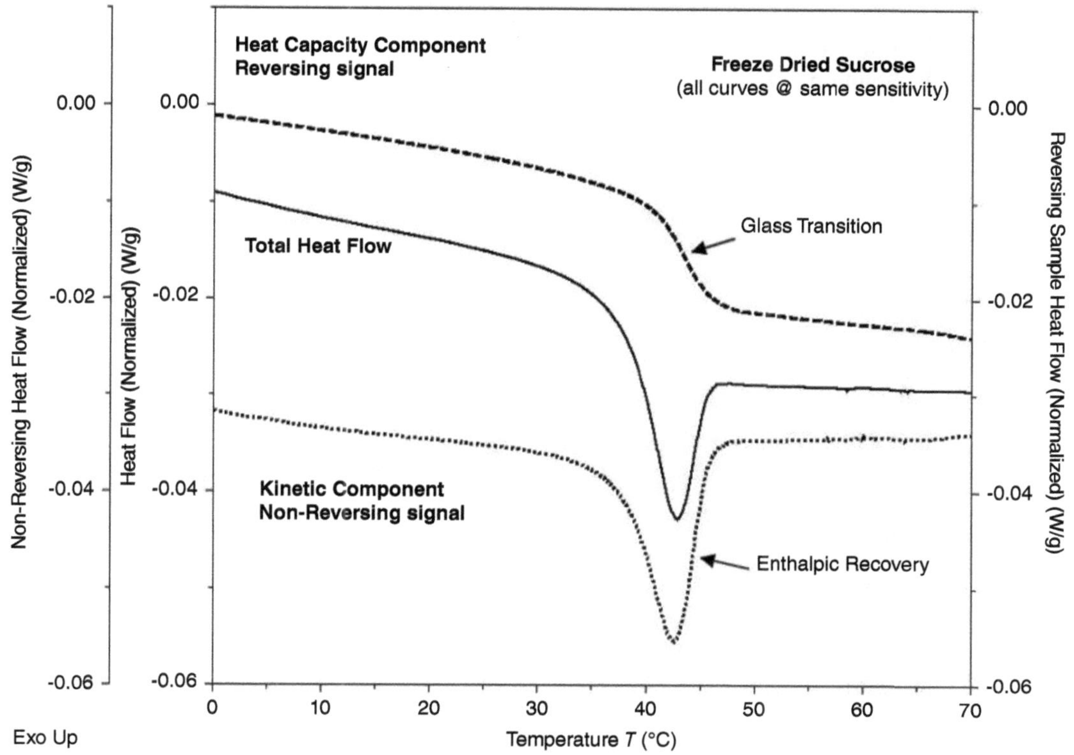

Fig. 30.10 Modulated temperature DSC (MTDSC) data for a sample of freeze dried amorphous sucrose. The sample was scanned in a hermetically sealed pan using the modulation conditions shown in Fig. 30.9. The total heat flow signal would be difficult to analyze because there are two overlapping events between 40 and 50 °C. However, the reversing and nonreversing MTDSC signals allow the two overlapping events to be separated, identified, and analyzed

is due to enthalpic recovery, a kinetic process caused by enthalpic relaxation during processing and storage, and is clearly seen in the **non-reversing heat flow** signal. Use of MTDSC allows these events to be separated, identified, and analyzed. Additional examples of complex analyses unraveled by using MTDSC can be found in Morrow et al. [11], Lee et al. [13], and Obas [16].

30.4 UNDERSTANDING THE EFFECT OF STRUCTURE ON PHYSICAL PROPERTIES

As illustrated in this book, there are numerous analytical techniques used to characterize food products. The challenge for food scientists is to develop an understanding of the relationship between analytical measurements and end-use product properties, such as texture, flow, stickiness, and rigidity. It is important to comprehend that even small changes in structure (occurring at the molecular level) can produce substantial changes in physical properties (observed at the macroscopic level). The goal of this section is to illustrate this concept with a familiar food material – gelatin.

30.4.1 Connecting Changes in Gelatin Structure as Probed by DSC to Its Physical Properties

Gelatin is a protein extracted from collagen, a connective tissue found in most animal parts, including bones, hides, and skin. As derived, it is amorphous, colorless, tasteless, and always contains some amount of water. A rather unique feature of gelatin is the wide range of functional properties it can provide to a variety of food and pharmaceutical products, such as jello, luncheon meats, gummy bears, marshmallows, and soft and hard capsules, over a wide range of water content from 95 to 1% by weight [17]. A rather unique structural feature of collagen, which is retained by gelatin, is the ability to form a triple helix that is stabilized by hydrogen bonding [18]. Formation of the triple helix at the molecular level results in the observation of a gel structure at the macroscopic level as the warm aqueous gelatin solution is cooled (termed a coil-to-helix or a sol–gel transition). Creation of triple helix structure upon cooling results in lower mobility and lower enthalpy molecules. If enthalpy is lowered, then the transition from solution to gel must be exothermic, as seen in Fig. 30.11 (top curve) for a solution containing 10% gelatin by weight in water. If the gelled sample is heated,

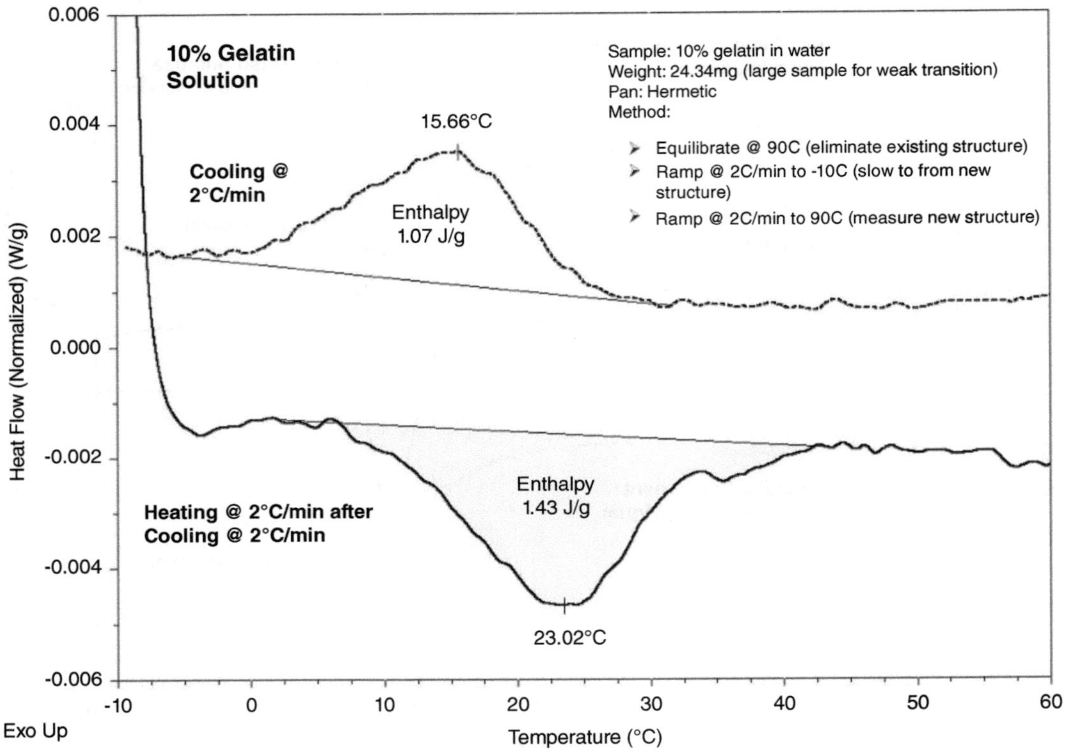

Fig. 30.11 DSC thermogram of cooling and subsequent heating of a 10% gelatin by weight in water solution. The cooling curve shows the exothermic transition from a solution to a gel (structure formation, loss of heat) and the subsequent heating curve shows the endothermic transition from a gel to a solution (structure loss, absorption of heat)

then the structure transitions from a gel to the higher enthalpy solution state (termed a helix-to-coil or a gel–sol transition). In order to increase in enthalpy, the sample must absorb heat, which results in an endothermic event, also observed in Fig. 30.11 (bottom curve). Formation of triple helix structure is a reversible process (termed thermoreversible), in which structure forms during cooling (protein renaturation) and is lost during heating (protein denaturation). Formation of structure is a kinetic (time and temperature dependent) process and, therefore, changes with factors such as concentration and cooling rate.

As can be observed in Fig. 30.11, the peak temperature and enthalpy are greater for the heating peak (23.02 °C and 1.43 J/g) compared to the cooling peak (15.66 °C and 1.07 J/g). The heating peak temperature is higher because the denaturation process is a kinetic event. The time dependence results in the temperature changing faster than the process can follow, causing a shift to higher temperatures when heating, and conversely to lower temperatures when cooling. The heating peak enthalpy is larger because the temperature is above the glass transition temperature (Tg is below the −90 °C limit of the cooling accessory), and there is enough molecular mobility for additional structure to form as the sample is slowly heated.

Below is an experiment that you can perform in the lab or at home to help you connect the observable changes in the physical properties of a gelatin gummy bear to underlying structural changes as probed by thermal analysis.

30.4.2 Gelatin Lab or Home Experiment

Materials needed for the experiment are listed below and shown in Fig. 30.12 (top left):

1. Two gummy bears of the same color from the same package containing gelatin as the only gelling agent (as determined by the ingredient list on the back of the package)
2. A small, sealable plastic food storage bag
3. A cup of hot water

Experimental Steps

1. Place one gummy in the bottom of a plastic bag and squeeze out the air. Seal the bag so that the moisture content of the gummy is not changed during the experiment. This is similar to the use of hermetically sealed pans to prevent water loss in the DSC experiment.
2. Heat a cup of water to boiling (near 100 °C) in a microwave or on the stove.
3. Dip the part of the bag containing the gummy into the hot water for one minute, being careful not to touch the hot water with your hand. Note: Heating the gummy in a

Fig. 30.12 Materials needed for the lab or home experiment are shown on the top left and resultant heated and cooled gummy in the sealed bag compared to the original gummy on the top right. Compared to the freshly heated and cooled gummy (bottom left), the 72 hr at room temperature gummy (bottom right) is now relatively easy to remove from the bag and has a texture similar to that of the original gummy bear. Note: To enhance the macroscopic visual observation of the texture in the bottom left and right images, 6 gummies were used, instead of one

sealed bag in boiling water evenly heats the gummy and prevents moisture loss, allowing for a direct comparison between the original and heated gummy samples. Heating the gummy in the bag in a microwave (uneven heating) or heating the gummy directly in a microwave (water loss) would confound the comparison.

4. Remove the bag from the water and allow the gummy to cool to room temperature (approximately 15 mins) in the sealed bag (Fig. 30.12, top right).
5. Once cooled, open the bag and compare the texture/feel of the heated and cooled gummy with the original gummy. Note: The gummy will be too soft to easily remove from the bag, so do your best to evaluate the material in the bag.
6. Using the annotated DSC method, thermograms, and associated data provided in Fig. 30.13, can you explain the observed differences in physical properties between the original and heated gummies? Record your observations and explanations to discuss with others.

Observations and Explanations

The original purchased gummy is firm, rubbery, and chewy and readily retains its external bear shape at room temperature. The DSC gummy thermogram (Fig. 30.13, first heat) shows a large glass transition near −30 °C, which is the gummy transitioning from the glassy to the rubbery state. The location of this transition confirms that the gummy is an amorphous material and is in the rubbery state at room temperature, and even below freezing, as the gummy *Tg* midpoint is approximately −27 °C. To test this, place a gummy in the freezer (approximately −18 °C) for around 4 hours; upon removal the gummy is firmer than at room temperature, but is still quite rubbery and chewable. Note that the second heat in Fig. 30.13 shows virtually the same glass transition event. One last feature to notice in the thermogram is the absence of an ice melting endothermic peak. Due to its extensive interaction with the large concentration of solids in the gummy (approximately 82%), the water is not able to freeze (termed unfreezable water) and, thus, there is no ice melting peak [16].

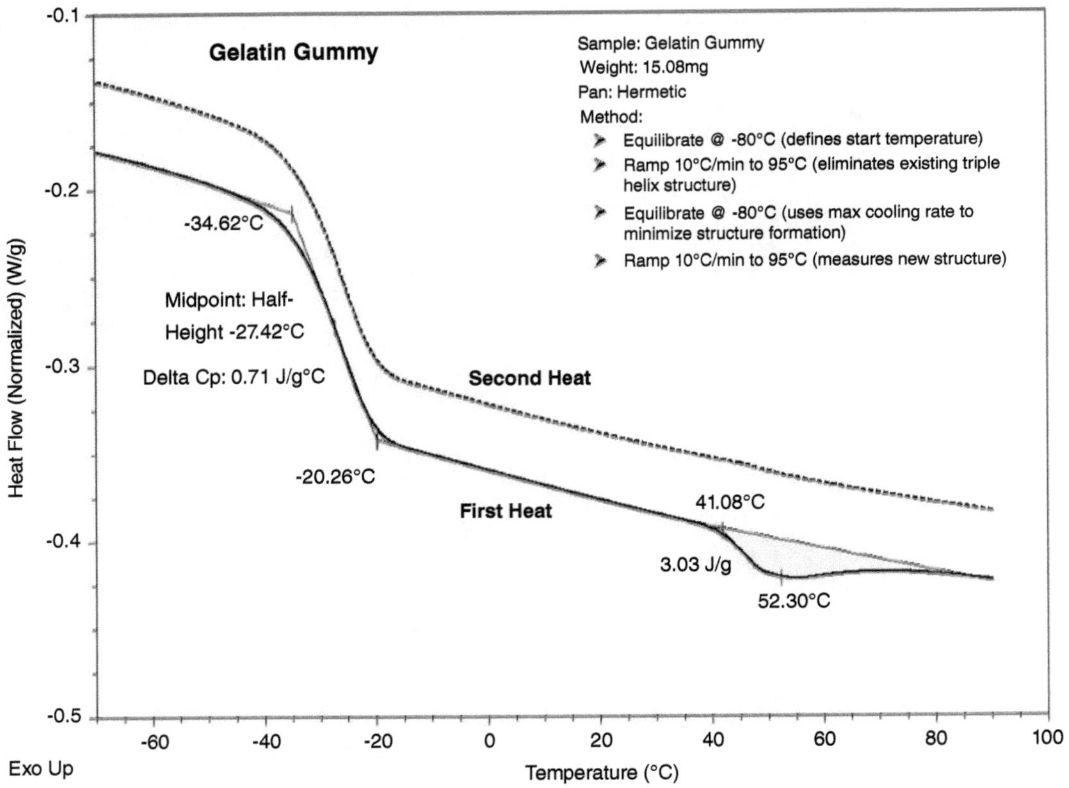

Fig. 30.13 DSC thermogram for a heat-cool-heat experiment on a gummy bear containing gelatin as the only gelling agent. Only material from the inside of the gummy was used, as the outside is coated with carnauba and beeswax to add shine and help prevent the gummies from sticking together. The first heat shows a *Tg* midpoint of approximately −27 °C and a small endothermic peak (3.03 J/g) near 50 °C. At the end of the first heat, the sample was rapidly cooled to −80 °C prior to a second heat to 95 °C. The second heat shows an essentially identical glass transition, but no endothermic event at 50 °C, indicating that the triple helix structure did not reform during cooling

The first heat in Fig. 30.13 contains an additional transition (endothermic peak) near 50 °C. It is small (3.03 J/g) and in the expected temperature range for denaturation of the gelatin triple helix structure. The first heat ended at 95 °C and the sample was then rapidly cooled to −80 °C prior to a second heat to 95 °C. The second heat is essentially identical to the first heat except for the transition near 50 °C. The absence of the 50 °C transition indicates that the gelatin triple helix structure does not immediately reform during cooling. With only about 18% moisture content (wet basis) in the sample, the kinetics of renaturation is too slow for the structure to form so quickly.

When comparing the physical properties of the two gummies, gummy texture is much more important than just the external shape. Though the heated gummy has the same glass transition event (Fig. 30.13, second heat), it no longer has the gelatin triple helix structure and associated hydrogen bonds. Thus, the heated and cooled gummy is a sticky, viscous, elastic, and shapeless mass, which is typical of a moist, amorphous material at a temperature above its glass transition. Even though the gelatin triple helix transition is small (3.03 J/g), it has a very significant effect on the physical properties of the gummy.

But wait, the thermal behavior of the gummy story is not done yet, since research shows that over time at room temperature (72 hr at 22 °C) the gelatin triple helix structure will reform (renature) to about 70% of its original amount of structure [16]. The decrease in structure compared to the original gummy is mainly attributable to the higher moisture content of the hot mix (25–30% moisture content, wet basis) from which the gummy was first formed compared to the final product gummy moisture content. To explore this renaturation of triple-helix structure on the macroscopic scale, heat a third gummy in a sealed bag and allow it to cool to room temperature and, after approximately 72 hr at room temperature, evaluate the gummy to determine whether it has regained a majority of its original textural attributes. Compared to the freshly heated and cooled gummy (Fig. 30.12, bottom left), you will find that the 72 hr at room temperature gummy (Fig. 30.12, bottom right) is now relatively easy to remove from the bag and has a texture similar to that of the original gummy bear! All that the gummy needed was time to reform its triple helix. For additional details regarding the thermal and kinetic behavior of gelatin gummies, see Obas [16] and Obas et al. [19].

30.5 SUMMARY

Thermal analysis is a series of laboratory techniques that measures physical and chemical properties of materials as a function of temperature and time. In a thermal analysis experiment, temperature is either held constant (isothermal) or programmed to increase or decrease at a linear rate. Since temperature and time are controlled in all food preparation processes, thermal analysis instruments can simulate these processes on a very small scale (milligrams) and measure the response of the material. The most frequently used techniques include TGA and DSC.

The utility of thermal analysis to the food scientist is due to the fact that product properties (functionality) at a specific temperature are dependent on the structure of the components at that temperature. Therefore, it is necessary to measure structure as a function of temperature. Structure can be defined as amorphous (no molecular order), crystalline, or semi-crystalline, with most food products having both amorphous and crystalline components. The challenge for food scientists is to develop an understanding of the relationship between analytical measurements and end-use physical properties.

Amorphous structure is characterized by analysis of the glass transition, which involves a significant change in the material's heat capacity and molecular mobility. Crystalline structure melts as temperature increases, creating an endothermic peak that provides information on the quantity of crystalline structure and its melting temperature. This chapter has illustrated application of thermal analysis to a variety of food materials.

30.6 STUDY QUESTIONS

1. Explain the differences between amorphous and crystalline structures and the thermal transitions they experience.
2. Design an experiment that would help you determine if the endotherm in the DSC data you collected is due to a thermodynamic or kinetic event.
3. In the TGA data you collected, for a new whole grain bread you are developing, there were two step changes in weight, one around 100 °C (212 °F) and one around 250 °C (482 °F).
   (a) Identify the probable cause of each step change.
   (b) Design a potential experiment to determine if you were correct in your identification.
4. You have two brands of caramels, one that is hard and one that is chewy.
   (a) What structural transition would you expect to detect in the DSC data for each type of caramel?
   (b) How would the DSC data be useful in understanding the differences in texture of the two caramel types?
5. A DSC "heat-cool-heat "experiment is commonly carried out when you have a food material with an unknown thermal history.
   (a) Explain what it means for a food material to have an unknown thermal history?
   (b) Describe the difference between the first and second heats for the gelatin gummy in Fig. 30.13.
6. Assume that you have been asked to develop a recipe for a crisp cookie based on your knowledge of material properties. To develop the recipe, you need to answer the following questions:
   (a) Should the finished cookie be in the glassy or rubbery state at room temperature? Explain your choice.
   (b) In selecting TGA experimental conditions, would you use a sealed or unsealed sample pan and why?
   (c) In selecting DSC experimental conditions, would you typically use a sealed or unsealed pan and why?
   (d) If you were to run a TMDSC experiment, which TMDSC signal would contain the glass transition and why?
   (e) While performing a DSC or TMDSC experiment, an amorphous component in the sample begins to crystallize. Would this cause an increase or decrease in the measured heat capacity?
   (f) You find out that high-fructose corn syrup (HFCS) is less expensive than sucrose, so to decrease the cost of your cookie formulation you make the cookies with HFCS. However, the cookies are no longer crisp, but chewy.
      1. Using what you learned from Sect. 30.4, explain what may have happened to cause the texture of the cookies to change? Be sure to relate the macroscopic level change in physical properties to the molecular level changes in structure.
      2. You also notice that the cookies are much browner when using HFCS compared to sucrose. Though not directly a thermal property, explain why the HFSC cookies are darker than the sucrose containing cookies. (Chap. 19)
      3. Disappointed in the results of your HFCS cookie, you unintentionally leave the HFCS cookie on the counter. After a few days, you notice it and find that it is now crispier. What may have happened to cause this increase in crispness? What thermal experiment could you do to test your hypothesis? (Chap. 15)

REFERENCES

1. MacNaughtan, B, Farhat IA (2008) Thermal methods in the study of foods and food ingredients (Chap. 9). In: Gabbott, P. (ed) Principles and Applications of Thermal Analysis. Blackwell Publishing, Oxford, UK, pp. 330–409

2. Raemy A (2003) Behavior of foods studied by thermal analysis. J Therm Anal Calorim 71:273–278
3. Ievolella J, Wang M, Slade L, Levine H (2003) Application of thermal analysis to cookie, cracker, and pretzel manufacturing, Ch. 2. In: Kaletunc G, Breslauer KJ (eds) Characterization of cereals and flours: properties, analysis, and applications. CRC Press, Boca Raton, FL, pp. 37–63
4. Kaletunc, G. (ed.). (2009) Calorimetry in food processing: Analysis and design of food systems. Wiley-Blackwell, Ames, Iowa. p. 412
5. Slade L, Levine H (1988) Non-equilibrium behavior of small carbohydrate-water systems. Pure Appl Chem 60(12):1841–1864
6. Slade L, Levine H (1991) Beyond water activity: recent advances based on an alternative approach to the assessment of food quality and safety. Crit Rev Food Sci Nutr 30(2-3): 115–360
7. Schmidt SJ (2004) Water and solids mobility in foods. Advances in Food and Nutrition Research, vol 48. Academic Press, London, UK, pp 1–101
8. Aguilera JM, Lillford PJ (2008) Food materials science: principles and practice. Springer, New York, p 616
9. Bhandari B, Roos YH (2012) Food material science and engineering. John Wiley & Sons, New York, NY. p. 414
10. Obas FL, Schmidt SJ (2022) Beyond the candy thermometer: Importance of glass transition to the stages of sugar cooking, Manufacturing Confectioner, August: 19–31
11. Morrow EA, Terban MW, Thomas LC, Gray DL, Bowman MJ, Billinge SJL, Schmidt SJ (2019) Effect of amorphization method on the physicochemical properties of amorphous sucrose. J. Food Eng. 243:125–141
12. Schmidt SJ (2012) Exploring the sucrose-water state diagram: Applications to hard candy cooking and confection quality and stability. Manufacturing Confectioner, January: 79–89
13. Lee JW, Thomas LC and Schmidt SJ (2011) Investigation of the heating rate dependency associated with the loss of crystalline structure in sucrose, glucose, and fructose using a thermal analysis approach (Part I). J Ag Food Chem (59):684–701
14. Morrow EA, Terban MW, Lee JW, Thomas LC, Billinge SJL, Schmidt SJ (2019). Investigation of thermal decomposition as a critical factor inhibiting cold crystallization in amorphous sucrose prepared by melt-quenching. J. Food Eng. 261:87–99
15. Thomas L (2006) Modulated DSC technology manual. TA Instruments, New Castle, DE
16. Obas F-L. (2023) Impact of formulation and processing conditions on the material properties and sensory characteristics of confectionery products and model systems across the stages of sugar cooking. Ph.D. Thesis, University of Illinois at Urbana-Champaign, Illinois
17. Gelatin Manufactures Institute of America (2012) Gelatin Handbook. Retrieved from: https://nitta-gelatin.com/wp-content/uploads/2018/02/GMIA_Gelatin-Handbook.pdf
18. Brodsky B, Werkmeister JA, Ramshaw JAM (2005) Collagens and gelatins. Biopolymers Online, 119–128. https://doi.org/10.1002/3527600035.bpol8006
19. Obas F-L, Lee S-Y, Thomas LC, Schmidt SJ (2021) Characterization and categorization of commercial confectionary gels through Napping-ultra Flash Profile (UFP) and Hierarchical Clustering Analysis. J Food Sci 86:2655–2670

Color Analysis

31

M. Monica Giusti, Belén Gordillo,
and M. Lourdes González-Miret

Contents

31.1 INTRODUCTION

Color is one of the main sensory attributes that determine food quality and acceptance along flavor and texture. Color, as the first characteristic perceived and often the only criterion attracting our attention, has the greatest influence on consumer's preferences. Color is a physical property, yet it is directly related to the chemical composition of foods and it is also an indirect indicator of many other aspects of food quality. In the food industry, the analysis of color allows us to estimate the levels of bioactive pigments in fruits and vegetables, to determine their ripening state, and to help us discriminate cultivars, varieties, and food types that have different compositions. Color also can be used to classify food quality grades or to assess safety. Certain color variations, such as those occurring on the surface of apples or meat, can be a sign of physical, chemical, or microbiological damage, allowing for their rapid detection. Color also influences our perception of other sensory attributes (taste/flavor/texture) having an important impact on consumers' expecta-

M. M. Giusti (✉)
Department of Food Science and Technology, The Ohio State University, Columbus, OH, USA
e-mail: giusti.6@osu.edu

B. Gordillo · M. L. González-Miret
Food Colour and Quality Laboratory, Facultad de Farmacia, Universidad de Sevilla, Sevilla, Spain
e-mail: bgordillo@us.es; miret@us.es

B. P. Ismail, S. S. Nielsen (eds.), *Nielsen's Food Analysis*, Food Science Text Series,
https://doi.org/10.1007/978-3-031-50643-7_31

tions and buying decisions. Hence, for food corporations and scientists who establish quality control specifications, it is essential to ensure that food products maintain an acceptable or desirable color over their shelf life.

While subjective visual assessment and the use of visual color standards are still used in the food industry, an objective evaluation of color is preferable for both research and industrial applications, where instrumental color measurements are extensively employed to provide repeatable data that correspond to how the eye sees color.

Colorimetry is the science of color measurement [1], which allows defining any color in terms of mathematical coordinates to replace subjective visual responses with an objective numerical system. Various color-ordering systems and color spaces have been developed to better correlate to the visual perception of color. This chapter provides a brief description of the human physiology of vision and presents the basics for understanding how color of food products should be measured.

31.2 PHYSIOLOGICAL BASIS OF COLOR

Color can be defined as the mental response experienced by an individual when radiant energy within the visible spectrum (380–770 nm) falls upon the retina of the human eye [2]. When light strikes an object, it can be absorbed, reflected, and/or scattered, depending on the surface characteristics and chemical composition of that object. Selective absorption of certain wavelengths of visible light is the primary basis for its color. The human eye perceives the reflected, scattered, or transmitted light. This optical stimulus is sent to the brain which processes the information, translates it, and interprets it as a "color".

Thus, color perception is considered a psychophysical concept, related to the psychology of the observer, the physiology of vision, and the spectral radiant energy of a light source.

Figure 31.1 is a simplified diagram of the human eye. Light enters the eye through the cornea and is focused on the **retina**, which contains two types of photoreceptor cells: cones and rods [3]. **Cones,** highly concentrated in the fovea, are responsible for color vision at high levels of illumination (**"photopic"** vision). People with typical color vision have three types of cones (ρ,γ, and β receptors) sensitive to red (long wavelengths), green (medium wavelengths), and blue (short wavelengths) light, respectively (Fig. 31.2). All colors can be produced by combining the stimuli of the three types of cones. The deficiency or absence of any of the three receptors results in color vision deficiency. **Rods,** widely distributed in the retina, have no color discrimination and are responsible for lightness discrimination at low-intensity light (**"scotopic"** vision). Under low levels of illumination rods work and cones do not; hence, the color discrimination is limited.

The brain's interpretation of optical signals is a complex phenomenon and is influenced by a variety of psychological aspects. According to the "Color Opponent theory" [4], signals are sent to the brain through three channels, and color is perceived as pairs of opposites: light/dark, red/green and yellow/blue. One channel (brightness signal) carries out achromatic information indicating darkness and lightness. The other two channels (hue signals) carry out color information indicating red vs. green and blue vs. yellow (Fig. 31.3).

Color is three dimensional, and to define or specify any color, three attributes are necessary [5]: (1) **hue**, the aspect of visual sensation that we instinctively think of as color (e.g., red, blue, green); (2) **lightness or luminosity**, which represents lightness and darkness; and (3) **chroma** or **saturation,** which indicates color intensity. The human eye has an extraordinary ability to discriminate color attributes, being able to appreciate lightness differences of 1% or hue differences of about 2–3 nm between two monochromatic colors with equal lightness. Humans are able to distinguish

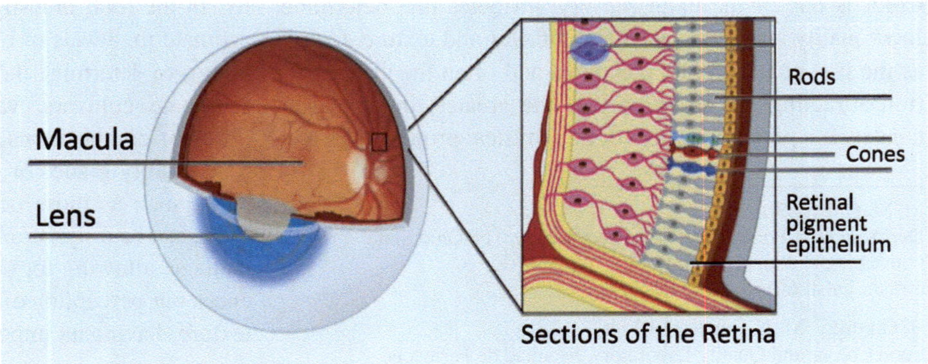

Fig. 31.1 The anatomy of the human eye. (Adapted from https://doi.org/10.1007/s11818-019-0025-x)

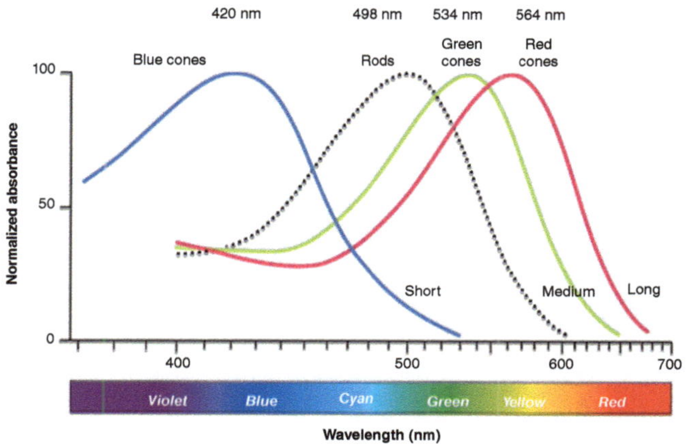

Fig. 31.2 Spectral sensitivity curves of the three types of cones and rods comprising photopic and scotopic vision. (Adapted from http://cnx.org/content/col11496/1.6/)

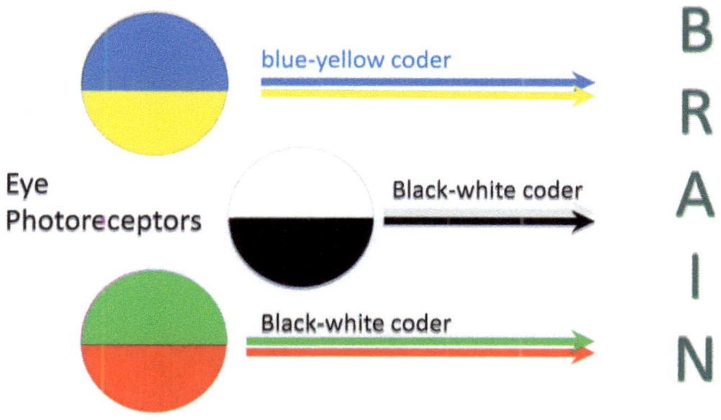

Fig. 31.3 The opponent colors theory

more than seven million different chromatic stimuli (about 200 hues, 450 lightness values, and about 80 saturation levels). However, humans have very poor color memory and cannot accurately recall colors previously observed [6, 7]. This factor, together with the possible differences among individuals caused by physiological (thresholds), psychological (motivation, tiredness), or pathological (color blindness) factors, make visual perception of color inherently subjective, hence the need for objective measurement of color.

31.3 COLOR SPECIFICATION AND ORDER SYSTEMS

Verbal, visual matching, and instrumental methods for describing and specifying color have been extensively developed. Below we describe some relevant to the food industry.

31.3.1 Visual Systems: The Munsell Color System

Color-order systems based on visual comparison with systematically organized color standards (linear scales, chromatic circles, color triangles, or color solids) have been developed through the years to select, measure, and compare colors [6, 8]. Color standards for several food products are available (honey or olive oil, for example) [9]. Visual methods are simple, convenient, and easy to understand; however, they may be more difficult to control and prone to subjectivity.

The first visual color system to define color in three separate, independent, and uniform dimensions, based on rigorous measurements of the responses of the human visual system, was the **Munsell** system. This system was developed in 1905 by Albert H. Munsell, an artist and educator. In the Munsell system, a color is specified with three attributes: hue, value, and chroma, represented cylindrically in three dimensions as an irregular color solid (Fig. 31.4):

1. The **hue** scale varies horizontally and is divided into 100 perceptually equal parts according to 10 shades, each of the following 5 reference colors: red (R), yellow (Y), green (G), blue (B), and purple (P); plus 5 mixed color pairs, green-yellow (GY), yellow-red (YR), red-purple (RP), purple-blue (PB), and blue-green (BG).
2. **Value** varies vertically and describes the lightness and darkness scale, from white to gray to black (10 designates absolute white and 0 designates absolute black).
3. **Chroma** varies radially from the center of the color solid and represents the "purity" or intensity of a color, as the extent of difference from a gray of the same value. It is designated in increasing numbers starting with 0 (neutral grey) to a maximum chroma that differs considerably for different hues. For example, R5 has a maximum chroma of 12 while Y5 has a maximum of 6.

In the **Munsell notation**, hue is listed first and designated by a number (from 1 to 10) and letter combination (the ten defined hues). Value follows with a number from 0 to 10 followed by a slash mark, which is followed by a number for chroma. For example, a color designated as 5R 7/10 indicates a hue 5R, a value of 7, and a chroma of 10.

One of Munsell's goals was to develop a system based on equal visual perception, with equal steps of perception for each of the coordinates. For example, the difference in value between 2 and 3 is visually equivalent to the difference between 5 and 6, which applies to the other coordinates as well. The Munsell system's visual linearity undoubtedly contributes to its success in many different fields.

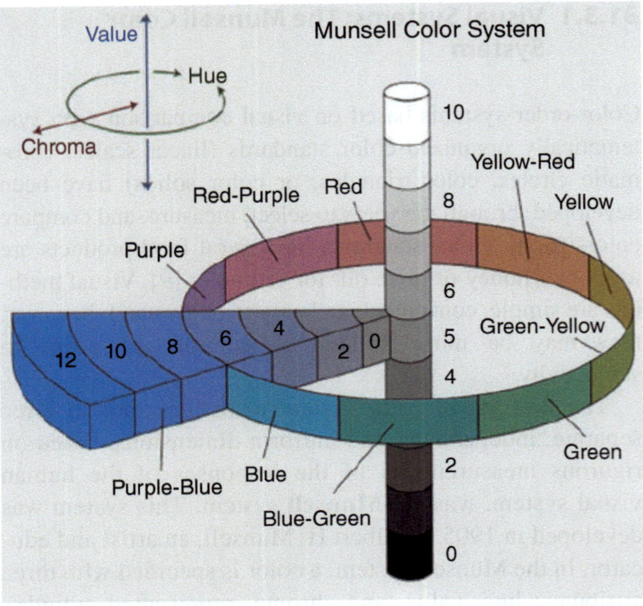

Fig. 31.4 The Munsell color system, showing: a circle of hues at value 5 chroma 6; the neutral values from 0 to 10; and the chromas of purple-blue (5 PB) at value 5. (Source, Wikipedia: https://en.wikipedia.org/wiki/Munsell_color_system)

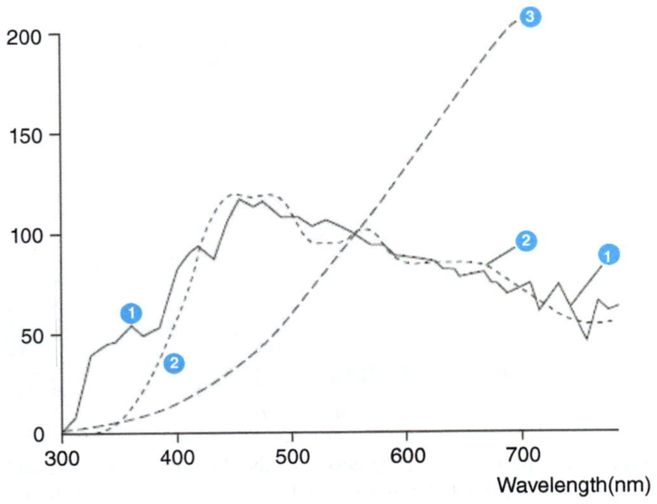

Fig. 31.5 The spectral power distribution curves of three standard CIE illuminants: (1) illuminant D_{65}, average daylight including ultraviolet wavelength, (2) illuminant C, average daylight (not including ultraviolet wavelength region), and (3) illuminant A, incandescent light. (Courtesy of Konica Minolta Sensing Americas, Inc., Ramsey, NJ)

31.3.2 Instrumental Systems: The CIE Color Specification Systems

The **CIE (Commission Internationale de l'Eclairage**, or the International Commission on Illumination) is the main international organization concerned with color, and it is accepted as the authority on the subject matter of light and lighting. The CIE described the basis for color measurement [10, 11] and introduced the concepts of "standard illumination" and "standard observers" that are used in color measurement worldwide.

Objects may appear to have different colors when viewed under different lights. **Metamerism** occurs when two objects appear to have the same color under one light source but exhibit different colors under another source. Hence, standard conditions of illumination must be defined to measure color. **Standard illuminants** for color measurement have been established by the CIE as theoretical power distribution of visible light that resemble specific light sources and that can be used for color calculations. Figure 31.5 shows the spectral power distribution curves of three CIE standard illuminants, A, C, and D_{65}. **Illuminant A,** adopted in 1931, represents an incandescent light bulb (2856 Kelvin). **Illuminant C** was also adopted in 1931 and represents overcast daylight (6774 Kelvin), while **Illuminant D65**, adopted in 1965, also represents average daylight but includes the ultraviolet wavelength region (6504 Kelvin). D65 is the most used illuminant since the CIE recommends its use when there are no prior requisites for using another illuminant.

Any color sensation can be matched by mixing three colored lights. W.D. Wright in 1928 and J. Guild in 1931 conducted independent experiments in which people with normal color vision visually matched spectral (single wavelength) light by mixing different amounts of three primary lights (red, green, and blue) using rheostats (Fig. 31.6), and repeating the process until covering the entire visible spectrum. The **field of view** for these experiments was **2°** (like viewing a dime at an arm's length), with the belief that the cones were within a 2° arc of the fovea (the greatest visual acuity area of retina). The red, green, and blue response factors were averaged and mathematically converted to *x*, *y*, and *z* functions (**color matching functions**) that quantify the red, green, and blue cone sensitivity of the average human observer. These observer functions were standardized and adopted by CIE in 1931 as the **CIE 2° Standard Observer**. These curves provide

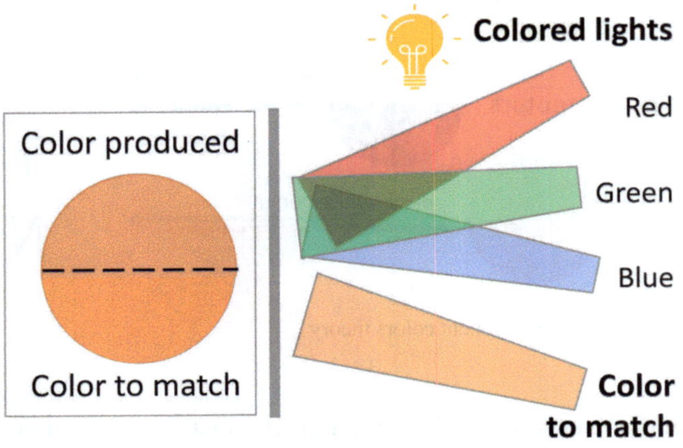

Fig. 31.6 Additive color matching. Diagram showing three projectors focused on the upper half of a circle to reproduce the color to be measured, presented on the lower half

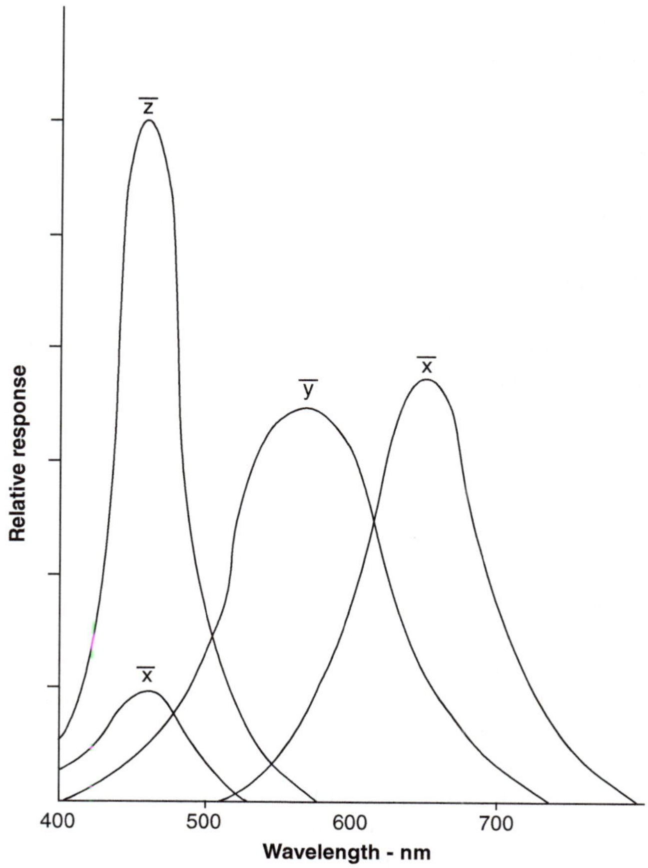

Fig. 31.7 Standard observer curves showing the relationship between the red (x), green (y) and blue (z) cone sensitivity and the visible spectrum

human sensory response factors (Fig. 31.7) needed to convert the spectral transmission or reflectance curve of any object to three numerical values. Later, it was established that most visual assessments are done with a greater field of view, 10° field, that encompasses the area in the fovea that contains the rod receptors, responsible for lightness discrimination. New functions were developed and adopted by CIE in 1964 as the **10° Supplementary Standard Observer,** which better correlates to visual assessments. Color data obtained by using the 2° observer will differ from that obtained using the 10° standard observer, but many color software packages have the capabilities to convert data between the two.

31.3.3 The CIE Tristimulus System and the Tristimulus Values

With the adoption of standard observer functions and standard illuminants, it became possible to convert the spectral transmission or reflectance curve of any object to three numerical values (within a 3-D color space), known as the **CIE tristimulus values:** ***X*, *Y*, and *Z*,** based on mathematical calculations [10]. Tristimulus Y is known as **luminosity**

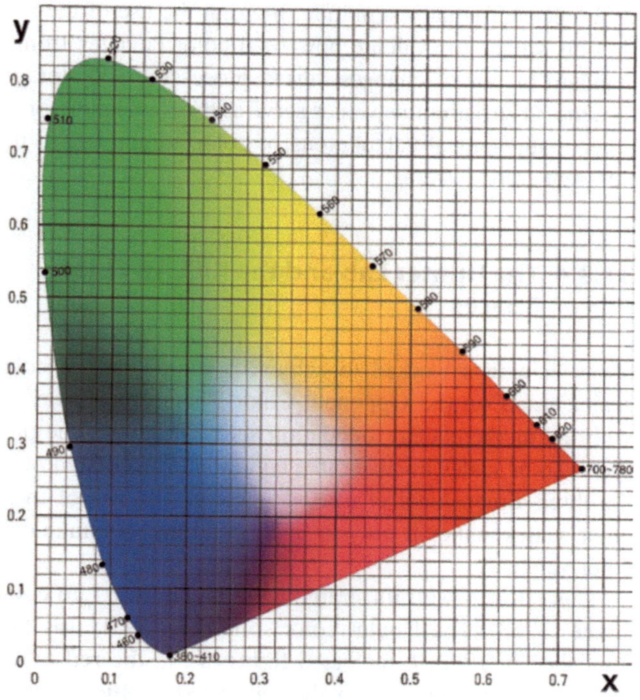

Fig. 31.8 The 1931 x, y chromaticity diagram. (Courtesy of Konica Minolta Sensing Americas, Inc., Ramsey, NJ)

(luminous reflectance or transmittance). In the 1931, CIE color space each color is defined by the Y value and the two chromaticity coordinates, being $x = X/(X + Y + Z)$ and $y = Y/(X + Y + Z)$ [12]. Figure 31.8 shows the CIE 1931 **chromaticity diagram** where x vs y are plotted to give the horseshoe-shaped locus. Spectral colors lie around the perimeter (named *spectrum locus*), and white light (illuminant D_{65}) has the coordinates $x = 0.314$, $y = 0.331$. Drawing a line from the coordinates for white light through the object coordinates to the edge gives the **dominant wavelength,** λ_d, which is analogous to hue in the Munsell system. The distance from the white light coordinates to the object coordinates, relative to the distance from the white light coordinates to λ_d, is described as **% purity**, and is analogous to chroma in the Munsell system. The tristimulus value ***Y*** is the **luminosity** and is analogous to "value" in the Munsell system.

Manual calculation of XYZ tristimulus values from reflectance/transmission spectra is a tedious operation. Modern colorimetric spectrophotometers measure the light reflected or transmitted from an object, and a processor multiplies spectral data by standard illuminant and standard observer functions to give the XYZ tristimulus values. The XYZ values do not easily relate to observed color, and the distribution of colors in the x, y chromaticity diagram is very nonuniform [see Fig. 31.8: the wavelength spacing in the green region (500–540 nm) is much larger than that in the red (600–700 nm) or blue (380–480 nm) regions]. This means that equal changes in x, y, or Y do not correspond to perceived differences of equal magnitude, limiting the use of the tristimulus values to evaluate how food color deviates from a standard or how it changes during processing and storage. Yet, the *XYZ* tristimulus values serve as the foundation for all other mathematical calculations of color parameters in the different color systems.

31.3.4 The Hunter Lab and the CIELAB Color Spaces

Many scientists worked to achieve a more uniform color scale, based on the Opponent-Color Theory, and working with the mathematical foundation set by the CIE. The proposed color scales involve three coordinates, consistent with the three-dimensional character of color. The use of these color scales makes it easy to understand and communicate color values.

This effort is well illustrated with the **Hunter Lab color space** (1958). The Hunter color solid (Fig. 31.9) was first published in 1942; *L* indicated **lightness**, *a*, the **red** (+) or **green** (−) coordinate; and *b*, the **yellow** (+) or **blue** (−) coordinate. The Hunter Lab color space was widely adopted by the food industry, and it is very effective for measuring color differences. The *Lab* system was subsequently improved by the CIE to give an even more uniform color spacing.

In 1976, the CIE officially adopted and recommended a more perceptually linear uniform color space obtained from a nonlinear transformation of *XYZ*: the CIE 1976 ($L^*a^*b^*$) or **CIELAB.** The CIELAB system is based on the opponent color model of human vision, like the Hunter Lab, and defines color based on the same three color coordinates: L^*, a^*, and b^*. It is important to note that the asterisk superscript was introduced to differentiate CIELAB parameters from Hunter Lab. Coordinate $\boldsymbol{L^*}$ represents **lightness**, ranging from black ($L^* = 0$) to white ($L^* = 100$). Coordinate $\boldsymbol{a^*}$ represents the red-green component (positive and negative values, respectively) of a color, and coordinate $\boldsymbol{b^*}$ represents the yellow-blue component (positive and negative values, respectively) of a color. L^*, a^*, and b^* serve, hence, to describe any color as three numerical data and to define its location in the three-dimensional CIELAB color space, in which the chromatic a^* and b^* axes are orthogonal to the achromatic lightness axis L^*.

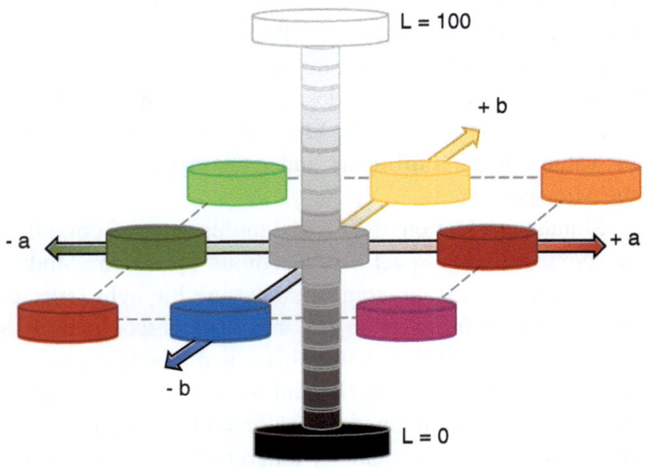

Fig. 31.9 The Hunter Lab color solid (adapted). (Courtesy of Hunter Lab, Reston, VA)

The rectangular a^*, b^* axes can be in turn converted into two polar coordinates, hue (h_{ab}) and chroma (C^*_{ab}), according to the formulae:

$$h_{ab} = \arctan\left(\frac{b^*}{a^*}\right) \tag{31.1}$$

$$C^*_{ab} = \sqrt{\left(a^*\right)^2 + \left(b^*\right)^2} \tag{31.2}$$

where:

h_{ab} = reddish, greenish, bluish, etc., character and values of 0° (or 360°), 90°, 180°, and 270° that indicate red, yellow, green, and blue, respectively

C^*_{ab} = the intensity of the color that represents the distance from the coordinate's origin ($a^* = 0$, $b^* = 0$) to a determined point within the CIELAB space for any L^* value

Figure 31.10 shows the CIELAB color space.

Each individual parameter (L^*, a^*, and b^*) provides only partial information about the color of an object, and objects sharing similar value in one coordinate could have very different colors. Figure 31.11 shows plots for three hypothetical objects having the following a^*b^* coordinates: $a^* = 12$ and $b^* = 8$; $a^* = 12$ and $b^* = 4$; $a^* = 12$ and $b^* = -4$. While all objects have identical a^* values, their colors range from purplish red (h_{ab} = 342°) to red (h_{ab} = 18°) to orange (h_{ab} = 34°). A combination of lightness (L^*), hue angle (h_{ab}), and chroma (C^*_{ab}) is usually easier to correlate to human perception. Chroma will increase with increasing pigment concentration and then decrease as the sample becomes darker. Thus, it is possible for two samples, one light and one dark, to have the same hue angle and the same chroma. They will readily be distinguished, however, because of their different L^* values.

The CIELAB color space has obvious parallels with previous color solid spaces such as Munsell (describing value, chroma, and hue) and Hunter Lab (describing *L*, *a*, *b* coordinates), which have been widely adopted by the food industry. However, the current CIE recommendation is the use of CIELAB, and it is important to highlight that color data from different systems are not comparable. In general, CIELAB has replaced the Hunter space for industrial appli-

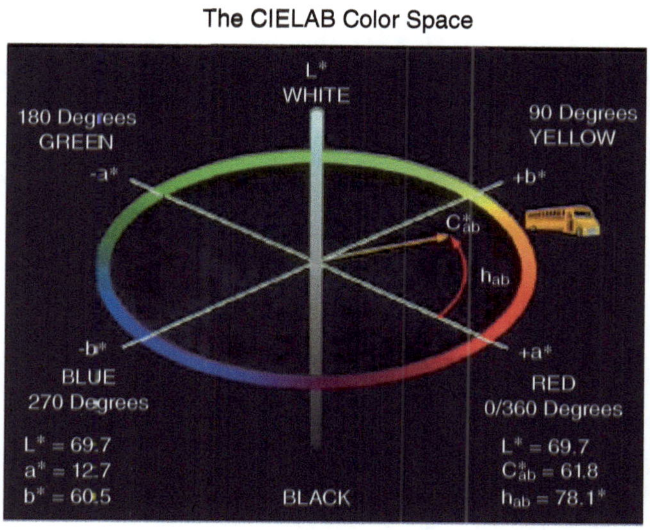

Fig. 31.10 The CIELAB color space showing the location of "school bus yellow". (Courtesy of HunterLab, Reston, VA)

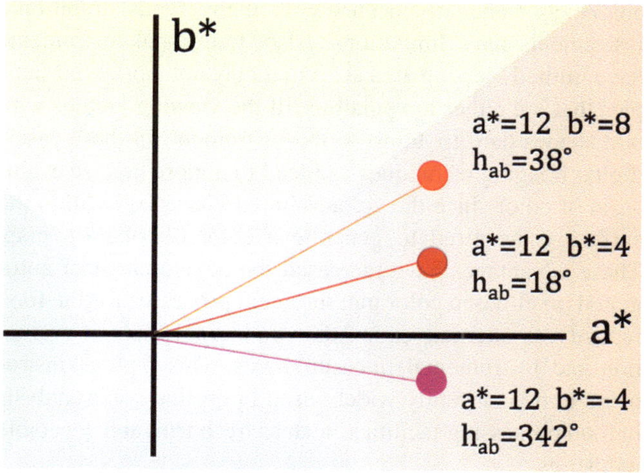

Fig. 31.11 Plots of *a**and *b** for three hypothetical samples

cations although *Lab* is still in use in some food companies, since updating standardized methods can be a costly endeavor.

31.4 INSTRUMENTAL MEASUREMENT OF COLOR

Color is the only sensory attribute for which instrumental methods are available to replace a sensory measurement with sufficient precision. In color measurement, the selection of the right **instrument** is very important. Likewise, **illuminant**, **viewing angle (2° or 10°)** and color space for data presentation (**XYZ**, ***Lab***, **CIELAB**) must be chosen. These should be specified in technical reports and/or research publications. Illuminant D65, 10° viewing angle, and CIELAB are appropriate for most food applications. Nowadays, instrumental color measurements can be conducted by using traditional instruments (colorimeters or spectrophotometers) or advanced computer vision methods (digital imaging techniques).

31.4.1 Colorimeters vs Spectrophotometers for Color Measurement

Colorimeters and spectrophotometers are two types of instruments that can be used to measure, analyze, and communicate color. Both types of instruments can provide a quantitative measurement that represents the way the average human eye sees the color of an object, yet they work in different ways, and each is best suited to different types of measurements.

Colorimeters use filters to obtain the tristimulus values (*X*, *Y*, *Z*) optically. These values can then be mathematically converted to other color systems. A colorimeter does not provide full-range spectral data and are used in applications that require fast results with less analytic depth. Colorimeters allow for color measurements with high stability, precision, ruggedness, ease of using, portability, online process control, and specialization for specific commodities. However, colorimeters do not have a high degree of accuracy with respect to identifying or matching colors. Colorimeters can vary with respect to operating in transmission or reflectance mode and can have different sizes of sample viewing area.

Spectrophotometers are instruments that quantitatively measure the transmission or reflection of light. There are different types of spectrophotometers specifically designed for the measurement of color, also referred to as spectrocolorimeters. Spectral data is then used to calculate the *X*, *Y*, *Z* values mathematically, in accordance with the CIE definitions depending on the illuminant, the measurement geometry, and the observer [6]. Spectrophotometers are precise, accurate, and the more sophisticated color measurement instrument available for accurately identifying the color of an object and are valuable tools for quality control and color formulation. Spectrophotometers are often used in research for which high analytical levels and accuracy of results are needed.

31.4.2 Reflection and Transmission Measurements

The measured values depend on geometrical relationships between the measuring instrument and the sample. Similarly, the visual assessment of colored samples is affected by the geometry of illumination and vision. The correlation between measured values and visual perception depends on the degree to which the geometry of the measurement simulates the geometrical conditions of viewing. The **instrument geometry** refers to the **arrangement of the light source**, **sample placement**, and **detector**. The angle of light incidence will play an important role, and the light may be diffuse (scatters from all sides) or direct (coming from a particular direction). The CIE recognizes different instrument geometries [10]:

1. The normal geometry **0°/180°** is used for **spectrophotometric transmission measurements**, which involves transparent samples. In this geometry, direct light illuminates at 0° and passes through the sample, being measured at 180°.
2. In the case of reflecting materials, **spectrophotometric reflection measurements** are indicated. Geometries are **45°/0°** when the specimen is illuminated at 45° and measured at 0° and, the inverse, **0°/45°** when the specimen is illuminated at 0° and measured at 45°. Instruments with these geometries can only measure reflectance (Fig. 31.12).
3. On the other hand, the diffuse sphere geometry has a white-coated **integrating sphere** that uses diffuse light for illuminating the sample, with a **d:8°** geometry (diffuse illumination, 8° angle of viewing). Instruments with the sphere geometry will often allow for measurements of reflectance and transmittance and can collect data with the specular component (Fig. 31.13) either included (SCI) or excluded (SCE). These instruments are very versatile and particularly helpful for color measurement of translucent and opaque samples and also can measure the amount of light scattering, turbidity or haze in liquid samples, and the amount of gloss in solid samples.

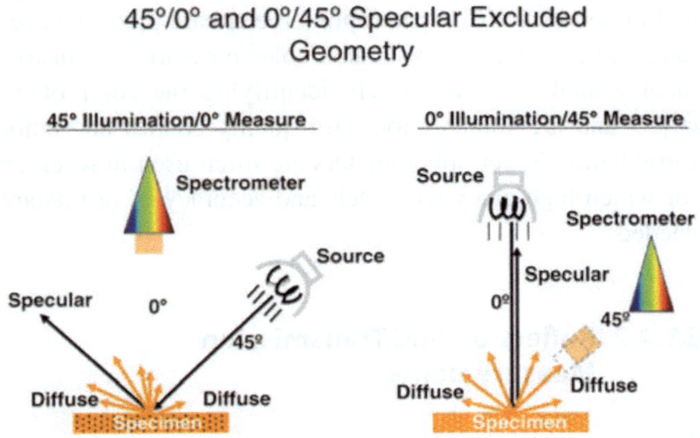

Fig. 31.12 CIE standardized geometries for 45°/0° and 0°/45° instruments. (Courtesy of HunterLab, Reston, VA)

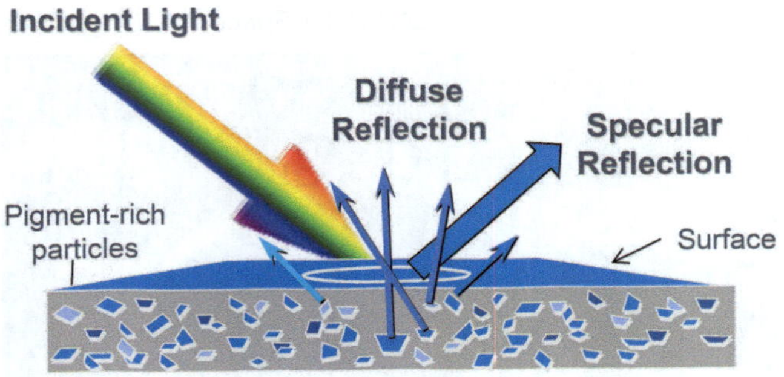

Fig. 31.13 Interaction of light with an object. (Adapted from Loughry [4], used with permission)

31.4.3 Advanced Optical Systems: Digital Imaging Techniques

Digital imaging techniques have increased in popularity over the last decades. For samples that have nonhomogeneous colors, as frequently occurs with many foods, traditional instruments have limitations. Also, traditional instruments have limited viewing area sizes that may not work with samples that are either too small to fill the viewing area or with odd shapes hard to adjust to the instrument. In these cases, digital imaging techniques can lead to a more precise evaluation of color since the color value of each pixel within the surface is acquired to generate a color distribution map. These advantages have increased the development of automated pixel-based color measurement processes for the food industry as alternatives to other traditional (human evaluation and instrumental) measurements. This type of instrumentation is currently widely used in on-line color analysis in food processing facilities, such as fresh fruit and vegetable industries.

Computer vision is the science that develops theoretical and algorithmic basis to automatically extract and analyze useful information about an object. It is possible to extract quantitative color information from digital images by using image processing and analysis [13]. Every different color present in the image can be accounted for in one or more pixels. Moreover, it allows measuring not only color but also other features related to appearance, such as texture or heterogeneity, which do not vary the color but affect how the human eye perceives it [14].

A computer vision system includes (1) a controlled illumination system, (2) a charge-coupled device (CCD), (3) a frame grabber that converts the analogue image from the camera into a digitized one, and (4) a computer with the suitable software for image processing and the interpretation of results [15]. The CIELAB coordinates are obtained by image analysis from the RGB data of the calibrated camera. A segmentation criterion (thresholding or edge-detection as the

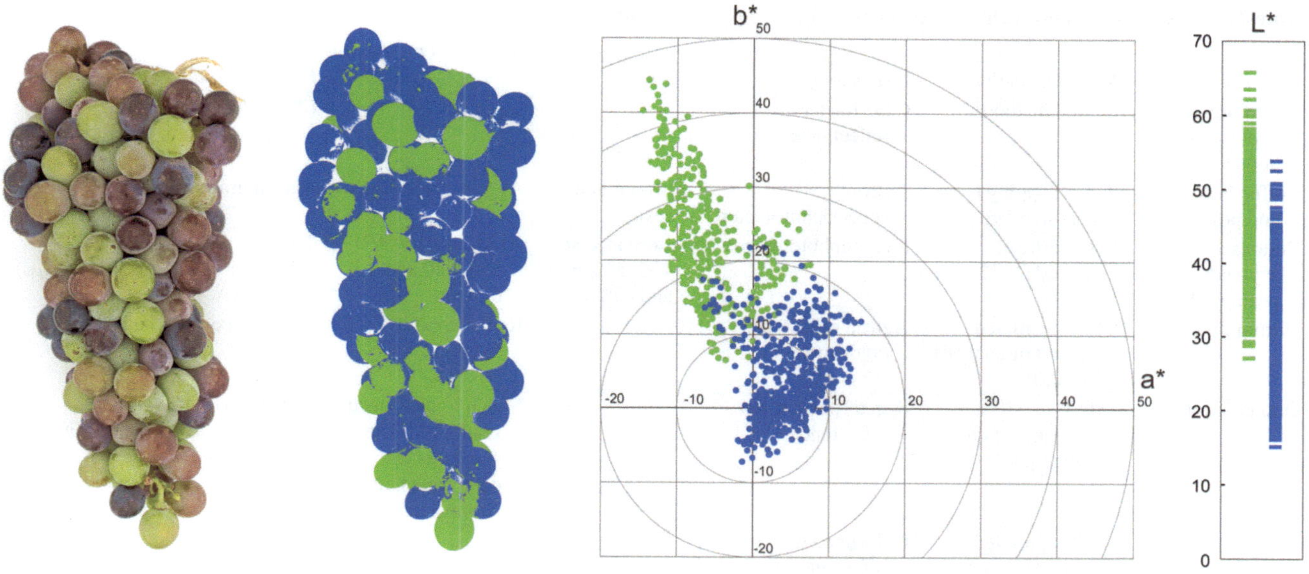

Fig. 31.14 CIELAB coordinates obtained by image analysis from a grape cluster during *veraison*. (Courtesy of Rodríguez-Pulido FJ, Food Color and Quality Lab., Seville, Spain)

main ones) is applied for calculating color only from pixels with analytical information, which simplify and/or change the representation of an image into something more meaningful and easier to analyze (Fig. 31.14).

31.5 COLOR MEASUREMENTS IN THE FOOD INDUSTRY

Color is one of the defining characteristics of any food product, and as such, it is a critical quality parameter monitored by most food industries during the quality control processes, as well as during distribution, and storage.

31.5.1 Color Measurements as a Tool for Quality Control

In the food industry, color measurements are widely used as valuable tools for quality control monitoring, process validation, commercial grading, and to ensure that products meet applicable regulations and safety standards. Color measurements offer numerous advantages for quality control purposes such as obtaining a great amount of quantifiable, reliable, and objective data in a rapid, repeatable, effective, consistent, noninvasive, and relatively economic manner. Currently, there are many automated colorimetric methods that allow for quickly assessing the consistency and uniformity of food products for real-time inspection and quality control In general, color measurements are simple and effective for quality control. Some examples of how color analysis is performed as part of quality control in food industries are shown in Table 31.1. The color quality standards might follow procedures carefully developed and validated in-house, or might be derived from official methods or approved methods recommended by a commodity association [such as the American Association of Cereal Chemists (AACC), or the American Oil Chemists Society (AOCS), or a regulatory agency (such as the USDA)]. In the case of baked products (e.g., breads, cookies, and goods), instrumental color measurement is an essential part of analyzing the quality for food manufacturers. Reflectance spectrophotometers are ideally suited to measuring the color of baked products due to their ability to objectively quantify chromatic information and translate it into applicable color spaces and indexes. On the other hand, the AOCS provides guidelines for the use of a Tintometer Color scale for the evaluation of oil and fats in refineries. The AOCS-Tintometer® Color Scale is a modified red and yellow version of the Lovibond® RYBN Color scale used for fats, oils, and derivatives. Other food products such as orange juice, tomato products, or peanut butter are also commercially graded by comparing their color with the USDA Color Standards as points of reference.

31.5.2 Color Differences and Color Tolerances for Industrial Applications

In both industrial and research applications, the interest in measuring the color of the products often lies on **acceptability** data, in which a sample is judged as commercially acceptable or not against a standard, or on how they change from batch to batch, year to year, or during processing and storage.

Table 31.1 Examples of instrumental color analysis for food quality control

Food product	*Reference method*	*Instrument*	*Raw ingredient testing*	*In-process production testing*	*Finished product testing*
Baked breads, cookies and goods	Internal methods	Colorimeter, reflectance			L^*, a^*, b^*
Coffee beans and powders	SCAA cupping roast level standard	Agtron spectrophotometer	Agtron grade scale	Agtron grade scale	Agtron grade scale
Cottonseed, soybean, and peanut oil	AOCS official method Cc 13c-50	Spectrophotometer	Absorbance at 460, 550, 620 and 670 nm	Absorbance at 460, 550, 620 and 670 nm	Color index = $1.29(A_{460}) + 69.7(A_{550}) + 41.2(A_{620}) - 56.4(A_{670})$
Durum flour	AACC approved method of analysis 14-30.01	Agtron reflectance colorimeter		Reflectance at 546 nm	Reflectance at 546 nm
Orange juice	USDA grade standards for orange juice	Spectrophotometer reflectance		Match to calibrated grade A-C color standards to determine Citrus Number score	Match to USDA 1–6 color standards to determine citrus number score
Pasta	AACC approved method of analysis 14-22.01	Colorimeter, reflectance			*L* and *b* values
Peanut butter	USDA grade standards for peanut butter	Hunter colorimeter, reflectance		Match to calibrated grade A-B color standards	Match to calibrated grade A-B color standards
Refined fats, oils, and derivatives	AOCS official method Cc 13b-45 the Wesson method Cc 8d-55 Cc 13j-97	AOCS-Tintometer color	R, Y, B, N	R, Y, B, N	R, Y, B, N, ratio Y/R
Tomato products (paste, sauce, ketchup and juice)	USDA grade standards for tomato products	Colorimeter, reflectance	Match to calibrated grade A-C color standards	Match to calibrated grade A-C color standards	Match to calibrated grade A-C color standards
Wheat flour semolina pasta products	AACC approved method of analysis 14-50.01	Spectrophotometer		Absorbance at 435.8 nm	Absorbance at 435.8 nm

Abbreviations: *AACC* American Association of Cereal Chemists, *AOCS* American Oil Chemists Society, *SCAA* Specialty Coffee Association of America, *USDA* United States Department of Agriculture

A single number is often desired in the industry for establishing pass/fail acceptability limits. A **color-difference formulae** allows for quantitation of the magnitude of color differences visually perceived with a single number, simplifying three numerical data (chromatic stimulus). The *CIE 1976 L*a*b** (CIELAB) *color difference* (ΔE^*_{ab}) between two color stimuli (i.e., a food and the standard) is calculated as the Euclidean distance between the points representing them in the space, according to the formula:

$$\Delta E^*_{ab} = \left[(\Delta L^*)^2 + (\Delta a^*)^2 + (\Delta b^*)^2 \right]^{1/2} \quad (31.3)$$

where:

$\Delta L^* = L^*_{sample} - L^*_{standard}$,
$\Delta a^* = a^*_{sample} - a^*_{standard}$,
$\Delta b^* = b^*_{sample} - b^*_{standard}$.

A limitation of ΔE^*_{ab} is that the single number will only indicate the magnitude of color difference, but not the trend of the color variation. Thus, samples with identical ΔE^*_{ab} values with respect to a standard will not necessarily have the same color and visual appearance. The differences in individual color attributes, lightness (ΔL^*), chroma (ΔC^*_{ab}), and hue (Δh) can be complementarily evaluated to assess the trend of the changes:

- $\Delta L^* = L^*_{sample} - L^*_{standard}$. Positive values indicate that the sample is lighter than the standard and darker for negative values.
- $\Delta C^*_{ab} = C^*_{ab\ sample} - C^*_{ab\ standard}$. Positive values indicate that the sample is more intense or saturated than the standard and the contrary for negative values.
- $\Delta h_{ab} = h_{ab\ sample} - h_{ab\ standard}$. In this case, positive or negative values indicate a different hue (not "more than" or "less than").

The visual **color tolerance** (color difference threshold) is the limit of the color difference between a sample and the standard for acceptance. However, in most practical applications, color differences above the threshold may be considered "acceptable differences" or "color tolerances". Thus, it has been proposed to use ranges of "rigorous" color tolerance to ΔE^*_{ab} values between 1.1 and 2.8 CIELAB units, "normal" tolerance between 2.8 and 5.6 CIELAB units, and "loose" tolerance above 5 CIELAB units [17].

Tolerance values create an area around the standard that is considered acceptable, while color outside is rejected. ΔE^*_{ab} values can be used for tolerances when individual attributes also are evaluated. A diagram showing acceptable tolerances based on ΔL^*, ΔC^*_{ab}, and Δh_{ab} numbers is shown in Fig. 31.15. The elliptical shape of the solid arises since tolerances for Δh_{ab} are considerably narrower than for ΔC^*_{ab} and ΔL^*.

Based on CIELAB, more advanced color difference formulae have been developed (CMC formula, CIE94 formula, DIN99 formula, and CIEDE2000 formula), aimed to correct the nonuniformity of the CIELAB color space for **small color** differences [18, 19]. These formulae are usually non-Euclidean, and color differences cannot be represented as simple distances in an ordinary three-dimensional space.

A simpler method to establish tolerances is through the use of color charts, chips, or pictures to serve as their color standards and provide a range of acceptable colors. A few chips can cover the relevant range of colors for a specific application and are usually easy to use, easy to handle, and inexpensive. Colored chips, pictures, or illustrations can be used to classify apples, determine peach maturity, or grading of bananas. There are USDA color scales for canned products, frozen peas and peanut butter, and color cards are used to determine grades of walnuts, coffee beans, and cocoa beans, among others. These tools clearly establish the limits that should not be exceeded for a product or process to be under control.

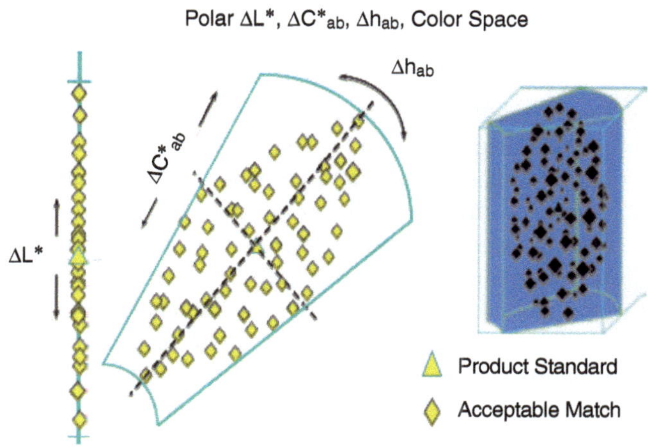

Fig. 31.15 Diagram showing acceptable ΔL^*, ΔC^*_{ab}, and Δh_{ab} tolerance limits for a product

31.5.3 Color Measurements to Monitor Shelf-Life

Monitoring the degradation of foods throughout their shelf life is crucial to ensure their nutritional and sensory value and, therefore, their success in the market. Some of these processes can cause color changes, often before detectable chemical changes occur. Therefore, color measurements are frequently used as a valuable method to quickly assess quality defects and to predict or estimate the shelf life of many food products.

Monitoring fruit maturation (ripening acceleration in climacteric fruit or de-greening process in non-climacteric ones) has economic importance since it allows for anticipating the selling period and improving the appearance of the fruit. Also, some changes in color (loss of green color in vegetables due to chlorophyll degradation, or loss of color in red/blue fruits due to anthocyanins degradation) are related to loss of quality and can lead to consumer rejection. Consumers often associate color with safety, and a product that changes color over time may be rejected by consumers as unsafe. Color changes over time can serve as indicators of enzymatic activity, or nonenzymatic browning. For these cases, color monitoring is a fast and effective way to assess shelf life. Table 31.2 presents examples of indices based on color parameters that have been applied to evaluate and control the quality characteristics of foods.

31.5.4 Practical Considerations for Sample Preparation and Presentation

For color to occur, three elements are needed: (1) a colored object, (2) visible radiation (light) that illuminates the object, and (3) an observer. The choice of sample preparation, sample presentation, instrument choice and instrument settings are issues that must be chosen with care in color measurement. Each commodity or sample will present its own peculiarities when it comes to measuring color and appearance.

The interaction of light with a sample is illustrated in Fig. 31.15. Light for which the angle of reflection is equal to the angle of incidence is described as **specular light**. Smooth polished surfaces will appear **glossy** because of the high degree of **specular reflection**. Rough surfaces will have a great diffuse reflection and a dull or matte appearance. Selective absorption of light will result in the appearance of color. **Opaque** samples will **reflect** light. **Transparent** samples will primarily **transmit** light, and **translucent** samples will both **reflect** and **transmit** light.

Table 31.2 Representative examples of how CIELAB parameters are used to monitor shelf-life of different food products

Food application	*Colorimetric index*	*Predicted parameter*	*References*
Browning	h_{ab}	Enzymatic browning in dates	Al-Amrani et al., 2020 https://doi.org/doi.org/10.1155/2020/8380461
	ΔE^*_{ab}	Browning degree in apples	Shimizu et al., 2021 https://doi.org/doi.org/10.21273/JASHS05027-20
Eggs freshness	L^*	Carotenoid degradation in fresh eggs	Yüceer and Caner, 2021 https://doi.org/doi.org/10.1111/jfpp.15261
Fruits and vegetables de-greening and ripening	CCI = 1000 $a^*/(L^*b^*)$	Ripening stages in orange fruit	Cubero et al., 2018 https://doi.org/doi.org/10.1016/j.biosystemseng.2017.12.012
	L^*, a^*, b^* RCI = 200/(1-a^*/b^*)	Ripening stages in banana fruit	Mendoza and Aguilera, 2004 https://doi.org/doi.org/10.1111/j.1365-2621.2004.tb09932.x Santos Gomes et al., 2013 https://doi.org/doi.org/10.1016/j.scienta.2012.11.014
	L^*, a^*, b^*	Ripening stages in papaya fruit	Santos Pereira et al., 2018 https://doi.org/doi.org/10.1016/j.compag.2017.12.029
	L^*, h_{ab}	Ripening stages in reed beet fruit	Montes-Lora et al., 2018 https://doi.org/doi.org/10.1007/s11130-018-0678-x
	a^*, b^*, C^*_{ab}	Ripening stages in grapes	Rodríguez-Pulido et al., 2012 https://doi.org/doi.org/10.1016/j.compag.2012.01.004
Green peas color discoloration	a^*, a^*/b^* ratio, h_{ab}	Chlorophyll degradation	Koca et al., 2007 https://doi.org/doi.org/10.1016/j.foodchem.2005.09.079
Meat freshness	L^*, a^*, b^*, C^*_{ab}, ΔE^*_{ab}	Freshness classification	Taheri-Garavand et al., 2019 https://doi.org/doi.org/10.1016/j.meatsci.2019.06.002
Prediction of bioactive compounds	L^*, a^*, (a^*/b^*) ratio, $(a^*/b^*)^2$ ratio	Lycopene in tomato fruit	Goisser et al., 2020 https://doi.org/10.1016/j.postharvbio.2020.111232 Arias et al., 2000 https://doi.org/10.1021/jf990974e
	L^*, a^*, b^* L^*, C^*_{ab}, h_{ab}	Vitamin A in orange juice	Meléndez-Martínez et al., 2007 https://doi.org/10.1021/jf0635412
	a^*, b^*, C^*_{ab}	Carotenoids in orange juice	Meléndez-Martínez et al., 2011 https://doi.org/doi.org/10.1016/j.foodchem.2010.11.167
	a^*, C^*_{ab}	Anthocyanin in strawberry	Hernanz et al., 2008 https://doi.org/doi.org/10.1021/jf073389j
	L^*, C^*_{ab}, h_{ab}	Total anthocyanins in red wines	Gomez-Míguez and Heredia, 2004 https://doi.org/doi.org/10.1021/jf049570z
	C^*_{ab}, b^*	Mineral content in honeys	González-Miret et al., 2005 https://doi.org/doi.org/10.1021/jf048207p
	L^*, C^*_{ab}, b^*	Chlorophylls and carotenoids in olive oil	Moyano et al., 2008 https://doi.org/doi.org/10.1016/j.foodres.2008.03.006

For color data to be useful, the numbers must be consistent and repeatable. **Sampling** of the product must be done so that it is representative and prepared so that it represents the product's color characteristics. Many food samples are partially transmitting and partially reflecting. Rather than being uniform, they may be mottled or highly variable in color. The number of readings that need to be taken for acceptable repeatability depends on the sample's nature. Another problem is that often the only instrument available is not ideal for the sample. Gordon Leggett [16] provides some practical tips and a systematic protocol for consistent color measurement of different food categories.

1. *Transparent samples: Transparent liquids* should be measured using a clear glass or plastic **cell** (not absorbing the visible light). Cell path length is selected based on color intensity, also considering the measurement (transmission or reflection). There are analytical cells of 1, 2, 5, 10 and even 20- and 50-mm path length available in the market for the different samples and measurements. A cell with distilled water or a noncolored liquid with refractive index similar to the sample can be used as a blank to negate the effects of cell and solvent. For clear transparent liquids, a single measurement using a **viewing area** of 15 mm diameter or greater may be sufficient for good repeatability. For hazy transparent liquids, two to four readings with replacement of the liquid between readings is necessary to get acceptable repeatability when using a 10 mm path length cell and a sphere instrument. Similarly, the color of solid samples that are transparent, such as plastic films for packaging, should be measured by transmittance.

2. *Translucent samples.* Liquids and gels with high solid content or suspended particles are translucent rather than transparent. They can be measured by transmission using a very thin 2-mm path length cell or measured in reflectance. Here, it is necessary to control the thickness of the sample so that it is effectively opaque. The use of large viewing area can also help to obtain more reproducible data.
3. *Opaque foods:* The color of opaque foods is measured by reflectance, and different samples may require different sample preparation or presentation to the instrument. Solid foods vary with respect to size, geometry, and uniformity. Instruments with viewing area of 25 mm diameter or larger are helpful for area-averaging no-uniform color. For powders, two readings with replacement of the powder between readings may be sufficient, but for flakes, chunks, and large particulates, a large field of view with three to six readings and sample replacement between readings is recommended. For example, instruments with a viewing area of ~150 mm diameter or more, such as the Agtron spectrophotometer, are particularly helpful for solids with heterogeneous color and shape, such as potato chips and other snack foods (see Table 31.1 for additional examples). For these kind of samples, digital image analysis may also be very helpful.

31.6 SUMMARY

Color is one of the most critical quality food attributes affecting consumer purchase decisions. Therefore, it is imperative for the food industry to carefully standardize, monitor and control the color of foods, and minimize batch to batch variability. For color to exist, an object, light, and an observer are needed. Hence, when evaluating color, all three elements must be considered, regardless if the measurement will be made visually or instrumentally. The Munsell system is a visual system that designates color in terms of its three dimensions: hue, value, and chroma (hue value/chroma). It correlates well to the way the humans see color and served as the basis for many other color matching systems.

Standardization of illuminants and the sensitivity of the human eyes (standard observer) allowed for the development of color-matching functions used to calculate numerical *XYZ* tristimulus values to represent a particular color. This system, known as the in the CIE tristimulus system, does not have equivalent visual spacing, a disadvantage when measuring how a sample differs from a standard or changes during processing and storage. Color-order systems with more uniform color spaces were later developed to facilitate the evaluation of color differences. These include the Hunter Lab system and the CIELAB color space, which have been widely adopted by the food industry. There are many instruments available for industrial and research applications that are rugged, easy to standardize, and user friendly. They vary with respect to presentation of sample, size of viewing area, portability, and the ability to measure by transmittance or reflectance. Many food samples may be partial transmitting and partial reflecting, nonuniform, and of varying size and shape. Thus, sample preparation and presentation are important to get measurements that are repeatable and that correspond to visual appearance. In both industrial and research applications, the interest in measuring the color of the products often lies on acceptability data and visual color tolerances for which color difference formula can be applied.

31.7 STUDY QUESTIONS

1. The color of a food product is reported as 4 PB 8/2. What color system was used to report this color? What do the different numbers and letters mean?
2. Using a calculator, determine hue angle and chroma for the following sets of a^*, b^* data: $a^* = +12$ and $b^* = +8$, $a^* = +12$ and $b^* = +4$, $a^* = 12$ and $b^* = -4$.
3. If one wants to use a colorimeter to measure the amount of browning in maple syrup, what indices would you expect to correspond well with visual assessment?
4. How sensitive is the human eye to perceive color differences? What are the human capabilities for color perception and color memory?
5. What type of instrument would you recommend to measure the color of a clarified juice sample? And what could be recommended for hazy samples?
6. What strategy would you recommend to obtain representative color measurement of irregular samples such as potato chips that do not have uniform shape or color?
7. How could you measure the color of a food product if you do not have an instrument available?

Acknowledgments The authors of this chapter thank Ronald E. Wrolstad, emeritus Distinguished Professor, Department of Food Science and Technology, Oregon State University, Corvallis, OR, USA, for his contributions to previous editions of this chapter. The chapter authors also thank Indarpal Singh of Conagra Brands for developing the initial version of Table 31.1 in this chapter.

REFERENCES

1. Loughrey K (2005) Overview of color analysis. Unit F5.1. In: Wrolstad RE, Acree TE, Decker EA, Penner MH, Reid DS, Schwartz SJ, Shoemaker CF, Smith D, Sporns P. (eds) Handbook of food analytical chemistry-pigments, colorants, flavors, texture, and bioactive food components. Wiley, NY

2. Berns RS (2000) Billmeyer and Saltzman's principles of color technology, 3rd edn. Wiley, NY
3. Campbell NA, Reece JB, Mitchell LG (1999) Biology, 5th edn. Benjamin/Cummings, Addison Wesley Longman, Menlo Park, CA
4. Loughry K (2000) The measurement of color, Chap 13. In: Francis FJ, Lauro GJ (eds) Natural food colorants. Marcel Dekker, NY
5. Wyszecki G, Stiles WS (1982). Color Science. Concepts and methods. Quantitative data and formulae, 2nd edn. John Wiley & Sons, Inc, NY.
6. Hutchings JB (1999) Food color and appearance, 2nd edn. Aspen Publishers, Gaithersburg, MD
7. Bartleson CJ (1960) Memory colors of familiar objects. J Opt Soc Am 50:73-77
8. Caivano, J.L., Nemcsics, A. (2022). Color Order Systems. In: Shamey, R. (eds) Encyclopedia of Color Science and Technology. Springer, Berlin, Heidelberg.
9. X-rite (2015) Munsell products. Available from https://www.xrite.com/top_munsell.aspx?action=products Accessed 18 Dec. 2015
10. Commision Internationale de l'Eclairage (2018). CIE 015: Technical Report. Colorimetry. 4th edn. Vienna, Austria.
11. Melgosa, M. (2022) Some recent advances in color science. Journal of Physics: Conference Series. 2407. 012030
12. MacDougall DB (2002). Colour in food. Improving quality. Woodhead Publishing. Cambridge, UK
13. Wu D and Sun DW (2013) Colour measurements by computer vision for food quality control. A review. Trends Food Sci Technol 29:5–20
14. Rodríguez-Pulido FJ, Gordillo B, González-Miret ML and Heredia FJ (2013). Analysis of food appearance properties by computer vision applying ellipsoids to colour data. Comput Electron Agric 99:108–115
15. Rodríguez-Pulido FJ, Gómez-Robledo L, Melgosa M., Gordillo B, González-Miret ML and Heredia FJ (2012). Ripeness estimation of grape berries and seeds by image analysis. Comput Electron Agric 82:128–133
16. Leggett GJ (2008) Color measurement techniques for food products, chap 2. In: Culver CA, Wrolstad RE (eds) Color quality of fresh and processed foods. ACS Symposium Series No. 983, American Chemical Society, Washington DC
17. Martínez JA, Melgosa M, Pérez MM, Hita E and Negueruela AI (2001). Visual and instrumental color evaluation in red wines. Food Sci Tech Int 7(5):439–444
18. Luo R (2002). Development of colour-difference formulae. Rev. Prog. Color., 32:28–39
19. Melgosa M, Alman DH, Grosman M, Gómez-Robledo L, Trémeau A, Cui G, García PA, Vázquez D, Li C and Luo MR (2013). Practical demonstration of the CIEDE2000 corrections to CIELAB using a small set of sample pairs. Color Res App 38(6):391–468

Food Microstructure Techniques

32

Jinping Dong

Contents

ABBREVIATIONS

2D	Two dimensional
3D	Three dimensional
AFM	Atomic force microscopy
CAT	Computerized axial tomography
CCD	Charge-coupled device
CLSM	Confocal laser scanning microscopy
CMC	Critical micelle concentration
CT	Computed tomography
DIC	Differential interference contrast
EDS	Energy-dispersive spectroscopy
EM	Electron microscopy
E-SEM	Environmental scanning electron microscopy
FPA	Focal plane array
FTIR	Fourier transform infrared
HSI	Hyperspectral imaging
IR	Infrared
LM	Light microscopy
SEM	Scanning electron microscopy
TEM	Transmission electron microscopy
XRD	X-ray diffraction

J. Dong (✉)
Global Core R&D, Cargill Research and Development Center, Cargill, Inc., Plymouth, MN, USA
e-mail: jinping_dong@cargill.com

B. P. Ismail, S. S. Nielsen (eds.), *Nielsen's Food Analysis*, Food Science Text Series,
https://doi.org/10.1007/978-3-031-50643-7_32

32.1 INTRODUCTION

One of the trends in food research and manufacturing is to design and produce the food with desired functionalities such as more appealing textures, longer shelf-life, improved health benefit, and friendly labeling. The prevailing approach to achieve this goal is to understand and therefore to control food structures, especially at the microscopic scale. The study of food microstructures typically involves three aspects – visualization, identification, and quantification – with each requiring the aid of different tools.

Food is a complex and often heterogeneous system. Both fresh (typically dominated with cellular structure) and processed (microscopic domains formed by mixed ingredients) foods contain structures that cannot be directly seen with naked eyes. Human eyes are capable of seeing things down to about 1 μm with proper lighting (compared to ~20–50 μm being the diameter of a human hair). Many biological cells are a few micrometers in size, but they are not usually visible to our eyes because good illumination is not always available. They appear big and clear under a regular light microscope. Some salad dressing emulsions have oil or water droplets well below 1 μm in diameter. Elemental plant fibers have a diameter of only a few nanometers. To be able to observe these systems requires high resolving power microscopes such as a scanning electron microscope (SEM).

Showing the morphology of the food with the help of microscopes is only one part of the microstructure elucidation. Identifying the distribution of ingredients with different chemical and physical characteristics gives another insight, and this often requires a chemical imaging tool [e.g., Fourier transform infrared (FTIR), Raman, fluorescence, or confocal laser scanning microscope] or a physical structural tool for understanding molecular arrangement or crystallinity (e.g., X-ray diffractometer).

Modern microscopy and chemical imaging tools allow users to not only determine food morphology and ingredient distribution but also quantify dimension parameters, concentrations, fractions, and kinetic constants. Using X-ray computed tomography, quantification can be done in 3D with nondestructive imaging. Microscale interactions and forces also can be probed among the food ingredients with the help of instruments such as an atomic force microscope. Values obtained from quantitative measurements are the last piece of the puzzle to solve food microstructures.

All the microstructure techniques listed above will be briefly introduced in this chapter. Many of them were not initially invented for food applications but have been borrowed by food researchers to understand food and to study the relationship between structure and its functionality. This chapter is not intended to be comprehensive in either breadth or depth regarding these techniques but rather to provide an overview, with references for more details. Note that the chapter will not cover measuring particle size and shape (see Chap. 5, Sect. 5.5.2.3) or color (see Chap. 31) and will refer to a number of spectroscopy chapters (Chaps. 7–10) as they relate to techniques for characterizing food microstructure. Because there are so many acronyms associated with instrumentation to evaluate food microstructure, a listing of acronyms used in this chapter is given.

32.2 MICROSCOPY

32.2.1 Introduction

One of the most frequently used family of instrumental techniques for analyzing foods and food microstructures is **imaging**. In modern terms, imaging is often regarded as picture-taking with a camera, but historically, it was always referred to as microscopy. **Microscopy** is the art and science of using microscopes as scientific investigative tools. Since the original development of light microscopes [1, 2], drastic improvements have been achieved in our ability to see in magnification and to differentiate in feature contrast. Example imaging agents include light (photons), X-rays (high-energy photons), electrons, ultrasound, microwaves, radio waves, etc. Most of the imaging agents are based in, and are a part of, the electromagnetic spectrum. Each of these techniques is subdivided into various imaging methods.

Microscopes were invented to visualize objects that cannot be seen by human eyes. How small of an object that can be clearly identified with the microscope determines the resolving power of the instrument. **Resolution** is the ability to distinguish or resolve two small points, which are very close together, as two separate entities. Factors that affect a microscope's resolution include the properties of the imaging agent (e.g., wavelength of the light) and the focusing power of the instrument (e.g., numerical aperture of the objective for light microscope). The **theoretical resolution limit** for a given microscope's primary or objective lens is given by a simple equation below:

$$R = \lambda / 2NA \qquad (32.1)$$

where:

R = resolution (theoretical resolution limit, minimum distance of the two adjacent objects)

λ = wavelength of the visualizing agent (e.g., light or electron)

NA = numerical aperture of the lens [proportional to the refractive index and sin (θ), where θ is the half angle of the incidence of the incoming light to the lens]

Inherent to any lens are imperfections, referred to as **aberrations**, which can cause images to appear distorted, out of focus, with colored fringes, etc. Correction of optical aberrations includes improvement of lens fabrication and grinding techniques, optimization of glass formulations, application of antireflection coatings, control of optical pathways, and combination of multiple lens elements.

Misalignment of the optical illumination is another factor that impacts a lens' optimum resolving power. Specific procedures and the care from the microscopist are needed to achieve perfect alignment and focus of the light beam to give uniform and bright illumination of the specimen. The alignment procedure is called **Köhler alignment**, named after August Köhler, a German physicist and microscopist [3].

32.2.2 Light Microscopy

32.2.2.1 Introduction

Light microscopy, as referred to in its name, employs light (or photons) as the imaging agent and magnifying lenses to visualize objects that cannot be seen with naked eyes. Light can be either reflected off of or transmitted through the sample and is then directed through the objective lens to the eye piece(s). Depending on the instrument design, there are two major categories of LMs, namely, stereo and compound microscopes. **Stereo** is a word related to **parallax**, implying two light beams reflecting off the samples and arriving at your two eyes separately. This allows the visualization from two different angles on the surface, resulting in a 3D interpretation of the view from the brain. Therefore, a stereo microscope is also called a dissecting microscope. The stereo microscope typically has low magnification (2–100×) but has long working distance and large viewing depth, which makes it easy for visualizing large and odd-shaped specimen. **A compound** microscope often has multiple lenses working together to achieve a clear, in-focus, magnified image. It usually works in higher magnification range (40–1200×), with the combination of the lenses from objectives and eyepieces.

32.2.2.2 Contrasting Modes

In LM, there are many ways to manipulate the light both before it reaches the specimen and after it interacts with the specimen to highlight certain features. Many of these are commonly used in the food industry and constitute what are called **contrasting** or **imaging modes** of light microscopy. A partial list of imaging modes includes bright-field, dark-field, phase contrast, birefringence (or cross polarization), differential interference contrast (DIC), and oblique lighting [4]. With the exception of bright-field mode (the simplest lighting mode), each of these alternant modes requires additional special fixtures to be attached to the microscope to accomplish the desired effect. These contrasting modes do not usually increase resolution per sé, but they can allow one to distinguish structures that may not be easily evident using a regular illumination. Starch is a classic example of a food component that can be examined using any of the imaging modes listed (Fig. 32.1).

32.2.2.3 Fluorescence Microscopy

Fluorescence is the light emitted from an atom, a molecule, or a material excited to the electronically active state (see Chap. 7, Sect. 7.3). The wavelength or energy of the excitation light is characteristic to the chemical bonds within a molecule or to the chemical/physical state of a material. By applying proper light sources and optical filters, fluorescence can be used to give unique contrast in light microscopy. Optical filters are added to a bright light source to select target wavelength for excitation, or **excitation spectrum**, and fluorescence filters are added after the sample to capture **emission spectrum**. Fluorescent stains, which contain functional groups that fluoresce upon excitation (called fluorochrome), also can be added to a sample to contrast components that they have strong affinity with. Staining can be done positively (e.g., staining the target structure) or negatively (e.g., staining nontarget structure). Many food materials contain natural fluorochrome. When a proper excitation wavelength is selected, bright color from fluorescence emission can be easily visualized under the light microscope. Figure 32.2 shows the aleurone layer within a wheat kernel that auto-fluoresces bright blue, making that particular layer of cells become obvious to the viewer, over any other cell types within the wheat kernel.

32.2.2.4 Histology

Histology, which refers to sample staining combined with light microscopy [5], is often used in biology and medicine to study cells and tissues. Samples are typically sectioned in some ways to create relatively thin slices with even thickness. This procedure is known as **microtomy** and usually involves a microtome or other mechanical cutting tools [6]. Histology has been applied in food research for a long time owing to its unique contrasting capability. The numerous dyes available on the market [7–9] are each designed to stain different materials different colors. Dye molecules interact with food ingredients mostly through physical interactions, which make the affiliated ingredients appear with the color of the dye itself. However, the dye-ingredient interaction sometimes can shift the color spectrum. For example, although common cornstarch stains blue with iodine, waxy cornstarch stains red (Fig. 32.3). This is the result of amylopectin dominating the starch in the waxy species, as opposed to amylose in the common species. Iodine is referred to as a **metachromatic stain**, meaning one dye can stain different materials in different colors. Other specialty dyes,

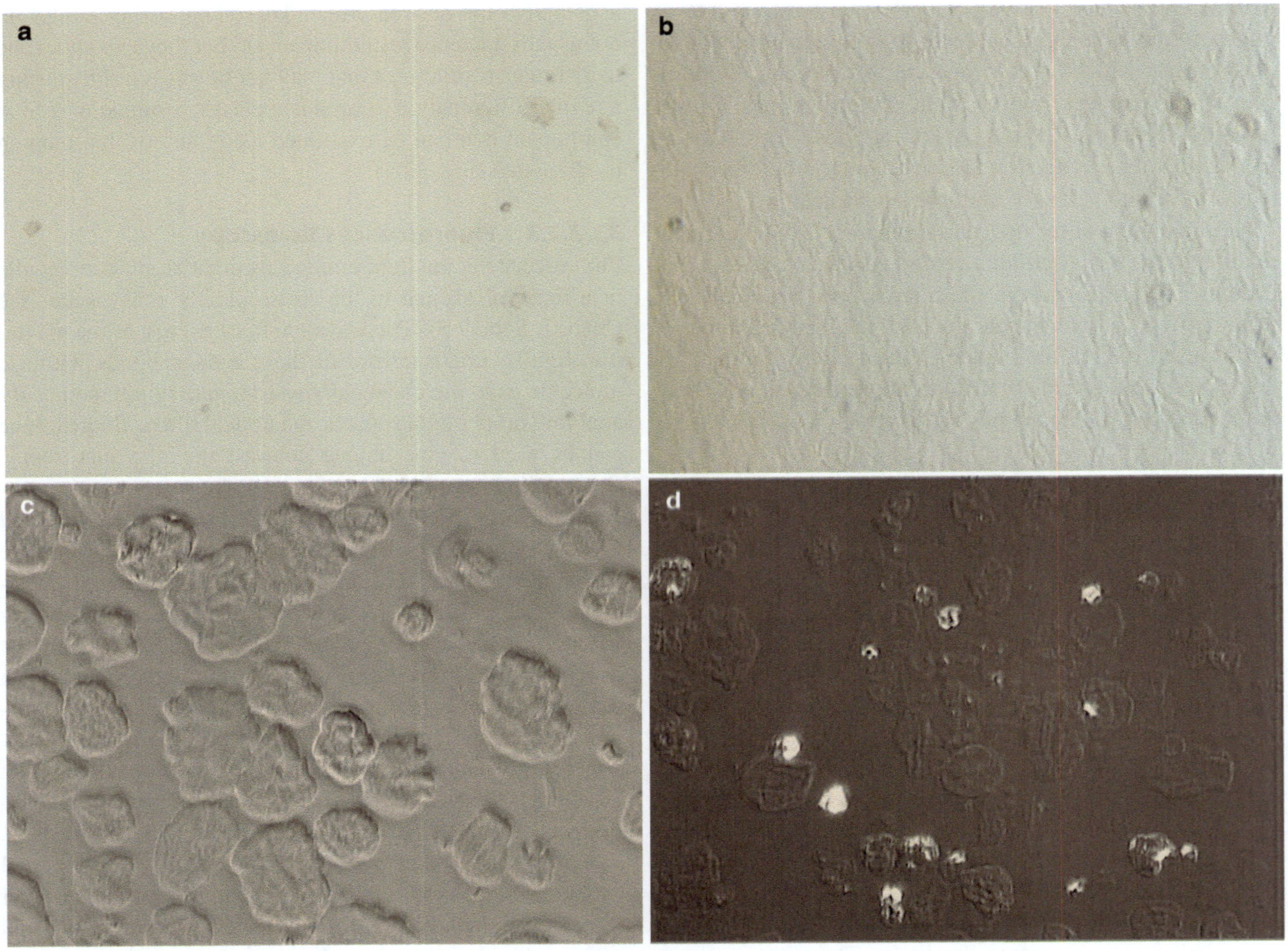

Fig. 32.1 Light microscopy of starch with different imaging modes. (**a**) Bright field image of cooked out starch (all or most of the crystallinity is gone). Low contrast associated with this type of sample makes it nearly impossible to see structure. (**b**) Oblique lighting. (**c**) Phase contrast. (**d**) Cross polarized image of *partially* cooked out starch. Bright granules are not cooked out and are retaining their crystallinity, or at least a portion of it

Fig. 32.2 Particles of wheat fiber demonstrating various colors of auto-fluorescence

called **polychromatic stains,** are combinations of dyes that do not react with each other but will stain food ingredients into different colors.

32.2.3 Electron Microscopy

Electron microscopy (EM), which uses electrons as the imaging agent, is of two common types: **scanning electron microscopy** (SEM) and **transmission electron microscopy** (TEM) [10]. SEM finds more frequent applications in the food industry, as described further below. Although TEM may provide an order of magnitude of higher resolution, it is not commonly used in food research mainly because it is very time consuming and requires delicate sample preparation (e.g., extremely thin sections of the sample material, typically 60–80 nm).

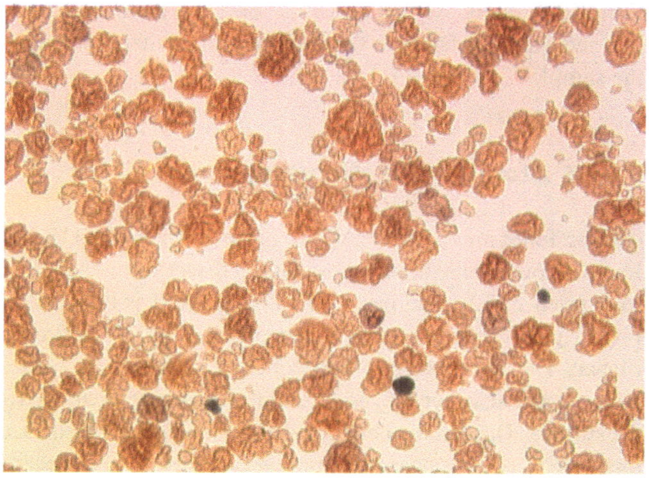

Fig. 32.3 Partially cooked out waxy cornstarch, stained with iodine. *Red* color is due to the amylopectin content of the starch. A few *dark*, common cornstarch granules are also seen in this image, which stain *blue* rather than red because they contain amylose rather than amylopectin. Iodine, being a metachromatic stain, can stain these various types of starch different colors, which aids in starch identification

The five main ways electron microscopes (EMs) differ from light microscopes (LMs) are (1) the imaging agent (electron vs. photon); (2) their resolving power (electrons have up to 100,000 times shorter wavelength than photons, capable of resolving individual atoms); (3) their magnifying power (wide range from 20× to 1,000,000×); (4) LMs can provide images with visible colors to the eyes, while EMs do not give contrast in color, although false colors can be applied to SEM images to differentiate brightness levels; and (5) EMs work in high vacuum, while LMs tend to work in ambient conditions.

SEMs employ an electron beam to scan across the surface of interest. After interacting with the materials, electrons come out in several different forms. The **secondary electrons** are produced by inelastic collisions of the probe electrons (**primary electrons**) with sample electrons. Sample electrons are ejected at slower speeds than probe electrons, and with lower energy levels. They are collected at the detector for quality imaging with good depth perception. This is the primary imaging mode of SEMs. **Backscattered electrons** (BSE) involve imaging with a high-energy probe electron interacting with the sample elastically. They arrive at the detector with nearly the same speed and energy as they had in the electron probe, resulting in an image that is material dependent. Dense materials, like metals, appear bright in the BSE-SEM image, while less dense materials appear darker (e.g., carbon-based biologicals). BSE imaging is proven to be more valuable than secondary electron imaging because of the details it can provide and because it is less subject to specimen charging (Fig. 32.4).

During SEM imaging, electrons accumulate on the surface (**surface charging**), especially where it is low in con-

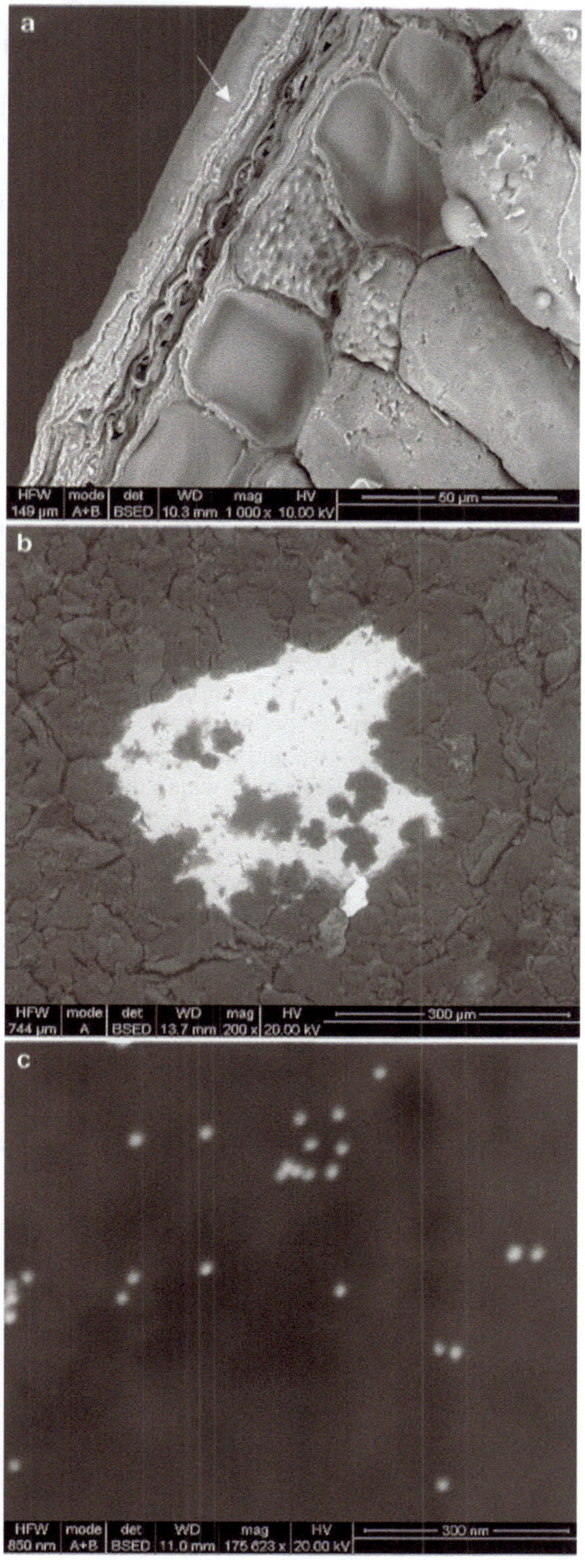

Fig. 32.4 BSE-SEM imaging of various materials. (**a**) A fracture through a seed in cross section. The seed coat is evident at the arrow. Cellular structure dominates the tissue below the seed coat. 1000×. (**b**) A contaminating particle (white) within a pharmaceutical product. 200×. (**c**) 20 nm gold particles (white dots) used to label and identify molecular structure within a biological system 175,000×

ductivity. Applying a thin metal or conductive coating to the specimen surface has been an essential sample preparation step. However, low vacuum and **environmental SEM** (E-SEM) modes use water vapor or other gases within the specimen chamber to help diminish sample charging artifacts. Water vapor can, under the right conditions, also help preserve tender biological samples (including many foods and food ingredients) within the otherwise very high vacuum condition of an SEM.

32.2.4 Energy-Dispersive X-Ray Spectroscopy

When scanning electrons of SEM interact with the atoms of the specimen, it may cause electrons to migrate in between electron shells, from which an X-ray emission could be induced. The energies of the X-ray are characteristic to the atomic structures, allowing identification of the elements through an energy-dispersive spectrometer, hence the name of this technique [**energy dispersive spectroscope** (EDS)]. The peaks from the EDS spectrum (X-ray energy in KeV in the X-axis and total X-ray counts in the Y-axis) represent specific elements from a periodic table (Fig. 32.5a). Additionally, both qualitative and quantitative analyses are possible for whatever elements exist within the sample.

Since the SEM works in a scanning fashion, restoring its electron probe in a predictable and reproducible way, we can choose specific elements from the spectrum and record a 2D array of element dots, from which an elemental map can be generated to show distribution of compositions (Fig. 32.5b). From EDS maps, we can know not only the elemental identity and quantity but also the distribution of that element in the sample.

32.2.5 Atomic Force Microscopy

Atomic force microscopy (AFM), invented in 1986 [11], has found applications in most scientific disciplines, including food science [12]. The basic principle of AFM is depicted in Fig. 32.6. A very sharp tip is attached to the end of a soft **cantilever**. The tip is brought into contact to the sample, and a piezoelectric **scanner** moves the cantilever to raster scan across the surface. The cantilever deflects (up or down) as the surface topography or tip-surface interaction changes. A **laser beam** bounced off the back of the cantilever moves up and down (or left and right if there is lateral deformation of the cantilever) on a **position-sensitive photodiode detector** (PSPD) with the cantilever deflection. The photoelectric signal is then recorded and transferred into a colored map with contrast corresponding to the cantilever deflection or surface height change. Very often, to avoid large forces from the cantilever/tip damaging the scanned surface, a constant force mode is enabled in which a **feedback** signal is sent to the piezo scanner to compensate for the height change of the sample, thus maintaining a constant tip-to-sample interaction. Depending on the sharpness of the tip (**radius of curvature**) and the **spring constant** of the cantilever (usually in the range of 0.1 to a few nN/nm), the resolution of the acquired image can be in the sub-0.1 nm range, which is sufficient to resolve atoms on a surface.

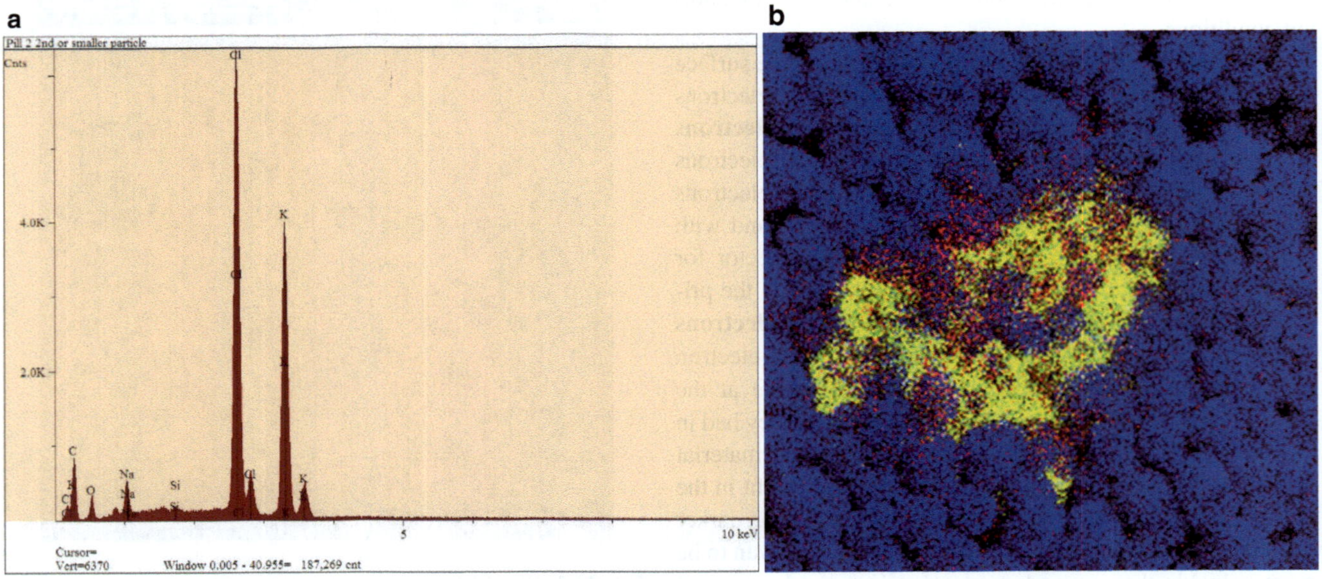

Fig. 32.5 EDS spectrum and EDS mapping. (**a**) An EDS elemental spectrum, with the various peaks labeled for the elements they represent. (**b**) An x-ray dot map (or x-ray-based elemental image) of an organic powder displaying a contaminating material. *Blue* represents where carbon is found in the sample. Powder particles are detected in approximate shape within the blue field. The contamination is seen as *yellow* (represents the element sodium) and *red* (represents the element chlorine). *Black* represents x-ray shadowing (meaning none of those x-ray are making it to the detector)

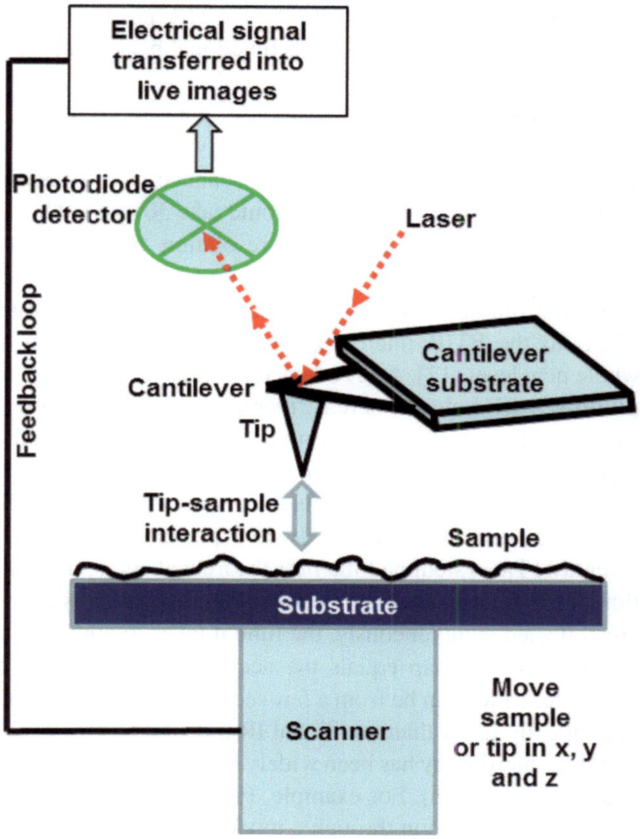

Fig. 32.6 Schematic of a typical AFM instrument

Although proving to have similar minimum resolution to electron microscopy, AFM finds many advantages over EM and other microscopy techniques. First, AFM measures the true height from the sample surface, which is needed to generate a 3D surface profile. Second, AFM is not limited by the environment it operates in. Conditions such as vacuum, ambient air, different humidity, elevated or depressed temperature, and even in liquid medium all work well for AFM. Especially in liquid media, study of biological macromolecules and living organisms is possible [13]. Third, because the tip physically touches the sample, a variety of interactions and microscale forces can be measured [14].

Imagine when a tip is brought close to the surface, forces between the tip and surface constantly change as the tip gets closer (repulsive, long range electrostatic, etc.), touches the surface (capillary, van der Waals attraction, etc.), indents into the surface (various nanomechanical, etc.), and breaks away from the surface (adhesive, etc.). As the tip scans laterally, frictional forces also can be detected by quantifying the torsion of the cantilever.

AFM was first introduced to the food science field in 1993 with research work focusing on monitoring food protein changes. Since then, AFM applications have expanded to many areas of food science and technology, such as qualitative imaging of polysaccharides and proteins to understand their conformation and organization in certain environments, quantitative structure analysis on complex food systems (e.g., mechanical strength of food gels) to correlate with functionality, probing molecular interactions (e.g., interaction between protein and surfactant as co-emulsifiers), and molecular manipulations to observe the reactions among food macromolecules [15].

32.3 CHEMICAL IMAGING

32.3.1 Introduction

Chemical imaging is a set of analytical techniques that can generate a contrast-based image to show the distribution of the chemical or molecular composition of an object (surface or bulk). Chemical images are usually created from a **data cube** as shown in Fig. 32.7. A full spectrum (e.g., from FTIR or Raman) containing chemical signatures of all components, at a single point, is acquired. Then, either the sample is moved through a motorized mechanical or piezoelectric stage or the probing beam [e.g., focused **infrared** (IR) light for **Fourier transform infrared** (FTIR), visible laser for Raman, or electrons for SEM] scans to the next point of interest, at which a second spectrum is collected. Many points are collected on one line, and many lines of data are acquired in a raster scanning fashion. After a 2D array of spectra is saved, various data processing algorithms are applied to each individual spectrum to extract needed information. Figure 32.7 shows a typical Raman spectrum, with all peaks corresponding to different functional groups, under certain vibrational modes. The peak inside the blue box is the OH stretching vibration from water in this case, and its area is proportional to the concentration of water at the current measurement location. The colored image is created by integrating this peak, for all collected spectra, with the contrast level proportional to the integrated value. If other components are of interest, integration of the corresponding functional peaks from the spectra can be easily applied to generate as many chemical images as possible [16].

32.3.2 Fourier Transform Infrared Microscopy

Chemical imaging with FTIR microscopy employs IR light (mid- or near-infrared) as the incident radiation (refer to Chap. 8, Sect. 8.2 for the basic introduction of IR spectroscopy). Because of the potential absorption of IR by optical components along the beam path, no transmission type of lenses is used in FTIR microscopy (like a regular light microscope, as described in Sect. 32.2.2). Spherical or parabolic mirrors are used to focus the light, and the same optical resolution limits apply as described in Eq. 32.1. Therefore, if imaging in mid-IR (wavelength of 2.5–50 μm), FTIR microscopy can provide chemical maps with theoretical spatial resolution of less than 1.5 μm. However, in practical applications, the resolution limit of FTIR microscopy is close to 5–10 μm when selecting functional peaks with different wavelength from the spectra.

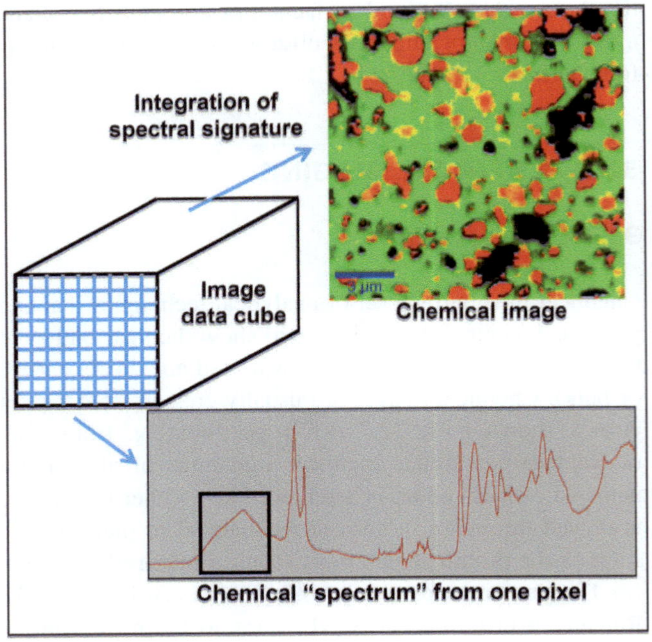

Fig. 32.7 Schematic of chemical imaging

Most FTIR microscopes adopt a point-to-point scan design, meaning the detector records only one spectrum at a time when the sample is being raster scanned. A full chemical map (e.g., 100 × 100 spectra) could take 30 min – 2 h (or longer depending on the spectral resolution and pixel settings), including the time of scanning and data recording. In 1995, the introduction of the **focal plane array detector** (FPA) to the FTIR microscope brought IR imaging to a whole new level [17]. A FPA detector contains a 2D array of **photo-sensitive elements** (e.g., mercury cadmium telluride, MCT), with each capable of capturing a complete but totally separate IR spectrum. The detector array can be anywhere from 16 × 16 to 128 × 128 elements, enabling chemical image acquisition of up to 16,384 pixels. The Digilab Stingray FPA system had the best theoretical spatial resolution of 5.5 μm using a 36× objective [17]. Because all spectra are collected simultaneously, the time it takes to complete a whole chemical map equals the acquisition of one single spectrum, which can be from a few seconds to a few minutes, dramatically faster than traditional IR imaging.

FTIR microscopy has been widely applied in food science and technology [16]. For example, Fig. 32.8 shows a study on moisture migration through a model sugar film, mimicking a cereal coating [18].

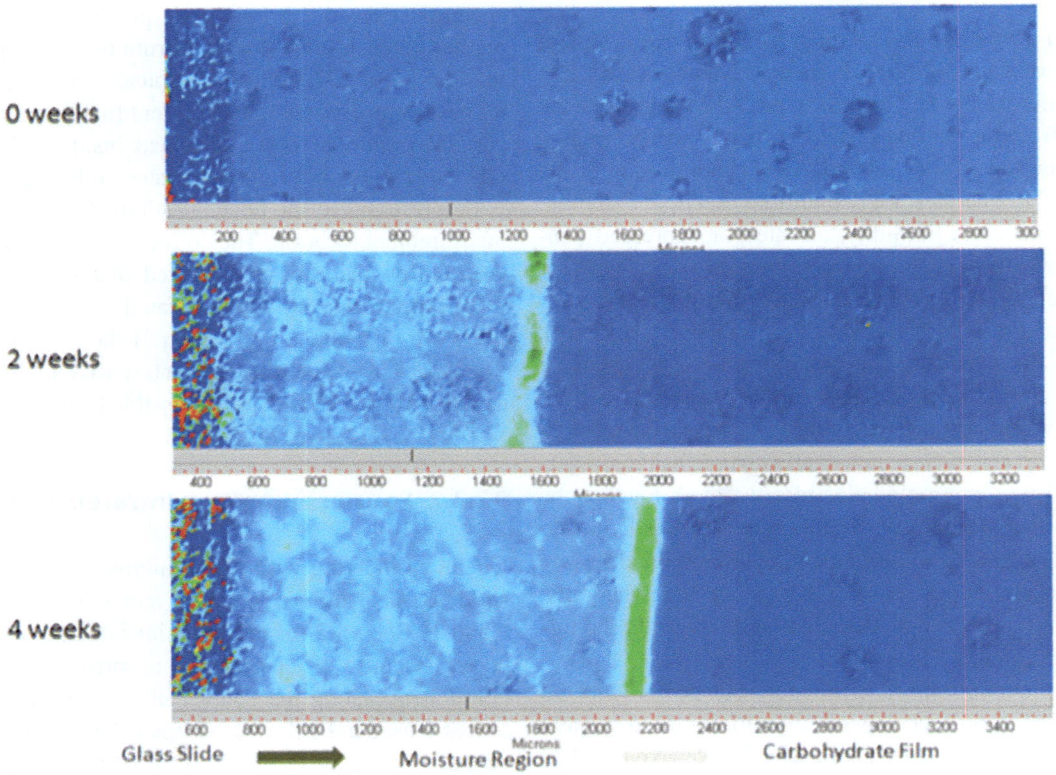

Fig. 32.8 FT-NIR mapping of moisture sorption in a model sugar film

32.3.3 Confocal Raman Microscopy

As introduced in Chap. 8 (Sect. 8.5), Raman scattering is intrinsically weak. For Raman microscopy, because the available radiation is focused into a much smaller area, to avoid thermal damage in the sample from the laser, the effective Raman scattering is even weaker, making Raman microscopy a much less appealing technique. A great leap was seen recently in Raman microscopy technology with the development of ultrahigh throughput spectrometers and high quantum efficiency **charge coupled devices** (CCDs) detectors. A high-resolution Raman image (e.g., a 200 × 200 pixel scan or a 40,000 spectral collection) could be acquired within a few minutes or even faster, comparing to a few hours with the old generation Raman microscope [19].

As described in Chap. 8, Raman scattering is independent of excitation wavelength, meaning visible lasers can be employed for Raman experiments, which makes the design of a Raman microscope much easier than FTIR. A regular light microscope can be readily equipped with introducing an incident laser and a mechanism to collect back scattering to enable Raman measurement.

Because visible light (much shorter wavelength than IR) can be used for Raman microscopy, image resolution can be much higher than FTIR imaging. With a green laser (e.g., 532 nm) and an oil immersion objective (e.g., NA = 1.3), the resolution limit of a Raman image can be in the 200–300 nm range.

Confocal Raman microscopy has been used extensively in food research [20–22]. Probing emulsifier efficiency at the oil/water interface, interactions between protein and lipid in cheeses, and distribution of starch in gluten matrix in bread, etc. are just a few examples demonstrating its advantages of high spatial resolution and no need for contrasting additives. However, as described in Chap. 8, fluorescence is often troublesome in image acquisition, especially in food systems with natural ingredients. Using a longer wavelength laser (e.g., 785 nm), defocusing of the incident beam, and applying preimaging bleaching are common ways to reduce fluorescence, but image resolution may be deteriorated.

32.3.4 Confocal Laser Scanning Microscopy

Confocal laser scanning microscopy (CLSM) is one of the four types of **confocal microscopy** techniques (the other three are spinning-disk, micro-lens enhanced, and programmable array microscopes), which provide high-resolution chemical images based on fluorescence emission from the sample [23].

Figure 32.9 shows a typical setup of a confocal microscope. The incident light (a laser or white light source) enters the microscope objective through a dichroic mirror and is focused on the sample through the objective lens. The reflected or backscattered light goes back through the lens and converges at the detector surface, through which the radiation signal is recorded. With a confocal setup, a light blocking plate with a small pinhole is placed right in front of the detector at the converging point to still allow the back-focused light to reach the detector. Imaging light also may be reflected back from locations below (e.g., green lines) or above the focal plane, but it will converge before or after the pinhole, which will be blocked by the pinhole plate and is not able to reach the detector.

With this configuration, only light coming from the focal plane can be recorded by the detector. No background radiation from below or above the focal plane reaches the detector, which gives the recorded images a much sharper appearance. However, one limitation of the confocal setup is that the field of view is significantly smaller due to the size of the pinhole. Therefore, a scanning mechanism, typically through a set of vibrating mirrors, is often incorporated to provide images with a large field of view. The mirrors vibrate through piezoelectric elements, capable of providing scanning speed of 1800 Hz (lines/s) or faster. The entire focusing system (anything above the sample) is usually capable of moving in the vertical direction, allowing image acquisition

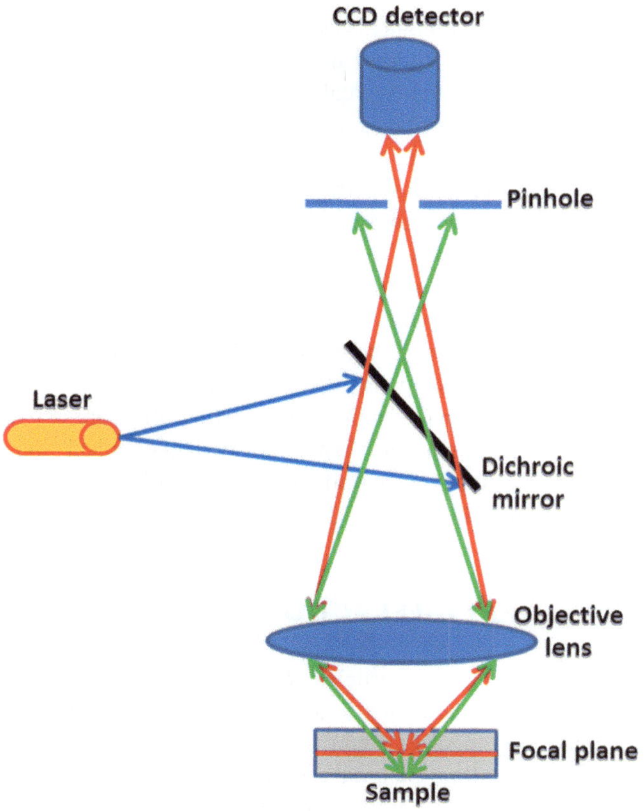

Fig. 32.9 Schematic of a confocal microscope

at different focal planes. A 3D image then can be generated by stacking images acquired from many focal planes.

The contrast of CLSM images is based on the fluorescence intensity from the laser's illuminating point. As described in Sect. 32.2.5, there are many ways to label the target object to give fluorescence emission. Samples are often labeled with multiple fluorophores to allow co-localization studies of a complex system.

CLSM has wide applications in biological and medical science disciplines but also finds popular applications in foods (e.g., imaging fat crystal structures, dairy products, food emulsions, food gels, plant materials, etc. [24]).

32.3.5 Hyperspectral Imaging

Hyperspectral imaging (HSI), a relatively new technology for the food industry, was initially developed in 1970s for military use on drones, airplanes, or satellites. It was soon adopted by geologist to survey earth morphology and composition and monitor environmental conditions. Application in agriculture to assess crop health was popularized in the last decade. In recent years, the food industry has shown increasing interests in HSI, mainly for foreign material identification in processing lines, nutritional evaluation of raw ingredients, and food safety related monitoring (e.g., bacteria growth) [25].

Hyperspectral imaging is essentially chemical imaging. It employs the same concept as explained in Sect. 32.3.1. A hyperspectral camera, typically equipped with an element array detector, either takes a snapshot or raster-scans over a target to acquire an image. Each pixel in the image contains a full electromagnetic spectrum that arises from ultraviolet, visible, infrared, or fluorescence radiation. By analyzing the individual spectrum to extract the desired chemical signature, a colored image can be rendered to illustrate the distribution of the component.

The major difference between a HSI system and a chemical imaging microscope is the processing software. Because real-time image analysis and feature recognition are essential for HSI cameras, it often requires mathematical algorithms such as machine learning that enable fast and complex computation and modeling [26].

32.4 X-RAY DIFFRACTION

Atoms, molecules, or particles arrange differently in a solid material. When they pack in an ordered fashion, with symmetrical and repetitive pattern, the solid is usually called a **crystal**. Crystalline materials are rigid, have fixed melting point, can be cleaved along a definite plane, and have anisotropic physical properties such as electrical conductivity, refractive index, and thermal expansion. In contrast, **amorphous** materials have short range, or no order of molecular alignment. They are less rigid, usually do not have sharp melting points, and have isotropic physical properties.

Within a three-dimensional crystalline structure, a volume element that repeats its geometry and orientation in all directions can usually be identified as a **unit cell**, with the smallest volume often considered as the representative unit (Fig. 32.10a). The unit cell can be defined with six parameters, with three being the edge length (e.g., Fig. 32.10b) and three being the angles in between the edges (denoted as α, β, and γ, not shown). For a given crystal, when these six parameters are measured, the unit cell, or the crystal structure, can be elucidated [27].

The structural elements (atoms, molecules, or particles) in a unit cell form parallel planes (dotted lines in Fig. 32.10c), which intersect the unit cell in many ways. When electromagnetic radiation such as X-rays strikes the crystals, they are reflected by these planes. Assuming the incident X-ray has parallel beams (e.g., Fig. 32.10c, beam 1 and 2), to have a constructive interference between the beams reflecting off of the crystal planes, the additional distance that beam 2 travels ($2 \times d \times \sin\theta$, where d is the spacing of crystal planes, and θ is the incident angle of incoming radiation) has to equal $n \times \lambda$ (n is an integer and λ is the wavelength of the radiation):

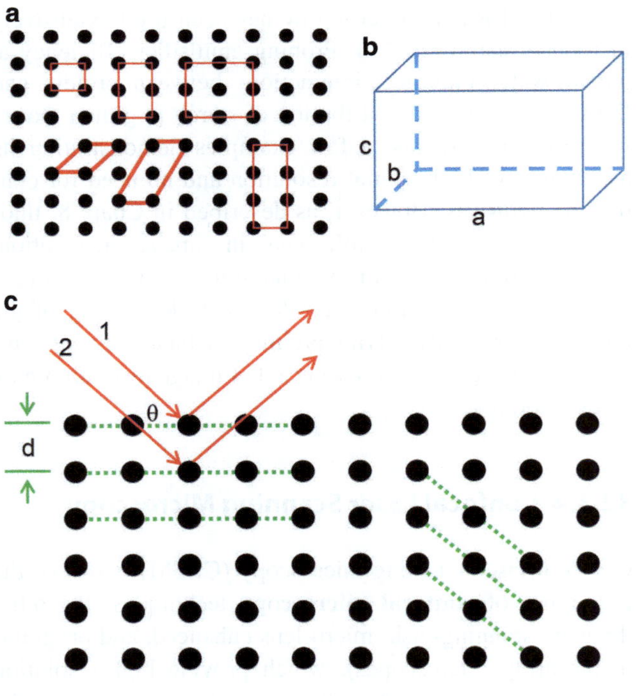

Fig. 32.10 Crystal unit cell and x-ray diffraction. (**a**) Repeating volume elements, or unit cells in a crystal, (**b**) Unit cell dimension parameters, (**c**) Parallel planes intersecting a unit cell, and reflection of x-rays off the planes

$$n\lambda = 2d\sin\theta \quad (32.2)$$

Equation (32.2) is called **Bragg's equation**. Under this condition, the reflected beams add up to give a diffraction pattern. Those with destructive interference are subtracted from each other, which produces much weaker radiation at the detector. With λ known from the X-ray source, and θ measured from X-ray/sample geometry, d can be obtained for a certain set of crystal planes. However, for a single crystal, within the scan range of θ, not all crystal planes can give **constructive diffraction**. A polycrystalline sample (powdered crystals) is usually used as an alternative to capture all diffraction patterns/peaks to resolve the complete crystal structure, and the technique is therefore referred to as **powder X-ray diffraction**.

A simple schematic of an **X-ray diffractometer** (XRD) is depicted in Fig. 32.11a. The detector usually scans along the arc to vary 2θ, and the XRD spectrum is generated by plotting it with X-ray intensity as a function of 2θ [28].

X-ray diffraction patterns are characteristic of unit cell types and other structural properties of the crystals (e.g., lattice parameters, phase identity, crystallinity, composition, etc.). Figure 32.11b is a typical XRD spectrum for crystalline cellulose, which is isolated from plant cell wall materials (from cotton linter in this case).

X-ray diffraction is a powerful tool for food research. Wide applications are found in carbohydrates (e.g., sugar, polysaccharides, etc.), and lipid/fats areas since these materials can easily form crystalline structures under application conditions [29, 30].

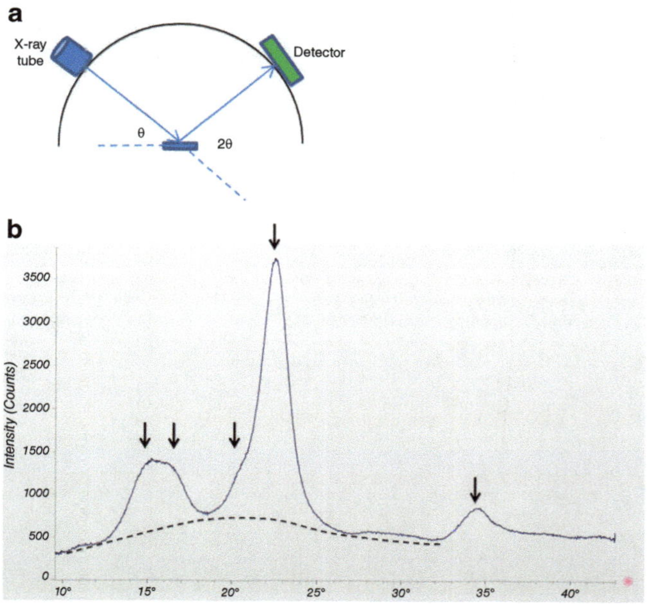

Fig. 32.11 Schematic of powder diffraction. (**a**) Simple schematic of an x-ray diffractometer. (**b**) XRD spectrum of crystalline cellulose. Dotted line shows the residual amorphous phase

32.5 TOMOGRAPHY

32.5.1 Introduction

The word **tomography** is based on the two Greek words, *tomos* (meaning slice or section) and *graphe* (meaning to draw or write). The process of tomography, then, is the reconstructing of the original material, or specimen, from either a series of physical sections (through microtomy or sequential removal from the bulk) of that specimen or from a series of "optical" sections. Though tomography was originally done methodically by hand, the general practice now is to use computer assistance, hence, the term **computer-assisted tomography** [31]. Tomography can be done using any of several possible mechanisms, such as confocal light microscopy, electron microscopy, radio frequency waves (MRI), or X-ray microscopy.

32.5.2 X-Ray Computed Tomography

X-ray computed tomography (CT) [older term is **computerized axial tomography** (CAT)] employs X-rays as the imaging agent to generate a 3D digital image of the specimen. As the X-ray strikes through the object, a 2D shadow image is collected on the detector, with contrast proportional to the X-ray attenuation or the physical density of the sample. The sample is rotated in single axis, and multiple images are collected with the orientation. A special computing algorithm (digital geometry processing) is applied to the series of 2D images to construct a 3D image that shows the inside of the specimen.

CT is both qualitative and quantitative. CT images can be rendered in both surface and volume modes. Segmentation can be applied to the images to render structures with similar density. The resolution of CT is mostly dependent on the X-ray beam size and the detector parameters. With a microscopic CT system, image resolution can be achieved in the range of a few micrometers.

CT is predominantly used in medical field but has found more and more applications in food research, e.g., to visualize fillings inside a chocolate cookie, to quantify air cell size and cell wall thickness of a gluten-free bread, to monitor bubble formation of an icing, and to study salt dissolution in a cookie dough.

32.6 CASE STUDIES

32.6.1 Fat Blends

One trend in fat–/oil-related food research is to reduce the amount of saturated fat. Saturated fat forms unique crystal structures at room temperature, which not only act as the framework of the blends to give desired physical functionality but also provide a network to hold liquid oil. Reducing or replacing the saturated fat, but keeping the same functionality, demands a thorough understanding of the fat crystal structure. In addition to ingredients, processing conditions may also be adjusted to allow manipulation of fat crystal formation.

In a fat blend case study, high oleic oil was the liquid fraction and palm stearin was the solid fraction. The ratio of the two was varied from 90:10 to 80:20. Structuring agents, which affect crystal formation and growth, were added in small amounts, 2–4% (e.g., monoglyceride, wax, lecithin, etc.). Functional properties such as oil-holding capacity, mechanical strength (e.g., Young's modulus), and plasticity were evaluated through different measurements. Crystallization was characterized by DSC, and crystal polymorphs were measured by XRD. Microstructure was imaged by confocal Raman microscopy and CLSM. Figure 32.12 shows some CLSM images of the fat blends with varying processing conditions and compositions. Drastic differences were observed in crystal structures (Nile red-stained liquid oil, with dark contrast corresponding to fat crystals), which correlated well with other functionality parameters.

32.6.2 Food Emulsions

To make a stable emulsion, emulsifiers with amphiphilic nature are added to prevent liquid droplets from coalescence. Common food emulsifiers are lecithin (from soy or egg yolk), mono- or diglycerides, and proteins. When concentration in a liquid is higher than its critical micelle concentration (CMC), lecithin molecules start to aggregate. The size and packing structure of the aggregates are indications of its packing ability at the oil/water interface when being used as emulsifiers. Figure 32.13 shows the microstructure characterization of emulsions prepared in a study comparing the performance of two different plant-source lecithins. TEM images show that one type of lecithin formed spherical micelles (smaller radius of curvature) and the other formed worm-like micelles. Light microscopy suggested significant differences in droplet size and morphology. CLSM shows that the emulsion on the left had lecithin nicely ordered at the

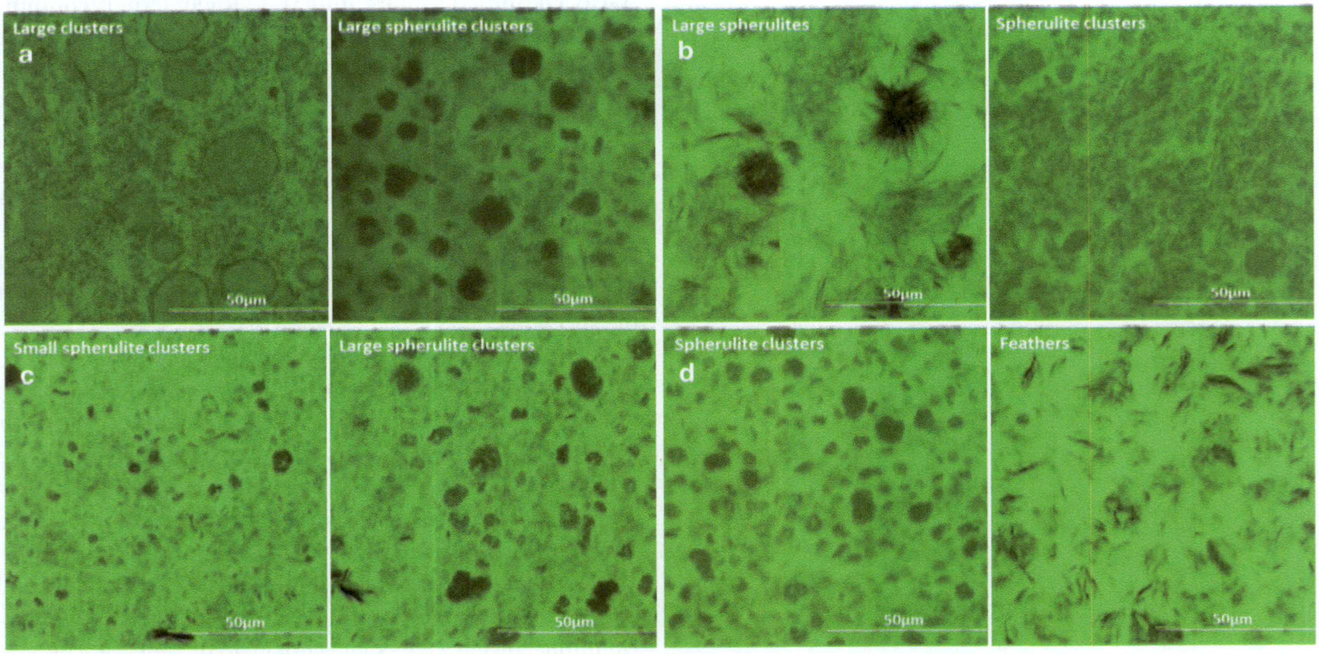

Fig. 32.12 CLSM of fat crystal structures. (**a**) Varying cooling rate. (**b**) Different structuring agent. (**c**) Varying liquid/saturated fat ratio. (**d**) Same structuring agent but with varying concentration

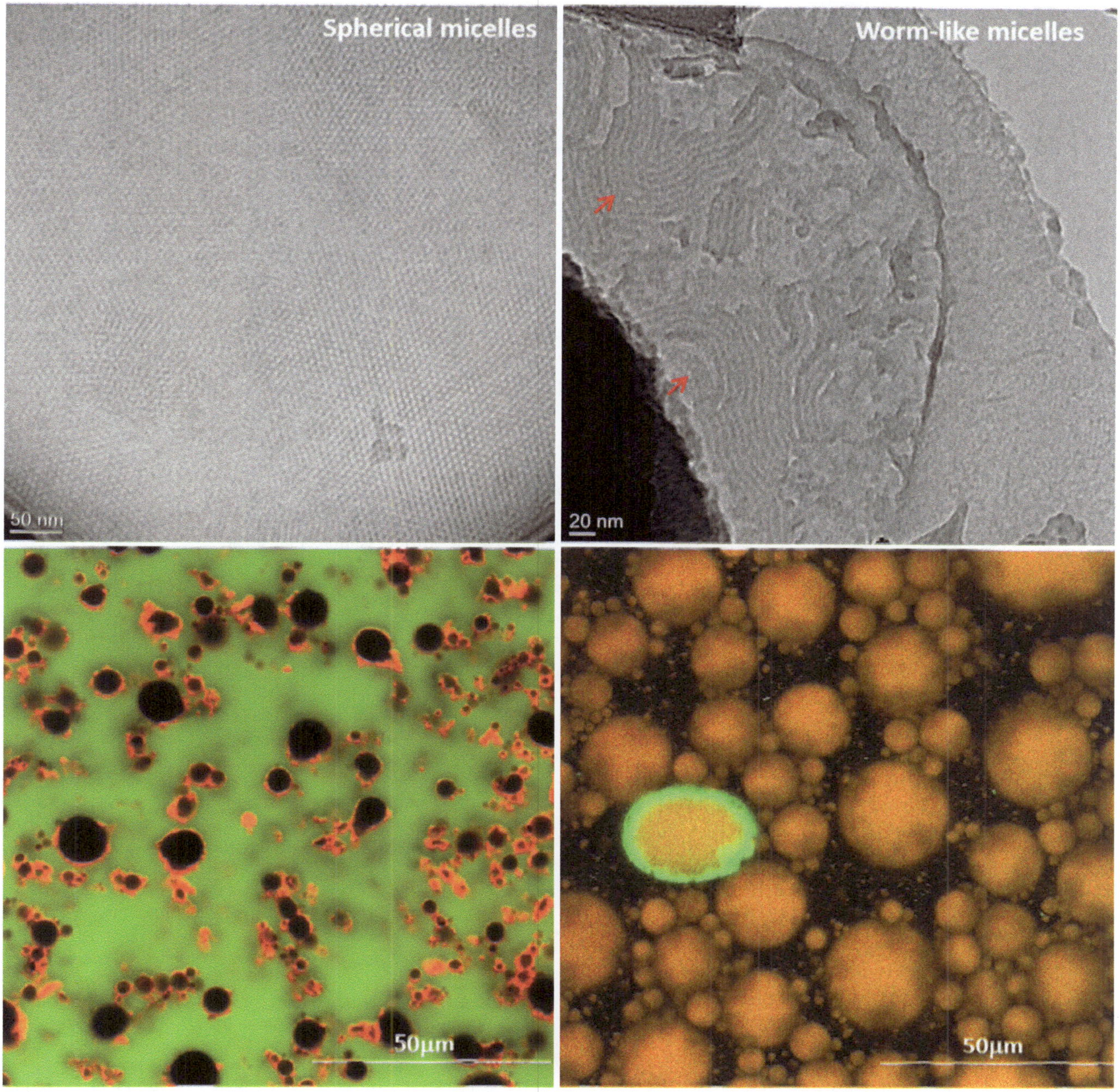

Fig. 32.13 Imaging characterization of emulsions with different lecithin: *Top row*, TEM; *Bottom row*, CLSM

interface of the water droplets, while the one on the right did not form the water-in-oil emulsion as desired. Lecithin stayed with the oil in the droplet instead of stabilizing at the interface. Emulsion stability measurement (through centrifuge), particle size measurement [through time-domain nuclear magnetic resonance (TD-NMR) (Chap. 10, Sect. 10.3) and particle size analyzer (Chap. 5, Sect. 5.5.2.3)], and conductivity measurement gave similar microstructure predictions.

32.7 SUMMARY

The future of food research and food production resides in controlling and manipulating food structures to give desired functionality. This chapter briefly introduced some direct and indirect microstructure characterization tools that were mostly invented or developed in and for other science disciplines. They have been borrowed by food scientists and manufacturers for many years to help better understand food

Table 32.1 Key characteristics of tools to examine food microstructure

Microstructure tool	*Imaging agent*	*Typical resolution*	*Key information provided*
Light microscope	Visible light	~200 nm	Morphology. Composition with histology
Fluorescence microscope	Visible light	~200 nm	Composition
SEM	Electrons	Sub-nm	Surface morphology
EDS	X-rays	Sub-nm	Elemental composition. Elemental distribution
AFM	Sharp tip	Sub-nm	Surface morphology; interfacial forces
FTIR microscope	IR light	5–10 mm	Chemical distribution
Confocal Raman microscope	Visible laser	~200 nm	Chemical distribution
CLSM	Visible laser	~200 nm	Chemical distribution
XRD	X-rays	20–50 μm	Crystallinity
CT	X-rays	5–10 μm	3D morphology

systems. Some techniques have become essential tools in food science and technology. More detailed discussions of each technique can be easily found in the references cited in the text. Table 32.1 summarizes some key characteristics of these techniques.

One must be aware that none of the techniques included in Table 32.1 are universal in their applications. Foods can be very complex systems, with a very broad spectrum of types and modifications. Very often, combinations of several scientific techniques are needed to grasp the full picture of the structure of just one food type. The purpose of learning food microstructure is to be able to control and design new structures. Progress is continuously being made in this developing field, and huge opportunities are just ahead of us.

32.8 STUDY QUESTIONS

1. What is the definition of "resolution" with regard to microscopy? How is the resolution of an optical microscope determined?
2. What is the main reason that electron microscopes can work at much higher magnifications (or resolution) than light microscopes?
3. Why do we apply a conductive coating to samples examined by an SEM? Why is a conductive coating usually not needed when examining samples using either a low vacuum SEM or an environmental SEM?
4. When using energy-dispersive X-ray spectroscopy (EDS), what carry the information telling us the elemental content of the sample, and where do they come from?
5. FT-Raman is often an option when sample has strong fluorescence. Why is it not used in Raman microscope?
6. What is "tomography" and what advantages does it have over other imaging techniques? How does computer-assisted X-ray tomography create a 3D image?
7. You are working on a project to reduce the sugar content in gummy candies. You replaced some sugar with high-intensity sweetener, and added some starch to try to maintain the elastic property. However, the candy does not look and taste like the normal gummy candy. You are asked to understand the candy's microstructure. What would be your experimental plan to do the characterization?

Acknowledgments The author of this chapter thanks Var L. St. Jeor (formerly with Global Food Research, Cargill Research and Development Center, Cargill, Inc., Plymouth, MN, USA) for his contribution to the previous edition of the book.

REFERENCES

1. Van Helden A, Dupre S, van Gent R, Zuidervaart H (eds) (2010) The origins of the telescope. KNAW Press (now Aksant Academic Publishers), Amsterdam, Netherland
2. James PJ, Thorpe N (1994) Ancient inventions. Random House Publishing, London, UK
3. Delly JG (1988) Photography through the microscope. Eastman Kodak Co., Rochester, NY
4. Bradbury S (1998) Introduction to Light Microscopy. Bios Scientific Publishers Ltd, Oxford, UK
5. Green FJ (1991) The Sigma-Aldrich handbook of stains, dyes & indicators. Aldrich Chemical Co., Inc., St. Louis, MO
6. Peter G (1964) Handbook of basic microtechnique, 3rd edn. McGraw Hill, New York
7. Clark G (1981) Staining procedures. Williams & Wilkins, Essex, UK
8. Lillie RD (1977) H. J. Conn's biological stains, 9th edn. Williams & Wilkins Co., Reprinted by Sigma Chemical Co., St. Louis, MO
9. Hergert W, Wriedt T (2012) The Mie theory basics and applications (Chapter 2 authored by Wriedt T), Springer, New York
10. Goodhew PJ, Humphreys J, Beanland R (2000) Electron Microscopy and Analysis 3rd Edition. CRC Press, Florida
11. Binnig G, Quate CF, Gerber CH (1986) Atomic force microscope, Physical Review Letters 56:930
12. Haugstad G (2012) Atomic force microscopy: understanding basic modes and advanced applications. Wiley, New York
13. Baró AM, Reifenberger RG (2012) Atomic force microscopy in liquid: biological applications. Wiley, New York
14. Sarid D (1994) Scanning force microscopy: with applications to electric, magnetic, and atomic forces. Oxford University Press, Oxford, UK
15. Yang H., Wang Y, Lai S, An H, Li Y, Chen F (2007) Application of atomic force microscopy as a nanotechnology tool in food science. J Food Sci 72:R65
16. Sasic S, Ozaki Y (2011) Raman, infrared, and near-infrared chemical imaging. Wiley, New York

17. Lewis EN, Treado PJ, Reeder RC, Story GM, Dowrey AE, Marcott C, Levin IW (1995) Fourier transform spectroscopic imaging using an infrared focal-plane array detector, Anal Chem 67:3377
18. Nowakowski CM, Aimutis WR, Helstad S, Elmore DL, Muroski A (2015) Mapping moisture sorption through carbohydrate composite glass with Fourier transform near-infrared (FT-NIR) hyperspectral imaging. Food Biophysics 10:207
19. Dieing T, Hollricher O, Toporski J (2011) Confocal Raman microscopy. Springer, New York
20. Thygesen LG, Lokke MM, Micklander E, Engelsen SB (2003) Vibrational microspectroscopy of food. Raman vs. FT-IR. Trends Food Sci & Tech 14:50
21. Roeffaers MBJ, Zhang X, Freudiger CW, Saar BG, van Ruijver M, van Dalen G, Xiao C, Xie XS (2011) Label-free imaging of biomolecules in food products using stimulated Raman microscopy. J Biomed Optics 16(2):021118
22. Gierlinger N, Schwanninger M (2007) The potential of Raman microscopy and Raman imaging in plant research. Spectroscopy 21:69
23. Jerome WG, Price RL (2018) Basic Confocal Microscopy. Springer, New York
24. Lagali N (ed) (2013) Confocal laser microscopy: principles and applications in medicine, biology, and sciences. InTech (open access), Rijeka, Croatia
25. Khan MJ, Khan HS, Yousaf A, Khurshid K, and Abbas A (2018) Modern Trends in Hyperspectral Image Analysis: A Review. IEEE Access 6:14118
26. Paoletti ME, Haut JM, Plaza J, Plaza A (2019) Deep Learning Classifiers for Hyperspectral Imaging: A Review. ISPRS Journal of Photogrammetry and Remote Sensing 158:279
27. Yoshio W, Eiichiro M, Kozo S (2011) X-ray diffraction crystallography. Springer, New York
28. Jenkins R, Snyder R (1996) Introduction to x-ray powder diffractometry. Wiley-Interscience, New York
29. Putaux JL (2005) Morpholody and structure of crystalline polysaccharides: some recent studies. Macromolecular Symposia 229:66
30. Idziak SHJ (2012) Powder x-ray diffraction of triglycerides in the study of polymorphism, ch. 3. In: Marangoni AG (ed) Structure-function analysis of edible fats, AOCS Press, Urbana, IL
31. Seeram E (2022) Computed Tomography, 5th Edition. Elsevier, Cambridge, MA

Part VII

Analysis of Objectionable Matter and Constituents

Analysis of Food Contaminants, Residues, and Chemical Constituents of Concern

33

B. Pam Ismail and S. Suzanne Nielsen

Contents

B. P. Ismail (✉)
Department of Food Science and Nutrition, University of Minnesota, St. Paul, MN, USA
e-mail: bismailm@umn.edu

S. S. Nielsen
Department of Food Science, Purdue University, West Lafayette, IN, USA
e-mail: nielsens@purdue.edu

33.1 INTRODUCTION: CURRENT AND EMERGING FOOD HAZARDS

The food chain that starts with farmers and ends with consumers can be complex (Fig. 33.1, reference [1]). Pesticide treatment, agricultural bioengineering (i.e., genetic

B. P. Ismail, S. S. Nielsen (eds.), *Nielsen's Food Analysis*, Food Science Text Series,
https://doi.org/10.1007/978-3-031-50643-7_33

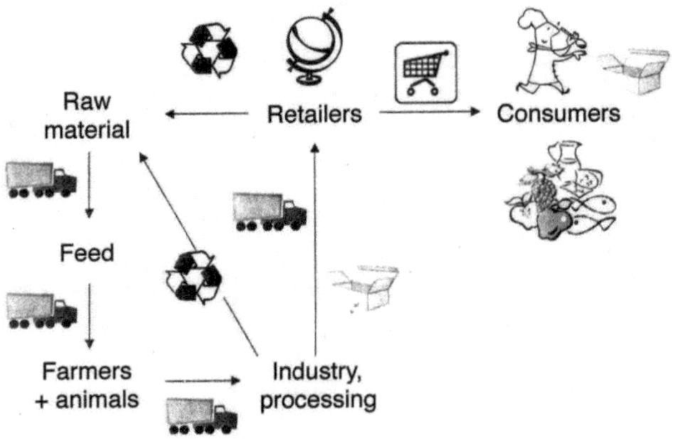

Fig. 33.1 An illustration of the food chain. (From Nielen and Marvin [1], with permission)

modification), veterinary drug administration (e.g., antibiotics, growth hormones), environmental and storage conditions, processing, transportation, economic gain practices, use of food additives, and/or choice of packaging material may lead to contamination with or introduction (intentionally and non-intentionally) of hazardous substances. Legislation and regulation to ensure food quality and safety are in place and continue to develop to protect stakeholders, namely farmers, consumers, and industry [2–4, references within].

Contaminants/food hazards (including extraneous matter, Chap. 34) often have threshold levels (**tolerance levels**) below which no adverse effects are observed. The **US Environmental Protection Agency** (EPA) establishes these tolerance levels, and the **Food and Drug Administration** (FDA) and **US Department of Agriculture** (USDA) enforce them. However, food safety incidents, microbial, biological, and chemical in nature, continue to occur. The **Rapid Alert System for Food and Feed** (RASFF) reported a total of 3862 alerts/notifications in 2020, a value that is 22% higher than what was reported in 2019 [5]. Microbial contamination ranked first in the top ten hazards, while pesticides ranked second. Notifications are generally categorized as follows: microbial, chemical, mycotoxins, and other hazards. Microbial contaminants include molds, viruses, and bacteria (discussion and methods of analysis for this category are beyond the scope of this chapter). Within the chemical category, the most frequently reported hazards include pesticides (e.g., ethylene oxide, omethoate, and isophenfos-methyl), veterinary drugs including antibiotics (e.g., β-lactam and chloramphenicol), and growth hormones (e.g., estradiol and progesterone, allergens, and heavy metals such as mercury, lead, and cadmium). Examples of common chemical hazards include fraud and food adulterants (e.g., melamine), packaging chemicals (e.g., bisphenol A and 4-methylbenzophenone), degradation metabolites (e.g., acrylamide, heterocyclic amines, and furan), naturally occurring substances (e.g., cyanide and solanine), and other chemical contaminants (e.g., 3-monochloropropane-1,2-diol, benzene, and perchlorate). An emerging chemical concern include nanomaterials used in packaging for functions such as antibacterial activity and oxygen scavenging ability [6]. Another category that can be of concern includes the genetically modified organisms (GMO) and their products. Introduction and usages of GMO in food products resulted in the development of legal requirements of safety and labeling.

Given the extent of concerns with contaminants, there is a strong need for adequate and reliable methods of detection and analysis to ensure food quality, safety, and fair trade. There are several well-established and reliable methods of analysis to detect certain food hazards. Development and validation of methods to detect and analyze emerging food hazards is a continuous effort. This chapter will cover some of the screening methods and quantitative methods that are commonly used for the detection and quantification of several food hazards. The focus in this chapter on chemical contamination is intended to compliment that of Chap. 35 on Food Forensic Investigations.

33.2 ANALYTICAL APPROACH

As with the analysis of any food constituent, there is an array of methodological approaches and techniques to choose from for the analysis of food hazards. Multiple factors influence the method of choice as listed in Chap. 1, Table 1.5, including in this case complexity of the food matrix, characteristics of the analyte (e.g., polarity, hydrophobicity, volatility, thermal stability, and chemical reactivity), and suspected level of the contaminant. The objectives for the analysis of contaminants can vary in complexity from mere detection of several suspected contaminants belonging to the same family to the determination of the exact level of a particular contaminant or, in more complex cases, detection of unknown adulterants. The general trend for regulatory institutions and industry is to implement inexpensive and rapid **screening methods**. However, depending on the objective of the analysis, **quantitative methods** that require sophisticated equipment might be needed. In this case, the industry may choose to send their samples to specialized laboratories. Once a method of analysis is chosen, appropriate consideration with regard to sampling and sample preparation needs to be made. Most importantly, lab safety needs to be considered as with any method of analysis with acute consideration of handling substances that might be contaminated with chemical hazards. Appropriate safety measures should be taken during sampling, sample preparation, and analysis of any food constituent. Personal protective equipment and laboratory design are unique for handling hazardous or hazard containing substances. For safety measures, the readers are directed to the EPA guide on safety related to analysis of toxins [7].

33.2.1 Choice of Analytical Method

The complexity of the food matrix (see Chap. 1, Sect. 1.4.4) and the characteristics of the analyte significantly influence the choice of extraction, separation, detection, and quantification techniques, as will be discussed in subsequent sections. Accuracy, precision, specificity, and sensitivity of the analytical method (see Chap. 4) are also important considerations. There are some official methods for the analysis of contaminants and residues, others are validated, and some are being developed and validated. Methods for the analysis of food hazards can be either qualitative, semiquantitative, or quantitative. A summary of analyses for some of the main contaminants is given in Table 33.1.

Table 33.1 Summary of analyses for contaminants, residues, and compounds of concern in foods

Contaminant	*Quantitative*	*Semiquantitative or qualitative (screening methods)*
Allergens	ELISA PCR LC-MS	LFS
Antibiotics	HPLC (mostly) GC Immunoassays	Microbial growth inhibition Receptor assays Enzyme-substrate assays Immunoassays
Genetically modified organisms (GMOs)	PCR (mostly) ELISA	LFS
Heavy metals	AES ICP ICP-MS	XRF
Mycotoxins	HPLC (mostly) GC Capillary electrophoresis Immunoassays (mostly)	TLC Immunoassay
Nitrites	Colorimetric methods Ion chromatography	Ion-selective electrode
Pesticides	Multiresidue (MRMs) GC (mostly) HPLC Single residue (SRMs) GC (mostly) HPLC Immunoassay	TLC Enzyme inhibition Immunoassay
Sulfites	Monier-Williams method Ion chromatography Enzyme method HPLC	"Ripper method"

ELISA enzyme-linked immunosorbent assay, *GC* gas chromatography, *HPLC* high-performance liquid chromatography, *ICP-MS* Inductively coupled plasma – mass spectrometry, *LC-MS* liquid chromatography-mass spectrometry, *LFS* lateral flow strip, *PCR* polymerase chain reaction, *TLC* thin-layer chromatography, *XRF* x-ray fluorescence

33.2.1.1 Qualitative or Semiquantitative Methods

Qualitative and **semiquantitative** methods, also known as **screening methods**, are usually used to assay a large number of samples for the presence of one or more contaminants belonging to the same family (e.g., pesticides Sect. 33.3.3.2.1, mycotoxins Sect. 33.4.3.1.1, and antibiotic residues Sect. 33.5.2.1). These methods are fast, of low cost, simple, and robust; they are often less sensitive to small changes in experimental and/or environmental conditions. While qualitative methods detect the presence of certain contaminants at or above a specified threshold, semiquantitative methods provide an estimate of the concentration of a detected contaminant. These methods include techniques such as **thin-layer chromatography** (TLC), **enzyme inhibition**, and **immunoassays** (see later sections of this chapter).

33.2.1.2 Quantitative Methods

For simultaneous quantitation and structural identification of chemical food contaminants and residues, **gas chromatography** (GC; Chap. 14) and **high-performance liquid chromatography** (HPLC; Chap. 13) are the two main analytical methodologies employed. The combination of GC or HPLC with mass spectrometry (MS) allowed for sensitive detections at very low concentrations. For example, GC–MS is commonly used for the simultaneous analysis of a large number of pesticides [8]. Advancement in LC–MS allowed its use for the analysis of antibiotics [9, 10], pesticides [11, 12], and mycotoxins [13].

Immunoassays (Chap. 27) are rapid, simple, and cost-effective means for the detection and quantification of both single and multiple contaminants or residues, such as pesticides, antibiotics, and mycotoxins. Of the various immunoassay techniques, **enzyme-linked immunosorbent assay** (ELISA) (Sect. 27.3.2) and **lateral flow strip** (LFS) tests (Sect. 27.3.4) provide a sensitive detection of toxic analytes, whereas **immunoaffinity chromatography** (Sect. 27.4) is used for concentration and cleanup of the analyte of interest. One drawback of immunoassays is that the antibodies used in the assay may have **cross-reactivity** (affinity) for related chemical structures.

33.2.2 Sample Preparation

33.2.2.1 Introduction

Food samples are usually too dilute (e.g., beverages) or too complex (e.g., cereals) for direct analysis of trace contaminants and residues. Therefore, sample preparation, including homogenization, extraction, fractionation/cleanup, concentration, and/or derivatization (for GC analysis), normally precedes the analysis of food contaminants and residues (Fig. 33.2). Analysts are faced with continuous decrease of

Sampling	Collection Storage
Sample preparation	Extraction Fractionation/Clean-up Concentration Derivatisation
Chromatographic analysis	Separation
Data Handling	Identification Quantification
Reporting	

Fig. 33.2 Flow diagram for the analysis of contaminants and residues. (From Sandra [14], with permission)

legislative limits for food contaminants and residues, resulting in the need for more sensitive, precise, and accurate measurements. Sample preparation techniques for the analysis of food contaminants and residues are continuously improved to guarantee high recovery and reproducibility (see reference [15] as an example). Fortunately, there also have been significant advances in analytical methods and equipment. The advanced technology of MS (see Chap. 11), specifically, their enhanced "recognition features" when used in tandem, allows them to take over the "selectivity" of classical sample preparation methods. However, for quantification purposes, cleaner extracts are preferred. In both target and **multiresidue** methods, isotope dilution and addition of an internal standard provide the most accurate results, compensating for sample matrix effects and ion suppression in MS.

Sampling and sample preparation for contaminants analysis are labor intensive and time consuming but are essential prerequisites for acquiring accurate analytical data. The stakes are higher for sample preparation errors than analytical errors in making a safety decision of a possibly contaminated sample. Obtaining a sample that is representative of the level of trace substances in a heterogeneous mixture is not an easy task. Sampling for the analysis of a particular contaminant or residue will be briefly discussed in relevant sections of this chapter.

33.2.2.2 Sample Homogenization

Food contaminants and residues in many cases are unequally distributed in a food system. For instance, most pesticides are not translocated within plants and are expected to be located on the surface of fresh produce. Thus, removal of inedible parts (outer skins, stems, and cores), when applicable, and homogenization are necessary for reliable and accurate analysis. Homogenization can be achieved by chopping or grinding, followed by blending and mixing. Apart from homogenization, grinding reduces the structural features of the sample and enhances extraction efficiency. Contaminating the sample or exposing it to unnecessary heat, which can cause volatilization or degradation of contaminants or residues, should be avoided. Sometimes, cryogrinding (see Chap. 5, Sect. 5.5.4) is preferred for soft samples and to avoid thermal degradation.

33.2.2.3 Extraction and Cleanup

33.2.2.3.1 Introduction

Almost all food samples, with the exception of samples directly soluble in an organic solvent (e.g., vegetable oil), require a solvent extraction step (Chap. 12, Sect. 12.2; Chap. 14, Sect. 14.2.2.4) to isolate the target analytes from the matrix. Defatting of lipid-rich food matrices is often required prior to the extraction of target contaminants. Extraction of contaminants and residues is traditionally done by solubilizing them in a suitable organic solvent, generally acetonitrile or acetone. **Anhydrous salt** (NaCl or Na_2SO_4) can be added to absorb water. In some cases, water is added so that the crude extract can be purified using a subsequent partitioning step with a second, water-immiscible solvent. When the extraction process is complete, the solvent is separated from insoluble solids by filtration.

Often, the crude sample extract is purified and concentrated before the separation/determination steps. The objective of cleanup is to separate the target analytes from coextracts that can interfere with their detection. Often, the preliminary cleanup step is followed by a preparative chromatography step (see Chap. 12 for basic information on chromatography). Cleanup of crude water-acetone or water-acetonitrile extract, for example, can be done by partitioning with a relatively nonpolar organic solvent. The residues then can be further purified by column chromatography (either adsorption or size-exclusion chromatography). Separate fractions of column eluate can be analyzed. The choice of packing material is both analyte and matrix dependent.

Extraction and purification techniques have been developed based on the characteristics of the target analyte(s), which can be categorized into three classes: (1) **volatile compounds** (VC) that can be analyzed via GC, after derivatization, if needed, (2) **semi-volatile compounds** (SVC) that are GC amendable (thermally stable), and (3) the **nonvolatile** (thermally labile) **compounds** (NVC), which are mostly analyzed by HPLC after extraction. Some of the widely used extraction techniques will be covered briefly in subsequent sections.

33.2.2.3.2 Solid-Phase Microextraction

All three classes of compounds listed above can be extracted using **solid-phase microextraction** (SPME), which is a type of **solid-phase extraction** (SPE) (Chap. 14, Sect. 14.2.2.5). Headspace (see Sect. 14.2.2.2) sampling can be performed for the analysis of VC, including pesticides, furan, and residues from packaging materials. Headspace sampling can be done by immersing the SPME polymer-coated fibers into the headspace. SPME also can be used for SVCs and NVCs, with the polymer-coated fiber directly immersed in the aqueous samples or placed inside a hollow cellulose membrane.

33.2.2.3.3 QuEChERS

The **QuEChERS method** [16, 17], which stands for quick, easy, cheap, effective, rugged, and safe, is a dispersive SPE (DSPE) technique that is by far the best method to extract multiple pesticide residues (Sect. 33.3.2) (see Chap. 14, Sect. 14.2.2.5 for SPE). A single sample preparation method and one chromatography method linked to MS (preferably in tandem) to determine as many pesticides as possible are very much desired. The advantages of the QuEChERS method (outlined in Fig. 33.3) include speed, ease, minimal solvent use, and lower cost as compared to conventional SPE. The types of adsorbents and solvents, as well as the pH and polarity of the solvents, can be adjusted based on the sample matrix and types of analytes. QuEChERS kits are commercially available (e.g., Sigma Aldrich/Supelco, Restek, and United Chemical Technologies).

33.2.2.3.4 Energized-Dispersive Guided Extraction

The **energized-dispersive guided extraction (EDGE)** is a relatively new extraction method developed by CEM Corporation in 2017 and offers a more effective extraction than QuEChERS for difficult matrices. EDGE is a method that combines pressurized liquid extraction with DSPE and is more automated than QuEChERS and is simpler than other extraction systems. For details on how this system works, the readers are directed to the CEM website. For the analysis of pesticides [18], it was shown to be 4.5 times faster than accelerated solvent extraction (Chap. 17, Sect. 17.4.4).

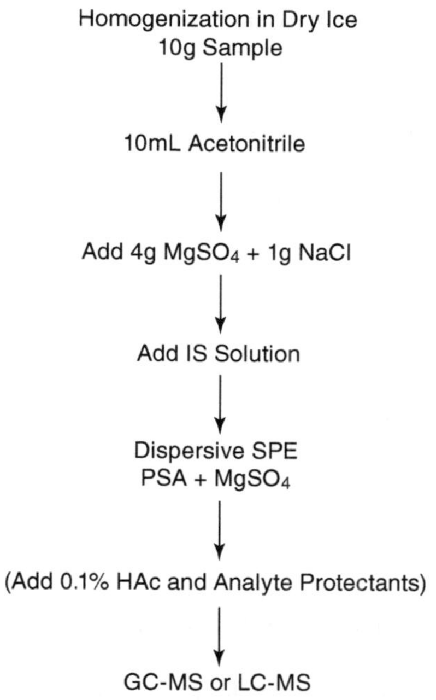

Fig. 33.3 Flow diagram for the QuEChERS procedure. (From Sandra [14], with permission)

33.2.2.3.5 Microwave-Assisted Solvent Extraction

Microwave-assisted solvent extraction (MASE), which utilizes electromagnetic radiation to desorb organics from their solid matrix, can decrease significantly the extraction time and the quantity of solvent required to efficiently extract target analytes (see reference [19] for example application). The efficiency of MASE is attributed to the elevated temperatures that exceed the boiling point of the solvent(s) and the rapid transfer of the analyte(s) to the solvent phase. Commercial systems are available that incorporate the capacity for simultaneously extracting multiple samples within closed, lined (perfluoroalkoxy), pressurized vessels (up to 8–12), using microwave absorbing solvents. Disadvantages of MASE include lack of selectivity and loss of thermolabile analytes.

33.2.2.4 Derivatization

The chemical structure of analytes may need to be modified to become suitable for separation and detection by specific chromatographic techniques and detected via available and/or desirable detectors. The process of structural modification through multiple chemical reactions is known as **derivatization** (see Chap. 12, Sect. 12.3.4.1.2; Chap. 13, Sect. 13.2.4.7; Chap. 14, Sect. 14.2.3). Many types of derivatization reactions have been used in the analysis of contaminants and residues, such as selected pesticide residues as discussed in reference [20] and the antibiotic nitrofurans as discussed in reference [14].

33.3 PESTICIDE RESIDUE ANALYSIS

33.3.1 Introduction

A pesticide is any substance or a mixture of substances formulated to destroy or control "pests" including weeds, microorganisms (e.g., fungi or bacteria), insects, and even mammals. Currently, there are over 1300 pesticides registered by the EPA [21], main classes of which include: (1) **herbicides** [for weed control, e.g., triazine (e.g., atrazine)], (2) **insecticides** [e.g., organochloride (OC) (e.g., dichlorodiphenyltrichloroethane (DDT)), organophosphates (OP) (e.g., malathion, dimethoate, omethoate), and methylcarbamates (e.g., aldicarb)], and (3) **fungicides** [e.g., phthalimide (e.g., captan)]. Other types of pesticides may include acaricides, molluscicides, nematicides, pheromones, plant growth regulators, repellents, and rodenticides. Information about active ingredients in pesticides are provided by the EPA [22].

Pesticide residues may occur in food as a result of the direct application to a crop or a farm animal or as a postharvest treatment of food commodities. Pesticide residues also can occur in meat, milk, and eggs as a result of the farm animal consumption of feed from treated crops. Additionally, pesticide residues can occur in food from environmental contamination and spray drift.

If pesticides were not applied, an estimated one-third of the crop production would be lost. However, pesticides may have adverse effects on human health, including, but not limited to, cancer, acute neurologic toxicity, chronic neurodevelopment impairment, and dysfunction of the immune, reproductive, and endocrine systems. Accordingly, there are strict regulations for pesticide registration and use all over the world.

Risk assessment studies, required by the EPA, are done to determine the nature and extent of toxic effects and to establish the level at which no adverse effects are observed [**no observed adverse effect level** (NOAEL)]. Based on risk assessment studies, **tolerance levels** [or **maximum residue level** (MRL)] of pesticides (the active ingredients as well as toxic metabolites, and transformation products) in food are set and enforced by government agencies, as mentioned in Sect. 33.1, to protect stakeholders and regulate international trade. In fact, tolerance levels must be established prior to registration. In general, the tolerance levels in foods range between 0.01 and 10 mg/kg, depending on the commodity and the pesticide used. Low tolerance levels have necessitated the development of more accurate and sensitive analytical methods to meet the requirements in food.

33.3.2 Types of Analytical Methods

Several factors make the analysis of pesticide residues in food a complex task. These factors include the complexity of the food matrix, the possibility of having pesticide levels as low as pictogram or femtogram amounts, and the considerable differences in the physical and chemical properties of pesticides.

Analytical methods employed for the analysis of pesticides can be categorized into either **single-residue methods** (SRM) or **multiple-residue methods** (MRM). SRMs are designed to measure a single analyte and, often, its toxic metabolites. The majority of SRMs have been developed for purposes of registration and setting tolerance levels, or for investigating the metabolism and environmental fate of a specific pesticide. Sampling, extraction, purification, and determination in SRMs are optimized for the target pesticide. The SRMs that have undergone EPA review are found in Volume II of the ***Pesticide Analytical Manual*** (*PAM* II) [23]. For purposes of monitoring quality and safety, given the large number of pesticides and the considerable differences in their physical and chemical characteristics (e.g., acidic, basic, or neutral, polar or nonpolar, and volatiles or nonvolatiles), MRMs can determine various pesticides in a single run and are much preferred. Many of the MRMs are found in Volume I of the ***Pesticide Analytical Manual*** (*PAM* I) [24]. AOAC International [25] also has developed an MRM for pesticide residues, the AOAC Pesticide Screen (AOAC Method 970.52). The identification and quantification in MRMs currently used by the FDA and USDA are based on GC and HPLC analysis. Reference [26] includes detailed descriptions of different MRM extraction and cleanup protocols specific for pesticide analysis.

33.3.3 Analytical Techniques Used for the Detection, Identification, and/or Quantification

A wide range of analytical techniques can be used in pesticide analysis. This section highlights some of the current analytical techniques used for the detection and determination of pesticide residues.

33.3.3.1 Biochemical Techniques

Biochemical techniques, such as enzyme inhibition assays and immunoassays, are widely used for the detection of pesticides. Commercially available test kits include enzyme inhibition assays. The principle of these assays is based on the inhibition of a specific enzyme, essential for vital functions in insects, by pesticides present in the sample. If no pesticides are present, the enzyme will be active and act on a substrate to cause a change in color. If no change in color occurs, then the test is positive, and further confirmation can be performed following more sophisticated analysis, such as HPLC and GC, to identify and quantify the specific pesticides present.

While enzyme inhibition assays are mostly used as screening methods, with limited sensitivity and selectivity, immunoassays can be tailored to particular purposes ranging from simple screening tests (field portable [27]) to quantitative laboratory tests. Additionally, extensive cleanup of extracts is not necessary for immunoassays, unless cross-reactivity exists. ELISA (Chap. 27, Sect. 27.3.2) accounts for almost 90% of the immunoassays used for pesticide residue analysis.

33.3.3.2 Chromatographic Techniques

33.3.3.2.1 Thin-Layer Chromatography

Thin-layer chromatography (TLC) (Chap. 12, Sect. 12.3.4.1.2) can be used for screening purposes in the analysis of pesticides. TLC can be used as a semiquantitative method that precedes more accurate detection and quantification by GC or LC. An example application is the detection and estimation of pesticides that inhibit insect enzymes such as cholinesterases. Once a crude extract is separated by TLC, the plate is sprayed with a solution containing the enzyme(s), followed by a solution containing a specific substrate, which releases a colored product. The lack of color change indicates enzyme inhibition, due to the presence of pesticide residues, and the zone of inhibition is proportional to the quantity of pesticide present.

33.3.3.2.2 Gas Chromatography

In general, GC is the preferred method for the determination of volatile and thermally stable pesticides, such as the OC and OP classes. Choice of columns and detectors is made based on the nature of the pesticides. For example, 5% diphenyl, 95% dimethylpolysiloxane stationary phase columns are commonly used in MRMs.

Pesticides often contain heteroatoms, such as O, S, N, Cl, Br, and F, in a single molecule. Therefore, element-selective detectors are often used, such as a flame photometric detector (FPD), which is suited for the detection of P-containing compounds (see Chap. 14, Sect. 14.3.5.4) such as OP pesticides in various crops, without extensive cleanup required. For the determination of OC, the electron capture detector (ECD, see Chap. 14, Sect. 14.3.5.3) is used extensively due to its high sensitivity to organic halogen compounds. Identification in conventional GC analysis is highly dependent on retention time, which is not an absolute confirmation of identity, due to matrix interferences. Coupling of fused capillary columns to MS detection enhances not only the confirmation process but also the quantitative determination (see Sect. 33.3.3.3).

33.3.3.2.3 High-Performance Liquid Chromatography

Development of HPLC analysis for the separation and detection of pesticides became a necessity as the number of pesticides with poor volatility, relatively high polarity, and thermal instability increased. Classes such as N-methyl carbamate (NMC), urea herbicides, benzoylurea insecticides, and benzimidazole fungicides are typically analyzed by HPLC. These compounds are often analyzed by reversed-phase chromatography (Chap. 13, Sect. 13.3.3) with C18 or C8 columns, followed by UV absorption, fluorescence, or MS detection (Chaps. 7 and 11). Utilization of MS detection widened the scope of HPLC analysis of pesticides. LC–MS is one of the most powerful techniques for the analysis of polar, ionic, and thermally labile pesticides (see Sect. 33.3.3.3.2).

33.3.3.3 Mass Spectrometry Detection

The subsequent sections will provide sample applications of GC–MS and LC–MS used in pesticide residue analysis. References [11, 13, 18, 26, 28–31] describe more application examples and details on method development.

33.3.3.3.1 Gas Chromatography-Mass Spectrometry

The most common GC–MS technique for the analysis of pesticide residue involves single quadrupole instruments with electron impact (EI) ionization. The selected ion monitoring (SIM) option enhances selectivity, increases sensitivity, and minimizes interferences from co-extracted compounds. Ion-trap detectors (ITD) are also used in the analysis of pesticide residues. Using ITD, the analysis in full-scan mode provides higher sensitivity as compared to single quadrupole analysis and allows for confirmation by library searches (NIST library spectrum search). Additionally, the ITD enables tandem MS analysis (MS/MS) by means of collision-induced dissociation (CID). The use of tandem MS improves selectivity and significantly reduces background without loss of identification capability, thus enabling the analysis of pesticides at trace levels in the presence of many interfering compounds. Figure 33.4 provides an illustration of enhanced compound identity confirmation upon the use of

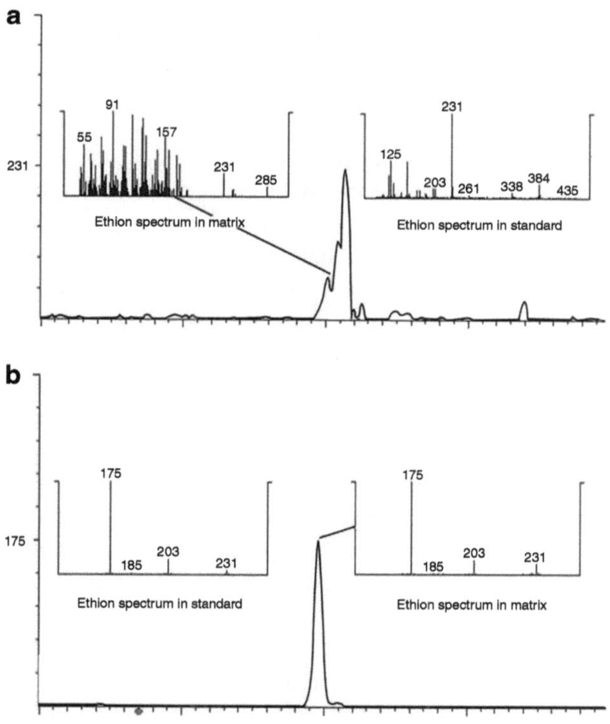

Fig. 33.4 Pesticide residue analysis in vegetable extract using (**a**) electron impact (EI) full scan, showing noisy baseline and no spectral match; vs. (**b**) EI MS/MS showing clean baseline, symmetrical peaks, and excellent spectral match. (Used with permission from Varian Inc., Palo Alto, CA)

MS/MS vs. only MS. For the determination of multiclass residues, triple quadrupole (QqQ) MS is powerful and fast analytical tool, requiring minimum sample preparation. The chromatogram separation becomes less important when the analysis is carried out by GC/QqQ-MS/MS, since the analyzer is able to monitor simultaneously a large number of co-eluting compounds. Reliable quantitation and confirmation can be easily achieved, even at trace concentration levels. Time-of-flight (TOF) MS instruments also are used for the simultaneous analysis of multiclass pesticides [18, 30].

33.3.3.3.2 High-Performance Liquid Chromatography-Mass Spectrometry

LC coupled with tandem MS analysis, using atmospheric-pressure ionization (API), atmospheric-pressure chemical ionization (APCI), and electrospray ionization (ESI), has been developed for many pesticides such as OP, carbamate, and sulfonylurea pesticides [29]. Because of HPLC solvent interferences, single quadrupole instruments are not as widely used in LC-MS as compared to GC-MS for pesticide analysis. When using LC–MS for pesticide residue analysis, QqQ are the most widely used mass analyzers. LC/TOF-MS has also gained popularity due to its high speed, sensitivity, and selectivity [31].

33.4 MYCOTOXIN ANALYSIS

33.4.1 Introduction

Molds, which are filamentous fungi, can develop on food commodities and produce various types of chemical toxins, collectively known as **mycotoxins.** The main producers of mycotoxins are fungal species belonging to the genera *Aspergillus*, *Fusarium,* and *Penicillin.* Crops can be directly infected with fungal growth and subsequent mycotoxin contamination as a result of environmental factors such as temperature, humidity, weather fluctuations, mechanical damage of crops, and pest attack. In addition, plant stress due to extreme soil dryness or lack of a balanced nutrient absorption can induce fungal growth. Fungal infection and mycotoxin contamination can occur at any stage of the food chain (Sect. 33.1). The United Nation's Food and Agriculture Organization estimated that up to 25% of the world's total crops per year are affected with "unacceptable" levels of mycotoxins [32].

Mycotoxin contamination may occur in food as a result of the direct mold infection of plant-origin commodities, such as cereals, dried fruit, spices, grape, coffee, cocoa, and fruit juices (especially apple based). Mycotoxins also can occur in milk, eggs, and, to a minor extent, in meat, as a result of the farm animal consumption of feed from contaminated crops. Additionally, humans and animals can be exposed to mycotoxins through heavily contaminated dust, as in the case of harbors and warehouses.

More than 300 mycotoxins, belonging to various chemical classes, are known. However, the major classes of mycotoxin with a toxicological impact on human health include **aflatoxins** (B1, B2, M1, M2, G1, and G2), **ochratoxins** (e.g., ochratoxin A, OTA), **trichothecenes** [e.g., deoxynivalenol (DON), T2, and HT-2], **fumonisins** (FBs, e.g., FB1, FB2, and FB3), **patulin** (a mycotoxin that occurs mainly in apples and apple products), and **zearalenone** (ZEA). The chemical classification and occurrence of mycotoxins (the major as well as minor ones) in raw commodities and processed foods is described in reference [32]. The toxic effects of the aforementioned mycotoxins include, but are not limited to, genotoxicity, carcinogenicity, mutagenicity, and immunotoxicity. Genotoxic compounds have a probability of inducing an effect at any dose; therefore, no threshold dose should be considered and they should not be present in food. For example, there is no postulated safe dose for aflatoxin B1, which is genotoxic and is the most potent naturally occurring carcinogenic substance known. However, to provide risk managers with necessary basis for making decisions, threshold doses and "safe" **total daily intake** or **total weekly intake** (TDI or TWI) have been set for other mycotoxins (Table 33.2). Based on the threshold doses, tolerance levels

Table 33.2 Total daily intake (TDI)/total weekly intake (TWI) for the major mycotoxins

Mycotoxin[a]	*TDI (ng/kg bw per day)*	*Organization*[b]
OTA	4	Health Canada (1989, 1996)
	5	Nordic Council (1991)
	5	EU (1998)
	14	JECFA (1996, 2001)
	120 (TWI)	EFSA (2006 [81])
FBs	2000	EU (2000)
	2000	JECFA (2000)
	400	Health Canada (2001)
DON	3000	Health Canada (1985)
	1000	Health Canada (2001)
	1000	Nordic Council (1998)
	1000	EU (1999)
	1000	JECFA (2000)
ZEA	100	Health Canada (1987)
	100	Nordic Council (1998)
	500	JECFA (2000)
	200	EU (2000)
Patulin	400	JECFA (1996)
	400	EU (2000)
	400	Health Canada (1996)

From Brera et al. [32], used with permission

[a]*OTA* ochratoxin, *FBs* fumonisins, *DON* deoxynivalenol, *ZEA* zearalenone

[b]*EU* European Union, *JECFA* Joint FAO/WHO Expert Committee on Food Additives

for mycotoxin were set in the USA (e.g., 0–35 μg/kg for aflatoxins, 2.5–50 μg/kg for OTA, 300–2000 μg/kg for DON, and 5–50 μg/kg for patulin).

While molds can be destroyed by natural causes or processing of food, mycotoxins can survive. To limit human exposure to mycotoxins, a key critical control point is avoiding the processing of raw materials with unacceptable mycotoxin levels. Therefore, implementing periodic testing is a necessity, including the adoption of reliable sampling procedures and validated methods of analysis. Organizations such as the AOAC International, American Oil Chemists' Society (AOCS), AACC International (now called Cereals and Grains Association), and the International Union of Pure and Applied Chemistry (IUPAC) have method validation programs for mycotoxin analysis. References [33–38] provide detailed information on mycotoxin occurrence, health effects, control, sampling and sample preparation, and analysis.

33.4.2 Sampling

Compared to the analysis of other types of residues, the sampling step in the analysis of mycotoxins is by far the largest contributor to the total error. A heterogeneous distribution (e.g., 0.1% of the lot is usually highly contaminated) may contribute to an overall detected level above the tolerance limit. In addition, an incorrect sampling protocol may cause false negative, leading to undesirable health, economic, and trade impacts. Therefore, an appropriate sampling plan is needed to ensure that the concentration in a sample is the same as that in the whole lot.

The Commission Regulation 401/2006 [38] provided specifications for the sampling of regulated mycotoxins in various food commodities including cereals, dried fruits, nuts, spices, milk and derived products, fruit juice, and solid apple products. Also, AOAC International, AOCS, and the FDA in collaboration with the USDA have developed detailed sampling plans for separate commodities.

33.4.3 Detection and Determination

Sample preparation steps (discussed in Sect. 33.2.2) that are specific for mycotoxin analysis have been developed. Shaking or blending often is used for mycotoxin extraction, while SPE often is used for cleanup, especially when multi-mycotoxin analysis is needed. For a specific mycotoxin, immunoaffinity LC (IAC) is used for cleanup.

Tremendous effort has been made to develop and optimize qualitative and quantitative methods for mycotoxin analysis. A list of official methods of mycotoxin analysis is included in reference [32] and others in references [33–35, 37]. Mycotoxin analysis kits, both for cleanup (for GC, HPLC, or TLC analysis) and for detection (e.g., ELISA, LFS, IAC), are commercially available. The sections below briefly discuss some of the current analytical techniques used for the detection and determination of mycotoxins in food.

33.4.3.1 Rapid Methods of Detection

33.4.3.1.1 TLC

A large number of TLC methods, with detection limits reaching 2 ng/g, for the analysis of mycotoxins have been accepted by AOAC International, including DON in barley and wheat, aflatoxin in peanuts and corn, aflatoxin M1 in milk and cheeses, OTA in barley and green coffee, and ZEA in corn. When results are positive, confirmatory and more sensitive quantitative analysis follows. The overall performance of TLC for the analysis of mycotoxin is improved when used in combination with IAC. Additionally, a microcomputer interfaced with a fluorodensitometer and semiconductor-based detection improves the data handling.

33.4.3.1.2 Immunoassays

The most common types of immunoassays used for the analysis of mycotoxins are ELISA (see Chap. 27) and **fluorescence polarization immunoassay** (FPIA). Since all mycotoxins are relatively small molecules (MW < 1000), competitive ELISA is used (see Chap. 27, Sect. 27.3.2). In FPIA, fluorescein-labeled mycotoxin competes with unlabeled mycotoxin analyte in a sample for binding with the antibodies. FPIA has gained popularity since it involves no coating of the plate and less analysis time as compared to ELISA. Results of FPIA were comparable to that of ELISA for the analysis of DON, ZEA, and OTA [32].

Immuno-based testing methods, including membrane-based immunoassays, LFS assays, and biosensors, also have been developed for online control or field testing of mycotoxins. Membrane-based immunoassays are based on the principle of direct competitive ELISA, with the antimycotoxin antibody being coated on the membrane surface. For example, test kits based on membrane-based immunoassays have been validated for OTA determination in wheat, rye, maize, and barley [38]. LFS (see Chap. 27, Sect. 27.4) is capable of detecting many mycotoxins simultaneously. Biosensors, on the other hand, are compact analytical devices that use biological components, such as nucleic acids, enzymes, antibodies, or cells, associated with a transduction system. The transduction system processes the signal produced by the interaction between the target molecule and the biological component. The use of biosensors for the detection of several mycotoxins, such as aflatoxins DON and OTA, is rapidly increasing [32].

33.4.3.2 Quantitative and Confirmative Chemical Methods

33.4.3.2.1 HPLC

For quantitative determination, HPLC is the methodology of choice for most mycotoxins, specifically, aflatoxins, DON, OTA, ZEA, FBs, T-2 and HT-2 toxins, and patulin. Aflatoxins, OTA, and ZEA exhibit a native fluorescence and are detected directly by fluorescence detector. Pre- or postcolumn derivatization (with trifluoroacetic acid, iodine, or bromine) is required to enhance the fluorescence detection of some aflatoxins (i.e., AFB1 and AFG1), whereas precolumn derivatization with the OPA reagent is required for fluorescence detection of FBs. However, direct UV detection is used for DON and patulin. Reversed-phase chromatographic separation is normally employed for multimycotoxin analysis. A multifunctional column, which consists of a mixture of reversed-phase, size exclusion, and ion-exchange stationary phases, also can be used (e.g., AOAC Method 49.2.19A for aflatoxin analysis). Methods using IAC-HPLC are reported for the determination of aflatoxins in different foods [39]. Validated HPLC methods for the analysis of various mycotoxins are periodically reported in Methods Committee Reports published in *Journal of AOAC International.*

As is the case with pesticide residue analysis, coupling of HPLC with MS analysis, especially LC–MS/MS, provides greater sensitivity and selectivity and allows for simultaneous analysis of multiclass mycotoxins. Additionally, the use of LC–MS/MS allows for the detection of conjugated mycotoxins (i.e., the toxin bound to a polar compound such as glucose or another sugar), which are referred to as "masked" or "modified" mycotoxins because they escape routine detection. A method has been developed for the determination of 39 mycotoxins, including conjugated DON, FBs, ZEA, and aflatoxins, in wheat and maize, using LC-ESI-triple quadrupole MS, without cleanup [40].

33.4.3.2.2 GC

GC is not widely used for the detection of mycotoxins, except in the case of trichothecenes. Trichothecenes do not strongly absorb in the UV-Vis range and are nonfluorescent; therefore, GC methods were developed for their determination. Capillary column GC is commonly employed for the simultaneous detection of different trichothecenes, e.g., DON, T2 toxin, and HT-2 toxin, using trifluoroacetyl, heptafluorobutyry, or trimethylsilyl derivatization coupled with electron capture detection. GC is often linked to MS for peak confirmation. GC–MS also can be used for the confirmation of patulin in apple juice [32]. Validated and accepted methods for the determination of trichothecenes using GC are reported as AOAC Official Methods and by the American Society of Brewing Chemists.

33.4.3.2.3 Capillary Electrophoresis

Capillary electrophoresis (CE, see Chap. 24, Sect. 24.2.6.3), generally used as a chromatographic technique, can be employed to separate mycotoxins from matrix components using electrical potential. Methods are available for the determination of patulin in apple juice [41], simultaneous determination of ochratoxins (A and B) and aflatoxins [42], and determination of ZEA using cyclodextrins (for enhancing the native fluorescence) [43].

33.4.3.3 Other Methods of Analysis

Other methods for the detection of mycotoxins have been reported, such as **near-infrared** (NIR) and **mid-infrared** (mid-IR) spectroscopy, especially **Fourier transform infrared** (FTIR) spectroscopy (a type of MIR) (see Chap. 8, Sect. 8.3.1.2). FTIR spectroscopy was used for the detection of mycotoxins in corn and peanuts spectra [44]. Rapid and accurate identification of mycotoxigenic fungi and their mycotoxins (e.g., FB1 and DON) was achieved using MIR and NIR spectroscopy [45, 46]. Calibration for these methods is based on reference HPLC or GC methods.

33.5 ANTIBIOTIC RESIDUE ANALYSIS

33.5.1 Introduction

Animals intended for human consumption may not only be given drugs (e.g., antibiotics, antifungals, tranquilizers, and anti-inflammatory drugs) at therapeutic levels to combat diseases but also some countries allow use of drugs (mainly antibiotics) at subtherapeutic levels to reduce the incidence of infectious diseases and for weight gain advantages. The **Center for Veterinary Medicine** (CVM), a branch of the FDA, regulates the manufacture, distribution, administration, and withdrawal period (time between the last drug treatment to the animal and the slaughter or use of milk or eggs by humans) of drugs given to animals, including animals from which human foods are derived. Most drug residues of concern for human food are antibiotic residues and thus are the focus of this section. Some of the major families/types of antibiotics, referred to in later sections, include β-lactams, sulfa drugs, cephalosporins, tetracyclines, and chloramphenicol.

Residual levels of any antibiotics given to animals used for human consumption are of concern for a variety of reasons, including the fact that some consumers are allergic to certain antibiotics, and the possibility that some microbes may develop antibiotic resistance due to constant exposure. Some antibiotics can be carcinogenic as well, such as nitrofuran compounds [47]. Also, on a practical level for dairy products such as cheese made with starter cultures, antibiotic residues would reduce the intended microbial growth and therefore reduce acid production. For all these reasons and more, the FDA has strict regulations on antibiotic residues in human food. Due to these regulations, antibiotic-contaminated meat and milk (including milk products) are considered to be adulterated. The FSIS monitors antibiotic residues in meat products from cattle, swine, lamb, goat, and poultry [48].

33.5.2 Detection and Determination

Often, procedures such as defatting, protein hydrolysis (in case of meat or egg samples), protein precipitation (in case of dairy samples), and aqueous wash (in case of honey to remove excess sugar) precede the extraction of antibiotic residues. For the extraction of many antibiotics, liquid–liquid extraction and SPE are commonly used, followed by a partial purification step often using ion-exchange cleanup systems that take advantage of the acid/base character of the antibiotics. Sample preparation steps, including extraction and isolation (as discussed in Sect. 33.2.2) specific for drug residues, are reviewed in reference [49].

Tolerance levels are established for some antibiotics, while others have a zero tolerance (as is the case for nitrofurans and chloramphenicol). Chloramphenicol, for instance, is an antibiotic of considerable current concern in the USA, European Union, and other countries. It is used in some parts of the world in producing shrimp and has been found in imported seafood products (e.g., shrimp, crayfish, and crab). Because of the adverse health effects on humans, the FDA has banned the use of chloramphenicol in animals raised for food production and set a zero tolerance in human food (21 CFR 522.390 [9]). Therefore, the analytical methods need to be as sensitive and selective as possible.

A wide variety of analytical methods have been developed and optimized for the analysis of antibiotics, categorized as screening or determinative and confirmatory. Reference [9] lists the methods commonly used in the analysis of several antibiotics, and reference [50] gives the process used by the AOAC to validate test kits. One excellent resource for antibiotic test methods, specific for milk and dairy products, is reference [51]. The sections below briefly discuss some of the analytical techniques used for the detection and determination of antibiotics in susceptible food commodities.

33.5.2.1 Screening Methods

Some of the major categories for rapid screening assays, some of which are quantitative, are: (1) microbial growth inhibition, (2) receptor assays, (3) enzyme-substrate assays, and (4) immunoassays. Some screening methods are specific to individual antibiotics, some are specific to a class of antibiotics, and some have no specificity. Screening assays for antibiotic residues in test samples initially relied mostly on inhibition of microbial growth, but now many use other principles for detection.

In **microbial growth inhibition** assays, turbidity, zone of inhibition, or acid production can be measured. In a **turbidity assay**, an indicator organism growing in a clear liquid culture will cause an increased turbidity; growth is inhibited, so turbidity is reduced if antibiotics are present. In a **zone of inhibition assay**, the test material diffuses through an agar-based nutrient medium that has been uniformly inoculated with spores of a susceptible organism. Any antibiotics present in the test material will inhibit the germination and growth of the organism, creating clear zones. In the **acid production assays**, the acid produced when microbes grow causes a color change in the medium. No color change means that the test sample contained an inhibitory substance. While microbial growth inhibition assays are more time consuming than many newer screening tests, they are inexpensive and applicable to testing large numbers of sample and provide some sensitivity to multiple antibiotic categories [51]. AOAC Official Methods include a nonspecific microbiological method for antibiotics and numerous microbiological methods for specific antibiotics [25].

An example of a **receptor assay** is the Charm II® test (Charm Science, Lawrence, MA), which has different versions, designed to detect different groups of antibiotics. The assay involves a competition between labeled antibiotics (labeled using ^{14}C or ^{3}H, depending on the specific type of Charm II® test system) and antibiotic residues in a milk sample for a limited number of specific binding sites on the surface of bacteria added to the test sample. The greater the concentration of antibiotic residue in the sample, the less radiolabeled tracer will become bound to the microorganism. The method can be applied to milk, certain dairy products, honey, and meat.

Enzyme substrate assays measure the inhibition of an enzyme acting on a substrate, caused by the presence of an antibiotic. An example used for testing raw milk is the Penzyme® III commercial kit (Neogen, Lansing, MI), which is specific for β-lactam antibiotics, and inhibits D,D-carboxypeptidase on an equimolar basis. When this enzyme acts on a specific substrate, it causes release of D-alanine, which can be measured in additional steps of the assay resulting in a color change [52].

Immunoassays, namely, ELISA and LFS (see Chap. 27, Sects. 27.3.3 and 27.3.4, respectively), also can be used for screening of antibiotic residues. The Charm ROSA (Rapid One Step Assay) MRL assay, intended for milk and cream testing, is an example that uses a lateral flow strip. Another example of an immunoassay marketed for testing milk is the

SNAP® kit (IDEXX Laboratories, Inc., Westbrook, ME). The assay is based on a competition between residual antibiotics in a milk sample and enzyme-labeled antibiotics in the test kit. The enzyme acts on a substrate to cause a color change; any antibiotics in the milk will result in a decrease in color development. Competitive ELISA can be used for the detection of a specific antibody, for example, the Veratox® assay (Neogen Corporation, Lansing, MI) used to detect chloramphenicol.

33.5.2.2 Determinative and Confirmatory Methods

Quantitative determination of antibiotic residues in food products follows the same general steps as for other trace analytes. After sample preparation steps, the partially purified extract is subjected to chromatographic separation, detection, and quantification. The most commonly used chromatography system is HPLC (mostly using reversed-phase separation mode) coupled with UV, fluorescence, chemiluminescence, or postcolumn reaction detectors. For confirmatory and identification purposes at trace levels, LC–MS and LC–MS/MS are increasingly used in the analysis of antibiotics [10, 20]. The FDA provides regulatory LC–MS/MS methods to detect fluoroquinolones in honey [52]. For fluoroquinolones in salmon, catfish, and shrimp, a LC method with fluorescence detection has been used [53], with the confirmatory LC/MS method provided by FDA.

Additionally, LC–MS/MS has been used for confirmation of β-lactam residues in milk [54]. The detection of 14 different types of sulfonamides, in milk [55] and in condensed milk and soft cheeses [56], at levels below 10 ng/mL, also was achieved using LC-MS/MS. LC-MS/MS has been compared with ultra-high-performance liquid chromatography/quadrupole time-of-flight MS (UHPLC/Q-Tof MS) for analysis of macrolide antibiotic residues in a variety of foods [57]. LC–MS/MS gave a lower limit of detection and better precision, but the UPLC/Q-Tof MS provided better confirmation of positive findings (See Chap. 13, Sect. 13.2.3.3 for UHPLC.)

33.6 ANALYSIS OF GMOS

33.6.1 Introduction

Agriculturally important plants may be genetically modified by the insertion of DNA from a different organism (**transgene**) into the plant's genome, conferring novel traits that it would not have otherwise, such as herbicide tolerance or insect resistance. The modified plant is termed a **genetically modified organism**, or **GMO**. GMO production relates to a number of food crops, including the following: corn, soybeans, cotton, canola (a cultivar of rapeseed), rice, sugar beets, Granny Smith apples, and papaya. Although protection from pests and herbicide tolerance are the most common GMO traits, other GMO crops include plants that have been modified to improve postharvest quality (e.g., reduce browning rate in cut apples, by reducing the content of polyphenol oxidase) or enhance the nutritional makeup of the food (e.g., golden rice with enhanced vitamin A precursor).

Despite the prevalence of GMO crops, there are concerns about limiting crop variation (i.e., monocropping); unintended impacts on other plants, insects, wildlife, and nearby communities; and possible allergies. The debate about the use of GMO crops has resulted in government regulation for application and labeling. There are specific guidelines put forth by various governments regarding the analyses used. Therefore, even producers must be able to accurately test ingredients and other products.

Many companies produce high-quality, easy-to-use test kits, many of which are specific for a certain GMO protein or gene. These kits generally fall into two categories, based on the methodology: **polymerase chain reaction** (PCR) kits that specifically amplify the DNA of the GMO gene (specific or shared by many GMOs) so that it can be detected and **immunoassays** (ELISA and LFS) that are specific for the proteins. Further readings on the GMO and GMO detection topics are listed in the references [58–60].

33.6.2 DNA Methods

Detection of the transgene DNA is an effective method of testing for GMO material in a sample. The analysis involves three distinct steps: (1) **extraction** of the DNA from the sample, (2) **amplification** of the DNA by PCR, and (3) **identification and quantification** of the amplified DNA (note: for **real-time quantitative PCR analyses**, amplification, and detection/quantification occur at the same time). While all three steps are important, the PCR amplification is critical to the specificity and success of the analysis.

33.6.2.1 DNA Extraction

Application of extreme levels of sheer and heat can damage the DNA, rendering the subsequent PCR and detection ineffective. Extraction of the DNA from the food matrix, therefore, should be performed prior to significant processing of the material, preferably on raw ingredients. DNA extraction protocols differ somewhat, but all include disrupting the matrix to release the DNA, usually by grinding the sample to a fine powder. This is followed by dispersal of the ground material into an extraction solution and removal of unwanted components. For example, lipids may be removed by solvent extraction and protein by the addition of a protease. A final step may involve the precipitation of the DNA with cold alcohol, such as ethanol or isopropanol.

33.6.2.2 PCR Amplification

PCR is a cyclic method that exponentially increases the copy number of a specific DNA sequence, by means of enzymatic replication. It uses thermal cycling to alternately replicate the target sequence and then melt the DNA into single strands in order to repeat the process. The method is based on the use of two synthetic DNA fragments that are complementary to opposite ends of the target sequence. These fragments are termed the **primers**, which are commonly 18 to 35 bases in length, and they can be produced only when the target sequence is known. If a general identification of any GMO material is desired, then the primers used would be complementary to the promoter sequence, which is common to all transgenic crop plants that are normally grown for industrial food production. If a specific GMO product is to be identified, then the primers would consist of a sequence that includes the transgene DNA and plant DNA. This is necessary to avoid the detection of DNA that originates from bacteria that may have been on or in the plants.

In addition to the specific primers, the PCR mixture also includes a heat-stable **DNA polymerase**, such as Taq polymerase (from *Thermus aquaticus*); the **nucleotide bases** that comprise DNA in the form of **deoxynucleoside triphosphates** (dNTPs); and a buffer solution to maintain the optimal conditions for the reactions. All of these components are present in the commercial kits. The reaction vials are then placed in a **thermal cycler**. Once the PCR system is started, the mixture is first put through a high-temperature cycle to melt the DNA (separate it into single strands), and then a lower temperature cycle to allow the primers to anneal to the single-stranded target DNA, and finally an intermediate-temperature cycle to allow the DNA polymerase to synthesize a new DNA strand complementary to the target strand by adding dNTPs, starting at the primers. The process is then repeated, usually for 30–50 cycles, which is sufficient to produce millions of copies of the DNA.

33.6.2.3 DNA Analysis

When sufficient DNA has been generated by PCR, the sample can be analyzed by agarose gel electrophoresis. The sample and standards migrate through the gel, and after the run, the gel is stained and the presence and abundance of DNA can be identified, by comparison to the position and degree of staining of the standards.

Some of the commercial test kits contain reagents to specifically label the double-stranded DNA with tags, such as fluorescein. These kits are meant to be used with special PCR equipment that deposits the finished, labeled mixture into a capillary where the intensity of the fluorescence, which is proportional to the abundance of DNA, is determined by high sensitivity fluorescence spectroscopy, rendering the electrophoresis step unnecessary. In this case, the specificity is contingent on the sequence of the primers, which directly determine the nature of the double-stranded DNA present. These kits also include standards and other reagents necessary for accurate quantification of the target DNA abundance in the sample. If real-time PCR is used, the fluorescence is read in the reaction tube after each cycle of amplification, yielding a curve containing multiple data points rather than a single data point for each sample.

33.6.3 Protein Methods

The protein methods for GMOs are immunoassays, primarily ELISA (Chap. 27, Sect. 27.3.2) and LFS (Chap. 27, Sect. 27.3.4), generally used for testing raw agricultural products and not processed products. Both qualitative and quantitative ELISA test kits for GMOs are commercially available, with the latter having typical detection limits of 0.01–0.1%. The LFSs are generally less sensitive (0.1–1.0%), but their speed and ease of use make them ideal for testing in the growing field, at storage areas, and at points of transport.

33.7 ALLERGEN ANALYSIS

33.7.1 Introduction

Food allergens are food proteins that trigger an allergic response. Symptoms of an allergic response include hives, face and tongue swelling, and difficulty breathing. The reaction can include the severe and life-threatening allergic reaction called anaphylactic shock. It is important to note that food allergy, which triggers an immune system reaction, is distinct from other adverse responses to food, such as food intolerance (e.g., lactose intolerance), pharmacologic reactions (due mainly to food additives such as sulfites and benzoate), and toxin-mediated reactions (due to residues such as pesticides and mycotoxins).

A significant percentage of the population has food allergies, and the prevalence is rising. Over 90% of the food allergic reactions in the USA are caused by the nine most common allergenic foods, referred to as **"the big nine"**: milk, eggs, fish, crustacean shellfish, tree nuts, peanuts, wheat, soybeans, and sesame [61]. In the USA, the Food Allergen Labeling and Consumer Protection Act of 2004 targets these food allergens [62]. It applies to all foods regulated by the FDA (both domestic and imported), requiring that labels list all ingredients by their common names and identify the source of all ingredients derived from the eight most common food allergens. Food allergies are an issue around the world; numerous other countries have or are considering labeling regulations. Since there is no cure for food allergies, strict avoidance of food allergens is the only effective action. All of these facts point to the importance of both screening and quantitative methods for analysis of allergenic foods.

Methods available for the detection of food allergens are mainly based on protein or DNA detection, as discussed in subsequent sections. A commercially available rapid screening method that is not based on protein or DNA detection is the swab and adenosine triphosphate (ATP)-sensitive detection method. This method is based on the detection of ATP present on the surface of multiple allergenic foods, e.g., peanut butter, whole egg, soybeans, and milk. It is used mainly to prevent food allergen cross contact during cleaning of processing equipment. Test kits for food allergen analysis (protein or DNA based and others) are commercially available. A review of food allergen analytical methods is given in reference [63].

33.7.2 Protein Methods

33.7.2.1 General Considerations

Similar to the analysis of many other food constituents of concern present in small amounts, sampling adequacy and the detection limits are of concern with analysis for allergens. Another concern is the adequate extraction of the different allergens. Unlike the trace analytes of concern mentioned thus far, since the analytes are proteins, the extraction solution is normally a buffer at various pHs and salt concentration. Extraction buffers differ in their ability to extract food allergens, with some solutions extracting the same allergens in different concentrations, while others cannot extract all the allergens. For example, phosphate buffer fails to extract the major peanut allergen (Ara h 3); however, the efficiency of extracting this allergen is greatly enhanced upon the addition of salt (Fig. 33.5). Additionally, it is crucial that the extraction solution is compatible with the assay used (e.g., immunoassay) and does not alter the chemical structure of the analyte. The choice of extraction procedure should also take into consideration the food processing conditions employed. Upon processing, the solubility of the proteins can be reduced due to denaturation and aggregation, resulting in reduced protein recovery [64]. Therefore, to obtain reliable and accurate results, it is crucial to select the right extraction solution for the target analyte(s).

33.7.2.2 Protein-Based Analytical Techniques

Classical protein-based methods used for the analysis of food allergens usually involve antibody-based assays (immunoassays) due to their specificity and sensitivity. There are commercially available test kits to detect and quantitate allergens. Immunoassay-based methods used for the analysis of food allergens include ELISA (Chap. 27, Sect. 27.3.2), LFS (Chap. 27, Sect. 27.3.4), Western blot (Chap. 27, Sect. 27.3.3.1), biosensor immunoassays (antibodies immobilized on a biosensor chip), and dot immunoblotting (Chap. 27, Sect. 27.3.3.2) [65].

Western blot and dot immunoblotting are mostly used for qualitative and screening purposes. The most commonly used immunoassays for the quantitative analysis of food allergens are ELISA methods, using a competitive or, more commonly, sandwich format (see Chap. 27, Sect. 27.3.2.3). The competitive ELISA is used for the small protein allergens, with a molecular weight less than 5 KDa. Numerous sandwich and competitive ELISA methods have been developed for several food allergens [63]. Lateral flow test strips

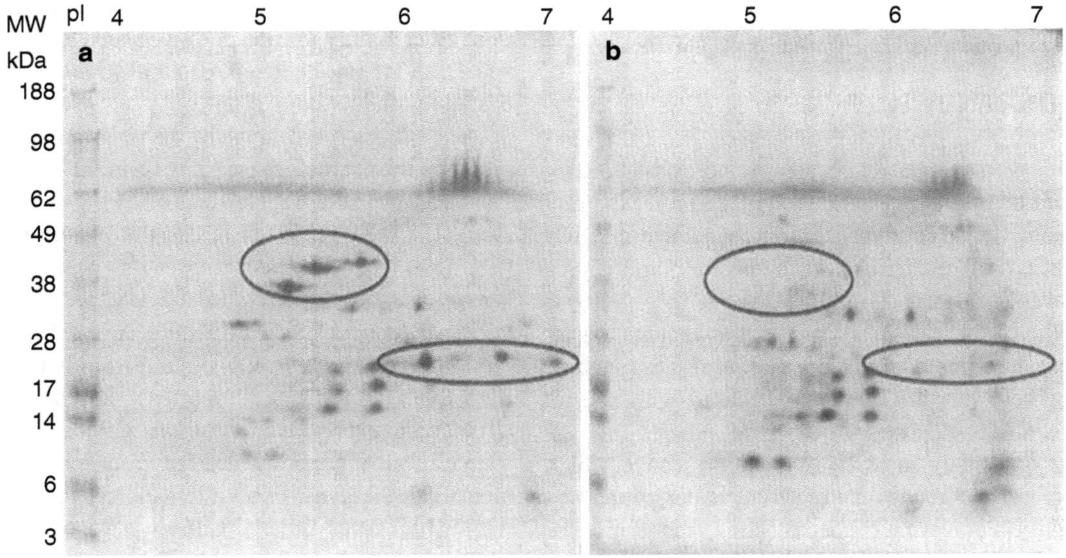

Fig. 33.5 Evaluation, by two-dimensional electrophoresis, of the ability of (**a**) high salt buffer and (**b**) phosphate buffer to extract the peanut allergen Ara h 3 from raw peanuts. (From Westphal [64], with permission)

are useful for screening purposes because they are very rapid and inexpensive and do not require instrumentation.

There are several drawbacks when using immunoassays, including cross-reactivity of antibodies with other proteins, food matrix interferences, structural changes due to food processing, destruction of epitopes upon adsorption onto solid matrices, and variability between manufactures due to use of different extraction buffers and antibodies. Often times, 1-D or 2-D electrophoresis is used to isolate target protein to avoid cross-reactivity. In spite of the drawbacks, immunoassays are most commonly used for the detection of allergens. However, advanced proteomic techniques are evolving including mass spectrometry-based proteomic methods, most notably LC coupled with tandem MS, capable of detecting and quantifying multiple allergens simultaneously [63].

33.7.3 DNA Methods

There are several advantages and disadvantages for the use of DNA-based methods in the analysis of food allergens. DNA-based methods do not target the allergen in the sample; therefore, the detection of the allergen-encoding DNA does not always correlate with the presence of the allergen, especially when the food has been fortified with purified protein. Upon processing, for example, production of protein isolates, protein, and DNA could be separated resulting in false conclusions regarding the presence of the allergen in the sample. Regardless of these disadvantages, DNA-based methods are very specific and sensitive techniques with the advantage that targeted DNA is less affected by several processing and extraction conditions as compared to proteins. The choice to use DNA-based methods is dependent on the type of sample being analyzed.

DNA-based methods involve the extraction of the DNA (see Sect. 33.6.2.1) followed by amplification by PCR using a thermostable polymerase (see Sect. 33.6.2.2). The amplified sample is then visualized by fluorescence staining or by Southern blotting following agarose gel electrophoresis. This procedure normally provides qualitative data, or semi-quantitative data if internal standards were used. Quantification can be achieved if real-time PCR (see Sect. 33.6.2.3 and reference [63]) or PCR-ELISA was used. PCR-ELISA method involves linking the amplified DNA fragment of an allergenic food to a specific protein-labeled DNA probe, which then is coupled with a specific enzyme-labeled antibody. Quantification of the DNA is based on the enzyme-substrate color producing reaction. Reference [57] includes a list of commercially available real-time PCR and PCR-ELISA test kits; for the analysis of allergens, readers are referred to reference [63].

33.8 ANALYSIS OF OTHER CHEMICAL CONTAMINANTS AND UNDESIRABLE CONSTITUENTS

33.8.1 Introduction

Pesticide residues, mycotoxins, antibiotic residues, GMOs, and allergens in foods have been of concern for many years. However, in any given time period, there are numerous additional chemical hazards that must be addressed. Many of these substances fall into the following categories: (1) banned (e.g., coumarin) or allowed in some countries but not others, (2) legally limited (e.g., sulfites, benzene), (3) intentional contaminants (e.g., melamine), (4) approved for use but of concern (e.g., monosodium glutamate), or (5) natural constituents of concern (e.g., acrylamide, furan). This section would be very lengthy to cover in detail the screening and quantitative methods for many of these chemical hazards. Instead, only a summary (Table 33.3) is given of select current and emerging hazards, with their major method(s) of analysis [66, 67]. To reach the limit of detection (LOD) desired and to positively identify the compounds, many of the major analyses rely on gas or liquid chromatography, with mass spectrometry. Described in more detail below is the analysis for two compounds of concern, sulfites and nitrates/nitrites, which are not as commonly analyzed by chromatography as by other methods. There are legal limits in foods for both of these compounds, so they are monitored regularly in select foods for quality control purposes.

33.8.2 Sulfites

Although sulfites are classified as allergens, they are covered in this separate section because the nature of the chemical compound, symptoms, and methods of analysis are quite different than for other allergens. Sulfites and sulfiting agents are a group of chemical compounds that include sulfur dioxide (SO_2), sulfurous acid (HSO_3), and the following inorganic sulfite salts that can liberate SO_2: sodium (Na) and potassium (K) sulfite, Na and K bisulfite, and Na and K metabisulfite [68, 69]. In some foods they occur naturally, but in other foods, they are added for a variety of reasons, including preventing microbial growth and browning. Sulfites naturally occur to some extent in all wines but are commonly added to stop fermentation at the appropriate time and to prevent spoilage and oxidation. Dried fruits and vegetable products are sometimes treated with sulfites to reduce browning. Shrimp, lobster, and related crustaceans can be treated with sulfite to prevent "blackspot." Some consumers are highly intolerant to sulfite residues in food, most commonly resulting in asthma attacks [69]. Therefore, the

Table 33.3 Select compounds of concern and their methods of analysis

Compound	*Nature of compound and reason in food*	*Reason of concern*	*Major foods/ ingredients identified*	*Major methods*
Acrylamide	Formed in carbohydrate-rich foods cooked at high temperature	Neurotoxin and carcinogen, and other health risks	Fried and oven-based foods high in carbohydrates	LC-MS/MS
Arsenic, cadmium, lead, and mercury	Occur in environment from industrial pollution and accumulate in plants through soil and water uptake	Toxic metals that can affect brain development in children	Drinking water, agricultural foods, and fish	ICP-OES, ICP-MS
Benzene	Can form at low levels in some beverages containing both ascorbic acid and benzoate salts	Carcinogen	Soft drinks	Headspace GC-MS
Bisphenol A (BPA)	Used to make polycarbonate plastic, but can leach from plastic	Can mimic body's own hormones. Associated with multiple health problems. Banned or amounts are limited by various countries	Foods/beverages in polycarbonate plastic	GC-MS, GC-MS/MS, LC-MS, LC-MS/MS
Cyanide	Naturally found in certain seeds, fruit stones, and cassava roots	Many cyanides are highly toxic to humans	Apple seeds, cherry pits, cassava	UV colorimetry (after reflux distillation)
Furans	Volatile liquid that seems to form during traditional heat treatments	Carcinogen	Variety of foods	Headspace GC-MS
Heterocyclic amines	Formed in meat products heated at high temperatures, with creatine being a necessary precursor with free radicals formed as a result of the Maillard reaction initiating their production	Neurotoxin, mutagen, and carcinogen	Grilled/fried meat and fish to well-done stage	Reversed-phase LC coupled with fluorescence detection, LC-MS
Lectins	Toxins naturally occurring in kidney beans. Boiling destroys toxins and makes beans safe to eat	Consumption of a single raw beans can cause severe gastrointestinal distress	Raw kidney beans (not a concern in cooked beans)	LC-MS/MS
Melamine and cyanuric acid	A trimer of cyanamide (with 66% N) that has been added illegally to increase apparent protein content of foods and ingredients	Causes kidney failure; Legal limits have been set	Added previously to wheat gluten, milk, and infant formula	HPLC-MS/MS (GC-MS for screening)
4-Methylbenzophenone (4MBP)	Metabolite of a chemical component of ink used for food packaging	Concern for health risk with long exposure	Boxes of cereal	LC-MS/MS, GC-MS
4-Methylimidazole	Formed via Maillard browning during normal cooking of certain foods. Formed in manufacturing of caramel coloring	Some concern about whether it is a possible carcinogen to humans	Roasted meats and coffee, caramel coloring, beverages with caramel coloring	LC-MS/MS (ion chromatography), GC
3-Monochloropropane 1,2-diol (3-MCPD) and glycidyl esters (GE)	Formed when proteins (soy) ae hydrolyzed by heat and food-grade acids to create hydrolyzed vegetable protein	Carcinogen	Soups, savory snacks, and gravy mixes flavored with acid-hydrolyzed vegetable protein	GC-MS, LC-MS, SCF-MS/MS
Nitrosamines	Produced from nitrites in the presence of proteins upon processing at high temperatures, as in frying	Carcinogen	Cured meats, primarily cooked bacon, beer, some cheeses, nonfat dry milk, fish	GC-MS for volatile nitrosamines, LC-MS for nonvolatile nitrosamines
Perchlorate	Component of rocket fuel and formed naturally	Can interfere with iodide uptake into the thyroid gland, leading to hypothyroidism	Bottled water, milk, lettuce	LC-MS/MS (ion chromatograph)

Table 33.3 (continued)

Compound	*Nature of compound and reason in food*	*Reason of concern*	*Major foods/ ingredients identified*	*Major methods*
Per- and polyfluoralkyl Substances (PFAS)	Used in packaging and cookware for grease and oil resistance or non-stick	Compounds are persistent in the environment, can bioaccumulate in tissue and associated with serious health effects	Drinking water, paperboard packaging, and Teflon coated cookware	LC-MS/MS
Pyrrolizidine Alkaloids	Alkaloids found in flowering weeds introduced into food through poor agricultural practices	Toxins cause liver failure	Teas, Herbs, Spices and Honey	LC-MS/MS
Solanine	A glycoalkaloid found in nightshade plants including potatoes that helps protect the plant, but causes bitter taste when at certain level in potato tuber. Peeling potatoes decreases content	Toxic to humans above certain levels	Potatoes	LC-MS/MS, HPTLC, GC

GC gas chromatography-mass spectrometry, *GC-MS/MS* gas chromatrography-tandem mass spectrometery (MS/MS), *HPLC* high performance liquid chromatography, *ICP* inductively couple plasma, *ICP-MS* inductively coupled plasma-mass spectrometry, *ICP-OES* inductively couple plasma-optical emission spectroscopy, *LC* liquid chromatography, *LC-MS* liquid chromatography-mass spectrometry, *LC-MS/MS* liquid chromatography- tandem mass spectrometry, *UV* Ultraviolet spectroscopy

FDA forbids sulfites from being added to foods intended for raw consumption (e.g., salad bar foods) and requires the phrase "contains sulfites" on the label of foods that contain greater than 10 ppm sulfites, whether naturally occurring or added during manufacturing [21 CFR 101.100 (a)(4)] [70]. Many countries besides the USA have set strict limits on the residual levels of sulfites in various foods.

Reactions of sulfites with other food components make analysis challenging, often resulting in decreased levels during storage. Most methods of analysis detect free forms of sulfite plus some bound forms. However, until recently, none of the available methods, alone, measured all forms of sulfite in foods very quantitatively, which includes free inorganic sulfite plus the many sulfite bound forms. It is not known which forms of sulfite cause the adverse responses in sulfite-sensitive consumers, so the focus has been on measuring as much as possible of the residual sulfite, both free and bound forms [71].

The FDA recently published a method to determine sulfites in food using liquid chromatography-tandem mass spectrometry (LC–MS/MS) [71]. This method measures free sulfite plus a reproducible portion of bound sulfites in food. The method has been validated for a wide variety of fruit and vegetable products, juices, and shrimp. Foods of high and low moisture content require some modifications in the procedure. Time will tell if this new sophisticated method for sulfite analysis replaces other long-time methods described below.

One of the long-time, quantitative methods for sulfite analysis of foods is the **Monier-Williams procedure** (AOAC Method 990.28) [25], which measures "total" SO_2 (actually, free sulfite plus reproducible portion of bound sulfites, such as carbonyl addition products). The FDA refers to this method in regulations for labeling sulfite-containing foods. In this method, the test sample is heated with HCl, converting sulfite to SO_2. Nitrogen gas bubbled through the sample sweeps SO_2 through a condenser and a hydrogen peroxide solution, oxidizing SO_2 to H_2SO_4. The sulfite content of the sample is directly related to the amount of H_2SO_4 generated, measured by either a gravimetric or turbidimetric procedure.

Other analytical methods for sulfites include the following:

1. The **"Ripper" method**, long used by the wine industry as a rapid screening method, compared to the more time-consuming Monier–Williams method; sulfite is titrated with an iodide-iodate solution, using a starch endpoint indicator; measures "free" SO_2 (APHA Standard Method 4500 – SO_3^{-2} B) [72].
2. Enzymatic method; sulfite is oxidized to sulfate, generating hydrogen peroxide that is further reacted with NADH-peroxidase, to produce NAD which is measured by absorption at 340 nm (Sect. 26.3.1.3.1).
3. Ion chromatography, using amperometric detector (AOAC Method 990.31) [25].
4. HPLC, with ultraviolet [73] or fluorometric [74] detection.

33.8.3 Nitrates/Nitrites

Sodium nitrate and sodium nitrite are both food preservatives commonly used to cure processed meat products, but there are legal limits on the level of sodium nitrite in the finished product. Nitrates (NO_3) and nitrites (NO_2) are very similar chemically, and sodium nitrate is readily converted to sodium nitrite. Sodium nitrate is naturally occurring in some vegetables, and it converts to sodium nitrite when it comes in contact with your saliva. The naturally occurring high levels of nitrates in celery and Swiss chard make powders of them common ingredients in some cured meat products. Both sodium nitrate and sodium nitrite give cured meat an appealing pinkish-reddish color and prevents it from turning brown. More importantly, sodium nitrite works with sodium chloride to prevent the growth of *Clostridium botulinum*. However, when products with high levels of residual nitrites are cooked at high temperatures (e.g., frying of bacon), the nitrites react with amines naturally present in the meat to form nitrosamines, which are suspected carcinogens. Also, high levels of nitrates or nitrites in the diet may induce the disorder methemoglobinemia, which can be fatal due to the reduced ability for oxygen transport in the blood. In addition to the residual nitrates in processed meat products, there is concern about the levels of nitrates and nitrites in drinking water, and in the water and soil used to grown vegetables, especially when consumed by infants and young children [75, 76]. Health concerns have led to the FDA-regulated limits of 10 ppm nitrates, 1 ppm nitrite, and 10 ppm total nitrates and nitrites in bottled water (21 CFR 165.110) [77].

During the curing of processed meat products, processors ensure the legal limits of sodium nitrite in the finished product largely by carefully monitoring the amount of sodium nitrate used in the formulation of the product. The levels of nitrates/nitrites may be monitored throughout the production system with relatively rapid tests for nitrates/nitrites. There are both official methods and more rapid rapids of analysis used for testing both processed meat products and water, as described below.

Three of the AOAC International official methods [25] for nitrates/nitrites include the following:

1. **Xylenol Method** (AOAC Method 935.48, Nitrates and Nitrites in Meat)

   In the xylenol method, the sample, with sulfuric acid added, is treated with 2,4-xylenol to produce 6-nitro-2,4-xylenol, which is distilled into a water-isopropanol alcohol-ammonia hydroxide mixture. The ammonia salt of 6-nitro-2,4-xylenol is a yellow color measured at 450 nm, compared to a nitrate N standard curve.
2. **Colorimetric Method** (AOAC 973.31, Nitrites in Cured Meat)

   In the colorimetric method, sodium nitrite is extracted from the sample then reacted with two reagents, sulfanilamide (sulfa) and naphthylethylenediamine (NEDA). These compounds react with nitrite to produce a purple dye, which is proportional to the concentration of nitrite ions.
3. **Ion Chromatographic Method** (AOAC 993.30, Inorganic Anions in Water)

   This method is intended for measuring nitrate-N, nitrite-N, and several other inorganic anions in drinking water, but is also used within the meat industry with meat extracts as the sample. Anions in the sample are separated using an ion chromatography system that includes a guard column, separator column, and suppressor device and quantitated using a conductivity detector.

Rapid methods of analysis include ion-selective electrodes (ISEs) and test strips. Regarding ISE, both nitrate and nitrite ISEs are commercially available for testing food products (liquid samples) (see Chap. 21, Sect. 21.3.4 for details of ISEs). Regarding the test strips, various companies make test strips intended for checking the nitrate content of water. For example, AquaChek nitrite/nitrate test strips (Hach, Loveland, CO) are able to test for either nitrates or nitrites, or both, using two pad areas on the test strip, with one area measuring nitrates and the other pad area measuring nitrites. The nitrate test area contains a combination of chemicals to reduce nitrates to nitrites. Nitrites, at an acid pH, react with sulfanilic acid to form a diazonium compound, which couples with an indicator to produce a pink color. The color intensity is proportional to the concentration of nitrate. Color blocks for interpretation of results are provided for both nitrate N (0–50 ppm) and nitrite N (0–3 ppm).

33.9 SUMMARY

Consumer concerns and government regulations focused on the safety of foods dictate the need for analysis of various food contaminants, residues, and chemical constituents of concern. These compounds include pesticide residues, mycotoxins, antibiotic residues, GMOs, allergens, food adulterants, packaging material hazardous chemicals, naturally occurring substances formed during processing, environmental contaminants, and certain other chemicals. Both rapid screening methods and more time-consuming quantitative methods are required to meet the needs of industry and government, in an effort to ensure a safe and reliable food supply. A positive result from a screening method usually leads to further testing to confirm and quantify the presence of the compound of concern. Sampling and sample preparation can be a significant challenge due to the low levels of the analytes and the complexity of food matrices. Sample preparation often includes homogenization, extraction, and cleanup and sometimes requires derivatization. Screening methods increasingly utilize immune-based techniques, such as ELISA, LFS, immunosensors, and immunoaffinity chromatography columns. Some immunoassays can be considered quantitative, rather than just screening methods. Other screening methods commonly used include enzyme inhibition assays, thin-layer chromatography, and inhibition of microbial growth. While GC is a common chromatographic technique used for the quantitative analysis of some pesticides, for many other compounds of concern covered in this chapter, the predominant chromatographic method is HPLC. Both GC and HPLC analysis are now commonly coupled with mass spectrometry detection, often using MS tandem systems. The testing for GMOs and allergens typically involves either protein-based methods (e.g., immunoassays) or DNA methods using PCR. Work continues to improve and develop various methods of analysis for chemical residues and compounds of concern, focusing largely on speed, cost, and reliability of screening methods and detection limits of quantitative methods.

33.10 STUDY QUESTIONS

1. Explain the importance of each of the following steps in sample preparation for the analysis of contaminants and residues of concern:
   (a) Grinding/homogenization
   (b) Extraction
   (c) Cleanup/purification
   (d) Derivatization
2. In the analysis of contaminants and residues of concern, compare and contrast:
   (a) GC vs. LC analysis
   (b) MS vs. MS/MS analysis
   (c) LC with fluorescence or UV detection vs. LC-MS
   (d) TLC vs. automated chromatography (LC or GC)
   (e) Microplate ELISA vs. LFS
3. The "tolerance level" for residues of the pesticide chlorpyrifos on corn grain is 0.05 ppm.
   (a) What is meant by "tolerance level"?
   (b) What federal agency sets that tolerance level?
   (c) What federal agency enforces that tolerance level?
   (d) What is "ppm" equivalent to in terms of (a) weight per volume, and (b) weight per weight units commonly used to express concentration?

   wt/vol: wt/wt:

   (e) In Volumes I and II of the *Pesticide Analytical Manual,* you find described "multiresidue" methods and "single-residue" methods. You also have found numerous screening methods. Which one of these three types of methods (i.e., multiresidue, single residue, and screening) would you use to ensure your compliance with the tolerance level for this pesticide? Briefly explain the nature of this type of method, and explain why you chose this method over the other two types of methods.
4. Mycotoxins are of potential concern in corn, especially in certain growing and storage conditions.
   (a) Sampling is a major contributor to error in the analysis for mycotoxins. Why is sampling for mycotoxin analysis such a challenge?
   (b) Identify the most commonly used quantitative chromatographic method for mycotoxins. Justify the preference of this method.
5. Regarding antibiotic residues, briefly explain the following:
   (a) How might these get into foods?
   (b) What types of foods most likely contain them?
   (c) Why are these antibiotic residues a problem?
   (d) Why have techniques such as immunoassays largely replaced microbial growth inhibition assays for screening antibiotic residues?
   (e) What method is most commonly used for accurate quantitative determination and confirmation?
6. You want to identify GMOs in a soybean field where you do not have access to a sophisticated analytical laboratory. What food constituent would you be analyzing, and which analytical technique would you use? Explain the principle of this technique.
7. You are analyzing unwanted chemical components in the foods/raw materials listed below. Identify one possible unwanted chemical component that would likely be of concern with this specific food, and state an appropriate method for quantitative analysis. Also give one appropriate screening method for each chemical component identified. (Note: For each food, give a different quantitative likely unwanted chemical component and a different method of analysis.)

	Likely unwanted chemical component	Quantitative analysis method	Screening method
(a) Oats			
(b) Peanuts			
(c) Milk			
(d) Wine			
(e) Cured meat			

8. Describe the DNA-based method and identify the two protein-based methods for GMO detection and quantitation. Describe the strengths and weaknesses of each.
9. For the analysis of food allergens, protein-based or DNA-based methods are commonly used.
   (a) Give an example of when you would choose a protein-based method over a DNA-based method. Justify your choice and provide the principle of the method of choice.

(b) Give an example of when you would choose a DNA-based method over a protein-based method. Justify your choice and provide the principle of the method of choice.

10. Regarding the compounds described in Table 33.3:
    (a) Identify five compounds associated with the heat of cooking/frying of foods.
    (b) Identify two compounds typically analyzed by GC (vs. LC) and explain the preference for GC methods with these compounds.
    (c) Identify two compounds associated with packaging material.
    (d) Identify a compound that is an economic adulterant.
    (e) Identify two compounds naturally present in specific foods, but toxic to humans at certain levels.
11. Regarding sulfites, explain how these differs from the food allergens described in Sect. 33.7 in the response of sensitive humans, how they differ in the nature of quantitative methods, and why quantitative determination is relatively difficult for sulfites compared to food allergens.

Acknowledgments The authors thank Indarpal Singh (Analytical & Applied Science, Conagra Brands) for his input on Tables 33.1 and 33.3 in this chapter.

REFERENCES

1. Nielen MWF, Marvin HJP (2008) Challenges in chemical food contaminants and residue analysis, Ch. 1. In: Pico Y (ed) Food contaminants and residue analysis. Comprehensive analytical chemistry, volume 51. Elsevier, Oxford, UK
2. Arvanitoyannis IS (2008) International regulations on food contaminants and residues, Ch. 2. In: Pico Y (ed) Food contaminants and residue analysis. Comprehensive analytical chemistry, volume 51. Elsevier, Oxford, UK
3. FDA (2023) Chemical Contaminants & Pesticides. https://www.fda.gov/food/chemical-contaminants-pesticides
4. FDA (2023) Food Chemical Safety. https://www.fda.gov/food/food-ingredients-packaging/food-chemical-safety
5. European Union (2020) Rapid alert system for food and feed. Preliminary annual report. https://food.ec.europa.eu/system/files/2021-08/rasff_pub_annual-report_2020.pdf
6. Alweera A, Khursheed N, Samra F, Anjum Z, Younis K (2022) Application of nanotechnology in food packaging: Pros and Cons, J Agric Food Res 7:100270
7. US Environmental Protection Agency (Accessed Sep 2023) Safety and Waste Management for SAM Biotoxin Methods https://www.epa.gov/esam/safety-and-waste-management-sam-biotoxin-methods
8. Lee J, Kim L, Shin Y, Lee J, Lee Jiho, Kim E, Moon J, Kim J (2017) Rapid and Simultaneous Analysis of 360 Pesticides in Brown Rice, Spinach, Orange, and Potato Using Microbore GC-MS/MS. J Agric Food Chem, 65:3253–3430
9. Turnipseed SB, Andersen WC (2008) Veterinary drug residues, Ch. 10. In: Pico Y (ed) Food contaminants and residue analysis. Comprehensive analytical chemistry, volume 51. Elsevier, Oxford, UK
10. Díaz-Bao M, Barreiro R, Miranda JM, Cepeda A, Regal P (2015) Fast HPLC-MS/MS method for determining penicillin antibiotics in infant formulas using molecularly imprinted solid-phase extraction. J Anal Methods Chem 2015:1–8
11. Sannino A (2008) Pesticide Residues, Ch. 9. In: Pico Y (ed) Food contaminants and residue analysis. Comprehensive analytical chemistry, volume 51. Elsevier, Oxford, UK
12. Stachniuk A, Fornal E (2016) Liquid chromatography-mass spectrometry in the analysis of pesticide residues in food. Food Anal Methods 9:1654–1665
13. De Santis B, Debegnach F, Gregori E, Russo S, Marchegiani F, Moracci G, Brera C (2017) Development of a LC-MS/MS method for the multi-mycotoxin determination in composite cereal-based samples, Toxins (Basel) 9:169
14. Sandra P, David F, Vanhoenacker G (2008) Advanced sample preparation techniques for the analysis of food contaminants and residues, ch. 5. In: Pico Y (ed) Food contaminants and residue analysis. Comprehensive analytical chemistry, volume 51. Elsevier, Oxford, UK
15. Lehotay SJ, Cook JM (2015) Sampling and sample processing in pesticide and residue analysis. J Agric Food Chem 63:4395–4404
16. Anastassiades M, Lehotay SJ, Stajnbaher D, Schenck FJ (2003) Fast and easy multiresidue method employing acetonitrile extraction/partitioning and "dispersive solid-phase extraction" for the determination of pesticide residues in produce. J AOAC Int 86:412–431
17. Niell S, Cesio V, Hepperle J, Doerk D, Kirsch L, Kolberg D, Scherbaum E, Anastassiades M, Heinzen H (2014) QuEChERS-based method for the multiresidue analysis of pesticides in beeswax by LC-MS/MS and GC×GC-TOF. J Agric Food Chem 62:3675–3683
18. Kinross AD, Hageman KJ, Doucette WJ, Foster AL (2020) Comparison of Accelerated Solvent Extraction (ASE) and Energized Dispersive Guided Extraction (EDGE) for the analysis of pesticides in leaves. J Chroma A, 1627:461414
19. Jiao, Z, Guo Z, Zhang S, Chen H (2015) Microwave-assisted micro-solid-phase extraction for analysis of tetracycline antibiotic in environmental samples. Int J Environ An Ch 95:82–91
20. Merkle S, Kleeberg K, Fritsche J (2015) Recent Developments and Applications of Solid Phase Microextraction (SPME) in Food and Environmental Analysis—A Review. Chromatography, 2:293–381
21. EPA (2022) Pesticide Product Information System. https://www.epa.gov/ingredients-used-pesticide-products/pesticide-product-information-system-ppis
22. EPA Basic Information about Pesticide Ingredients (2023) https://www.epa.gov/ingredients-used-pesticide-products/basic-information-about-pesticide-ingredients
23. FDA (2002) Pesticide analytical manual volume II (updated January, 2002). http://www.fda.gov/food/foodscienceresearch/laboratorymethods/ucm113710.htm
24. FDA (1999) Pesticide analytical manual volume I (PAM), 3rd edn (1994, updated October, 1999). http://www.fda.gov/food/foodscienceresearch/laboratorymethods/ucm111455.htm
25. AOAC International (2023) Official methods of analysis (on line), 22nd edn. Official methods of analysis, 22nd edn. AOAC International, Gaithersburg, MD
26. Tadeo JL (2008) Analysis of pesticides in food and environmental samples. CRC, Taylor and Francis Group, Boca Ranton, FL
27. Lee NA, Kennedy IR (2001) Environmental monitoring of pesticides by immunoanalytical techniques: Validation, current status, and future perspectives. J AOAC Int 84:1393–1406
28. Soderberg D (2005) Committee on residues and related topics; pesticides and other chemical contaminants. General referee reports. J AOAC Int 88:331–345
29. Pico Y, Blasco C, Font G (2004) Environmental and food applications of LC tandem mass spectrometry in pesticide-residues analysis: an overview. Mass Spectrom Rev 23:45–85

30. Patel K, Fussell RJ, Goodall DM, Keelye BJ (2004) Evaluation of large volume-difficult matrix introduction-gas chromatography-time of flight-mass spectrometry (LV-DMI-GC-TOF-MS) for the determination of pesticides in fruit-based baby foods. Food Addit Contam 21:658–669
31. Gilbert-López B, García-Reyes JF, Ortega-Barrales P, Molina-Díaz A, Fernández-Alba AR (2007) Analyses of pesticide residues in fruit-based baby food by liquid chromatography/electrospray ionization time-of-flight mass spectrometry. Rapid Commun Mass Spectrom 21(13):2059–2071
32. Brera C, De Santis B, Debegnach F, Miraglia M (2008) Mycotoxins, ch. 12. In: Pico Y (ed) Food contaminants and residue analysis. Comprehensive analytical chemistry, volume 51. Elsevier, Oxford, UK
33. Pitt JI, Wild CP, Baan RA, Gelderblom WCA, Miller JD, Riley RT, Wu F (2012) Improving public health through mycotoxin control. International Agency for Research on Cancer Publication No 158, Geneva, Switzerland
34. Miller JD, Schaafsma AW, Bhatnagar D, Bondy G, Carbone I, Harris LJ, Harrison G, Munkvold GP, Oswald IP, Pestka JJ, Sharpe L, Sumarah MW, Tittlemier SA, Zhou T (2014) Mycotoxins that affect the North America agri-food sector:state of the art and directions for the future. World Mycotoxin Journal 7:63–82
35. Rahmani A, Jinap S, Soleimany F (2009) Qualitative and quantitative analysis of mycotoxins. Comprehensive Reviews in Food Science and Food Safety 8:202–251
36. European Union (2022) Methods of sampling and analysis for the control of levels of certain contaminants in foodstuffs. https://eur-lex.europa.eu/EN/legal-content/summary/methods-of-sampling-and-analysis-for-the-control-of-levels-of-certain-contaminants-in-foodstuffs.html
37. Siantar DP, Trucksess MW, Scott PM, Herman EM (eds) (2008) Food contaminants: mycotoxins and food allergens. ACS symposium series 1001. American Chemical Society, Washington, DC
38. Commission Regulation (EC) N 401/2006 of 23 February 2006. Laying down the methods of sampling and analysis for the official control of the levels of mycotoxins in food stuff
39. Agriopoulou S, Stamatelopoulou E, Varzakas T (2020) Advances in Analysis and Detection of Major Mycotoxins in Foods. Foods, 9:518
40. Sulyok M, Berthiller F, Krska R, Schuhmacher R (2006) Development and validation of a liquid chromatography/tandem mass spectrometric method for the determination of 39 mycotoxins in wheat and maize. Rapid Commun Mass Spectrom 20:2649–2659
41. Tsao R, Zhou T (2000) Micellar electrokinetic capillary electrophoresis for rapid analysis of patulin in apple cider. J Agric Food Chem 48:5231–5235
42. Peña R, Alcaraz MC, Arce L, Ríos A, Valcárcel M (2002) Screening of aflatoxins in feed samples using a flow system coupled to capillary electrophoresis. J Chromatogr A 967:303–314
43. Maragos CM, Appell M (2007) Capillary electrophoresis of the mycotoxin zearalenone using cyclodextrin-enhanced fluorescence. J Chromatogr A 1143:252–257
44. Kos G, Sieger M, McMullin D, Zahradnik C, Sulyok M, Oner T, Mizaikoff B, Krska R (2016) A novel chemometric classification for ftir spectra of mycotoxin-contaminated maize and peanuts at regulatory limits, Food Addit Contam Part A, 33:1596–1607.
45. Berardo N, Pisacane V, Battilani P, Scandolara A, Pietri A, Marocco A (2005) Rapid detection of kernel rots and mycotoxin in maize by near-infrared reflectance spectroscopy. J Agric Food Chem 53:8128–8134
46. Pettersson H, Åberg L (2003) Near infrared spectroscopy for determination of mycotoxin in cereals. Food Control 14:229–232
47. De la Calle MB, Anklam E (2005) Semicarbazide: occurrence in food products and state-of-the-art in analytical methods used for its determination. Anal Bioanaly Chem 382:968–977
48. United States Department of Agriculture Red book archives. USDA Food Safety Inspection Service. http://www.fsis.usda.gov/wps/portal/fsis/topics/data-collection-and-reports/chemistry/red-books/archive
49. Chen J, Ying G, Deng W (2019) Antibiotic Residues in Food: Extraction, Analysis, and Human Health Concerns. J Agric Food Chem, 67:7569–7586
50. AOAC (2023) The AOAC International rapid methods validation process. http://jornades.uab.cat/workshopmrama/sites/jornades.uab.cat.workshopmrama/files/AOAC_RI.pdf
51. Bulthaus M (2004) Detection of Antibiotic/Drug Residues in Milk and Dairy Products, ch. 12. In: Wehr HM, Frank JF (eds) Standard methods for the examination of dairy products, 17th edn. American Public Health Association, Washington, DC
52. FDA (2006) Preparation and LC/MS/MS analysis of honey for fluoroquinolone residues. 29 Sept, 2006 (last updated 11/01/2017). http://www.fda.gov/Food/FoodScienceResearch/LaboratoryMethods/ucm071495.htm
53. Roybal J, Walker CC, Pfenning AP, Turnipseed SB, Storey JM, Gonzales SA, Hurlbut JA (2002) Concurrent determination of four fluoroquinolones in catfish, shrimp, and salmon by liquid chromatography with fluorescence detection. J AOAC Int, 85:1293–1301
54. Holstege DM, Puschner B, Whitehead G, Galey FD (2002) Screening and mass spectral confirmation of B-lactam antibiotic residues in milk using LC-MS/MS. J Agric Food Chem 50(2):406–411
55. Cavalier C, Curini R, Di Corcia A, Nazzari M, Samperi R (2003) A simple and sensitive liquid chromatography-mass spectrometry confirmatory method for analyzing sulfonamide antibacterials in milk and egg. J Agric Food Chem 51:558–566
56. Clark SB, Turnipseed SB, Madson MR (2005) Confirmation of sulfamethazine, sulfathiazole, and sulfadimethoxine residues in condensed milk and soft-cheese products by liquid chromatography/tandem mass spectrometry. J AOAC Int 88:736–743
57. Wang J, Leung D (2007) Analyses of macrolide antibiotic residues I eggs, raw milk, and honey using both ultra-performance liquid chromatography/quadrupole time-of-flight mass spectrometry and high-performance liquid chromatography/tandem mass spectrometry. Rapid Commun Mass Spectrom 21(19):3213–3122
58. Ahmed FE (2004) Testing of genetically modified organisms in foods. CRC, Boca Raton, FL
59. Heller KJ (2003) Genetically engineered food: methods and detection. Wiley-VCH, Weinheim, Germany
60. Jackson JF, Linskens HF (2009) Testing for genetic manipulation in plants (molecular methods of plant analysis). Springer, New York
61. FDA (2023) Food allergies: What you need to know. (Last updated 04/12/2023). https://www.fda.gov/food/buy-store-serve-safe-food/food-allergies-what-you-need-know
62. FDA (2004) Food allergen labeling and consumer protection act of 2004 (Public Law 108–282, Title II). (Last updated 11/29/2022). http://www.fda.gov/Food/GuidanceRegulation/GuidanceDocumentsRegulatoryInformation/Allergens/ucm106187.htm
63. Prado M, Ortea I, Vial S, Rivas J, Calo-Mata P, Velázquez J (2015) Advanced DNA- and protein-based methods for the detection and investigation of food allergens. Crit Rev Food Sci Nutr. https://doi.org/10.1080/10408398.2013.873767
64. Westphal CD (2008) Improvement of immunoassays for the detection of food allergens, ch. 29. In: Siantar DP, Trucksess MW, Scott PM, Herman EM (eds) Food contaminants: mycotoxins and food allergens. ACS symposium series 1001, American Chemical Society, Washington, DC
65. Yman IM, Eriksson A, Johansson MA, Hellenas K-E (2006) Food allergen detection with biosensor immunoassays. J AOAC Int 89(3):856–861

66. FSIS (2023) Chemical residues and contaminants. https://www.fsis.usda.gov/science-data/data-sets-visualizations/chemical-residues-and-contaminants
67. FDA (2023) Food program compendium of analytical laboratory methods: Chemical analytical manual. https://www.fda.gov/food/laboratory-methods-food/foods-program-compendium-analytical-laboratory-methodsCAM
68. Taylor SL, Bush RK, Nordlee JA (2003) Sulfites, ch. 24. In: Metcalfe DD, Simon RA (eds) Food allergy: adverse reactions to foods and food additives, 3rd edn. Blackwell, Malden, MA, p 324–341
69. Bush RK, Montalbano MM (2014) Asthna and Food Additives, ch. 27. In: Metcalfe DD, Sampson HA, Simon RA, Lack G (eds) Food allergy: adverse reactions to foods and food additives, 5th edn. Wiley-Blackwell, Hoboken, NJ, p 361–374
70. Anonymous (2016) Code of federal regulations. Food; exemptions from labeling. 21 CFR 101.100 (a) (4). US Government Printing Office, Washington, DC
71. FDA (2021) Determination of sulfites in food using liquid chromatography-tandem mass spectrometry (LC-MS/MS). FDA Foods program compendium of analytical laboratory methods: Chemical analytical manual (CAM). Method C-004.03. Sept 15, 2021. https://www.fda.gov/media/114411/download
72. Eaton AD, Clesceri LS, Rice EW, Greenberg AE (eds) (2005) Standard methods for the examination of water and wastewater, 21st edn., Method 4500 – SO_3^{-2} B. American Public Health Association, Washington, DC
73. McFeeters RF, Barish AO (2003) Sulfite analysis of fruits and vegetables by high-performance liquid chromatography (HPLC) with ultraviolet spectrophotometric detection. J Agric Food Chem 51:1513–1517
74. Chung SWC, Chan BTP, Chan ACM (2008) Determination of free and reversibly-bound sulfite in selected foods by high-performance liquid chromatography with fluorometric detection. J AOAC Int 91:98–102
75. Sindelar JJ, Miklowski AL (2011) Sodium nitrite in processed meat and poultry meats: a review of curing and examining the risk/benefit of its use. American Meat Science Association 3:1–14
76. Katan MB (2009) Nitrate in foods: harmful or healthy? Amer J Clin Nutr 90:11–12
77. Anonymous (2023) Code of federal regulations. Bottled water. 21 CFR 165.110. US Government Printing Office, Washington, DC

34 Analysis for Extraneous Matter

Hulya Dogan and Bhadriraju Subramanyam

Contents

H. Dogan (✉) · B. Subramanyam
Department of Grain Science and Industry, Kansas State University, Manhattan, KS, USA
e-mail: dogan@ksu.edu; sbhadrir@ksu.edu

34.1 INTRODUCTION

Analysis for extraneous matter is an important element both in the selection of raw materials for food manufacturing and for monitoring the quality of processed foods. The presence

B. P. Ismail, S. S. Nielsen (eds.), *Nielsen's Food Analysis*, Food Science Text Series,
https://doi.org/10.1007/978-3-031-50643-7_34

of extraneous material in a food product is unappealing and can pose a serious health hazard to the consumer. It also represents lack of good manufacturing practices and sanitary conditions in production, storage, or distribution. The presence of extraneous materials in the product ingredients may render the final product adulterated and not suitable for human food.

34.1.1 Federal Food, Drug, and Cosmetic Act

The **Federal Food, Drug, and Cosmetic Act** (FD&C Act) of 1938 with amendments administered and enforced by the US Food and Drug Administration (FDA) [1] defines a food as **adulterated** "if it consists in whole or in part of any filthy, putrid, or decomposed substance, or if it is otherwise unfit for food [Section 402 (21 USC 342) (a)(3)]; or if it has been prepared, packed, or held under unsanitary conditions whereby it may have become contaminated with filth, or whereby it may have been rendered injurious to health" [Section 402 (21 USC 342) (a)(4)]. The filthy, putrid, or decomposed substances referred to in the law include the extraneous matter addressed in this chapter. In addition, extraneous matter includes adulterants that may be encountered in processing systems, such as lubricants, metal particles, or other contaminants (animate or inanimate), that may be introduced into a food intentionally or because of a poorly operated food processing system. These aspects are not covered in this chapter.

34.1.2 Good Manufacturing Practices

The current good manufacturing practices (cGMPs) were published in 1969 by the Food and Drug Administration (FDA) (21 CFR Part 117) to provide guidance for compliance with the FD&C Act [2] (see also Chap. 2). That regulation provides guidelines for operating a food processing facility in compliance with Section 402 (a)(4), revised in 2015. Currently, the cGMPs are amended to make the compliance guidelines more risk-based under the Food Safety Modernization Act of 2011. Paramount to complying with the FD&C Act and cGMPs is the thorough inspection of raw materials and routine monitoring of food processing operations to ensure protection of the consuming public from harmful or filthy food products.

34.1.3 Defect Action Levels

Most of our foods are made from or consist in part of ingredients that are obtained from plants or animals and are mechanically stored, handled, and transported in large quantities. It would be virtually impossible to keep those materials completely free of various forms of contaminants. In recognition of that, the FDA [3] has established **defect action levels** (DALs) that reflect current maximum levels for natural or unavoidable defects in food for human use that present no health hazard. They reflect the maximum levels that are considered unavoidable under good manufacturing practices. They apply mainly to contaminants that are unavoidably carried over from raw agricultural commodities into the food processing system. The manner in which foods are manufactured may lead to their contamination with extraneous materials if strict controls in processing are not maintained. This latter type of contamination leads to food safety issues and DALs are not used to determine compliance. Other actionable levels of contaminants may be found in the FDA *Compliance Policy Guide (CPG) Manual* [4].

DALs are published in the FDA handbook and represent limits at which the food product is considered "adulterated" and subject to enforcement action. If there is no published DAL for a product, FDA evaluates and decides on a case-by-case basis. FDA's technical and regulatory experts in filth and extraneous materials use a variety of criteria to determine the significance and regulatory impact of the findings. The most current information of FDA laws and regulations relevant to extraneous matter, including cGMPs, DALs, and CPGs, can be found on the Internet:

Federal Food, Drug, and Cosmetic Act (FD&C Act)	https://www.fda.gov/regulatory-information/laws-enforced-fda/federal-food-drug-and-cosmetic-act-fdc-act
Current Good Manufacturing Practice, Hazard Analysis, and Risk-Based Preventive Controls for Human	https://www.fda.gov/food/food-safety-modernization-act-fsma/fsma-final-rule-preventive-controls-human-food
Food Defect Action Levels (DALs)	https://www.fda.gov/food/ingredients-additives-gras-packaging-guidance-documents-regulatory-information/food-defect-levels-handbook
Compliance Policy Guidance (CPG)	https://www.fda.gov/inspections-compliance-enforcement-and-criminal-investigations/compliance-manuals/manual-compliance-policy-guides

34.1.4 Purposes of Analyses

The major purposes for conducting analyses for extraneous matter in foods are to ensure the protection of the consuming public from harmful or filthy food products, to meet regulatory requirements of the FD&C Act Sections 402 (a)(3) and 402 (a)(4), and to comply with DALs.

34.2 GENERAL CONSIDERATIONS

34.2.1 Definition of Terms

Terms used by **AOAC International** (AOAC Method 970.66) to classify or characterize various types of extraneous materials are defined as follows:

- **Extraneous Materials.** Any foreign matter in product associated with objectionable conditions or practices in production, storage, or distribution; included are various classes of filth, decomposed material (decayed tissues due to parasitic or nonparasitic causes), and miscellaneous matter such as sand and soil, glass, rust, or other foreign substances. Bacterial counts are not included.
- **Filth.** Any objectionable matter contributed by animal contamination, such as rodent, insect, or bird matter, or any other objectionable matter contributed by unsanitary conditions.
- **Heavy Filth.** Heavier material separated from products by sedimentation based on different densities of filth, food particles, and immersion liquids. Examples of such filth are sand, soil, insect and rodent excreta pellets and pellet fragments, and some animal excreta pellets.
- **Light Filth.** Lighter filth particles that are oleophilic and are separated from product by floating them in an oil-aqueous liquid mixture. Examples are insect fragments, whole insects, rodent hairs and fragments, and feather barbules.
- **Sieved Filth.** Filth particles of specific size ranges separated quantitatively from product by use of selected sieve mesh sizes.

34.2.2 Diagnostic Characteristics of Filth

There are certain qualities characteristic to extraneous materials that serve as proof of presence of foreign or objectionable matter in food. Examples include specific diagnostic characteristics of **molds** (i.e., parallel hyphal walls, septation, granular appearance of cell contents, branching of hyphae, blunt ends of hyphal filaments, nonrefracted appearance of hyphae); diagnostic characteristics of **insect fragments** (i.e., recognizable shape, form, or surface sculpture, an articulation or joint, setae or setal pits, sutures), **rodent hairs** (i.e., pigment patterns and structural features), and **feather barbules** (i.e., structural features); diagnostic characteristics of **insect-damaged kernels** (IDK) and packaging materials; and chemical identification of **animal urine** and **excrement**. These diagnostic characteristics are outlined by AOAC International (formerly Association of Official Analytical Chemists) for positive identification of extraneous matter or filth [5].

The **AACC International** (AACCI) (formerly American Association of Cereal Chemists, AACC) publishes a methods book that includes a section on extraneous matter, containing descriptive material helpful in identifying insect and rodent contaminants [6]. Several microscopic and radiographic illustrations are provided by the AACCI as authentic reference materials to help analysts to identify filth. AACCI Method 28-95.01, "Insect, Rodent Hair, and Radiographic Illustrations," provides a series of colored pictures representative of insect fragments commonly found in cereal products and pictorial examples of rodent hair structure.

Kurtz and Harris [7] provided a virtual parts catalog of insect fragments with a series of micrographs. Gentry et al. [8], in an updated version of the Kurtz and Harris publication, include colored micrographs of common insect fragments. Also included in AACCI Method 28-95.01 are radiographic examples of grain kernels that contain internal insect infestation. AACCI Method 28-21.02, "X-ray Examination for Internal Insect Infestation," provides an outline of the apparatus and procedure for X-ray examination of internal insect infestation in grain [9].

34.3 OFFICIAL AND APPROVED METHODS

There are various laboratory methods for separating (isolating) extraneous materials from foods and for identifying and enumerating them. The FDA and the AOAC International have published reference articles, books, and methods on the analysis of extraneous materials. The most authoritative source, and that generally considered official by the FDA, is the *Official Methods of Analysis of AOAC International*, Chap. 16, "Extraneous Materials: Isolation" [5]. This chapter includes methods for extraneous matter isolation in various food categories (Table 34.1). The AOAC International "Extraneous Materials: Isolation" chapter contains a subchapter dealing with **molds.** This includes identification of molds and methods for isolation of molds in fruits and fruit products and vegetables and vegetable products.

The AACCI [6] has established methods for isolating and identifying extraneous matter in cereal grains and their products (AACCI Method 28, listed in Table 34.2). In most instances, the AACCI methods are based on FDA or AOAC methods, but the format is slightly different. The AACCI presents each procedure in an outline form that includes the scope, apparatus, and reagents required and the procedure in itemized steps, while the AOAC methods use a narrative paragraph form (Table 34.2).

A valuable resource on analysis for extraneous matter is *Principles of Food Analysis for Filth,*

Table 34.1 Official methods of AOAC International for analysis of extraneous materials

Section	*Title*
16.	Extraneous materials: isolation
16.1	General
16.2	Beverages and beverage materials
16.3	Dairy products
16.4	Nuts and nut products
16.5	Grains and their products
16.6	Baked goods
16.7	Breakfast cereals
16.8	Eggs and egg products
16.9	Poultry, meat, and fish and other marine products
16.10	Fruits and fruit products
16.11	Snack food products
16.12	Sugars and sugar products
16.13	Vegetables and vegetable products
16.14	Spices and other condiments
16.15	Miscellaneous
16.16	Animal excretions
16.17	Mold
16.18	Fruits and fruit products
16.19	Vegetables and vegetable products

Decomposition and Foreign Matter (FDA Technical Bulletin No. 1) [10]. The FDA *Training Manual for Analytical Entomology in the Food Industry* [11] is prepared to facilitate the orientation of food analysts to the basic techniques they will need for filth analysis. A recent, more advanced resource is *Fundamentals of Microanalytical Entomology: A Practical Guide to Detecting and Identifying Filth in Foods* [12]. Most chapter authors of this resource are, or have been, FDA personnel "involved in the forensic aspect of piecing together the etiological puzzles of how insect filth gets into processed food products" [12]. The authors share their experience gained in gathering and developing evidence used to document violations of the law that the FDA is mandated to enforce.

34.4 BASIC ANALYSIS

Various methods for isolation of extraneous matter are suggested in Sect. 34.3, which defines different types of filth: (1) separation on the basis of differences in **density**, **affinity for oleophilic solvents**, or **particle size**; (2) **diagnostic characteristics** for identification of filth; and (3) **chemical identification** of contaminants. Since all methods of analysis for extraneous matter for all categories of food cannot be discussed in this chapter, only the underlying principles of the methods are summarized below. Readers may need to refer to the specific AOAC methods cited for detailed instructions of the procedures.

Table 34.2 Approved methods of the AACC International for analysis of extraneous materials

Number	*Title*
28	Extraneous matter
28-01.01	Apparatus or materials for extraneous matter methods
28-02.01	Reagents for extraneous matter methods
28-03.02	Special techniques for extraneous matter methods
28-06.01	Cinder and sand particles in farina–counting method
28-07.01	Cinder and sand particles in farina–gravimetric method
28-10.02	Macroscopic examination of external contamination in whole grains
28-19.01	External filth and internal insect infestation in whole corn
28-20.02	Microscopic examination of external contamination in whole grains
28-21.02	X-ray examination for internal insect infestation
28-22.02	Cracking-flotation test for internal insects in whole grains
28-30.02	Macroscopic examination of materials hard to hydrate
28-31.02	Pancreatin sieving method, for insect and rodent filth in materials hard to hydrate
28-32.02	Sieving method, for materials hard to hydrate
28-33.02	Pancreatin nonsieving method for insect and rodent filth in materials easy to hydrate
28-40.01	Acid hydrolysis method for insect fragments and rodent hairs–wheat-soy blend
28-41.03	Acid hydrolysis method for extracting insect fragments and rodent hairs–light filth in white flour
28-43.01	Glass plate method, for insect excreta
28-44.01	Iodine method, for insect eggs in flour
28-50.01	Decantation method, for rodent excreta
28-51.02	Flotation method, for insect and rodent filth
28-60.02	Tween-versene method, for insect fragments and rodent hairs in rye flour
28-70.01	Defatting-digestion method, for insect fragments and rodent hairs
28-75.02	Sieving method, for light filth in starch
28-80.01	Flotation method, for insect and rodent filth in popped popcorn
28-85.01	Ultraviolet light examination, for rodent urine
28-86.01	Xanthydrol test, for urea
28-87.01	Urease-bromothymol blue test paper, for urea
28-93.01	Direction of insect penetration into food packaging
28-95.01	Insect, rodent hair, and radiographic illustrations

The AOAC and the AACCI methods for analysis of extraneous materials involve the use of one or more of the following basic methods: filtration, sieving, wet sieving, gravimetry, sedimentation/flotation, cracking flotation, heat, acid or enzyme digestion, macroscopic and microscopic methods, and mold counts.

34.4.1 Sieving Method

Separation is based on difference in particle size of product and contaminant using **standard test sieves** (Fig. 34.1). For instance, the insects are (larger) separated from spices (smaller) using a 20-mesh sieve, and wheat grains (larger) are separated from insects (smaller) using a 10- or 12-mesh sieve. Then the contaminant is identified using a **widefield stereomicroscope**.

34.4.2 Sedimentation Method

Separation is based on different densities of product, contaminant, and immersion fluid. Specific gravity of immersion solution (carbon tetrachloride/chloroform) allows heavier shell, sand, glass, metal, or excreta contaminants to settle; less dense product floats. Apparent lower **specific gravity** of internally infested wheat kernels, for instance, allows them to float, while sound wheat kernels settle in 1.27 specific gravity solution. Contaminants are then identified using a microscope.

Analysis of high-fat-containing samples such as nuts requires defatting using petroleum ether prior to filth analysis (AOAC Method 968.33A). The chloroform and chloroform/carbon tetrachloride solvents allow pieces of shell, sand, and soil to settle at the bottom of the beaker on the basis of specific gravity and cause the defatted nut meats to float and be decanted. Essentially, the same procedure is suggested to isolate pieces of rodent excreta from corn grits, rye and wheat meal, whole wheat flour, farina, and semolina in AOAC Method 941.16A. It should be noted that the use of the more toxic solvents such as carbon tetrachloride, chloroform, and petroleum ether is avoided in most contemporary analytical methods.

Fig. 34.1 Fisher US Standard test sieve. Range of mesh sizes is available for various particle size separations

34.4.3 Flotation Methods

Flotation methods are designed to isolate microscopic filth by floating the filth upward, typically in an **oil/water-phased system**. Insect fragments, mites, and hairs are lipophilic and like to be in the oil phase, thus they float to the surface with the oils. Plant tissues and most related tissues are hydrophilic, and they tend to stay in the water phase. Therefore, separation is based on the principle of affinity for oleophilic solvents. Gravity further helps this process, and larger particles sink. To accomplish the separation of filth from food, a number of solution systems are used to ensure that the majority of the product sinks, while the oils with trapped filth float. The oil phase is trapped off with a **Wildman trap flask** (Fig. 34.2), filtered, collected on a filter paper, and examined microscopically to determine the amount and kinds of filth present [13].

Flotation is a common method used to determine insect fragments, rodent hairs, and other forms of light filth in wheat flour (AOAC Method 972.32). The acid digestion is used to break down the starch in the flour and allows the other flour constituents to more cleanly separate from the dilute acid solution. Although the AOAC method calls for digestion by autoclaving, AACCI Method 28-41.03 provides for an alternative hotplate digestion, which might be more convenient for some laboratories. The oleophilic property of insect fragments, rodent hairs, and feather barbules allows them to be coated by the mineral oil and trapped in the oil layer for separation and collection on ruled filter paper. The heavier sediments of the digestion are washed and drained from the funnel. Fragments and rodent hairs are reported on the basis of 50 g of flour.

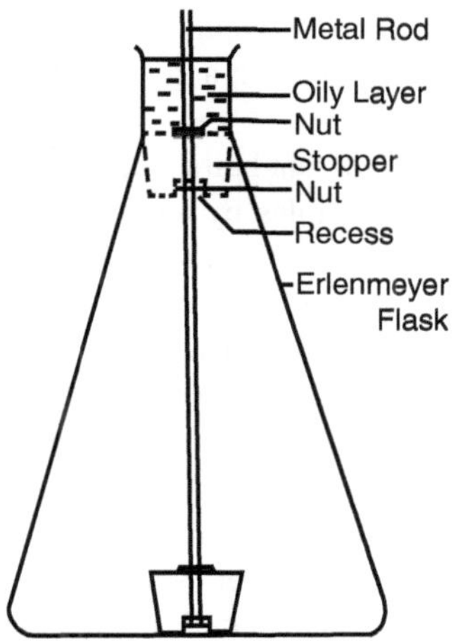

Fig. 34.2 Wildman trap flask. Stopper on shaft is lifted up to neck of flask to trap off floating layer. (Adapted from [5], AOAC Method 945.75. Extraneous Materials in Products)

34.4.3.1 Cracking-Flotation Method

Internal infesting insects (such as in grains) can be determined using an oleophilic method. First, any external insects are removed by sieving. Grain sample is coarsely cracked to free insects from kernels. Cracked grain sample is digested in 3–5% HCl solution and sieved with water to remove hydrolyzed starch and acid. The sample is transferred to a Wildman trap flask and boiled in 40% alcohol solution to deaerate. Tween 80 (polyoxyethylenesorbitan monooleate) and Na_4EDTA (tetrasodium salt of ethylenediaminetetraacetic acid) solutions are added to cause light bran particles to remain in solution during oil extraction of insect material. Light mineral oil is added to the solution to form a floating layer in which insect material is attracted due to its oleophilic nature. The oil layer is filtered through a ruled paper to collect the contaminating insects. The filter paper is examined microscopically.

34.4.3.2 Light Filth Flotation Method

Oleophilic filth is defined as light filth. Examples of light filth include insects, insect fragments, hairs, and feather barbules which can be detected in a food product by separating them from the food in the oil phase of an oil-aqueous mixture. The analysis of light filth is accomplished through a series of steps, starting with a pretreatment that removes fats, oils, soluble solids, and fine particulate matter to enhance the wettability of the food. The second step requires mixing the food with a water and oil mixture. The food will remain in the aqueous phase and the light filth will rise to the top with the oil phase. In the third step, the extract with filth elements is poured onto a ruled filter paper using a filter flask and funnel (Fig. 34.3) and examined line by line under a stereomicroscope (Fig. 34.4). After identification and enumeration, the results are reported to provide the following information: (i) whole or equivalent insects (adults, pupae, maggots, larvae, cast skins); (ii) insect fragments, identified; (iii) insect fragments, unidentified; (iv) aphids, scale insects, mites, spiders, etc. and their fragments; and (v) rodent hairs (state the length of the hairs).

Fig. 34.3 Filter flask and funnel. Funnel has a collar (partially raised) that holds ruled filter paper in place on the funnel base for trapping filth for examination. Suction is applied with a water aspirator. (www.whatman.com)

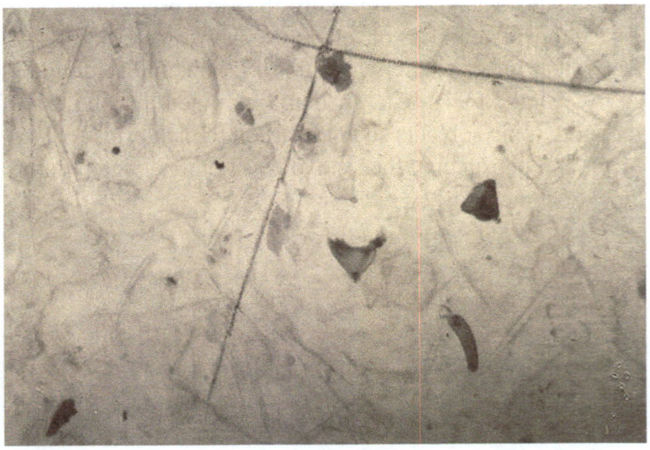

Fig. 34.4 Microscopic view of insect fragments and rodent hairs on a filter paper

Insect fragments, rodent hairs, and other light filth can be isolated from flour samples by an acid hydrolysis method. Sediment products of the digestion are allowed to settle out in a separatory funnel and are drained away. The remaining oil layer is filtered through a ruled filter paper and the contaminants identified microscopically. However, certain grain products such as whole wheat flour contain amounts of bran particles that may result in excessive amounts of material being trapped in the oil layer, making it difficult to identify particles of filth. The "Tween-Versene Method for Insect Fragments and Rodent Hairs in Rye Flour" (AACCI Method 28-60.02) utilizes two chemical agents that tend to suppress bran accumulation in the heptane recovery layer. Tween 80 (polyoxyethylene sorbitan monooleate) is a nonionic agent that appears to have certain surface-active properties that make it a useful adjunct to Na_4EDTA (tetrasodium salt of ethylenediaminetetraacetic acid). In the presence of Tween 80, Na_4EDTA appears to be a depressor for food materials (such as bran and other light plant matter), which otherwise tend to float. It has been suggested that the chelating properties of Na_4EDTA may result in its adsorption onto the surfaces of food particles along with the surfactant Tween 80, thereby preventing an attraction of food particles to oils used to isolate light filth. By preventing plant material from being collected in the heptane layer that is trapped off, contaminants such as oleophilic insect parts (exoskeleton) that are contained in the separating oil are much easier to distinguish and identify. AACCI Method 28-95.01 provides a description of insect fragments and rodent hair characteristics with illustrations [6]. Tilocca et al. [14] found the floatation test to be a reliable method for detecting insect fragments in semolina and tomato products. Plastic fragments were also isolated from grated bread by using the floatation test procedure [14]. The light filth method yielded an average recovery rate of 78–84% of rat hairs and 97–99% of insect fragments when tofu was spiked at levels of 5–15 and 5–30, respectively [15].

34.4.4 Subjectivity of Methods

Insect parts, rodent hairs, and feather barbules in food products are generally reported as the total number of filth elements counted of each kind encountered per sample unit. They are identified on the basis of objective criteria. However, identifying insect fragments is not a simple task. Training and supervised practice are required to achieve competence and consistency. Some fragments are easily identified on the basis of structural shape and form. Mandibles, for example, are quite distinctive in their shape and configuration; certain species of insects can be determined on the basis of this one structure. In other instances, fragments may be mere chips of insect cuticle that have neither distinctive shape nor form but can be identified as being of insect origin if they have one or more of the characteristics given in Sect. 34.2.2. Experienced analysts should rarely misinterpret fragments.

Isolation of extraneous material from a food product so that it can be identified and enumerated can be a very simple procedure or one that requires a series of several rather involved steps. In the process of isolating fragments from flour by the acid hydrolysis method, for instance, the sample is transferred from the digestion container to the separatory container and then to the filter paper for identification and enumeration. At each of those transfers, there is an opportunity for loss of fragments. Although the analyst may have made every effort to maintain the isolation "quantitative," there are opportunities for error. Both fragment loss and analyst variation are minimized by common use of standard methods and procedures and by proper training and supervised practice.

Another concern involves the significance of **insect fragment counts** (as well as particles of sand, pieces of rodent excreta, rodent hairs, etc.) in relation to **fragment or particle size**. Fragment counts are reported on a numerical basis; they do not reflect the total contaminant biomass that is present. A small fragment is counted the same as a large fragment. The size of the fragment may be a reflection of the process to which a common raw material (e.g., wheat) has been subjected; a more vigorous process produces more and smaller fragments than a less vigorous process. The state of insects may also be a factor. Dead (dried) forms produce greater numbers of fragments than live forms. These factors have been of concern to food processors for some time and have prompted the search for more objective means of determining insect contamination.

34.5 OTHER TECHNIQUES

34.5.1 Overview

Methods described in Sect. 34.3 are directed primarily at routine quality control efforts to determine if the level of natural or unavoidable defects is below the defect action level. To a certain extent, those routine methods can be used to identify the source of contaminants in processed foods. For example, the identification of certain insect fragments can indicate infestation in the raw commodity rather than in the processing system. However, other more sophisticated techniques offer opportunities to pinpoint the nature and source of other contaminants that may exist unavoidably or due to mistakes, accidents, material or equipment failures, or intentional adulteration.

The detection of insects in stored grain and the quantitation of insect parts present in grain products represent a serious and continuing problem for the grain industry. Approved methods of detection primarily involve visual and microscopic inspection and X-ray analysis which require trained personnel and are time consuming, difficult to standardize, and expensive. The assays for insect contamination are preferred to be highly specific, sensitive, rapid, and inexpensive. Moreover, it ideally should be employable by persons having minimal training, particularly in nonlaboratory settings such as at grain elevator and processing sites.

There are several attempts to develop rapid and efficient methods including the use of nuclear magnetic resonance, sound amplification, and infrared spectrometry as alternatives to presently used chemical techniques mentioned in preceding sections. Most of these techniques are expensive and challenging due to difficulty in quantification and identification of specific infestations. Immunological assays, which have found widespread use in clinical diagnostic settings and also in home use, have been explored to detect insect contamination. These methods are described below as they relate to detecting an infestation.

34.5.2 X-Ray Radiography

X-ray radiography is widely used as a test reference method [16]. Grain processors use it as a means of inspecting wheat for internal insect infestation, which is the main the source of insect fragments in processed cereal products (Fig. 34.5). The existing X-ray techniques enable the classification of at least four stages of insect development by measuring the area occupied by the insect, and an accurate classification is also possible based on visible insect morphology [17]. The use of real-time digital imaging instead of X-ray radiographs to discriminate the infested kernels significantly shortened the X-ray procedures. Conventional film observations give, however, better accuracy (3% error rate) than the digital images (11.7%) for infestation by third-larval instars, while the error is less than 1% for both methods with a more advanced stage of larval development [18].

Fig. 34.5 X-ray radiograph of infested wheat. (**a**) Lesser grain borer pupae. (**b**) Rice weevil pupae. (Courtesy of Moses Khamis)

34.5.3 X-Ray Microtomography

X-ray microtomography (XMT) is an emerging 3-D imaging technique that operates on the same basic principles as medical computed tomography (CT) scanners but has much higher resolution (see also Chap. 32, Sect. 32.5). It is very effective in characterizing various internal structural features, which are not possible with conventional 2-D imaging methods (Chap. 32, Sect. 32.5.2). Conventional imaging techniques such as light microscopy, scanning electron microscopy (SEM), and digital video imaging have some limitations: they are destructive in nature, as sample preparation involves cutting to expose the cross section to be viewed. High-resolution XMT has a wide range of applications in science and engineering for which accurate 3-D imaging of internal structure of objects is crucial. XMT is able to capture several features of the internal structures of grain kernels, which are not possible with the conventional imaging methods [19]. Figure 34.6 is demonstration of shadow images taken at several step angles during scanning. A typical scan creates 200–400 of those images that are then used to create axial, sagittal, and coronal views shown in Fig. 34.7.

34.5.4 Electrical Conductance Method

The **electrical conductance** method is based on monitoring the conductance signals for each single kernel during milling in a single kernel characteristics system (SKCS), which is commonly used for wheat hardness determination [17]. This method is highly accurate for detecting older developmental stages of insects: the percentage of properly classified cases for small, medium, and large larvae and pupae are 24.5, 62.2, 87.5, and 88.6, respectively. The accuracy of this method depends also on insect species (rice weevil and lesser grain borer) and wheat type (soft or hard red winter wheat). This approach is not a viable option for identification of insect eggs, undeveloped larvae, and deceased insects in grains with low moisture content due to poor conductivity [20]. A separate study reported detection of maize weevil infestation during various stages of development in popcorn kernels using a conductive roller mill [21].

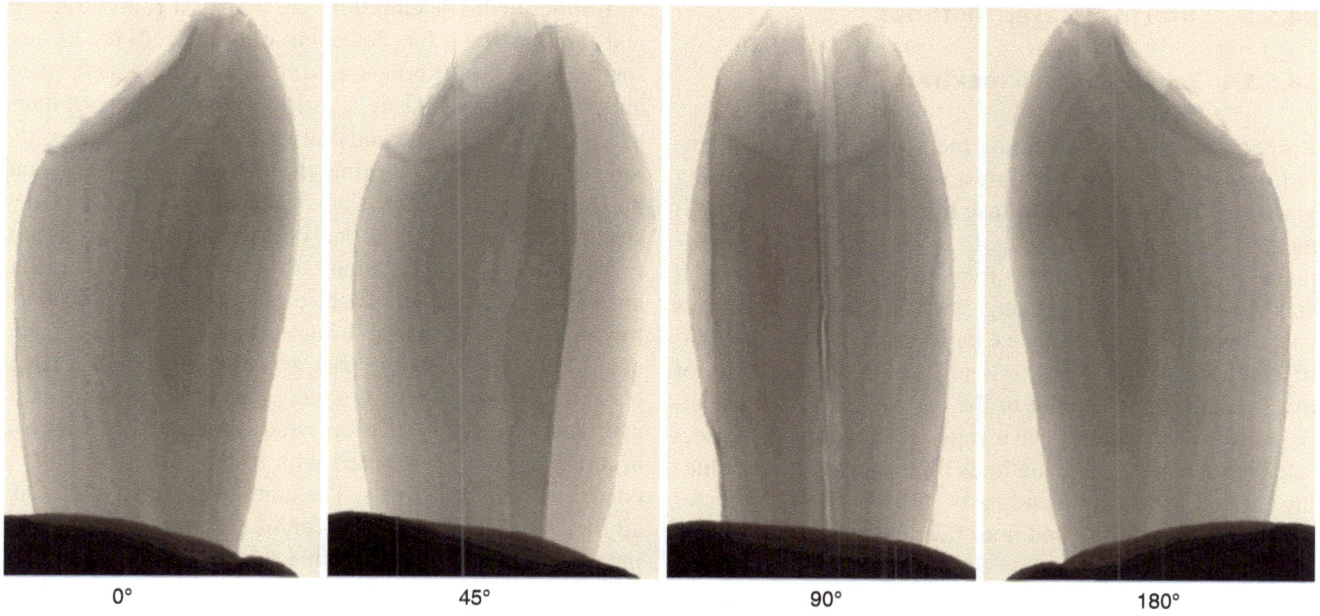

Fig. 34.6 Shadow images of a wheat kernel at various rotation angles

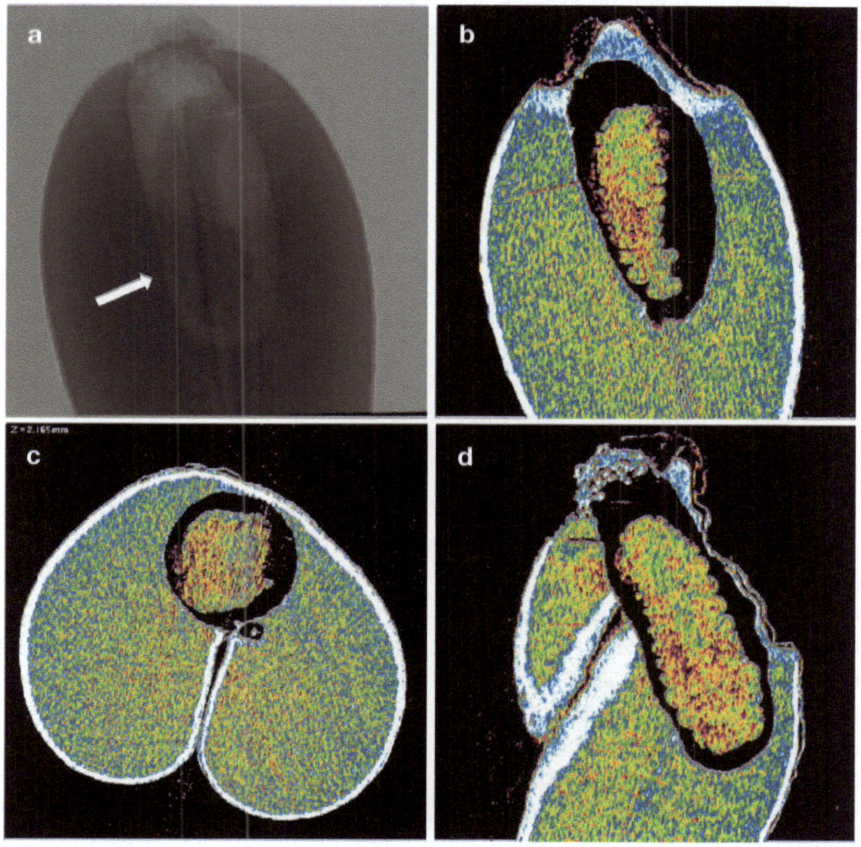

Fig. 34.7 X-ray microtomography (XMT) images. (**a**) Shadow image, (**b**) sagittal view, (**c**) axial view, and (**d**) coronal view of lesser grain borer in a wheat kernel

34.5.5 Olfactory-Based Methods

34.5.5.1 Solid-Phase Microextraction (SPME)

Recently, odor-based detection methods are gaining popularity in detecting insect infestation and for evaluating grain quality. In the **solid-phase microextraction** (SPME) method (see Chap. 14, Sect. 14.2.2.4), the volatile compounds vaporized from the food samples are condensed and then evaluated by using gas chromatography-mass spectrometry (GC–MS) (see Chap. 11, Sect. 11.4) to quantify the volatiles [22]. The efficiency and the sensitivity of this method are attributed to the extraction time and temperature. Many studies have reported detection of insect infestation using SPME methods for insect species, including the red flour beetle and rusty grain beetle [23], lesser grain borer [24], confused flour beetle, and granary weevil [25].

34.5.5.2 Electronic Nose (E-nose)

Electronic nose encompasses various types of sensors and operates on the fundamental principle of electronic aroma detection [26]. The system consists of an odor sensor, data preprocessor, and an interpretation system. The sensor detects the volatile compounds in the headspace of the food samples and responds by changing the electrical properties and differentiates volatiles from a predefined database [27] (see Chap. 14, Sect. 14.3.5; Chap 35, Sect. 35.5.2.1). Using E-nose, rapid and automatic detection of insect infestation in grain samples is possible [28]. One of the key considerations for the detection test by E-nose is to select appropriate sensors specific to volatile organic compounds (VOCs). A few studies have reported success in detecting insect infestation, identifying insect species, and predicting insect population in grain samples [27, 29]. However, these methods show promise; further investigation is required to assess their reliability at lower insect densities and to ensure high accuracy in identifying insect species.

34.5.6 Impact-Acoustic Emission

Acoustic detection is a very promising method for early detection of insects in bulk grain. Acoustic identification of insects is based on their ability to generate sound by eating, flying, egg laying, or locomotion [30]. Despite the fact that it has been a known pest detection method in storage facilities for a long time, reliability, and efficacy of acoustic pest detection have been improved only recently as a result of the development of improved acoustic devices and signal processing methods [30].

Impact-acoustic emissions are used as a nondestructive, real-time method for detection of damaged grains and shelled nuts [31]. Kernels are impacted onto a steel plate, and the resulting acoustic signal is analyzed to detect damage using different methods: modeling of the signal in the time domain, computing time-domain signal variances and maximums in short-time windows, analysis of the frequency spectrum magnitudes, and analysis of a derivative spectrum. Features were used as inputs to a stepwise discriminant analysis routine, which selected a small subset of features for accurate classification using a neural network. Pearson et al. [31] reported that impact-acoustic emission is a feasible and promising method for detection of IDK, sprout damage, and scab damage. More study is needed to improve accuracy on kernels infested with insects that have not yet emerged from the kernels. The computational cost of classifying a kernel using this technique is very low, allowing inspection of large numbers of wheat kernels very rapidly, ~40 kernels/s. Grain inspectors usually use a 100 g (3000 kernel) sample to inspect for IDK. This takes an inspector approximately 20 min to analyze manually but can be accomplished in about 75 s with an acoustic system.

According to official standards for wheat grain, a sample as a whole is considered infested when it contains two or more live weevils, or one live weevil and one or more other live insects injurious to stored grain, or two or more other live insects injurious to stored grain [32]. Eliopoulos et al. [30] evaluated the efficacy of bioacoustics in detecting the presence of adult beetles inside the grain mass using a piezoelectric sensor and an acoustic emission amplifier. The accuracy of this system has been reported to be 72–100% in detecting 1–2 insects per kilogram of hard wheat, which is the standard threshold for classifying a grain mass "clean" or "infested." Also, the detection threshold of this system is very low, 0.1–0.5 insects per kilogram of grain, which is not possible with many other detection methods [30].

34.5.7 Microscopy Techniques

Microscopy techniques including light microscopy, fluorescence microscopy, and scanning electron microscopy (SEM) are used to study the structure/function relationships of food but also can be applied to questions of extraneous matter (see also Chap. 32, Sect. 32.2). For example, SEM with energy-dispersive spectroscopy (EDS) can be used to determine the nature of metals in products that may be due to equipment failure or intentional adulteration due to tampering [33]. Light microscopy in a polarized mode can be used to distinguish between plastics, glass, and other fiber or crystalline contaminants [34].

34.5.8 Near-Infrared Spectroscopy

Near-infrared spectroscopy (NIRS) (see also Chap. 8, Sect. 8.4) is a relatively fast, accurate, and economical technique available to the grain industry for compositional analysis such as water, oil, fiber, starch, and protein in grains and seeds. It also has relatively recent applications in analysis of extraneous matter. NIRS has been used to: (1) identify several coleopteran species [35], (2) detect parasitized weevils in wheat kernels [36], and (3) detect external and internal insect infestation in wheat [37–39]. The near-infrared (NIR) technique has proven to be a reliable method for rapid assessment of immature and adult fragments of rice weevil in flour obtained from infested wheat kernels [40]. Berardo et al. [41] reported that NIRS predicts the percentage of *Fusarium verticillioides* infection in maize kernels and the content of ergosterol and fumonisin B1 in meals. (*Fusarium verticillioides* is the most common fungal species infecting maize; it generates the mycotoxin, fumonisin B1.) In the same way, promising results were obtained when NIRS methodology was applied to detect scab-damaged kernels [42] and estimate deoxynivalenol, ergosterol, and fumonisin in single kernels of wheat [35] and corn [43].

Near-infrared spectroscopy used with a single-kernel characterization system is able to detect later stages of internal insect infestation in wheat with a 95% confidence [39]. In contrast to other procedures, this system is capable of being automated and incorporated into the current grain inspection process. NIRS also has been compared with the current standard insect fragment flotation method for its ability to detect insect fragments in flour [44]. Fragment counts with both techniques were correlated; however, the flotation method was more sensitive below the FDA DAL of 75 insect fragments per 50 g of wheat flour. NIR spectroscopy was able to predict accurately whether flour samples contained less than or more than 130 fragment/50 grams.

Hyperspectral imaging is a recent development that has proven to be effective in grain quality inspection. It is an integration of conventional imaging and spectrometry, which involves acquiring images at more than 20 wavelengths within the electromagnetic spectrum [45, 46]. Since it is a complex indirect method, hyperspectral data should be confirmed for homogeneity before statistical modeling through various preprocessing techniques, as described in detail by Ravikanth et al. [45]. In the area of cereal grain quality, hyperspectral imaging has been successfully used to identify fungus-damaged kernels [46–48], insect-damaged kernels [49, 50], defective wheat [42, 51], and wheat classes [52, 53]; to detect contaminants from different origins (animals, other cereal grains, botanical impurities, other contaminants) in cereal grains; [54], and to quantify ergot bodies in cereals [55].

34.5.9 Artificial Intelligence-Based Techniques

Artificial intelligence (AI) has witnessed significant advances in recent years with the utilization of high-performance computing and big data technologies [56]. The utilization of computer-assisted techniques in beetle and food safety investigations has been established over an extended period [57–61]. Among these methods, **deep learning** (DL) has emerged as a promising solution for image processing, object detection, and recognition. Specifically, **convolutional neural networks** (CNNs) are capable of directly incorporating the raw image as input, allowing integration of the feature construction step within the learning process. This integration occurs through the adjustment of CNN parameters and connections based on the error derived from a specific training dataset [62]. A CNN trained on a dataset of beetle elytra images was employed for the detection and identification of 15 species of stored-product insects [63]. The findings revealed that this approach achieved an accuracy of over 83% in cross-validation, suggesting the potential for even greater performance when applied to a real fragment recognition model encompassing varying sizes and shapes. A similar approach was undertaken by utilizing microscopic images of elytra to develop a dataset for detection and identification of 15 stored-product insect beetles. The artificial neural network provided about 80% accuracy to differentiate species [64]. A recent study showed that CNN models with improved image quality datasets provided about 90% accuracy in prediction of 27 different pantry beetles [65].

34.5.10 Molecular-Based Methods

To date, only a limited number of **molecular-based approaches** have been devised for the detection of insect contaminants beetle and moth species contaminating food products especially grains [66, 67]. In the field of insect identification, DNA-based methods predominantly rely on amplifying the DNA barcode region of the mitochondrial cytochrome c oxidase subunit I (COI) [68, 69]. A multiplex PCR-based method was employed to detect two insect species commonly found in stored products [70]. The study was conducted by spiking insect fragments into whole wheat flour at a concentration of 1%. The results indicated the effectiveness of PCR method for identifying insect species. However, the researchers suggested the need for additional investigations to extend the applicability of insect identification at lower spike levels, thus aligning with real-world scenarios. Another method of identification is the microarray chip method which has received considerable attention due to its high throughput, parallelism, and sensitivity [71]. Few studies have used microarray-based methods for identification of different psocid (which are tiny insects that live in moist environments generally feeding on fungi or mold; they feed on the germ of damaged and broken kernels) species and reported promising results [71–73].

34.5.11 Enzyme-Linked Immunosorbent Assays

To develop an optimal immunological assay for an insect contamination of foodstuffs, antibodies are required that are directed against an insect-specific antigen, preferably protein, likely to be present in any life stage of the contaminating insect or in insect remains. Antigens and antibodies are two key parts of any immunoassay (see also Chap. 27).

For an immunoassay with broad specificity, it is required to use an insect-specific protein such as myosin. Myosin is ubiquitous in insects; it is present in large quantities in adult insect tissue and is also present in appreciable quantities in other life stages [74]. An **enzyme-linked immunosorbent assay** (ELISA) method has been developed to measure quantitatively the amount of insect material in a sample [74]. It is also possible to develop an immunoassay specific for a particular species of insect contamination using antibodies having a unique species specificity. Kitto et al. [75] developed such techniques (patented in 1992) for detecting the amount of insect contamination in foodstuffs. The method comprises the following steps:

1. Preparing an aqueous solution or suspension of a homogenized grain sample
2. Substantially affixing at least a portion of solution or suspension to a solid surface
3. Applying to solid surface a specifically binding insect antigen (or antibody) and enzyme to form an antibody–enzyme conjugate, resulting in formation of a colored product when the enzyme reacts with a substrate
4. Washing unbound conjugate from the solid surface
5. Incubating the solid surface with an enzyme substrate under conditions allowing colored product to be formed when enzyme is present
6. Correlating amounts of color formed with an amount of insect contamination

Recent research [76] showed that the myosin in fourth instars of the lesser grain borer developing within kernels of wheat degraded within the first 2 weeks when larvae were killed with phosphine, a fumigant commonly used to manage insect infestations in stored grain. Myosin degradation resulted in underestimating insect fragment estimates by about 58%.

34.6 COMPARISON OF METHODS

A number of methods that have been developed to detect insects in commodity samples (Table 34.3) are described here in general terms:

1. Density separation based on infested kernels being lighter weight and floating in a liquid
2. Staining kernels to detect weevil egg plugs
3. Detection of carbon dioxide or uric acid produced by the internally feeding insects
4. Detection of insects hidden inside kernels using near-infrared spectroscopy (NIRS)
5. Detection by olfactory based methods
6. Detection by use of nuclear magnetic resonance (NMR)
7. Detection by X-ray images and digital image analysis techniques
8. Acoustical sensors to hear sounds from insects feeding inside kernels
9. Artificial intelligence to identify insect species by deep learning
10. Identification of insects using molecular methods by DNA bar coding
11. Enzyme-linked immunosorbent assays (ELISA) to detect myosin in insect muscle

Some of the recent methods have been developed by adapting the single-kernel characterization system (SKCS), computed tomography (CT), acoustic-impact emissions, and use of an electrical conductive roller mill [17, 77, 78].

The choice of method depends on several factors: (i) type of infestation (inside or outside food grains, in the surrounding premises or inside bulk grain), (ii) required level of inspection (macroscopic vs. microscopic, qualitative vs. quantitative), (iii) availability of equipment and facilities, and (iv) required sensitivity [79]. Most of the methods aim to detect the presence of live insects directly or indirectly. External insects are detected by visual inspection, sampling, sieving, and heat-extraction methods, while internal (hidden) insects are detected by radiography, staining techniques to identify egg plugs, and near-infrared and fragment count methods. Determination of uric acid or CO_2 level serves as an indirect way of detecting and estimating internally feeding insects, and these methods may be suitable if infestations are restricted to one insect species. Depending on some storage conditions, grain may contain molds and insects, and in such cases CO_2 produced by molds may interfere in accurately detecting or estimating insects. Olfaction based methods can be mostly effective at higher insect densities as the efficacy is dependent based on the volatile organic compound (VOCs) secretion by insects. Artificial Intelligence using deep learning has shown 90% accuracy but needs further exploration to confirm their efficacy at larger scale and

Table 34.3 Insect detection methods applicable for commodity samples

Test method	*Applicability*	*Comments*
Visual inspection	Whole grains, milled products	Qualitative. Only high-level infestation detected
Sampling and sieving	Whole grains, milled products	Commonly practiced. Hidden infestation not detected
Heat extraction	Whole grains	Adults and larvae detected
Acoustics	Whole grains	Feeding sounds: Active stages detected Impact-acoustic emissions: Nondestructive. Real time. Detect insect, sprout, and scab damage
Breeding out	Whole grains	Time consuming
Imaging techniques		
X-ray method	Whole grains	Nondestructive. Highly accurate. Able to detect both live and dead insects inside grain kernels. Cannot detect insect eggs. Prohibitive capital cost
Near-infrared spectroscopy	Whole grains, milled products	Rapid. Sensitive. Can be automated. No sample preparation. Cannot detect low levels of infestation. Sensitive to moisture content. Calibration of equipment is complex and frequent
Nuclear magnetic resonance	Whole grains	Less sensitive
Serological techniques	Whole grains, milled products	Highly sensitive. Species specific. Shows infestation from unknown past to till date
Uric acid determination	Whole grains, milled products	Shows infestation from unknown past to till date
CO_2 analysis	Whole grains	Simple. Time consuming. Indicates current level of infestation. Not suitable for grains having >15% moisture
Specific gravity methods	Whole grains	Simple and quick. Not suitable for oats and maize
Cracking and flotation method	Whole grains	Variable results noted
Fragment count	Whole grains, milled products	Highly variable results noted. Shows infestation from unknown past to till date
Staining techniques		
Egg plugs	Whole grains	Specific for *Sitophilus* spp.
Ninhydrin method	Whole grains	Eggs and early larvae not indicated
Olfactory method		
Solid Phase Microextraction	Whole grains, milled products	Dependent upon insect volatile secretion, and effectiveness at lower insect densities unknown.
Electronic-Nose	Whole grains, milled products	Dependent upon insect volatile secretion, and effectiveness at lower insect densities unknown.
Artificial Intelligence	Whole grains	Have shown 90% accuracy for identification of insect species, and needs further exploration.
Molecular methods	Whole grains	Not reported to effective at lower spiked levels of contaminants.

Adapted from Rajendran [79]

detecting different insect stages. Fragment count and ELISA methods can be used for the detection of both living and dead insects. In general, problems encountered with these detection methods are that the most accurate methods, such as X-ray and computed tomography (CT), are laborious and expensive, while rapid, automated methods may not be suitable for detecting eggs and young larvae [76].

34.7 ISOLATION PRINCIPLES APPLIED TO FOOD PROCESSING

Examination of stored-product insects often requires extracting them from the commodity. An intensive summary of literature survey on insect extraction and detection methods can be found in reference [80]. Isolation principles, such as particle size and density, discussed in the preceding sections are designed to identify extraneous materials in finished food products, monitoring quality, and compliance with DALs. In addition, some of these principles of isolation are used in a **proactive** way during processing to prevent extraneous matter from being incorporated into finished food products.

Wheat that contains hidden internal insect infestation is the primary source of insect fragments in processed cereal products. The current DAL for internal insect infestation in wheat is 32 **IDK** per 100 g of wheat [3]. IDK are those **visually** determined to have insect tunneling or emergence holes. Most processors rely on much lower levels of IDK (≤6 IDK/100 g) to produce flour that meets customer tolerances and the FDA's DAL for insect fragments in flour. In addition, to prevent adulteration of flour with filth, entoleters, and infestation destroyers in the milling process break up insect-damaged kernels and these broken kernels along with the insect fragments are aspirated out of the milling stream. As previously indicated, **X-ray radiography** is used by some as a means for selecting grains for processing or for research purposes to age-grade internally developing stages of stored-grain insects. More recently, NIR spectroscopy has provided a new tool for assessing internal insect infestation in wheat. By selectively milling only wheat that has minimal or no evidence of internal insect infestation, grain processors can effectively limit insect fragments in their products. In like manner, bakers and other users of processed grain products can selectively monitor for insect fragments in their raw materials using one of the approved methods for extraction and enumeration of fragments or by sending samples to a private laboratory for fragment analysis.

Most food processing systems that deal with agricultural products generally apply some type of cleaning operations as an initial step. In flour milling, for example, wheat is passed through a system called the "cleaning house," which consists of a series of machines that apply the principles of particle size and density separation. Sieves remove contaminants larger than wheat kernels as well as finer contaminants such as sand. In addition, air (aspiration) is used to remove plant material that is lighter than the grain. Current equipment to remove stones and other dense materials the same size as grain kernels uses air passed upward through an inclined, tilted table. This causes the grain to "float" off the side of the table and the heavier material to continue and "tail" over the end of the table. In earlier systems, grains were passed through washers in which water separated the grain from heavier material (such as stones) much like fluming of potatoes or fruit does. Impact with rotating disks and steel pegs (entoleters and infestation destroyers) or grinding operations are used prior to milling to break open kernels of wheat containing internal insect infestation. As a means of reducing insect fragments in the finished product, this process is followed by aspiration to lift out any light insect contaminants released in the operation.

As a final step in wheat milling, flour is generally passed through sieves fine enough to remove insect eggs and any other contaminants that might be present. This is to assure that when flour leaves the mill, it is free of any viable form of insect contamination [81]. Where flour is used in large quantities, such as commercial bakeries, prior to use, flour is again sieved to ensure that no contamination has occurred in transport and storage of the flour.

Metal contamination has been a major concern of all food processors. Although metal detection methods are not specifically among the isolation techniques represented in AOAC International or AACCI methods manuals, they serve the purpose of isolating contaminants from food products. Magnets of various types have been used on raw materials and processing systems to prevent the passage of metal into handling and processing equipment where both equipment damage and product contamination are concerns. **Metal detectors** are employed in many food processing operations and on finished product packaging lines to detect ferrous and nonferrous metal fragments and to prevent contaminated products from entering consumer food channels.

Recent X-ray technology suggests that X-rays may have an advantage over other methods for detecting metal and that they can also be used to detect glass, wood, plastic, and bone chips in foods. Detection of these extraneous materials also can be automated with rejection systems in packaging lines [82].

34.8 SUMMARY

Extraneous matter in raw ingredients and in processed foods might be unavoidable in the array of foods that are stored, handled, processed, and transported. DALs are established for amounts considered unavoidable and of no health hazard. A variety of methods are available to isolate extraneous matter from foods. Those methods largely prescribed by AOAC International employ a series of physical and chemical means to separate the extraneous material for identification and enumeration. Major concerns in the analysis of food products for extraneous matter are the objectivity of methods and the availability of adequately trained analysts. Some "principles" of isolation are applied in a proactive way in food processing operations.

Currently available methods (both macroscopic and microscopic) show varying degrees of efficiency in analysis of extraneous matter and filth in foods. Some techniques are time consuming, require trained personnel, and are difficult to implement in real time. Some techniques have not been found feasible to be implemented in food inspection systems because of their cost, unreliability, and the varying degrees of success obtained in detecting infestations. Macroscopic and microscopic procedures for characterizing defects in foods tend to supplement each other and together provide a comprehensive evaluation of defects in the product. It is important that the analyst realizes the close association of complementary methods for use as a joint approach in solving analytical problems.

Examining products for extraneous matter is quite different than performing product evaluations for chemical and microbiological contaminants. Typically, the extraneous matter becomes more difficult to detect following its introduction to the process; thus, implementation of comprehensive preventative measures are the ultimate cost-effective safety measures.

34.9 STUDY QUESTIONS

1. Indicate why the FDA has established DALs.
2. Explain why practicing cGMPs has no impact on DALs.
3. List three major reasons for conducting analysis for extraneous matter in foods.

4. What two resources provide methods for separating extraneous matter from cereal grains and their products?
5. There are several basic principles involved in separating (isolating) extraneous matter from foods. List five of these principles and give an example of each principle.
6. Briefly describe the major constraint(s) to currently accepted methods for analyses of extraneous matter in foods.
7. Explain how some of the more recent analytical techniques can assist in identifying sources of extraneous matter in foods.
8. What are some likely sources of error with the various analytical methods?

Acknowledgments We thank Mr. Manivannan Selladurai, Ph.D. Candidate, Department of Grain Science and Industry, Kansas State University, Manhattan, Kansas, for his invaluable assistance in updating information in this chapter. This article is contribution number 24-014-E of the Kansas Agricultural Experiment Station.

REFERENCES

1. FDA (2022) Federal Food Drug and Cosmetic Act, as amended. In: Compilation of food and drug laws. Section 201. Authenticated US government information. Accessed from https://www.govinfo.gov/content/pkg/COMPS-973/pdf/COMPS-973.pdf.
2. FDA (2015) Current good manufacturing practice in manufacturing, packing, or holding human food. Part 110, Title 21: food and drugs. In: Code of federal regulations. Office of the Federal Register National Archives and Records Administration, Washington, DC
3. FDA (2018) The food Defect Action Levels - current levels for natural or unavoidable defects for human use that present no health hazard. Department of Health and Human Services, Food and Drug Administration. Washington, DC. Accessed from https://www.fda.gov/food/ingredients-additives-gras-packaging-guidance-documents-regulatory-information/food-defect-levels-handbook
4. FDA (2016) Compliance policy guide manual. Food and Drug Administration, Office of Regulatory Affairs, Washington, DC
5. AOAC International (2016) Extraneous Materials: isolation. In: Official methods of analysis, 20th edn., 2016 (online). AOAC International, Rockville, MD
6. AACC International (2010) AACCI method 28 extraneous matter. In: Approved methods of the American association of cereal chemists, 11th edn., AACC International, St. Paul, MN
7. Kurtz OL, Harris KL (1962) Micro-Analytical entomology for food sanitation control. Association of Official Analytical Chemists, Washington, DC
8. Gentry JW, Harris KL (1991) Microanalytical entomology for food sanitation control, vols 1 and 2. Association of Official Analytical Chemists, Melbourne, FL
9. AACC International (2010) X-ray examination for internal insect infestation, AACCI method 28–21.02. In: Approved methods of the American association of cereal chemists, 11th edn. AACC International, St. Paul, MN
10. FDA (1981) Principles of food analysis for filth, decomposition, and foreign matter. FDA technical bulletin no. 1, Gorham JR (ed), Association of Official Analytical Chemists, Arlington, VA
11. FDA (1978) Training manual for analytical entomology in the food industry. FDA technical bulletin no. 2, Gorham JR (ed), Association of Official Analytical Chemists, Arlington, VA
12. Olsen AR (ed) (1995) Fundamentals of microanalytical entomology - a practical guide to detecting and identifying filth in foods. CRC, Boca Raton, FL
13. FDA (1998) Introduction and apparatus for macroanalytical methods. In: FDA technical bulletin number 5, macroanalytical procedures manual (MPM), FDA, Washington, DC
14. Tilocca MG, Caneglias E, Vodret B, Mancuso MR, Zimmardi A, Manno C, Schiavo, M R (2015) Analysis of foreign matter in foodstuffs using the light filth test: report 2012-2013. Ital J Food Saf 4(3):4504
15. Nakashbma MJ, Hajoway JM, Keizer H, McGary ED, Senff WA, Silva ML, Spurlock LF, Stenske MA, Torn A, Wheeler M (1993) Extraction of light filth from tofu: collaborative study. J AOAC Intl 76(1):50–52
16. Pedersen JR (1992) Insects: identification, damage, and detection, ch. 12. In: Sauer DB (ed) Storage of cereal grains and their products. American Association of Cereal Chemists, St. Paul, MN, pp. 635–689
17. Pearson TC, Brabec DL, Schwartz CR (2003) Automated detection of internal insect infestations in whole wheat kernels using a Perten SKCS 4100. Appl Eng Agric 19:727–733
18. Haff RP, Slaughter DC (2004) Real-time X-ray inspection of wheat for infestation by the granary weevil, *Sitophilus granarius* (L.). Trans ASAE 47:531–537
19. Dogan H (2007) Non-destructive Imaging of agricultural products using X-ray microtomography. Proc Microsc Microanal Conf, 13(2):512–513
20. Pearson T, Brabec DL (2007) Detection of wheat kernels with hidden insect infestations with an electrically conductive roller mill. Appl Eng Agri 23(5):639
21. Brabec D, Dowell F, Campbell J, West M (2017) Detection of internally infested popcorn using electrically conductive roller mills. J Stored Prod Res 70:37–43
22. Banga KS, Kotwaliwale N, Mohapatra D, Giri SK (2018) Techniques for insect detection in stored food grains: An overview. Food Control 94:167–76.
23. Senthilkumar T, Jayas DS, White NDG, Freund MS, Shafai C, Thomson DJ (2012) Characterization of volatile organic compounds released by granivorous insects in stored wheat. J Stored Prod Res, 48:91–96
24. Niu Y, Hardy G, Agarwal M, Hua L, Ren Y (2016) Characterization of volatiles *Tribolium castaneum* (H.) in flour using solid phase micro extraction–gas chromatography mass spectrometry (SPME–GCMS). Food Sci Hum Well 5(1):24–29
25. Abuelnnor N, Jones PRH, Ratcliffe NM, de Lacy Costello B, Spencer-Phillips PT (2010) Investigation of the semiochemicals of confused flour beetle *Tribolium confusum* Jaquelin du Val and grain weevil *Sitophilus granarius* (L.) in stored wheat grain and flour. Integr Pest Manag Rev 7(2):105–121
26. Wilson AD (2013) Diverse applications of electronic-nose technologies in agriculture and forestry. Sensors 13(2):2295–2348.
27. Wu J, Jayas DS, Zhang Q, White NDG, & York RK (2013). Feasibility of the application of electronic nose technology to detect insect infestation in wheat. Can Biosyst Eng 55:3.1–3.9
28. Zhang H, Wang J, Tian X, Yu H, Yu Y (2007) Optimization of sensor array and detection of stored duration of wheat by electronic nose. J Food Eng 82(4):403–408
29. Stetter JR, Findlay Jr MW, Schroeder KM, Yue C, Penrose, WR, (1993) Quality classification of grain using a sensor array and pattern recognition. Anal Chim Acta 284:1–11
30. Eliopoulos PA, Potamitis I, Kontodimas DC, Givropoulou EG (2015) Detection of adult beetles inside the stored wheat mass based on their acoustic emissions. J Econ Entomol 108(6):1–7
31. Pearson TC, Cetin AE, Tewfik AH, Haff RP (2007) Feasibility of impact-acoustic emissions for detection of damaged wheat kernels. Digital Signal Process 17:617–633

32. FGIS (2015) Official United States standards for grain, 7 CFR Part 810. Federal Grain Inspection Service. USDA, Washington, DC.
33. Goldstein JI, Newbury DE, Michael JR, Ritchie NW, Scott JHJ, Joy DC (2017) Scanning electron microscopy and X-ray microanalysis. New York:Springer.
34. McCrone WC, Delly JG (1973) The particle atlas, 2nd edn., Ann Arbor Science, Ann Arbor, MI
35. Dowell FE, Ram MS, Seitz LM (1999) Predicting scab, vomitoxin, and ergosterol in single wheat kernels using near-infrared spectroscopy. Cereal Chem 76(4):573–576
36. Baker JE, Dowell FE, Throne JE (1999) Detection of parasitized rice weevils in wheat kernels with near-infrared spectroscopy. Biol Control 16:88–90
37. Johnson JB (2020) An overview of near-infrared spectroscopy (NIRS) for the detection of insect pests in stored grains. J Stored Prod Res 86:101558.
38. Ghaedian AR, Wehling RL (1997) Discrimination of sound and granary-weevil-larva-infested wheat kernels by near-infrared diffuse reflectance spectroscopy. J AOAC Int 80: 997–1005
39. Dowell FD, Throne JE, Baker JE (1998) Automated nondestructive detection of internal insect infestation of wheat kernels by using near-infrared reflectance spectroscopy. J Econ Entomol 91:899–904
40. Toews MD, Perez-Mendoza J, Throne JE, Dowell FE, Maghirang E, Arthur FH, Campbell JF (2014) Rapid assessment of insect fragments in flour milled from wheat infested with known densities of immature and adult *Sitophilus oryzae* (Coleoptera: Curculionidae), J Econ Entomol 100(5):1714–23.
41. Berardo N, Pisacane V, Battilani P, Scandolara A, Pietro A, Marocco A (2005) Rapid detection of kernel rots and mycotoxins in maize by near-infrared reflectance spectroscopy. J Agri Food Chem 53:8128–8134
42. Delwiche SR, Hareland GA (2004) Detection of scab-damaged hard red spring wheat kernels by near-infrared reflectance. Cereal Chem 81(5):643–649
43. Dowell FE, Pearson TC, Maghirang EB, Xie F, Wicklow DT (2002) Reflectance and transmittance spectroscopy applied to detecting fumonisin in single corn kernels infected with *Fusarium verticillioides*. Cereal Chem 79(2):222–226
44. Perez-Mendoza P, Throne JE, Dowell FE, Baker JE (2003) Detection of insect fragments in wheat flour by near-infrared spectroscopy. J Stored Prod Res 39:305–312
45. Ravikanth L, Singh CB, Jayas, DS, White NDG (2015) Classification of contaminants from wheat using near-infrared hyperspectral imaging, BiosystEng 135:73–86
46. Bauriegel E, Giebel A, Geyer M, Schmidt U, Herppich WB (2011) Early detection of Fusarium infection in wheat using hyperspectral imaging. Comput Electron Agri 75(2):304–312
47. Choudhary R, Mahesh S, Paliwal J, Jayas DS (2009) Identification of wheat classes using wavelet features from near infrared hyperspectral images of bulk samples. Biosyst Eng 102(2):115–127
48. Singh CB, Jayas DS, Paliwal J, White NDG (2007) Fungal detection in wheat using near-infrared hyperspectral imaging. Trans ASABE 50(6):2171–2176
49. Kaliramesh S, Chelladurai V, Jayas DS, Alagusundaram K, White N, Fields P (2013) Detection of infestation by *Callosobruchus maculatus* in mungbean using near-infrared hyperspectral imaging. J Stored Prod Res 52:107–111
50. Singh CB, Jayas, DS, Paliwal, J, White, NDG (2009) Detection of insect-damaged wheat kernels using near-infrared hyperspectral imaging. J Stored Prod Res 45(3):151–158
51. Singh CB, Jayas DS, Paliwal J, White NDG (2010) Detection of midge-damaged wheat kernels using short-wave near-infrared hyperspectral and digital color imaging. Biosyst Eng 105(3):380–387
52. Mahesh S, Manickavasagan A, Jayas DS, Paliwal J, White NDG (2008) Feasibility of near-infrared hyperspectral imaging to differentiate Canadian wheat classes. Biosyst Eng 101(1):50–57
53. Mahesh S, Jayas DS, Paliwal J, White NDG (2011) Identification of wheat classes at different moisture levels using near-infrared hyperspectral images of bulk samples. Sens Instru Food QualSaf 5(1):1–9
54. Pierna JAF, Vermeulen P, Amand O, Tossens A, Dardenne P, Baeten V (2012) NIR hyperspectral imaging spectroscopy and chemometrics for the detection of undesirable substances in food and feed. Chemom Intell Lab Syst 117:233–239
55. Vermeulen P, Pierna JAF, Van Egmond HP, Dardenne P, Baeten V (2011) Online detection and quantification of ergot bodies in cereals using infrared hyperspectral imaging. Food Additives and Contaminants. Part A, Chemistry, Analysis, Control, Exposure and Risk Assessment 29(2):232–240
56. Othman S, Mavani NR, Hussain MA, Abd Rahman N, Ali JM (2023) Artificial intelligence-based techniques for adulteration and defect detections in food and agricultural industry: A review, J Agri Food Res 25:100590.
57. Boyer DM, Lipman Y, St. Clair E, Puente J, Patel BA, Funkhouser T, Jernvall J, Daubechies I (2011) Algorithms to automatically quantify the geometric similarity of anatomical surfaces, Proc Nat Acad Sci 108(45):18221–6.
58. Daly HV, Hoelmer KI, Norman P, Allen T (1982) Computer-assisted measurement and identification of honey bees (Hymenoptera: Apidae), Ann Entomol Soc Am 75(6):591–4
59. Gaston KJ, O'Neill MA (2004) Automated species identification: why not?, Philosophical Transactions of the Royal Society of London. Series B: Biological Sciences. 359(1444):655–67.
60. Larios N, Deng H, Zhang W, Sarpola M, Yuen J, Paasch R, Moldenke A, Lytle DA, Correa SR, Mortensen EN, Shapiro LG (2008) Automated insect identification through concatenated histograms of local appearance features: feature vector generation and region detection for deformable objects. Mach Vis Appl 19:105–23.
61. LeCun Y, Bengio Y, Hinton G. (2015) Deep learning. Nature 521(7553):436–44.
62. Weeks PJ, Gauld ID, Gaston KJ, O'Neill MA(1997) Automating the identification of insects: a new solution to an old problem. Bull Entomol Res 87(2):203–11.
63. Wu L, Liu Z, Bera T, Ding H, Langley DA, Jenkins-Barnes A, Furlanello C, Maggio V, Tong W, Xu J (2019) A deep learning model to recognize food contaminating beetle species based on elytra fragments. Comput Electron Agric 166:105002.
64. Park SI, Bisgin H, Ding H, Semey HG, Langley DA, Tong W, Xu J (2016). Species identification of food contaminating beetles by recognizing patterns in microscopic images of elytra fragments. PLoS One, 11(6):e0157940.
65. Bisgin H, Bera T, Wu L, Ding H, Bisgin N, Liu Z, Pava-Ripoll M, Barnes A, Campbell JF, Vyas H, Furlanello C (2022) Accurate species identification of food-contaminating beetles with quality-improved elytral images and deep learning. Front Artif Intell 5:952424
66. Negi A, Anandharaj A, Kalakandan S, Rajamani MA (2021) Molecular approach for the detection and quantification of *Tribolium castaneum* (Herbst) infestation in stored wheat flour, Food Technol Biotechnol 59(1):112–21.
67. Solà M, Riudavets J, Agustí N (2018) Detection and identification of five common internal grain insect pests by multiplex PCR. Food Cont 84:246–54.
68. Folmer O, Black M, Hoeh W, Lutz R, Vrijenhoek R (1994) DNA primers for amplification of mitochondrial cytochrome c oxidase subunit I from diverse metazoan invertebrates. Mol Mar Biol Biotechnol 9:7881515.

69. Hebert PD, Cywinska A, Ball SL, DeWaard JR (2003) Biological identifications through DNA barcodes. Proc Royal Soc London. Series B: Biol Sc 270(1512):313–21.
70. Pava-Ripoll M, Miller AK, Ziobro GC (2023) Development of A Multiplex Polymerase Chain Reaction (PCR) Assay for the Potential Detection of Insect Contaminants in Food. J Food Prot 20:100120.
71. Liu LJ, Pang AH, Feng SQ, Cui BY, Zhao ZH, Kučerová Z, Stejskal V, Opit G, Aulicky R, Cao Y, Li FJ (2017) Molecular identification of ten species of stored-product psocids through microarray method based on ITS2 rDNA. Sci Rep 7(1):1–9.
72. Wei DD, Yuan ML, Wang ZY, Wang D, Wang BJ, Dou W, Wang JJ (2011) Sequence analysis of the ribosomal internal transcribed spacers region in psocids (Psocoptera: Liposcelididae) for phylogenetic inference and species discrimination. J Econ Entomol 104(5):1720–9.
73. Zhao ZH, Cui BY, Li ZH, Jiang F, Yang QQ, Kučerová Z, Stejskal V, Opit G, Cao Y, Li FJ (2016) The establishment of species-specific primers for the molecular identification of ten stored-product psocids based on ITS2 rDNA. Sci Rep 6(1):1–8.
74. Quinn FA, Burkholder WE, Kitto GB (1992) Immunological technique for measuring insect contamination of grain. J Econ Entomol 85:1463–1470
75. Kitto GB, Quinn FA, Burkholder W (1992) Techniques for detecting insect contamination of foodstuffs, US Patent 5118610
76. Atui MB, Flin PW, Lazzari SMN, Lazzari FA (2007) Detection of *Rhyzopertha dominica* larvae in stored wheat using ELISA: the impact of myosin degradation following fumigation. J Stored Prod Res, 43:156–159
77. Toews MD, Pearson TC, Campbell JF (2006) Imaging and automated detection of *Sitophilus oryzae* (Coleoptera: Curculionidae) pupae in hard red winter wheat. J Econ Entomol 99(2):583–592
78. Pearson TC, Brabec DL (2007) Detection of wheat kernels with hidden insect infestations with an electrically conductive roller mill. Appl Eng Agric 23(5):639–645
79. Rajendran S (2005) Detection of insect infestation in stored foods. In: Taylor SL (ed) Advances in food and nutrition research, volume 49. Elsevier Academic, UK, pp. 163–232
80. Hagstrum DW, Subramanyam Bh (2006) Fundamentals of stored product entomology. AACC International, St. Paul, MN
81. Mills R, Pedersen J (1990) A flour mill sanitation manual. Eagan, St. Paul, MN
82. FMC Food Tech (2001) X-ray technology. Solut 2:20, 21

Food Forensic Investigation

35

William R. Aimutis, Michael A. Mortenson, and Jinping Dong

Contents

35.1 INTRODUCTION

Modern food manufacturing has evolved to consistently offer food products to consumers that are safe and free from microbial contamination and foreign materials, as well as having a pleasing odor and flavor. However, since most commercial food products are produced on machinery operating at a very high speed, there are incidences of something physically wearing or breaking and occasionally placing unwanted foreign material into a food product. Additionally, extraneous environmental odors may contaminate food raw materials or processed foods and be carried over into the finished product, causing an off-odor and/or taint in the product. There are also occasions when someone has deliberately and fraudulently added something to mixing vessels or to a

W. R. Aimutis (✉)
North Carolina Food Innovation Lab, North Carolina State University, Kannapolis, NC, USA
e-mail: bill_aimutis@ncsu.edu

M. A. Mortenson
Core R&D, Cargill Research and Development Center, Plymouth, MN, USA
e-mail: michael_mortenson@cargill.com

J. Dong
Core R&D, Cargill Research and Development Center, Cargill, Inc., Plymouth, MN, USA
e-mail: jinping_dong@cargill.com

B. P. Ismail, S. S. Nielsen (eds.), *Nielsen's Food Analysis*, Food Science Text Series,
https://doi.org/10.1007/978-3-031-50643-7_35

package to discredit a company or potentially cause physical harm to the consumer. This type of maleficence has been termed "**food tampering.**" Investigating the origins of all these mishaps necessitates special handling and testing of the affected products using food forensic investigation. Once contaminants have been identified in either the company's own laboratory or by an accredited outside third-party laboratory, further investigation must be conducted to identify where, when, and how the contamination occurred.

Food forensic investigation is a logical process for the investigation of root causes for a product that is perceived to be physically objectionable to a customer or consumer because it may have an off-odor/taint, have been contaminated by foreign material during processing or at some later time, or have been intentionally tampered with by an outside party. Products under these circumstances must be analyzed from a perspective different from understanding the quality or composition of a product. Food ingredients or finished products with off-odors/taint or containing foreign material contamination could be the subject of litigation as customers sue a supplier for damages as well as facing large fines from the government for not complying with rules and regulations. The food industry has not typically had to comply with such stringent rules for foreign material contamination in the past, and therefore most suppliers are ill-prepared to meet the legislated rules concerning this matter.

Many companies have traditionally relied on bulk testing of their products to meet established quality standards that were often guaranteed by a certificate of analysis (see Chap. 1, Sect. 1.2.4.3). However, bulk testing will miss the what, when, where, and how of foreign material contamination. Therefore, when a complaint is registered that a food is contaminated by foreign material or has an off-odor/taint, a food company must react with appropriate procedures.

Testing to identify the what, when, where, and how can be either handled by the implicated company or by an accredited third party. When a third party is used, it is still imperative for the implicated company to have standardized procedures for how the products will be received, documented, and delivered to the outside party for analysis. This chapter provides an introduction to the craft of food forensic investigation.

35.2 TYPICAL/ATYPICAL ISSUES REQUIRING FORENSIC ANALYSIS

35.2.1 Foreign Material Contamination

Foreign material or extraneous matter contamination in foods accounts for a large majority of the customer complaints received by food manufacturers, retail outlets, and regulatory authorities [1]. **Foreign material** is defined as anything perceived by the consumer as alien to the food. Consumer perception is important because not all foreign material will be deleterious to those consuming the material and may have originated from the food itself, such as bone fragments or crystals of salt, sugar, or minerals. These are often mistaken by the consumer for glass pieces and will trigger a negative response from consumers, resulting in customer complaints. The majority of foreign material issues can be broadly classified as intrinsic or extrinsic materials to a particular food. **Intrinsic foreign materials** are derived from the raw materials, usually plant or animal origin, used as starting material for an ingredient or processed food product that may have unknowingly contaminated the finished food product. For example, plant stems and leaves, pits and seeds, animal hair/fur, corn cobs, and bone chips are commonly identified from products that were investigated after a consumer complaint and are not foreign to the food being analyzed. These are often referred to as extraneous matter as outlined in Chap. 34. Most intrinsic foreign material complaints will be in solid foods and have entered the food product either at a primary raw material production site (fruit or vegetable processor, slaughter house, or food ingredient manufacturer) or during a consumer's preparation of the food [1]. The types of foods most often involved with a foreign material complaint are vegetables and vegetable products, followed by cereals and cereal products.

Extrinsic foreign material is much more varied and arises in a food usually inadvertently during processing from the farm to the package or often after it has been opened. Common examples of these materials include stones, broken glass, metal fragments and shavings, nuts and bolts, dental materials, human hair, plastics, string, packaging materials, insect parts, grease globules, wood chips and sawdust, gasket materials, or filter fibers. This chapter on food forensic investigation goes beyond typical analysis of extraneous matter but is limited to the causes making a food product physically objectionable to a consumer, i.e., not covering chemical contaminants, which are the subject of Chap. 33. Figure 35.1 shows a few examples of foreign materials found in food production facilities.

Food companies may also need to conduct a forensic analysis of products that are suspected to have been tampered with by outside parties with intent to discredit the company or to cause physical harm to consumers. In these cases, it is imperative to carefully handle received samples to identify the foreign material and how it may have found its way into the food. If foul play is suspected when a complaint is received, it should be a standard operating procedure (SOP) for the food company to contact local and national law enforcement agencies for assistance in analyzing the materials. Meticulous sample handling and documentation of the

Fig. 35.1 Representative foreign materials found in food production facilities

sample at all times is imperative due to the possibility or likelihood that the sample becomes part of a lawsuit or criminal investigation. It is not unusual to be advised of a lawsuit months after completing a seemingly routine investigation.

35.2.2 Food Authenticity and Adulteration

Our global food supply chain must be routinely characterized for authenticity to serve as a baseline in the unlikely event it is intentionally or unintentionally compromised in a fraudulent manner. **Food authenticity** refers to the unique properties of a food much like a human fingerprint. Advanced analytical and chemometric methods are important in establishing baseline biomarkers and concentrations for selected typical components of a food. This requires a well-conceived analytical testing strategy to identify the unique fingerprint of a food and catalogue this knowledge. An established database of this information allows investigation of potential food fraud and/or adulteration when a product seems untypical. The characterization of food fingerprints goes beyond quality aspects of a product but also can be important in relation to human health. Therefore, these factors are important for the food industry and public health and regulatory authorities. Extensive legislation in recent years has been codified in relation to food authenticity and food safety leading to strict and comprehensive regulations in both the United States (https://www.fda.gov/food/guidanceregulation/FSMA) and Europe (https://www.efsa.europa.eu/en/aboutefsa).

The analytical approach to food authentication and investigating food adulteration utilizes several targeted and untargeted methods because not all authenticity aspects can be identified using a single analytical method. A **targeted analytical method** is employed to directly link to an authenticity marker (e.g., DNA molecule or chemical compound) or a parameter associated with a chemical compound like its stable isotope content [2]. **Untargeted analytical methods** measure a wide range of analytes or biomarkers, establishing a fingerprint of the food or ingredient being analyzed. Unfortunately, many of these methods do not measure only a single parameter but multiple parameters that generate large quantities of data to fingerprint a formulated food or food ingredient. Ultimately this requires very large databases to compare against and complex statistics to confirm if a food authenticity profile is typical or untypical. Our choice of words here are very specific. **Untypical** means the sample being analyzed does not have the exact fingerprint of authentic (or **typical**) samples in the database. Occasionally, rare varieties of a food can be encountered during an analysis that have not been deposited in the database. One should be cautious of labeling a food as adulterated until several different analyzes confirm multiple untypical characteristics.

Several methods described earlier in this textbook are frequently used to fingerprint foods. Ideally noninvasive methods like electronic nose (this chapter), infrared and Ramen spectroscopy (Chap. 8), and hyperspectral imaging (Chap. 32) combined with chemometric analysis provide a significant amount of information to characterize a food. However, this is often not adequate, so more invasive or targeted analyzes must be done to qualitatively identify chemical analytes in a food and then quantify the concentrations of particular molecular species. These techniques are important for establishing the fingerprint of a food but also if it has been adulterated to discriminate the likely adulterant. These methods include GC–MS (Chap. 11, Sect. 11.4), LC-MS (Chap. 11, Sect. 11.5), NIR, FTIR, and fluorescence spectroscopy (Chap. 8), NMR (Chap. 10), HPLC (Chap. 13), and inductively coupled plasma-mass spectrometry (ICP-MS) (Chap. 9). As mentioned earlier, honey and olive oil are foods frequently adulterated, and many of these methods have been employed to confirm adulteration. As these databases continue to be established and further populated, it is imperative standardized approaches be used for fingerprinting and analysis.

Every food has typical specific metabolites and groups of metabolites that provide discriminatory patterns which can be varied by a response to environmental, genetic, or human alterations (adulteration). For food authentication, several *ohmics* technologies (genomics, proteomics, and metabolomics) have become an important discipline for assessing safety and authenticity [3]. This untargeted approach to characterization can provide information to discriminate and predict differences between food samples that allow statistical models to predict typical food fingerprints.

There are several emerging ambient mass spectrometry methods that will be important for fingerprinting foods in the future, including desorption electrospray ionization (DESI), direct analysis in real time (DART), and atmospheric pressure solid analysis probe (ASAP) [4]. Sensor technologies are rapidly advancing, and these will enable handheld detection devices soon to qualitatively detect food adulteration. One such method is DNAFoil that quickly analyzes if a meat product from one species is contaminated with another species [5].

Food fraud is the deliberate and intentional substitution, misrepresentation, addition, or tampering of food, food ingredients, or packaging for economic gain. **Food adulteration** is often interchangeably used with food fraud. Food fraud has become much more prevalent during the last decade as prices for some crops and ingredients have risen. Unscrupulous companies will qualify ingredients with food processors by sending pure ingredients for initial analysis and process efficacy but will deliver fraudulent material after they make the qualification and sale. For example, as olive oil prices have risen, some global suppliers began intentionally adding canola oil (which is much less expensive and clean flavored) to consciously be price competitive. Unfortunately, these oils were represented as 100% olive oil so the act of diluting the oil is fraud. Recent estimates indicate greater than 70% of the olive oil sold in the United States (US) is fraudulently contaminated with other oils [6]. Some other common fraudulent foods are the following: milk, honey, saffron, orange juice, coffee, apple juice, tea, fish, and black pepper [6]. Fortunately, many of the same analytical methods described elsewhere in this book and in this chapter are useful to identify fraudulent products. US Pharmacopeia (USP) maintains an online database as a repository for food ingredient fraud reports and associated analytical detection methods [7]. USP also includes in the *Food Chemicals Codex* appendix, *Food Fraud Mitigation Guidance* [8].

At the writing of this chapter, there has not been an international effort to standardize food forensic methods for sample handling and testing, despite new regulations being enacted internationally that require food companies to conduct forensic analysis and account for its origin. For example, the US Food and Drug Administration has enacted the *Food Safety Modernization Act* [9], and the Canadian Food Inspection Agency has a similar law, *Safe Food for Canadians Act* [10]. The scientific bases for analytical methods used to date have been well accepted, but there is still a greater need to identify the range of scientific needs to continue this field's development. Food forensic investigation has additional specialized needs because of the demand for "proof" and "evidence" in the context of legal proceedings or law enforcement. These needs may relate more to the quality of the *process* by which material is collected and analyzed as opposed to the science and technology employed.

35.3 ESSENTIAL ELEMENTS OF FOOD FORENSIC TEAMS

35.3.1 Nature of Food Forensic Teams

Food forensic investigation is a multidisciplined approach drawing on knowledge from a number of fields including biology, metallurgy, crystallography, and forensic science. An individual responsible for forensic analysis should have investigative experience especially in identifying foreign material contamination or off-odors/flavors, broad food industry experience (including knowledge about food harvesting, supply chain logistics, agricultural harvesting, and food processing unit operations), and a strong intuition toward detective work. Although many scientific tools and techniques have emerged to conduct food analysis and forensic investigation, the most powerful tools investigators have are their observation skills and the ability to ask the right questions to put the puzzle together.

Many food companies are forming food forensic teams, much like their product recall teams, to handle these instances. As shown above, many intracompany disciplines (including a corporate legal representative) and responsibilities will need to participate in an investigation. Documentation development and training of individuals in following established processes are important to maintain a constant state of readiness to conduct a forensic investigation [11]. Fortunately, these investigations are not routine, but it is still an expectation of regulatory agencies that skilled personnel are available to be responsible for identifying and finding the source of contamination. This will require personnel who have the necessary basic education, experience, and skills. Proactive companies will allow their food forensic personnel to routinely practice their art when not under the pressure of an investigation so that when the need arises, they are comfortable with the standardized procedures they will need to follow. Furthermore, continuing education for forensic personnel should be a facet of the overall food forensic program. Proficiency testing of investigators should be part of a company's quality assurance operating procedures [12].

As standard operating procedures (SOPs) are documented by a company for its food forensic team and program, these must include approaches for: (1) planning and design, (2) documentation, (3) protocols for quality assurance and quality control, and (4) sampling and logistics (Table 35.1). Each of these topics is covered in sections that follow.

The following description is a real-life example demonstrating how a forensics team worked together to address an urgent issue at a food production facility:

A cleaning team discovered an electronic device was left in the cooking oven and was severely burned (Fig. 35.2). The Food Safety team was immediately notified and by working with the Operations team, it was found that the device was accidently left in the oven the previous night when staff was conducting routine maintenance inside the oven. This incident put an entire day of finished product (more than 15 batches or > 50,000 pounds) at risk. After analyzing the components of the device, the Quality Control team identified possible contaminants include volatile compounds released from the burning of the plastics and the fume particles from explosion of the lithium battery. The Chemical Characterization team compared selected batches with the control using liquid and gas chromatography coupled with nontargeted mass spectrometry. Elemental analysis with ICP was also performed. Chemical differences from certain time

Table 35.1 Important points to consider in a codified food forensic plan for sampling, collection, and analysis

Develop a mechanism for quickly formulating a "consensus" analytical plan when a new sample (or set of samples) arises
Keep and update a set of standard operating procedures and validation data for analyzing case samples
Maintain a set of documented guidelines, requirements, and procedures for sample preparation for each analytical procedure
Maintain approved procedures for handling and storing samples
Develop standardized methods for data analysis, reporting, and presentation
Maintain reliable logistic plans for sending and receiving samples
Maintain clear and secure lines of communication for data, information, and discussion
Develop a mechanism for formulating an on-the-fly validation plan for a new procedure

Fig. 35.2 Burnt electronic device found inside an oven

point of the day were identified. The Sensory team was then brought over to evaluate the batches that did not show signs of chemical contamination. The Regulatory team made the final assessment based on all the test results. It was determined that the incident happened during early production hours and batches after 12:15 pm posed no health risk. Through the work of the entire forensics team, a large amount of finished products were saved and those with safety risk were discarded.

35.3.2 Planning and Design

Planning and design means that the company has considered and documented the exact steps to be followed when it has discovered or been alerted to a problem necessitating activation of the food forensic team. Development of a very high-level schematic of the process flow is a good starting point (see an example in Fig. 35.3). Once this is codified, the next steps are to go into more depth on every step of the process and detail the questions on what, where, when, and how. This document may contain a set of questions that need to be answered before proceeding in the standardized process. An example of the initial steps and questions to initiate an investigation is presented in Fig. 35.4.

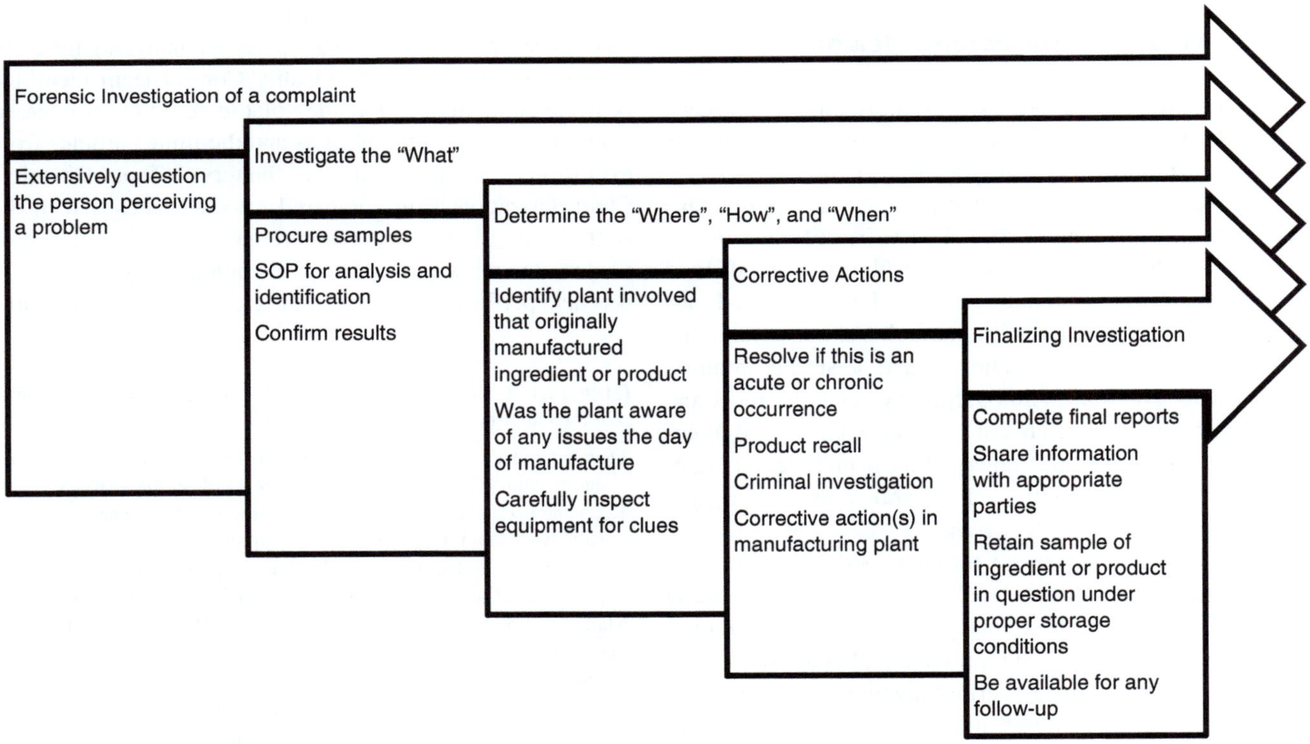

Fig. 35.3 Generalized process flow of a food forensic investigation

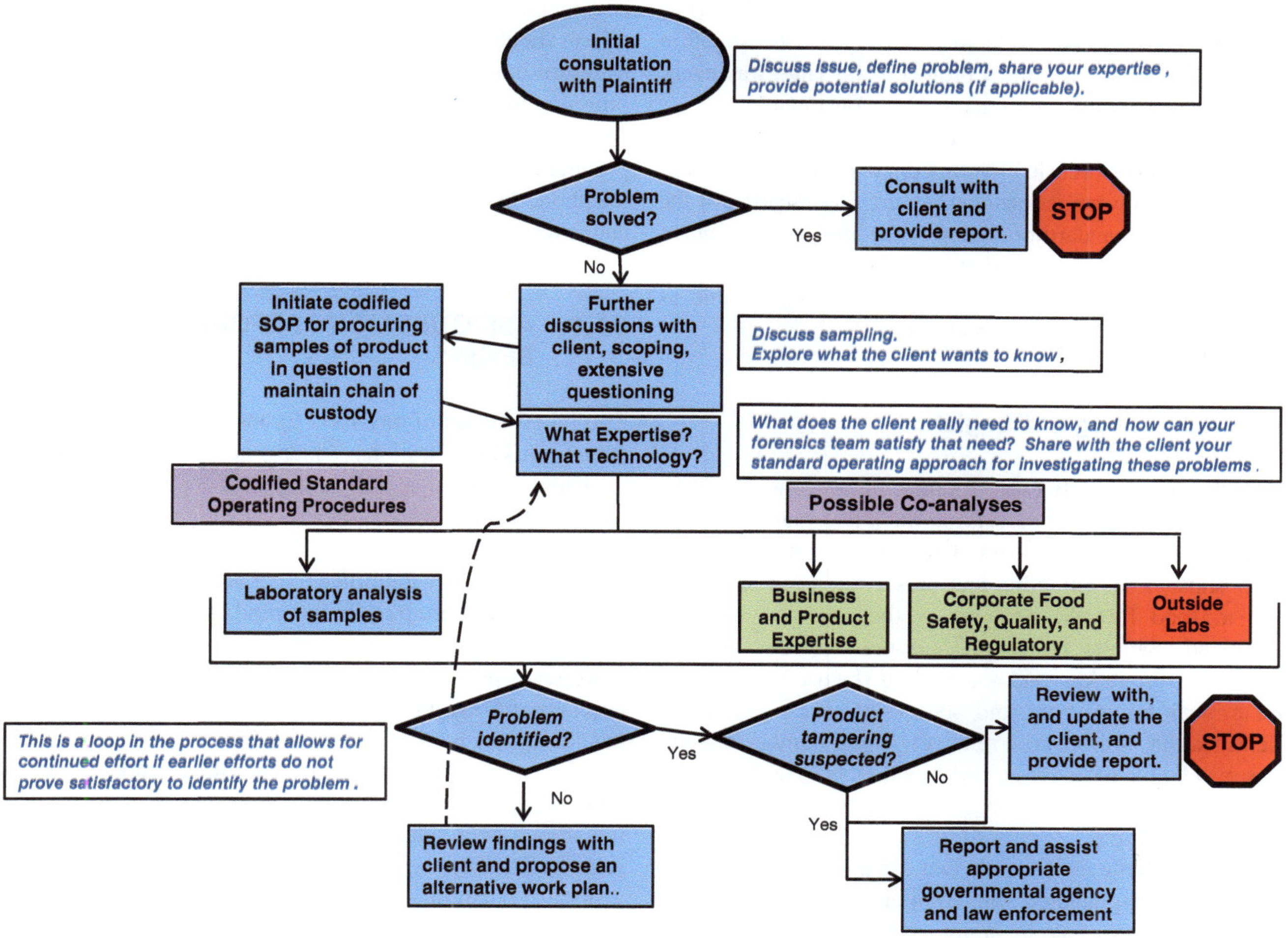

Fig. 35.4 Decision tree used in a food forensic investigation

35.3.3 Documentation

Documentation refers to a collection of documents and records [13]. A food forensic laboratory that has proactively written all of its SOPs, policies, and instructions into a codified form has met the guidelines for quality management [14]. Documents should explain in a written or graphical form what a food forensic laboratory plans to do and how it will be accomplished, as well as instructing scientists how to perform tasks. Codified procedures have been reviewed and accepted by a company's leadership team. A good forensic laboratory will have all of its documents in as complete a form as possible before they are needed in an investigation. Once an investigation is initiated, the documents will provide guidance, explanations, and instructions about how to operate. Records are recorded information about an activity and, thus, do not exist until after the activity has been performed. It is important to maintain records (including who conducted the work) from all stages of a forensic investigation for a successful outcome when a regulatory agency audits.

35.3.4 Quality Assurance and Quality Control

SOPs for quality assurance and quality control are also important for an accused company to be able to defend itself in litigation cases. Yet, this is an area that is often either taken for granted or ignored. **Quality assurance** (QA) is different than **quality control** (QC). In a forensic investigation program, QA encompasses all the activities undertaken by the laboratory to assure itself (and its customers) that reliable and accurate testing will be undertaken at all times according to the codified standard operating procedures [11]. QC in a forensic laboratory are the activities undertaken to confirm that test results are accurate and reliable [11].

QA programs also should be developed for laboratory consumables, such as stock solutions, laboratory standards, and carrier gases, all of which can impact results being generated in a laboratory. Therefore, food forensic teams developing their protocols must think in advance as to what they may need for a number of different circumstances, and accordingly codify where to purchase these materials so they are on an approved list of suppliers that have developed a reputation for quality and are willing to certify the materials purchased. Scientists conducting a forensic investigation must be disciplined to follow protocols the same way every time, including using only preapproved consumables.

Critical components of the QA program are presented in Table 35.1. Food forensic investigators will encounter exceptions in many investigations that require modification of the SOP. Codified standard operating procedures are not meant to cover every possible situation, and it is therefore important to also have in place procedures to help make adjustments as an outlying situation develops (Table 35.1).

35.3.5 Sampling and Logistics

The first step in nearly every forensic investigation involves collecting samples and preserving them in a manner to prevent or at least minimize degradation or contamination from other substances. Sampling and preservation are as important to food forensic analysis as the scientific methods employed. Sample collection and preservation for food forensic investigations and potentially to be used as evidence are critical for efficient and successful investigation and identification. Moreover, a relatively small number of samples may be the basis for an investigation, but yet they could be highly significant in being able to identify the contaminant and its source.

Food is a biological sample that can undergo degradation if not properly handled, including how it is stored and shipped to the forensic laboratory. Any mishandling during this phase could have ramifications in ultimately identifying what the problem is and how it got in the food in the first place. Sampling and logistics refer to the process or procedures used to procure a sample for testing and how a sample is shipped to the forensic laboratory. The fewer number of people who handle a contaminated sample, the less likely a sample will be corrupted or damaged in some manner. This is referred to as the **chain of custody** in criminal forensic cases but is also applicable to food forensic investigation. A "chain of custody" is a system of unbroken control of a sample. It involves isolating a sample and packaging it securely while documenting carefully what has been done. The person carrying out this work formally takes control (custody) of the sample and is responsible for its safe keeping until it is transferred and signed over to the custody of the next person in the chain and so on.

Sampling and logistics is a fairly small part of the overall process and often is not under the manufacturer's control but can be particularly important for maintaining chain of custody. It would be somewhat unreasonable to expect a company to send its food forensic personnel to the offended customer to prepare the samples and return them to the food forensic laboratory, but in the case of a suspected tampering case that might be required or at least considered. Minimally, a form should be included with any packaging materials sent to return the product in question, and the method for shipping product back to the manufacturer should be indicated. This form should be completed by the person returning the product and include information similar to those asked in a telephone interview as outlined below but should also indicate anyone who has contacted the product in question.

35.4 ASK QUESTIONS BEFORE ANALYSIS BEGINS

The initial stages of any food forensic investigation should start by asking questions – many questions! The types of questions being asked will be different depending where the problem was discovered – at a food manufacturing site or by a consumer (Table 35.2). If the problem is perceived by a consumer, it is very important to get details quickly to record them while they are fresh in the mind. Additionally, speed is important to differentiate if this is a once-off problem or more widespread. If the problem is widespread, it may be necessary for a company to require a product recall to protect the consuming public. If a consumer perceived the problem, start by communicating directly with the plaintiff by asking about what was observed or perceived. This line of questioning should be conducted by one individual, preferably the scientist leading the investigation. Many cases of foreign material contamination can be traced back to consumers

Table 35.2 Examples of questions to ask a person perceiving a problem product

Questions for food manufacturer	*Questions for a consumer*
Problem description? Details?	What did you perceive as a problem?
What do we know about the problem?	How many members of your family perceived a problem?
Chronic, sporadic, acute?	Where was the product purchased?
Urgency? Do you suspect this is dangerous to the public or is a tampered product?	Lot code of the product?
Is a customer (consumer) involved?	How was the product prepared with details?
Description of samples to be analyzed? Get more details	Was anyone injured by the problem?
What are the lot numbers of ingredients that may be involved?	Do you still have any of the product remaining?
What is the extent or breadth of the problem?	Would you carefully follow instructions to return it to our investigation team?
Have samples been obtained? Who collected the samples, and how?	How much sample is available?
Where were the samples collected?	Who collected the samples, and how?
How much sample is available?	Where were the samples collected?
Is there a control?	
Can we get access to your manufacturing records for the period of time the product in question was manufactured?	

themselves, and they should be questioned about how they have handled the product from the time they purchased it until they perceived a problem (usually when they are consuming the product). An investigator cannot be timid about asking very difficult questions to the plaintiff without seemingly placing blame. Details gathered during early stages of an investigation are often important in identifying where to begin looking for answers, i.e., where the contaminant came from initially and how it entered the ingredient or processed food. Finally, it is imperative that answers all questions be honestly captured and recorded in written detail since they may be important later if a company needs to defend itself in a lawsuit or if there is criminal prosecution for product tampering.

35.5 ANALYZING "PROBLEM SAMPLES"

When the sample is received in the food forensic laboratory, it is important to document all the details. Photographs and written descriptions should be taken in every step from the preopening of the package to the contents inside. Forensic laboratories will then follow their preestablished decision trees (examples are in Figs. 35.5 and 35.6) to initiate analysis and decide on appropriate methods (Table 35.3). Many food ingredient companies and processors also retain limited quantities of materials from each production lot in the event they need to conduct follow-up analysis. These samples also should be readied for further inspection and/or analysis during a food forensic investigation to compare to suspected products.

35.5.1 Foreign Material Contamination

35.5.1.1 Introduction

The majority of issues to be investigated can be handled by personnel routinely conducting extraneous matter analysis (Chap. 34), but inevitably there will be investigations that must be handled by internal or external specialists due to the nature of the contaminant or testing and instrumentation required. A forensic analyst should be familiar with identifying the most commonly encountered foreign materials (Table 35.4).

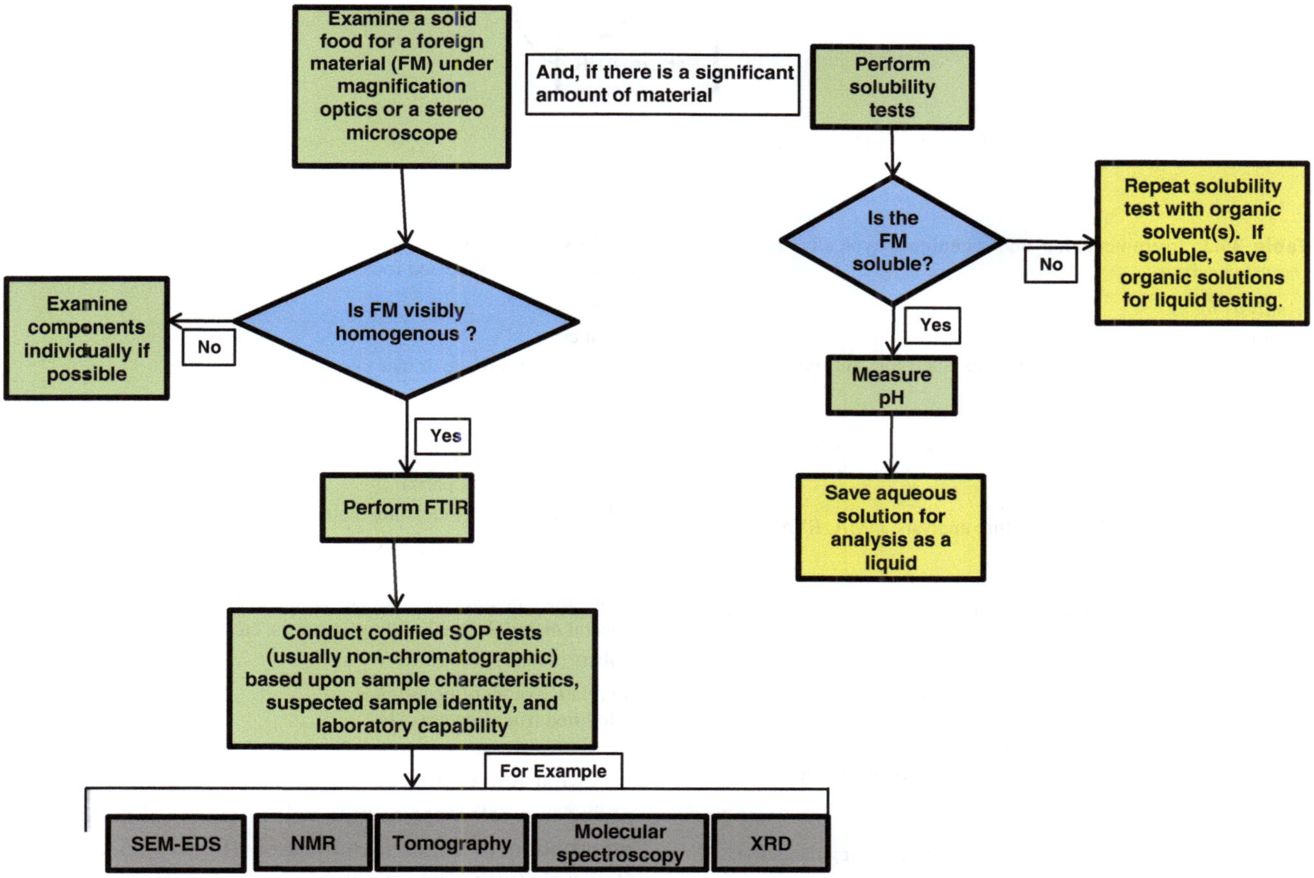

Fig. 35.5 Flowchart for analysis of an unknown solid sample

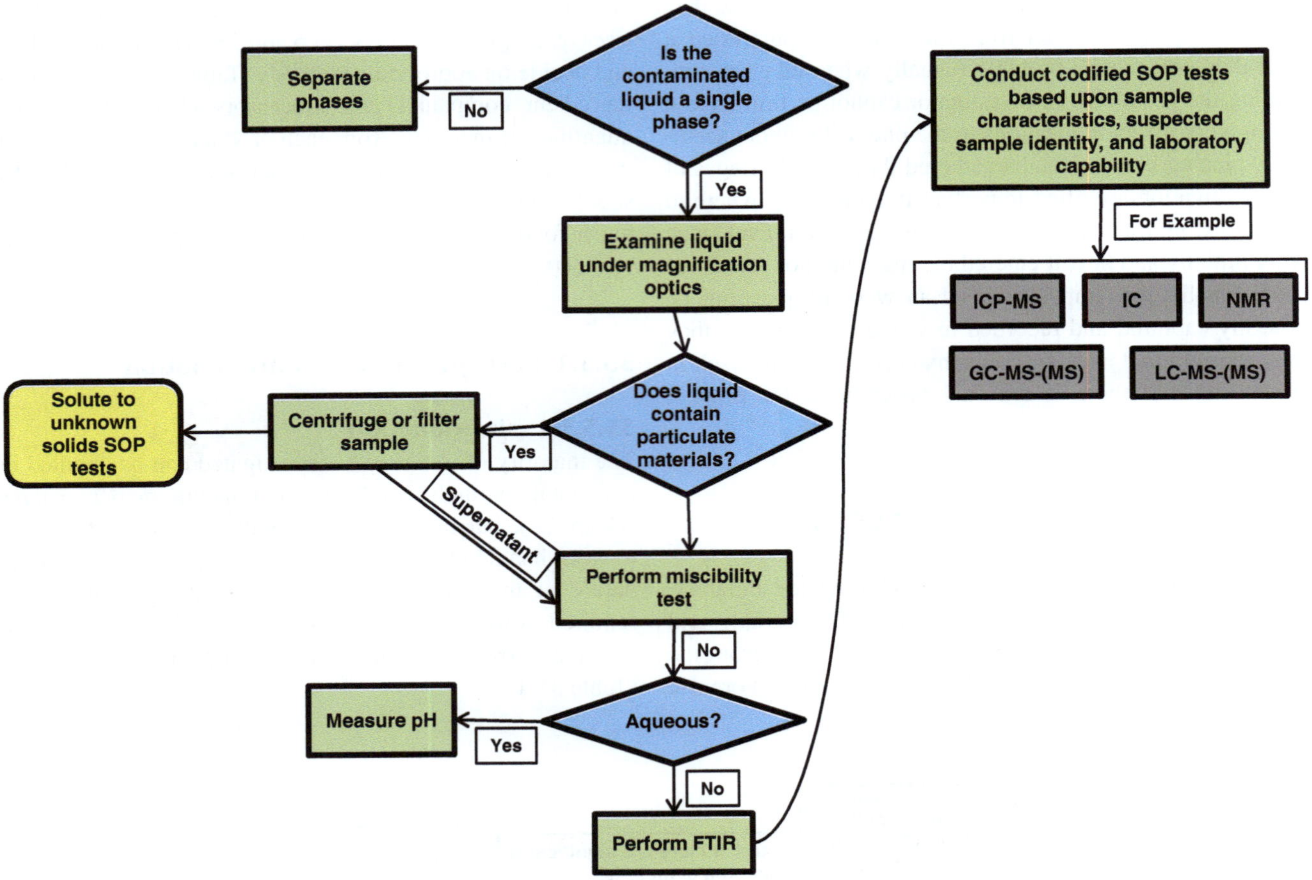

Fig. 35.6 Flowchart for analysis of an unknown liquid sample

Table 35.3 Instrumentation and techniques typically used in food forensic analysis

Analysis focus
Lipids analysis Extractions, composition, volatiles, shelf life, antioxidants, unknown ID, mass spectrometry
Spectroscopy FTIT, Raman, EDS, XRF, μ-XRF, fluorescence, chemical and elemental imaging, ICP
Thermal and mechanical DSC, moisture isotherms, texture analysis, DMA, RVA, rheometers
Liquid chromatography HPLC, UHPLC, ion chromatography (IC), mass spectroscopy, mass analysis, fractionation
Gas chromatography (GC) 1D & 2D, flame ionization, olfactory, mass spectroscopy
Microscopy Light microscopy, SEM, confocal, x-ray microscopy, histological staining, fluorescence, immunostaining, thin sectioning, TEM
X-ray tomography 3-D imaging, surface analysis, nondestructive segmentation, ingredient ID, and localization

Table 35.4 Foreign materials typically identified in food ingredients, raw foods, and processed foods[a, b]

Glass
Plastics
Animal origin (bone or hair usually)
Metals
Extraneous vegetative matter (including wood)
Minerals (usually as crystals)
Arthropods (insect parts)
Chemicals (usually cleaning materials)
Food material (scorch particles, ingredient clumps that are undissolved, specks, etc.)
Yeast flocculation (especially in carbonated beverages)
Dental materials (teeth, fillings, crowns, etc.)
Fibers (usually from filters)

[a]In relative order of occurrence
[b]Modified from Edwards [1]

Visual observation of a contaminated product and comparing to a reference sample will often reveal quite a bit of information to begin identification (Figs. 35.5 and 35.6). However, further investigation often involves understanding the chemical, elemental, and physical/structural characteristics of the contaminant. One should be mindful of the sample preparation procedures for some of these analytical techniques and consider use of nondestructive techniques to preserve physical integrity of the object.

35.5.1.2 Chemical Composition

When analyzing a foreign material (FM), the first thing one wants to determine is the chemical composition, or the molecular makeup of the object. If the FM can be isolated and purified, nuclear magnetic resonance (NMR) (Chap. 10) and mass spectrometry (MS) (Chap. 11) are ideal tools to elucidate the molecular structure. For example, when an industrial lubricant was spilled into a fryer, mass spectrometry quickly identified the chemical composition of the FM and confirmed the source.

Sometimes purifying the contaminant is not possible. Characterizing the chemical functionality may be sufficient to determine composition of the FM. Spectroscopy techniques such as Fourier-transform infrared (FTIR) and Raman spectroscopy (Chap. 8) can identify the chemical bonds in a molecule, e.g., the sharp peak near 1742 cm^{-1} in a FTIR spectrum comes from the ester bond in lipids, and the two broad peaks near 1635 and 1560 cm^{-1} are typical protein amide I and II groups. There are numerous spectral libraries one can utilize to find the chemical fingerprints of unknowns.

Very often, reference or suspicious source materials are available. Then one only needs to confirm that the FM has the chemical match to one of these materials. High-performance liquid chromatography (HPLC) (Chap. 13) or gas chromatography (GC) (Chap. 14) and FTIR (Chap. 8) are used regularly to do the comparison.

35.5.1.3 Elemental Composition

Most food materials are organic compounds, meaning they are derived from living organisms, and typically composed of covalent chemical bonds. Many foreign objects in food forensics are inorganic and contain ionic bonds. Methods mentioned in Sect. 35.5.1.2 may not work on materials such as salts (e.g., NaCl, Na_2CO_3, etc.), metals (e.g., iron, aluminum, etc.), and minerals (e.g., sand, glass, ceramic, etc.). The following are techniques used often in forensics to identify the elemental composition, or the atomic makeup of the foreign materials.

Inductively coupled plasma (ICP) and other atomic spectroscopies introduced in Chap. 9 only work for homogeneous liquid samples. Therefore, foreign materials need to be acid-digested into a soluble solution before injection into the instrument. This sometimes is problematic for minerals and stainless steel.

X-ray fluorescence (XRF) employs high-energy X-rays to excite fluorescence emission from the target material. The wavelength, or the energy of fluorescence, is characteristic to the atom and can be used to identify the material's elemental structure [15]. Laser-induced breakdown spectroscopy (LIBS) uses a high-energy laser pulse to excite electrons from the atoms. As the electronic energy decays, a photon emission is generated with the wavelength characteristic to the atomic number [16]. Both XRF and LIBS can focus on a microscopic spot for FM detection and handheld scanners are commercially available. Energy-dispersive spectroscopy (EDS) (Chap. 32, Sect. 32.2.4) is part of a scanning electron microscope (SEM) and capable of analyzing the elemental content of an object as small as a few nanometers. Because the SEM sample chamber is under vacuum, EDS cannot be used on liquids.

35.5.1.4 Physical and Structural Identification

Plastic foreign materials such as polyethylene may have very different hardness because of the difference in polymer chain length. Gelatinized starch becomes brittle over time because of the retrogradation, or recrystallization, of the amylose molecules. Two packaging films can have the same composition but be manufactured by different processes (e.g., extrusion vs. casting), resulting in different microscopic surface patterns. These mechanical and morphological properties are sometimes critical information in FM identification.

X-ray diffraction (Chap. 32, Sect. 32.4) is used not only to characterize crystallinity of materials but also to differentiate polymorphs. Rheometers and texture analyzers (Chap. 29) can be used to evaluate the stiffness, hardness, elasticity, and viscoelastic properties of an FM and compare with potential candidate references.

Microscopy (Chap. 32, Sect. 32.2) is another essential tool in FM structural identification. In addition to assisting in early assessment on sample preparation and identification pathways, microscopy also provides detailed surface morphology useful for comparing samples with the same composition. Sometimes, structural differences are not obvious on the surface of FMs; X-ray microtomography (Chap. 32, Sect. 32.5) is then employed to reveal a 3-D image.

35.5.2 Off-Flavors and Taints

At some point during their careers, all food professionals will experience an ingredient or product exhibiting an off-flavor or a taint. **Off-flavors** result from food degradation or reaction of the ingredient or product, whereas **taints** result from an external contamination [17–19]. The identification and remediation of off-flavors and taints are a blend of science (e.g., sensory, analytical chemistry, food chemistry, food processing, packaging, transportation, etc.) and art (e.g., tactful communication, creative extraction and isolation techniques, application of previous experiences, etc.). Given these numerous facets, the remainder of this section focuses on a practical strategy a food professional might take when presented with an off-flavor or taint. The strategy does not include technical details or specific methodology; for this information, the reader is referred to earlier chapters in this textbook and to the works of others skilled in the field of off-flavors and taints [18, 20–24].

35.5.2.1 Strategy and Supportive Examples

As mentioned earlier in this chapter, forensic teams often follow a high-level schematic to direct process flow. A schematic for elucidation of off-flavors and taints is presented in Fig. 35.7. This flowchart begins with initial consultation and ends with rigorous instrumental analysis; some example techniques are referenced throughout the figure. One underlying theme throughout the schematic is the use of sensory evaluation to guide the process. There are several reasons to use traditional sensory testing procedures for off-flavors and taints (e.g., determining the nature of the problem, threshold, diagnostic, quality control, shelf life, etc.). However, for this strategy, it is assumed the problem is established, and sensory (visual and smelling) evaluation is used to confirm the causative stimuli through the entire forensics process (from consultation, through analysis, and finally linkage to the cause). If the forensic scientists do not experience the off-flavor or taint, there is no chance of identifying the cause of the problem. The following examples illustrate how the forensic scientist's olfaction or taste system might be the most important tool in the strategy.

Consider a "smoky" off-aroma in a food product reported by a client during an initial telephone consultation. To the scientist, the descriptor "smoky" meant "fire" and the logical thought process pointed toward a taint issue. However, the scientist knew the complex nature of "smoky" descriptors and requested a sample for simple sensory evaluation. After smelling the sample, the odor described as "smoky" by the customer was clearly "toasted bread" to the scientist and not "fire" as originally thought. This confusion was cleared by the scientist smelling a sample of the affected product and ultimately guided the investigation toward an off-flavor (volatiles from Maillard browning reaction) instead of a taint (absorbed volatiles from pyrolysis). In this example, the problem was not appropriately defined until the scientist's experience was applied to the sample.

Now, consider the analysis portion of Fig. 35.7. The sensory character of the off-flavor or taint must be considered for any analytical approach chosen. For example, the scientist may have a good idea of a compound that is "musty" in odor character (e.g., in potato products) and wants instrumental confirmation (e.g., GC with retention index or MS). The scientist also should use a **GC-O** technique (i.e., gas chromatography with an olfactory detector outlet) (see Chap. 14, Sect. 14.3.5) on the problem sample and compare data to a compound database containing chromatographic and odor character information (e.g., Flavornet [25]) and, whenever possible, a chemical standard. In other cases of off-flavors or taint, sensory can serve to guide the forensic scientist through challenging odor isolation experiments. The scientist may begin the process with a simple headspace analysis. A GC-MS chromatogram may not show a peak at the same time the offending aroma is detected in a corresponding GC-O aromagram; this is not uncommon since the human nose is more sensitive than MS for certain compounds. From this experience, the scientist learns the compound of interest appears to be captured by the headspace technique but it might require larger sample sizes to boost MS signal. Perhaps the headspace technique might be abandoned for a more rigorous approach (e.g., solvent extraction and/or distillation); sensory evaluation can be used throughout this process as well. There are several different solvents used for aroma extraction, but these have compound biases [18]. Smelling a few drops of an extract from a sniff strip might give early insight into appropriate solvent choice; if the off-aroma is present, the solvent choice appears to be appropriate. Additionally, if the scientist decided to use a solvent extraction followed by high-vacuum distillation to remove nonvolatile components, it is a good idea to use a sniff strip to make sure the offending aroma is present before spending the time on distillation and subsequent solvent concentration.

35.5.2.2 Off-Odors and Taint Determination

Identifying a problem molecule (or set of molecules) is the first step in determining the origin of an off-flavor or taint. As challenging as identification can be, linkage of the offending molecule(s) to a mechanism of formation or point of contamination could prove more difficult than identification. An off-flavor or taint may begin at any point of the food production process. A number of examples illustrate some possibilities. The raw material may create a problem from the beginning of the supply chain (e.g., feedy, barny, cowy, weedy compounds in milk). Improper handling of materials prior to the production facility may accelerate detrimental reactions (e.g., corn germ sitting in a hot rail car accelerates lipid oxidation). The composition of a food matrix may make it more susceptible to detrimental reactions before, during, and after processing (e.g., lipid oxidation or Maillard browning). The processing environment (e.g., residual cleanser) or unit operations (e.g., microbial outgrowth in pipes) may taint product. Inks from packaging materials may penetrate into the product. Environmental exposure during transportation may taint the product (e.g., diesel fuel/exhaust contamination, mustiness from halogenated anisoles on pallets, or vola-

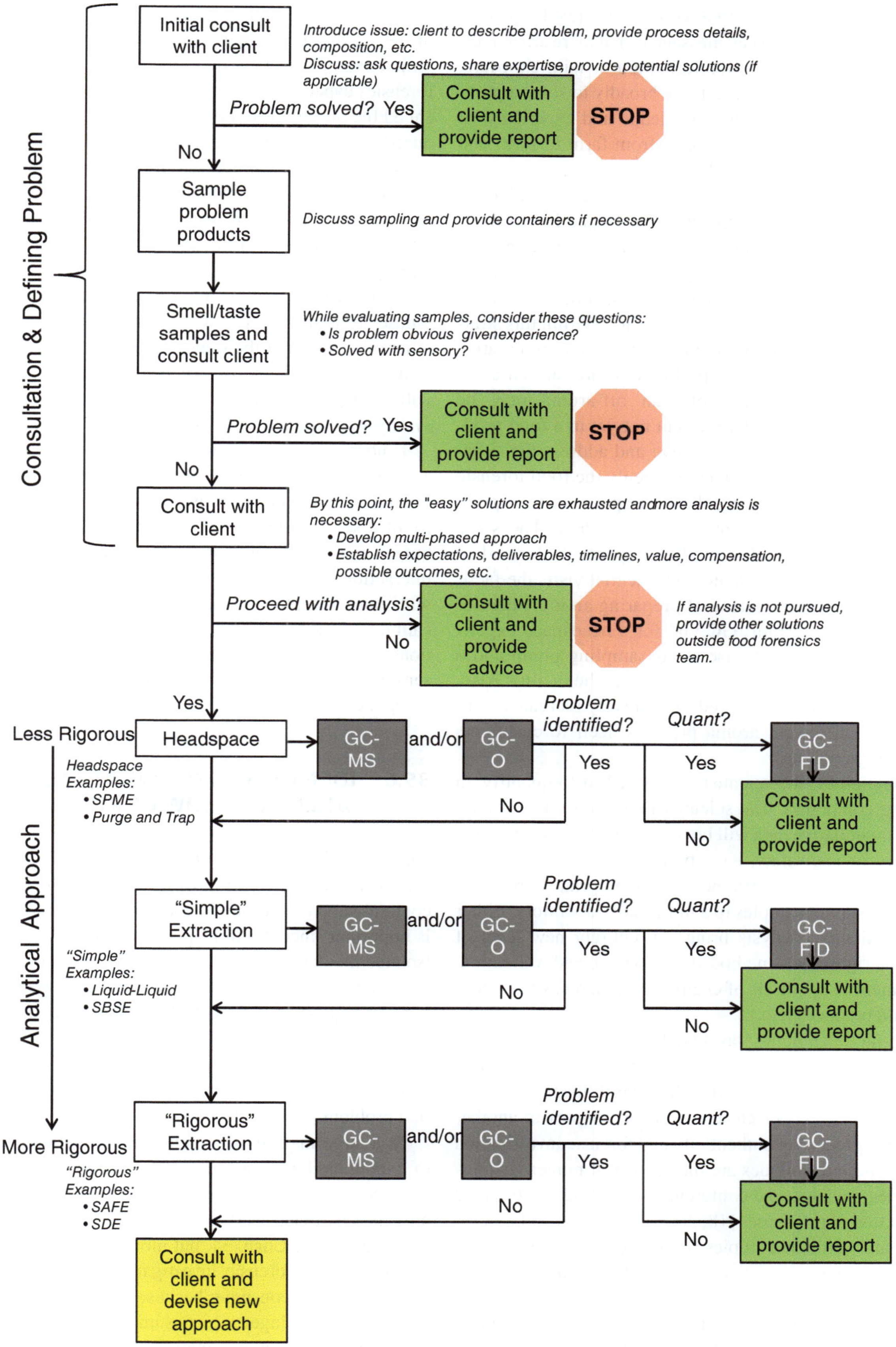

Fig. 35.7 Decision tree for off-flavor/taint forensic investigation

tile organic compounds (VOCs) from a freshly painted shipping container). The key message is that there are numerous possibilities for off-flavor formation or taint exposure, and the forensic scientist must think broadly to solve a problem. They must tap into the knowledge and experience of colleagues involved in the process (from farm to fork) and from experts in the field [18, 20–23].

Established volatiles isolation, separation, and identification techniques are covered in Chap. 14 and in other references [18, 22, 23, 25]. For efficiency and cost saving, the off-flavor and taint strategy begins by using the simplest approaches then going to more rigorous, time- and labor-intensive techniques only when necessary. Within this strategy, established isolation techniques may require creative modifications to capture the problem before any chromatography occurs. For example, an off-aroma may be detected in a production facility but not yet in a food product; the client wants to be proactive and address the situation before it taints the product. Suddenly the food forensic scientist is now considering the best approach to capture the problem from an entire production plant. The same strategy applies but with a modified approach. Instead of receiving a sample to evaluate, the scientist visits the facility and smells the air. Instead of preparing an ingredient or food product in a 20-mL GC vial for a headspace experiment, the scientist might use a gas sampling pump and a number of adsorbent traps to sample air in the facility. Also, the forensic scientist may need to creatively recreate a situation in which the off-aroma presents itself to customers (e.g., during cooking).

In some cases, the scientist will be asked to quantify an offending compound. The scientist must carefully consider how the quantitative data will be used (e.g., general curiosity, regulatory issues, safety, legal proceedings, etc.). Then the scientist will need to determine if it is appropriate to run tests in house or submit samples to a third-party laboratory. Either way, quantitative analysis brings an entirely new level of analytical rigor (i.e., method development and validation, procurement or synthesis of chemical standards, strict documentation practices, etc.).

Thus far, techniques presented for off-odor and taint determination are based largely on sensory and hyphenated gas chromatography techniques. It is worth mentioning that artificial sense (i.e., electronic nose or tongue) is another potential approach for authenticity and forensic investigation of food. These techniques are the subject of numerous journal articles, reviews, and conferences. The reader is directed to a selection of references [18, 26–34] for reviews of e-nose/tongue and affiliated topics (e.g., sensor technologies, applied statistics, data science, computing, machine learning, etc.).

There are several reported uses of e-nose in food applications including evaluations of raw material validation, product quality control, fruit/vegetable maturity, shelf-life, degree of freshness, and product authentication [26, 29, 32]. A simplified example of how an e-nose might be used in a food forensics capacity follows: the vapor phase of several different edible oil samples of the same botanical source are introduced to an assembled array of chemical sensors. Sensor response signals are processed into a "fingerprint" (not individual odor-active compounds) for each particular sample and are combined with data from correlating sensory and other supportive instrumental techniques and processed (e.g., pattern recognition, neural net, multivariate analysis, PCA, machine learning, etc.) into a database. Then this e-nose "training" is used as a foundation to compare against other oil samples of the same botanical source. Products not fitting within a range of acceptable sample "fingerprints" will be flagged as pass/fail to whatever criteria the e-nose system was "trained" [18, 30].

E-noses are appealing for their reported ease of use (once a system is "trained"), very little sample preparation, and fast run times. However, one must carefully consider the sensitivity of a sensor array (limits of detection) in comparison to human noses (odorant threshold), indiscriminate chemical interaction with sensors compared to biological interactions with odor receptors in human response, and complicated calibration and data analysis. Also, more practical limitations of e-nose may be frequent calibration, poisoning of sensors, and unintended sensor interactions with food components.

35.6 IDENTIFYING THE WHAT, WHERE, WHEN, AND HOW OF AN ISSUE

Once the unknown foreign material or off-odor is identified, a second phase begins to investigate where, when, and how it contaminated a food ingredient or processed food. This step is important since it will help a company decide if this is an isolated incident; if more widespread, the company may initiate a large-scale recall. If public safety is jeopardized, there is never a question if a recall should be initiated. Identifying what went wrong may require close examination of each step in the manufacturing process, but with experience the investigator will often know where to begin looking for the source of a problem [35] (Table 35.5).

In a large majority of forensic investigations, the source of contamination by a foreign material has originated from the consumer's handling of the food, and the culprit is usually glass particles [1] (Fig. 35.8a). Extraneous matter analysis is covered in Chap. 34, but more details are offered here in relation to a forensic investigation. Glass is particularly bothersome to a consumer because perceptually it can cause great harm to an ingesting consumer. This point emphasizes the need to carefully document conversations with the con-

Table 35.5 Other issues addressed by food forensic analysts: requests to determine what went wrong

Black specks
Broken emulsions
Cause of hysteresis
Crystallization
Degradation products or byproducts
Flock issues
Freeze/thaw damage
Ingredient-ingredient interactions
Missing ingredients
Off color
Poor product functionality
Poor product performance
Processing damage
Processing equipment cleaning issues
Quality of cook
Sediments
Starch analysis (cook, application)
Substituted ingredients
Undercook or overcook
Viscosity issues
Water migration issues
Water activity
Wrong ingredient ID

sumer in the initial investigatory stages, including how they prepared the food. Most glass chips originate from the rim of glasses or casserole dishes used in preparation. Occasionally glass shelves in retailer display cases, such as in a delicatessen, may chip and contaminate a finished or processed food product. Other objects can be mistaken by the consumer as glass fragments. These include calcium lactate crystals from cheeses, magnesium ammonium phosphate (clear and colorless crystalline material commonly known as struvite) from canned fish products like tuna or salmon, and rock salt (sometimes from the consumer's own salt grinder). Another common complaint area is metal from hard or very chewy foods that the consumer finds during chewing. Often times the forensic laboratory will have identified these as metal fragments characteristic of dental appliances including amalgam fillings and root canal posts. Occasionally, a tooth chip also will be identified. These are almost always from the oral cavity of the consumer (usually one with a preexisting dental problem) [1]. However, on occasion, there are instances of contamination by animal teeth, either coming with the meat or from a family pet. It is not unknown for dogs to have restorative dentistry!

Some contaminants will be easily identified as to where they originated since they may be part of the intrinsic materials used to manufacture a food ingredient or processed food. Extrinsic material contaminants will require more investigative work, usually at the plant that manufactured the implicated ingredient or processed food.

Food forensic investigators will need to work closely with operations, sanitation, engineering, and maintenance leaders who oversee unit operations of an implicated plant. The initial investigation in identifying the unknown contaminant may provide clues as to where to begin looking for its origin. For example, metal shavings (Fig. 35.8b) that indicate stainless steel contamination (Fig. 35.9) suggest looking initially at the processing equipment itself for scores etched in the metal or an item broken in the processor. Wires contaminating food often can be traced back to milling and sieving equipment. Gasket material is easily traced back to every gasket in the process (Fig. 35.8c). Contaminants identified as grease globules or oil spots will indicate an investigator should look at every joint and seal that is lubricated. Fine fibers (Fig. 35.8d) are often traced back to filters used in the process. Plastic or string in a contaminated product is often traced back to early staging and premixing steps before processing begins when ingredients are being emptied from sacks into a mixing vessel. Plastic also can contaminate a product when a package is being manufactured onsite immediately prior to filling, such as in some bottling operations.

A forensic investigator will need to determine, after identifying the contaminant, if it makes sense that it could have gotten into the food at some point from the time it was harvested and processed until it left its manufacturing facility. If it does not make sense, you should suspect product tampering and engage the proper authorities including a regulatory body and law enforcement. For example, if ground glass is identified in a product and there is no glass anywhere in the process, you should become speculative. However, do not overlook the fact that glass can originate from light bulbs, light fixtures, or gauge lenses near the processor. The same concerns should be raised if the identified contaminant is a foreign material with no relationship to food processing. Product tampering is often associated with disgruntled employees or individuals upset with a company for various reasons. Fortunately, product tampering to cause widespread disease or harm to consumers is rare.

35.7 INTERPRETING AND REPORTING DATA

The goals for forensic investigations are to construct explanations as to how a food material was contaminated by foreign materials or developed an off-odor/off-flavor by uncovering origins of the stated problem using accepted scientific protocols and then to understand how it happened. Ultimately, discoveries made in this process will identify steps that can be taken to minimize this happening again in the future. The process of sharing data and results requires careful record keeping throughout the process which when reviewed by others can be replicated and validated.

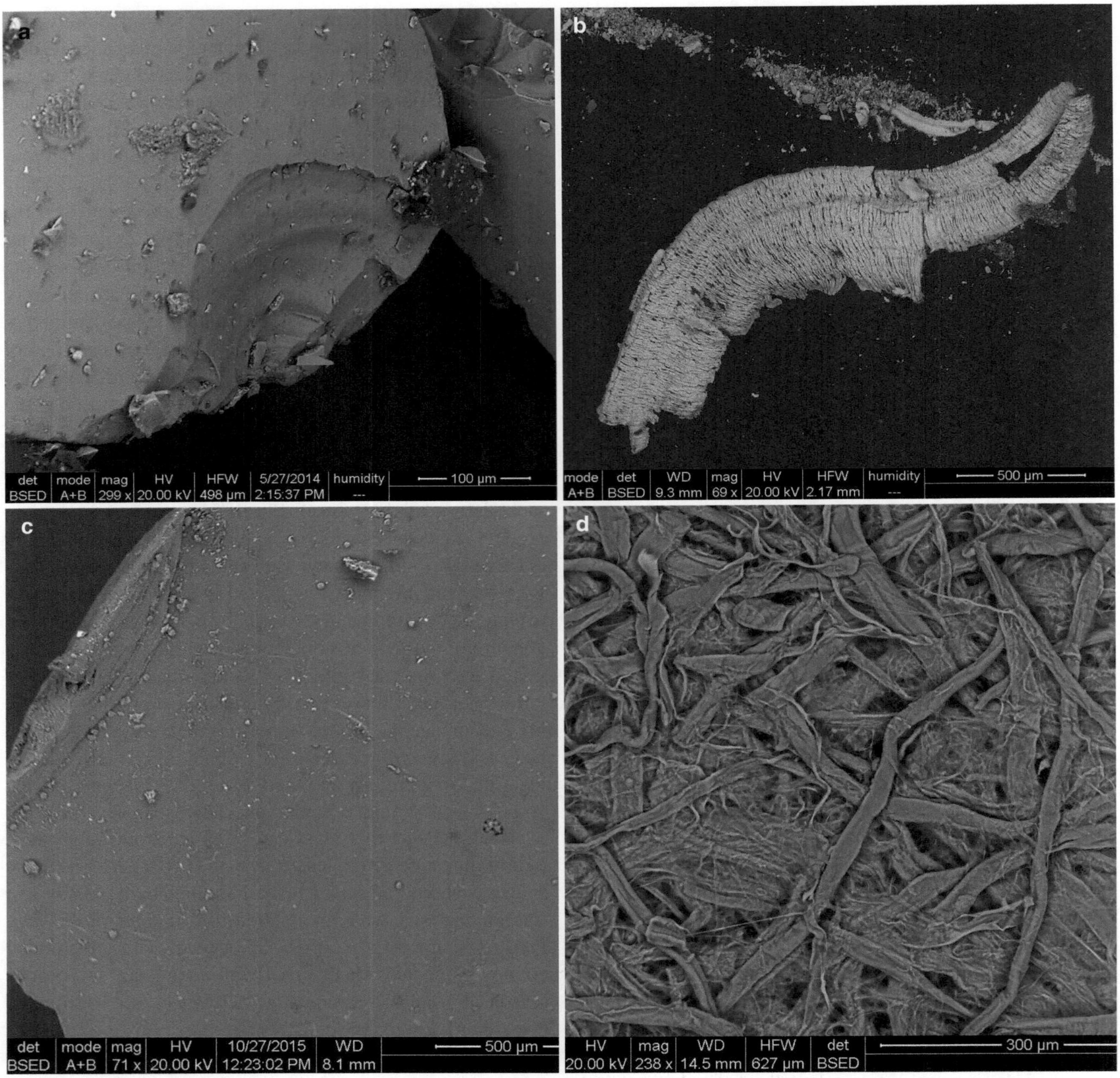

Fig. 35.8 Scanning electron micrographs of commonly identified food foreign material contaminants. (**a**) Glass shard isolated from a processed food product (magnification 299×). (**b**) Stainless steel metal fragment isolated from flour after passing through a sieve (magnification 69×). (**c**) Polymeric gasket material recovered from a processed food product (magnification 71×). (**d**) Cellulosic fibers from in-process filters recovered from corn syrup (magnification 238×)

Reported data must be restricted to exactly what is known, and an investigator should not speculate or guess in reporting results. This is especially important in the fast pace of a potential product recall or product tampering event when there will be intense pressure being exerted to deliver results that may be inaccurate and/or incomplete. Wrongfully reported data could result in falsely reassuring the public or causing unnecessary panic [13]. Food forensic teams should have a single point of contact to the customer, regulatory, or law enforcement to discuss results to avoid confusion or improper release of information.

A forensic investigator's role is not finished at reporting data. Samples will need to be properly stored, and chain of control protocols must be followed. An analyst may need to testify in a lawsuit or criminal proceeding, offer interpretation of results including those that are not conclusive, and recommend any changes to SOPs [14].

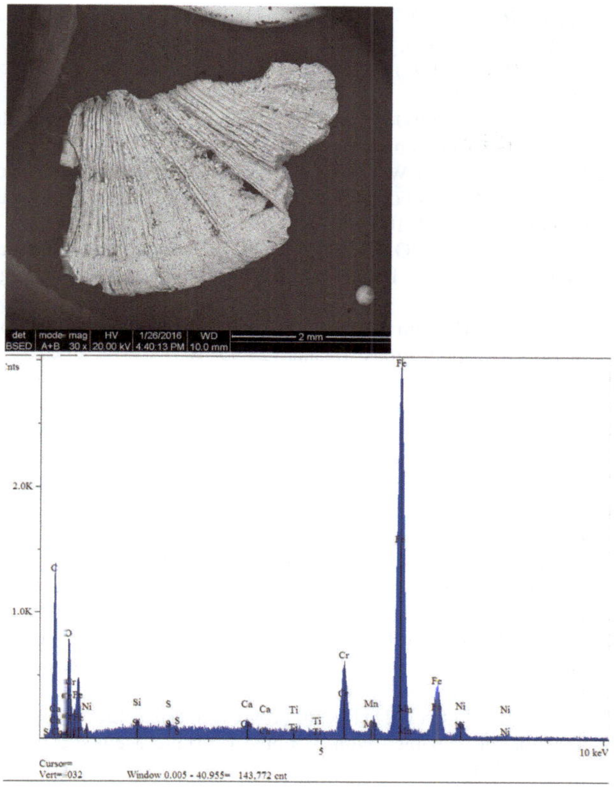

Fig. 35.9 Scanning electron microscopy (SEM) micrograph and energy-dispersive x-ray spectrometry (EDS) spectrum of a metallic foreign material contamination

35.8 SUMMARY

Much of the food we consume daily comes from crops and livestock that are further processed by food companies using a broad range of processes and production systems. Seemingly unexplainable circumstances do happen in manufacturing environments and can go undetected until a consumer bites into a piece of food and notices a problem. The importance of food forensic investigation is becoming greater given the thorough investigation of problem situations required by new regulatory requirements instituted globally and given the damage to a company's reputation caused by such problems.

Food forensic investigation requires experienced individuals to analyze samples that have something wrong with them either caused by an error in processing or by deliberate acts of individuals trying to discredit a company or cause physical harm to the consumer. When a product is reported with a problem, it is important to follow codified SOPs to analyze the product to identify what the problem is, where and when it occurred in the supply chain, and how it occurred, so corrective action can be taken. Various specialized nondestructive and destructive techniques are critical to investigate such problems of foreign matter contamination. Sensitive instrumental techniques and simple sensory evaluation are critical to identify contaminants causing off-flavors/odors and taints. Food forensic tools and experienced personnel are critical to determine the root cause of both product failures and food tampering.

35.9 STUDY QUESTIONS

1. What is the difference between forensic quality control and forensic quality assurance?
2. Explain what is meant by what, where, when, and how in relation to a food forensic investigation.
3. Explain why it is important for a food forensic investigator to be familiar with unit operations in manufacturing plants.
4. What are potential sources of glass as foreign material contaminants in a food plant that does not use glass packaging?
5. What is meant by chain of custody in a food forensic investigation?
6. The phrase "codified standard operating procedures" was used throughout this chapter. Explain what is meant by that phrase?
7. When you receive a package with samples of a suspect product, what should be your first observations? How will you record these?
8. What is "taint"?
9. A consumer files a complaint that a piece of glass was found in cheese manufactured from your plant. Draw a decision tree outlining the steps you will need to take in a forensic investigation to identify what the customer actually perceived.

REFERENCES

1. Edwards MC, Stringer MF (2007) The breakdowns in food group. Observations on patterns in foreign material investigations. Food Control 18:773–82
2. Donarski J, Camin F (2019) Sampling guidelines for building and curating food authenticity databases, Trends Food Sci Technol https://doi.org/10.1016/j.tifs.2019.02.109
3. Medina S, Pereira JA (2019) Food fingerprints – A valuable tool to monitor food authenticity and safety Food Chem 278:144–162
4. Black C, Chevaillier OP (2016) The current and potential applications of ambient mass spectrometry in detecting food fraud. Trends Anal Chem https://doi.org/10.1016/j.trac.2016.06.005
5. El Sheikha, AF (2019) DNAFoil: Novel technology for the rapid detection of food adulteration Trends Food Sci Technol 86:544–552
6. Hsieh D (2015) Food fraud: a criminal activity implementing preventative measures that increase difficulty in carrying out the crime. Food Qual Safety 2(2):18–23
7. U.S. Pharmacopeia: USP's Food Fraud Database Available from: http://www.usp.org/food-ingredients/food-fraud-database

8. USP – Food Chemicals Codex, Appendix XVII; Food Fraud Mitigation Guidance. Available from: http://www.usp.org/sites/default/files/usp_pdf/EN/fcc/food-fraud-mitigation-guidance.pdf
9. United States Food and Drug Administration (Internet). Washington D.C.: Food Safety Modernization Act. Available from: http://www.fda.gov/Food/GuidanceRegulation/FSMA/.
10. Canadian Food Agency (Internet). Quebec: Safe Food for Canadians Act. Available from: http://www.inspection.gc.ca/food/.
11. Levy S, Bergman P, Frank A (1999) Quality assurance in forensic evidence. Accredited Quality Assurance 4:253–55
12. Meek T (2015) Engineering defensibility in food labs. Food Quality and Safety 21(4):35–37
13. Pyzdek T, Keller, P (2013) The Handbook for Quality Management: A Complete Guide to Operational Excellence, 2nd edn, McGraw-Hill Publishers, London, UK
14. Magnusun ML, Satzger RD, Alcarez A, Brewer J, Fetterolf D, Harper M, Hrynchuk R, McNally MF, Montgomery M., Nottingham E, Peterson J, Rickenbach M, Seidel JL, Wolnik K (2012) Guidelines for the identification of unknown samples for laboratories performing forensic analysis for chemical terrorism. J Forensic Sci 57:636–642
15. Potts PJ, Webb PC (1992) X-ray fluorescence spectrometry. Journal of Geochemical Exploration 44:251–296
16. Miziolek AW, Palleschi V, Schechter I (2006) Laser Induced Breakdown Spectroscopy. Cambridge University Press, Cambridge, UK
17. Kilcast D (1996) Sensory evaluation of taints and off-flavours. In: Saxby MJ (Ed) Food Taints and Off-flavors. Chapman and Hall, London, UK, p. 1–40
18. Reineccius G. (2006) Flavor chemistry and technology. Taylor & Francis, Boca Raton, FL, p. 161–200
19. Baigrie B (2003) Introduction. In: Baigrie B (Ed) Taints and off-flavours in food, Woodhead, Boca Raton, FL, p. 1–4
20. Saxby MJ (Ed) (1996) Food taints and off-flavors. Chapman and Hall, London, UK
21. Baigrie B (Ed) (2003) Taints and off-flavours in food. Woodhead, Boca Raton, FL
22. Marsili R (Ed) (2007) Sensory-directed flavor analysis. Taylor & Francis, Boca Raton, FL
23. Marsili R (Ed) (1997) Techniques for analyzing food aroma. Marcel Dekker, New York
24. Acree T, Arn H (2004) Flavornet. Available from: http://www.flavornet.org/flavornet.html
25. Maarse H, Grosch W (1996) Analysis of taints and off-flavours. In: Saxby MJ (Ed) Food taints and off-flavors. Chapman and Hall, London, UK, p. 72–106
26. Schaller E, Bosset JO, Escher, F (1998) 'Electronic noses' and their application to food. Lebensmittel-Wissenschaft und -Technologie 31(4):305–316
27. Gutierrez J, Horrillo MC (2014) Advances in artificial olfaction: Sensors and applications. Talanta 124:95–105
28. Capelli L, Sironi S, Del Rosso R (2014) Electronic noses for environmental monitoring applications. Sensors 14(11):19979–20007
29. Sliwinska M, Wisniewska P, Dymerski T, Namiesnik J, Wardencki W (2014) Food analysis using artificial senses. Journal of Agricultural and Food Chemistry 62(7):1423–1448
30. Persaud K (2016) Electronic Noses and Tongues in the Food Industry. In: Rodríguez-Méndez, ML (Ed) Electronic Noses and Tongues in Food Science. Academic Press, p. 1–12.
31. Zhang X, Cheng J, Wu L, Mei Y, Jaffrezic-Renault N, Guo Z (2018) An overview of an artificial nose system. Talanta 184:93–102
32. Shi H, Zhang M, Adhikari B (2018) Advances of electronic nose and its application in fresh foods: A review. Critical Reviews in Food Science and Nutrition 58(16):2700–2710
33. Ye Z, Liu Y, Li Q (2021) Recent Progress in Smart Electronic Nose Technologies Enabled with Machine Learning Methods. Sensors 21(22):7620–7641
34. Ranbir MK, Singh G, Singh J, Kaur N, Singh N (2022) Machine Learning-based Analytical Systems: Food Forensics. ACS Omega 7(51):47518–47535
35. Stringer MF, Hall MN (2007) The breakdowns in food group. A generic model of the integrated food supply chain to aid the investigation of food safety breakdowns. Food Control 18:755–765

Index

B. P. Ismail, S. S. Nielsen (eds.), *Nielsen's Food Analysis*, Food Science Text Series,
https://doi.org/10.1007/978-3-031-50643-7

M

Q

T

U

V

W

X

Z

MIX
Papier aus verantwortungsvollen Quellen
Paper from responsible sources
FSC® C105338

If you have any concerns about our products,
you can contact us on
ProductSafety@springernature.com

In case Publisher is established outside the EU,
the EU authorized representative is:
Springer Nature Customer Service Center GmbH
Europaplatz 3, 69115 Heidelberg, Germany

Printed by Libri Plureos GmbH
in Hamburg, Germany